Geometry

($A =$ area, $B =$ area of base, $C =$ circumference, $S =$ lateral area or surface area, $V =$ volume)

1. Triangle

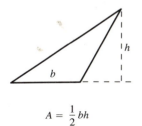

$$A = \frac{1}{2}bh$$

2. Similar Triangles

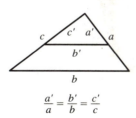

$$\frac{a'}{a} = \frac{b'}{b} = \frac{c'}{c}$$

3. Pythagorean Theorem

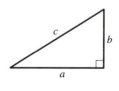

$$a^2 + b^2 = c^2$$

4. Parallelogram

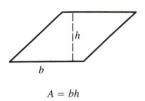

$$A = bh$$

5. Trapezoid

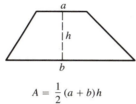

$$A = \frac{1}{2}(a + b)h$$

6. Circle

$$A = \pi r^2, \quad C = 2\pi r$$

7. Any Cylinder or Prism with Parallel Bases

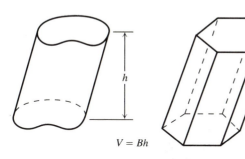

$$V = Bh$$

8. Right Circular Cylinder

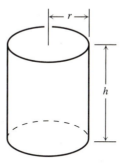

$$V = \pi r^2 h, \quad S = 2\pi rh$$

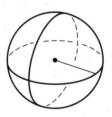

9. Any Cone or Pyramid

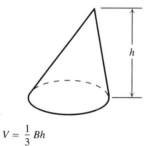

$$V = \frac{1}{3}Bh$$

10. Right Circular Cone

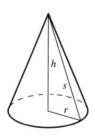

$$V = \frac{1}{3}\pi r^2 h, \quad S = \pi rs$$

11. Sphere

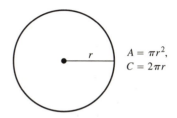

$$V = \frac{4}{3}\pi r^3, \quad S = 4\pi r^2$$

9TH EDITION

Calculus and Analytic Geometry

PART II

9TH EDITION

Calculus and Analytic Geometry

PART II

George B. Thomas, Jr.
Massachusetts Institute of Technology

Ross L. Finney

With the collaboration of
Maurice D. Weir
Naval Postgraduate School

 Addison-Wesley Publishing Company

Reading, Massachusetts • Menlo Park, California • New York
Don Mills, Ontario • Wokingham, England • Amsterdam
Bonn • Sydney • Singapore • Tokyo • Madrid
San Juan • Milan • Paris

Acquisitions Editor	Laurie Rosatone	*Production Editorial Services*	Barbara Pendergast
Development Editor	Marianne Lepp	*Art Editors*	Susan London-Payne, Connie Hulse
Managing Editor	Karen Guardino	*Copy Editor*	Barbara Flanagan
Senior Production Supervisor	Jennifer Bagdigian	*Proofreader*	Joyce Grandy
Senior Marketing Manager	Andrew Fisher	*Text Design*	Martha Podren, Podren Design;
Marketing Coordinator	Benjamin Rivera		Geri Davis, Quadrata, Inc.
Prepress Buying Manager	Sarah McCracken	*Cover Design*	Marshall Henrichs
Art Buyer	Joseph Vetere	*Cover Photo*	John Lund/Tony Stone Worldwide
Senior Manufacturing Manager	Roy Logan	*Composition*	TSI Graphics, Inc.
Manufacturing Coordinator	Evelyn Beaton	*Technical Illustration*	Tech Graphics

Reprinted with corrections, March, 1996.

Photo Credits: 633, 722, 875, 899, From *PSSC Physics 2/e,* 1965; D.C. Heath & Co. with Education Development Center, Inc., Newton, MA. Reprinted with permission **872,** AP/Wide World Photos **889,** © 1994 Nelson L. Max, University of California/Biological Photo Service; Graphic by Alfred Gray **938,** ND Roger-Viollet **1068,** NASA/Jet Propulsion Laboratory

Library of Congress Cataloging-in-Publication Data
Thomas, George Brinton, 1914–
      Calculus and analytic geometry / George B. Thomas, Ross L. Finney.
      —9th ed.
      p. cm.
      Includes index.
      ISBN 0-201-53176-3
      1. Calculus. 2. Geometry, Analytic. I. Finney, Ross L.
II. Title.
QA303.T42 1996
515'. 15—dc20
                                                    94-30543
                                                   CIP

3 4 5 6 7 8 9 10   VH   989796

Contents

CAS Explorations and Projects
(Listed by chapter and section)

To the Instructor

This Is a Substantial Revision

Throughout the 40 years that it has been in print, Thomas/Finney has been used to support a variety of teaching methods from traditional to experimental. In response to the many exciting currents in teaching calculus in the 1990s, the new edition builds on the traditional strengths of the book—excellent exercises, sound mathematics, variety in applications—to produce a flexible text that contains all the elements needed to teach the many different kinds of courses that exist today.

- The exercises have been reorganized to facilitate assigning a subset of the material in a section.
- The grapher explorations, all accessible with any graphing calculator, many suitable for in-class and group work, have been expanded.
- New Computer Algebra System (CAS) explorations and projects that require a CAS have been included. Some of these can be done quickly while others require several hours. All are suitable for either individual or group work. You will find a list of CAS exercise topics following the Table of Contents.
- Technology Connection notes appear throughout the text suggesting experiments students might do with a grapher to supplement their understanding of a given topic. These notes are meant to encourage students to think of their grapher as a casually available tool, like a pencil.
- We have revised the entire first semester and large portions of the second and third semesters to provide what we believe is a cleaner, more visual, and more accessible book.

With all these changes, we have not compromised our belief that the fundamental goal of a calculus book is to prepare students to enter the scientific community.

Students Will Find Even More Support for Creative Problem Solving

Throughout this book, we have included examples and discussions that encourage students to think visually and numerically. Almost every exercise set has easy to mid-level exercises that require students to think analytically or to generate and interpret graphs as a tool for understanding mathematical or real-world relationships.

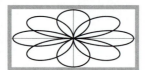

$r_1 = \sin 2\theta$ and $r_2 = \cos 2\theta$ graphed together.

762

Technology   *Finding Intersections*   The *simultaneous mode* of a graphing utility gives new meaning to the *simultaneous solution* of a pair of polar co-ordinate equations. A simultaneous solution occurs only where the two graphs "collide" while they are being drawn simultaneously and not where one graph intersects the other at a point that had been illuminated earlier. The distinction is particularly important in the areas of traffic control or missile defense. For example, in traffic control the only issue is whether two aircraft are in the same place at the same time. The question of whether the curves the craft follow intersect is unimportant.

To illustrate, graph the polar equations

$$r = \cos 2\theta \qquad \text{and} \qquad r = \sin 2\theta$$

in simultaneous mode with $0 \le \theta < 2\pi$, θ Step $= 0.1$, and view dimensions [xmin, xmax] $= [-1, 1]$ by [ymin, ymax] $= [-1, 1]$. *While the graphs are being drawn on the screen,* count the number of times the two graphs illuminate a single pixel simultaneously. Explain why these points of intersection of the two graphs correspond to simultaneous solutions of the equations. (You may find it helpful to slow down the graphing by making θ Step smaller, say 0.05, for example.) In how many points total do the graphs actually intersect?

37. *Safe and effective dosage.* The concentration in the blood re-sulting from a single dose of a drug normally decreases with time as the drug is eliminated from the body. Doses may there-fore need to be repeated periodically to keep the concentration from dropping below some particular level. One model for the ef-fect of repeated doses gives the residual concentration just before the $(n + 1)$st dose as

$$R_n = C_0 e^{-kt_0} + C_0 e^{-2kt_0} + \cdots + C_0 e^{-nkt_0},$$

where $C_0 =$ the change in concentration achievable by a single dose (mg/ml), $k =$ the *elimination constant* (h^{-1}), and $t_0 =$ time between doses (h). See Fig. 8.22.

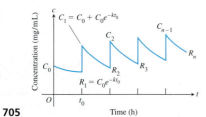

705

Many sections also contain a few more challenging problems to extend the range of the mathematically curious.

This volume has more than 1000 figures to appeal to the students' geometric intuition. Drawing lessons aid students with difficult 3-dimensional sketches, enhancing their ability to think in 3-space. In this edition we have increased the use of visualization internal to the discus-sion. The burden of exposition is shared by art in the body of the text when we feel that pictures and text together will convey ideas better than words alone.

Throughout the text, students are asked to experiment, investigate, and explain. Writing exercises are placed throughout the text. In addi-tion, each chapter end contains a list of questions that ask students to re-view and summarize what they have learned. Many of these exercises make good writing assignments.

Students Will Master Techniques

Problem-Solving Strategies   We believe that the students learn best when procedural techniques are laid out as clearly as possible. To this end we have revisited the summaries of the steps used to solve problems, adding some where necessary, deleting some where a thought process rather than a tech-nique was at issue, and making each one clear and useful. As always, we are espe-cially careful that examples in the text follow the steps outlined by the discussion.

Exercises   Every exercise set has been reviewed and revised. Exercises are now *grouped by topic,* with special sections for grapher explorations. Many sections also have a set of Computer Algebra System (CAS) Explorations and Projects, a new fea-ture for this edition. Within each group, the exercises are graded and paired. Within this framework, the exercises generally follow the order of presentation of the text.

Exercises that require a calculator or computer are identified by icons: calcu-lator exercise, graphing utility (such as graphing calculator) exercise, and Computer Algebra System exercise.

How to Integrate in Spherical Coordinates

To evaluate

$$\iiint_D f(\rho, \phi, \theta)\ dV$$

over a region D in space in spherical coordinates, integrating first with respect to ρ, then with respect to ϕ, and finally with respect to θ, take the following steps.

1. *A sketch.* Sketch the region D along with its projection R on the xy-plane. Label the surfaces that bound D.
2. *The ρ-limits of integration.* Draw a ray M from the origin making an angle ϕ with the positive z-axis. Also draw the projection of M on the xy-plane (call the projection L). The ray L makes an angle θ with the positive x-axis. As ρ increases, M enters D at $\rho = g_1(\phi, \theta)$ and leaves at $\rho = g_2(\phi, \theta)$. These are the ρ-limits of integration.
3. *The ϕ-limits of integration.* For any given θ, the angle ϕ that M makes with the z-axis runs from $\phi = \phi_{min}$ to $\phi = \phi_{max}$. These are the ϕ-limits of integration.
4. *The θ-limits of integration.* The ray L sweeps over R as θ runs from α to β. These are the θ-limits of integration. The integral is

$$\iiint_D f(\rho, \phi, \theta)\ dV = \int_{\theta=\alpha}^{\theta=\beta} \int_{\phi=\phi_{min}}^{\phi=\phi_{max}} \int_{\rho=g_1(\phi,\theta)}^{\rho=g_2(\phi,\theta)} f(\rho, \phi, \theta)\, \rho^2 \sin \phi\, d\rho\, d\phi\, d\theta.\quad (5)$$

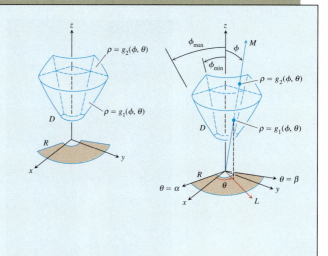

1042

Within the exercise sets, we have practice exercises, exercises that encourage critical thinking, more challenging exercises (in subsections marked "Applications and Theory"), and exercises that require writing in English about concepts. Writing exercises are placed both throughout the exercise sets, and in an end-of-chapter feature called "Questions to Guide Your Review."

How to Test a Power Series for Convergence

Step 1: Use the Ratio Test (or nth-Root Test) to find the interval where the series converges absolutely. Ordinarily, this is an open interval

$$|x - a| < R \qquad \text{or} \qquad a - R < x < a + R.$$

Step 2: If the interval of absolute convergence is finite, test for convergence or divergence at each endpoint, as in Examples 3(a) and (b). Use a Comparison Test, the Integral Test, or the Alternating Series Test.

Step 3: If the interval of absolute convergence is $a - R < x < a + R$, the series diverges for $|x - a| > R$ (it does not even converge conditionally), because the nth term does not approach zero for those values of x.

666

Chapter End   At the end of each chapter are three features with questions that summarize the chapter in different ways.

Questions to Guide Your Review ask students to think about concepts and verbalize their understanding without trying to calculate numeric answers. These are, as always, suitable for writing exercises.

Practice Exercises provide a review of the techniques, ideas, and key applications.

Additional Exercises—Theory, Examples, Applications supply challenging applications and theoretic problems that deepen the understanding of mathematical ideas.

Applications, Technology, History—Features That Bring Calculus to Life

Applications and Examples   It has been a hallmark of this book through the years that we illustrate applications of calculus with real data based on already familiar situations or situations students are likely to encounter soon. Throughout the text, we cite sources for the data and/or articles from which the applications are drawn, helping students understand that calculus is a current, dynamic field. Most of these appli-

EXAMPLE 3   To open the 1992 Summer Olympics in Barcelona, bronze medalist archer Antonio Rebollo lit the Olympic torch with a flaming arrow (Fig. 11.9).

Suppose that Rebollo wanted the arrow to reach its maximum height exactly 4 ft above the center of the cauldron (Fig. 11.10).

a) If he shot the arrow at a height of 6 ft above ground level 30 yd from the 70-ft-high cauldron, express y_{max} in terms of the initial speed v_0 and firing angle α.

b) If $y_{max} = 74$ ft (Fig. 11.10), use the results of part (a) to find the value of $v_0 \sin \alpha$.

c) When the arrow reaches y_{max}, the horizontal distance traveled to the center of the cauldron is $x = 90$ ft. Use this fact to find the value of $v_0 \cos \alpha$.

d) Find the initial firing angle of the arrow.

11.9 Spanish archer Antonio Rebollo lights the Olympic torch in Barcelona with a flaming arrow.

872

cations are directed toward science and engineering, but there are many from other fields as well.

Technology: Graphing Calculator and Computer Algebra Systems Explorations   Wherever appropriate, the text presents calculator exercises that explore numerical patterns and/or graphing calculator exercises that ask students to generate and interpret graphs as a tool to understanding mathematical or real-world relationships. Many of these exercises are suitable for classroom demonstration or for group work by students in or out of class.

Computer Algebra System (CAS) exercises have been added to every chapter. These exercises, more than 100 in all, have been tested on both Mathematica and Maple. A full list of CAS exercise topics follows the Table of Contents.

Notes appear throughout the text encouraging students to explore with graphers.

History   Any student is enriched by seeing the human side of mathematics. As in earlier editions, we feature history boxes that describe origins of ideas and entertaining quirks of mathematics history.

The Many Faces of This Book

Mathematics Is a Formal and Beautiful Language   A good part of the beauty of the calculus lies in the fact that it is a stunning creation of the human mind. As in previous editions we have been careful to say only what is true and mathematically sound. In this edition we reviewed every definition, theorem, corollary, and proof for clarity and mathematical correctness.

Even Better Suited to Be the Reference Text in a Reform Course   Whether calculus is taught by a traditional lecture or entirely in labs with individual and group learning which focuses on numeric and graphical experimentation, ideas and techniques need to be articulated clearly. This book provides the exercises for computer and grapher experiments and group learning and, in a traditional format, the summation of the lesson—the formal statement of the mathematics and the clear presentation of the technique.

Students Will Learn from This Book for Many Years to Come   We provide far more material than any one instructor would want to teach. We do this intentionally. Students can continue to learn calculus from this book long after the class has ended. It provides an accessible review of the calculus a student has already studied. It is a resource for the working engineer or scientist.

Content Features of Part II of the Ninth Edition

Sequences and Series

- The introduction to sequences has been spread over two sections (Sections 8.1, 8.2), providing more time for this idea.
- Chapter 8 has been reorganized to allow one section per lecture. (See the Table of Contents.)
- Taylor series (Section 8.9) are introduced by exploring the question When can a function be expressed as a power series and, if it can, what must the coefficients be?
- Power series are applied to solve differential equations and initial value problems (Section 8.11).

Conic Sections

- The geometry of conic sections is treated in Section 9.1.
- Eccentricity is covered separately in Section 9.2, where it is used to classify the conics.

Multivariate Calculus

- The treatments of curvature and torsion and the **TNB** frame in Chapter 11 (Vector-Valued Functions and Motion in Space) has been simplified and unified.
- In Chapter 12 (Multivariable Functions and Partial Derivatives), improved tree diagrams cover more cases in partial derivatives, and a simplified treatment of directional derivatives provides better introductory motivation.

- Chapter 13 (Multiple Integrals) offers illustrated strategies for finding limits of integration in double and triple integrals in various coordinate systems.

- Chapter 14 (Integration in Vector Fields) has been reorganized to place all the material on line integrals before the material on surface integrals.

- There is now a section (Section 14.6) on integration over parametrized surfaces.

Supplements for the Instructor

OmniTest[3] in DOS-Based Format   This easy-to-use software is developed exclusively for Addison-Wesley by ips Publishing, a leader in computerized testing and assessment. Among its features are the following.

- **DOS interface is easy to learn and operate.** The windows look-alike interface makes it easy to choose and control the items as well as the format for each test.

- **You can easily create make-up exams, customized homework assignments, and multiple test forms to prevent plagiarism.** OmniTest[3] is algorithm driven—meaning the program can automatically insert new numbers into the same equation—creating hundreds of variations of that equation. The numbers are constrained to keep answers reasonable. This allows you to create a virtually endless supply of parallel versions of the same test. This new version of OmniTest also allows you to "lock in" the values shown in the model problem, if you wish.

- **Test items are keyed by section** to the text. Within the section, you can select questions that test individual objectives from that section.

- **You can enter your own questions** by way of OmniTest[3]'s sophisticated editor—complete with mathematical notation.

Instructor's Solutions Manual   by Maurice D. Weir (Naval Postgraduate School). This two-volume supplement contains the worked-out solutions for *all* the exercises in the text.

Answer Book   contains short answers to most exercises in the text.

Supplements for the Instructor and Student

Student Study Guide   by Maurice D. Weir (Naval Postgraduate School). Organized to correspond with the text, this workbook in a semiprogrammed format increases student proficiency with study tips and additional practice.

Student Solutions Manual   by Maurice D. Weir (Naval Postgraduate School). This manual is designed for the student and contains carefully worked-out solutions to all of the odd-numbered exercises in the text.

Differential Equations Primer   A short, supplementary manual containing approximately a chapter's-worth of material. Available should the instructor choose to cover this material within the calculus sequence.

Technology-Related Supplements

Analyzer* This program is a tool for exploring functions in calculus and many other disciplines. It can graph a function of a single variable and overlay graphs of other functions. It can differentiate, integrate, or iterate a function. It can find roots, maxima and minima, and inflection points, as well as vertical asymptotes. In addition, Analyzer* can compose functions, graph polar and parametric equations, make families of curves, and make animated sequences with changing parameters. It exploits the unique flexibility of the Macintosh wherever possible, allowing input to be either numeric (from the keyboard) or graphic (with a mouse). Analyzer* runs on Macintosh II, Plus, or better.

The Calculus Explorer Consisting of 27 programs ranging from functions to vector fields, this software enables the instructor and student to use the computer as an "electronic chalkboard." The Explorer is highly interactive and allows for manipulation of variables and equations to provide graphical visualization of mathematical relationships that are not intuitively obvious. The Explorer provides user-friendly operation through an easy-to-use menu-driven system, extensive on-line documentation, superior graphics capability, and fast operation. An accompanying manual includes sections covering each program, with appropriate examples and exercises. Available for IBM PC/compatibles.

InSight A calculus demonstration software program that enhances understanding of calculus concepts graphically. The program consists of ten simulations. Each presents an application and takes the user through the solution visually. The format is interactive. Available for IBM PC/compatibles.

Laboratories for Calculus I Using Mathematica By Margaret Höft, The University of Michigan-Dearborn. An inexpensive collection of *Mathematica* lab experiments consisting of material usually covered in the first term of the calculus sequence.

Math Explorations Series Each manual provides problems and explorations in calculus. Intended for self-paced and laboratory settings, these books are an excellent complement to the text.

Exploring Calculus with a Graphing Calculator, Second Edition, by Charlene E. Beckmann and Ted Sundstrom of Grand Valley State University.

Exploring Calculus with Mathematica, by James K. Finch and Millianne Lehmann of the University of San Francisco.

Exploring Calculus with Derive, by David C. Arney of the United States Military Academy at West Point.

Exploring Calculus with Maple, by Mark H. Holmes, Joseph G. Ecker, William E. Boyce, and William L. Seigmann of Rensselaer Polytechnic Institute.

Exploring Calculus with Analyzer*, by Richard E. Sours of Wilkes University.

Exploring Calculus with the IBM PC Version 2.0, by John B. Fraleigh and Lewis I. Pakula of the University of Rhode Island.

Acknowledgments

We would like to express our thanks for the many valuable contributions of the people who reviewed this book as it developed through its various stages:

Manuscript Reviewers

Erol Barbut, *University of Idaho*
Neil E. Berger, *University of Illinois at Chicago*
George Bradley, *Duquesne University*
Thomas R. Caplinger, *Memphis State University*
Curtis L. Card, *Black Hills State University*
James C. Chesla, *Grand Rapids Community College*
P.M. Dearing, *Clemson University*
Maureen H. Fenrick, *Mankato State University*
Stuart Goldenberg, *CA Polytechnic State University*
Johnny L. Henderson, *Auburn University*
James V. Herod, *Georgia Institute of Technology*
Paul Hess, *Grand Rapids Community College*
Alice J. Kelly, Santa Clara University
Jeuel G. LaTorre, *Clemson University*
Pamela Lowry, *Lawrence Technological University*
John E. Martin, III, *Santa Rosa Junior College*
James Martino, *Johns Hopkins University*
James R. McKinney, *California State Polytechnic University*
Jeff Morgan, *Texas A & M University*
F. J. Papp, *University of Michigan—Dearborn*
Peter Ross, *Santa Clara University*
Rouben Rostamian, *University of Maryland—Baltimore County*
William L. Siegmann, *Rensselaer Polytechnic Institute*
John R. Smart, *University of Wisconsin—Madison*
Dennis C. Smolarski, S. J., *Santa Clara University*
Bobby N. Winters, *Pittsburgh State University*

Technology Notes Reviewers

Lynn Kamstra Ipina, *University of Wyoming*
Robert Flagg, *University of Southern Maine*
Jeffrey Stephen Fox, *University of Colorado at Boulder*
James Martino, *Johns Hopkins University*
Carl W. Morris, *University of Missouri—Columbia*
Robert G. Stein, *California State University—San Bernardino*

Accuracy Checkers

Steven R. Finch, *Massachusetts Bay Community College*
Paul R. Lorczak, *MathSoft, Inc.*
John R. Martin, *Tarrant County Junior College*
Jeffrey D. Oldham, *Stanford University*

Exercises

In addition, we thank the following people who reviewed the exercise sets for content and balance and contributed many of the interesting new exercises:

> Jennifer Earles Szydlik, *University of Wisconsin—Madison*
> Aparna W. Higgins, *University of Dayton*
> William Higgins, *Wittenberg University*
> Leonard F. Klosinski, *Santa Clara University*
> David Mann, *Naval Postgraduate School*
> Kirby C. Smith, *Texas A & M University*

Kirby Smith was also a pre-revision reviewer and we wish to thank him for his many helpful suggestions.

We would like to express our appreciation to David Canright, Naval Postgraduate School, for his advice and his contributions to the CAS exercise sets, and Gladwin Bartel, at Otero Junior College, for many helpful suggestions.

Answers

We would like to thank Cynthia Hutcherson for providing answers for exercises in some of the chapters in this edition. We also appreciate the work of an outstanding team of graduate students at Stanford University, who checked every answer in the text for accuracy: Miguel Abreu, David Cardon, Tanya Kalich, Jeffrey D. Oldham, and Julie Roskies. Jeffrey D. Oldham also tested all the CAS exercises, and we thank him for his many helpful suggestions.

Other Contributors

We are particularly grateful to Maurice D. Weir, Naval Postgraduate School, who shared his teaching ideas throughout the preparation of this book. He produced the final exercise sets and wrote most of the CAS exercises for this edition. We appreciate his constant encouragement and thoughtful advice.

We thank Richard A. Askey, University of Wisconsin—Madison, David McKay, Oregon State University, and Richard G. Montgomery, Southern Oregon State College, for sharing their teaching ideas for this edition.

We are also grateful to Erich Laurence Hauenstein, College of DuPage, for generously providing an improved treatment of chaos in Newton's method, and to Robert Carlson, University of Colorado, Colorado Springs, for improving the exposition in the section on relative rates of growth of functions.

To the Student

What Is Calculus?

Calculus is the mathematics of motion and change. Where there is motion or growth, where variable forces are at work producing acceleration, calculus is the right mathematics to apply. This was true in the beginnings of the subject, and it is true today.

Calculus was first invented to meet the mathematical needs of the scientists of the sixteenth and seventeenth centuries, needs that were mainly mechanical in nature. Differential calculus dealt with the problem of calculating rates of change. It enabled people to define slopes of curves, to calculate velocities and accelerations of moving bodies, to find firing angles that would give cannons their greatest range, and to predict the times when planets would be closest together or farthest apart. Integral calculus dealt with the problem of determining a function from information about its rate of change. It enabled people to calculate the future location of a body from its present position and a knowledge of the forces acting on it, to find the areas of irregular regions in the plane, to measure the lengths of curves, and to find the volumes and masses of arbitrary solids.

Today, calculus and its extensions in mathematical analysis are far reaching indeed, and the physicists, mathematicians, and astronomers who first invented the subject would surely be amazed and delighted, as we hope you will be, to see what a profusion of problems it solves and what a range of fields now use it in the mathematical models that bring understanding about the universe and the world around us. The goal of this edition is to present a modern view of calculus enhanced by the use of technology.

How to Learn Calculus

Learning calculus is not the same as learning arithmetic, algebra, and geometry. In those subjects, you learn primarily how to calculate with numbers, how to simplify algebraic expressions and calculate with variables, and how to reason about points, lines, and figures in the plane. Calculus involves those techniques and skills but develops others as well, with greater precision and at a deeper level. Calculus introduces so many new concepts and computational operations, in fact, that you will no longer be able to learn everything you need in class. You will have to learn a fair amount on your own or by working with other students. What should you do to learn?

1. Read the text. You will not be able to learn all the meanings and connections you need just by attempting the exercises. You will need to read relevant

passages in the book and work through examples step by step. Speed reading will not work here. You are reading and searching for detail in a step-by-step logical fashion. This kind of reading, required by any deep and technical content, takes attention, patience, and practice.

2. Do the homework, keeping the following principles in mind.
   a) Sketch diagrams whenever possible.
   b) Write your solutions in a connected step-by-step logical fashion, as if you were explaining to someone else.
   c) Think about why each exercise is there. Why was it assigned? How is it related to the other assigned exercises?

3. Use your calculator and computer whenever possible. Complete as many grapher and CAS (Computer Algebra System) exercises as you can, *even if they are not assigned*. Graphs provide insight and visual representations of important concepts and relationships. Numbers can reveal important patterns. A CAS gives you the freedom to explore realistic problems and examples that involve calculations that are too difficult or lengthy to do by hand.

4. Try on your own to write short descriptions of the key points each time you complete a section of the text. If you succeed, you probably understand the material. If you do not, you will know where there is a gap in your understanding.

Learning calculus is a process—it does not come all at once. Be patient, persevere, ask questions, discuss ideas and work with classmates, and seek help when you need it, right away. The rewards of learning calculus will be very satisfying, both intellectually and professionally.

G.B.T., Jr., *State College, PA*
R.L.F., *Monterey, CA*

Infinite Series

OVERVIEW In this chapter we develop a remarkable formula that enables us to express many functions as "infinite polynomials" and at the same time tells how much error we will incur if we truncate those polynomials to make them finite. In addition to providing effective polynomial approximations of differentiable functions, these infinite polynomials (called power series) have many other uses. They provide an efficient way to evaluate nonelementary integrals and they solve differential equations that give insight into heat flow, vibration, chemical diffusion, and signal transmission. What you will learn here sets the stage for the roles played by series of functions of all kinds in science and mathematics.

8.1 Limits of Sequences of Numbers

Informally, a sequence is an ordered list of things, but in this chapter the things will usually be numbers. We have seen sequences before, such as the sequence $x_0, x_1, \ldots, x_n, \ldots$ of numbers generated by Newton's method and the sequence $c_1, c_2, \ldots, c_n, \ldots$ of polygons that define Helga von Koch's snowflake. These sequences have limits, but many equally important sequences do not.

Definitions and Notation

We can list the integer multiples of 3 by assigning each multiple a position:

$$
\begin{array}{llllll}
\text{Domain:} & 1 & 2 & 3 \ldots n \ldots \\
& \downarrow & \downarrow & \downarrow & \downarrow \\
\text{Range:} & 3 & 6 & 9 & 3n
\end{array}
$$

The first number is 3, the second 6, the third 9, and so on. The assignment is a function that assigns $3n$ to the nth place. And that is the basic idea for constructing sequences. There is a function that tells us where each item is to be placed.

Definition

An **infinite sequence** (or **sequence**) of numbers is a function whose domain is the set of integers greater than or equal to some integer n_0.

Usually n_0 is 1 and the domain of the sequence is the set of positive integers. But sometimes we want to start sequences elsewhere. We take $n_0 = 0$ when we begin Newton's method. We might take $n_0 = 3$ if we were defining a sequence of n-sided polygons.

Sequences are defined the way other functions are, some typical rules being

$$a(n) = \sqrt{n}, \qquad a(n) = (-1)^{n+1}\frac{1}{n}, \qquad a(n) = \frac{n-1}{n}$$

(Example 1 and Fig. 8.1).

To indicate that the domains are sets of integers, we use a letter like n from the middle of the alphabet for the independent variable, instead of the x, y, z, and t used widely in other contexts. The formulas in the defining rules, however, like those above, are often valid for domains larger than the set of positive integers. This can be an advantage, as we will see.

The number $a(n)$ is the **nth term** of the sequence, or the **term with index n.** If $a(n) = (n-1)/n$, we have

First term	Second term	Third term		nth term
$a(1) = 0$	$a(2) = \dfrac{1}{2},$	$a(3) = \dfrac{2}{3},$	$\dots,$	$a(n) = \dfrac{n-1}{n}.$

When we use the subscript notation a_n for $a(n)$, the sequence is written

$$a_1 = 0, \qquad a_2 = \frac{1}{2}, \qquad a_3 = \frac{2}{3}, \qquad \dots, \qquad a_n = \frac{n-1}{n}.$$

To describe sequences, we often write the first few terms as well as a formula for the nth term.

EXAMPLE 1

We write	For the sequence whose defining rule is
$1, \sqrt{2}, \sqrt{3}, \sqrt{4}, \dots, \sqrt{n}, \dots$	$a_n = \sqrt{n}$
$1, \dfrac{1}{2}, \dfrac{1}{3}, \dots, \dfrac{1}{n}, \dots$	$a_n = \dfrac{1}{n}$
$1, -\dfrac{1}{2}, \dfrac{1}{3}, -\dfrac{1}{4}, \dots, (-1)^{n+1}\dfrac{1}{n}, \dots$	$a_n = (-1)^{n+1}\dfrac{1}{n}$
$0, \dfrac{1}{2}, \dfrac{2}{3}, \dfrac{3}{4}, \dots, \dfrac{n-1}{n}, \dots$	$a_n = \dfrac{n-1}{n}$
$0, -\dfrac{1}{2}, \dfrac{2}{3}, -\dfrac{3}{4}, \dots, (-1)^{n+1}\left(\dfrac{n-1}{n}\right), \dots$	$a_n = (-1)^{n+1}\left(\dfrac{n-1}{n}\right)$
$3, 3, 3, \dots, 3, \dots$	$a_n = 3$

Notation We refer to the sequence whose nth term is a_n with the notation $\{a_n\}$ ("the sequence a sub n"). The second sequence in Example 1 is $\{1/n\}$ ("the sequence 1 over n"); the last sequence is $\{3\}$ ("the constant sequence 3").

8.1 The sequences of Example 1 are graphed here in two different ways: by plotting the numbers a_n on a horizontal axis and by plotting the points (n, a_n) in the coordinate plane.

The terms $a_n = \sqrt{n}$ eventually surpass every integer, so the sequence $\{a_n\}$ diverges, . . .

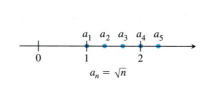

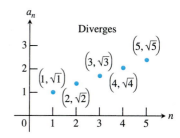

. . . but the terms $a_n = 1/n$ decrease steadily and get arbitrarily close to 0 as n increases, so the sequence $\{a_n\}$ converges to 0.

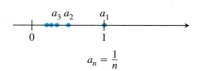

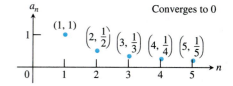

The terms $a_n = (-1)^{n+1}(1/n)$ alternate in sign but still converge to 0.

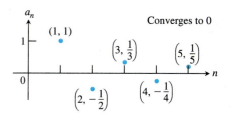

The terms $a_n = (n-1)/n$ approach 1 steadily and get arbitrarily close as n increases, so the sequence $\{a_n\}$ converges to 1.

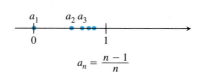

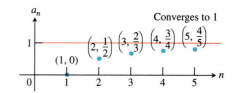

The terms $a_n = (-1)^{n+1}[(n-1)/n]$ alternate in sign. The positive terms approach 1. But the negative terms approach -1 as n increases, so the sequence $\{a_n\}$ diverges.

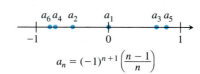

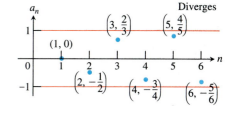

The terms in the sequence of constants $a_n = 3$ have the same value regardless of n, so the sequence $\{a_n\}$ converges to 3.

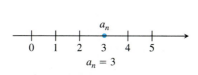

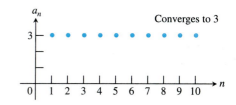

Convergence and Divergence

As Fig. 8.1 shows, the sequences of Example 1 do not behave the same way. The sequences $\{1/n\}$, $\{(-1)^{n+1}(1/n)\}$, and $\{(n-1)/n\}$ each seem to approach a single limiting value as n increases, and $\{3\}$ is at a limiting value from the very first. On the other hand, terms of $\{(-1)^{n+1}(n-1)/n\}$ seem to accumulate near two different values, -1 and 1, while the terms of $\{\sqrt{n}\,\}$ become increasingly large and do not accumulate anywhere.

To distinguish sequences that approach a unique limiting value L, as n increases, from those that do not, we say that the former sequences *converge*, according to the following definition.

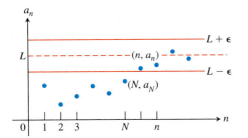

8.2 $a_n \to L$ if $y = L$ is a horizontal asymptote of the sequence of points $\{(n, a_n)\}$. In this figure, all the a_n's after a_N lie within ϵ of L.

Definitions

The sequence $\{a_n\}$ **converges** to the number L if to every positive number ϵ there corresponds an integer N such that for all n,

$$n > N \quad \Rightarrow \quad |a_n - L| < \epsilon.$$

If no such number L exists, we say that $\{a_n\}$ **diverges**.

If $\{a_n\}$ converges to L, we write $\lim_{n\to\infty} a_n = L$, or simply $a_n \to L$, and call L the **limit** of the sequence (Fig. 8.2).

EXAMPLE 2 *Testing the definition*

Show that

a) $\displaystyle \lim_{n\to\infty} \frac{1}{n} = 0$ b) $\displaystyle \lim_{n\to\infty} k = k$ (any constant k)

Solution

a) Let $\epsilon > 0$ be given. We must show that there exists an integer N such that for all n,

$$n > N \quad \Rightarrow \quad \left| \frac{1}{n} - 0 \right| < \epsilon.$$

This implication will hold if $(1/n) < \epsilon$ or $n > 1/\epsilon$. If N is any integer greater than $1/\epsilon$, the implication will hold for all $n > N$. This proves that $\lim_{n\to\infty}(1/n) = 0$.

b) Let $\epsilon > 0$ be given. We must show that there exists an integer N such that for all n,

$$n > N \quad \Rightarrow \quad |k - k| < \epsilon.$$

Since $k - k = 0$, we can use any positive integer for N and the implication will hold. This proves that $\lim_{n\to\infty} k = k$ for any constant k. ☐

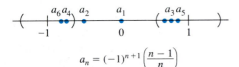

$$a_n = (-1)^{n+1}\left(\frac{n-1}{n}\right)$$

Neither the ϵ-interval about 1 nor the ϵ-interval about -1 contains a complete tail of the sequence.

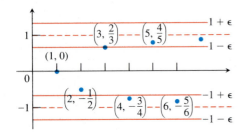

8.3 The sequence $\{(-1)^{n+1}[(n-1)/n]\}$ diverges.

EXAMPLE 3 Show that $\{(-1)^{n+1}[(n-1)/n]\}$ diverges.

Solution Take a positive ϵ smaller than 1 so that the bands shown in Fig. 8.3 about the lines $y = 1$ and $y = -1$ do not overlap. Any $\epsilon < 1$ will do. Convergence to 1 would require every point of the graph beyond a certain index N to lie inside the upper band, but this will never happen. As soon as a point (n, a_n) lies in the upper band, every alternate point starting with $(n + 1, a_{n+1})$ will lie in the lower band. Hence the sequence cannot converge to 1. Likewise, it cannot converge to -1. On the other hand, because the terms of the sequence get alternately closer to 1 and -1, they never accumulate near any other value. Therefore, the sequence diverges. ❑

The behavior of $\{(-1)^{n+1}[(n - 1)/n]\}$ is qualitatively different from that of $\{\sqrt{n}\}$, which diverges because it outgrows every real number L. To describe the behavior of $\{\sqrt{n}\}$ we write

$$\lim_{n \to \infty} (\sqrt{n}) = \infty.$$

In speaking of infinity as a limit of a sequence $\{a_n\}$, we do not mean that the difference between a_n and infinity becomes small as n increases. We mean that a_n becomes numerically large as n increases.

Recursive Definitions

Recursion formulas arise regularly in computer programs and numerical routines for solving differential equations.

So far, we have calculated each a_n directly from the value of n. But sequences are often defined **recursively** by giving

1. The value(s) of the initial term or terms, and
2. A rule, called a **recursion formula,** for calculating any later term from terms that precede it.

Factorial notation

The notation $n!$ ("n factorial") means the product $1 \cdot 2 \cdot 3 \cdot \cdots \cdot n$ of the integers from 1 to n. Notice that $(n + 1)! = (n + 1) \cdot n!$. Thus, $4! = 1 \cdot 2 \cdot 3 \cdot 4 = 24$ and $5! = 1 \cdot 2 \cdot 3 \cdot 4 \cdot 5 = 5 \cdot 4! = 120$. We define 0! to be 1. Factorials grow even faster than exponentials, as the following table suggests.

n	e^n (rounded)	$n!$
1	3	1
5	148	120
10	22,026	3,628,800
20	4.9×10^8	2.4×10^{18}

EXAMPLE 4   *Sequences constructed recursively*

a) The statements $a_1 = 1$ and $a_n = a_{n-1} + 1$ define the sequence $1, 2, 3, \ldots, n, \ldots$ of positive integers. With $a_1 = 1$, we have $a_2 = a_1 + 1 = 2$, $a_3 = a_2 + 1 = 3$, and so on.

b) The statements $a_1 = 1$ and $a_n = n \cdot a_{n-1}$ define the sequence $1, 2, 6, 24, \ldots, n!, \ldots$ of factorials. With $a_1 = 1$, we have $a_2 = 2 \cdot a_1 = 2$, $a_3 = 3 \cdot a_2 = 6$, $a_4 = 4 \cdot a_3 = 24$, and so on.

c) The statements $a_1 = 1$, $a_2 = 1$, and $a_{n+1} = a_n + a_{n-1}$ define the sequence $1, 1, 2, 3, 5, \ldots$ of **Fibonacci numbers.** With $a_1 = 1$ and $a_2 = 1$, we have $a_3 = 1 + 1 = 2$, $a_4 = 2 + 1 = 3$, $a_5 = 3 + 2 = 5$, and so on.

d) As we can see by applying Newton's method, the statements $x_0 = 1$ and $x_{n+1} = x_n - [(\sin x_n - x_n^2)/(\cos x_n - 2x_n)]$ define a sequence that converges to a solution of the equation $\sin x - x^2 = 0$. ❑

Subsequences

If the terms of one sequence appear in another sequence in their given order, we call the first sequence a **subsequence** of the second.

EXAMPLE 5   *Subsequences of the sequence of positive integers*

a) The subsequence of even integers: $2, 4, 6, \ldots, 2n, \ldots$
b) The subsequence of odd integers: $1, 3, 5, \ldots, 2n - 1, \ldots$
c) The subsequence of primes: $2, 3, 5, 7, 11, \ldots$ ❑

Subsequences are important for two reasons:

1. If a sequence $\{a_n\}$ converges to L, then all of its subsequences converge to L. If we know that a sequence converges, it may be quicker to find or estimate its limit by examining a particular subsequence.

2. If any subsequence of a sequence $\{a_n\}$ diverges, or if two subsequences have different limits, then $\{a_n\}$ diverges. For example, the sequence $\{(-1)^n\}$ diverges because the subsequence $-1, -1, -1, \ldots$ of odd numbered terms converges to -1 while the subsequence $1, 1, 1, \ldots$ of even numbered terms converges to 1, a different limit.

Subsequences also provide a new way to view convergence. A **tail** of a sequence is a subsequence that consists of all terms of the sequence from some index N on. In other words, a tail is one of the sets $\{a_n \mid n \geq N\}$. Another way to say that $a_n \to L$ is to say that every ϵ-interval about L contains a tail of the sequence.

> The convergence or divergence of a sequence has nothing to do with how the sequence begins. It depends only on how the tails behave.

Bounded Nondecreasing Sequences

> **Definition**
>
> A sequence $\{a_n\}$ with the property that $a_n \leq a_{n+1}$ for all n is called a **nondecreasing sequence.**

EXAMPLE 6   *Nondecreasing sequences*

a) The sequence $1, 2, 3, \ldots, n, \ldots$ of natural numbers

b) The sequence $\dfrac{1}{2}, \dfrac{2}{3}, \dfrac{3}{4}, \ldots, \dfrac{n}{n+1}, \ldots$

c) The constant sequence $\{3\}$

There are two kinds of nondecreasing sequences—those whose terms increase beyond any finite bound and those whose terms do not.

> **Definitions**
>
> A sequence $\{a_n\}$ is **bounded from above** if there exists a number M such that $a_n \leq M$ for all n. The number M is an **upper bound** for $\{a_n\}$. If M is an upper bound for $\{a_n\}$ but no number less than M is an upper bound for $\{a_n\}$, then M is the **least upper bound** for $\{a_n\}$.

EXAMPLE 7

a) The sequence $1, 2, 3, \ldots, n, \ldots$ has no upper bound.

b) The sequence $\dfrac{1}{2}, \dfrac{2}{3}, \dfrac{3}{4}, \ldots, \dfrac{n}{n+1}, \ldots$ is bounded above by $M = 1$.
   No number less than 1 is an upper bound for the sequence, so 1 is the least upper bound (Exercise 47).

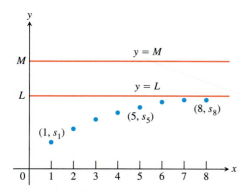

8.4 If the terms of a nondecreasing sequence have an upper bound M, they have a limit $L \le M$.

A nondecreasing sequence that is bounded from above always has a least upper bound. This fact is a consequence of the completeness property of real numbers but we will not prove it here. Instead, we will prove that if L is the least upper bound, then the sequence converges to L.

Suppose we plot the points $(1, s_1), (2, s_2), \ldots, (n, s_n), \ldots$ in the xy-plane. If M is an upper bound of the sequence, all these points will lie on or below the line $y = M$ (Fig. 8.4). The line $y = L$ is the lowest such line. None of the points (n, s_n) lies above $y = L$, but some do lie above any lower line $y = L - \epsilon$, if ϵ is a positive number. The sequence converges to L because

a) $s_n \le L$ for *all* values of n and

b) given any $\epsilon > 0$, there exists at least one integer N for which $s_N > L - \epsilon$.

The fact that $\{s_n\}$ is nondecreasing tells us further that

$$s_n \ge s_N > L - \epsilon \qquad \text{for all } n \ge N.$$

Thus, *all* the numbers s_n beyond the Nth number lie within ϵ of L. This is precisely the condition for L to be the limit of the sequence s_n.

The facts for nondecreasing sequences are summarized in the following theorem. A similar result holds for nonincreasing sequences (Exercise 41).

> **Theorem 1**
> **The Nondecreasing Sequence Theorem**
> A nondecreasing sequence of real numbers converges if and only if it is bounded from above. If a nondecreasing sequence converges, it converges to its least upper bound.

Exercises 8.1

Finding Terms of a Sequence

Each of Exercises 1–6 gives a formula for the nth term a_n of a sequence $\{a_n\}$. Find the values of $a_1, a_2, a_3,$ and a_4.

1. $a_n = \dfrac{1 - n}{n^2}$

2. $a_n = \dfrac{1}{n!}$

3. $a_n = \dfrac{(-1)^{n+1}}{2n - 1}$

4. $a_n = 2 + (-1)^n$

5. $a_n = \dfrac{2^n}{2^{n+1}}$

6. $a_n = \dfrac{2^n - 1}{2^n}$

Each of Exercises 7–12 gives the first term or two of a sequence along with a recursion formula for the remaining terms. Write out the first ten terms of the sequence.

7. $a_1 = 1, \quad a_{n+1} = a_n + (1/2^n)$

8. $a_1 = 1, \quad a_{n+1} = a_n/(n + 1)$

9. $a_1 = 2, \quad a_{n+1} = (-1)^{n+1} a_n/2$

10. $a_1 = -2, \quad a_{n+1} = n a_n/(n + 1)$

11. $a_1 = a_2 = 1, \quad a_{n+2} = a_{n+1} + a_n$

12. $a_1 = 2, \quad a_2 = -1, \quad a_{n+2} = a_{n+1}/a_n$

Finding a Sequence's Formula

In Exercises 13–22, find a formula for the nth term of the sequence.

13. The sequence $1, -1, 1, -1, 1, \ldots$ 1's with alternating signs

14. The sequence $-1, 1, -1, 1, -1, \ldots$ 1's with alternating signs

15. The sequence $1, -4, 9, -16, 25, \ldots$ Squares of the positive integers, with alternating signs

16. The sequence $1, -\dfrac{1}{4}, \dfrac{1}{9}, -\dfrac{1}{16}, \dfrac{1}{25}, \ldots$ Reciprocals of squares of the positive integers, with alternating signs

17. The sequence $0, 3, 8, 15, 24, \ldots$ Squares of the positive integers diminished by 1

18. The sequence $-3, -2, -1, 0, 1, \ldots$ Integers beginning with -3

19. The sequence $1, 5, 9, 13, 17, \ldots$ Every other odd positive integer

20. The sequence $2, 6, 10, 14, 18, \ldots$ Every other even positive integer

21. The sequence $1, 0, 1, 0, 1, \ldots$ Alternating 1's and 0's

22. The sequence $0, 1, 1, 2, 2, 3, 3, 4, \ldots$ Each positive integer repeated

Calculator Explorations of Limits

In Exercises 23–26, experiment with a calculator to find a value of N that will make the inequality hold for all $n > N$. Assuming that the inequality is the one from the formal definition of the limit of a sequence, what sequence is being considered in each case and what is its limit?

23. $|\sqrt[n]{0.5} - 1| < 10^{-3}$ 24. $|\sqrt[n]{n} - 1| < 10^{-3}$

25. $(0.9)^n < 10^{-3}$ 26. $2^n/n! < 10^{-7}$

27. *Sequences generated by Newton's method.* Newton's method, applied to a differentiable function $f(x)$, begins with a starting value x_0 and constructs from it a sequence of numbers $\{x_n\}$ that under favorable circumstances converges to a zero of f. The recursion formula for the sequence is

$$x_{n+1} = x_n - \frac{f(x_n)}{f'(x_n)}.$$

   a) Show that the recursion formula for $f(x) = x^2 - a$, $a > 0$, can be written as $x_{n+1} = (x_n + a/x_n)/2$.

   b) Starting with $x_0 = 1$ and $a = 3$, calculate successive terms of the sequence until the display begins to repeat. What number is being approximated? Explain.

28. (*Continuation of Exercise 27.*) Repeat part (b) of Exercise 27 with $a = 2$ in place of $a = 3$.

29. *A recursive definition of $\pi/2$.* If you start with $x_1 = 1$ and define the subsequent terms of $\{x_n\}$ by the rule $x_n = x_{n-1} + \cos x_{n-1}$, you generate a sequence that converges rapidly to $\pi/2$. (a) Try it. (b) Use the accompanying figure to explain why the convergence is so rapid.

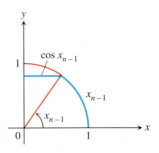

30. According to a front-page article in the December 15, 1992, issue of *The Wall Street Journal,* Ford Motor Company now uses about $7\frac{1}{4}$ hours of labor to produce stampings for the average vehicle, down from an estimated 15 hours in 1980. The Japanese need only about $3\frac{1}{2}$ hours.

    Ford's improvement since 1980 represents an average decrease of 6% per year. If that rate continues, then n years from now Ford will use about

$$S_n = 7.25(0.94)^n$$

hours of labor to produce stampings for the average vehicle. Assuming that the Japanese continue to spend $3\frac{1}{2}$ hours per vehicle, how many more years will it take Ford to catch up? Find out two ways:

   a) Find the first term of the sequence $\{S_n\}$ that is less than or equal to 3.5.

   b) **GRAPHER** Graph $f(x) = 7.25(0.94)^x$ and use TRACE to find where the graph crosses the line $y = 3.5$.

Theory and Examples

In Exercises 31–34, determine if the sequence is nondecreasing and if it is bounded from above.

31. $a_n = \dfrac{3n + 1}{n + 1}$ 32. $a_n = \dfrac{(2n + 3)!}{(n + 1)!}$

33. $a_n = \dfrac{2^n 3^n}{n!}$ 34. $a_n = 2 - \dfrac{2}{n} - \dfrac{1}{2^n}$

Which of the sequences in Exercises 35–40 converge, and which diverge? Give reasons for your answers.

35. $a_n = 1 - \dfrac{1}{n}$ 36. $a_n = n - \dfrac{1}{n}$

37. $a_n = \dfrac{2^n - 1}{2^n}$ 38. $a_n = \dfrac{2^n - 1}{3^n}$

39. $a_n = ((-1)^n + 1)\left(\dfrac{n + 1}{n}\right)$

40. The first term of a sequence is $x_1 = \cos(1)$. The next terms are $x_2 = x_1$ or $\cos(2)$, whichever is larger; and $x_3 = x_2$ or $\cos(3)$, whichever is larger (farther to the right). In general,

$$x_{n+1} = \max\{x_n, \cos(n + 1)\}.$$

41. *Nonincreasing sequences.* A sequence of numbers $\{a_n\}$ in which $a_n \geq a_{n+1}$ for every n is called a **nonincreasing sequence.** A sequence $\{a_n\}$ is **bounded from below** if there is a number M with $M \leq a_n$ for every n. Such a number M is called a **lower bound** for the sequence. Deduce from Theorem 1 that a nonincreasing sequence that is bounded from below converges and that a nonincreasing sequence that is not bounded from below diverges.

(*Continuation of Exercise 41.*) Using the conclusion of Exercise 41, determine which of the sequences in Exercises 42–46 converge and which diverge.

42. $a_n = \dfrac{n+1}{n}$

43. $a_n = \dfrac{1 + \sqrt{2n}}{\sqrt{n}}$

44. $a_n = \dfrac{1 - 4^n}{2^n}$

45. $a_n = \dfrac{4^{n+1} + 3^n}{4^n}$

46. $a_1 = 1, \quad a_{n+1} = 2a_n - 3$

47. *The sequence $\{n/(n+1)\}$ has a least upper bound of 1.* Show that if M is a number less than 1, then the terms of $\{n/(n+1)\}$ eventually exceed M. That is, if $M < 1$ there is an integer N such that $n/(n+1) > M$ whenever $n > N$. Since $n/(n+1) < 1$ for every n, this proves that 1 is a least upper bound for $\{n/(n+1)\}$.

48. *Uniqueness of least upper bounds.* Show that if M_1 and M_2 are least upper bounds for the sequence $\{a_n\}$, then $M_1 = M_2$. That is, a sequence cannot have two different least upper bounds.

49. Is it true that a sequence $\{a_n\}$ of positive numbers must converge if it is bounded from above? Give reasons for your answer.

50. Prove that if $\{a_n\}$ is a convergent sequence, then to every positive number ϵ there corresponds an integer N such that for all m and n,

$$m > N \quad \text{and} \quad n > N \implies |a_m - a_n| < \epsilon.$$

51. *Uniqueness of limits.* Prove that limits of sequences are unique. That is, show that if L_1 and L_2 are numbers such that $a_n \to L_1$ and $a_n \to L_2$, then $L_1 = L_2$.

52. *Limits and subsequences.* Prove that if two subsequences of a sequence $\{a_n\}$ have different limits $L_1 \neq L_2$, then $\{a_n\}$ diverges.

53. For a sequence $\{a_n\}$ the terms of even index are denoted by a_{2k} and the terms of odd index by a_{2k+1}. Prove that if $a_{2k} \to L$ and $a_{2k+1} \to L$, then $a_n \to L$.

54. Prove that a sequence $\{a_n\}$ converges to 0 if and only if the sequence of absolute values $\{|a_n|\}$ converges to 0.

✿ CAS Explorations and Projects

Use a CAS to perform the following steps for the sequences in Exercises 55–66.

a) Calculate and then plot the first 25 terms of the sequence. Does the sequence appear to be bounded from above or below? Does it appear to converge or diverge? If it does converge, what is the limit L?

b) If the sequence converges, find an integer N such that $|a_n - L| \leq 0.01$ for $n \geq N$. How far in the sequence do you have to get for the terms to lie within 0.0001 of L?

55. $a_n = \sqrt[n]{n}$

56. $a_n = \left(1 + \dfrac{0.5}{n}\right)^n$

57. $a_1 = 1, \quad a_{n+1} = a_n + \dfrac{1}{5^n}$

58. $a_1 = 1, \quad a_{n+1} = a_n + (-2)^n$

59. $a_n = \sin n$

60. $a_n = n \sin \dfrac{1}{n}$

61. $a_n = \dfrac{\sin n}{n}$

62. $a_n = \dfrac{\ln n}{n}$

63. $a_n = (0.9999)^n$

64. $a_n = 123456^{1/n}$

65. $a_n = \dfrac{8^n}{n!}$

66. $a_n = \dfrac{n^{41}}{19^n}$

67. *Compound interest, deposits, and withdrawals.* If you invest an amount of money A_0 at a fixed annual interest rate r compounded m times per year, and if the constant amount b is added to the account at the end of each compounding period (or taken from the account if $b < 0$), then the amount you have after $n+1$ compounding periods is

$$A_{n+1} = \left(1 + \frac{r}{m}\right) A_n + b. \tag{1}$$

a) If $A_0 = 1000$, $r = 0.02015$, $m = 12$, and $b = 50$, calculate and plot the first 100 points (n, A_n). How much money is in your account at the end of 5 years? Does $\{A_n\}$ converge? Is $\{A_n\}$ bounded?

b) Repeat part (a) with $A_0 = 5000$, $r = 0.0589$, $m = 12$, and $b = -50$.

c) If you invest 5000 dollars in a certificate of deposit (CD) that pays 4.5% annually, compounded quarterly, and you make no further investments in the CD, approximately how many years will it take before you have 20,000 dollars? What if the CD earns 6.25%?

d) It can be shown that for any $k \geq 0$, the sequence defined recursively by Eq. (1) satisfies the relation

$$A_k = \left(1 + \frac{r}{m}\right)^k \left(A_0 + \frac{mb}{r}\right) - \frac{mb}{r}. \tag{2}$$

For the values of the constants A_0, r, m, and b given in part (a), validate this assertion by comparing the values of the first 50 terms of both sequences. Then show by direct substitution that the terms in Eq. (2) satisfy the recursion formula (1).

68. *Logistic difference equation.* The recursive relation

$$a_{n+1} = r a_n (1 - a_n)$$

is called the **logistic difference equation,** and when the initial value a_0 is given the equation defines the **logistic sequence** $\{a_n\}$. Throughout this exercise we choose a_0 in the interval $0 < a_0 < 1$, say $a_0 = 0.3$.

a) Choose $r = 3/4$. Calculate and plot the points (n, a_n) for the first 100 terms in the sequence. Does it appear to converge? What do you guess is the limit? Does the limit seem to depend on your choice of a_0?

b) Choose several values of r in the interval $1 < r < 3$ and repeat the procedures in part (a). Be sure to choose some points near the endpoints of the interval. Describe the behavior of the sequences you observe in your plots.

c) Now examine the behavior of the sequence for values of r near the endpoints of the interval $3 < r < 3.45$. The transition value $r = 3$ is called a **bifurcation value** and the new behavior of the sequence in the interval is called an **attracting 2-cycle.** Explain why this reasonably describes the behavior.

d) Next explore the behavior for r values near the endpoints of each of the intervals $3.45 < r < 3.54$ and $3.54 < r < 3.55$. Plot the first 200 terms of the sequences. Describe in your own words the behavior observed in your plots for each interval. Among how many values does the sequence appear to oscillate for each interval? The values $r = 3.45$ and $r = 3.54$ (rounded to 2 decimal places) are also called bifurcation values because the behavior of the sequence changes as r crosses over those values.

e) The situation gets even more interesting. There is actually an increasing sequence of bifurcation values $3 < 3.45 < 3.54 < \cdots < c_n < c_{n+1} \cdots$ such that for $c_n < r < c_{n+1}$ the logistic sequence $\{a_n\}$ eventually oscillates steadily among 2^n values, called an **attracting 2^n-cycle.** Moreover, the bifurcation sequence $\{c_n\}$ is bounded above by 3.57 (so it converges). If you choose a value of $r < 3.57$ you will observe a 2^n-cycle of some sort. Choose $r = 3.5695$ and plot 300 points.

f) Let us see what happens when $r > 3.57$. Choose $r = 3.65$ and calculate and plot the first 300 terms of $\{a_n\}$. Observe how the terms wander around in an unpredictable, chaotic fashion. You cannot predict the value of a_{n+1} from the value of a_n.

g) For $r = 3.65$ choose two starting values of a_0 that are close together, say, $a_0 = 0.3$ and $a_0 = 0.301$. Calculate and plot the first 300 values of the sequences determined by each starting value. Compare the behaviors observed in your plots. How far out do you go before the corresponding terms of your two sequences appear to depart from each other? Repeat the exploration for $r = 3.75$. Can you see how the plots look different depending on your choice of a_0? We say that the logistic sequence is **sensitive to the initial condition** a_0.

8.2 Theorems for Calculating Limits of Sequences

The study of limits would be cumbersome if we had to answer every question about convergence by applying the definition. Fortunately, three theorems make this largely unnecessary. The first is a version of Theorem 1, Section 1.2.

Theorem 2

Let $\{a_n\}$ and $\{b_n\}$ be sequences of real numbers and let A and B be real numbers. The following rules hold if $\lim_{n\to\infty} a_n = A$ and $\lim_{n\to\infty} b_n = B$.

1. *Sum Rule:* $\lim_{n\to\infty} (a_n + b_n) = A + B$
2. *Difference Rule:* $\lim_{n\to\infty} (a_n - b_n) = A - B$
3. *Product Rule:* $\lim_{n\to\infty} (a_n \cdot b_n) = A \cdot B$
4. *Constant Multiple Rule:* $\lim_{n\to\infty} (k \cdot b_n) = k \cdot B$ (Any number k)
5. *Quotient Rule:* $\lim_{n\to\infty} \dfrac{a_n}{b_n} = \dfrac{A}{B}$ if $B \neq 0$

EXAMPLE 1   By combining Theorem 2 with the limit results in Example 2 of the preceding section, we have

$$\lim_{n\to\infty} \left(-\frac{1}{n} \right) = -1 \cdot \lim_{n\to\infty} \frac{1}{n} = -1 \cdot 0 = 0$$

$$\lim_{n\to\infty} \left(\frac{n-1}{n} \right) = \lim_{n\to\infty} \left(1 - \frac{1}{n} \right) = \lim_{n\to\infty} 1 - \lim_{n\to\infty} \frac{1}{n} = 1 - 0 = 1$$

$$\lim_{n\to\infty} \frac{5}{n^2} = 5 \cdot \lim_{n\to\infty} \frac{1}{n} \cdot \lim_{n\to\infty} \frac{1}{n} = 5 \cdot 0 \cdot 0 = 0$$

$$\lim_{n\to\infty} \frac{4 - 7n^6}{n^6 + 3} = \lim_{n\to\infty} \frac{(4/n^6) - 7}{1 + (3/n^6)} = \frac{0 - 7}{1 + 0} = -7.$$

One consequence of Theorem 2 is that every nonzero multiple of a divergent sequence $\{a_n\}$ diverges. For suppose, to the contrary, that $\{ca_n\}$ converges for some number $c \neq 0$. Then, by taking $k = 1/c$ in the Constant Multiple Rule in Theorem 2, we see that the sequence

$$\left\{ \frac{1}{c} \cdot ca_n \right\} = \{a_n\}$$

converges. Thus, $\{ca_n\}$ cannot converge unless $\{a_n\}$ also converges. If $\{a_n\}$ does not converge, then $\{ca_n\}$ does not converge.

The next theorem is the sequence version of the Sandwich Theorem in Section 1.2.

Theorem 3
The Sandwich Theorem for Sequences

Let $\{a_n\}$, $\{b_n\}$, and $\{c_n\}$ be sequences of real numbers. If $a_n \leq b_n \leq c_n$ holds for all n beyond some index N, and if $\lim_{n \to \infty} a_n = \lim_{n \to \infty} c_n = L$, then $\lim_{n \to \infty} b_n = L$ also.

An immediate consequence of Theorem 3 is that, if $|b_n| \leq c_n$ and $c_n \to 0$, then $b_n \to 0$ because $-c_n \leq b_n \leq c_n$. We use this fact in the next example.

EXAMPLE 2   Since $1/n \to 0$, we know that

a) $\dfrac{\cos n}{n} \to 0$ because $\left| \dfrac{\cos n}{n} \right| = \dfrac{|\cos n|}{n} \leq \dfrac{1}{n}$;

b) $\dfrac{1}{2^n} \to 0$ because $\dfrac{1}{2^n} \leq \dfrac{1}{n}$;

c) $(-1)^n \dfrac{1}{n} \to 0$ because $\left| (-1)^n \dfrac{1}{n} \right| \leq \dfrac{1}{n}$.   ❑

The application of Theorems 2 and 3 is broadened by a theorem stating that applying a continuous function to a convergent sequence produces a convergent sequence. We state the theorem without proof.

Theorem 4
The Continuous Function Theorem for Sequences

Let $\{a_n\}$ be a sequence of real numbers. If $a_n \to L$ and if f is a function that is continuous at L and defined at all a_n, then $f(a_n) \to f(L)$.

EXAMPLE 3   Show that $\sqrt{(n+1)/n} \to 1$.

Solution   We know that $(n+1)/n \to 1$. Taking $f(x) = \sqrt{x}$ and $L = 1$ in Theorem 4 gives $\sqrt{(n+1)/n} \to \sqrt{1} = 1$.   ❑

Technology *The Sequence $\{2^{1/n}\}$* What happens if you enter 2 in your calculator and take square roots repeatedly? The numbers form a sequence that appears to converge to 1, as suggested in the accompanying table. Try it for yourself.

n	$2^{1/n}$
2	1.4142 13562
4	1.1892 07115
8	1.0905 07733
64	1.0108 89286
256	1.0027 11275
1024	1.0006 77131
16384	1.0000 42307

What is happening in the table above? The sequence $\{1/n\}$ converges to 0. By taking $a_n = 1/n$, $f(x) = 2^x$, and $L = 0$ in Theorem 4, we see that $2^{1/n} = f(1/n) \to f(L) = 2^0 = 1$. Since the successive square roots of 2 form a subsequence $2^{1/2}, 2^{1/4}, 2^{1/8}, \ldots$ of $\{2^{1/n}\}$, the square roots must converge to 1 also (Fig. 8.5).

Using l'Hôpital's Rule

The next theorem enables us to use l'Hôpital's rule to find the limits of some sequences.

> ### Theorem 5
> Suppose that $f(x)$ is a function defined for all $x \geq n_0$ and that $\{a_n\}$ is a sequence of real numbers such that $a_n = f(n)$ for $n \geq n_0$. Then
> $$\lim_{x \to \infty} f(x) = L \quad \Rightarrow \quad \lim_{n \to \infty} a_n = L.$$

Proof Suppose that $\lim_{x \to \infty} f(x) = L$. Then for each positive number ϵ there is a number M such that for all x,
$$x > M \quad \Rightarrow \quad |f(x) - L| < \epsilon.$$

Let N be an integer greater than M and greater than or equal to n_0. Then
$$n > N \quad \Rightarrow \quad a_n = f(n) \quad \text{and} \quad |a_n - L| = |f(n) - L| < \epsilon. \quad \square$$

EXAMPLE 4 Show that $\lim_{n \to \infty} (\ln n)/n = 0$.

Solution The function $(\ln x)/x$ is defined for all $x \geq 1$ and agrees with the given sequence at positive integers. Therefore, by Theorem 5, $\lim_{n \to \infty} (\ln n)/n$ will equal $\lim_{x \to \infty} (\ln x)/x$ if the latter exists. A single application of l'Hôpital's rule shows that
$$\lim_{x \to \infty} \frac{\ln x}{x} = \lim_{x \to \infty} \frac{1/x}{1} = \frac{0}{1} = 0.$$

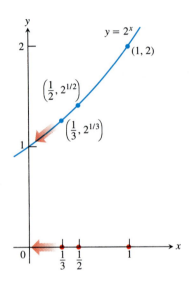

8.5 As $n \to \infty$, $1/n \to 0$ and $2^{1/n} \to 2^0$.

We conclude that $\lim_{n \to \infty} (\ln n)/n = 0$.   ❑

When we use l'Hôpital's rule to find the limit of a sequence, we often treat n as a continuous real variable and differentiate directly with respect to n. This saves us from having to rewrite the formula for a_n as we did in Example 4.

EXAMPLE 5   Find $\lim_{n \to \infty} (2^n/5n)$.

Solution   By l'Hôpital's rule,

$$\lim_{n \to \infty} \frac{2^n}{5n} = \lim_{n \to \infty} \frac{2^n \cdot \ln 2}{5}$$

$$= \infty.$$ ❑

Limits That Arise Frequently

The limits in Table 8.1 arise frequently. The first limit is from Example 4. The next two can be proved by taking logarithms and applying Theorem 4 (Exercises 71 and 72). The remaining proofs can be found in Appendix 6.

Table 8.1

1. $\lim_{n \to \infty} \dfrac{\ln n}{n} = 0$
2. $\lim_{n \to \infty} \sqrt[n]{n} = 1$
3. $\lim_{n \to \infty} x^{1/n} = 1 \quad (x > 0)$
4. $\lim_{n \to \infty} x^n = 0 \quad (|x| < 1)$
5. $\lim_{n \to \infty} \left(1 + \dfrac{x}{n}\right)^n = e^x \quad$ (Any x)
6. $\lim_{n \to \infty} \dfrac{x^n}{n!} = 0 \quad$ (Any x)

In formulas (3)–(6), x remains fixed as $n \to \infty$.

EXAMPLE 6   *Limits from Table 8.1*

1. $\dfrac{\ln (n^2)}{n} = \dfrac{2 \ln n}{n} \to 2 \cdot 0 = 0$ Formula 1

2. $\sqrt[n]{n^2} = n^{2/n} = (n^{1/n})^2 \to (1)^2 = 1$ Formula 2

3. $\sqrt[n]{3n} = 3^{1/n}(n^{1/n}) \to 1 \cdot 1 = 1$ Formula 3 with $x = 3$, and Formula 2

4. $\left(-\dfrac{1}{2}\right)^n \to 0$ Formula 4 with $x = -\dfrac{1}{2}$

5. $\left(\dfrac{n - 2}{n}\right)^n = \left(1 + \dfrac{-2}{n}\right)^n \to e^{-2}$ Formula 5 with $x = -2$

6. $\dfrac{100^n}{n!} \to 0$ Formula 6 with $x = 100$ ❑

EXAMPLE 7   Does the sequence whose nth term is

$$a_n = \left(\frac{n + 1}{n - 1}\right)^n$$

converge? If so, find $\lim_{n \to \infty} a_n$.

Solution   The limit leads to the indeterminate form 1^∞. We can apply l'Hôpital's rule if we first change the form to $\infty \cdot 0$ by taking the natural logarithm of a_n:

$$\ln a_n = \ln \left(\frac{n + 1}{n - 1}\right)^n$$

$$= n \ln \left(\frac{n + 1}{n - 1}\right).$$

Then,

$$\lim_{n\to\infty} \ln a_n = \lim_{n\to\infty} n \ln\left(\frac{n+1}{n-1}\right) \qquad \infty \cdot 0$$

$$= \lim_{n\to\infty} \frac{\ln\left(\dfrac{n+1}{n-1}\right)}{1/n} \qquad \frac{0}{0}$$

$$= \lim_{n\to\infty} \frac{-2/(n^2-1)}{-1/n^2} \qquad \text{l'Hôpital's rule}$$

$$= \lim_{n\to\infty} \frac{2n^2}{n^2-1} = 2.$$

Since $\ln a_n \to 2$, and $f(x) = e^x$ is continuous, Theorem 4 tells us that

$$a_n = e^{\ln a_n} \to e^2.$$

The sequence $\{a_n\}$ converges to e^2. ❑

✳ Picard's Method for Finding Roots

The problem of solving the equation

$$f(x) = 0 \tag{1}$$

is equivalent to that of solving the equation

$$g(x) = f(x) + x = x, \tag{2}$$

obtained by adding x to both sides of Eq. (1). By this simple change, we cast Eq. (1) into a form that may render it solvable on a computer by a powerful method called **Picard's method** (after the French mathematician Charles Émile Picard, 1856–1941).

If the domain of g contains the range of g, we can start with a point x_0 in the domain and apply g repeatedly to get

$$x_1 = g(x_0), \qquad x_2 = g(x_1), \qquad x_3 = g(x_2), \qquad \dots \tag{3}$$

Under simple restrictions that we will describe shortly, the sequence generated by the recursion formula $x_{n+1} = g(x_n)$ will converge to a point x for which $g(x) = x$. This point solves the equation $f(x) = 0$ because

$$f(x) = g(x) - x = x - x = 0. \tag{4}$$

A point x for which $g(x) = x$ is a **fixed point** of g. We see in Eq. (4) that the fixed points of g are precisely the roots of f.

EXAMPLE 8   *Testing the method*

Solve the equation

$$\frac{1}{4}x + 3 = x.$$

Solution   By algebra, we know that the solution is $x = 4$. To apply Picard's method, we take

$$g(x) = \frac{1}{4}x + 3,$$

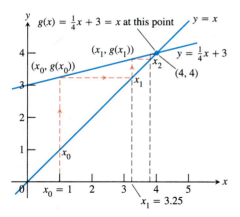

8.6 The Picard solution of the equation $g(x) = (1/4)x + 3 = x$ (Example 8).

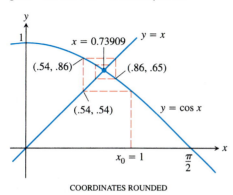

COORDINATES ROUNDED

8.7 The solution of $\cos x = x$ by Picard's method starting at $x_0 = 1$ (Example 9).

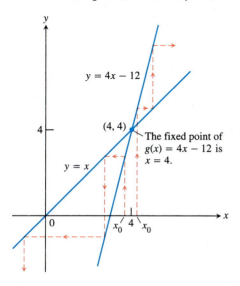

8.8 Applying the Picard method to $g(x) = 4x - 12$ will not find the fixed point unless x_0 is the fixed point 4 itself (Example 10).

choose a starting point, say $x_0 = 1$, and calculate the initial terms of the sequence $x_{n+1} = g(x_n)$. Table 8.2 lists the results. In 10 steps, the solution of the original equation is found with an error of magnitude less than 3×10^{-6}.

Figure 8.6 shows the geometry of the solution. We start with $x_0 = 1$ and calculate the first value $g(x_0)$. This becomes the second x-value x_1. The second y-value $g(x_1)$ becomes the third x-value x_2, and so on. The process is shown as a path (called the *iteration path*) that starts at $x_0 = 1$, moves up to $(x_0, g(x_0)) = (x_0, x_1)$, over to (x_1, x_1), up to $(x_1, g(x_1))$, and so on. The path converges to the point where the graph of g meets the line $y = x$. This is the point where $g(x) = x$. ❏

Table 8.2 Successive iterates of $g(x) = (1/4)x + 3$, starting with $x_0 = 1$

x_n	$x_{n+1} = g(x_n) = (1/4)x_n + 3$
$x_0 = 1$	$x_1 = g(x_0) = (1/4)(1) + 3 = 3.25$
$x_1 = 3.25$	$x_2 = g(x_1) = (1/4)(3.25) + 3 = 3.8125$
$x_2 = 3.8125$	$x_3 = g(x_2) = 3.9531\,25$
$x_3 = 3.9531\,25$	$x_4 = 3.9882\,8125$
$\vdots$	$x_5 = 3.9970\,70313$
	$x_6 = 3.9992\,67578$
	$x_7 = 3.9998\,16895$
	$x_8 = 3.9999\,54224$
	$x_9 = 3.9999\,88556$
	$x_{10} = 3.9999\,97139$
	$\vdots$

EXAMPLE 9 Solve the equation $\cos x = x$.

Solution We take $g(x) = \cos x$, choose $x_0 = 1$ as a starting value, and use the recursion formula $x_{n+1} = g(x_n)$ to find

$$x_0 = 1, \qquad x_1 = \cos 1, \qquad x_2 = \cos(x_1), \ldots.$$

We can approximate the first 50 terms or so on a calculator in radian mode by entering 1 and taking the cosine repeatedly. The display stops changing when $\cos x = x$ to the number of decimal places in the display.

Try it for yourself. As you continue to take the cosine, the successive approximations lie alternately above and below the fixed point $x = 0.739085133\ldots$.

Figure 8.7 shows that the values oscillate this way because the path of the procedure spirals around the fixed point. ❏

EXAMPLE 10 Picard's method will not solve the equation

$$g(x) = 4x - 12 = x.$$

As Fig. 8.8 shows, any choice of x_0 except $x_0 = 4$, the solution itself, generates a divergent sequence that moves away from the solution. ❏

The difficulty in Example 10 can be traced to the fact that the slope of the line $y = 4x - 12$ exceeds 1, the slope of the line $y = x$. Conversely, the process worked in Example 8 because the slope of the line $y = (1/4)x + 3$ was numerically less than 1. A theorem from advanced calculus tells us that if $g'(x)$ is continuous on a closed interval I whose interior contains a solution of the equation $g(x) = x$, and if $|g'(x)| < 1$ on I, then any choice of x_0 in the interior of I will lead to the solution. (See the introduction to Exercises 83 and 84 about what to do if $|g'(x)| > 1$.)

Exercises 8.2

Finding Limits

Which of the sequences $\{a_n\}$ in Exercises 1–62 converge, and which diverge? Find the limit of each convergent sequence.

1. $a_n = 2 + (0.1)^n$

2. $a_n = \dfrac{n + (-1)^n}{n}$

3. $a_n = \dfrac{1 - 2n}{1 + 2n}$

4. $a_n = \dfrac{2n + 1}{1 - 3\sqrt{n}}$

5. $a_n = \dfrac{1 - 5n^4}{n^4 + 8n^3}$

6. $a_n = \dfrac{n + 3}{n^2 + 5n + 6}$

7. $a_n = \dfrac{n^2 - 2n + 1}{n - 1}$

8. $a_n = \dfrac{1 - n^3}{70 - 4n^2}$

9. $a_n = 1 + (-1)^n$

10. $a_n = (-1)^n \left(1 - \dfrac{1}{n}\right)$

11. $a_n = \left(\dfrac{n + 1}{2n}\right)\left(1 - \dfrac{1}{n}\right)$

12. $a_n = \left(2 - \dfrac{1}{2^n}\right)\left(3 + \dfrac{1}{2^n}\right)$

13. $a_n = \dfrac{(-1)^{n+1}}{2n - 1}$

14. $a_n = \left(-\dfrac{1}{2}\right)^n$

15. $a_n = \sqrt{\dfrac{2n}{n + 1}}$

16. $a_n = \dfrac{1}{(0.9)^n}$

17. $a_n = \sin\left(\dfrac{\pi}{2} + \dfrac{1}{n}\right)$

18. $a_n = n\pi \cos(n\pi)$

19. $a_n = \dfrac{\sin n}{n}$

20. $a_n = \dfrac{\sin^2 n}{2^n}$

21. $a_n = \dfrac{n}{2^n}$

22. $a_n = \dfrac{3^n}{n^3}$

23. $a_n = \dfrac{\ln(n + 1)}{\sqrt{n}}$

24. $a_n = \dfrac{\ln n}{\ln 2n}$

25. $a_n = 8^{1/n}$

26. $a_n = (0.03)^{1/n}$

27. $a_n = \left(1 + \dfrac{7}{n}\right)^n$

28. $a_n = \left(1 - \dfrac{1}{n}\right)^n$

29. $a_n = \sqrt[n]{10n}$

30. $a_n = \sqrt[n]{n^2}$

31. $a_n = \left(\dfrac{3}{n}\right)^{1/n}$

32. $a_n = (n + 4)^{1/(n+4)}$

33. $a_n = \dfrac{\ln n}{n^{1/n}}$

34. $a_n = \ln n - \ln(n + 1)$

35. $a_n = \sqrt[n]{4^n n}$

36. $a_n = \sqrt[n]{3^{2n+1}}$

37. $a_n = \dfrac{n!}{n^n}$ (*Hint:* Compare with $1/n$.)

38. $a_n = \dfrac{(-4)^n}{n!}$

39. $a_n = \dfrac{n!}{10^{6n}}$

40. $a_n = \dfrac{n!}{2^n \cdot 3^n}$

41. $a_n = \left(\dfrac{1}{n}\right)^{1/(\ln n)}$

42. $a_n = \ln\left(1 + \dfrac{1}{n}\right)^n$

43. $a_n = \left(\dfrac{3n + 1}{3n - 1}\right)^n$

44. $a_n = \left(\dfrac{n}{n + 1}\right)^n$

45. $a_n = \left(\dfrac{x^n}{2n + 1}\right)^{1/n}$, $\quad x > 0$

46. $a_n = \left(1 - \dfrac{1}{n^2}\right)^n$

47. $a_n = \dfrac{3^n \cdot 6^n}{2^{-n} \cdot n!}$

48. $a_n = \dfrac{(10/11)^n}{(9/10)^n + (11/12)^n}$

49. $a_n = \tanh n$

50. $a_n = \sinh(\ln n)$

51. $a_n = \dfrac{n^2}{2n - 1} \sin \dfrac{1}{n}$

52. $a_n = n\left(1 - \cos \dfrac{1}{n}\right)$

53. $a_n = \tan^{-1} n$

54. $a_n = \dfrac{1}{\sqrt{n}} \tan^{-1} n$

55. $a_n = \left(\dfrac{1}{3}\right)^n + \dfrac{1}{\sqrt{2^n}}$

56. $a_n = \sqrt[n]{n^2 + n}$

57. $a_n = \dfrac{(\ln n)^{200}}{n}$

58. $a_n = \dfrac{(\ln n)^5}{\sqrt{n}}$

59. $a_n = n - \sqrt{n^2 - n}$

60. $a_n = \dfrac{1}{\sqrt{n^2 - 1} - \sqrt{n^2 + n}}$

61. $a_n = \dfrac{1}{n}\displaystyle\int_1^n \dfrac{1}{x}\,dx$

62. $a_n = \displaystyle\int_1^n \dfrac{1}{x^p}\,dx, \quad p > 1$

Theory and Examples

63. The first term of a sequence is $x_1 = 1$. Each succeeding term is the sum of all those that come before it:

$$x_{n+1} = x_1 + x_2 + \cdots + x_n.$$

Write out enough early terms of the sequence to deduce a general formula for x_n that holds for $n \geq 2$.

64. A sequence of rational numbers is described as follows:

$$\frac{1}{1}, \frac{3}{2}, \frac{7}{5}, \frac{17}{12}, \cdots, \frac{a}{b}, \frac{a+2b}{a+b}, \cdots.$$

Here the numerators form one sequence, the denominators form a second sequence, and their ratios form a third sequence. Let x_n and y_n be, respectively, the numerator and the denominator of the nth fraction $r_n = x_n/y_n$.

a) Verify that $x_1^2 - 2y_1^2 = -1$, $x_2^2 - 2y_2^2 = +1$ and, more generally, that if $a^2 - 2b^2 = -1$ or $+1$, then

$$(a + 2b)^2 - 2(a + b)^2 = +1 \quad \text{or} \quad -1,$$

respectively.

b) The fractions $r_n = x_n/y_n$ approach a limit as n increases. What is that limit? (*Hint:* Use part (a) to show that $r_n^2 - 2 = \pm(1/y_n)^2$ and that y_n is not less than n.)

65. *Newton's method.* The following sequences come from the recursion formula for Newton's method,

$$x_{n+1} = x_n - \frac{f(x_n)}{f'(x_n)}.$$

Do the sequences converge? If so, to what value? In each case, begin by identifying the function f that generates the sequence.

a) $x_0 = 1, \quad x_{n+1} = x_n - \dfrac{x_n^2 - 2}{2x_n} = \dfrac{x_n}{2} + \dfrac{1}{x_n}$

b) $x_0 = 1, \quad x_{n+1} = x_n - \dfrac{\tan x_n - 1}{\sec^2 x_n}$

c) $x_0 = 1, \quad x_{n+1} = x_n - 1$

66. a) Suppose that $f(x)$ is differentiable for all x in $[0, 1]$ and that $f(0) = 0$. Define the sequence $\{a_n\}$ by the rule $a_n = nf(1/n)$. Show that $\lim_{n\to\infty} a_n = f'(0)$.

Use the result in part (a) to find the limits of the following sequences $\{a_n\}$.

b) $a_n = n \tan^{-1} \dfrac{1}{n}$
      c) $a_n = n(e^{1/n} - 1)$

d) $a_n = n \ln \left(1 + \dfrac{2}{n}\right)$

67. *Pythagorean triples.* A triple of positive integers a, b, and c is called a **Pythagorean triple** if $a^2 + b^2 = c^2$. Let a be an odd positive integer and let

$$b = \left\lfloor \frac{a^2}{2} \right\rfloor \quad \text{and} \quad c = \left\lceil \frac{a^2}{2} \right\rceil$$

be, respectively, the integer floor and ceiling for $a^2/2$.

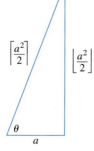

a) Show that $a^2 + b^2 = c^2$. (*Hint:* Let $a = 2n + 1$ and express b and c in terms of n.)

b) By direct calculation, or by appealing to the figure here, find

$$\lim_{a \to \infty} \frac{\left\lfloor \dfrac{a^2}{2} \right\rfloor}{\left\lceil \dfrac{a^2}{2} \right\rceil}.$$

68. *The nth root of n!*

a) Show that $\lim_{n\to\infty} (2n\pi)^{1/(2n)} = 1$ and hence, using Stirling's approximation (Chapter 7, Additional Exercise 50a), that

$$\sqrt[n]{n!} \approx \frac{n}{e} \quad \text{for large values of } n.$$

b) CALCULATOR   Test the approximation in (a) for $n = 40$, $50, 60, \ldots$, as far as your calculator will allow.

69. a) Assuming that $\lim_{n\to\infty} (1/n^c) = 0$ if c is any positive constant, show that

$$\lim_{n\to\infty} \frac{\ln n}{n^c} = 0$$

if c is any positive constant.

b) Prove that $\lim_{n\to\infty} (1/n^c) = 0$ if c is any positive constant. (*Hint:* If $\epsilon = 0.001$ and $c = 0.04$, how large should N be to ensure that $|1/n^c - 0| < \epsilon$ if $n > N$?)

70. *The zipper theorem.* Prove the "zipper theorem" for sequences: If $\{a_n\}$ and $\{b_n\}$ both converge to L, then the sequence

$$a_1, b_1, a_2, b_2, \ldots, a_n, b_n, \ldots$$

converges to L.

71. Prove that $\lim_{n\to\infty} \sqrt[n]{n} = 1$.

72. Prove that $\lim_{n\to\infty} x^{1/n} = 1$, $(x > 0)$.

73. Prove Theorem 3.

74. Prove Theorem 4.

❋ Picard's Method

CALCULATOR   Use Picard's method to solve the equations in Exercises 75–80.

75. $\sqrt{x} = x$
      76. $x^2 = x$

77. $\cos x + x = 0$
      78. $\cos x = x + 1$

79. $x - \sin x = 0.1$

80. $\sqrt{x} = 4 - \sqrt{1 + x}$ (*Hint:* Square both sides first.)

81. Solving the equation $\sqrt{x} = x$ by Picard's method finds the solution $x = 1$ but not the solution $x = 0$. Why? (*Hint:* Graph $y = x$ and $y = \sqrt{x}$ together.)

82. Solving the equation $x^2 = x$ by Picard's method with $|x_0| \neq 1$ can find the solution $x = 0$ but not the solution $x = 1$. Why? (*Hint:* Graph $y = x^2$ and $y = x$ together.)

Slope greater than 1. Example 10 showed that we cannot apply Picard's method to find a fixed point of $g(x) = 4x - 12$. But we can apply the method to find a fixed point of $g^{-1}(x) = (1/4)x + 3$ because the derivative of g^{-1} is $1/4$, whose value is less than 1 in magnitude on any interval. In Example 8, we found the fixed point of g^{-1} to be $x = 4$. Now notice that 4 is also a fixed point of g, since

$$g(4) = 4(4) - 12 = 4.$$

In finding the fixed point of g^{-1}, we found the fixed point of g.

A function and its inverse always have the same fixed points. The graphs of the functions are symmetric about the line $y = x$ and therefore intersect the line at the same points.

We now see that the application of Picard's method is quite broad. For suppose g is one-to-one, with a continuous first derivative whose magnitude is greater than 1 on a closed interval I whose interior contains a fixed point of g. Then the derivative of g^{-1}, being the reciprocal of g', has magnitude less than 1 on I. Picard's method applied to g^{-1} on I will find the fixed point of g. As cases in point, find the fixed points of the functions in Exercises 83 and 84.

83. $g(x) = 2x + 3$

84. $g(x) = 1 - 4x$

8.3 Infinite Series

In mathematics and science we often write functions as infinite polynomials, such as

$$\frac{1}{1-x} = 1 + x + x^2 + x^3 + \cdots + x^n + \cdots, \qquad |x| < 1,$$

(we will see the importance of doing so as the chapter continues). For any allowable value of x, we evaluate the polynomial as an infinite sum of constants, a sum we call an *infinite series*. The goal of this section and the next four is to familiarize ourselves with infinite series.

Series and Partial Sums

We begin by asking how to assign meaning to an expression like

$$1 + \frac{1}{2} + \frac{1}{4} + \frac{1}{8} + \frac{1}{16} + \cdots.$$

The way to do so is not to try to add all the terms at once (we cannot) but rather to add the terms one at a time from the beginning and look for a pattern in how these partial sums grow.

Partial sum		Value
first:	$s_1 = 1$	$2 - 1$
second:	$s_2 = 1 + \dfrac{1}{2}$	$2 - \dfrac{1}{2}$
third:	$s_3 = 1 + \dfrac{1}{2} + \dfrac{1}{4}$	$2 - \dfrac{1}{4}$
$\vdots$	$\vdots$	$\vdots$
nth:	$s_n = 1 + \dfrac{1}{2} + \dfrac{1}{4} + \cdots + \dfrac{1}{2^{n-1}}$	$2 - \dfrac{1}{2^{n-1}}$

Indeed there is a pattern. The partial sums form a sequence whose nth term is

$$s_n = 2 - \frac{1}{2^{n-1}}.$$

This sequence converges to 2 because $\lim_{n \to \infty} (1/2^n) = 0$. We say

"the sum of the infinite series $1 + \dfrac{1}{2} + \dfrac{1}{4} + \cdots + \dfrac{1}{2^{n-1}} + \cdots$ is 2."

Is the sum of any finite number of terms in this series equal to 2? No. Can we actually add an infinite number of terms one by one? No. But we can still define their sum by defining it to be the limit of the sequence of partial sums as $n \to \infty$, in this case 2 (Fig. 8.9). Our knowledge of sequences and limits enables us to break away from the confines of finite sums.

8.9 As the lengths 1, 1/2, 1/4, 1/8, ... are added one by one, the sum approaches 2.

Definitions

Given a sequence of numbers $\{a_n\}$, an expression of the form

$$a_1 + a_2 + a_3 + \cdots + a_n + \cdots$$

is an **infinite series.** The number a_n is the **nth term** of the series. The sequence $\{s_n\}$ defined by

$$s_1 = a_1$$

$$s_2 = a_1 + a_2$$

$$\vdots$$

$$s_n = a_1 + a_2 + \cdots + a_n = \sum_{k=1}^{n} a_k$$

$$\vdots$$

is the **sequence of partial sums** of the series, the number s_n being the **nth partial sum.** If the sequence of partial sums converges to a limit L, we say that the series **converges** and that its **sum** is L. In this case, we also write

$$a_1 + a_2 + \cdots + a_n + \cdots = \sum_{n=1}^{\infty} a_n = L.$$

If the sequence of partial sums of the series does not converge, we say that the series **diverges.**

When we begin to study a given series $a_1 + a_2 + \cdots + a_n + \cdots$, we might not know whether it converges or diverges. In either case, it is convenient to use sigma notation to write the series as

$$\sum_{n=1}^{\infty} a_n, \qquad \sum_{k=1}^{\infty} a_k, \qquad \text{or} \qquad \sum a_n$$

A useful shorthand when summation from 1 to ∞ is understood

Geometric Series

Geometric series are series of the form

$$a + ar + ar^2 + \cdots + ar^{n-1} + \cdots = \sum_{n=1}^{\infty} ar^{n-1} \qquad (1)$$

in which a and r are fixed real numbers and $a \neq 0$. The **ratio** r can be positive, as in

$$1 + \frac{1}{2} + \frac{1}{4} + \cdots + \left(\frac{1}{2}\right)^{n-1} + \cdots,$$

or negative, as in

$$1 - \frac{1}{3} + \frac{1}{9} - \cdots + \left(-\frac{1}{3}\right)^{n-1} + \cdots.$$

If $r = 1$, the nth partial sum of the series in (1) is

$$s_n = a + a(1) + a(1)^2 + \cdots + a(1)^{n-1} = na,$$

and the series diverges because $\lim_{n \to \infty} s_n = \pm \infty$, depending on the sign of a. If $r = -1$, the series diverges because the nth partial sums alternate between a and 0. If $|r| \neq 1$, we can determine the convergence or divergence of the series in the following way:

$$s_n = a + ar + ar^2 + \cdots + ar^{n-1}$$

$$rs_n = ar + ar^2 + \cdots + ar^{n-1} + ar^n \qquad \text{Multiply } s_n \text{ by } r.$$

$$s_n - rs_n = a - ar^n \qquad \begin{array}{l}\text{Subtract } rs_n \text{ from } s_n.\\ \text{Most of the terms on}\\ \text{the right cancel.}\end{array}$$

$$s_n(1 - r) = a(1 - r^n) \qquad \text{Factor.}$$

$$s_n = \frac{a(1 - r^n)}{1 - r}, \qquad (r \neq 1). \qquad \begin{array}{l}\text{We can solve for}\\ s_n \text{ if } r \neq 1.\end{array}$$

If $|r| < 1$, then $r^n \to 0$ as $n \to \infty$ (as in Section 8.2) and $s_n \to a/(1 - r)$. If $|r| > 1$, then $|r^n| \to \infty$ and the series diverges.

Equation (2) holds *only* if the summation begins with $n = 1$.

> If $|r| < 1$, the geometric series $a + ar + ar^2 + \cdots + ar^{n-1} + \cdots$ converges to $a/(1 - r)$:
>
> $$\sum_{n=1}^{\infty} ar^{n-1} = \frac{a}{1 - r}, \qquad |r| < 1. \qquad (2)$$
>
> If $|r| \geq 1$, the series diverges.

EXAMPLE 1   The geometric series with $a = 1/9$ and $r = 1/3$ is

$$\frac{1}{9} + \frac{1}{27} + \frac{1}{81} + \cdots = \sum_{n=1}^{\infty} \frac{1}{9}\left(\frac{1}{3}\right)^{n-1} = \frac{1/9}{1 - (1/3)} = \frac{1}{6}.$$

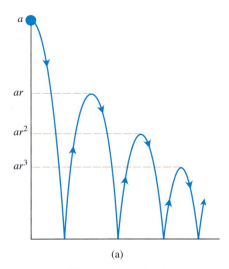

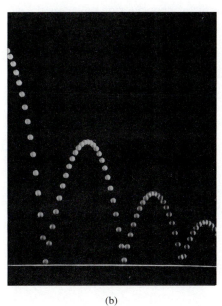

8.10 (a) Example 3 shows how to use a geometric series to calculate the total vertical distance traveled by a bouncing ball if the height of each rebound is reduced by the factor r. (b) A stroboscopic photo of a bouncing ball.

EXAMPLE 2 The series

$$\sum_{n=1}^{\infty} \frac{(-1)^n 5}{4^n} = -\frac{5}{4} + \frac{5}{16} - \frac{5}{64} + \cdots$$

is a geometric series with $a = -5/4$ and $r = -1/4$. It converges to

$$\frac{a}{1-r} = \frac{-5/4}{1 + (1/4)} = -1.$$

EXAMPLE 3 You drop a ball from a meters above a flat surface. Each time the ball hits the surface after falling a distance h, it rebounds a distance rh, where r is positive but less than 1. Find the total distance the ball travels up and down (Fig. 8.10).

Solution The total distance is

$$s = a + \underbrace{2ar + 2ar^2 + 2ar^3 + \cdots}_{\text{This sum is } 2ar/(1-r).} = a + \frac{2ar}{1-r} = a\frac{1+r}{1-r}.$$

If $a = 6$ m and $r = 2/3$, for instance, the distance is

$$s = 6\frac{1 + (2/3)}{1 - (2/3)} = 6\left(\frac{5/3}{1/3}\right) = 30\,\text{m}.$$

EXAMPLE 4 *Repeating decimals*

Express the repeating decimal 5.23 23 23 ... as the ratio of two integers.

Solution

$$5.23\ 23\ 23\ldots = 5 + \frac{23}{100} + \frac{23}{(100)^2} + \frac{23}{(100)^3} + \cdots$$

$$= 5 + \frac{23}{100}\underbrace{\left(1 + \frac{1}{100} + \left(\frac{1}{100}\right)^2 + \cdots\right)}_{1/(1-0.01)} \qquad a = 1,\ r = 1/100$$

$$= 5 + \frac{23}{100}\left(\frac{1}{0.99}\right) = 5 + \frac{23}{99} = \frac{518}{99}$$

Telescoping Series

Unfortunately, formulas like the one for the sum of a convergent geometric series are rare and we usually have to settle for an estimate of a series' sum (more about this later). The next example, however, is another case in which we can find the sum exactly.

EXAMPLE 5 Find the sum of the series $\displaystyle\sum_{n=1}^{\infty} \frac{1}{n(n+1)}$.

Solution We look for a pattern in the sequence of partial sums that might lead to

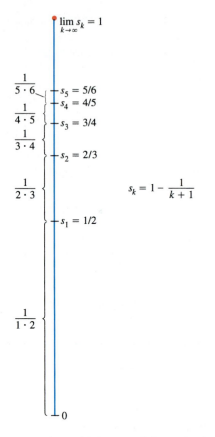

$\lim_{k\to\infty} s_k = 1$

$\frac{1}{5\cdot 6}$ — $s_5 = 5/6$
— $s_4 = 4/5$
$\frac{1}{4\cdot 5}$ — $s_3 = 3/4$
$\frac{1}{3\cdot 4}$ — $s_2 = 2/3$

$\frac{1}{2\cdot 3}$ — $s_k = 1 - \frac{1}{k+1}$

— $s_1 = 1/2$

$\frac{1}{1\cdot 2}$

0

8.11 The partial sums of the series in Example 5.

a formula for s_k. The key, as in the integration

$$\int \frac{dx}{x(x+1)} = \int \frac{dx}{x} - \int \frac{dx}{x+1},$$

is partial fractions. The observation that

$$\frac{1}{k(k+1)} = \frac{1}{k} - \frac{1}{k+1} \tag{3}$$

permits us to write the partial sum

$$\sum_{n=1}^{k} \frac{1}{n(n+1)} = \frac{1}{1\cdot 2} + \frac{1}{2\cdot 3} + \cdots + \frac{1}{k\cdot(k+1)}$$

as

$$s_k = \left(\frac{1}{1} - \frac{1}{2}\right) + \left(\frac{1}{2} - \frac{1}{3}\right) + \cdots + \left(\frac{1}{k} - \frac{1}{k+1}\right). \tag{4}$$

Removing parentheses and canceling the terms of opposite sign collapses the sum to

$$s_k = 1 - \frac{1}{k+1}. \tag{5}$$

We now see that $s_k \to 1$ as $k \to \infty$. The series converges, and its sum is 1 (Fig. 8.11).

$$\sum_{n=1}^{\infty} \frac{1}{n(n+1)} = 1.$$

❑

Divergent Series

Geometric series with $|r| \geq 1$ are not the only series to diverge.

EXAMPLE 6   The series

$$\sum_{n=1}^{\infty} n^2 = 1 + 4 + 9 + \cdots + n^2 + \cdots$$

diverges because the partial sums grow beyond every number L. After $n = 1$, the partial sum $s_n = 1 + 4 + 9 + \cdots + n^2$ is greater than n^2.   ❑

EXAMPLE 7   The series

$$\sum_{n=1}^{\infty} \frac{n+1}{n} = \frac{2}{1} + \frac{3}{2} + \frac{4}{3} + \cdots + \frac{n+1}{n} + \cdots$$

diverges because the partial sums eventually outgrow every preassigned number. Each term is greater than 1, so the sum of n terms is greater than n.   ❑

The nth-Term Test for Divergence

Observe that $\lim_{n\to\infty} a_n$ must equal zero if the series $\sum_{n=1}^{\infty} a_n$ converges. To see why, let S represent the series' sum and $s_n = a_1 + a_2 + \cdots + a_n$ the nth partial sum. When n is large, both s_n and s_{n-1} are close to S, so their difference, a_n, is close to zero. More formally,

$$a_n = s_n - s_{n-1} \quad \rightarrow \quad S - S = 0. \qquad \text{\color{blue}Difference Rule for sequences}$$

Caution

Theorem 6 *does not say* that $\sum_{n=1}^{\infty} a_n$ converges if $a_n \rightarrow 0$. It is possible for a series to diverge when $a_n \rightarrow 0$.

Theorem 6

If $\displaystyle\sum_{n=1}^{\infty} a_n$ converges, then $a_n \rightarrow 0$.

Theorem 6 leads to a test for detecting the kind of divergence that occurred in Examples 6–8.

The nth-Term Test for Divergence

$\displaystyle\sum_{n=1}^{\infty} a_n$ diverges if $\displaystyle\lim_{n\to\infty} a_n$ fails to exist or is different from zero.

EXAMPLE 8   In applying the nth-Term Test, we can see that

a) $\displaystyle\sum_{n=1}^{\infty} n^2$ diverges because $n^2 \rightarrow \infty$

b) $\displaystyle\sum_{n=1}^{\infty} \frac{n+1}{n}$ diverges because $\dfrac{n+1}{n} \rightarrow 1$

c) $\displaystyle\sum_{n=1}^{\infty} (-1)^{n+1}$ diverges because $\lim_{n\to\infty} (-1)^{n+1}$ does not exist

d) $\displaystyle\sum_{n=1}^{\infty} \frac{-n}{2n+5}$ diverges because $\lim_{n\to\infty} \dfrac{-n}{2n+5} = -\dfrac{1}{2} \neq 0$.   ❑

EXAMPLE 9   *$a_n \rightarrow 0$ but the series diverges*

The series

$$1 + \underbrace{\frac{1}{2} + \frac{1}{2}}_{\text{2 terms}} + \underbrace{\frac{1}{4} + \frac{1}{4} + \frac{1}{4} + \frac{1}{4}}_{\text{4 terms}} + \cdots + \underbrace{\frac{1}{2^n} + \frac{1}{2^n} + \cdots + \frac{1}{2^n}}_{2^n \text{ terms}} + \cdots$$

diverges even though its terms form a sequence that converges to 0.   ❑

Combining Series

Whenever we have two convergent series, we can add them term by term, subtract them term by term, or multiply them by constants to make new convergent series.

Theorem 7

If $\sum a_n = A$ and $\sum b_n = B$ are convergent series, then

1. *Sum Rule:* $\qquad\qquad\qquad \sum (a_n + b_n) = \sum a_n + \sum b_n = A + B$

2. *Difference Rule:* $\qquad\qquad \sum (a_n - b_n) = \sum a_n - \sum b_n = A - B$

3. *Constant Multiple Rule:* $\qquad \sum k a_n = k \sum a_n = kA \qquad$ (Any number k).

Proof The three rules for series follow from the analogous rules for sequences in Theorem 2, Section 8.2. To prove the Sum Rule for series, let

$$A_n = a_1 + a_2 + \cdots + a_n, \quad B_n = b_1 + b_2 + \cdots + b_n.$$

Then the partial sums of $\sum (a_n + b_n)$ are

$$S_n = (a_1 + b_1) + (a_2 + b_2) + \cdots + (a_n + b_n)$$
$$= (a_1 + \cdots + a_n) + (b_1 + \cdots + b_n)$$
$$= A_n + B_n.$$

Since $A_n \to A$ and $B_n \to B$, we have $S_n \to A + B$ by the Sum Rule for sequences. The proof of the Difference Rule is similar.

To prove the Constant Multiple Rule for series, observe that the partial sums of $\sum k a_n$ form the sequence

$$S_n = k a_1 + k a_2 + \cdots + k a_n = k(a_1 + a_2 + \cdots + a_n) = kA_n,$$

which converges to kA by the Constant Multiple Rule for sequences. $\qquad \square$

As corollaries of Theorem 7, we have

1. Every nonzero constant multiple of a divergent series diverges.

2. If $\sum a_n$ converges and $\sum b_n$ diverges, then $\sum (a_n + b_n)$ and $\sum (a_n - b_n)$ both diverge.

We omit the proofs.

EXAMPLE 10 Find the sums of the following series.

a) $\displaystyle \sum_{n=1}^{\infty} \frac{3^{n-1} - 1}{6^{n-1}} = \sum_{n=1}^{\infty} \left(\frac{1}{2^{n-1}} - \frac{1}{6^{n-1}} \right)$

$\qquad\qquad\qquad = \displaystyle \sum_{n=1}^{\infty} \frac{1}{2^{n-1}} - \sum_{n=1}^{\infty} \frac{1}{6^{n-1}} \qquad$ Difference Rule

$\qquad\qquad\qquad = \dfrac{1}{1 - (1/2)} - \dfrac{1}{1 - (1/6)} \qquad$ Geometric series with $a = 1$ and $r = 1/2, 1/6$

$\qquad\qquad\qquad = 2 - \dfrac{6}{5}$

$\qquad\qquad\qquad = \dfrac{4}{5}$

b) $\displaystyle\sum_{n=1}^{\infty} \frac{4}{2^{n-1}} = 4 \sum_{n=1}^{\infty} \frac{1}{2^{n-1}}$ Constant Multiple Rule

$\displaystyle = 4 \left(\frac{1}{1 - (1/2)} \right)$ Geometric series with $a = 1$, $r = 1/2$

$\displaystyle = 8$ ❏

Adding or Deleting Terms

We can always add a finite number of terms to a series or delete a finite number of terms without altering the series' convergence or divergence, although in the case of convergence this will usually change the sum. If $\sum_{n=1}^{\infty} a_n$ converges, then $\sum_{n=k}^{\infty} a_n$ converges for any $k > 1$ and

$$\sum_{n=1}^{\infty} a_n = a_1 + a_2 + \cdots + a_{k-1} + \sum_{n=k}^{\infty} a_n. \tag{6}$$

Conversely, if $\sum_{n=k}^{\infty} a_n$ converges for any $k > 1$, then $\sum_{n=1}^{\infty} a_n$ converges. Thus,

$$\sum_{n=1}^{\infty} \frac{1}{5^n} = \frac{1}{5} + \frac{1}{25} + \frac{1}{125} + \sum_{n=4}^{\infty} \frac{1}{5^n} \tag{7}$$

and

$$\sum_{n=4}^{\infty} \frac{1}{5^n} = \left(\sum_{n=1}^{\infty} \frac{1}{5^n} \right) - \frac{1}{5} - \frac{1}{25} - \frac{1}{125}. \tag{8}$$

Reindexing

As long as we preserve the order of its terms, we can reindex any series without altering its convergence. To raise the starting value of the index h units, replace the n in the formula for a_n by $n - h$:

$$\sum_{n=1}^{\infty} a_n = \sum_{n=1+h}^{\infty} a_{n-h} = a_1 + a_2 + a_3 + \cdots .$$

To lower the starting value of the index h units, replace the n in the formula for a_n by $n + h$:

$$\sum_{n=1}^{\infty} a_n = \sum_{n=1-h}^{\infty} a_{n+h} = a_1 + a_2 + a_3 + \cdots .$$

It works like a horizontal shift.

EXAMPLE 11   We can write the geometric series that starts with

$$1 + \frac{1}{2} + \frac{1}{4} + \cdots$$

as

$$\sum_{n=0}^{\infty} \frac{1}{2^n}, \qquad \sum_{n=5}^{\infty} \frac{1}{2^{n-5}}, \qquad \text{or even} \qquad \sum_{n=-4}^{\infty} \frac{1}{2^{n+4}}.$$

The partial sums remain the same no matter what indexing we choose.   ❏

We usually give preference to indexings that lead to simple expressions.

Exercises 8.3

Finding nth Partial Sums

In Exercises 1–6, find a formula for the nth partial sum of each series and use it to find the series' sum if the series converges.

1. $2 + \dfrac{2}{3} + \dfrac{2}{9} + \dfrac{2}{27} + \cdots + \dfrac{2}{3^{n-1}} + \cdots$

2. $\dfrac{9}{100} + \dfrac{9}{100^2} + \dfrac{9}{100^3} + \cdots + \dfrac{9}{100^n} + \cdots$

3. $1 - \dfrac{1}{2} + \dfrac{1}{4} - \dfrac{1}{8} + \cdots + (-1)^{n-1}\dfrac{1}{2^{n-1}} + \cdots$

4. $1 - 2 + 4 - 8 + \cdots + (-1)^{n-1}2^{n-1} + \cdots$

5. $\dfrac{1}{2 \cdot 3} + \dfrac{1}{3 \cdot 4} + \dfrac{1}{4 \cdot 5} + \cdots + \dfrac{1}{(n+1)(n+2)} + \cdots$

6. $\dfrac{5}{1 \cdot 2} + \dfrac{5}{2 \cdot 3} + \dfrac{5}{3 \cdot 4} + \cdots + \dfrac{5}{n(n+1)} + \cdots$

Series with Geometric Terms

In Exercises 7–14, write out the first few terms of each series to show how the series starts. Then find the sum of the series.

7. $\displaystyle\sum_{n=0}^{\infty} \dfrac{(-1)^n}{4^n}$

8. $\displaystyle\sum_{n=2}^{\infty} \dfrac{1}{4^n}$

9. $\displaystyle\sum_{n=1}^{\infty} \dfrac{7}{4^n}$

10. $\displaystyle\sum_{n=0}^{\infty} (-1)^n \dfrac{5}{4^n}$

11. $\displaystyle\sum_{n=0}^{\infty} \left(\dfrac{5}{2^n} + \dfrac{1}{3^n}\right)$

12. $\displaystyle\sum_{n=0}^{\infty} \left(\dfrac{5}{2^n} - \dfrac{1}{3^n}\right)$

13. $\displaystyle\sum_{n=0}^{\infty} \left(\dfrac{1}{2^n} + \dfrac{(-1)^n}{5^n}\right)$

14. $\displaystyle\sum_{n=0}^{\infty} \left(\dfrac{2^{n+1}}{5^n}\right)$

Telescoping Series

Use partial fractions to find the sum of each series in Exercises 15–22.

15. $\displaystyle\sum_{n=1}^{\infty} \dfrac{4}{(4n-3)(4n+1)}$

16. $\displaystyle\sum_{n=1}^{\infty} \dfrac{6}{(2n-1)(2n+1)}$

17. $\displaystyle\sum_{n=1}^{\infty} \dfrac{40n}{(2n-1)^2(2n+1)^2}$

18. $\displaystyle\sum_{n=1}^{\infty} \dfrac{2n+1}{n^2(n+1)^2}$

19. $\displaystyle\sum_{n=1}^{\infty} \left(\dfrac{1}{\sqrt{n}} - \dfrac{1}{\sqrt{n+1}}\right)$

20. $\displaystyle\sum_{n=1}^{\infty} \left(\dfrac{1}{2^{1/n}} - \dfrac{1}{2^{1/(n+1)}}\right)$

21. $\displaystyle\sum_{n=1}^{\infty} \left(\dfrac{1}{\ln(n+2)} - \dfrac{1}{\ln(n+1)}\right)$

22. $\displaystyle\sum_{n=1}^{\infty} (\tan^{-1}(n) - \tan^{-1}(n+1))$

Convergence or Divergence

Which series in Exercises 23–40 converge, and which diverge? Give reasons for your answers. If a series converges, find its sum.

23. $\displaystyle\sum_{n=0}^{\infty} \left(\dfrac{1}{\sqrt{2}}\right)^n$

24. $\displaystyle\sum_{n=0}^{\infty} (\sqrt{2})^n$

25. $\displaystyle\sum_{n=1}^{\infty} (-1)^{n+1}\dfrac{3}{2^n}$

26. $\displaystyle\sum_{n=1}^{\infty} (-1)^{n+1}n$

27. $\displaystyle\sum_{n=0}^{\infty} \cos n\pi$

28. $\displaystyle\sum_{n=0}^{\infty} \dfrac{\cos n\pi}{5^n}$

29. $\displaystyle\sum_{n=0}^{\infty} e^{-2n}$

30. $\displaystyle\sum_{n=1}^{\infty} \ln \dfrac{1}{n}$

31. $\displaystyle\sum_{n=1}^{\infty} \dfrac{2}{10^n}$

32. $\displaystyle\sum_{n=0}^{\infty} \dfrac{1}{x^n}, \quad |x| > 1$

33. $\displaystyle\sum_{n=0}^{\infty} \dfrac{2^n - 1}{3^n}$

34. $\displaystyle\sum_{n=1}^{\infty} \left(1 - \dfrac{1}{n}\right)^n$

35. $\displaystyle\sum_{n=0}^{\infty} \dfrac{n!}{1000^n}$

36. $\displaystyle\sum_{n=1}^{\infty} \dfrac{n^n}{n!}$

37. $\displaystyle\sum_{n=1}^{\infty} \ln \left(\dfrac{n}{n+1}\right)$

38. $\displaystyle\sum_{n=1}^{\infty} \ln \left(\dfrac{n}{2n+1}\right)$

39. $\displaystyle\sum_{n=0}^{\infty} \left(\dfrac{e}{\pi}\right)^n$

40. $\displaystyle\sum_{n=0}^{\infty} \dfrac{e^{n\pi}}{\pi^{ne}}$

Geometric Series

In each of the geometric series in Exercises 41–44, write out the first few terms of the series to find a and r, and find the sum of the series. Then express the inequality $|r| < 1$ in terms of x and find the values of x for which the inequality holds and the series converges.

41. $\displaystyle\sum_{n=0}^{\infty} (-1)^n x^n$

42. $\displaystyle\sum_{n=0}^{\infty} (-1)^n x^{2n}$

43. $\displaystyle\sum_{n=0}^{\infty} 3\left(\dfrac{x-1}{2}\right)^n$

44. $\displaystyle\sum_{n=0}^{\infty} \dfrac{(-1)^n}{2}\left(\dfrac{1}{3 + \sin x}\right)^n$

In Exercises 45–50, find the values of x for which the given geometric series converges. Also, find the sum of the series (as a function of x) for those values of x.

45. $\displaystyle\sum_{n=0}^{\infty} 2^n x^n$

46. $\displaystyle\sum_{n=0}^{\infty} (-1)^n x^{-2n}$

47. $\displaystyle\sum_{n=0}^{\infty} (-1)^n (x+1)^n$

48. $\displaystyle\sum_{n=0}^{\infty} \left(-\dfrac{1}{2}\right)^n (x-3)^n$

49. $\displaystyle\sum_{n=0}^{\infty} \sin^n x$

50. $\displaystyle\sum_{n=0}^{\infty} (\ln x)^n$

Repeating Decimals

Express each of the numbers in Exercises 51–58 as the ratio of two integers.

51. $0.\overline{23} = 0.23\ 23\ 23\ \ldots$

52. $0.\overline{234} = 0.234\ 234\ 234\ \ldots$

53. $0.\overline{7} = 0.7777\ldots$

54. $0.\overline{d} = 0.dddd\ldots,$ where d is a digit

55. $0.0\overline{6} = 0.06666\ldots$

56. $1.\overline{414} = 1.414\ 414\ 414\ \ldots$

57. $1.24\overline{123} = 1.24\ 123\ 123\ 123\ \ldots$

58. $3.\overline{142857} = 3.142857\ 142857\ \ldots$

Theory and Examples

59. The series in Exercise 5 can also be written as

$$\sum_{n=1}^{\infty} \frac{1}{(n+1)(n+2)} \quad \text{and} \quad \sum_{n=-1}^{\infty} \frac{1}{(n+3)(n+4)}.$$

Write it as a sum beginning with (a) $n = -2$, (b) $n = 0$, (c) $n = 5$.

60. The series in Exercise 6 can also be written as

$$\sum_{n=1}^{\infty} \frac{5}{n(n+1)} \quad \text{and} \quad \sum_{n=0}^{\infty} \frac{5}{(n+1)(n+2)}.$$

Write it as a sum beginning with (a) $n = -1$, (b) $n = 3$, (c) $n = 20$.

61. Make up an infinite series of nonzero terms whose sum is

a) 1 **b)** -3 **c)** 0.

Can you make an infinite series of nonzero terms that converges to any number you want? Explain.

62. Make up an example of two divergent infinite series whose term-by-term sum converges.

63. Show by example that $\sum(a_n/b_n)$ may diverge even though $\sum a_n$ and $\sum b_n$ converge and no b_n equals 0.

64. Find convergent geometric series $A = \sum a_n$ and $B = \sum b_n$ that illustrate the fact that $\sum a_n b_n$ may converge without being equal to AB.

65. Show by example that $\sum(a_n/b_n)$ may converge to something other than A/B even when $A = \sum a_n$, $B = \sum b_n \neq 0$, and no b_n equals 0.

66. If $\sum a_n$ converges and $a_n > 0$ for all n, can anything be said about $\sum(1/a_n)$? Give reasons for your answer.

67. What happens if you add a finite number of terms to a divergent series or delete a finite number of terms from a divergent series? Give reasons for your answer.

68. If $\sum a_n$ converges and $\sum b_n$ diverges, can anything be said about their term-by-term sum $\sum(a_n + b_n)$? Give reasons for your answer.

69. Make up a geometric series $\sum ar^{n-1}$ that converges to the number 5 if

a) $a = 2$ **b)** $a = 13/2.$

70. Find the value of b for which

$$1 + e^b + e^{2b} + e^{3b} + \cdots = 9.$$

71. For what values of r does the infinite series

$$1 + 2r + r^2 + 2r^3 + r^4 + 2r^5 + r^6 + \cdots$$

converge? Find the sum of the series when it converges.

72. Show that the error $(L - s_n)$ obtained by replacing a convergent geometric series with one of its partial sums s_n is $ar^n/(1-r)$.

73. A ball is dropped from a height of 4 m. Each time it strikes the pavement after falling from a height of h meters it rebounds to a height of $0.75h$ meters. Find the total distance the ball travels up and down.

74. (*Continuation of Exercise 73.*) Find the total number of seconds the ball in Exercise 73 is traveling. (*Hint:* The formula $s = 4.9t^2$ gives $t = \sqrt{s/4.9}$.)

75. The accompanying figure shows the first five of a sequence of squares. The outermost square has an area of $4\ \text{m}^2$. Each of the other squares is obtained by joining the midpoints of the sides of the squares before it. Find the sum of the areas of all the squares.

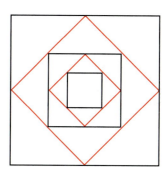

76. The accompanying figure shows the first three rows and part of the fourth row of a sequence of rows of semicircles. There are 2^n semicircles in the nth row, each of radius $1/2^n$. Find the sum of the areas of all the semicircles.

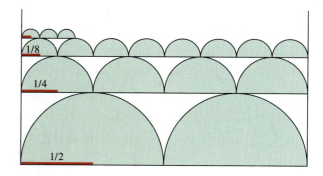

77. *Helga von Koch's snowflake curve.* Helga von Koch's snowflake (p. 167) is a curve of infinite length that encloses a region of finite area. To see why this is so, suppose the curve is generated by starting with an equilateral triangle whose sides have length 1.

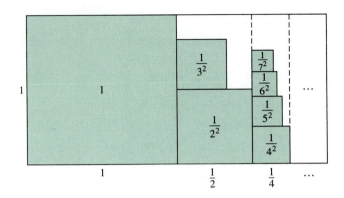

   a) Find the length L_n of the nth curve C_n and show that $\lim_{n \to \infty} L_n = \infty$.

   b) Find the area A_n of the region enclosed by C_n and calculate $\lim_{n \to \infty} A_n$.

78. The accompanying figure provides an informal proof that $\sum_{n=1}^{\infty} (1/n^2)$ is less than 2. Explain what is going on. (Source: "Convergence with Pictures" by P. J. Rippon, *American Mathematical Monthly*, Vol. 93, No. 6, 1986, pp. 476–78.)

<table>
<tr><td>**8.4**</td><td></td></tr>
</table>

The Integral Test for Series of Nonnegative Terms

Given a series $\sum a_n$, we have two questions:

1. Does the series converge?
2. If it converges, what is its sum?

Much of the rest of this chapter is devoted to the first question. But as a practical matter, the second question is just as important, and we will return to it later.

    In this section and the next two, we study series that do not have negative terms. The reason for this restriction is that the partial sums of these series form nondecreasing sequences, and nondecreasing sequences that are bounded from above always converge (Theorem 1, Section 8.1). To show that a series of nonnegative terms converges, we need only show that its partial sums are bounded from above.

    It may at first seem to be a drawback that this approach establishes the fact of convergence without producing the sum of the series in question. Surely it would be better to compute sums of series directly from formulas for their partial sums. But in most cases such formulas are not available, and in their absence we have to turn instead to the two-step procedure of first establishing convergence and then approximating the sum.

Nondecreasing Partial Sums

Suppose that $\sum_{n=1}^{\infty} a_n$ is an infinite series with $a_n \geq 0$ for all n. Then each partial sum is greater than or equal to its predecessor because $s_{n+1} = s_n + a_n$:

$$s_1 \leq s_2 \leq s_3 \leq \cdots \leq s_n \leq s_{n+1} \leq \cdots .$$

Since the partial sums form a nondecreasing sequence, the Nondecreasing Sequence Theorem (Theorem 1, Section 8.1) tells us that the series will converge if and only if the partial sums are bounded from above.

Corollary of Theorem 1

A series $\sum_{n=1}^{\infty} a_n$ of nonnegative terms converges if and only if its partial sums are bounded from above.

Caution

Notice that the nth-Term Test for divergence does not detect the divergence of the harmonic series. The nth term, $1/n$, goes to zero, but the series still diverges.

Nicole Oresme (1320–1382)

The argument we use to show the divergence of the harmonic series was devised by the French theologian, mathematician, physicist, and bishop Nicole Oresme (pronounced "orem"). Oresme was a vigorous opponent of astrology, a dynamic preacher, an adviser of princes, a friend of King Charles V, a popularizer of science, and a skillful translator of Latin into French.

Oresme did not believe in Albert of Saxony's generally accepted model of free fall (Chapter 6, Additional Exercise 26) but preferred Aristotle's constant-acceleration model, the model that became popular among Oxford scholars in the 1330s and that Galileo eventually used three hundred years later.

EXAMPLE 1   *The harmonic series*

The series

$$\sum_{n=1}^{\infty} \frac{1}{n} = 1 + \frac{1}{2} + \frac{1}{3} + \cdots + \frac{1}{n} + \cdots$$

is called the **harmonic series**. It diverges because there is no upper bound for its partial sums. To see why, group the terms of the series in the following way:

$$1 + \frac{1}{2} + \underbrace{\left(\frac{1}{3} + \frac{1}{4}\right)}_{> \frac{2}{4} = \frac{1}{2}} + \underbrace{\left(\frac{1}{5} + \frac{1}{6} + \frac{1}{7} + \frac{1}{8}\right)}_{> \frac{4}{8} = \frac{1}{2}} + \underbrace{\left(\frac{1}{9} + \frac{1}{10} + \cdots + \frac{1}{16}\right)}_{> \frac{8}{16} = \frac{1}{2}} + \cdots.$$

The sum of the first two terms is 1.5. The sum of the next two terms is $1/3 + 1/4$, which is greater than $1/4 + 1/4 = 1/2$. The sum of the next four terms is $1/5 + 1/6 + 1/7 + 1/8$, which is greater than $1/8 + 1/8 + 1/8 + 1/8 = 1/2$. The sum of the next eight terms is $1/9 + 1/10 + 1/11 + 1/12 + 1/13 + 1/14 + 1/15 + 1/16$, which is greater than $8/16 = 1/2$. The sum of the next 16 terms is greater than $16/32 = 1/2$, and so on. In general, the sum of 2^n terms ending with $1/2^{n+1}$ is greater than $2^n/2^{n+1} = 1/2$. The sequence of partial sums is not bounded from above: If $n = 2^k$, the partial sum s_n is greater than $k/2$. The harmonic series diverges. ❑

The Integral Test

We introduce the Integral Test with a series that is related to the harmonic series, but whose nth term is $1/n^2$ instead of $1/n$.

EXAMPLE 2   Does the following series converge?

$$\sum_{n=1}^{\infty} \frac{1}{n^2} = 1 + \frac{1}{4} + \frac{1}{9} + \frac{1}{16} + \cdots + \frac{1}{n^2} + \cdots \tag{1}$$

Solution   We determine the convergence of $\sum_{n=1}^{\infty}(1/n^2)$ by comparing it with $\int_{1}^{\infty}(1/x^2)\,dx$. To carry out the comparison, we think of the terms of the series as values of the function $f(x) = 1/x^2$ and interpret these values as the areas of rectangles under the curve $y = 1/x^2$.

As Fig. 8.12 shows,

$$s_n = \frac{1}{1^2} + \frac{1}{2^2} + \frac{1}{3^2} + \cdots + \frac{1}{n^2}$$

$$= f(1) + f(2) + f(3) + \cdots + f(n)$$

$$< f(1) + \int_{1}^{n} \frac{1}{x^2}\,dx$$

$$< 1 + \int_{1}^{\infty} \frac{1}{x^2}\,dx$$

$$< 1 + 1 = 2. \qquad \text{As in Section 7.6, Example 8,}$$
$$\int_{1}^{\infty}(1/x^2)\,dx = 1.$$

Thus the partial sums of $\sum_{n=1}^{\infty} 1/n^2$ are bounded from above (by 2) and the series converges. The sum of the series is known to be $\pi^2/6 \approx 1.64493$. ❑

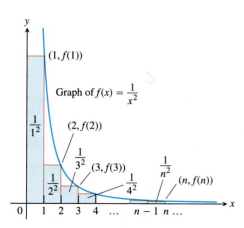

$(1, f(1))$

Graph of $f(x) = \dfrac{1}{x^2}$

$\dfrac{1}{1^2}$ $(2, f(2))$

$\dfrac{1}{2^2}$ $\dfrac{1}{3^2}$ $(3, f(3))$ $\dfrac{1}{4^2}$ $\dfrac{1}{n^2}$ $(n, f(n))$

$0 \quad 1 \quad 2 \quad 3 \quad 4 \quad \cdots \quad n-1 \; n \; \cdots$

8.12 Figure for the area comparisons in Example 2.

The Integral Test

Let $\{a_n\}$ be a sequence of positive terms. Suppose that $a_n = f(n)$, where f is a continuous, positive, decreasing function of x for all $x \geq N$ (N a positive integer). Then the series $\sum_{n=N}^{\infty} a_n$ and the integral $\int_N^{\infty} f(x)\, dx$ both converge or both diverge.

Proof We establish the test for the case $N = 1$. The proof for general N is similar.

We start with the assumption that f is a decreasing function with $f(n) = a_n$ for every n. This leads us to observe that the rectangles in Fig. 8.13(a), which have areas $a_1, a_2, \ldots, a_n$, collectively enclose more area than that under the curve $y = f(x)$ from $x = 1$ to $x = n + 1$. That is,

$$\int_1^{n+1} f(x)\, dx \leq a_1 + a_2 + \cdots + a_n.$$

Caution

The series and integral need not have the same value in the convergent case. As we saw in Example 2, $\sum_{n=1}^{\infty} (1/n^2) = \pi^2/6$ while $\int_1^{\infty} (1/x^2)\, dx = 1$.

In Fig. 8.13(b) the rectangles have been faced to the left instead of to the right. If we momentarily disregard the first rectangle, of area a_1, we see that

$$a_2 + a_3 + \cdots + a_n \leq \int_1^n f(x)\, dx.$$

If we include a_1, we have

$$a_1 + a_2 + \cdots + a_n \leq a_1 + \int_1^n f(x)\, dx.$$

Combining these results gives

$$\int_1^{n+1} f(x)\, dx \leq a_1 + a_2 + \cdots + a_n \leq a_1 + \int_1^n f(x)\, dx. \tag{2}$$

If $\int_1^{\infty} f(x)\, dx$ is finite, the right-hand inequality shows that $\sum a_n$ is finite. If $\int_1^{\infty} f(x)\, dx$ is infinite, the left-hand inequality shows that $\sum a_n$ is infinite. Hence the series and the integral are both finite or both infinite. ❏

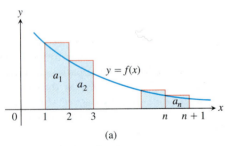

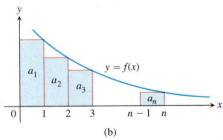

8.13 Subject to the conditions of the Integral Test, the series $\sum_{n=1}^{\infty} a_n$ and the integral $\int_1^{\infty} f(x)\, dx$ both converge or both diverge.

EXAMPLE 3 *The p-series.* Show that the **p-series**

$$\sum_{n=1}^{\infty} \frac{1}{n^p} = \frac{1}{1^p} + \frac{1}{2^p} + \frac{1}{3^p} + \cdots + \frac{1}{n^p} + \cdots \tag{3}$$

(p a real constant) converges if $p > 1$, and diverges if $p \leq 1$.

Solution If $p > 1$, then $f(x) = 1/x^p$ is a positive decreasing function of x. Since

$$\int_1^{\infty} \frac{1}{x^p}\, dx = \int_1^{\infty} x^{-p}\, dx = \lim_{b \to \infty} \left[\frac{x^{-p+1}}{-p+1} \right]_1^b$$

$$= \frac{1}{1-p} \lim_{b \to \infty} \left(\frac{1}{b^{p-1}} - 1 \right)$$

$$= \frac{1}{1-p} (0 - 1) = \frac{1}{p-1}, \qquad \begin{array}{l} b^{p-1} \to \infty \text{ as } b \to \infty \\ \text{because } p - 1 > 0. \end{array}$$

the series converges by the Integral Test.

If $p < 1$, then $1 - p > 0$ and

$$\int_1^\infty \frac{1}{x^p}\, dx = \frac{1}{1-p} \lim_{b\to\infty} (b^{1-p} - 1) = \infty.$$

The series diverges by the Integral Test.
If $p = 1$, we have the (divergent) harmonic series

$$1 + \frac{1}{2} + \frac{1}{3} + \cdots + \frac{1}{n} + \cdots.$$

We have convergence for $p > 1$ but divergence for every other value of p.

❏

Exercises 8.4

Determining Convergence or Divergence

Which of the series in Exercises 1–30 converge, and which diverge? Give reasons for your answers. (When you check an answer, remember that there may be more than one way to determine the series' convergence or divergence.)

1. $\sum_{n=1}^{\infty} \frac{1}{10^n}$

2. $\sum_{n=1}^{\infty} e^{-n}$

3. $\sum_{n=1}^{\infty} \frac{n}{n+1}$

4. $\sum_{n=1}^{\infty} \frac{5}{n+1}$

5. $\sum_{n=1}^{\infty} \frac{3}{\sqrt{n}}$

6. $\sum_{n=1}^{\infty} \frac{-2}{n\sqrt{n}}$

7. $\sum_{n=1}^{\infty} -\frac{1}{8^n}$

8. $\sum_{n=1}^{\infty} \frac{-8}{n}$

9. $\sum_{n=2}^{\infty} \frac{\ln n}{n}$

10. $\sum_{n=2}^{\infty} \frac{\ln n}{\sqrt{n}}$

11. $\sum_{n=1}^{\infty} \frac{2^n}{3^n}$

12. $\sum_{n=1}^{\infty} \frac{5^n}{4^n + 3}$

13. $\sum_{n=0}^{\infty} \frac{-2}{n+1}$

14. $\sum_{n=1}^{\infty} \frac{1}{2n-1}$

15. $\sum_{n=1}^{\infty} \frac{2^n}{n+1}$

16. $\sum_{n=1}^{\infty} \frac{1}{\sqrt{n}(\sqrt{n}+1)}$

17. $\sum_{n=2}^{\infty} \frac{\sqrt{n}}{\ln n}$

18. $\sum_{n=1}^{\infty} \left(1 + \frac{1}{n}\right)^n$

19. $\sum_{n=1}^{\infty} \frac{1}{(\ln 2)^n}$

20. $\sum_{n=1}^{\infty} \frac{1}{(\ln 3)^n}$

21. $\sum_{n=3}^{\infty} \frac{(1/n)}{(\ln n)\sqrt{\ln^2 n - 1}}$

22. $\sum_{n=1}^{\infty} \frac{1}{n(1 + \ln^2 n)}$

23. $\sum_{n=1}^{\infty} n \sin \frac{1}{n}$

24. $\sum_{n=1}^{\infty} n \tan \frac{1}{n}$

25. $\sum_{n=1}^{\infty} \frac{e^n}{1 + e^{2n}}$

26. $\sum_{n=1}^{\infty} \frac{2}{1 + e^n}$

27. $\sum_{n=1}^{\infty} \frac{8 \tan^{-1} n}{1 + n^2}$

28. $\sum_{n=1}^{\infty} \frac{n}{n^2 + 1}$

29. $\sum_{n=1}^{\infty} \text{sech}\, n$

30. $\sum_{n=1}^{\infty} \text{sech}^2 n$

Theory and Examples

For what values of a, if any, do the series in Exercises 31 and 32 converge?

31. $\sum_{n=1}^{\infty} \left(\frac{a}{n+2} - \frac{1}{n+4}\right)$

32. $\sum_{n=3}^{\infty} \left(\frac{1}{n-1} - \frac{2a}{n+1}\right)$

33. a) Draw illustrations like those in Figs. 8.12 and 8.13 to show that the partial sums of the harmonic series satisfy the inequalities

$$\ln (n+1) = \int_1^{n+1} \frac{1}{x}\, dx \le 1 + \frac{1}{2} + \cdots + \frac{1}{n}$$

$$\le 1 + \int_1^n \frac{1}{x}\, dx = 1 + \ln n.$$

b) There is absolutely no empirical evidence for the divergence of the harmonic series even though we know it diverges. The partial sums just grow too slowly. To see what we mean, suppose you had started with $s_1 = 1$ the day the universe was formed, 13 billion years ago, and added a new term every *second*. About how large would the partial sum s_n be today, assuming a 365-day year?

34. Are there any values of x for which $\sum_{n=1}^{\infty}(1/(nx))$ converges? Give reasons for your answer.

35. Is it true that if $\sum_{n=1}^{\infty} a_n$ is a divergent series of positive numbers then there is also a divergent series $\sum_{n=1}^{\infty} b_n$ of positive numbers with $b_n < a_n$ for every n? Is there a "smallest" divergent series of positive numbers? Give reasons for your answers.

36. (*Continuation of Exercise 35*) Is there a "largest" convergent series of positive numbers? Explain.

37. *The Cauchy condensation test.* The Cauchy condensation test says: Let $\{a_n\}$ be a nonincreasing sequence ($a_n \ge a_{n+1}$ for all n) of positive terms that converges to 0. Then $\sum a_n$ converges if and only if $\sum 2^n a_{2^n}$ converges. For example, $\sum(1/n)$ diverges because $\sum 2^n \cdot (1/2^n) = \sum 1$ diverges. Show why the test works.

38. Use the Cauchy condensation test from Exercise 37 to show that

    a) $\displaystyle\sum_{n=2}^{\infty} \frac{1}{n \ln n}$ diverges;

    b) $\displaystyle\sum_{n=1}^{\infty} \frac{1}{n^p}$ converges if $p > 1$ and diverges if $p \le 1$.

39. *Logarithmic p-series*

    a) Show that

$$\int_{2}^{\infty} \frac{dx}{x(\ln x)^p} \quad (p \text{ a positive constant})$$

    converges if and only if $p > 1$.

    b) What implications does the fact in (a) have for the convergence of the series

$$\sum_{n=2}^{\infty} \frac{1}{n(\ln n)^p}?$$

    Give reasons for your answer.

40. (*Continuation of Exercise 39.*) Use the result in Exercise 39 to determine which of the following series converge and which diverge. Support your answer in each case.

    a) $\displaystyle\sum_{n=2}^{\infty} \frac{1}{n(\ln n)}$ **b)** $\displaystyle\sum_{n=2}^{\infty} \frac{1}{n(\ln n)^{1.01}}$

    c) $\displaystyle\sum_{n=2}^{\infty} \frac{1}{n \ln (n^3)}$ **d)** $\displaystyle\sum_{n=2}^{\infty} \frac{1}{n(\ln n)^3}$

41. *Euler's constant.* Graphs like those in Fig. 8.13 suggest that as n increases there is little change in the difference between the sum

$$1 + \frac{1}{2} + \cdots + \frac{1}{n}$$

and the integral

$$\ln n = \int_{1}^{n} \frac{1}{x} dx.$$

To explore this idea, carry out the following steps.

a) By taking $f(x) = 1/x$ in inequality (2), show that

$$\ln (n + 1) \le 1 + \frac{1}{2} + \cdots + \frac{1}{n} \le 1 + \ln n$$

or

$$0 < \ln (n + 1) - \ln n \le 1 + \frac{1}{2} + \cdots + \frac{1}{n} - \ln n \le 1.$$

Thus, the sequence

$$a_n = 1 + \frac{1}{2} + \cdots + \frac{1}{n} - \ln n$$

is bounded from below and from above.

b) Show that

$$\frac{1}{n+1} < \int_{n}^{n+1} \frac{1}{x} dx = \ln (n + 1) - \ln n,$$

and use this result to show that the sequence $\{a_n\}$ in part (a) is decreasing.

Since a decreasing sequence that is bounded from below converges (Exercise 41 in Section 8.1), the numbers a_n defined in (a) converge:

$$1 + \frac{1}{2} + \cdots + \frac{1}{n} - \ln n \to \gamma.$$

The number γ, whose value is 0.5772 ..., is called *Euler's constant*. In contrast to other special numbers like π and e, no other expression with a simple law of formulation has ever been found for γ.

42. Use the integral test to show that

$$\sum_{n=0}^{\infty} e^{-n^2}$$

converges.

<table>
<tr><td>8.5</td><td></td></tr>
</table>

Comparison Tests for Series of Nonnegative Terms

The key question in using Corollary 1 in the preceding section is how to determine in any particular instance whether the s_n's are bounded from above. Sometimes we can establish this by showing that each s_n is less than or equal to the corresponding partial sum of a series already known to converge.

EXAMPLE 1   The series

$$\sum_{n=0}^{\infty} \frac{1}{n!} = 1 + \frac{1}{1!} + \frac{1}{2!} + \frac{1}{3!} + \cdots \qquad (1)$$

converges because its terms are all positive and less than or equal to the corresponding terms of

$$1 + \sum_{n=0}^{\infty} \frac{1}{2^n} = 1 + 1 + \frac{1}{2} + \frac{1}{2^2} + \cdots. \tag{2}$$

To see how this relationship leads to an upper bound for the partial sums of $\sum_{n=0}^{\infty}(1/(n!))$, let

$$s_n = 1 + \frac{1}{1!} + \frac{1}{2!} + \cdots + \frac{1}{n!}$$

and observe that, for each n,

$$s_n \le 1 + 1 + \frac{1}{2} + \frac{1}{2^2} + \cdots + \frac{1}{2^{n-1}} < 1 + \sum_{n=0}^{\infty} \frac{1}{2^n} = 1 + \frac{1}{1 - (1/2)} = 3.$$

Thus the partial sums of $\sum_{n=0}^{\infty}(1/(n!))$ are all less than 3, so $\sum_{n=0}^{\infty}(1/(n!))$ converges.

The fact that 3 is an upper bound for the partial sums of $\sum_{n=0}^{\infty}(1/(n!))$ does not mean that the series converges to 3. As we will see in Section 8.10, the series converges to e. ❏

The Direct Comparison Test

We established the convergence in Example 1 by comparing the terms of the given series with the terms of a series known to converge. This idea can be pursued further to yield a number of tests known as *comparison tests*.

Direct Comparison Test for Series of Nonnegative Terms

Let $\sum a_n$ be a series with no negative terms.

a) $\sum a_n$ converges if there is a convergent series $\sum c_n$ with $a_n \le c_n$ for all $n > N$, for some integer N.

b) $\sum a_n$ diverges if there is a divergent series of nonnegative terms $\sum d_n$ with $a_n \ge d_n$ for all $n > N$, for some integer N.

Proof In part (a), the partial sums of $\sum a_n$ are bounded above by

$$M = a_1 + a_2 + \cdots + a_n + \sum_{n=N+1}^{\infty} c_n.$$

They therefore form a nondecreasing sequence with a limit $L \le M$.

In part (b), the partial sums of $\sum a_n$ are not bounded from above. If they were, the partial sums for $\sum d_n$ would be bounded by

$$M' = d_1 + d_2 + \cdots + d_N + \sum_{n=N+1}^{\infty} a_n$$

and $\sum d_n$ would have to converge instead of diverge. ❏

To apply the Direct Comparison Test to a series, we need not include the early terms of the series. We can start the test with any index N provided we include all the terms of the series being tested from there on.

EXAMPLE 2   Does the following series converge?

$$5 + \frac{2}{3} + 1 + \frac{1}{7} + \frac{1}{2} + \frac{1}{3!} + \frac{1}{4!} + \cdots + \frac{1}{k!} + \cdots$$

Solution   We ignore the first four terms and compare the remaining terms with those of the convergent geometric series $\sum_{n=1}^{\infty} 1/2^n$. We see that

$$\frac{1}{2} + \frac{1}{3!} + \frac{1}{4!} + \cdots \leq \frac{1}{2} + \frac{1}{4} + \frac{1}{8} + \cdots.$$

Therefore, the original series converges by the Direct Comparison Test.   ❑

To apply the Direct Comparison Test, we need to have on hand a list of series whose convergence or divergence we know. Here is what we know so far:

Convergent series	**Divergent series**
Geometric series with $\lvert r \rvert < 1$	Geometric series with $\lvert r \rvert \geq 1$
Telescoping series like $\displaystyle\sum_{n=1}^{\infty} \frac{1}{n(n+1)}$	The harmonic series $\displaystyle\sum_{n=1}^{\infty} \frac{1}{n}$
The series $\displaystyle\sum_{n=0}^{\infty} \frac{1}{n!}$	Any series $\sum a_n$ for which $\lim_{n\to\infty} a_n$ does not exist or $\lim_{n\to\infty} a_n \neq 0$
The p-series $\displaystyle\sum_{n=1}^{\infty} \frac{1}{n^p}$ with $p > 1$	The p-series $\displaystyle\sum_{n=1}^{\infty} \frac{1}{n^p}$ with $p \leq 1$

The Limit Comparison Test

We now introduce a comparison test that is particularly handy for series in which a_n is a rational function of n.

Suppose we wanted to investigate the convergence of the series

a) $\displaystyle\sum_{n=2}^{\infty} \frac{2n}{n^2 - n + 1}$ b) $\displaystyle\sum_{n=2}^{\infty} \frac{8n^3 + 100n^2 + 1000}{2n^6 - n + 5}.$

In determining convergence or divergence, only the tails matter. And when n is very large, the highest powers in the numerator and denominator matter the most. So in (a), we might reason this way: For n large,

$$a_n = \frac{2n}{n^2 - n + 1}$$

behaves like $2n/n^2 = 2/n$. Since $\sum 1/n$ diverges, we expect $\sum a_n$ to diverge, too.

In (b) we might reason that for n large

$$a_n = \frac{8n^3 + 100n^2 + 1000}{2n^6 - n + 5}$$

will behave approximately like $(8n^3)/(2n^6) = 4/n^3$. Since $\sum 4/n^3$ converges (it is 4 times a convergent p-series), we expect $\sum a_n$ to converge, too.

Our expectations about $\sum a_n$ in each case are correct, as the following test shows.

Limit Comparison Test

Suppose that $a_n > 0$ and $b_n > 0$ for all $n \geq N$ (N an integer).

1. If $\lim\limits_{n \to \infty} \dfrac{a_n}{b_n} = c > 0$, then $\sum a_n$ and $\sum b_n$ both converge or both diverge.

2. If $\lim\limits_{n \to \infty} \dfrac{a_n}{b_n} = 0$ and $\sum b_n$ converges, then $\sum a_n$ converges.

3. If $\lim\limits_{n \to \infty} \dfrac{a_n}{b_n} = \infty$ and $\sum b_n$ diverges, then $\sum a_n$ diverges.

Proof We will prove part (1). Parts (2) and (3) are left as Exercises 37 (a) and (b). Since $c/2 > 0$, there exists an integer N such that for all n

$$n > N \Rightarrow \left| \frac{a_n}{b_n} - c \right| < \frac{c}{2}.$$

Limit definition with $\epsilon = c/2$, $L = c$, and a_n replaced by a_n/b_n

Thus, for $n > N$,

$$-\frac{c}{2} < \frac{a_n}{b_n} - c < \frac{c}{2},$$

$$\frac{c}{2} < \frac{a_n}{b_n} < \frac{3c}{2},$$

$$\left(\frac{c}{2}\right) b_n < a_n < \left(\frac{3c}{2}\right) b_n.$$

If $\sum b_n$ converges, then $\sum (3c/2)b_n$ converges and $\sum a_n$ converges by the Direct Comparison Test. If $\sum b_n$ diverges, then $\sum (c/2)b_n$ diverges and $\sum a_n$ diverges by the Direct Comparison Test. ∎

EXAMPLE 3 Which of the following series converge, and which diverge?

a) $\dfrac{3}{4} + \dfrac{5}{9} + \dfrac{7}{16} + \dfrac{9}{25} + \cdots = \sum\limits_{n=1}^{\infty} \dfrac{2n+1}{(n+1)^2} = \sum\limits_{n=1}^{\infty} \dfrac{2n+1}{n^2 + 2n + 1}$

b) $\dfrac{1}{1} + \dfrac{1}{3} + \dfrac{1}{7} + \dfrac{1}{15} + \cdots = \sum\limits_{n=1}^{\infty} \dfrac{1}{2^n - 1}$

c) $\dfrac{1 + 2 \ln 2}{9} + \dfrac{1 + 3 \ln 3}{14} + \dfrac{1 + 4 \ln 4}{21} + \cdots = \sum\limits_{n=2}^{\infty} \dfrac{1 + n \ln n}{n^2 + 5}$

Solution

We could just as well have taken $b_n = 2/n$ but $1/n$ is simpler.

a) Let $a_n = (2n+1)/(n^2 + 2n + 1)$. For n large, we expect a_n to behave like $2n/n^2 = 2/n$, so we let $b_n = 1/n$. Since

$$\sum_{n=1}^{\infty} b_n = \sum_{n=1}^{\infty} \frac{1}{n} \quad \text{diverges}$$

and

$$\lim_{n \to \infty} \frac{a_n}{b_n} = \lim_{n \to \infty} \frac{2n^2 + n}{n^2 + 2n + 1} = 2,$$

$\sum a_n$ diverges by part 1 of the Limit Comparison Test.

b) Let $a_n = 1/(2^n - 1)$. For n large, we expect a_n to behave like $1/2^n$, so we let $b_n = 1/2^n$. Since

$$\sum_{n=1}^{\infty} b_n = \sum_{n=1}^{\infty} \frac{1}{2^n} \text{ converges}$$

and

$$\lim_{n \to \infty} \frac{a_n}{b_n} = \lim_{n \to \infty} \frac{2^n}{2^n - 1}$$

$$= \lim_{n \to \infty} \frac{1}{1 - (1/2^n)}$$

$$= 1,$$

$\sum a_n$ converges by part 1 of the Limit Comparison Test.

c) Let $a_n = (1 + n \ln n)/(n^2 + 5)$. For n large, we expect a_n to behave like $(n \ln n)/n^2 = (\ln n)/n$, which is greater than $1/n$ for $n \geq 3$, so we take $b_n = 1/n$. Since

$$\sum_{n=2}^{\infty} b_n = \sum_{n=2}^{\infty} \frac{1}{n} \text{ diverges}$$

and

$$\lim_{n \to \infty} \frac{a_n}{b_n} = \lim_{n \to \infty} \frac{n + n^2 \ln n}{n^2 + 5}$$

$$= \infty,$$

$\sum a_n$ diverges by part 3 of the Limit Comparison Test. ❏

EXAMPLE 4 Does $\displaystyle\sum_{n=1}^{\infty} \frac{\ln n}{n^{3/2}}$ converge?

Solution Because $\ln n$ grows more slowly than n^c for any positive constant c (Section 8.2, Exercise 69), we would expect to have

$$\frac{\ln n}{n^{3/2}} < \frac{n^{1/4}}{n^{3/2}} = \frac{1}{n^{5/4}}$$

for n sufficiently large. Indeed, taking $a_n = (\ln n)/n^{3/2}$ and $b_n = 1/n^{5/4}$, we have

$$\lim_{n \to \infty} \frac{a_n}{b_n} = \lim_{n \to \infty} \frac{\ln n}{n^{1/4}}$$

$$= \lim_{n \to \infty} \frac{1/n}{(1/4)\, n^{-3/4}} \qquad \text{l'Hôpital's rule}$$

$$= \lim_{n \to \infty} \frac{4}{n^{1/4}} = 0.$$

Since $\sum b_n = \sum (1/n^{5/4})$ (a p-series with $p > 1$) converges, $\sum a_n$ converges by part 2 of the Limit Comparison Test. ❏

Exercises 8.5

Determining Convergence or Divergence

Which of the series in Exercises 1–36 converge, and which diverge? Give reasons for your answers.

1. $\sum_{n=1}^{\infty} \dfrac{1}{2\sqrt{n} + \sqrt[3]{n}}$

2. $\sum_{n=1}^{\infty} \dfrac{3}{n + \sqrt{n}}$

3. $\sum_{n=1}^{\infty} \dfrac{\sin^2 n}{2^n}$

4. $\sum_{n=1}^{\infty} \dfrac{1 + \cos n}{n^2}$

5. $\sum_{n=1}^{\infty} \dfrac{2n}{3n - 1}$

6. $\sum_{n=1}^{\infty} \dfrac{n + 1}{n^2 \sqrt{n}}$

7. $\sum_{n=1}^{\infty} \left(\dfrac{n}{3n + 1} \right)^n$

8. $\sum_{n=1}^{\infty} \dfrac{1}{\sqrt{n^3 + 2}}$

9. $\sum_{n=3}^{\infty} \dfrac{1}{\ln (\ln n)}$

10. $\sum_{n=2}^{\infty} \dfrac{1}{(\ln n)^2}$

11. $\sum_{n=1}^{\infty} \dfrac{(\ln n)^2}{n^3}$

12. $\sum_{n=1}^{\infty} \dfrac{(\ln n)^3}{n^3}$

13. $\sum_{n=2}^{\infty} \dfrac{1}{\sqrt{n} \ln n}$

14. $\sum_{n=1}^{\infty} \dfrac{(\ln n)^2}{n^{3/2}}$

15. $\sum_{n=1}^{\infty} \dfrac{1}{1 + \ln n}$

16. $\sum_{n=1}^{\infty} \dfrac{1}{(1 + \ln n)^2}$

17. $\sum_{n=2}^{\infty} \dfrac{\ln (n + 1)}{n + 1}$

18. $\sum_{n=1}^{\infty} \dfrac{1}{(1 + \ln^2 n)}$

19. $\sum_{n=2}^{\infty} \dfrac{1}{n\sqrt{n^2 - 1}}$

20. $\sum_{n=1}^{\infty} \dfrac{\sqrt{n}}{n^2 + 1}$

21. $\sum_{n=1}^{\infty} \dfrac{1 - n}{n2^n}$

22. $\sum_{n=1}^{\infty} \dfrac{n + 2^n}{n^2 2^n}$

23. $\sum_{n=1}^{\infty} \dfrac{1}{3^{n-1} + 1}$

24. $\sum_{n=1}^{\infty} \dfrac{3^{n-1} + 1}{3^n}$

25. $\sum_{n=1}^{\infty} \sin \dfrac{1}{n}$

26. $\sum_{n=1}^{\infty} \tan \dfrac{1}{n}$

27. $\sum_{n=1}^{\infty} \dfrac{10n + 1}{n(n + 1)(n + 2)}$

28. $\sum_{n=3}^{\infty} \dfrac{5n^3 - 3n}{n^2(n - 2)(n^2 + 5)}$

29. $\sum_{n=1}^{\infty} \dfrac{\tan^{-1} n}{n^{1.1}}$

30. $\sum_{n=1}^{\infty} \dfrac{\sec^{-1} n}{n^{1.3}}$

31. $\sum_{n=1}^{\infty} \dfrac{\coth n}{n^2}$

32. $\sum_{n=1}^{\infty} \dfrac{\tanh n}{n^2}$

33. $\sum_{n=1}^{\infty} \dfrac{1}{n \sqrt[n]{n}}$

34. $\sum_{n=1}^{\infty} \dfrac{\sqrt[n]{n}}{n^2}$

35. $\sum_{n=1}^{\infty} \dfrac{1}{1 + 2 + 3 + \cdots + n}$

36. $\sum_{n=1}^{\infty} \dfrac{1}{1 + 2^2 + 3^2 + \cdots + n^2}$

Theory and Examples

37. Prove (a) Part 2 and (b) Part 3 of the Limit Comparison Test.

38. If $\sum_{n=1}^{\infty} a_n$ is a convergent series of nonnegative numbers, can anything be said about $\sum_{n=1}^{\infty} (a_n/n)$? Explain.

39. Suppose that $a_n > 0$ and $b_n > 0$ for $n \geq N$ (N an integer). If $\lim_{n \to \infty} (a_n/b_n) = \infty$ and $\sum a_n$ converges, can anything be said about $\sum b_n$? Give reasons for your answer.

40. Prove that if $\sum a_n$ is a convergent series of nonnegative terms, then $\sum a_n^2$ converges.

CAS Exploration and Project

41. It is not yet known whether the series

$$\sum_{n=1}^{\infty} \dfrac{1}{n^3 \sin^2 n}$$

converges or diverges. Use a CAS to explore the behavior of the series by performing the following steps.

   a) Define the sequence of partial sums

   $$s_k = \sum_{n=1}^{k} \dfrac{1}{n^3 \sin^2 n}.$$

   What happens when you try to find the limit of s_k as $k \to \infty$? Does your CAS find a closed form answer for this limit?

   b) Plot the first 100 points (k, s_k) for the sequence of partial sums. Do they appear to converge? What would you estimate the limit to be?

   c) Next plot the first 200 points (k, s_k). Discuss the behavior in your own words.

   d) Plot the first 400 points (k, s_k). What happens when $k = 355$? Calculate the number 355/113. Explain from your calculation what happened at $k = 355$. For what values of k would you guess this behavior might occur again?

   You will find an interesting discussion of this series in Chapter 72 of *Mazes for the Mind* by Clifford A. Pickover, St. Martin's Press, Inc., New York, 1992.

8.6

The Ratio and Root Tests for Series of Nonnegative Terms

Convergence tests that depend on comparing series with integrals or other series are called *extrinsic* tests. They are useful, but there are reasons to look for tests that do not require comparison. As a practical matter, we may not be able to find the series or functions we need to make a comparison work. And, in principle, all the information about a given series should be contained in its own terms. We therefore turn our attention to *intrinsic* tests—tests that depend only on the series at hand.

The Ratio Test

The first intrinsic test, the Ratio Test, measures the rate of growth (or decline) of a series by examining the ratio a_{n+1}/a_n. For a geometric series $\sum ar^n$, this rate is a constant $((ar^{n+1})/(ar^n) = r)$, and the series converges if and only if its ratio is less than 1 in absolute value. But even if the ratio is not constant, we may be able to find a geometric series for comparison, as in Example 1.

The series in Example 1 converges rapidly, as the following computer data suggest.

n	s_n
5	1.5492 06349
10	1.5702 89085
15	1.5707 83080
20	1.5707 95964
25	1.5707 96317
30	1.5707 96327
35	1.5707 96327

EXAMPLE 1   Let $a_1 = 1$ and let $a_{n+1} = \dfrac{n}{2n+1} a_n$ for all n. Does the series $\sum a_n$ converge?

Solution   We begin by writing a few terms of the series:

$$a_1 = 1, \qquad a_2 = \frac{1}{3}a_1 = \frac{1}{3}, \qquad a_3 = \frac{2}{5}a_2 = \frac{1 \cdot 2}{3 \cdot 5}, \qquad a_4 = \frac{3}{7}a_3 = \frac{1 \cdot 2 \cdot 3}{3 \cdot 5 \cdot 7}.$$

Each term is somewhat less than 1/2 the term before it, because $n/(2n+1)$ is less than 1/2. Therefore the terms of the series are less than or equal to the terms of the geometric series

$$1 + \left(\frac{1}{2}\right) + \left(\frac{1}{2}\right)^2 + \cdots + \left(\frac{1}{2}\right)^{n-1} + \cdots,$$

which converges to 2. So our series also converges, and its sum is less than 2. The table in the margin shows how quickly the series converges to its known limit, $\pi/2$.   ☐

In proving the Ratio Test, we will make a comparison with an appropriate geometric series as in Example 1, but when we *apply* the test there is no need for comparison.

The Ratio Test

Let $\sum a_n$ be a series with positive terms, and suppose that

$$\lim_{n \to \infty} \frac{a_{n+1}}{a_n} = \rho.$$

Then

a)   the series *converges* if $\rho < 1$,
b)   the series *diverges* if $\rho > 1$ or ρ is infinite,
c)   the test is *inconclusive* if $\rho = 1$.

Proof

a) $\rho < 1$. Let r be a number between ρ and 1. Then the number $\epsilon = r - \rho$ is positive. Since

$$\frac{a_{n+1}}{a_n} \to \rho,$$

a_{n+1}/a_n must lie within ϵ of ρ when n is large enough, say for all $n \geq N$. In particular,

$$\frac{a_{n+1}}{a_n} < \rho + \epsilon = r, \qquad \text{when } n \geq N.$$

That is,

$$a_{N+1} < ra_N,$$

$$a_{N+2} < ra_{N+1} < r^2 a_N,$$

$$a_{N+3} < ra_{N+2} < r^3 a_N,$$

$$\vdots$$

$$a_{N+m} < ra_{N+m-1} < r^m a_N.$$

These inequalities show that the terms of our series, after the Nth term, approach zero more rapidly than the terms in a geometric series with ratio $r < 1$. More precisely, consider the series $\sum c_n$, where $c_n = a_n$ for $n = 1, 2, \ldots, N$ and $c_{N+1} = ra_N, c_{N+2} = r^2 a_N, \ldots, c_{N+m} = r^m a_N, \ldots$. Now $a_n \le c_n$ for all n, and

$$\sum_{n=1}^{\infty} c_n = a_1 + a_2 + \cdots + a_{N-1} + a_N + ra_N + r^2 a_N + \cdots$$

$$= a_1 + a_2 + \cdots + a_{N-1} + a_N(1 + r + r^2 + \cdots).$$

The geometric series $1 + r + r^2 + \cdots$ converges because $|r| < 1$, so $\sum c_n$ converges. Since $a_n \le c_n$, $\sum a_n$ also converges.

b) $1 < \rho \le \infty$. From some index M on,

$$\frac{a_{n+1}}{a_n} > 1 \qquad \text{and} \qquad a_M < a_{M+1} < a_{M+2} < \cdots.$$

The terms of the series do not approach zero as n becomes infinite, and the series diverges by the nth-Term Test.

c) $\rho = 1$. The two series

$$\sum_{n=1}^{\infty} \frac{1}{n} \qquad \text{and} \qquad \sum_{n=1}^{\infty} \frac{1}{n^2}$$

show that some other test for convergence must be used when $\rho = 1$.

For $\sum_{n=1}^{\infty} \frac{1}{n}$: $\dfrac{a_{n+1}}{a_n} = \dfrac{1/(n+1)}{1/n} = \dfrac{n}{n+1} \to 1.$

For $\sum_{n=1}^{\infty} \frac{1}{n^2}$: $\dfrac{a_{n+1}}{a_n} = \dfrac{1/(n+1)^2}{1/n^2} = \left(\dfrac{n}{n+1}\right)^2 \to 1^2 = 1.$

In both cases $\rho = 1$, yet the first series diverges while the second converges.   ❑

The Ratio Test is often effective when the terms of a series contain factorials of expressions involving n or expressions raised to the nth power.

EXAMPLE 2   Investigate the convergence of the following series.

a) $\displaystyle\sum_{n=0}^{\infty} \frac{2^n + 5}{3^n}$

b) $\displaystyle\sum_{n=1}^{\infty} \frac{(2n)!}{n!n!}$

c) $\displaystyle\sum_{n=1}^{\infty} \frac{4^n n!n!}{(2n)!}$

Solution

a) For the series $\sum_{n=0}^{\infty}(2^n + 5)/3^n$,

$$\frac{a_{n+1}}{a_n} = \frac{(2^{n+1} + 5)/3^{n+1}}{(2^n + 5)/3^n} = \frac{1}{3} \cdot \frac{2^{n+1} + 5}{2^n + 5} = \frac{1}{3} \cdot \left(\frac{2 + 5 \cdot 2^{-n}}{1 + 5 \cdot 2^{-n}}\right) \to \frac{1}{3} \cdot \frac{2}{1} = \frac{2}{3}.$$

The series converges because $\rho = 2/3$ is less than 1.

This does *not* mean that 2/3 is the sum of the series. In fact,

$$\sum_{n=0}^{\infty} \frac{2^n + 5}{3^n} = \sum_{n=0}^{\infty} \left(\frac{2}{3}\right)^n + \sum_{n=0}^{\infty} \frac{5}{3^n} = \frac{1}{1 - (2/3)} + \frac{5}{1 - (1/3)} = \frac{21}{2}.$$

b) If $a_n = \dfrac{(2n)!}{n!n!}$, then $a_{n+1} = \dfrac{(2n + 2)!}{(n + 1)!(n + 1)!}$ and

$$\frac{a_{n+1}}{a_n} = \frac{n!n!(2n + 2)(2n + 1)(2n)!}{(n + 1)!(n + 1)!(2n)!}$$

$$= \frac{(2n + 2)(2n + 1)}{(n + 1)(n + 1)} = \frac{4n + 2}{n + 1} \to 4.$$

The series diverges because $\rho = 4$ is greater than 1.

c) If $a_n = 4^n n!n!/(2n)!$, then

$$\frac{a_{n+1}}{a_n} = \frac{4^{n+1}(n + 1)!(n + 1)!}{(2n + 2)(2n + 1)(2n)!} \cdot \frac{(2n)!}{4^n n!n!}$$

$$= \frac{4(n + 1)(n + 1)}{(2n + 2)(2n + 1)} = \frac{2(n + 1)}{2n + 1} \to 1.$$

Because the limit is $\rho = 1$, we cannot decide from the Ratio Test whether the series converges. However, when we notice that $a_{n+1}/a_n = (2n + 2)/(2n + 1)$, we conclude that a_{n+1} is always greater than a_n because $(2n + 2)/(2n + 1)$ is always greater than 1. Therefore, all terms are greater than or equal to $a_1 = 2$, and the nth term does not approach zero as $n \to \infty$. The series diverges. ❑

The *n*th-Root Test

The convergence tests we have so far for $\sum a_n$ work best when the formula for a_n is relatively simple. But consider the following.

EXAMPLE 3   Let $a_n = \begin{cases} n/2^n, & n \text{ odd} \\ 1/2^n, & n \text{ even.} \end{cases}$ Does $\sum a_n$ converge?

Solution   We write out several terms of the series:

$$\sum_{n=1}^{\infty} a_n = \frac{1}{2^1} + \frac{1}{2^2} + \frac{3}{2^3} + \frac{1}{2^4} + \frac{5}{2^5} + \frac{1}{2^6} + \frac{7}{2^7} + \cdots$$

$$= \frac{1}{2} + \frac{1}{4} + \frac{3}{8} + \frac{1}{16} + \frac{5}{32} + \frac{1}{64} + \frac{7}{128} + \cdots.$$

Clearly, this is not a geometric series. The nth term approaches zero as $n \to \infty$, so we do not know if the series diverges. The Integral Test does not look promising. The Ratio Test produces

$$\frac{a_{n+1}}{a_n} = \begin{cases} \dfrac{1}{2n}, & n \text{ odd} \\ \dfrac{n + 1}{2}, & n \text{ even.} \end{cases}$$

As $n \to \infty$, the ratio is alternately small and large and has no limit.

A test that will answer the question (the series converges) is the nth-Root Test.

❑

The nth-Root Test

Let $\sum a_n$ be a series with $a_n \geq 0$ for $n \geq N$, and suppose that

$$\lim_{n \to \infty} \sqrt[n]{a_n} = \rho.$$

Then

a) the series *converges* if $\rho < 1$,
b) the series *diverges* if $\rho > 1$ or ρ is infinite,
c) the test is *inconclusive* if $\rho = 1$.

Proof

a) $\rho < 1.$ Choose an $\epsilon > 0$ so small that $\rho + \epsilon < 1$. Since $\sqrt[n]{a_n} \to \rho$, the terms $\sqrt[n]{a_n}$ eventually get closer than ϵ to ρ. In other words, there exists an index $M \geq N$ such that

$$\sqrt[n]{a_n} < \rho + \epsilon \qquad \text{when } n \geq M.$$

Then it is also true that

$$a_n < (\rho + \epsilon)^n \qquad \text{for } n \geq M.$$

Now, $\sum_{n=M}^{\infty} (\rho + \epsilon)^n$, a geometric series with ratio $(\rho + \epsilon) < 1$, converges. By comparison, $\sum_{n=M}^{\infty} a_n$ converges, from which it follows that

$$\sum_{n=1}^{\infty} a_n = a_1 + \cdots + a_{M-1} + \sum_{n=M}^{\infty} a_n$$

converges.

b) $1 < \rho \leq \infty.$ For all indices beyond some integer M, we have $\sqrt[n]{a_n} > 1$, so that $a_n > 1$ for $n > M$. The terms of the series do not converge to zero. The series diverges by the nth-Term Test.

c) $\rho = 1.$ The series $\sum_{n=1}^{\infty}(1/n)$ and $\sum_{n=1}^{\infty}(1/n^2)$ show that the test is not conclusive when $\rho = 1$. The first series diverges and the second converges, but in both cases $\sqrt[n]{a_n} \to 1$. ❏

EXAMPLE 3 (continued) Let $a_n = \begin{cases} n/2^n, & n \text{ odd} \\ 1/2^n, & n \text{ even}. \end{cases}$ Does $\sum a_n$ converge?

Solution We apply the nth-Root Test, finding that

$$\sqrt[n]{a_n} = \begin{cases} \sqrt[n]{n}/2, & n \text{ odd} \\ 1/2, & n \text{ even}. \end{cases}$$

Therefore,

$$\frac{1}{2} \leq \sqrt[n]{a_n} \leq \frac{\sqrt[n]{n}}{2}.$$

Since $\sqrt[n]{n} \to 1$ (Section 8.2, Table 8.1), we have $\lim_{n \to \infty} \sqrt[n]{a_n} = 1/2$ by the Sandwich Theorem. The limit is less than 1, so the series converges by the nth-Root Test.

❏

EXAMPLE 4 Which of the following series converges, and which diverges?

a) $\displaystyle\sum_{n=1}^{\infty} \frac{n^2}{2^n}$ b) $\displaystyle\sum_{n=1}^{\infty} \frac{2^n}{n^2}$

Solution

a) $\displaystyle\sum_{n=1}^{\infty} \frac{n^2}{2^n}$ converges because $\sqrt[n]{\dfrac{n^2}{2^n}} = \dfrac{\sqrt[n]{n^2}}{\sqrt[n]{2^n}} = \dfrac{\left(\sqrt[n]{n}\right)^2}{2} \to \dfrac{1}{2} < 1.$

b) $\displaystyle\sum_{n=1}^{\infty} \frac{2^n}{n^2}$ diverges because $\sqrt[n]{\dfrac{2^n}{n^2}} = \dfrac{2}{\left(\sqrt[n]{n}\right)^2} \to \dfrac{2}{1} > 1.$

Exercises 8.6

Determining Convergence or Divergence

Which of the series in Exercises 1–26 converge, and which diverge? Give reasons for your answers. (When checking your answers, remember there may be more than one way to determine a series' convergence or divergence.)

1. $\displaystyle\sum_{n=1}^{\infty} \frac{n^{\sqrt{2}}}{2^n}$

2. $\displaystyle\sum_{n=1}^{\infty} n^2 e^{-n}$

3. $\displaystyle\sum_{n=1}^{\infty} n!\, e^{-n}$

4. $\displaystyle\sum_{n=1}^{\infty} \frac{n!}{10^n}$

5. $\displaystyle\sum_{n=1}^{\infty} \frac{n^{10}}{10^n}$

6. $\displaystyle\sum_{n=1}^{\infty} \left(\frac{n-2}{n}\right)^n$

7. $\displaystyle\sum_{n=1}^{\infty} \frac{2+(-1)^n}{1.25^n}$

8. $\displaystyle\sum_{n=1}^{\infty} \frac{(-2)^n}{3^n}$

9. $\displaystyle\sum_{n=1}^{\infty} \left(1-\frac{3}{n}\right)^n$

10. $\displaystyle\sum_{n=1}^{\infty} \left(1-\frac{1}{3n}\right)^n$

11. $\displaystyle\sum_{n=1}^{\infty} \frac{\ln n}{n^3}$

12. $\displaystyle\sum_{n=1}^{\infty} \frac{(\ln n)^n}{n^n}$

13. $\displaystyle\sum_{n=1}^{\infty} \left(\frac{1}{n}-\frac{1}{n^2}\right)$

14. $\displaystyle\sum_{n=1}^{\infty} \left(\frac{1}{n}-\frac{1}{n^2}\right)^n$

15. $\displaystyle\sum_{n=1}^{\infty} \frac{\ln n}{n}$

16. $\displaystyle\sum_{n=1}^{\infty} \frac{n \ln n}{2^n}$

17. $\displaystyle\sum_{n=1}^{\infty} \frac{(n+1)(n+2)}{n!}$

18. $\displaystyle\sum_{n=1}^{\infty} e^{-n}(n^3)$

19. $\displaystyle\sum_{n=1}^{\infty} \frac{(n+3)!}{3!\,n!\,3^n}$

20. $\displaystyle\sum_{n=1}^{\infty} \frac{n 2^n (n+1)!}{3^n n!}$

21. $\displaystyle\sum_{n=1}^{\infty} \frac{n!}{(2n+1)!}$

22. $\displaystyle\sum_{n=1}^{\infty} \frac{n!}{n^n}$

23. $\displaystyle\sum_{n=2}^{\infty} \frac{n}{(\ln n)^n}$

24. $\displaystyle\sum_{n=2}^{\infty} \frac{n}{(\ln n)^{(n/2)}}$

25. $\displaystyle\sum_{n=1}^{\infty} \frac{n! \ln n}{n(n+2)!}$

26. $\displaystyle\sum_{n=1}^{\infty} \frac{3^n}{n^3 2^n}$

Which of the series $\sum_{n=1}^{\infty} a_n$ defined by the formulas in Exercises 27–38 converge, and which diverge? Give reasons for your answers.

27. $a_1 = 2, \quad a_{n+1} = \dfrac{1+\sin n}{n}\, a_n$

28. $a_1 = 1, \quad a_{n+1} = \dfrac{1+\tan^{-1} n}{n}\, a_n$

29. $a_1 = \dfrac{1}{3}, \quad a_{n+1} = \dfrac{3n-1}{2n+5}\, a_n$

30. $a_1 = 3, \quad a_{n+1} = \dfrac{n}{n+1}\, a_n$

31. $a_1 = 2, \quad a_{n+1} = \dfrac{2}{n}\, a_n$

32. $a_1 = 5, \quad a_{n+1} = \dfrac{\sqrt[n]{n}}{2}\, a_n$

33. $a_1 = 1, \quad a_{n+1} = \dfrac{1+\ln n}{n}\, a_n$

34. $a_1 = \dfrac{1}{2}, \quad a_{n+1} = \dfrac{n+\ln n}{n+10}\, a_n$

35. $a_1 = \dfrac{1}{3}, \quad a_{n+1} = \sqrt[n]{a_n}$

36. $a_1 = \dfrac{1}{2}, \quad a_{n+1} = (a_n)^{n+1}$

37. $a_n = \dfrac{2^n n! n!}{(2n)!}$

38. $a_n = \dfrac{(3n)!}{n!(n+1)!(n+2)!}$

Which of the series in Exercises 39–44 converge, and which diverge? Give reasons for your answers.

39. $\displaystyle\sum_{n=1}^{\infty} \frac{(n!)^n}{(n^n)^2}$

40. $\displaystyle\sum_{n=1}^{\infty} \frac{(n!)^n}{n^{(n^2)}}$

41. $\displaystyle\sum_{n=1}^{\infty} \frac{n^n}{2^{(n^2)}}$

42. $\displaystyle\sum_{n=1}^{\infty} \frac{n^n}{(2^n)^2}$

43. $\displaystyle\sum_{n=1}^{\infty} \frac{1 \cdot 3 \cdot \cdots \cdot (2n-1)}{4^n 2^n n!}$

44. $\displaystyle\sum_{n=1}^{\infty} \frac{1 \cdot 3 \cdot \cdots \cdot (2n-1)}{[2 \cdot 4 \cdot \cdots \cdot (2n)](3^n + 1)}$

Theory and Examples

45. Neither the Ratio nor the nth-Root Test helps with p-series. Try them on

$$\sum_{n=1}^{\infty} \frac{1}{n^p}$$

and show that both tests fail to provide information about convergence.

46. Show that neither the Ratio Test nor the nth-Root Test provides information about the convergence of

$$\sum_{n=2}^{\infty} \frac{1}{(\ln n)^p} \qquad (p \text{ constant}).$$

47. Let $a_n = \begin{cases} n/2^n & \text{if } n \text{ is a prime number} \\ 1/2^n & \text{otherwise.} \end{cases}$

Does $\sum a_n$ converge? Give reasons for your answer.

8.7 Alternating Series, Absolute and Conditional Convergence

A series in which the terms are alternately positive and negative is an **alternating series.**

Here are three examples:

$$1 - \frac{1}{2} + \frac{1}{3} - \frac{1}{4} + \frac{1}{5} - \cdots + \frac{(-1)^{n+1}}{n} + \cdots \tag{1}$$

$$-2 + 1 - \frac{1}{2} + \frac{1}{4} - \frac{1}{8} + \cdots + \frac{(-1)^n 4}{2^n} + \cdots \tag{2}$$

$$1 - 2 + 3 - 4 + 5 - 6 + \cdots + (-1)^{n+1} n + \cdots \tag{3}$$

Series (1), called the **alternating harmonic series,** converges, as we will see in a moment. Series (2), a geometric series with ratio $r = -1/2$, converges to $-2/[1 + (1/2)] = -4/3$. Series (3) diverges because the nth term does not approach zero.

We prove the convergence of the alternating harmonic series by applying the Alternating Series Test.

Theorem 8
The Alternating Series Test (Leibniz's Theorem)
The series

$$\sum_{n=1}^{\infty} (-1)^{n+1} u_n = u_1 - u_2 + u_3 - u_4 + \cdots$$

converges if all three of the following conditions are satisfied:

1. The u_n's are all positive.
2. $u_n \geq u_{n+1}$ for all $n \geq N$, for some integer N,
3. $u_n \to 0$.

Proof If n is an even integer, say $n = 2m$, then the sum of the first n terms is

$$s_{2m} = (u_1 - u_2) + (u_3 - u_4) + \cdots + (u_{2m-1} - u_{2m})$$

$$= u_1 - (u_2 - u_3) - (u_4 - u_5) - \cdots - (u_{2m-2} - u_{2m-1}) - u_{2m}.$$

The first equality shows that s_{2m} is the sum of m nonnegative terms, since each term in parentheses is positive or zero. Hence $s_{2m+2} \geq s_{2m}$, and the sequence $\{s_{2m}\}$ is nondecreasing. The second equality shows that $s_{2m} \leq u_1$. Since $\{s_{2m}\}$ is nondecreasing and bounded from above, it has a limit, say

$$\lim_{m \to \infty} s_{2m} = L. \tag{4}$$

If n is an odd integer, say $n = 2m + 1$, then the sum of the first n terms is $s_{2m+1} = s_{2m} + u_{2m+1}$. Since $u_n \to 0$,

$$\lim_{m \to \infty} u_{2m+1} = 0$$

and, as $m \to \infty$,

$$s_{2m+1} = s_{2m} + u_{2m+1} \to L + 0 = L. \tag{5}$$

Combining the results of (4) and (5) gives $\lim_{n \to \infty} s_n = L$ (Section 8.1, Exercise 53).
❏

EXAMPLE 1   The alternating harmonic series

$$\sum_{n=1}^{\infty} (-1)^{n+1} \frac{1}{n} = 1 - \frac{1}{2} + \frac{1}{3} - \frac{1}{4} + \cdots$$

satisfies the three requirements of Theorem 8 with $N = 1$; it therefore converges.
❏

A graphical interpretation of the partial sums (Fig. 8.14) shows how an alternating series converges to its limit L when the three conditions of Theorem 8 are satisfied with $N = 1$. (Exercise 63 asks you to picture the case $N > 1$.) Starting from the origin of the x-axis, we lay off the positive distance $s_1 = u_1$. To find the point corresponding to $s_2 = u_1 - u_2$, we back up a distance equal to u_2. Since $u_2 \leq u_1$, we do not back up any farther than the origin. We continue in this seesaw fashion, backing up or going forward as the signs in the series demand. But for $n \geq N$, each forward or backward step is shorter than (or at most the same size as) the preceding step, because $u_{n+1} \leq u_n$. And since the nth term approaches zero as n increases, the size of step we take forward or backward gets smaller and smaller. We oscillate across the limit L, and the amplitude of oscillation approaches zero. The limit L lies between any two successive sums s_n and s_{n+1} and hence differs from s_n by an amount less than u_{n+1}.

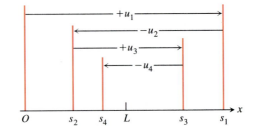

8.14 The partial sums of an alternating series that satisfies the hypotheses of Theorem 8 for $N = 1$ straddle the limit from the beginning.

Because

$$|L - s_n| < u_{n+1} \qquad \text{for } n \geq N,$$

we can make useful estimates of the sums of convergent alternating series.

Theorem 9

The Alternating Series Estimation Theorem

If the alternating series $\sum_{n=1}^{\infty}(-1)^{n+1}u_n$ satisfies the three conditions of Theorem 8, then for $n \geq N$,

$$s_n = u_1 - u_2 + \cdots + (-1)^{n+1}u_n$$

approximates the sum L of the series with an error whose absolute value is less than u_{n+1}, the numerical value of the first unused term. Furthermore, the remainder, $L - s_n$, has the same sign as the first unused term.

We leave the verification of the sign of the remainder for Exercise 53.

EXAMPLE 2   We try Theorem 9 on a series whose sum we know:

$$\sum_{n=0}^{\infty}(-1)^n\frac{1}{2^n} = 1 - \frac{1}{2} + \frac{1}{4} - \frac{1}{8} + \frac{1}{16} - \frac{1}{32} + \frac{1}{64} - \frac{1}{128} + \frac{1}{256} - \cdots.$$

The theorem says that if we truncate the series after the eighth term, we throw away a total that is positive and less than 1/256. The sum of the first eight terms is 0.6640 625. The sum of the series is

$$\frac{1}{1 - (-1/2)} = \frac{1}{3/2} = \frac{2}{3}.$$

The difference, $(2/3) - 0.6640\ 625 = 0.0026\ 04166\ 6\ldots$, is positive and less than $(1/256) = 0.0039\ 0625$. ◻

Absolute Convergence

Definition

A series $\sum a_n$ **converges absolutely** (is **absolutely convergent**) if the corresponding series of absolute values, $\sum |a_n|$, converges.

The geometric series

$$1 - \frac{1}{2} + \frac{1}{4} - \frac{1}{8} + \cdots$$

converges absolutely because the corresponding series of absolute values

$$1 + \frac{1}{2} + \frac{1}{4} + \frac{1}{8} + \cdots$$

converges. The alternating harmonic series does not converge absolutely. The corresponding series of absolute values is the (divergent) harmonic series.

Definition

A series that converges but does not converge absolutely **converges conditionally.**

The alternating harmonic series converges conditionally.

Absolute convergence is important for two reasons. First, we have good tests for convergence of series of positive terms. Second, if a series converges absolutely, then it converges. That is the thrust of the next theorem.

Caution

We can rephrase Theorem 10 to say that *every absolutely convergent series converges.* However, the converse statement is false: Many convergent series do not converge absolutely.

Theorem 10

The Absolute Convergence Test

If $\sum_{n=1}^{\infty} |a_n|$ converges, then $\sum_{n=1}^{\infty} a_n$ converges.

Proof For each n,

$$-|a_n| \le a_n \le |a_n|, \qquad \text{so} \qquad 0 \le a_n + |a_n| \le 2|a_n|.$$

If $\sum_{n=1}^{\infty} |a_n|$ converges, then $\sum_{n=1}^{\infty} 2|a_n|$ converges and, by the Direct Comparison Test, the nonnegative series $\sum_{n=1}^{\infty} (a_n + |a_n|)$ converges. The equality $a_n = (a_n + |a_n|) - |a_n|$ now lets us express $\sum_{n=1}^{\infty} a_n$ as the difference of two convergent series:

$$\sum_{n=1}^{\infty} a_n = \sum_{n=1}^{\infty} (a_n + |a_n| - |a_n|) = \sum_{n=1}^{\infty} (a_n + |a_n|) - \sum_{n=1}^{\infty} |a_n|.$$

Therefore, $\sum_{n=1}^{\infty} a_n$ converges. ❑

EXAMPLE 3 For $\sum_{n=1}^{\infty} (-1)^{n+1} \dfrac{1}{n^2} = 1 - \dfrac{1}{4} + \dfrac{1}{9} - \dfrac{1}{16} + \cdots$, the corresponding series of absolute values is the convergent series

$$\sum_{n=1}^{\infty} \frac{1}{n^2} = 1 + \frac{1}{4} + \frac{1}{9} + \frac{1}{16} + \cdots.$$

The original series converges because it converges absolutely. ❑

EXAMPLE 4 For $\sum_{n=1}^{\infty} \dfrac{\sin n}{n^2} = \dfrac{\sin 1}{1} + \dfrac{\sin 2}{4} + \dfrac{\sin 3}{9} + \cdots$, the corresponding series of absolute values is

$$\sum_{n=1}^{\infty} \left| \frac{\sin n}{n^2} \right| = \frac{|\sin 1|}{1} + \frac{|\sin 2|}{4} + \cdots,$$

which converges by comparison with $\sum_{n=1}^{\infty} (1/n^2)$ because $|\sin n| \le 1$ for every n. The original series converges absolutely; therefore it converges. ❑

EXAMPLE 5 *Alternating p-series*

If p is a positive constant, the sequence $\{1/n^p\}$ is a decreasing sequence with limit zero. Therefore the alternating p-series

$$\sum_{n=1}^{\infty} \frac{(-1)^{n-1}}{n^p} = 1 - \frac{1}{2^p} + \frac{1}{3^p} - \frac{1}{4^p} + \cdots, \qquad p > 0$$

converges.

If $p > 1$, the series converges absolutely. If $0 < p \le 1$, the series converges conditionally.

Conditional convergence: $\qquad 1 - \dfrac{1}{\sqrt{2}} + \dfrac{1}{\sqrt{3}} - \dfrac{1}{\sqrt{4}} + \cdots$

Absolute convergence: $\qquad 1 - \dfrac{1}{2^{3/2}} + \dfrac{1}{3^{3/2}} - \dfrac{1}{4^{3/2}} + \cdots$ ❑

Rearranging Series

Theorem 11

The Rearrangement Theorem for Absolutely Convergent Series

If $\sum_{n=1}^{\infty} a_n$ converges absolutely, and $b_1, b_2, \ldots, b_n, \ldots$ is any arrangement of the sequence $\{a_n\}$, then $\sum b_n$ converges absolutely and

$$\sum_{n=1}^{\infty} b_n = \sum_{n=1}^{\infty} a_n.$$

(For an outline of the proof, see Exercise 60.)

EXAMPLE 6 As we saw in Example 3, the series

$$1 - \frac{1}{4} + \frac{1}{9} - \frac{1}{16} + \cdots + (-1)^{n-1}\frac{1}{n^2} + \cdots$$

converges absolutely. A possible rearrangement of the terms of the series might start with a positive term, then two negative terms, then three positive terms, then four negative terms, and so on: After k terms of one sign, take $k + 1$ terms of the opposite sign. The first ten terms of such a series look like this:

$$1 - \frac{1}{4} - \frac{1}{16} + \frac{1}{9} + \frac{1}{25} + \frac{1}{49} - \frac{1}{36} - \frac{1}{64} - \frac{1}{100} - \frac{1}{144} + \cdots.$$

The Rearrangement Theorem says that both series converge to the same value. In this example, if we had the second series to begin with, we would probably be glad to exchange it for the first, if we knew that we could. We can do even better: The sum of either series is also equal to

$$\sum_{n=1}^{\infty} \frac{1}{(2n-1)^2} - \sum_{n=1}^{\infty} \frac{1}{(2n)^2}.$$

(See Exercise 61.) ❑

Caution

If we rearrange infinitely many terms of a conditionally convergent series, we can get results that are far different from the sum of the original series.

The kind of behavior illustrated by this example is typical of what can happen with any conditionally convergent series. Moral: Add the terms of a conditionally convergent series in the order given.

EXAMPLE 7 *Rearranging the alternating harmonic series*

The alternating harmonic series

$$\frac{1}{1} - \frac{1}{2} + \frac{1}{3} - \frac{1}{4} + \frac{1}{5} - \frac{1}{6} + \frac{1}{7} - \frac{1}{8} + \frac{1}{9} - \frac{1}{10} + \frac{1}{11} - \cdots$$

can be rearranged to diverge or to reach any preassigned sum.

a) *Rearranging $\sum_{n=1}^{\infty} (-1)^{n+1}/n$ to diverge.* The series of terms $\sum [1/(2n - 1)]$ diverges to $+\infty$ and the series of terms $\sum(-1/2n)$ diverges to $-\infty$. No matter how far out in the sequence of odd-numbered terms we begin, we can always add enough positive terms to get an arbitrarily large sum. Similarly, with the negative terms, no matter how far out we start, we can add enough consecutive even-numbered terms to get a negative sum of arbitrarily large absolute value. If we wished to do so, we could start adding odd-numbered terms until we had a sum greater than $+3$, say, and then follow that with enough consecutive negative terms to make the new total less than -4. We could then add enough positive terms to make the total greater than $+5$ and follow with consecutive unused negative terms to make a new total less than -6, and so on. In this way, we could make the swings arbitrarily large in either direction.

Flowchart 8.1 Procedure for Determining Convergence

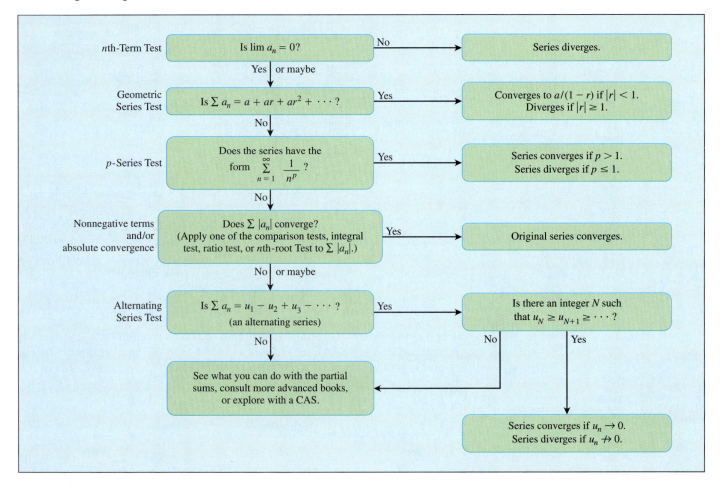

b) *Rearranging* $\sum_{n=1}^{\infty}(-1)^{n+1}/n$ *to converge to* 1. Another possibility is to focus on a particular limit. Suppose we try to get sums that converge to 1. We start with the first term, 1/1, and then subtract 1/2. Next we add 1/3 and 1/5, which brings the total back to 1 or above. Then we add consecutive negative terms until the total is less than 1. We continue in this manner: When the sum is less than 1, add positive terms until the total is 1 or more; then subtract (add negative) terms until the total is again less than 1. This process can be continued indefinitely. Because both the odd-numbered terms and the even-numbered terms of the original series approach zero as $n \to \infty$, the amount by which our partial sums exceed 1 or fall below it approaches zero. So the new series converges to 1. The rearranged series starts like this:

$$\frac{1}{1} - \frac{1}{2} + \frac{1}{3} + \frac{1}{5} - \frac{1}{4} + \frac{1}{7} + \frac{1}{9} - \frac{1}{6} + \frac{1}{11} + \frac{1}{13} - \frac{1}{8} + \frac{1}{15} + \frac{1}{17} - \frac{1}{10}$$

$$+ \frac{1}{19} + \frac{1}{21} - \frac{1}{12} + \frac{1}{23} + \frac{1}{25} - \frac{1}{14} + \frac{1}{27} - \frac{1}{16} + \cdots$$

❏

Exercises 8.7

Determining Convergence or Divergence

Which of the alternating series in Exercises 1–10 converge, and which diverge? Give reasons for your answers.

1. $\sum_{n=1}^{\infty}(-1)^{n+1}\frac{1}{n^2}$

2. $\sum_{n=1}^{\infty}(-1)^{n+1}\frac{1}{n^{3/2}}$

3. $\sum_{n=1}^{\infty}(-1)^{n+1}\left(\frac{n}{10}\right)^n$

4. $\sum_{n=1}^{\infty}(-1)^{n+1}\frac{10^n}{n^{10}}$

5. $\sum_{n=2}^{\infty}(-1)^{n+1}\frac{1}{\ln n}$

6. $\sum_{n=1}^{\infty}(-1)^{n+1}\frac{\ln n}{n}$

7. $\sum_{n=2}^{\infty}(-1)^{n+1}\frac{\ln n}{\ln n^2}$

8. $\sum_{n=1}^{\infty}(-1)^n \ln\left(1+\frac{1}{n}\right)$

9. $\sum_{n=1}^{\infty}(-1)^{n+1}\frac{\sqrt{n}+1}{n+1}$

10. $\sum_{n=1}^{\infty}(-1)^{n+1}\frac{3\sqrt{n}+1}{\sqrt{n}+1}$

Absolute Convergence

Which of the series in Exercises 11–44 converge absolutely, which converge, and which diverge? Give reasons for your answers.

11. $\sum_{n=1}^{\infty}(-1)^{n+1}(0.1)^n$

12. $\sum_{n=1}^{\infty}(-1)^{n+1}\frac{(0.1)^n}{n}$

13. $\sum_{n=1}^{\infty}(-1)^n\frac{1}{\sqrt{n}}$

14. $\sum_{n=1}^{\infty}\frac{(-1)^n}{1+\sqrt{n}}$

15. $\sum_{n=1}^{\infty}(-1)^{n+1}\frac{n}{n^3+1}$

16. $\sum_{n=1}^{\infty}(-1)^{n+1}\frac{n!}{2^n}$

17. $\sum_{n=1}^{\infty}(-1)^n\frac{1}{n+3}$

18. $\sum_{n=1}^{\infty}(-1)^n\frac{\sin n}{n^2}$

19. $\sum_{n=1}^{\infty}(-1)^{n+1}\frac{3+n}{5+n}$

20. $\sum_{n=2}^{\infty}(-1)^n\frac{1}{\ln(n^3)}$

21. $\sum_{n=1}^{\infty}(-1)^{n+1}\frac{1+n}{n^2}$

22. $\sum_{n=1}^{\infty}\frac{(-2)^{n+1}}{n+5^n}$

23. $\sum_{n=1}^{\infty}(-1)^n n^2(2/3)^n$

24. $\sum_{n=1}^{\infty}(-1)^{n+1}(\sqrt[n]{10})$

25. $\sum_{n=1}^{\infty}(-1)^n\frac{\tan^{-1}n}{n^2+1}$

26. $\sum_{n=2}^{\infty}(-1)^{n+1}\frac{1}{n\ln n}$

27. $\sum_{n=1}^{\infty}(-1)^n\frac{n}{n+1}$

28. $\sum_{n=1}^{\infty}(-1)^n\frac{\ln n}{n-\ln n}$

29. $\sum_{n=1}^{\infty}\frac{(-100)^n}{n!}$

30. $\sum_{n=1}^{\infty}(-5)^{-n}$

31. $\sum_{n=1}^{\infty}\frac{(-1)^{n-1}}{n^2+2n+1}$

32. $\sum_{n=2}^{\infty}(-1)^n\left(\frac{\ln n}{\ln n^2}\right)^n$

33. $\sum_{n=1}^{\infty}\frac{\cos n\pi}{n\sqrt{n}}$

34. $\sum_{n=1}^{\infty}\frac{\cos n\pi}{n}$

35. $\sum_{n=1}^{\infty}\frac{(-1)^n(n+1)^n}{(2n)^n}$

36. $\sum_{n=1}^{\infty}\frac{(-1)^{n+1}(n!)^2}{(2n)!}$

37. $\sum_{n=1}^{\infty}(-1)^n\frac{(2n)!}{2^n n! n}$

38. $\sum_{n=1}^{\infty}(-1)^n\frac{(n!)^2 3^n}{(2n+1)!}$

39. $\sum_{n=1}^{\infty}(-1)^n(\sqrt{n+1}-\sqrt{n})$

40. $\sum_{n=1}^{\infty}(-1)^n(\sqrt{n^2+n}-n)$

41. $\sum_{n=1}^{\infty}(-1)^n\left(\sqrt{n+\sqrt{n}}-\sqrt{n}\right)$

42. $\sum_{n=1}^{\infty}\frac{(-1)^n}{\sqrt{n}+\sqrt{n+1}}$

43. $\sum_{n=1}^{\infty} (-1)^n \operatorname{sech} n$

44. $\sum_{n=1}^{\infty} (-1)^n \operatorname{csch} n$

Error Estimation

In Exercises 45–48, estimate the magnitude of the error involved in using the sum of the first four terms to approximate the sum of the entire series.

45. $\sum_{n=1}^{\infty} (-1)^{n+1} \dfrac{1}{n}$ It can be shown that the sum is ln 2.

46. $\sum_{n=1}^{\infty} (-1)^{n+1} \dfrac{1}{10^n}$

47. $\sum_{n=1}^{\infty} (-1)^{n+1} \dfrac{(0.01)^n}{n}$ As you will see in Section 8.8, the sum is ln (1.01).

48. $\dfrac{1}{1+t} = \sum_{n=0}^{\infty} (-1)^n t^n, \quad 0 < t < 1$

CALCULATOR Approximate the sums in Exercises 49 and 50 with an error of magnitude less than 5×10^{-6}.

49. $\sum_{n=0}^{\infty} (-1)^n \dfrac{1}{(2n)!}$ As you will see in Section 8.10, the sum is cos 1, the cosine of 1 radian.

50. $\sum_{n=0}^{\infty} (-1)^n \dfrac{1}{n!}$ As you will see in Section 8.10, the sum is e^{-1}.

Theory and Examples

51. a)   The series

$$\frac{1}{3} - \frac{1}{2} + \frac{1}{9} - \frac{1}{4} + \frac{1}{27} - \frac{1}{8} + \cdots + \frac{1}{3^n} - \frac{1}{2^n} + \cdots$$

does not meet one of the conditions of Theorem 8. Which one?

b)   Find the sum of the series in (a).

52. CALCULATOR The limit L of an alternating series that satisfies the conditions of Theorem 8 lies between the values of any two consecutive partial sums. This suggests using the average

$$\frac{s_n + s_{n+1}}{2} = s_n + \frac{1}{2}(-1)^{n+2} a_{n+1}$$

to estimate L. Compute

$$s_{20} + \frac{1}{2} \cdot \frac{1}{21}$$

as an approximation to the sum of the alternating harmonic series. The exact sum is ln 2 = 0.6931....

53. *The sign of the remainder of an alternating series that satisfies the conditions of Theorem 8.* Prove the assertion in Theorem 9 that whenever an alternating series satisfying the conditions of Theorem 8 is approximated with one of its partial sums, then the remainder (sum of the unused terms) has the same sign as the first unused term. (*Hint:* Group the remainder's terms in consecutive pairs.)

54. Show that the sum of the first $2n$ terms of the series

$$1 - \frac{1}{2} + \frac{1}{2} - \frac{1}{3} + \frac{1}{3} - \frac{1}{4} + \frac{1}{4} - \frac{1}{5} + \frac{1}{5} - \frac{1}{6} + \cdots$$

is the same as the sum of the first n terms of the series

$$\frac{1}{1 \cdot 2} + \frac{1}{2 \cdot 3} + \frac{1}{3 \cdot 4} + \frac{1}{4 \cdot 5} + \frac{1}{5 \cdot 6} + \cdots.$$

Do these series converge? What is the sum of the first $2n + 1$ terms of the first series? If the series converge, what is their sum?

55. Show that if $\sum_{n=1}^{\infty} a_n$ diverges, then $\sum_{n=1}^{\infty} |a_n|$ diverges.

56. Show that if $\sum_{n=1}^{\infty} a_n$ converges absolutely, then

$$\left| \sum_{n=1}^{\infty} a_n \right| \le \sum_{n=1}^{\infty} |a_n|.$$

57. Show that if $\sum_{n=1}^{\infty} a_n$ and $\sum_{n=1}^{\infty} b_n$ both converge absolutely, then so does

a) $\sum_{n=1}^{\infty} (a_n + b_n)$ **b)** $\sum_{n=1}^{\infty} (a_n - b_n)$

c) $\sum_{n=1}^{\infty} k a_n$ (k any number)

58. Show by example that $\sum_{n=1}^{\infty} a_n b_n$ may diverge even if $\sum_{n=1}^{\infty} a_n$ and $\sum_{n=1}^{\infty} b_n$ both converge.

59. CALCULATOR In Example 7, suppose the goal is to arrange the terms to get a new series that converges to $-1/2$. Start the new arrangement with the first negative term, which is $-1/2$. Whenever you have a sum that is less than or equal to $-1/2$, start introducing positive terms, taken in order, until the new total is greater than $-1/2$. Then add negative terms until the total is less than or equal to $-1/2$ again. Continue this process until your partial sums have been above the target at least three times and finish at or below it. If s_n is the sum of the first n terms of your new series, plot the points (n, s_n) to illustrate how the sums are behaving.

60. *Outline of the proof of the Rearrangement Theorem (Theorem 11).*

a)   Let ϵ be a positive real number, let $L = \sum_{n=1}^{\infty} a_n$, and let $s_k = \sum_{n=1}^{k} a_n$. Show that for some index N_1 and for some index $N_2 \ge N_1$,

$$\sum_{n=N_1}^{\infty} |a_n| < \frac{\epsilon}{2} \quad \text{and} \quad |s_{N_2} - L| < \frac{\epsilon}{2}.$$

Since all the terms $a_1, a_2, \ldots, a_{N_2}$ appear somewhere in the sequence $\{b_n\}$, there is an index $N_3 \ge N_2$ such that if $n \ge N_3$, then $(\sum_{k=1}^{n} b_k) - s_{N_2}$ is at most a sum of terms a_m with $m \ge N_1$. Therefore, if $n \ge N_3$,

$$\left| \sum_{k=1}^{n} b_k - L \right| \le \left| \sum_{k=1}^{n} b_k - s_{N_2} \right| + |s_{N_2} - L|$$

$$\le \sum_{k=N_1}^{\infty} |a_k| + |s_{N_2} - L| < \epsilon.$$

b) The argument in (a) shows that if $\sum_{n=1}^{\infty} a_n$ converges absolutely then $\sum_{n=1}^{\infty} b_n$ converges and $\sum_{n=1}^{\infty} b_n = \sum_{n=1}^{\infty} a_n$. Now show that because $\sum_{n=1}^{\infty} |a_n|$ converges, $\sum_{n=1}^{\infty} |b_n|$ converges to $\sum_{n=1}^{\infty} |a_n|$.

61. *Unzipping absolutely convergent series.*

a) Show that if $\sum_{n=1}^{\infty} |a_n|$ converges and

$$b_n = \begin{cases} a_n & \text{if } a_n \geq 0 \\ 0 & \text{if } a_n < 0, \end{cases}$$

then $\sum_{n=1}^{\infty} b_n$ converges.

b) Use the results in (a) to show likewise that if $\sum_{n=1}^{\infty} |a_n|$ converges and

$$c_n = \begin{cases} 0 & \text{if } a_n \geq 0 \\ a_n & \text{if } a_n < 0, \end{cases}$$

then $\sum_{n=1}^{\infty} c_n$ converges.

In other words, if a series converges absolutely, its positive terms form a convergent series, and so do its negative terms. Furthermore,

$$\sum_{n=1}^{\infty} a_n = \sum_{n=1}^{\infty} b_n + \sum_{n=1}^{\infty} c_n$$

because $b_n = (a_n + |a_n|)/2$ and $c_n = (a_n - |a_n|)/2$.

62. What is wrong here:

Multiply both sides of the alternating harmonic series

$$S = 1 - \frac{1}{2} + \frac{1}{3} - \frac{1}{4} + \frac{1}{5} - \frac{1}{6} +$$

$$\frac{1}{7} - \frac{1}{8} + \frac{1}{9} - \frac{1}{10} + \frac{1}{11} - \frac{1}{12} + \cdots$$

by 2 to get

$$2S = 2 - 1 +$$

$$\frac{2}{3} - \frac{1}{2} + \frac{2}{5} - \frac{1}{3} + \frac{2}{7} - \frac{1}{4} + \frac{2}{9} - \frac{1}{5} + \frac{2}{11} - \frac{1}{6} + \cdots.$$

Collect terms with the same denominator, as the arrows indicate, to arrive at

$$2S = 1 - \frac{1}{2} + \frac{1}{3} - \frac{1}{4} + \frac{1}{5} - \frac{1}{6} + \cdots.$$

The series on the right-hand side of this equation is the series we started with. Therefore, $2S = S$, and dividing by S gives $2 = 1$. (Source: "Riemann's Rearrangement Theorem" by Stewart Galanor, *Mathematics Teacher,* Vol. 80, No. 8, 1987, pp. 675–81.)

63. Draw a figure similar to Fig. 8.14 to illustrate the convergence of the series in Theorem 8 when $N > 1$.

8.8

Power Series

Now that we can test infinite series for convergence we can study the infinite polynomials mentioned at the beginning of Section 8.3. We call these polynomials power series because they are defined as infinite series of powers of some variable, in our case x. Like polynomials, power series can be added, subtracted, multiplied, differentiated, and integrated to give new power series.

Power Series and Convergence

We begin with the formal definition.

Equation (1) is the special case obtained by taking $a = 0$ in Eq. (2).

Definition

A **power series about $x = 0$** is a series of the form

$$\sum_{n=0}^{\infty} c_n x^n = c_0 + c_1 x + c_2 x^2 + \cdots + c_n x^n + \cdots. \tag{1}$$

A **power series about $x = a$** is a series of the form

$$\sum_{n=0}^{\infty} c_n (x - a)^n = c_0 + c_1(x - a) + c_2(x - a)^2 + \cdots + c_n(x - a)^n + \cdots \tag{2}$$

in which the **center** a and the **coefficients** $c_0, c_1, c_2, \ldots, c_n, \ldots$ are constants.

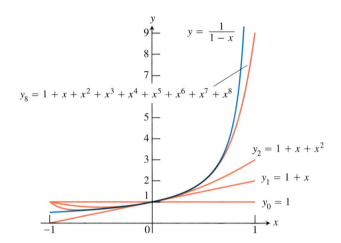

8.15 The graphs of $f(x) = 1/(1 - x)$ and four of its polynomial approximations (Example 1).

EXAMPLE 1   Taking all the coefficients to be 1 in Eq. (1) gives the geometric power series

$$\sum_{n=0}^{\infty} x^n = 1 + x + x^2 + \cdots + x^n + \cdots.$$

This is the geometric series with first term 1 and ratio x. It converges to $1/(1 - x)$ for $|x| < 1$. We express this fact by writing

$$\frac{1}{1 - x} = 1 + x + x^2 + \cdots + x^n + \cdots, \qquad -1 < x < 1. \tag{3}$$

$\square$

Up to now, we have used Eq. (3) as a formula for the sum of the series on the right. We now change the focus: We think of the partial sums of the series on the right as polynomials $P_n(x)$ that approximate the function on the left. For values of x near zero, we need take only a few terms of the series to get a good approximation. As we move toward $x = 1$, or -1, we must take more terms. Figure 8.15 shows the graphs of $f(x) = 1/(1 - x)$, and the approximating polynomials $y_n = P_n(x)$ for $n = 0, 1, 2,$ and 8.

EXAMPLE 2   The power series

$$1 - \frac{1}{2}(x - 2) + \frac{1}{4}(x - 2)^2 + \cdots + \left(-\frac{1}{2}\right)^n (x - 2)^n + \cdots \tag{4}$$

matches Eq. (2) with $a = 2$, $c_0 = 1$, $c_1 = -1/2$, $c_2 = 1/4, \ldots, c_n = (-1/2)^n$. This is a geometric series with first term 1 and ratio $r = -\dfrac{x - 2}{2}$. The series converges for $\left| \dfrac{x - 2}{2} \right| < 1$ or $0 < x < 4$. The sum is

$$\frac{1}{1 - r} = \frac{1}{1 + \dfrac{x - 2}{2}} = \frac{2}{x},$$

so

$$\frac{2}{x} = 1 - \frac{(x - 2)}{2} + \frac{(x - 2)^2}{4} - \cdots + \left(-\frac{1}{2}\right)^n (x - 2)^n + \cdots, \qquad 0 < x < 4.$$

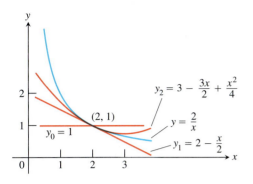

8.16 The graphs of $f(x) = 2/x$ and its first three polynomial approximations (Example 2).

Series (4) generates useful polynomial approximations of $f(x) = 2/x$ for values of x near 2:

$$P_0(x) = 1$$

$$P_1(x) = 1 - \frac{1}{2}(x - 2) = 2 - \frac{x}{2}$$

$$P_2(x) = 1 - \frac{1}{2}(x - 2) + \frac{1}{4}(x - 2)^2 = 3 - \frac{3x}{2} + \frac{x^2}{4},$$

and so on (Fig. 8.16).

EXAMPLE 3 For what values of x do the following power series converge?

a) $\displaystyle\sum_{n=1}^{\infty} (-1)^{n-1} \frac{x^n}{n} = x - \frac{x^2}{2} + \frac{x^3}{3} - \cdots$

b) $\displaystyle\sum_{n=1}^{\infty} (-1)^{n-1} \frac{x^{2n-1}}{2n-1} = x - \frac{x^3}{3} + \frac{x^5}{5} - \cdots$

c) $\displaystyle\sum_{n=0}^{\infty} \frac{x^n}{n!} = 1 + x + \frac{x^2}{2!} + \frac{x^3}{3!} + \cdots$

d) $\displaystyle\sum_{n=0}^{\infty} n!\, x^n = 1 + x + 2!\, x^2 + 3!\, x^3 + \cdots$

Solution Apply the Ratio Test to the series $\sum |u_n|$, where u_n is the nth term of the series in question.

a) $\left| \dfrac{u_{n+1}}{u_n} \right| = \dfrac{n}{n+1} |x| \to |x|.$

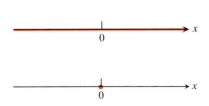

The series converges absolutely for $|x| < 1$. It diverges if $|x| > 1$ because the nth term does not converge to zero. At $x = 1$, we get the alternating harmonic series $1 - 1/2 + 1/3 - 1/4 + \cdots$, which converges. At $x = -1$ we get $-1 - 1/2 - 1/3 - 1/4 - \cdots$, the negative of the harmonic series; it diverges. Series (a) converges for $-1 < x \le 1$ and diverges elsewhere.

b) $\left| \dfrac{u_{n+1}}{u_n} \right| = \dfrac{2n-1}{2n+1} x^2 \to x^2.$

The series converges absolutely for $x^2 < 1$. It diverges for $x^2 > 1$ because the nth term does not converge to zero. At $x = 1$ the series becomes $1 - 1/3 + 1/5 - 1/7 + \cdots$, which converges by the Alternating Series Theorem. It also converges at $x = -1$ because it is again an alternating series that satisfies the conditions for convergence. The value at $x = -1$ is the negative of the value at $x = 1$. Series (b) converges for $-1 \le x \le 1$ and diverges elsewhere.

c) $\left| \dfrac{u_{n+1}}{u_n} \right| = \left| \dfrac{x^{n+1}}{(n+1)!} \cdot \dfrac{n!}{x^n} \right| = \dfrac{|x|}{n+1} \to 0$ for every x.

The series converges absolutely for all x.

d) $\left| \dfrac{u_{n+1}}{u_n} \right| = \left| \dfrac{(n+1)!\, x^{n+1}}{n!\, x^n} \right| = (n+1)|x| \to \infty$ unless $x = 0$.

The series diverges for all values of x except $x = 0$.

Example 3 illustrates how we usually test a power series for convergence, and the possible results.

How to Test a Power Series for Convergence

Step 1: Use the Ratio Test (or nth-Root Test) to find the interval where the series converges absolutely. Ordinarily, this is an open interval

$$|x - a| < R \qquad \text{or} \qquad a - R < x < a + R.$$

Step 2: If the interval of absolute convergence is finite, test for convergence or divergence at each endpoint, as in Examples 3(a) and (b). Use a Comparison Test, the Integral Test, or the Alternating Series Test.

Step 3: If the interval of absolute convergence is $a - R < x < a + R$, the series diverges for $|x - a| > R$ (it does not even converge conditionally), because the nth term does not approach zero for those values of x.

To simplify the notation, Theorem 12 deals with the convergence of series of the form $\sum a_n x^n$. For series of the form $\sum a_n (x - a)^n$ we can replace $x - a$ by x' and apply the results to the series $\sum a_n (x')^n$.

Theorem 12

The Convergence Theorem for Power Series

If $\displaystyle\sum_{n=0}^{\infty} a_n x^n = a_0 + a_1 x + a_2 x^2 + \cdots$

converges for $x = c \neq 0$, then it converges absolutely for all $|x| < |c|$. If the series diverges for $x = d$, then it diverges for all $|x| > |d|$.

Proof Suppose the series $\sum_{n=0}^{\infty} a_n c^n$ converges. Then $\lim_{n \to \infty} a_n c^n = 0$. Hence, there is an integer N such that $|a_n c^n| < 1$ for all $n \geq N$. That is,

$$|a_n| < \frac{1}{|c|^n} \qquad \text{for } n \geq N. \tag{5}$$

Now take any x such that $|x| < |c|$ and consider

$$|a_0| + |a_1 x| + \cdots + |a_{N-1} x^{N-1}| + |a_N x^N| + |a_{N+1} x^{N+1}| + \cdots.$$

There are only a finite number of terms prior to $|a_N x^N|$, and their sum is finite. Starting with $|a_N x^N|$ and beyond, the terms are less than

$$\left|\frac{x}{c}\right|^N + \left|\frac{x}{c}\right|^{N+1} + \left|\frac{x}{c}\right|^{N+2} + \cdots \tag{6}$$

because of (5). But the series in (6) is a geometric series with ratio $r = |x/c|$, which is less than 1, since $|x| < |c|$. Hence the series (6) converges, so the original series converges absolutely. This proves the first half of the theorem.

The second half of the theorem follows from the first. If the series diverges at $x = d$ and converges at a value x_0 with $|x_0| > |d|$, we may take $c = x_0$ in the first half of the theorem and conclude that the series converges absolutely at d. But the series cannot converge absolutely and diverge at one and the same time. Hence, if it diverges at d, it diverges for all $|x| > |d|$. ∎

The Radius and Interval of Convergence

The examples we have looked at, and the theorem we just proved, lead to the conclusion that a power series behaves in one of the following three ways.

Possible Behavior of $\sum c_n(x - a)^n$

1. There is a positive number R such that the series diverges for $|x - a| > R$ but converges absolutely for $|x - a| < R$. The series may or may not converge at either of the endpoints $x = a - R$ and $x = a + R$.

2. The series converges absolutely for every x ($R = \infty$).

3. The series converges at $x = a$ and diverges elsewhere ($R = 0$).

In case 1, the set of points at which the series converges is a finite interval, called the **interval of convergence.** We know from the examples that the interval can be open, half-open, or closed, depending on the particular series. But no matter which kind of interval it is, R is called the **radius of convergence** of the series, and $a + R$ is the least upper bound of the set of points at which the series converges. The convergence is absolute at every point in the interior of the interval. If a power series converges absolutely for all values of x, we say that its **radius of convergence is infinite.** If it converges only at $x = a$, the **radius of convergence is zero.**

Term-by-Term Differentiation

A theorem from advanced calculus says that a power series can be differentiated term by term at each interior point of its interval of convergence.

A word of caution

Term-by-term differentiation might not work for other kinds of series. For example, the trigonometric series

$$\sum_{n=1}^{\infty} \frac{\sin (n! \, x)}{n^2}$$

converges for all x. But if we differentiate term by term we get the series

$$\sum_{n=1}^{\infty} \frac{n! \cos (n! \, x)}{n^2},$$

which diverges for all x.

Theorem 13
The Term-by-Term Differentiation Theorem

If $\sum c_n(x - a)^n$ converges for $a - R < x < a + R$ for some $R > 0$, it defines a function f:

$$f(x) = \sum_{n=0}^{\infty} c_n(x - a)^n, \qquad a - R < x < a + R.$$

Such a function f has derivatives of all orders inside the interval of convergence. We can obtain the derivatives by differentiating the original series term by term:

$$f'(x) = \sum_{n=1}^{\infty} nc_n(x - a)^{n-1}$$

$$f''(x) = \sum_{n=2}^{\infty} n(n - 1)c_n(x - a)^{n-2},$$

and so on. Each of these derived series converges at every interior point of the interval of convergence of the original series.

EXAMPLE 4   Find series for $f'(x)$ and $f''(x)$ if

$$f(x) = \frac{1}{1-x} = 1 + x + x^2 + x^3 + x^4 + \cdots + x^n + \cdots$$

$$= \sum_{n=0}^{\infty} x^n, \quad -1 < x < 1$$

Solution

$$f'(x) = \frac{1}{(1-x)^2} = 1 + 2x + 3x^2 + 4x^3 + \cdots + nx^{n-1} + \cdots$$

$$= \sum_{n=1}^{\infty} nx^{n-1}, \quad -1 < x < 1$$

$$f''(x) = \frac{2}{(1-x)^3} = 2 + 6x + 12x^2 + \cdots + n(n-1)x^{n-2} + \cdots$$

$$= \sum_{n=2}^{\infty} n(n-1)x^{n-2}, \quad -1 < x < 1$$

Term-by-Term Integration

Another advanced theorem states that a power series can be integrated term by term throughout its interval of convergence.

> **Theorem 14**
> **The Term-by-Term Integration Theorem**
> Suppose that
> $$f(x) = \sum_{n=0}^{\infty} c_n(x-a)^n$$
> converges for $a - R < x < a + R \ (R > 0)$. Then
> $$\sum_{n=0}^{\infty} c_n(x-a)^{n+1}/(n+1)$$
> converges for $a - R < x < a + R$ and
> $$\int f(x)\,dx = \sum_{n=0}^{\infty} c_n \frac{(x-a)^{n+1}}{n+1} + C$$
> for $a - R < x < a + R$.

EXAMPLE 5   *A series for* $\tan^{-1} x$, $-1 \leq x \leq 1$

Identify the function

$$f(x) = x - \frac{x^3}{3} + \frac{x^5}{5} - \cdots, \quad -1 \leq x \leq 1.$$

Solution   We differentiate the original series term by term and get

$$f'(x) = 1 - x^2 + x^4 - x^6 + \cdots, \quad -1 < x < 1.$$

This is a geometric series with first term 1 and ratio $-x^2$, so

$$f'(x) = \frac{1}{1 - (-x^2)} = \frac{1}{1 + x^2}.$$

We can now integrate $f'(x) = 1/(1 + x^2)$ to get

$$\int f'(x)\,dx = \int \frac{dx}{1+x^2} = \tan^{-1} x + C.$$

The series for $f(x)$ is zero when $x = 0$, so $C = 0$. Hence

$$f(x) = x - \frac{x^3}{3} + \frac{x^5}{5} - \frac{x^7}{7} + \cdots = \tan^{-1} x, \qquad -1 < x < 1. \qquad (7)$$

In Section 8.11, we will see that the series also converges to $\tan^{-1} x$ at $x = \pm 1$.

❏

EXAMPLE 6   *A series for* $\ln(1 + x)$, $-1 < x \leq 1$

The series

$$\frac{1}{1+t} = 1 - t + t^2 - t^3 + \cdots$$

converges on the open interval $-1 < t < 1$. Therefore,

$$\ln(1 + x) = \int_0^x \frac{1}{1+t}\,dt = t - \frac{t^2}{2} + \frac{t^3}{3} - \frac{t^4}{4} + \cdots \Bigg]_0^x$$

$$= x - \frac{x^2}{2} + \frac{x^3}{3} - \frac{x^4}{4} + \cdots, \qquad -1 < x < 1.$$

It can also be shown that the series converges at $x = 1$ to the number $\ln 2$, but that was not guaranteed by the theorem.

❏

Notice that the original series in Example 5 converges at both endpoints of the original interval of convergence, but Theorem 13 can guarantee the convergence of the differentiated series only inside the interval.

Technology   *Study of Series*   Series are in many ways analogous to integrals. Just as the number of functions with explicit antiderivatives in terms of elementary functions is small compared to the number of integrable functions, the number of power series in x that agree with explicit elementary functions on x-intervals is small compared to the number of power series that converge on some x-interval. Graphing utilities can aid in the study of such series in much the same way that numerical integration aids in the study of definite integrals. The ability to study power series at particular values of x is built into most Computer Algebra Systems.

If a series converges rapidly enough, CAS exploration might give us an idea of the sum. For instance, in calculating the early partial sums of the series $\sum_{n=1}^{\infty}[1/(2^{n-1})]$ (Section 8.5, Example 3b), Maple returns $S_n = 1.6066\,95152$ for $31 \leq n \leq 200$. This suggests that the sum of the series is $1.6066\,95152$ to 10 digits. Indeed,

$$\sum_{n=201}^{\infty} \frac{1}{2^n - 1} = \sum_{n=201}^{\infty} \frac{1}{2^{n-1}(2 - (1/2^{n-1}))} < \sum_{n=201}^{\infty} \frac{1}{2^{n-1}} = \frac{1}{2^{199}} < 1.25 \times 10^{-60}.$$

The remainder after 200 terms is negligible.

However, CAS and calculator exploration cannot do much for us if the series converges or diverges very slowly, and indeed can be downright misleading. For example, try calculating the partial sums of the series $\sum_{n=1}^{\infty}[1/(10^{10}n)]$. The terms are tiny in comparison to the numbers we normally work with and the partial sums, even for hundreds of terms, are miniscule. We might well be fooled into thinking that the series converges. In fact, it diverges, as we can see by writing it as $(1/10^{10})\sum_{n=1}^{\infty}(1/n)$.

We will know better how to interpret numerical results after studying error estimates in Section 8.10.

Multiplication of Power Series

Still another advanced theorem states that absolutely converging power series can be multiplied the way we multiply polynomials.

Theorem 15

The Series Multiplication Theorem for Power Series

If $A(x) = \sum_{n=0}^{\infty} a_n x^n$ and $B(x) = \sum_{n=0}^{\infty} b_n x^n$ converge absolutely for $|x| < R$, and

$$c_n = a_0 b_n + a_1 b_{n-1} + a_2 b_{n-2} + \cdots + a_{n-1} b_1 + a_n b_0 = \sum_{k=0}^{n} a_k b_{n-k},$$

then $\sum_{n=0}^{\infty} c_n x^n$ converges absolutely to $A(x)B(x)$ for $|x| < R$:

$$\left(\sum_{n=0}^{\infty} a_n x^n \right) \cdot \left(\sum_{n=0}^{\infty} b_n x^n \right) = \sum_{n=0}^{\infty} c_n x^n.$$

EXAMPLE 7   Multiply the geometric series

$$\sum_{n=0}^{\infty} x^n = 1 + x + x^2 + \cdots + x^n + \cdots = \frac{1}{1-x}, \qquad \text{for } |x| < 1,$$

by itself to get a power series for $1/(1-x)^2$, for $|x| < 1$.

Solution   Let

$$A(x) = \sum_{n=0}^{\infty} a_n x^n = 1 + x + x^2 + \cdots + x^n + \cdots = 1/(1-x)$$

$$B(x) = \sum_{n=0}^{\infty} b_n x^n = 1 + x + x^2 + \cdots + x^n + \cdots = 1/(1-x)$$

and

$$c_n = \underbrace{a_0 b_n + a_1 b_{n-1} + \cdots + a_k b_{n-k} + \cdots + a_n b_0}_{n+1 \text{ terms}}$$

$$= \underbrace{1 + 1 + \cdots + 1}_{n+1 \text{ ones}} = n+1.$$

Then, by the Series Multiplication Theorem,

$$A(x) \cdot B(x) = \sum_{n=0}^{\infty} c_n x^n = \sum_{n=0}^{\infty} (n+1) x^n$$

$$= 1 + 2x + 3x^2 + 4x^3 + \cdots + (n+1)x^n + \cdots$$

is the series for $1/(1-x)^2$. The series all converge absolutely for $|x| < 1$.

Notice that Example 4 gives the same answer because

$$\frac{d}{dx} \left(\frac{1}{1-x} \right) = \frac{1}{(1-x)^2}.$$

Exercises 8.8

Intervals of Convergence

In Exercises 1–32, (a) find the series' radius and interval of convergence. For what values of x does the series converge (b) absolutely, (c) conditionally?

1. $\displaystyle\sum_{n=0}^{\infty} x^n$

2. $\displaystyle\sum_{n=0}^{\infty} (x+5)^n$

3. $\displaystyle\sum_{n=0}^{\infty} (-1)^n (4x+1)^n$

4. $\displaystyle\sum_{n=1}^{\infty} \frac{(3x-2)^n}{n}$

5. $\displaystyle\sum_{n=0}^{\infty} \frac{(x-2)^n}{10^n}$

6. $\displaystyle\sum_{n=0}^{\infty} (2x)^n$

7. $\displaystyle\sum_{n=0}^{\infty} \frac{nx^n}{n+2}$

8. $\displaystyle\sum_{n=1}^{\infty} \frac{(-1)^n (x+2)^n}{n}$

9. $\displaystyle\sum_{n=1}^{\infty} \frac{x^n}{n\sqrt{n}\,3^n}$

10. $\displaystyle\sum_{n=1}^{\infty} \frac{(x-1)^n}{\sqrt{n}}$

11. $\displaystyle\sum_{n=0}^{\infty} \frac{(-1)^n x^n}{n!}$

12. $\displaystyle\sum_{n=0}^{\infty} \frac{3^n x^n}{n!}$

13. $\displaystyle\sum_{n=0}^{\infty} \frac{x^{2n+1}}{n!}$

14. $\displaystyle\sum_{n=0}^{\infty} \frac{(2x+3)^{2n+1}}{n!}$

15. $\displaystyle\sum_{n=0}^{\infty} \frac{x^n}{\sqrt{n^2+3}}$

16. $\displaystyle\sum_{n=0}^{\infty} \frac{(-1)^n x^n}{\sqrt{n^2+3}}$

17. $\displaystyle\sum_{n=0}^{\infty} \frac{n(x+3)^n}{5^n}$

18. $\displaystyle\sum_{n=0}^{\infty} \frac{nx^n}{4^n(n^2+1)}$

19. $\displaystyle\sum_{n=0}^{\infty} \frac{\sqrt{n}\,x^n}{3^n}$

20. $\displaystyle\sum_{n=1}^{\infty} \sqrt[n]{n}\,(2x+5)^n$

21. $\displaystyle\sum_{n=1}^{\infty} \left(1+\frac{1}{n}\right)^n x^n$

22. $\displaystyle\sum_{n=1}^{\infty} (\ln n)\, x^n$

23. $\displaystyle\sum_{n=1}^{\infty} n^n x^n$

24. $\displaystyle\sum_{n=0}^{\infty} n!(x-4)^n$

25. $\displaystyle\sum_{n=1}^{\infty} \frac{(-1)^{n+1}(x+2)^n}{n2^n}$

26. $\displaystyle\sum_{n=0}^{\infty} (-2)^n(n+1)(x-1)^n$

27. $\displaystyle\sum_{n=2}^{\infty} \frac{x^n}{n(\ln n)^2}$ $\left(\begin{array}{l}\text{Get the information you need}\\ \text{about } \sum 1/(n(\ln n)^2) \text{ from}\\ \text{Section 8.4, Exercise 39.}\end{array}\right)$

28. $\displaystyle\sum_{n=2}^{\infty} \frac{x^n}{n\ln n}$ $\left(\begin{array}{l}\text{Get the information you need about}\\ \sum 1/(n\ln n) \text{ from}\\ \text{Section 8.4, Exercise 38.}\end{array}\right)$

29. $\displaystyle\sum_{n=1}^{\infty} \frac{(4x-5)^{2n+1}}{n^{3/2}}$

30. $\displaystyle\sum_{n=1}^{\infty} \frac{(3x+1)^{n+1}}{2n+2}$

31. $\displaystyle\sum_{n=1}^{\infty} \frac{(x+\pi)^n}{\sqrt{n}}$

32. $\displaystyle\sum_{n=0}^{\infty} \frac{(x-\sqrt{2})^{2n+1}}{2^n}$

In Exercises 33–38, find the series' interval of convergence and, within this interval, the sum of the series as a function of x.

33. $\displaystyle\sum_{n=0}^{\infty} \frac{(x-1)^{2n}}{4^n}$

34. $\displaystyle\sum_{n=0}^{\infty} \frac{(x+1)^{2n}}{9^n}$

35. $\displaystyle\sum_{n=0}^{\infty} \left(\frac{\sqrt{x}}{2}-1\right)^n$

36. $\displaystyle\sum_{n=0}^{\infty} (\ln x)^n$

37. $\displaystyle\sum_{n=0}^{\infty} \left(\frac{x^2+1}{3}\right)^n$

38. $\displaystyle\sum_{n=0}^{\infty} \left(\frac{x^2-1}{2}\right)^n$

Theory and Examples

39. For what values of x does the series

$$1 - \frac{1}{2}(x-3) + \frac{1}{4}(x-3)^2 + \cdots + \left(-\frac{1}{2}\right)^n (x-3)^n + \cdots$$

converge? What is its sum? What series do you get if you differentiate the given series term by term? For what values of x does the new series converge? What is its sum?

40. If you integrate the series in Exercise 39 term by term, what new series do you get? For what values of x does the new series converge, and what is another name for its sum?

41. The series

$$\sin x = x - \frac{x^3}{3!} + \frac{x^5}{5!} - \frac{x^7}{7!} + \frac{x^9}{9!} - \frac{x^{11}}{11!} + \cdots$$

converges to $\sin x$ for all x.

a) Find the first six terms of a series for $\cos x$. For what values of x should the series converge?

b) By replacing x by $2x$ in the series for $\sin x$, find a series that converges to $\sin 2x$ for all x.

c) Using the result in (a) and series multiplication, calculate the first six terms of a series for $2\sin x \cos x$. Compare your answer with the answer in (b).

42. The series

$$e^x = 1 + x + \frac{x^2}{2!} + \frac{x^3}{3!} + \frac{x^4}{4!} + \frac{x^5}{5!} + \cdots$$

converges to e^x for all x.

a) Find a series for $(d/dx)e^x$. Do you get the series for e^x? Explain your answer.

b) Find a series for $\int e^x dx$. Do you get the series for e^x? Explain your answer.

c) Replace x by $-x$ in the series for e^x to find a series that converges to e^{-x} for all x. Then multiply the series for e^x and e^{-x} to find the first six terms of a series for $e^{-x} \cdot e^x$.

43. The series

$$\tan x = x + \frac{x^3}{3} + \frac{2x^5}{15} + \frac{17x^7}{315} + \frac{62x^9}{2835} + \cdots$$

converges to $\tan x$ for $-\pi/2 < x < \pi/2$.

a) Find the first five terms of the series for $\ln|\sec x|$. For what values of x should the series converge?

b) Find the first five terms of the series for $\sec^2 x$. For what values of x should this series converge?

c) Check your result in (b) by squaring the series given for $\sec x$ in Exercise 44.

44. The series for

$$\sec x = 1 + \frac{x^2}{2} + \frac{5}{24}x^4 + \frac{61}{720}x^6 + \frac{277}{8064}x^8 + \cdots$$

converges to $\sec x$ for $-\pi/2 < x < \pi/2$.

a) Find the first five terms of a power series for the function $\ln|\sec x + \tan x|$. For what values of x should the series converge?

b) Find the first four terms of a series for $\sec x \tan x$. For what values of x should the series converge?

c) Check your result in (b) by multiplying the series for $\sec x$ by the series given for $\tan x$ in Exercise 43.

45. *Uniqueness of convergent power series*

a) Show that if two power series $\sum_{n=0}^{\infty} a_n x^n$ and $\sum_{n=0}^{\infty} b_n x^n$ are convergent and equal for all values of x in an open interval $(-c, c)$, then $a_n = b_n$ for every n. (*Hint:* Let $f(x) = \sum_{n=0}^{\infty} a_n x^n = \sum_{n=0}^{\infty} b_n x^n$. Differentiate term by term to show that a_n and b_n both equal $f^{(n)}(0)/(n!)$.)

b) Show that if $\sum_{n=0}^{\infty} a_n x^n = 0$ for all x in an open interval $(-c, c)$, then $a_n = 0$ for every n.

46. *The sum of the series* $\sum_{n=0}^{\infty} (n^2/2^n)$. To find the sum of this series, express $1/(1 - x)$ as a geometric series, differentiate both sides of the resulting equation with respect to x, multiply both sides of the result by x, differentiate again, multiply by x again, and set x equal to $1/2$. What do you get? (Source: David E. Dobbs' letter to the editor, *Illinois Mathematics Teacher,* Vol. 33, Issue 4, 1982, p. 27.)

47. *Convergence at endpoints.* Show by examples that the convergence of a power series at an endpoint of its interval of convergence may be either conditional or absolute.

48. Make up a power series whose interval of convergence is

a) $(-3, 3)$ **b)** $(-2, 0)$ **c)** $(1, 5)$.

8.9 Taylor and Maclaurin Series

This section shows how functions that are infinitely differentiable generate power series called Taylor series. In many cases, these series can provide useful polynomial approximations of the generating functions.

Series Representations

We know that within its interval of convergence the sum of a power series is a continuous function with derivatives of all orders. But what about the other way around? If a function $f(x)$ has derivatives of all orders on an interval I, can it be expressed as a power series on I? And if it can, what will its coefficients be?

We can answer the last question readily if we assume that $f(x)$ is the sum of a power series

$$f(x) = \sum_{n=0}^{\infty} a_n (x - a)^n$$

$$= a_0 + a_1(x - a) + a_2(x - a)^2 + \cdots + a_n(x - a)^n + \cdots$$

with a positive radius of convergence. By repeated term-by-term differentiation within the interval of convergence I we obtain

$$f'(x) = a_1 + 2a_2(x - a) + 3a_3(x - a)^2 + \cdots + na_n(x - a)^{n-1} + \cdots$$

$$f''(x) = 1 \cdot 2a_2 + 2 \cdot 3a_3(x - a) + 3 \cdot 4a_4(x - a)^2 + \cdots$$

$$f'''(x) = 1 \cdot 2 \cdot 3a_3 + 2 \cdot 3 \cdot 4a_4(x - a) + 3 \cdot 4 \cdot 5a_5(x - a)^2 + \cdots,$$

with the nth derivative, for all n, being

$$f^{(n)}(x) = n!\, a_n +\ \text{a sum of terms with } (x - a) \text{ as a factor.}$$

Since these equations all hold at $x = a$, we have

$$f'(a) = a_1,$$

$$f''(a) = 1 \cdot 2a_2,$$

$$f'''(a) = 1 \cdot 2 \cdot 3a_3,$$

and, in general,

$$f^{(n)}(a) = n! \, a_n.$$

These formulas reveal a marvelous pattern in the coefficients of any power series $\sum_{n=0}^{\infty} a_n(x - a)^n$ that converges to the values of f on I ("represents f on I," we say). If there *is* such a series (still an open question), then there is only one such series and its nth coefficient is

$$a_n = \frac{f^{(n)}(a)}{n!}.$$

If f has a series representation, then the series must be

$$f(x) = f(a) + f'(a)(x - a) + \frac{f''(a)}{2!}(x - a)^2$$

$$+ \cdots + \frac{f^{(n)}(a)}{n!}(x - a)^n + \cdots. \tag{1}$$

But if we start with an arbitrary function f that is infinitely differentiable on an interval I centered at $x = a$ and use it to generate the series in Eq. (1), will the series then converge to $f(x)$ at each x in the interior of I? The answer is maybe—for some functions it will but for other functions it will not, as we will see.

Taylor and Maclaurin Series

Definitions
Let f be a function with derivatives of all orders throughout some interval containing a as an interior point. Then the **Taylor series generated by f at $x = a$** is

$$\sum_{k=0}^{\infty} \frac{f^{(k)}(a)}{k!}(x - a)^k = f(a) + f'(a)(x - a) + \frac{f''(a)}{2!}(x - a)^2$$

$$+ \cdots + \frac{f^{(n)}(a)}{n!}(x - a)^n + \cdots.$$

The **Maclaurin series generated by f** is

$$\sum_{k=0}^{\infty} \frac{f^{(k)}(0)}{k!}x^k = f(0) + f'(0)x + \frac{f''(0)}{2!}x^2 + \cdots + \frac{f^{(n)}(0)}{n!}x^n + \cdots,$$

the Taylor series generated by f at $x = 0$.

EXAMPLE 1   Find the Taylor series generated by $f(x) = 1/x$ at $a = 2$. Where, if anywhere, does the series converge to $1/x$?

Solution We need to find $f(2), f'(2), f''(2), \ldots$. Taking derivatives we get

$$f(x) = x^{-1}, \qquad\qquad f(2) = 2^{-1} = \frac{1}{2},$$

$$f'(x) = -x^{-2}, \qquad\qquad f'(2) = -\frac{1}{2^2},$$

$$f''(x) = 2!\, x^{-3}, \qquad\qquad \frac{f''(2)}{2!} = 2^{-3} = \frac{1}{2^3},$$

$$f'''(x) = -3!\, x^{-4}, \qquad\qquad \frac{f'''(2)}{3!} = -\frac{1}{2^4},$$

$$\vdots \qquad\qquad\qquad\qquad \vdots$$

$$f^{(n)}(x) = (-1)^n n!\, x^{-(n+1)}, \qquad \frac{f^{(n)}(2)}{n!} = \frac{(-1)^n}{2^{n+1}}.$$

The Taylor series is

$$f(2) + f'(2)(x-2) + \frac{f''(2)}{2!}(x-2)^2 + \cdots + \frac{f^{(n)}}{n!}(x-2)^n + \cdots$$

$$= \frac{1}{2} - \frac{(x-2)}{2^2} + \frac{(x-2)^2}{2^3} - \cdots + (-1)^n \frac{(x-2)^n}{2^{n+1}} + \cdots.$$

This is a geometric series with first term 1/2 and ratio $r = -(x-2)/2$. It converges absolutely for $|x - 2| < 2$ and its sum is

$$\frac{1/2}{1 + (x-2)/2} = \frac{1}{2 + (x-2)} = \frac{1}{x}.$$

In this example the Taylor series generated by $f(x) = 1/x$ at $a = 2$ converges to $1/x$ for $|x - 2| < 2$ or $0 < x < 4$. ❑

Taylor Polynomials

The linearization of a differentiable function f at a point a is the polynomial

$$P_1(x) = f(a) + f'(a)(x - a).$$

If f has derivatives of higher order at a, then it has higher order polynomial approximations as well, one for each available derivative. These polynomials are called the Taylor polynomials of f.

We speak of a Taylor polynomial of *order n* rather than *degree n* because $f^{(n)}(a)$ may be zero. The first two Taylor polynomials of $\cos x$ at $x = 0$, for example, are $P_0(x) = 1$ and $P_1(x) = 1$. The first order polynomial has degree zero, not one.

Definition

Let f be a function with derivatives of order k for $k = 1, 2, \ldots, N$ in some interval containing a as an interior point. Then for any integer n from 0 through N, the **Taylor polynomial of order n** generated by f at $x = a$ is the polynomial

$$P_n(x) = f(a) + f'(a)(x - a) + \frac{f''(a)}{2!}(x - a)^2 + \cdots$$

$$+ \frac{f^{(k)}(a)}{k!}(x - a)^k + \cdots + \frac{f^{(n)}(a)}{n!}(x - a)^n.$$

Just as the linearization of f at $x = a$ provides the best linear approximation of f in the neighborhood of a, the higher order Taylor polynomials provide the best polynomial approximations of their respective degrees. (See Exercise 32.)

EXAMPLE 2   Find the Taylor series and the Taylor polynomials generated by $f(x) = e^x$ at $x = 0$.

Solution   Since

$$f(x) = e^x, \qquad f'(x) = e^x, \qquad \cdots, \qquad f^{(n)}(x) = e^x, \qquad \cdots,$$

we have

$$f(0) = e^0 = 1, \qquad f'(0) = 1, \qquad \cdots, \qquad f^{(n)}(0) = 1, \qquad \cdots.$$

The Taylor series generated by f at $x = 0$ is

$$f(0) + f'(0)x + \frac{f''(0)}{2!}x^2 + \cdots + \frac{f^{(n)}(0)}{n!}x^n + \cdots$$

$$= 1 + x + \frac{x^2}{2} + \cdots + \frac{x^n}{n!} + \cdots$$

$$= \sum_{k=0}^{\infty} \frac{x^k}{k!}.$$

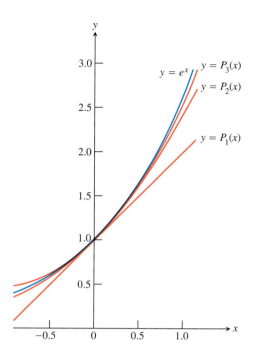

8.17 The graph of $f(x) = e^x$ and its Taylor polynomials

$P_1(x) = 1 + x,$

$P_2(x) = 1 + x + (x^2/2!),$ and

$P_3(x) = 1 + x + (x^2/2!) + (x^3/3!).$

Notice the very close agreement near the center $x = 0$.

By definition, this is also the Maclaurin series for e^x. In Section 8.10 we will see that the series converges to e^x at every x.

The Taylor polynomial of order n at $x = 0$ is

$$P_n(x) = 1 + x + \frac{x^2}{2} + \cdots + \frac{x^n}{n!}.$$

See Fig. 8.17.   ❑

EXAMPLE 3   Find the Taylor series and Taylor polynomials generated by $f(x) = \cos x$ at $x = 0$.

Solution   The cosine and its derivatives are

$$\begin{aligned} f(x) &= \cos x & f'(x) &= -\sin x, \\ f''(x) &= -\cos x, & f^{(3)}(x) &= \sin x, \\ &\ \ \vdots & &\ \ \vdots \\ f^{(2n)}(x) &= (-1)^n \cos x, & f^{(2n+1)}(x) &= (-1)^{n+1} \sin x. \end{aligned}$$

At $x = 0$, the cosines are 1 and the sines are 0, so

$$f^{(2n)}(0) = (-1)^n, \qquad f^{(2n+1)}(0) = 0.$$

The Taylor series generated by f at 0 is

$$f(0) + f'(0)x + \frac{f''(0)}{2!}x^2 + \frac{f'''(0)}{3!}x^3 + \cdots + \frac{f^{(n)}(0)}{n!}x^n + \cdots$$

$$= 1 + 0 \cdot x - \frac{x^2}{2!} + 0 \cdot x^3 + \frac{x^4}{4!} + \cdots + (-1)^n \frac{x^{2n}}{(2n)!} + \cdots$$

$$= \sum_{n=0}^{\infty} \frac{(-1)^n x^{2n}}{(2n)!}.$$

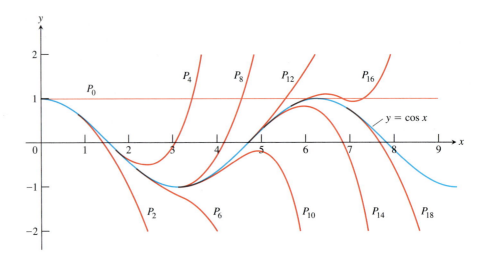

8.18 The polynomials

$$P_{2n}(x) = \sum_{k=0}^{n} [(-1)^k x^{2k}/(2k)!]$$

converge to cos x as $n \to \infty$. We can deduce the behavior of cos x arbitrarily far away solely from knowing the values of the cosine and its derivatives at $x = 0$.

Infinitely differentiable functions that are represented by their Taylor series only at isolated points are, in practice, very rare.

Who invented Taylor series?

Brook Taylor (1685–1731) did not invent Taylor series, and Maclaurin series were not developed by Colin Maclaurin (1698–1746). James Gregory was already working with Taylor series when Taylor was only a few years old, and he published the Maclaurin series for tan x, sec x, $\tan^{-1}x$, and $\sec^{-1}x$ ten years before Maclaurin was born. Nicolaus Mercator discovered the Maclaurin series for ln $(1 + x)$ at about the same time.

Taylor was unaware of Gregory's work when he published his book *Methodus incrementorum directa et inversa* in 1715, containing what we now call Taylor series. Maclaurin quoted Taylor's work in a calculus book he wrote in 1742. The book popularized series representations of functions and although Maclaurin never claimed to have discovered them, Taylor series centered at $x = 0$ became known as Maclaurin series. History evened things up in the end. Maclaurin, a brilliant mathematician, was the original discoverer of the rule for solving systems of equations that we call Cramer's rule.

By definition, this is also the Maclaurin series for cos x. In Section 8.10, we will see that the series converges to cos x at every x.

Because $f^{(2n+1)}(0) = 0$, the Taylor polynomials of orders $2n$ and $2n + 1$ are identical:

$$P_{2n}(x) = P_{2n+1}(x) = 1 - \frac{x^2}{2!} + \frac{x^4}{4!} - \cdots + (-1)^n \frac{x^{2n}}{(2n)!}.$$

Figure 8.18 shows how well these polynomials approximate $f(x) = \cos x$ near $x = 0$. Only the right-hand portions of the graphs are given because the graphs are symmetric about the y-axis. ❑

EXAMPLE 4   *A function f whose Taylor series converges at every x but converges to f(x) only at x = 0*

It can be shown (though not easily) that

$$f(x) = \begin{cases} 0, & x = 0 \\ e^{-1/x^2}, & x \neq 0 \end{cases}$$

(Fig. 8.19) has derivatives of all orders at $x = 0$ and that $f^{(n)}(0) = 0$ for all n. This means that the Taylor series generated by f at $x = 0$ is

$$f(0) + f'(0)x + \frac{f''(0)}{2!}x^2 + \cdots + \frac{f^{(n)}(0)}{n!}x^n + \cdots$$

$$= 0 + 0 \cdot x + 0 \cdot x^2 + \cdots + 0 \cdot x^n + \cdots$$

$$= 0 + 0 + \cdots + 0 + \cdots.$$

The series converges for every x (its sum is 0) but converges to $f(x)$ only at $x = 0$. ❑

Two questions still remain.

1.   For what values of x can we normally expect a Taylor series to converge to its generating function?

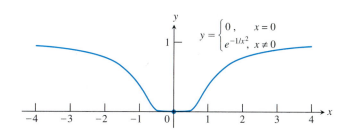

8.19 The graph of the continuous extension of $y = e^{-1/x^2}$ is so flat at the origin that all of its derivatives there are zero (Example 4).

2. How accurately do a function's Taylor polynomials approximate the function on a given interval?

The answers are provided by a theorem of Taylor in the next section.

Exercises 8.9

Finding Taylor Polynomials

In Exercises 1–8, find the Taylor polynomials of orders 0, 1, 2, and 3 generated by f at a.

1. $f(x) = \ln x, \quad a = 1$

2. $f(x) = \ln(1 + x), \quad a = 0$

3. $f(x) = 1/x, \quad a = 2$

4. $f(x) = 1/(x + 2), \quad a = 0$

5. $f(x) = \sin x, \quad a = \pi/4$

6. $f(x) = \cos x, \quad a = \pi/4$

7. $f(x) = \sqrt{x}, \quad a = 4$

8. $f(x) = \sqrt{x + 4}, \quad a = 0$

Finding Maclaurin Series

Find the Maclaurin series for the functions in Exercises 9–20.

9. e^{-x}

10. $e^{x/2}$

11. $\dfrac{1}{1 + x}$

12. $\dfrac{1}{1 - x}$

13. $\sin 3x$

14. $\sin \dfrac{x}{2}$

15. $7\cos(-x)$

16. $5\cos \pi x$

17. $\cosh x = \dfrac{e^x + e^{-x}}{2}$

18. $\sinh x = \dfrac{e^x - e^{-x}}{2}$

19. $x^4 - 2x^3 - 5x + 4$

20. $(x + 1)^2$

Finding Taylor Series

In Exercises 21–28, find the Taylor series generated by f at $x = a$.

21. $f(x) = x^3 - 2x + 4, \quad a = 2$

22. $f(x) = 2x^3 + x^2 + 3x - 8, \quad a = 1$

23. $f(x) = x^4 + x^2 + 1, \quad a = -2$

24. $f(x) = 3x^5 - x^4 + 2x^3 + x^2 - 2, \quad a = -1$

25. $f(x) = 1/x^2, \quad a = 1$

26. $f(x) = x/(1 - x), \quad a = 0$

27. $f(x) = e^x, \quad a = 2$

28. $f(x) = 2^x, \quad a = 1$

Theory and Examples

29. Use the Taylor series generated by e^x at $x = a$ to show that

$$e^x = e^a \left[1 + (x - a) + \frac{(x - a)^2}{2!} + \cdots \right].$$

30. *(Continuation of Exercise 29.)* Find the Taylor series generated by e^x at $x = 1$. Compare your answer with the formula in Exercise 29.

31. Let $f(x)$ have derivatives through order n at $x = a$. Show that the Taylor polynomial of order n and its first n derivatives have the same values that f and its first n derivatives have at $x = a$.

32. *Of all polynomials of degree $\leq n$, the Taylor polynomial of order n gives the best approximation.* Suppose that $f(x)$ is differentiable on an interval centered at $x = a$ and that $g(x) = b_0 + b_1(x - a) + \cdots + b_n(x - a)^n$ is a polynomial of degree n with constant coefficients $b_0, \cdots, b_n$. Let $E(x) = f(x) - g(x)$. Show that if we impose on g the conditions

a) $E(a) = 0$ The approximation error is zero at $x = a$.

b) $\displaystyle\lim_{x \to a} \frac{E(x)}{(x - a)^n} = 0,$ The error is negligible when compared to $(x - a)^n$.

then

$$g(x) = f(a) + f'(a)(x - a) + \frac{f''(a)}{2!}(x - a)^2 + \cdots$$
$$+ \frac{f^{(n)}(a)}{n!}(x - a)^n.$$

Thus, the Taylor polynomial $P_n(x)$ is the only polynomial of degree less than or equal to n whose error is both zero at $x = a$ and negligible when compared with $(x - a)^n$.

Quadratic Approximations

The Taylor polynomial of order 2 generated by a twice-differentiable function $f(x)$ at $x = a$ is called the **quadratic approximation** of f at $x = a$. In Exercises 33–38, find the (a) linearization (Taylor polynomial of order 1) and (b) quadratic approximation of f at $x = 0$.

33. $f(x) = \ln(\cos x)$ **34.** $f(x) = e^{\sin x}$

35. $f(x) = 1/\sqrt{1 - x^2}$ **36.** $f(x) = \cosh x$

37. $f(x) = \sin x$ **38.** $f(x) = \tan x$

8.10

Convergence of Taylor Series; Error Estimates

This section addresses the two questions left unanswered by Section 8.9:

1. When does a Taylor series converge to its generating function?
2. How accurately do a function's Taylor polynomials approximate the function on a given interval?

Taylor's Theorem

We answer these questions with the following theorem.

Theorem 16

Taylor's Theorem

If f and its first n derivatives $f', f'', \ldots, f^{(n)}$ are continuous on $[a, b]$ or on $[b, a]$, and $f^{(n)}$ is differentiable on (a, b) or on (b, a), then there exists a number c between a and b such that

$$f(b) = f(a) + f'(a)(b - a) + \frac{f''(a)}{2!}(b - a)^2 + \cdots$$
$$+ \frac{f^{(n)}(a)}{n!}(b - a)^n + \frac{f^{(n+1)}(c)}{(n + 1)!}(b - a)^{n+1}.$$

Taylor's theorem is a generalization of the Mean Value Theorem (Exercise 39). There is a proof of Taylor's theorem at the end of this section.

When we apply Taylor's theorem, we usually want to hold a fixed and treat b as an independent variable. Taylor's formula is easier to use in circumstances like these if we change b to x. Here is how the theorem reads with this change.

Corollary to Taylor's Theorem
Taylor's Formula

If f has derivatives of all orders in an open interval I containing a, then for each positive integer n and for each x in I,

$$f(x) = f(a) + f'(a)(x-a) + \frac{f''(a)}{2!}(x-a)^2 + \cdots$$

$$+ \frac{f^{(n)}(a)}{n!}(x-a)^n + R_n(x), \qquad \text{(1)}$$

where

$$R_n(x) = \frac{f^{(n+1)}(c)}{(n+1)!}(x-a)^{n+1} \qquad \text{for some } c \text{ between } a \text{ and } x. \quad \text{(2)}$$

When we state Taylor's theorem this way, it says that for each x in I,

$$f(x) = P_n(x) + R_n(x).$$

Pause for a moment to think about how remarkable this equation is. For any value of n we want, the equation gives both a polynomial approximation of f of that order and a formula for the error involved in using that approximation over the interval I.

Equation (1) is called **Taylor's formula.** The function $R_n(x)$ is called the **remainder of order** n or the **error term** for the approximation of f by $P_n(x)$ over I. If $R_n(x) \to 0$ as $n \to \infty$ for all x in I, we say that the Taylor series generated by f at $x = a$ **converges** to f on I, and we write

$$f(x) = \sum_{k=0}^{\infty} \frac{f^{(k)}(a)}{k!}(x-a)^k.$$

EXAMPLE 1   *The Maclaurin series for e^x*

Show that the Taylor series generated by $f(x) = e^x$ at $x = 0$ converges to $f(x)$ for every real value of x.

Solution   The function has derivatives of all orders throughout the interval $I = (-\infty, \infty)$. Equations (1) and (2) with $f(x) = e^x$ and $a = 0$ give

$$e^x = 1 + x + \frac{x^2}{2!} + \cdots + \frac{x^n}{n!} + R_n(x) \qquad \text{Polynomial from Section 8.9, Example 2}$$

and

$$R_n(x) = \frac{e^c}{(n+1)!}x^{n+1} \qquad \text{for some } c \text{ between } 0 \text{ and } x.$$

Since e^x is an increasing function of x, e^c lies between $e^0 = 1$ and e^x. When x is negative, so is c, and $e^c < 1$. When x is zero, $e^x = 1$ and $R_n(x) = 0$. When x is positive, so is c, and $e^c < e^x$. Thus,

$$|R_n(x)| \le \frac{|x|^{n+1}}{(n+1)!} \qquad \text{when } x \le 0,$$

and

$$|R_n(x)| < e^x \frac{x^{n+1}}{(n+1)!} \qquad \text{when } x > 0.$$

Finally, because

$$\lim_{n \to \infty} \frac{x^{n+1}}{(n+1)!} = 0 \qquad \text{for every } x, \qquad \text{Section 8.2}$$

$\lim_{n \to \infty} R_n(x) = 0$, and the series converges to e^x for every x.

$$e^x = \sum_{k=0}^{\infty} \frac{x^k}{k!} = 1 + x + \frac{x^2}{2!} + \cdots + \frac{x^k}{k!} + \cdots.$$

Estimating the Remainder

It is often possible to estimate $R_n(x)$ as we did in Example 1. This method of estimation is so convenient that we state it as a theorem for future reference.

Theorem 17
The Remainder Estimation Theorem

If there are positive constants M and r such that $|f^{(n+1)}(t)| \leq M r^{n+1}$ for all t between a and x, inclusive, then the remainder term $R_n(x)$ in Taylor's theorem satisfies the inequality

$$|R_n(x)| \leq M \frac{r^{n+1}|x - a|^{n+1}}{(n+1)!}.$$

If these conditions hold for every n and all the other conditions of Taylor's theorem are satisfied by f, then the series converges to $f(x)$.

In the simplest examples, we can take $r = 1$ provided f and all its derivatives are bounded in magnitude by some constant M. In other cases, we may need to consider r. For example, if $f(x) = 2 \cos(3x)$, each time we differentiate we get a factor of 3 and r needs to be greater than 1. In this particular case, we can take $r = 3$ along with $M = 2$.

We are now ready to look at some examples of how the Remainder Estimation Theorem and Taylor's theorem can be used together to settle questions of convergence. As you will see, they can also be used to determine the accuracy with which a function is approximated by one of its Taylor polynomials.

EXAMPLE 2   *The Maclaurin series for* sin *x*

Show that the Maclaurin series for $\sin x$ converges to $\sin x$ for all x.

Solution The function and its derivatives are

$$f(x) = \sin x, \quad f'(x) = \cos x,$$
$$f''(x) = -\sin x, \quad f'''(x) = -\cos x,$$
$$\vdots \qquad\qquad \vdots$$
$$f^{(2k)}(x) = (-1)^k \sin x, \quad f^{(2k+1)}(x) = (-1)^k \cos x,$$

so

$$f^{(2k)}(0) = 0 \quad \text{and} \quad f^{(2k+1)}(0) = (-1)^k.$$

The series has only odd-powered terms and, for $n = 2k + 1$, Taylor's theorem gives

$$\sin x = x - \frac{x^3}{3!} + \frac{x^5}{5!} - \cdots + \frac{(-1)^k x^{2k+1}}{(2k+1)!} + R_{2k+1}(x).$$

All the derivatives of $\sin x$ have absolute values less than or equal to 1, so we can apply the Remainder Estimation Theorem with $M = 1$ and $r = 1$ to obtain

$$|R_{2k+1}(x)| \leq 1 \cdot \frac{|x|^{2k+2}}{(2k+2)!}.$$

Since $(|x|^{2k+2}/(2k+2)!) \to 0$ as $k \to \infty$, whatever the value of x, $R_{2k+1}(x) \to 0$, and the Maclaurin series for $\sin x$ converges to $\sin x$ for every x.

$$\sin x = \sum_{k=0}^{\infty} \frac{(-1)^k x^{2k+1}}{(2k+1)!} = x - \frac{x^3}{3!} + \frac{x^5}{5!} - \frac{x^7}{7!} + \cdots. \qquad (3)$$

EXAMPLE 3 *The Maclaurin series for cos x*

Show that the Maclaurin series for $\cos x$ converges to $\cos x$ for every value of x.

Solution We add the remainder term to the Taylor polynomial for $\cos x$ (Section 8.9, Example 3) to obtain Taylor's formula for $\cos x$ with $n = 2k$:

$$\cos x = 1 - \frac{x^2}{2!} + \frac{x^4}{4!} - \cdots + (-1)^k \frac{x^{2k}}{(2k)!} + R_{2k}(x).$$

Because the derivatives of the cosine have absolute value less than or equal to 1, the Remainder Estimation Theorem with $M = 1$ and $r = 1$ gives

$$|R_{2k}(x)| \leq 1 \cdot \frac{|x|^{2k+1}}{(2k+1)!}.$$

For every value of x, $R_{2k} \to 0$ as $k \to \infty$. Therefore, the series converges to $\cos x$ for every value of x.

$$\cos x = \sum_{k=0}^{\infty} \frac{(-1)^k x^{2k}}{(2k)!} = 1 - \frac{x^2}{2!} + \frac{x^4}{4!} - \frac{x^6}{6!} + \cdots. \qquad (4)$$

EXAMPLE 4   *Finding a Maclaurin series by substitution*

Find the Maclaurin series for $\cos 2x$.

Solution   We can find the Maclaurin series for $\cos 2x$ by substituting $2x$ for x in the Maclaurin series for $\cos x$:

$$\cos 2x = \sum_{k=0}^{\infty} \frac{(-1)^k (2x)^{2k}}{(2k)!} = 1 - \frac{(2x)^2}{2!} + \frac{(2x)^4}{4!} - \frac{(2x)^6}{6!} + \cdots \qquad \text{Eq. (4) with } 2x \text{ for } x$$

$$= 1 - \frac{2^2 x^2}{2!} + \frac{2^4 x^4}{4!} - \frac{2^6 x^6}{6!} + \cdots$$

$$= \sum_{k=0}^{\infty} (-1)^k \frac{2^{2k} x^{2k}}{(2k)!}.$$

Eq. (4) holds for $-\infty < x < \infty$, implying that it holds for $-\infty < 2x < \infty$, so the newly created series converges for all x. Exercise 45 explains why the series is in fact the Maclaurin series for $\cos 2x$. ❑

EXAMPLE 5   *Finding a Maclaurin series by multiplication*

Find the Maclaurin series for $x \sin x$.

Solution   We can find the Maclaurin series for $x \sin x$ by multiplying the Maclaurin series for $\sin x$ (Eq. 3) by x:

$$x \sin x = x \left(x - \frac{x^3}{3!} + \frac{x^5}{5!} - \frac{x^7}{7!} + \cdots \right)$$

$$= x^2 - \frac{x^4}{3!} + \frac{x^6}{5!} - \frac{x^8}{7!} + \cdots.$$

The new series converges for all x because the series for $\sin x$ converges for all x. Exercise 45 explains why the series is the Maclaurin series for $x \sin x$. ❑

Truncation Error

The Maclaurin series for e^x converges to e^x for all x. But we still need to decide how many terms to use to approximate e^x to a given degree of accuracy. We get this information from the Remainder Estimation Theorem.

EXAMPLE 6   Calculate e with an error of less than 10^{-6}.

Solution   We can use the result of Example 1 with $x = 1$ to write

$$e = 1 + 1 + \frac{1}{2!} + \cdots + \frac{1}{n!} + R_n(1),$$

with

$$R_n(1) = e^c \frac{1}{(n+1)!} \qquad \text{for some } c \text{ between } 0 \text{ and } 1.$$

For the purposes of this example, we assume that we know that $e < 3$. Hence, we are certain that

$$\frac{1}{(n+1)!} < R_n(1) < \frac{3}{(n+1)!}$$

because $1 < e^c < 3$ for $0 < c < 1$.

By experiment we find that $1/9! > 10^{-6}$, while $3/10! < 10^{-6}$. Thus we should take $(n + 1)$ to be at least 10, or n to be at least 9. With an error of less than 10^{-6},

$$e = 1 + 1 + \frac{1}{2} + \frac{1}{3!} + \cdots + \frac{1}{9!} \approx 2.7182\ 82.$$ ◻

EXAMPLE 7   For what values of x can we replace $\sin x$ by $x - (x^3/3!)$ with an error of magnitude no greater than 3×10^{-4}?

Solution   Here we can take advantage of the fact that the Maclaurin series for $\sin x$ is an alternating series for every nonzero value of x. According to the Alternating Series Estimation Theorem (Section 8.7), the error in truncating

$$\sin x = x - \frac{x^3}{3!} + \frac{x^5}{5!} - \cdots$$

after $(x^3/3!)$ is no greater than

$$\left| \frac{x^5}{5!} \right| = \frac{|x|^5}{120}.$$

Therefore the error will be less than or equal to 3×10^{-4} if

$$\frac{|x|^5}{120} < 3 \times 10^{-4} \qquad \text{or} \qquad |x| < \sqrt[5]{360 \times 10^{-4}} \approx 0.514. \qquad \text{\color{blue}Rounded down, to be safe}$$

The Alternating Series Estimation Theorem tells us something that the Remainder Estimation Theorem does not: namely, that the estimate $x - (x^3/3!)$ for $\sin x$ is an underestimate when x is positive because then $x^5/120$ is positive.

Figure 8.20 shows the graph of $\sin x$, along with the graphs of a number of its approximating Taylor polynomials. The graph of $P_3(x) = x - (x^3/3!)$ is almost indistinguishable from the sine curve when $-1 \le x \le 1$.

8.20 The polynomials

$$P_{2n+1}(x) = \sum_{k=0}^{n} \frac{(-1)^k x^{2k+1}}{(2k+1)!}$$

converge to $\sin x$ as $n \to \infty$.

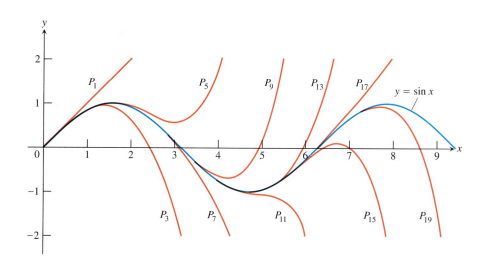

You might wonder how the estimate given by the Remainder Estimation Theorem compares with the one just obtained from the Alternating Series Estimation Theorem. If we write

$$\sin x = x - \frac{x^3}{3!} + R_3,$$

then the Remainder Estimation Theorem gives

$$|R_3| \leq 1 \cdot \frac{|x|^4}{4!} = \frac{|x|^4}{24},$$

which is not as good. But if we recognize that $x - (x^3/3!) = 0 + x + 0x^2 - (x^3/3!) + 0x^4$ is the Taylor polynomial of order 4 as well as of order 3, then

$$\sin x = x - \frac{x^3}{3!} + 0 + R_4,$$

and the Remainder Estimation Theorem with $M = r = 1$ gives

$$|R_4| \leq 1 \cdot \frac{|x|^5}{5!} = \frac{|x|^5}{120}.$$

This is what we had from the Alternating Series Estimation Theorem. ❑

Combining Taylor Series

On the intersection of their intervals of convergence, Taylor series can be added, subtracted, and multiplied by constants, and the results are once again Taylor series. The Taylor series for $f(x) + g(x)$ is the sum of the Taylor series for $f(x)$ and $g(x)$ because the nth derivative of $f + g$ is $f^{(n)} + g^{(n)}$, and so on. Thus we obtain the Maclaurin series for $(1 + \cos 2x)/2$ by adding 1 to the Maclaurin series for $\cos 2x$ and dividing the combined results by 2, and the Maclaurin series for $\sin x + \cos x$ is the term-by-term sum of the Maclaurin series for $\sin x$ and $\cos x$.

✳ Euler's Formula

As you may recall, a complex number is a number of the form $a + bi$, where a and b are real numbers and $i = \sqrt{-1}$. If we substitute $x = i\theta$ (θ real) in the Maclaurin series for e^x and use the relations

$$i^2 = -1, \qquad i^3 = i^2 i = -i, \qquad i^4 = i^2 i^2 = 1, \qquad i^5 = i^4 i = i,$$

and so on, to simplify the result, we obtain

$$e^{i\theta} = 1 + \frac{i\theta}{1!} + \frac{i^2\theta^2}{2!} + \frac{i^3\theta^3}{3!} + \frac{i^4\theta^4}{4!} + \frac{i^5\theta^5}{5!} + \frac{i^6\theta^6}{6!} + \cdots$$

$$= \left(1 - \frac{\theta^2}{2!} + \frac{\theta^4}{4!} - \frac{\theta^6}{6!} + \cdots\right) + i\left(\theta - \frac{\theta^3}{3!} + \frac{\theta^5}{5!} - \cdots\right) = \cos\theta + i\sin\theta.$$

This does not *prove* that $e^{i\theta} = \cos\theta + i\sin\theta$ because we have not yet defined what it means to raise e to an imaginary power. But it does say how to define $e^{i\theta}$ to be consistent with other things we know.

One of the amazing consequences of Euler's formula is the equation

$$e^{i\pi} = -1.$$

When written in the form $e^{i\pi} + 1 = 0$, this equation combines the five most important constants in mathematics.

Definition

For any real number θ, $e^{i\theta} = \cos\theta + i\sin\theta$.

(5)

Equation (5), called **Euler's formula,** enables us to define e^{a+bi} to be $e^a \cdot e^{bi}$ for any complex number $a + bi$.

A Proof of Taylor's Theorem

We prove Taylor's theorem assuming $a < b$. The proof for $a > b$ is nearly the same.

The Taylor polynomial

$$P_n(x) = f(a) + f'(a)(x-a) + \frac{f''(a)}{2!}(x-a)^2 + \cdots + \frac{f^{(n)}(a)}{n!}(x-a)^n$$

and its first n derivatives match the function f and its first n derivatives at $x = a$. We do not disturb that matching if we add another term of the form $K(x-a)^{n+1}$, where K is any constant, because such a term and its first n derivatives are all equal to zero at $x = a$. The new function

$$\phi_n(x) = P_n(x) + K(x-a)^{n+1}$$

and its first n derivatives still agree with f and its first n derivatives at $x = a$.

We now choose the particular value of K that makes the curve $y = \phi_n(x)$ agree with the original curve $y = f(x)$ at $x = b$. In symbols,

$$f(b) = P_n(b) + K(b-a)^{n+1}, \quad \text{or} \quad K = \frac{f(b) - P_n(b)}{(b-a)^{n+1}}. \tag{6}$$

With K defined by Eq. (6), the function

$$F(x) = f(x) - \phi_n(x)$$

measures the difference between the original function f and the approximating function ϕ_n for each x in $[a, b]$.

We now use Rolle's theorem (Section 3.2). First, because $F(a) = F(b) = 0$ and both F and F' are continuous on $[a, b]$, we know that

$$F'(c_1) = 0 \quad \text{for some } c_1 \text{ in } (a, b).$$

Next, because $F'(a) = F'(c_1) = 0$ and both F' and F'' are continuous on $[a, c_1]$, we know that

$$F''(c_2) = 0 \quad \text{for some } c_2 \text{ in } (a, c_1).$$

Rolle's theorem, applied successively to F'', F''', $\ldots$, $F^{(n-1)}$ implies the existence of

$$c_3 \quad \text{in } (a, c_2) \quad \text{such that } F'''(c_3) = 0,$$
$$c_4 \quad \text{in } (a, c_3) \quad \text{such that } F^{(4)}(c_4) = 0,$$
$$\vdots$$
$$c_n \quad \text{in } (a, c_{n-1}) \quad \text{such that } F^{(n)}(c_n) = 0.$$

Finally, because $F^{(n)}$ is continuous on $[a, c_n]$ and differentiable on (a, c_n), and $F^{(n)}(a) = F^{(n)}(c_n) = 0$, Rolle's theorem implies that there is a number c_{n+1} in (a, c_n) such that

$$F^{(n+1)}(c_{n+1}) = 0. \tag{7}$$

If we differentiate $F(x) = f(x) - P_n(x) - K(x-a)^{n+1}$ a total of $n+1$ times, we get

$$F^{(n+1)}(x) = f^{(n+1)}(x) - 0 - (n+1)!K. \qquad (8)$$

Equations (7) and (8) together give

$$K = \frac{f^{(n+1)}(c)}{(n+1)!} \quad \text{for some number } c = c_{n+1} \text{ in } (a, b). \qquad (9)$$

Equations (6) and (9) give

$$f(b) = P_n(b) + \frac{f^{(n+1)}(c)}{(n+1)!}(b-a)^{n+1}.$$

This concludes the proof. ❏

Exercises 8.10

Maclaurin Series by Substitution

Use substitution (as in Example 4) to find the Maclaurin series of the functions in Exercises 1–6.

1. e^{-5x} **2.** $e^{-x/2}$ **3.** $5 \sin(-x)$

4. $\sin\left(\dfrac{\pi x}{2}\right)$ **5.** $\cos \sqrt{x}$ **6.** $\cos(x^{3/2}/\sqrt{2})$

More Maclaurin Series

Find Maclaurin series for the functions in Exercises 7–18.

7. xe^x **8.** $x^2 \sin x$ **9.** $\dfrac{x^2}{2} - 1 + \cos x$

10. $\sin x - x + \dfrac{x^3}{3!}$ **11.** $x \cos \pi x$ **12.** $x^2 \cos(x^2)$

13. $\cos^2 x$ (*Hint:* $\cos^2 x = (1 + \cos 2x)/2$.)

14. $\sin^2 x$ **15.** $\dfrac{x^2}{1 - 2x}$ **16.** $x \ln(1 + 2x)$

17. $\dfrac{1}{(1-x)^2}$ **18.** $\dfrac{2}{(1-x)^3}$

Error Estimates

19. For approximately what values of x can you replace $\sin x$ by $x - (x^3/6)$ with an error of magnitude no greater than 5×10^{-4}? Give reasons for your answer.

20. If $\cos x$ is replaced by $1 - (x^2/2)$ and $|x| < 0.5$, what estimate can be made of the error? Does $1 - (x^2/2)$ tend to be too large, or too small? Give reasons for your answer.

21. How close is the approximation $\sin x = x$ when $|x| < 10^{-3}$? For which of these values of x is $x < \sin x$?

22. The estimate $\sqrt{1+x} = 1 + (x/2)$ is used when x is small. Estimate the error when $|x| < 0.01$.

23. The approximation $e^x = 1 + x + (x^2/2)$ is used when x is small. Use the Remainder Estimation Theorem to estimate the error when $|x| < 0.1$.

24. (*Continuation of Exercise 23.*) When $x < 0$, the series for e^x is an alternating series. Use the Alternating Series Estimation Theorem to estimate the error that results from replacing e^x by $1 + x + (x^2/2)$ when $-0.1 < x < 0$. Compare your estimate with the one you obtained in Exercise 23.

25. Estimate the error in the approximation $\sinh x = x + (x^3/3!)$ when $|x| < 0.5$. (*Hint:* Use R_4, not R_3.)

26. When $0 \le h \le 0.01$, show that e^h may be replaced by $1 + h$ with an error of magnitude no greater than 0.6% of h. Use $e^{0.01} = 1.01$.

27. For what positive values of x can you replace $\ln(1+x)$ by x with an error of magnitude no greater than 1% of the value of x?

28. You plan to estimate $\pi/4$ by evaluating the Maclaurin series for $\tan^{-1} x$ at $x = 1$. Use the Alternating Series Estimation Theorem to determine how many terms of the series you would have to add to be sure the estimate is good to 2 decimal places.

29. a) Use the Maclaurin series for $\sin x$ and the Alternating Series Estimation Theorem to show that

$$1 - \frac{x^2}{6} < \frac{\sin x}{x} < 1, \quad x \ne 0.$$

b) GRAPHER Graph $f(x) = (\sin x)/x$ together with the functions $y = 1 - (x^2/6)$ and $y = 1$ for $-5 \le x \le 5$. Comment on the relationships among the graphs.

30. a) Use the Maclaurin series for $\cos x$ and the Alternating Series Estimation Theorem to show that

$$\frac{1}{2} - \frac{x^2}{24} < \frac{1 - \cos x}{x^2} < \frac{1}{2}, \quad x \ne 0.$$

(This is the inequality in Section 1.2, Exercise 46.)

b)   **GRAPHER** Graph $f(x) = (1 - \cos x)/x^2$ together with $y = (1/2) - (x^2/24)$ and $y = 1/2$ for $-9 \leq x \leq 9$. Comment on the relationships among the graphs.

Finding and Identifying Maclaurin Series

Each of the series in Exercises 31–34 is the value of the Maclaurin series of a function $f(x)$ at some point. What function and what point? What is the sum of the series?

31. $(0.1) - \dfrac{(0.1)^3}{3!} + \dfrac{(0.1)^5}{5!} - \cdots + \dfrac{(-1)^k (0.1)^{2k+1}}{(2k+1)!} + \cdots$

32. $1 - \dfrac{\pi^2}{4^2 \cdot 2!} + \dfrac{\pi^4}{4^4 \cdot 4!} - \cdots + \dfrac{(-1)^k (\pi)^{2k}}{4^{2k} \cdot (2k)!} + \cdots$

33. $\dfrac{\pi}{3} - \dfrac{\pi^3}{3^3 \cdot 3} + \dfrac{\pi^5}{3^5 \cdot 5} - \cdots + \dfrac{(-1)^k \pi^{2k+1}}{3^{2k+1}(2k+1)} + \cdots$

34. $\pi - \dfrac{\pi^2}{2} + \dfrac{\pi^3}{3} - \cdots + (-1)^{k-1}\dfrac{\pi^k}{k} + \cdots$

35. Multiply the Maclaurin series for e^x and $\sin x$ together to find the first five nonzero terms of the Maclaurin series for $e^x \sin x$.

36. Multiply the Maclaurin series for e^x and $\cos x$ together to find the first five nonzero terms of the Maclaurin series for $e^x \cos x$.

37. Use the identity $\sin^2 x = (1 - \cos 2x)/2$ to obtain the Maclaurin series for $\sin^2 x$. Then differentiate this series to obtain the Maclaurin series for $2 \sin x \cos x$. Check that this is the series for $\sin 2x$.

38. (*Continuation of Exercise 37.*) Use the identity $\cos^2 x = \cos 2x + \sin^2 x$ to obtain a power series for $\cos^2 x$.

Theory and Examples

39. *Taylor's theorem and the Mean Value Theorem.* Explain how the Mean Value Theorem (Section 3.2, Theorem 4) is a special case of Taylor's theorem.

40. *Linearizations at inflection points* (*Continuation of Section 3.7, Exercise 63*). Show that if the graph of a twice-differentiable function $f(x)$ has an inflection point at $x = a$, then the linearization of f at $x = a$ is also the quadratic approximation of f at $x = a$. This explains why tangent lines fit so well at inflection points.

41. *The (second) second derivative test.* Use the equation

$$f(x) = f(a) + f'(a)(x - a) + \frac{f''(c_2)}{2}(x - a)^2$$

to establish the following test.

   Let f have continuous first and second derivatives and suppose that $f'(a) = 0$. Then

a) f has a local maximum at a if $f'' \leq 0$ throughout an interval whose interior contains a;

b) f has a local minimum at a if $f'' \geq 0$ throughout an interval whose interior contains a.

42. *A cubic approximation.* Use Taylor's formula with $a = 0$ and $n = 3$ to find the standard cubic approximation of $f(x) = 1/(1 - x)$ at $x = 0$. Give an upper bound for the magnitude of the error in the approximation when $|x| \leq 0.1$.

43. a)   Use Taylor's formula with $n = 2$ to find the quadratic approximation of $f(x) = (1 + x)^k$ at $x = 0$ (k a constant).

b)   If $k = 3$, for approximately what values of x in the interval $[0, 1]$ will the error in the quadratic approximation be less than $1/100$?

44. *Improving approximations to π.*

a)   Let P be an approximation of π accurate to n decimals. Show that $P + \sin P$ gives an approximation correct to $3n$ decimals. (*Hint:* Let $P = \pi + x$.)

b)   Try it with a calculator.

45. *The Maclaurin series generated by $f(x) = \sum_{n=0}^{\infty} a_n x^n$ is $\sum_{n=0}^{\infty} a_n x^n$.* A function defined by a power series $\sum_{n=0}^{\infty} a_n x^n$ with a radius of convergence $c > 0$ has a Maclaurin series that converges to the function at every point of $(-c, c)$. Show this by showing that the Maclaurin series generated by $f(x) = \sum_{n=0}^{\infty} a_n x^n$ is the series $\sum_{n=0}^{\infty} a_n x^n$ itself.

   An immediate consequence of this is that series like

$$x \sin x = x^2 - \frac{x^4}{3!} + \frac{x^6}{5!} - \frac{x^8}{7!} + \cdots$$

and

$$x^2 e^x = x^2 + x^3 + \frac{x^4}{2!} + \frac{x^5}{3!} + \cdots,$$

obtained by multiplying Maclaurin series by powers of x, as well as series obtained by integration and differentiation of convergent power series, are themselves the Maclaurin series generated by the functions they represent.

46. *Maclaurin series for even functions and odd functions* (*Continuation of Section 8.8, Exercise 45*). Suppose that $f(x) = \sum_{n=0}^{\infty} a_n x^n$ converges for all x in an open interval $(-c, c)$. Show that

a)   If f is even, then $a_1 = a_3 = a_5 = \cdots = 0$, i.e., the series for f contains only even powers of x.

b)   If f is odd, then $a_0 = a_2 = a_4 = \cdots = 0$, i.e., the series for f contains only odd powers of x.

47. *Taylor polynomials of periodic functions*

a)   Show that every continuous periodic function $f(x)$, $-\infty < x < \infty$, is bounded in magnitude by showing that there exists a positive constant M such that $|f(x)| \leq M$ for all x.

b)   Show that the graph of every Taylor polynomial of positive degree generated by $f(x) = \cos x$ must eventually move away from the graph of $\cos x$ as $|x|$ increases. You can see this in Fig. 8.18. The Taylor polynomials of $\sin x$ behave in a similar way (Fig. 8.20).

48. **GRAPHER**

a)   Graph the curves $y = (1/3) - (x^2)/5$ and $y = (x - \tan^{-1} x)/x^3$ together with the line $y = 1/3$.

b)   Use a Maclaurin series to explain what you see. What is

$$\lim_{x \to 0} \frac{x - \tan^{-1} x}{x^3} ?$$

Euler's Formula

49. Use Eq. (5) to write the following powers of e in the form $a + bi$.

a) $e^{-i\pi}$ **b)** $e^{i\pi/4}$ **c)** $e^{-i\pi/2}$

50. *Euler's identities.* Use Eq. (5) to show that

$$\cos\theta = \frac{e^{i\theta} + e^{-i\theta}}{2} \quad \text{and} \quad \sin\theta = \frac{e^{i\theta} - e^{-i\theta}}{2i}.$$

51. Establish the equations in Exercise 50 by combining the formal Maclaurin series for $e^{i\theta}$ and $e^{-i\theta}$.

52. Show that

a) $\cosh i\theta = \cos\theta,$ **b)** $\sinh i\theta = i\sin\theta.$

53. By multiplying the Maclaurin series for e^x and $\sin x$, find the terms through x^5 of the Maclaurin series for $e^x \sin x$. This series is the imaginary part of the series for

$$e^x \cdot e^{ix} = e^{(1+i)x}.$$

Use this fact to check your answer. For what values of x should the series for $e^x \sin x$ converge?

54. When a and b are real, we define $e^{(a+ib)x}$ with the equation

$$e^{(a+ib)x} = e^{ax} \cdot e^{ibx} = e^{ax}(\cos bx + i\sin bx).$$

Differentiate the right-hand side of this equation to show that

$$\frac{d}{dx} e^{(a+ib)x} = (a + ib)e^{(a+ib)x}.$$

Thus the familiar rule $(d/dx)e^{kx} = ke^{kx}$ holds for k complex as well as real.

55. Use the definition of $e^{i\theta}$ to show that for any real numbers $\theta, \theta_1,$ and $\theta_2,$

a) $e^{i\theta_1} e^{i\theta_2} = e^{i(\theta_1 + \theta_2)},$

b) $e^{-i\theta} = 1/e^{i\theta}.$

56. Two complex numbers $a + ib$ and $c + id$ are equal if and only if $a = c$ and $b = d$. Use this fact to evaluate

$$\int e^{ax} \cos bx \, dx \quad \text{and} \quad \int e^{ax} \sin bx \, dx$$

from

$$\int e^{(a+ib)x} dx = \frac{a - ib}{a^2 + b^2} e^{(a+ib)x} + C,$$

where $C = C_1 + iC_2$ is a complex constant of integration.

✪ CAS Explorations and Projects—Linear, Quadratic, and Cubic Approximations

Taylor's formula with $n = 1$ and $a = 0$ gives the linearization of a function at $x = 0$. With $n = 2$ and $n = 3$ we obtain the standard quadratic and cubic approximations. In these exercises we explore the errors associated with these approximations. We seek answers to two questions:

a) For what values of x can the function be replaced by each approximation with an error less than 10^{-2}?

b) What is the maximum error we could expect if we replace the function by each approximation over the specified interval?

Using a CAS, perform the following steps to aid in answering questions (a) and (b) for the functions and intervals in Exercises 57–62.

Step 1: Plot the function over the specified interval.

Step 2: Find the Taylor polynomials $P_1(x)$, $P_2(x)$, and $P_3(x)$ at $x = 0$.

Step 3: Calculate the $(n + 1)$st derivative $f^{(n+1)}(c)$ associated with the remainder term for each Taylor polynomial. Plot the derivative as a function of c over the specified interval and estimate its maximum absolute value, M.

Step 4: Calculate the remainder $R_n(x)$ for each polynomial. Using the estimate M from step 3 in place of $f^{(n+1)}(c)$, plot $R_n(x)$ over the specified interval. Then estimate the values of x that answer question (a).

Step 5: Compare your estimated error with the actual error $E_n(x) = |f(x) - P_n(x)|$ by plotting $E_n(x)$ over the specified interval. This will help answer question (b).

Step 6: Graph the function and its three Taylor approximations together. Discuss the graphs in relation to the information discovered in steps 4 and 5.

57. $f(x) = \dfrac{1}{\sqrt{1+x}}, \quad |x| \le \dfrac{3}{4}$

58. $f(x) = (1+x)^{3/2}, \quad -\dfrac{1}{2} \le x \le 2$

59. $f(x) = \dfrac{x}{x^2 + 1}, \quad |x| \le 2$

60. $f(x) = (\cos x)(\sin 2x), \quad |x| \le 2$

61. $f(x) = e^{-x} \cos 2x, \quad |x| \le 1$

62. $f(x) = e^{x/3} \sin 2x, \quad |x| \le 2$

8.11

Applications of Power Series

This section introduces the binomial series for estimating powers and roots and shows how series are sometimes used to approximate the solution of an initial value problem, to evaluate nonelementary integrals, and to evaluate limits that lead to indeterminate forms. We provide a self-contained derivation of the Maclaurin series for $\tan^{-1} x$ and conclude with a reference table of frequently used series.

The Binomial Series for Powers and Roots

The Maclaurin series generated by $f(x) = (1 + x)^m$, when m is constant, is

$$1 + mx + \frac{m(m-1)}{2!} x^2 + \frac{m(m-1)(m-2)}{3!} x^3 + \cdots$$

$$+ \frac{m(m-1)(m-2)\cdots(m-k+1)}{k!} x^k + \cdots. \quad (1)$$

This series, called the **binomial series,** converges absolutely for $|x| < 1$. To derive the series, we first list the function and its derivatives:

$$f(x) = (1 + x)^m$$

$$f'(x) = m(1 + x)^{m-1}$$

$$f''(x) = m(m - 1)(1 + x)^{m-2}$$

$$f'''(x) = m(m - 1)(m - 2)(1 + x)^{m-3}$$

$$\vdots$$

$$f^{(k)}(x) = m(m - 1)(m - 2)\cdots(m - k + 1)(1 + x)^{m-k}.$$

We then evaluate these at $x = 0$ and substitute into the Maclaurin series formula to obtain the series in (1).

If m is an integer greater than or equal to zero, the series stops after $(m + 1)$ terms because the coefficients from $k = m + 1$ on are zero.

If m is not a positive integer or zero, the series is infinite and converges for $|x| < 1$. To see why, let u_k be the term involving x^k. Then apply the Ratio Test for absolute convergence to see that

$$\left| \frac{u_{k+1}}{u_k} \right| = \left| \frac{m - k}{k + 1} x \right| \to |x| \quad \text{as } k \to \infty.$$

Our derivation of the binomial series shows only that it is generated by $(1 + x)^m$ and converges for $|x| < 1$. The derivation does not show that the series converges to $(1 + x)^m$. It does, but we assume that part without proof.

For $-1 < x < 1$,

$$(1 + x)^m = 1 + \sum_{k=1}^{\infty} \binom{m}{k} x^k, \quad (2)$$

where we define

$$\binom{m}{1} = m, \quad \binom{m}{2} = \frac{m(m-1)}{2!},$$

and

$$\binom{m}{k} = \frac{m(m-1)(m-2)\cdots(m-k+1)}{k!} \quad \text{for } k \geq 3.$$

EXAMPLE 1   If $m = -1$,

$$\binom{-1}{1} = -1, \qquad \binom{-1}{2} = \frac{-1(-2)}{2!} = 1,$$

and

$$\binom{-1}{k} = \frac{-1(-2)(-3)\cdots(-1-k+1)}{k!} = (-1)^k \left(\frac{k!}{k!}\right) = (-1)^k.$$

With these coefficient values, Eq. (2) becomes the geometric series

$$(1+x)^{-1} = 1 + \sum_{k=1}^{\infty} (-1)^k x^k = 1 - x + x^2 - x^3 + \cdots + (-1)^k x^k + \cdots.$$

EXAMPLE 2   We know from Section 3.7, Example 1, that $\sqrt{1-x} \approx 1 + (x/2)$ for $|x|$ small. With $m = 1/2$, the binomial series gives quadratic and higher order approximations as well, along with error estimates that come from the Alternating Series Estimation Theorem:

$$(1+x)^{1/2} = 1 + \frac{x}{2} + \frac{\left(\frac{1}{2}\right)\left(-\frac{1}{2}\right)}{2!} x^2 + \frac{\left(\frac{1}{2}\right)\left(-\frac{1}{2}\right)\left(-\frac{3}{2}\right)}{3!} x^3$$

$$+ \frac{\left(\frac{1}{2}\right)\left(-\frac{1}{2}\right)\left(-\frac{3}{2}\right)\left(-\frac{5}{2}\right)}{4!} x^4 + \cdots$$

$$= 1 + \frac{x}{2} - \frac{x^2}{8} + \frac{x^3}{16} - \frac{5x^4}{128} + \cdots.$$

Substitution for x gives still other approximations. For example,

$$\sqrt{1-x^2} \approx 1 - \frac{x^2}{2} - \frac{x^4}{8} \qquad \text{for } |x^2| \text{ small}$$

$$\sqrt{1-\frac{1}{x}} \approx 1 - \frac{1}{2x} - \frac{1}{8x^2} \qquad \text{for } \left|\frac{1}{x}\right| \text{ small, i.e., } |x| \text{ large.}$$

Power Series Solutions of Differential Equations and Initial Value Problems

When we cannot find a relatively simple expression for the solution of an initial value problem or differential equation, we try to get information about the solution in other ways. One way is to try to find a power series representation for the solution. If we can do so, we immediately have a source of polynomial approximations of the solution, which may be all that we really need. The first example (Example 3) deals with a first order linear differential equation that could be solved with the methods of Section 6.11. The example shows how, not knowing this, we can solve the equation with power series. The second example (Example 4) deals with an equation that cannot be solved by previous methods.

EXAMPLE 3   Solve the initial value problem

$$y' - y = x, \qquad y(0) = 1.$$

Solution   We assume that there is a solution of the form

$$y = a_0 + a_1 x + a_2 x^2 + \cdots + a_{n-1} x^{n-1} + a_n x^n + \cdots. \tag{3}$$

Our goal is to find values for the coefficients a_k that make the series and its first derivative

$$y' = a_1 + 2a_2 x + 3a_3 x^2 + \cdots + n a_n x^{n-1} + \cdots \tag{4}$$

satisfy the given differential equation and initial condition. The series $y' - y$ is the difference of the series in Eqs. (3) and (4):

$$y' - y = (a_1 - a_0) + (2a_2 - a_1)x + (3a_3 - a_2)x^2 + \cdots$$

$$+ (n a_n - a_{n-1})x^{n-1} + \cdots. \tag{5}$$

If y is to satisfy the equation $y' - y = x$, the series in (5) must equal x. Since power series representations are unique, as you saw if you did Exercise 45 in Section 8.8, the coefficients in Eq. (5) must satisfy the equations

$$a_1 - a_0 = 0 \qquad \text{Constant terms}$$

$$2a_2 - a_1 = 1 \qquad \text{Coefficients of } x$$

$$3a_3 - a_2 = 0 \qquad \text{Coefficients of } x^2$$

$$\vdots \qquad\qquad \vdots$$

$$n a_n - a_{n-1} = 0 \qquad \text{Coefficients of } x^{n-1}$$

$$\vdots \qquad\qquad \vdots$$

We can also see from Eq. (3) that $y = a_0$ when $x = 0$, so that $a_0 = 1$ (this being the initial condition). Putting it all together, we have

$$a_0 = 1, \qquad a_1 = a_0 = 1, \qquad a_2 = \frac{1 + a_1}{2} = \frac{1 + 1}{2} = \frac{2}{2},$$

$$a_3 = \frac{a_2}{3} = \frac{2}{3 \cdot 2} = \frac{2}{3!}, \quad \cdots, \quad a_n = \frac{a_{n-1}}{n} = \frac{2}{n!}, \quad \cdots$$

Substituting these coefficient values into the equation for y (Eq. 3) gives

$$y = 1 + x + 2 \cdot \frac{x^2}{2!} + 2 \cdot \frac{x^3}{3!} + \cdots + 2 \cdot \frac{x^n}{n!} + \cdots$$

$$= 1 + x + 2 \underbrace{\left(\frac{x^2}{2!} + \frac{x^3}{3!} + \cdots + \frac{x^n}{n!} + \cdots \right)}_{\text{the Maclaurin series for } e^x - 1 - x}$$

$$= 1 + x + 2(e^x - 1 - x) = 2e^x - 1 - x.$$

The solution of the initial value problem is $y = 2e^x - 1 - x$.

As a check, we see that

$$y(0) = 2e^0 - 1 - 0 = 2 - 1 = 1$$

and

$$y' - y = (2e^x - 1) - (2e^x - 1 - x) = x.$$ ❑

EXAMPLE 4   Find a power series solution for

$$y'' + x^2 y = 0. \tag{6}$$

Solution   We assume that there is a solution of the form

$$y = a_0 + a_1 x + a_2 x^2 + \cdots + a_n x^n + \cdots, \tag{7}$$

and find what the coefficients a_k have to be to make the series and its second derivative

$$y'' = 2a_2 + 3 \cdot 2a_3 x + \cdots + n(n-1)a_n x^{n-2} + \cdots \tag{8}$$

satisfy Eq. (6). The series for $x^2 y$ is x^2 times the right-hand side of Eq. (7):

$$x^2 y = a_0 x^2 + a_1 x^3 + a_2 x^4 + \cdots + a_n x^{n+2} + \cdots. \tag{9}$$

The series for $y'' + x^2 y$ is the sum of the series in Eqs. (8) and (9):

$$\begin{aligned} y'' + x^2 y = 2a_2 + 6a_3 x + (12a_4 + a_0)x^2 + (20a_5 + a_1)x^3 \\ + \cdots + (n(n-1)a_n + a_{n-4})x^{n-2} + \cdots. \end{aligned} \tag{10}$$

Notice that the coefficient of x^{n-2} in Eq. (9) is a_{n-4}. If y and its second derivative y'' are to satisfy Eq. (6), the coefficients of the individual powers of x on the right-hand side of Eq. (10) must all be zero:

$$2a_2 = 0, \quad 6a_3 = 0, \quad 12a_4 + a_0 = 0, \quad 20a_5 + a_1 = 0, \tag{11}$$

and for all $n \geq 4$,

$$n(n-1)a_n + a_{n-4} = 0. \tag{12}$$

We can see from Eq. (7) that

$$a_0 = y(0), \quad a_1 = y'(0).$$

In other words, the first two coefficients of the series are the values of y and y' at $x = 0$. The equations in (11) and the recursion formula in (12) enable us to evaluate all the other coefficients in terms of a_0 and a_1.

The first two of Eqs. (11) give

$$a_2 = 0, \quad a_3 = 0.$$

Equation (12) shows that if $a_{n-4} = 0$, then $a_n = 0$; so we conclude that

$$a_6 = 0, \quad a_7 = 0, \quad a_{10} = 0, \quad a_{11} = 0,$$

and whenever $n = 4k + 2$ or $4k + 3$, a_n is zero. For the other coefficients we have

$$a_n = \frac{-a_{n-4}}{n(n-1)}$$

so that

$$a_4 = \frac{-a_0}{4 \cdot 3}, \quad a_8 = \frac{-a_4}{8 \cdot 7} = \frac{a_0}{3 \cdot 4 \cdot 7 \cdot 8}$$

$$a_{12} = \frac{-a_8}{11 \cdot 12} = \frac{-a_0}{3 \cdot 4 \cdot 7 \cdot 8 \cdot 11 \cdot 12}$$

and

$$a_5 = \frac{-a_1}{5 \cdot 4}, \qquad a_9 = \frac{-a_5}{9 \cdot 8} = \frac{a_1}{4 \cdot 5 \cdot 8 \cdot 9}$$

$$a_{13} = \frac{-a_9}{12 \cdot 13} = \frac{-a_1}{4 \cdot 5 \cdot 8 \cdot 9 \cdot 12 \cdot 13}.$$

The answer is best expressed as the sum of two separate series—one multiplied by a_0, the other by a_1:

$$y = a_0 \left(1 - \frac{x^4}{3 \cdot 4} + \frac{x^8}{3 \cdot 4 \cdot 7 \cdot 8} - \frac{x^{12}}{3 \cdot 4 \cdot 7 \cdot 8 \cdot 11 \cdot 12} + \cdots \right)$$

$$+ a_1 \left(x - \frac{x^5}{4 \cdot 5} + \frac{x^9}{4 \cdot 5 \cdot 8 \cdot 9} - \frac{x^{13}}{4 \cdot 5 \cdot 8 \cdot 9 \cdot 12 \cdot 13} + \cdots \right).$$

Both series converge absolutely for all x, as is readily seen by the ratio test. ❏

Evaluating Nonelementary Integrals

Maclaurin series can be used to express nonelementary integrals in terms of series.

Integrals like $\int \sin x^2 \, dx$ arise in the study of the diffraction of light.

EXAMPLE 5   Express $\int \sin x^2 \, dx$ as a power series.

Solution   From the series for $\sin x$ we obtain

$$\sin x^2 = x^2 - \frac{x^6}{3!} + \frac{x^{10}}{5!} - \frac{x^{14}}{7!} + \frac{x^{18}}{9!} - \cdots.$$

Therefore,

$$\int \sin x^2 \, dx = C + \frac{x^3}{3} - \frac{x^7}{7 \cdot 3!} + \frac{x^{11}}{11 \cdot 5!} - \frac{x^{15}}{15 \cdot 7!} + \frac{x^{19}}{19 \cdot 9!} - \cdots.$$ ❏

EXAMPLE 6   Estimate $\int_0^1 \sin x^2 \, dx$ with an error of less than 0.001.

Solution   From the indefinite integral in Example 5,

$$\int_0^1 \sin x^2 \, dx = \frac{1}{3} - \frac{1}{7 \cdot 3!} + \frac{1}{11 \cdot 5!} - \frac{1}{15 \cdot 7!} + \frac{1}{19 \cdot 9!} - \cdots.$$

The series alternates, and we find by experiment that

$$\frac{1}{11 \cdot 5!} \approx 0.0007\,6$$

is the first term to be numerically less than 0.001. The sum of the preceding two terms gives

$$\int_0^1 \sin x^2 \, dx \approx \frac{1}{3} - \frac{1}{42} \approx 0.310.$$

With two more terms we could estimate

$$\int_0^1 \sin x^2 \, dx \approx 0.3102\,68$$

with an error of less than 10^{-6}. With only one term beyond that we have

$$\int_0^1 \sin x^2 \, dx \approx \frac{1}{3} - \frac{1}{42} + \frac{1}{1320} - \frac{1}{75600} + \frac{1}{6894720} \approx 0.3102\,68303,$$

with an error of about 1.08×10^{-9}. To guarantee this accuracy with the error formula for the trapezoidal rule would require using about 8,000 subintervals. ☐

Arctangents

In Section 8.8, Example 5, we found a series for $\tan^{-1} x$ by differentiating to get

$$\frac{d}{dx} \tan^{-1} x = \frac{1}{1 + x^2} = 1 - x^2 + x^4 - x^6 + \cdots$$

and integrating to get

$$\tan^{-1} x = x - \frac{x^3}{3} + \frac{x^5}{5} - \frac{x^7}{7} + \cdots .$$

However, we did not prove the term-by-term integration theorem on which this conclusion depended. We now derive the series again by integrating both sides of the finite formula

$$\frac{1}{1 + t^2} = 1 - t^2 + t^4 - t^6 + \cdots + (-1)^n t^{2n} + \frac{(-1)^{n+1} t^{2n+2}}{1 + t^2}, \qquad (13)$$

in which the last term comes from adding the remaining terms as a geometric series with first term $a = (-1)^{n+1} t^{2n+2}$ and ratio $r = -t^2$. Integrating both sides of Eq. (13) from $t = 0$ to $t = x$ gives

$$\tan^{-1} x = x - \frac{x^3}{3} + \frac{x^5}{5} - \frac{x^7}{7} + \cdots + (-1)^n \frac{x^{2n+1}}{2n + 1} + R(n, x),$$

where

$$R(n, x) = \int_0^x \frac{(-1)^{n+1} t^{2n+2}}{1 + t^2} \, dt.$$

The denominator of the integrand is greater than or equal to 1; hence

$$|R(n, x)| \leq \int_0^{|x|} t^{2n+2} \, dt = \frac{|x|^{2n+3}}{2n + 3}.$$

We take this route instead of finding the Maclaurin series directly because the formulas for the higher order derivatives of $\tan^{-1} x$ are unmanageable.

If $|x| \leq 1$, the right side of this inequality approaches zero as $n \to \infty$. Therefore $\lim_{n \to \infty} R(n, x) = 0$ if $|x| \leq 1$ and

$$\tan^{-1} x = \sum_{n=0}^{\infty} \frac{(-1)^n x^{2n+1}}{2n + 1}, \qquad |x| \leq 1.$$

$$\tan^{-1} x = x - \frac{x^3}{3} + \frac{x^5}{5} - \frac{x^7}{7} + \cdots, \qquad |x| \leq 1 \qquad (14)$$

When we put $x = 1$ in Eq. (14), we get **Leibniz's formula:**

$$\frac{\pi}{4} = 1 - \frac{1}{3} + \frac{1}{5} - \frac{1}{7} + \frac{1}{9} - \cdots + \frac{(-1)^n}{2n + 1} + \cdots .$$

This series converges too slowly to be a useful source of decimal approximations of π. It is better to use a formula like

$$\pi = 48 \tan^{-1} \frac{1}{18} + 32 \tan^{-1} \frac{1}{57} - 20 \tan^{-1} \frac{1}{239},$$

which uses values of x closer to zero.

Evaluating Indeterminate Forms

We can sometimes evaluate indeterminate forms by expressing the functions involved as Taylor series.

EXAMPLE 7   Evaluate $\lim\limits_{x \to 1} \dfrac{\ln x}{x - 1}$.

Solution   We represent $\ln x$ as a Taylor series in powers of $x - 1$. This can be accomplished by calculating the Taylor series generated by $\ln x$ at $x = 1$ directly or by replacing x by $x - 1$ in the series for $\ln x$ in Section 8.8, Example 6. Either way, we obtain

$$\ln x = (x - 1) - \frac{1}{2}(x - 1)^2 + \cdots,$$

from which we find that

$$\lim_{x \to 1} \frac{\ln x}{x - 1} = \lim_{x \to 1} \left(1 - \frac{1}{2}(x - 1) + \cdots \right) = 1.$$ ❑

EXAMPLE 8   Evaluate $\lim\limits_{x \to 0} \dfrac{\sin x - \tan x}{x^3}$.

Solution   The Maclaurin series for $\sin x$ and $\tan x$, to terms in x^5, are

$$\sin x = x - \frac{x^3}{3!} + \frac{x^5}{5!} - \cdots, \qquad \tan x = x + \frac{x^3}{3} + \frac{2x^5}{15} + \cdots.$$

Hence,

$$\sin x - \tan x = -\frac{x^3}{2} - \frac{x^5}{8} - \cdots = x^3 \left(-\frac{1}{2} - \frac{x^2}{8} - \cdots \right)$$

and

$$\lim_{x \to 0} \frac{\sin x - \tan x}{x^3} = \lim_{x \to 0} \left(-\frac{1}{2} - \frac{x^2}{8} - \cdots \right)$$

$$= -\frac{1}{2}.$$ ❑

If we apply series to calculate $\lim_{x \to 0}((1/\sin x) - (1/x))$, we not only find the limit successfully but also discover an approximation formula for $\csc x$.

EXAMPLE 9   Find $\lim\limits_{x \to 0} \left(\dfrac{1}{\sin x} - \dfrac{1}{x} \right)$.

Solution

$$\frac{1}{\sin x} - \frac{1}{x} = \frac{x - \sin x}{x \sin x} = \frac{x - \left(x - \dfrac{x^3}{3!} + \dfrac{x^5}{5!} - \cdots\right)}{x \cdot \left(x - \dfrac{x^3}{3!} + \dfrac{x^5}{5!} - \cdots\right)}$$

$$= \frac{x^3 \left(\dfrac{1}{3!} - \dfrac{x^2}{5!} + \cdots\right)}{x^2 \left(1 - \dfrac{x^2}{3!} + \cdots\right)} = x \frac{\dfrac{1}{3!} - \dfrac{x^2}{5!} + \cdots}{1 - \dfrac{x^2}{3!} + \cdots}.$$

Therefore,

$$\lim_{x \to 0} \left(\frac{1}{\sin x} - \frac{1}{x}\right) = \lim_{x \to 0} \left(x \frac{\dfrac{1}{3!} - \dfrac{x^2}{5!} + \cdots}{1 - \dfrac{x^2}{3!} + \cdots}\right) = 0.$$

From the quotient on the right, we can see that if $|x|$ is small, then

$$\frac{1}{\sin x} - \frac{1}{x} \approx x \cdot \frac{1}{3!} = \frac{x}{6} \qquad \text{or} \qquad \csc x \approx \frac{1}{x} + \frac{x}{6}.$$

Frequently Used Maclaurin Series

$$\frac{1}{1 - x} = 1 + x + x^2 + \cdots + x^n + \cdots = \sum_{n=0}^{\infty} x^n, \qquad |x| < 1$$

$$\frac{1}{1 + x} = 1 - x + x^2 - \cdots + (-x)^n + \cdots = \sum_{n=0}^{\infty} (-1)^n x^n, \qquad |x| < 1$$

$$e^x = 1 + x + \frac{x^2}{2!} + \cdots + \frac{x^n}{n!} + \cdots = \sum_{n=0}^{\infty} \frac{x^n}{n!}, \qquad |x| < \infty$$

$$\sin x = x - \frac{x^3}{3!} + \frac{x^5}{5!} - \cdots + (-1)^n \frac{x^{2n+1}}{(2n+1)!} + \cdots = \sum_{n=0}^{\infty} \frac{(-1)^n x^{2n+1}}{(2n+1)!}, \qquad |x| < \infty$$

$$\cos x = 1 - \frac{x^2}{2!} + \frac{x^4}{4!} - \cdots + (-1)^n \frac{x^{2n}}{(2n)!} + \cdots = \sum_{n=0}^{\infty} \frac{(-1)^n x^{2n}}{(2n)!}, \qquad |x| < \infty$$

$$\ln(1 + x) = x - \frac{x^2}{2!} + \frac{x^3}{3} - \cdots + (-1)^{n-1} \frac{x^n}{n} + \cdots = \sum_{n=1}^{\infty} \frac{(-1)^{n-1} x^n}{n}, \qquad -1 < x \leq 1$$

$$\ln \frac{1+x}{1-x} = 2 \tanh^{-1} x = 2\left(x + \frac{x^3}{3} + \frac{x^5}{5} + \cdots + \frac{x^{2n+1}}{2n+1} + \cdots\right) = 2 \sum_{n=0}^{\infty} \frac{x^{2n+1}}{2n+1}, \qquad |x| < 1$$

$$\tan^{-1} x = x - \frac{x^3}{3} + \frac{x^5}{5} - \cdots + (-1)^n \frac{x^{2n+1}}{2n+1} + \cdots = \sum_{n=0}^{\infty} \frac{(-1)^n x^{2n+1}}{2n+1}, \qquad |x| \leq 1$$

(Continued)

Binomial Series

$$(1 + x)^m = 1 + mx + \frac{m(m - 1)x^2}{2!} + \frac{m(m - 1)(m - 2)x^3}{3!} + \cdots + \frac{m(m - 1)(m - 2) \cdots (m - k + 1)x^k}{k!} + \cdots$$

$$= 1 + \sum_{k=1}^{\infty} \binom{m}{k} x^k, \qquad |x| < 1,$$

where

$$\binom{m}{1} = m, \qquad \binom{m}{2} = \frac{m(m - 1)}{2!}, \qquad \binom{m}{k} = \frac{m(m - 1) \cdots (m - k + 1)}{k!} \qquad \text{for } k \geq 3.$$

Note: To write the binomial series compactly, it is customary to define $\binom{m}{0}$ to be 1 and to take $x^0 = 1$ (even in the usually excluded case where $x = 0$), yielding $(1 + x)^m = \sum_{k=0}^{\infty} \binom{m}{k} x^k$. If m is a *positive integer,* the series terminates at x^m and the result converges for all x.

Exercises 8.11

Binomial Series

Find the first four terms of the binomial series for the functions in Exercises 1–10.

1. $(1 + x)^{1/2}$ **2.** $(1 + x)^{1/3}$ **3.** $(1 - x)^{-1/2}$

4. $(1 - 2x)^{1/2}$ **5.** $\left(1 + \frac{x}{2}\right)^{-2}$ **6.** $\left(1 - \frac{x}{2}\right)^{-2}$

7. $(1 + x^3)^{-1/2}$ **8.** $(1 + x^2)^{-1/3}$

9. $\left(1 + \frac{1}{x}\right)^{1/2}$ **10.** $\left(1 - \frac{2}{x}\right)^{1/3}$

Find the binomial series for the functions in Exercises 11–14.

11. $(1 + x)^4$ **12.** $(1 + x^2)^3$

13. $(1 - 2x)^3$ **14.** $\left(1 - \frac{x}{2}\right)^4$

Initial Value Problems

Find series solutions for the initial value problems in Exercises 15–32.

15. $y' + y = 0, \quad y(0) = 1$ **16.** $y' - 2y = 0, \quad y(0) = 1$

17. $y' - y = 1, \quad y(0) = 0$ **18.** $y' + y = 1, \quad y(0) = 2$

19. $y' - y = x, \quad y(0) = 0$ **20.** $y' + y = 2x, \quad y(0) = -1$

21. $y' - xy = 0, \quad y(0) = 1$ **22.** $y' - x^2y = 0, \quad y(0) = 1$

23. $(1 - x)y' - y = 0, \quad y(0) = 2$

24. $(1 + x^2)y' + 2xy = 0, \quad y(0) = 3$

25. $y'' - y = 0, \quad y'(0) = 1$ and $y(0) = 0$

26. $y'' + y = 0, \quad y'(0) = 0$ and $y(0) = 1$

27. $y'' + y = x, \quad y'(0) = 1$ and $y(0) = 2$

28. $y'' - y = x, \quad y'(0) = 2$ and $y(0) = -1$

29. $y'' - y = -x, \quad y'(2) = -2$ and $y(2) = 0$

30. $y'' - x^2y = 0, \quad y'(0) = b$ and $y(0) = a$

31. $y'' + x^2y = x, \quad y'(0) = b$ and $y(0) = a$

32. $y'' - 2y' + y = 0, \quad y'(0) = 1$ and $y(0) = 0$

Approximations and Nonelementary Integrals

⊞ **CALCULATOR** In Exercises 33–36, use series to estimate the integrals' values with an error of magnitude less than 10^{-3}. (The answer section gives the integrals' values rounded to 5 decimal places.)

33. $\displaystyle\int_0^{0.2} \sin x^2 \, dx$ **34.** $\displaystyle\int_0^{0.2} \frac{e^{-x} - 1}{x} \, dx$

35. $\displaystyle\int_0^{0.1} \frac{1}{\sqrt{1 + x^4}} \, dx$ **36.** $\displaystyle\int_0^{0.25} \sqrt[3]{1 + x^2} \, dx$

⊞ **CALCULATOR** Use series to approximate the values of the integrals in Exercises 37–40 with an error of magnitude less than 10^{-8}. (The answer section gives the integrals' values rounded to 10 decimal places.)

37. $\displaystyle\int_0^{0.1} \frac{\sin x}{x} \, dx$ **38.** $\displaystyle\int_0^{0.1} e^{-x^2} \, dx$

39. $\displaystyle\int_0^{0.1} \sqrt{1 + x^4} \, dx$ **40.** $\displaystyle\int_0^{1} \frac{1 - \cos x}{x^2} \, dx$

41. Estimate the error if $\cos t^2$ is approximated by $1 - \frac{t^4}{2} + \frac{t^8}{4!}$ in the integral $\int_0^1 \cos t^2 \, dt$.

42. Estimate the error if $\cos \sqrt{t}$ is approximated by $1 - \dfrac{t}{2} + \dfrac{t^2}{4!} - \dfrac{t^3}{6!}$ in the integral $\int_0^1 \cos \sqrt{t}\, dt$.

In Exercises 43–46, find a polynomial that will approximate $F(x)$ throughout the given interval with an error of magnitude less than 10^{-3}.

43. $F(x) = \displaystyle\int_0^x \sin t^2 \, dt, \quad [0, 1]$

44. $F(x) = \displaystyle\int_0^x t^2 e^{-t^2} \, dt, \quad [0, 1]$

45. $F(x) = \displaystyle\int_0^x \tan^{-1} t \, dt, \quad$ a) $[0, 0.5]$ b) $[0, 1]$

46. $F(x) = \displaystyle\int_0^x \dfrac{\ln(1+t)}{t} \, dt, \quad$ a) $[0, 0.5]$ b) $[0, 1]$

Indeterminate Forms

Use series to evaluate the limits in Exercises 47–56.

47. $\displaystyle\lim_{x \to 0} \dfrac{e^x - (1 + x)}{x^2}$

48. $\displaystyle\lim_{x \to 0} \dfrac{e^x - e^{-x}}{x}$

49. $\displaystyle\lim_{t \to 0} \dfrac{1 - \cos t - (t^2/2)}{t^4}$

50. $\displaystyle\lim_{\theta \to 0} \dfrac{\sin \theta - \theta + (\theta^3/6)}{\theta^5}$

51. $\displaystyle\lim_{y \to 0} \dfrac{y - \tan^{-1} y}{y^3}$

52. $\displaystyle\lim_{y \to 0} \dfrac{\tan^{-1} y - \sin y}{y^3 \cos y}$

53. $\displaystyle\lim_{x \to \infty} x^2 (e^{-1/x^2} - 1)$

54. $\displaystyle\lim_{x \to \infty} (x + 1) \sin \dfrac{1}{x + 1}$

55. $\displaystyle\lim_{x \to 0} \dfrac{\ln(1 + x^2)}{1 - \cos x}$

56. $\displaystyle\lim_{x \to 2} \dfrac{x^2 - 4}{\ln(x - 1)}$

Theory and Examples

57. Replace x by $-x$ in the Maclaurin series for $\ln(1 + x)$ to obtain a series for $\ln(1 - x)$. Then subtract this from the Maclaurin series for $\ln(1 + x)$ to show that for $|x| < 1$,

$$\ln \dfrac{1 + x}{1 - x} = 2\left(x + \dfrac{x^3}{3} + \dfrac{x^5}{5} + \cdots\right).$$

58. How many terms of the Maclaurin series for $\ln(1 + x)$ should you add to be sure of calculating $\ln(1.1)$ with an error of magnitude less than 10^{-8}? Give reasons for your answer.

59. According to the Alternating Series Estimation Theorem, how many terms of the Maclaurin series for $\tan^{-1} 1$ would you have to add to be sure of finding $\pi/4$ with an error of magnitude less than 10^{-3}? Give reasons for your answer.

60. Show that the Maclaurin series for $f(x) = \tan^{-1} x$ diverges for $|x| > 1$.

61. CALCULATOR About how many terms of the Maclaurin series for $\tan^{-1} x$ would you have to use to evaluate each term on the right-hand side of the equation

$$\pi = 48 \tan^{-1} \dfrac{1}{18} + 32 \tan^{-1} \dfrac{1}{57} - 20 \tan^{-1} \dfrac{1}{239}$$

with an error of magnitude less than 10^{-6}? In contrast, the convergence of $\sum_{n=1}^{\infty} (1/n^2)$ to $\pi^2/6$ is so slow that even 50 terms will not yield two-place accuracy.

62. Integrate the first three nonzero terms of the Maclaurin series for $\tan t$ from 0 to x to obtain the first three nonzero terms of the Maclaurin series for $\ln \sec x$.

63. a) Use the binomial series and the fact that

$$\dfrac{d}{dx} \sin^{-1} x = (1 - x^2)^{-1/2}$$

to generate the first four nonzero terms of the Maclaurin series for $\sin^{-1} x$. What is the radius of convergence?

b) Use your result in (a) to find the first five nonzero terms of the Maclaurin series for $\cos^{-1} x$.

64. a) Find the first four nonzero terms of the Maclaurin series for

$$\sinh^{-1} x = \int_0^x \dfrac{dt}{\sqrt{1 + t^2}}.$$

b) CALCULATOR Use the first *three* terms of the series in (a) to estimate $\sinh^{-1} 0.25$. Give an upper bound for the magnitude of the estimation error.

65. Obtain the Maclaurin series for $1/(1 + x)^2$ from the series for $-1/(1 + x)$.

66. Use the Maclaurin series for $1/(1 - x^2)$ to obtain a series for $2x/(1 - x^2)^2$.

67. CAS The English mathematician Wallis discovered the formula

$$\dfrac{\pi}{4} = \dfrac{2 \cdot 4 \cdot 4 \cdot 6 \cdot 6 \cdot 8 \cdot \cdots}{3 \cdot 3 \cdot 5 \cdot 5 \cdot 7 \cdot 7 \cdot \cdots}.$$

Find π to 2 decimal places with this formula.

68. CALCULATOR Construct a table of natural logarithms $\ln n$ for $n = 1, 2, 3, \ldots, 10$ by using the formula in Exercise 57, but taking advantage of the relationships $\ln 4 = 2 \ln 2$, $\ln 6 = \ln 2 + \ln 3$, $\ln 8 = 3 \ln 2$, $\ln 9 = 2 \ln 3$, and $\ln 10 = \ln 2 + \ln 5$ to reduce the job to the calculation of relatively few logarithms by series. Start by using the following values for x in Exercise 57:

$$\dfrac{1}{3}, \quad \dfrac{1}{5}, \quad \dfrac{1}{9}, \quad \dfrac{1}{13}.$$

69. Integrate the binomial series for $(1 - x^2)^{-1/2}$ to show that for $|x| < 1$,

$$\sin^{-1} x = x + \sum_{n=1}^{\infty} \dfrac{1 \cdot 3 \cdot 5 \cdot \cdots \cdot (2n - 1)}{2 \cdot 4 \cdot 6 \cdot \cdots \cdot (2n)} \dfrac{x^{2n+1}}{2n + 1}.$$

70. *Series for $\tan^{-1} x$ for $|x| > 1$.* Derive the series

$$\tan^{-1} x = \dfrac{\pi}{2} - \dfrac{1}{x} + \dfrac{1}{3x^3} - \dfrac{1}{5x^5} + \cdots, \quad x > 1$$

$$\tan^{-1} x = -\dfrac{\pi}{2} - \dfrac{1}{x} + \dfrac{1}{3x^3} - \dfrac{1}{5x^5} + \cdots, \quad x < -1,$$

by integrating the series

$$\dfrac{1}{1 + t^2} = \dfrac{1}{t^2} \cdot \dfrac{1}{1 + (1/t^2)} = \dfrac{1}{t^2} - \dfrac{1}{t^4} + \dfrac{1}{t^6} - \dfrac{1}{t^8} + \cdots$$

in the first case from x to ∞ and in the second case from $-\infty$ to x.   )

71. *The value of $\sum_{n=1}^{\infty} \tan^{-1}(2/n^2)$*

    a) Use the formula for the tangent of the difference of two angles to show that

$$\tan\left(\tan^{-1}(n+1) - \tan^{-1}(n-1)\right) = \frac{2}{n^2}$$

CHAPTER   **8**   QUESTIONS TO GUIDE YOUR REVIEW

1. What is an infinite sequence? What does it mean for such a sequence to converge? to diverge? Give examples.

2. What uses can be found for subsequences? Give examples.

3. What is a nondecreasing sequence? Under what circumstances does such a sequence have a limit? Give examples.

4. What theorems are available for calculating limits of sequences? Give examples.

5. What theorem sometimes enables us to use l'Hôpital's rule to calculate the limit of a sequence? Give an example.

6. What six sequence limits are likely to arise when you work with sequences and series?

7. What is Picard's method for solving the equation $f(x) = 0$? Give an example.

8. What is an infinite series? What does it mean for such a series to converge? to diverge? Give examples.

9. What is a geometric series? When does such a series converge? diverge? When it does converge, what is its sum? Give examples.

10. Besides geometric series, what other convergent and divergent series do you know?

11. What is the nth-Term Test for Divergence? What is the idea behind the test?

12. What can be said about term-by-term sums and differences of convergent series? about constant multiples of convergent and divergent series?

13. What happens if you add a finite number of terms to a convergent series? a divergent series? What happens if you delete a finite number of terms from a convergent series? a divergent series?

14. How do you reindex a series? Why might you want to do this?

15. Under what circumstances will an infinite series of nonnegative terms converge? diverge? Why study series of nonnegative terms?

16. What is the Integral Test? What is the reasoning behind it? Give an example of its use.

17. When do p-series converge? diverge? How do you know? Give examples of convergent and divergent p-series.

18. What are the Direct Comparison Test and the Limit Comparison Test? What is the reasoning behind these tests? Give examples of their use.

19. What are the Ratio and Root Tests? Do they always give you the information you need to determine convergence or divergence? Give examples.

20. What is an alternating series? What theorem is available for determining the convergence of such a series?

21. How can you estimate the error involved in approximating the sum of an alternating series with one of the series' partial sums? What is the reasoning behind the estimate?

22. What is absolute convergence? conditional convergence? How are the two related?

23. What do you know about rearranging the terms of an absolutely convergent series? of a conditionally convergent series? Give examples.

24. What is a power series? How do you test a power series for convergence? What are the possible outcomes?

25. What are the basic facts about

    a) term-by-term differentiation of power series?

    b) term-by-term integration of power series?

    c) multiplication of power series?

Give examples.

26. What is the Taylor series generated by a function $f(x)$ at a point $x = a$? What information do you need about f to construct the series? Give an example.

27. What is a Maclaurin series?

28. Does a Taylor series always converge to its generating function? Explain.

29. What are Taylor polynomials? Of what use are they?

30. What is Taylor's formula? What does it say about the errors involved in using Taylor polynomials to approximate functions? In particular, what does Taylor's formula say about the error in a linearization? a quadratic approximation?

31. What is the binomial series? On what interval does it converge? How is it used?

32. How can you sometimes use power series to solve initial value problems?

33. How can you sometimes use power series to estimate the values of nonelementary definite integrals?

34. What are the Maclaurin series for $1/(1 - x)$, $1/(1 + x)$, e^x, $\sin x$, $\cos x$, $\ln(1 + x)$, $\ln[(1 + x)/(1 - x)]$, and $\tan^{-1} x$? How do you estimate the errors involved in replacing these series with their partial sums?

CHAPTER 8 PRACTICE EXERCISES

Convergent or Divergent Sequences

Which of the sequences whose nth terms appear in Exercises 1–18 converge, and which diverge? Find the limit of each convergent sequence.

1. $a_n = 1 + \dfrac{(-1)^n}{n}$

2. $a_n = \dfrac{1 - (-1)^n}{\sqrt{n}}$

3. $a_n = \dfrac{1 - 2^n}{2^n}$

4. $a_n = 1 + (0.9)^n$

5. $a_n = \sin \dfrac{n\pi}{2}$

6. $a_n = \sin n\pi$

7. $a_n = \dfrac{\ln(n^2)}{n}$

8. $a_n = \dfrac{\ln(2n + 1)}{n}$

9. $a_n = \dfrac{n + \ln n}{n}$

10. $a_n = \dfrac{\ln(2n^3 + 1)}{n}$

11. $a_n = \left(\dfrac{n - 5}{n}\right)^n$

12. $a_n = \left(1 + \dfrac{1}{n}\right)^{-n}$

13. $a_n = \sqrt[n]{\dfrac{3^n}{n}}$

14. $a_n = \left(\dfrac{3}{n}\right)^{1/n}$

15. $a_n = n(2^{1/n} - 1)$

16. $a_n = \sqrt[n]{2n + 1}$

17. $a_n = \dfrac{(n + 1)!}{n!}$

18. $a_n = \dfrac{(-4)^n}{n!}$

Convergent Series

Find the sums of the series in Exercises 19–24.

19. $\displaystyle\sum_{n=3}^{\infty} \dfrac{1}{(2n - 3)(2n - 1)}$

20. $\displaystyle\sum_{n=2}^{\infty} \dfrac{-2}{n(n + 1)}$

21. $\displaystyle\sum_{n=1}^{\infty} \dfrac{9}{(3n - 1)(3n + 2)}$

22. $\displaystyle\sum_{n=3}^{\infty} \dfrac{-8}{(4n - 3)(4n + 1)}$

23. $\displaystyle\sum_{n=0}^{\infty} e^{-n}$

24. $\displaystyle\sum_{n=1}^{\infty} (-1)^n \dfrac{3}{4^n}$

Convergent or Divergent Series

Which of the series in Exercises 25–40 converge absolutely, which converge conditionally, and which diverge? Give reasons for your answers.

25. $\displaystyle\sum_{n=1}^{\infty} \dfrac{1}{\sqrt{n}}$

26. $\displaystyle\sum_{n=1}^{\infty} \dfrac{-5}{n}$

27. $\displaystyle\sum_{n=1}^{\infty} \dfrac{(-1)^n}{\sqrt{n}}$

28. $\displaystyle\sum_{n=1}^{\infty} \dfrac{1}{2n^3}$

29. $\displaystyle\sum_{n=1}^{\infty} \dfrac{(-1)^n}{\ln(n + 1)}$

30. $\displaystyle\sum_{n=2}^{\infty} \dfrac{1}{n(\ln n)^2}$

31. $\displaystyle\sum_{n=1}^{\infty} \dfrac{\ln n}{n^3}$

32. $\displaystyle\sum_{n=3}^{\infty} \dfrac{\ln n}{\ln(\ln n)}$

33. $\displaystyle\sum_{n=1}^{\infty} \dfrac{(-1)^n}{n\sqrt{n^2 + 1}}$

34. $\displaystyle\sum_{n=1}^{\infty} \dfrac{(-1)^n 3n^2}{n^3 + 1}$

35. $\displaystyle\sum_{n=1}^{\infty} \dfrac{n + 1}{n!}$

36. $\displaystyle\sum_{n=1}^{\infty} \dfrac{(-1)^n(n^2 + 1)}{2n^2 + n - 1}$

37. $\displaystyle\sum_{n=1}^{\infty} \dfrac{(-3)^n}{n!}$

38. $\displaystyle\sum_{n=1}^{\infty} \dfrac{2^n 3^n}{n^n}$

39. $\displaystyle\sum_{n=1}^{\infty} \dfrac{1}{\sqrt{n(n + 1)(n + 2)}}$

40. $\displaystyle\sum_{n=2}^{\infty} \dfrac{1}{n\sqrt{n^2 - 1}}$

Power Series

In Exercises 41–50, (a) find the series' radius and interval of convergence. Then identify the values of x for which the series converges (b) absolutely and (c) conditionally.

41. $\displaystyle\sum_{n=1}^{\infty} \dfrac{(x + 4)^n}{n \, 3^n}$

42. $\displaystyle\sum_{n=1}^{\infty} \dfrac{(x - 1)^{2n-2}}{(2n - 1)!}$

43. $\displaystyle\sum_{n=1}^{\infty} \dfrac{(-1)^{n-1}(3x - 1)^n}{n^2}$

44. $\displaystyle\sum_{n=0}^{\infty} \dfrac{(n + 1)(2x + 1)^n}{(2n + 1)2^n}$

45. $\displaystyle\sum_{n=1}^{\infty} \dfrac{x^n}{n^n}$

46. $\displaystyle\sum_{n=1}^{\infty} \dfrac{x^n}{\sqrt{n}}$

47. $\displaystyle\sum_{n=0}^{\infty} \frac{(n+1)\,x^{2n-1}}{3^n}$

48. $\displaystyle\sum_{n=0}^{\infty} \frac{(-1)^n(x-1)^{2n+1}}{2n+1}$

49. $\displaystyle\sum_{n=1}^{\infty} (\operatorname{csch} n)\, x^n$

50. $\displaystyle\sum_{n=1}^{\infty} (\coth n)\, x^n$

Maclaurin Series

Each of the series in Exercises 51–56 is the value of the Maclaurin series of a function $f(x)$ at a particular point. What function and what point? What is the sum of the series?

51. $1 - \dfrac{1}{4} + \dfrac{1}{16} - \cdots + (-1)^n \dfrac{1}{4^n} + \cdots$

52. $\dfrac{2}{3} - \dfrac{4}{18} + \dfrac{8}{81} - \cdots + (-1)^{n-1} \dfrac{2^n}{n3^n} + \cdots$

53. $\pi - \dfrac{\pi^3}{3!} + \dfrac{\pi^5}{5!} - \cdots + (-1)^n \dfrac{\pi^{2n+1}}{(2n+1)!} + \cdots$

54. $1 - \dfrac{\pi^2}{9\cdot 2!} + \dfrac{\pi^4}{81\cdot 4!} - \cdots + (-1)^n \dfrac{\pi^{2n}}{3^{2n}(2n)!} + \cdots$

55. $1 + \ln 2 + \dfrac{(\ln 2)^2}{2!} + \cdots + \dfrac{(\ln 2)^n}{n!} + \cdots$

56. $\dfrac{1}{\sqrt{3}} - \dfrac{1}{9\sqrt{3}} + \dfrac{1}{45\sqrt{3}} - \cdots$

$\qquad + (-1)^{n-1} \dfrac{1}{(2n-1)(\sqrt{3})^{2n-1}} + \cdots$

Find Maclaurin series for the functions in Exercises 57–64.

57. $\dfrac{1}{1-2x}$

58. $\dfrac{1}{1+x^3}$

59. $\sin \pi x$

60. $\sin \dfrac{2x}{3}$

61. $\cos (x^{5/2})$

62. $\cos \sqrt{5x}$

63. $e^{(\pi x/2)}$

64. e^{-x^2}

Taylor Series

In Exercises 65–68, find the first four nonzero terms of the Taylor series generated by f at $x = a$.

65. $f(x) = \sqrt{3+x^2}$ at $x = -1$

66. $f(x) = 1/(1-x)$ at $x = 2$

67. $f(x) = 1/(x+1)$ at $x = 3$

68. $f(x) = 1/x$ at $x = a > 0$

Initial Value Problems

Use power series to solve the initial value problems in Exercises 69–76.

69. $y' + y = 0, \quad y(0) = -1$

70. $y' - y = 0, \quad y(0) = -3$

71. $y' + 2y = 0, \quad y(0) = 3$

72. $y' + y = 1, \quad y(0) = 0$

73. $y' - y = 3x, \quad y(0) = -1$

74. $y' + y = x, \quad y(0) = 0$

75. $y' - y = x, \quad y(0) = 1$

76. $y' - y = -x, \quad y(0) = 2$

Nonelementary Integrals

Use series to approximate the values of the integrals in Exercises 77–80 with an error of magnitude less than 10^{-8}. (The answer section gives the integrals' values rounded to 10 decimal places.)

77. $\displaystyle\int_0^{1/2} e^{-x^3}\, dx$

78. $\displaystyle\int_0^1 x\, \sin(x^3)\, dx$

79. $\displaystyle\int_0^{1/2} \frac{\tan^{-1} x}{x}\, dx$

80. $\displaystyle\int_0^{1/64} \frac{\tan^{-1} x}{\sqrt{x}}\, dx$

Indeterminate Forms

In Exercises 81–86:

  a) Use power series to evaluate the limit.

  b) GRAPHER Then use a grapher to support your calculation.

81. $\displaystyle\lim_{x\to 0} \frac{7\sin x}{e^{2x}-1}$

82. $\displaystyle\lim_{\theta\to 0} \frac{e^\theta - e^{-\theta} - 2\theta}{\theta - \sin \theta}$

83. $\displaystyle\lim_{t\to 0} \left(\frac{1}{2-2\cos t} - \frac{1}{t^2} \right)$

84. $\displaystyle\lim_{h\to 0} \frac{(\sin h)/h - \cos h}{h^2}$

85. $\displaystyle\lim_{z\to 0} \frac{1-\cos^2 z}{\ln(1-z) + \sin z}$

86. $\displaystyle\lim_{y\to 0} \frac{y^2}{\cos y - \cosh y}$

87. Use a series representation of $\sin 3x$ to find values of r and s for which

$$\lim_{x\to 0} \left(\frac{\sin 3x}{x^3} + \frac{r}{x^2} + s \right) = 0.$$

88. a) Show that the approximation $\csc x \approx 1/x + x/6$ in Section 8.11, Example 9, leads to the approximation $\sin x \approx 6x/(6+x^2)$.

  b) GRAPHER EXPLORATION Compare the accuracies of the approximations $\sin x \approx x$ and $\sin x \approx 6x/(6+x^2)$ by comparing the graphs of $f(x) = \sin x - x$ and $g(x) = \sin x - (6x/(6+x^2))$. Describe what you find.

Theory and Examples

89. a) Show that the series

$$\sum_{n=1}^{\infty} \left(\sin \frac{1}{2n} - \sin \frac{1}{2n+1} \right)$$

converges.

  b) CALCULATOR Estimate the magnitude of the error involved in using the sum of the sines through $n = 20$ to approximate the sum of the series. Is the approximation too large, or too small? Give reasons for your answer.

90. a) Show that the series $\displaystyle\sum_{n=1}^{\infty} \left(\tan \frac{1}{2n} - \tan \frac{1}{2n+1} \right)$ converges.

  b) CALCULATOR Estimate the magnitude of the error in using the sum of the tangents through $-\tan(1/41)$ to approximate the sum of the series. Is the approximation too large, or too small? Give reasons for your answer.

91. Find the radius of convergence of the series

$$\sum_{n=1}^{\infty} \frac{2 \cdot 5 \cdot 8 \cdot \cdots \cdot (3n-1)}{2 \cdot 4 \cdot 6 \cdot \cdots \cdot (2n)} x^n.$$

92. Find the radius of convergence of the series

$$\sum_{n=1}^{\infty} \frac{3 \cdot 5 \cdot 7 \cdot \cdots \cdot (2n+1)}{4 \cdot 9 \cdot 14 \cdot \cdots \cdot (5n-1)} (x-1)^n.$$

93. Find a closed-form formula for the nth partial sum of the series $\sum_{n=2}^{\infty} \ln\left(1 - (1/n^2)\right)$ and use it to determine the convergence or divergence of the series.

94. Evaluate $\sum_{k=2}^{\infty} \left(1/(k^2 - 1)\right)$ by finding the limit as $n \to \infty$ of the series' nth partial sum.

95. a) Find the interval of convergence of the series

$$y = 1 + \frac{1}{6} x^3 + \frac{1}{180} x^6 + \cdots$$

$$+ \frac{1 \cdot 4 \cdot 7 \cdot \cdots \cdot (3n-2)}{(3n)!} x^{3n} + \cdots.$$

b) Show that the function defined by the series satisfies a differential equation of the form

$$\frac{d^2 y}{dx^2} = x^a y + b$$

and find the values of the constants a and b.

96. a) Find the Maclaurin series for the function $x^2/(1 + x)$.

b) Does the series converge at $x = 1$? Explain.

97. If $\sum_{n=1}^{\infty} a_n$ and $\sum_{n=1}^{\infty} b_n$ are convergent series of nonnegative numbers, can anything be said about $\sum_{n=1}^{\infty} a_n b_n$? Give reasons for your answer.

98. If $\sum_{n=1}^{\infty} a_n$ and $\sum_{n=1}^{\infty} b_n$ are divergent series of nonnegative numbers, can anything be said about $\sum_{n=1}^{\infty} a_n b_n$? Give reasons for your answer.

99. Prove that the sequence $\{x_n\}$ and the series $\sum_{k=1}^{\infty} (x_{k+1} - x_k)$ both converge or both diverge.

100. Prove that $\sum_{n=1}^{\infty} (a_n/(1 + a_n))$ converges if $a_n > 0$ for all n and $\sum_{n=1}^{\infty} a_n$ converges.

101. (*Continuation of Section 3.8, Exercise 25.*) If you did Exercise 25 in Section 3.8, you saw that in practice Newton's method stopped too far from the root of $f(x) = (x - 1)^{40}$ to give a useful estimate of its value, $x = 1$. Prove that nevertheless, for any starting value $x_0 \neq 1$, the sequence $x_0, x_1, x_2, \ldots, x_n, \ldots$ of approximations generated by Newton's method really does converge to 1.

102. a) Suppose that $a_1, a_2, a_3, \ldots, a_n$ are positive numbers satisfying the following conditions:

i) $a_1 \geq a_2 \geq a_3 \geq \cdots$;

ii)   the series $a_2 + a_4 + a_8 + a_{16} + \cdots$ diverges.

Show that the series

$$\frac{a_1}{1} + \frac{a_2}{2} + \frac{a_3}{3} + \cdots$$

diverges.

b) Use the result in (a) to show that

$$1 + \sum_{n=2}^{\infty} \frac{1}{n \ln n}$$

diverges.

103. Suppose you wish to obtain a quick estimate for the value of $\int_0^1 x^2 e^x \, dx$. There are several ways to do this.

a) Use the trapezoidal rule with $n = 2$ to estimate $\int_0^1 x^2 e^x \, dx$.

b) Write out the first three nonzero terms of the Maclaurin series for $x^2 e^x$ to obtain the fourth Maclaurin polynomial $P(x)$ for $x^2 e^x$. Use $\int_0^1 P(x) \, dx$ to obtain another estimate for $\int_0^1 x^2 e^x \, dx$.

c) The second derivative of $f(x) = x^2 e^x$ is positive for all $x > 0$. Explain why this enables you to conclude that the trapezoidal rule estimate obtained in (a) is too large. (*Hint:* What does the second derivative tell you about the graph of a function? How does this relate to the trapezoidal approximation of the area under this graph?)

d) All the derivatives of $f(x) = x^2 e^x$ are positive for $x > 0$. Explain why this enables you to conclude that all Maclaurin polynomial approximations to $f(x)$ for x in [0, 1] will be too small. (*Hint:* $f(x) = P_n(x) + R_n(x)$.)

e) Use integration by parts to evaluate $\int_0^1 x^2 e^x \, dx$.

Convergence or Divergence

Which of the series $\sum_{n=1}^{\infty} a_n$ defined by the formulas in Exercises 1–4 converge, and which diverge? Give reasons for your answers.

1. $\displaystyle\sum_{n=1}^{\infty} \frac{1}{(3n-2)^{n+(1/2)}}$

2. $\displaystyle\sum_{n=1}^{\infty} \frac{(\tan^{-1} n)^2}{n^2 + 1}$

3. $\displaystyle\sum_{n=1}^{\infty} (-1)^n \tanh n$

4. $\displaystyle\sum_{n=2}^{\infty} \frac{\log_n(n!)}{n^3}$

Which of the series $\sum_{n=1}^{\infty} a_n$ defined by the formulas in Exercises 5–8 converge, and which diverge? Give reasons for your answers.

5. $a_1 = 1, \quad a_{n+1} = \dfrac{n(n+1)}{(n+2)(n+3)} a_n$

   (*Hint:* Write out several terms, see which factors cancel, and then generalize.)

6. $a_1 = a_2 = 7, \quad a_{n+1} = \dfrac{n}{(n-1)(n+1)} a_n$ if $n \geq 2$

7. $a_1 = a_2 = 1, \quad a_{n+1} = \dfrac{1}{1 + a_n}$ if $n \geq 2$

8. $a_n = 1/3^n$ if n is odd, $\quad a_n = n/3^n$ if n is even

Choosing Centers for Taylor Series

Taylor's formula

$$f(x) = f(a) + f'(a)(x-a) + \frac{f''(a)}{2!}(x-a)^2 + \cdots$$

$$+ \frac{f^{(n)}(a)}{n!}(x-a)^n + \frac{f^{(n+1)}(c)}{(n+1)!}(x-a)^{n+1}$$

expresses the value of f at x in terms of the values of f and its derivatives at $x = a$. In numerical computations, we therefore need f to be a point where we know the values of f and its derivatives. We also need a to be close enough to the values of f we are interested in to make $(x-a)^{n+1}$ so small we can neglect the remainder.

In Exercises 9–14, what Taylor series would you choose to represent the function near the given value of x? (There may be more than one good answer.) Write out the first four nonzero terms of the series you choose.

9. $\cos x \quad$ near $\quad x = 1$

10. $\sin x \quad$ near $\quad x = 6.3$

11. $e^x \quad$ near $\quad x = 0.4$

12. $\ln x \quad$ near $\quad x = 1.3$

13. $\cos x \quad$ near $\quad x = 69$

14. $\tan^{-1} x \quad$ near $\quad x = 2$

Theory and Examples

15. Let a and b be constants with $0 < a < b$. Does the sequence $\{(a^n + b^n)^{1/n}\}$ converge? If it does converge, what is the limit?

16. Find the sum of the infinite series

$$1 + \frac{2}{10} + \frac{3}{10^2} + \frac{7}{10^3} + \frac{2}{10^4} + \frac{3}{10^5} + \frac{7}{10^6} + \frac{2}{10^7} + \frac{3}{10^8} + \frac{7}{10^9} + \cdots.$$

17. Evaluate

$$\sum_{n=0}^{\infty} \int_n^{n+1} \frac{1}{1 + x^2} \, dx.$$

18. Find all values of x for which

$$\sum_{n=1}^{\infty} \frac{nx^n}{(n+1)(2x+1)^n}$$

converges absolutely.

19. *Generalizing Euler's constant.* Figure 8.21 shows the graph of a positive twice-differentiable decreasing function f whose second derivative is positive on $(0, \infty)$. For each n, the number A_n is the area of the lunar region between the curve and the line segment joining the points $(n, f(n))$ and $(n+1, f(n+1))$.

a) Use the figure to show that $\sum_{n=1}^{\infty} A_n < (1/2)(f(1) - f(2))$.

b) Then show the existence of

$$\lim_{n \to \infty} \left[\sum_{k=1}^{n} f(k) - \frac{1}{2}(f(1) + f(n)) - \int_1^n f(x) \, dx \right].$$

c) Then show the existence of

$$\lim_{n \to \infty} \left[\sum_{k=1}^{n} f(k) - \int_1^n f(x) \, dx \right].$$

If $f(x) = 1/x$, the limit in (c) is Euler's constant (Section 8.4, Exercise 41). (Source: "Convergence with Pictures" by P. J. Rippon, *American Mathematical Monthly*, Vol. 93, No. 6, 1986, pp. 476–78.)

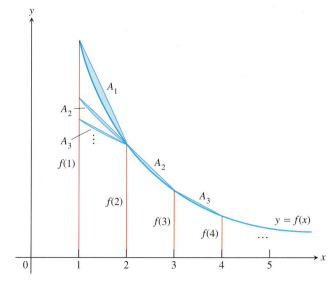

8.21 The figure for Exercise 19.

20. This exercise refers to the "right side up" equilateral triangle with sides of length $2b$ in the accompanying figure. "Upside down" equilateral triangles are removed from the original triangle as the sequence of pictures suggests. The sum of the areas removed from the original triangle forms an infinite series.

a) Find this infinite series.

b) Find the sum of this infinite series and hence find the total area removed from the original triangle.

c) Is every point on the original triangle removed? Explain why or why not.

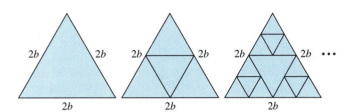

✪ 21. CAS EXPLORATION

a) Does the value of

$$\lim_{n \to \infty} \left(1 - \frac{\cos (a/n)}{n} \right)^n, \quad a \text{ constant},$$

appear to depend on the value of a? If so, how?

b) Does the value of

$$\lim_{n \to \infty} \left(1 - \frac{\cos (a/n)}{bn} \right)^n, \quad a \text{ and } b \text{ constant}, b \neq 0,$$

appear to depend on the value of b? If so, how?

c) Use calculus to confirm your findings in (a) and (b).

22. Show that if $\sum_{n=1}^{\infty} a_n$ converges, then

$$\sum_{n=1}^{\infty} \left(\frac{1 + \sin (a_n)}{2} \right)^n$$

converges.

23. Find a value for the constant b that will make the radius of convergence of the power series

$$\sum_{n=2}^{\infty} \frac{b^n x^n}{\ln n}$$

equal to 5.

24. How do you know that the functions $\sin x$, $\ln x$, and e^x are not polynomials? Give reasons for your answer.

25. Find the value of a for which the limit

$$\lim_{x \to 0} \frac{\sin (ax) - \sin x - x}{x^3}$$

is finite and evaluate the limit.

26. Find values of a and b for which

$$\lim_{x \to 0} \frac{\cos (ax) - b}{2x^2} = -1.$$

27. *Raabe's (or Gauss's) test.* The following test, which we state without proof, is an extension of the Ratio Test.

Raabe's test: If $\sum_{n=1}^{\infty} u_n$ is a series of positive constants and there exist constants C, K, and N such that

$$\frac{u_n}{u_{n+1}} = 1 + \frac{C}{n} + \frac{f(n)}{n^2}, \tag{1}$$

where $|f(n)| < K$ for $n \geq N$, then $\sum_{n=1}^{\infty} u_n$ converges if $C > 1$ and diverges if $C \leq 1$.

Show that the results of Raabe's test agree with what you know about the series $\sum_{n=1}^{\infty} (1/n^2)$ and $\sum_{n=1}^{\infty} (1/n)$.

28. (*Continuation of Exercise 27.*) Suppose that the terms of $\sum_{n=1}^{\infty} u_n$ are defined recursively by the formulas

$$u_1 = 1, \quad u_{n+1} = \frac{(2n-1)^2}{(2n)(2n+1)} u_n.$$

Apply Raabe's test to determine whether the series converges.

29. If $\sum_{n=1}^{\infty} a_n$ converges, and if $a_n \neq 1$ and $a_n > 0$ for all n,

a) Show that $\sum_{n=1}^{\infty} a_n^2$ converges.

b) Does $\sum_{n=1}^{\infty} a_n/(1 - a_n)$ converge? Explain.

30. (*Continuation of Exercise 29.*) If $\sum_{n=1}^{\infty} a_n$ converges, and if $1 > a_n > 0$ for all n, show that $\sum_{n=1}^{\infty} \ln (1 - a_n)$ converges. (*Hint:* First show that $|\ln (1 - a_n)| \leq a_n/(1 - a_n)$.)

31. *Nicole Oresme's theorem.* Prove Nicole Oresme's theorem that

$$1 + \frac{1}{2} \cdot 2 + \frac{1}{4} \cdot 3 + \cdots + \frac{n}{2^{n-1}} + \cdots = 4.$$

(*Hint*: Differentiate both sides of the equation $1/(1 - x) = 1 + \sum_{n=1}^{\infty} x^n$.)

32. a) Show that

$$\sum_{n=1}^{\infty} \frac{n(n+1)}{x^n} = \frac{2x^2}{(x-1)^3}$$

for $|x| > 1$ by differentiating the identity

$$\sum_{n=1}^{\infty} x^{n+1} = \frac{x^2}{1 - x}$$

twice, multiplying the result by x, and then replacing x by $1/x$.

b) CALCULATOR Use part (a) to find the real solution greater than 1 of the equation

$$x = \sum_{n=1}^{\infty} \frac{n(n+1)}{x^n}.$$

33. *A fast estimate of $\pi/2$.* As you saw if you did Exercise 29 in Section 8.1, the sequence generated by starting with $x_0 = 1$

and applying the recursion formula $x_{n+1} = x_n + \cos x_n$ converges rapidly to $\pi/2$. To explain the speed of the convergence, let $\epsilon_n = (\pi/2) - x_n$. (See the accompanying figure.) Then

$$\epsilon_{n+1} = \frac{\pi}{2} - x_n - \cos x_n$$

$$= \epsilon_n - \cos\left(\frac{\pi}{2} - \epsilon_n\right)$$

$$= \epsilon_n - \sin \epsilon_n$$

$$= \frac{1}{3!}(\epsilon_n)^3 - \frac{1}{5!}(\epsilon_n)^5 + \cdots.$$

Use this equality to show that

$$0 < \epsilon_{n+1} < \frac{1}{6}(\epsilon_n)^3.$$

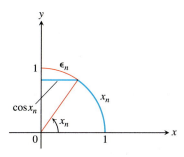

34. If $\sum_{n=1}^{\infty} a_n$ is a convergent series of positive numbers, can anything be said about the convergence of $\sum_{n=1}^{\infty} \ln(1 + a_n)$? Give reasons for your answer.

35. *Quality control*

   a) Differentiate the series

   $$\frac{1}{1-x} = 1 + x + x^2 + \cdots + x^n + \cdots$$

   to obtain a series for $1/(1-x)^2$.

   b) In one throw of two dice, the probability of getting a roll of 7 is $p = 1/6$. If you throw the dice repeatedly, the probability that a 7 will appear for the first time at the nth throw is $q^{n-1}p$, where $q = 1 - p = 5/6$. The expected number of throws until a 7 first appears is $\sum_{n=1}^{\infty} nq^{n-1}p$. Find the sum of this series.

   c) As an engineer applying statistical control to an industrial operation, you inspect items taken at random from the assembly line. You classify each sampled item as either "good" or "bad." If the probability of an item's being good is p and of an item's being bad is $q = 1 - p$, the probability that the first bad item found is the nth one inspected is $p^{n-1}q$. The average number inspected up to and including the first bad item found is $\sum_{n=1}^{\infty} np^{n-1}q$. Evaluate this sum, assuming $0 < p < 1$.

36. *Expected value.* Suppose that a random variable X may assume the values $1, 2, 3, \ldots$, with probabilities $p_1, p_2, p_3, \ldots$, where p_k is the probability that X equals k ($k = 1, 2, 3, \ldots$). Suppose also that $p_k \geq 0$ and that $\sum_{k=1}^{\infty} p_k = 1$. The **expected value** of X,

denoted by $E(X)$, is the number $\sum_{k=1}^{\infty} kp_k$, provided the series converges. In each of the following cases, show that $\sum_{k=1}^{\infty} p_k = 1$ and find $E(X)$ if it exists. (*Hint:* See Exercise 35.)

   a) $p_k = 2^{-k}$ **b)** $p_k = \dfrac{5^{k-1}}{6^k}$

   c) $p_k = \dfrac{1}{k(k+1)} = \dfrac{1}{k} - \dfrac{1}{k+1}$

37. *Safe and effective dosage.* The concentration in the blood resulting from a single dose of a drug normally decreases with time as the drug is eliminated from the body. Doses may therefore need to be repeated periodically to keep the concentration from dropping below some particular level. One model for the effect of repeated doses gives the residual concentration just before the $(n + 1)$st dose as

$$R_n = C_0\,e^{-kt_0} + C_0\,e^{-2kt_0} + \cdots + C_0\,e^{-nkt_0},$$

where $C_0 = $ the change in concentration achievable by a single dose (mg/ml), $k = $ the *elimination constant* (h^{-1}), and $t_0 = $ time between doses (h). See Fig. 8.22.

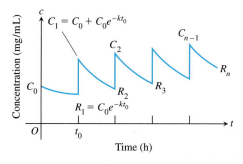

8.22 One possible effect of repeated doses on the concentration of a drug in the bloodstream.

   a) Write R_n in closed form as a single fraction, and find $R = \lim_{n\to\infty} R_n$.

   b) Calculate R_1 and R_{10} for $C_0 = 1$ mg/ml, $k = 0.1$ h^{-1}, and $t_0 = 10$ h. How good an estimate of R is R_{10}?

   c) If $k = 0.01$ h^{-1} and $t_0 = 10$ h, find the smallest n such that $R_n > (1/2)R$.

(Source: *Prescribing Safe and Effective Dosage*, B. Horelick and S. Koont, COMAP, Inc., Lexington, MA.)

38. (*Continuation of Exercise 37.*) If a drug is known to be ineffective below a concentration C_L and harmful above some higher concentration C_H, one needs to find values of C_0 and t_0 that will produce a concentration that is safe (not above C_H) but effective (not below C_L). See Fig. 8.23. We therefore want to find values for C_0 and t_0 for which

$$R = C_L \quad \text{and} \quad C_0 + R = C_H.$$

Thus $C_0 = C_H - C_L$. When these values are substituted in the

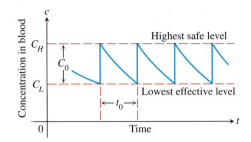

8.23 Safe and effective concentrations of a drug. C_0 is the change in concentration produced by one dose; t_0 is the time between doses.

equation for R obtained in part (a) of Exercise 37, the resulting equation simplifies to

$$t_0 = \frac{1}{k} \ln \frac{C_H}{C_L}.$$

To reach an effective level rapidly, one might administer a "loading" dose that would produce a concentration of C_H mg/ml. This could be followed every t_0 hours by a dose that raises the concentration by $C_0 = C_H - C_L$ mg/ml.

a) Verify the preceding equation for t_0.

b) If $k = 0.05\,\text{h}^{-1}$ and the highest safe concentration is e times the lowest effective concentration, find the length of time between doses that will assure safe and effective concentrations.

c) Given $C_H = 2$ mg/ml, $C_L = 0.5$ mg/ml, and $k = 0.02\,\text{h}^{-1}$, determine a scheme for administering the drug.

d) Suppose that $k = 0.2\,\text{h}^{-1}$ and that the smallest effective concentration is 0.03 mg/ml. A single dose that produces a concentration of 0.1 mg/ml is administered. About how long will the drug remain effective?

39. *An infinite product.* The infinite product

$$\prod_{n=1}^{\infty} (1 + a_n) = (1 + a_1)(1 + a_2)(1 + a_3) \cdots$$

is said to converge if the series

$$\sum_{n=1}^{\infty} \ln (1 + a_n),$$

obtained by taking the natural logarithm of the product, converges. Prove that the product converges if $a_n > -1$ for every n and if $\sum_{n=1}^{\infty} |a_n|$ converges. (*Hint:* Show that

$$|\ln (1 + a_n)| \leq \frac{|a_n|}{1 - |a_n|} \leq 2 |a_n|$$

when $|a_n| < 1/2$.)

40. If p is a constant, show that the series

$$1 + \sum_{n=3}^{\infty} \frac{1}{n \cdot \ln n \cdot [\ln (\ln n)]^p}$$

(a) converges if $p > 1$, (b) diverges if $p \leq 1$. In general, if $f_1(x) = x$, $f_{n+1}(x) = \ln (f_n(x))$, and n takes on the values 1,

2, 3, $\ldots$, we find that $f_2(x) = \ln x$, $f_3(x) = \ln (\ln x)$, and so on. If $f_n(a) > 1$, then

$$\int_a^{\infty} \frac{dx}{f_1(x) f_2(x) \cdots f_n(x)(f_{n+1}(x))^p}$$

converges if $p > 1$ and diverges if $p \leq 1$.

41. a) Prove the following theorem: If $\{c_n\}$ is a sequence of numbers such that every sum $t_n = \sum_{k=1}^{n} c_k$ is bounded, then the series $\sum_{n=1}^{\infty} c_n/n$ converges and is equal to $\sum_{n=1}^{\infty} t_n/(n(n+1))$.

Outline of proof: Replace c_1 by t_1 and c_n by $t_n - t_{n-1}$ for $n \geq 2$. If $s_{2n+1} = \sum_{k=1}^{2n+1} c_k/k$, show that

$$s_{2n+1} = t_1 \left(1 - \frac{1}{2}\right) + t_2 \left(\frac{1}{2} - \frac{1}{3}\right)$$

$$+ \cdots + t_{2n} \left(\frac{1}{2n} - \frac{1}{2n+1}\right) + \frac{t_{2n+1}}{2n+1}$$

$$= \sum_{k=1}^{2n} \frac{t_k}{k(k+1)} + \frac{t_{2n+1}}{2n+1}.$$

Because $|t_k| < M$ for some constant M, the series

$$\sum_{k=1}^{\infty} \frac{t_k}{k(k+1)}$$

converges absolutely and s_{2n+1} has a limit as $n \to \infty$. Finally, if $s_{2n} = \sum_{k=1}^{2n} c_k/k$, then $s_{2n+1} - s_{2n} = c_{2n+1}/(2n+1)$ approaches zero as $n \to \infty$ because $|c_{2n+1}| = |t_{2n+1} - t_{2n}| < 2M$. Hence the sequence of partial sums of the series $\sum c_k/k$ converges and the limit is $\sum_{k=1}^{\infty} t_k/(k(k+1))$.

b) Show how the foregoing theorem applies to the alternating harmonic series

$$1 - \frac{1}{2} + \frac{1}{3} - \frac{1}{4} + \frac{1}{5} - \frac{1}{6} + \cdots.$$

c) Show that the series

$$1 - \frac{1}{2} - \frac{1}{3} + \frac{1}{4} + \frac{1}{5} - \frac{1}{6} - \frac{1}{7} + \cdots$$

converges. (After the first term, the signs are two negative, two positive, two negative, two positive, and so on in that pattern.)

42. *The convergence of $\sum_{n=1}^{\infty} [(-1)^{n-1} x^n]/n$ to $\ln(1+x)$ for $-1 < x \leq 1$*

a) Show by long division or otherwise that

$$\frac{1}{1+t} = 1 - t + t^2 - t^3 + \cdots + (-1)^n t^n + \frac{(-1)^{n+1} t^{n+1}}{1+t}.$$

b) By integrating the equation of part (a) with respect to t from 0 to x, show that

$$\ln (1+x) = x - \frac{x^2}{2} + \frac{x^3}{3} - \frac{x^4}{4} + \cdots$$

$$+ (-1)^n \frac{x^{n+1}}{n+1} + R_{n+1}$$

where

$$R_{n+1} = (-1)^{n+1} \int_0^x \frac{t^{n+1}}{1+t} \, dt.$$

c) If $x \geq 0$, show that

$$|R_{n+1}| \leq \int_0^x t^{n+1} \, dt = \frac{x^{n+2}}{n+2}.$$

$\Big($ *Hint*: As t varies from 0 to x,

$$1 + t \geq 1 \quad \text{and} \quad t^{n+1}/(1+t) \leq t^{n+1},$$

and

$$\left| \int_0^x f(t) \, dt \right| \leq \int_0^x \left| f(t) \right| dt. \Big)$$

d) If $-1 < x < 0$, show that

$$\left| R_{n+1} \right| \leq \left| \int_0^x \frac{t^{n+1}}{1-|x|} \, dt \right| = \frac{|x|^{n+2}}{(n+2)(1-|x|)}.$$

$\Big($ *Hint*: If $x < t \leq 0$, then $|1 + t| \geq 1 - |x|$ and

$$\left| \frac{t^{n+1}}{1+t} \right| \leq \frac{|t|^{n+1}}{1-|x|}. \Big)$$

e) Use the foregoing results to prove that the series

$$x - \frac{x^2}{2} + \frac{x^3}{3} - \frac{x^4}{4} + \cdots + \frac{(-1)^n x^{n+1}}{n+1} + \cdots$$

converges to $\ln(1+x)$ for $-1 < x \leq 1$.

Conic Sections, Parametrized Curves, and Polar Coordinates

OVERVIEW The study of motion has been important since ancient times, and calculus provides the mathematics we need to describe it. In this chapter, we extend our ability to analyze motion by showing how to track the position of a moving body as a function of time. We begin with equations for conic sections, since these are the paths traveled by planets, satellites, and other bodies (even electrons) whose motions are driven by inverse square forces. As we will see in Chapter 11, once we know that the path of a moving body is a conic section, we immediately have information about the body's velocity and the force that drives it. Planetary motion is best described with the help of polar coordinates (another of Newton's inventions, although James-Jakob-Jacques Bernoulli (1655–1705) usually gets the credit), so we also investigate curves, derivatives, and integrals in this new coordinate system.

9.1 Conic Sections and Quadratic Equations

This section shows how the conic sections from Greek geometry are described today as the graphs of quadratic equations in the coordinate plane. The Greeks of Plato's time described these curves as the curves formed by cutting a double cone with a plane (Fig. 9.1, on the following page); hence the name *conic section*.

Circles

Definitions

A **circle** is the set of points in a plane whose distance from a given fixed point in the plane is constant. The fixed point is the **center** of the circle; the constant distance is the **radius**.

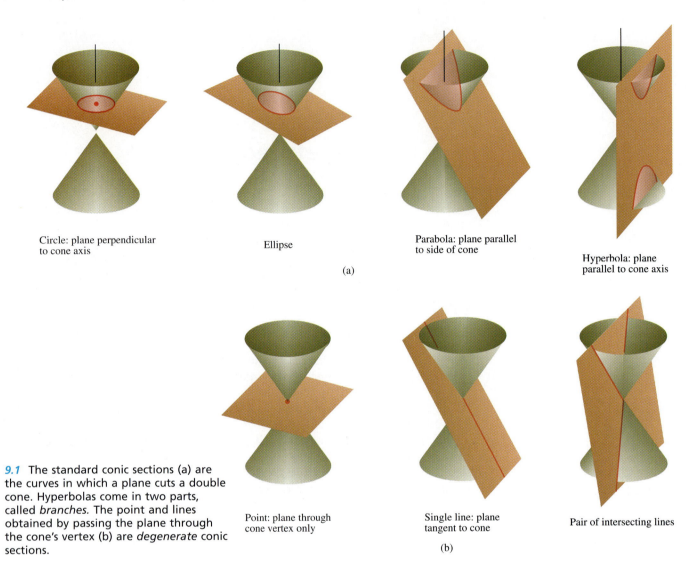

Circle: plane perpendicular
to cone axis

Ellipse

Parabola: plane parallel
to side of cone

Hyperbola: plane
parallel to cone axis

(a)

9.1 The standard conic sections (a) are
the curves in which a plane cuts a double
cone. Hyperbolas come in two parts,
called *branches.* The point and lines
obtained by passing the plane through
the cone's vertex (b) are *degenerate* conic
sections.

Point: plane through
cone vertex only

Single line: plane
tangent to cone

Pair of intersecting lines

(b)

The standard-form equations for circles, derived in Preliminaries, Section 4, from
the distance formula $d = \sqrt{(x_2 - x_1)^2 + (y_2 - y_1)^2}$, are these:

Circles

Circle of radius a centered
at the origin:

$$x^2 + y^2 = a^2$$

Circle of radius a centered
at the point (h, k):

$$(x - h)^2 + (y - k)^2 = a^2$$

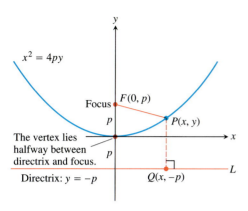

9.2 The parabola $x^2 = 4py$.

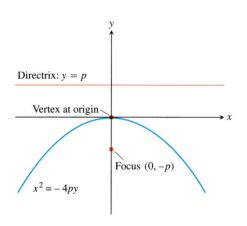

9.3 The parabola $x^2 = -4py$.

Parabolas

Definitions

A set that consists of all the points in a plane equidistant from a given fixed point and a given fixed line in the plane is a **parabola.** The fixed point is the **focus** of the parabola. The fixed line is the **directrix.**

If the focus F lies on the directrix L, the parabola is the line through F perpendicular to L. We consider this to be a degenerate case and assume henceforth that F does not lie on L.

A parabola has its simplest equation when its focus and directrix straddle one of the coordinate axes. For example, suppose that the focus lies at the point $F(0, p)$ on the positive y-axis and that the directrix is the line $y = -p$ (Fig. 9.2). In the notation of the figure, a point $P(x, y)$ lies on the parabola if and only if $PF = PQ$. From the distance formula,

$$PF = \sqrt{(x - 0)^2 + (y - p)^2} = \sqrt{x^2 + (y - p)^2}$$

$$PQ = \sqrt{(x - x)^2 + (y - (-p))^2} = \sqrt{(y + p)^2}.$$

When we equate these expressions, square, and simplify, we get

$$y = \frac{x^2}{4p} \qquad \text{or} \qquad x^2 = 4py. \qquad \text{Standard form} \qquad (1)$$

These equations reveal the parabola's symmetry about the y-axis. We call the y-axis the **axis** of the parabola (short for "axis of symmetry").

The point where a parabola crosses its axis is the **vertex.** The vertex of the parabola $x^2 = 4py$ lies at the origin (Fig. 9.2). The positive number p is the parabola's **focal length.**

If the parabola opens downward, with its focus at $(0, -p)$ and its directrix the line $y = p$, then Eqs. (1) become

$$y = -\frac{x^2}{4p} \qquad \text{and} \qquad x^2 = -4py$$

(Fig. 9.3). We obtain similar equations for parabolas opening to the right or to the left (Fig. 9.4, on the following page, and Table 9.1).

Table 9.1 Standard-form equations for parabolas with vertices at the origin ($p > 0$)

Equation	Focus	Directrix	Axis	Opens
$x^2 = 4py$	$(0, p)$	$y = -p$	y-axis	Up
$x^2 = -4py$	$(0, -p)$	$y = p$	y-axis	Down
$y^2 = 4px$	$(p, 0)$	$x = -p$	x-axis	To the right
$y^2 = -4px$	$(-p, 0)$	$x = p$	x-axis	To the left

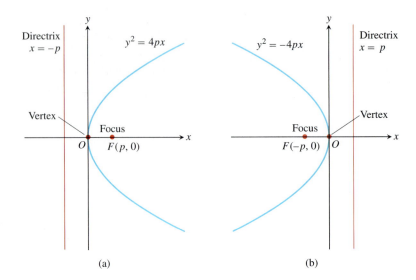

9.4 (a) The parabola $y^2 = 4px$. (b) The parabola $y^2 = -4px$.

(a)   (b)

EXAMPLE 1   Find the focus and directrix of the parabola $y^2 = 10x$.

Solution   We find the value of p in the standard equation $y^2 = 4px$:

$$4p = 10, \quad \text{so} \quad p = \frac{10}{4} = \frac{5}{2}.$$

Then we find the focus and directrix for this value of p:

Focus: $(p, 0) = \left(\frac{5}{2}, 0\right)$

Directrix: $x = -p \quad \text{or} \quad x = -\frac{5}{2}.$

The horizontal and vertical shift formulas in Preliminaries, Section 4, can be applied to the equations in Table 9.1 to give equations for a variety of parabolas in other locations (see Exercises 39, 40, and 45–48).

Ellipses

> **Definitions**
> An **ellipse** is the set of points in a plane whose distances from two fixed points in the plane have a constant sum. The two fixed points are the **foci** of the ellipse.

The quickest way to construct an ellipse uses the definition. Put a loop of string around two tacks F_1 and F_2, pull the string taut with a pencil point P, and move the pencil around to trace a closed curve (Fig. 9.5). The curve is an ellipse because the sum $PF_1 + PF_2$, being the length of the loop minus the distance between the tacks, remains constant. The ellipse's foci lie at F_1 and F_2.

9.5 How to draw an ellipse.

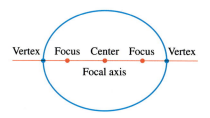

9.6 Points on the focal axis of an ellipse.

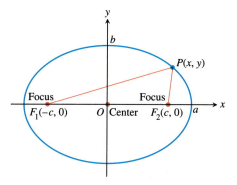

9.7 The ellipse defined by the equation $PF_1 + PF_2 = 2a$ is the graph of the equation $(x^2/a^2) + (y^2/b^2) = 1$.

Definitions
The line through the foci of an ellipse is the ellipse's **focal axis.** The point on the axis halfway between the foci is the **center.** The points where the focal axis and ellipse cross are the ellipse's **vertices** (Fig. 9.6).

If the foci are $F_1(-c, 0)$ and $F_2(c, 0)$ (Fig. 9.7), and $PF_1 + PF_2$ is denoted by $2a$, then the coordinates of a point P on the ellipse satisfy the equation

$$\sqrt{(x + c)^2 + y^2} + \sqrt{(x - c)^2 + y^2} = 2a.$$

To simplify this equation, we move the second radical to the right-hand side, square, isolate the remaining radical, and square again, obtaining

$$\frac{x^2}{a^2} + \frac{y^2}{a^2 - c^2} = 1. \tag{2}$$

Since $PF_1 + PF_2$ is greater than the length F_1F_2 (triangle inequality for triangle PF_1F_2), the number $2a$ is greater than $2c$. Accordingly, $a > c$ and the number $a^2 - c^2$ in Eq. (2) is positive.

The algebraic steps leading to Eq. (2) can be reversed to show that every point P whose coordinates satisfy an equation of this form with $0 < c < a$ also satisfies the equation $PF_1 + PF_2 = 2a$. A point therefore lies on the ellipse if and only if its coordinates satisfy Eq. (2).

If

$$b = \sqrt{a^2 - c^2}, \tag{3}$$

then $a^2 - c^2 = b^2$ and Eq. (2) takes the form

$$\frac{x^2}{a^2} + \frac{y^2}{b^2} = 1. \tag{4}$$

Equation (4) reveals that this ellipse is symmetric with respect to the origin and both coordinate axes. It lies inside the rectangle bounded by the lines $x = \pm a$ and $y = \pm b$. It crosses the axes at the points $(\pm a, 0)$ and $(0, \pm b)$. The tangents at these points are perpendicular to the axes because

$$\frac{dy}{dx} = -\frac{b^2 x}{a^2 y} \qquad \text{Obtained from Eq. (4) by implicit differentiation}$$

is zero if $x = 0$ and infinite if $y = 0$.

The Major and Minor Axes of an Ellipse

The **major axis** of the ellipse in Eq. (4) is the line segment of length $2a$ joining the points $(\pm a, 0)$. The **minor axis** is the line segment of length $2b$ joining the points $(0, \pm b)$. The number a itself is the **semimajor axis,** the number b the **semiminor axis.** The number c, found from Eq. (3) as

$$c = \sqrt{a^2 - b^2},$$

is the **center-to-focus distance** of the ellipse.

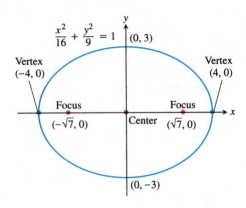

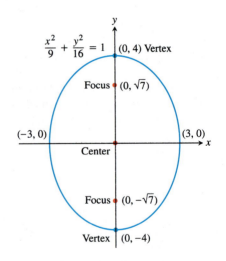

9.8 Major axis horizontal (Example 2).

9.9 Major axis vertical (Example 3).

EXAMPLE 2   *Major axis horizontal*

The ellipse

$$\frac{x^2}{16} + \frac{y^2}{9} = 1 \tag{5}$$

(Fig. 9.8) has

Semimajor axis: $a = \sqrt{16} = 4,$ Semiminor axis: $b = \sqrt{9} = 3$

Center-to-focus distance: $c = \sqrt{16 - 9} = \sqrt{7}$

Foci: $(\pm c, 0) = (\pm \sqrt{7}, 0)$

Vertices: $(\pm a, 0) = (\pm 4, 0)$

Center: $(0, 0).$ ❑

EXAMPLE 3   *Major axis vertical*

The ellipse

$$\frac{x^2}{9} + \frac{y^2}{16} = 1, \tag{6}$$

obtained by interchanging x and y in Eq. (5), has its major axis vertical instead of horizontal (Fig. 9.9). With a^2 still equal to 16 and b^2 equal to 9, we have

Semimajor axis: $a = \sqrt{16} = 4,$ Semiminor axis: $b = \sqrt{9} = 3$

Center-to-focus distance: $c = \sqrt{16 - 9} = \sqrt{7}$

Foci: $(0, \pm c) = (0, \pm \sqrt{7})$

Vertices: $(0, \pm a) = (0, \pm 4)$

Center: $(0, 0).$ ❑

There is never any cause for confusion in analyzing equations like (5) and (6). We simply find the intercepts on the coordinate axes; then we know which way the major axis runs because it is the longer of the two axes. The center always lies at the origin and the foci lie on the major axis.

Standard-Form Equations for Ellipses Centered at the Origin

Foci on the x-axis: $\dfrac{x^2}{a^2} + \dfrac{y^2}{b^2} = 1$ $(a > b)$

　Center-to-focus distance: $c = \sqrt{a^2 - b^2}$
　Foci: $(\pm c, 0)$
　Vertices: $(\pm a, 0)$

Foci on the y-axis: $\dfrac{x^2}{b^2} + \dfrac{y^2}{a^2} = 1$ $(a > b)$

　Center-to-focus distance: $c = \sqrt{a^2 - b^2}$
　Foci: $(0, \pm c)$
　Vertices: $(0, \pm a)$

In each case, a is the semimajor axis and b is the semiminor axis.

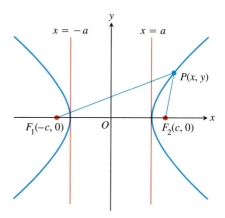

9.10 Hyperbolas have two branches. For points on the right-hand branch of the hyperbola shown here, $PF_1 - PF_2 = 2a$. For points on the left-hand branch, $PF_2 - PF_1 = 2a$.

Hyperbolas

Definitions

A **hyperbola** is the set of points in a plane whose distances from two fixed points in the plane have a constant difference. The two fixed points are the **foci** of the hyperbola.

If the foci are $F_1(-c, 0)$ and $F_2(c, 0)$ (Fig. 9.10) and the constant difference is $2a$, then a point (x, y) lies on the hyperbola if and only if

$$\sqrt{(x + c)^2 + y^2} - \sqrt{(x - c)^2 + y^2} = \pm 2a. \tag{7}$$

To simplify this equation, we move the second radical to the right-hand side, square, isolate the remaining radical, and square again, obtaining

$$\frac{x^2}{a^2} + \frac{y^2}{a^2 - c^2} = 1. \tag{8}$$

So far, this looks just like the equation for an ellipse. But now $a^2 - c^2$ is negative because $2a$, being the difference of two sides of triangle PF_1F_2, is less than $2c$, the third side.

The algebraic steps leading to Eq. (8) can be reversed to show that every point P whose coordinates satisfy an equation of this form with $0 < a < c$ also satisfies Eq. (7). A point therefore lies on the hyperbola if and only if its coordinates satisfy Eq. (8).

If we let b denote the positive square root of $c^2 - a^2$,

$$b = \sqrt{c^2 - a^2}, \tag{9}$$

then $a^2 - c^2 = -b^2$ and Eq. (8) takes the more compact form

$$\frac{x^2}{a^2} - \frac{y^2}{b^2} = 1. \tag{10}$$

The differences between Eq. (10) and the equation for an ellipse (Eq. 4) are the minus sign and the new relation

$$c^2 = a^2 + b^2. \qquad \text{From Eq. (9)}$$

Like the ellipse, the hyperbola is symmetric with respect to the origin and coordinate axes. It crosses the x-axis at the points $(\pm a, 0)$. The tangents at these points are vertical because

$$\frac{dy}{dx} = \frac{b^2 x}{a^2 y} \qquad \text{\textcolor{blue}{Obtained from Eq. (10) by implicit differentiation}}$$

is infinite when $y = 0$. The hyperbola has no y-intercepts; in fact, no part of the curve lies between the lines $x = -a$ and $x = a$.

Definitions

The line through the foci of a hyperbola is the **focal axis.** The point on the axis halfway between the foci is the hyperbola's **center.** The points where the focal axis and hyperbola cross are the **vertices** (Fig. 9.11).

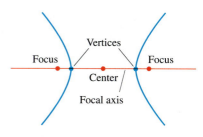

9.11 Points on the focal axis of a hyperbola.

Asymptotes of Hyperbolas—Graphing

The hyperbola

$$\frac{x^2}{a^2} - \frac{y^2}{b^2} = 1 \qquad\qquad (11)$$

has two asymptotes, the lines

$$y = \pm \frac{b}{a} x.$$

The asymptotes give us the guidance we need to graph hyperbolas quickly. (See the drawing lesson.) The fastest way to find the equations of the asymptotes is to replace the 1 in Eq. (11) by 0 and solve the new equation for y:

$$\underbrace{\frac{x^2}{a^2} - \frac{y^2}{b^2} = 1}_{\text{hyperbola}} \quad \Rightarrow \quad \underbrace{\frac{x^2}{a^2} - \frac{y^2}{b^2} = 0}_{\text{0 for 1}} \quad \Rightarrow \quad \underbrace{y = \pm\frac{b}{a}x.}_{\text{asymptotes}}$$

Standard-Form Equations for Hyperbolas Centered at the Origin

Foci on the x-axis: $\dfrac{x^2}{a^2} - \dfrac{y^2}{b^2} = 1$ *Foci on the y-axis:* $\dfrac{y^2}{a^2} - \dfrac{x^2}{b^2} = 1$

  Center-to-focus distance: $c = \sqrt{a^2 + b^2}$ Center-to-focus distance: $c = \sqrt{a^2 + b^2}$

  Foci: $(\pm c, 0)$ Foci: $(0, \pm c)$

  Vertices: $(\pm a, 0)$ Vertices: $(0, \pm a)$

  Asymptotes: $\dfrac{x^2}{a^2} - \dfrac{y^2}{b^2} = 0$ or $y = \pm\dfrac{b}{a}x$ Asymptotes: $\dfrac{y^2}{a^2} - \dfrac{x^2}{b^2} = 0$ or $y = \pm\dfrac{a}{b}x$

Notice the difference in the asymptote equations (b/a in the first, a/b in the second).

DRAWING LESSON

How to Graph the Hyperbola $\dfrac{x^2}{a^2} - \dfrac{y^2}{b^2} = 1$

1   Mark the points $(\pm a, 0)$ and $(0, \pm b)$ with line segments and complete the rectangle they determine.

2   Sketch the asymptotes by extending the rectangle's diagonals.

3   Use the rectangle and asymptotes to guide your drawing.

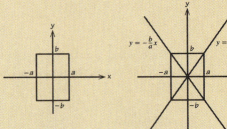 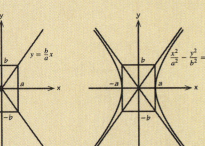

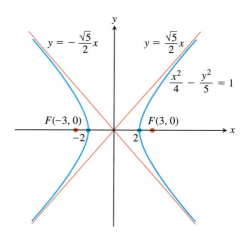

9.12 The hyperbola in Example 4.

EXAMPLE 4   *Foci on the x-axis*

The equation

$$\frac{x^2}{4} - \frac{y^2}{5} = 1 \tag{12}$$

is Eq. (10) with $a^2 = 4$ and $b^2 = 5$ (Fig. 9.12). We have

Center-to-focus distance: $c = \sqrt{a^2 + b^2} = \sqrt{4 + 5} = 3$

Foci: $(\pm c, 0) = (\pm 3, 0)$,   Vertices: $(\pm a, 0) = (\pm 2, 0)$

Center: $(0, 0)$

Asymptotes: $\dfrac{x^2}{4} - \dfrac{y^2}{5} = 0$ or $y = \pm \dfrac{\sqrt{5}}{2} x$.   ☐

EXAMPLE 5   *Foci on the y-axis*

The hyperbola

$$\frac{y^2}{4} - \frac{x^2}{5} = 1,$$

obtained by interchanging x and y in Eq. (12), has its vertices on the y-axis instead of the x-axis (Fig. 9.13). With a^2 still equal to 4 and b^2 equal to 5, we have

Center-to-focus distance: $c = \sqrt{a^2 + b^2} = \sqrt{4 + 5} = 3$

Foci: $(0, \pm c) = (0, \pm 3)$,   Vertices: $(0, \pm a) = (0, \pm 2)$

Center: $(0, 0)$

Asymptotes: $\dfrac{y^2}{4} - \dfrac{x^2}{5} = 0$ or $y = \pm \dfrac{2}{\sqrt{5}} x$.

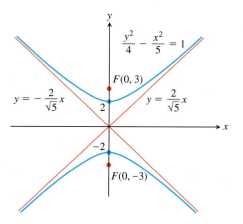

9.13 The hyperbola in Example 5.   ☐

Reflective Properties

The chief applications of parabolas involve their use as reflectors of light and radio waves. Rays originating at a parabola's focus are reflected out of the parabola parallel to the parabola's axis (Fig. 9.14, on the following page, and Exercise 90). This property is used by flashlight, headlight, and spotlight reflectors and by microwave broadcast antennas to direct radiation from point sources into narrow beams. Conversely, electromagnetic waves arriving parallel to a parabolic reflector's axis are directed

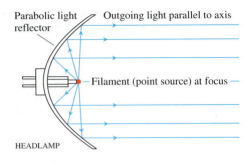

HEADLAMP

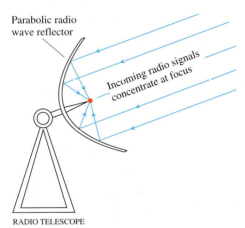

RADIO TELESCOPE

9.14 Two of the many uses of parabolic reflectors.

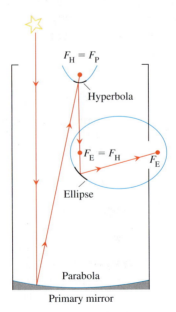

9.16 Schematic drawing of a reflecting telescope.

toward the reflector's focus. This property is used to intensify signals picked up by radio telescopes and television satellite dishes, to focus arriving light in telescopes, and to concentrate sunlight in solar heaters.

If an ellipse is revolved about its major axis to generate a surface (the surface is called an *ellipsoid*) and the interior is silvered to produce a mirror, light from one focus will be reflected to the other focus (Fig. 9.15). Ellipsoids reflect sound the same way, and this property is used to construct *whispering galleries,* rooms in which a person standing at one focus can hear a whisper from the other focus. Statuary Hall in the U.S. Capitol building is a whispering gallery. Ellipsoids also appear in instruments used to study aircraft noise in wind tunnels (sound at one focus can be received at the other focus with relatively little interference from other sources).

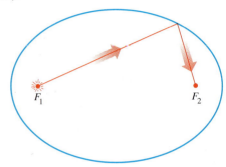

9.15 An elliptical mirror (shown here in profile) reflects light from one focus to the other.

Light directed toward one focus of a hyperbolic mirror is reflected toward the other focus. This property of hyperbolas is combined with the reflective properties of parabolas and ellipses in designing modern telescopes. In Fig. 9.16 starlight reflects off a primary parabolic mirror toward the mirror's focus F_P. It is then reflected by a small hyperbolic mirror, whose focus is $F_H = F_P$, toward the second focus of the hyperbola, $F_E = F_H$. Since this focus is shared by an ellipse, the light is reflected by the elliptical mirror to the ellipse's second focus to be seen by an observer.

As recent experience with NASA's Hubble space telescope shows, the mirrors have to be nearly perfect to focus properly. The aberration that caused the malfunction in Hubble's primary mirror (now corrected with additional mirrors) amounted to about half a wavelength of visible light, no more than 1/50 the width of a human hair.

Other Applications

Water pipes are sometimes designed with elliptical cross sections to allow for expansion when the water freezes. The triggering mechanisms in some lasers are elliptical, and stones on a beach become more and more elliptical as they are ground down by waves. There are also applications of ellipses to fossil formation. The ellipsolith, once thought to be a separate species, is now known to be an elliptically deformed nautilus.

Hyperbolic paths arise in Einstein's theory of relativity and form the basis for the (unrelated) LORAN radio navigation system. (LORAN is short for "long range navigation.") Hyperbolas also form the basis for a new system the Burlington Northern Railroad developed for using synchronized electronic signals from satellites to track freight trains. Computers aboard Burlington Northern locomotives in Minnesota have been able to track trains to within one mile per hour of their speed and to within 150 feet of their actual location.

Exercises 9.1

Identifying Graphs

Match the parabolas in Exercises 1–4 with the following equations:

$$x^2 = 2y, \quad x^2 = -6y, \quad y^2 = 8x, \quad y^2 = -4x.$$

Then find the parabola's focus and directrix.

1.

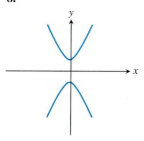

2.

3.

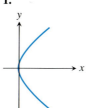

4.

Match each conic section in Exercises 5–8 with one of these equations:

$$\frac{x^2}{4} + \frac{y^2}{9} = 1, \quad \frac{x^2}{2} + y^2 = 1,$$

$$\frac{y^2}{4} - x^2 = 1, \quad \frac{x^2}{4} - \frac{y^2}{9} = 1.$$

Then find the conic section's foci and vertices. If the conic section is a hyperbola, find its asymptotes as well.

5.

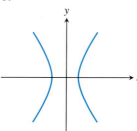

6.

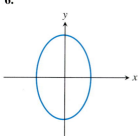

7.

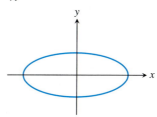

8.

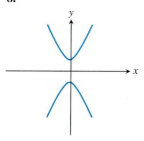

Parabolas

Exercises 9–16 give equations of parabolas. Find each parabola's focus and directrix. Then sketch the parabola. Include the focus and directrix in your sketch.

9. $y^2 = 12x$ **10.** $x^2 = 6y$ **11.** $x^2 = -8y$

12. $y^2 = -2x$ **13.** $y = 4x^2$ **14.** $y = -8x^2$

15. $x = -3y^2$ **16.** $x = 2y^2$

Ellipses

Exercises 17–24 give equations for ellipses. Put each equation in standard form. Then sketch the ellipse. Include the foci in your sketch.

17. $16x^2 + 25y^2 = 400$ **18.** $7x^2 + 16y^2 = 112$

19. $2x^2 + y^2 = 2$ **20.** $2x^2 + y^2 = 4$

21. $3x^2 + 2y^2 = 6$ **22.** $9x^2 + 10y^2 = 90$

23. $6x^2 + 9y^2 = 54$ **24.** $169x^2 + 25y^2 = 4225$

Exercises 25 and 26 give information about the foci and vertices of ellipses centered at the origin of the xy-plane. In each case, find the ellipse's standard-form equation from the given information.

25. Foci: $(\pm\sqrt{2}, 0)$ **26.** Foci: $(0, \pm 4)$
Vertices: $(\pm 2, 0)$ Vertices: $(0, \pm 5)$

Hyperbolas

Exercises 27–34 give equations for hyperbolas. Put each equation in standard form and find the hyperbola's asymptotes. Then sketch the hyperbola. Include the asymptotes and foci in your sketch.

27. $x^2 - y^2 = 1$ **28.** $9x^2 - 16y^2 = 144$

29. $y^2 - x^2 = 8$ **30.** $y^2 - x^2 = 4$

31. $8x^2 - 2y^2 = 16$ **32.** $y^2 - 3x^2 = 3$

33. $8y^2 - 2x^2 = 16$ **34.** $64x^2 - 36y^2 = 2304$

Exercises 35–38 give information about the foci, vertices, and asymptotes of hyperbolas centered at the origin of the xy-plane. In each case, find the hyperbola's standard-form equation from the information given.

35. Foci: $(0, \pm\sqrt{2})$ **36.** Foci: $(\pm 2, 0)$
Asymptotes: $y = \pm x$
Asymptotes: $y = \pm \dfrac{1}{\sqrt{3}} x$

37. Vertices: $(\pm 3, 0)$ **38.** Vertices: $(0, \pm 2)$
Asymptotes: $y = \pm \dfrac{4}{3} x$ Asymptotes: $y = \pm \dfrac{1}{2} x$

Shifting Conic Sections

39. The parabola $y^2 = 8x$ is shifted down 2 units and right 1 unit to generate the parabola $(y + 2)^2 = 8(x - 1)$. (a) Find the new parabola's vertex, focus, and directrix. (b) Plot the new vertex, focus, and directrix, and sketch in the parabola.

40. The parabola $x^2 = -4y$ is shifted left 1 unit and up 3 units to generate the parabola $(x + 1)^2 = -4(y - 3)$. (a) Find the new parabola's vertex, focus, and directrix. (b) Plot the new vertex, focus, and directrix, and sketch in the parabola.

41. The ellipse $(x^2/16) + (y^2/9) = 1$ is shifted 4 units to the right and 3 units up to generate the ellipse

$$\frac{(x - 4)^2}{16} + \frac{(y - 3)^2}{9} = 1.$$

(a) Find the foci, vertices, and center of the new ellipse. (b) Plot the new foci, vertices, and center, and sketch in the new ellipse.

42. The ellipse $(x^2/9) + (y^2/25) = 1$ is shifted 3 units to the left and 2 units down to generate the ellipse

$$\frac{(x + 3)^2}{9} + \frac{(y + 2)^2}{25} = 1.$$

(a) Find the foci, vertices, and center of the new ellipse. (b) Plot the new foci, vertices, and center, and sketch in the new ellipse.

43. The hyperbola $(x^2/16) - (y^2/9) = 1$ is shifted 2 units to the right to generate the hyperbola

$$\frac{(x - 2)^2}{16} - \frac{y^2}{9} = 1.$$

(a) Find the center, foci, vertices, and asymptotes of the new hyperbola. (b) Plot the new center, foci, vertices, and asymptotes, and sketch in the hyperbola.

44. The hyperbola $(y^2/4) - (x^2/5) = 1$ is shifted 2 units down to generate the hyperbola

$$\frac{(y + 2)^2}{4} - \frac{x^2}{5} = 1.$$

(a) Find the center, foci, vertices, and asymptotes of the new hyperbola. (b) Plot the new center, foci, vertices, and asymptotes, and sketch in the hyperbola.

Exercises 45–48 give equations for parabolas and tell how many units up or down and to the right or left each parabola is to be shifted. Find an equation for the new parabola, and find the new vertex, focus, and directrix.

45. $y^2 = 4x$,   left 2, down 3

46. $y^2 = -12x$,   right 4, up 3

47. $x^2 = 8y$,   right 1, down 7

48. $x^2 = 6y$,   left 3, down 2

Exercises 49–52 give equations for ellipses and tell how many units up or down and to the right or left each ellipse is to be shifted. Find an equation for the new ellipse, and find the new foci, vertices, and center.

49. $\dfrac{x^2}{6} + \dfrac{y^2}{9} = 1$,   left 2, down 1

50. $\dfrac{x^2}{2} + y^2 = 1$,   right 3, up 4

51. $\dfrac{x^2}{3} + \dfrac{y^2}{2} = 1$,   right 2, up 3

52. $\dfrac{x^2}{16} + \dfrac{y^2}{25} = 1$,   left 4, down 5

Exercises 53–56 give equations for hyperbolas and tell how many units up or down and to the right or left each hyperbola is to be shifted. Find an equation for the new hyperbola, and find the new center, foci, vertices, and asymptotes.

53. $\dfrac{x^2}{4} - \dfrac{y^2}{5} = 1$,   right 2, up 2

54. $\dfrac{x^2}{16} - \dfrac{y^2}{9} = 1$,   left 5, down 1

55. $y^2 - x^2 = 1$,   left 1, down 1

56. $\dfrac{y^2}{3} - x^2 = 1$,   right 1, up 3

Find the center, foci, vertices, asymptotes, and radius, as appropriate, of the conic sections in Exercises 57–68.

57. $x^2 + 4x + y^2 = 12$

58. $2x^2 + 2y^2 - 28x + 12y + 114 = 0$

59. $x^2 + 2x + 4y - 3 = 0$

60. $y^2 - 4y - 8x - 12 = 0$

61. $x^2 + 5y^2 + 4x = 1$

62. $9x^2 + 6y^2 + 36y = 0$

63. $x^2 + 2y^2 - 2x - 4y = -1$

64. $4x^2 + y^2 + 8x - 2y = -1$

65. $x^2 - y^2 - 2x + 4y = 4$

66. $x^2 - y^2 + 4x - 6y = 6$

67. $2x^2 - y^2 + 6y = 3$

68. $y^2 - 4x^2 + 16x = 24$

Inequalities

Sketch the regions in the xy-plane whose coordinates satisfy the inequalities or pairs of inequalities in Exercises 69–74.

69. $9x^2 + 16y^2 \leq 144$

70. $x^2 + y^2 \geq 1$ and $4x^2 + y^2 \leq 4$

71. $x^2 + 4y^2 \geq 4$ and $4x^2 + 9y^2 \leq 36$

72. $(x^2 + y^2 - 4)(x^2 + 9y^2 - 9) \leq 0$

73. $4y^2 - x^2 \geq 4$

74. $|x^2 - y^2| \leq 1$

Theory and Examples

75. *Archimedes' formula for the volume of a parabolic solid.*
The region enclosed by the parabola $y = (4h/b^2)x^2$ and the line $y = h$ is revolved about the y-axis to generate the solid shown here. Show that the volume of the solid is $3/2$ the volume of the corresponding cone.

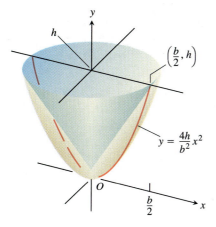

$$y = \frac{4h}{b^2}x^2$$

76. *Suspension bridge cables hang in parabolas.* The suspension bridge cable shown here supports a uniform load of w pounds per horizontal foot. It can be shown that if H is the horizontal tension of the cable at the origin, then the curve of the cable satisfies the equation

$$\frac{dy}{dx} = \frac{w}{H}x.$$

Show that the cable hangs in a parabola by solving this differential equation subject to the initial condition that $y = 0$ when $x = 0$.

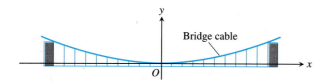

Bridge cable

77. Find an equation for the circle through the points $(1, 0)$, $(0, 1)$, and $(2, 2)$.

78. Find an equation for the circle through the points $(2, 3)$, $(3, 2)$, and $(-4, 3)$.

79. Find an equation for the circle centered at $(-2, 1)$ that passes through the point $(1, 3)$. Is the point $(1.1, 2.8)$ inside, outside, or on the circle?

80. Find equations for the tangents to the circle $(x - 2)^2 + (y - 1)^2 = 5$ at the points where the circle crosses the coordinate axes. (*Hint:* Use implicit differentiation.)

81. If lines are drawn parallel to the coordinate axes through a point P on the parabola $y^2 = kx, k > 0$, the parabola partitions the rectangular region bounded by these lines and the coordinate axes into two smaller regions, A and B.

a) If the two smaller regions are revolved about the y-axis, show that they generate solids whose volumes have the ratio 4:1.

b) What is the ratio of the volumes generated by revolving the regions about the x-axis?

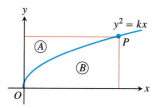

82. Show that the tangents to the curve $y^2 = 4px$ from any point on the line $x = -p$ are perpendicular.

83. Find the dimensions of the rectangle of largest area that can be inscribed in the ellipse $x^2 + 4y^2 = 4$ with its sides parallel to the coordinate axes. What is the area of the rectangle?

84. Find the volume of the solid generated by revolving the region enclosed by the ellipse $9x^2 + 4y^2 = 36$ about the (a) x-axis, (b) y-axis.

85. The "triangular" region in the first quadrant bounded by the x-axis, the line $x = 4$, and the hyperbola $9x^2 - 4y^2 = 36$ is revolved about the x-axis to generate a solid. Find the volume of the solid.

86. The region bounded on the left by the y-axis, on the right by the hyperbola $x^2 - y^2 = 1$, and above and below by the lines $y = \pm 3$ is revolved about the y-axis to generate a solid. Find the volume of the solid.

87. Find the centroid of the region that is bounded below by the x-axis and above by the ellipse $(x^2/9) + (y^2/16) = 1$.

88. The curve $y = \sqrt{x^2 + 1}, 0 \le x \le \sqrt{2}$, which is part of the upper branch of the hyperbola $y^2 - x^2 = 1$, is revolved about the x-axis to generate a surface. Find the area of the surface.

89. The circular waves in the photograph here were made by touching the surface of a ripple tank, first at A and then at B. As the waves expanded, their point of intersection appeared to trace a hyperbola. Did it really do that? To find out, we can model the waves with circles centered at A and B.

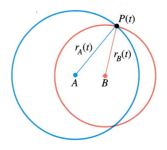

At time t, the point P is $r_A(t)$ units from A and $r_B(t)$ units from B. Since the radii of the circles increase at a constant rate, the rate at which the waves are traveling is

$$\frac{dr_A}{dt} = \frac{dr_B}{dt}.$$

Conclude from this equation that $r_A - r_B$ has a constant value, so that P must lie on a hyperbola with foci at A and B.

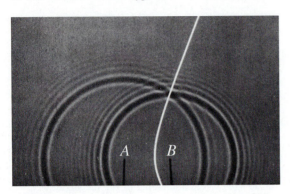

The expanding waves in Exercise 89.

90. *The reflective property of parabolas.* The figure here shows a typical point $P(x_0, y_0)$ on the parabola $y^2 = 4px$. The line L is tangent to the parabola at P. The parabola's focus lies at $F(p, 0)$. The ray L' extending from P to the right is parallel to the x-axis. We show that light from F to P will be reflected out along L' by showing that β equals α. Establish this equality by taking the following steps.

a) Show that $\tan \beta = 2p/y_0$.
b) Show that $\tan \phi = y_0/(x_0 - p)$.
c) Use the identity

$$\tan \alpha = \frac{\tan \phi - \tan \beta}{1 + \tan \phi \tan \beta}$$

to show that $\tan \alpha = 2p/y_0$.

Since α and β are both acute, $\tan \beta = \tan \alpha$ implies $\beta = \alpha$.

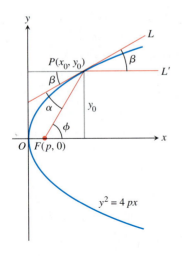

91. *How the astronomer Kepler used string to draw parabolas.* Kepler's method for drawing a parabola (with more modern tools) requires a string the length of a T square and a table whose edge can serve as the parabola's directrix. Pin one end of the string to the point where you want the focus to be and the other end to the upper end of the T square. Then, holding the string taut against the T square with a pencil, slide the T square along the table's edge. As the T square moves, the pencil will trace a parabola. Why?

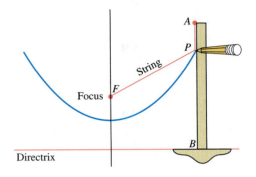

92. *Construction of a hyperbola.* The following diagrams appeared (unlabeled) in Ernest J. Eckert, "Constructions Without Words," *Mathematics Magazine,* Vol. 66, No. 2, April 1993, p. 113. Explain the constructions.

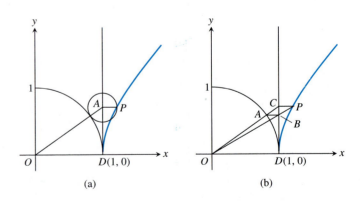

93. *The width of a parabola at the focus.* Show that the number $4p$ is the **width** of the parabola $x^2 = 4py$ $(p > 0)$ at the focus by showing that the line $y = p$ cuts the parabola at points that are $4p$ units apart.

94. *The asymptotes of $(x^2/a^2) - (y^2/b^2) = 1$.* Show that the vertical distance between the line $y = (b/a)x$ and the upper half of the right-hand branch $y = (b/a)\sqrt{x^2 - a^2}$ of the hyperbola $(x^2/a^2) - (y^2/b^2) = 1$ approaches 0 by showing that

$$\lim_{x \to \infty}\left(\frac{b}{a}x - \frac{b}{a}\sqrt{x^2 - a^2}\right) = \frac{b}{a}\lim_{x \to \infty}\left(x - \sqrt{x^2 - a^2}\right) = 0.$$

Similar results hold for the remaining portions of the hyperbola and the lines $y = \pm(b/a)x$.

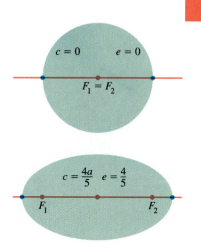

$c = 0$ $e = 0$

$F_1 = F_2$

$c = \dfrac{4a}{5}$ $e = \dfrac{4}{5}$

F_1 F_2

F_1 $c = a$ $e = 1$ F_2

9.17 The ellipse changes from a circle to a line segment as c increases from 0 to a.

Table 9.2 Eccentricities of planetary orbits

Mercury	0.21	Saturn	0.06
Venus	0.01	Uranus	0.05
Earth	0.02	Neptune	0.01
Mars	0.09	Pluto	0.25
Jupiter	0.05		

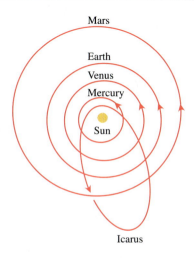

Mars

Earth

Venus

Mercury

Sun

Icarus

9.18 The orbit of the asteroid Icarus is highly eccentric. Earth's orbit is so nearly circular that its foci lie inside the sun.

Classifying Conic Sections by Eccentricity

We now show how to associate with each conic section a number called the conic section's eccentricity. The eccentricity reveals the conic section's type (circle, ellipse, parabola, or hyperbola) and, in the case of ellipses and hyperbolas, describes the conic section's general proportions.

Eccentricity

Although the center-to-focus distance c does not appear in the equation

$$\frac{x^2}{a^2} + \frac{y^2}{b^2} = 1, \qquad (a > b)$$

for an ellipse, we can still determine c from the equation $c = \sqrt{a^2 - b^2}$. If we fix a and vary c over the interval $0 \leq c \leq a$, the resulting ellipses will vary in shape (Fig. 9.17). They are circles if $c = 0$ (so that $a = b$) and flatten as c increases. If $c = a$, the foci and vertices overlap and the ellipse degenerates into a line segment.

We use the ratio of c to a to describe the various shapes the ellipse can take. We call this ratio the ellipse's eccentricity.

Definition

The **eccentricity** of the ellipse $(x^2/a^2) + (y^2/b^2) = 1 \ (a > b)$ is

$$e = \frac{c}{a} = \frac{\sqrt{a^2 - b^2}}{a}.$$

The planets in the solar system revolve around the sun in elliptical orbits with the sun at one focus. Most of the orbits are nearly circular, as can be seen from the eccentricities in Table 9.2. Pluto has a fairly eccentric orbit, with $e = 0.25$, as does Mercury, with $e = 0.21$. Other members of the solar system have orbits that are even more eccentric. Icarus, an asteroid about 1 mile wide that revolves around the sun every 409 Earth days, has an orbital eccentricity of 0.83 (Fig. 9.18).

EXAMPLE 1 The orbit of Halley's comet is an ellipse 36.18 astronomical units long by 9.12 astronomical units wide. (One *astronomical unit* [AU] is 149,597,870 km, the semimajor axis of Earth's orbit.) Its eccentricity is

$$e = \frac{\sqrt{a^2 - b^2}}{a} = \frac{\sqrt{(36.18/2)^2 - (9.12/2)^2}}{(1/2)(36.18)} = \frac{\sqrt{(18.09)^2 - (4.56)^2}}{18.09} \approx 0.97. \ \square$$

Whereas a parabola has one focus and one directrix, each ellipse has two foci and two directrices. These are the lines perpendicular to the major axis at distances $\pm a/e$ from the center. The parabola has the property that

$$PF = 1 \cdot PD \tag{1}$$

for any point P on it, where F is the focus and D is the point nearest P on the directrix. For an ellipse, it can be shown that the equations that replace (1) are

$$PF_1 = e \cdot PD_1, \qquad PF_2 = e \cdot PD_2. \tag{2}$$

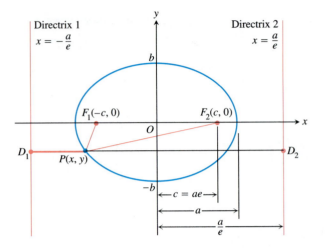

9.19 The foci and directrices of the ellipse $(x^2/a^2) + (y^2/b^2) = 1$. Directrix 1 corresponds to focus F_1, and directrix 2 to focus F_2.

Here, e is the eccentricity, P is any point on the ellipse, F_1 and F_2 are the foci, and D_1 and D_2 are the points on the directrices nearest P (Fig. 9.19).

In each equation in (2) the directrix and focus must correspond; that is, if we use the distance from P to F_1, we must also use the distance from P to the directrix at the same end of the ellipse. The directrix $x = -a/e$ corresponds to $F_1(-c, 0)$, and the directrix $x = a/e$ corresponds to $F_2(c, 0)$.

The eccentricity of a hyperbola is also $e = c/a$, only in this case c equals $\sqrt{a^2 + b^2}$ instead of $\sqrt{a^2 - b^2}$. In contrast to the eccentricity of an ellipse, the eccentricity of a hyperbola is always greater than 1.

Definition

The **eccentricity** of the hyperbola $(x^2/a^2) - (y^2/b^2) = 1$ is

$$e = \frac{c}{a} = \frac{\sqrt{a^2 + b^2}}{a}.$$

In both ellipse and hyperbola, the eccentricity is the ratio of the distance between the foci to the distance between the vertices (because $c/a = 2c/2a$).

$$\text{Eccentricity} = \frac{\text{distance between foci}}{\text{distance between vertices}}$$

In an ellipse, the foci are closer together than the vertices and the ratio is less than 1. In a hyperbola, the foci are farther apart than the vertices and the ratio is greater than 1.

EXAMPLE 2   Locate the vertices of an ellipse of eccentricity 0.8 whose foci lie at the points $(0, \pm 7)$.

Halley's comet

Edmund Halley (1656–1742; pronounced "*haw*-ley"), British biologist, geologist, sea captain, pirate, spy, Antarctic voyager, astronomer, adviser on fortifications, company founder and director, and the author of the first actuarial mortality tables, was also the mathematician who pushed and harried Newton into writing his *Principia*. Despite his accomplishments, Halley is known today chiefly as the man who calculated the orbit of the great comet of 1682: "wherefore if according to what we have already said [the comet] should return again about the year 1758, candid posterity will not refuse to acknowledge that this was first discovered by an Englishman." Indeed, candid posterity did not refuse—ever since the comet's return in 1758, it has been known as Halley's comet.

Last seen rounding the sun during the winter and spring of 1985–86, the comet is due to return in the year 2062. A recent study indicates that the comet has made about 2000 cycles so far with about the same number to go before the sun erodes it away completely.

Solution Since $e = c/a$, the vertices are the points $(0, \pm a)$ where

$$a = \frac{c}{e} = \frac{7}{0.8} = 8.75,$$

or $(0, \pm 8.75)$.

EXAMPLE 3   Find the eccentricity of the hyperbola $9x^2 - 16y^2 = 144$.

Solution We divide both sides of the hyperbola's equation by 144 to put it in standard form, obtaining

$$\frac{9x^2}{144} - \frac{16y^2}{144} = 1 \quad \text{and} \quad \frac{x^2}{16} - \frac{y^2}{9} = 1.$$

With $a^2 = 16$ and $b^2 = 9$, we find that $c = \sqrt{a^2 + b^2} = \sqrt{16 + 9} = 5$, so

$$e = \frac{c}{a} = \frac{5}{4}.$$

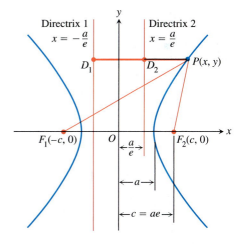

Directrix 1
$x = -\dfrac{a}{e}$

Directrix 2
$x = \dfrac{a}{e}$

D_1 D_2 $P(x, y)$

$F_1(-c, 0)$ O $\leftarrow \dfrac{a}{e} \rightarrow$ $F_2(c, 0)$

$\leftarrow a \rightarrow$

$\leftarrow c = ae \rightarrow$

9.20 The foci and directrices of the hyperbola $(x^2/a^2) - (y^2/b^2) = 1$. No matter where P lies on the hyperbola, $PF_1 = e \cdot PD_1$ and $PF_2 = e \cdot PD_2$.

As with the ellipse, it can be shown that the lines $x = \pm a/e$ act as directrices for the hyperbola and that

$$PF_1 = e \cdot PD_1 \quad \text{and} \quad PF_2 = e \cdot PD_2. \tag{3}$$

Here P is any point on the hyperbola, F_1 and F_2 are the foci, and D_1 and D_2 are the points nearest P on the directrices (Fig. 9.20).

To complete the picture, we define the eccentricity of a parabola to be $e = 1$. Equations (1) – (3) then have the common form $PF = e \cdot PD$.

Definition

The **eccentricity** of a parabola is $e = 1$.

The "focus–directrix" equation $PF = e \cdot PD$ unites the parabola, ellipse, and hyperbola in the following way. Suppose that the distance PF of a point P from a fixed point F (the focus) is a constant multiple of its distance from a fixed line (the directrix). That is, suppose

$$PF = e \cdot PD, \tag{4}$$

where e is the constant of proportionality. Then the path traced by P is

a) a *parabola* if $e = 1$,
b) an *ellipse* of eccentricity e if $e < 1$, and
c) a *hyperbola* of eccentricity e if $e > 1$.

Equation (4) may not look like much to get excited about. There are no coordinates in it and when we try to translate it into coordinate form it translates in different ways, depending on the size of e. At least, that is what happens in Cartesian coordinates. However, in polar coordinates, as we will see in Section 9.8, the equation $PF = e \cdot PD$ translates into a single equation regardless of the value of e, an equation so simple that it has been the equation of choice of astronomers and space scientists for nearly 300 years.

Given the focus and corresponding directrix of a hyperbola centered at the origin and with foci on the x-axis, we can use the dimensions shown in Fig. 9.20 to find e. Knowing e, we can derive a Cartesian equation for the hyperbola from the equation $PF = e \cdot PD$, as in the next example. We can find equations for ellipses centered at the origin and with foci on the x-axis in a similar way, using the dimensions shown in Fig. 9.19.

EXAMPLE 4   Find a Cartesian equation for the hyperbola centered at the origin that has a focus at (3, 0) and the line $x = 1$ as the corresponding directrix.

Solution   We first use the dimensions shown in Fig. 9.20 to find the hyperbola's eccentricity. The focus is

$$(c, 0) = (3, 0), \quad \text{so} \quad c = 3.$$

The directrix is the line

$$x = \frac{a}{e} = 1, \quad \text{so} \quad a = e.$$

When combined with the equation $e = c/a$ that defines eccentricity, these results give

$$e = \frac{c}{a} = \frac{3}{e}, \quad \text{so} \quad e^2 = 3 \quad \text{and} \quad e = \sqrt{3}.$$

Knowing e, we can now derive the equation we want from the equation $PF = e \cdot PD$. In the notation of Fig. 9.21, we have

$$PF = e \cdot PD \qquad \text{Eq. (4)}$$
$$\sqrt{(x-3)^2 + (y-0)^2} = \sqrt{3}\,|x - 1| \qquad e = \sqrt{3}$$
$$x^2 - 6x + 9 + y^2 = 3(x^2 - 2x + 1)$$
$$2x^2 - y^2 = 6$$
$$\frac{x^2}{3} - \frac{y^2}{6} = 1.$$

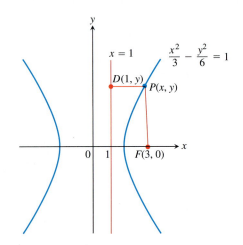

9.21 The hyperbola in Example 4.

Exercises 9.2

Ellipses

In Exercises 1–8, find the eccentricity of the ellipse. Then find and graph the ellipse's foci and directrices.

1. $16x^2 + 25y^2 = 400$

2. $7x^2 + 16y^2 = 112$

3. $2x^2 + y^2 = 2$

4. $2x^2 + y^2 = 4$

5. $3x^2 + 2y^2 = 6$

6. $9x^2 + 10y^2 = 90$

7. $6x^2 + 9y^2 = 54$

8. $169x^2 + 25y^2 = 4225$

Exercises 9–12 give the foci or vertices and the eccentricities of ellipses centered at the origin of the xy-plane. In each case, find the ellipse's standard-form equation.

9. Foci: (0, ±3)
   Eccentricity: 0.5

10. Foci: (±8, 0)
   Eccentricity: 0.2

11. Vertices: (0, ±70)
   Eccentricity: 0.1

12. Vertices: (±10, 0)
   Eccentricity: 0.24

Exercises 13–16 give foci and corresponding directrices of ellipses centered at the origin of the xy-plane. In each case, use the dimensions in Fig. 9.19 to find the eccentricity of the ellipse. Then find the ellipse's standard-form equation.

13. Focus: $(\sqrt{5}, 0)$

Directrix: $x = \dfrac{9}{\sqrt{5}}$

14. Focus: $(4, 0)$

Directrix: $x = \dfrac{16}{3}$

15. Focus: $(-4, 0)$

Directrix: $x = -16$

16. Focus: $(-\sqrt{2}, 0)$

Directrix: $x = -2\sqrt{2}$

17. Draw an ellipse of eccentricity 4/5. Explain your procedure.

18. Draw the orbit of Pluto (eccentricity 0.25) to scale. Explain your procedure.

19. The endpoints of the major and minor axes of an ellipse are $(1, 1)$, $(3, 4)$, $(1, 7)$, and $(-1, 4)$. Sketch the ellipse, give its equation in standard form, and find its foci, eccentricity, and directrices.

20. Find an equation for the ellipse of eccentricity 2/3 that has the line $x = 9$ as a directrix and the point $(4, 0)$ as the corresponding focus.

21. What values of the constants a, b, and c make the ellipse

$$4x^2 + y^2 + ax + by + c = 0$$

lie tangent to the x-axis at the origin and pass through the point $(-1, 2)$? What is the eccentricity of the ellipse?

22. *The reflective property of ellipses.* An ellipse is revolved about its major axis to generate an ellipsoid. The inner surface of the ellipsoid is silvered to make a mirror. Show that a ray of light emanating from one focus will be reflected to the other focus. Sound waves also follow such paths, and this property is used in constructing "whispering galleries." (*Hint:* Place the ellipse in standard position in the xy-plane and show that the lines from a point P on the ellipse to the two foci make congruent angles with the tangent to the ellipse at P.)

Hyperbolas

In Exercises 23–30, find the eccentricity of the hyperbola. Then find and graph the hyperbola's foci and directrices.

23. $x^2 - y^2 = 1$

24. $9x^2 - 16y^2 = 144$

25. $y^2 - x^2 = 8$

26. $y^2 - x^2 = 4$

27. $8x^2 - 2y^2 = 16$

28. $y^2 - 3x^2 = 3$

29. $8y^2 - 2x^2 = 16$

30. $64x^2 - 36y^2 = 2304$

Exercises 31–34 give the eccentricities and the vertices or foci of hyperbolas centered at the origin of the xy-plane. In each case, find the hyperbola's standard-form equation.

31. Eccentricity: 3
Vertices: $(0, \pm 1)$

32. Eccentricity: 2
Vertices: $(\pm 2, 0)$

33. Eccentricity: 3
Foci: $(\pm 3, 0)$

34. Eccentricity: 1.25
Foci: $(0, \pm 5)$

Exercises 35–38 give foci and corresponding directrices of hyperbolas centered at the origin of the xy-plane. In each case, find the hyperbola's eccentricity. Then find the hyperbola's standard-form equation.

35. Focus: $(4, 0)$
Directrix: $x = 2$

36. Focus: $(\sqrt{10}, 0)$
Directrix: $x = \sqrt{2}$

37. Focus: $(-2, 0)$

Directrix: $x = -\dfrac{1}{2}$

38. Focus: $(-6, 0)$
Directrix: $x = -2$

39. A hyperbola of eccentricity 3/2 has one focus at $(1, -3)$. The corresponding directrix is the line $y = 2$. Find an equation for the hyperbola.

40. *The effect of eccentricity on a hyperbola's shape.* What happens to the graph of a hyperbola as its eccentricity increases? To find out, rewrite the equation $(x^2/a^2) - (y^2/b^2) = 1$ in terms of a and e instead of a and b. Graph the hyperbola for various values of e and describe what you find.

41. *The reflective property of hyperbolas.* Show that a ray of light directed toward one focus of a hyperbolic mirror, as in the accompanying figure, is reflected toward the other focus. (*Hint:* Show that the tangent to the hyperbola at P bisects the angle made by segments PF_1 and PF_2.)

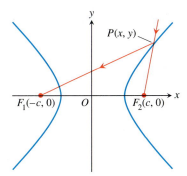

42. *A confocal ellipse and hyperbola.* Show that an ellipse and a hyperbola that have the same foci A and B, as in the accompanying figure, cross at right angles at their point of intersection. (*Hint:* A ray of light from focus A that met the hyperbola at P would be reflected from the hyperbola as if it came directly from B (Exercise 41). The same ray would be reflected off the ellipse to pass through B (Exercise 22).)

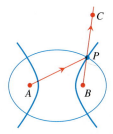

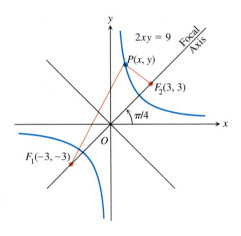

9.22 The focal axis of the hyperbola $2xy = 9$ makes an angle of $\pi/4$ radians with the positive x-axis.

9.3

Quadratic Equations and Rotations

In this section, we examine one of the most amazing results in analytic geometry, which is that the Cartesian graph of any equation

$$Ax^2 + Bxy + Cy^2 + Dx + Ey + F = 0, \tag{1}$$

in which A, B, and C are not all zero, is nearly always a conic section. The exceptions are the cases in which there is no graph at all or the graph consists of two parallel lines. It is conventional to call all graphs of Eq. (1), curved or not, **quadratic curves.**

The Cross Product Term

You may have noticed that the term Bxy did not appear in the equations for the conic sections in Section 9.1. This happened because the axes of the conic sections ran parallel to (in fact, coincided with) the coordinate axes.

To see what happens when the parallelism is absent, let us write an equation for a hyperbola with $a = 3$ and foci at $F_1(-3, -3)$ and $F_2(3, 3)$ (Fig. 9.22). The equation $|PF_1 - PF_2| = 2a$ becomes $|PF_1 - PF_2| = 2(3) = 6$ and

$$\sqrt{(x+3)^2 + (y+3)^2} - \sqrt{(x-3)^2 + (y-3)^2} = \pm 6.$$

When we transpose one radical, square, solve for the radical that still appears, and square again, the equation reduces to

$$2xy = 9, \tag{2}$$

a case of Eq. (1) in which the cross-product term is present. The asymptotes of the hyperbola in Eq. (2) are the x- and y-axes, and the focal axis makes an angle of $\pi/4$ radians with the positive x-axis. As in this example, the cross product term is present in Eq. (1) only when the axes of the conic are tilted.

Rotating the Coordinate Axes to Eliminate the Cross Product Term

To eliminate the xy-term from the equation of a conic, we rotate the coordinate axes to eliminate the "tilt" in the axes of the conic. The equations for the rotations we use are derived in the following way. In the notation of Fig. 9.23, which shows a counterclockwise rotation about the origin through an angle α,

$$x = OM = OP\cos(\theta + \alpha) = OP\cos\theta\cos\alpha - OP\sin\theta\sin\alpha$$

$$y = MP = OP\sin(\theta + \alpha) = OP\cos\theta\sin\alpha + OP\sin\theta\cos\alpha. \tag{3}$$

Since

$$OP\cos\theta = OM' = x'$$

and

$$OP\sin\theta = M'P = y',$$

the equations in (3) reduce to the following.

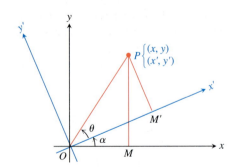

9.23 A counterclockwise rotation through angle α about the origin.

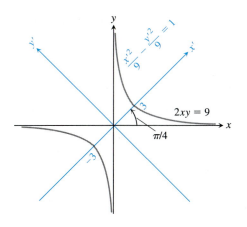

9.24 The hyperbola in Example 1 (x' and y' are the new coordinates).

Equations for Rotating Coordinate Axes

$$x = x' \cos \alpha - y' \sin \alpha$$

$$y = x' \sin \alpha + y' \cos \alpha \tag{4}$$

EXAMPLE 1 The x- and y-axes are rotated through an angle of $\pi/4$ radians about the origin. Find an equation for the hyperbola $2xy = 9$ in the new coordinates.

Solution Since $\cos \pi/4 = \sin \pi/4 = 1/\sqrt{2}$, we substitute

$$x = \frac{x' - y'}{\sqrt{2}}, \qquad y = \frac{x' + y'}{\sqrt{2}}$$

from Eqs. (4) into the equation $2xy = 9$ and obtain

$$2 \left(\frac{x' - y'}{\sqrt{2}} \right) \left(\frac{x' + y'}{\sqrt{2}} \right) = 9$$

$$x'^2 - y'^2 = 9$$

$$\frac{x'^2}{9} - \frac{y'^2}{9} = 1.$$

See Fig. 9.24.

If we apply Eqs. (4) to the quadratic equation (1), we obtain a new quadratic equation

$$A' x'^2 + B' x' y' + C' y'^2 + D' x' + E' y' + F' = 0. \tag{5}$$

The new and old coefficients are related by the equations

$$A' = A \cos^2 \alpha + B \cos \alpha \sin \alpha + C \sin^2 \alpha$$

$$B' = B \cos 2\alpha + (C - A) \sin 2\alpha$$

$$C' = A \sin^2 \alpha - B \sin \alpha \cos \alpha + C \cos^2 \alpha \tag{6}$$

$$D' = D \cos \alpha + E \sin \alpha$$

$$E' = -D \sin \alpha + E \cos \alpha$$

$$F' = F.$$

These equations show, among other things, that if we start with an equation for a curve in which the cross product term is present ($B \neq 0$), we can find a rotation angle α that produces an equation in which no cross product term appears ($B' = 0$). To find α, we set $B' = 0$ in the second equation in (6) and solve the resulting equation,

$$B \cos 2\alpha + (C - A) \sin 2\alpha = 0,$$

for α. In practice, this means determining α from one of the two equations

$$\cot 2\alpha = \frac{A - C}{B} \qquad \text{or} \qquad \tan 2\alpha = \frac{B}{A - C}. \tag{7}$$

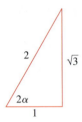

9.25 This triangle identifies $2\alpha = \cot^{-1}(1/\sqrt{3})$ as $\pi/3$ (Example 2).

EXAMPLE 2 The coordinate axes are to be rotated through an angle α to produce an equation for the curve

$$2x^2 + \sqrt{3}\,xy + y^2 - 10 = 0$$

that has no cross product term. Find α and the new equation. Identify the curve.

Solution The equation $2x^2 + \sqrt{3}\,xy + y^2 - 10 = 0$ has $A = 2$, $B = \sqrt{3}$, and $C = 1$. We substitute these values into Eq. (7) to find α:

$$\cot 2\alpha = \frac{A - C}{B} = \frac{2 - 1}{\sqrt{3}} = \frac{1}{\sqrt{3}}.$$

From the right triangle in Fig. 9.25, we see that one appropriate choice of angle is $2\alpha = \pi/3$, so we take $\alpha = \pi/6$. Substituting $\alpha = \pi/6$, $A = 2$, $B = \sqrt{3}$, $C = 1$, $D = E = 0$, and $F = -10$ into Eqs. (6) gives

$$A' = \frac{5}{2}, \qquad B' = 0, \qquad C' = \frac{1}{2}, \qquad D' = E' = 0, \qquad F' = -10.$$

Equation (5) then gives

$$\frac{5}{2}x'^2 + \frac{1}{2}y'^2 - 10 = 0, \qquad \text{or} \qquad \frac{x'^2}{4} + \frac{y'^2}{20} = 1.$$

The curve is an ellipse with foci on the new y'-axis (Fig. 9.26). ❏

Possible Graphs of Quadratic Equations

We now return to the graph of the general quadratic equation.

Since axes can always be rotated to eliminate the cross product term, there is no loss of generality in assuming that this has been done and that our equation has the form

$$Ax^2 + Cy^2 + Dx + Ey + F = 0. \tag{8}$$

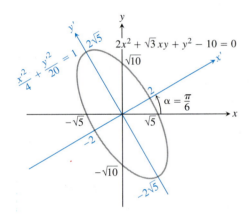

9.26 The conic section in Example 2.

Equation (8) represents

a) a *circle* if $A = C \neq 0$ (special cases: the graph is a point or there is no graph at all);

b) a *parabola* if Eq. (8) is quadratic in one variable and linear in the other;

c) an *ellipse* if A and C are both positive or both negative (special cases: circles, a single point or no graph at all);

d) a *hyperbola* if A and C have opposite signs (special case: a pair of intersecting lines);

e) a *straight line* if A and C are zero and at least one of D and E is different from zero;

f) *one or two straight lines* if the left-hand side of Eq. (8) can be factored into the product of two linear factors.

See Table 9.3 (on page 732) for examples.

The Discriminant Test

We do not need to eliminate the xy-term from the equation

$$Ax^2 + Bxy + Cy^2 + Dx + Ey + F = 0 \tag{9}$$

to tell what kind of conic section the equation represents. If this is the only information we want, we can apply the following test instead.

As we have seen, if $B \neq 0$, then rotating the coordinate axes through an angle α that satisfies the equation

$$\cot 2\alpha = \frac{A - C}{B} \tag{10}$$

will change Eq. (9) into an equivalent form

$$A' x'^2 + C' y'^2 + D' x' + E' y' + F' = 0 \tag{11}$$

without a cross product term.

Now, the graph of Eq. (11) is a (real or degenerate)

a) *parabola* if A' or $C' = 0$; that is, if $A' C' = 0$;
b) *ellipse* if A' and C' have the same sign; that is, if $A' C' > 0$;
c) *hyperbola* if A' and C' have opposite signs; that is, if $A' C' < 0$.

It can also be verified from Eqs. (6) that for any rotation of axes,

$$B^2 - 4AC = B'^2 - 4A' C'. \tag{12}$$

This means that the quantity $B^2 - 4AC$ is not changed by a rotation. But when we rotate through the angle α given by Eq. (10), B' becomes zero, so

$$B^2 - 4AC = -4A' C'.$$

Since the curve is a parabola if $A' C' = 0$, an ellipse if $A' C' > 0$, and a hyperbola if $A' C' < 0$, the curve must be a parabola if $B^2 - 4AC = 0$, an ellipse if $B^2 - 4AC < 0$, and a hyperbola if $B^2 - 4AC > 0$. The number $B^2 - 4AC$ is called the **discriminant** of Eq. (9).

The Discriminant Test

With the understanding that occasional degenerate cases may arise, the quadratic curve $Ax^2 + Bxy + Cy^2 + Dx + Ey + F = 0$ is

a) a **parabola** if $B^2 - 4AC = 0$,
b) an **ellipse** if $B^2 - 4AC < 0$,
c) a **hyperbola** if $B^2 - 4AC > 0$.

EXAMPLE 3

a) $3x^2 - 6xy + 3y^2 + 2x - 7 = 0$ represents a parabola because

$$B^2 - 4AC = (-6)^2 - 4 \cdot 3 \cdot 3 = 36 - 36 = 0.$$

b) $x^2 + xy + y^2 - 1 = 0$ represents an ellipse because

$$B^2 - 4AC = (1)^2 - 4 \cdot 1 \cdot 1 = -3 < 0.$$

c) $xy - y^2 - 5y + 1 = 0$ represents a hyperbola because

$$B^2 - 4AC = (1)^2 - 4(0)(-1) = 1 > 0. \qquad \square$$

Table 9.3 Examples of quadratic curves

$Ax^2 + Bxy + Cy^2 + Dx + Ey + F = 0$								
	A	B	C	D	E	F	Equation	Remarks
Circle	1		1			−4	$x^2 + y^2 = 4$	$A = C;\ F < 0$
Parabola			1	−9			$y^2 = 9x$	Quadratic in y, linear in x
Ellipse	4		9			−36	$4x^2 + 9y^2 = 36$	A, C have same sign, $A \neq C;\ F < 0$
Hyperbola	1		−1			−1	$x^2 - y^2 = 1$	A, C have opposite signs
One line (still a conic section)	1						$x^2 = 0$	y-axis
Intersecting lines (still a conic section)		1		1	−1	−1	$xy + x - y - 1 = 0$	Factors to $(x - 1)(y + 1) = 0$, so $x = 1,\ y = -1$
Parallel lines (not a conic section)	1			−3		2	$x^2 - 3x + 2 = 0$	Factors to $(x - 1)(x - 2) = 0$, so $x = 1,\ x = 2$
Point	1		1				$x^2 + y^2 = 0$	The origin
No graph	1					1	$x^2 = -1$	No graph

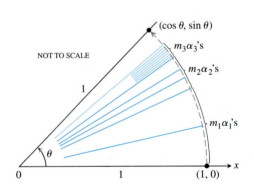

NOT TO SCALE

9.27 To calculate the sine and cosine of an angle θ between 0 and 2π, the calculator rotates the point (1, 0) to an appropriate location on the unit circle and displays the resulting coordinates.

Technology *How Calculators Use Rotations to Evaluate Sines and Cosines*
Some calculators use rotations to calculate sines and cosines of arbitrary angles. The procedure goes something like this: The calculator has, stored,

1. ten angles or so, say

    $$\alpha_1 = \sin^{-1}(10^{-1}), \qquad \alpha_2 = \sin^{-1}(10^{-2}), \qquad \ldots, \qquad \alpha_{10} = \sin^{-1}(10^{-10}),$$

    and

2. twenty numbers, the sines and cosines of the angles $\alpha_1, \alpha_2, \ldots, \alpha_{10}$.

To calculate the sine and cosine of an arbitrary angle θ, we enter θ (in radians) into the calculator. The calculator subtracts or adds multiples of 2π to θ to replace θ by the angle between 0 and 2π that has the same sine and cosine as θ (we continue to call the angle θ). The calculator then "writes" θ as a sum of multiples of α_1 (as many as possible without overshooting) plus multiples of α_2 (again, as many as possible), and so on, working its way to α_{10}. This gives

$$\theta \approx m_1\alpha_1 + m_2\alpha_2 + \cdots + m_{10}\alpha_{10}.$$

The calculator then rotates the point (1, 0) through m_1 copies of α_1 (through α_1, m_1 times in succession), plus m_2 copies of α_2, and so on, finishing off with m_{10} copies of α_{10} (Fig. 9.27). The coordinates of the final position of (1, 0) on the unit circle are the values the calculator gives for $(\cos \theta, \sin \theta)$.

Exercises 9.3

Using the Discriminant

Use the discriminant $B^2 - 4AC$ to decide whether the equations in Exercises 1–16 represent parabolas, ellipses, or hyperbolas.

1. $x^2 - 3xy + y^2 - x = 0$
2. $3x^2 - 18xy + 27y^2 - 5x + 7y = -4$
3. $3x^2 - 7xy + \sqrt{17}\, y^2 = 1$
4. $2x^2 - \sqrt{15}\, xy + 2y^2 + x + y = 0$
5. $x^2 + 2xy + y^2 + 2x - y + 2 = 0$
6. $2x^2 - y^2 + 4xy - 2x + 3y = 6$
7. $x^2 + 4xy + 4y^2 - 3x = 6$
8. $x^2 + y^2 + 3x - 2y = 10$
9. $xy + y^2 - 3x = 5$
10. $3x^2 + 6xy + 3y^2 - 4x + 5y = 12$
11. $3x^2 - 5xy + 2y^2 - 7x - 14y = -1$
12. $2x^2 - 4.9xy + 3y^2 - 4x = 7$
13. $x^2 - 3xy + 3y^2 + 6y = 7$
14. $25x^2 + 21xy + 4y^2 - 350x = 0$
15. $6x^2 + 3xy + 2y^2 + 17y + 2 = 0$
16. $3x^2 + 12xy + 12y^2 + 435x - 9y + 72 = 0$

Rotating Coordinate Axes

In Exercises 17–26, rotate the coordinate axes to change the given equation into an equation that has no cross product (xy) term. Then identify the graph of the equation. (The new equations will vary with the size and direction of the rotation you use.)

17. $xy = 2$
18. $x^2 + xy + y^2 = 1$
19. $3x^2 + 2\sqrt{3}\, xy + y^2 - 8x + 8\sqrt{3}\, y = 0$
20. $x^2 - \sqrt{3}\, xy + 2y^2 = 1$
21. $x^2 - 2xy + y^2 = 2$
22. $3x^2 - 2\sqrt{3}\, xy + y^2 = 1$
23. $\sqrt{2}\, x^2 + 2\sqrt{2}\, xy + \sqrt{2}\, y^2 - 8x + 8y = 0$
24. $xy - y - x + 1 = 0$
25. $3x^2 + 2xy + 3y^2 = 19$
26. $3x^2 + 4\sqrt{3}\, xy - y^2 = 7$

27. Find the sine and cosine of an angle through which the coordinate axes can be rotated to eliminate the cross product term from the equation

$$14x^2 + 16xy + 2y^2 - 10x + 26,370\, y - 17 = 0.$$

Do not carry out the rotation.

28. Find the sine and cosine of an angle through which the coordinate axes can be rotated to eliminate the cross product term from the equation

$$4x^2 - 4xy + y^2 - 8\sqrt{5}\, x - 16\sqrt{5}\, y = 0.$$

Do not carry out the rotation.

Calculator

The conic sections in Exercises 17–26 were chosen to have rotation angles that were "nice" in the sense that once we knew $\cot 2\alpha$ or $\tan 2\alpha$ we could identify 2α and find $\sin \alpha$ and $\cos \alpha$ from familiar triangles. The conic sections encountered in practice may not have such nice rotation angles, and we may have to use a calculator to determine α from the value of $\cot 2\alpha$ or $\tan 2\alpha$.

In Exercises 29–34, use a calculator to find an angle α through which the coordinate axes can be rotated to change the given equation into a quadratic equation that has no cross product term. Then find $\sin \alpha$ and $\cos \alpha$ to 2 decimal places and use Eqs. (6) to find the coefficients of the new equation to the nearest decimal place. In each case, say whether the conic section is an ellipse, a hyperbola, or a parabola.

29. $x^2 - xy + 3y^2 + x - y - 3 = 0$
30. $2x^2 + xy - 3y^2 + 3x - 7 = 0$
31. $x^2 - 4xy + 4y^2 - 5 = 0$
32. $2x^2 - 12xy + 18y^2 - 49 = 0$
33. $3x^2 + 5xy + 2y^2 - 8y - 1 = 0$
34. $2x^2 + 7xy + 9y^2 + 20x - 86 = 0$

Theory and Examples

35. What effect does a $90°$ rotation about the origin have on the equations of the following conic sections? Give the new equation in each case.

   a)   The ellipse $(x^2/a^2) + (y^2/b^2) = 1$ $(a > b)$
   b)   The hyperbola $(x^2/a^2) - (y^2/b^2) = 1$
   c)   The circle $x^2 + y^2 = a^2$
   d)   The line $y = mx$
   e)   The line $y = mx + b$

36. What effect does a $180°$ rotation about the origin have on the equations of the following conic sections? Give the new equation in each case.

   a)   The ellipse $(x^2/a^2) + (y^2/b^2) = 1$ $(a > b)$
   b)   The hyperbola $(x^2/a^2) - (y^2/b^2) = 1$
   c)   The circle $x^2 + y^2 = a^2$
   d)   The line $y = mx$
   e)   The line $y = mx + b$

37. *The Hyperbola $xy = a$.* The hyperbola $xy = 1$ is one of many hyperbolas of the form $xy = a$ that appear in science and mathematics.

   a) Rotate the coordinate axes through an angle of $45°$ to change the equation $xy = 1$ into an equation with no xy-term. What is the new equation?

   b) Do the same for the equation $xy = a$.

38. Find the eccentricity of the hyperbola $xy = 2$.

39. Can anything be said about the graph of the equation $Ax^2 + Bxy + Cy^2 + Dx + Ey + F = 0$ if $AC < 0$? Give reasons for your answer.

40. Does any nondegenerate conic section $Ax^2 + Bxy + Cy^2 + Dx + Ey + F = 0$ have all of the following properties?

   a) It is symmetric with respect to the origin.

   b) It passes through the point $(1, 0)$.

   c) It is tangent to the line $y = 1$ at the point $(-2, 1)$.

   Give reasons for your answer.

41. Show that the equation $x^2 + y^2 = a^2$ becomes $x'^2 + y'^2 = a^2$ for every choice of the angle α in the rotation equations (4).

42. Show that rotating the axes through an angle of $\pi/4$ radians will eliminate the xy-term from Eq. (1) whenever $A = C$.

43. **a)** Decide whether the equation

$$x^2 + 4xy + 4y^2 + 6x + 12y + 9 = 0$$

   represents an ellipse, a parabola, or a hyperbola.

   b) Show that the graph of the equation in (a) is the line $2y = -x - 3$.

44. **a)** Decide whether the conic section with equation

$$9x^2 + 6xy + y^2 - 12x - 4y + 4 = 0$$

   represents a parabola, an ellipse, or a hyperbola.

   b) Show that the graph of the equation in (a) is the line $y = -3x + 2$.

45. **a)** What kind of conic section is the curve $xy + 2x - y = 0$?

   b) Solve the equation $xy + 2x - y = 0$ for y and sketch the curve as the graph of a rational function of x.

   c) Find equations for the lines parallel to the line $y = -2x$ that are normal to the curve. Add the lines to your sketch.

46. Prove or find counterexamples to the following statements about the graph of $Ax^2 + Bxy + Cy^2 + Dx + Ey + F = 0$.

   a) If $AC > 0$, the graph is an ellipse.

   b) If $AC > 0$, the graph is a hyperbola.

   c) If $AC < 0$, the graph is a hyperbola.

47. *A nice area formula for ellipses.* When $B^2 - 4AC$ is negative, the equation

$$Ax^2 + Bxy + Cy^2 = 1$$

represents an ellipse. If the ellipse's semi-axes are a and b, its area is πab (a standard formula). Show that the area is also given by the formula $2\pi / \sqrt{4AC - B^2}$. (*Hint:* Rotate the coordinate axes to eliminate the xy-term and apply Eq. (12) to the new equation.)

48. *Other invariants.* We describe the fact that $B'^2 - 4A'C'$ equals $B^2 - 4AC$ after a rotation about the origin by saying that the discriminant of a quadratic equation is an **invariant** of the equation. Use Eqs. (6) to show that the numbers (a) $A + C$ and (b) $D^2 + E^2$ are also invariants, in the sense that

$$A' + C' = A + C \quad \text{and} \quad D'^2 + E'^2 = D^2 + E^2.$$

We can use these equalities to check against numerical errors when we rotate axes. They can also be helpful in shortening the work required to find values for the new coefficients.

49. *A proof that $B'^2 - 4A'C' = B^2 - 4AC$.* Use Eqs. (6) to show that $B'^2 - 4A'C' = B^2 - 4AC$ for any rotation of axes about the origin. The calculation works out nicely but requires patience.

9.4   Parametrizations of Plane Curves

When the path of a particle moving in the plane looks like the curve in Fig. 9.28, we cannot hope to describe it with a Cartesian formula that expresses y directly in terms of x or x directly in terms of y. Instead, we express each of the particle's coordinates as a function of time t and describe the path with a pair of equations, $x = f(t)$ and $y = g(t)$. For studying motion, equations like these are preferable to a Cartesian formula because they tell us the particle's position at any time t.

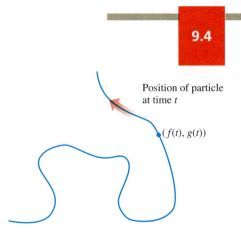

Position of particle at time t

$(f(t), g(t))$

9.28 The path traced by a particle moving in the *xy*-plane is not always the graph of a function of x or a function of y.

Definitions

If x and y are given as continuous functions

$$x = f(t), \qquad y = g(t)$$

over an interval of t-values, then the set of points $(x, y) = (f(t), g(t))$ defined by these equations is a **curve** in the coordinate plane. The equations are **parametric equations** for the curve. The variable t is a **parameter** for the curve and its domain I is the **parameter interval.** If I is a closed interval, $a \leq t \leq b$, the point $(f(a), g(a))$ is the **initial point** of the curve and $(f(b), g(b))$ is the **terminal point** of the curve. When we give parametric equations and a parameter interval for a curve in the plane, we say that we have **parametrized** the curve. The equations and interval constitute a **parametrization** of the curve.

In many applications t denotes time, but it might instead denote an angle (as in some of the following examples) or the distance a particle has traveled along its path from its starting point (as it sometimes will when we later study motion).

EXAMPLE 1   *The circle $x^2 + y^2 = 1$*

The equations and parameter interval

$$x = \cos t, \qquad y = \sin t, \qquad 0 \leq t \leq 2\pi,$$

describe the position $P(x, y)$ of a particle that moves counterclockwise around the circle $x^2 + y^2 = 1$ as t increases (Fig. 9.29).

We know that the point lies on this circle for every value of t because

$$x^2 + y^2 = \cos^2 t + \sin^2 t = 1.$$

But how much of the circle does the point $P(x, y)$ actually traverse?

To find out, we track the motion as t runs from 0 to 2π. The parameter t is the radian measure of the angle that radius OP makes with the positive x-axis. The particle starts at $(1, 0)$, moves up and to the left as t approaches $\pi/2$, and continues around the circle to stop again at $(1, 0)$ when $t = 2\pi$. The particle traces the circle exactly once. ❑

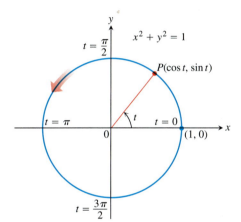

9.29 The equations $x = \cos t, y = \sin t$ describe motion on the circle $x^2 + y^2 = 1$. The arrow shows the direction of increasing t (Example 1).

EXAMPLE 2   *A semicircle*

The equations and parameter interval

$$x = \cos t, \qquad y = -\sin t, \qquad 0 \leq t \leq \pi,$$

describe the position $P(x, y)$ of a particle that moves clockwise around the circle $x^2 + y^2 = 1$ as t increases from 0 to π.

We know that the point P lies on this circle for all t because its coordinates satisfy the circle's equation. How much of the circle does the particle traverse? To find out, we track the motion as t runs from 0 to π. As in Example 1, the particle starts at $(1, 0)$. But now as t increases, y becomes negative, decreasing to -1 when $t = \pi/2$ and then increasing back to 0 as t approaches π. The motion stops at $t = \pi$ with only the lower half of the circle covered (Fig. 9.30). ❑

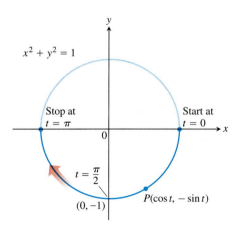

9.30 The point $P(\cos t, -\sin t)$ moves clockwise as t increases from 0 to π (Example 2).

EXAMPLE 3   *Half a parabola*

The position $P(x, y)$ of a particle moving in the xy-plane is given by the equations

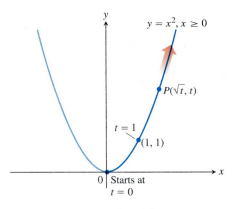

9.31 The equations $x = \sqrt{t}, y = t$ and interval $t \geq 0$ describe the motion of a particle that traces the right-hand half of the parabola $y = x^2$ (Example 3).

and parameter interval

$$x = \sqrt{t}, \qquad y = t, \qquad t \geq 0.$$

Identify the path traced by the particle and describe the motion.

Solution We try to identify the path by eliminating t between the equations $x = \sqrt{t}$ and $y = t$. With any luck, this will produce a recognizable algebraic relation between x and y. We find that

$$y = t = \left(\sqrt{t}\right)^2 = x^2.$$

This means that the particle's position coordinates satisfy the equation $y = x^2$, so the particle moves along the parabola $y = x^2$.

It would be a mistake, however, to conclude that the particle's path is the entire parabola $y = x^2$— it is only half the parabola. The particle's x-coordinate is never negative. The particle starts at $(0, 0)$ when $t = 0$ and rises into the first quadrant as t increases (Fig. 9.31). ☐

EXAMPLE 4   *An entire parabola*

The position $P(x, y)$ of a particle moving in the xy-plane is given by the equations and parameter interval

$$x = t, \qquad y = t^2, \qquad -\infty < t < \infty.$$

Identify the particle's path and describe the motion.

Solution We identify the path by eliminating t between the equations $x = t$ and $y = t^2$, obtaining

$$y = (t)^2 = x^2.$$

The particle's position coordinates satisfy the equation $y = x^2$, so the particle moves along this curve.

In contrast to Example 3, the particle now traverses the entire parabola. As t increases from $-\infty$ to ∞, the particle comes down the left-hand side, passes through the origin, and moves up the right-hand side (Fig. 9.32).

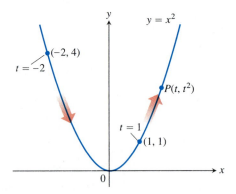

9.32 The path defined by $x = t$, $y = t^2, -\infty < t < \infty$ is the entire parabola $y = x^2$ (Example 4).

As Example 4 illustrates, any curve $y = f(x)$ has the parametrization $x = t$, $y = f(t)$. This is so simple we usually do not use it, but the point of view is occasionally helpful.

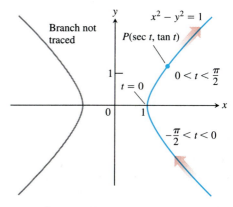

9.33 The ellipse in Example 5, drawn for $a > b$. The coordinates of P are $x = a \cos t$, $y = b \sin t$.

EXAMPLE 5 *A parametrization of the ellipse $x^2/a^2 + y^2/b^2 = 1$*

Describe the motion of a particle whose position $P(x, y)$ at time t is given by

$$x = a \cos t, \qquad y = b \sin t, \qquad 0 \le t \le 2\pi.$$

Solution We find a Cartesian equation for the particle's coordinates by eliminating t between the equations

$$\cos t = \frac{x}{a}, \qquad \sin t = \frac{y}{b}.$$

We accomplish this with the identity $\cos^2 t + \sin^2 t = 1$, which yields

$$\left(\frac{x}{a}\right)^2 + \left(\frac{y}{b}\right)^2 = 1, \qquad \text{or} \qquad \frac{x^2}{a^2} + \frac{y^2}{b^2} = 1.$$

The particle's coordinates (x, y) satisfy the equation $(x^2/a^2) + (y^2/b^2) = 1$, so the particle moves along this ellipse. When $t = 0$, the particle's coordinates are

$$x = a \cos(0) = a, \qquad y = b \sin(0) = 0,$$

so the motion starts at $(a, 0)$. As t increases, the particle rises and moves toward the left, moving counterclockwise. It traverses the ellipse once, returning to its starting position $(a, 0)$ at time $t = 2\pi$ (Fig. 9.33). ❑

EXAMPLE 6 *A parametrization of the circle $x^2 + y^2 = a^2$*

The equations and parameter interval

$$x = a \cos t, \qquad y = a \sin t, \qquad 0 \le t \le 2\pi,$$

obtained by taking $b = a$ in Example 5, describe the circle $x^2 + y^2 = a^2$. ❑

EXAMPLE 7 *A parametrization of the right-hand branch of the hyperbola $x^2 - y^2 = 1$*

Describe the motion of the particle whose position $P(x, y)$ at time t is given by

$$x = \sec t, \qquad y = \tan t, \qquad -\frac{\pi}{2} < t < \frac{\pi}{2}.$$

Solution We find a Cartesian equation for the coordinates of P by eliminating t between the equations

$$\sec t = x, \qquad \tan t = y.$$

We accomplish this with the identity $\sec^2 t - \tan^2 t = 1$, which yields

$$x^2 - y^2 = 1.$$

Since the particle's coordinates (x, y) satisfy the equation $x^2 - y^2 = 1$, the motion takes place somewhere on this hyperbola. As t runs between $-\pi/2$ and $\pi/2$, $x = \sec t$ remains positive and $y = \tan t$ runs between $-\infty$ and ∞, so P traverses the hyperbola's right-hand branch. It comes in along the branch's lower half as $t \to 0^-$, reaches $(1, 0)$ at $t = 0$, and moves out into the first quadrant as t increases toward $\pi/2$ (Fig. 9.34). ❑

9.34 The equations $x = \sec t$, $y = \tan t$ and interval $-\pi/2 < t < \pi/2$ describe the right-hand branch of the hyperbola $x^2 - y^2 = 1$ (Example 7).

Huygen's clock

The problem with a pendulum clock whose bob swings in a circular arc is that the frequency of the swing depends on the amplitude of the swing. The wider the swing, the longer it takes the bob to return to center.

This does not happen if the bob can be made to swing in a cycloid. In 1673, Christiaan Huygens (1629–1695), the Dutch mathematician, physicist, and astronomer who discovered the rings of Saturn, driven by a need to make accurate determinations of longitude at sea, designed a pendulum clock whose bob would swing in a cycloid. He hung the bob from a fine wire constrained by guards that caused it to draw up as it swung away from center. How were the guards shaped? They were cycloids, too.

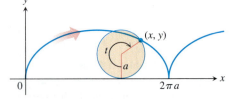

EXAMPLE 8   *Cycloids*

A wheel of radius a rolls along a horizontal straight line. Find parametric equations for the path traced by a point P on the wheel's circumference. The path is called a **cycloid.**

Solution   We take the line to be the x-axis, mark a point P on the wheel, start the wheel with P at the origin, and roll the wheel to the right. As parameter, we use the angle t through which the wheel turns, measured in radians. Figure 9.35 shows the wheel a short while later, when its base lies at units from the origin. The wheel's center C lies at (at, a) and the coordinates of P are

$$x = at + a \cos \theta, \qquad y = a + a \sin \theta.$$

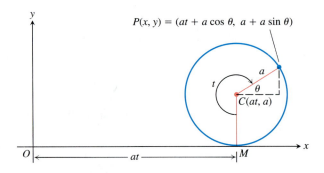

9.35 The position of $P(x, y)$ on the rolling wheel at angle t (Example 8).

To express θ in terms of t, we observe that $t + \theta = 3\pi/2$, so that

$$\theta = \frac{3\pi}{2} - t.$$

This makes

$$\cos \theta = \cos \left(\frac{3\pi}{2} - t \right) = -\sin t, \qquad \sin \theta = \sin \left(\frac{3\pi}{2} - t \right) = -\cos t.$$

The equations we seek are

$$x = at - a \sin t, \qquad y = a - a \cos t.$$

These are usually written with the a factored out:

$$x = a(t - \sin t), \qquad y = a(1 - \cos t). \tag{1}$$

Figure 9.36 shows the first arch of the cycloid and part of the next.   ❑

✳ Brachistochrones and Tautochrones

If we turn Fig. 9.36 upside down, Eqs. (1) still apply and the resulting curve (Fig. 9.37) has two interesting physical properties. The first relates to the origin O and the point B at the bottom of the first arch. Among all smooth curves joining these points, the cycloid is the curve along which a frictionless bead, subject only to the force of gravity, will slide from O to B the fastest. This makes the cycloid a **brachistochrone** ("brah-*kiss*-toe-krone"), or shortest time curve for these points. The second property is that even if you start the bead partway down the curve

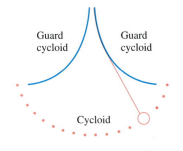

9.36 The cycloid $x = a(t - \sin t)$, $y = a(1 - \cos t)$, for $t \geq 0$.

The witch of Agnesi

Although l'Hôpital wrote the first text on differential calculus, the first text to include differential and integral calculus along with analytic geometry, infinite series, and differential equations was written in the 1740s by the Italian mathematician Maria Gaetana Agnesi (1718–1799). Agnesi, a gifted scholar and linguist whose Latin essay defending higher education for women was published when she was only nine years old, was a well-published scientist by age 20 and an honorary faculty member of the University of Bologna by age 30.

Today, Agnesi is remembered chiefly for a bell-shaped curve called *the witch of Agnesi.* This name, found only in English texts, is the result of a mistranslation. Agnesi's own name for the curve was *versiera* or "turning curve." John Colson, a noted Cambridge mathematician who felt Agnesi's text so important that he learned Italian to translate it "for the benefit of British youth" (he particularly had in mind young women, for whom he hoped Agnesi would be a role model), probably confused versiera with *avversiera,* which means "wife of the devil" and translates into "witch." You can find out more about the witch by doing Exercise 29.

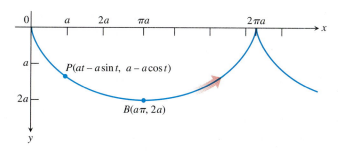

9.37 To study motion along an upside-down cycloid under the influence of gravity, we turn Fig. 9.36 upside down. This points the *y*-axis in the direction of the gravitational force and makes the downward *y*-coordinates positive. The equations and parameter interval for the cycloid are still

$$x = a(t - \sin t),$$
$$y = a(1 - \cos t), \quad t \geq 0.$$

The arrow shows the direction of increasing *t*.

toward B, it will still take the bead the same amount of time to reach B. This makes the cycloid a **tautochrone** ("*taw*-toe-krone"), or same-time curve for O and B.

Are there any other brachistochrones joining O and B, or is the cycloid the only one? We can formulate this as a mathematical question in the following way. At the start, the kinetic energy of the bead is zero, since its velocity is zero. The work done by gravity in moving the bead from $(0, 0)$ to any other point (x, y) in the plane is mgy, and this must equal the change in kinetic energy. That is,

$$mgy = \frac{1}{2}mv^2 - \frac{1}{2}m(0)^2.$$

Thus, the velocity of the bead when it reaches (x, y) has to be

$$v = \sqrt{2gy}.$$

That is,

$$\frac{ds}{dt} = \sqrt{2gy} \qquad \text{\textit{ds} is the arc length differential along the bead's path.}$$

or

$$dt = \frac{ds}{\sqrt{2gy}} = \frac{\sqrt{1 + (dy/dx)^2}\, dx}{\sqrt{2gy}}.$$

The time T_f it takes the bead to slide along a particular path $y = f(x)$ from O to $B(a\pi, 2a)$ is

$$T_f = \int_{x=0}^{x=a\pi} \sqrt{\frac{1 + (dy/dx)^2}{2gy}}\, dx. \tag{2}$$

What curves $y = f(x)$, if any, minimize the value of this integral?

At first sight, we might guess that the straight line joining O and B would give the shortest time, but perhaps not. There might be some advantage in having the bead fall vertically at first to build up its velocity faster. With a higher velocity, the bead could travel a longer path and still reach B first. Indeed, this is the right idea.

The solution, from a branch of mathematics known as the calculus of variations, is that the original cycloid from O to B is the one and only brachistochrone for O and B.

While the solution of the brachistrochrone problem is beyond our present reach, we can still show why the cycloid is a tautochrone. For the cycloid, Eq. (2) takes the form

$$T_{\text{cycloid}} = \int_{x=0}^{x=a\pi} \sqrt{\frac{dx^2 + dy^2}{2gy}}$$

$$= \int_{t=0}^{t=\pi} \sqrt{\frac{a^2(2 - 2\cos t)}{2ga(1 - \cos t)}}\, dt$$

From Eqs. (1),
$dx = a(1 - \cos t)\, dt,$
$dy = a \sin t\, dt,$ and
$y = a(1 - \cos t)$

$$= \int_0^\pi \sqrt{\frac{a}{g}}\, dt = \pi \sqrt{\frac{a}{g}}.$$

Thus, the amount of time it takes the frictionless bead to slide down the cycloid to B after it is released from rest at O is $\pi\sqrt{a/g}$.

Suppose that instead of starting the bead at O we start it at some lower point on the cycloid, a point (x_0, y_0) corresponding to the parameter value $t_0 > 0$. The bead's velocity at any later point (x, y) on the cycloid is

$$v = \sqrt{2g\,(y - y_0)} = \sqrt{2ga\,(\cos t_0 - \cos t)}. \qquad y = a(1 - \cos t)$$

Accordingly, the time required for the bead to slide from (x_0, y_0) down to B is

$$T = \int_{t_0}^\pi \sqrt{\frac{a^2(2 - 2\cos t)}{2ga\,(\cos t_0 - \cos t)}}\, dt = \sqrt{\frac{a}{g}} \int_{t_0}^\pi \sqrt{\frac{1 - \cos t}{\cos t_0 - \cos t}}\, dt$$

$$= \sqrt{\frac{a}{g}} \int_{t_0}^\pi \sqrt{\frac{2\sin^2(t/2)}{(2\cos^2(t_0/2) - 1) - (2\cos^2(t/2) - 1)}}\, dt$$

$$= \sqrt{\frac{a}{g}} \int_{t_0}^\pi \frac{\sin(t/2)\, dt}{\sqrt{\cos^2(t_0/2) - \cos^2(t/2)}}$$

$$= \sqrt{\frac{a}{g}} \int_{t=t_0}^{t=\pi} \frac{-2\, du}{\sqrt{a^2 - u^2}}$$

$u = \cos(t/2)$
$-2du = \sin(t/2)\, dt$
$c = \cos(t_0/2)$

$$= 2\sqrt{\frac{a}{g}} \left[-\sin^{-1} \frac{u}{c} \right]_{t=t_0}^{t=\pi}$$

$$= 2\sqrt{\frac{a}{g}} \left[-\sin^{-1} \frac{\cos(t/2)}{\cos(t_0/2)} \right]_{t_0}^{\pi}$$

$$= 2\sqrt{\frac{a}{g}} (-\sin^{-1} 0 + \sin^{-1} 1) = \pi \sqrt{\frac{a}{g}}.$$

This is precisely the time it takes the bead to slide to B from O. It takes the bead the same amount of time to reach B no matter where it starts. Beads starting simultaneously from O, A, and C in Fig. 9.38, for instance, will all reach B at the same time.

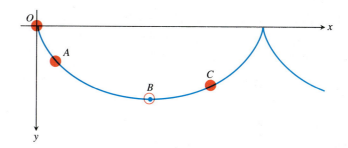

9.38 Beads released simultaneously on the cycloid at O, A, and C will reach B at the same time.

Standard Parametrizations

Circle $x^2 + y^2 = a^2$:

$$x = a \cos t$$

$$y = a \sin t$$

$$0 \leq t \leq 2\pi$$

Ellipse $\dfrac{x^2}{a^2} + \dfrac{y^2}{b^2} = 1$:

$$x = a \cos t$$

$$y = b \sin t$$

$$0 \leq t \leq 2\pi$$

Cycloid generated by a circle of radius a:

$$x = a(t - \sin t), \qquad y = a(1 - \cos t)$$

Exercises 9.4

Finding Cartesian Equations from Parametric Equations

Exercises 1–24 give parametric equations and parameter intervals for the motion of a particle in the xy-plane. Identify the particle's path by finding a Cartesian equation for it. Graph the Cartesian equation. (The graphs will vary with the equation used.) Indicate the portion of the graph traced by the particle and the direction of motion.

1. $x = \cos t, \quad y = \sin t, \quad 0 \leq t \leq \pi$

2. $x = \cos 2t, \quad y = \sin 2t, \quad 0 \leq t \leq \pi$

3. $x = \sin (2\pi (1 - t)), \quad y = \cos (2\pi (1 - t)), \quad 0 \leq t \leq 1$

4. $x = \cos (\pi - t), \quad y = \sin (\pi - t), \quad 0 \leq t \leq \pi$

5. $x = 4 \cos t, \quad y = 2 \sin t, \quad 0 \leq t \leq 2\pi$

6. $x = 4 \sin t, \quad y = 2 \cos t, \quad 0 \leq t \leq \pi$

7. $x = 4 \cos t, \quad y = 5 \sin t, \quad 0 \leq t \leq \pi$

8. $x = 4 \sin t, \quad y = 5 \cos t, \quad 0 \leq t \leq 2\pi$

9. $x = 3t, \quad y = 9t^2, \quad -\infty < t < \infty$

10. $x = -\sqrt{t}, \quad y = t, \quad t \geq 0$

11. $x = t, \quad y = \sqrt{t}, \quad t \geq 0$

12. $x = \sec^2 t - 1, \quad y = \tan t, \quad -\pi/2 < t < \pi/2$

13. $x = -\sec t, \quad y = \tan t, \quad -\pi/2 < t < \pi/2$

14. $x = \csc t, \quad y = \cot t, \quad 0 < t < \pi$

15. $x = 2t - 5, \quad y = 4t - 7, \quad -\infty < t < \infty$

16. $x = 1 - t, \quad y = 1 + t, \quad -\infty < t < \infty$

17. $x = t, \quad y = 1 - t, \quad 0 \leq t \leq 1$

18. $x = 3 - 3t, \quad y = 2t, \quad 0 \leq t \leq 1$

19. $x = t, \quad y = \sqrt{1 - t^2}, \quad -1 \leq t \leq 0$

20. $x = t, \quad y = \sqrt{4 - t^2}, \quad 0 \leq t \leq 2$

21. $x = t^2, \quad y = \sqrt{t^4 + 1}, \quad t \geq 0$

22. $x = \sqrt{t + 1}, \quad y = \sqrt{t}, \quad t \geq 0$

23. $x = -\cosh t, \quad y = \sinh t, \quad -\infty < t < \infty$

24. $x = 2 \sinh t, \quad y = 2 \cosh t, \quad -\infty < t < \infty$

Determining Parametric Equations

25. Find parametric equations and a parameter interval for the motion of a particle that starts at $(a, 0)$ and traces the circle $x^2 + y^2 = a^2$

a) once clockwise, **b)** once counterclockwise,
c) twice clockwise, **d)** twice counterclockwise.

(There are many ways to do these, so your answers may not be the same as the ones in the back of the book.)

26. Find parametric equations and a parameter interval for the motion of a particle that starts at $(a, 0)$ and traces the ellipse $(x^2/a^2) + (y^2/b^2) = 1$

a) once clockwise, **b)** once counterclockwise,
c) twice clockwise, **d)** twice counterclockwise.

(As in Exercise 25, there are many correct answers.)

27. Find parametric equations for the semicircle

$$x^2 + y^2 = a^2, \quad y > 0,$$

using as parameter the slope $t = dy/dx$ of the tangent to the curve at (x, y).

28. Find parametric equations for the circle

$$x^2 + y^2 = a^2,$$

using as parameter the arc length s measured counterclockwise from the point $(a, 0)$ to the point (x, y).

29. *The witch of Maria Agnesi.* The bell-shaped witch of Maria Agnesi can be constructed in the following way. Start with a circle of radius 1, centered at the point $(0, 1)$, as shown in the accompanying figure. Choose a point A on the line $y = 2$ and connect it to the origin with a line segment. Call the point where the segment crosses the circle B. Let P be the point where the vertical line through A crosses the horizontal line through B. The witch is the curve traced by P as A moves along the line $y = 2$. Find parametric equations and a parameter interval for the witch by expressing the coordinates of P in terms of t, the radian measure of the angle that segment OA makes with the positive x-axis. The following equalities (which you may assume) will help.

a) $x = AQ$ **b)** $y = 2 - AB \sin t$
c) $AB \cdot OA = (AQ)^2$

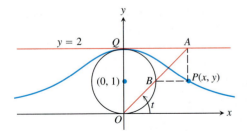

30. *The involute of a circle.* If a string wound around a fixed circle is unwound while held taut in the plane of the circle, its end P

traces an *involute* of the circle. In Fig. 9.39, the circle in question is the circle $x^2 + y^2 = 1$ and the tracing point starts at $(1, 0)$. The unwound portion of the string is tangent to the circle at Q, and t is the radian measure of the angle from the positive x-axis to segment OQ. Derive parametric equations for the involute by expressing the coordinates x and y of P in terms of t for $t \geq 0$.

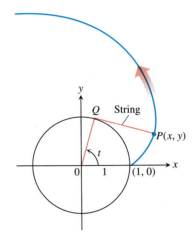

9.39 The involute of a circle of radius 1 (Exercise30.)

31. *Parametrizations of lines in the plane* (Fig. 9.40).

a) Show that the equations and parameter interval

$$x = x_0 + (x_1 - x_0)t, \quad y = y_0 + (y_1 - y_0)t, \quad -\infty < t < \infty,$$

describe the line through the points (x_0, y_0) and (x_1, y_1).

b) Using the same parameter interval, write parametric equations for the line through a point (x_1, y_1) and the origin.

c) Using the same parameter interval, write parametric equations for the line through $(-1, 0)$ and $(0, 1)$.

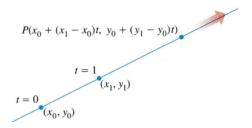

9.40 The line in Exercise 31. The arrow shows the direction of increasing t.

32. *The trammel of Archimedes.* The mechanical system pictured here is called the trammel of Archimedes. It consists of a rigid bar of length L, one end attached to a roller that rolls along the y-axis. At a fixed distance R from this end, the bar is attached to a second roller on the x-axis. Let P be the point at the free end of the bar and let θ be the angle the bar makes with the positive x-axis.

a) Find parametric equations for the path of P in terms of the parameter θ.

b) Find an equation in x and y whose graph is the path of P, and identify this path.

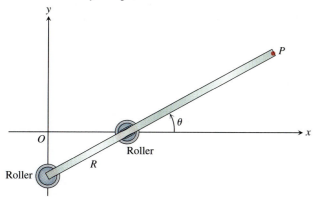

33. *Hypocycloids.* When a circle rolls on the inside of a fixed circle, any point P on the circumference of the rolling circle describes a *hypocycloid*. Let the fixed circle be $x^2 + y^2 = a^2$, let the radius of the rolling circle be b, and let the initial position of the tracing point P be $A(a, 0)$. Find parametric equations for the hypocycloid, using as the parameter the angle θ from the positive x-axis to the line joining the circles' centers. In particular, if $b = a/4$, as in the accompanying figure, show that the hypocycloid is the astroid

$$x = a \cos^3 \theta, \quad y = a \sin^3 \theta.$$

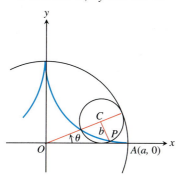

34. *More about hypocycloids.* The accompanying figure shows a circle of radius a tangent to the inside of a circle of radius $2a$. The point P, shown as the point of tangency in the figure, is attached to the smaller circle. What path does P trace as the smaller circle rolls around the inside of the larger circle?

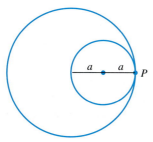

35. As the point N moves along the line $y = a$ in the accompanying figure, P moves in such a way that $OP = MN$. Find parametric equations for the coordinates of P as functions of the angle t that the line ON makes with the positive y-axis.

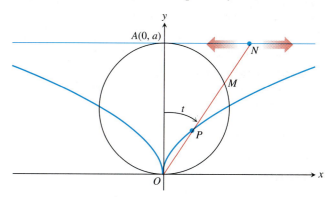

36. *Trochoids.* A wheel of radius a rolls along a horizontal straight line without slipping. Find parametric equations for the curve traced out by a point P on a spoke of the wheel b units from its center. As parameter, use the angle θ through which the wheel turns. The curve is called a **trochoid,** which is a cycloid when $b = a$.

Distance Using Parametric Equations

37. Find the point on the parabola $x = t$, $y = t^2$, $-\infty < t < \infty$, closest to the point $(2, 1/2)$. (*Hint:* Minimize the square of the distance as a function of t.)

38. Find the point on the ellipse $x = 2 \cos t$, $y = \sin t$, $0 \le t \le 2\pi$ closest to the point $(3/4, 0)$. (*Hint:* Minimize the square of the distance as a function of t.)

▦ Grapher Explorations

If you have a parametric equation grapher, graph the following equations over the given intervals.

39. *Ellipse.* $x = 4 \cos t$, $y = 2 \sin t$, over

a) $0 \le t \le 2\pi$ **b)** $0 \le t \le \pi$

c) $-\pi/2 \le t \le \pi/2$.

40. *Hyperbola branch.* $x = \sec t$ (enter as $1/\cos(t)$), $y = \tan t$ (enter as $\sin(t)/\cos(t)$), over

a) $-1.5 \le t \le 1.5$ **b)** $-0.5 \le t \le 0.5$

c) $-0.1 \le t \le 0.1$.

41. *Parabola.* $x = 2t + 3$, $y = t^2 - 1$, $-2 \le t \le 2$

42. *Cycloid.* $x = t - \sin t$, $y = 1 - \cos t$, over

a) $0 \le t \le 2\pi$ **b)** $0 \le t \le 4\pi$

c) $\pi \le t \le 3\pi$.

43. *Astroid.* $x = \cos^3 t$, $y = \sin^3 t$, over

a) $0 \le t \le 2\pi$ **b)** $-\pi/2 \le t \le \pi/2$.

44. *A nice curve (a deltoid)*

$$x = 2 \cos t + \cos 2t, \quad y = 2 \sin t - \sin 2t, \quad 0 \le t \le 2\pi$$

What happens if you replace 2 with -2 in the equations for x and y? Graph the new equations and find out.

45. *An even nicer curve*

$$x = 3 \cos t + \cos 3t, \quad y = 3 \sin t - \sin 3t, \quad 0 \le t \le 2\pi$$

What happens if you replace 3 with -3 in the equations for x and y? Graph the new equations and find out.

46. *Projectile motion.* Graph

$$x = (64 \cos \alpha) t, \quad y = -16t^2 + (64 \sin \alpha) t, \quad 0 \le t \le 4 \sin \alpha$$

for the following firing angles.

a) $\alpha = \pi/4$ **b)** $\alpha = \pi/6$ **c)** $\alpha = \pi/3$
d) $\alpha = \pi/2$ (watch out—here it comes!)

47. *Three beautiful curves*

a) *Epicycloid:*

$$x = 9 \cos t - \cos 9t, \quad y = 9 \sin t - \sin 9t, \quad 0 \le t \le 2\pi$$

b) *Hypocycloid:*

$$x = 8 \cos t + 2 \cos 4t, \quad y = 8 \sin t - 2 \sin 4t, \quad 0 \le t \le 2\pi$$

c) *Hypotrochoid:*

$$x = \cos t + 5 \cos 3t, \quad y = 6 \cos t - 5 \sin 3t, \quad 0 \le t \le 2\pi$$

48. *More beautiful curves*

a) $x = 6 \cos t + 5 \cos 3t, \quad y = 6 \sin t - 5 \sin 3t,$
$0 \le t \le 2\pi$

b) $x = 6 \cos 2t + 5 \cos 6t, \quad y = 6 \sin 2t - 5 \sin 6t,$
$0 \le t \le \pi$

c) $x = 6 \cos t + 5 \cos 3t, \quad y = 6 \sin 2t - 5 \sin 3t,$
$0 \le t \le 2\pi$

d) $x = 6 \cos 2t + 5 \cos 6t, \quad y = 6 \sin 4t - 5 \sin 6t,$
$0 \le t \le \pi$

9.5   Calculus with Parametrized Curves

This section shows how to find slopes, lengths, and surface areas associated with parametrized curves.

Slopes of Parametrized Curves

> **Definitions**
>
> A parametrized curve $x = f(t)$, $y = g(t)$ is **differentiable at $t = t_0$** if f and g are differentiable at $t = t_0$. The curve is **differentiable** if it is differentiable at every parameter value. The curve is **smooth** if f' and g' are continuous and not simultaneously zero.

At a point on a differentiable parametrized curve where y is also a differentiable function of x, the derivatives dx/dt, dy/dt, and dy/dx are related by the Chain Rule equation

$$\frac{dy}{dt} = \frac{dy}{dx} \frac{dx}{dt}.$$

If $dx/dt \ne 0$, we may divide both sides of this equation by dx/dt to solve for dy/dx.

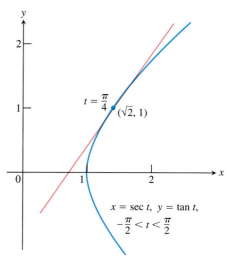

9.41 The hyperbola branch in Example 1.

Formula for Finding dy/dx from dy/dt and dx/dt ($dx/dt \neq 0$)

$$\frac{dy}{dx} = \frac{dy/dt}{dx/dt} \qquad (1)$$

EXAMPLE 1   Find the tangent to the right-hand hyperbola branch

$$x = \sec t, \qquad y = \tan t, \qquad -\frac{\pi}{2} < t < \frac{\pi}{2},$$

at the point $(\sqrt{2}, 1)$, where $t = \pi/4$ (Fig. 9.41).

Solution   The slope of the curve at t is

$$\frac{dy}{dx} = \frac{dy/dt}{dx/dt} = \frac{\sec^2 t}{\sec t \tan t} = \frac{\sec t}{\tan t}. \qquad \text{Eq. (1)}$$

Setting t equal to $\pi/4$ gives

$$\left. \frac{dy}{dx} \right|_{t=\pi/4} = \frac{\sec (\pi/4)}{\tan (\pi/4)}$$

$$= \frac{\sqrt{2}}{1} = \sqrt{2}.$$

The point–slope equation of the tangent is

$$y - y_0 = m \, (x - x_0)$$
$$y - 1 = \sqrt{2} \, (x - \sqrt{2})$$
$$y = \sqrt{2}x - 2 + 1$$
$$y = \sqrt{2}x - 1. \qquad \square$$

The Parametric Formula for d^2y/dx^2

If the parametric equations for a curve define y as a twice-differentiable function of x, we may calculate d^2y/dx^2 as a function of t in the following way:

$$\frac{d^2y}{dx^2} = \frac{d}{dx} \, (y') = \frac{dy'/dt}{dx/dt}. \qquad \text{Eq. (1) with } y \text{ replaced by } y'$$

Notice the lack of symmetry in Eq. (2). To find d^2y/dx^2, we divide the derivative of y' by the derivative of x, not by the derivative of x'.

How to Express d^2y/dx^2 in Terms of t

Step 1: Express $y' = dy/dx$ in terms of t.

Step 2: Find dy'/dt.

Step 3: Divide dy'/dt by dx/dt. The quotient is d^2y/dx^2.

Formula for Finding d^2y/dx^2 from $y' = dy/dx$ and dx/dt ($dx/dt \neq 0$)

$$\frac{d^2y}{dx^2} = \frac{dy'/dt}{dx/dt} \qquad (2)$$

EXAMPLE 2   Find d^2y/dx^2 if $x = t - t^2$ and $y = t - t^3$.

Solution

Step 1: *Express y' in terms of t:*

$$y' = \frac{dy}{dx} = \frac{dy/dt}{dx/dt} = \frac{1 - 3t^2}{1 - 2t} \qquad \text{Eq. (1) with } x = t - t^2, \; y = t - t^3$$

Step 2: *Differentiate y' with respect to t:*

$$\frac{dy'}{dt} = \frac{d}{dt}\left(\frac{1 - 3t^2}{1 - 2t}\right)$$

$$= \frac{2 - 6t + 6t^2}{(1 - 2t)^2}$$

Step 3: *Divide dy'/dt by dx/dt.* Since

$$\frac{dx}{dt} = \frac{d}{dt}(t - t^2) = 1 - 2t, \qquad x = t - t^2$$

we have

$$\frac{d^2y}{dx^2} = \frac{dy'/dt}{dx/dt} \qquad \text{Eq. (2)}$$

$$= \frac{2 - 6t + 6t^2}{(1 - 2t)^2} \cdot \frac{1}{1 - 2t}$$

$$= \frac{2 - 6t + 6t^2}{(1 - 2t)^3}.$$ ❑

Lengths of Parametrized Curves. Centroids

We find an integral for the length of a smooth curve $x = f(t), y = g(t), a \le t \le b$, by rewriting the integral $L = \int ds$ from Section 5.5 in the following way:

$$L = \int_{t=a}^{t=b} ds = \int_a^b \sqrt{dx^2 + dy^2}$$

$$= \int_a^b \sqrt{\left(\frac{(dx)^2}{(dt)^2} + \frac{(dy)^2}{(dt)^2}\right)} \, dt^2 = \int_a^b \sqrt{\left(\frac{dx}{dt}\right)^2 + \left(\frac{dy}{dt}\right)^2} \, dt.$$

The only requirement besides the continuity of the integrand is that the point $P(x, y) = P(f(t), g(t))$ not trace any portion of the curve more than once as t moves from a to b.

Length

If a smooth curve $x = f(t), y = g(t), a \le t \le b$, is traversed exactly once as t increases from a to b, the curve's length is

$$L = \int_a^b \sqrt{\left(\frac{dx}{dt}\right)^2 + \left(\frac{dy}{dt}\right)^2} \, dt. \qquad (3)$$

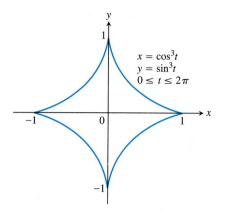

9.42 The astroid in Example 3.

The length formulas in Section 5.5 are special cases of Eq. (3) (Exercises 35 and 36).

What if there are two different parametrizations for a curve whose length we want to find—does it matter which one we use? The answer, from advanced calculus, is no, as long as the parametrization we choose meets the conditions preceding Eq. (3).

EXAMPLE 3 Find the length of the astroid (Fig. 9.42)

$$x = \cos^3 t, \qquad y = \sin^3 t, \qquad 0 \le t \le 2\pi.$$

Solution Because of the curve's symmetry with respect to the coordinate axes, its length is four times the length of the first-quadrant portion. We have

$$x = \cos^3 t, \qquad y = \sin^3 t$$

$$\left(\frac{dx}{dt}\right)^2 = [3\cos^2 t(-\sin t)]^2 = 9\cos^4 t \sin^2 t$$

$$\left(\frac{dy}{dt}\right)^2 = [3\sin^2 t(\cos t)]^2 = 9\sin^4 t \cos^2 t$$

$$\sqrt{\left(\frac{dx}{dt}\right)^2 + \left(\frac{dy}{dt}\right)^2} = \sqrt{9\cos^2 t \sin^2 t \underbrace{(\cos^2 t + \sin^2 t)}_{1}}$$

$$= \sqrt{9\cos^2 t \sin^2 t}$$

$$= 3|\cos t \sin t|$$

$$= 3\cos t \sin t. \qquad \text{\small $\cos t \sin t \ge 0$ for $0 \le t \le \pi/2$}$$

Therefore,

$$\text{Length of first-quadrant portion} = \int_0^{\pi/2} 3\cos t \sin t \, dt$$

$$= \frac{3}{2} \int_0^{\pi/2} \sin 2t \, dt \qquad \text{\small $\cos t \sin t = $} \atop \text{\small $(1/2)\sin 2t$}$$

$$= -\frac{3}{4} \cos 2t \Big]_0^{\pi/2} = \frac{3}{2}.$$

The length of the astroid is four times this: $4(3/2) = 6.$ ∎

EXAMPLE 4 Find the centroid of the first-quadrant arc of the astroid in Example 3.

Solution We take the curve's density to be $\delta = 1$ and calculate the curve's mass and moments about the coordinate axes as we did at the end of Section 5.7.

The distribution of mass is symmetric about the line $y = x$, so $\bar{x} = \bar{y}$. A typical segment of the curve (Fig. 9.43) has mass

$$dm = 1 \cdot ds = \sqrt{\left(\frac{dx}{dt}\right)^2 + \left(\frac{dy}{dt}\right)^2} \, dt = 3\cos t \sin t \, dt. \qquad \text{\small From} \atop \text{\small Example 3}$$

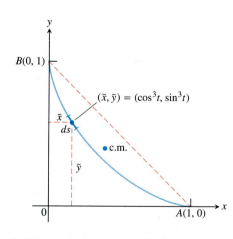

9.43 A typical segment of the arc in Example 4. The centroid (c.m.) of the curve lies about a third of the way toward chord *AB*.

The curve's mass is

$$M = \int_0^{\pi/2} dm = \int_0^{\pi/2} 3\cos t \, \sin t \, dt = \frac{3}{2}.$$

Again from Example 3

The curve's moment about the x-axis is

$$M_x = \int \tilde{y} \, dm = \int_0^{\pi/2} \sin^3 t \cdot 3 \cos t \sin t \, dt$$

$$= 3 \int_0^{\pi/2} \sin^4 t \cos t \, dt = 3 \cdot \frac{\sin^5 t}{5} \Big]_0^{\pi/2} = \frac{3}{5}.$$

Hence,

$$\overline{y} = \frac{M_x}{M} = \frac{3/5}{3/2} = \frac{2}{5}.$$

The centroid is the point (2/5, 2/5) (Fig. 9.43).

The Area of a Surface of Revolution

For smooth parametrized curves, the length formula in Eq. (3) leads to the following formulas for surfaces of revolution. The derivations are similar to the derivations of the Cartesian formulas in Section 5.6.

Surface Area

If a smooth curve $x = f(t)$, $y = g(t)$, $a \leq t \leq b$, is traversed exactly once as t increases from a to b, then the areas of the surfaces generated by revolving the curve about the coordinate axes are as follows.

1. Revolution about the x-axis $(y \geq 0)$:
$$S = \int_a^b 2\pi y \sqrt{\left(\frac{dx}{dt}\right)^2 + \left(\frac{dy}{dt}\right)^2} \, dt \qquad (4)$$

2. Revolution about the y-axis $(x \geq 0)$:
$$S = \int_a^b 2\pi x \sqrt{\left(\frac{dx}{dt}\right)^2 + \left(\frac{dy}{dt}\right)^2} \, dt \qquad (5)$$

As with length, we can calculate surface area from any convenient parametrization that meets the stated criteria.

EXAMPLE 5   The standard parametrization of the circle of radius 1 centered at the point (0, 1) in the xy-plane is

$$x = \cos t, \qquad y = 1 + \sin t, \qquad 0 \leq t \leq 2\pi.$$

Use this parametrization to find the area of the surface swept out by revolving the circle about the x-axis (Fig. 9.44).

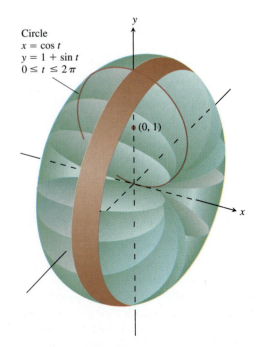

Circle
$x = \cos t$
$y = 1 + \sin t$
$0 \leq t \leq 2\pi$

$(0, 1)$

9.44 The surface in Example 5.

Solution We evaluate the formula

$$S = \int_a^b 2\pi\, y \sqrt{\left(\frac{dx}{dt}\right)^2 + \left(\frac{dy}{dt}\right)^2}\, dt \qquad \text{Eq. (4) for revolution about the } x\text{-axis}$$

$$= \int_0^{2\pi} 2\pi\, (1 + \sin t) \sqrt{\underbrace{(-\sin t)^2 + (\cos t)^2}_{1}}\, dt$$

$$= 2\pi \int_0^{2\pi} (1 + \sin t)\, dt$$

$$= 2\pi \Big[t - \cos t \Big]_0^{2\pi} = 4\pi^2.$$

Exercises 9.5

Tangents to Parametrized Curves

In Exercises 1–12, find an equation for the line tangent to the curve at the point defined by the given value of t. Also, find the value of $d^2 y/dx^2$ at this point.

1. $x = 2 \cos t, \quad y = 2 \sin t, \quad t = \pi/4$

2. $x = \sin 2\pi t, \quad y = \cos 2\pi t, \quad t = -1/6$

3. $x = 4 \sin t, \quad y = 2 \cos t, \quad t = \pi/4$

4. $x = \cos t, \quad y = \sqrt{3} \cos t, \quad t = 2\pi/3$

5. $x = t, \quad y = \sqrt{t}, \quad t = 1/4$

6. $x = \sec^2 t - 1, \quad y = \tan t, \quad t = -\pi/4$

7. $x = \sec t, \quad y = \tan t, \quad t = \pi/6$

8. $x = -\sqrt{t+1}, \quad y = \sqrt{3t}, \quad t = 3$

9. $x = 2t^2 + 3, \quad y = t^4, \quad t = -1$

10. $x = 1/t, \quad y = -2 + \ln t, \quad t = 1$

11. $x = t - \sin t, \quad y = 1 - \cos t, \quad t = \pi/3$

12. $x = \cos t, \quad y = 1 + \sin t, \quad t = \pi/2$

Implicitly Defined Parametrizations

Assuming that the equations in Exercises 13–16 define x and y implicitly as differentiable functions $x = f(t), y = g(t)$, find the slope of the curve $x = f(t), y = g(t)$ at the given value of t.

13. $x^2 - 2tx + 2t^2 = 4, \quad 2y^3 - 3t^2 = 4, \quad t = 2$

14. $x = \sqrt{5 - \sqrt{t}}, \quad y(t - 1) = \ln y, \quad t = 1$

15. $x + 2x^{3/2} = t^2 + t, \quad y\sqrt{t+1} + 2t\sqrt{y} = 4, \quad t = 0$

16. $x \sin t + 2x = t, \quad t \sin t - 2t = y, \quad t = \pi$

Lengths of Curves

Find the lengths of the curves in Exercises 17–22.

17. $x = \cos t, \quad y = t + \sin t, \quad 0 \le t \le \pi$

18. $x = t^3, \quad y = 3t^2/2, \quad 0 \le t \le \sqrt{3}$

19. $x = t^2/2, \quad y = (2t + 1)^{3/2}/3, \quad 0 \le t \le 4$

20. $x = (2t + 3)^{3/2}/3, \quad y = t + t^2/2, \quad 0 \le t \le 3$

21. $x = 8 \cos t + 8t \sin t$
    $y = 8 \sin t - 8t \cos t,$
    $0 \le t \le \pi/2$

22. $x = \ln (\sec t + \tan t) - \sin t$
    $y = \cos t, \quad 0 \le t \le \pi/3$

Surface Area

Find the areas of the surfaces generated by revolving the curves in Exercises 23–26 about the indicated axes.

23. $x = \cos t, \quad y = 2 + \sin t, \quad 0 \le t \le 2\pi; \quad x$-axis

24. $x = (2/3)t^{3/2}, \quad y = 2\sqrt{t}, \quad 0 \le t \le \sqrt{3}; \quad y$-axis

25. $x = t + \sqrt{2}, \quad y = (t^2/2) + \sqrt{2}\,t, \quad -\sqrt{2} \le t \le \sqrt{2}; \quad y$-axis

26. $x = \ln (\sec t + \tan t) - \sin t, \quad y = \cos t, \quad 0 \le t \le \pi/3;$
    x-axis

27. *A cone frustum.* The line segment joining the points $(0, 1)$ and $(2, 2)$ is revolved about the x-axis to generate a frustum of a cone. Find the surface area of the frustum using the parametrization $x = 2t, y = t + 1, 0 \le t \le 1$. Check your result with the geometry formula: Area $= \pi\, (r_1 + r_2)$(slant height).

28. *A cone.* The line segment joining the origin to the point (h, r) is revolved about the x-axis to generate a cone of height h and base radius r. Find the cone's surface area with the parametric equations $x = ht, y = rt, 0 \le t \le 1$. Check your result with the geometry formula: Area $= \pi r$(slant height).

Centroids

29. a) Find the coordinates of the centroid of the curve
$$x = \cos t + t \sin t, \quad y = \sin t - t \cos t, \quad 0 \le t \le \pi/2.$$

b) CALCULATOR The curve is a portion of the involute in Fig. 9.39. Sketch the curve. Find the centroid's coordinates to the nearest tenth and add the centroid to your sketch.

30. a) Find the coordinates of the centroid of the curve
$$x = e^t \cos t, \quad y = e^t \sin t, \quad 0 \le t \le \pi.$$

b) CALCULATOR Sketch the curve. Find the centroid's coordinates to the nearest tenth and add the centroid to your sketch.

31. a) Find the coordinates of the centroid of the curve
$$x = \cos t, \quad y = t + \sin t, \quad 0 \le t \le \pi.$$

b) Sketch the curve and add the centroid to your sketch.

32. INTEGRAL EVALUATOR Most centroid calculations for curves are done with a calculator or computer that has an integral evaluation program. As a case in point, find, to the nearest hundredth, the coordinates of the centroid of the curve
$$x = t^3, \quad y = 3t^2/2, \quad 0 \le t \le \sqrt{3}.$$

Theory and Examples

33. *Length is independent of parametrization.* To illustrate the fact that the numbers we get for length do not depend on the way we parametrize our curves (except for the mild restrictions mentioned earlier), calculate the length of the semicircle $y = \sqrt{1 - x^2}$ with these two different parametrizations:

a) $x = \cos 2t, \quad y = \sin 2t, \quad 0 \le t \le \pi/2$

b) $x = \sin \pi t, \quad y = \cos \pi t, \quad -1/2 \le t \le 1/2$

34. *Elliptic integrals.* The length of the ellipse
$$x = a \cos t, \quad y = b \sin t, \quad 0 \le t \le 2\pi$$
turns out to be
$$\text{Length} = 4a \int_0^{\pi/2} \sqrt{1 - e^2 \cos^2 t}\, dt,$$
where e is the ellipse's eccentricity. The integral in this formula, called an *elliptic integral*, is nonelementary except when $e = 0$ or 1.

a) CALCULATOR Use the trapezoidal rule with $n = 10$ to estimate the length of the ellipse when $a = 1$ and $e = 1/2$.

b) Use the fact that the absolute value of the second derivative of $f(t) = \sqrt{1 - e^2 \cos^2 t}$ is less than 1 to find an upper bound for the error in the estimate you obtained in (a).

35. As mentioned in Section 9.4, the graph of a function $y = f(x)$ over an interval $[a, b]$ automatically has the parametrization
$$x = x, \quad y = f(x), \quad a \le x \le b.$$
The parameter, in this case, is x itself.

Show that for this parametrization the parametric length formula
$$L = \int_a^b \sqrt{\left(\frac{dx}{dt}\right)^2 + \left(\frac{dy}{dt}\right)^2}\, dt$$
reduces to the Cartesian formula
$$L = \int_a^b \sqrt{1 + \left(\frac{dy}{dx}\right)^2}\, dx$$
derived in Section 5.5. This will show that the Cartesian formula is a special case of the parametric formula.

36. (*Continuation of Exercise 35.*) Show that the Cartesian formula
$$L = \int_c^d \sqrt{1 + \left(\frac{dx}{dy}\right)^2}\, dy$$
for the length of the curve $x = g(y), c \le y \le d$ (Section 5.5, Eq. 3), is a special case of the parametric length formula
$$L = \int_a^b \sqrt{\left(\frac{dx}{dt}\right)^2 + \left(\frac{dy}{dt}\right)^2}\, dt.$$

37. Find the area under one arch of the cycloid
$$x = a(\theta - \sin \theta), \quad y = a(1 - \cos \theta).$$
(*Hint:* Use $dx = (dx/d\theta)\, d\theta$.)

38. Find the length of one arch of the cycloid
$$x = a(\theta - \sin \theta), \quad y = a(1 - \cos \theta).$$

39. Find the area of the surface generated by revolving one arch of the cycloid $x = \theta - \sin \theta, y = 1 - \cos \theta$ about the x-axis.

40. Find the volume swept out by revolving the region bounded by the x-axis and one arch of the cycloid $x = \theta - \sin \theta, y = 1 - \cos \theta$ about the x-axis. (*Hint:* $dV = \pi y^2 \, dx = \pi y^2 \, (dx/d\theta)\, d\theta$.)

Grapher Explorations

The curves in Exercises 41 and 42 are called *Bowditch curves* or *Lissajous figures*. In each case, find the point in the interior of the first quadrant where the tangent to the curve is horizontal, and find the equations of the two tangents at the origin.

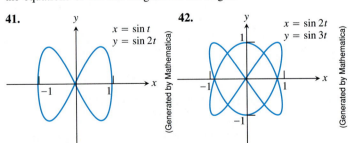

Graph the parametric curves in Exercises 43–49 over parameter intervals of your choice. The curves are Bowditch curves (Lissajous figures), the general formula being
$$x = a \sin(mt + d), \quad y = b \sin nt,$$
with m and n integers.

43. $x = \sin 2t, \quad y = \sin t$

44. $x = \sin 3t, \quad y = \sin 4t$

45. $x = \sin t, \quad y = \sin 4t$

46. $x = \sin t, \quad y = \sin 5t$

47. $x = \sin 3t, \quad y = \sin 5t$

48. $x = \sin(3t + \pi/2), \quad y = \sin 5t$

49. $x = \sin(3t + \pi/4), \quad y = \sin 5t$

✦ CAS Explorations and Projects

Use a CAS to perform the following steps on the parametrized curves in Exercises 50–55.

a) Plot the curve for the given interval of t values.

b) Find dy/dx and d^2y/dx^2 at the point t_0.

c) Find an equation for the tangent line to the curve at the point defined by the given value t_0. Plot the curve together with the tangent line on a single graph.

d) Find the length of the curve over the interval.

50. $x = \dfrac{1}{3}t^3, \quad y = \dfrac{1}{2}t^2, \quad 0 \le t \le 1, \quad t_0 = 1/2$

51. $x = 2t^3 - 16t^2 + 25t + 5, \quad y = t^2 + t - 3, \quad 0 \le t \le 6,$
$t_0 = 3/2$

52. $x = e^t - t^2, \quad y = t + e^{-t}, \quad -1 \le t \le 2, \quad t_0 = 1$

53. $x = t - \cos t, \quad y = 1 + \sin t, \quad -\pi \le t \le \pi, \quad t_0 = \pi/4$

54. $x = e^t + \sin 2t, \quad y = e^t + \cos(t^2), \quad -\sqrt{2}\pi \le t \le \pi/4,$
$t_0 = -\pi/4$

55. $x = e^t \cos t, \quad y = e^t \sin t, \quad 0 \le t \le \pi, \quad t_0 = \pi/2$

The equations in Exercises 56 and 57 define x and y implicitly as differentiable functions of t. Use a CAS to perform the following steps:

a) Solve the first equation for x and the second equation for y to find $x = f(t)$ and $y = g(t)$.

b) Find the slope of the curve $x = f(t)$ and $y = g(t)$ at t_0.

c) Find an equation for the tangent line to the curve at the point defined by t_0.

d) Plot the curve together with the tangent line over the specified interval of t-values.

56. $x^2 - 2tx + 3t^2 = 4, \quad y^3 - 2t^2 = 7, \quad -1 \le t \le 2, \quad t_0 = 1$

57. $x^2 \cos t + 2x = t, \quad t \sin t + 2\sqrt{y} = y, \quad -2\pi \le t \le 2\pi,$
$t_0 = -\pi/4$

9.6

Polar Coordinates

In this section, we study polar coordinates and their relation to Cartesian coordinates. While a point in the plane has just one pair of Cartesian coordinates, it has infinitely many pairs of polar coordinates. This has interesting consequences for graphing, as we will see in the next section.

Definition of Polar Coordinates

To define polar coordinates, we first fix an **origin** O (called the **pole**) and an **initial ray** from O (Fig. 9.45). Then each point P can be located by assigning to it a **polar coordinate pair** (r, θ) in which r gives the directed distance from O to P and θ gives the directed angle from the initial ray to ray OP.

9.45 To define polar coordinates for the plane, we start with an origin, called the pole, and an initial ray.

Polar Coordinates	(1)
$P(r, \theta)$	
Directed distance from O to P — Directed angle from initial ray to OP	

As in trigonometry, θ is positive when measured counterclockwise and negative when measured clockwise. The angle associated with a given point is not unique.

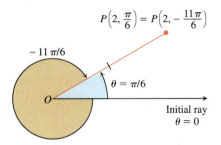

9.46 Polar coordinates are not unique.

For instance, the point 2 units from the origin along the ray $\theta = \pi/6$ has polar coordinates $r = 2, \theta = \pi/6$. It also has coordinates $r = 2, \theta = -11\pi/6$ (Fig. 9.46).

Negative Values of *r*

There are occasions when we wish to allow r to be negative. That is why we use directed distance in (1). The point $P(2, 7\pi/6)$ can be reached by turning $7\pi/6$ rad counterclockwise from the initial ray and going forward 2 units (Fig. 9.47). It can also be reached by turning $\pi/6$ rad counterclockwise from the initial ray and going *backward* 2 units. So the point also has polar coordinates $r = -2, \theta = \pi/6$.

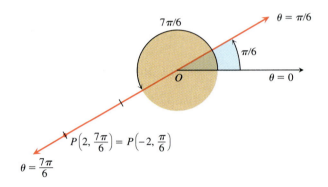

9.47 Polar coordinates can have negative *r*-values.

EXAMPLE 1 Find all the polar coordinates of the point $P(2, \pi/6)$.

Solution We sketch the initial ray of the coordinate system, draw the ray from the origin that makes an angle of $\pi/6$ rad with the initial ray, and mark the point $(2, \pi/6)$ (Fig. 9.48). We then find the angles for the other coordinate pairs of P in which $r = 2$ and $r = -2$.

For $r = 2$, the complete list of angles is

$$\frac{\pi}{6}, \quad \frac{\pi}{6} \pm 2\pi, \quad \frac{\pi}{6} \pm 4\pi, \quad \frac{\pi}{6} \pm 6\pi, \quad \dots .$$

For $r = -2$, the angles are

$$-\frac{5\pi}{6}, \quad -\frac{5\pi}{6} \pm 2\pi, \quad -\frac{5\pi}{6} \pm 4\pi, \quad -\frac{5\pi}{6} \pm 6\pi, \quad \dots .$$

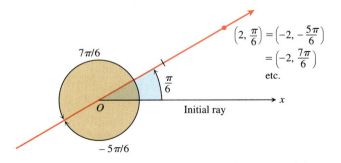

9.48 The point $P(2, \pi/6)$ has infinitely many polar coordinate pairs.

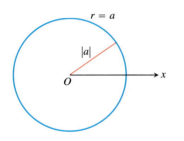

9.49 The polar equation for this circle is $r = a$.

The corresponding coordinate pairs of P are

$$\left(2, \frac{\pi}{6} + 2n\pi\right), \quad n = 0, \pm 1, \pm 2, \ldots$$

and

$$\left(-2, -\frac{5\pi}{6} + 2n\pi\right), \quad n = 0, \pm 1, \pm 2, \ldots.$$

When $n = 0$, the formulas give $(2, \pi/6)$ and $(-2, -5\pi/6)$. When $n = 1$, they give $(2, 13\pi/6)$ and $(-2, 7\pi/6)$, and so on. ❑

Elementary Coordinate Equations and Inequalities

If we hold r fixed at a constant value $r = a \neq 0$, the point $P(r, \theta)$ will lie $|a|$ units from the origin O. As θ varies over any interval of length 2π, P then traces a circle of radius $|a|$ centered at O (Fig. 9.49).

If we hold θ fixed at a constant value $\theta = \theta_0$ and let r vary between $-\infty$ and ∞, the point $P(r, \theta)$ traces the line through O that makes an angle of measure θ_0 with the initial ray.

(a)

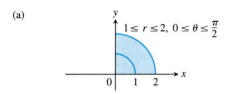

Equation	Graph
$r = a$	Circle of radius $|a|$ centered at O
$\theta = \theta_0$	Line through O making an angle θ_0 with the initial ray

(b)

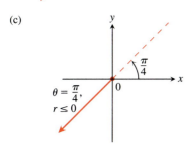

(c)

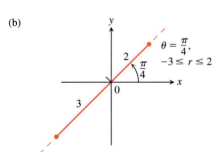

EXAMPLE 2

a) $r = 1$ and $r = -1$ are equations for the circle of radius 1 centered at O.

b) $\theta = \pi/6$, $\theta = 7\pi/6$, and $\theta = -5\pi/6$ are equations for the line in Fig. 9.48. ❑

Equations of the form $r = a$ and $\theta = \theta_0$ can be combined to define regions, segments, and rays.

EXAMPLE 3 Graph the sets of points whose polar coordinates satisfy the following conditions.

a) $1 \le r \le 2$ and $0 \le \theta \le \dfrac{\pi}{2}$

b) $-3 \le r \le 2$ and $\theta = \dfrac{\pi}{4}$

c) $r \le 0$ and $\theta = \dfrac{\pi}{4}$

d) $\dfrac{2\pi}{3} \le \theta \le \dfrac{5\pi}{6}$ (no restriction on r)

(d)

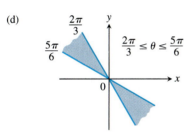

9.50 The graphs of typical inequalities in r and θ (Example 3).

Solution The graphs are shown in Fig. 9.50. ❑

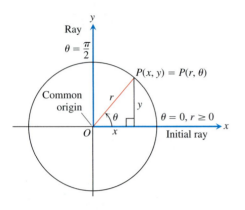

9.51 The usual way to relate polar and Cartesian coordinates.

Cartesian Versus Polar Coordinates

When we use both polar and Cartesian coordinates in a plane, we place the two origins together and take the initial polar ray as the positive x-axis. The ray $\theta = \pi/2$, $r > 0$, becomes the positive y-axis (Fig. 9.51). The two coordinate systems are then related by the following equations.

Equations Relating Polar and Cartesian Coordinates

$$x = r \cos \theta, \qquad y = r \sin \theta, \qquad x^2 + y^2 = r^2, \qquad \frac{y}{x} = \tan \theta \quad (2)$$

We use Eqs. (2) to rewrite polar equations in Cartesian form and vice versa.

EXAMPLE 4

Polar equation	Cartesian equivalent
$r \cos \theta = 2$	$x = 2$
$r^2 \cos \theta \sin \theta = 4$	$xy = 4$
$r^2 \cos^2 \theta - r^2 \sin^2 \theta = 1$	$x^2 - y^2 = 1$
$r = 1 + 2r \cos \theta$	$y^2 - 3x^2 - 4x - 1 = 0$
$r = 1 - \cos \theta$	$x^4 + y^4 + 2x^2y^2 + 2x^3 + 2xy^2 - y^2 = 0$

With some curves, we are better off with polar coordinates; with others, we aren't. ❏

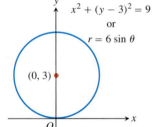

9.52 The circle in Example 5.

EXAMPLE 5   Find a polar equation for the circle $x^2 + (y - 3)^2 = 9$ (Fig. 9.52).

Solution

$$
\begin{aligned}
x^2 + y^2 - 6y + 9 &= 9 && \text{Expand } (y - 3)^2. \\
x^2 + y^2 - 6y &= 0 && \text{The 9's cancel.} \\
r^2 - 6r \sin \theta &= 0 && x^2 + y^2 = r^2 \\
r = 0 \quad \text{or} \quad r - 6 \sin \theta &= 0 \\
r &= 6 \sin \theta && \text{Includes both possibilities}
\end{aligned}
$$

We will say more about polar equations of conic sections in Section 9.8. ❏

EXAMPLE 6   Replace the following polar equations by equivalent Cartesian equations, and identify their graphs.

a) $r \cos \theta = -4$

b) $r^2 = 4r \cos \theta$

c) $r = \dfrac{4}{2 \cos \theta - \sin \theta}$

Solution   We use the substitutions $r \cos \theta = x$, $r \sin \theta = y$, $r^2 = x^2 + y^2$.

a) $r \cos \theta = -4$

The Cartesian equation: $r \cos \theta = -4$
$$x = -4$$
The graph: Vertical line through $x = -4$ on the x-axis

b) $r^2 = 4r \cos \theta$

The Cartesian equation: $r^2 = 4r \cos \theta$
$$x^2 + y^2 = 4x$$
$$x^2 - 4x + y^2 = 0$$
$$x^2 - 4x + 4 + y^2 = 4 \qquad \text{Completing the square}$$
$$(x - 2)^2 + y^2 = 4$$

The graph: Circle, radius 2, center $(h, k) = (2, 0)$

c) $r = \dfrac{4}{2 \cos \theta - \sin \theta}$

The Cartesian equation: $r(2 \cos \theta - \sin \theta) = 4$
$$2r \cos \theta - r \sin \theta = 4$$
$$2x - y = 4$$
$$y = 2x - 4$$

The graph: Line, slope $m = 2$, y-intercept $b = -4$

Exercises 9.6

Polar Coordinate Pairs

1. Which polar coordinate pairs label the same point?

   a) $(3, 0)$ **b)** $(-3, 0)$ **c)** $(2, 2\pi/3)$
   d) $(2, 7\pi/3)$ **e)** $(-3, \pi)$ **f)** $(2, \pi/3)$
   g) $(-3, 2\pi)$ **h)** $(-2, -\pi/3)$

2. Which polar coordinate pairs label the same point?

   a) $(-2, \pi/3)$ **b)** $(2, -\pi/3)$ **c)** (r, θ)
   d) $(r, \theta + \pi)$ **e)** $(-r, \theta)$ **f)** $(2, -2\pi/3)$
   g) $(-r, \theta + \pi)$ **h)** $(-2, 2\pi/3)$

3. Plot the following points (given in polar coordinates). Then find all the polar coordinates of each point.

   a) $(2, \pi/2)$ **b)** $(2, 0)$
   c) $(-2, \pi/2)$ **d)** $(-2, 0)$

4. Plot the following points (given in polar coordinates). Then find all the polar coordinates of each point.

   a) $(3, \pi/4)$ **b)** $(-3, \pi/4)$
   c) $(3, -\pi/4)$ **d)** $(-3, -\pi/4)$

Polar to Cartesian Coordinates

5. Find the Cartesian coordinates of the points in Exercise 1.

6. Find the Cartesian coordinates of the following points (given in polar coordinates).

   a) $\left(\sqrt{2}, \pi/4\right)$ **b)** $(1, 0)$

   c) $(0, \pi/2)$ **d)** $\left(-\sqrt{2}, \pi/4\right)$

   e) $(-3, 5\pi/6)$ **f)** $(5, \tan^{-1}(4/3))$

   g) $(-1, 7\pi)$ **h)** $\left(2\sqrt{3}, 2\pi/3\right)$

Graphing Polar Equations and Inequalities

Graph the sets of points whose polar coordinates satisfy the equations and inequalities in Exercises 7–22.

7. $r = 2$ 8. $0 \le r \le 2$

9. $r \ge 1$ 10. $1 \le r \le 2$

11. $0 \le \theta \le \pi/6, \quad r \ge 0$ 12. $\theta = 2\pi/3, \quad r \le -2$

13. $\theta = \pi/3, \quad -1 \le r \le 3$

14. $\theta = 11\pi/4, \quad r \geq -1$

15. $\theta = \pi/2, \quad r \geq 0$

16. $\theta = \pi/2, \quad r \leq 0$

17. $0 \leq \theta \leq \pi, \quad r = 1$

18. $0 \leq \theta \leq \pi, \quad r = -1$

19. $\pi/4 \leq \theta \leq 3\pi/4, \quad 0 \leq r \leq 1$

20. $-\pi/4 \leq \theta \leq \pi/4, \quad -1 \leq r \leq 1$

21. $-\pi/2 \leq \theta \leq \pi/2, \quad 1 \leq r \leq 2$

22. $0 \leq \theta \leq \pi/2, \quad 1 \leq |r| \leq 2$

Polar to Cartesian Equations

Replace the polar equations in Exercises 23–48 by equivalent Cartesian equations. Then describe or identify the graph.

23. $r \cos \theta = 2$

24. $r \sin \theta = -1$

25. $r \sin \theta = 0$

26. $r \cos \theta = 0$

27. $r = 4 \csc \theta$

28. $r = -3 \sec \theta$

29. $r \cos \theta + r \sin \theta = 1$

30. $r \sin \theta = r \cos \theta$

31. $r^2 = 1$

32. $r^2 = 4r \sin \theta$

33. $r = \dfrac{5}{\sin \theta - 2 \cos \theta}$

34. $r^2 \sin 2\theta = 2$

35. $r = \cot \theta \csc \theta$

36. $r = 4 \tan \theta \sec \theta$

37. $r = \csc \theta \, e^{r \cos \theta}$

38. $r \sin \theta = \ln r + \ln \cos \theta$

39. $r^2 + 2r^2 \cos \theta \sin \theta = 1$

40. $\cos^2 \theta = \sin^2 \theta$

41. $r^2 = -4r \cos \theta$

42. $r^2 = -6r \sin \theta$

43. $r = 8 \sin \theta$

44. $r = 3 \cos \theta$

45. $r = 2 \cos \theta + 2 \sin \theta$

46. $r = 2 \cos \theta - \sin \theta$

47. $r \sin \left(\theta + \dfrac{\pi}{6} \right) = 2$

48. $r \sin \left(\dfrac{2\pi}{3} - \theta \right) = 5$

Cartesian to Polar Equations

Replace the Cartesian equations in Exercises 49–62 by equivalent polar equations.

49. $x = 7$

50. $y = 1$

51. $x = y$

52. $x - y = 3$

53. $x^2 + y^2 = 4$

54. $x^2 - y^2 = 1$

55. $\dfrac{x^2}{9} + \dfrac{y^2}{4} = 1$

56. $xy = 2$

57. $y^2 = 4x$

58. $x^2 + xy + y^2 = 1$

59. $x^2 + (y - 2)^2 = 4$

60. $(x - 5)^2 + y^2 = 25$

61. $(x - 3)^2 + (y + 1)^2 = 4$

62. $(x + 2)^2 + (y - 5)^2 = 16$

Theory and Examples

63. Find all polar coordinates of the origin.

64. *Vertical and horizontal lines*

   a) Show that every vertical line in the xy-plane has a polar equation of the form $r = a \sec \theta$.

   b) Find the analogous polar equation for horizontal lines in the xy-plane.

| 9.7 | # Graphing in Polar Coordinates |

This section describes techniques for graphing equations in polar coordinates.

Symmetry

Figure 9.53 illustrates the standard polar coordinate tests for symmetry.

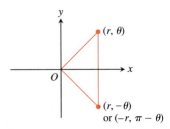

About the x-axis
(a)

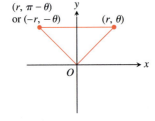

About the y-axis
(b)

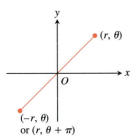

About the origin
(c)

9.53 Three tests for symmetry.

> **Symmetry Tests for Polar Graphs**
>
> 1. *Symmetry about the x-axis:* If the point (r, θ) lies on the graph, the point $(r, -\theta)$ or $(-r, \pi - \theta)$ lies on the graph (Fig. 9.53a).
> 2. *Symmetry about the y-axis:* If the point (r, θ) lies on the graph, the point $(r, \pi - \theta)$ or $(-r, -\theta)$ lies on the graph (Fig. 9.53b).
> 3. *Symmetry about the origin:* If the point (r, θ) lies on the graph, the point $(-r, \theta)$ or $(r, \theta + \pi)$ lies on the graph (Fig. 9.53c).

Slope

The slope of a polar curve $r = f(\theta)$ is given by dy/dx, not by $r' = df/d\theta$. To see why, think of the graph of f as the graph of the parametric equations

$$x = r \cos \theta = f(\theta) \cos \theta, \qquad y = r \sin \theta = f(\theta) \sin \theta.$$

If f is a differentiable function of θ, then so are x and y and, when $dx/d\theta \neq 0$, we can calculate dy/dx from the parametric formula

$$\frac{dy}{dx} = \frac{dy/d\theta}{dx/d\theta} \qquad \text{Section 9.5, Eq. (1) with } t = \theta$$

$$= \frac{\dfrac{d}{d\theta}(f(\theta) \cdot \sin \theta)}{\dfrac{d}{d\theta}(f(\theta) \cdot \cos \theta)}$$

$$= \frac{\dfrac{df}{d\theta} \sin \theta + f(\theta) \cos \theta}{\dfrac{df}{d\theta} \cos \theta - f(\theta) \sin \theta} \qquad \text{Product Rule for Derivatives}$$

> **Slope of the Curve $r = f(\theta)$**
>
> $$\left. \frac{dy}{dx} \right|_{(r,\theta)} = \frac{f'(\theta) \sin \theta + f(\theta) \cos \theta}{f'(\theta) \cos \theta - f(\theta) \sin \theta}, \qquad (1)$$
>
> provided $dx/d\theta \neq 0$ at (r, θ).

If the curve $r = f(\theta)$ passes through the origin at $\theta = \theta_0$, then $f(\theta_0) = 0$, and Eq. (1) gives

$$\left. \frac{dy}{dx} \right|_{(0, \theta_0)} = \frac{f'(\theta_0) \sin \theta_0}{f'(\theta_0) \cos \theta_0} = \tan \theta_0.$$

If the graph of $r = f(\theta)$ passes through the origin at the value $\theta = \theta_0$, the slope of the curve there is $\tan \theta_0$. The reason we say "slope at $(0, \theta_0)$" and not just "slope at the origin" is that a polar curve may pass through the origin more than once, with different slopes at different θ-values. This is not the case in our first example, however.

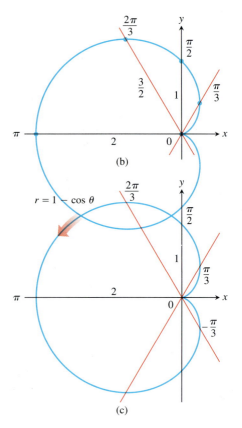

θ	$r = 1 - \cos \theta$
0	0
$\frac{\pi}{3}$	$\frac{1}{2}$
$\frac{\pi}{2}$	1
$\frac{2\pi}{3}$	$\frac{3}{2}$
π	2

(a)

(b)

$r = 1 - \cos \theta$

(c)

9.54 The steps in graphing the cardioid $r = 1 - \cos \theta$ (Example 1). The arrow shows the direction of increasing θ.

EXAMPLE 1   *A cardioid*

Graph the curve $r = 1 - \cos \theta$.

Solution   The curve is symmetric about the x-axis because

$$(r, \theta) \text{ on the graph} \;\Rightarrow\; r = 1 - \cos \theta$$
$$\Rightarrow\; r = 1 - \cos(-\theta) \qquad \cos \theta = \cos(-\theta)$$
$$\Rightarrow\; (r, -\theta) \text{ on the graph.}$$

As θ increases from 0 to π, $\cos \theta$ decreases from 1 to -1, and $r = 1 - \cos \theta$ increases from a minimum value of 0 to a maximum value of 2. As θ continues on from π to 2π, $\cos \theta$ increases from -1 back to 1 and r decreases from 2 back to 0. The curve starts to repeat when $\theta = 2\pi$ because the cosine has period 2π.

The curve leaves the origin with slope $\tan(0) = 0$ and returns to the origin with slope $\tan(2\pi) = 0$.

We make a table of values from $\theta = 0$ to $\theta = \pi$, plot the points, draw a smooth curve through them with a horizontal tangent at the origin, and reflect the curve across the x-axis to complete the graph (Fig. 9.54). The curve is called a *cardioid* because of its heart shape. Cardioid shapes appear in the cams that direct the even layering of thread on bobbins and reels, and in the signal-strength pattern of certain radio antennae. ❏

EXAMPLE 2   Graph the curve $r^2 = 4 \cos \theta$.

Solution   The equation $r^2 = 4 \cos \theta$ requires $\cos \theta \geq 0$, so we get the entire graph by running θ from $-\pi/2$ to $\pi/2$. The curve is symmetric about the x-axis because

$$(r, \theta) \text{ on the graph} \;\Rightarrow\; r^2 = 4 \cos \theta$$
$$\Rightarrow\; r^2 = 4 \cos(-\theta) \qquad \cos \theta = \cos(-\theta)$$
$$\Rightarrow\; (r, -\theta) \text{ on the graph.}$$

The curve is also symmetric about the origin because

$$(r, \theta) \text{ on the graph} \;\Rightarrow\; r^2 = 4 \cos \theta$$
$$\Rightarrow\; (-r)^2 = 4 \cos \theta$$
$$\Rightarrow\; (-r, \theta) \text{ on the graph.}$$

Together, these two symmetries imply symmetry about the y-axis.

The curve passes through the origin when $\theta = -\pi/2$ and $\theta = \pi/2$. It has a vertical tangent both times because $\tan \theta$ is infinite.

For each value of θ in the interval between $-\pi/2$ and $\pi/2$, the formula $r^2 = 4 \cos \theta$ gives two values of r:

$$r = \pm 2\sqrt{\cos \theta}.$$

We make a short table of values, plot the corresponding points, and use information about symmetry and tangents to guide us in connecting the points with a smooth curve (Fig. 9.55). ❏

θ	$\cos \theta$	$r = \pm 2\sqrt{\cos \theta}$
0	1	± 2
$\pm\dfrac{\pi}{6}$	$\dfrac{\sqrt{3}}{2}$	± 1.9
$\pm\dfrac{\pi}{4}$	$\dfrac{1}{\sqrt{2}}$	± 1.7
$\pm\dfrac{\pi}{3}$	$\dfrac{1}{2}$	± 1.4
$\pm\dfrac{\pi}{2}$	0	0

(a)

(b)

Loop for $r = -2\sqrt{\cos \theta}$, $-\dfrac{\pi}{2} \le \theta \le \dfrac{\pi}{2}$ Loop for $r = 2\sqrt{\cos \theta}$, $-\dfrac{\pi}{2} \le \theta \le \dfrac{\pi}{2}$

9.55 The graph of $r^2 = 4 \cos \theta$ (Example 2). The arrows show the direction of increasing θ. The values of r in the table are rounded.

Steps for Faster Graphing

Step 1: First graph $r = f(\theta)$ in the *Cartesian* $r\theta$-plane (that is, plot the values of θ on a horizontal axis and the corresponding values of r along a vertical axis).

Step 2: Then use the Cartesian graph as a "table" and guide to sketch the *polar* coordinate graph.

Faster Graphing

One way to graph a polar equation $r = f(\theta)$ is to make a table of (r, θ) values, plot the corresponding points, and connect them in order of increasing θ. This can work well if there are enough points to reveal all the loops and dimples in the graph. Here we describe another method of graphing that is usually quicker and more reliable. The steps are listed at left.

This method is better than simple point plotting because the Cartesian graph, even when hastily drawn, shows at a glance where r is positive, negative, and nonexistent, as well as where r is increasing and decreasing. As examples, we graph $r = 1 + \cos(\theta/2)$ and $r^2 = \sin 2\theta$.

EXAMPLE 3   Graph the curve

$$r = 1 + \cos \frac{\theta}{2}.$$

Solution We first graph r as a function of θ in the Cartesian $r\theta$-plane. Since the cosine has period 2π, we must let θ run from 0 to 4π to produce the entire graph (Fig. 9.56a, on the following page). The arrows from the θ-axis to the curve give radii for graphing $r = 1 + \cos(\theta/2)$ in the polar plane (Fig. 9.56b, on the following page). ❑

EXAMPLE 4   *A lemniscate*

Graph the curve $r^2 = \sin 2\theta$.

Solution Here we begin by plotting r^2 (not r) as a function of θ in the Cartesian $r^2\theta$-plane, treating r^2 as a variable that may have negative as well as positive values (Fig. 9.57a, on the following page). We pass from there to the graph of $r = \pm\sqrt{\sin 2\theta}$ in the $r\theta$-plane (Fig. 9.57b, on the following page) and then draw the polar graph (Fig. 9.57c, on the following page). The graph in Fig. 9.57(b) "covers" the final polar graph in Fig. 9.57(c) twice. We could have managed with either loop alone, with the two upper halves, or with the two lower halves. The double covering does no harm, however, and we learn a little more about the behavior of the function this way. ❑

DRAWING LESSON
How to Use Cartesian Graphs to Draw Polar Graphs

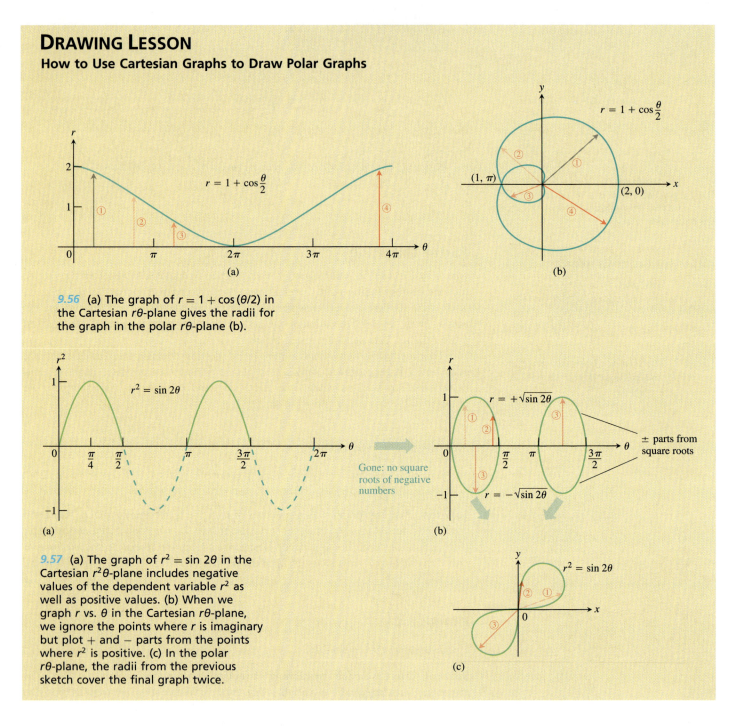

9.56 (a) The graph of $r = 1 + \cos(\theta/2)$ in the Cartesian $r\theta$-plane gives the radii for the graph in the polar $r\theta$-plane (b).

9.57 (a) The graph of $r^2 = \sin 2\theta$ in the Cartesian $r^2\theta$-plane includes negative values of the dependent variable r^2 as well as positive values. (b) When we graph r vs. θ in the Cartesian $r\theta$-plane, we ignore the points where r is imaginary but plot $+$ and $-$ parts from the points where r^2 is positive. (c) In the polar $r\theta$-plane, the radii from the previous sketch cover the final graph twice.

Finding Points Where Polar Graphs Intersect

The fact that we can represent a point in different ways in polar coordinates makes extra care necessary in deciding when a point lies on the graph of a polar equation and in determining the points in which polar graphs intersect. The problem is that a point of intersection may satisfy the equation of one curve with polar coordinates that are different from the ones with which it satisfies the equation of another curve.

Thus, solving the equations of two curves simultaneously may not identify all their points of intersection. The only sure way to identify all the points of intersection is to graph the equations.

EXAMPLE 5   *Deceptive coordinates*

Show that the point $(2, \pi/2)$ lies on the curve $r = 2 \cos 2\theta$.

Solution   It may seem at first that the point $(2, \pi/2)$ does not lie on the curve because substituting the given coordinates into the equation gives

$$2 = 2 \cos 2 \left(\frac{\pi}{2}\right) = 2 \cos \pi = -2,$$

which is not a true equality. The magnitude is right, but the sign is wrong. This suggests looking for a pair of coordinates for the given point in which r is negative, for example, $(-2, -(\pi/2))$. If we try these in the equation $r = 2 \cos 2\theta$, we find

$$-2 = 2 \cos 2 \left(-\frac{\pi}{2}\right) = 2(-1) = -2,$$

and the equation is satisfied. The point $(2, \pi/2)$ does lie on the curve.   ❏

EXAMPLE 6   *Elusive intersection points*

Find the points of intersection of the curves

$$r^2 = 4 \cos \theta \qquad \text{and} \qquad r = 1 - \cos \theta.$$

Solution   In Cartesian coordinates, we can always find the points where two curves cross by solving their equations simultaneously. In polar coordinates, the story is different. Simultaneous solution may reveal some intersection points without revealing others. In this example, simultaneous solution reveals only two of the four intersection points. The others are found by graphing. (Also, see Exercise 49.)

If we substitute $\cos \theta = r^2/4$ in the equation $r = 1 - \cos \theta$, we get

$$r = 1 - \cos \theta = 1 - \frac{r^2}{4}$$

$$4r = 4 - r^2$$

$$r^2 + 4r - 4 = 0$$

$$r = -2 \pm 2\sqrt{2}. \qquad \text{\textcolor{blue}{Quadratic formula}}$$

The value $r = -2 - 2\sqrt{2}$ has too large an absolute value to belong to either curve. The values of θ corresponding to $r = -2 + 2\sqrt{2}$ are

$$\theta = \cos^{-1}(1 - r) \qquad \text{\textcolor{blue}{From } } r = 1 - \cos \theta$$

$$= \cos^{-1}\left(1 - \left(2\sqrt{2} - 2\right)\right) \qquad \text{\textcolor{blue}{Set } } r = 2\sqrt{2} - 2.$$

$$= \cos^{-1}\left(3 - 2\sqrt{2}\right)$$

$$= \pm 80°. \qquad \text{\textcolor{blue}{Rounded to the nearest degree}}$$

We have thus identified two intersection points: $(r, \theta) = (2\sqrt{2} - 2, \pm 80°)$.

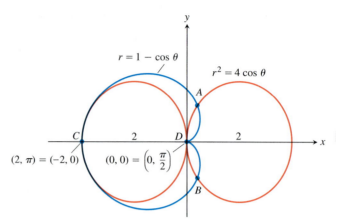

9.58 The four points of intersection of the curves $r = 1 - \cos \theta$ and $r^2 = 4 \cos \theta$ (Example 6). Only A and B were found by simultaneous solution. The other two were disclosed by graphing.

If we graph the equations $r^2 = 4 \cos \theta$ and $r = 1 - \cos \theta$ together (Fig. 9.58), as we can now do by combining the graphs in Figs. 9.54 and 9.55, we see that the curves also intersect at the point $(2, \pi)$ and the origin. Why weren't the r-values of these points revealed by the simultaneous solution? The answer is that the points $(0, 0)$ and $(2, \pi)$ are not on the curves "simultaneously." They are not reached at the same value of θ. On the curve $r = 1 - \cos \theta$, the point $(2, \pi)$ is reached when $\theta = \pi$. On the curve $r^2 = 4 \cos \theta$, it is reached when $\theta = 0$, where it is identified not by the coordinates $(2, \pi)$, which do not satisfy the equation, but by the coordinates $(-2, 0)$, which do. Similarly, the cardioid reaches the origin when $\theta = 0$, but the curve $r^2 = 4 \cos \theta$ reaches the origin when $\theta = \pi/2$. ❑

Technology *Finding Intersections* The *simultaneous mode* of a graphing utility gives new meaning to the *simultaneous solution* of a pair of polar coordinate equations. A simultaneous solution occurs only where the two graphs "collide" while they are being drawn simultaneously and not where one graph intersects the other at a point that had been illuminated earlier. The distinction is particularly important in the areas of traffic control or missile defense. For example, in traffic control the only issue is whether two aircraft are in the same place at the same time. The question of whether the curves the craft follow intersect is unimportant.

To illustrate, graph the polar equations

$$r = \cos 2\theta \qquad \text{and} \qquad r = \sin 2\theta$$

in simultaneous mode with $0 \le \theta < 2\pi$, θ Step $= 0.1$, and view dimensions [xmin, xmax] $= [-1, 1]$ by [ymin, ymax] $= [-1, 1]$. *While the graphs are being drawn on the screen,* count the number of times the two graphs illuminate a single pixel simultaneously. Explain why these points of intersection of the two graphs correspond to simultaneous solutions of the equations. (You may find it helpful to slow down the graphing by making θ Step smaller, say 0.05, for example.) In how many points total do the graphs actually intersect?

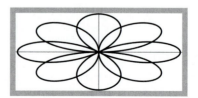

$r_1 = \sin 2\theta$ and $r_2 = \cos 2\theta$ graphed together.

Exercises 9.7

Symmetries and Polar Graphs

Identify the symmetries of the curves in Exercises 1–12. Then sketch the curves.

1. $r = 1 + \cos\theta$ **2.** $r = 2 - 2\cos\theta$

3. $r = 1 - \sin\theta$ **4.** $r = 1 + \sin\theta$

5. $r = 2 + \sin\theta$ **6.** $r = 1 + 2\sin\theta$

7. $r = \sin(\theta/2)$ **8.** $r = \cos(\theta/2)$

9. $r^2 = \cos\theta$ **10.** $r^2 = \sin\theta$

11. $r^2 = -\sin\theta$ **12.** $r^2 = -\cos\theta$

Graph the lemniscates in Exercises 13–16. What symmetries do these curves have?

13. $r^2 = 4\cos 2\theta$ **14.** $r^2 = 4\sin 2\theta$

15. $r^2 = -\sin 2\theta$ **16.** $r^2 = -\cos 2\theta$

Slopes of Polar Curves

Use Eq. (1) to find the slopes of the curves in Exercises 17–20 at the given points. Sketch the curves along with their tangents at these points.

17. *Cardioid.* $r = -1 + \cos\theta$; $\theta = \pm\pi/2$

18. *Cardioid.* $r = -1 + \sin\theta$; $\theta = 0, \pi$

19. *Four-leaved rose.* $r = \sin 2\theta$; $\theta = \pm\pi/4, \pm 3\pi/4$

20. *Four-leaved rose.* $r = \cos 2\theta$; $\theta = 0, \pm\pi/2, \pi$

Limaçons

Graph the limaçons in Exercises 21–24. Limaçon ("*lee*-ma-sahn") is Old French for "snail." You will understand the name when you graph the limaçons in Exercise 21. Equations for limaçons have the form $r = a \pm b\cos\theta$ or $r = a \pm b\sin\theta$. There are four basic shapes.

21. *Limaçons with an inner loop*

   a) $r = \dfrac{1}{2} + \cos\theta$ **b)** $r = \dfrac{1}{2} + \sin\theta$

22. *Cardioids*

   a) $r = 1 - \cos\theta$ **b)** $r = -1 + \sin\theta$

23. *Dimpled limaçons*

   a) $r = \dfrac{3}{2} + \cos\theta$ **b)** $r = \dfrac{3}{2} - \sin\theta$

24. *Oval limaçons*

   a) $r = 2 + \cos\theta$ **b)** $r = -2 + \sin\theta$

Graphing Polar Inequalities

25. Sketch the region defined by the inequalities $-1 \le r \le 2$ and $-\pi/2 \le \theta \le \pi/2$.

26. Sketch the region defined by the inequalities $0 \le r \le 2\sec\theta$ and $-\pi/4 \le \theta \le \pi/4$.

In Exercises 27 and 28, sketch the region defined by the inequality.

27. $0 \le r \le 2 - 2\cos\theta$ **28.** $0 \le r^2 \le \cos\theta$

Intersections

29. Show that the point $(2, 3\pi/4)$ lies on the curve $r = 2\sin 2\theta$.

30. Show that $(1/2, 3\pi/2)$ lies on the curve $r = -\sin(\theta/3)$.

Find the points of intersection of the pairs of curves in Exercises 31–38.

31. $r = 1 + \cos\theta$, $r = 1 - \cos\theta$

32. $r = 1 + \sin\theta$, $r = 1 - \sin\theta$

33. $r = 2\sin\theta$, $r = 2\sin 2\theta$

34. $r = \cos\theta$, $r = 1 - \cos\theta$

35. $r = \sqrt{2}$, $r^2 = 4\sin\theta$

36. $r^2 = \sqrt{2}\sin\theta$, $r^2 = \sqrt{2}\cos\theta$

37. $r = 1$, $r^2 = 2\sin 2\theta$

38. $r^2 = \sqrt{2}\cos 2\theta$, $r^2 = \sqrt{2}\sin 2\theta$

GRAPHER Find the points of intersection of the pairs of curves in Exercises 39–42.

39. $r^2 = \sin 2\theta$, $r^2 = \cos 2\theta$

40. $r = 1 + \cos\dfrac{\theta}{2}$, $r = 1 - \sin\dfrac{\theta}{2}$

41. $r = 1$, $r = 2\sin 2\theta$

42. $r = 1$, $r^2 = 2\sin 2\theta$

Grapher Explorations

43. Which of the following has the same graph as $r = 1 - \cos\theta$?

   a) $r = -1 - \cos\theta$
   b) $r = 1 + \cos\theta$

Confirm your answer with algebra.

44. Which of the following has the same graph as $r = \cos 2\theta$?

   a) $r = -\sin(2\theta + \pi/2)$
   b) $r = -\cos(\theta/2)$

Confirm your answer with algebra.

45. *A rose within a rose.* Graph the equation $r = 1 - 2\sin 3\theta$.

46. *The nephroid of Freeth.* Graph the nephroid of Freeth:

$$r = 1 + 2\sin\frac{\theta}{2}.$$

47. *Roses.* Graph the roses $r = \cos m\theta$ for $m = 1/3, 2, 3,$ and 7.

48. *Spirals.* Polar coordinates are just the thing for defining spirals. Graph the following spirals.

a) $r = \theta$
b) $r = -\theta$
c) *A logarithmic spiral:* $r = e^{\theta/10}$
d) *A hyperbolic spiral:* $r = 8/\theta$
e) *An equilateral hyperbola:* $r = \pm 10/\sqrt{\theta}$

(Use different colors for the two branches.)

Theory and Examples

49. (*Continuation of Example 6.*) The simultaneous solution of the equations

$$r^2 = 4 \cos \theta \tag{2}$$

$$r = 1 - \cos \theta \tag{3}$$

in the text did not reveal the points $(0, 0)$ and $(2, \pi)$ in which their graphs intersected.

a) We could have found the point $(2, \pi)$, however, by replacing the (r, θ) in Eq. (2) by the equivalent $(-r, \theta + \pi)$ to obtain

$$r^2 = 4 \cos \theta$$

$$(-r)^2 = 4 \cos (\theta + \pi) \tag{4}$$

$$r^2 = -4 \cos \theta.$$

Solve Eqs. (3) and (4) simultaneously to show that $(2, \pi)$ is a common solution. (This will still not reveal that the graphs intersect at $(0, 0)$.)

b) The origin is still a special case. (It often is.) Here is one way to handle it: Set $r = 0$ in Eqs. (2) and (3) and solve each equation for a corresponding value of θ. Since $(0, \theta)$ is the origin for *any* θ, this will show that both curves pass through the origin even if they do so for different θ-values.

50. If a curve has any two of the symmetries listed at the beginning of the section, can anything be said about its having or not having the third symmetry? Give reasons for your answer.

*51. Find the maximum width of the petal of the four-leaved rose $r = \cos 2\theta$, which lies along the x-axis.

*52. Find the maximum height above the x-axis of the cardioid $r = 2 (1 + \cos \theta)$.

| 9.8 | **Polar Equations for Conic Sections** |

Polar coordinates are important in astronomy and astronautical engineering because the ellipses, parabolas, and hyperbolas along which satellites, moons, planets, and comets move can all be described with a single relatively simple coordinate equation. We develop that equation here.

Lines

Suppose the perpendicular from the origin to line L meets L at the point $P_0(r_0, \theta_0)$, with $r_0 \geq 0$ (Fig. 9.59). Then, if $P(r, \theta)$ is any other point on L, the points P, P_0, and O are the vertices of a right triangle, from which we can read the relation

$$\frac{r_0}{r} = \cos (\theta - \theta_0)$$

or

$$r \cos (\theta - \theta_0) = r_0.$$

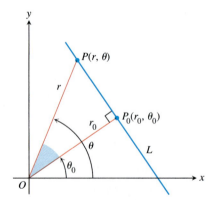

9.59 We can obtain a polar equation for line L by reading the relation $r_0 / r = \cos (\theta - \theta_0)$ from triangle $OP_0 P$.

The Standard Polar Equation for Lines

If the point $P_0(r_0, \theta_0)$ is the foot of the perpendicular from the origin to the line L, and $r_0 \geq 0$, then an equation for L is

$$r \cos (\theta - \theta_0) = r_0.$$

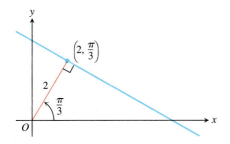

9.60 The standard polar equation of this line is

$$r \cos \left(\theta - \frac{\pi}{3} \right) = 2$$

(Example 1).

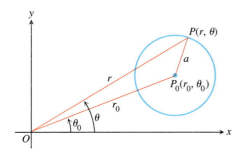

9.61 We can get a polar equation for this circle by applying the Law of Cosines to triangle OP_0P.

EXAMPLE 1   Use the identity $\cos (A - B) = \cos A \cos B + \sin A \sin B$ to find a Cartesian equation for the line in Fig. 9.60.

Solution

$$r \cos \left(\theta - \frac{\pi}{3} \right) = 2$$

$$r \left(\cos \theta \cos \frac{\pi}{3} + \sin \theta \sin \frac{\pi}{3} \right) = 2$$

$$\frac{1}{2} r \cos \theta + \frac{\sqrt{3}}{2} r \sin \theta = 2$$

$$\frac{1}{2} x + \frac{\sqrt{3}}{2} y = 2$$

$$x + \sqrt{3} y = 4 \qquad \Box$$

Circles

To find a polar equation for the circle of radius a centered at $P_0(r_0, \theta_0)$, we let $P(r, \theta)$ be a point on the circle and apply the Law of Cosines to triangle OP_0P (Fig. 9.61). This gives

$$a^2 = r_0^2 + r^2 - 2r_0r \cos (\theta - \theta_0). \qquad (1)$$

If the circle passes through the origin, then $r_0 = a$ and Eq. (1) simplifies to

$$a^2 = a^2 + r^2 - 2ar \cos (\theta - \theta_0) \qquad \text{Eq. (1) with } r_0 = a$$

$$r^2 = 2ar \cos (\theta - \theta_0)$$

$$r = 2a \cos (\theta - \theta_0). \qquad (2)$$

If the circle's center lies on the positive x-axis, $\theta_0 = 0$ and Eq. (2) becomes

$$r = 2a \cos \theta. \qquad (3)$$

If the center lies on the positive y-axis, $\theta = \pi/2$, $\cos (\theta - \pi/2) = \sin \theta$, and Eq. (2) becomes

$$r = 2a \sin \theta. \qquad (4)$$

Equations for circles through the origin centered on the negative x- and y-axes can be obtained from Eqs. (3) and (4) by replacing r with $-r$.

Polar Equations for Circles Through the Origin Centered on the x- and y-axes, Radius a

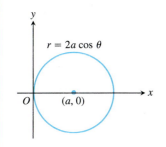

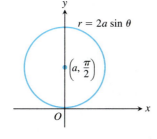

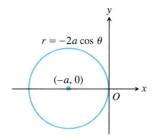

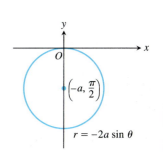

EXAMPLE 2   *Circles through the origin*

Radius	Center (polar coordinates)	Equation
3	$(3, 0)$	$r = 6 \cos \theta$
2	$(2, \pi/2)$	$r = 4 \sin \theta$
1/2	$(-1/2, 0)$	$r = -\cos \theta$
1	$(-1, \pi/2)$	$r = -2 \sin \theta$

Ellipses, Parabolas, and Hyperbolas Unified

To find polar equations for ellipses, parabolas, and hyperbolas, we place one focus at the origin and the corresponding directrix to the right of the origin along the vertical line $x = k$ (Fig. 9.62). This makes

$$PF = r$$

and

$$PD = k - FB = k - r \cos \theta.$$

The conic's focus–directrix equation $PF = e \cdot PD$ then becomes

$$r = e(k - r \cos \theta),$$

which can be solved for r to obtain

$$r = \frac{ke}{1 + e \cos \theta}. \tag{5}$$

This equation represents an ellipse if $0 < e < 1$, a parabola if $e = 1$, and a hyperbola if $e > 1$. And there we have it—ellipses, parabolas, and hyperbolas all with the same basic equation.

9.62 If a conic section is put in this position, then $PF = r$ and $PD = k - r \cos \theta$.

EXAMPLE 3   *Typical conics from Eq. (5)*

$$e = \frac{1}{2}: \quad \text{ellipse} \quad r = \frac{k}{2 + \cos \theta}$$

$$e = 1: \quad \text{parabola} \quad r = \frac{k}{1 + \cos \theta}$$

$$e = 2: \quad \text{hyperbola} \quad r = \frac{2k}{1 + 2 \cos \theta}$$

You may see variations of Eq. (5) from time to time, depending on the location of the directrix. If the directrix is the line $x = -k$ to the left of the origin (the origin is still a focus), we replace Eq. (5) by

$$r = \frac{ke}{1 - e \cos \theta}.$$

Table 9.4 Equations for conic sections ($e > 0$)

A.
$$r = \frac{ke}{1 + e \cos \theta}$$

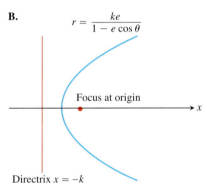

B.
$$r = \frac{ke}{1 - e \cos \theta}$$

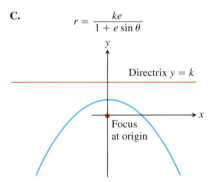

C.
$$r = \frac{ke}{1 + e \sin \theta}$$

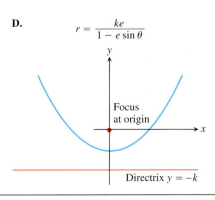

D.
$$r = \frac{ke}{1 - e \sin \theta}$$

The denominator now has a $(-)$ instead of a $(+)$. If the directrix is either of the lines $y = k$ or $y = -k$, the equations we get have sines in them instead of cosines, as shown in Table 9.4.

EXAMPLE 4   Find an equation for the hyperbola with eccentricity 3/2 and directrix $x = 2$.

Solution   We use Eq. (A) in Table 9.4 with $k = 2$ and $e = 3/2$ to get

$$r = \frac{2(3/2)}{1 + (3/2) \cos \theta} \qquad \text{or} \qquad r = \frac{6}{2 + 3 \cos \theta}.$$

EXAMPLE 5   Find the directrix of the parabola

$$r = \frac{25}{10 + 10 \cos \theta}.$$

Solution   We divide the numerator and denominator by 10 to put the equation in standard form:

$$r = \frac{5/2}{1 + \cos \theta}.$$

This is the equation

$$r = \frac{ke}{1 + e \cos \theta}$$

with $k = 5/2$ and $e = 1$. The equation of the directrix is $x = 5/2$.

From the ellipse diagram in Fig. 9.63, we see that k is related to the eccentricity e and the semimajor axis a by the equation

$$k = \frac{a}{e} - ea. \tag{6}$$

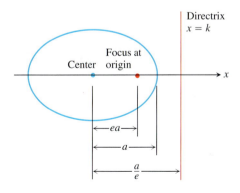

9.63 In an ellipse with semimajor axis a, the focus–directrix distance is $k = (a/e) - ea$, so $ke = a(1 - e^2)$.

From this, we find that $ke = a(1 - e^2)$. Replacing ke in Eq. (5) by $a(1 - e^2)$ gives the standard polar equation for an ellipse.

Ellipse with Eccentricity *e* and Semimajor Axis *a*

$$r = \frac{a(1 - e^2)}{1 + e \cos \theta} \qquad (7)$$

Notice that when $e = 0$, Eq. (7) becomes $r = a$, which represents a circle.
Equation (7) is the starting point for calculating planetary orbits.

EXAMPLE 6 Find a polar equation for an ellipse with semimajor axis 39.44 AU (astronomical units) and eccentricity 0.25. This is the approximate size of Pluto's orbit around the sun.

Solution We use Eq. (7) with $a = 39.44$ and $e = 0.25$ to find

$$r = \frac{39.44(1 - (0.25)^2)}{1 + 0.25 \cos \theta} = \frac{147.9}{4 + \cos \theta}.$$

At its point of closest approach (perihelion), Pluto is

$$r = \frac{147.9}{4 + 1} = 29.58 \text{ AU}$$

from the sun. At its most distant point (aphelion), Pluto is

$$r = \frac{147.9}{4 - 1} = 49.3 \text{ AU}$$

from the sun (Fig. 9.64).

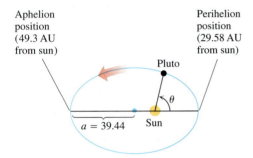

Aphelion position (49.3 AU from sun)

Perihelion position (29.58 AU from sun)

9.64 The orbit of Pluto (Example 6).

EXAMPLE 7 Find the distance from one focus of the ellipse in Example 6 to the associated directrix.

Solution We use Eq. (6) with $a = 39.44$ and $e = 0.25$ to find

$$k = 39.44 \left(\frac{1}{0.25} - 0.25 \right) = 147.9 \text{ AU}.$$

Exercises 9.8

Lines

Find polar and Cartesian equations for the lines in Exercises 1–4.

1.

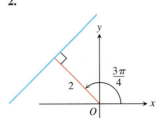

2.

3.

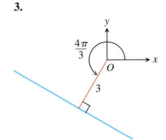

4.

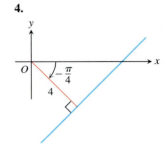

Sketch the lines in Exercises 5–8 and find Cartesian equations for them.

5. $r \cos\left(\theta - \frac{\pi}{4}\right) = \sqrt{2}$

6. $r \cos\left(\theta + \frac{3\pi}{4}\right) = 1$

7. $r \cos\left(\theta - \frac{2\pi}{3}\right) = 3$

8. $r \cos\left(\theta + \frac{\pi}{3}\right) = 2$

Find a polar equation in the form $r \cos(\theta - \theta_0) = r_0$ for each of the lines in Exercises 9–12.

9. $\sqrt{2}x + \sqrt{2}y = 6$

10. $\sqrt{3}x - y = 1$

11. $y = -5$

12. $x = -4$

Circles

Find polar equations for the circles in Exercises 13–16.

13.

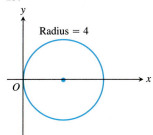

Radius = 4

14.

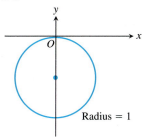

Radius = 1

15.

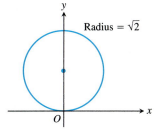

Radius = $\sqrt{2}$

16.

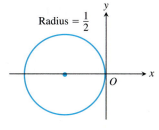

Radius = $\frac{1}{2}$

Sketch the circles in Exercises 17–20. Give polar coordinates for their centers and identify their radii.

17. $r = 4 \cos\theta$

18. $r = 6 \sin\theta$

19. $r = -2 \cos\theta$

20. $r = -8 \sin\theta$

Find polar equations for the circles in Exercises 21–28. Sketch each circle in the coordinate plane and label it with both its Cartesian and polar equations.

21. $(x - 6)^2 + y^2 = 36$

22. $(x + 2)^2 + y^2 = 4$

23. $x^2 + (y - 5)^2 = 25$

24. $x^2 + (y + 7)^2 = 49$

25. $x^2 + 2x + y^2 = 0$

26. $x^2 - 16x + y^2 = 0$

27. $x^2 + y^2 + y = 0$

28. $x^2 + y^2 - \frac{4}{3}y = 0$

Conic Sections from Eccentricities and Directrices

Exercises 29–36 give the eccentricities of conic sections with one focus at the origin, along with the directrix corresponding to that focus. Find a polar equation for each conic section.

29. $e = 1$, $x = 2$

30. $e = 1$, $y = 2$

31. $e = 5$, $y = -6$

32. $e = 2$, $x = 4$

33. $e = 1/2$, $x = 1$

34. $e = 1/4$, $x = -2$

35. $e = 1/5$, $y = -10$

36. $e = 1/3$, $y = 6$

Parabolas and Ellipses

Sketch the parabolas and ellipses in Exercises 37–44. Include the directrix that corresponds to the focus at the origin. Label the vertices with appropriate polar coordinates. Label the centers of the ellipses as well.

37. $r = \dfrac{1}{1 + \cos\theta}$

38. $r = \dfrac{6}{2 + \cos\theta}$

39. $r = \dfrac{25}{10 - 5\cos\theta}$

40. $r = \dfrac{4}{2 - 2\cos\theta}$

41. $r = \dfrac{400}{16 + 8\sin\theta}$

42. $r = \dfrac{12}{3 + 3\sin\theta}$

43. $r = \dfrac{8}{2 - 2\sin\theta}$

44. $r = \dfrac{4}{2 - \sin\theta}$

Graphing Inequalities

Sketch the regions defined by the inequalities in Exercises 45 and 46.

45. $0 \le r \le 2\cos\theta$

46. $-3\cos\theta \le r \le 0$

▓ Grapher Explorations

Graph the lines and conic sections in Exercises 47–56.

47. $r = 3\sec(\theta - \pi/3)$

48. $r = 4\sec(\theta + \pi/6)$

49. $r = 4\sin\theta$

50. $r = -2\cos\theta$

51. $r = 8/(4 + \cos\theta)$

52. $r = 8/(4 + \sin\theta)$

53. $r = 1/(1 - \sin\theta)$

54. $r = 1/(1 + \cos\theta)$

55. $r = 1/(1 + 2\sin\theta)$

56. $r = 1/(1 + 2\cos\theta)$

Theory and Examples

57. *Perihelion and aphelion.* A planet travels about its sun in an ellipse whose semimajor axis has length a.

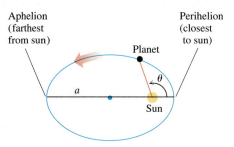

Aphelion (farthest from sun)

Perihelion (closest to sun)

Planet

Sun

a) Show that $r = a(1 - e)$ when the planet is closest to the sun and that $r = a(1 + e)$ when the planet is farthest from the sun.

b) Use the data in the table below to find how close each planet in our solar system comes to the sun and how far away each planet gets from the sun.

58. *Planetary orbits.* In Example 6, we found a polar equation for the orbit of Pluto. Use the data in the table below to find polar equations for the orbits of the other planets.

Planet	Semimajor axis (astronomical units)	Eccentricity
Mercury	0.3871	0.2056
Venus	0.7233	0.0068
Earth	1.000	0.0167
Mars	1.524	0.0934
Jupiter	5.203	0.0484
Saturn	9.539	0.0543
Uranus	19.18	0.0460
Neptune	30.06	0.0082
Pluto	39.44	0.2481

59. a) Find Cartesian equations for the curves $r = 4 \sin \theta$ and $r = \sqrt{3} \sec \theta$.

b) Sketch the curves together and label their points of intersection in both Cartesian and polar coordinates.

60. Repeat Exercise 59 for $r = 8 \cos \theta$ and $r = 2 \sec \theta$.

61. Find a polar equation for the parabola with focus (0, 0) and directrix $r \cos \theta = 4$.

62. Find a polar equation for the parabola with focus $(0, 0)$ and directrix $r \cos (\theta - \pi/2) = 2$.

63. a) *The space engineer's formula for eccentricity.* The space engineer's formula for the eccentricity of an elliptical orbit is

$$e = \frac{r_{max} - r_{min}}{r_{max} + r_{min}},$$

where r is the distance from the space vehicle to the attracting focus of the ellipse along which it travels. Why does the formula work?

b) *Drawing ellipses with string.* You have a string with a knot in each end that can be pinned to a drawing board. The string is 10 in. long from the center of one knot to the center of the other. How far apart should the pins be to

use the method illustrated in Fig. 9.5 (Section 9.1) to draw an ellipse of eccentricity 0.2? The resulting ellipse would resemble the orbit of Mercury.

64. *Halley's comet* (See Section 9.2, Example 1.)

a) Write an equation for the orbit of Halley's comet in a coordinate system in which the sun lies at the origin and the other focus lies on the negative x-axis, scaled in astronomical units.

b) How close does the comet come to the sun in astronomical units? in kilometers?

c) What is the farthest the comet gets from the sun in astronomical units? in kilometers?

In Exercises 65–68, find a polar equation for the given curve. In each case, sketch a typical curve.

65. $x^2 + y^2 - 2 ay = 0$

66. $y^2 = 4ax + 4a^2$

67. $x \cos \alpha + y \sin \alpha = p$ $(\alpha, p$ constant)

68. $(x^2 + y^2)^2 + 2 ax(x^2 + y^2) - a^2 y^2 = 0$

✿ CAS Explorations and Projects

69. Use a CAS to plot the polar equation

$$r = \frac{ke}{1 + e \cos \theta}$$

for various values of k and e, $-\pi \leq \theta \leq \pi$. Answer the following questions.

a) Take $k = -2$. Describe what happens to the plots as you take e to be 3/4, 1, and 5/4. Repeat for $k = 2$.

b) Take $k = -1$. Describe what happens to the plots as you take e to be 7/6, 5/4, 4/3, 3/2, 2, 3, 5, 10, and 20. Repeat for $e = 1/2, 1/3, 1/4, 1/10$, and 1/20.

c) Now keep $e > 0$ fixed and describe what happens as you take k to be $-1, -2, -3, -4$, and -5. Be sure to look at graphs for parabolas, ellipses, and hyperbolas.

70. Use a CAS to plot the polar ellipse

$$r = \frac{a(1 - e^2)}{1 + e \cos \theta}$$

for various values of $a > 0$ and $0 < e < 1$, $-\pi \leq \theta \leq \pi$.

a) Take $e = 9/10$. Describe what happens to the plots as you let a equal 1, 3/2, 2, 3, 5, and 10. Repeat with $e = 1/4$.

b) Take $a = 2$. Describe what happens as you take e to be 9/10, 8/10, 7/10, ..., 1/10, 1/20, and 1/50.

9.9

Integration in Polar Coordinates

This section shows how to calculate areas of plane regions, lengths of curves, and areas of surfaces of revolution in polar coordinates.

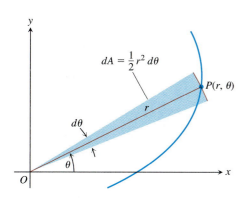

9.65 To derive a formula for the area of region *OTS*, we approximate the region with fan-shaped circular sectors.

Area in the Plane

The region OTS in Fig. 9.65 is bounded by the rays $\theta = \alpha$ and $\theta = \beta$ and the curve $r = f(\theta)$. We approximate the region with n nonoverlapping fan-shaped circular sectors based on a partition P of angle TOS. The typical sector has radius $r_k = f(\theta_k)$ and central angle of radian measure $\Delta\theta_k$. Its area is

$$A_k = \frac{1}{2} r_k^2 \Delta\theta_k = \frac{1}{2}\big(f(\theta_k)\big)^2 \Delta\theta_k.$$

The area of region OTS is approximately

$$\sum_{k=1}^{n} A_k = \sum_{k=1}^{n} \frac{1}{2}\big(f(\theta_k)\big)^2 \Delta\theta_k.$$

If f is continuous, we expect the approximations to improve as $\|P\| \to 0$, and we are led to the following formula for the region's area:

$$A = \lim_{\|P\| \to 0} \sum_{k=1}^{n} \frac{1}{2}\big(f(\theta_k)\big)^2 \Delta\theta_k$$

$$= \int_\alpha^\beta \frac{1}{2}\big(f(\theta)\big)^2 \, d\theta.$$

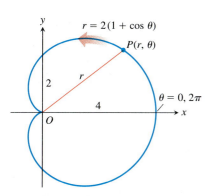

9.66 The area differential dA.

> **Area of the Fan-shaped Region Between the Origin and the Curve**
> $r = f(\theta), \quad \alpha \le \theta \le \beta$
>
> $$A = \int_\alpha^\beta \frac{1}{2} r^2 \, d\theta.$$
>
> This is the integral of the **area differential** (Fig. 9.66)
>
> $$dA = \frac{1}{2} r^2 \, d\theta.$$

EXAMPLE 1   Find the area of the region in the plane enclosed by the cardioid $r = 2(1 + \cos\theta)$.

Solution   We graph the cardioid (Fig. 9.67) and determine that the radius OP sweeps out the region exactly once as θ runs from 0 to 2π. The area is therefore

$$\int_{\theta=0}^{\theta=2\pi} \frac{1}{2} r^2 \, d\theta = \int_0^{2\pi} \frac{1}{2} \cdot 4(1 + \cos\theta)^2 \, d\theta$$

$$= \int_0^{2\pi} 2(1 + 2\cos\theta + \cos^2\theta) \, d\theta$$

$$= \int_0^{2\pi} \left(2 + 4\cos\theta + 2\,\frac{1+\cos 2\theta}{2}\right) d\theta$$

$$= \int_0^{2\pi} (3 + 4\cos\theta + \cos 2\theta) \, d\theta$$

$$= \left[3\theta + 4\sin\theta + \frac{\sin 2\theta}{2}\right]_0^{2\pi} = 6\pi - 0 = 6\pi.$$

9.67 The cardioid in Example 1.

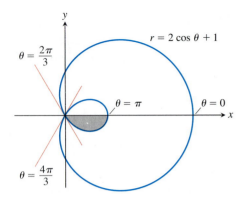

9.68 The limaçon in Example 2.

EXAMPLE 2 Find the area inside the smaller loop of the limaçon

$$r = 2 \cos \theta + 1.$$

Solution After sketching the curve (Fig. 9.68), we see that the smaller loop is traced out by the point (r, θ) as θ increases from $\theta = 2\pi/3$ to $\theta = 4\pi/3$. Since the curve is symmetric about the x-axis (the equation is unaltered when we replace θ by $-\theta$), we may calculate the area of the shaded half of the inner loop by integrating from $\theta = 2\pi/3$ to $\theta = \pi$. The area we seek will be twice the resulting integral:

$$A = 2 \int_{2\pi/3}^{\pi} \frac{1}{2} r^2 \, d\theta = \int_{2\pi/3}^{\pi} r^2 \, d\theta.$$

Since

$$r^2 = (2 \cos \theta + 1)^2 = 4 \cos^2 \theta + 4 \cos \theta + 1$$
$$= 4 \cdot \frac{1 + \cos 2\theta}{2} + 4 \cos \theta + 1$$
$$= 2 + 2 \cos 2\theta + 4 \cos \theta + 1$$
$$= 3 + 2 \cos 2\theta + 4 \cos \theta,$$

we have

$$A = \int_{2\pi/3}^{\pi} (3 + 2 \cos 2\theta + 4 \cos \theta) \, d\theta$$
$$= \left[3\theta + \sin 2\theta + 4 \sin \theta \right]_{2\pi/3}^{\pi}$$
$$= (3\pi) - \left(2\pi - \frac{\sqrt{3}}{2} + 4 \cdot \frac{\sqrt{3}}{2} \right)$$
$$= \pi - \frac{3\sqrt{3}}{2}.$$

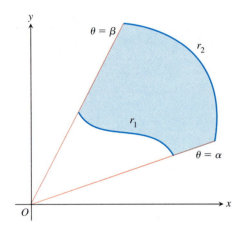

9.69 The area of the shaded region is calculated by subtracting the area of the region between r_1 and the origin from the area of the region between r_2 and the origin.

To find the area of a region like the one in Fig. 9.69, which lies between two polar curves $r_1 = r_1 (\theta)$ and $r_2 = r_2(\theta)$ from $\theta = \alpha$ to $\theta = \beta$, we subtract the integral of $(1/2)r_1^2 \, d\theta$ from the integral of $(1/2)r_2^2 \, d\theta$. This leads to the following formula.

Area of the Region $0 \le r_1(\theta) \le r \le r_2(\theta), \quad \alpha \le \theta \le \beta$

$$A = \int_{\alpha}^{\beta} \frac{1}{2} r_2^2 \, d\theta - \int_{\alpha}^{\beta} \frac{1}{2} r_1^2 \, d\theta = \int_{\alpha}^{\beta} \frac{1}{2} (r_2^2 - r_1^2) \, d\theta \qquad (1)$$

EXAMPLE 3 Find the area of the region that lies inside the circle $r = 1$ and outside the cardioid $r = 1 - \cos \theta$.

Solution We sketch the region to determine its boundaries and find the limits of integration (Fig. 9.70). The outer curve is $r_2 = 1$, the inner curve is $r_1 = 1 - \cos \theta$,

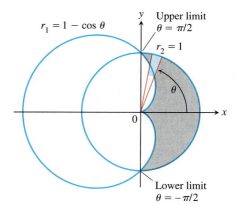

9.70 The region and limits of integration in Example 3.

and θ runs from $-\pi/2$ to $\pi/2$. The area, from Eq. (1), is

$$A = \int_{-\pi/2}^{\pi/2} \frac{1}{2}(r_2^2 - r_1^2)\,d\theta$$

$$= 2\int_0^{\pi/2} \frac{1}{2}(r_2^2 - r_1^2)\,d\theta \qquad \text{Symmetry}$$

$$= \int_0^{\pi/2} (1 - (1 - 2\cos\theta + \cos^2\theta))\,d\theta$$

$$= \int_0^{\pi/2} (2\cos\theta - \cos^2\theta)\,d\theta = \int_0^{\pi/2} \left(2\cos\theta - \frac{1 + \cos 2\theta}{2}\right)\,d\theta$$

$$= \left[2\sin\theta - \frac{\theta}{2} - \frac{\sin 2\theta}{4}\right]_0^{\pi/2} = 2 - \frac{\pi}{4}. \qquad \square$$

The Length of a Curve

We can obtain a polar coordinate formula for the length of a curve $r = f(\theta)$, $\alpha \le \theta \le \beta$, by parametrizing the curve as

$$x = r\cos\theta = f(\theta)\cos\theta, \qquad y = r\sin\theta = f(\theta)\sin\theta, \qquad \alpha \le \theta \le \beta. \qquad (2)$$

The parametric length formula, Eq. (3) from Section 9.5, then gives the length as

$$L = \int_\alpha^\beta \sqrt{\left(\frac{dx}{d\theta}\right)^2 + \left(\frac{dy}{d\theta}\right)^2}\,d\theta.$$

This equation becomes

$$L = \int_\alpha^\beta \sqrt{r^2 + \left(\frac{dr}{d\theta}\right)^2}\,d\theta$$

when Eqs. (2) are substituted for x and y (Exercise 33).

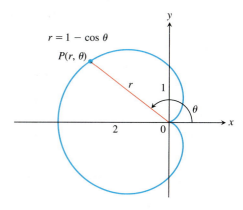

9.71 Example 4 calculates the length of this cardioid.

> **Length of a Curve**
>
> If $r = f(\theta)$ has a continuous first derivative for $\alpha \le \theta \le \beta$ and if the point $P(r, \theta)$ traces the curve $r = f(\theta)$ exactly once as θ runs from α to β, then the length of the curve is
>
> $$L = \int_\alpha^\beta \sqrt{r^2 + \left(\frac{dr}{d\theta}\right)^2}\,d\theta. \qquad (3)$$

EXAMPLE 4   Find the length of the cardioid $r = 1 - \cos\theta$.

Solution   We sketch the cardioid to determine the limits of integration (Fig. 9.71). The point $P(r, \theta)$ traces the curve once, counterclockwise as θ runs from 0 to 2π, so these are the values we take for α and β.

With

$$r = 1 - \cos\theta, \qquad \frac{dr}{d\theta} = \sin\theta,$$

we have

$$r^2 + \left(\frac{dr}{d\theta}\right)^2 = (1 - \cos\theta)^2 + (\sin\theta)^2$$

$$= 1 - 2\cos\theta + \underbrace{\cos^2\theta + \sin^2\theta}_{1} = 2 - 2\cos\theta$$

and

$$L = \int_\alpha^\beta \sqrt{r^2 + \left(\frac{dr}{d\theta}\right)^2}\, d\theta = \int_0^{2\pi} \sqrt{2 - 2\cos\theta}\, d\theta$$

$$= \int_0^{2\pi} \sqrt{4\sin^2\frac{\theta}{2}}\, d\theta \qquad 1 - \cos\theta = 2\sin^2\frac{\theta}{2}$$

$$= \int_0^{2\pi} 2\left|\sin\frac{\theta}{2}\right|\, d\theta$$

$$= \int_0^{2\pi} 2\sin\frac{\theta}{2}\, d\theta \qquad \sin\frac{\theta}{2} \geq 0 \quad \text{for} \quad 0 \leq \theta \leq 2\pi$$

$$= \left[-4\cos\frac{\theta}{2}\right]_0^{2\pi} = 4 + 4 = 8.$$

The Area of a Surface of Revolution

To derive polar coordinate formulas for the area of a surface of revolution, we parametrize the curve $r = f(\theta)$, $\alpha \leq \theta \leq \beta$, with Eqs. (2) and apply the surface area equations in Section 9.5.

Area of a Surface of Revolution

If $r = f(\theta)$ has a continuous first derivative for $\alpha \leq \theta \leq \beta$ and if the point $P(r, \theta)$ traces the curve $r = f(\theta)$ exactly once as θ runs from α to β, then the areas of the surfaces generated by revolving the curve about the x- and y-axes are given by the following formulas:

1. Revolution about the x-axis ($y \geq 0$):
$$S = \int_\alpha^\beta 2\pi r \sin\theta \sqrt{r^2 + \left(\frac{dr}{d\theta}\right)^2}\, d\theta \qquad (4)$$

2. Revolution about the y-axis ($x \geq 0$):
$$S = \int_\alpha^\beta 2\pi r \cos\theta \sqrt{r^2 + \left(\frac{dr}{d\theta}\right)^2}\, d\theta \qquad (5)$$

EXAMPLE 5   Find the area of the surface generated by revolving the right-hand loop of the lemniscate $r^2 = \cos 2\theta$ about the y-axis.

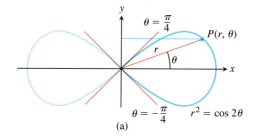

$\theta = \dfrac{\pi}{4}$

$P(r, \theta)$

r

θ

x

$\theta = -\dfrac{\pi}{4}$ $r^2 = \cos 2\theta$

(a)

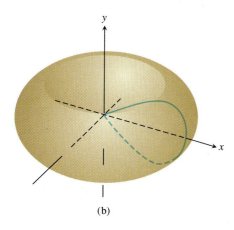

(b)

9.72 The right-hand half of a lemniscate (a) is revolved about the y-axis to generate a surface (b), whose area is calculated in Example 5.

Solution   We sketch the loop to determine the limits of integration (Fig. 9.72). The point $P(r, \theta)$ traces the curve once, counterclockwise as θ runs from $-\pi/4$ to $\pi/4$, so these are the values we take for α and β.

We evaluate the area integrand in Eq. (4) in stages. First,

$$2\pi r \cos\theta \sqrt{r^2 + \left(\frac{dr}{d\theta}\right)^2} = 2\pi \cos\theta \sqrt{r^4 + \left(r\frac{dr}{d\theta}\right)^2}. \tag{6}$$

Next, $r^2 = \cos 2\theta$, so

$$2r\frac{dr}{d\theta} = -2\sin 2\theta$$

$$r\frac{dr}{d\theta} = -\sin 2\theta$$

$$\left(r\frac{dr}{d\theta}\right)^2 = \sin^2 2\theta.$$

Finally, $r^4 = (r^2)^2 = \cos^2 2\theta$, so the square root on the right-hand side of Eq. (6) simplifies to

$$\sqrt{r^4 + \left(r\frac{dr}{d\theta}\right)^2} = \sqrt{\cos^2 2\theta + \sin^2 2\theta} = 1.$$

All together, we have

$$S = \int_\alpha^\beta 2\pi r \cos\theta \sqrt{r^2 + \left(\frac{dr}{d\theta}\right)^2}\, d\theta \qquad \text{Eq. (4)}$$

$$= \int_{-\pi/4}^{\pi/4} 2\pi \cos\theta \cdot (1)\, d\theta$$

$$= 2\pi \left[\sin\theta \right]_{-\pi/4}^{\pi/4}$$

$$= 2\pi \left[\frac{\sqrt{2}}{2} + \frac{\sqrt{2}}{2} \right] = 2\pi \sqrt{2}.$$

Exercises 9.9

Areas Inside Polar Curves

Find the areas of the regions in Exercises 1–6.

1. Inside the oval limaçon $r = 4 + 2\cos\theta$

2. Inside the cardioid $r = a(1 + \cos\theta), \quad a > 0$

3. Inside one leaf of the four-leaved rose $r = \cos 2\theta$

4. Inside the lemniscate $r^2 = 2a^2 \cos 2\theta, \quad a > 0$

5. Inside one loop of the lemniscate $r^2 = 4\sin 2\theta$

6. Inside the six-leaved rose $r^2 = 2\sin 3\theta$

Areas Shared by Polar Regions

Find the areas of the regions in Exercises 7–16.

7. Shared by the circles $r = 2\cos\theta$ and $r = 2\sin\theta$

8. Shared by the circles $r = 1$ and $r = 2\sin\theta$

9. Shared by the circle $r = 2$ and the cardioid $r = 2(1 - \cos\theta)$

10. Shared by the cardioids $r = 2(1 + \cos\theta)$ and $r = 2(1 - \cos\theta)$

11. Inside the lemniscate $r^2 = 6\cos 2\theta$ and outside the circle $r = \sqrt{3}$

12. Inside the circle $r = 3a \cos \theta$ and outside the cardioid $r = a(1 + \cos \theta)$, $a > 0$

13. Inside the circle $r = -2 \cos \theta$ and outside the circle $r = 1$

14. a) Inside the outer loop of the limaçon $r = 2 \cos \theta + 1$ (See Fig. 9.68.)

    b) Inside the outer loop and outside the inner loop of the limaçon $r = 2 \cos \theta + 1$

15. Inside the circle $r = 6$ above the line $r = 3 \csc \theta$

16. Inside the lemniscate $r^2 = 6 \cos 2\theta$ to the right of the line $r = (3/2) \sec \theta$

17. a) Find the area of the shaded region in Fig. 9.73.

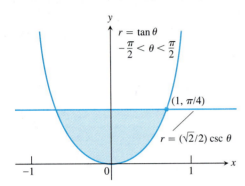

9.73 The region in Exercise 17.

    b) It looks as if the graph of $r = \tan \theta$, $-\pi/2 < \theta < \pi/2$, could be asymptotic to the lines $x = 1$ and $x = -1$. Is it? Give reasons for your answer.

18. The area of the region that lies inside the cardioid curve $r = \cos \theta + 1$ and outside the circle $r = \cos \theta$ is not

$$\frac{1}{2} \int_0^{2\pi} [(\cos \theta + 1)^2 - \cos^2 \theta] \, d\theta = \pi.$$

Why not? What *is* the area? Give reasons for your answers.

Lengths of Polar Curves

Find the lengths of the curves in Exercises 19–27.

19. The spiral $r = \theta^2$, $\quad 0 \leq \theta \leq \sqrt{5}$

20. The spiral $r = e^\theta / \sqrt{2}$, $\quad 0 \leq \theta \leq \pi$

21. The cardioid $r = 1 + \cos \theta$

22. The curve $r = a \sin^2 (\theta/2)$, $\quad 0 \leq \theta \leq \pi$, $\quad a > 0$

23. The parabolic segment $r = 6/(1 + \cos \theta)$, $\quad 0 \leq \theta \leq \pi/2$

24. The parabolic segment $r = 2/(1 - \cos \theta)$, $\quad \pi/2 \leq \theta \leq \pi$

25. The curve $r = \cos^3 (\theta/3)$, $\quad 0 \leq \theta \leq \pi/4$

26. The curve $r = \sqrt{1 + \sin 2\theta}$, $\quad 0 \leq \theta \leq \pi \sqrt{2}$

27. The curve $r = \sqrt{1 + \cos 2\theta}$, $\quad 0 \leq \theta \leq \pi \sqrt{2}$

28. *Circumferences of circles.* As usual, when faced with a new formula, it is a good idea to try it on familiar objects to be sure it gives results consistent with past experience. Use the length formula in Eq. (3) to calculate the circumferences of the following circles ($a > 0$):

  a) $r = a$ **b)** $r = a \cos \theta$ **c)** $r = a \sin \theta$

Surface Area

Find the areas of the surfaces generated by revolving the curves in Exercises 29–32 about the indicated axes.

29. $r = \sqrt{\cos 2\theta}$, $\quad 0 \leq \theta \leq \pi/4$, $\quad y$-axis

30. $r = \sqrt{2} e^{\theta/2}$, $\quad 0 \leq \theta \leq \pi/2$, $\quad x$-axis

31. $r^2 = \cos 2\theta$, $\quad x$-axis

32. $r = 2a \cos \theta$, $a > 0$, $\quad y$-axis

Theory and Examples

33. *The length of the curve $r = f(\theta)$, $\alpha \leq \theta \leq \beta$.* Assuming that the necessary derivatives are continuous, show how the substitutions

$$x = f(\theta) \cos \theta, \quad y = f(\theta) \sin \theta$$

(Eqs. 2 in the text) transform

$$L = \int_\alpha^\beta \sqrt{\left(\frac{dx}{d\theta}\right)^2 + \left(\frac{dy}{d\theta}\right)^2} \, d\theta$$

into

$$L = \int_\alpha^\beta \sqrt{r^2 + \left(\frac{dr}{d\theta}\right)^2} \, d\theta.$$

34. *Average value.* If f is continuous, the average value of the polar coordinate r over the curve $r = f(\theta)$, $\alpha \leq \theta \leq \beta$, with respect to θ is given by the formula

$$r_{av} = \frac{1}{\beta - \alpha} \int_\alpha^\beta f(\theta) \, d\theta.$$

Use this formula to find the average value of r with respect to θ over the following curves ($a > 0$).

  a) The cardioid $r = a(1 - \cos \theta)$

  b) The circle $r = a$

  c) The circle $r = a \cos \theta$, $\quad -\pi/2 \leq \theta \leq \pi/2$

35. *$r = f(\theta)$ vs. $r = 2 f(\theta)$.* Can anything be said about the relative lengths of the curves $r = f(\theta)$, $\alpha \leq \theta \leq \beta$, and $r = 2 f(\theta)$, $\alpha \leq \theta \leq \beta$? Give reasons for your answer.

36. *$r = f(\theta)$ vs. $r = 2 f(\theta)$.* The curves $r = f(\theta)$, $\alpha \leq \theta \leq \beta$, and $r = 2 f(\theta)$, $\alpha \leq \theta \leq \beta$, are revolved about the x-axis to generate surfaces. Can anything be said about the relative areas of these surfaces? Give reasons for your answer.

Centroids of Fan-Shaped Regions

Since the centroid of a triangle is located on each median, two-thirds of the way from the vertex to the opposite base, the lever arm for the moment about the x-axis of the thin triangular region in Fig. 9.74 is about $(2/3)r \sin \theta$. Similarly, the lever arm for the moment of the triangular region about the y-axis is about $(2/3)r \cos \theta$. These approximations improve as $\Delta\theta \to 0$ and lead to the following formulas

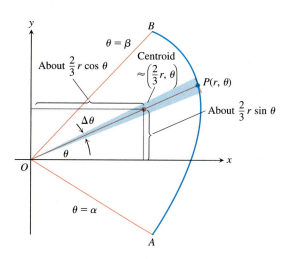

for the coordinates of the centroid of region AOB:

$$\bar{x} = \frac{\displaystyle\int \frac{2}{3}r\cos\theta \cdot \frac{1}{2}r^2\,d\theta}{\displaystyle\int \frac{1}{2}r^2\,d\theta} = \frac{\displaystyle\frac{2}{3}\int r^3\cos\theta\,d\theta}{\displaystyle\int r^2\,d\theta},$$

$$\bar{y} = \frac{\displaystyle\int \frac{2}{3}r\sin\theta \cdot \frac{1}{2}r^2\,d\theta}{\displaystyle\int \frac{1}{2}r^2\,d\theta} = \frac{\displaystyle\frac{2}{3}\int r^3\sin\theta\,d\theta}{\displaystyle\int r^2\,d\theta},$$

with limits $\theta = \alpha$ to $\theta = \beta$ on all integrals.

37. Find the centroid of the region enclosed by the cardioid $r = a(1 + \cos\theta)$.

38. Find the centroid of the semicircular region $0 \le r \le a$, $0 \le \theta \le \pi$.

9.74 The moment of the thin triangular sector about the x-axis is approximately

$$\frac{2}{3}r\sin\theta\,dA = \frac{2}{3}r\sin\theta \cdot \frac{1}{2}r^2 d\theta = \frac{1}{3}r^3\sin\theta\,d\theta.$$

Chapter 9   Questions to Guide Your Review

1. What is a parabola? What are the Cartesian equations for parabolas whose vertices lie at the origin and whose foci lie on the coordinate axes? How can you find the focus and directrix of such a parabola from its equation?

2. What is an ellipse? What are the Cartesian equations for ellipses centered at the origin with foci on one of the coordinate axes? How can you find the foci, vertices, and directrices of such an ellipse from its equation?

3. What is a hyperbola? What are the Cartesian equations for hyperbolas centered at the origin with foci on one of the coordinate axes? How can you find the foci, vertices, and directrices of such an ellipse from its equation?

4. What is the eccentricity of a conic section? How can you classify conic sections by eccentricity? How are an ellipse's shape and eccentricity related?

5. Explain the equation $PF = e \cdot PD$.

6. What is a quadratic curve in the xy-plane? Give examples of degenerate and nondegenerate quadratic curves.

7. How can you find a Cartesian coordinate system in which the new equation for a conic section in the plane has no xy-term? Give an example.

8. How can you tell what kind of graph to expect from a quadratic equation in x and y? Give examples.

9. What is a parametrized curve in the xy-plane? If you find a Cartesian equation for the path of a particle whose motion in the plane is described parametrically, what kind of match can you expect between the Cartesian equation's graph and the path of motion? Give examples.

10. What are some typical parametrizations for conic sections?

11. What is a cycloid? What are typical parametric equations for cycloids? What physical properties account for the importance of cycloids?

12. What is the formula for the slope dy/dx of a parametrized curve $x = f(t), y = g(t)$? When does the formula apply? When can you expect to be able to find d^2y/dx^2 as well? Give examples.

13. How do you find the length of a smooth parametrized curve $x = f(t), y = g(t), a \le t \le b$? What does smoothness have to do with length? What else do you need to know about the parametrization in order to find the curve's length? Give examples.

14. Under what conditions can you find the area of the surface generated by revolving a curve $x = f(t), y = g(t), a \le t \le b$, about the x-axis? the y-axis? Give examples.

15. How do you find the centroid of a smooth parametrized curve $x = f(t), y = g(t), a \le t \le b$? Give an example.

16. What are polar coordinates? What equations relate polar coordinates to Cartesian coordinates? Why might you want to change from one coordinate system to the other?

17. What consequence does the lack of uniqueness of polar coordinates have for graphing? Give an example.

18. How do you graph equations in polar coordinates? Include in your discussion symmetry, slope, behavior at the origin, and the use of Cartesian graphs. Give examples.

19. What are the standard equations for lines and conic sections in polar coordinates? Give examples.

20. How do you find the area of a region $0 \leq r_1(\theta) \leq r \leq r_2(\theta)$, $\alpha \leq \theta \leq \beta$, in the polar coordinate plane? Give examples.

21. Under what conditions can you find the length of a curve $r = f(\theta), \alpha \leq \theta \leq \beta$, in the polar coordinate plane? Give an example of a typical calculation.

22. Under what conditions can you find the area of the surface generated by revolving a curve $r = f(\theta), \alpha \leq \theta \leq \beta$, about the x-axis? the y-axis? Give examples of typical calculations.

CHAPTER 9 PRACTICE EXERCISES

Graphing Conic Sections

Sketch the parabolas in Exercises 1–4. Include the focus and directrix in each sketch.

1. $x^2 = -4y$ **2.** $x^2 = 2y$

3. $y^2 = 3x$ **4.** $y^2 = -(8/3)x$

Find the eccentricities of the ellipses and hyperbolas in Exercises 5–8. Sketch each conic section. Include the foci, vertices, and asymptotes (as appropriate) in your sketch.

5. $16x^2 + 7y^2 = 112$ **6.** $x^2 + 2y^2 = 4$

7. $3x^2 - y^2 = 3$ **8.** $5y^2 - 4x^2 = 20$

Shifting Conic Sections

Exercises 9–14 give equations for conic sections and tell how many units up or down and to the right or left each curve is to be shifted. Find an equation for the new conic section and find the new foci, vertices, centers, and asymptotes, as appropriate. If the curve is a parabola, find the new directrix as well.

9. $x^2 = -12y$,   right 2, up 3

10. $y^2 = 10x$,   left 1/2, down 1

11. $\dfrac{x^2}{9} + \dfrac{y^2}{25} = 1$,   left 3, down 5

12. $\dfrac{x^2}{169} + \dfrac{y^2}{144} = 1$,   right 5, up 12

13. $\dfrac{y^2}{8} - \dfrac{x^2}{2} = 1$,   right 2, up $2\sqrt{2}$

14. $\dfrac{x^2}{36} - \dfrac{y^2}{64} = 1$,   left 10, down 3

Identifying Conic Sections

Identify the conic sections in Exercises 15–22 and find their foci, vertices, centers, and asymptotes (as appropriate). If the curve is a parabola, find its directrix as well.

15. $x^2 - 4x - 4y^2 = 0$ **16.** $4x^2 - y^2 + 4y = 8$

17. $y^2 - 2y + 16x = -49$ **18.** $x^2 - 2x + 8y = -17$

19. $9x^2 + 16y^2 + 54x - 64y = -1$

20. $25x^2 + 9y^2 - 100x + 54y = 44$

21. $x^2 + y^2 - 2x - 2y = 0$

22. $x^2 + y^2 + 4x + 2y = 1$

Using the Discriminant

What conic sections or degenerate cases do the equations in Exercises 23–28 represent? Give a reason for your answer in each case.

23. $x^2 + xy + y^2 + x + y + 1 = 0$

24. $x^2 + 4xy + 4y^2 + x + y + 1 = 0$

25. $x^2 + 3xy + 2y^2 + x + y + 1 = 0$

26. $x^2 + 2xy - 2y^2 + x + y + 1 = 0$

27. $x^2 - 2xy + y^2 = 0$

28. $x^2 - 3xy + 4y^2 = 0$

Rotating Conic Sections

Identify the conic sections in Exercises 29–32. Then rotate the coordinate axes to find a new equation for the conic section that has no cross product term. (The new equations will vary with the size and direction of the rotations used.)

29. $2x^2 + xy + 2y^2 - 15 = 0$ **30.** $3x^2 + 2xy + 3y^2 = 19$

31. $x^2 + 2\sqrt{3}xy - y^2 + 4 = 0$

32. $x^2 - 3xy + y^2 = 5$

Identifying Parametric Equations in the Plane

Exercises 33–38 give parametric equations and parameter intervals for the motion of a particle in the xy-plane. Identify the particle's path by finding a Cartesian equation for it. Graph the Cartesian equation and indicate the direction of motion and the portion traced by the particle.

33. $x = t/2, \quad y = t + 1; \quad -\infty < t < \infty$

34. $x = \sqrt{t}, \quad y = 1 - \sqrt{t}; \quad t \geq 0$

35. $x = (1/2) \tan t, \quad y = (1/2) \sec t; \quad -\pi/2 < t < \pi/2$

36. $x = -2 \cos t, \quad y = 2 \sin t; \quad 0 \leq t \leq \pi$

37. $x = -\cos t, \quad y = \cos^2 t; \quad 0 \leq t \leq \pi$

38. $x = 4 \cos t, \quad y = 9 \sin t; \quad 0 \leq t \leq 2\pi$

Finding Parametric Equations and Tangent Lines

39. Find parametric equations and a parameter interval for the motion of a particle in the xy-plane that traces the ellipse $16x^2 + 9y^2 = 144$ once counterclockwise. (There are many ways to do this, so your answer may not be the same as the one in the back of the book.)

40. Find parametric equations and a parameter interval for the motion of a particle that starts at the point $(-2, 0)$ in the xy-plane and traces the circle $x^2 + y^2 = 4$ three times clockwise. (There are many ways to do this.)

In Exercises 41 and 42, find an equation for the line in the xy-plane that is tangent to the curve at the point corresponding to the given value of t. Also, find the value of d^2y/dx^2 at this point.

41. $x = (1/2) \tan t, \quad y = (1/2) \sec t; \quad t = \pi/3$

42. $x = 1 + 1/t^2, \quad y = 1 - 3/t; \quad t = 2$

Lengths of Parametrized Curves

Find the lengths of the curves in Exercises 43 and 44.

43. $x = e^{2t} - \dfrac{t}{8}, \quad y = e^t; \quad 0 \leq t \leq \ln 2$

44. The enclosed loop in Fig. 9.75.

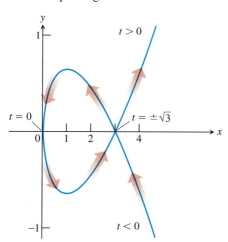

9.75 Exercise 44 refers to the curve $x = t^2, y = (t^3/3) - t$ shown here. The loop starts at $t = -\sqrt{3}$ and ends at $t = \sqrt{3}$.

Surface Areas

Find the areas of the surfaces generated by revolving the curves in Exercises 45 and 46 about the indicated axes.

45. $x = t^2/2, \quad y = 2t, \quad 0 \leq t \leq \sqrt{5}; \quad x$-axis

46. $x = t^2 + 1/(2t), \quad y = 4\sqrt{t}, \quad 1/\sqrt{2} \leq t \leq 1; \quad y$-axis

Graphs in the Polar Plane

Sketch the regions defined by the polar coordinate inequalities in Exercises 47 and 48.

47. $0 \leq r \leq 6 \cos \theta$

48. $-4 \sin \theta \leq r \leq 0$

Match each graph in Exercises 49–56 with the appropriate equation (a)–(l). There are more equations than graphs, so some equations will not be matched.

a) $r = \cos 2\theta$ **b)** $r \cos \theta = 1$

c) $r = \dfrac{6}{1 - 2 \cos \theta}$ **d)** $r = \sin 2\theta$

e) $r = \theta$ **f)** $r^2 = \cos 2\theta$

g) $r = 1 + \cos \theta$ **h)** $r = 1 - \sin \theta$

i) $r = \dfrac{2}{1 - \cos \theta}$ **j)** $r^2 = \sin 2\theta$

k) $r = -\sin \theta$ **l)** $r = 2 \cos \theta + 1$

49. Four-leaved rose

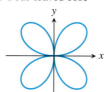

50. Spiral

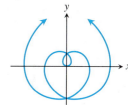

51. Limaçon

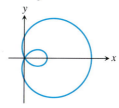

52. Lemniscate

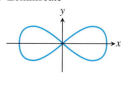

53. Circle

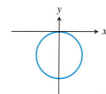

54. Cardioid

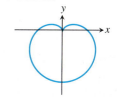

55. Parabola

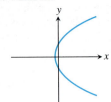

56. Lemniscate

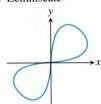

Intersections of Graphs in the Polar Plane

Find the points of intersection of the curves given by the polar coordinate equations in Exercises 57–64.

57. $r = \sin \theta, \quad r = 1 + \sin \theta$

58. $r = \cos \theta, \quad r = 1 - \cos \theta$

59. $r = 1 + \cos \theta, \quad r = 1 - \cos \theta$

60. $r = 1 + \sin \theta, \quad r = 1 - \sin \theta$

61. $r = 1 + \sin \theta, \quad r = -1 + \sin \theta$

62. $r = 1 + \cos \theta, \quad r = -1 + \cos \theta$

63. $r = \sec \theta, \quad r = 2 \sin \theta$

64. $r = -2 \csc \theta, \quad r = -4 \cos \theta$

Tangent Lines in the Polar Plane

In Exercises 65 and 66, find equations for the lines that are tangent to the polar coordinate curves at the origin.

65. The lemniscate $r^2 = \cos 2\theta$

66. The limaçon $r = 2 \cos \theta + 1$

67. Find polar coordinate equations for the lines that are tangent to the tips of the petals of the four-leaved rose $r = \sin 2\theta$.

68. Find polar coordinate equations for the lines that are tangent to the cardioid $r = 1 + \sin \theta$ at the points where it crosses the x-axis.

Polar to Cartesian Equations

Sketch the lines in Exercises 69–74. Also, find a Cartesian equation for each line.

69. $r \cos \left(\theta + \dfrac{\pi}{3} \right) = 2 \sqrt{3}$

70. $r \cos \left(\theta - \dfrac{3\pi}{4} \right) = \dfrac{\sqrt{2}}{2}$

71. $r = 2 \sec \theta$

72. $r = -\sqrt{2} \sec \theta$

73. $r = -(3/2) \csc \theta$

74. $r = \left(3\sqrt{3} \right) \csc \theta$

Find Cartesian equations for the circles in Exercises 75–78. Sketch each circle in the coordinate plane and label it with both its Cartesian and polar equations.

75. $r = -4 \sin \theta$

76. $r = 3\sqrt{3} \sin \theta$

77. $r = 2\sqrt{2} \cos \theta$

78. $r = -6 \cos \theta$

Cartesian to Polar Equations

Find polar equations for the circles in Exercises 79–82. Sketch each circle in the coordinate plane and label it with both its Cartesian and polar equations.

79. $x^2 + y^2 + 5y = 0$

80. $x^2 + y^2 - 2y = 0$

81. $x^2 + y^2 - 3x = 0$

82. $x^2 + y^2 + 4x = 0$

Conic Sections in Polar Coordinates

Sketch the conic sections whose polar coordinate equations are given in Exercises 83–86. Give polar coordinates for the vertices and, in the case of ellipses, for the centers as well.

83. $r = \dfrac{2}{1 + \cos \theta}$

84. $r = \dfrac{8}{2 + \cos \theta}$

85. $r = \dfrac{6}{1 - 2 \cos \theta}$

86. $r = \dfrac{12}{3 + \sin \theta}$

Exercises 87–90 give the eccentricities of conic sections with one focus at the origin of the polar coordinate plane, along with the directrix for that focus. Find a polar equation for each conic section.

87. $e = 2, \quad r \cos \theta = 2$

88. $e = 1, \quad r \cos \theta = -4$

89. $e = 1/2, \quad r \sin \theta = 2$

90. $e = 1/3, \quad r \sin \theta = -6$

Area, Length, and Surface Area in the Polar Plane

Find the areas of the regions in the polar coordinate plane described in Exercises 91–94.

91. Enclosed by the limaçon $r = 2 - \cos \theta$

92. Enclosed by one leaf of the three-leaved rose $r = \sin 3\theta$

93. Inside the "figure eight" $r = 1 + \cos 2\theta$ and outside the circle $r = 1$

94. Inside the cardioid $r = 2(1 + \sin \theta)$ and outside the circle $r = 2 \sin \theta$

Find the lengths of the curves given by the polar coordinate equations in Exercises 95–98.

95. $r = -1 + \cos \theta$

96. $r = 2 \sin \theta + 2 \cos \theta, \quad 0 \le \theta \le \pi/2$

97. $r = 8 \sin^3(\theta/3), \quad 0 \le \theta \le \pi/4$

98. $r = \sqrt{1 + \cos 2\theta}, \quad -\pi/2 \le \theta \le \pi/2$

Find the areas of the surfaces generated by revolving the polar coordinate curves in Exercises 99 and 100 about the indicated axes.

99. $r = \sqrt{\cos 2\theta}, \quad 0 \le \theta \le \pi/4, \quad x\text{-axis}$

100. $r^2 = \sin 2\theta, \quad y\text{-axis}$

Theory and Examples

101. Find the volume of the solid generated by revolving the region enclosed by the ellipse $9x^2 + 4y^2 = 36$ about (a) the x-axis, (b) the y-axis.

102. The "triangular" region in the first quadrant bounded by the x-axis, the line $x = 4$, and the hyperbola $9x^2 - 4y^2 = 36$ is revolved about the x-axis to generate a solid. Find the volume of the solid.

103. A ripple tank is made by bending a strip of tin around the perimeter of an ellipse for the wall of the tank and soldering a flat bottom onto this. An inch or two of water is put in the tank and you drop a marble into it, right at one focus of the ellipse. Ripples radiate outward through the water, reflect from the strip around the edge of the tank, and a few seconds later a drop of water spurts up at the second focus. Why?

104. *LORAN.* A radio signal was sent simultaneously from towers A and B, located several hundred miles apart on the northern California coast. A ship offshore received the signal from A 1400 microseconds before receiving the signal from B. Assuming that the signals traveled at the rate of 980 ft/microsecond, what can be said about the location of the ship relative to the two towers?

105. On a level plane, at the same instant, you hear the sound of a rifle and that of the bullet hitting the target. What can be said about your location relative to the rifle and target?

106. *Archimedes spirals.* The graph of an equation of the form $r = a\theta$, where a is a nonzero constant, is called an **Archimedes spiral**. Is there anything special about the widths between the successive turns of such a spiral?

107. a) Show that the equations $x = r \cos \theta, y = r \sin \theta$ transform the polar equation

$$r = \frac{k}{1 + e \cos \theta}$$

into the Cartesian equation

$$(1 - e^2)x^2 + y^2 + 2kex - k^2 = 0.$$

b) Then apply the criteria of Section 9.3 to show that

$$e = 0 \quad \Rightarrow \quad \text{circle}$$
$$0 < e < 1 \quad \Rightarrow \quad \text{ellipse}$$
$$e = 1 \quad \Rightarrow \quad \text{parabola}$$
$$e > 1 \quad \Rightarrow \quad \text{hyperbola.}$$

108. *A satellite orbit.* A satellite is in an orbit that passes over the North and South Poles of the earth. When it is over the South Pole it is at the highest point of its orbit, 1000 miles above the earth's surface. Above the North Pole it is at the lowest point of its orbit, 300 miles above the earth's surface.

a) Assuming that the orbit is an ellipse with one focus at the center of the earth, find its eccentricity. (Take the diameter of the earth to be 8000 miles.)

b) Using the north–south axis of the earth as the x-axis and the center of the earth as origin, find a polar equation for the orbit.

The Angle Between the Radius Vector and the Tangent Line to a Polar Coordinate Curve

In Cartesian coordinates, when we want to discuss the direction of a curve at a point, we use the angle ϕ measured counterclockwise from the positive x-axis to the tangent line. In polar coordinates, it is more convenient to calculate the angle ψ from the *radius vector* to the tangent line (Fig. 9.76). The angle ϕ can then be calculated from the relation

$$\phi = \theta + \psi, \tag{1}$$

which comes from applying the exterior angle theorem to the triangle in Fig. 9.76.

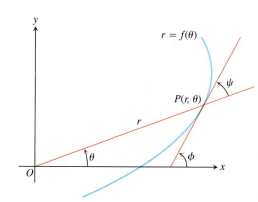

9.76 The angle ψ between the tangent line and the radius vector.

Suppose the equation of the curve is given in the form $r = f(\theta)$, where $f(\theta)$ is a differentiable function of θ. Then

$$x = r \cos \theta \quad \text{and} \quad y = r \sin \theta \tag{2}$$

are differentiable functions of θ with

$$\frac{dx}{d\theta} = -r \sin \theta + \cos \theta \, \frac{dr}{d\theta},$$
$$\frac{dy}{d\theta} = r \cos \theta + \sin \theta \, \frac{dr}{d\theta}. \tag{3}$$

Since $\psi = \phi - \theta$ from (1),

$$\tan \psi = \tan(\phi - \theta) = \frac{\tan \phi - \tan \theta}{1 + \tan \phi \tan \theta}.$$

Furthermore,

$$\tan \phi = \frac{dy}{dx} = \frac{dy/d\theta}{dx/d\theta}$$

because $\tan \phi$ is the slope of the curve at P. Also,

$$\tan \theta = \frac{y}{x}.$$

Hence

$$\tan \psi = \frac{\dfrac{dy/d\theta}{dx/d\theta} - \dfrac{y}{x}}{1 + \dfrac{y}{x}\dfrac{dy/d\theta}{dx/d\theta}}$$

(4)

$$= \frac{x\dfrac{dy}{d\theta} - y\dfrac{dx}{d\theta}}{x\dfrac{dx}{d\theta} + y\dfrac{dy}{d\theta}}.$$

The numerator in the last expression in Eq. (4) is found from Eqs. (2) and (3) to be

$$x\frac{dy}{d\theta} - y\frac{dx}{d\theta} = r^2.$$

Similarily, the denominator is

$$x\frac{dx}{d\theta} + y\frac{dy}{d\theta} = r\frac{dr}{d\theta}.$$

When we substitute these into Eq. (4), we obtain

$$\tan \psi = \frac{r}{dr/d\theta}.$$

(5)

This is the equation we use for finding ψ as a function of θ.

109. Show, by reference to a figure, that the angle β between the tangents to two curves at a point of intersection may be found from the formula

$$\tan \beta = \frac{\tan \psi_2 - \tan \psi_1}{1 + \tan \psi_2 \tan \psi_1}.$$

(6)

When will the two curves intersect at right angles?

110. Find the value of $\tan \psi$ for the curve $r = \sin^4(\theta/4)$.

111. Find the angle between the radius vector to the curve $r = 2a \sin 3\theta$ and its tangent when $\theta = \pi/6$.

112. a) GRAPHER Graph the hyperbolic spiral $r\theta = 1$. What appears to happen to ψ as the spiral winds in around the origin?

b) Confirm your finding in (a) analytically.

113. The circles $r = \sqrt{3} \cos \theta$ and $r = \sin \theta$ intersect at the point $(\sqrt{3}/2, \pi/3)$. Show that their tangents are perpendicular there.

114. Sketch the cardioid $r = a(1 + \cos \theta)$ and circle $r = 3a \cos \theta$ in one diagram and find the angle between their tangents at the point of intersection that lies in the first quadrant.

115. Find the points of intersection of the parabolas

$$r = \frac{1}{1 - \cos \theta} \quad \text{and} \quad r = \frac{3}{1 + \cos \theta}$$

and the angles between their tangents at these points.

116. Find points on the cardioid $r = a(1 + \cos \theta)$ where the tangent line is (a) horizontal, (b) vertical.

117. Show that parabolas $r = a/(1 + \cos \theta)$ and $r = b/(1 - \cos \theta)$ are orthogonal at each point of intersection $(ab \neq 0)$.

118. Find the angle at which the cardioid $r = a(1 - \cos \theta)$ crosses the ray $\theta = \pi/2$.

119. Find the angle between the line $r = 3 \sec \theta$ and the cardioid $r = 4(1 + \cos \theta)$ at one of their intersections.

120. Find the slope of the tangent line to the curve $r = a \tan(\theta/2)$ at $\theta = \pi/2$.

121. Find the angle at which the parabolas $r = 1/(1 - \cos \theta)$ and $r = 1/(1 - \sin \theta)$ intersect in the first quadrant.

122. The equation $r^2 = 2 \csc 2\theta$ represents a curve in polar coordinates.

a) Sketch the curve.

b) Find an equivalent Cartesian equation for the curve.

c) Find the angle at which the curve intersects the ray $\theta = \pi/4$.

123. Suppose that the angle ψ from the radius vector to the tangent line of the curve $r = f(\theta)$ has the constant value α.

a) Show that the area bounded by the curve and two rays $\theta = \theta_1, \theta = \theta_2$, is proportional to $r_2^2 - r_1^2$, where (r_1, θ_1) and (r_2, θ_2) are polar coordinates of the ends of the arc of the curve between these rays. Find the factor of proportionality.

b) Show that the length of the arc of the curve in part (a) is proportional to $r_2 - r_1$, and find the proportionality constant.

124. Let P be a point on the hyperbola $r^2 \sin 2\theta = 2a^2$. Show that the triangle formed by OP, the tangent at P, and the initial line is isosceles.

Finding Conic Sections

1. Find an equation for the parabola with focus $(4, 0)$ and directrix $x = 3$. Sketch the parabola together with its vertex, focus, and directrix.

2. Find the vertex, focus, and directrix of the parabola

$$x^2 - 6x - 12y + 9 = 0.$$

3. Find an equation for the curve traced by the point $P(x, y)$ if the distance from P to the vertex of the parabola $x^2 = 4y$ is twice the distance from P to the focus. Identify the curve.

4. A line segment of length $a + b$ runs from the x-axis to the y-axis. The point P on the segment lies a units from one end and b units from the other end. Show that P traces an ellipse as the ends of the segment slide along the axes.

5. The vertices of an ellipse of eccentricity 0.5 lie at the points $(0, \pm 2)$. Where do the foci lie?

6. Find an equation for the ellipse of eccentricity 2/3 that has the line $x = 2$ as a directrix and the point $(4, 0)$ as the corresponding focus.

7. One focus of a hyperbola lies at the point $(0, -7)$ and the corresponding directrix is the line $y = -1$. Find an equation for the hyperbola if its eccentricity is (a) 2, (b) 5.

8. Find an equation for the hyperbola with foci $(0, -2)$ and $(0, 2)$ that passes through the point $(12, 7)$.

Orthogonal Curves

Two curves are said to be **orthogonal** if their tangents cross at right angles at every point where the curves intersect. Exercises 9–12 are about orthogonal conic sections.

9. Sketch the curves $xy = 2$ and $x^2 - y^2 = 3$ together and show that they are orthogonal.

10. Sketch the curves $y^2 = 4x + 4$ and $y^2 = 64 - 16x$ together and show that they are orthogonal.

11. Show that the curves $2x^2 + 3y^2 = a^2$ and $ky^2 = x^3$ are orthogonal for all values of the constants a and k ($a \neq 0, k \neq 0$). Sketch the four curves corresponding to $a = 2, a = 4, k = 1/2, k = -2$ in one diagram.

12. Show that the parabolas

$$y^2 = 4a(a - x), \quad a > 0$$

and

$$y^2 = 4b(b + x), \quad b > 0$$

have a common focus, the same for any a and b. Show that the parabolas intersect at the points $(a - b, \pm 2\sqrt{ab})$ and that each a-parabola is orthogonal to every b-parabola. By varying a and b, we obtain two families of confocal parabolas. Each family is said to be a set of **orthogonal trajectories** of the other family (Fig. 9.77).

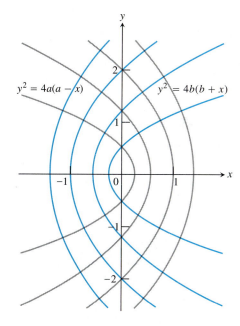

9.77 Confocal parabolas in Exercise 12.

Tangents to Conic Sections

13. *Constructing tangents to parabolas.* Show that the tangent to the parabola $y^2 = 4px$ at the point $P(x_1, y_1) \neq (0, 0)$ on the parabola meets the axis of symmetry x_1 units to the left of the vertex. This provides an accurate way to construct a tangent to the parabola at any point other than the origin (where we already have the y-axis): Mark the point $P(x_1, y_1)$ in question, drop a perpendicular from P to the x-axis, measure $2x_1$ units to the left, mark that point, and draw a line from there through P.

14. Show that no tangent can be drawn from the origin to the hyperbola $x^2 - y^2 = 1$. (*Hint:* If the tangent to a curve at a point $P(x, y)$ on the curve passes through the origin, then the slope of the curve at P is y/x.)

15. Show that any tangent to the hyperbola $xy = a^2$ makes a triangle of area $2a^2$ with the hyperbola's asymptotes.

16. a)   Show that the line

$$b^2 xx_1 + a^2 yy_1 - a^2 b^2 = 0$$

is tangent to the ellipse $b^2 x^2 + a^2 y^2 - a^2 b^2 = 0$ at the point (x_1, y_1) on the ellipse.

b)   Show that the line

$$b^2 xx_1 - a^2 yy_1 - a^2 b^2 = 0$$

is tangent to the hyperbola $b^2 x^2 - a^2 y^2 - a^2 b^2 = 0$ at the point (x_1, y_1) on the hyperbola.

c)   Show that the tangent to the conic section

$$Ax^2 + B xy + C y^2 + Dx + Ey + F = 0$$

at a point (x_1, y_1) on it has an equation that may be written in the form

$$Axx_1 + B\left(\frac{x_1 y + xy_1}{2}\right) + C yy_1 + D\left(\frac{x + x_1}{2}\right)$$

$$+ E\left(\frac{y + y_1}{2}\right) + F = 0.$$

Equations and Inequalities

What points in the xy-plane satisfy the equations and inequalities in Exercises 17–24? Draw a figure for each exercise.

17. $(x^2 - y^2 - 1)(x^2 + y^2 - 25)(x^2 + 4y^2 - 4) = 0$

18. $(x + y)(x^2 + y^2 - 1) = 0$

19. $(x^2/9) + (y^2/16) \le 1$

20. $(x^2/9) - (y^2/16) \le 1$

21. $(9x^2 + 4y^2 - 36)(4x^2 + 9y^2 - 16) \le 0$

22. $(9x^2 + 4y^2 - 36)(4x^2 + 9y^2 - 16) > 0$

23. $x^4 - (y^2 - 9)^2 = 0$

24. $x^2 + xy + y^2 < 3$

Parametric Equations

25. *Epicycloids.* When a circle rolls externally along the circumference of a second, fixed circle, any point P on the circumference of the rolling circle describes an *epicycloid*, as shown here. Let the fixed circle have its center at the origin O and have radius a.

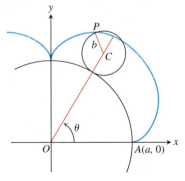

Let the radius of the rolling circle be b and let the initial position of the tracing point P be $A(a, 0)$. Find parametric equations for the epicycloid, using as the parameter the angle θ from the positive x-axis to the line through the circles' centers.

26. Find parametric equations and a Cartesian equation for the curve traced by the point $P(x, y)$ if its coordinates satisfy the differential equations

$$\frac{dx}{dt} = -2y, \quad \frac{dy}{dt} = \cos t,$$

subject to the conditions that $x = 3$ and $y = 0$ when $t = 0$. Identify the curve.

27. *Pythagorean triples.* Suppose that the coordinates of a particle $P(x, y)$ moving in the plane are

$$x = \frac{1 - t^2}{1 + t^2} \quad \text{and} \quad y = \frac{2t}{1 + t^2}$$

for $-\infty < t < \infty$. Show that $x^2 + y^2 = 1$ and hence that the motion takes place on the unit circle. What one point of the circle is not covered by the motion? Sketch the circle and indicate the direction of motion for increasing t. For what values of t does $(x, y) = (0, -1)?\ (1, 0)?\ (0, 1)?$

From $x^2 + y^2 = 1$, we obtain

$$(t^2 - 1)^2 + (2t)^2 = (t^2 + 1)^2,$$

an equation of interest in number theory because it generates *Pythagorean triples* of integers. When t is an integer greater than 1, $a = t^2 - 1$, $b = 2t$, and $c = t^2 + 1$ are positive integers that satisfy the equation $a^2 + b^2 = c^2$.

28. a)   Find the centroid of the region enclosed by the x-axis and the cycloid arch

$$x = a (t - \sin t), \quad y = a (1 - \cos t), \quad 0 \le t \le 2\pi.$$

b)   Find the first moments about the coordinate axes of the curve

$$x = (2/3) t^{3/2}, \quad y = 2 \sqrt{t}, \quad 0 \le t \le \sqrt{3}.$$

Polar Coordinates

29. a)   Find an equation in polar coordinates for the curve

$$x = e^{2t} \cos t, \quad y = e^{2t} \sin t, \quad -\infty < t < \infty.$$

b)   Find the length of the curve from $t = 0$ to $t = 2\pi$.

30. Find the length of the curve $r = 2 \sin^3(\theta/3), 0 \le \theta \le 3\pi$, in the polar coordinate plane.

31. Find the area of the surface generated by revolving the first-quadrant portion of the cardioid $r = 1 + \cos \theta$ about the x-axis. (*Hint:* Use the identities $1 + \cos \theta = 2 \cos^2 (\theta/2)$ and $\sin \theta = 2 \sin (\theta/2) \cos(\theta/2)$ to simplify the integral.)

32. Sketch the regions enclosed by the curves $r = 2a \cos^2 (\theta/2)$ and $r = 2a \sin^2 (\theta/2), a > 0$, in the polar coordinate plane and find the area of the portion of the plane they have in common.

Exercises 33–36 give the eccentricities of conic sections with one focus at the origin of the polar coordinate plane, along with the directrix for that focus. Find a polar equation for each conic section.

33. $e = 2$, $r \cos \theta = 2$

34. $e = 1$, $r \cos \theta = -4$

35. $e = 1/2$, $r \sin \theta = 2$

36. $e = 1/3$, $r \sin \theta = -6$

Theory and Examples

37. A rope with a ring in one end is looped over two pegs in a horizontal line. The free end, after being passed through the ring, has a weight suspended from it to make the rope hang taut. If the rope slips freely over the pegs and through the ring, the weight will descend as far as possible. Assume that the length of the rope is at least four times as great as the distance between the pegs and that the configuration of the rope is symmetric with respect to the line of the vertical part of the rope.

a)   Find the angle formed at the bottom of the loop (Fig. 9.78).

b)   Show that for each fixed position of the ring on the rope, the possible locations of the ring in space lie on an ellipse with foci at the pegs.

c)   Justify the original symmetry assumption by combining the result in (b) with the assumption that the rope and weight will take a rest position of minimal potential energy.

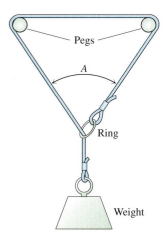

9.78 Exercise 37 asks how large the angle A will be when the frictionless rope shown here is pulled tight by the weight.

38. Two radar stations lie 20 km apart along an east–west line. A low-flying plane traveling from west to east is known to have a speed of v_0 km/sec. At $t = 0$ a signal is sent from the station at $(-10, 0)$, bounces off the plane, and is received at $(10, 0)$ $30/c$

seconds later (c is the velocity of the signal). When $t = 10/v_0$, another signal is sent out from the station at $(-10, 0)$, reflects off the plane, and is once again received $30/c$ seconds later by the other station. Find the position of the plane when it reflects the second signal under the assumption that v_0 is much less than c.

39. A comet moves in a parabolic orbit with the sun at the focus. When the comet is 4×10^7 miles from the sun, the line from the comet to the sun makes a $60°$ angle with the orbit's axis, as shown here. How close will the comet come to the sun?

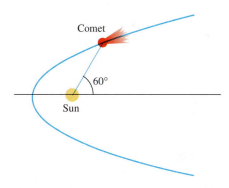

40. Find the points on the parabola $x = 2t$, $y = t^2$, $-\infty < t < \infty$, closest to the point $(0, 3)$.

41. Find the eccentricity of the ellipse $x^2 + xy + y^2 = 1$ to the nearest hundredth.

42. Find the eccentricity of the hyperbola $xy = 1$.

43. Is the curve $\sqrt{x} + \sqrt{y} = 1$ part of a conic section? If so, what kind of conic section? If not, why not?

44. Show that the curve $2xy - \sqrt{2}\, y + 2 = 0$ is a hyperbola. Find the hyperbola's center, vertices, foci, axes, and asymptotes.

45. Find a polar coordinate equation for

a)   the parabola with focus at the origin and vertex at $(a, \pi/4)$;

b)   the ellipse with foci at the origin and $(2, 0)$ and one vertex at $(4, 0)$;

c)   the hyperbola with one focus at the origin, center at $(2, \pi/2)$, and a vertex at $(1, \pi/2)$.

46. Any line through the origin will intersect the ellipse $r = 3/(2 + \cos \theta)$ in two points P_1 and P_2. Let d_1 be the distance between P_1 and the origin and let d_2 be the distance between P_2 and the origin. Compute $(1/d_1) + (1/d_2)$.

47. *Generating a cardioid with circles.* Cardioids are special epicycloids (Exercise 25). Show that if you roll a circle of radius a about another circle of radius a in the polar coordinate plane, as in Fig. 9.79, the original point of contact P will trace a cardioid. (*Hint:* Start by showing that angles OBC and PAD both have measure θ.)

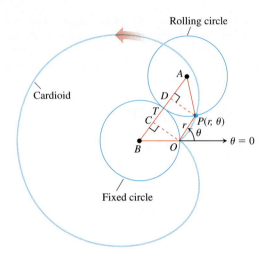

9.79 As the circle centered at A rolls around the circle centered at B, the point P traces a cardioid (Exercise 47).

48. *A bifold closet door.* A bifold closet door consists of two one-foot-wide panels, hinged at point P. The outside bottom corner of one panel rests on a pivot at O (see the accompanying figure). The outside bottom corner of the other panel, denoted by Q, slides along a straight track, shown in the figure as a portion of the x-axis. Assume that as Q moves back and forth, the bottom of the door rubs against a thick carpet. What shape will the door sweep out on the surface of the carpet?

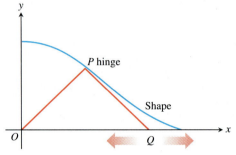

49. GRAPHER EXPLORATION Graph the curve $r = \cos 5\theta + n \cos \theta$, $0 \le \theta \le \pi$ for integers $n = -5$ (heart) to $n = 5$ (bell). (Source: *The College Mathematics Journal*, Vol. 25, No. 1, Jan. 1994.)

Vectors and Analytic Geometry in Space

OVERVIEW   This chapter introduces vectors and three-dimensional coordinate systems. Just as the coordinate plane is the natural place to study functions of a single variable, coordinate space is the place to study functions of two variables (or more). We establish coordinates in space by adding a third axis that measures distance above and below the xy-plane. This builds on what we already know without forcing us to start over again.

10.1 Vectors in the Plane

Some of the things we measure are determined by their magnitudes. To record mass, length, or time, for example, we need only write down a number and name an appropriate unit of measure. But we need more information to describe a force, displacement, or velocity. To describe a force, we need to record the direction in which it acts as well as how large it is. To describe a body's displacement, we have to say in what direction it moves as well as how far. To describe a body's velocity, we have to know where the body is headed as well as how fast it is going.

Quantities that have direction as well as magnitude are usually represented by arrows that point in the direction of the action and whose lengths give the magnitude of the action in terms of a suitably chosen unit.

When we discuss these arrows abstractly, we think of them as directed line segments and we call them *vectors*.

> **Definitions**
>
> A **vector** in the plane is a directed line segment. Two vectors are **equal** or **the same** if they have the same length and direction.

Thus, the arrows we use when we draw vectors are understood to represent the same vector if they have the same length, are parallel, and point in the same direction.

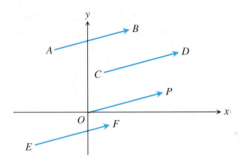

10.1 Arrows with the same length and direction represent the same vector (Example 1).

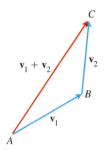

10.2 Scalar multiples of **v**.

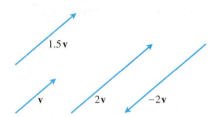

10.3 The sum of $\mathbf{v}_1$ and $\mathbf{v}_2$.

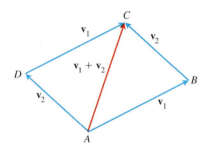

10.4 The Parallelogram Law of Addition. Quadrilateral *ABCD* is a parallelogram because opposite sides have equal lengths. The law was used by Aristotle to describe the combined action of two forces.

In print, vectors are usually described with single boldface roman letters, as in **v** ("vector vee"). The vector defined by the directed line segment from point A to point B is written as $\overrightarrow{AB}$ ("vector *ab*").

EXAMPLE 1   The four arrows in Fig. 10.1 have the same length and direction. They therefore represent the same vector, and we write

$$\overrightarrow{AB} = \overrightarrow{CD} = \overrightarrow{OP} = \overrightarrow{EF}.$$ ❑

Scalars and Scalar Multiples

We multiply a vector by a positive real number by multiplying its length by the number (Fig. 10.2). To multiply a vector by 2, we double its length. To multiply a vector by 1.5, we increase its length by 50%, and so on. We multiply a vector by a negative number by reversing the vector's direction and multiplying the length by the number's absolute value.

If c is a nonzero real number and **v** a vector, the direction of $c\mathbf{v}$ agrees with that of **v** if c is positive and is opposite to that of **v** if c is negative. Since real numbers work like scaling factors in this context, we call them **scalars** and call multiples like $c\mathbf{v}$ **scalar multiples** of **v**.

To include multiplication by zero, we adopt the convention that multiplying a vector by zero produces the **zero vector 0,** consisting of points that are degenerate line segments of zero length. Unlike other vectors, the vector **0** has no direction.

Geometric Addition: The Parallelogram Law

Two nonzero vectors $\mathbf{v}_1$ and $\mathbf{v}_2$ can be added geometrically by drawing a representative of $\mathbf{v}_1$, say from A to B as in Fig. 10.3, and then a representative of $\mathbf{v}_2$ starting from the terminal point B of $\mathbf{v}_1$. In Fig. 10.3, $\mathbf{v}_2 = \overrightarrow{BC}$. The sum $\mathbf{v}_1 + \mathbf{v}_2$ is then the vector represented by the arrow from the initial point A of $\mathbf{v}_1$ to the terminal point C of $\mathbf{v}_2$. That is, if

$$\mathbf{v}_1 = \overrightarrow{AB} \qquad \text{and} \qquad \mathbf{v}_2 = \overrightarrow{BC},$$

then

$$\mathbf{v}_1 + \mathbf{v}_2 = \overrightarrow{AB} + \overrightarrow{BC} = \overrightarrow{AC}.$$

This description of addition is sometimes called the **Parallelogram Law** of addition because $\mathbf{v}_1 + \mathbf{v}_2$ is given by the diagonal of the parallelogram determined by $\mathbf{v}_1$ and $\mathbf{v}_2$ (Fig. 10.4).

Components

Two vectors are said to be **parallel** if they are nonzero scalar multiples of one another or, equivalently, if the line segments representing them are parallel.

Whenever a vector **v** can be written as a sum

$$\mathbf{v} = \mathbf{v}_1 + \mathbf{v}_2$$

of two *nonparallel* vectors, the vectors $\mathbf{v}_1$ and $\mathbf{v}_2$ are said to be **components** of **v**. We also say that we have **represented** or **resolved v** in terms of $\mathbf{v}_1$ and $\mathbf{v}_2$.

The most common algebra of vectors is based on representing each vector in terms of components parallel to the Cartesian coordinate axes and writing each

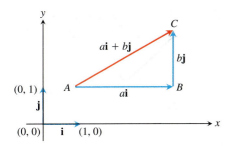

10.5 The basic vectors **i** and **j**. Any vector $\overrightarrow{AC}$ in the plane can be expressed as a scalar multiple of **i** plus a scalar multiple of **j**.

component as an appropriate multiple of a **basic** vector of length l. The basic vector in the positive x-direction is the vector **i** determined by the directed line segment that runs from (0, 0) to (1, 0). The basic vector in the positive y-direction is the vector **j** determined by the directed line segment from (0, 0) to (0, 1). Then $a\,\mathbf{i}$, a being a scalar, represents a vector of length $|a|$ parallel to the x-axis, pointing to the right if $a > 0$ and to the left if $a < 0$. Similarly, $b\,\mathbf{j}$ is a vector of length $|b|$ parallel to the y-axis, pointing up if $b > 0$ and down if $b < 0$. Figure 10.5 shows a vector $\mathbf{v} = \overrightarrow{AC}$ resolved into its **i**- and **j**-components as the sum

$$\mathbf{v} = a\,\mathbf{i} + b\,\mathbf{j}.$$

Definitions

If $\mathbf{v} = a\,\mathbf{i} + b\,\mathbf{j}$, the vectors $a\,\mathbf{i}$ and $b\,\mathbf{j}$ are the **vector components of v in the directions of i and j.** The numbers a and b are the **scalar components of v in the directions of i and j.**

Components enable us to define the equality of vectors algebraically.

Definition

Equality of Vectors (Algebraic Definition)

$$a\,\mathbf{i} + b\,\mathbf{j} = a'\,\mathbf{i} + b'\,\mathbf{j} \qquad \Leftrightarrow \qquad a = a' \text{ and } b = b' \qquad (1)$$

Two vectors are equal if and only if their scalar components in the directions of **i** and **j** are identical.

Algebraic Addition

Vectors may be added algebraically by adding their corresponding scalar components, as shown in Fig. 10.6.

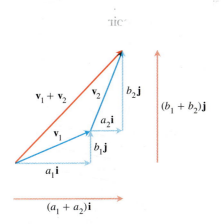

10.6 If $\mathbf{v}_1 = a_1\,\mathbf{i} + b_1\,\mathbf{j}$ and $\mathbf{v}_2 = a_2\,\mathbf{i} + b_2\,\mathbf{j}$, then $\mathbf{v}_1 + \mathbf{v}_2 = (a_1 + a_2)\,\mathbf{i} + (b_1 + b_2)\,\mathbf{j}$.

If $\mathbf{v}_1 = a_1\,\mathbf{i} + b_1\,\mathbf{j}$ and $\mathbf{v}_2 = a_2\,\mathbf{i} + b_2\,\mathbf{j}$, then

$$\mathbf{v}_1 + \mathbf{v}_2 = (a_1 + a_2)\,\mathbf{i} + (b_1 + b_2)\,\mathbf{j}. \qquad (2)$$

EXAMPLE 2

$$(2\,\mathbf{i} - 4\,\mathbf{j}) + (5\,\mathbf{i} + 3\,\mathbf{j}) = (2 + 5)\,\mathbf{i} + (-4 + 3)\,\mathbf{j} = 7\,\mathbf{i} - \mathbf{j} \qquad \square$$

Subtraction

The negative of a vector $\mathbf{v}$ is the vector $-\mathbf{v} = (-1)\,\mathbf{v}$. It has the same length as $\mathbf{v}$ but points in the opposite direction. To subtract a vector $\mathbf{v}_2$ from a vector $\mathbf{v}_1$, we add $-\mathbf{v}_2$ to $\mathbf{v}_1$. This can be done geometrically by drawing $-\mathbf{v}_2$ from the tip of $\mathbf{v}_1$

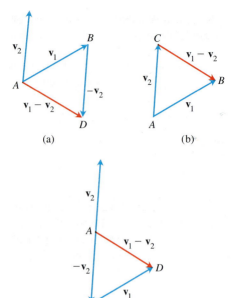

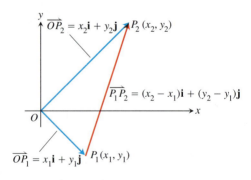

10.7 Three ways to draw $\mathbf{v}_1 - \mathbf{v}_2$ (there are others): (a) as $\mathbf{v}_1 + (-\mathbf{v}_2)$; (b) as the vector from the tip of $\mathbf{v}_2$ to the tip of $\mathbf{v}_1$; and (c) as $-\mathbf{v}_2 + \mathbf{v}_1$.

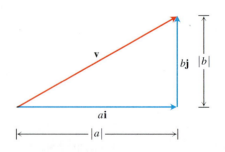

10.8 $\overrightarrow{P_1P_2} = (x_2 - x_1)\mathbf{i} + (y_2 - y_1)\mathbf{j}$

10.9 The length of $\mathbf{v}$ is $\sqrt{|a|^2 + |b|^2} = \sqrt{a^2 + b^2}$.

and then drawing the vector from the initial point of $\mathbf{v}_1$ to the tip of $-\mathbf{v}_2$, as shown in Fig. 10.7(a), where

$$\overrightarrow{AD} = \overrightarrow{AB} + \overrightarrow{BD} = \mathbf{v}_1 + (-\mathbf{v}_2) = \mathbf{v}_1 - \mathbf{v}_2.$$

Another way to draw $\mathbf{v}_1 - \mathbf{v}_2$ is to draw $\mathbf{v}_1$ and $\mathbf{v}_2$ with a common initial point and then draw $\mathbf{v}_1 - \mathbf{v}_2$ as the vector from the tip of $\mathbf{v}_2$ to the tip of $\mathbf{v}_1$. This is illustrated in Fig. 10.7(b), where

$$\overrightarrow{CB} = \overrightarrow{CA} + \overrightarrow{AB} = -\mathbf{v}_2 + \mathbf{v}_1 = \mathbf{v}_1 - \mathbf{v}_2.$$

Still another way is to draw $\mathbf{v}_1$ from the tip of $-\mathbf{v}_2$ (Fig. 10.7c).
In terms of components, vector subtraction follows the algebraic law

$$\mathbf{v}_1 - \mathbf{v}_2 = (a_1 - a_2)\mathbf{i} + (b_1 - b_2)\mathbf{j}, \tag{3}$$

which says that corresponding scalar components are subtracted.

EXAMPLE 3

$$(6\mathbf{i} + 2\mathbf{j}) - (3\mathbf{i} - 5\mathbf{j}) = (6 - 3)\mathbf{i} + (2 - (-5))\mathbf{j} = 3\mathbf{i} + 7\mathbf{j} \qquad \square$$

We find the components of the vector from a point $P_1(x_1, y_1)$ to a point $P_2(x_2, y_2)$ by subtracting the components of $\overrightarrow{OP}_1 = x_1\mathbf{i} + y_1\mathbf{j}$ from the components of $\overrightarrow{OP}_2 = x_2\mathbf{i} + y_2\mathbf{j}$ (Fig. 10.8).

The vector from $P_1(x_1, y_1)$ to $P_2(x_2, y_2)$ is

$$\overrightarrow{P_1P_2} = (x_2 - x_1)\mathbf{i} + (y_2 - y_1)\mathbf{j}. \tag{4}$$

EXAMPLE 4 The vector from $P_1(3, 4)$ to $P_2(5, 1)$ is

$$\overrightarrow{P_1P_2} = (5 - 3)\mathbf{i} + (1 - 4)\mathbf{j} = 2\mathbf{i} - 3\mathbf{j}. \qquad \square$$

Magnitude

The **magnitude** or **length** of $\mathbf{v} = a\mathbf{i} + b\mathbf{j}$ is $|\mathbf{v}| = \sqrt{a^2 + b^2}$. We arrive at this number by applying the Pythagorean theorem to the right triangle determined by $\mathbf{v}$ and its two vector components (Fig. 10.9). The bars in $|\mathbf{v}|$ (read "the magnitude of $\mathbf{v}$" or "the length of $\mathbf{v}$") are the same bars we use for absolute values.

The magnitude or length of $\mathbf{v} = a\mathbf{i} + b\mathbf{j}$ is $|\mathbf{v}| = \sqrt{a^2 + b^2}$. $\tag{5}$

EXAMPLE 5 You push a loaded supermarket cart by applying a 20-lb force $\mathbf{F}$ that makes a 30° angle with the horizontal. Resolve $\mathbf{F}$ into its horizontal and vertical components. (The horizontal component is the effective force in the direction of motion. The vertical component just adds weight to the cart.)

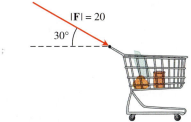

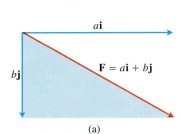

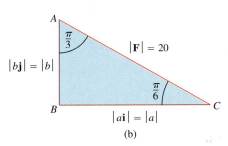

10.10 (a) The horizontal and vertical components of the vector **F** in Example 5. (b) The right triangle determined by **F** and its components.

Solution   We draw a vector triangle for $\mathbf{F} = a\,\mathbf{i} + b\,\mathbf{j}$ and its vector components along with the right triangle determined by their magnitudes (Fig. 10.10). The triangle is a 30-60-90 triangle, so $|a| = 10\sqrt{3}$ and $|b| = 10$. The horizontal component of **F** is $10\sqrt{3}\,\mathbf{i}$. The vertical component is $-10\,\mathbf{j}$ (negative because it points down). That is, $\mathbf{F} = 10\sqrt{3}\,\mathbf{i} - 10\,\mathbf{j}$.   ❑

Scalar Multiplication

Scalar multiplication can be accomplished component by component.

> If c is a scalar and $\mathbf{v} = a\,\mathbf{i} + b\,\mathbf{j}$ is a vector, then
> $$c\,\mathbf{v} = c(a\,\mathbf{i} + b\,\mathbf{j}) = (ca)\,\mathbf{i} + (cb)\,\mathbf{j}. \tag{6}$$

The length of $c\,\mathbf{v}$ is $|c|$ times the length of **v**:

$$
\begin{aligned}
|c\,\mathbf{v}| &= |(ca)\,\mathbf{i} + (cb)\,\mathbf{j}| && \text{Eq. (6)}\\
&= \sqrt{(ca)^2 + (cb)^2} && \text{Eq. (5) with } ca \text{ and } cb \text{ in place of } a \text{ and } b\\
&= \sqrt{c^2(a^2 + b^2)}\\
&= \sqrt{c^2}\sqrt{a^2 + b^2}\\
&= |c||\mathbf{v}|.
\end{aligned}
$$

> If c is a scalar and **v** is a vector, then $|c\,\mathbf{v}| = |c||\mathbf{v}|$.

EXAMPLE 6   If $c = -2$ and $\mathbf{v} = -3\,\mathbf{i} + 4\,\mathbf{j}$, then

$$|\mathbf{v}| = |-3\,\mathbf{i} + 4\,\mathbf{j}| = \sqrt{(-3)^2 + (4)^2} = \sqrt{9 + 16} = \sqrt{25} = 5$$

$$|-2\,\mathbf{v}| = |(-2)(-3\,\mathbf{i} + 4\,\mathbf{j})| = |6\,\mathbf{i} - 8\,\mathbf{j}| = \sqrt{(6)^2 + (-8)^2} = \sqrt{36 + 64}$$

$$= \sqrt{100} = 10 = |-2|5 = |c||\mathbf{v}|.$$ ❑

The Zero Vector

In terms of components, the zero vector is the vector

$$\mathbf{0} = 0\,\mathbf{i} + 0\,\mathbf{j}.$$

It is the only vector whose length is zero, as we can see from the fact that

$$|a\,\mathbf{i} + b\,\mathbf{j}| = \sqrt{a^2 + b^2} = 0 \quad \Leftrightarrow \quad a = b = 0.$$

Unit Vectors

Any vector whose length is 1 is a **unit vector.** The vectors $\mathbf{i}$ and $\mathbf{j}$ are unit vectors.

$$|\mathbf{i}| = |1\,\mathbf{i} + 0\,\mathbf{j}| = \sqrt{1^2 + 0^2} = 1, \qquad |\mathbf{j}| = |0\,\mathbf{i} + 1\,\mathbf{j}| = \sqrt{0^2 + 1^2} = 1$$

If $\mathbf{u}$ is the unit vector obtained by rotating $\mathbf{i}$ through an angle θ in the positive direction, then $\mathbf{u}$ has a horizontal component $\cos\theta$ and vertical component $\sin\theta$ (Fig. 10.11), so that

$$\mathbf{u} = (\cos\theta)\,\mathbf{i} + (\sin\theta)\,\mathbf{j}. \qquad (7)$$

As θ varies from 0 to 2π, the point P in Fig. 10.11 traces the circle $x^2 + y^2 = 1$ counterclockwise. This takes in all possible directions, so *Eq. (7) gives every unit vector in the plane.*

Length vs. Direction

If $\mathbf{v} \neq \mathbf{0},$ then

$$\left|\frac{\mathbf{v}}{|\mathbf{v}|}\right| = \left|\frac{1}{|\mathbf{v}|}\mathbf{v}\right| = \frac{1}{|\mathbf{v}|}|\mathbf{v}| = 1,$$

so $\mathbf{v}/|\mathbf{v}|$ is a unit vector in the direction of $\mathbf{v}$. We can therefore express $\mathbf{v}$ in terms of its two important features, length and direction, by writing $\mathbf{v} = |\mathbf{v}|(\mathbf{v}/|\mathbf{v}|)$.

If $\mathbf{v} \neq \mathbf{0},$ then

1. $\dfrac{\mathbf{v}}{|\mathbf{v}|}$ is a unit vector in the direction of $\mathbf{v}$;

2.   the equation $\mathbf{v} = |\mathbf{v}|\dfrac{\mathbf{v}}{|\mathbf{v}|}$ expresses $\mathbf{v}$ in terms of its length and direction.

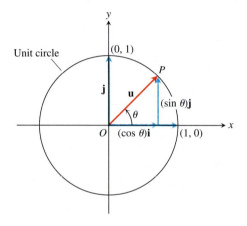

In handwritten work it is common to denote unit vectors with small "hats," as in $\hat{u}$ (pronounced "u hat"). In hat notation, $\mathbf{i}$ and $\mathbf{j}$ become $\hat{\mathbf{i}}$ and $\hat{\mathbf{j}}$.

10.11 The unit vector that makes an angle of measure θ with the positive x-axis. Every unit vector in the plane has the form

$$\mathbf{u} = (\cos\theta)\,\mathbf{i} + (\sin\theta)\,\mathbf{j}$$

for some θ.

It is common in applied fields that use vectors a great deal to refer to the vector $\mathbf{v}/|\mathbf{v}|$ itself as the direction of $\mathbf{v}$. The equation $\mathbf{v} = |\mathbf{v}|(\mathbf{v}/|\mathbf{v}|)$ is then said to express $\mathbf{v}$ as a *product* of its length and direction.

EXAMPLE 7   Express $\mathbf{v} = 3\,\mathbf{i} - 4\,\mathbf{j}$ as a product of its length and direction.

Solution   Length of $\mathbf{v}$: $|\mathbf{v}| = \sqrt{(3)^2 + (-4)^2} = \sqrt{9 + 16} = 5$

Direction of $\mathbf{v}$: $\dfrac{\mathbf{v}}{|\mathbf{v}|} = \dfrac{3\,\mathbf{i} - 4\,\mathbf{j}}{5} = \dfrac{3}{5}\mathbf{i} - \dfrac{4}{5}\mathbf{j}$

$$\mathbf{v} = 3\,\mathbf{i} - 4\,\mathbf{j} = 5\underbrace{\left(\frac{3}{5}\mathbf{i} - \frac{4}{5}\mathbf{j}\right)}$$

length   direction

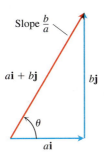

10.12 If $a \neq 0$, the vector $a\,\mathbf{i} + b\,\mathbf{j}$ has slope $b/a = \tan\theta$.

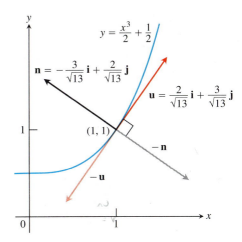

10.13 The unit tangent and normal vectors at the point $(1, 1)$ on the curve $y = (x^3/2) + 1/2$.

If $\mathbf{v} = a\,\mathbf{i} + b\,\mathbf{j}$, then $\mathbf{p} = -b\,\mathbf{i} + a\,\mathbf{j}$ and $\mathbf{q} = b\,\mathbf{i} - a\,\mathbf{j}$ are perpendicular to $\mathbf{v}$ because their slopes are both $-a/b$, the negative reciprocal of $\mathbf{v}$'s slope.

Slopes, Tangents, and Normals

A vector is **parallel** to a line if the segments that represent the vector are parallel to the line. The **slope** of a vector that is not vertical is the slope shared by the lines parallel to the vector. Thus, if $a \neq 0$, the vector $\mathbf{v} = a\,\mathbf{i} + b\,\mathbf{j}$ has a well-defined slope, which can be calculated from the components of $\mathbf{v}$ as the number b/a (Fig. 10.12).

A vector is **tangent** or **normal** to a curve at a point if it is parallel or normal to the line that is tangent to the curve at that point. The next example shows how to find such vectors.

EXAMPLE 8   Find unit vectors tangent and normal to the curve

$$y = \frac{x^3}{2} + \frac{1}{2}$$

at the point $(1, 1)$.

Solution   We find the unit vectors that are parallel and normal to the curve's tangent line at $(1, 1)$ (Fig. 10.13).

The slope of the line tangent to the curve at $(1, 1)$ is

$$y' = \frac{3x^2}{2}\bigg|_{x=1} = \frac{3}{2}.$$

We look for a unit vector with this slope. The vector $\mathbf{v} = 2\,\mathbf{i} + 3\,\mathbf{j}$ has slope $3/2$, as does every nonzero multiple of $\mathbf{v}$. To find a multiple of $\mathbf{v}$ that is a unit vector, we divide $\mathbf{v}$ by

$$|\mathbf{v}| = \sqrt{2^2 + 3^2} = \sqrt{13},$$

obtaining

$$\mathbf{u} = \frac{\mathbf{v}}{|\mathbf{v}|} = \frac{2}{\sqrt{13}}\,\mathbf{i} + \frac{3}{\sqrt{13}}\,\mathbf{j}.$$

The vector $\mathbf{u}$ is tangent to the curve at $(1, 1)$ because it has the same direction as $\mathbf{v}$. Of course,

$$-\mathbf{u} = -\frac{2}{\sqrt{13}}\,\mathbf{i} - \frac{3}{\sqrt{13}}\,\mathbf{j},$$

which points in the opposite direction, is also tangent to the curve at $(1, 1)$. Without some additional requirement, there is no reason to prefer one of these vectors to the other.

To find unit vectors normal to the curve at $(1, 1)$, we look for unit vectors whose slopes are the negative reciprocal of the slope of $\mathbf{u}$. This is quickly done by interchanging the scalar components of $\mathbf{u}$ and changing the sign of one of them. We obtain

$$\mathbf{n} = -\frac{3}{\sqrt{13}}\,\mathbf{i} + \frac{2}{\sqrt{13}}\,\mathbf{j} \quad \text{and} \quad -\mathbf{n} = \frac{3}{\sqrt{13}}\,\mathbf{i} - \frac{2}{\sqrt{13}}\,\mathbf{j}.$$

Again, either one will do. The vectors have opposite directions but both are normal to the curve at $(1, 1)$.   ◻

Exercises 10.1

Geometry and Calculation

1. The vectors **A**, **B**, and **C** in the figure here lie in a plane. Copy them on a sheet of paper. Then, by arranging vectors head to tail, as in Figs. 10.3 and 10.6, sketch

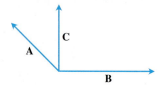

a) **A** + **B**

b) **A** + **B** + **C**

c) **A** − 2**B**

d) $\frac{1}{2}$**A** − **C**

2. The vectors **A**, **B**, and **C** in the figure here lie in a plane. Copy them on a sheet of paper. Then, by arranging vectors head to tail, as in Figs. 10.3 and 10.6, sketch

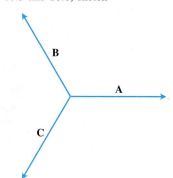

a) **A** − **B**

b) **A** + **B** + **C**

c) 2 **A** − $\frac{1}{2}$ **B**

d) **A** − (**B** − **C**)

Let **A** = 2 **i** − 7 **j**, **B** = **i** + 6 **j**, and **C** = $\sqrt{3}$ **i** − π **j**. Write each of the vectors in Exercises 3–6 in the form a **i** + b **j**.

3. **A** + 2 **B**

4. **A** + **B** − **C**

5. 3 **A** − $\frac{1}{\pi}$ **C**

6. 2 **A** − 3 **B** + 32 **j**

7. Vectors **u**, **v**, and **w** are determined by the sides of triangle ABC as shown.

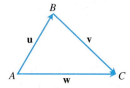

a) Express **w** in terms of **u** and **v**.

b) Express **v** in terms of **u** and **w**.

8. Vectors **u** and **w** are determined by the sides of triangle ABC as shown, and P is the midpoint of side BC. Express **a** in terms of **u** and **w**.

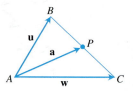

Express the vectors in Exercises 9–16 in the form a **i** + b **j** and sketch them as arrows in the coordinate plane beginning at the origin.

9. $\overrightarrow{P_1 P_2}$ if P_1 is the point (5, 7) and P_2 is the point (2, 9)

10. $\overrightarrow{P_1 P_2}$ if P_1 is the point (1, 2) and P_2 is the point (−3, 5)

11. $\overrightarrow{AB}$ if A is the point (−5, 3) and B is the point (−10, 8)

12. $\overrightarrow{AB}$ if A is the point (−7, −8) and B is the point (6, 11)

13. $\overrightarrow{P_1 P_2}$ if P_1 is the point (1, 3) and P_2 is the point (2, −1)

14. $\overrightarrow{P_3 P_4}$ if P_3 is the point (1, 3) and P_4 is the midpoint of the line segment $P_1 P_2$ joining $P_1(2, -1)$ and $P_2(-4, 3)$

15. The sum of the vectors $\overrightarrow{AB}$ and $\overrightarrow{CD}$, given the four points $A(1, -1)$, $B(2, 0)$, $C(-1, 3)$, and $D(-2, 2)$

16. The vector from the point A to the origin where $\overrightarrow{AB} = 4$ **i** − 2 **j** and B is the point (−2, 5)

17. Given the vector $\overrightarrow{AB} = 3$ **i** − **j** and A is the point (2, 9), find the point B.

18. Given the vector $\overrightarrow{PQ} = -6$ **i** − 4 **j** and Q is the point (3, 3), find the point P.

Unit Vectors

Sketch the vectors in Exercises 19–22 and express each vector in the form a **i** + b **j**.

19. The unit vectors **u** = (cos θ) **i** + (sin θ) **j** for $\theta = \pi/6$ and $\theta = 2\pi/3$. Include the circle $x^2 + y^2 = 1$ in your sketch.

20. The unit vectors **u** = (cos θ) **i** + (sin θ) **j** for $\theta = -\pi/4$ and $\theta = -3\pi/4$. Include the circle $x^2 + y^2 = 1$ in your sketch.

21. The unit vector obtained by rotating **j** counterclockwise $3\pi/4$ rad about the origin

22. The unit vector obtained by rotating **j** clockwise $2\pi/3$ rad about the origin

For the vectors in Exercises 23 and 24, find unit vectors **u** = (cos θ) **i** + (sin θ) **j** in the same direction.

23. 6 **i** − 8 **j**

24. − **i** + 3 **j**

In Exercises 25–28, find the unit vectors that are tangent and normal to the curve at the given point (four vectors in all). Then sketch the vectors and curve together.

25. $y = x^2$, $(2, 4)$

26. $x^2 + 2y^2 = 6$, $(2, 1)$

27. $y = \tan^{-1} x$, $(1, \pi/4)$

28. $y = \sum_{n=0}^{\infty} \dfrac{x^n}{n!}$, $(0, 1)$

In Exercises 29–32, find the unit vectors that are tangent and normal to the curve at the given point (four vectors in all).

29. $3x^2 + 8xy + 2y^2 - 3 = 0$, $(1, 0)$

30. $x^2 - 6xy + 8y^2 - 2x - 1 = 0$, $(1, 1)$

31. $y = \int_0^x \sqrt{3 + t^4} \, dt$, $(0, 0)$

32. $y = \int_e^x \ln(\ln t) \, dt$, $(e, 0)$

Length and Direction

In Exercises 33 and 34, express each vector as a product of its length and direction.

33. $5\mathbf{i} + 12\mathbf{j}$

34. $2\mathbf{i} - 3\mathbf{j}$

35. Find the unit vectors that are parallel to the vector $3\mathbf{i} - 4\mathbf{j}$ (two vectors in all).

36. Find a vector of length 2 whose direction is the opposite of the direction of the vector $\mathbf{A} = -\mathbf{i} + 2\mathbf{j}$. How many such vectors are there?

37. Show that $\mathbf{A} = 3\mathbf{i} + 6\mathbf{j}$ and $\mathbf{B} = -\mathbf{i} - 2\mathbf{j}$ have opposite directions. Sketch $\mathbf{A}$ and $\mathbf{B}$ together.

38. Show that $\mathbf{A} = 3\mathbf{i} + 6\mathbf{j}$ and $\mathbf{B} = (1/2)\mathbf{i} + \mathbf{j}$ have the same direction.

Theory and Applications

39. You are pulling on a suitcase with a force $\mathbf{F}$ (pictured here) whose magnitude is $|\mathbf{F}| = 10$ lb. Find the x- and y-components of $\mathbf{F}$.

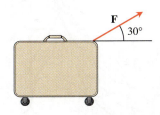

40. A kite string exerts a 12-lb pull ($|\mathbf{F}| = 12$) on a kite and makes a 45° angle with the horizontal. Find the horizontal and vertical components of $\mathbf{F}$.

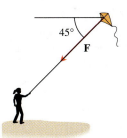

41. Let $\mathbf{A} = 2\mathbf{i} + \mathbf{j}$, $\mathbf{B} = \mathbf{i} + \mathbf{j}$, and $\mathbf{C} = \mathbf{i} - \mathbf{j}$. Find scalars α, β, such that $\mathbf{A} = \alpha\mathbf{B} + \beta\mathbf{C}$.

42. Let $\mathbf{A} = \mathbf{i} - 2\mathbf{j}$, $\mathbf{B} = 2\mathbf{i} + 3\mathbf{j}$, and $\mathbf{C} = \mathbf{i} + \mathbf{j}$. Write $\mathbf{A} = \mathbf{A}_1 + \mathbf{A}_2$ where $\mathbf{A}_1$ is parallel to $\mathbf{B}$ and $\mathbf{A}_2$ is parallel to $\mathbf{C}$. (See Exercise 41.)

43. A bird flies from its nest 5 km in the direction 60° north of east, where it stops to rest on a tree. It then flies 10 km in the direction due southeast to land atop a telephone pole. Place an xy-coordinate system so that the origin is the bird's nest, the x-axis points east, and the y-axis points north.

  a) At what point is the tree located?

  b) At what point is the telephone pole?

44. A bird flies from its nest 7 km in the direction northeast, where it stops to rest on a tree. It then flies 8 km in the direction 30° south of west to land atop a telephone pole. Place an xy-coordinate system so that the origin is the bird's nest, the x-axis points east, and the y-axis points north.

  a) At what point is the tree located?

  b) At what point is the telephone pole?

45. Let $\mathbf{v}$ be a vector in the plane not parallel to the y-axis. How is the slope of $-\mathbf{v}$ related to the slope of $\mathbf{v}$? Give reasons for your answer.

10.2

Cartesian (Rectangular) Coordinates and Vectors in Space

Our goal now is to describe the three-dimensional Cartesian coordinate system and learn our way around in space. This means defining distance, practicing with the arithmetic of vectors in space (the rules are the same as in the plane but with an extra term), and making connections between sets of points and equations and inequalities.

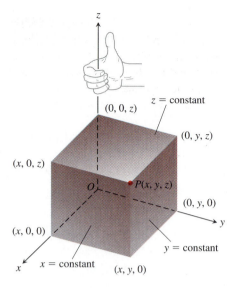

10.14 The Cartesian coordinate system is right-handed.

Cartesian Coordinates

To locate points in space, we use three mutually perpendicular coordinate axes, arranged as in Fig. 10.14. The axes Ox, Oy, and Oz shown there make a *right-handed* coordinate frame. When you hold your right hand so that the fingers curl from the positive x-axis toward the positive y-axis, your thumb points along the positive z-axis.

The Cartesian coordinates (x, y, z) of a point P in space are the numbers at which the planes through P perpendicular to the axes cut the axes.

Points on the x-axis have y- and z-coordinates equal to zero. That is, they have coordinates of the form $(x, 0, 0)$. Similarly, points on the y-axis have coordinates of the form $(0, y, 0)$. Points on the z-axis have coordinates of the form $(0, 0, z)$.

The points in a plane perpendicular to the x-axis all have the same x-coordinate, this being the number at which that plane cuts the x-axis. The y- and z-coordinates can be any numbers. Similarly, the points in a plane perpendicular to the y-axis have a common y-coordinate and the points in a plane perpendicular to the z-axis have a common z-coordinate. To write equations for these planes, we name the common coordinate's value. The plane $x = 2$ is the plane perpendicular to the x-axis at $x = 2$. The plane $y = 3$ is the plane perpendicular to the y-axis at $y = 3$. The plane $z = 5$ is the plane perpendicular to the z-axis at $z = 5$. Figure 10.15 shows the planes $x = 2$, $y = 3$, and $z = 5$, together with their intersection point $(2, 3, 5)$.

The planes $x = 2$ and $y = 3$ in Fig. 10.15 intersect in a line parallel to the z-axis. This line is described by the *pair* of equations $x = 2$, $y = 3$. A point (x, y, z) lies on the line if and only if $x = 2$ and $y = 3$. Similarly, the line of intersection of the planes $y = 3$ and $z = 5$ is described by the equation pair $y = 3$, $z = 5$. This line runs parallel to the x-axis. The line of intersection of the planes $x = 2$ and $z = 5$, parallel to the y-axis, is described by the equation pair $x = 2$, $z = 5$.

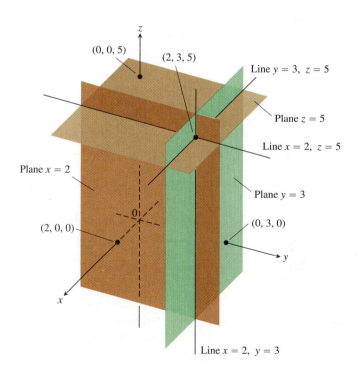

10.15 The planes $x = 2$, $y = 3$, and $z = 5$ determine three lines through the point $(2, 3, 5)$.

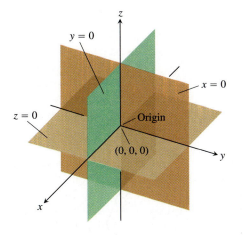

10.16 The planes $x = 0$, $y = 0$, and $z = 0$ divide space into eight octants.

The planes determined by the coordinate axes are the **xy-plane,** whose standard equation is $z = 0$; the **yz-plane,** whose standard equation is $x = 0$; and the **xz-plane,** whose standard equation is $y = 0$. They meet at the **origin** $(0, 0, 0)$ (Fig. 10.16).

The three **coordinate planes** $x = 0$, $y = 0$, and $z = 0$ divide space into eight cells called **octants.** The octant in which the point coordinates are all positive is called the **first octant;** there is no conventional numbering for the other seven octants.

Cartesian coordinates for space are also called **rectangular coordinates** because the axes that define them meet at right angles.

In the following examples, we match coordinate equations and inequalities with the sets of points they define in space.

EXAMPLE 1

Defining equations and inequalities	Verbal description
$z \geq 0$	The half-space consisting of the points on and above the x-plane.
$x = -3$	The plane perpendicular to the x-axis at $x = -3$. This plane lies parallel to the yz-plane and 3 units behind it.
$z = 0, x \leq 0, y \geq 0$	The second quadrant of the xy-plane.
$x \geq 0, y \geq 0, z \geq 0$	The first octant.
$-1 \leq y \leq 1$	The slab between the planes $y = -1$ and $y = 1$ (planes included).
$y = -2, z = 2$	The line in which the planes $y = -2$ and $z = 2$ intersect. Alternatively, the line through the point $(0, -2, 2)$ parallel to the x-axis.

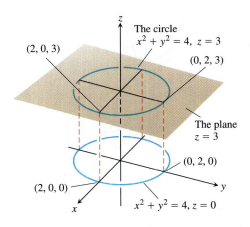

10.17 The circle $x^2 + y^2 = 4$, $z = 3$.

EXAMPLE 2   What points $P(x, y, z)$ satisfy the equations

$$x^2 + y^2 = 4 \qquad \text{and} \qquad z = 3?$$

Solution   The points lie in the horizontal plane $z = 3$ and, in this plane, make up the circle $x^2 + y^2 = 4$. We call this set of points "the circle $x^2 + y^2 = 4$ in the plane $z = 3$" or, more simply, "the circle $x^2 + y^2 = 4$, $z = 3$" (Fig. 10.17).

Vectors in Space

The sets of equivalent directed line segments that we use to represent forces, displacements, and velocities in space are called vectors, just as in the plane. The same rules of addition, subtraction, and scalar multiplication apply.

The vectors represented by the directed line segments from the origin to the points $(1, 0, 0)$, $(0, 1, 0)$, and $(0, 0, 1)$ are the **basic vectors** (Fig. 10.18, on the following page). We denote them by $\mathbf{i}$, $\mathbf{j}$, and $\mathbf{k}$. The **position vector r** from the origin O to the typical point $P(x, y, z)$ is

$$\mathbf{r} = \overrightarrow{OP} = x\,\mathbf{i} + y\,\mathbf{j} + z\,\mathbf{k}. \qquad (1)$$

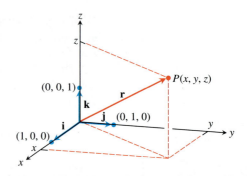

10.18 The position vector of a point in space.

> **Definition**
>
> **Addition and Subtraction for Vectors in Space**
>
> For any vectors $\mathbf{A} = a_1\,\mathbf{i} + a_2\,\mathbf{j} + a_3\,\mathbf{k}$ and $\mathbf{B} = b_1\,\mathbf{i} + b_2\,\mathbf{j} + b_3\,\mathbf{k}$,
>
> $$\mathbf{A} + \mathbf{B} = (a_1 + b_1)\,\mathbf{i} + (a_2 + b_2)\,\mathbf{j} + (a_3 + b_3)\,\mathbf{k}$$
>
> $$\mathbf{A} - \mathbf{B} = (a_1 - b_1)\,\mathbf{i} + (a_2 - b_2)\,\mathbf{j} + (a_3 - b_3)\,\mathbf{k}.$$

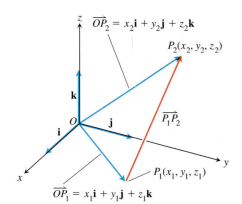

10.19 The vector from P_1 to P_2 is $\overrightarrow{P_1P_2}$ $= (x_2 - x_1)\,\mathbf{i} + (y_2 - y_1)\,\mathbf{j} + (z_2 - z_1)\,\mathbf{k}$.

The Vector Between Two Points

We can express the vector $\overrightarrow{P_1P_2}$ from the point $P_1(x_1, y_1, z_1)$ to the point $P_2(x_2, y_2, z_2)$ in terms of the coordinates of P_1 and P_2 because (Fig. 10.19)

$$\overrightarrow{P_1P_2} = \overrightarrow{OP_2} - \overrightarrow{OP_1}$$

$$= (x_2\,\mathbf{i} + y_2\,\mathbf{j} + z_2\,\mathbf{k}) - (x_1\,\mathbf{i} + y_1\,\mathbf{j} + z_1\,\mathbf{k}) \qquad (2)$$

$$= (x_2 - x_1)\,\mathbf{i} + (y_2 - y_1)\,\mathbf{j} + (z_2 - z_1)\,\mathbf{k}.$$

> The vector from $P_1(x_1, y_1, z_1)$ to $P_2(x_2, y_2, z_2)$ is
>
> $$\overrightarrow{P_1P_2} = (x_2 - x_1)\,\mathbf{i} + (y_2 - y_1)\,\mathbf{j} + (z_2 - z_1)\,\mathbf{k}. \qquad (3)$$

Magnitude

As always, the important features of a vector are its magnitude and direction. We find a formula for the magnitude (length) of $a_1\,\mathbf{i} + a_2\,\mathbf{j} + a_3\,\mathbf{k}$ by applying the Pythagorean theorem to the right triangles in Fig. 10.20. From triangle ABC,

$$|\overrightarrow{AC}| = \sqrt{a_1{}^2 + a_2{}^2}$$

and from triangle ACD,

$$|a_1\,\mathbf{i} + a_2\,\mathbf{j} + a_3\,\mathbf{k}| = |\overrightarrow{AD}| = \sqrt{|\overrightarrow{AC}|^2 + |\overrightarrow{CD}|^2} = \sqrt{a_1{}^2 + a_2{}^2 + a_3{}^2}.$$

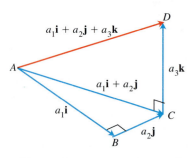

10.20 We find the length of $\overrightarrow{AD}$ by applying the Pythagorean theorem to the right triangles ABC and ACD.

> The **magnitude (length)** of $\mathbf{A} = a_1\,\mathbf{i} + a_2\,\mathbf{j} + a_3\,\mathbf{k}$ is
>
> $$|\mathbf{A}| = |a_1\,\mathbf{i} + a_2\,\mathbf{j} + a_3\,\mathbf{k}| = \sqrt{a_1{}^2 + a_2{}^2 + a_3{}^2}. \qquad (4)$$

Scalar Multiplication

> ### Definition
> If c is a scalar and $\mathbf{A} = a_1\,\mathbf{i} + a_2\,\mathbf{j} + a_3\,\mathbf{k}$ is a vector, then
> $$c\,\mathbf{A} = (ca_1)\,\mathbf{i} + (ca_2)\,\mathbf{j} + (ca_3)\,\mathbf{k}.$$

EXAMPLE 3   The length of $\mathbf{A} = \mathbf{i} - 2\,\mathbf{j} + 3\,\mathbf{k}$ is

$$|\mathbf{A}| = \sqrt{(1)^2 + (-2)^2 + (3)^2} = \sqrt{1 + 4 + 9} = \sqrt{14}.\qquad\square$$

If we multiply $\mathbf{A} = a_1\mathbf{i} + a_2\mathbf{j} + a_3\mathbf{k}$ by a scalar c, the length of $c\mathbf{A}$ is $|c|$ times the length of $\mathbf{A}$, as in the plane. The reason is the same, as well:

$$c\,\mathbf{A} = ca_1\,\mathbf{i} + ca_2\,\mathbf{j} + ca_3\,\mathbf{k},$$

$$|c\,\mathbf{A}| = \sqrt{(ca_1)^2 + (ca_2)^2 + (ca_3)^2} = \sqrt{c^2a_1{}^2 + c^2a_2{}^2 + c^2a_3{}^2} \qquad (5)$$

$$= |c|\sqrt{a_1{}^2 + a_2{}^2 + a_3{}^2} = |c||\mathbf{A}|.$$

EXAMPLE 4   If $\mathbf{A}$ is the vector of Example 3, then the length of

$$2\,\mathbf{A} = 2(\mathbf{i} - 2\,\mathbf{j} + 3\,\mathbf{k}) = 2\,\mathbf{i} - 4\,\mathbf{j} + 6\,\mathbf{k}$$

is

$$\sqrt{(2)^2 + (-4)^2 + (6)^2} = \sqrt{4 + 16 + 36} = \sqrt{56}$$

$$= \sqrt{4 \cdot 14} = 2\sqrt{14} = 2|\mathbf{A}|.\qquad\square$$

The Zero Vector

The **zero vector** in space is the vector $\mathbf{0} = 0\,\mathbf{i} + 0\,\mathbf{j} + 0\,\mathbf{k}$. As in the plane, $\mathbf{0}$ has zero length, and no direction.

Unit Vectors

A **unit vector** in space is a vector of length 1. The basic vectors are unit vectors because

$$|\mathbf{i}| = |1\,\mathbf{i} + 0\,\mathbf{j} + 0\,\mathbf{k}| = \sqrt{1^2 + 0^2 + 0^2} = 1,$$

$$|\mathbf{j}| = |0\,\mathbf{i} + 1\,\mathbf{j} + 0\,\mathbf{k}| = \sqrt{0^2 + 1^2 + 0^2} = 1,$$

$$|\mathbf{k}| = |0\,\mathbf{i} + 0\,\mathbf{j} + 1\,\mathbf{k}| = \sqrt{0^2 + 0^2 + 1^2} = 1.$$

Magnitude and Direction

If $\mathbf{A} \neq \mathbf{0}$, then $\mathbf{A}/|\mathbf{A}|$ is a unit vector in the direction of $\mathbf{A}$ and we can use the equation

$$\mathbf{A} = |\mathbf{A}|\ \frac{\mathbf{A}}{|\mathbf{A}|} \qquad (6)$$

to express $\mathbf{A}$ as a product of its magnitude and direction.

EXAMPLE 5   Express $\mathbf{A} = \mathbf{i} - 2\mathbf{j} + 3\mathbf{k}$ as a product of its magnitude and direction.

Solution

$$\mathbf{A} = |\mathbf{A}| \cdot \frac{\mathbf{A}}{|\mathbf{A}|} \qquad \text{Eq. (6)}$$

$$= \sqrt{14} \cdot \frac{\mathbf{i} - 2\mathbf{j} + 3\mathbf{k}}{\sqrt{14}} \qquad \text{From Example 3}$$

$$= \sqrt{14}\left(\frac{1}{\sqrt{14}}\mathbf{i} - \frac{2}{\sqrt{14}}\mathbf{j} + \frac{3}{\sqrt{14}}\mathbf{k}\right) = (\text{length of } \mathbf{A}) \cdot (\text{direction of } \mathbf{A})$$

EXAMPLE 6   Find a unit vector $\mathbf{u}$ in the direction of the vector from $P_1(1, 0, 1)$ to $P_2(3, 2, 0)$.

Solution   We divide $\overrightarrow{P_1 P_2}$ by its length:

$$\overrightarrow{P_1 P_2} = (3 - 1)\mathbf{i} + (2 - 0)\mathbf{j} + (0 - 1)\mathbf{k} = 2\mathbf{i} + 2\mathbf{j} - \mathbf{k}$$

$$|\overrightarrow{P_1 P_2}| = \sqrt{(2)^2 + (2)^2 + (-1)^2} = \sqrt{4 + 4 + 1} = \sqrt{9} = 3$$

$$\mathbf{u} = \frac{\overrightarrow{P_1 P_2}}{|\overrightarrow{P_1 P_2}|} = \frac{2\mathbf{i} + 2\mathbf{j} - \mathbf{k}}{3} = \frac{2}{3}\mathbf{i} + \frac{2}{3}\mathbf{j} - \frac{1}{3}\mathbf{k}.$$

EXAMPLE 7   Find a vector 6 units long in the direction of $\mathbf{A} = 2\mathbf{i} + 2\mathbf{j} - \mathbf{k}$.

Solution   The vector we want is

$$6\frac{\mathbf{A}}{|\mathbf{A}|} = 6\frac{2\mathbf{i} + 2\mathbf{j} - \mathbf{k}}{\sqrt{2^2 + 2^2 + (-1)^2}} = 6\frac{2\mathbf{i} + 2\mathbf{j} - \mathbf{k}}{3} = 4\mathbf{i} + 4\mathbf{j} - 2\mathbf{k}.$$

Distance in Space

The distance between two points P_1 and P_2 in space is the length of $\overrightarrow{P_1 P_2}$.

The Distance Between $P_1(x_1, y_1, z_1)$ and $P_2(x_2, y_2, z_2)$

$$|\overrightarrow{P_1 P_2}| = \sqrt{(x_2 - x_1)^2 + (y_2 - y_1)^2 + (z_2 - z_1)^2} \qquad (7)$$

EXAMPLE 8   The distance between $P_1(2, 1, 5)$ and $P_2(-2, 3, 0)$ is

$$|\overrightarrow{P_1 P_2}| = \sqrt{(-2 - 2)^2 + (3 - 1)^2 + (0 - 5)^2}$$

$$= \sqrt{16 + 4 + 25}$$

$$= \sqrt{45} = 3\sqrt{5}.$$

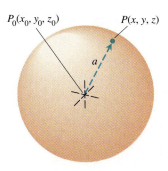

10.21 The standard equation of the sphere of radius a centered at (x_0, y_0, z_0) is

$$(x - x_0)^2 + (y - y_0)^2 + (z - z_0)^2 = a^2.$$

Spheres

We use Eq. (7) to write equations for spheres in space (Fig. 10.21). A point $P(x, y, z)$ lies on the sphere of radius a centered at $P_0(x_0, y_0, z_0)$ precisely when $|\overrightarrow{P_0P}| = a$ or

$$(x - x_0)^2 + (y - y_0)^2 + (z - z_0)^2 = a^2.$$

The Standard Equation for the Sphere of Radius a and Center (x_0, y_0, z_0)

$$(x - x_0)^2 + (y - y_0)^2 + (z - z_0)^2 = a^2 \qquad (8)$$

EXAMPLE 9   Find the center and radius of the sphere

$$x^2 + y^2 + z^2 + 3x - 4z + 1 = 0.$$

Solution   We find the center and radius of a sphere the way we find the center and radius of a circle: Complete the squares on the x-, y-, and z-terms as necessary and write each quadratic as a squared linear expression. Then, from the equation in standard form, read off the center and radius. For the sphere here we have

$$x^2 + y^2 + z^2 + 3x - 4z + 1 = 0$$

$$(x^2 + 3x \quad) + y^2 + (z^2 - 4z \quad) = -1$$

$$\left(x^2 + 3x + \left(\frac{3}{2}\right)^2\right) + y^2 + \left(z^2 - 4z + \left(\frac{-4}{2}\right)^2\right) = -1 + \left(\frac{3}{2}\right)^2 + \left(\frac{-4}{2}\right)^2$$

$$\left(x + \frac{3}{2}\right)^2 + y^2 + (z - 2)^2 = -1 + \frac{9}{4} + 4 = \frac{21}{4}.$$

This is Eq. (8) with $x_0 = -3/2$, $y_0 = 0$, $z_0 = 2$, and $a = \sqrt{21}/2$. The center is $(-3/2, 0, 2)$. The radius is $\sqrt{21}/2$.   ❑

EXAMPLE 10   *Sets bounded by spheres or portions of spheres*

Defining equations and inequalities	Description
a) $x^2 + y^2 + z^2 < 4$	The interior of the sphere $x^2 + y^2 + z^2 = 4$.
b) $x^2 + y^2 + z^2 \leq 4$	The solid ball bounded by the sphere $x^2 + y^2 + z^2 = 4$. Alternatively, the sphere $x^2 + y^2 + z^2 = 4$ together with its interior.
c) $x^2 + y^2 + z^2 > 4$	The exterior of the sphere $x^2 + y^2 + z^2 = 4$.
d) $x^2 + y^2 + z^2 = 4,$ $z \leq 0$	The lower hemisphere cut from the sphere $x^2 + y^2 + z^2 = 4$ by the xy-plane (the plane $z = 0$). ❑

DRAWING LESSON
How to Draw Three-dimensional Objects to Look Three-dimensional

1   Break lines. When one line passes behind another, break it to show that it doesn't touch and that part of it is hidden.

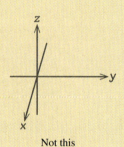

Intersecting   *CD* behind *AB*   *AB* behind *CD*

2   Make the angle between the positive *x*-axis and the positive *y*-axis large enough.

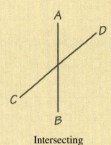   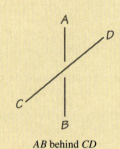

This   Not this

3   Draw planes parallel to the coordinate planes as if they were rectangles with sides parallel to the coordinate axes.

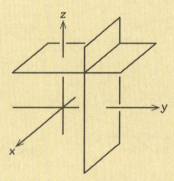

4   Dash or omit hidden portions of lines. Don't let the line touch the boundary of the parallelogram that represents the plane, unless the line lies in the plane.

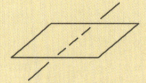   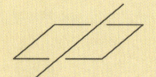   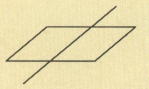

Line below plane   Line above plane   Line *in* plane

5 **Spheres: Draw the sphere first (outline and equator); draw axes, if any, later. Use line breaks and dashed lines.**

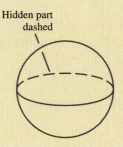

Hidden part dashed

Sphere first

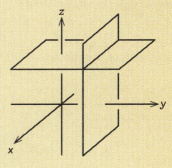

Break

A contact dot sometimes helps

Break

Axes later

6 **A general rule for perspective: Draw the object as if it lies some distance away, below, and to your left.**

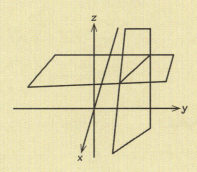

Advice ignored

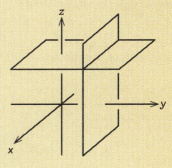

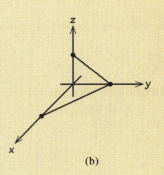

Advice followed

7 **To draw a plane that crosses all three coordinate axes, follow the steps shown here: (a) Sketch the axes and mark the intercepts. (b) Connect the intercepts to form two sides of a parallelogram. (c) Complete the parallelogram and enlarge it by drawing lines parallel to its sides. (d) Darken the exposed parts, break hidden lines, and, if desired, dash hidden portions of the axes. You may wish to erase the smaller parallelogram at this point.**

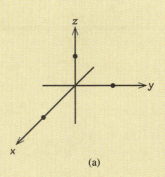

(a)

(b)

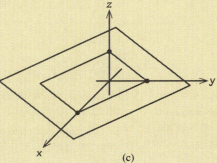

(c)

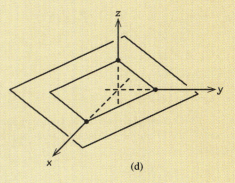

(d)

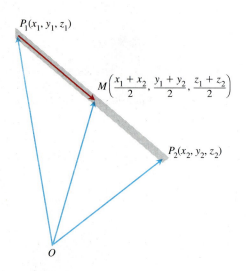

10.22 The coordinates of the midpoint are the averages of the coordinates of P_1 and P_2.

Midpoints

The coordinates of the midpoint of a line segment are found by averaging.

> The midpoint M of the line segment joining points $P_1(x_1, y_1, z_1)$ and $P_2(x_2, y_2, z_2)$ is the point
> $$\left(\frac{x_1 + x_2}{2}, \frac{y_1 + y_2}{2}, \frac{z_1 + z_2}{2}\right).$$

To see why, observe (Fig. 10.22) that

$$\overrightarrow{OM} = \overrightarrow{OP_1} + \frac{1}{2}(\overrightarrow{P_1P_2}) = \overrightarrow{OP_1} + \frac{1}{2}(\overrightarrow{OP_2} - \overrightarrow{OP_1})$$

$$= \frac{1}{2}(\overrightarrow{OP_1} + \overrightarrow{OP_2})$$

$$= \frac{x_1 + x_2}{2}\mathbf{i} + \frac{y_1 + y_2}{2}\mathbf{j} + \frac{z_1 + z_2}{2}\mathbf{k}.$$

EXAMPLE 11 The midpoint of the segment joining $P_1(3, -2, 0)$ and $P_2(7, 4, 4)$ is

$$\left(\frac{3 + 7}{2}, \frac{-2 + 4}{2}, \frac{0 + 4}{2}\right) = (5, 1, 2).$$

Exercises 10.2

Sets, Equations, and Inequalities

In Exercises 1–12, give a geometric description of the set of points in space whose coordinates satisfy the given pairs of equations.

1. $x = 2, \quad y = 3$
2. $x = -1, \quad z = 0$
3. $y = 0, \quad z = 0$
4. $x = 1, \quad y = 0$
5. $x^2 + y^2 = 4, \quad z = 0$
6. $x^2 + y^2 = 4, \quad z = -2$
7. $x^2 + z^2 = 4, \quad y = 0$
8. $y^2 + z^2 = 1, \quad x = 0$
9. $x^2 + y^2 + z^2 = 1, \quad x = 0$
10. $x^2 + y^2 + z^2 = 25, \quad y = -4$
11. $x^2 + y^2 + (z + 3)^2 = 25, \quad z = 0$
12. $x^2 + (y - 1)^2 + z^2 = 4, \quad y = 0$

In Exercises 13–18, describe the sets of points in space whose coordinates satisfy the given inequalities or combinations of equations and inequalities.

13. **a)** $x \geq 0, \quad y \geq 0, \quad z = 0$
    b) $x \geq 0, \quad y \leq 0, \quad z = 0$

14. **a)** $0 \leq x \leq 1$
    b) $0 \leq x \leq 1, \quad 0 \leq y \leq 1$
    c) $0 \leq x \leq 1, \quad 0 \leq y \leq 1, \quad 0 \leq z \leq 1$

15. **a)** $x^2 + y^2 + z^2 \leq 1$
    b) $x^2 + y^2 + z^2 > 1$

16. **a)** $x^2 + y^2 \leq 1, \quad z = 0$
    b) $x^2 + y^2 \leq 1, \quad z = 3$
    c) $x^2 + y^2 \leq 1, \quad$ no restriction on z

17. **a)** $x^2 + y^2 + z^2 = 1, \quad z \geq 0$
    b) $x^2 + y^2 + z^2 \leq 1, \quad z \geq 0$

18. **a)** $x = y, \quad z = 0$
    b) $x = y, \quad$ no restriction on z

In Exercises 19–28, describe the given set with a single equation or with a pair of equations.

19. The plane perpendicular to the
    a) x-axis at $(3, 0, 0)$
    b) y-axis at $(0, -1, 0)$
    c) z-axis at $(0, 0, -2)$

20. The plane through the point $(3, -1, 2)$ perpendicular to the

    a) x-axis   **b)** y-axis   **c)** z-axis

21. The plane through the point $(3, -1, 1)$ parallel to the

    a) xy-plane   **b)** yz-plane   **c)** xz-plane

22. The circle of radius 2 centered at $(0, 0, 0)$ and lying in the

    a) xy-plane   **b)** yz-plane   **c)** xz-plane

23. The circle of radius 2 centered at $(0, 2, 0)$ and lying in the

    a) xy-plane   **b)** yz-plane   **c)** plane $y = 2$

24. The circle of radius 1 centered at $(-3, 4, 1)$ and lying in a plane parallel to the

    a) xy-plane   **b)** yz-plane   **c)** xz-plane

25. The line through the point $(1, 3, -1)$ parallel to the

    a) x-axis   **b)** y-axis   **c)** z-axis

26. The set of points in space equidistant from the origin and the point $(0, 2, 0)$

27. The circle in which the plane through the point $(1, 1, 3)$ perpendicular to the z-axis meets the sphere of radius 5 centered at the origin

28. The set of points in space that lie 2 units from the point $(0, 0, 1)$ and, at the same time, 2 units from the point $(0, 0, -1)$

Write inequalities to describe the sets in Exercises 29–34.

29. The slab bounded by the planes $z = 0$ and $z = 1$ (planes included)

30. The solid cube in the first octant bounded by the coordinate planes and the planes $x = 2$, $y = 2$, and $z = 2$

31. The half-space consisting of the points on and below the xy-plane

32. The upper hemisphere of the sphere of radius 1 centered at the origin

33. The (a) interior and (b) exterior of the sphere of radius 1 centered at the point $(1, 1, 1)$

34. The closed region bounded by the spheres of radius 1 and radius 2 centered at the origin. (*Closed* means the spheres are to be included. Had we wanted the spheres left out, we would have asked for the *open* region bounded by the spheres. This is analogous to the way we use *closed* and *open* to describe intervals: *closed* means endpoints included, *open* means endpoints left out. Closed sets include boundaries; open sets leave them out.)

Length and Direction

In Exercises 35–44, express each vector as a product of its length and direction.

35. $2\mathbf{i} + \mathbf{j} - 2\mathbf{k}$ **36.** $3\mathbf{i} - 6\mathbf{j} + 2\mathbf{k}$

37. $\mathbf{i} + 4\mathbf{j} - 8\mathbf{k}$ **38.** $9\mathbf{i} - 2\mathbf{j} + 6\mathbf{k}$

39. $5\mathbf{k}$ **40.** $-4\mathbf{j}$

41. $\dfrac{3}{5}\mathbf{i} + \dfrac{4}{5}\mathbf{k}$ **42.** $\dfrac{1}{\sqrt{2}}\mathbf{i} - \dfrac{1}{\sqrt{2}}\mathbf{k}$

43. $\dfrac{1}{\sqrt{6}}\mathbf{i} - \dfrac{1}{\sqrt{6}}\mathbf{j} - \dfrac{1}{\sqrt{6}}\mathbf{k}$ **44.** $\dfrac{\mathbf{i}}{\sqrt{3}} + \dfrac{\mathbf{j}}{\sqrt{3}} + \dfrac{\mathbf{k}}{\sqrt{3}}$

45. Find the vectors whose lengths and directions are given. Try to do the calculations without writing.

	Length	Direction
a)	2	$\mathbf{i}$
b)	$\sqrt{3}$	$-\mathbf{k}$
c)	$\dfrac{1}{2}$	$\dfrac{3}{5}\mathbf{j} + \dfrac{4}{5}\mathbf{k}$
d)	7	$\dfrac{6}{7}\mathbf{i} - \dfrac{2}{7}\mathbf{j} + \dfrac{3}{7}\mathbf{k}$

46. Find the vectors whose lengths and directions are given. Try to do the calculations without writing.

	Length	Direction
a)	7	$-\mathbf{j}$
b)	$\sqrt{2}$	$-\dfrac{3}{5}\mathbf{i} - \dfrac{4}{5}\mathbf{k}$
c)	$\dfrac{13}{12}$	$\dfrac{3}{13}\mathbf{i} - \dfrac{4}{13}\mathbf{j} - \dfrac{12}{13}\mathbf{k}$
d)	$a > 0$	$\dfrac{1}{\sqrt{2}}\mathbf{i} + \dfrac{1}{\sqrt{3}}\mathbf{j} - \dfrac{1}{\sqrt{6}}\mathbf{k}$

47. Find a vector of magnitude 7 in the direction of $\mathbf{A} = 12\mathbf{i} - 5\mathbf{k}$.

48. Find a vector $\sqrt{5}$ units long in the direction of $\mathbf{A} = \mathbf{i} + \mathbf{j} + \mathbf{k}$.

49. Find a vector 5 units long in the direction opposite to the direction of $\mathbf{A} = 2\mathbf{i} - 3\mathbf{j} + 6\mathbf{k}$.

50. Find a vector of magnitude 3 in the direction opposite to the direction of $\mathbf{A} = (1/2)\mathbf{i} - (1/2)\mathbf{j} - (1/2)\mathbf{k}$.

Vectors Determined by Points; Midpoints and Distance

In Exercises 51–56, find

    a) the distance between points P_1 and P_2,

    b) the direction of $\overrightarrow{P_1 P_2}$,

    c) the midpoint of line segment $P_1 P_2$.

51. $P_1(1, 1, 1)$, $P_2(3, 3, 0)$

52. $P_1(-1, 1, 5)$, $P_2(2, 5, 0)$

53. $P_1(1, 4, 5)$, $P_2(4, -2, 7)$

54. $P_1(3, 4, 5)$, $P_2(2, 3, 4)$

55. $P_1(0, 0, 0)$, $P_2(2, -2, -2)$

56. $P_1(5, 3, -2)$, $P_2(0, 0, 0)$

57. If $\overrightarrow{AB} = \mathbf{i} + 4\,\mathbf{j} - 2\,\mathbf{k}$ and B is the point $(5, 1, 3)$, find A.

58. If $\overrightarrow{AB} = -7\,\mathbf{i} + 3\,\mathbf{j} + 8\,\mathbf{k}$ and A is the point $(-2, -3, 6)$, find B.

Spheres

Find the centers and radii of the spheres in Exercises 59–62.

59. $(x + 2)^2 + y^2 + (z - 2)^2 = 8$

60. $\left(x + \frac{1}{2}\right)^2 + \left(y + \frac{1}{2}\right)^2 + \left(z + \frac{1}{2}\right)^2 = \frac{21}{4}$

61. $\left(x - \sqrt{2}\right)^2 + \left(y - \sqrt{2}\right)^2 + \left(z + \sqrt{2}\right)^2 = 2$

62. $x^2 + \left(y + \frac{1}{3}\right)^2 + \left(z - \frac{1}{3}\right)^2 = \frac{29}{9}$

Find equations for the spheres whose centers and radii are given in Exercises 63–66.

	Center	Radius
63.	(1, 2, 3)	$\sqrt{14}$
64.	(0, −1, 5)	2
65.	(−2, 0, 0)	$\sqrt{3}$
66.	(0, −7, 0)	7

Find the centers and radii of the spheres in Exercises 67–70.

67. $x^2 + y^2 + z^2 + 4x - 4z = 0$

68. $x^2 + y^2 + z^2 - 6y + 8z = 0$

69. $2x^2 + 2y^2 + 2z^2 + x + y + z = 9$

70. $3x^2 + 3y^2 + 3z^2 + 2y - 2z = 9$

71. Find a formula for the distance from the point $P(x, y, z)$ to the

   a) x-axis   **b)** y-axis   **c)** z-axis

72. Find a formula for the distance from the point $P(x, y, z)$ to the

   a) xy-plane   **b)** yz-plane   **c)** xz-plane

Geometry with Vectors

73. Suppose that A, B, and C are the corner points of the thin triangular plate of constant density shown here.

   a) Find the vector from C to the midpoint M of side AB.

   b) Find the vector from C to the point that lies two-thirds of the way from C to M on the median CM.

   c) Find the coordinates of the point in which the medians of $\triangle ABC$ intersect. According to Exercise 31, Section 5.7, this point is the plate's center of mass.

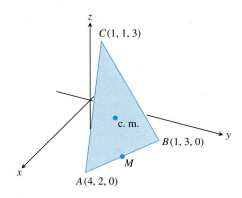

74. Find the vector from the origin to the point of intersection of the medians of the triangle whose vertices are

$$A(1, -1, 2), \quad B(2, 1, 3), \quad \text{and} \quad C(-1, 2, -1).$$

75. Let $ABCD$ be a general, not necessarily planar, quadrilateral in space. Show that the two segments joining the midpoints of opposite sides of $ABCD$ bisect each other. (*Hint:* Show that the segments have the same midpoint.)

76. Vectors are drawn from the center of a regular n-sided polygon in the plane to the vertices of the polygon. Show that the sum of the vectors is zero. (*Hint:* What happens to the sum if you rotate the polygon about its center?)

77. Suppose that A, B, and C are vertices of a triangle and that a, b, and c are, respectively, the midpoints of the opposite sides. Show that $\overrightarrow{Aa} + \overrightarrow{Bb} + \overrightarrow{Cc} = 0$.

10.3

Dot Products

We now introduce the dot product, the first of two methods we will learn for multiplying vectors together. Dot products are also called *scalar* products because the multiplication results in a scalar, not a vector. The products in the next section are vectors.

Scalar Products

When two nonzero vectors **A** and **B** are placed so their initial points coincide, they form an angle θ of measure $0 \leq \theta \leq \pi$ (Fig. 10.23). This angle is called the **angle** between **A** and **B**.

10.23 The angle between **A** and **B**.

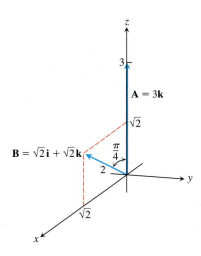

10.24 The vectors in Example 1.

In words, $\mathbf{A} \cdot \mathbf{B}$ is the length of $\mathbf{A}$ times the length of $\mathbf{B}$ times the cosine of the angle between $\mathbf{A}$ and $\mathbf{B}$.

EXAMPLE 1   If $\mathbf{A} = 3\,\mathbf{k}$ and $\mathbf{B} = \sqrt{2}\,\mathbf{i} + \sqrt{2}\,\mathbf{k}$ (Fig. 10.24), then

$$\mathbf{A} \cdot \mathbf{B} = |\mathbf{A}||\mathbf{B}| \cos \theta = (3)(2) \cos \frac{\pi}{4} = 6 \cdot \frac{\sqrt{2}}{2} = 3\sqrt{2}. \qquad \square$$

Since the sign of $\mathbf{A} \cdot \mathbf{B}$ is determined by $\cos \theta$, the scalar product is positive if the angle between the vectors is acute, negative if the angle is obtuse. (We look at right angles in a moment.)

Since the angle a vector $\mathbf{A}$ makes with itself is zero, and $\cos 0 = 1$,

$$\mathbf{A} \cdot \mathbf{A} = |\mathbf{A}||\mathbf{A}| \cos 0 = |\mathbf{A}||\mathbf{A}|(1) = |\mathbf{A}|^2, \qquad \text{or} \qquad |\mathbf{A}| = \sqrt{\mathbf{A} \cdot \mathbf{A}}. \tag{2}$$

Calculation

To calculate $\mathbf{A} \cdot \mathbf{B}$ from the components of $\mathbf{A}$ and $\mathbf{B}$ in a Cartesian coordinate system with unit vectors $\mathbf{i}, \mathbf{j},$ and $\mathbf{k}$, we let

$$\mathbf{A} = a_1\,\mathbf{i} + a_2\,\mathbf{j} + a_3\,\mathbf{k},$$
$$\mathbf{B} = b_1\,\mathbf{i} + b_2\,\mathbf{j} + b_3\,\mathbf{k},$$

and

$$\mathbf{C} = \mathbf{B} - \mathbf{A} = (b_1 - a_1)\,\mathbf{i} + (b_2 - a_2)\,\mathbf{j} + (b_3 - a_3)\,\mathbf{k}.$$

The law of cosines for the triangle whose sides represent $\mathbf{A}, \mathbf{B},$ and $\mathbf{C}$ (Fig. 10.25) is

$$|\mathbf{C}|^2 = |\mathbf{A}|^2 + |\mathbf{B}|^2 - 2|\mathbf{A}||\mathbf{B}| \cos \theta,$$

$$|\mathbf{A}||\mathbf{B}| \cos \theta = \frac{|\mathbf{A}|^2 + |\mathbf{B}|^2 - |\mathbf{C}|^2}{2}.$$

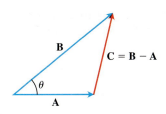

10.25 We obtain Eq. (3) by applying the law of cosines to a triangle whose sides represent $\mathbf{A}, \mathbf{B},$ and $\mathbf{C} = \mathbf{B} - \mathbf{A}$.

The left side of this equation is $\mathbf{A} \cdot \mathbf{B}$. We evaluate the right side by squaring the components of $\mathbf{A}, \mathbf{B},$ and $\mathbf{C}$ (Eq. 4, Section 10.2). The resulting cancellations give

$$\mathbf{A} \cdot \mathbf{B} = a_1 b_1 + a_2 b_2 + a_3 b_3. \tag{3}$$

Thus, to find the scalar product of two given vectors we multiply their corresponding $\mathbf{i}$-, $\mathbf{j}$-, and $\mathbf{k}$- components and add the results.

Solving Eq. (1) for θ gives a formula for finding angles between vectors.

The angle between two nonzero vectors **A** and **B** is

$$\theta = \cos^{-1}\left(\frac{\mathbf{A} \cdot \mathbf{B}}{|\mathbf{A}||\mathbf{B}|}\right). \tag{4}$$

Since the values of the arc cosine lie in $[0, \pi]$, Eq. (4) automatically gives the angle made by **A** and **B.**

EXAMPLE 2   Find the angle between $\mathbf{A} = \mathbf{i} - 2\mathbf{j} - 2\mathbf{k}$ and $\mathbf{B} = 6\mathbf{i} + 3\mathbf{j} + 2\mathbf{k}$.

Solution   We use Eq. (4):

$$\mathbf{A} \cdot \mathbf{B} = (1)(6) + (-2)(3) + (-2)(2) = 6 - 6 - 4 = -4$$

$$|\mathbf{A}| = \sqrt{(1)^2 + (-2)^2 + (-2)^2} = \sqrt{9} = 3$$

$$|\mathbf{B}| = \sqrt{(6)^2 + (3)^2 + (2)^2} = \sqrt{49} = 7$$

$$\theta = \cos^{-1}\left(\frac{\mathbf{A} \cdot \mathbf{B}}{|\mathbf{A}||\mathbf{B}|}\right) \qquad \text{Eq. (4)}$$

$$= \cos^{-1}\left(\frac{-4}{(3)(7)}\right) = \cos^{-1}\left(-\frac{4}{21}\right) \approx 1.76 \text{ rad} \qquad \text{About } 101°$$

❏

Laws of the Dot Product

From the equation $\mathbf{A} \cdot \mathbf{B} = a_1b_1 + a_2b_2 + a_3b_3$, we can see right away that

---

$$\mathbf{A} \cdot \mathbf{B} = \mathbf{B} \cdot \mathbf{A}. \tag{5}$$

In other words, the dot product is commutative. We can also see from Eq. (3) that if c is any number (or scalar), then

---

$$(c\,\mathbf{A}) \cdot \mathbf{B} = \mathbf{A} \cdot (c\,\mathbf{B}) = c\,(\mathbf{A} \cdot \mathbf{B}). \tag{6}$$

If $\mathbf{C} = c_1\mathbf{i} + c_2\mathbf{j} + c_3\mathbf{k}$ is any third vector, then

$$\mathbf{A} \cdot (\mathbf{B} + \mathbf{C}) = a_1(b_1 + c_1) + a_2(b_2 + c_2) + a_3(b_3 + c_3)$$

$$= (a_1b_1 + a_2b_2 + a_3b_3) + (a_1c_1 + a_2c_2 + a_3c_3)$$

$$= \mathbf{A} \cdot \mathbf{B} + \mathbf{A} \cdot \mathbf{C}.$$

Hence dot products obey the distributive law:

$$\mathbf{A} \cdot (\mathbf{B} + \mathbf{C}) = \mathbf{A} \cdot \mathbf{B} + \mathbf{A} \cdot \mathbf{C}. \tag{7}$$

If we combine this with the commutative law, Eq. (5), it is also evident that

$$(\mathbf{A} + \mathbf{B}) \cdot \mathbf{C} = \mathbf{A} \cdot \mathbf{C} + \mathbf{B} \cdot \mathbf{C}. \tag{8}$$

Equations (7) and (8) together permit us to multiply sums of vectors by the familiar laws of algebra. For example,

$$(\mathbf{A} + \mathbf{B}) \cdot (\mathbf{C} + \mathbf{D}) = \mathbf{A} \cdot \mathbf{C} + \mathbf{A} \cdot \mathbf{D} + \mathbf{B} \cdot \mathbf{C} + \mathbf{B} \cdot \mathbf{D}. \tag{9}$$

Perpendicular (Orthogonal) Vectors

Two nonzero vectors **A** and **B** are perpendicular or **orthogonal** if the angle between them is $\pi/2$. For such vectors, we automatically have $\mathbf{A} \cdot \mathbf{B} = 0$ because $\cos(\pi/2) = 0$. The converse is also true. If **A** and **B** are nonzero vectors with $\mathbf{A} \cdot \mathbf{B} = |\mathbf{A}||\mathbf{B}| \cos\theta = 0$, then $\cos\theta = 0$ and $\theta = \cos^{-1} 0 = \pi/2$.

> Nonzero vectors **A** and **B** are perpendicular (orthogonal) if and only if $\mathbf{A} \cdot \mathbf{B} = 0$.

EXAMPLE 3 $\mathbf{A} = 3\mathbf{i} - 2\mathbf{j} + \mathbf{k}$ and $\mathbf{B} = 2\mathbf{j} + 4\mathbf{k}$ are orthogonal because

$$\mathbf{A} \cdot \mathbf{B} = (3)(0) + (-2)(2) + (1)(4) = 0. \qquad \square$$

Vector Projections

The **vector projection** of $\mathbf{B} = \overrightarrow{PQ}$ onto a nonzero vector $\mathbf{A} = \overrightarrow{PS}$ (Fig. 10.26) is the vector $\overrightarrow{PR}$ determined by dropping a perpendicular from Q to the line PS. The notation for this vector is

$$\text{proj}_{\mathbf{A}}\,\mathbf{B} \qquad (\text{"the vector projection of } \mathbf{B} \text{ onto } \mathbf{A}\text{"}).$$

If **B** represents a force, then $\text{proj}_{\mathbf{A}}\,\mathbf{B}$ represents the effective force in the direction of **A** (Fig. 10.27).

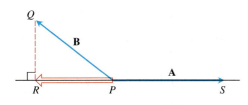

10.26 The vector projection of **B** onto **A**.

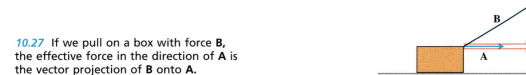

10.27 If we pull on a box with force **B**, the effective force in the direction of **A** is the vector projection of **B** onto **A**.

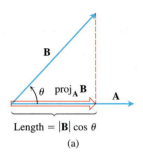

Length = $|\mathbf{B}| \cos \theta$
(a)

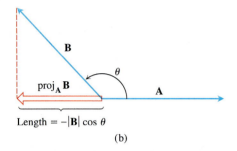

Length = $-|\mathbf{B}| \cos \theta$
(b)

10.28 The length of proj$_A$ **B** is (a) $|\mathbf{B}| \cos \theta$ if $\cos \theta \geq 0$ and (b) $-|\mathbf{B}| \cos \theta$ if $\cos \theta < 0$.

If the angle θ between **A** and **B** is acute, proj$_A$ **B** has length $|\mathbf{B}| \cos \theta$ and direction $\mathbf{A}/|\mathbf{A}|$(Fig. 10.28). If θ is obtuse, $\cos \theta < 0$ and proj$_A$ **B** has length $-|\mathbf{B}| \cos \theta$ and direction $-\mathbf{A}/|\mathbf{A}|$. In any case,

$$\text{proj}_A\,\mathbf{B} = (|\mathbf{B}| \cos \theta)\frac{\mathbf{A}}{|\mathbf{A}|}$$

$$= \left(\frac{\mathbf{A} \cdot \mathbf{B}}{|\mathbf{A}|}\right)\frac{\mathbf{A}}{|\mathbf{A}|}$$

$$= \left(\mathbf{B} \cdot \frac{\mathbf{A}}{|\mathbf{A}|}\right)\frac{\mathbf{A}}{|\mathbf{A}|}.$$

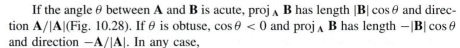

$$|\mathbf{B}| \cos \theta = \frac{|\mathbf{A}||\mathbf{B}| \cos \theta}{|\mathbf{A}|}$$
$$= \frac{\mathbf{A} \cdot \mathbf{B}}{|\mathbf{A}|}$$

$$\boxed{\text{proj}_A\,\mathbf{B} = \left(\mathbf{B} \cdot \frac{\mathbf{A}}{|\mathbf{A}|}\right)\frac{\mathbf{A}}{|\mathbf{A}|} = \left(\frac{\mathbf{B} \cdot \mathbf{A}}{\mathbf{A} \cdot \mathbf{A}}\right)\mathbf{A}} \qquad (10)$$

The number $|\mathbf{B}| \cos \theta$ is called the **scalar component of B in the direction of A.** Since

$$|\mathbf{B}| \cos \theta = \mathbf{B} \cdot \frac{\mathbf{A}}{|\mathbf{A}|}, \qquad (11)$$

we can find the scalar component by "dotting" **B** with the direction of **A**. Equation (10) says that the vector projection of **B** onto **A** is the scalar component of **B** in the direction of **A** times the direction of **A**.

While the first part of Eq. (10) describes the effect of **B** in the direction of **A**, the second part is better for calculation because it avoids square roots.

EXAMPLE 4   Find the vector projection of $\mathbf{B} = 6\,\mathbf{i} + 3\,\mathbf{j} + 2\,\mathbf{k}$ onto $\mathbf{A} = \mathbf{i} - 2\,\mathbf{j} - 2\,\mathbf{k}$ and the scalar component of **B** in the direction of **A**.

Solution   We find proj$_A$ **B** from Eq. (10):

$$\text{proj}_A\,\mathbf{B} = \frac{\mathbf{B} \cdot \mathbf{A}}{\mathbf{A} \cdot \mathbf{A}}\mathbf{A} = \frac{6 - 6 - 4}{1 + 4 + 4}(\mathbf{i} - 2\,\mathbf{j} - 2\,\mathbf{k})$$

$$= -\frac{4}{9}(\mathbf{i} - 2\,\mathbf{j} - 2\,\mathbf{k}) = -\frac{4}{9}\mathbf{i} + \frac{8}{9}\mathbf{j} + \frac{8}{9}\mathbf{k}.$$

We find the scalar component of **B** in the direction of **A** from Eq. (11):

$$|\mathbf{B}| \cos \theta = \mathbf{B} \cdot \frac{\mathbf{A}}{|\mathbf{A}|} = (6\,\mathbf{i} + 3\,\mathbf{j} + 2\,\mathbf{k}) \cdot \left(\frac{1}{3}\mathbf{i} - \frac{2}{3}\mathbf{j} - \frac{2}{3}\mathbf{k}\right)$$

$$= 2 - 2 - \frac{4}{3} = -\frac{4}{3}. \qquad \square$$

Writing a Vector as a Sum of Orthogonal Vectors

In mechanics, we often need to express a vector **B** as a sum of a vector parallel to a vector **A** and a vector orthogonal to **A**. We can accomplish this with the equation

$$\mathbf{B} = \text{proj}_A\,\mathbf{B} + (\mathbf{B} - \text{proj}_A\,\mathbf{B}), \qquad (12)$$

shown in Fig. 10.29.

10.29 Writing **B** as the sum of vectors parallel and orthogonal to **A**.

Where vectors came from

Although Aristotle used vectors to describe the effects of forces, the idea of resolving vectors into geometric components parallel to the coordinate axes came from Descartes. The algebra of vectors we use today was developed simultaneously and independently in the 1870s by Josiah Willard Gibbs (1839–1903), a mathematical physicist at Yale University, and by the English mathematical physicist Oliver Heaviside (1850–1925), the Heaviside of Heaviside layer fame. The works of Gibbs and Heaviside grew out of more complicated mathematical theories developed some years earlier by the Irish mathematician William Hamilton (1805–1865) and the German linguist, physicist, and geometer Hermann Grassman (1809–1877).

How to Write B as a Vector Parallel to A Plus a Vector Orthogonal to A

$$\mathbf{B} = \text{proj}_{\mathbf{A}}\,\mathbf{B} + (\mathbf{B} - \text{proj}_{\mathbf{A}}\,\mathbf{B})$$

$$= \underbrace{\left(\frac{\mathbf{B}\cdot\mathbf{A}}{\mathbf{A}\cdot\mathbf{A}}\right)\mathbf{A}}_{\text{parallel to }\mathbf{A}} + \underbrace{\left(\mathbf{B} - \left(\frac{\mathbf{B}\cdot\mathbf{A}}{\mathbf{A}\cdot\mathbf{A}}\right)\mathbf{A}\right)}_{\text{orthogonal to }\mathbf{A}} \tag{13}$$

EXAMPLE 5   Express $\mathbf{B} = 2\,\mathbf{i} + \mathbf{j} - 3\,\mathbf{k}$ as the sum of a vector parallel to $\mathbf{A} = 3\,\mathbf{i} - \mathbf{j}$ and a vector orthogonal to $\mathbf{A}$.

Solution   We use Eq. (13). With

$$\mathbf{A}\cdot\mathbf{B} = 6 - 1 = 5 \quad\text{and}\quad \mathbf{A}\cdot\mathbf{A} = 9 + 1 = 10,$$

Eq. (13) gives

$$\mathbf{B} = \frac{\mathbf{B}\cdot\mathbf{A}}{\mathbf{A}\cdot\mathbf{A}}\mathbf{A} + \left(\mathbf{B} - \frac{\mathbf{B}\cdot\mathbf{A}}{\mathbf{A}\cdot\mathbf{A}}\mathbf{A}\right) = \frac{5}{10}(3\,\mathbf{i} - \mathbf{j}) + \left(2\,\mathbf{i} + \mathbf{j} - 3\,\mathbf{k} - \frac{5}{10}(3\,\mathbf{i} - \mathbf{j})\right)$$

$$= \left(\frac{3}{2}\,\mathbf{i} - \frac{1}{2}\,\mathbf{j}\right) + \left(\frac{1}{2}\,\mathbf{i} + \frac{3}{2}\,\mathbf{j} - 3\,\mathbf{k}\right).$$

Check: The first vector in the sum is parallel to $\mathbf{A}$ because it is $(1/2)\,\mathbf{A}$. The second vector in the sum is orthogonal to $\mathbf{A}$ because

$$\left(\frac{1}{2}\,\mathbf{i} + \frac{3}{2}\,\mathbf{j} - 3\,\mathbf{k}\right)\cdot(3\,\mathbf{i} - \mathbf{j}) = \frac{3}{2} - \frac{3}{2} = 0.$$

Work

In Section 5.8, we calculated the work done by a constant force of magnitude F in moving an object through a distance d as $W = Fd$. That formula holds only if the force is directed along the line of motion. If a force $\mathbf{F}$ moving an object through a displacement $\mathbf{D} = \overrightarrow{PQ}$ has some other direction, the work is performed by the component of $\mathbf{F}$ in the direction of $\mathbf{D}$. If θ is the angle between $\mathbf{F}$ and $\mathbf{D}$ (Fig. 10.30), then

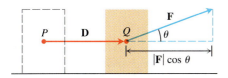

10.30 The work done by a constant force **F** during a displacement **D** is $(|\mathbf{F}|\cos\theta)|\mathbf{D}|$.

$$\text{Work} = \binom{\text{scalar component of }\mathbf{F}}{\text{in the direction of }\mathbf{D}}\binom{}{\text{length of }\mathbf{D}}$$

$$= (|\mathbf{F}|\cos\theta)\,|\mathbf{D}|$$

$$= \mathbf{F}\cdot\mathbf{D}$$

Definition

The **work** done by a constant force $\mathbf{F}$ acting through a displacement $\mathbf{D} = \overrightarrow{PQ}$ is

$$W = \mathbf{F}\cdot\mathbf{D} = |\mathbf{F}||\mathbf{D}|\cos\theta, \tag{14}$$

where θ is the angle between $\mathbf{F}$ and $\mathbf{D}$.

Work

The standard units of work are the foot-pound and newton-meter, both force-distance units. The newton-meter is usually called a *joule*. For more about the notion of work and the units involved, read the first few paragraphs of Section 5.8.

EXAMPLE 6   If $|\mathbf{F}| = 40$ N (newtons), $|\mathbf{D}| = 3$ m, and $\theta = 60°$, the work done by $\mathbf{F}$ in acting from P to Q is

$$\text{Work} = |\mathbf{F}||\mathbf{D}|\cos\theta \qquad \text{Eq. (14)}$$

$$= (40)(3)\cos 60° \qquad \text{Given values}$$

$$= (120)(1/2)$$

$$= 60 \text{ J (joules)}$$

We will encounter more interesting work problems in Chapter 14 when we can find the work done by a variable force along a path in space.

Exercises 10.3

Calculations

In Exercises 1–10, find

a) $\mathbf{A} \cdot \mathbf{B}$, $|\mathbf{A}|$, $|\mathbf{B}|$;
b) the cosine of the angle between $\mathbf{A}$ and $\mathbf{B}$;
c) the scalar component of $\mathbf{B}$ in the direction of $\mathbf{A}$;
d) the vector $\text{proj}_\mathbf{A}\mathbf{B}$.

1. $\mathbf{A} = 2\mathbf{i} - 4\mathbf{j} + \sqrt{5}\,\mathbf{k}$, $\mathbf{B} = -2\mathbf{i} + 4\mathbf{j} - \sqrt{5}\,\mathbf{k}$

2. $\mathbf{A} = (3/5)\mathbf{i} + (4/5)\mathbf{k}$, $\mathbf{B} = 5\mathbf{i} + 12\mathbf{j}$

3. $\mathbf{A} = 10\mathbf{i} + 11\mathbf{j} - 2\mathbf{k}$, $\mathbf{B} = 3\mathbf{j} + 4\mathbf{k}$

4. $\mathbf{A} = 2\mathbf{i} + 10\mathbf{j} - 11\mathbf{k}$, $\mathbf{B} = 2\mathbf{i} + 2\mathbf{j} + \mathbf{k}$

5. $\mathbf{A} = -2\mathbf{i} + 7\mathbf{j}$, $\mathbf{B} = \mathbf{k}$

6. $\mathbf{A} = \dfrac{1}{\sqrt{2}}\mathbf{i} + \dfrac{1}{\sqrt{3}}\mathbf{j} + \dfrac{1}{\sqrt{6}}\mathbf{k}$, $\mathbf{B} = \dfrac{1}{\sqrt{2}}\mathbf{j} - \mathbf{k}$

7. $\mathbf{A} = 5\mathbf{j} - 3\mathbf{k}$, $\mathbf{B} = \mathbf{i} + \mathbf{j} + \mathbf{k}$

8. $\mathbf{A} = \mathbf{i} + \mathbf{k}$, $\mathbf{B} = \mathbf{i} + \mathbf{j} + \mathbf{k}$

9. $\mathbf{A} = -\mathbf{i} + \mathbf{j}$, $\mathbf{B} = \sqrt{2}\mathbf{i} + \sqrt{3}\mathbf{j} + 2\mathbf{k}$

10. $\mathbf{A} = -5\mathbf{i} + \mathbf{j}$, $\mathbf{B} = 2\mathbf{i} + \sqrt{17}\mathbf{j} + 10\mathbf{k}$

11. Write $\mathbf{B} = 3\mathbf{j} + 4\mathbf{k}$ as the sum of a vector parallel to $\mathbf{A} = \mathbf{i} + \mathbf{j}$ and a vector orthogonal to $\mathbf{A}$.

12. Write $\mathbf{B} = \mathbf{j} + \mathbf{k}$ as the sum of a vector parallel to $\mathbf{A} = \mathbf{i} + \mathbf{j}$ and a vector orthogonal to $\mathbf{A}$.

13. Write $\mathbf{B} = 8\mathbf{i} + 4\mathbf{j} - 12\mathbf{k}$ as the sum of a vector parallel to $\mathbf{A} = \mathbf{i} + 2\mathbf{j} - \mathbf{k}$ and a vector orthogonal to $\mathbf{A}$.

14. $\mathbf{B} = \mathbf{i} + (\mathbf{j} + \mathbf{k})$ is already the sum of a vector parallel to $\mathbf{i}$ and a vector orthogonal to $\mathbf{i}$. If you use Eq. (13) with $\mathbf{A} = \mathbf{i}$, do you get $\mathbf{B}_1 = \mathbf{i}$ and $\mathbf{B}_2 = \mathbf{j} + \mathbf{k}$? Try it and find out.

Geometry

15. *Sums and differences.* In the accompanying figure, it looks as

if $\mathbf{v}_1 + \mathbf{v}_2$ and $\mathbf{v}_1 - \mathbf{v}_2$ are orthogonal. Is this mere coincidence, or are there circumstances under which we may expect the sum of two vectors to be orthogonal to their difference? Give reasons for your answer.

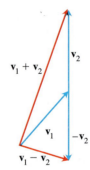

16. Suppose that AB is the diameter of a circle with center O and that C is a point on one of the two arcs joining A and B. Show that $\overrightarrow{CA}$ and $\overrightarrow{CB}$ are orthogonal.

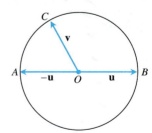

17. Show that the diagonals of a rhombus (parallelogram with sides of equal length) are perpendicular.

18. Show that squares are the only rectangles with perpendicular diagonals.

19. Prove the fact, often exploited by carpenters, that a parallelogram is a rectangle if and only if its diagonals are equal in length.

20. Show that the indicated diagonal of the parallelogram determined by vectors **u** and **v** bisects the angle between **u** and **v** if $|\mathbf{u}| = |\mathbf{v}|$.

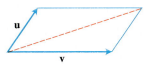

21. The accompanying figure shows a pyramid $OABCD$ with a square base whose sides are 1 unit long. The pyramid's height is also 1 unit, and the point D stands directly above the midpoint of the diagonal OB. Find the angle between $\overrightarrow{OB}$ and $\overrightarrow{OD}$.

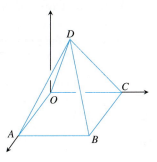

22. *Direction angles and direction cosines.* The direction angles α, β, and γ of a vector $\mathbf{v} = a\,\mathbf{i} + b\,\mathbf{j} + c\,\mathbf{k}$ are defined as follows:

α is the angle between **v** and the positive x-axis $(0 \le \alpha \le \pi)$,

β is the angle between **v** and the positive y-axis $(0 \le \beta \le \pi)$,

γ is the angle between **v** and the positive z-axis $(0 \le \gamma \le \pi)$.

a) Show that

$$\cos\alpha = \frac{a}{|\mathbf{v}|}, \quad \cos\beta = \frac{b}{|\mathbf{v}|}, \quad \cos\gamma = \frac{c}{|\mathbf{v}|},$$

and $\cos^2\alpha + \cos^2\beta + \cos^2\gamma = 1$. These cosines are called the direction cosines of **v**.

b) *Unit vectors are built from direction cosines.* Show that if $\mathbf{v} = a\,\mathbf{i} + b\,\mathbf{j} + c\,\mathbf{k}$ is a unit vector then a, b, and c are the direction cosines of **v**.

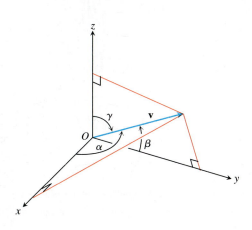

Angles Between Vectors

CALCULATOR Find the angles between the vectors in Exercises 23–26 to the nearest hundredth of a radian.

23. $\mathbf{A} = 2\mathbf{i} + \mathbf{j}, \quad \mathbf{B} = \mathbf{i} + 2\mathbf{j} - \mathbf{k}$

24. $\mathbf{A} = 2\mathbf{i} - 2\mathbf{j} + \mathbf{k}, \quad \mathbf{B} = 3\mathbf{i} + 4\mathbf{k}$

25. $\mathbf{A} = \sqrt{3}\,\mathbf{i} - 7\mathbf{j}, \quad \mathbf{B} = \sqrt{3}\,\mathbf{i} + \mathbf{j} - 2\mathbf{k}$

26. $\mathbf{A} = \mathbf{i} + \sqrt{2}\,\mathbf{j} - \sqrt{2}\,\mathbf{k}, \quad \mathbf{B} = -\mathbf{i} + \mathbf{j} + \mathbf{k}$

CALCULATOR Find the angles in Exercises 27–29 to the nearest hundredth of a radian.

27. The interior angles of the triangle ABC whose vertices are $A(-1, 0, 2)$, $B(2, 1, -1)$, and $C(1, -2, 2)$

28. The angle between $\mathbf{A} = 2\mathbf{i} + 2\mathbf{j} + \mathbf{k}$ and $\mathbf{B} = 2\mathbf{i} + 10\mathbf{j} - 11\mathbf{k}$

29. The angle between the diagonal of a cube and the diagonal of one of its faces. (*Hint:* Use a cube whose edges represent **i, j,** and **k.**)

30. **CALCULATOR** A water main is to be constructed with a 20% grade in the north direction and a 10% grade in the east direction. Determine the angle θ required in the water main for the turn from north to east. (See Preliminaries, Section 2, Exercise 46.)

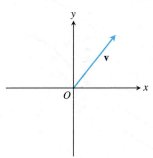

Theory and Examples

31. a) Use the fact that $\mathbf{u} \cdot \mathbf{v} = |\mathbf{u}||\mathbf{v}| \cos\theta$ to show that the inequality $|\mathbf{u} \cdot \mathbf{v}| \le |\mathbf{u}||\mathbf{v}|$ for any vectors **u** and **v**.

b) Under what circumstances, if any, does $|\mathbf{u} \cdot \mathbf{v}|$ equal $|\mathbf{u}||\mathbf{v}|$? Give reasons for your answer.

32. Copy the axes and vector shown here. Then shade in the points (x, y) for which $(x\,\mathbf{i} + y\,\mathbf{j}) \cdot \mathbf{v} \le 0$. Justify your answer.

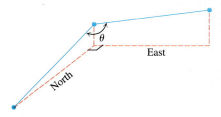

33. If $\mathbf{u}_1$ and $\mathbf{u}_2$ are orthogonal unit vectors and $\mathbf{v} = a\,\mathbf{u}_1 + b\,\mathbf{u}_2$, find $\mathbf{v} \cdot \mathbf{u}_1$.

34. *Cancellation in dot products.* In real-number multiplication, if $ab_1 = ab_2$ and a is not zero, we can cancel the a and conclude that $b_1 = b_2$. Does the same rule hold for vector multiplication: If $\mathbf{A} \cdot \mathbf{B}_1 = \mathbf{A} \cdot \mathbf{B}_2$ and $\mathbf{A} \neq 0$, can you conclude that $\mathbf{B}_1 = \mathbf{B}_2$? Give reasons for your answer.

35. Suppose $\mathbf{A}$, $\mathbf{B}$, and $\mathbf{C}$ are mutually orthogonal vectors. Let $\mathbf{D} = 5\mathbf{A} - 6\mathbf{B} + 3\mathbf{C}$.

  a) If $\mathbf{A}$, $\mathbf{B}$, and $\mathbf{C}$ are unit vectors, find $|\mathbf{D}|$, the magnitude of $\mathbf{D}$.

  b) If $|\mathbf{A}| = 2$, $|\mathbf{B}| = 3$, and $|\mathbf{C}| = 4$, find $|\mathbf{D}|$.

36. Suppose $\mathbf{A}$, $\mathbf{B}$, and $\mathbf{C}$ are mutually orthogonal unit vectors. If $\mathbf{D}$ is a vector such that $\mathbf{D} = \alpha\mathbf{A} + \beta\mathbf{B} + \gamma\mathbf{C}$ where α, β, and γ are scalars, prove that $\alpha = \mathbf{D} \cdot \mathbf{A}$, $\beta = \mathbf{D} \cdot \mathbf{B}$, and $\gamma = \mathbf{D} \cdot \mathbf{C}$.

Work

37. Find the work done by a force $\mathbf{F} = 5\mathbf{k}$ (magnitude 5 N) in moving an object along the line from the origin to the point $(1, 1, 1)$ (distance in meters).

38. The union Pacific's *Big Boy* locomotive could pull 6000-ton trains with a tractive effort (pull) of 602,148 N (135,375 lb). At this level of effort, about how much work did *Big Boy* do on the (approximately straight) 605-km journey from San Francisco to Los Angeles?

39. How much work does it take to slide a crate 20 m along a loading dock by pulling on it with a 200-N force at an angle of $30°$ from the horizontal?

40. The wind passing over a boat's sail exerted a 1000-lb magnitude force $\mathbf{F}$ as shown here. How much work did the wind perform in moving the boat forward 1 mi? Answer in foot-pounds.

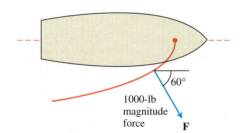

1000-lb
magnitude
force   **F**

Equations for Lines in the Plane

41. Show that the vector $\mathbf{v} = a\mathbf{i} + b\mathbf{j}$ is perpendicular to the line $ax + by = c$ by establishing that the slope of $\mathbf{v}$ is the negative reciprocal of the slope of the given line.

42. Show that the vector $\mathbf{v} = a\mathbf{i} + b\mathbf{j}$ is parallel to the line $bx - ay = c$ by establishing that the slope of the line segment representing $\mathbf{v}$ is the same as the slope of the given line.

In Exercises 43–46, use the result of Exercise 41 to find an equation for the line through P perpendicular to $\mathbf{v}$. Then sketch the line. Include $\mathbf{v}$ in your sketch *as a vector starting at the origin.*

43. $P(2, 1)$, $\mathbf{v} = \mathbf{i} + 2\mathbf{j}$

44. $P(-1, 2)$, $\mathbf{v} = -2\mathbf{i} - \mathbf{j}$

45. $P(-2, -7)$, $\mathbf{v} = -2\mathbf{i} + \mathbf{j}$

46. $P(11, 10)$, $\mathbf{v} = 2\mathbf{i} - 3\mathbf{j}$

In Exercises 47–50, use the result of Exercise 42 to find an equation for the line through P parallel to $\mathbf{v}$. Then sketch the line. Include $\mathbf{v}$ in your sketch *as a vector starting at the origin.*

47. $P(-2, 1)$, $\mathbf{v} = \mathbf{i} - \mathbf{j}$

48. $P(0, -2)$, $\mathbf{v} = 2\mathbf{i} + 3\mathbf{j}$

49. $P(1, 2)$, $\mathbf{v} = -\mathbf{i} - 2\mathbf{j}$

50. $P(1, 3)$, $\mathbf{v} = 3\mathbf{i} - 2\mathbf{j}$

Angles Between Lines in the Plane

The acute angle between intersecting lines that do not cross at right angles is the same as the angle determined by vectors normal to the lines or by the vectors parallel to the lines.

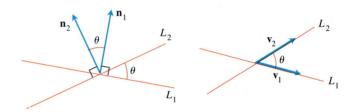

Use this fact and the results of Exercise 41 or 42 to find the acute angles between the lines in Exercises 51–54.

51. $3x + y = 5$, $2x - y = 4$

52. $y = \sqrt{3}x - 1$, $y = -\sqrt{3}x + 2$

53. $\sqrt{3}x - y = -2$, $x - \sqrt{3}y = 1$

54. $x + \sqrt{3}y = 1$, $(1 - \sqrt{3})x + (1 + \sqrt{3})y = 8$

CALCULATOR In Exercises 55 and 56, find the acute angle between the lines to the nearest hundredth of a radian.

55. $3x - 4y = 3$, $x - y = 7$

56. $12x + 5y = 1$, $2x - 2y = 3$

Angles Between Differentiable Curves

The angles between two differentiable curves at a point of intersection are the angles between the curves' tangent lines at these points. Find the angles between the curves in Exercises 57–60. (You will not need a calculator.)

57. $y = (3/2) - x^2$, $y = x^2$ (two points of intersection)

58. $x = (3/4) - y^2$, $x = y^2 - (3/4)$ (two points of intersection)

59. $y = x^3$, $x = y^2$ (two points of intersection)

60. $y = -x^2$, $y = \sqrt[3]{x}$ (two points of intersection)

10.4 Cross Products

In studying lines in the plane, we needed to describe how a line was tilting, and we did so with the notions of slope and angle of inclination. In space, we need to be able to describe how a plane is tilting. We accomplish this by multiplying two vectors in the plane together to get a third vector perpendicular to the plane. The direction of this third vector tells us the "inclination" of the plane. The product we use to multiply the vectors together is the *vector* or *cross product,* the second of the two vector multiplication methods we study in calculus.

Cross products are widely used to describe the effects of forces in studies of electricity, magnetism, fluid flows, and orbital mechanics. This section presents the mathematical properties that account for the use of cross products in these fields.

The Cross Product of Two Vectors in Space

We start with two nonzero vectors **A** and **B** in space. If **A** and **B** are not parallel, they determine a plane. We select a unit vector **n** perpendicular to the plane by the **right-hand rule.** This means we choose **n** to be the unit (normal) vector that points the way your right thumb points when your fingers curl through the angle θ from **A** to **B** (Fig. 10.31). We then define the **vector product A × B** ("A cross **B**") to be the *vector*

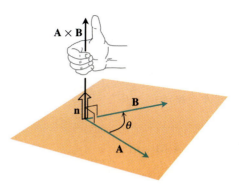

10.31 The construction of **A** × **B**.

> ### Definition
>
> $$\mathbf{A} \times \mathbf{B} = (|\mathbf{A}||\mathbf{B}|\sin\theta)\,\mathbf{n} \qquad (1)$$

The vector **A** × **B** is orthogonal to both **A** and **B** because it is a scalar multiple of **n.** The vector product of **A** and **B** is often called the **cross product** of **A** and **B** because of the cross in the notation **A** × **B.**

Since the sines of 0 and π are both zero in Eq. (1), it makes sense to define the cross product of two parallel nonzero vectors to be **0.**

If one or both of **A** and **B** are zero, we also define **A** × **B** to be zero. This way, the cross product of two vectors **A** and **B** is zero if and only if **A** and **B** are parallel or one or both of them are zero.

> Nonzero vectors **A** and **B** are parallel if and only if **A** × **B** = **0.**

A x B vs. B x A

Reversing the order of the factors in a nonzero cross product reverses the direction of the product. When the fingers of our right hand curl through the angle θ from **B** to **A,** our thumb points the opposite way and the unit vector we choose in forming **B** × **A** is the negative of the one we choose in forming **A** × **B** (Fig. 10.32). Thus,

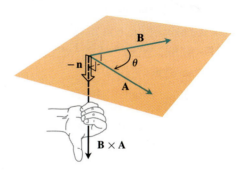

10.32 The construction of **B** × **A**.

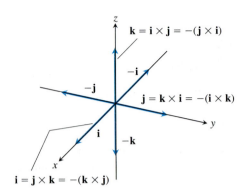

10.33 The pairwise cross products of **i**, **j**, and **k**.

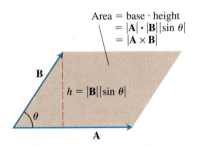

10.34 The parallelogram determined by **A** and **B**.

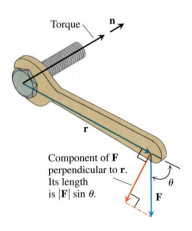

10.35 The torque vector describes the tendency of the force **F** to drive the bolt forward.

for all vectors **A** and **B**,

$$\mathbf{B} \times \mathbf{A} = -(\mathbf{A} \times \mathbf{B}). \qquad (2)$$

Unlike the dot product, the cross product is not commutative.

When we apply the definition to calculate the pairwise cross products of **i**, **j**, and **k**, we find (Fig. 10.33)

$$\mathbf{i} \times \mathbf{j} = -(\mathbf{j} \times \mathbf{i}) = \mathbf{k}$$

$$\mathbf{j} \times \mathbf{k} = -(\mathbf{k} \times \mathbf{j}) = \mathbf{i} \qquad (3)$$

$$\mathbf{k} \times \mathbf{i} = -(\mathbf{i} \times \mathbf{k}) = \mathbf{j}$$

Diagram for recalling these products

and

$$\mathbf{i} \times \mathbf{i} = \mathbf{j} \times \mathbf{j} = \mathbf{k} \times \mathbf{k} = \mathbf{0}.$$

| A x B | Is the Area of a Parallelogram

Because **n** is a unit vector, the magnitude of **A** × **B** is

$$|\mathbf{A} \times \mathbf{B}| = |\mathbf{A}||\mathbf{B}|| \sin \theta ||\mathbf{n}| = |\mathbf{A}||\mathbf{B}| \sin \theta. \qquad (4)$$

This is the area of the parallelogram determined by **A** and **B** (Fig. 10.34), $|\mathbf{A}|$ being the base of the parallelogram and $|\mathbf{B}|| \sin \theta|$ the height.

Torque

When we turn a bolt by applying a force **F** to a wrench (Fig. 10.35), the torque we produce acts along the axis of the bolt to drive the bolt forward. The magnitude of the torque depends on how far out on the wrench the force is applied and on how much of the force is perpendicular to the wrench at the point of application. The number we use to measure the torque's magnitude is the product of the length of the lever arm **r** and the scalar component of **F** perpendicular to **r**. In the notation of Fig. 10.35,

Magnitude of torque vector $= |\mathbf{r}||\mathbf{F}| \sin \theta,$

or $|\mathbf{r} \times \mathbf{F}|$. If we let **n** be a unit vector along the axis of the bolt in the direction of the torque, then a complete description of the torque vector is **r** × **F**, or

Torque vector $= (|\mathbf{r}||\mathbf{F}| \sin \theta) \, \mathbf{n}.$

Recall that we defined **A** × **B** to be **0** when **A** and **B** are parallel. This is consistent with the torque interpretation as well. If the force **F** in Fig. 10.35 is parallel to the wrench, meaning that we are trying to turn the bolt by pushing or pulling along the line of the wrench's handle, the torque produced is zero.

EXAMPLE 1   The magnitude of the torque exerted by force **F** about the pivot

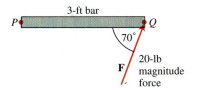

3-ft bar

P Q

70°

F   20-lb magnitude force

10.36 The magnitude of the torque exerted by **F** and P is about 56.4 ft-lb (Example 1).

point P in Fig. 10.36 is

$$|\overrightarrow{PQ} \times \mathbf{F}| = |\overrightarrow{PQ}||\mathbf{F}| \sin 70° \qquad \text{Eq. (4)}$$

$$\approx (3)(20)(0.94)$$

$$\approx 56.4 \text{ ft-lb.} \qquad \Box$$

The Associative and Distributive Laws

As a rule, cross-product multiplication is *not associative* because $(\mathbf{A} \times \mathbf{B}) \times \mathbf{C}$ lies in the plane of **A** and **B** whereas $\mathbf{A} \times (\mathbf{B} \times \mathbf{C})$ lies in the plane of **B** and **C**. However, the following laws do hold.

Scalar Distributive Law

$$(r\mathbf{A}) \times (s\mathbf{B}) = (rs)(\mathbf{A} \times \mathbf{B}) \tag{5}$$

Vector Distributive Laws

$$\mathbf{A} \times (\mathbf{B} + \mathbf{C}) = \mathbf{A} \times \mathbf{B} + \mathbf{A} \times \mathbf{C} \tag{6}$$

$$(\mathbf{B} + \mathbf{C}) \times \mathbf{A} = \mathbf{B} \times \mathbf{A} + \mathbf{C} \times \mathbf{A} \tag{7}$$

As a special case of Eq. (5) we also have

$$(-\mathbf{A}) \times \mathbf{B} = \mathbf{A} \times (-\mathbf{B}) = -(\mathbf{A} \times \mathbf{B}). \tag{8}$$

The Scalar Distributive Law can be verified by applying Eq. (1) to the products on both sides of Eq. (5) and comparing the results. The Vector Distributive Law in Eq. (6) is not easy to prove. We will assume it here and leave the proof to Appendix 7. Equation (7) follows from Eq. (6): Multiply both sides of Eq. (6) by -1 and reverse the orders of the products.

The Determinant Formula for A x B

Our next objective is to calculate $\mathbf{A} \times \mathbf{B}$ from the components of **A** and **B** relative to a Cartesian coordinate system.

Suppose

$$\mathbf{A} = a_1 \mathbf{i} + a_2 \mathbf{j} + a_3 \mathbf{k}, \qquad \mathbf{B} = b_1 \mathbf{i} + b_2 \mathbf{j} + b_3 \mathbf{k}.$$

Then the distributive laws and the rules for multiplying **i, j,** and **k** tell us that

$$\mathbf{A} \times \mathbf{B} = (a_1 \mathbf{i} + a_2 \mathbf{j} + a_3 \mathbf{k}) \times (b_1 \mathbf{i} + b_2 \mathbf{j} + b_3 \mathbf{k})$$

$$= a_1 b_1 \mathbf{i} \times \mathbf{i} + a_1 b_2 \mathbf{i} \times \mathbf{j} + a_1 b_3 \mathbf{i} \times \mathbf{k}$$

$$+ a_2 b_1 \mathbf{j} \times \mathbf{i} + a_2 b_2 \mathbf{j} \times \mathbf{j} + a_2 b_3 \mathbf{j} \times \mathbf{k}$$

$$+ a_3 b_1 \mathbf{k} \times \mathbf{i} + a_3 b_2 \mathbf{k} \times \mathbf{j} + a_3 b_3 \mathbf{k} \times \mathbf{k}$$

$$= (a_2 b_3 - a_3 b_2) \mathbf{i} - (a_1 b_3 - a_3 b_1) \mathbf{j} + (a_1 b_2 - a_2 b_1) \mathbf{k}.$$

Determinants

(For more information, see Appendix 8.)

$$\begin{vmatrix} a & b \\ c & d \end{vmatrix} = ad - bc$$

EXAMPLE

$$\begin{vmatrix} 2 & 1 \\ -4 & 3 \end{vmatrix} = (2)(3) - (1)(-4)$$

$$= 6 + 4 = 10 \qquad \square$$

$$\begin{vmatrix} a_1 & a_2 & a_3 \\ b_1 & b_2 & b_3 \\ c_1 & c_2 & c_3 \end{vmatrix}$$

$$= a_1 \begin{vmatrix} b_2 & b_3 \\ c_2 & c_3 \end{vmatrix} - a_2 \begin{vmatrix} b_1 & b_3 \\ c_1 & c_3 \end{vmatrix} + a_3 \begin{vmatrix} b_1 & b_2 \\ c_1 & c_2 \end{vmatrix}$$

EXAMPLE

$$\begin{vmatrix} -5 & 3 & 1 \\ 2 & 1 & 1 \\ -4 & 3 & 1 \end{vmatrix}$$

$$= (-5) \begin{vmatrix} 1 & 1 \\ 3 & 1 \end{vmatrix} - (3) \begin{vmatrix} 2 & 1 \\ -4 & 1 \end{vmatrix}$$

$$+ (1) \begin{vmatrix} 2 & 1 \\ -4 & 3 \end{vmatrix}$$

$$= -5(1 - 3) - 3(2 + 4) + 1(6 + 4)$$

$$= 10 - 18 + 10 = 2 \qquad \square$$

The terms in the last line are the same as the terms in the expansion of the symbolic determinant

$$\begin{vmatrix} \mathbf{i} & \mathbf{j} & \mathbf{k} \\ a_1 & a_2 & a_3 \\ b_1 & b_2 & b_3 \end{vmatrix}.$$

We therefore have the following rule.

If $\mathbf{A} = a_1\,\mathbf{i} + a_2\,\mathbf{j} + a_3\,\mathbf{k}$ and $\mathbf{B} = b_1\,\mathbf{i} + b_2\,\mathbf{j} + b_3\,\mathbf{k}$, then

$$\mathbf{A} \times \mathbf{B} = \begin{vmatrix} \mathbf{i} & \mathbf{j} & \mathbf{k} \\ a_1 & a_2 & a_3 \\ b_1 & b_2 & b_3 \end{vmatrix}. \tag{9}$$

EXAMPLE 2

Find $\mathbf{A} \times \mathbf{B}$ and $\mathbf{B} \times \mathbf{A}$ if

$$\mathbf{A} = 2\,\mathbf{i} + \mathbf{j} + \mathbf{k}, \qquad \mathbf{B} = -4\,\mathbf{i} + 3\,\mathbf{j} + \mathbf{k}.$$

Solution

$$\mathbf{A} \times \mathbf{B} = \begin{vmatrix} \mathbf{i} & \mathbf{j} & \mathbf{k} \\ 2 & 1 & 1 \\ -4 & 3 & 1 \end{vmatrix} = \begin{vmatrix} 1 & 1 \\ 3 & 1 \end{vmatrix}\mathbf{i} - \begin{vmatrix} 2 & 1 \\ -4 & 1 \end{vmatrix}\mathbf{j} + \begin{vmatrix} 2 & 1 \\ -4 & 3 \end{vmatrix}\mathbf{k} \qquad \text{Eq. (9)}$$

$$= -2\,\mathbf{i} - 6\,\mathbf{j} + 10\,\mathbf{k}$$

$$\mathbf{B} \times \mathbf{A} = -(\mathbf{A} \times \mathbf{B}) = 2\,\mathbf{i} + 6\,\mathbf{j} - 10\,\mathbf{k} \qquad \text{Eq. (2)}$$

$$\square$$

EXAMPLE 3

Find a vector perpendicular to the plane of $P(1, -1, 0)$, $Q(2, 1, -1)$, and $R(-1, 1, 2)$.

Solution The vector $\overrightarrow{PQ} \times \overrightarrow{PR}$ is perpendicular to the plane because it is perpendicular to both vectors. In terms of components,

$$\overrightarrow{PQ} = (2 - 1)\,\mathbf{i} + (1 + 1)\,\mathbf{j} + (-1 - 0)\,\mathbf{k} = \mathbf{i} + 2\,\mathbf{j} - \mathbf{k}$$

$$\overrightarrow{PR} = (-1 - 1)\,\mathbf{i} + (1 + 1)\,\mathbf{j} + (2 - 0)\,\mathbf{k} = -2\,\mathbf{i} + 2\,\mathbf{j} + 2\,\mathbf{k}$$

$$\overrightarrow{PQ} \times \overrightarrow{PR} = \begin{vmatrix} \mathbf{i} & \mathbf{j} & \mathbf{k} \\ 1 & 2 & -1 \\ -2 & 2 & 2 \end{vmatrix} = \begin{vmatrix} 2 & -1 \\ 2 & 2 \end{vmatrix}\mathbf{i} - \begin{vmatrix} 1 & -1 \\ -2 & 2 \end{vmatrix}\mathbf{j} + \begin{vmatrix} 1 & 2 \\ -2 & 2 \end{vmatrix}\mathbf{k}$$

$$= 6\,\mathbf{i} + 6\,\mathbf{k}. \qquad \square$$

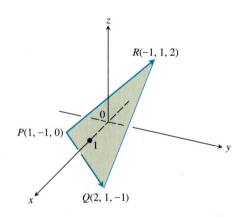

10.37 The area of triangle PQR is half of $|\overrightarrow{PQ} \times \overrightarrow{PR}|$ (Example 4).

EXAMPLE 4

Find the area of the triangle with vertices $P(1, -1, 0)$, $Q(2, 1, -1)$, and $R(-1, 1, 2)$ (Fig. 10.37).

Solution   The area of the parallelogram determined by P, Q, and R is

$$|\overrightarrow{PQ} \times \overrightarrow{PR}| = |6\,\mathbf{i} + 6\,\mathbf{k}| \qquad \text{Values from Example 3}$$

$$= \sqrt{(6)^2 + (6)^2} = \sqrt{2 \cdot 36} = 6\sqrt{2}.$$

The triangle's area is half of this, or $3\sqrt{2}$. ❑

EXAMPLE 5   Find a unit vector perpendicular to the plane of $P(1, -1, 0)$, $Q(2, 1, -1)$, and $R(-1, 1, 2)$.

Solution   Since $\overrightarrow{PQ} \times \overrightarrow{PR}$ is perpendicular to the plane, its direction $\mathbf{n}$ is a unit vector perpendicular to the plane. Taking values from Examples 3 and 4, we have

$$\mathbf{n} = \frac{\overrightarrow{PQ} \times \overrightarrow{PR}}{|\overrightarrow{PQ} \times \overrightarrow{PR}|} = \frac{6\,\mathbf{i} + 6\,\mathbf{k}}{6\sqrt{2}} = \frac{1}{\sqrt{2}}\,\mathbf{i} + \frac{1}{\sqrt{2}}\,\mathbf{k}.$$ ❑

The Triple Scalar or Box Product

The product $(\mathbf{A} \times \mathbf{B}) \cdot \mathbf{C}$ is called the **triple scalar product** of $\mathbf{A}$, $\mathbf{B}$, and $\mathbf{C}$ (in that order). As you can see from the formula

$$|(\mathbf{A} \times \mathbf{B}) \cdot \mathbf{C}| = |\mathbf{A} \times \mathbf{B}||\mathbf{C}||\cos\theta|, \tag{10}$$

the absolute value of the product is the volume of the parallelepiped (parallelogram-sided box) determined by $\mathbf{A}$, $\mathbf{B}$, and $\mathbf{C}$ (Fig. 10.38). The number $|\mathbf{A} \times \mathbf{B}|$ is the area of the base parallelogram. The number $|\mathbf{C}||\cos\theta|$ is the parallelepiped's height. Because of this geometry, $(\mathbf{A} \times \mathbf{B}) \cdot \mathbf{C}$ is also called the **box product** of $\mathbf{A}$, $\mathbf{B}$, and $\mathbf{C}$.

By treating the planes of $\mathbf{B}$ and $\mathbf{C}$ and of $\mathbf{C}$ and $\mathbf{A}$ as the base planes of the parallelepiped determined by $\mathbf{A}$, $\mathbf{B}$, and $\mathbf{C}$, we see that

> The dot and cross may be interchanged in a triple scalar product without altering its value.

$$(\mathbf{A} \times \mathbf{B}) \cdot \mathbf{C} = (\mathbf{B} \times \mathbf{C}) \cdot \mathbf{A} = (\mathbf{C} \times \mathbf{A}) \cdot \mathbf{B}. \tag{11}$$

Since the dot product is commutative, Eq. (11) also gives

$$(\mathbf{A} \times \mathbf{B}) \cdot \mathbf{C} = \mathbf{A} \cdot (\mathbf{B} \times \mathbf{C}). \tag{12}$$

$$\text{Volume} = \text{area of base} \cdot \text{height}$$
$$= |\mathbf{A} \times \mathbf{B}| \cdot |\mathbf{C}||\cos\theta|$$
$$= |(\mathbf{A} \times \mathbf{B}) \cdot \mathbf{C}|$$

10.38 The number $|(\mathbf{A} \times \mathbf{B}) \cdot \mathbf{C}|$ is the volume of a parallelepiped.

The triple scalar product can be evaluated as a determinant:

$$\mathbf{A} \cdot (\mathbf{B} \times \mathbf{C}) = \mathbf{A} \cdot \left[\begin{vmatrix} b_2 & b_3 \\ c_2 & c_3 \end{vmatrix} \mathbf{i} - \begin{vmatrix} b_1 & b_3 \\ c_1 & c_3 \end{vmatrix} \mathbf{j} + \begin{vmatrix} b_1 & b_2 \\ c_1 & c_2 \end{vmatrix} \mathbf{k} \right]$$

$$= a_1 \begin{vmatrix} b_2 & b_3 \\ c_2 & c_3 \end{vmatrix} - a_2 \begin{vmatrix} b_1 & b_3 \\ c_1 & c_3 \end{vmatrix} + a_3 \begin{vmatrix} b_1 & b_2 \\ c_1 & c_2 \end{vmatrix}$$

$$= \begin{vmatrix} a_1 & a_2 & a_3 \\ b_1 & b_2 & b_3 \\ c_1 & c_2 & c_3 \end{vmatrix}$$

$$\mathbf{A} \cdot (\mathbf{B} \times \mathbf{C}) = (\mathbf{A} \times \mathbf{B}) \cdot \mathbf{C} = \begin{vmatrix} a_1 & a_2 & a_3 \\ b_1 & b_2 & b_3 \\ c_1 & c_2 & c_3 \end{vmatrix} \qquad (13)$$

EXAMPLE 6   Find the volume of the box (parallelepiped) determined by $\mathbf{A} = \mathbf{i} + 2\mathbf{j} - \mathbf{k}$, $\mathbf{B} = -2\mathbf{i} + 3\mathbf{k}$, and $\mathbf{C} = 7\mathbf{j} - 4\mathbf{k}$.

Solution

$$\mathbf{A} \cdot (\mathbf{B} \times \mathbf{C}) = \begin{vmatrix} 1 & 2 & -1 \\ -2 & 0 & 3 \\ 0 & 7 & -4 \end{vmatrix} = \begin{vmatrix} 0 & 3 \\ 7 & -4 \end{vmatrix} - 2 \begin{vmatrix} -2 & 3 \\ 0 & -4 \end{vmatrix} - \begin{vmatrix} -2 & 0 \\ 0 & 7 \end{vmatrix}$$

$$= -21 - 16 + 14 = -23$$

The volume is $|\mathbf{A} \cdot (\mathbf{B} \times \mathbf{C})| = 23$.   ❑

Exercises 10.4

Calculations

In Exercises 1–8, find the length and direction (when defined) of $\mathbf{A} \times \mathbf{B}$ and $\mathbf{B} \times \mathbf{A}$.

1. $\mathbf{A} = 2\mathbf{i} - 2\mathbf{j} - \mathbf{k}$, $\mathbf{B} = \mathbf{i} - \mathbf{k}$

2. $\mathbf{A} = 2\mathbf{i} + 3\mathbf{j}$, $\mathbf{B} = -\mathbf{i} + \mathbf{j}$

3. $\mathbf{A} = 2\mathbf{i} - 2\mathbf{j} + 4\mathbf{k}$, $\mathbf{B} = -\mathbf{i} + \mathbf{j} - 2\mathbf{k}$

4. $\mathbf{A} = \mathbf{i} + \mathbf{j} - \mathbf{k}$, $\mathbf{B} = 0$

5. $\mathbf{A} = 2\mathbf{i}$, $\mathbf{B} = -3\mathbf{j}$

6. $\mathbf{A} = \mathbf{i} \times \mathbf{j}$, $\mathbf{B} = \mathbf{j} \times \mathbf{k}$

7. $\mathbf{A} = -8\mathbf{i} - 2\mathbf{j} - 4\mathbf{k}$, $\mathbf{B} = 2\mathbf{i} + 2\mathbf{j} + \mathbf{k}$

8. $\mathbf{A} = \dfrac{3}{2}\mathbf{i} - \dfrac{1}{2}\mathbf{j} + \mathbf{k}$, $\mathbf{B} = \mathbf{i} + \mathbf{j} + 2\mathbf{k}$

In Exercises 9–14, sketch the coordinate axes and then include the vectors $\mathbf{A}$, $\mathbf{B}$, and $\mathbf{A} \times \mathbf{B}$ as vectors starting at the origin.

9. $\mathbf{A} = \mathbf{i}$, $\mathbf{B} = \mathbf{j}$

10. $\mathbf{A} = \mathbf{i} - \mathbf{k}$, $\mathbf{B} = \mathbf{j}$

11. $\mathbf{A} = \mathbf{i} - \mathbf{k}$, $\mathbf{B} = \mathbf{j} + \mathbf{k}$

12. $\mathbf{A} = 2\mathbf{i} - \mathbf{j}$, $\mathbf{B} = \mathbf{i} + 2\mathbf{j}$

13. $\mathbf{A} = \mathbf{i} + \mathbf{j}$, $\mathbf{B} = \mathbf{i} - \mathbf{j}$

14. $\mathbf{A} = \mathbf{j} + 2\mathbf{k}$, $\mathbf{B} = \mathbf{i}$

In Exercises 15–18:

a) Find the area of the triangle determined by the points P, Q, and R.

b) Find a unit vector perpendicular to plane PQR.

15. $P(1, -1, 2)$, $Q(2, 0, -1)$, $R(0, 2, 1)$

16. $P(1, 1, 1)$, $Q(2, 1, 3)$, $R(3, -1, 1)$

17. $P(2, -2, 1)$, $Q(3, -1, 2)$, $R(3, -1, 1)$

18. $P(-2, 2, 0)$, $Q(0, 1, -1)$, $R(-1, 2, -2)$

19. Let $\mathbf{A} = 5\mathbf{i} - \mathbf{j} + \mathbf{k}$, $\mathbf{B} = \mathbf{j} - 5\mathbf{k}$, $\mathbf{C} = -15\mathbf{i} + 3\mathbf{j} - 3\mathbf{k}$. Which vectors, if any, are (a) perpendicular, (b) parallel? Give reasons for your answers.

20. Let $\mathbf{A} = \mathbf{i} + 2\mathbf{j} - \mathbf{k}$, $\mathbf{B} = -\mathbf{i} + \mathbf{j} + \mathbf{k}$, $\mathbf{C} = \mathbf{i} + \mathbf{k}$, $\mathbf{D} = -(\pi/2)\mathbf{i} - \pi\mathbf{j} + (\pi/2)\mathbf{k}$. Which vectors, if any, are (a) perpendicular, (b) parallel? Give reasons for your answers.

In Exercises 21 and 22, find the magnitude of the torque exerted by $\mathbf{F}$ on the bolt at P if $|\overrightarrow{PQ}| = 8$ in. and $|\mathbf{F}| = 30$ lb. Answer in foot-pounds.

21.

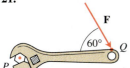

22.

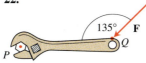

In Exercises 23–26, verify that $(\mathbf{A} \times \mathbf{B}) \cdot \mathbf{C} = (\mathbf{B} \times \mathbf{C}) \cdot \mathbf{A} = (\mathbf{C} \times \mathbf{A}) \cdot \mathbf{B}$ and find the volume of the parallelepiped (box) determined by $\mathbf{A}$, $\mathbf{B}$, and $\mathbf{C}$.

	A	**B**	**C**
23.	$2\mathbf{i}$	$2\mathbf{j}$	$2\mathbf{k}$
24.	$\mathbf{i} - \mathbf{j} + \mathbf{k}$	$2\mathbf{i} + \mathbf{j} - 2\mathbf{k}$	$-\mathbf{i} + 2\mathbf{j} - \mathbf{k}$
25.	$2\mathbf{i} + \mathbf{j}$	$2\mathbf{i} - \mathbf{j} + \mathbf{k}$	$\mathbf{i} + 2\mathbf{k}$
26.	$\mathbf{i} + \mathbf{j} - 2\mathbf{k}$	$-\mathbf{i} - \mathbf{k}$	$2\mathbf{i} + 4\mathbf{j} - 2\mathbf{k}$

Theory and Examples

27. Which of the following are *always true* and which are *not always true?* Give reasons for your answers.

a) $|\mathbf{A}| = \sqrt{\mathbf{A} \cdot \mathbf{A}}$

b) $\mathbf{A} \cdot \mathbf{A} = |\mathbf{A}|$

c) $\mathbf{A} \times \mathbf{0} = \mathbf{0} \times \mathbf{A} = \mathbf{0}$

d) $\mathbf{A} \times (-\mathbf{A}) = \mathbf{0}$

e) $\mathbf{A} \times \mathbf{B} = \mathbf{B} \times \mathbf{A}$

f) $\mathbf{A} \times (\mathbf{B} + \mathbf{C}) = \mathbf{A} \times \mathbf{B} + \mathbf{A} \times \mathbf{C}$

g) $(\mathbf{A} \times \mathbf{B}) \cdot \mathbf{B} = 0$

h) $(\mathbf{A} \times \mathbf{B}) \cdot \mathbf{C} = \mathbf{A} \cdot (\mathbf{B} \times \mathbf{C})$

28. Which of the following are *always true* and which are *not always true?* Give reasons for your answers.

a) $\mathbf{A} \cdot \mathbf{B} = \mathbf{B} \cdot \mathbf{A}$

b) $\mathbf{A} \times \mathbf{B} = -(\mathbf{B} \times \mathbf{A})$

c) $(-\mathbf{A}) \times \mathbf{B} = -(\mathbf{A} \times \mathbf{B})$

d) $(c\mathbf{A}) \cdot \mathbf{B} = \mathbf{A} \cdot (c\mathbf{B}) = c(\mathbf{A} \cdot \mathbf{B})$ (any number c)

e) $c(\mathbf{A} \times \mathbf{B}) = (c\mathbf{A}) \times \mathbf{B} = \mathbf{A} \times (c\mathbf{B})$ (any number c)

f) $\mathbf{A} \cdot \mathbf{A} = |\mathbf{A}|^2$

g) $(\mathbf{A} \times \mathbf{A}) \cdot \mathbf{A} = 0$

h) $(\mathbf{A} \times \mathbf{B}) \cdot \mathbf{A} = \mathbf{B} \cdot (\mathbf{A} \times \mathbf{B})$

29. Given nonzero vectors $\mathbf{A}$, $\mathbf{B}$, and $\mathbf{C}$, use dot-product and cross-product notation, as appropriate, to describe the following.

a)   The vector projection of $\mathbf{A}$ onto $\mathbf{B}$

b)   A vector orthogonal to $\mathbf{A}$ and $\mathbf{B}$

c)   A vector orthogonal to $\mathbf{A} \times \mathbf{B}$ and $\mathbf{C}$

d)   The volume of the parallelepiped determined by $\mathbf{A}$, $\mathbf{B}$, and $\mathbf{C}$

30. Given nonzero vectors $\mathbf{A}$, $\mathbf{B}$, and $\mathbf{C}$, use dot-product and cross-product notation to describe the following.

a)   A vector orthogonal to $\mathbf{A} \times \mathbf{B}$ and $\mathbf{A} \times \mathbf{C}$

b)   A vector orthogonal to $\mathbf{A} + \mathbf{B}$ and $\mathbf{A} - \mathbf{B}$

c)   A vector of length $|\mathbf{A}|$ in the direction of $\mathbf{B}$

d)   The area of the parallelogram determined by $\mathbf{A}$ and $\mathbf{C}$

31. Let $\mathbf{A}$, $\mathbf{B}$, and $\mathbf{C}$ be vectors. Which of the following make sense, and which do not? Give reasons for your answers.

a) $(\mathbf{A} \times \mathbf{B}) \cdot \mathbf{C}$

b) $\mathbf{A} \times (\mathbf{B} \cdot \mathbf{C})$

c) $\mathbf{A} \times (\mathbf{B} \times \mathbf{C})$

d) $\mathbf{A} \cdot (\mathbf{B} \cdot \mathbf{C})$

32. Show that except in degenerate cases $(\mathbf{A} \times \mathbf{B}) \times \mathbf{C}$ lies in the plane of $\mathbf{A}$ and $\mathbf{B}$ while $\mathbf{A} \times (\mathbf{B} \times \mathbf{C})$ lies in the plane of $\mathbf{B}$ and $\mathbf{C}$. What *are* the degenerate cases?

33. *Cancellation in cross products.* If $\mathbf{A} \times \mathbf{B} = \mathbf{A} \times \mathbf{C}$ and $\mathbf{A} \neq \mathbf{0}$, then does $\mathbf{B} = \mathbf{C}$? Give reasons for your answer.

34. *Double cancellation.* If $\mathbf{A} \neq \mathbf{0}$, and if $\mathbf{A} \times \mathbf{B} = \mathbf{A} \times \mathbf{C}$ and $\mathbf{A} \cdot \mathbf{B} = \mathbf{A} \cdot \mathbf{C}$, then does $\mathbf{B} = \mathbf{C}$? Give reasons for your answers.

Area in the Plane

Find the areas of the parallelograms whose vertices are given in Exercises 35–38.

35. $A(1, 0)$, $B(0, 1)$, $C(-1, 0)$, $D(0, -1)$

36. $A(0, 0)$, $B(7, 3)$, $C(9, 8)$, $D(2, 5)$

37. $A(-1, 2)$, $B(2, 0)$, $C(7, 1)$, $D(4, 3)$

38. $A(-6, 0)$, $B(1, -4)$, $C(3, 1)$, $D(-4, 5)$

Find the areas of the triangles whose vertices are given in Exercises 39–42.

39. $A(0, 0)$, $B(-2, 3)$, $C(3, 1)$

40. $A(-1, -1)$, $B(3, 3)$, $C(2, 1)$

41. $A(-5, 3)$, $B(1, -2)$, $C(6, -2)$

42. $A(-6, 0)$, $B(10, -5)$, $C(-2, 4)$

43. Find a formula for the area of the triangle in the xy-plane with vertices at $(0, 0)$, (a_1, a_2), and (b_1, b_2). Explain your work.

44. Find a concise formula for the area of a triangle with vertices (a_1, a_2), (b_1, b_2), and (c_1, c_2).

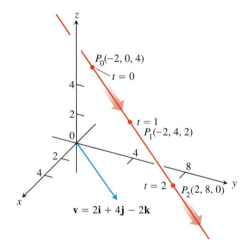

10.39 A point P lies on the line through P_0 parallel to $\mathbf{v}$ if and only if $\overrightarrow{P_0P}$ is a scalar multiple of $\mathbf{v}$.

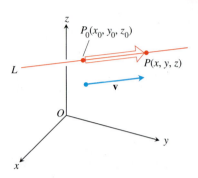

10.40 Selected points and parameter values on the line $x = -2 + 2t$, $y = 4t$, $z = 4 - 2t$. The arrows show the direction of increasing t (Example 1).

10.5 Lines and Planes in Space

This section shows how to use scalar and vector products to write equations for lines, line segments, and planes in space.

Lines and Line Segments in Space

Suppose L is a line in space passing through a point $P_0(x_0, y_0, z_0)$ parallel to a vector $\mathbf{v} = A\,\mathbf{i} + B\,\mathbf{j} + C\,\mathbf{k}$. Then L is the set of all points $P(x, y, z)$ for which $\overrightarrow{P_0P}$ is parallel to $\mathbf{v}$ (Fig. 10.39). That is, P lies on L if and only if $\overrightarrow{P_0P}$ is a scalar multiple of $\mathbf{v}$.

Vector Equation for the Line Through $P_0(x_0, y_0, z_0)$ Parallel to v

$$\overrightarrow{P_0P} = t\,\mathbf{v}, \qquad -\infty < t < \infty \qquad (1)$$

Equating the corresponding components of the two sides of Eq. (1) gives three scalar equations involving the parameter t:

$$(x - x_0)\,\mathbf{i} + (y - y_0)\,\mathbf{j} + (z - z_0)\,\mathbf{k} = t(A\,\mathbf{i} + B\,\mathbf{j} + C\,\mathbf{k}), \qquad \text{Eq. (1) expanded}$$

$$x - x_0 = tA, \qquad y - y_0 = tB, \qquad z - z_0 = tC. \qquad \text{Components equated}$$

When rearranged, these equations give us the standard parametrization of the line for the parameter interval $-\infty < t < \infty$.

Standard Parametrization of the Line Through $P_0(x_0, y_0, z_0)$ Parallel to $\mathbf{v} = A\,\mathbf{i} + B\,\mathbf{j} + C\,\mathbf{k}$

$$x = x_0 + tA, \qquad y = y_0 + tB, \qquad z = z_0 + tC, \qquad -\infty < t < \infty \qquad (2)$$

EXAMPLE 1 Find parametric equations for the line through $(-2, 0, 4)$ parallel to $\mathbf{v} = 2\,\mathbf{i} + 4\,\mathbf{j} - 2\,\mathbf{k}$ (Fig. 10.40).

Solution With $P_0(x_0, y_0, z_0)$ equal to $(-2, 0, 4)$ and $A\,\mathbf{i} + B\,\mathbf{j} + C\,\mathbf{k}$ equal to $2\,\mathbf{i} + 4\,\mathbf{j} - 2\,\mathbf{k}$, Eqs. (2) become

$$x = -2 + 2t, \qquad y = 4t, \qquad z = 4 - 2t. \qquad \square$$

EXAMPLE 2 Find parametric equations for the line through $P(-3, 2, -3)$ and $Q(1, -1, 4)$.

Solution The vector

$$\overrightarrow{PQ} = (1 - (-3))\,\mathbf{i} + (-1 - 2)\,\mathbf{j} + (4 - (-3))\,\mathbf{k}$$

$$= 4\,\mathbf{i} - 3\,\mathbf{j} + 7\,\mathbf{k}$$

is parallel to the line, and Eqs. (2) with $(x_0, y_0, z_0) = (-3, 2, -3)$ give

$$x = -3 + 4t, \qquad y = 2 - 3t, \qquad z = -3 + 7t.$$

We could have chosen $Q(1, -1, 4)$ as the "base point" and written

$$x = 1 + 4t, \qquad y = -1 - 3t, \qquad z = 4 + 7t.$$

These equations serve as well as the first; they simply place you at a different point for a given value of t.   ❑

To parametrize a line segment joining two points, we first parametrize the line through the points. We then find the t-values for the endpoints and restrict t to lie in the closed interval bounded by these values. The line equations together with this added restriction parametrize the segment.

EXAMPLE 3   Parametrize the line segment joining the points $P(-3, 2, -3)$ and $Q(1, -1, 4)$ (Fig. 10.41).

Solution   We begin with equations for the line through P and Q, taking them, in this case, from Example 2:

$$x = -3 + 4t, \qquad y = 2 - 3t, \qquad z = -3 + 7t. \tag{3}$$

We observe that the point

$$(x, y, z) = (-3 + 4t, 2 - 3t, -3 + 7t)$$

passes through $P(-3, 2, -3)$ at $t = 0$ and $Q(1, -1, 4)$ at $t = 1$. We add the restriction $0 \le t \le 1$ to Eqs. (3) to parametrize the segment:

$$x = -3 + 4t, \qquad y = 2 - 3t, \qquad z = -3 + 7t, \qquad 0 \le t \le 1. \quad ❑$$

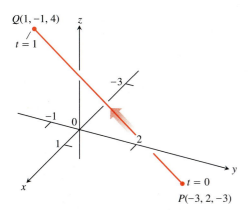

Notice that parametrizations are not unique.

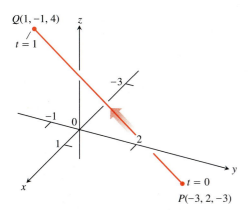

10.41 Example 3 derives a parametrization of line segment PQ. The arrow shows the direction of increasing t.

The Distance from a Point to a Line in Space

To find the distance from a point S to a line that passes through a point P parallel to a vector $\mathbf{v}$, we find the length of the component of $\overrightarrow{PS}$ normal to the line (Fig. 10.42). In the notation of the figure, the length is $|\overrightarrow{PS}| \sin\theta$, which is $|\overrightarrow{PS} \times \mathbf{v}|/|\mathbf{v}|$.

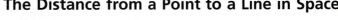

> **Distance from a Point S to a Line Through P Parallel to v**
>
> $$d = \frac{|\overrightarrow{PS} \times \mathbf{v}|}{|\mathbf{v}|} \tag{4}$$

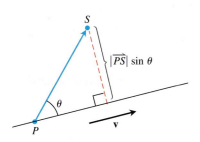

10.42 The distance from S to the line through P parallel to $\mathbf{v}$ is $|\overrightarrow{PS}| \sin\theta$, where θ is the angle between $\overrightarrow{PS}$ and $\mathbf{v}$.

EXAMPLE 4   Find the distance from the point $S(1, 1, 5)$ to the line

$$L: \qquad x = 1 + t, \quad y = 3 - t, \quad z = 2t.$$

Solution   We see from the equations for L that L passes through $P(1, 3, 0)$ parallel to $\mathbf{v} = \mathbf{i} - \mathbf{j} + 2\mathbf{k}$. With

$$\overrightarrow{PS} = (1 - 1)\mathbf{i} + (1 - 3)\mathbf{j} + (5 - 0)\mathbf{k} = -2\mathbf{j} + 5\mathbf{k}$$

and

$$\overrightarrow{PS} \times \mathbf{v} = \begin{vmatrix} \mathbf{i} & \mathbf{j} & \mathbf{k} \\ 0 & -2 & 5 \\ 1 & -1 & 2 \end{vmatrix} = \mathbf{i} + 5\mathbf{j} + 2\mathbf{k},$$

Eq. (4) gives

$$d = \frac{|\overrightarrow{PS} \times \mathbf{v}|}{|\mathbf{v}|} = \frac{\sqrt{1 + 25 + 4}}{\sqrt{1 + 1 + 4}} = \frac{\sqrt{30}}{\sqrt{6}} = \sqrt{5}. \qquad \square$$

Equations for Planes in Space

Suppose plane M passes through a point $P_0(x_0, y_0, z_0)$ and is normal (perpendicular) to the nonzero vector $\mathbf{n} = A\mathbf{i} + B\mathbf{j} + C\mathbf{k}$. Then M is the set of all points $P(x, y, z)$ for which $\overrightarrow{P_0P}$ is orthogonal to $\mathbf{n}$ (Fig. 10.43). That is, P lies on M if and only if $\mathbf{n} \cdot \overrightarrow{P_0P} = 0$. This equation is equivalent to

$$(A\mathbf{i} + B\mathbf{j} + C\mathbf{k}) \cdot [(x - x_0)\mathbf{i} + (y - y_0)\mathbf{j} + (z - z_0)\mathbf{k}] = 0$$

or

$$A(x - x_0) + B(y - y_0) + C(z - z_0) = 0.$$

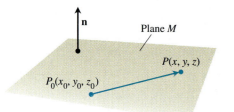

10.43 The standard equation for a plane in space is defined in terms of a vector normal to the plane: A point P lies in the plane through P_0 normal to $\mathbf{n}$ if and only if $\mathbf{n} \cdot \overrightarrow{P_0P} = 0$.

Plane Through $P_0(x_0, y_0, z_0)$ Normal to $\mathbf{n} = A\mathbf{i} + B\mathbf{j} + C\mathbf{k}$

Vector equation:	$\mathbf{n} \cdot \overrightarrow{P_0P} = 0$	(5)
Component equation:	$A(x - x_0) + B(y - y_0) + C(z - z_0) = 0$	(6)

EXAMPLE 5   Find an equation for the plane through $P_0(-3, 0, 7)$ perpendicular to $\mathbf{n} = 5\mathbf{i} + 2\mathbf{j} - \mathbf{k}$.

Solution

$$5(x - (-3)) + 2(y - 0) + (-1)(z - 7) = 0 \qquad \text{Eq. (5) with the given } \mathbf{n} \text{ and } P_0$$

$$5x + 15 + 2y - z + 7 = 0$$

$$5x + 2y - z = -22 \qquad \square$$

Notice in Example 5 how the components of $\mathbf{n} = 5\mathbf{i} + 2\mathbf{j} - \mathbf{k}$ became the coefficients of x, y, and z in the equation $5x + 2y - z = -22$. This was no accident, because Eq. (6), when rearranged, takes the form

$$Ax + By + Cz = D,$$

where $D = Ax_0 + By_0 + Cz_0$.

$A\mathbf{i} + B\mathbf{j} + C\mathbf{k}$ is normal to the plane $Ax + By + Cz = D$.

EXAMPLE 6   *A plane determined by three points*

Find the plane through $A(0, 0, 1)$, $B(2, 0, 0)$, and $C(0, 3, 0)$.

Solution We find a vector normal to the plane and use it with one of the points (it does not matter which) to write an equation for the plane.

The cross product

$$\overrightarrow{AB} \times \overrightarrow{AC} = \begin{vmatrix} \mathbf{i} & \mathbf{j} & \mathbf{k} \\ 2 & 0 & -1 \\ 0 & 3 & -1 \end{vmatrix} = 3\mathbf{i} + 2\mathbf{j} + 6\mathbf{k}$$

is normal to the plane. We substitute the components of this vector and the coordinates of the point $(0, 0, 1)$ into Eq. (6) to get

$$3(x - 0) + 2(y - 0) + 6(z - 1) = 0,$$

$$3x + 2y + 6z = 6. \qquad \square$$

EXAMPLE 7   *The intersection of a line and a plane*

Find the point where the line

$$x = \frac{8}{3} + 2t, \qquad y = -2t, \qquad z = 1 + t$$

intersects the plane $3x + 2y + 6z = 6$.

Solution The point

$$\left(\frac{8}{3} + 2t, -2t, 1 + t \right)$$

lies in the plane if its coordinates satisfy the equation of the plane; that is, if

$$3\left(\frac{8}{3} + 2t \right) + 2(-2t) + 6(1 + t) = 6$$

$$8 + 6t - 4t + 6 + 6t = 6$$

$$8t = -8$$

$$t = -1.$$

The point of intersection is

$$(x, y, z)\big|_{t=-1} = \left(\frac{8}{3} - 2, 2, 1 - 1 \right) = \left(\frac{2}{3}, 2, 0 \right). \qquad \square$$

EXAMPLE 8   *The distance from a point to a plane*

Find the distance from $S(1, 1, 3)$ to the plane $3x + 2y + 6z = 6$.

Solution We find a point P in the plane and calculate the length of the vector projection of $\overrightarrow{PS}$ onto a vector $\mathbf{n}$ normal to the plane (Fig. 10.44, on the following page).

The coefficients in the equation $3x + 2y + 6z = 6$ give

$$\mathbf{n} = 3\mathbf{i} + 2\mathbf{j} + 6\mathbf{k}.$$

10.44 The distance from S to the plane is the length of the vector projection of $\overrightarrow{PS}$ onto $\mathbf{n}$ (Example 8).

The points on the plane easiest to find from the plane's equation are the intercepts. If we take P to be the y-intercept $(0, 3, 0)$, then

$$\overrightarrow{PS} = (1 - 0)\,\mathbf{i} + (1 - 3)\,\mathbf{j} + (3 - 0)\,\mathbf{k}$$

$$= \mathbf{i} - 2\,\mathbf{j} + 3\,\mathbf{k},$$

$$|\mathbf{n}| = \sqrt{(3)^2 + (2)^2 + (6)^2} = \sqrt{49} = 7.$$

The distance from S to the plane is

$$d = \left| \overrightarrow{PS} \cdot \frac{\mathbf{n}}{|\mathbf{n}|} \right| \qquad \text{length of proj}_{\mathbf{n}}\ \overrightarrow{PS}$$

$$= \left| (\mathbf{i} - 2\,\mathbf{j} + 3\,\mathbf{k}) \cdot \left(\frac{3}{7}\mathbf{i} + \frac{2}{7}\mathbf{j} + \frac{6}{7}\mathbf{k} \right) \right|$$

$$= \left| \frac{3}{7} - \frac{4}{7} + \frac{18}{7} \right| = \frac{17}{7}.$$

> **How to Find the Distance from a Point to a Plane**
>
> To find the distance from a point S to a plane $Ax + By + Cz = D$, find
>
> 1. a point P on the plane,
> 2. $\overrightarrow{PS}$,
> 3. the direction of $\mathbf{n} = A\,\mathbf{i} + B\,\mathbf{j} + C\,\mathbf{k}$.
>
> Then calculate the distance as
>
> $$d = \left| \overrightarrow{PS} \cdot \frac{\mathbf{n}}{|\mathbf{n}|} \right|.$$

Angles Between Planes; Lines of Intersection

The angle between two intersecting planes is defined to be the (acute) angle determined by their normal vectors (Fig. 10.45).

EXAMPLE 9   Find the angle between the planes $3x - 6y - 2z = 15$ and $2x + y - 2z = 5$.

Solution   The vectors

$$\mathbf{n}_1 = 3\,\mathbf{i} - 6\,\mathbf{j} - 2\,\mathbf{k}, \qquad \mathbf{n}_2 = 2\,\mathbf{i} + \mathbf{j} - 2\,\mathbf{k}$$

are normals to the planes. The angle between them is

$$\theta = \cos^{-1}\left(\frac{\mathbf{n}_1 \cdot \mathbf{n}_2}{|\mathbf{n}_1||\mathbf{n}_2|} \right) \qquad \text{Eq. (4), Section 10.3}$$

$$= \cos^{-1}\left(\frac{4}{21} \right)$$

$$\approx 1.38 \text{ rad} \qquad \text{About } 79°$$

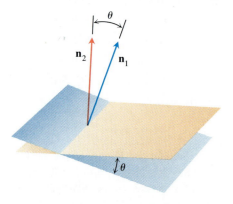

10.45 The angle between two planes is obtained from the angle between their normals.

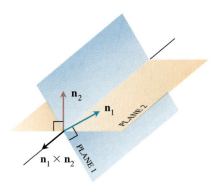

10.46 How the line of intersection of two planes is related to the planes' normal vectors (Example 10).

EXAMPLE 10   Find a vector parallel to the line of intersection of the planes $3x - 6y - 2z = 15$ and $2x + y - 2z = 5$.

Solution   The line of intersection of two planes is perpendicular to the planes' normal vectors $\mathbf{n}_1$ and $\mathbf{n}_2$ (Fig. 10.46), and therefore parallel to $\mathbf{n}_1 \times \mathbf{n}_2$. Turning this around, $\mathbf{n}_1 \times \mathbf{n}_2$ is a vector parallel to the planes' line of intersection. In our case,

$$\mathbf{n}_1 \times \mathbf{n}_2 = \begin{vmatrix} \mathbf{i} & \mathbf{j} & \mathbf{k} \\ 3 & -6 & -2 \\ 2 & 1 & -2 \end{vmatrix} = 14\,\mathbf{i} + 2\,\mathbf{j} + 15\,\mathbf{k}.$$

Any nonzero scalar multiple of $\mathbf{n}_1 \times \mathbf{n}_2$ will do as well.   ❑

EXAMPLE 11   Find parametric equations for the line in which the planes $3x - 6y - 2z = 15$ and $2x + y - 2z = 5$ intersect.

Solution   We find a vector parallel to the line and a point on the line and use Eqs. (2).

Example 10 identifies $\mathbf{v} = 14\,\mathbf{i} + 2\,\mathbf{j} + 15\,\mathbf{k}$ as a vector parallel to the line. To find a point on the line, we can take any point common to the two planes. Substituting $z = 0$ in the plane equations and solving for x and y simultaneously identifies one of these points as $(3, -1, 0)$. The line is

$$x = 3 + 14t, \qquad y = -1 + 2t, \qquad z = 15t.$$ ❑

Exercises 10.5

Lines and Line Segments

Find parametric equations for the lines in Exercises 1–12.

1. The line through the point $P(3, -4, -1)$ parallel to the vector $\mathbf{i} + \mathbf{j} + \mathbf{k}$

2. The line through $P(1, 2, -1)$ and $Q(-1, 0, 1)$

3. The line through $P(-2, 0, 3)$ and $Q(3, 5, -2)$

4. The line through $P(1, 2, 0)$ and $Q(1, 1, -1)$

5. The line through the origin parallel to the vector $2\mathbf{j} + \mathbf{k}$

6. The line through the point $(3, -2, 1)$ parallel to the line $x = 1 + 2t, y = 2 - t, z = 3t$

7. The line through $(1, 1, 1)$ parallel to the z-axis

8. The line through $(2, 4, 5)$ perpendicular to the plane $3x + 7y - 5z = 21$

9. The line through $(0, -7, 0)$ perpendicular to the plane $x + 2y + 2z = 13$

10. The line through $(2, 3, 0)$ perpendicular to the vectors $\mathbf{A} = \mathbf{i} + 2\mathbf{j} + 3\mathbf{k}$ and $\mathbf{B} = 3\mathbf{i} + 4\mathbf{j} + 5\mathbf{k}$

11. The x-axis

12. The z-axis

Find parametrizations for the line segments joining the points in Exercises 13–20. Draw coordinate axes and sketch each segment, indicating the direction of increasing t for your parametrization.

13. $(0, 0, 0)$, $(1, 1, 3/2)$

14. $(0, 0, 0)$, $(1, 0, 0)$

15. $(1, 0, 0)$, $(1, 1, 0)$

16. $(1, 1, 0)$, $(1, 1, 1)$

17. $(0, 1, 1)$, $(0, -1, 1)$

18. $(0, 2, 0)$, $(3, 0, 0)$

19. $(2, 0, 2)$, $(0, 2, 0)$

20. $(1, 0, -1)$, $(0, 3, 0)$

Planes

Find equations for the planes in Exercises 21–26.

21. The plane through $P_0(0, 2, -1)$ normal to $\mathbf{n} = 3\mathbf{i} - 2\mathbf{j} - \mathbf{k}$

22. The plane through $(1, -1, 3)$ parallel to the plane $3x + y + z = 7$

23. The plane through $(1, 1, -1)$, $(2, 0, 2)$, and $(0, -2, 1)$

24. The plane through $(2, 4, 5)$, $(1, 5, 7)$, and $(-1, 6, 8)$

25. The plane through $P_0(2, 4, 5)$ perpendicular to the line
$$x = 5 + t, \quad y = 1 + 3t, \quad z = 4t$$

26. The plane through $A(1, -2, 1)$ perpendicular to the vector from the origin to A

27. Find the point of intersection of the lines $x = 2t + 1, y = 3t + 2, z = 4t + 3$, and $x = s + 2, y = 2s + 4, z = -4s - 1$, and then find the plane determined by these lines.

28. Find the point of intersection of the lines $x = t, y = -t + 2, z = t + 1$, and $x = 2s + 2, y = s + 3, z = 5s + 6$, and then find the plane determined by these lines.

In Exercises 29 and 30, find the plane determined by the intersecting lines.

29. $L1$: $x = -1 + t, \ y = 2 + t, \ z = 1 - t, \ -\infty < t < \infty$
$L2$: $x = 1 - 4s, \ y = 1 + 2s, \ z = 2 - 2s, \ -\infty < s < \infty$

30. $L1$: $x = t, \ y = 3 - 3t, \ z = -2 - t, \ -\infty < t < \infty$
$L2$: $x = 1 + s, \ y = 4 + s, \ z = -1 + s, \ -\infty < s < \infty$

31. Find a plane through $P_0(2, 1, -1)$ and perpendicular to the line of intersection of the planes $2x + y - z = 3, x + 2y + z = 2$.

32. Find a plane through the points $P_1(1, 2, 3), P_2(3, 2, 1)$ and perpendicular to the plane $4x - y + 2z = 7$.

Distances

In Exercises 33–38, find the distance from the point to the line.

33. $(0, 0, 12)$; $x = 4t, \quad y = -2t, \quad z = 2t$

34. $(0, 0, 0)$; $x = 5 + 3t, \quad y = 5 + 4t, \quad z = -3 - 5t$

35. $(2, 1, 3)$; $x = 2 + 2t, \quad y = 1 + 6t, \quad z = 3$

36. $(2, 1, -1)$; $x = 2t, \quad y = 1 + 2t, \quad z = 2t$

37. $(3, -1, 4)$; $x = 4 - t, \quad y = 3 + 2t, \quad z = -5 + 3t$

38. $(-1, 4, 3)$; $x = 10 + 4t, \quad y = -3, \quad z = 4t$

In Exercises 39–44, find the distance from the point to the plane.

39. $(2, -3, 4)$, $x + 2y + 2z = 13$

40. $(0, 0, 0)$, $3x + 2y + 6z = 6$

41. $(0, 1, 1)$, $4y + 3z = -12$

42. $(2, 2, 3)$, $2x + y + 2z = 4$

43. $(0, -1, 0)$, $2x + y + 2z = 4$

44. $(1, 0, -1)$, $-4x + y + z = 4$

45. Find the distance from the plane $x + 2y + 6z = 1$ to the plane $x + 2y + 6z = 10$.

46. Find the distance from the line $x = 2 + t, y = 1 + t, z = -(1/2) - (1/2)t$ to the plane $x + 2y + 6z = 10$.

Angles

Find the angles between the planes in Exercises 47 and 48. (You will not need a calculator.)

47. $x + y = 1, \quad 2x + y - 2z = 2$

48. $5x + y - z = 10, \quad x - 2y + 3z = -1$

CALCULATOR Use a calculator to find the acute angles between the planes in Exercises 49–52 to the nearest hundredth of a radian.

49. $2x + 2y + 2z = 3, \quad 2x - 2y - z = 5$

50. $x + y + z = 1, \quad z = 0$ (the xy-plane)

51. $2x + 2y - z = 3, \quad x + 2y + z = 2$

52. $4y + 3z = -12, \quad 3x + 2y + 6z = 6$

Intersecting Lines and Planes

In Exercises 53–56, find the point in which the line meets the plane.

53. $x = 1 - t, \quad y = 3t, \quad z = 1 + t; \quad 2x - y + 3z = 6$

54. $x = 2, \quad y = 3 + 2t, \quad z = -2 - 2t; \quad 6x + 3y - 4z = -12$

55. $x = 1 + 2t, \quad y = 1 + 5t, \quad z = 3t; \quad x + y + z = 2$

56. $x = -1 + 3t, \quad y = -2, \quad z = 5t; \quad 2x - 3z = 7$

Find parametrizations for the lines in which the planes in Exercises 57–60 intersect.

57. $x + y + z = 1, \quad x + y = 2$

58. $3x - 6y - 2z = 3, \quad 2x + y - 2z = 2$

59. $x - 2y + 4z = 2, \quad x + y - 2z = 5$

60. $5x - 2y = 11, \quad 4y - 5z = -17$

Given two lines in space, either they are parallel, or they intersect, or they are skew (imagine, for example, the flight paths of two planes in the sky). Exercises 61 and 62 each give three lines. In each exercise, determine whether the lines, taken two at a time, are parallel, intersect, or are skew. If they intersect, find the point of intersection.

61. $L1$: $x = 3 + 2t, y = -1 + 4t, z = 2 - t, -\infty < t < \infty$
$L2$: $x = 1 + 4s, y = 1 + 2s, z = -3 + 4s, -\infty < s < \infty$
$L3$: $x = 3 + 2r, y = 2 + r, z = -2 + 2r, -\infty < r < \infty$

62. $L1$: $x = 1 + 2t, y = -1 - t, z = 3t, -\infty < t < \infty$
$L2$: $x = 2 - s, y = 3s, z = 1 + s, -\infty < s < \infty$
$L3$: $x = 5 + 2r, y = 1 - r, z = 8 + 3r, -\infty < r < \infty$

Theory and Examples

63. Use Eqs. (2) to generate a parametrization of the line through $P(2, -4, 7)$ parallel to $\mathbf{v}_1 = 2\mathbf{i} - \mathbf{j} + 3\mathbf{k}$. Then generate another parametrization of the line using the point $P_2(-2, -2, 1)$ and the vector $\mathbf{v}_2 = -\mathbf{i} + (1/2)\mathbf{j} - (3/2)\mathbf{k}$.

64. Use Eq. (6) to generate an equation for the plane through $P_1(4, 1, 5)$ normal to $\mathbf{n}_1 = \mathbf{i} - 2\mathbf{j} + \mathbf{k}$. Then generate another equation for the same plane using the point $P_2(3, -2, 0)$ and the normal vector $\mathbf{n}_2 = -\sqrt{2}\mathbf{i} + 2\sqrt{2}\mathbf{j} - \sqrt{2}\mathbf{k}$.

65. Find the points in which the line $x = 1 + 2t, y = -1 - t, z = 3t$ meets the coordinate planes. Describe the reasoning behind your answer.

66. Find equations for the line in the plane $z = 3$ that makes an angle of $\pi/6$ rad with $\mathbf{i}$ and an angle of $\pi/3$ rad with $\mathbf{j}$. Describe the reasoning behind your answer.

67. Is the line $x = 1 - 2t$, $y = 2 + 5t$, $z = -3t$ parallel to the plane $2x + y - z = 8$? Give reasons for your answer.

68. How can you tell when two planes $A_1x + B_1y + C_1z = D_1$ and $A_2x + B_2y + C_2z = D_2$ are parallel? perpendicular? Give reasons for your answer.

69. Find two different planes whose intersection is the line $x = 1 + t$, $y = 2 - t$, $z = 3 + 2t$. Write equations for each plane in the form $Ax + By + Cz = D$.

70. Find a plane through the origin that meets the plane M: $2x + 3y + z = 12$ in a right angle. How do you know that your plane is perpendicular to M?

71. For any nonzero numbers a, b, and c, the graph of $(x/a) + (y/b) + (z/c) = 1$ is a plane. Which planes have an equation of this form?

72. Suppose L_1 and L_2 are disjoint (nonintersecting) nonparallel lines. Is it possible for a nonzero vector to be perpendicular to both L_1 and L_2? Give reasons for your answer.

Computer Graphics

73. *Perspective in computer graphics.* In computer graphics and perspective drawing we need to represent objects seen by the eye in space as images on a two-dimensional plane. Suppose the eye is at $E(x_0, 0, 0)$ as shown here and that we want to represent a point $P_1(x_1, y_1, z_1)$ as a point on the yz-plane. We do this by projecting P_1 onto the plane with a ray from E. The point P_1 will be portrayed as the point $P(0, y, z)$. The problem for us as graphics designers is to find y and z given E and P_1.

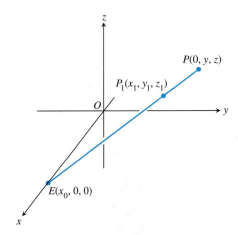

a) Write a vector equation that holds between $\overrightarrow{EP}$ and $\overrightarrow{EP_1}$. Use the equation to express y and z in terms of x_0, x_1, y_1, and z_1.

b) Test the formulas obtained for y and z in part (a) by investigating their behavior at $x_1 = 0$ and $x_1 = x_0$ and by seeing what happens as $x_0 \to \infty$.

74. *Hidden lines.* Here is another typical problem in computer graphics. Your eye is at $(4, 0, 0)$. You are looking at a triangular plate whose vertices are at $(1, 0, 1)$, $(1, 1, 0)$, and $(-2, 2, 2)$. The line segment from $(1, 0, 0)$ to $(0, 2, 2)$ passes through the plate. What portion of the line segment is hidden from your view by the plate? (This is an exercise in finding intersections of lines and planes.)

| 10.6 | # Cylinders and Quadric Surfaces |

In the calculus of functions of a single variable, we began with lines and used our knowledge of lines to study curves in the plane. We investigated tangents and found that, when highly magnified, differentiable curves were effectively linear. Among the curves of special interest were the conic sections, curves defined by second-degree equations in x and y.

To study the calculus of functions of more than one variable, we begin much the same way. We start with planes and use our knowledge of planes to study surfaces in space. Among the surfaces of special interest are cylinders and quadric surfaces, surfaces defined by second-degree equations in x, y, and z. We described the planes in Section 10.5. In this section, we describe the surfaces.

Cylinders

A **cylinder** is the surface composed of all the lines that (1) lie parallel to a given line in space and (2) pass through a given plane curve. The curve is a **generating curve** for the cylinder (Fig. 10.47, on the following page). In solid geometry, where *cylinder* means *circular cylinder,* the generating curves are circles, but now we allow generating curves of any kind. The cylinder in our first example is generated by a parabola.

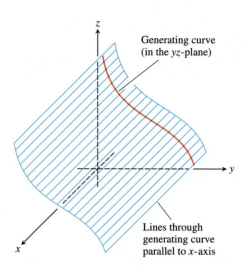

10.47 A cylinder and generating curve.

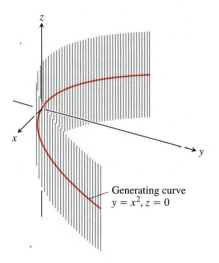

10.48 The cylinder of lines passing through the parabola $y = x^2$ in the xy-plane parallel to the z-axis (Example 1).

When graphing a cylinder or other surface by hand or analyzing one generated by a computer, it helps to look at the curves formed by intersecting the surface with planes parallel to the coordinate planes. These curves are called **cross sections** or **traces.**

EXAMPLE 1   *The parabolic cylinder $y = x^2$*

Find an equation for the cylinder made by the lines parallel to the z-axis that pass through the parabola $y = x^2$, $z = 0$ (Fig. 10.48).

Solution   Suppose that the point $P_0(x_0, x_0{}^2, 0)$ lies on the parabola $y = x^2$ in the xy-plane. Then, for any value of z, the point $Q(x_0, x_0{}^2, z)$ will lie on the cylinder because it lies on the line $x = x_0$, $y = x_0{}^2$ through P_0 parallel to the z-axis. Conversely, any point $Q(x_0, x_0{}^2, z)$ whose y-coordinate is the square of its x-coordinate lies on the cylinder because it lies on the line $x = x_0$, $y = x_0{}^2$ through P_0 parallel to the z-axis (Fig. 10.49).

Regardless of the value of z, therefore, the points on the surface are the points whose coordinates satisfy the equation $y = x^2$. This makes $y = x^2$ an equation for the cylinder. Because of this, we call the cylinder "the cylinder $y = x^2$."

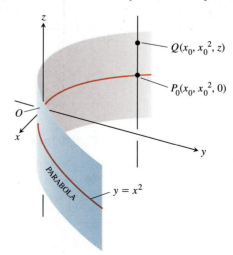

10.49 Every point of the cylinder in Fig. 10.48 has coordinates of the form $(x_0, x_0{}^2, z)$. We call the cylinder "the cylinder $y = x^2$."

DRAWING LESSON
How to Draw Cylinders Parallel to the Coordinate Axes

$$x^2 + y^2 = 1 \qquad\qquad z = y^2$$

1 Sketch all three coordinate axes *very lightly.*

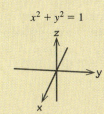 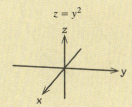

2 Sketch the trace of the cylinder in the coordinate plane of the two variables that appear in the cylinder's equation. Sketch *very lightly.*

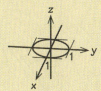 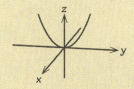

3 Sketch traces in parallel planes on either side (again, *lightly*).

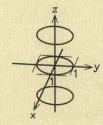 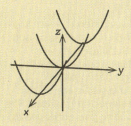

4 Add parallel outer edges to give the shape definition.

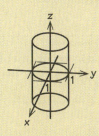 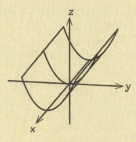

5 If more definition is required, darken the parts of the lines that are exposed to view. Leave the hidden parts light. Use line breaks when you can.

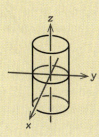 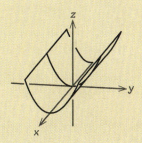

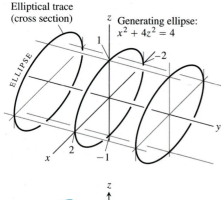

Elliptical trace (cross section)

Generating ellipse: $x^2 + 4z^2 = 4$

ELLIPSE

As Example 1 suggests, any curve $f(x, y) = c$ in the xy-plane defines a cylinder parallel to the z-axis whose equation is also $f(x, y) = c$. The equation $x^2 + y^2 = 1$ defines the circular cylinder made by the lines parallel to the z-axis that pass through the circle $x^2 + y^2 = 1$ in the xy-plane. The equation $x^2 + 4y^2 = 9$ defines the elliptical cylinder made by the lines parallel to the z-axis that pass through the ellipse $x^2 + 4y^2 = 9$ in the xy-plane.

In a similar way, any curve $g(x, z) = c$ in the xz-plane defines a cylinder parallel to the y-axis whose space equation is also $g(x, z) = c$ (Fig. 10.50). Any curve $h(y, z) = c$ defines a cylinder parallel to the x-axis whose space equation is also $h(y, z) = c$ (Fig. 10.51).

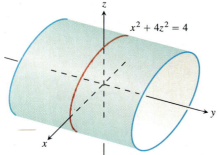

$x^2 + 4z^2 = 4$

10.50 The elliptic cylinder $x^2 + 4z^2 = 4$ is made of lines parallel to the y-axis and passing through the ellipse $x^2 + 4z^2 = 4$ in the xz-plane. The cross sections or "traces" of the cylinder in planes perpendicular to the y-axis are ellipses congruent to the generating ellipse. The cylinder extends along the entire y-axis.

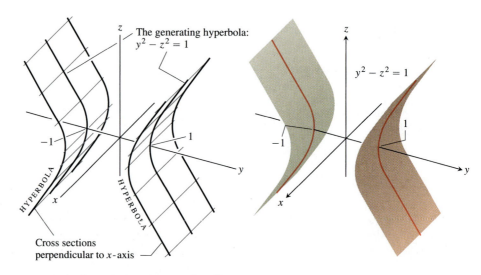

The generating hyperbola: $y^2 - z^2 = 1$

HYPERBOLA

HYPERBOLA

Cross sections perpendicular to x-axis

$y^2 - z^2 = 1$

10.51 The hyperbolic cylinder $y^2 - z^2 = 1$ is made of lines parallel to the x-axis and passing through the hyperbola $y^2 - z^2 = 1$ in the yz-plane. The cross sections of the cylinder in planes perpendicular to the x-axis are hyperbolas congruent to the generating hyperbola.

> An equation in any two of the three Cartesian coordinates defines a cylinder parallel to the axis of the third coordinate.

Quadric Surfaces

A **quadric surface** is the graph in space of a second-degree equation in x, y, and z. The most general form is

$$Ax^2 + By^2 + Cz^2 + Dxy + Eyz + Fxz + Gx + Hy + Jz + K = 0,$$

where A, B, C, and so on are constants, but the equation can be simplified by translation and rotation, as in the two-dimensional case in Section 9.3. We will study only the simpler equations. Although the definition did not require it, the cylinders in Figs. 10.49–10.51 were also examples of quadric surfaces. We now examine ellipsoids (these include spheres), paraboloids, cones, and hyperboloids.

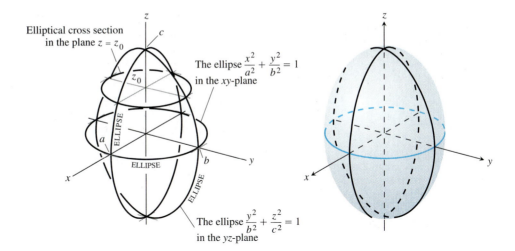

10.52 The ellipsoid

$$\frac{x^2}{a^2} + \frac{y^2}{b^2} + \frac{z^2}{c^2} = 1$$

in Example 2.

EXAMPLE 2   The **ellipsoid**

$$\frac{x^2}{a^2} + \frac{y^2}{b^2} + \frac{z^2}{c^2} = 1 \tag{1}$$

(Fig. 10.52) cuts the coordinate axes at $(\pm a, 0, 0)$, $(0, \pm b, 0)$, and $(0, 0, \pm c)$. It lies within the rectangular box defined by the inequalities $|x| \le a$, $|y| \le b$, and $|z| \le c$. The surface is symmetric with respect to each of the coordinate planes because the variables in the defining equation are squared.

The curves in which the three coordinate planes cut the surface are ellipses. For example,

$$\frac{x^2}{a^2} + \frac{y^2}{b^2} = 1 \qquad \text{when} \qquad z = 0.$$

The section cut from the surface by the plane $z = z_0$, $|z_0| < c$, is the ellipse

$$\frac{x^2}{a^2(1 - (z_0/c)^2)} + \frac{y^2}{b^2(1 - (z_0/c)^2)} = 1. \tag{2}$$

If any two of the semiaxes a, b, and c are equal, the surface is an **ellipsoid of revolution.** If all three are equal, the surface is a sphere. ❑

Technology   *Visualizing in Space*   A Computer Algebra System (CAS) or other computer graphing utility can help in visualizing surfaces in space. It can draw traces in different planes with far more patience than most people can muster. Many computer graphing systems can rotate the figure so you can see it as if it were a physical model you could turn in your hand. Hidden-line algorithms (see Exercise 74, Section 10.5) are used to block out portions of the surface that you would not see from your current viewing angle. Often a CAS will require surfaces to be entered in parametric form, as discussed in Section 14.6 (see also CAS Exercises 57–60 in Section 12.1). Sometimes you may have to manipulate the grid mesh to see all portions of a surface.

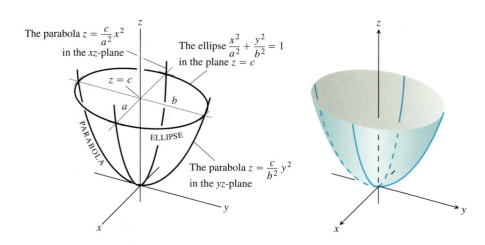

10.53 The elliptic paraboloid $(x^2/a^2) + (y^2/b^2) = z/c$ in Example 3, shown for $c > 0$. The cross sections perpendicular to the z-axis above the xy-plane are ellipses. The cross sections in the planes that contain the z-axis are parabolas.

In the figure: The parabola $z = \dfrac{c}{a^2}x^2$ in the xz-plane. The ellipse $\dfrac{x^2}{a^2} + \dfrac{y^2}{b^2} = 1$ in the plane $z = c$. The parabola $z = \dfrac{c}{b^2}y^2$ in the yz-plane.

(a)

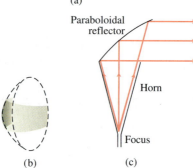

Paraboloidal reflector

Horn

Focus

(b)   (c)

10.54 Many antennas are shaped like pieces of paraboloids of revolution. (a) Radio telescopes use the same principles as optical telescopes. (b) A "rectangular-cut" radar reflector. (c) The profile of a horn antenna in a microwave radio link.

EXAMPLE 3   The **elliptic paraboloid**

$$\frac{x^2}{a^2} + \frac{y^2}{b^2} = \frac{z}{c} \tag{3}$$

is symmetric with respect to the planes $x = 0$ and $y = 0$ (Fig. 10.53). The only intercept on the axes is the origin. Except for this point, the surface lies above or entirely below the xy-plane, depending on the sign of c. The sections cut by the coordinate planes are

$x = 0$:   the parabola $z = \dfrac{c}{b^2}y^2$

$y = 0$:   the parabola $z = \dfrac{c}{a^2}x^2$ $\qquad(4)$

$z = 0$:   the point $(0, 0, 0)$.

Each plane $z = z_0$ above the xy-plane cuts the surface in the ellipse

$$\frac{x^2}{a^2} + \frac{y^2}{b^2} = \frac{z_0}{c}. \qquad \square$$

EXAMPLE 4   The **circular paraboloid** or **paraboloid of revolution**

$$\frac{x^2}{a^2} + \frac{y^2}{a^2} = \frac{z}{c} \tag{5}$$

is obtained by taking $b = a$ in Eq. (3) for the elliptic paraboloid. The cross sections of the surface by planes perpendicular to the z-axis are circles centered on the z-axis. The cross sections by planes containing the z-axis are congruent parabolas with a common focus at the point $(0, 0, a^2/4c)$.

Shapes cut from circular paraboloids are used for antennas in radio telescopes, satellite trackers, and microwave radio links (Fig. 10.54). $\qquad \square$

EXAMPLE 5   The **elliptic cone**

$$\frac{x^2}{a^2} + \frac{y^2}{b^2} = \frac{z^2}{c^2} \tag{6}$$

10.55 The elliptic cone $(x^2/a^2) + (y^2/b^2) = (z^2/c^2)$ in Example 5. Planes perpendicular to the z-axis cut the cone in ellipses above and below the xy-plane. Vertical planes that contain the z-axis cut it in pairs of intersecting lines.

is symmetric with respect to the three coordinate planes (Fig. 10.55). The sections cut by the coordinate planes are

$x = 0$:   the lines $z = \pm \dfrac{c}{b} y$ (7)

$y = 0$:   the lines $z = \pm \dfrac{c}{a} x$ (8)

$z = 0$:   the point $(0, 0, 0)$.

The sections cut by planes $z = z_0$ above and below the xy-plane are ellipses whose centers lie on the z-axis and whose vertices lie on the lines in Eqs. (7) and (8).

If $a = b$, the cone is a right circular cone. ❏

EXAMPLE 6   The **hyperboloid of one sheet**

$$\frac{x^2}{a^2} + \frac{y^2}{b^2} - \frac{z^2}{c^2} = 1 \tag{9}$$

is symmetric with respect to each of the three coordinate planes (Fig. 10.56, on the following page). The sections cut out by the coordinate planes are

$x = 0$:   the hyperbola $\dfrac{y^2}{b^2} - \dfrac{z^2}{c^2} = 1$

$y = 0$:   the hyperbola $\dfrac{x^2}{a^2} - \dfrac{z^2}{c^2} = 1$ (10)

$z = 0$:   the ellipse $\dfrac{x^2}{a^2} + \dfrac{y^2}{b^2} = 1.$

The plane $z = z_0$ cuts the surface in an ellipse with center on the z-axis and vertices on one of the hyperbolas in (10).

10.56 The hyperboloid $(x^2/a^2) + (y^2/b^2) - (z^2/c^2) = 1$ in Example 6. Planes perpendicular to the z-axis cut it in ellipses. Vertical planes containing the z-axis cut it in hyperbolas.

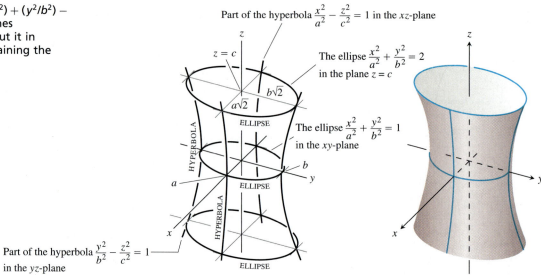

The surface is connected, meaning that it is possible to travel from one point on it to any other without leaving the surface. For this reason, it is said to have *one* sheet, in contrast to the hyperboloid in the next example, which has two sheets.

If $a = b$, the hyperboloid is a surface of revolution.

EXAMPLE 7 The **hyperboloid of two sheets**

$$\frac{z^2}{c^2} - \frac{x^2}{a^2} - \frac{y^2}{b^2} = 1 \tag{11}$$

is symmetric with respect to the three coordinate planes (Fig. 10.57). The plane $z = 0$ does not intersect the surface; in fact, for a horizontal plane to intersect the

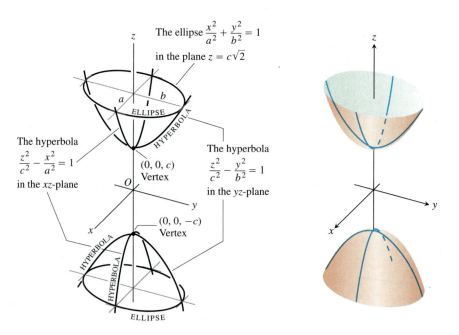

10.57 The hyperboloid $(z^2/c^2) - (x^2/a^2) - (y^2/b^2) = 1$ in Example 7. Planes perpendicular to the z-axis above and below the vertices cut it in ellipses. Vertical planes containing the z-axis cut it in hyperbolas.

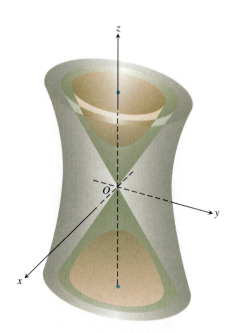

10.58 Both hyperboloids are asymptotic to the cone (Example 7).

surface, we must have $|z| \geq c$. The hyperbolic sections

$$x = 0: \quad \frac{z^2}{c^2} - \frac{y^2}{b^2} = 1, \qquad y = 0: \quad \frac{z^2}{c^2} - \frac{x^2}{a^2} = 1,$$

have their vertices and foci on the z-axis. The surface is separated into two portions, one above the plane $z = c$ and the other below the plane $z = -c$. This accounts for its name.

Equations (9) and (11) have different numbers of negative terms. The number in each case is the same as the number of sheets of the hyperboloid. If we replace the 1 on the right side of either Eq. (9) or Eq. (11) by 0, we obtain the equation

$$\frac{x^2}{a^2} + \frac{y^2}{b^2} = \frac{z^2}{c^2}$$

for an elliptic cone (Eq. 6). The hyperboloids are asymptotic to this cone (Fig. 10.58) in the same way that the hyperbolas

$$\frac{x^2}{a^2} - \frac{y^2}{b^2} = \pm 1$$

are asymptotic to the lines

$$\frac{x^2}{a^2} - \frac{y^2}{b^2} = 0$$

in the xy-plane. ◻

EXAMPLE 8   The **hyperbolic paraboloid**

$$\frac{y^2}{b^2} - \frac{x^2}{a^2} = \frac{z}{c}, \qquad c > 0 \tag{12}$$

has symmetry with respect to the planes $x = 0$ and $y = 0$ (Fig. 10.59). The sections in these planes are

$$x = 0: \quad \text{the parabola } z = \frac{c}{b^2} y^2, \tag{13}$$

$$y = 0: \quad \text{the parabola } z = -\frac{c}{a^2} x^2. \tag{14}$$

10.59 The hyperbolic paraboloid $(y^2/b^2) - (x^2/a^2) = z/c$, $c > 0$ in Example 8. The cross sections in planes perpendicular to the z-axis above and below the xy-plane are hyperbolas. The cross sections in planes perpendicular to the other axes are parabolas.

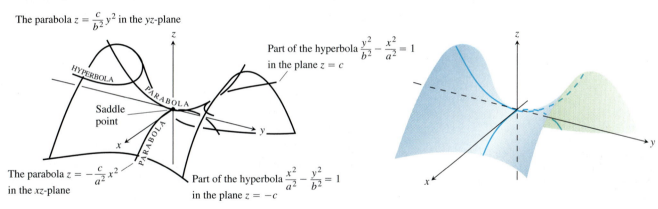

The parabola $z = \frac{c}{b^2} y^2$ in the yz-plane

HYPERBOLA

PARABOLA

Saddle point

The parabola $z = -\frac{c}{a^2} x^2$ in the xz-plane

Part of the hyperbola $\frac{y^2}{b^2} - \frac{x^2}{a^2} = 1$ in the plane $z = c$

Part of the hyperbola $\frac{x^2}{a^2} - \frac{y^2}{b^2} = 1$ in the plane $z = -c$

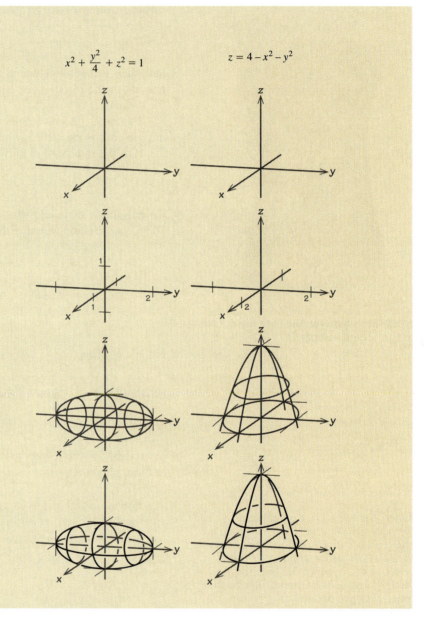

$$x^2 + \frac{y^2}{4} + z^2 = 1 \qquad\qquad z = 4 - x^2 - y^2$$

1 Lightly sketch the three coordinate axes.

2 Decide on a scale and mark the intercepts on the axes.

3 Sketch cross sections in the coordinate planes and in a few parallel planes, but don't clutter the picture. Use tangent lines as guides.

4 If more is required, darken the parts exposed to view. Leave the rest light. Use line breaks when you can.

In the plane $x = 0$, the parabola opens upward from the origin. The parabola in the plane $y = 0$ opens downward.

If we cut the surface by a plane $z = z_0 > 0$, the section is a hyperbola,

$$\frac{y^2}{b^2} - \frac{x^2}{a^2} = \frac{z_0}{c}, \tag{15}$$

with its focal axis parallel to the y-axis and its vertices on the parabola in (13). If z_0 is negative, the focal axis is parallel to the x-axis and the vertices lie on the parabola in (14).

Near the origin, the surface is shaped like a saddle. To a person traveling along the surface in the yz-plane, the origin looks like a minimum. To a person traveling in the xz-plane, the origin looks like a maximum. Such a point is called a **minimax** or **saddle point** of a surface.   ❑

Liquid Mirror Telescopes

When a circular pan of liquid is rotated about its vertical axis, the surface of the liquid does not stay flat. Instead, it assumes the shape of a paraboloid of revolution, exactly what is needed for the primary mirror of a reflecting telescope. At the turn of the century, attempts to make reliable mirrors with revolving mercury failed because of surface ripples and focus losses caused by variations in the speed of rotation. Today, these difficulties can be overcome with synchronous motors driven by oscillator-stabilized power supplies and checked against constant clocks.

Using the same idea, astronomers at the Steward Observatory's Mirror Laboratory in Tucson, Arizona, have used a large heated spinning turntable to cast borosilicate glass blanks for lightweight mirrors. Spincast mirrors cost less than traditionally cast mirrors and can be made larger. Their shorter focal lengths also allow them to be installed in compact telescope frames that are less expensive to house and less likely to flex in strong winds.

Exercises 10.6

Matching Equations with Surfaces

In Exercises 1–12, match the equation with the surface it defines. Also, identify each surface by type (paraboloid, ellipsoid, etc.) The surfaces are labeled (a)–(l).

1. $x^2 + y^2 + 4z^2 = 10$

2. $z^2 + 4y^2 - 4x^2 = 4$

3. $9y^2 + z^2 = 16$

4. $y^2 + z^2 = x^2$

5. $x = y^2 - z^2$

6. $x = -y^2 - z^2$

7. $x^2 + 2z^2 = 8$

8. $z^2 + x^2 - y^2 = 1$

9. $x = z^2 - y^2$

10. $z = -4x^2 - y^2$

11. $x^2 + 4z^2 = y^2$

12. $9x^2 + 4y^2 + 2z^2 = 36$

a)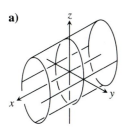

b)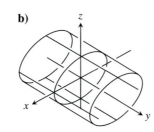

c)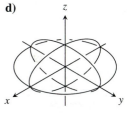

d)

e)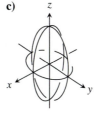

f)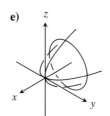

g)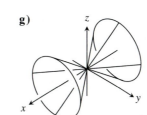

h)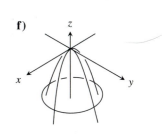

i)

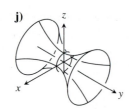

j)

k)

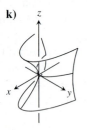

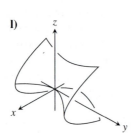

l)

Drawing

Sketch the surfaces in Exercises 13–76.

CYLINDERS

13. $x^2 + y^2 = 4$

14. $x^2 + z^2 = 4$

15. $z = y^2 - 1$

16. $x = y^2$

17. $x^2 + 4z^2 = 16$

18. $4x^2 + y^2 = 36$

19. $z^2 - y^2 = 1$

20. $yz = 1$

ELLIPSOIDS

21. $9x^2 + y^2 + z^2 = 9$

22. $4x^2 + 4y^2 + z^2 = 16$

23. $4x^2 + 9y^2 + 4z^2 = 36$

24. $9x^2 + 4y^2 + 36z^2 = 36$

PARABOLOIDS

25. $z = x^2 + 4y^2$

26. $z = x^2 + 9y^2$

27. $z = 8 - x^2 - y^2$

28. $z = 18 - x^2 - 9y^2$

29. $x = 4 - 4y^2 - z^2$

30. $y = 1 - x^2 - z^2$

CONES

31. $x^2 + y^2 = z^2$

32. $y^2 + z^2 = x^2$

33. $4x^2 + 9z^2 = 9y^2$

34. $9x^2 + 4y^2 = 36z^2$

HYPERBOLOIDS

35. $x^2 + y^2 - z^2 = 1$

36. $y^2 + z^2 - x^2 = 1$

37. $(y^2/4) + (z^2/9) - (x^2/4) = 1$

38. $(x^2/4) + (y^2/4) - (z^2/9) = 1$

39. $z^2 - x^2 - y^2 = 1$

40. $(y^2/4) - (x^2/4) - z^2 = 1$

41. $x^2 - y^2 - (z^2/4) = 1$

42. $(x^2/4) - y^2 - (z^2/4) = 1$

HYPERBOLIC PARABOLOIDS

43. $y^2 - x^2 = z$

44. $x^2 - y^2 = z$

ASSORTED

45. $x^2 + y^2 + z^2 = 4$

46. $4x^2 + 4y^2 = z^2$

47. $z = 1 + y^2 - x^2$

48. $y^2 - z^2 = 4$

49. $y = -(x^2 + z^2)$

50. $z^2 - 4x^2 - 4y^2 = 4$

51. $16x^2 + 4y^2 = 1$

52. $z = x^2 + y^2 + 1$

53. $x^2 + y^2 - z^2 = 4$

54. $x = 4 - y^2$

55. $x^2 + z^2 = y$

56. $z^2 - (x^2/4) - y^2 = 1$

57. $x^2 + z^2 = 1$

58. $4x^2 + 4y^2 + z^2 = 4$

59. $16y^2 + 9z^2 = 4x^2$

60. $z = x^2 - y^2 - 1$

61. $9x^2 + 4y^2 + z^2 = 36$

62. $4x^2 + 9z^2 = y^2$

63. $x^2 + y^2 - 16z^2 = 16$

64. $z^2 + 4y^2 = 9$

65. $z = -(x^2 + y^2)$

66. $y^2 - x^2 - z^2 = 1$

67. $x^2 - 4y^2 = 1$

68. $z = 4x^2 + y^2 - 4$

69. $4y^2 + z^2 - 4x^2 = 4$

70. $z = 1 - x^2$

71. $x^2 + y^2 = z$

72. $(x^2/4) + y^2 - z^2 = 1$

73. $yz = 1$

74. $36x^2 + 9y^2 + 4z^2 = 36$

75. $9x^2 + 16y^2 = 4z^2$

76. $4z^2 - x^2 - y^2 = 4$

Theory and Examples

77. a) Express the area A of the cross section cut from the ellipsoid

$$x^2 + \frac{y^2}{4} + \frac{z^2}{9} = 1$$

by the plane $z = c$ as a function of c. (The area of an ellipse with semiaxes a and b is πab.)

b) Use slices perpendicular to the z-axis to find the volume of the ellipsoid in (a).

c) Now find the volume of the ellipsoid

$$\frac{x^2}{a^2} + \frac{y^2}{b^2} + \frac{z^2}{c^2} = 1.$$

Does your formula give the volume of a sphere of radius a if $a = b = c$?

78. The barrel shown here is shaped like an ellipsoid with equal pieces cut from the ends by planes perpendicular to the z-axis. The cross sections perpendicular to the z-axis are circular. The barrel is $2h$ units high, its midsection radius is R, and its end radii are both r. Find a formula for the barrel's volume. Then check two things. First, suppose the sides of the barrel are straightened to turn the barrel into a cylinder of radius R and height $2h$. Does your formula give the cylinder's volume? Second, suppose $r = 0$

and $h = R$ so the barrel is a sphere. Does your formula give the sphere's volume?

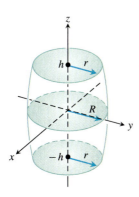

79. Show that the volume of the segment cut from the paraboloid

$$\frac{x^2}{a^2} + \frac{y^2}{b^2} = \frac{z}{c}$$

by the plane $z = h$ equals half the segment's base times its altitude. (Figure 10.53 shows the segment for the special case $h = c$.)

80. a) Find the volume of the solid bounded by the hyperboloid

$$\frac{x^2}{a^2} + \frac{y^2}{b^2} - \frac{z^2}{c^2} = 1$$

and the planes $z = 0$ and $z = h, h > 0$.

b) Express your answer in (a) in terms of h and the areas A_0 and A_h of the regions cut by the hyperboloid from the planes $z = 0$ and $z = h$.

c) Show that the volume in (a) is also given by the formula

$$V = \frac{h}{6}\left(A_0 + 4A_m + A_h\right),$$

where A_m is the area of the region cut by the hyperboloid from the plane $z = h/2$.

81. If the hyperbolic paraboloid $(y^2/b^2) - (x^2/a^2) = z/c$ is cut by the plane $y = y_1$, the resulting curve is a parabola. Find its vertex and focus.

82. Suppose you set $z = 0$ in the equation

$$A x^2 + B y^2 + C z^2 + D xy + E yz + F xz + G x + H y + J z + K = 0$$

to obtain a curve in the xy-plane. What will the curve be like? Give reasons for your answer.

83. Every time we found the trace of a quadric surface in a plane parallel to one of the coordinate planes, it turned out to be a conic section. Was this mere coincidence? Did it have to happen? Give reasons for your answer.

84. Suppose you intersect a quadric surface with a plane that is *not* parallel to one of the coordinate planes. What will the trace in the plane be like? Give reasons for your answer.

▦ Computer Grapher Explorations

Plot the surfaces in Exercises 85–88 over the indicated domains. If you can, rotate the surface into different viewing positions.

85. $z = y^2, \quad -2 \le x \le 2, \quad -0.5 \le y \le 2$

86. $z = 1 - y^2, \quad -2 \le x \le 2, \quad -2 \le y \le 2$

87. $z = x^2 + y^2, \quad -3 \le x \le 3, \quad -3 \le y \le 3$

88. $z = x^2 + 2y^2 \quad$ over

    a) $-3 \le x \le 3, \quad -3 \le y \le 3$
    b) $-1 \le x \le 1, \quad -2 \le y \le 3$
    c) $-2 \le x \le 2, \quad -2 \le y \le 2$
    d) $-2 \le x \le 2, \quad -1 \le y \le 1$

✿ CAS Explorations and Projects

Use a CAS to plot the surfaces in Exercises 89–94. Identify the type of quadric surface from your graph.

89. $\dfrac{x^2}{9} + \dfrac{y^2}{36} = 1 - \dfrac{z^2}{25}$ **90.** $\dfrac{x^2}{9} - \dfrac{z^2}{9} = 1 - \dfrac{y^2}{16}$

91. $5x^2 = z^2 - 3y^2$ **92.** $\dfrac{y^2}{16} = 1 - \dfrac{x^2}{9} + z$

93. $\dfrac{x^2}{9} - 1 = \dfrac{y^2}{16} + \dfrac{z^2}{2}$ **94.** $y - \sqrt{4 - z^2} = 0$

10.7   Cylindrical and Spherical Coordinates

This section introduces two new coordinate systems for space: the cylindrical coordinate system and the spherical coordinate system. Cylindrical coordinates simplify the equations of cylinders. Spherical coordinates simplify the equations of spheres and cones. We will use cylindrical coordinates to study planetary motion in Section 11.5.

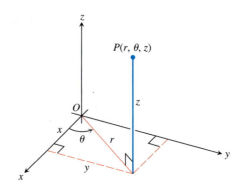

10.60 The cylindrical coordinates of a point in space are r, θ, and z.

Cylindrical Coordinates

We obtain cylindrical coordinates for space by combining polar coordinates in the xy-plane with the usual z-axis. This assigns to every point in space one or more coordinate triples of the form (r, θ, z), as shown in Fig. 10.60.

Definition

Cylindrical coordinates represent a point P in space by ordered triples (r, θ, z) in which

1. r and θ are polar coordinates for the vertical projection of P on the xy-plane,

2. z is the rectangular vertical coordinate.

The values of x, y, r, and θ in rectangular and cylindrical coordinates are related by the usual equations.

Equations Relating Rectangular (x, y, z) and Cylindrical (r, θ, z) Coordinates

$$x = r \cos \theta, \qquad y = r \sin \theta, \qquad z = z$$
$$r^2 = x^2 + y^2, \qquad \tan \theta = y/x$$

(1)

In cylindrical coordinates, the equation $r = a$ describes not just a circle in the xy-plane but an entire cylinder about the z-axis (Fig. 10.61). The z-axis is given by $r = 0$. The equation $\theta = \theta_0$ describes the plane that contains the z-axis and makes an angle θ_0 with the positive x-axis. And, just as in rectangular coordinates, the equation $z = z_0$ describes a plane perpendicular to the z-axis.

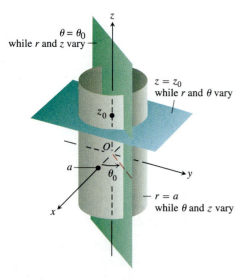

10.61 Constant-coordinate equations in cylindrical coordinates yield cylinders and planes.

EXAMPLE 1 What points satisfy the equations

$$r = 2, \qquad \theta = \frac{\pi}{4}?$$

Along this line, z varies while r and θ have the constant values $r = 2$ and $\theta = \pi/4$.

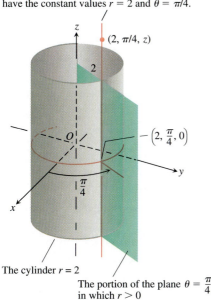

10.62 The points whose first two cylindrical coordinates are $r = 2$ and $\theta = \pi/4$ form a line parallel to the z-axis (Example 1).

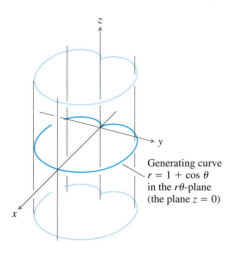

10.63 The cylindrical coordinate equation $r = 1 + \cos\theta$ defines a cylinder in space whose cross sections perpendicular to the z-axis are cardioids (Example 2).

10.64 The cylinder in Example 5.

Solution These points make up the line in which the cylinder $r = 2$ cuts the portion of the plane $\theta = \pi/4$ where r is positive (Fig. 10.62). This is the line through the point $(2, \pi/4, 0)$ parallel to the z-axis. ❏

EXAMPLE 2 Sketch the surface $r = 1 + \cos\theta$.

Solution The equation involves only r and θ; the coordinate variable z is missing. Therefore, the surface is a cylinder of lines that pass through the cardioid $r = 1 + \cos\theta$ in the $r\theta$-plane and lie parallel to the z-axis. The rules for sketching the cylinder are the same as always: sketch the x-, y-, and z-axes, draw a few perpendicular cross sections, connect the cross sections with parallel lines, and darken the exposed parts (Fig. 10.63). ❏

EXAMPLE 3 Find a Cartesian equation for the surface $z = r^2$ and identify the surface.

Solution From Eqs. (1) we have $z = r^2 = x^2 + y^2$. The surface is the circular paraboloid $x^2 + y^2 = z$. ❏

EXAMPLE 4 Find an equation for the circular cylinder $4x^2 + 4y^2 = 9$ in cylindrical coordinates.

Solution The cylinder consists of the points whose distance from the z-axis is $\sqrt{x^2 + y^2} = 3/2$. The corresponding equation in cylindrical coordinates is $r = 3/2$. ❏

EXAMPLE 5 Find an equation for the cylinder $x^2 + (y - 3)^2 = 9$ in cylindrical coordinates (Fig. 10.64).

Solution The equation for the cylinder in cylindrical coordinates is the same as the polar equation for the cylinder's base in the xy-plane:

$$r = 6\sin\theta.$$ ❏

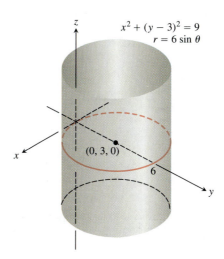

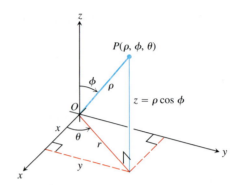

10.65 The spherical coordinates ρ, ϕ, and θ and their relation to x, y, z, and r.

A few books give spherical coordinates in the order (ρ, θ, ϕ), with θ and ϕ reversed. In some cases, you may also find r being used for ρ. Watch out for this when you read elsewhere.

Spherical Coordinates

Spherical coordinates locate points in space with angles and a distance, as shown in Fig. 10.65.

The first coordinate, $\rho = |\overrightarrow{OP}|$, is the point's distance from the origin. Unlike r, *the variable ρ is never negative.* The second coordinate, ϕ, is the angle $\overrightarrow{OP}$ makes with the positive z-axis. It is required to lie in the interval $[0, \pi]$. The third coordinate is the angle θ as measured in cylindrical coordinates.

Definition

Spherical coordinates represent a point P in space by ordered triples (ρ, ϕ, θ) in which

1. ρ is the distance from P to the origin,
2. ϕ is the angle $\overrightarrow{OP}$ makes with the positive z-axis ($0 \le \phi \le \pi$),
3. θ is the angle from cylindrical coordinates.

The equation $\rho = a$ describes the sphere of radius a centered at the origin (Fig. 10.66). The equation $\phi = \phi_0$ describes a single cone whose vertex lies at the origin and whose axis lies along the z-axis. (We broaden our interpretation to include the xy-plane as the cone $\phi = \pi/2$.) If ϕ_0 is greater than $\pi/2$, the cone $\phi = \phi_0$ opens downward.

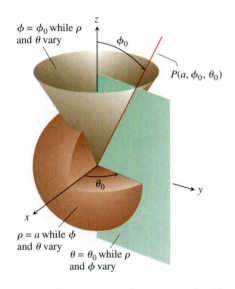

$\phi = \phi_0$ while ρ and θ vary

$P(a, \phi_0, \theta_0)$

$\rho = a$ while ϕ and θ vary

$\theta = \theta_0$ while ρ and ϕ vary

10.66 Constant-coordinate equations in spherical coordinates yield spheres, single cones, and half-planes.

Equations Relating Spherical Coordinates to Cartesian and Cylindrical Coordinates

$$r = \rho \sin\phi, \qquad x = r\cos\theta = \rho\sin\phi\cos\theta,$$

$$z = \rho\cos\phi, \qquad y = r\sin\theta = \rho\sin\phi\sin\theta, \qquad (2)$$

$$\rho = \sqrt{x^2 + y^2 + z^2} = \sqrt{r^2 + z^2}$$

EXAMPLE 6 Find a spherical coordinate equation for the sphere

$$x^2 + y^2 + (z-1)^2 = 1.$$

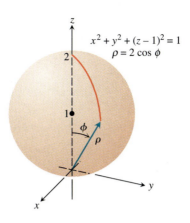

10.67 The sphere in Example 6.

Solution We use Eqs. (2) to substitute for x, y, and z:

$$x^2 + y^2 + (z-1)^2 = 1$$

$$\rho^2 \sin^2\phi \cos^2\theta + \rho^2 \sin^2\phi \sin^2\theta + (\rho\cos\phi - 1)^2 = 1 \qquad \text{Eqs. (2)}$$

$$\rho^2 \sin^2\phi \underbrace{(\cos^2\theta + \sin^2\theta)}_{1} + \rho^2 \cos^2\phi - 2\rho\cos\phi + 1 = 1$$

$$\rho^2 \underbrace{(\sin^2\phi + \cos^2\phi)}_{1} = 2\rho\cos\phi$$

$$\rho^2 = 2\rho\cos\phi$$

$$\rho = 2\cos\phi$$

See Fig. 10.67. ❏

EXAMPLE 7 Find a spherical coordinate equation for the cone $z = \sqrt{x^2 + y^2}$ (Fig. 10.68).

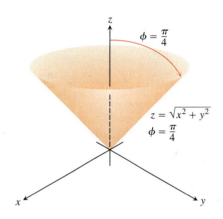

10.68 The cone in Example 7.

Solution 1 Use geometry. The cone is symmetric with respect to the z-axis and cuts the first quadrant of the yz-plane along the line $z = y$. The angle between the cone and the positive z-axis is therefore $\pi/4$ radians. The cone consists of the points whose spherical coordinates have ϕ equal to $\pi/4$, so its equation is $\phi = \pi/4$.

Solution 2 Use algebra. If we use Eqs. (2) to substitute for x, y, and z we obtain the same result:

$$z = \sqrt{x^2 + y^2}$$

$$\rho\cos\phi = \sqrt{\rho^2 \sin^2\phi} \qquad \text{Example 6}$$

$$\rho\cos\phi = \rho\sin\phi \qquad \rho \geq 0, \ \sin\phi \geq 0$$

$$\cos\phi = \sin\phi$$

$$\phi = \frac{\pi}{4} \qquad 0 \leq \phi \leq \pi$$

❏

Exercises 10.7

Converting Point Coordinates

The following table gives the coordinates of specific points in space in one of three coordinate systems. In Exercises 1–10, find coordinates for each point in the other two systems. There may be more than one right answer because points in cylindrical and spherical coordinates may have more than one coordinate triple.

Rectangular (x, y, z)	Cylindrical (r, θ, z)	Spherical (ρ, ϕ, θ)
1. $(0, 0, 0)$		
2. $(1, 0, 0)$		
3. $(0, 1, 0)$		
4. $(0, 0, 1)$		
5.	$(1, 0, 0)$	
6.	$(\sqrt{2}, 0, 1)$	
7.	$(1, \pi/2, 1)$	
8.		$(\sqrt{3}, \pi/3, -\pi/2)$
9.		$(2\sqrt{2}, \pi/2, 3\pi/2)$
10.		$(\sqrt{2}, \pi, 2\pi/2)$

Converting Equations and Inequalities; Associated Figures

In Exercises 11–36, translate the equations and inequalities from the given coordinate system (rectangular, cylindrical, spherical) into equations and inequalities in the other two systems. Also, identify the figure being defined.

11. $r = 0$

12. $x^2 + y^2 = 5$

13. $z = 0$

14. $z = -2$

15. $z = \sqrt{x^2 + y^2}, \quad z \le 1$

16. $z = \sqrt{x^2 + y^2}, \quad 1 \le z \le 2$

17. $\rho \sin \phi \cos \theta = 0$

18. $\tan^2 \phi = 1$

19. $x^2 + y^2 + z^2 = 4$

20. $x^2 + y^2 + \left(z - \dfrac{1}{2}\right)^2 = \dfrac{1}{4}$

21. $\rho = 5 \cos \phi$

22. $\rho = -6 \cos \phi$

23. $r = \csc \theta$

24. $r = -3 \sec \theta$

25. $\rho = \sqrt{2} \sec \phi$

26. $\rho = 9 \csc \phi$

27. $x^2 + y^2 + (z - 1)^2 = 1, \quad z \le 1$

28. $r^2 + z^2 = 4, \quad z \le -\sqrt{2}$

29. $\rho = 3, \quad \pi/3 \le \phi \le 2\pi/3$

30. $x^2 + y^2 + z^2 = 3, \quad 0 \le z \le \sqrt{3}/2$

31. $z = 4 - 4r^2, \quad 0 \le r \le 1$

32. $z = 4 - r, \quad 0 \le r \le 4$

33. $\phi = 3\pi/4, \quad 0 \le \rho \le \sqrt{2}$

34. $\phi = \pi/2, \quad 0 \le \rho \le \sqrt{7}$

35. $z + r^2 \cos 2\theta = 0$

36. $z^2 - r^2 = 1$

37. Find the rectangular coordinates of the center of the sphere
$$r^2 + z^2 = 4r \cos \theta + 6r \sin \theta + 2z.$$

38. Find the rectangular coordinates of the center of the sphere
$$\rho = 2 \sin \phi \ (\cos \theta - 2 \sin \theta).$$

Sets Defined by Equations

In Exercises 39–42, describe the set of points whose cylindrical coordinates satisfy the given equation. Sketch each set.

39. $r = -2 \sin \theta$

40. $r = 2 \cos \theta$

41. $r = 1 - \cos \theta$

42. $r = 1 + \sin \theta$

In Exercises 43 and 44, describe the set of points in space whose spherical coordinates satisfy the given equation. Sketch each set.

43. $\rho = 1 - \cos \phi$

44. $\rho = 1 + \cos \phi$

Theory and Examples

45. *Horizontal planes in cylindrical and spherical coordinates.* (a) Show that the plane whose equation is $z = c \ (c \ne 0)$ in rectangular and cylindrical coordinates has the equation $\rho = c \sec \phi$ in spherical coordinates. (b) Find an equation for the *xy*-plane in spherical coordinates.

46. *Vertical circular cylinders in spherical coordinates.* Find an equation of the form $\rho = f(\phi)$ for the cylinder $x^2 + y^2 = a^2$.

47. *Vertical planes in cylindrical coordinates.* (a) Show that planes perpendicular to the *x*-axis have equations of the form $r = a \sec \theta$ in cylindrical coordinates. (b) Show that planes perpendicular to the *y*-axis have equations of the form $r = b \csc \theta$.

48. *(Continuation of Exercise 47.)* Find an equation of the form $r = f(\theta)$ in cylindrical coordinates for the plane $ax + by = c, c \ne 0$.

49. *Symmetry.* What symmetry will you find in a surface that has an equation of the form $r = f(z)$ in cylindrical coordinates? Give reasons for your answer.

50. *Symmetry.* What symmetry will you find in a surface that has an equation of the form $\rho = f(\phi)$ in spherical coordinates? Give reasons for your answer.

✳ CAS Explorations and Projects

Use a CAS to perform the following steps in Exercises 51 and 52.

a) Solve the given equation for r (cylindrical coordinates) or ρ (spherical coordinates). Choose the expression with the positive square root and simplify it.

b) Plot r as a function of θ and z, or ρ as a function of θ and ϕ, as appropriate. Use the specified ranges of the variables for your plots. (Other ranges can present difficulties for a computer grapher in producing the complete surface. Experiment with other ranges later if you wish.)

c) From your plot in (b), estimate the center and radius of the sphere you graphed to the nearest integer value.

d) Convert the given equation from cylindrical or spherical coordinates to rectangular coordinates. Simplify the results. You may need to complete the final simplification by hand to obtain an equation for the sphere in the factored form $(x - x_0)^2 + (y - y_0)^2 + (z - z_0)^2 = a^2$. Compare this result against your estimates from (c).

e) Plot the implicit equation obtained in (d). How does it compare with the plot produced in (b)? Can you explain any discrepancies from the way in which a computer grapher plots surfaces?

51. $r^2 + z^2 = 2r(\cos\theta + \sin\theta) + 2, \quad \dfrac{\pi}{4} \le \theta \le \dfrac{9\pi}{4}, \quad -2 \le z \le 2$

52. $\rho^2 = 2\rho(\cos\theta \sin\phi - \cos\phi) + 2, \quad 0 \le \theta \le 2\pi, \quad 0 \le \phi \le \pi$

CHAPTER 10 QUESTIONS TO GUIDE YOUR REVIEW

1. When do directed line segments (in the plane or space) represent the same vector?

2. How are vectors added and subtracted geometrically? algebraically?

3. How do you find a vector's magnitude and direction?

4. If a vector is multiplied by a positive scalar, how is the result related to the original vector? What if the scalar is zero? negative?

5. Define the *dot product* (*scalar product*) of two vectors. Which algebraic laws (commutative, associative, distributive, cancellation) are satisfied by dot products, and which, if any, are not? Give examples. When is the dot product of two vectors equal to zero?

6. What geometric or physical interpretations do dot products have? Give examples.

7. What is the vector projection of a vector **B** onto a vector **A**? How do you write **B** as the sum of a vector parallel to **A** and a vector orthogonal to **A**?

8. Define the *cross product* (*vector product*) of two vectors. Which algebraic laws (commutative, associative, distributive, cancellation) are satisfied by cross products, and which are not? Give examples. When is the cross product of two vectors equal to zero?

9. What geometric or physical interpretations do cross products have? Give examples.

10. What is the determinant formula for calculating the cross product of two vectors relative to the Cartesian **i, j, k**-coordinate system? Use it in an example.

11. How do you find equations for lines, line segments, and planes in space? Give examples. Can you express a line in space by a single equation? A plane?

12. How do you find the distance from a point to a line in space? From a point to a plane? Give examples.

13. What are box products? What significance do they have? How are they evaluated? Give an example.

14. How do you find equations for spheres in space? Give examples.

15. How do you find the intersection of two lines in space? a line and a plane? two planes? Give examples.

16. What is a cylinder? Give examples of equations that define cylinders in Cartesian coordinates; in cylindrical coordinates. What advice can you give about drawing cylinders?

17. What are quadric surfaces? Give examples of different kinds of ellipsoids, paraboloids, cones, and hyperboloids (equations and sketches). What advice can you give about sketching quadric surfaces?

18. How are cylindrical and spherical coordinates defined? Draw diagrams that show how cylindrical and spherical coordinates are related to Cartesian coordinates.

19. Are cylindrical and spherical coordinates for a point in space unique? Explain.

20. What sets have constant-coordinate equations (like $x = 1$ or $r = 1$ or $\phi = \pi/3$) in the three coordinate systems for space?

21. What equations are available for changing from one space coordinate system to another?

CHAPTER 10 PRACTICE EXERCISES

Vector Calculations

1. Draw the unit vectors $\mathbf{u} = (\cos\theta)\,\mathbf{i} + (\sin\theta)\,\mathbf{j}$ for $\theta = 0, \pi/2, 2\pi/3, 5\pi/4$, and $5\pi/3$, together with the coordinate axes and unit circle.

2. Find the unit vector obtained by rotating

   a) $\mathbf{i}$ clockwise through an angle $\pi/4$;

   b) $\mathbf{j}$ counterclockwise through an angle $2\pi/3$.

Express the vectors in Exercises 3–6 in terms of their lengths and directions.

3. $\sqrt{2}\,\mathbf{i} + \sqrt{2}\,\mathbf{j}$

4. $-\mathbf{i} - \mathbf{j}$

5. $2\,\mathbf{i} - 3\,\mathbf{j} + 6\,\mathbf{k}$

6. $\mathbf{i} + 2\,\mathbf{j} - \mathbf{k}$

7. Find a vector 2 units long in the direction of $\mathbf{A} = 4\,\mathbf{i} - \mathbf{j} + 4\mathbf{k}$.

8. Find a vector 5 units long in the direction opposite to the direction of $\mathbf{A} = (3/5)\,\mathbf{i} + (4/5)\mathbf{k}$.

9. The accompanying figure shows parallelogram $ABCD$ and the midpoint P of diagonal BD.

   a) Express $\overrightarrow{BD}$ in terms of $\overrightarrow{AB}$ and $\overrightarrow{AD}$.

   b) Express $\overrightarrow{AP}$ in terms of $\overrightarrow{AB}$ and $\overrightarrow{AD}$.

   c) Prove that P is also the midpoint of diagonal AC.

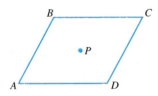

10. The points O, A, B, C, D, and E are vertices of the rectangular box shown here. Express $\overrightarrow{OD}$ and $\overrightarrow{OE}$ in terms of $\overrightarrow{OA}$, $\overrightarrow{OB}$, and $\overrightarrow{OC}$.

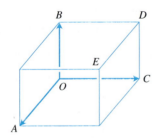

11. Copy the vectors $\mathbf{u}$ and $\mathbf{v}$ and sketch the vector projection of $\mathbf{v}$ onto $\mathbf{u}$.

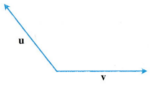

12. Express vectors $\mathbf{a}$, $\mathbf{b}$, and $\mathbf{c}$ in terms of $\mathbf{u}$ and $\mathbf{v}$.

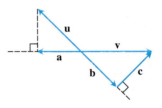

In Exercises 13 and 14, find $|\mathbf{A}|$, $|\mathbf{B}|$, $\mathbf{A} \cdot \mathbf{B}$, $\mathbf{B} \cdot \mathbf{A}$, $\mathbf{A} \times \mathbf{B}$, $\mathbf{B} \times \mathbf{A}$, $|\mathbf{A} \times \mathbf{B}|$, the angle between $\mathbf{A}$ and $\mathbf{B}$, the scalar component of $\mathbf{B}$ in the direction of $\mathbf{A}$, and the vector projection of $\mathbf{B}$ onto $\mathbf{A}$.

13. $\mathbf{A} = \mathbf{i} + \mathbf{j}$
    $\mathbf{B} = 2\,\mathbf{i} + \mathbf{j} - 2\,\mathbf{k}$

14. $\mathbf{A} = \mathbf{i} + \mathbf{j} + 2\,\mathbf{k}$
    $\mathbf{B} = -\mathbf{i} - \mathbf{k}$

In Exercises 15 and 16, write $\mathbf{B}$ as the sum of a vector parallel to $\mathbf{A}$ and a vector orthogonal to $\mathbf{A}$.

15. $\mathbf{A} = 2\,\mathbf{i} + \mathbf{j} - \mathbf{k}$,
    $\mathbf{B} = \mathbf{i} + \mathbf{j} - 5\,\mathbf{k}$

16. $\mathbf{A} = \mathbf{i} - 2\,\mathbf{j}$,
    $\mathbf{B} = \mathbf{i} + \mathbf{j} + \mathbf{k}$

In Exercises 17 and 18, draw coordinate axes and then sketch $\mathbf{A}$, $\mathbf{B}$, and $\mathbf{A} \times \mathbf{B}$ as vectors at the origin.

17. $\mathbf{A} = \mathbf{i}$, $\mathbf{B} = \mathbf{i} + \mathbf{j}$

18. $\mathbf{A} = \mathbf{i} - \mathbf{j}$, $\mathbf{B} = \mathbf{i} + \mathbf{j}$

In Exercises 19 and 20, find the unit vectors that are tangent and normal to the curve at point P.

19. $y = \tan x$, $P(\pi/4, 1)$

20. $x^2 + y^2 = 25$, $P(3, 4)$

21. For any vectors $\mathbf{A}$ and $\mathbf{B}$, show that $|\mathbf{A} + \mathbf{B}|^2 + |\mathbf{A} - \mathbf{B}|^2 = 2\,|\mathbf{A}|^2 + 2\,|\mathbf{B}|^2$.

22. Let ABC be the triangle determined by vectors $\mathbf{u}$ and $\mathbf{v}$.

    a) Express the area of $\triangle ABC$ in terms of $\mathbf{u}$ and $\mathbf{v}$.

    b) Express the triangle's altitude h in terms of $\mathbf{u}$ and $\mathbf{v}$.

    c) Find the area and altitude if $\mathbf{u} = \mathbf{i} - \mathbf{j} + \mathbf{k}$ and $\mathbf{v} = 2\,\mathbf{i} + \mathbf{k}$.

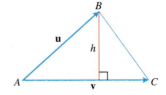

23. If $|\mathbf{v}| = 2$, $|\mathbf{w}| = 3$, and the angle between $\mathbf{v}$ and $\mathbf{w}$ is $\pi/3$, find $|\mathbf{v} - 2\mathbf{w}|$.

24. For what value or values of a will the vectors $\mathbf{u} = 2\mathbf{i} + 4\mathbf{j} - 5\mathbf{k}$ and $\mathbf{v} = -4\mathbf{i} - 8\mathbf{j} + a\mathbf{k}$ be parallel?

In Exercises 25 and 26, find (a) the area of the parallelogram determined by vectors $\mathbf{A}$ and $\mathbf{B}$, (b) the volume of the parallelepiped determined by the vectors $\mathbf{A}$, $\mathbf{B}$, and $\mathbf{C}$.

25. $\mathbf{A} = \mathbf{i} + \mathbf{j} - \mathbf{k}$, $\quad \mathbf{B} = 2\mathbf{i} + \mathbf{j} + \mathbf{k}$, $\quad \mathbf{C} = -\mathbf{i} - 2\mathbf{j} + 3\mathbf{k}$

26. $\mathbf{A} = \mathbf{i} + \mathbf{j}$, $\quad \mathbf{B} = \mathbf{j}$, $\quad \mathbf{C} = \mathbf{i} + \mathbf{j} + \mathbf{k}$

Lines, Planes, and Distances

27. Suppose that $\mathbf{n}$ is normal to a plane and that $\mathbf{v}$ is parallel to the plane. Describe how you would find a vector $\mathbf{u}$ that is both perpendicular to $\mathbf{v}$ and parallel to the plane.

28. Find a vector in the plane parallel to the line $ax + by = c$.

In Exercises 29 and 30, find the distance from the point to the line.

29. $(2, 2, 0)$; $x = -t$, $y = t$, $z = -1 + t$

30. $(0, 4, 1)$; $x = 2 + t$, $y = 2 + t$, $z = t$

31. Parametrize the line that passes through the point $(1, 2, 3)$ parallel to the vector $\mathbf{v} = -3\mathbf{i} + 7\mathbf{k}$.

32. Parametrize the line segment joining the points $P(1, 2, 0)$ and $Q(1, 3, -1)$.

In Exercises 33 and 34, find the distance from the point to the plane.

33. $(6, 0, -6)$, $x - y = 4$

34. $(3, 0, 10)$, $2x + 3y + z = 2$

35. Find an equation for the plane that passes through the point $(3, -2, 1)$ normal to the vector $\mathbf{n} = 2\mathbf{i} + \mathbf{j} + \mathbf{k}$.

36. Find an equation for the plane that passes through the point $(-1, 6, 0)$ perpendicular to the line $x = -1 + t$, $y = 6 - 2t$, $z = 3t$.

In Exercises 37 and 38, find an equation for the plane through points P, Q, and R.

37. $P(1, -1, 2)$, $Q(2, 1, 3)$, $R(-1, 2, -1)$

38. $P(1, 0, 0)$, $Q(0, 1, 0)$, $R(0, 0, 1)$

39. Find the points in which the line $x = 1 + 2t$, $y = -1 - t$, $z = 3t$ meets the three coordinate planes.

40. Find the point in which the line through the origin perpendicular to the plane $2x - y - z = 4$ meets the plane $3x - 5y + 2z = 6$.

41. Find the acute angle between the planes $x = 7$ and $x + y + \sqrt{2}z = -3$.

42. Find the acute angle between the planes $x + y = 1$ and $y + z = 1$.

43. Find parametric equations for the line in which the planes $x + 2y + z = 1$ and $x - y + 2z = -8$ intersect.

44. Show that the line in which the planes

$$x + 2y - 2z = 5 \quad \text{and} \quad 5x - 2y - z = 0$$

intersect is parallel to the line

$$x = -3 + 2t, \quad y = 3t, \quad z = 1 + 4t.$$

45. The planes $3x + 6z = 1$ and $2x + 2y - z = 3$ intersect in a line.

a) Show that the planes are orthogonal.
b) Find equations for the line of intersection.

46. Find an equation for the plane that passes through the point $(1, 2, 3)$ parallel to $\mathbf{u} = 2\mathbf{i} + 3\mathbf{j} + \mathbf{k}$ and $\mathbf{v} = \mathbf{i} - \mathbf{j} + 2\mathbf{k}$.

47. Is $\mathbf{v} = 2\mathbf{i} - 4\mathbf{j} + \mathbf{k}$ related in any special way to the plane $2x + y = 5$? Give reasons for your answer.

48. The equation $\mathbf{n} \cdot \overrightarrow{P_0 P} = 0$ represents the plane through P_0 normal to $\mathbf{n}$. What set does the inequality $\mathbf{n} \cdot \overrightarrow{P_0 P} > 0$ represent?

49. Find the distance from the point $P(1, 4, 0)$ to the plane through $A(0, 0, 0)$, $B(2, 0, -1)$ and $C(2, -1, 0)$.

50. Find the distance from the point $(2, 2, 3)$ to the plane $2x + 3y + 5z = 0$.

51. Find a vector parallel to the plane $2x - y - z = 4$ and orthogonal to $\mathbf{i} + \mathbf{j} + \mathbf{k}$.

52. Find a unit vector orthogonal to $\mathbf{A}$ in the plane of $\mathbf{B}$ and $\mathbf{C}$ if $\mathbf{A} = 2\mathbf{i} - \mathbf{j} + \mathbf{k}$, $\mathbf{B} = \mathbf{i} + 2\mathbf{j} + \mathbf{k}$, and $\mathbf{C} = \mathbf{i} + \mathbf{j} - 2\mathbf{k}$.

53. Find a vector of magnitude 2 parallel to the line of intersection of the planes $x + 2y + z - 1 = 0$ and $x - y + 2z + 7 = 0$.

54. Find the point in which the line through the origin perpendicular to the plane $2x - y - z = 4$ meets the plane $3x - 5y + 2z = 6$.

55. Find the point in which the line through $P(3, 2, 1)$ normal to the plane $2x - y + 2z = -2$ meets the plane.

56. What angle does the line of intersection of the planes $2x + y - z = 0$ and $x + y + 2z = 0$ make with the positive x-axis?

57. The line

$$L: \quad x = 3 + 2t, \quad y = 2t, \quad z = t$$

intersects the plane $x + 3y - z = -4$ in a point P. Find the coordinates of P and find equations for the line through P perpendicular to L.

58. Show that for every real number k the plane

$$x - 2y + z + 3 + k(2x - y - z + 1) = 0$$

contains the line of intersection of the planes

$$x - 2y + z + 3 = 0 \quad \text{and} \quad 2x - y - z + 1 = 0.$$

59. Find an equation for the plane through $A(-2, 0, -3)$ and $B(1, -2, 1)$ that lies parallel to the line through $C(-2, -13/5, 26/5)$ and $D(16/5, -13/5, 0)$.

60. Is the line $x = 1 + 2t$, $y = -2 + 3t$, $z = -5t$ related in any way to the plane $-4x - 6y + 10z = 9$? Give reasons for your answer.

61. Which of the following are equations for the plane through the points $P(1, 1, -1)$, $Q(3, 0, 2)$, and $R(-2, 1, 0)$?

a) $(2\mathbf{i} - 3\mathbf{j} + 3\mathbf{k}) \cdot ((x + 2)\mathbf{i} + (y - 1)\mathbf{j} + z\mathbf{k}) = 0$

b) $x = 3 - t$, $\quad y = -11t$, $\quad z = 2 - 3t$

c) $(x + 2) + 11(y - 1) = 3z$

d) $(2\mathbf{i} - 3\mathbf{j} + 3\mathbf{k}) \times ((x + 2)\mathbf{i} + (y - 1)\mathbf{j} + z\mathbf{k}) = \mathbf{0}$

e) $(2\mathbf{i} - \mathbf{j} + 3\mathbf{k}) \times (-3\mathbf{i} + \mathbf{k}) \cdot ((x + 2)\mathbf{i} + (y - 1)\mathbf{j} + z\mathbf{k})$ $= 0$

62. The parallelogram shown here has vertices at $A(2, -1, 4)$, $B(1, 0, -1)$, $C(1, 2, 3)$, and D. Find

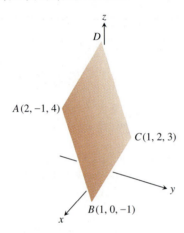

a) the coordinates of D,

b) the cosine of the interior angle at B,

c) the vector projection of $\overrightarrow{BA}$ onto $\overrightarrow{BC}$,

d) the area of the parallelogram,

e) an equation for the plane of the parallelogram,

f) the areas of the orthogonal projections of the parallelogram on the three coordinate planes.

63. *Distance between lines.* Find the distance between the line L_1 through the points $A(1, 0, -1)$ and $B(-1, 1, 0)$ and the line L_2 through the points $C(3, 1, -1)$ and $D(4, 5, -2)$. The distance is to be measured along the line perpendicular to the two lines. First find a vector $\mathbf{n}$ perpendicular to both lines. Then project $\overrightarrow{AC}$ onto $\mathbf{n}$.

64. *(Continuation of Exercise 63.)* Find the distance between the line through $A(4, 0, 2)$ and $B(2, 4, 1)$ and the line through $C(1, 3, 2)$ and $D(2, 2, 4)$.

Quadric Surfaces

Identify and sketch the surfaces in Exercises 65–76.

65. $x^2 + y^2 + z^2 = 4$

66. $x^2 + (y - 1)^2 + z^2 = 1$

67. $4x^2 + 4y^2 + z^2 = 4$

68. $36x^2 + 9y^2 + 4z^2 = 36$

69. $z = -(x^2 + y^2)$

70. $y = -(x^2 + z^2)$

71. $x^2 + y^2 = z^2$

72. $x^2 + z^2 = y^2$

73. $x^2 + y^2 - z^2 = 4$

74. $4y^2 + z^2 - 4x^2 = 4$

75. $y^2 - x^2 - z^2 = 1$

76. $z^2 - x^2 - y^2 = 1$

Coordinate Systems

The equations in Exercises 77–86 define sets both in the plane and in three-dimensional space. Identify both sets for each equation.

RECTANGULAR COORDINATES

77. $x = 0$

78. $x + y = 1$

79. $x^2 + y^2 = 4$

80. $x^2 + 4y^2 = 16$

81. $x = y^2$

82. $y^2 - x^2 = 1$

CYLINDRICAL COORDINATES

83. $r = 1 - \cos\theta$

84. $r = \sin\theta$

85. $r^2 = 2\cos 2\theta$

86. $r = \cos 2\theta$

Describe the sets defined by the spherical coordinate equations and inequalities in Exercises 87–92.

87. $\rho = 2$

88. $\theta = \pi/4$

89. $\phi = \pi/6$

90. $\rho = 1$, $\quad \phi = \pi/2$

91. $\rho = 1$, $\quad 0 \le \phi \le \pi/2$

92. $1 \le \rho \le 2$

The following table gives the coordinates of points in space in one of three coordinate systems. In Exercises 93–98, find coordinates for each point in the other two systems. There may be more than one right answer because cylindrical and spherical coordinates are not unique.

	Rectangular (x, y, z)	Cylindrical (r, θ, z)	Spherical (ρ, ϕ, θ)
93.		$(1, 0, 0)$	
94.		$\left(1, \dfrac{\pi}{2}, 0\right)$	
95.			$\left(\sqrt{2}, \dfrac{\pi}{4}, \dfrac{\pi}{2}\right)$
96.			$\left(2, \dfrac{5\pi}{6}, 0\right)$
97.	$(-1, 0, -1)$		
98.	$(0, -1, 1)$		

In Exercises 99–110, translate the equations from the given coordinate system (rectangular, cylindrical, spherical) into the other two systems. Identify the set of points in space defined by the equation.

RECTANGULAR

99. $z = 2$

100. $z = \sqrt{3x^2 + 3y^2}$

101. $x^2 + y^2 + (z + 1)^2 = 1$

102. $x^2 + y^2 + (z - 3)^2 = 9$

CYLINDRICAL

103. $z = r^2$

105. $r = 7 \sin \theta$

104. $z = |r|$

106. $r = 4 \cos \theta$

SPHERICAL

107. $\rho = 4$

109. $\phi = 3\pi / 4$

108. $\rho = \sqrt{3} \sec \phi$

110. $\rho \cos \phi + \rho^2 \sin^2 \phi = 1$

CHAPTER 10 ADDITIONAL EXERCISES–THEORY, EXAMPLES, APPLICATIONS

Applications

1. *Submarine hunting.* Two surface ships on maneuvers are trying to determine a submarine's course and speed to prepare for an aircraft intercept. As shown here, ship A is located at (4, 0, 0) while ship B is located at (0, 5, 0). All coordinates are given in thousands of feet. Ship A locates the submarine in the direction of the vector $2\mathbf{i} + 3\mathbf{j} - (1/3)\mathbf{k}$, and ship B locates it in the direction of the vector $18\mathbf{i} - 6\mathbf{j} - \mathbf{k}$. Four minutes ago, the submarine was located at $(2, -1, -1/3)$. The aircraft is due in 20 min. Assuming the submarine moves in a straight line at a constant speed, to what position should the surface ships direct the aircraft?

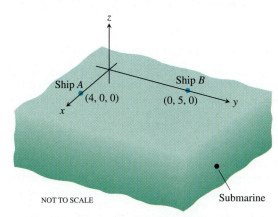

NOT TO SCALE

2. *A helicopter rescue.* Two helicopters, H_1 and H_2, are traveling together. At time $t = 0$ hours, they separate and follow different straight-line paths given by

$$H_1: \quad x = 6 + 40t, \quad y = -3 + 10t, \quad z = -3 + 2t$$
$$H_2: \quad x = 6 + 110t, \quad y = -3 + 4t, \quad z = -3 + t,$$

all coordinates measured in miles. Due to system malfunctions, H_2 stops its flight at (446, 13, 1) and, in a negligible amount of time, lands at (446, 13, 0). Two hours later, H_1 is advised of this fact and heads toward H_2 at 150 mph. How long will it take H_1 to reach H_2?

3. *Work.* Find the work done in pushing a car 250 m with a force of magnitude 160 N directed at an angle of $\pi / 6$ rad downward from the horizontal against the back of the car.

4. *Torque.* The operator's manual for the Toro® 21-in. lawnmower says "tighten the spark plug to 15 ft-lb (20.4 N · m)." If you are installing the plug with a 10.5-in. socket wrench that places the center of your hand 9 in. from the axis of the spark plug, about how hard should you pull? Answer in pounds.

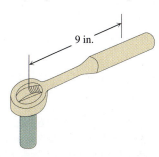

9 in.

Theory and Examples

5. Show that $|\mathbf{A} + \mathbf{B}| \le |\mathbf{A}| + |\mathbf{B}|$ for any vectors $\mathbf{A}$ and $\mathbf{B}$.

6. Suppose that vectors $\mathbf{A}$ and $\mathbf{B}$ are not parallel and that $\mathbf{A} = \mathbf{C} + \mathbf{D}$, where $\mathbf{C}$ is parallel to $\mathbf{B}$ and $\mathbf{D}$ is orthogonal to $\mathbf{B}$. Express $\mathbf{C}$ and $\mathbf{D}$ in terms of $\mathbf{A}$ and $\mathbf{B}$.

7. Show that $\mathbf{C} = |\mathbf{B}|\mathbf{A} + |\mathbf{A}|\mathbf{B}$ bisects the angle between $\mathbf{A}$ and $\mathbf{B}$.

8. Show that $|\mathbf{B}|\mathbf{A} + |\mathbf{A}|\mathbf{B}$ and $|\mathbf{B}|\mathbf{A} - |\mathbf{A}|\mathbf{B}$ are orthogonal.

9. *Dot multiplication is positive definite.* Show that dot multiplication of vectors is *positive definite;* that is, show that $\mathbf{A} \cdot \mathbf{A} \ge 0$ for every vector $\mathbf{A}$ and that $\mathbf{A} \cdot \mathbf{A} = 0$ if and only if $\mathbf{A} = \mathbf{0}$.

10. By forming the cross product of two appropriate vectors, derive the trigonometric identity

$$\sin(A - B) = \sin A \cos B - \cos A \sin B.$$

11. Use vectors to prove that

$$(a^2 + b^2)(c^2 + d^2) \ge (ac + bd)^2$$

for any four numbers $a, b, c,$ and d. (*Hint:* Let $\mathbf{A} = a\mathbf{i} + b\mathbf{j}$ and $\mathbf{B} = c\mathbf{i} + d\mathbf{j}$.)

12. In the figure here, D is the midpoint of side AB of triangle ABC, and E is one-third of the way between C and B. Use vectors to prove that F is the midpoint of line segment CD.

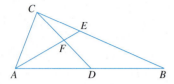

13. a) Show that

$$\begin{vmatrix} x_1 - x & y_1 - y & z_1 - z \\ x_2 - x & y_2 - y & z_2 - z \\ x_3 - x & y_3 - y & z_3 - z \end{vmatrix} = 0$$

is an equation for the plane through the three noncollinear points $P_1(x_1, y_1, z_1)$, $P_2(x_2, y_2, z_2)$, and $P_3(x_3, y_3, z_3)$.

b) What set of points in space is described by the equation

$$\begin{vmatrix} x & y & z & 1 \\ x_1 & y_1 & z_1 & 1 \\ x_2 & y_2 & z_2 & 1 \\ x_3 & y_3 & z_3 & 1 \end{vmatrix} = 0?$$

14. Show that the lines

$$x = a_1 s + b_1, \, y = a_2 s + b_2, \, z = a_3 s + b_3, \, -\infty < s < \infty,$$

and

$$x = c_1 t + d_1, \, y = c_2 t + d_2, \, z = c_3 t + d_3, \, -\infty < t < \infty,$$

intersect or are parallel if and only if

$$\begin{vmatrix} a_1 & c_1 & b_1 - d_1 \\ a_2 & c_2 & b_2 - d_2 \\ a_3 & c_3 & b_3 - d_3 \end{vmatrix} = 0.$$

15. Use vectors to show that the distance from $P_1(x_1, y_1)$ to the line $ax + by = c$ is

$$d = \frac{|ax_1 + by_1 - c|}{\sqrt{a^2 + b^2}}.$$

16. a) Use vectors to show that the distance from $P_1(x_1, y_1, z_1)$ to the plane $Ax + By + Cz = D$ is

$$d = \frac{|Ax_1 + By_1 + Cz_1 - D|}{\sqrt{A^2 + B^2 + C^2}}.$$

b) Find an equation for the sphere that is tangent to the planes $x + y + z = 3$ and $x + y + z = 9$ if the planes $2x - y = 0$ and $3x - z = 0$ pass through the center of the sphere.

17. a) Show that the distance between the parallel planes $Ax + By + Cz = D_1$ and $Ax + By + Cz = D_2$ is

$$d = \frac{|D_1 - D_2|}{|A\mathbf{i} + B\mathbf{j} + C\mathbf{k}|}. \tag{1}$$

b) Use Eq. (1) to find the distance between the planes $2x + 3y - z = 6$ and $2x + 3y - z = 12$.

c) Find an equation for the plane parallel to the plane $2x - y + 2z = -4$ if the point $(3, 2, -1)$ is equidistant from the two planes.

d) Write equations for the planes that lie parallel to and 5 units away from the plane $x - 2y + z = 3$.

18. Prove that four points A, B, C, and D are coplanar (lie in a common plane) if and only if $\vec{AD} \cdot (\vec{AB} \times \vec{BC}) = 0$.

19. *Triple vector products.* The **triple vector products** $(\mathbf{A} \times \mathbf{B}) \times \mathbf{C}$ and $\mathbf{A} \times (\mathbf{B} \times \mathbf{C})$ are usually not equal, although the formulas for evaluating them from components are similar:

$$(\mathbf{A} \times \mathbf{B}) \times \mathbf{C} = (\mathbf{A} \cdot \mathbf{C})\mathbf{B} - (\mathbf{B} \cdot \mathbf{C})\mathbf{A}. \tag{2}$$

$$\mathbf{A} \times (\mathbf{B} \times \mathbf{C}) = (\mathbf{A} \cdot \mathbf{C})\mathbf{B} - (\mathbf{A} \cdot \mathbf{B})\mathbf{C}. \tag{3}$$

Verify each formula for the following vectors by evaluating its two sides and comparing the results.

	A	**B**	**C**
a)	$2\mathbf{i}$	$2\mathbf{j}$	$2\mathbf{k}$
b)	$\mathbf{i} - \mathbf{j} + \mathbf{k}$	$2\mathbf{i} + \mathbf{j} - 2\mathbf{k}$	$-\mathbf{i} + 2\mathbf{j} - \mathbf{k}$
c)	$2\mathbf{i} + \mathbf{j}$	$2\mathbf{i} - \mathbf{j} + \mathbf{k}$	$\mathbf{i} + 2\mathbf{k}$
d)	$\mathbf{i} + \mathbf{j} - 2\mathbf{k}$	$-\mathbf{i} - \mathbf{k}$	$2\mathbf{i} + 4\mathbf{j} - 2\mathbf{k}$

20. Show that if $\mathbf{A}$, $\mathbf{B}$, $\mathbf{C}$, and $\mathbf{D}$ are any vectors, then

a) $\mathbf{A} \times (\mathbf{B} \times \mathbf{C}) + \mathbf{B} \times (\mathbf{C} \times \mathbf{A}) + \mathbf{C} \times (\mathbf{A} \times \mathbf{B}) = \mathbf{0}$,

b) $\mathbf{A} \times \mathbf{B} = (\mathbf{A} \cdot \mathbf{B} \times \mathbf{i})\mathbf{i} + (\mathbf{A} \cdot \mathbf{B} \times \mathbf{j})\mathbf{j} + (\mathbf{A} \cdot \mathbf{B} \times \mathbf{k})\mathbf{k}$,

c) $(\mathbf{A} \times \mathbf{B}) \cdot (\mathbf{C} \times \mathbf{D}) = \begin{vmatrix} \mathbf{A} \cdot \mathbf{C} & \mathbf{B} \cdot \mathbf{C} \\ \mathbf{A} \cdot \mathbf{D} & \mathbf{B} \cdot \mathbf{D} \end{vmatrix}$.

21. Prove or disprove the formula

$$\mathbf{A} \times (\mathbf{A} \times (\mathbf{A} \times \mathbf{B})) \cdot \mathbf{C} = -|\mathbf{A}|^2 \mathbf{A} \cdot \mathbf{B} \times \mathbf{C}.$$

22. *The projection of a vector on a plane.* Let P be a plane in space and let $\mathbf{v}$ be a vector. The vector projection of $\mathbf{v}$ onto the plane P, $\text{proj}_P \mathbf{v}$, can be defined informally as follows. Suppose the sun is shining so that its rays are normal to the plane P. Then $\text{proj}_P \mathbf{v}$ is the "shadow" of $\mathbf{v}$ onto P. If P is the plane $x + 2y + 6z = 6$ and $\mathbf{v} = \mathbf{i} + \mathbf{j} + \mathbf{k}$, find $\text{proj}_P \mathbf{v}$.

23. The accompanying figure shows nonzero vectors $\mathbf{v}$, $\mathbf{w}$, and $\mathbf{z}$, with $\mathbf{z}$ orthogonal to the line L, and $\mathbf{v}$ and $\mathbf{w}$ making equal angles β with L. Assuming $|\mathbf{v}| = |\mathbf{w}|$, find $\mathbf{w}$ in terms of $\mathbf{v}$ and $\mathbf{z}$.

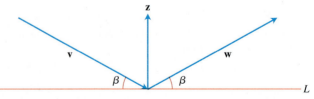

24. *The parabolic coordinate system.* This exercise introduces a new coordinate system for space, the **parabolic coordinate system.** A point P is determined by an ordered triple (α, β, γ) in which

i) $\alpha\beta$ is the square of the distance from P to the z-axis,

ii) $|\alpha - \beta|$ is twice the distance from P to the xy-plane, and P lies above the xy-plane if $\alpha - \beta > 0$ and below it if $\alpha - \beta < 0$,

iii) $\gamma = \theta$, where θ has the same meaning as in cylindrical and spherical coordinate systems, except that we restrict γ to lie in $[0, 2\pi]$.

a) What are the equations for changing from parabolic coordinates to Cartesian coordinates?

b) Why is "parabolic coordinate system" an appropriate name?

*25. *Point masses and gravitation.* In physics the law of gravitation says that if P and Q are (point) masses with mass M and m, respectively, then P is attracted to Q by the force

$$\mathbf{F} = \frac{GMm\,\mathbf{r}}{|\mathbf{r}|^3},$$

where $\mathbf{r}$ is the vector from P to Q and G is a constant (the gravitational constant). Moreover, if $Q_1, \ldots, Q_k$ are (point) masses with mass $m_1, \ldots, m_k$, respectively, then the force on P due to all the Q_i's is

$$\mathbf{F} = \sum_{i=1}^{k} \frac{GMm_i}{|\mathbf{r}_i|^3}\,\mathbf{r}_i,$$

where $\mathbf{r}_i$ is the vector from P to Q_i.

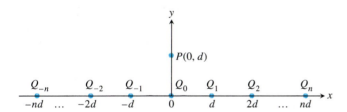

a) Let point P with mass M be located at the point $(0, d)$, $d > 0$, in the coordinate plane. For $i = -n, -n+1, \ldots, -1, 0, 1, \ldots, n$, let Q_i be located at the point $(id, 0)$ and have mass m. Find the magnitude of the gravitational force on P due to all the Q_i's.

b) Is the limit as $n \to \infty$ of the magnitude of the force on P finite? Why, or why not?

*26. *Relativistic sums.* Einstein's special theory of relativity roughly says that with respect to a reference frame (coordinate system) no material object can travel as fast as c, the speed of light. So, if $\overrightarrow{x}$ and $\overrightarrow{y}$ are two velocities such that $|\overrightarrow{x}| < c$ and $|\overrightarrow{y}| < c$, then the **relativistic sum** $\overrightarrow{x} \oplus \overrightarrow{y}$ of $\overrightarrow{x}$ and $\overrightarrow{y}$ must have length less than c. Einstein's special theory of relativity says

$$\overrightarrow{x} \oplus \overrightarrow{y} = \frac{\overrightarrow{x} + \overrightarrow{y}}{1 + \dfrac{\overrightarrow{x} \cdot \overrightarrow{y}}{c^2}} + \frac{1}{c^2} \cdot \frac{\gamma_x}{\gamma_x + 1} \cdot \frac{\overrightarrow{x} \times (\overrightarrow{x} \times \overrightarrow{y})}{1 + \dfrac{\overrightarrow{x} \cdot \overrightarrow{y}}{c^2}},$$

where

$$\gamma_x = \frac{1}{\sqrt{1 - \dfrac{\overrightarrow{x} \cdot \overrightarrow{x}}{c^2}}}.$$

It can be shown that if $|\overrightarrow{x}| < c$ and $|\overrightarrow{y}| < c$, then $|\overrightarrow{x} \oplus \overrightarrow{y}| < c$. This exercise deals with two special cases.

a) Prove that if $\overrightarrow{x}$ and $\overrightarrow{y}$ are orthogonal, $|\overrightarrow{x}| < c$, $|\overrightarrow{y}| < c$, then $|\overrightarrow{x} \oplus \overrightarrow{y}| < c$.

b) Prove that if $\overrightarrow{x}$ and $\overrightarrow{y}$ are parallel, $|\overrightarrow{x}| < c$, $|\overrightarrow{y}| < c$, then $|\overrightarrow{x} \oplus \overrightarrow{y}| < c$.

c) Compute $\lim_{c \to \infty} \overrightarrow{x} \oplus \overrightarrow{y}$.

Vector-Valued Functions and Motion in Space

OVERVIEW   When a body travels through space, the equations $x = f(t)$, $y = g(t)$, and $z = h(t)$ that give the body's coordinates as functions of time serve as parametric equations for the body's motion and path. With vector notation, we can condense these into a single equation $\mathbf{r}(t) = f(t)\,\mathbf{i} + g(t)\,\mathbf{j} + h(t)\,\mathbf{k}$ that gives the body's position as a vector function of time.

In this chapter, we show how to use calculus to study the paths, velocities, and accelerations of moving bodies. As we go along, we will see how our work answers the standard questions about the paths and motions of projectiles, planets, and satellites. In the final section, we use our new vector calculus to derive Kepler's laws of planetary motion from Newton's laws of motion and gravitation.

11.1 Vector-Valued Functions and Space Curves

To track a particle moving in space, we run a vector $\mathbf{r}$ from the origin to the particle (Fig. 11.1) and study the changes in $\mathbf{r}$. If the particle's position coordinates are twice-differentiable functions of time, then so is $\mathbf{r}$, and we can find the particle's velocity and acceleration vectors at any time by differentiating $\mathbf{r}$. Conversely, if we know either the particle's velocity vector or acceleration vector as a continuous function of time, and if we have enough information about the particle's initial velocity and position, we can find $\mathbf{r}$ as a function of time by integration.

Definitions

When a particle moves through space during a time interval I, we think of the particle's coordinates as functions defined on I:

$$x = f(t), \qquad y = g(t), \qquad z = h(t), \qquad t \,\varepsilon\, I. \tag{1}$$

The points $(x, y, z) = (f(t), g(t), h(t))$, $t \,\varepsilon\, I$, make up the **curve** in space that we call the particle's **path.** The equations and interval in (1) **parametrize** the curve. The vector

$$\mathbf{r}(t) = \overrightarrow{OP} = f(t)\,\mathbf{i} + g(t)\,\mathbf{j} + h(t)\,\mathbf{k}$$

from the origin to the particle's **position** $P(f(t), g(t), h(t))$ at time t is the particle's **position vector.** The functions f, g, and h are the **component functions**

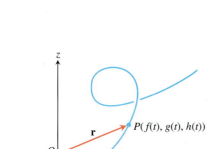

11.1 The position vector $\mathbf{r} = \overrightarrow{OP}$ of a particle moving through space is a function of time.

855

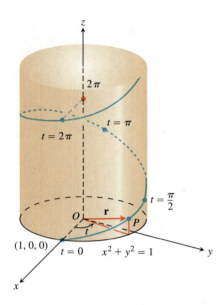

11.2 The upper half of the helix $\mathbf{r}(t) = (\cos t)\mathbf{i} + (\sin t)\mathbf{j} + t\mathbf{k}$.

(**components**) of the position vector. We think of the particle's path as the **curve traced by r** during the time interval I.

Equation (1) defines **r** as a vector function of the real variable t on the interval I. More generally, a **vector function** or **vector-valued function** on a domain set D is a rule that assigns a vector in space to each element in D. For now, the domains will be intervals of real numbers. Later, in Chapter 14, the domains will be regions in the plane or in space. Vector functions will then be called "vector fields."

We refer to real-valued functions as **scalar functions** to distinguish them from vector functions. The components of **r** are scalar functions of t. When we define a vector-valued function by giving its component functions, we assume the vector function's domain to be the common domain of the components.

EXAMPLE 1 *A Helix*

The vector function

$$\mathbf{r}(t) = (\cos t)\mathbf{i} + (\sin t)\mathbf{j} + t\mathbf{k}$$

is defined for all real values of t. The curve traced by **r** is a helix (from an old Greek word for "spiral") that winds around the circular cylinder $x^2 + y^2 = 1$ (Fig. 11.2). The curve lies on the cylinder because the **i**- and **j**-components of **r**, being the x- and y-coordinates of the tip of **r**, satisfy the cylinder's equation:

$$x^2 + y^2 = (\cos t)^2 + (\sin t)^2 = 1.$$

The curve rises as the **k**-component $z = t$ increases. Each time t increases by 2π, the curve completes one turn around the cylinder. The equations

$$x = \cos t, \qquad y = \sin t, \qquad z = t$$

parametrize the helix, the interval $-\infty < t < \infty$ being understood. You will find more helices in Fig. 11.3. ☐

Limits and Continuity

The way we define limits of vector-valued functions is similar to the way we define limits of real-valued functions.

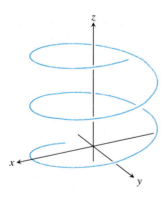

$\mathbf{r}(t) = (\cos t)\mathbf{i} + (\sin t)\mathbf{j} + t\mathbf{k}$

(Generated by Mathematica)

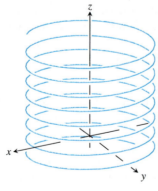

$\mathbf{r}(t) = (\cos t)\mathbf{i} + (\sin t)\mathbf{j} + 0.3t\mathbf{k}$

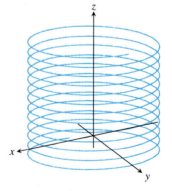

$\mathbf{r}(t) = (\cos 5t)\mathbf{i} + (\sin 5t)\mathbf{j} + t\mathbf{k}$

11.3 Helices drawn by computer.

Definition

Let $\mathbf{r}(t) = f(t)\,\mathbf{i} + g(t)\,\mathbf{j} + h(t)\,\mathbf{k}$ be a vector function and $\mathbf{L}$ a vector. We say that $\mathbf{r}$ has **limit** $\mathbf{L}$ as t approaches t_0 and write

$$\lim_{t \to t_0} \mathbf{r}(t) = \mathbf{L}$$

if, for every number $\epsilon > 0$, there exists a corresponding number $\delta > 0$ such that for all t

$$0 < |t - t_0| < \delta \quad \Rightarrow \quad |\mathbf{r}(t) - \mathbf{L}| < \epsilon.$$

If $\mathbf{L} = L_1\,\mathbf{i} + L_2\,\mathbf{j} + L_3\,\mathbf{k}$, then $\lim_{t \to t_0} \mathbf{r}(t) = \mathbf{L}$ precisely when

$$\lim_{t \to t_0} f(t) = L_1, \qquad \lim_{t \to t_0} g(t) = L_2, \qquad \text{and} \qquad \lim_{t \to t_0} h(t) = L_3.$$

The equation

$$\lim_{t \to t_0} \mathbf{r}(t) = \left(\lim_{t \to t_0} f(t)\right)\mathbf{i} + \left(\lim_{t \to t_0} g(t)\right)\mathbf{j} + \left(\lim_{t \to t_0} h(t)\right)\mathbf{k} \tag{2}$$

provides a practical way to calculate limits of vector functions.

EXAMPLE 2   If $\mathbf{r}(t) = (\cos t)\,\mathbf{i} + (\sin t)\,\mathbf{j} + t\,\mathbf{k}$, then

$$\lim_{t \to \pi/4} \mathbf{r}(t) = \left(\lim_{t \to \pi/4} \cos t\right)\mathbf{i} + \left(\lim_{t \to \pi/4} \sin t\right)\mathbf{j} + \left(\lim_{t \to \pi/4} t\right)\mathbf{k}$$

$$= \frac{\sqrt{2}}{2}\,\mathbf{i} + \frac{\sqrt{2}}{2}\,\mathbf{j} + \frac{\pi}{4}\,\mathbf{k}.$$

We define continuity for vector functions the same way we define continuity for scalar functions.

Definition

A vector function $\mathbf{r}(t)$ is **continuous at a point** $t = t_0$ in its domain if $\lim_{t \to t_0} \mathbf{r}(t) = \mathbf{r}(t_0)$. The function is **continuous** if it is continuous at every point in its domain.

Since limits can be expressed in terms of components, we can test vector functions for continuity by examining their components (Exercise 51).

Component Test for Continuity at a Point

The vector function $\mathbf{r}(t) = f(t)\,\mathbf{i} + g(t)\,\mathbf{j} + h(t)\,\mathbf{k}$ is continuous at $t = t_0$ if and only if f, g, and h are continuous at t_0.

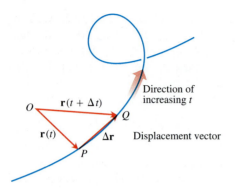

11.4 Between time t and time $t + \Delta t$, the particle moving along the path shown here undergoes the displacement $\overrightarrow{PQ} = \Delta\mathbf{r}$. The vector sum $\mathbf{r}(t) + \Delta\mathbf{r}$ gives the new position, $\mathbf{r}(t + \Delta t)$.

EXAMPLE 3

a) The function

$$\mathbf{r}(t) = (\cos t)\,\mathbf{i} + (\sin t)\,\mathbf{j} + t\,\mathbf{k}$$

is continuous because $\cos t$, $\sin t$, and t are continuous.

b) The function

$$\mathbf{g}(t) = (\cos t)\,\mathbf{i} + (\sin t)\,\mathbf{j} + \lfloor t \rfloor\,\mathbf{k}$$

is discontinuous at every integer.   ❑

Derivatives and Motion

Suppose that $\mathbf{r}(t) = f(t)\,\mathbf{i} + g(t)\,\mathbf{j} + h(t)\,\mathbf{k}$ is the position vector of a particle moving along a curve in space and that f, g, and h are differentiable functions of t. Then the difference between the particle's positions at time t and time $t + \Delta t$ is

$$\Delta\mathbf{r} = \mathbf{r}(t + \Delta t) - \mathbf{r}(t)$$

(Fig. 11.4). In terms of components,

$$\begin{aligned}
\Delta\mathbf{r} &= \mathbf{r}(t + \Delta t) - \mathbf{r}(t) \\
&= [f(t + \Delta t)\,\mathbf{i} + g(t + \Delta t)\,\mathbf{j} + h(t + \Delta t)\,\mathbf{k}] \\
&\quad -[f(t)\,\mathbf{i} + g(t)\,\mathbf{j} + h(t)\,\mathbf{k}] \\
&= [f(t + \Delta t) - f(t)]\,\mathbf{i} + [g(t + \Delta t) - g(t)]\,\mathbf{j} + [h(t + \Delta t) - h(t)]\,\mathbf{k}.
\end{aligned}$$

As Δt approaches zero, three things seem to happen simultaneously. First, Q approaches P along the curve. Second, the secant line PQ seems to approach a limiting position tangent to the curve at P. Third, the quotient $\Delta\mathbf{r}/\Delta t$ approaches the limit

$$\begin{aligned}
\lim_{\Delta t \to 0} \frac{\Delta\mathbf{r}}{\Delta t} &= \left[\lim_{\Delta t \to 0} \frac{f(t + \Delta t) - f(t)}{\Delta t}\right]\mathbf{i} + \left[\lim_{\Delta t \to 0} \frac{g(t + \Delta t) - g(t)}{\Delta t}\right]\mathbf{j} \\
&\quad + \left[\lim_{\Delta t \to 0} \frac{h(t + \Delta t) - h(t)}{\Delta t}\right]\mathbf{k} \\
&= \left[\frac{df}{dt}\right]\mathbf{i} + \left[\frac{dg}{dt}\right]\mathbf{j} + \left[\frac{dh}{dt}\right]\mathbf{k}.
\end{aligned}$$

We are therefore led by past experience to the following definitions.

> **Definitions**
>
> The vector function $\mathbf{r}(t) = f(t)\,\mathbf{i} + g(t)\,\mathbf{j} + h(t)\,\mathbf{k}$ is **differentiable at $\mathbf{t} = \mathbf{t_0}$** if f, g, and h are differentiable at t_0. Also, $\mathbf{r}$ is said to be **differentiable** if it is differentiable at every point of its domain. At any point t at which $\mathbf{r}$ is differentiable, its **derivative** is the vector
>
> $$\frac{d\mathbf{r}}{dt} = \lim_{\Delta t \to 0} \frac{\mathbf{r}(t + \Delta t) - \mathbf{r}(t)}{\Delta t} = \frac{df}{dt}\,\mathbf{i} + \frac{dg}{dt}\,\mathbf{j} + \frac{dh}{dt}\,\mathbf{k}.$$
>
> The curve traced by $\mathbf{r}$ is **smooth** if $d\mathbf{r}/dt$ is continuous and never $\mathbf{0}$, i.e., if f, g, and h have continuous first derivatives that are not simultaneously 0.

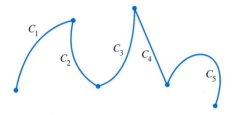

11.5 A piecewise smooth curve made up of five smooth curves connected end to end in continuous fashion.

The vector $d\mathbf{r}/dt$, when different from **0**, is also a vector *tangent* to the curve. The **tangent line** to the curve at a point $(f(t_0), g(t_0), h(t_0))$ is defined to be the line through the point parallel to $d\mathbf{r}/dt$ at $t = t_0$. We require $d\mathbf{r}/dt \neq \mathbf{0}$ for a smooth curve to make sure the curve has a continuously turning tangent at each point. On a smooth curve there are no sharp corners or cusps.

A curve that is made up of a finite number of smooth curves pieced together in a continuous fashion is called **piecewise smooth** (Fig. 11.5).

Look once again at Fig. 11.4. We drew the figure for Δt positive, so $\Delta \mathbf{r}$ points forward, in the direction of the motion. The vector $\Delta \mathbf{r}/\Delta t$ (not shown), having the same direction as $\Delta \mathbf{r}$, points forward too. Had Δt been negative, $\Delta \mathbf{r}$ would have pointed backward, against the direction of motion. The quotient $\Delta \mathbf{r}/\Delta t$, however, being a negative scalar multiple of $\Delta \mathbf{r}$, would once again have pointed forward. No matter how $\Delta \mathbf{r}$ points, $\Delta \mathbf{r}/\Delta t$ points forward and we expect the vector $d\mathbf{r}/dt = \lim_{\Delta t \to 0} \Delta \mathbf{r}/\Delta t$, when different from **0**, to do the same. This means that the derivative $d\mathbf{r}/dt$ is just what we want for modeling a particle's velocity. It points in the direction of motion and gives the rate of change of position with respect to time. For a smooth curve, the velocity is never zero; the particle does not stop or reverse direction.

> **Definitions**
>
> If **r** is the position vector of a particle moving along a smooth curve in space, then
>
> $$\mathbf{v}(t) = \frac{d\mathbf{r}}{dt}$$
>
> is the particle's **velocity vector**, tangent to the curve. At any time t, the direction of **v** is the **direction of motion**, the magnitude of **v** is the particle's **speed,** and the derivative $\mathbf{a} = d\mathbf{v}/dt$, when it exists, is the particle's **acceleration vector.** In short,
>
> 1. Velocity is the derivative of position: $\quad \mathbf{v} = \dfrac{d\mathbf{r}}{dt}.$
>
> 2. Speed is the magnitude of velocity: $\quad$ Speed $= |\mathbf{v}|.$
>
> 3. Acceleration is the derivative of velocity: $\quad \mathbf{a} = \dfrac{d\mathbf{v}}{dt} = \dfrac{d^2\mathbf{r}}{dt^2}.$
>
> 4. The vector $\mathbf{v}/|\mathbf{v}|$ is the direction of motion at time t.

We can express the velocity of a moving particle as the product of its speed and direction.

$$\text{Velocity} = |\mathbf{v}| \left(\frac{\mathbf{v}}{|\mathbf{v}|} \right) = (\text{speed}) \, (\text{direction})$$

EXAMPLE 4　　The vector

$$\mathbf{r}(t) = (3 \cos t)\,\mathbf{i} + (3 \sin t)\,\mathbf{j} + t^2\,\mathbf{k}$$

gives the position of a moving body at time t. Find the body's speed and direction when $t = 2$. At what times, if any, are the body's velocity and acceleration orthogonal?

Solution

$$\mathbf{r} = (3\cos t)\mathbf{i} + (3\sin t)\mathbf{j} + t^2\mathbf{k}$$

$$\mathbf{v} = \frac{d\mathbf{r}}{dt} = -(3\sin t)\mathbf{i} + (3\cos t)\mathbf{j} + 2t\mathbf{k}$$

$$\mathbf{a} = \frac{d^2\mathbf{r}}{dt^2} = -(3\cos t)\mathbf{i} - (3\sin t)\mathbf{j} + 2\mathbf{k}$$

At $t = 2$, the body's speed and direction are

Speed: $\qquad |\mathbf{v}(2)| = \sqrt{(-3\sin 2)^2 + (3\cos 2)^2 + (4)^2} = 5$

Direction: $\qquad \dfrac{\mathbf{v}(2)}{|\mathbf{v}(2)|} = -\left(\dfrac{3}{5}\sin 2\right)\mathbf{i} + \left(\dfrac{3}{5}\cos 2\right)\mathbf{j} + \dfrac{4}{5}\mathbf{k}.$

To find the times when $\mathbf{v}$ and $\mathbf{a}$ are orthogonal, we look for values of t for which

$$\mathbf{v} \cdot \mathbf{a} = 9\sin t\cos t - 9\cos t\sin t + 4t = 4t = 0.$$

The only value is $t = 0$. $\qquad\qquad\qquad\qquad\qquad\qquad\qquad\qquad$ ❏

Differentiation Rules

Because the derivatives of vector functions may be computed component by component, the rules for differentiating vector functions have the same form as the rules for differentiating scalar functions.

Differentiation Rules for Vector Functions

Constant Function Rule: $\qquad \dfrac{d}{dt}\mathbf{C} = \mathbf{0} \qquad$ (any constant vector $\mathbf{C}$)

If $\mathbf{u}$ and $\mathbf{v}$ are differentiable vector functions of t, then

Scalar Multiple Rules: $\qquad \dfrac{d}{dt}(c\,\mathbf{u}) = c\dfrac{d\mathbf{u}}{dt} \qquad$ (any number c)

$$\dfrac{d}{dt}(f\,\mathbf{u}) = \dfrac{df}{dt}\mathbf{u} + f\dfrac{d\mathbf{u}}{dt} \qquad \begin{array}{l}\text{(any differ-}\\ \text{entiable scalar}\\ \text{function } f(t))\end{array}$$

Sum Rule: $\qquad \dfrac{d}{dt}(\mathbf{u} + \mathbf{v}) = \dfrac{d\mathbf{u}}{dt} + \dfrac{d\mathbf{v}}{dt}$

Difference Rule: $\qquad \dfrac{d}{dt}(\mathbf{u} - \mathbf{v}) = \dfrac{d\mathbf{u}}{dt} - \dfrac{d\mathbf{v}}{dt}$

Dot Product Rule: $\qquad \dfrac{d}{dt}(\mathbf{u} \cdot \mathbf{v}) = \dfrac{d\mathbf{u}}{dt} \cdot \mathbf{v} + \mathbf{u} \cdot \dfrac{d\mathbf{v}}{dt}$

Cross Product Rule: $\qquad \dfrac{d}{dt}(\mathbf{u} \times \mathbf{v}) = \dfrac{d\mathbf{u}}{dt} \times \mathbf{v} + \mathbf{u} \times \dfrac{d\mathbf{v}}{dt}$

When you use the Cross Product Rule, remember to preserve the order of the factors. If $\mathbf{u}$ comes first on the left side of the equation, it must also come first on the right or the signs will be wrong.

As an algebraic convenience, we sometimes write the product of a scalar c and a vector $\mathbf{v}$ as $\mathbf{v}c$ instead of $c\mathbf{v}$. This permits us, for instance, to write the Chain Rule in a familiar form:

$$\frac{d\mathbf{r}}{ds} = \frac{d\mathbf{r}}{dt}\frac{dt}{ds}.$$

Chain Rule (Short Form): If $\mathbf{r}$ is a differentiable function of t and t is a differentiable function of s, then

$$\frac{d\mathbf{r}}{ds} = \frac{d\mathbf{r}}{dt}\frac{dt}{ds}.$$

We will prove the product rules and Chain Rule but leave the rules for constants, scalar multiples, sums, and differences as exercises.

Proof of the Dot Product Rule Suppose that

$$\mathbf{u} = u_1(t)\,\mathbf{i} + u_2(t)\,\mathbf{j} + u_3(t)\,\mathbf{k}$$

and

$$\mathbf{v} = v_1(t)\,\mathbf{i} + v_2(t)\,\mathbf{j} + v_3(t)\,\mathbf{k}.$$

Then

$$\frac{d}{dt}(\mathbf{u} \cdot \mathbf{v}) = \frac{d}{dt}(u_1v_1 + u_2v_2 + u_3v_3)$$

$$= \underbrace{u_1'v_1 + u_2'v_2 + u_3'v_3}_{\mathbf{u}' \,\cdot\, \mathbf{v}} + \underbrace{u_1v_1' + u_2v_2' + u_3v_3'}_{\mathbf{u} \,\cdot\, \mathbf{v}'}. \qquad \square$$

Proof of the Cross Product Rule We model the proof after the proof of the product rule for scalar functions. According to the definition of derivative,

$$\frac{d}{dt}(\mathbf{u} \times \mathbf{v}) = \lim_{h \to 0} \frac{\mathbf{u}(t+h) \times \mathbf{v}(t+h) - \mathbf{u}(t) \times \mathbf{v}(t)}{h}.$$

To change this fraction into an equivalent one that contains the difference quotients for the derivatives of $\mathbf{u}$ and $\mathbf{v}$, we subtract and add $\mathbf{u}(t) \times \mathbf{v}(t+h)$ in the numerator. Then

$$\frac{d}{dt}(\mathbf{u} \times \mathbf{v})$$

$$= \lim_{h \to 0} \frac{\mathbf{u}(t+h) \times \mathbf{v}(t+h) - \mathbf{u}(t) \times \mathbf{v}(t+h) + \mathbf{u}(t) \times \mathbf{v}(t+h) - \mathbf{u}(t) \times \mathbf{v}(t)}{h}$$

$$= \lim_{h \to 0} \left[\frac{\mathbf{u}(t+h) - \mathbf{u}(t)}{h} \times \mathbf{v}(t+h) + \mathbf{u}(t) \times \frac{\mathbf{v}(t+h) - \mathbf{v}(t)}{h} \right]$$

$$= \lim_{h \to 0} \frac{\mathbf{u}(t+h) - \mathbf{u}(t)}{h} \times \lim_{h \to 0} \mathbf{v}(t+h) + \lim_{h \to 0} \mathbf{u}(t) \times \lim_{h \to 0} \frac{\mathbf{v}(t+h) - \mathbf{v}(t)}{h}.$$

The last of these equalities holds because the limit of the cross product of two vector functions is the cross product of their limits if the latter exist (Exercise 52). As h approaches zero, $\mathbf{v}(t+h)$ approaches $\mathbf{v}(t)$ because $\mathbf{v}$, being differentiable at t, is continuous at t (Exercise 53). The two fractions approach the values of $d\mathbf{u}/dt$ and $d\mathbf{v}/dt$ at t. In short,

$$\frac{d}{dt}(\mathbf{u} \times \mathbf{v}) = \frac{d\mathbf{u}}{dt} \times \mathbf{v} + \mathbf{u} \times \frac{d\mathbf{v}}{dt}. \qquad \square$$

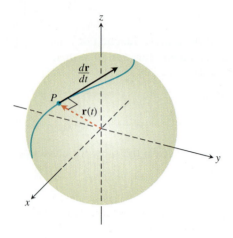

11.6 If a particle moves on a sphere in such a way that its position **r** is a differentiable function of time, then **r** · (d**r**/dt) = 0.

Proof of the Chain Rule Suppose that $\mathbf{r}(t) = f(t)\mathbf{i} + g(t)\mathbf{j} + h(t)\mathbf{k}$ is a differentiable vector function of t and that t is a differentiable scalar function of some other variable s. Then f, g, and h are differentiable functions of s, and the Chain Rule for differentiable real-valued functions gives

$$\frac{d\mathbf{r}}{ds} = \frac{df}{ds}\mathbf{i} + \frac{dg}{ds}\mathbf{j} + \frac{dh}{ds}\mathbf{k}$$

$$= \frac{df}{dt}\frac{dt}{ds}\mathbf{i} + \frac{dg}{dt}\frac{dt}{ds}\mathbf{j} + \frac{dh}{dt}\frac{dt}{ds}\mathbf{k}$$

$$= \left(\frac{df}{dt}\mathbf{i} + \frac{dg}{dt}\mathbf{j} + \frac{dh}{dt}\mathbf{k}\right)\frac{dt}{ds}$$

$$= \frac{d\mathbf{r}}{dt}\frac{dt}{ds}. \qquad \blacksquare$$

Vector Functions of Constant Length

When we track a particle moving on a sphere centered at the origin (Fig. 11.6), the position vector has a constant length equal to the radius of the sphere. The velocity vector $d\mathbf{r}/dt$, tangent to the path of motion, is tangent to the sphere and hence perpendicular to **r**. This is always the case for a differentiable vector function of constant length: The vector and its first derivative are orthogonal. With the length constant, the change in the function is a change in direction only, and direction changes take place at right angles.

We will use this observation repeatedly in Section 11.4.

If **u** is a differentiable vector function of t of constant length, then

$$\mathbf{u} \cdot \frac{d\mathbf{u}}{dt} = 0. \tag{3}$$

To see why Eq. (3) holds, suppose that **u** is a differentiable function of t and that $|\mathbf{u}|$ is constant. Then $\mathbf{u} \cdot \mathbf{u} = |\mathbf{u}|^2$ is constant and we may differentiate both sides of this equation to get

$$\frac{d}{dt}(\mathbf{u} \cdot \mathbf{u}) = \frac{d}{dt}(\text{constant}) = 0$$

$$\frac{d\mathbf{u}}{dt} \cdot \mathbf{u} + \mathbf{u} \cdot \frac{d\mathbf{u}}{dt} = 0 \qquad \text{Dot Product Rule with } \mathbf{v} = \mathbf{u}$$

$$2\mathbf{u} \cdot \frac{d\mathbf{u}}{dt} = 0 \qquad \text{Dot multiplication is commutative.}$$

$$\mathbf{u} \cdot \frac{d\mathbf{u}}{dt} = 0.$$

EXAMPLE 5 Show that

$$\mathbf{u}(t) = (\sin t)\mathbf{i} + (\cos t)\mathbf{j} + \sqrt{3}\,\mathbf{k}$$

has constant length and is orthogonal to its derivative.

Solution

$$\mathbf{u}(t) = (\sin t)\,\mathbf{i} + (\cos t)\,\mathbf{j} + \sqrt{3}\,\mathbf{k}$$

$$|\mathbf{u}(t)| = \sqrt{(\sin t)^2 + (\cos t)^2 + (\sqrt{3})^2} = \sqrt{1+3} = 2$$

$$\frac{d\mathbf{u}}{dt} = (\cos t)\,\mathbf{i} - (\sin t)\,\mathbf{j}$$

$$\mathbf{u} \cdot \frac{d\mathbf{u}}{dt} = \sin t \cos t - \sin t \cos t = 0$$

Integrals of Vector Functions

A differentiable vector function $\mathbf{R}(t)$ is an **antiderivative** of a vector function $\mathbf{r}(t)$ on an interval I if $d\mathbf{R}/dt = \mathbf{r}$ at each point of I. If $\mathbf{R}$ is an antiderivative of $\mathbf{r}$ on I, it can be shown, working one component at a time, that every antiderivative of $\mathbf{r}$ on I has the form $\mathbf{R} + \mathbf{C}$ for some constant vector $\mathbf{C}$ (Exercise 56). The set of all antiderivatives of $\mathbf{r}$ on I is the **indefinite integral** of $\mathbf{r}$ on I.

> **Definition**
>
> The **indefinite integral** of $\mathbf{r}$ with respect to t is the set of all antiderivatives of $\mathbf{r}$, denoted by $\int \mathbf{r}(t)\,dt$. If $\mathbf{R}$ is any antiderivative of $\mathbf{r}$, then
>
> $$\int \mathbf{r}(t)\,dt = \mathbf{R}(t) + \mathbf{C}.$$

The usual arithmetic rules for indefinite integrals apply.

EXAMPLE 6

$$\int ((\cos t)\,\mathbf{i} + \mathbf{j} - 2t\,\mathbf{k})\,dt$$

$$= \left(\int \cos t\, dt \right)\mathbf{i} + \left(\int dt \right)\mathbf{j} - \left(\int 2t\, dt \right)\mathbf{k} \qquad (4)$$

$$= (\sin t + C_1)\,\mathbf{i} + (t + C_2)\,\mathbf{j} - (t^2 + C_3)\,\mathbf{k} \qquad (5)$$

$$= (\sin t)\,\mathbf{i} + t\,\mathbf{j} - t^2\,\mathbf{k} + \mathbf{C} \qquad \mathbf{C} = C_1\,\mathbf{i} + C_2\,\mathbf{j} - C_3\,\mathbf{k}$$

As in the integration of scalar functions, we recommend that you skip the steps in (4) and (5) and go directly to the final form. Find an antiderivative for each component and add a constant vector at the end.

Definite integrals of vector functions are defined in terms of components.

> **Definition**
>
> If the components of $\mathbf{r}(t) = f(t)\,\mathbf{i} + g(t)\,\mathbf{j} + h(t)\,\mathbf{k}$ are integrable over $[a, b]$ then so is $\mathbf{r}$, and the **definite integral** of $\mathbf{r}$ from a to b is
>
> $$\int_a^b \mathbf{r}(t)\,dt = \left(\int_a^b f(t)\,dt \right)\mathbf{i} + \left(\int_a^b g(t)\,dt \right)\mathbf{j} + \left(\int_a^b h(t)\,dt \right)\mathbf{k}.$$

The usual arithmetic rules for definite integrals apply (Exercise 54).

EXAMPLE 7

$$\int_0^\pi ((\cos t)\,\mathbf{i} + \mathbf{j} - 2t\,\mathbf{k})\,dt = \left(\int_0^\pi \cos t\,dt\right)\mathbf{i} + \left(\int_0^\pi dt\right)\mathbf{j} - \left(\int_0^\pi 2t\,dt\right)\mathbf{k}$$

$$= \left[\sin t\right]_0^\pi \mathbf{i} + \left[t\right]_0^\pi \mathbf{j} - \left[t^2\right]_0^\pi \mathbf{k}$$

$$= [0 - 0]\,\mathbf{i} + [\pi - 0]\,\mathbf{j} - [\pi^2 - 0^2]\,\mathbf{k}$$

$$= \pi\,\mathbf{j} - \pi^2\,\mathbf{k} \qquad\qquad ❑$$

EXAMPLE 8   *Finding a particle's position function from its velocity function and initial position*

The velocity of a particle moving in space is

$$\frac{d\mathbf{r}}{dt} = (\cos t)\,\mathbf{i} - (\sin t)\,\mathbf{j} + \mathbf{k}.$$

Find the particle's position as a function of t if $\mathbf{r} = 2\,\mathbf{i} + \mathbf{k}$ when $t = 0$.

Solution   Our goal is to solve the initial value problem that consists of

The differential equation: $\dfrac{d\mathbf{r}}{dt} = (\cos t)\,\mathbf{i} - (\sin t)\,\mathbf{j} + \mathbf{k}$

The initial condition: $\mathbf{r}(0) = 2\,\mathbf{i} + \mathbf{k}$

Integrating both sides of the differential equation with respect to t gives

$$\mathbf{r}(t) = (\sin t)\,\mathbf{i} + (\cos t)\,\mathbf{j} + t\,\mathbf{k} + \mathbf{C}.$$

We then use the initial condition to find the right value for $\mathbf{C}$:

$$(\sin 0)\,\mathbf{i} + (\cos 0)\,\mathbf{j} + (0)\,\mathbf{k} + \mathbf{C} = 2\,\mathbf{i} + \mathbf{k} \qquad \mathbf{r}(0) = 2\,\mathbf{i} + \mathbf{k}$$

$$\mathbf{j} + \mathbf{C} = 2\,\mathbf{i} + \mathbf{k}$$

$$\mathbf{C} = 2\,\mathbf{i} - \mathbf{j} + \mathbf{k}.$$

The particle's position as a function of t is

$$\mathbf{r}(t) = (\sin t + 2)\,\mathbf{i} + (\cos t - 1)\,\mathbf{j} + (t + 1)\,\mathbf{k}.$$

To check (always a good idea), we can see from this formula that

$$\frac{d\mathbf{r}}{dt} = (\cos t + 0)\,\mathbf{i} + (-\sin t - 0)\,\mathbf{j} + (1 + 0)\,\mathbf{k}$$

$$= (\cos t)\,\mathbf{i} - (\sin t)\,\mathbf{j} + \mathbf{k}$$

and

$$\mathbf{r}(0) = (\sin 0 + 2)\,\mathbf{i} + (\cos 0 - 1)\,\mathbf{j} + (0 + 1)\,\mathbf{k}$$

$$= 2\,\mathbf{i} + \mathbf{k}. \qquad\qquad ❑$$

Exercises 11.1

Motion in the *xy*-plane

In Exercises 1–4, $\mathbf{r}(t)$ is the position of a particle in the *xy*-plane at time *t*. Find an equation in *x* and *y* whose graph is the path of the particle. Then find the particle's velocity and acceleration vectors at the given value of *t*.

1. $\mathbf{r}(t) = (t+1)\mathbf{i} + (t^2-1)\mathbf{j}, \quad t = 1$

2. $\mathbf{r}(t) = (t^2+1)\mathbf{i} + (2t-1)\mathbf{j}, \quad t = 1/2$

3. $\mathbf{r}(t) = e^t\mathbf{i} + \dfrac{2}{9}e^{2t}\mathbf{j}, \quad t = \ln 3$

4. $\mathbf{r}(t) = (\cos 2t)\mathbf{i} + (3\sin 2t)\mathbf{j}, \quad t = 0$

Exercises 5–8 give the position vectors of particles moving along various curves in the *xy*-plane. In each case, find the particle's velocity and acceleration vectors at the stated times and sketch them as vectors on the curve.

5. *Motion on the circle $x^2 + y^2 = 1$*

$$\mathbf{r}(t) = (\sin t)\mathbf{i} + (\cos t)\mathbf{j}; \quad t = \pi/4 \text{ and } \pi/2$$

6. *Motion on the circle $x^2 + y^2 = 16$*

$$\mathbf{r}(t) = \left(4\cos\frac{t}{2}\right)\mathbf{i} + \left(4\sin\frac{t}{2}\right)\mathbf{j}; \quad t = \pi \text{ and } 3\pi/2$$

7. *Motion on the cycloid $x = t - \sin t, \ y = 1 - \cos t$*

$$\mathbf{r}(t) = (t - \sin t)\mathbf{i} + (1 - \cos t)\mathbf{j}; \quad t = \pi \text{ and } 3\pi/2$$

8. *Motion on the parabola $y = x^2 + 1$*

$$\mathbf{r}(t) = t\mathbf{i} + (t^2+1)\mathbf{j}; \quad t = -1, 0, \text{ and } 1$$

Velocity and Acceleration in Space

In Exercises 9–14, $\mathbf{r}(t)$ is the position of a particle in space at time *t*. Find the particle's velocity and acceleration vectors. Then find the particle's speed and direction of motion at the given value of *t*. Write the particle's velocity at that time as the product of its speed and direction.

9. $\mathbf{r}(t) = (t+1)\mathbf{i} + (t^2-1)\mathbf{j} + 2t\mathbf{k}, \quad t = 1$

10. $\mathbf{r}(t) = (1+t)\mathbf{i} + \dfrac{t^2}{\sqrt{2}}\mathbf{j} + \dfrac{t^3}{3}\mathbf{k}, \quad t = 1$

11. $\mathbf{r}(t) = (2\cos t)\mathbf{i} + (3\sin t)\mathbf{j} + 4t\mathbf{k}, \quad t = \pi/2$

12. $\mathbf{r}(t) = (\sec t)\mathbf{i} + (\tan t)\mathbf{j} + \dfrac{4}{3}t\mathbf{k}, \quad t = \pi/6$

13. $\mathbf{r}(t) = (2\ln(t+1))\mathbf{i} + t^2\mathbf{j} + \dfrac{t^2}{2}\mathbf{k}, \quad t = 1$

14. $\mathbf{r}(t) = (e^{-t})\mathbf{i} + (2\cos 3t)\mathbf{j} + (2\sin 3t)\mathbf{k}, \quad t = 0$

In Exercises 15–18, $\mathbf{r}(t)$ is the position of a particle in space at time *t*. Find the angle between the velocity and acceleration vectors at time $t = 0$.

15. $\mathbf{r}(t) = (3t+1)\mathbf{i} + \sqrt{3}\,t\mathbf{j} + t^2\mathbf{k}$

16. $\mathbf{r}(t) = \left(\dfrac{\sqrt{2}}{2}t\right)\mathbf{i} + \left(\dfrac{\sqrt{2}}{2}t - 16t^2\right)\mathbf{j}$

17. $\mathbf{r}(t) = (\ln(t^2+1))\mathbf{i} + (\tan^{-1}t)\mathbf{j} + \sqrt{t^2+1}\,\mathbf{k}$

18. $\mathbf{r}(t) = \dfrac{4}{9}(1+t)^{3/2}\mathbf{i} + \dfrac{4}{9}(1-t)^{3/2}\mathbf{j} + \dfrac{1}{3}t\mathbf{k}$

In Exercises 19 and 20, $\mathbf{r}(t)$ is the position vector of a particle in space at time *t*. Find the time or times in the given time interval when the velocity and acceleration vectors are orthogonal.

19. $\mathbf{r}(t) = (t - \sin t)\mathbf{i} + (1 - \cos t)\mathbf{j}, \quad 0 \le t \le 2\pi$

20. $\mathbf{r}(t) = (\sin t)\mathbf{i} + t\mathbf{j} + (\cos t)\mathbf{k}, \quad t \ge 0$

Integrating Vector-valued Functions

Evaluate the integrals in Exercises 21–26.

21. $\displaystyle\int_0^1 [t^3\mathbf{i} + 7\mathbf{j} + (t+1)\mathbf{k}]\,dt$

22. $\displaystyle\int_1^2 \left[(6-6t)\mathbf{i} + 3\sqrt{t}\,\mathbf{j} + \left(\dfrac{4}{t^2}\right)\mathbf{k}\right]dt$

23. $\displaystyle\int_{-\pi/4}^{\pi/4} [(\sin t)\mathbf{i} + (1 + \cos t)\mathbf{j} + (\sec^2 t)\mathbf{k}]\,dt$

24. $\displaystyle\int_0^{\pi/3} [(\sec t \tan t)\mathbf{i} + (\tan t)\mathbf{j} + (2\sin t \cos t)\mathbf{k}]\,dt$

25. $\displaystyle\int_1^4 \left[\dfrac{1}{t}\mathbf{i} + \dfrac{1}{5-t}\mathbf{j} + \dfrac{1}{2t}\mathbf{k}\right]dt$

26. $\displaystyle\int_0^1 \left[\dfrac{2}{\sqrt{1-t^2}}\mathbf{i} + \dfrac{\sqrt{3}}{1+t^2}\mathbf{k}\right]dt$

Initial Value Problems for Vector–valued Functions

Solve the initial value problems in Exercises 27–32 for $\mathbf{r}$ as a vector function of *t*.

27. Differential equation: $\quad \dfrac{d\mathbf{r}}{dt} = -t\mathbf{i} - t\mathbf{j} - t\mathbf{k}$

Initial condition: $\quad \mathbf{r}(0) = \mathbf{i} + 2\mathbf{j} + 3\mathbf{k}$

28. Differential equation: $\quad \dfrac{d\mathbf{r}}{dt} = (180t)\mathbf{i} + (180t - 16t^2)\mathbf{j}$

Initial condition: $\quad \mathbf{r}(0) = 100\mathbf{j}$

29. Differential equation: $\quad \dfrac{d\mathbf{r}}{dt} = \dfrac{3}{2}(t+1)^{1/2}\mathbf{i} + e^{-t}\mathbf{j} + \dfrac{1}{t+1}\mathbf{k}$

Initial condition: $\quad \mathbf{r}(0) = \mathbf{k}$

30. Differential equation: $\dfrac{d\mathbf{r}}{dt} = (t^3 + 4t)\,\mathbf{i} + t\,\mathbf{j} + 2t^2\,\mathbf{k}$

Initial condition: $\mathbf{r}(0) = \mathbf{i} + \mathbf{j}$

31. Differential equation: $\dfrac{d^2\mathbf{r}}{dt^2} = -32\,\mathbf{k}$

Initial conditions: $\mathbf{r}(0) = 100\,\mathbf{k}$ and

$\dfrac{d\mathbf{r}}{dt}\bigg|_{t=0} = 8\,\mathbf{i} + 8\,\mathbf{j}$

32. Differential equation: $\dfrac{d^2\mathbf{r}}{dt^2} = -(\mathbf{i} + \mathbf{j} + \mathbf{k})$

Initial conditions: $\mathbf{r}(0) = 10\,\mathbf{i} + 10\,\mathbf{j} + 10\,\mathbf{k}$ and

$\dfrac{d\mathbf{r}}{dt}\bigg|_{t=0} = \mathbf{0}$

Tangent Lines to Smooth Curves

As mentioned in the text, the tangent line to a smooth curve $\mathbf{r}(t) = f(t)\,\mathbf{i} + g(t)\,\mathbf{j} + h(t)\,\mathbf{k}$ at $t = t_0$ is the line that passes through the point $(f(t_0), g(t_0), h(t_0))$ parallel to $\mathbf{v}(t_0)$, the curve's velocity vector at t_0. In Exercises 33–36, find parametric equations for the line that is tangent to the given curve at the given parameter value $t = t_0$.

33. $\mathbf{r}(t) = (\sin t)\,\mathbf{i} + (t^2 - \cos t)\,\mathbf{j} + e^t\,\mathbf{k}, \quad t_0 = 0$

34. $\mathbf{r}(t) = (2\sin t)\,\mathbf{i} + (2\cos t)\,\mathbf{j} + 5t\,\mathbf{k}, \quad t_0 = 4\pi$

35. $\mathbf{r}(t) = (a\sin t)\,\mathbf{i} + (a\cos t)\,\mathbf{j} + bt\,\mathbf{k}, \quad t_0 = 2\pi$

36. $\mathbf{r}(t) = (\cos t)\,\mathbf{i} + (\sin t)\,\mathbf{j} + (\sin 2t)\,\mathbf{k}, \quad t_0 = \dfrac{\pi}{2}$

Motion on Circular Paths

37. Each of the following equations (a)–(e) describes the motion of a particle having the same path, namely the unit circle $x^2 + y^2 = 1$. Although the path of each particle in (a)–(e) is the same, the behavior, or "dynamics," of each particle is different. For each particle, answer the following questions.

i) Does the particle have constant speed? If so, what is its constant speed?

ii) Is the particle's acceleration vector always orthogonal to its velocity vector?

iii) Does the particle move clockwise or counterclockwise around the circle?

iv) Does the particle begin at the point $(1, 0)$?

a) $\mathbf{r}(t) = (\cos t)\,\mathbf{i} + (\sin t)\,\mathbf{j}, \quad t \geq 0$

b) $\mathbf{r}(t) = \cos(2t)\,\mathbf{i} + \sin(2t)\,\mathbf{j}, \quad t \geq 0$

c) $\mathbf{r}(t) = \cos(t - \pi/2)\,\mathbf{i} + \sin(t - \pi/2)\,\mathbf{j}, \quad t \geq 0$

d) $\mathbf{r}(t) = (\cos t)\,\mathbf{i} - (\sin t)\,\mathbf{j}, \quad t \geq 0$

e) $\mathbf{r}(t) = \cos(t^2)\,\mathbf{i} + \sin(t^2)\,\mathbf{j}, \quad t \geq 0$

38. Show that the vector-valued function

$\mathbf{r}(t) = (2\,\mathbf{i} + 2\,\mathbf{j} + \mathbf{k})$

$\quad + \cos t \left(\dfrac{1}{\sqrt{2}}\,\mathbf{i} - \dfrac{1}{\sqrt{2}}\,\mathbf{j} \right) + \sin t \left(\dfrac{1}{\sqrt{3}}\,\mathbf{i} + \dfrac{1}{\sqrt{3}}\,\mathbf{j} + \dfrac{1}{\sqrt{3}}\,\mathbf{k} \right)$

describes the motion of a particle moving in the circle of radius 1 centered at the point $(2, 2, 1)$ and lying in the plane $x + y - 2z = 2$.

Motion along a Straight Line

39. At time $t = 0$, a particle is located at the point $(1, 2, 3)$. It travels in a straight line to the point $(4, 1, 4)$, has speed 2 at $(1, 2, 3)$ and constant acceleration $3\,\mathbf{i} - \mathbf{j} + \mathbf{k}$. Find an equation for the position vector $\mathbf{r}(t)$ of the particle at time t.

40. A particle traveling in a straight line is located at the point $(1, -1, 2)$ and has speed 2 at time $t = 0$. The particle moves toward the point $(3, 0, 3)$ with constant acceleration $2\,\mathbf{i} + \mathbf{j} + \mathbf{k}$. Find its position vector $\mathbf{r}(t)$ at time t.

Theory and Examples

41. A particle moves along the top of the parabola $y^2 = 2x$ from left to right at a constant speed of 5 units per second. Find the velocity of the particle as it moves through the point $(2, 2)$.

42. A particle moves on a cycloid in the xy-plane in such a way that its position at time t is

$$\mathbf{r}(t) = (t - \sin t)\,\mathbf{i} + (1 - \cos t)\,\mathbf{j}.$$

Find the maximum and minimum values of $|\mathbf{v}|$ and $|\mathbf{a}|$. (*Hint:* Find the extreme values of $|\mathbf{v}|^2$ and $|\mathbf{a}|^2$ first and take square roots later.)

43. A particle moves around the ellipse $(y/3)^2 + (z/2)^2 = 1$ in the yz plane in such a way that its position at time t is

$$\mathbf{r}(t) = (3\cos t)\,\mathbf{j} + (2\sin t)\,\mathbf{k}.$$

Find the maximum and minimum values of $|\mathbf{v}|$ and $|\mathbf{a}|$. (See the hint in Exercise 42.)

44. *A satellite in circular orbit.* A satellite of mass m is revolving at a constant speed v around a body of mass M (Earth, for example) in a circular orbit of radius r_0 (measured from the body's center of mass). Determine the satellite's orbital period T (the time to complete one full orbit), as follows:

a) Coordinatize the orbital plane by placing the origin at the body's center of mass, with the satellite on the x-axis at $t = 0$ and moving counterclockwise, as in the accompanying figure.

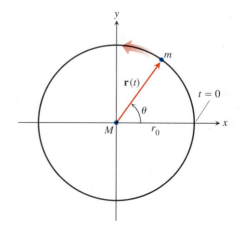

Let $\mathbf{r}(t)$ be the satellite's position vector at time t. Show that $\theta = vt/r_0$ and hence that

$$\mathbf{r}(t) = \left(r_0 \cos \frac{vt}{r_0}\right)\mathbf{i} + \left(r_0 \sin \frac{vt}{r_0}\right)\mathbf{j}.$$

b) Find the acceleration of the satellite.

c) According to Newton's Law of Gravitation, the gravitational force exerted on the satellite is directed toward M and is given by

$$\mathbf{F} = \left(-\frac{GmM}{r_0^2}\right)\frac{\mathbf{r}}{r_0},$$

where G is the universal constant of gravitation. Using Newton's second law, $\mathbf{F} = m\,\mathbf{a}$, show that $v^2 = GM/r_0$.

d) Show that the orbital period T satisfies $vT = 2\pi r_0$.

e) From parts (c) and (d), deduce that

$$T^2 = \frac{4\pi^2}{GM}r_0^3.$$

That is, the square of the period of a satellite in circular orbit is proportional to the cube of the radius from the orbital center.

45. Let $\mathbf{v}$ be a differentiable vector function of t. Show that if $\mathbf{v} \cdot (d\mathbf{v}/dt) = 0$ for all t, then $|\mathbf{v}|$ is constant.

46. *Derivatives of triple scalar products*

a) Show that if $\mathbf{u}$, $\mathbf{v}$, and $\mathbf{w}$ are differentiable vector functions of t, then

$$\frac{d}{dt}(\mathbf{u} \cdot \mathbf{v} \times \mathbf{w}) \tag{6}$$

$$= \frac{d\mathbf{u}}{dt} \cdot \mathbf{v} \times \mathbf{w} + \mathbf{u} \cdot \frac{d\mathbf{v}}{dt} \times \mathbf{w} + \mathbf{u} \cdot \mathbf{v} \times \frac{d\mathbf{w}}{dt}.$$

b) Show that Eq. (6) is equivalent to

$$\frac{d}{dt}\begin{vmatrix} u_1 & u_2 & u_3 \\ v_1 & v_2 & v_3 \\ w_1 & w_2 & w_3 \end{vmatrix} = \begin{vmatrix} \dfrac{du_1}{dt} & \dfrac{du_2}{dt} & \dfrac{du_3}{dt} \\ v_1 & v_2 & v_3 \\ w_1 & w_2 & w_3 \end{vmatrix} + \begin{vmatrix} u_1 & u_2 & u_3 \\ \dfrac{dv_1}{dt} & \dfrac{dv_2}{dt} & \dfrac{dv_3}{dt} \\ w_1 & w_2 & w_3 \end{vmatrix}$$

$$+ \begin{vmatrix} u_1 & u_2 & u_3 \\ v_1 & v_2 & v_3 \\ \dfrac{dw_1}{dt} & \dfrac{dw_2}{dt} & \dfrac{dw_3}{dt} \end{vmatrix}. \tag{7}$$

Equation (7) says that the derivative of a 3 by 3 determinant of differentiable functions is the sum of the three determinants obtained from the original by differentiating one row at a time. The result extends to determinants of any order.

47. *(Continuation of Exercise 46.)* Suppose that $\mathbf{r}(t) = f(t)\mathbf{i} + g(t)\mathbf{j} + h(t)\mathbf{k}$ and that f, g, and h have derivatives through order three. Use Eq. (6) or (7) to show that

$$\frac{d}{dt}\left(\mathbf{r} \cdot \frac{d\mathbf{r}}{dt} \times \frac{d^2\mathbf{r}}{dt^2}\right) = \mathbf{r} \cdot \left(\frac{d\mathbf{r}}{dt} \times \frac{d^3\mathbf{r}}{dt^3}\right). \tag{8}$$

(Hint: Differentiate on the left and look for vectors whose products are zero.)

48. *The Constant Function Rule.* Prove that if $\mathbf{u}$ is the vector function with the constant value $\mathbf{C}$, then $d\mathbf{u}/dt = \mathbf{0}$.

49. *The Scalar Multiple Rules*

a) Prove that if $\mathbf{u}$ is a differentiable function of t and c is any real number, then

$$\frac{d(c\mathbf{u})}{dt} = c\frac{d\mathbf{u}}{dt}.$$

b) Prove that if $\mathbf{u}$ is a differentiable function of t and f is a differentiable scalar function of t, then

$$\frac{d}{dt}(f\,\mathbf{u}) = \frac{df}{dt}\mathbf{u} + f\frac{d\mathbf{u}}{dt}.$$

50. *The Sum and Difference Rules.* Prove that if $\mathbf{u}$ and $\mathbf{v}$ are differentiable functions of t, then

$$\frac{d}{dt}(\mathbf{u} + \mathbf{v}) = \frac{d\mathbf{u}}{dt} + \frac{d\mathbf{v}}{dt}$$

and

$$\frac{d}{dt}(\mathbf{u} - \mathbf{v}) = \frac{d\mathbf{u}}{dt} - \frac{d\mathbf{v}}{dt}.$$

51. *The component test for continuity at a point.* Show that the vector function $\mathbf{r}$ defined by the rule $\mathbf{r}(t) = f(t)\mathbf{i} + g(t)\mathbf{j} + h(t)\mathbf{k}$ is continuous at $t = t_0$ if and only if f, g, and h are continuous at t_0.

52. *Limits of cross products of vector functions.* Suppose that $\mathbf{r}_1(t) = f_1(t)\mathbf{i} + f_2(t)\mathbf{j} + f_3(t)\mathbf{k}$, $\mathbf{r}_2(t) = g_1(t)\mathbf{i} + g_2(t)\mathbf{j} + g_3(t)\mathbf{k}$, $\lim_{t \to t_0}\mathbf{r}_1(t) = \mathbf{A}$, and $\lim_{t \to t_0}\mathbf{r}_2(t) = \mathbf{B}$. Use the determinant formula for cross products and the Limit Product Rule for scalar functions to show that

$$\lim_{t \to t_0}(\mathbf{r}_1(t) \times \mathbf{r}_2(t)) = \mathbf{A} \times \mathbf{B}.$$

53. *Differentiable vector functions are continuous.* Show that if $\mathbf{r}(t) = f(t)\mathbf{i} + g(t)\mathbf{j} + h(t)\mathbf{k}$ is differentiable at $t = t_0$, then it is continuous at t_0 as well.

54. Establish the following properties of integrable vector functions.

a) The *Constant Scalar Multiple Rule:*

$$\int_a^b k\,\mathbf{r}(t)\,dt = k\int_a^b \mathbf{r}(t)\,dt \quad \text{(any scalar } k\text{)}$$

The *Rule for Negatives,*

$$\int_a^b (-\mathbf{r}(t))\,dt = -\int_a^b \mathbf{r}(t)\,dt,$$

is obtained by taking $k = -1$.

b) The *Sum and Difference Rules:*

$$\int_a^b (\mathbf{r}_1(t) \pm \mathbf{r}_2(t))\,dt = \int_a^b \mathbf{r}_1(t)\,dt \pm \int_a^b \mathbf{r}_2(t)\,dt$$

c) The *Constant Vector Multiple Rules:*

$$\int_a^b \mathbf{C} \cdot \mathbf{r}(t)\,dt = \mathbf{C} \cdot \int_a^b \mathbf{r}(t)\,dt \quad \text{(any constant vector } \mathbf{C}\text{)}$$

and

$$\int_a^b \mathbf{C} \times \mathbf{r}(t)\, dt = \mathbf{C} \times \int_a^b \mathbf{r}(t)\, dt \quad \text{(any constant vector } \mathbf{C}\text{)}$$

55. *Products of scalar and vector functions.* Suppose that the scalar function $u(t)$ and the vector function $\mathbf{r}(t)$ are both defined for $a \le t \le b$.

a) Show that $u\,\mathbf{r}$ is continuous on $[a, b]$ if u and $\mathbf{r}$ are continuous on $[a, b]$.

b) If u and $\mathbf{r}$ are both differentiable on $[a, b]$, show that $u\,\mathbf{r}$ is differentiable on $[a, b]$ and that

$$\frac{d}{dt}(u\,\mathbf{r}) = u\,\frac{d\mathbf{r}}{dt} + \mathbf{r}\,\frac{du}{dt}.$$

56. *Antiderivatives of vector functions*

a) Use Corollary 2 of the Mean Value Theorem for scalar functions to show that if two vector functions $\mathbf{R}_1(t)$ and $\mathbf{R}_2(t)$ have identical derivatives on an interval I, then the functions differ by a constant vector value throughout I.

b) Use the result in (a) to show that if $\mathbf{R}(t)$ is any antiderivative of $\mathbf{r}(t)$ on I, then every other antiderivative of $\mathbf{r}$ on I equals $\mathbf{R}(t) + \mathbf{C}$ for some constant vector $\mathbf{C}$.

57. *The Fundamental Theorem of Calculus.* The Fundamental Theorem of Calculus for scalar functions of a real variable holds for vector functions of a real variable as well. Prove this by using the theorem for scalar functions to show first that if a vector function $\mathbf{r}(t)$ is continuous for $a \le t \le b$, then

$$\frac{d}{dt}\int_a^t \mathbf{r}(\tau)\, d\tau = \mathbf{r}(t)$$

at every point t of $[a, b]$. Then use the conclusion in part (b) of Exercise 56 to show that if $\mathbf{R}$ is any antiderivative of $\mathbf{r}$ on $[a, b]$ then

$$\int_a^b \mathbf{r}(t)\, dt = \mathbf{R}(b) - \mathbf{R}(a).$$

CAS Explorations and Projects

Use a CAS to perform the following steps in Exercises 58–61.

a) Plot the space curve traced out by the position vector $\mathbf{r}$.

b) Find the components of the velocity vector $d\mathbf{r}/dt$.

c) Evaluate $d\mathbf{r}/dt$ at the given point t_0 and determine the equation of the tangent line to the curve at $\mathbf{r}(t_0)$.

d) Plot the tangent line together with the curve over the given interval.

58. $\mathbf{r}(t) = (\sin t - t \cos t)\,\mathbf{i} + (\cos t + t \sin t)\,\mathbf{j} + t^2\,\mathbf{k}, \quad 0 \le t \le 6\pi,$ $t_0 = 3\pi/2$

59. $\mathbf{r}(t) = \sqrt{2}\,t\,\mathbf{i} + e^t\,\mathbf{j} + e^{-t}\,\mathbf{k}, \quad -2 \le t \le 3, \quad t_0 = 1$

60. $\mathbf{r}(t) = (\sin 2t)\,\mathbf{i} + (\ln (1+t))\,\mathbf{j} + t\,\mathbf{k}, \quad 0 \le t \le 4\pi, \quad t_0 = \pi/4$

61. $\mathbf{r}(t) = (\ln (t^2 + 2))\,\mathbf{i} + (\tan^{-1} 3t)\,\mathbf{j} + \sqrt{t^2 + 1}\,\mathbf{k}, \quad -3 \le t \le 5,$ $t_0 = 3$

In Exercises 62 and 63, you will explore graphically the behavior of the helix

$$\mathbf{r}(t) = (\cos at)\,\mathbf{i} + (\sin at)\,\mathbf{j} + bt\,\mathbf{k}$$

as you change the values of the constants a and b. Use a CAS to perform the steps in each exercise.

62. Set $b = 1$. Plot the helix $\mathbf{r}(t)$ together with the tangent line to the curve at $t = 3\pi/2$ for $a = 1, 2, 4,$ and 6 over the interval $0 \le t \le 4\pi$. Describe in your own words what happens to the graph of the helix and the position of the tangent line as a increases through these positive values.

63. Set $a = 1$. Plot the helix $\mathbf{r}(t)$ together with the tangent line to the curve at $t = 3\pi/2$ for $b = 1/4, 1/2, 2,$ and 4 over the interval $0 \le t \le 4\pi$. Describe in your own words what happens to the graph of the helix and the position of the tangent line as b increases through these positive values.

11.2

Modeling Projectile Motion

When we shoot a projectile into the air we usually want to know beforehand how far it will go (will it reach the target?), how high it will rise (will it clear the hill?), and when it will land (when do we get results?). We get this information from the direction and magnitude of the projectile's initial velocity vector, using Newton's Second Law of Motion.

The Vector and Parametric Equations for Ideal Projectile Motion

To derive equations for projectile motion, we assume that the projectile behaves like a particle moving in a vertical coordinate plane and that the only force acting on the projectile during its flight is the constant force of gravity, which always points

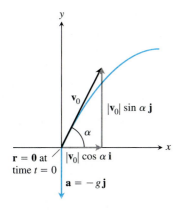

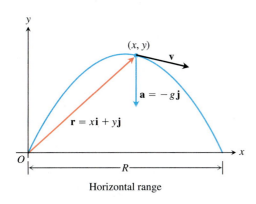

(a) Position, velocity, acceleration,
and launch angle at $t = 0$

(b) Position, velocity, and acceleration
at a later time t

11.7 The flight of an ideal projectile.

straight down. In practice, none of these assumptions really holds. The ground moves beneath the projectile as the earth turns, the air creates a frictional force that varies with the projectile's speed and altitude, and the force of gravity changes as the projectile moves along. All this must be taken into account by applying corrections to the predictions of the *ideal* equations we are about to derive. The corrections, however, are not the subject of this section.

We assume that our projectile is launched from the origin at time $t = 0$ into the first quadrant with an initial velocity $\mathbf{v}_0$ (Fig. 11.7). If $\mathbf{v}_0$ makes an angle α with the horizontal, then

$$\mathbf{v}_0 = (|\mathbf{v}_0| \cos \alpha)\,\mathbf{i} + (|\mathbf{v}_0| \sin \alpha)\,\mathbf{j}. \tag{1}$$

If we use the simpler notation v_0 for the initial speed $|\mathbf{v}_0|$, then

$$\mathbf{v}_0 = (v_0 \cos \alpha)\,\mathbf{i} + (v_0 \sin \alpha)\,\mathbf{j}. \tag{2}$$

The projectile's initial position is

$$\mathbf{r}_0 = 0\,\mathbf{i} + 0\,\mathbf{j} = \mathbf{0}. \tag{3}$$

Newton's Second Law of Motion says that the force acting on the projectile is its mass m times its acceleration, or $m(d^2\mathbf{r}/dt^2)$ if $\mathbf{r}$ is the projectile's position vector and t is time. If the force is solely the gravitational force $-mg\,\mathbf{j}$, then

$$m\frac{d^2\mathbf{r}}{dt^2} = -mg\,\mathbf{j} \quad \text{and} \quad \frac{d^2\mathbf{r}}{dt^2} = -g\,\mathbf{j}. \tag{4}$$

We find $\mathbf{r}$ as a function of t by solving the following initial value problem:

Differential equation: $\dfrac{d^2\mathbf{r}}{dt^2} = -g\,\mathbf{j}$

Initial conditions: $\mathbf{r} = \mathbf{r}_0$ and $\dfrac{d\mathbf{r}}{dt} = \mathbf{v}_0$ when $t = 0$

The first integration gives

$$\frac{d\mathbf{r}}{dt} = -(gt)\,\mathbf{j} + \mathbf{v}_0.$$

A second integration gives

$$\mathbf{r} = -\frac{1}{2}gt^2\,\mathbf{j} + \mathbf{v}_0 t + \mathbf{r}_0.$$

Substituting the values of $\mathbf{v}_0$ and $\mathbf{r}_0$ from Eqs. (2) and (3) gives

$$\mathbf{r} = -\frac{1}{2}gt^2\,\mathbf{j} + \underbrace{(v_0 \cos \alpha)t\,\mathbf{i} + (v_0 \sin \alpha)t\,\mathbf{j}}_{\mathbf{v}_0 t} + \mathbf{0}$$

or

$$\mathbf{r} = (v_0 \cos \alpha)t\,\mathbf{i} + \left((v_0 \sin \alpha)t - \frac{1}{2}gt^2\right)\mathbf{j}. \tag{5}$$

Equation (5) is the *vector equation* for ideal projectile motion. The angle α is the projectile's **launch angle (firing angle, angle of elevation)**, and v_0, as we said before, is the projectile's **initial speed.**

Equation (5) is equivalent to a pair of scalar equations,

$$x = (v_0 \cos \alpha)t, \qquad y = (v_0 \sin \alpha)t - \frac{1}{2}gt^2. \tag{6}$$

These are known as the *parametric equations* for ideal projectile motion. If time is measured in seconds and distance in meters, g is 9.8 m/sec² and Eqs. (6) give x and y in meters. With feet in place of meters, g is 32 ft/sec² and Eqs. (6) give x and y in feet.

EXAMPLE 1   A projectile is fired from the origin over horizontal ground at an initial speed of 500 m/sec at a launch angle of 60°. Where will the projectile be 10 sec later?

Solution   We use Eqs. (6) with $v_0 = 500$, $\alpha = 60°$, $g = 9.8$, and $t = 10$ to find the projectile's coordinates to the nearest meter 10 sec after firing:

$$x = (v_0 \cos \alpha)t = 500 \cdot \frac{1}{2} \cdot 10 = 2500 \text{ m}$$

$$y = (v_0 \sin \alpha)t - \frac{1}{2}gt^2$$

$$= 500 \cdot \frac{\sqrt{3}}{2} \cdot 10 - \frac{1}{2} \cdot 9.8 \cdot (10)^2$$

$$= 2500\sqrt{3} - 490$$

$$\approx 3840 \text{ m}.$$

Ten seconds after firing, the projectile is 3840 m in the air and 2500 m downrange.

Height, Flight Time, and Range

Equations (6) enable us to answer most questions about an ideal projectile fired from the origin.

The projectile reaches its highest point when its vertical velocity component is zero, that is, when

$$\frac{dy}{dt} = v_0 \sin \alpha - gt = 0, \qquad \text{or} \qquad t = \frac{v_0 \sin \alpha}{g}.$$

For this value of t, the value of y is

$$y_{\text{max}} = (v_0 \sin \alpha)\left(\frac{v_0 \sin \alpha}{g}\right) - \frac{1}{2} g \left(\frac{v_0 \sin \alpha}{g}\right)^2 = \frac{(v_0 \sin \alpha)^2}{2g}. \qquad (7)$$

To find when the projectile lands, when fired over horizontal ground, we set y equal to zero in Eqs. (6) and solve for t:

$$(v_0 \sin \alpha)t - \frac{1}{2}gt^2 = 0$$

$$t\left(v_0 \sin \alpha - \frac{1}{2}gt\right) = 0$$

$$t = 0, \qquad t = \frac{2v_0 \sin \alpha}{g}. \qquad (8)$$

Since 0 is the time the projectile is fired, $(2v_0 \sin \alpha)/g$ must be the time when the projectile strikes the ground.

To find the projectile's range R, the distance from the origin to the point of impact on horizontal ground, we find the value of x when $t = (2v_0 \sin \alpha)/g$:

$$x = (v_0 \cos \alpha)t$$

$$R = (v_0 \cos \alpha)\left(\frac{2v_0 \sin \alpha}{g}\right)$$

$$= \frac{v_0^2}{g}(2 \sin \alpha \cos \alpha) = \frac{v_0^2}{g} \sin 2\alpha. \qquad (9)$$

The range is largest when $\sin 2\alpha = 1$ or $\alpha = 45°$.

EXAMPLE 2   Find the maximum height, flight time, and range of a projectile fired from the origin over horizontal ground at an initial speed of 500 m/sec and a launch angle of 60° (same projectile as in Example 1).

Solution

Maximum height (Eq. 7): $y_{max} = \dfrac{(v_0 \sin \alpha)^2}{2g}$

$$= \frac{(500 \sin 60°)^2}{2(9.8)} \approx 9566 \text{ m}$$

Flight time (Eq. 8): $t = \dfrac{2v_0 \sin \alpha}{g}$

$$= \frac{2(500) \sin 60°}{9.8} \approx 88 \text{ sec}$$

Range (Eq. 9): $R = \dfrac{v_0^2}{g} \sin 2\alpha$

$$= \frac{(500)^2 \sin 120°}{9.8} \approx 22{,}092 \text{ m}$$

Ideal Trajectories Are Parabolic

It is often claimed that water from a hose traces a parabola in the air, but anyone who looks closely enough will see this is not so. The air slows the water down, and its forward progress is too slow at the end to match the rate at which it falls.

What is really being claimed is that ideal projectiles move along parabolas, and this we can see from Eqs. (6). If we substitute $t = x/(v_0 \cos \alpha)$ from the first equation into the second, we obtain the Cartesian coordinate equation

$$y = -\left(\frac{g}{2v_0^2 \cos^2 \alpha}\right)x^2 + (\tan \alpha)x.$$

This equation has the form $y = ax^2 + bx$, so its graph is a parabola.

Firing from (x_0, y_0)

If we fire our ideal projectile from the point (x_0, y_0) instead of the origin (Fig. 11.8), the equations that replace Eqs. (6) are

$$x = x_0 + (v_0 \cos \alpha)t, \qquad y = y_0 + (v_0 \sin \alpha)t - \frac{1}{2}gt^2, \qquad (10)$$

as you will be invited to show in Exercise 19.

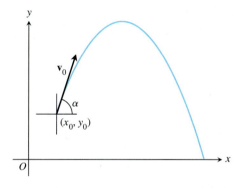

11.8 The path of a projectile fired from (x_0, y_0) with an initial velocity $\mathbf{v}_0$ at an angle of α degrees with the horizontal.

11.9 Spanish archer Antonio Rebollo lights the Olympic torch in Barcelona with a flaming arrow.

EXAMPLE 3   To open the 1992 Summer Olympics in Barcelona, bronze medalist archer Antonio Rebollo lit the Olympic torch with a flaming arrow (Fig. 11.9).

Suppose that Rebollo wanted the arrow to reach its maximum height exactly 4 ft above the center of the cauldron (Fig. 11.10).

a) If he shot the arrow at a height of 6 ft above ground level 30 yd from the 70-ft-high cauldron, express y_{max} in terms of the initial speed v_0 and firing angle α.

b) If $y_{max} = 74$ ft (Fig. 11.10), use the results of part (a) to find the value of $v_0 \sin \alpha$.

c) When the arrow reaches y_{max}, the horizontal distance traveled to the center of the cauldron is $x = 90$ ft. Use this fact to find the value of $v_0 \cos \alpha$.

d) Find the initial firing angle of the arrow.

Solution

a) We use a coordinate system in which the x-axis lies along the ground toward the left (to match the photograph in Fig. 11.9) and the coordinates of the flaming arrow at $t = 0$ are $x_0 = 0$ and $y_0 = 6$ (Fig. 11.10). We have

$$y = y_0 + (v_0 \sin \alpha)t - \frac{1}{2}gt^2 \qquad \text{Eqs. (10)}$$

$$= 6 + (v_0 \sin \alpha)t - \frac{1}{2}gt^2. \qquad y_0 = 6$$

We find the time when the arrow reaches its highest point by setting $dy/dt = 0$ and solving for t, obtaining

$$t = \frac{v_0 \sin \alpha}{g}.$$

For this value of t, the value of y is

$$y_{max} = 6 + (v_0 \sin \alpha)\left(\frac{v_0 \sin \alpha}{g}\right) - \frac{1}{2}g\left(\frac{v_0 \sin \alpha}{g}\right)^2 = 6 + \frac{(v_0 \sin \alpha)^2}{2g}.$$

b) Using $y_{max} = 74$ and $g = 32$, we see from part (a) that

$$74 = 6 + \frac{(v_0 \sin \alpha)^2}{(2)(32)}$$

or

$$v_0 \sin \alpha = \sqrt{(68)(64)}.$$

c) We substitute the time to reach y_{max} from part (a) and the horizontal distance $x = 90$ ft into Eqs. (10) to obtain

$$x = x_0 + (v_0 \cos \alpha)t \qquad \text{Eqs. (10)}$$

$$90 = 0 + (v_0 \cos \alpha)t \qquad x = 90, \quad x_0 = 0$$

$$= (v_0 \cos \alpha)\left(\frac{v_0 \sin \alpha}{g}\right). \qquad t = (v_0 \sin \alpha)/g$$

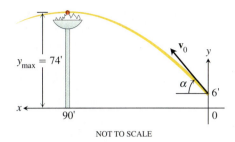

11.10 Ideal path of the arrow that lit the Olympic torch.

NOT TO SCALE

Solving this equation for $v_0 \cos \alpha$ and using the result from part (b), we have

$$v_0 \cos \alpha = \frac{90g}{v_0 \sin \alpha}$$

d) Parts (b) and (c) together tell us that

$$\begin{aligned}
\tan \alpha &= \frac{v_0 \sin \alpha}{v_0 \cos \alpha} \\
&= \frac{(v_0 \sin \alpha)^2}{90g} \\
&= \frac{(68)(64)}{(90)(32)} = \frac{68}{45}
\end{aligned}$$

or

$$\alpha \approx \tan^{-1}\left(\frac{68}{45}\right) \approx 57°.$$

This is Rebollo's firing angle.   ❑

Exercises 11.2

The projectiles in the following exercises are to be treated as ideal projectiles whose behavior is faithfully portrayed by the equations derived in the text. Most of the arithmetic, however, is realistic and is best done with a calculator. All launch angles are assumed to be measured from the horizontal. All projectiles are assumed to be fired from the origin over horizontal ground, unless stated otherwise.

1. A projectile is fired at a speed of 840 m/sec at an angle of 60°. How long will it take to get 21 km downrange?

2. Find the muzzle speed of a gun whose maximum range is 24.5 km.

3. A projectile is fired with an initial speed of 500 m/sec at an angle of elevation of 45°.

   a) When and how far away will the projectile strike?
   b) How high overhead will the projectile be when it is 5 km downrange?
   c) What is the highest the projectile will go?

4. A baseball is thrown from the stands 32 ft above the field at an angle of 30° up from horizontal. When and how far away will the ball strike the ground if its initial speed is 32 ft/sec?

5. An athlete throws a 16-lb shot at an angle of 45° to the horizontal from 6.5 ft above the ground at an initial speed of 44 ft/sec. How long after launch and how far from the inner edge of the stopboard does the shot land?

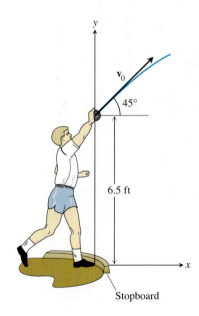

6. Because of its initial evaluation, the shot in Exercise 5 would have gone slightly farther if it had been launched at a 40° angle. How much farther? Answer in inches.

7. A spring gun at ground level fires a golf ball at an angle of 45°. The ball lands 10 m away. What was the ball's initial speed? For the same initial speed, find the two firing angles that make the range 6 m.

8. An electron in a TV tube is beamed horizontally at a speed of 5×10^6 m/sec toward the face of the tube 40 cm away. About how far will the electron drop before it hits?

9. Laboratory tests designed to find how far golf balls of different hardness go when hit with a driver showed that a 100-compression ball hit with a club head speed of 100 mph at a launch angle of 9° carried 248.8 yd. What was the launch speed of the ball? (It was more than 100 mph. At the same time the club head was moving forward, the compressed ball was kicking away from the club face, adding to the ball's forward speed.)

10. A human cannonball is to be fired with an initial speed of $v_0 = 80\sqrt{10}/3$ ft/sec. The circus performer (of the right caliber, naturally) hopes to land on a special cushion located 200 ft downrange. The circus is being held in a large room with a flat ceiling 75 ft high. Can the performer be fired to the cushion without striking the ceiling? If so, what should the cannon's angle of elevation be?

11. A golf ball leaves the ground at a 30° angle at a speed of 90 ft/sec. Will it clear the top of a 30-ft tree 135 ft away?

12. A golf ball is hit from the tee to a green elevated 45 ft above the tee with an initial speed of 116 ft/sec at an angle of elevation of 45°. Assuming that the pin, 369 ft downrange, does not get in the way, where will the ball land in relation to the pin?

13. In Moscow in 1987, Natalya Lisouskaya set a women's world record by putting an 8-lb 13-oz shot 73 ft 10 in. Assuming that she launched the shot at a 40° angle to the horizontal 6.5 ft above the ground, what was the shot's initial speed?

14. A baseball hit by a Boston Red Sox player at a 20° angle from 3 ft above the ground just cleared the left end of the "Green Monster," the left-field wall in Fenway Park (Fig. 11.11). This wall is 37 ft high and 315 ft from home plate. About how fast was the ball going? How long did it take the ball to reach the wall?

15. Show that a projectile fired at an angle of α degrees, $0 < \alpha < 90$, has the same range as a projectile fired at the same speed at an angle of $(90 - \alpha)$ degrees. (In models that take air resistance into account, this symmetry is lost.)

16. What two angles of elevation will enable a projectile to reach a target 16 km downrange on the same level as the gun if the projectile's initial speed is 400 m/sec?

17. Show that doubling a projectile's initial speed at a given launch angle multiplies its range by 4. By about what percentage should you increase the initial speed to double the height and range?

18. Show that a projectile attains three-quarters of its maximum height in half the time it takes to reach the maximum height.

19. Derive the equations

$$x = x_0 + (v_0 \cos \alpha)t$$

$$y = y_0 + (v_0 \sin \alpha)t - \frac{1}{2}gt^2$$

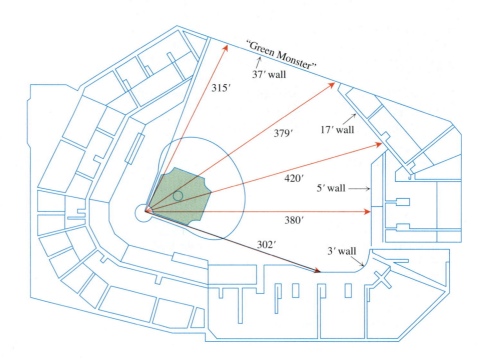

11.11 The "Green Monster," the left-field wall at Fenway Park in Boston (Exercise 14).

(Eqs. 10 in the text) by solving the following initial value problem for a vector **r** in the plane.

Differential equation: $\dfrac{d^2\mathbf{r}}{dt^2} = -g\,\mathbf{j}$

Initial conditions: $\mathbf{r} = x_0\,\mathbf{i} + y_0\,\mathbf{j}$

and

$\dfrac{d\mathbf{r}}{dt} = (v_0\cos\alpha)\,\mathbf{i} + (v_0\sin\alpha)\,\mathbf{j}$

when $t = 0$

20. Using the firing angle $\alpha = 57°$ in Example 3, find the speed with which the flaming arrow left Rebollo's bow. See Fig. 11.10.

21. The cauldron in Example 3 is 12 ft in diameter. Using Eqs. (10) and Example 3(c), find how long it takes the flaming arrow to cover the horizontal distance to the rim. How high is the arrow at this time?

22. The multiflash photograph shown below shows a model train engine moving at a constant speed on a straight track. As the engine moved along, a marble was fired into the air by a spring in the engine's smokestack. The marble, which continued to move with the same forward speed as the engine, rejoined the engine 1 sec after it was fired. Measure the angle the marble's path made with the horizontal and use the information to find how high the marble went and how fast the engine was moving.

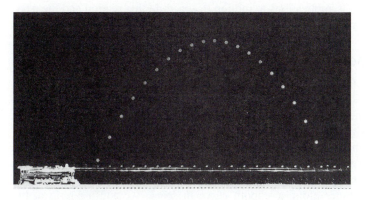

The train in Exercise 22.

23. Figure 11.12 shows an experiment with two marbles. Marble A was launched toward marble B with launch angle α and initial speed v_0. At the same instant, marble B was released to fall from rest $R\tan\alpha$ units directly above a spot R units downrange from A. The marbles were found to collide regardless of the value of v_0. Was this mere coincidence, or must this happen? Give reasons for your answer.

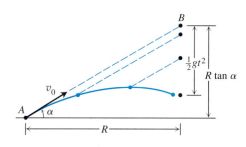

11.12 The marbles in Exercise 23.

24. An ideal projectile is launched straight down an inclined plane, as shown in profile in Fig. 11.13.

a) Show that the greatest downhill range is achieved when the initial velocity vector bisects angle AOR.

b) If the projectile were fired uphill instead of down, what launch angle would maximize its range? Give reasons for your answer.

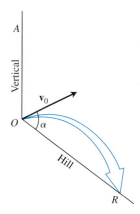

11.13 Maximum downhill range occurs when the velocity vector bisects angle AOR (Exercise 24).

25. An ideal projectile, launched from the origin into the first octant at time $t = 0$ with initial velocity $\mathbf{v}_0$, experiences a constant downward acceleration $\mathbf{a} = -g\mathbf{k}$ from gravity. Find the projectile's velocity and position as functions of t.

26. *Air resistance proportional to velocity.* If a projectile of mass m launched with initial velocity $\mathbf{v}_0$ encounters an air resistance proportional to its velocity, the total force $m(d^2\mathbf{r}/dt^2)$ on the projectile satisfies the equation

$$m\frac{d^2\mathbf{r}}{dt^2} = -mg\,\mathbf{j} - k\frac{d\mathbf{r}}{dt},$$

where k is the proportionality constant. Show that one integration of this equation gives

$$\frac{d\mathbf{r}}{dt} + \frac{k}{m}\mathbf{r} = \mathbf{v}_0 - gt\,\mathbf{j}.$$

Solve this equation. To do so, multiply both sides of the equation by $e^{(k/m)t}$. The left-hand side will then be the derivative of a product, with the result that both sides of the equation can now be integrated.

The function $e^{(k/m)t}$ is called an *integrating factor* for the differential equation because multiplying the equation by it makes the equation integrable.

27. For a projectile fired from the ground at launch angle α with initial speed v_0, consider α as a variable and v_0 as a fixed constant. For each α, $0 < \alpha \le \pi/2$, we obtain a parabolic trajectory (Fig 11.14). Show that the points in the plane that give the maximum heights of these parabolic trajectories all lie on the ellipse

$$x^2 + 4\left(y - \frac{v_0^2}{4g}\right)^2 = \frac{v_0^4}{4g^2},$$

where $x \ge 0$.

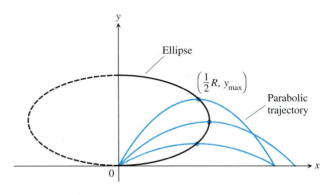

11.14 The parabolas in Exercise 27.

28. GRAPHER If you have access to a parametric equation grapher and have not yet done Exercise 46 in Section 9.4, do it now. It is about ideal projectile motion.

11.3

Arc Length and the Unit Tangent Vector **T**

As you can imagine, differentiable curves, especially those with continuous first and second derivatives, have been subjects of intense study, for their mathematical interest as well as their applications to motion in space. In this section and the next, we study some of the features that account for the importance of these curves.

Arc Length Along a Curve

One of the special features of smooth space curves is that they have a measurable length. This enables us to locate points along these curves by giving their directed distance s along the curve from some **base point,** the way we locate points on coordinate axes by giving their directed distance from the origin (Fig. 11.15). Time is the natural parameter for describing a moving body's velocity and acceleration, but s is the natural parameter for studying a curve's shape. Both parameters appear in analyses of space flight.

11.15 Smooth curves can be scaled like number lines, the coordinate of each point being its directed distance from a preselected base point. .

To measure distance along a smooth curve in space, we add a z-term to the formula we use for curves in the plane.

> ### Definition
> The **length** of a smooth curve $\mathbf{r}(t) = f(t)\,\mathbf{i} + g(t)\,\mathbf{j} + h(t)\,\mathbf{k}$, $a \le t \le b$, that is traced exactly once as t increases from $t = a$ to $t = b$ is
>
> $$L = \int_a^b \sqrt{\left(\frac{df}{dt}\right)^2 + \left(\frac{dg}{dt}\right)^2 + \left(\frac{dh}{dt}\right)^2}\,dt$$
>
> $$= \int_a^b \sqrt{\left(\frac{dx}{dt}\right)^2 + \left(\frac{dy}{dt}\right)^2 + \left(\frac{dz}{dt}\right)^2}\,dt. \qquad (1)$$

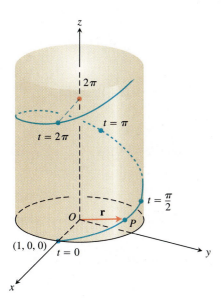

11.16 The helix $\mathbf{r}(t) = (\cos t)\mathbf{i} + (\sin t)\mathbf{j}$ $+ t\mathbf{k}$ in Example 1.

Just as for plane curves, we can calculate the length of a curve in space from any convenient parametrization that meets the stated conditions. Again, we omit the proof.

The square root in Eq. (1) is $|\mathbf{v}|$, the length of the velocity vector $d\mathbf{r}/dt$. This enables us to write the formula for length a shorter way.

Length Formula (Short Form)

$$L = \int_a^b |\mathbf{v}|\, dt \qquad (2)$$

EXAMPLE 1   Find the length of one turn of the helix

$$\mathbf{r}(t) = (\cos t)\mathbf{i} + (\sin t)\mathbf{j} + t\mathbf{k}.$$

Solution   The helix makes one full turn as t runs from 0 to 2π (Fig. 11.16). The length of this portion of the curve is

$$L = \int_a^b |\mathbf{v}|\, dt = \int_0^{2\pi} \sqrt{(-\sin t)^2 + (\cos t)^2 + (1)^2}\, dt$$

$$= \int_0^{2\pi} \sqrt{2}\, dt = 2\pi\sqrt{2}.$$

This is $\sqrt{2}$ times the length of the circle in the xy-plane over which the helix stands. ☐

We use the Greek letter τ ("tau") as the variable of integration in Eq. (3) because the letter t is already in use as the upper limit.

If we choose a base point $P(t_0)$ on a smooth curve C parametrized by t, each value of t determines a point $P(t) = (x(t), y(t), z(t))$ on C and a "directed distance"

$$s(t) = \int_{t_0}^t |\mathbf{v}(\tau)|\, d\tau, \qquad (3)$$

measured along C from the base point (Fig. 11.17). If $t > t_0$, $s(t)$ is the distance from $P(t_0)$ to $P(t)$. If $t < t_0$, $s(t)$ is the negative of the distance. Each value of s determines a point on C and this parametrizes C with respect to s. We call s an **arc length parameter** for the curve. The parameter's value increases in the direction of increasing t.

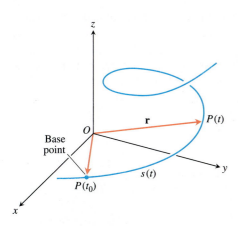

11.17 The directed distance along the curve from $P(t_0)$ to any point $P(t)$ is

$$s(t) = \int_{t_0}^t |\mathbf{v}(\tau)|\, d\tau.$$

Arc Length Parameter with Base Point $P(t_0)$

$$s(t) = \int_{t_0}^t \sqrt{[x'(\tau)]^2 + [y'(\tau)]^2 + [z'(\tau)]^2}\, d\tau = \int_{t_0}^t |\mathbf{v}(\tau)|\, d\tau \qquad (4)$$

EXAMPLE 2   If $t_0 = 0$, the arc length parameter along the helix

$$\mathbf{r}(t) = (\cos t)\mathbf{i} + (\sin t)\mathbf{j} + t\mathbf{k}$$

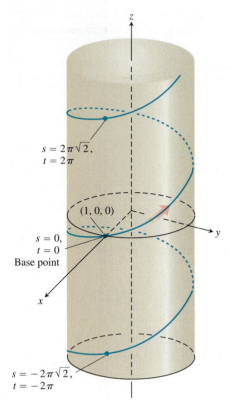

11.18 Arc length parameter values on the helix $\mathbf{r}(t) = (\cos t)\,\mathbf{i} + (\sin t)\,\mathbf{j} + t\,\mathbf{k}$ (Example 2).

from t_0 to t is

$$s(t) = \int_{t_0}^{t} |\mathbf{v}(\tau)|\, d\tau \qquad \text{Eq. (4)}$$

$$= \int_{0}^{t} \sqrt{2}\, d\tau \qquad \text{Value from Example 1}$$

$$= \sqrt{2}\, t.$$

Thus, $s(2\pi) = 2\pi\sqrt{2}$, $s(-2\pi) = -2\pi\sqrt{2}$, and so on (Fig. 11.18). ❏

EXAMPLE 3 *Distance Along a Line.*

Show that if $\mathbf{u} = u_1\,\mathbf{i} + u_2\,\mathbf{j} + u_3\,\mathbf{k}$ is a unit vector, then the directed distance along the line

$$\mathbf{r}(t) = (x_0 + tu_1)\,\mathbf{i} + (y_0 + tu_2)\,\mathbf{j} + (z_0 + tu_3)\,\mathbf{k}$$

from the point $P_0(x_0, y_0, z_0)$ where $t = 0$ is t itself.

Solution

$$\mathbf{v} = \frac{d}{dt}(x_0 + tu_1)\,\mathbf{i} + \frac{d}{dt}(y_0 + tu_2)\,\mathbf{j} + \frac{d}{dt}(z_0 + tu_3)\,\mathbf{k} = u_1\,\mathbf{i} + u_2\,\mathbf{j} + u_3\,\mathbf{k} = \mathbf{u},$$

so

$$s(t) = \int_{0}^{t} |\mathbf{v}|\, d\tau = \int_{0}^{t} |\mathbf{u}|\, d\tau = \int_{0}^{t} 1\, d\tau = t.$$

❏

Speed on a Smooth Curve

Since the derivatives beneath the radical in Eq. (4) are continuous (the curve is smooth), the Fundamental Theorem of Calculus tells us that s is a differentiable function of t with derivative

$$\frac{ds}{dt} = |\mathbf{v}(t)|. \tag{5}$$

As we expect, the speed with which the particle moves along its path is the magnitude of $\mathbf{v}$.

Notice that while the base point $P(t_0)$ plays a role in defining s in Eq. (4), it plays no role in Eq. (5). The rate at which a moving particle covers distance along its path has nothing to do with how far away the base point is.

Notice also that $ds/dt > 0$ since, by definition, $|\mathbf{v}|$ is never zero for a smooth curve. We see once again that s is an increasing function of t.

The Unit Tangent Vector **T**

Since $ds/dt > 0$ for the curves we are considering, s is one-to-one and has an inverse that gives t as a differentiable function of s (Section 6.1). The derivative of the inverse is

$$\frac{dt}{ds} = \frac{1}{ds/dt} = \frac{1}{|\mathbf{v}|}. \tag{6}$$

This makes $\mathbf{r}$ a differentiable function of s whose derivative can be calculated with

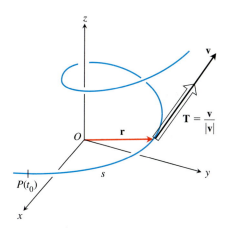

the Chain Rule to be

$$\frac{d\mathbf{r}}{ds} = \frac{d\mathbf{r}}{dt}\frac{dt}{ds} = \mathbf{v}\frac{1}{|\mathbf{v}|} = \frac{\mathbf{v}}{|\mathbf{v}|}. \tag{7}$$

Equation (7) says that $d\mathbf{r}/ds$ is a unit vector in the direction of $\mathbf{v}$. We call $d\mathbf{r}/ds$ the unit tangent vector of the curve traced by $\mathbf{r}$ and denote it by $\mathbf{T}$ (Fig. 11.19).

11.19 We find the unit tangent vector **T** by dividing **v** by |**v**|.

Definition

The **unit tangent vector** of a differentiable curve $\mathbf{r}(t)$ is

$$\mathbf{T} = \frac{d\mathbf{r}}{ds} = \frac{d\mathbf{r}/dt}{ds/dt} = \frac{\mathbf{v}}{|\mathbf{v}|}. \tag{8}$$

The unit tangent vector $\mathbf{T}$ is a differentiable function of t whenever $\mathbf{v}$ is a differentiable function of t. As we will see in the next section, $\mathbf{T}$ is one of three unit vectors in a traveling reference frame that is used to describe the motion of space vehicles and other bodies moving in three dimensions.

EXAMPLE 4   Find the unit tangent vector of the helix

$$\mathbf{r}(t) = (\cos t)\,\mathbf{i} + (\sin t)\,\mathbf{j} + t\,\mathbf{k}.$$

Solution

$$\mathbf{v} = (-\sin t)\,\mathbf{i} + (\cos t)\,\mathbf{j} + \mathbf{k}$$

$$|\mathbf{v}| = \sqrt{(-\sin t)^2 + (\cos t)^2 + (1)^2} = \sqrt{2}$$

$$\mathbf{T} = \frac{\mathbf{v}}{|\mathbf{v}|} = -\frac{\sin t}{\sqrt{2}}\,\mathbf{i} + \frac{\cos t}{\sqrt{2}}\,\mathbf{j} + \frac{1}{\sqrt{2}}\,\mathbf{k} \qquad \square$$

EXAMPLE 5   *The involute of a circle (Fig. 11.20)*

Find the unit tangent vector of the curve

$$\mathbf{r}(t) = (\cos t + t\sin t)\,\mathbf{i} + (\sin t - t\cos t)\,\mathbf{j}, \qquad t > 0.$$

Solution

$$\mathbf{v} = \frac{d\mathbf{r}}{dt} = (-\sin t + \sin t + t\cos t)\,\mathbf{i} + (\cos t - \cos t + t\sin t)\,\mathbf{j}$$

$$= (t\cos t)\,\mathbf{i} + (t\sin t)\,\mathbf{j}$$

$$|\mathbf{v}| = \sqrt{t^2\cos^2 t + t^2\sin^2 t} = \sqrt{t^2} = |t| = t \qquad {\scriptstyle |t| = t \text{ because } t > 0}$$

$$\mathbf{T} = \frac{\mathbf{v}}{|\mathbf{v}|} = \frac{\mathbf{v}}{t} = (\cos t)\,\mathbf{i} + (\sin t)\,\mathbf{j} \qquad \square$$

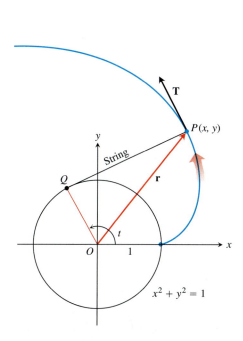

11.20 The *involute* of a circle is the path traced by the endpoint *P* of a string unwinding from a circle, here the unit circle in the *xy*-plane.

EXAMPLE 6   For the counterclockwise motion

$$\mathbf{r}(t) = (\cos t)\,\mathbf{i} + (\sin t)\,\mathbf{j}$$

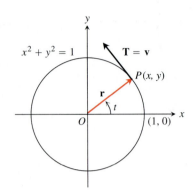

11.21 The motion $\mathbf{r}(t) = (\cos t)\,\mathbf{i} + (\sin t)\,\mathbf{j}$ (Example 6).

around the unit circle,

$$\mathbf{v} = (-\sin t)\,\mathbf{i} + (\cos t)\,\mathbf{j}$$

is already a unit vector, so $\mathbf{T} = \mathbf{v}$ (Fig. 11.21).

Exercises 11.3

In Exercises 1–8, find the curve's unit tangent vector. Also, find the length of the indicated portion of the curve.

1. $\mathbf{r}(t) = (2\cos t)\,\mathbf{i} + (2\sin t)\,\mathbf{j} + \sqrt{5}\,t\,\mathbf{k}, \quad 0 \le t \le \pi$

2. $\mathbf{r}(t) = (6\sin 2t)\,\mathbf{i} + (6\cos 2t)\,\mathbf{j} + 5t\,\mathbf{k}, \quad 0 \le t \le \pi$

3. $\mathbf{r}(t) = t\,\mathbf{i} + (2/3)t^{3/2}\,\mathbf{k}, \quad 0 \le t \le 8$

4. $\mathbf{r}(t) = (2 + t)\,\mathbf{i} - (t + 1)\,\mathbf{j} + t\,\mathbf{k}, \quad 0 \le t \le 3$

5. $\mathbf{r}(t) = (\cos^3 t)\,\mathbf{j} + (\sin^3 t)\,\mathbf{k}, \quad 0 \le t \le \pi/2$

6. $\mathbf{r}(t) = 6t^3\,\mathbf{i} - 2t^3\,\mathbf{j} - 3t^3\,\mathbf{k}, \quad 1 \le t \le 2$

7. $\mathbf{r}(t) = (t\cos t)\,\mathbf{i} + (t\sin t)\,\mathbf{j} + (2\sqrt{2}/3)t^{3/2}\,\mathbf{k}, \quad 0 \le t \le \pi$

8. $\mathbf{r}(t) = (t\sin t + \cos t)\,\mathbf{i} + (t\cos t - \sin t)\,\mathbf{j}, \quad \sqrt{2} \le t \le 2$

9. Find the point on the curve

$$\mathbf{r}(t) = (5\sin t)\,\mathbf{i} + (5\cos t)\,\mathbf{j} + 12t\,\mathbf{k}$$

at a distance 26π units along the curve from the origin in the direction of increasing arc length.

10. Find the point on the curve

$$\mathbf{r}(t) = (12\sin t)\,\mathbf{i} - (12\cos t)\,\mathbf{j} + 5t\,\mathbf{k}$$

at a distance 13π units along the curve from the origin in the direction opposite to the direction of increasing arc length.

In Exercises 11–14, find the arc length parameter along the curve from the point where $t = 0$ by evaluating the integral

$$s = \int_0^t |\mathbf{v}(\tau)|\,d\tau$$

from Eq. (3). Then find the length of the indicated portion of the curve.

11. $\mathbf{r}(t) = (4\cos t)\,\mathbf{i} + (4\sin t)\,\mathbf{j} + 3t\,\mathbf{k}, \quad 0 \le t \le \pi/2$

12. $\mathbf{r}(t) = (\cos t + t\sin t)\,\mathbf{i} + (\sin t - t\cos t)\,\mathbf{j}, \quad \pi/2 \le t \le \pi$

13. $\mathbf{r}(t) = (e^t \cos t)\,\mathbf{i} + (e^t \sin t)\,\mathbf{j} + e^t\,\mathbf{k}, \quad -\ln 4 \le t \le 0$

14. $\mathbf{r}(t) = (1 + 2t)\,\mathbf{i} + (1 + 3t)\,\mathbf{j} + (6 - 6t)\,\mathbf{k}, \quad -1 \le t \le 0$

15. Find the length of the curve

$$\mathbf{r}(t) = (\sqrt{2}\,t)\,\mathbf{i} + (\sqrt{2}\,t)\,\mathbf{j} + (1 - t^2)\,\mathbf{k}$$

from $(0, 0, 1)$ to $(\sqrt{2}, \sqrt{2}, 0)$.

16. The length $2\pi\sqrt{2}$ of the turn of the helix in Example 1 is also the length of the diagonal of a square 2π units on a side. Show how to obtain this square by cutting away and flattening a portion of the cylinder around which the helix winds.

17. a) Show that the curve $\mathbf{r}(t) = (\cos t)\,\mathbf{i} + (\sin t)\,\mathbf{j} + (1 - \cos t)\,\mathbf{k}$, $0 \le t \le 2\pi$, is an ellipse by showing that it is the intersection of a right circular cylinder and a plane. Find equations for the cylinder and plane.

b) Sketch the ellipse on the cylinder. Add to your sketch the unit tangent vectors at $t = 0, \pi/2, \pi$, and $3\pi/2$.

c) Show that the acceleration vector always lies parallel to the plane (orthogonal to a vector normal to the plane). Thus, if you draw the acceleration as a vector attached to the ellipse, it will lie in the plane of the ellipse. Add the acceleration vectors for $t = 0, \pi/2, \pi$, and $3\pi/2$ to your sketch.

d) Write an integral for the length of the ellipse. Do not try to evaluate the integral—it is nonelementary.

e)   NUMERICAL INTEGRATOR Estimate the length of the ellipse to two decimal places.

18. *Length is independent of parametrization.* To illustrate the fact that the length of a smooth space curve does not depend on the parametrization you use to compute it, calculate the length of one turn of the helix in Example 1 with the following parametrizations.

a) $\mathbf{r}(t) = (\cos 4t)\,\mathbf{i} + (\sin 4t)\,\mathbf{j} + 4t\,\mathbf{k}, \quad 0 \le t \le \pi/2$

b) $\mathbf{r}(t) = [\cos (t/2)]\,\mathbf{i} + [\sin (t/2)]\,\mathbf{j} + (t/2)\,\mathbf{k}, \quad 0 \le t \le 4\pi$

c) $\mathbf{r}(t) = (\cos t)\,\mathbf{i} - (\sin t)\,\mathbf{j} - t\,\mathbf{k}, \quad -2\pi \le t \le 0$

| **11.4** | # Curvature, Torsion, and the **TNB** Frame |

In this section we define a frame of mutually orthogonal unit vectors that always travels with a body moving along a curve in space (Fig. 11.22). The frame has three vectors. The first is **T**, the unit tangent vector. The second is **N**, the unit vector that gives the direction of $d\mathbf{T}/ds$. The third is $\mathbf{B} = \mathbf{T} \times \mathbf{N}$. These vectors and their derivatives, when available, give useful information about a vehicle's orientation in space and about how the vehicle's path turns and twists.

For example, $|d\mathbf{T}/ds|$ tells how much a vehicle's path turns to the left or right as it moves along; it is called the *curvature* of the vehicle's path. The number $-(d\mathbf{B}/ds) \cdot \mathbf{N}$ tells how much a vehicle's path rotates or twists out of its plane of motion as the vehicle moves along; it is called the *torsion* of the vehicle's path. Look at Fig. 11.22 again. If P is a train climbing up a curved track, the rate at which the headlight turns from side to side per unit distance is the curvature of the track. The rate at which the engine tends to twist out of the plane formed by **T** and **N** is the torsion.

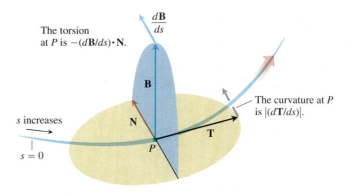

11.22 Every moving body travels with a **TNB** frame that characterizes the geometry of its path of motion.

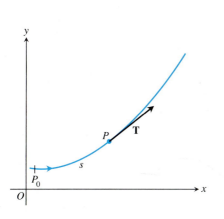

11.23 As P moves along the curve in the direction of increasing arc length, the unit tangent vector turns. The value of $|d\mathbf{T}/ds|$ at P is called the *curvature* of the curve at P.

The Curvature of a Plane Curve

As a particle moves along a smooth curve in the plane, $\mathbf{T} = d\mathbf{r}/ds$ turns as the curve bends. Since **T** is a unit vector, its length remains constant and only its direction changes as the particle moves along the curve. The rate at which **T** turns per unit of length along the curve is called the *curvature* (Fig. 11.23). The traditional symbol for the curvature function is the Greek letter κ ("kappa").

Definition

If $\mathbf{T}$ is the unit tangent vector of a smooth curve, the **curvature** function of the curve is

$$\kappa = \left| \frac{d\mathbf{T}}{ds} \right|.$$

If $|d\mathbf{T}/ds|$ is large, $\mathbf{T}$ turns sharply as the particle passes through P, and the curvature at P is large. If $|d\mathbf{T}/ds|$ is close to zero, $\mathbf{T}$ turns more slowly and the curvature at P is smaller. Testing the definition, we see in Examples 1 and 2 that the curvature is constant for straight lines and circles.

EXAMPLE 1   *The curvature of a straight line is zero*

On a straight line, the unit tangent vector $\mathbf{T}$ always points in the same direction, so its components are constants. Therefore $|d\mathbf{T}/ds| = |\mathbf{0}| = 0$ (Fig. 11.24).   ❑

EXAMPLE 2   *The curvature of a circle of radius a is 1/a*

To see why (Fig. 11.25), start with the parametrization

$$\mathbf{r}(\theta) = (a \cos \theta)\,\mathbf{i} + (a \sin \theta)\,\mathbf{j}$$

and substitute $\theta = s/a$ to parametrize in terms of arc length s.

$$\mathbf{r} = \left(a \cos \frac{s}{a} \right)\mathbf{i} + \left(a \sin \frac{s}{a} \right)\mathbf{j}.$$

Then

$$\mathbf{T} = \frac{d\mathbf{r}}{ds} = \left(-\sin \frac{s}{a} \right)\mathbf{i} + \left(\cos \frac{s}{a} \right)\mathbf{j}$$

and

$$\frac{d\mathbf{T}}{ds} = \left(-\frac{1}{a}\cos \frac{s}{a} \right)\mathbf{i} - \left(\frac{1}{a}\sin \frac{s}{a} \right)\mathbf{j}.$$

Hence, for any value of s,

$$\kappa = \left| \frac{d\mathbf{T}}{ds} \right|$$

$$= \sqrt{\frac{1}{a^2}\cos^2\left(\frac{s}{a}\right) + \frac{1}{a^2}\sin^2\left(\frac{s}{a}\right)}$$

$$= \frac{1}{\sqrt{a^2}} = \frac{1}{|a|} = \frac{1}{a}. \qquad \text{Since } a > 0, \ |a| = a.$$ ❑

11.24 Along a straight line, $\mathbf{T}$ always points in the same direction. The curvature, $|d\mathbf{T}/ds|$, is zero (Example 1).

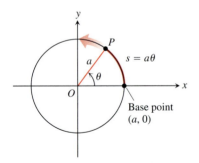

11.25 The point P has coordinates $(a \cos \theta, a \sin \theta) = (a \cos (s/a), \ a \sin (s/a))$ (Example 2).

The Principal Unit Normal Vector for Plane Curves

Since $\mathbf{T}$ has constant length, the vector $d\mathbf{T}/ds$ is orthogonal to $\mathbf{T}$ (Section 11.1). Therefore, if we divide $d\mathbf{T}/ds$ by the length κ, we obtain a *unit* vector orthogonal to $\mathbf{T}$ (Fig. 11.26).

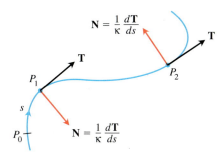

$$N = \frac{1}{\kappa} \frac{d\mathbf{T}}{ds}$$

11.26 The vector $d\mathbf{T}/ds$, normal to the curve, always points in the direction in which **T** is turning. The vector **N** is the direction of $d\mathbf{T}/ds$.

Definition

At a point where $\kappa \neq 0$, the **principal unit normal** vector for a curve in the plane is

$$\mathbf{N} = \frac{1}{\kappa} \frac{d\mathbf{T}}{ds}.$$

The vector $d\mathbf{T}/ds$ points in the direction in which **T** turns as the curve bends. Therefore, if we face in the direction of increasing arc length, the vector $d\mathbf{T}/ds$ points toward the right if **T** turns clockwise and toward the left if **T** turns counterclockwise. In other words, the principal normal vector **N** will point toward the concave side of the curve (Fig. 11.26). Exercise 10 illustrates what happens when $\kappa = 0$ at a point.

Because the arc length parameter for a smooth curve $\mathbf{r}(t) = f(t)\mathbf{i} + g(t)\mathbf{j}$ is defined with ds/dt positive, $ds/dt = |ds/dt|$, and the Chain Rule gives

$$\mathbf{N} = \frac{d\mathbf{T}/ds}{|d\mathbf{T}/ds|}$$

$$= \frac{(d\mathbf{T}/dt)(dt/ds)}{|d\mathbf{T}/dt||dt/ds|}$$

$$= \frac{d\mathbf{T}/dt}{|d\mathbf{T}/dt|}. \tag{1}$$

This formula enables us to find **N** without having to find κ and s first.

EXAMPLE 3   Find **T** and **N** for the circular motion

$$\mathbf{r}(t) = (\cos 2t)\mathbf{i} + (\sin 2t)\mathbf{j}.$$

Solution   We first find **T**:

$$\mathbf{v} = -(2\sin 2t)\mathbf{i} + (2\cos 2t)\mathbf{j},$$

$$|\mathbf{v}| = \sqrt{4\sin^2 2t + 4\cos^2 2t} = 2,$$

$$\mathbf{T} = \frac{\mathbf{v}}{|\mathbf{v}|}$$

$$= -(\sin 2t)\mathbf{i} + (\cos 2t)\mathbf{j}.$$

From this we find

$$\frac{d\mathbf{T}}{dt} = -(2\cos 2t)\mathbf{i} - (2\sin 2t)\mathbf{j},$$

$$\left|\frac{d\mathbf{T}}{dt}\right| = \sqrt{4\cos^2 2t + 4\sin^2 2t} = 2,$$

and

$$\mathbf{N} = \frac{d\mathbf{T}/dt}{|d\mathbf{T}/dt|}$$

$$= -(\cos 2t)\mathbf{i} - (\sin 2t)\mathbf{j}. \qquad \text{Eq. (1)} \qquad \square$$

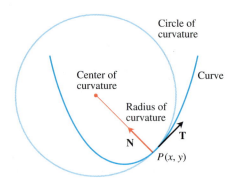

11.27 The osculating circle at $P(x, y)$ lies toward the inner side of the curve.

Circle of Curvature and Radius of Curvature

The **circle of curvature** or **osculating circle** at a point P on a plane curve where $\kappa \neq 0$ is the circle in the plane of the curve that

1. is tangent to the curve at P (has the same tangent line the curve has);
2. has the same curvature the curve has at P; and
3. lies toward the concave or inner side of the curve (as in Fig. 11.27).

The **radius of curvature** of the curve at P is the radius of the circle of curvature, which, according to Example 2, is

$$\text{Radius of curvature} = \rho = \frac{1}{\kappa}. \qquad (2)$$

To find ρ, we find κ and take the reciprocal. The **center of curvature** of the curve at P is the center of the circle of curvature.

Curvature and Normal Vectors for Space Curves

Just as it does for a curve in the plane, the arc length parameter s gives the unit tangent vector $\mathbf{T} = d\mathbf{r}/ds$ for a smooth curve in space. We again define the curvature to be

$$\kappa = \left| \frac{d\mathbf{T}}{ds} \right|. \qquad (3)$$

The vector $d\mathbf{T}/ds$ is orthogonal to $\mathbf{T}$ and we define the principal unit normal to be

$$\mathbf{N} = \frac{1}{\kappa} \frac{d\mathbf{T}}{ds} = \frac{d\mathbf{T}/dt}{|d\mathbf{T}/dt|}. \qquad (4)$$

EXAMPLE 4   Find the curvature for the helix (Fig. 11.28)

$$\mathbf{r}(t) = (a \cos t)\mathbf{i} + (a \sin t)\mathbf{j} + bt\,\mathbf{k}, \qquad a, b \geq 0, \ a^2 + b^2 \neq 0.$$

Solution   We calculate $\mathbf{T}$ from the velocity vector $\mathbf{v}$:

$$\mathbf{v} = -(a \sin t)\mathbf{i} + (a \cos t)\mathbf{j} + b\,\mathbf{k}$$

$$|\mathbf{v}| = \sqrt{a^2 \sin^2 t + a^2 \cos^2 t + b^2} = \sqrt{a^2 + b^2}$$

$$\mathbf{T} = \frac{\mathbf{v}}{|\mathbf{v}|} = \frac{1}{\sqrt{a^2 + b^2}}[-(a \sin t)\mathbf{i} + (a \cos t)\mathbf{j} + b\,\mathbf{k}].$$

Then, using the Chain Rule, we find $d\mathbf{T}/ds$ as

$$\frac{d\mathbf{T}}{ds} = \frac{d\mathbf{T}}{dt} \frac{dt}{ds} \qquad \color{teal}{\text{Chain Rule}}$$

$$= \frac{d\mathbf{T}}{dt} \cdot \frac{1}{|\mathbf{v}|} \qquad \color{teal}{\frac{ds}{dt} = |\mathbf{v}|, \text{ so } \frac{dt}{ds} = \frac{1}{|\mathbf{v}|}}$$

$$= \frac{1}{\sqrt{a^2 + b^2}}[-(a \cos t)\mathbf{i} - (a \sin t)\mathbf{j}] \cdot \left(\frac{1}{\sqrt{a^2 + b^2}} \right)$$

$$= \frac{a}{a^2 + b^2}[-(\cos t)\mathbf{i} - (\sin t)\mathbf{j}].$$

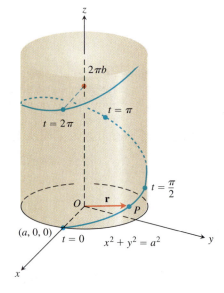

11.28 The helix $\mathbf{r}(t) = (a \cos t)\mathbf{i} + (a \sin t)\mathbf{j} + bt\,\mathbf{k}$, drawn with a and b positive and $t \geq 0$ (Example 4).

Therefore,

$$\kappa = \left| \frac{d\mathbf{T}}{ds} \right|$$

$$= \frac{a}{a^2 + b^2} \left| -(\cos t)\,\mathbf{i} - (\sin t)\,\mathbf{j} \right|$$

$$= \frac{a}{a^2 + b^2} \sqrt{(\cos t)^2 + (\sin t)^2} = \frac{a}{a^2 + b^2}. \tag{5}$$

From Eq. (5) we see that increasing b for a fixed a decreases the curvature. Decreasing a for a fixed b eventually decreases the curvature as well. Stretching a spring tends to straighten it.

If $b = 0$, the helix reduces to a circle of radius a and its curvature reduces to $1/a$, as it should. If $a = 0$, the helix becomes the z-axis, and its curvature reduces to 0, again as it should. ❏

EXAMPLE 5   Find **N** for the helix in Example 4.

Solution   We have

$$\frac{d\mathbf{T}}{dt} = -\frac{1}{\sqrt{a^2 + b^2}} [(a \cos t)\,\mathbf{i} + (a \sin t)\,\mathbf{j}] \qquad \text{Example 4}$$

$$\left| \frac{d\mathbf{T}}{dt} \right| = \frac{1}{\sqrt{a^2 + b^2}} \sqrt{a^2 \cos^2 t + a^2 \sin^2 t} = \frac{a}{\sqrt{a^2 + b^2}}$$

$$\mathbf{N} = \frac{d\mathbf{T}/dt}{|d\mathbf{T}/dt|} \qquad \text{Eq. (4)}$$

$$= -\frac{\sqrt{a^2 + b^2}}{a} \cdot \frac{1}{\sqrt{a^2 + b^2}} [(a \cos t)\,\mathbf{i} + (a \sin t)\,\mathbf{j}]$$

$$= -(\cos t)\,\mathbf{i} - (\sin t)\,\mathbf{j}. \qquad ❏$$

Torsion and the Binormal Vector

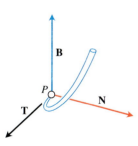

11.29 The vectors **T, N,** and **B** (in that order) make a right-handed frame of mutually orthogonal unit vectors in space. You can call it the **Frenet** ("fre-*nay*") **frame** (after Jean-Frédéric Frenet, 1816–1900), or you can call it the **TNB frame.**

The **binormal vector** of a curve in space is $\mathbf{B} = \mathbf{T} \times \mathbf{N}$, a unit vector orthogonal to both **T** and **N** (Fig. 11.29). Together **T, N,** and **B** define a moving right-handed vector frame that plays a significant role in calculating the flight paths of space vehicles.

How does $d\mathbf{B}/ds$ behave in relation to **T, N,** and **B**? From the rule for differentiating a cross product, we have

$$\frac{d\mathbf{B}}{ds} = \frac{d\mathbf{T}}{ds} \times \mathbf{N} + \mathbf{T} \times \frac{d\mathbf{N}}{ds}.$$

Since **N** is the direction of $d\mathbf{T}/ds$, $(d\mathbf{T}/ds) \times \mathbf{N} = \mathbf{0}$ and

$$\frac{d\mathbf{B}}{ds} = \mathbf{0} + \mathbf{T} \times \frac{d\mathbf{N}}{ds} = \mathbf{T} \times \frac{d\mathbf{N}}{ds}. \tag{6}$$

From this we see that $d\mathbf{B}/ds$ is orthogonal to **T** since a cross product is orthogonal to its factors.

Since $d\mathbf{B}/ds$ is also orthogonal to **B** (the latter has constant length), it follows that $d\mathbf{B}/ds$ is orthogonal to the plane of **B** and **T.** In other words, $d\mathbf{B}/ds$ is parallel to **N,** so $d\mathbf{B}/ds$ is a scalar multiple of **N.** In symbols,

$$\frac{d\mathbf{B}}{ds} = -\tau\,\mathbf{N}.$$

The minus sign in this equation is traditional. The scalar τ is called the torsion along the curve. Notice that

$$\frac{d\mathbf{B}}{ds} \cdot \mathbf{N} = -\tau\,\mathbf{N} \cdot \mathbf{N} = -\tau\,(1) = -\tau,$$

so that

$$\tau = -\frac{d\mathbf{B}}{ds} \cdot \mathbf{N}.$$

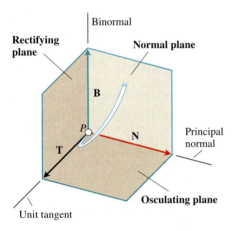

Binormal

Rectifying plane

Normal plane

B

P

N

Principal normal

T

Osculating plane

Unit tangent

11.30 The names of the three planes determined by **T, N,** and **B.**

Definition

Let $\mathbf{B} = \mathbf{T} \times \mathbf{N}$. The **torsion** function of a smooth curve is

$$\tau = -\frac{d\mathbf{B}}{ds} \cdot \mathbf{N}.$$

Unlike the curvature κ, which is never negative, the torsion τ may be positive, negative, or zero.

The three planes determined by **T, N,** and **B** are shown in Fig. 11.30. The curvature $\kappa = |d\mathbf{T}/ds|$ can be thought of as the rate at which the normal plane turns as the point P moves along the curve. Similarly, the torsion $\tau = -(d\mathbf{B}/ds) \cdot \mathbf{N}$ is the rate at which the osculating plane turns about **T** as P moves along the curve. Torsion measures how the curve twists.

The Tangential and Normal Components of Acceleration

When a body is accelerated by gravity, brakes, a combination of rocket motors, or whatever, we usually want to know how much of the acceleration acts to move the body straight ahead in the direction of motion, in the tangential direction **T.** We can find out if we use the Chain Rule to rewrite **v** as

$$\mathbf{v} = \frac{d\mathbf{r}}{dt} = \frac{d\mathbf{r}}{ds}\frac{ds}{dt} = \mathbf{T}\frac{ds}{dt}$$

and differentiate both ends of this string of equalities to get

$$\mathbf{a} = \frac{d\mathbf{v}}{dt} = \frac{d}{dt}\left(\mathbf{T}\frac{ds}{dt}\right) = \frac{d^2s}{dt^2}\mathbf{T} + \frac{ds}{dt}\frac{d\mathbf{T}}{dt}$$

$$= \frac{d^2s}{dt^2}\mathbf{T} + \frac{ds}{dt}\left(\frac{d\mathbf{T}}{ds}\frac{ds}{dt}\right) = \frac{d^2s}{dt^2}\mathbf{T} + \frac{ds}{dt}\left(\kappa\,\mathbf{N}\frac{ds}{dt}\right)$$

$$= \frac{d^2s}{dt^2}\mathbf{T} + \kappa\left(\frac{ds}{dt}\right)^2\mathbf{N}.$$

$$\mathbf{a} = a_{\mathrm{T}}\mathbf{T} + a_{\mathrm{N}}\mathbf{N}, \tag{7}$$

where

$$a_{\mathrm{T}} = \frac{d^2 s}{dt^2} = \frac{d}{dt}|\mathbf{v}| \quad \text{and} \quad a_{\mathrm{N}} = \kappa \left(\frac{ds}{dt}\right)^2 = \kappa |\mathbf{v}|^2 \tag{8}$$

are the **tangential** and **normal** scalar components of acceleration.

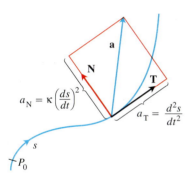

11.31 The tangential and normal components of acceleration. The acceleration **a** always lies in the plane of **T** and **N**, orthogonal to **B**.

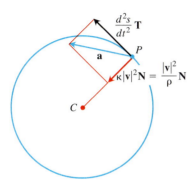

11.32 The tangential and normal components of the acceleration of a body that is speeding up as it moves counterclockwise around a circle of radius ρ.

Equation (7) is remarkable in that **B** does not appear. No matter how the path of the moving body we are watching may appear to twist and turn in space, the acceleration **a** *always lies in the plane of* **T** and **N** orthogonal to **B**. The equation also tells us exactly how much of the acceleration takes place tangent to the motion $(d^2 s/dt^2)$ and how much takes place normal to the motion $[\kappa(ds/dt)^2]$ (Fig. 11.31).

What information can we read from Eqs. (8)? By definition, acceleration **a** is the rate of change of velocity **v**, and in general both the length and direction of **v** change as a body moves along its path. The tangential component of acceleration a_{T} measures the rate of change of the *length* of **v** (that is, the change in the speed). The normal component of acceleration a_{N} measures the rate of change of the *direction* of **v**.

Notice that the normal scalar component of the acceleration is the curvature times the *square* of the speed. This explains why you have to hold on when your car makes a sharp (large κ), high-speed (large $|\mathbf{v}|$) turn. If you double the speed of your car, you will experience four times the normal component of acceleration for the same curvature.

If a body moves in a circle at a constant speed, $d^2 s/dt^2$ is zero and all the acceleration points along **N** toward the circle's center. If the body is speeding up or slowing down, **a** has a nonzero tangential component (Fig. 11.32).

To calculate a_{N} we usually use the formula $a_{\mathrm{N}} = \sqrt{|\mathbf{a}|^2 - a_{\mathrm{T}}^2}$, which comes from solving the equation $|\mathbf{a}|^2 = \mathbf{a} \cdot \mathbf{a} = a_{\mathrm{T}}^2 + a_{\mathrm{N}}^2$ for a_{N}. With this formula we can find a_{N} without having to calculate κ first.

$$a_{\mathrm{N}} = \sqrt{|\mathbf{a}|^2 - a_{\mathrm{T}}^2} \tag{9}$$

EXAMPLE 6   Without finding **T** and **N**, write the acceleration of the motion

$$\mathbf{r}(t) = (\cos t + t \sin t)\mathbf{i} + (\sin t - t \cos t)\mathbf{j}, \qquad t > 0$$

in the form $\mathbf{a} = a_{\mathrm{T}}\mathbf{T} + a_{\mathrm{N}}\mathbf{N}$. (The path of the motion is the involute of the circle in Fig. 11.33, on the following page.)

Solution   We use the first of Eqs. (8) to find a_{T}:

$$\mathbf{v} = (t \cos t)\mathbf{i} + (t \sin t)\mathbf{j} \qquad \text{Value from Section 11.3, Example 5}$$

$$|\mathbf{v}| = \sqrt{t^2 \cos^2 t + t^2 \sin^2 t} = \sqrt{t^2} = |t| = t \qquad t > 0$$

$$a_{\mathrm{T}} = \frac{d}{dt}|\mathbf{v}| = \frac{d}{dt}(t) = 1. \qquad \text{Eq. (8)}$$

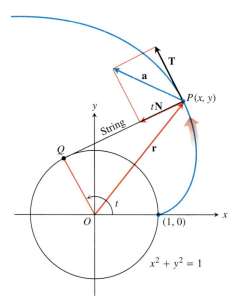

11.33 The tangential and normal components of the acceleration of the motion $r(t) = (\cos t + t \sin t)\,\mathbf{i} + (\sin t - t \cos t)\,\mathbf{j}$, for $t > 0$ (Example 6).

Knowing a_T, we use Eq. (9) to find a_N:

$$\mathbf{a} = (\cos t - t \sin t)\,\mathbf{i} + (\sin t + t \cos t)\,\mathbf{j}$$

$$|\mathbf{a}|^2 = t^2 + 1 \qquad \text{After some algebra}$$

$$a_N = \sqrt{|\mathbf{a}|^2 - a_T{}^2}$$

$$= \sqrt{(t^2 + 1) - (1)} = \sqrt{t^2} = t.$$

We then use Eq. (7) to find $\mathbf{a}$:

$$\mathbf{a} = a_T\,\mathbf{T} + a_N\,\mathbf{N} = (1)\,\mathbf{T} + (t)\,\mathbf{N} = \mathbf{T} + t\,\mathbf{N}.$$

See Fig. 11.33. ◻

Formulas for Computing Curvature and Torsion

We now give some easy-to-use formulas for computing the curvature and torsion of a smooth curve. From Eq. (7), we have

$$\mathbf{v} \times \mathbf{a} = \left(\frac{ds}{dt}\,\mathbf{T}\right) \times \left[\frac{d^2 s}{dt^2}\,\mathbf{T} + \kappa \left(\frac{ds}{dt}\right)^2 \mathbf{N}\right] \qquad \begin{array}{l}\text{From Section}\\ \text{11.3, Eq. (8),}\\ \mathbf{v} = d\mathbf{r}/dt =\\ (ds/dt)\,\mathbf{T}\end{array}$$

$$= \left(\frac{ds}{dt}\frac{d^2 s}{dt^2}\right)(\mathbf{T} \times \mathbf{T}) + \kappa \left(\frac{ds}{dt}\right)^3 (\mathbf{T} \times \mathbf{N})$$

$$= \kappa \left(\frac{ds}{dt}\right)^3 \mathbf{B}. \qquad \begin{array}{l}\mathbf{T} \times \mathbf{T} = \mathbf{0}\text{ and}\\ \mathbf{T} \times \mathbf{N} = \mathbf{B}\end{array}$$

It follows that

$$|\mathbf{v} \times \mathbf{a}| = \kappa \left|\frac{ds}{dt}\right|^3 |\mathbf{B}| = \kappa\,|\mathbf{v}|^3. \qquad \frac{ds}{dt} = |\mathbf{v}|\text{ and }|\mathbf{B}| = 1$$

Solving for κ gives the following formula.

A Vector Formula for Curvature

$$\kappa = \frac{|\mathbf{v} \times \mathbf{a}|}{|\mathbf{v}|^3} \qquad\qquad (10)$$

Equation (10) calculates the curvature, a geometric property of the curve, from the velocity and acceleration of any vector representation of the curve in which $|\mathbf{v}|$ is different from zero. Take a moment to think about how remarkable this really is: From any formula for motion along a curve, no matter how variable the motion may be (as long as $\mathbf{v}$ is never zero), we can calculate a physical property of the curve that seems to have nothing to do with the way the curve is traversed.

The most widely used formula for torsion, derived in more advanced texts, is

Newton's dot notation for derivatives

The dots in Eq. (11) denote differentiation with respect to t, one derivative for each dot. Thus, $\dot{x}$ ("x dot") means dx/dt, $\ddot{x}$ ("x double dot") means $d^2 x/dt^2$, and $\dddot{x}$ ("x triple dot") means $d^3 x/dt^3$. Similarly, $\dot{y} = dy/dt$ and so on.

$$\tau = \frac{\begin{vmatrix} \dot{x} & \dot{y} & \dot{z} \\ \ddot{x} & \ddot{y} & \ddot{z} \\ \dddot{x} & \dddot{y} & \dddot{z} \end{vmatrix}}{|\mathbf{v} \times \mathbf{a}|^2} \qquad (\text{if } \mathbf{v} \times \mathbf{a} \neq \mathbf{0}). \qquad (11)$$

This formula calculates the torsion directly from the derivatives of the component functions $x = f(t)$, $y = g(t)$, $z = h(t)$ that make up **r.** The determinant's first row comes from **v,** the second row comes from **a,** and the third row comes from $\dot{\mathbf{a}} = d\mathbf{a}/dt$.

EXAMPLE 7   Use Eqs. (10) and (11) to find κ and τ for the helix

$$\mathbf{r}(t) = (a\cos t)\,\mathbf{i} + (a\sin t)\,\mathbf{j} + bt\,\mathbf{k}, \qquad a, b \geq 0, \; a^2 + b^2 \neq 0.$$

Solution   We calculate the curvature with Eq. (10):

$$\mathbf{v} = -(a\sin t)\,\mathbf{i} + (a\cos t)\,\mathbf{j} + b\,\mathbf{k},$$

$$\mathbf{a} = -(a\cos t)\,\mathbf{i} - (a\sin t)\,\mathbf{j},$$

$$\mathbf{v} \times \mathbf{a} = \begin{vmatrix} \mathbf{i} & \mathbf{j} & \mathbf{k} \\ -a\sin t & a\cos t & b \\ -a\cos t & -a\sin t & 0 \end{vmatrix}$$

$$= (ab\sin t)\,\mathbf{i} - (ab\cos t)\,\mathbf{j} + a^2\,\mathbf{k},$$

$$\kappa = \frac{|\mathbf{v} \times \mathbf{a}|}{|\mathbf{v}|^3} = \frac{\sqrt{a^2 b^2 + a^4}}{(a^2 + b^2)^{3/2}} = \frac{a\sqrt{a^2 + b^2}}{(a^2 + b^2)^{3/2}} = \frac{a}{a^2 + b^2}. \tag{12}$$

Notice that Eq. (12) agrees with Eq. (5) in Example 4, where we calculated the curvature directly from its definition.

To evaluate Eq. (11) for the torsion, we find the entries in the determinant by differentiating **r** with respect to t. We already have **v** and **a,** and

$$\dot{\mathbf{a}} = \frac{d\mathbf{a}}{dt} = (a\sin t)\,\mathbf{i} - (a\cos t)\,\mathbf{j}.$$

Hence,

$$\tau = \frac{\begin{vmatrix} \dot{x} & \dot{y} & \dot{z} \\ \ddot{x} & \ddot{y} & \ddot{z} \\ \dddot{x} & \dddot{y} & \dddot{z} \end{vmatrix}}{|\mathbf{v} \times \mathbf{a}|^2} = \frac{\begin{vmatrix} -a\sin t & a\cos t & b \\ -a\cos t & -a\sin t & 0 \\ a\sin t & -a\cos t & 0 \end{vmatrix}}{(a\sqrt{a^2 + b^2}\,)^2} \qquad \begin{array}{l}\text{Value of}\\ |\mathbf{v} \times \mathbf{a}|\\ \text{from Eq. (12)}\end{array}$$

$$= \frac{b(a^2\cos^2 t + a^2\sin^2 t)}{a^2(a^2 + b^2)}$$

$$= \frac{b}{a^2 + b^2}. \tag{13}$$

From Eq. (13), we see that the torsion of a helix about a circular cylinder is constant. In fact, constant curvature and constant torsion characterize the helix among all curves in space.

The DNA molecule, the basic building block of life forms, is designed in the form of two helices winding around each other, a little like the rungs and sides of a twisted rope ladder (Fig. 11.34). Not only is the space occupied by the DNA molecule very much smaller than it would be if it were unraveled, but when the molecule is damaged the imperfect piece can be snipped out by a kind of molecular scissors (because the curvature and torsion functions are constant) and the DNA made right again.

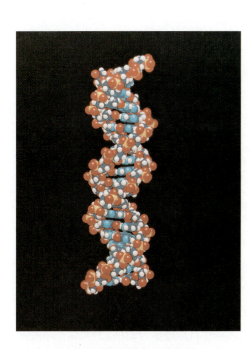

11.34 The helical shape of a DNA molecule is characterized by its constant curvature and torsion.

Formulas for Curves in Space

Unit tangent vector:	$\mathbf{T} = \dfrac{\mathbf{v}}{\|\mathbf{v}\|}$
Principal unit normal vector:	$\mathbf{N} = \dfrac{d\mathbf{T}/dt}{\|d\mathbf{T}/dt\|}$
Binormal vector:	$\mathbf{B} = \mathbf{T} \times \mathbf{N}$
Curvature:	$\kappa = \left| \dfrac{d\mathbf{T}}{ds} \right| = \dfrac{\|\mathbf{v} \times \mathbf{a}\|}{\|\mathbf{v}\|^3}$

Torsion:

$$\tau = -\frac{d\mathbf{B}}{ds} \cdot \mathbf{N} = \frac{\begin{vmatrix} \dot{x} & \dot{y} & \dot{z} \\ \ddot{x} & \ddot{y} & \ddot{z} \\ \dddot{x} & \dddot{y} & \dddot{z} \end{vmatrix}}{|\mathbf{v} \times \mathbf{a}|^2}$$

Tangential and normal scalar components of acceleration:

$$\mathbf{a} = a_{\mathrm{T}}\,\mathbf{T} + a_{\mathrm{N}}\,\mathbf{N}$$

$$a_{\mathrm{T}} = \frac{d}{dt}|\mathbf{v}|$$

$$a_{\mathrm{N}} = \kappa\,|\mathbf{v}|^2 = \sqrt{|\mathbf{a}|^2 - a_{\mathrm{T}}^2}$$

Exercises 11.4

Plane Curves

Find $\mathbf{T}$, $\mathbf{N}$, and κ for the plane curves in Exercises 1–4.

1. $\mathbf{r}(t) = t\,\mathbf{i} + (\ln \cos t)\,\mathbf{j}, \quad -\pi/2 < t < \pi/2$

2. $\mathbf{r}(t) = (\ln \sec t)\,\mathbf{i} + t\,\mathbf{j}, \quad -\pi/2 < t < \pi/2$

3. $\mathbf{r}(t) = (2t + 3)\,\mathbf{i} + (5 - t^2)\,\mathbf{j}$

4. $\mathbf{r}(t) = (\cos t + t \sin t)\,\mathbf{i} + (\sin t - t \cos t)\,\mathbf{j}, \quad t > 0$

In Exercises 5 and 6, write $\mathbf{a}$ in the form $\mathbf{a} = a_{\mathrm{T}}\,\mathbf{T} + a_{\mathrm{N}}\,\mathbf{N}$ without finding $\mathbf{T}$ and $\mathbf{N}$.

5. $\mathbf{r}(t) = (2t + 3)\,\mathbf{i} + (t^2 - 1)\,\mathbf{j}$

6. $\mathbf{r}(t) = \ln(t^2 + 1)\,\mathbf{i} + (t - 2\tan^{-1} t)\,\mathbf{j}$

7. *A formula for the curvature of the graph of a function in the xy-plane*

a) The graph $y = f(x)$ in the xy-plane automatically has the parametrization $x = x$, $y = f(x)$, and the vector formula $\mathbf{r}(x) = x\,\mathbf{i} + f(x)\,\mathbf{j}$. Use this formula to show that if f is a twice-differentiable function of x, then

$$\kappa(x) = \frac{|f''(x)|}{\left[1 + (f'(x))^2\right]^{3/2}}.$$

b) Use the formula for κ in (a) to find the curvature of $y = \ln(\cos x)$, $-\pi/2 < x < \pi/2$. Compare your answer with the answer in Exercise 1.

c) Show that the curvature is zero at a point of inflection.

8. *A formula for the curvature of a parametrized plane curve*

a) Show that the curvature of a smooth curve $\mathbf{r}(t) = f(t)\,\mathbf{i} + g(t)\,\mathbf{j}$ defined by twice-differentiable functions $x = f(t)$ and $y = g(t)$ is given by the formula

$$\kappa = \frac{|\dot{x}\ddot{y} - \dot{y}\ddot{x}|}{(\dot{x}^2 + \dot{y}^2)^{3/2}}.$$

Apply the formula to find the curvatures of the following curves.

b) $\mathbf{r}(t) = t\,\mathbf{i} + (\ln \sin t)\,\mathbf{j}, \quad 0 < t < \pi$

c) $\mathbf{r}(t) = [\tan^{-1}(\sinh t)]\,\mathbf{i} + (\ln \cosh t)\,\mathbf{j}$.

9. *Normals to plane curves*

a) Show that $\mathbf{n}(t) = -g'(t)\,\mathbf{i} + f'(t)\,\mathbf{j}$ and $-\mathbf{n}(t) = g'(t)\,\mathbf{i} - f'(t)\,\mathbf{j}$ are both normal to the curve $\mathbf{r}(t) = f(t)\,\mathbf{i} + g(t)\,\mathbf{j}$ at the point $(f(t), g(t))$.

To obtain $\mathbf{N}$ for a particular plane curve, we can choose the one of $\mathbf{n}$ or $-\mathbf{n}$ from part (a) that points toward the concave side of the curve, and make it into a unit vector. (See Fig. 11.26.) Apply this method to find $\mathbf{N}$ for the following curves.

b) $\mathbf{r}(t) = t\mathbf{i} + e^{2t}\mathbf{j}$

c) $\mathbf{r}(t) = \sqrt{4 - t^2}\,\mathbf{i} + t\mathbf{j}, \quad -2 \le t \le 2$

10. *(Continuation of Exercise 9.)*

a) Use the method of Exercise 9 to find $\mathbf{N}$ for the curve $\mathbf{r}(t) = t\mathbf{i} + (1/3)t^3\mathbf{j}$ when $t < 0$; when $t > 0$.

b) Calculate

$$\mathbf{N} = \frac{d\mathbf{T}/dt}{|d\mathbf{T}/dt|}, t \ne 0,$$

for the curve in part (a). Does $\mathbf{N}$ exist at $t = 0$? Graph the curve and explain what is happening to $\mathbf{N}$ as t passes from negative to positive values.

Space Curves

Find $\mathbf{T}$, $\mathbf{N}$, $\mathbf{B}$, κ, and τ for the space curves in Exercises 11–18.

11. $\mathbf{r}(t) = (3 \sin t)\mathbf{i} + (3 \cos t)\mathbf{j} + 4t\mathbf{k}$

12. $\mathbf{r}(t) = (\cos t + t \sin t)\mathbf{i} + (\sin t - t \cos t)\mathbf{j} + 3\mathbf{k}$

13. $\mathbf{r}(t) = (e^t \cos t)\mathbf{i} + (e^t \sin t)\mathbf{j} + 2\mathbf{k}$

14. $\mathbf{r}(t) = (6 \sin 2t)\mathbf{i} + (6 \cos 2t)\mathbf{j} + 5t\mathbf{k}$

15. $\mathbf{r}(t) = (t^3/3)\mathbf{i} + (t^2/2)\mathbf{j}, \quad t > 0$

16. $\mathbf{r}(t) = (\cos^3 t)\mathbf{i} + (\sin^3 t)\mathbf{j}, \quad 0 < t < \pi/2$

17. $\mathbf{r}(t) = t\mathbf{i} + (a \cosh(t/a))\mathbf{j}, \quad a > 0$

18. $\mathbf{r}(t) = (\cosh t)\mathbf{i} - (\sinh t)\mathbf{j} + t\mathbf{k}$

In Exercises 19 and 20, write $\mathbf{a}$ in the form $a_T\mathbf{T} + a_N\mathbf{N}$ without finding $\mathbf{T}$ and $\mathbf{N}$.

19. $\mathbf{r}(t) = (a \cos t)\mathbf{i} + (a \sin t)\mathbf{j} + bt\mathbf{k}$

20. $\mathbf{r}(t) = (1 + 3t)\mathbf{i} + (t - 2)\mathbf{j} - 3t\mathbf{k}$

In Exercises 21–24, write $\mathbf{a}$ in the form $\mathbf{a} = a_T\mathbf{T} + a_N\mathbf{N}$ at the given value of t without finding $\mathbf{T}$ and $\mathbf{N}$.

21. $\mathbf{r}(t) = (t + 1)\mathbf{i} + 2t\mathbf{j} + t^2\mathbf{k}, \quad t = 1$

22. $\mathbf{r}(t) = (t \cos t)\mathbf{i} + (t \sin t)\mathbf{j} + t^2\mathbf{k}, \quad t = 0$

23. $\mathbf{r}(t) = t^2\mathbf{i} + (t + (1/3)t^3)\mathbf{j} + (t - (1/3)t^3)\mathbf{k}, \quad t = 0$

24. $\mathbf{r}(t) = (e^t \cos t)\mathbf{i} + (e^t \sin t)\mathbf{j} + \sqrt{2}e^t\mathbf{k}, \quad t = 0$

In Exercises 25 and 26, find $\mathbf{r}$, $\mathbf{T}$, $\mathbf{N}$, and $\mathbf{B}$ at the given value of t. Then find equations for the osculating, normal, and rectifying planes at that value of t.

25. $\mathbf{r}(t) = (\cos t)\mathbf{i} + (\sin t)\mathbf{j} - \mathbf{k}, \quad t = \pi/4$

26. $\mathbf{r}(t) = (\cos t)\mathbf{i} + (\sin t)\mathbf{j} + t\mathbf{k}, \quad t = 0$

Physical Applications

27. The speedometer on your car reads a steady 35 mph. Could you be accelerating? Explain.

28. Can anything be said about the acceleration of a particle that is moving at a constant speed? Give reasons for your answer.

29. Can anything be said about the speed of a particle whose acceleration is always orthogonal to its velocity? Give reasons for your answer.

30. An object of mass m travels along the parabola $y = x^2$ with a constant speed of 10 units/sec. What is the force on the object due to its acceleration at $(0, 0)$? at $(2^{1/2}, 2)$? Write your answers in terms of $\mathbf{i}$ and $\mathbf{j}$. (Remember Newton's law, $\mathbf{F} = m\,\mathbf{a}$.)

31. The following is a quotation from an article in *The American Mathematical Monthly,* titled "Curvature in the Eighties" by Robert Osserman (October 1990, page 731):

> Curvature also plays a key role in physics. The magnitude of a force required to move an object at constant speed along a curved path is, according to Newton's laws, a constant multiple of the curvature of the trajectories.

Explain mathematically why the second sentence of the quotation is true.

32. Show that a moving particle will move in a straight line if the normal component of its acceleration is zero.

More on Curvature

33. Show that the parabola $y = ax^2, a \ne 0$, has its largest curvature at its vertex and has no minimum curvature. (*Note:* Since the curvature of a curve remains the same if the curve is translated or rotated, this result is true for any parabola.)

34. Show that the ellipse $x = a \cos t, y = b \sin t, a > b > 0$, has its largest curvature on its major axis and its smallest curvature on its minor axis. (As in Exercise 33, the same is true for any ellipse.)

35. *Maximizing the curvature of a helix.* In Example 4, we found the curvature of the helix $\mathbf{r}(t) = (a \cos t)\mathbf{i} + (a \sin t)\mathbf{j} + bt\mathbf{k}$ $(a, b \ge 0)$ to be $\kappa = a/(a^2 + b^2)$. What is the largest value κ can have for a given value of b? Give reasons for your answer.

36. *A sometime shortcut to curvature.* If you already know $|a_N|$ and $|\mathbf{v}|$, then the formula $a_N = \kappa|\mathbf{v}|^2$ gives a convenient way to find the curvature. Use it to find the curvature and radius of curvature of the curve

$$\mathbf{r}(t) = (\cos t + t \sin t)\mathbf{i} + (\sin t - t \cos t)\mathbf{j}, \quad t > 0.$$

(Take a_N and $|\mathbf{v}|$ from Example 6.)

37. Show that κ and τ are both zero for the line

$$\mathbf{r}(t) = (x_0 + At)\mathbf{i} + (y_0 + Bt)\mathbf{j} + (z_0 + Ct)\mathbf{k}.$$

38. *Total curvature.* We find the **total curvature** of the portion of a smooth curve that runs from $s = s_0$ to $s = s_1 > s_0$ by integrating κ from s_0 to s_1. If the curve has some other parameter, say t, then the total curvature is

$$K = \int_{s_0}^{s_1} \kappa\, ds = \int_{t_0}^{t_1} \kappa\, \frac{ds}{dt}\, dt = \int_{t_0}^{t_1} \kappa|\mathbf{v}|\,dt,$$

where t_0 and t_1 correspond to s_0 and s_1. Find the total curvature of the portion of the helix $\mathbf{r}(t) = (3 \cos t)\mathbf{i} + (3 \sin t)\mathbf{j} + t\mathbf{k}$, $0 \le t \le 4\pi$.

39. *(Continuation of Exercise 38.)* Find the total curvatures of the following curves.

    a) The involute of the unit circle: $\mathbf{r}(t) = (\cos t + t \sin t)\mathbf{i} + (\sin t - t \cos t)\mathbf{j}$, $a \le t \le b\, (a > 0)$. (Exercise 36 gives a convenient way to find κ. Use values from Example 6.)

    b) The parabola $y = x^2$, $-\infty < x < \infty$

40. **a)** Find an equation for the circle of curvature of the curve $\mathbf{r}(t) = t\mathbf{i} + (\sin t)\mathbf{j}$ at the point $(\pi/2, 1)$. (The curve parametrizes the graph of $y = \sin x$ in the xy-plane.)

    b) Find an equation for the circle of curvature of the curve $\mathbf{r}(t) = (2 \ln t)\mathbf{i} - [t + (1/t)]\mathbf{j}$, $e^{-2} \le t \le e^2$, at the point $(0, -2)$, where $t = 1$.

Theory and Examples

41. What can be said about the torsion of a (sufficiently differentiable) plane curve $\mathbf{r}(t) = f(t)\mathbf{i} + g(t)\mathbf{j}$? Give reasons for your answer.

42. *The torsion of a helix.* In Example 7, we found the torsion of the helix

$$\mathbf{r}(t) = (a \cos t)\mathbf{i} + (a \sin t)\mathbf{j} + bt\,\mathbf{k}, \quad a, b \ge 0$$

to be $\tau = b/(a^2 + b^2)$. What is the largest value τ can have for a given value of a? Give reasons for your answer.

43. *Differentiable curves with zero torsion lie in planes.* That a sufficiently differentiable curve with zero torsion lies in a plane is a special case of the fact that a particle whose velocity remains perpendicular to a fixed vector $\mathbf{C}$ moves in a plane perpendicular to $\mathbf{C}$. This, in turn, can be viewed as the solution of the following problem in calculus.

    Suppose $\mathbf{r}(t) = f(t)\mathbf{i} + g(t)\mathbf{j} + h(t)\mathbf{k}$ is twice differentiable for all t in an interval $[a, b]$, that $\mathbf{r} = 0$ when $t = a$, and that $\mathbf{v} \cdot \mathbf{k} = 0$ for all t in $[a, b]$. Then $h(t) = 0$ for all t in $[a, b]$.

    Solve this problem. (*Hint:* Start with $\mathbf{a} = d^2\mathbf{r}/dt^2$ and apply the initial conditions in reverse order.)

44. *A formula that calculates τ from $\mathbf{B}$ and $\mathbf{v}$.* If we start with the definition $\tau = -(d\mathbf{B}/ds) \cdot \mathbf{N}$ and apply the Chain Rule to rewrite $d\mathbf{B}/ds$ as

$$\frac{d\mathbf{B}}{ds} = \frac{d\mathbf{B}}{dt}\frac{dt}{ds} = \frac{d\mathbf{B}}{dt}\frac{1}{|\mathbf{v}|},$$

we arrive at the formula

$$\tau = -\frac{1}{|\mathbf{v}|}\left(\frac{d\mathbf{B}}{dt} \cdot \mathbf{N}\right).$$

The advantage of this formula over Eq. (11) is that it is easier to derive and state. The disadvantage is that it can take a lot of work to evaluate without a computer. Use the new formula to find the torsion of the helix in Example 7.

▦ Grapher Explorations

The formula

$$\kappa(x) = \frac{|f''(x)|}{\left[1 + (f'(x))^2\right]^{3/2}},$$

derived in Exercise 7, expresses the curvature $\kappa(x)$ of a twice-differentiable plane curve $y = f(x)$ as a function of x. Find the curvature function of each of the curves in Exercises 45–48. Then graph $f(x)$ together with $\kappa(x)$ over the given interval. You will find some surprises.

45. $y = x^2$, $-2 \le x \le 2$

46. $y = x^4/4$, $-2 \le x \le 2$

47. $y = \sin x$, $0 \le x \le 2\pi$

48. $y = e^x$, $-1 \le x \le 2$

✺ CAS Explorations and Projects—Circles of Curvature

In Exercises 49–56 you will use a CAS to explore the osculating circle at a point P on a plane curve where $\kappa \ne 0$. Use a CAS to perform the following steps:

    a) Plot the plane curve given in parametric or function form over the specified interval to see what it looks like.

    b) Calculate the curvature κ of the curve at the given value t_0 using the appropriate formula from Exercise 7 or 8. Use the parametrization $x = t$ and $y = f(t)$ if the curve is given as a function $y = f(x)$.

    c) Find the unit normal vector $\mathbf{N}$ at t_0. Notice that the signs of the components of $\mathbf{N}$ depend on whether the unit tangent vector $\mathbf{T}$ is turning clockwise or counterclockwise at $t = t_0$. (See Exercise 9.)

    d) If $\mathbf{C} = a\mathbf{i} + b\mathbf{j}$ is the vector from the origin to the center (a, b) of the osculating circle, find the center $\mathbf{C}$ from the vector equation

$$\mathbf{C} = \mathbf{r}(t_0) + \frac{1}{\kappa(t_0)}\mathbf{N}(t_0).$$

The point $P(x_0, y_0)$ on the curve is given by the position vector $\mathbf{r}(t_0)$.

    e) Plot implicitly the equation $(x - a)^2 + (y - b)^2 = 1/\kappa^2$ of the osculating circle. Then plot the curve and osculating circle together. You may need to experiment with the size of the viewing window, but be sure it is square.

49. $\mathbf{r}(t) = (3 \cos t)\mathbf{i} + (5 \sin t)\mathbf{j}$, $0 \le t \le 2\pi$, $t_0 = \pi/4$

50. $\mathbf{r}(t) = (\cos^3 t)\mathbf{i} + (\sin^3 t)\mathbf{j}$, $0 \le t \le 2\pi$, $t_0 = \pi/4$

51. $\mathbf{r}(t) = t^2\mathbf{i} + (t^3 - 3t)\mathbf{j}$, $-4 \le t \le 4$, $t_0 = 3/5$

52. $\mathbf{r}(t) = (t^3 - 2t^2 - t)\mathbf{i} + \dfrac{3t}{\sqrt{1 + t^2}}\mathbf{j}$, $-2 \le t \le 5$, $t_0 = 1$

53. $\mathbf{r}(t) = (2t - \sin t)\mathbf{i} + (2 - 2\cos t)\mathbf{j}$, $0 \le t \le 3\pi$, $t_0 = 3\pi/2$

54. $\mathbf{r}(t) = (e^{-t}\cos t)\mathbf{i} + (e^{-t}\sin t)\mathbf{j}$, $0 \le t \le 6\pi$, $t_0 = \pi/4$

55. $y = x^2 - x$, $-2 \le x \le 5$, $x_0 = 1$

56. $y = x(1 - x)^{2/5}$, $-1 \le x \le 2$, $x_0 = 1/2$

CAS Explorations and Projects—Curvature, Torsion, and the TNB Frame

Rounding the answers to four decimal places, use a CAS to find $\mathbf{v}$, $\mathbf{a}$, speed, $\mathbf{T}$, $\mathbf{N}$, $\mathbf{B}$, κ, τ, and the tangential and normal components of acceleration for the curves in Exercises 57–60 at the given values of t.

57. $\mathbf{r}(t) = (t \cos t)\mathbf{i} + (t \sin t)\mathbf{j} + t\mathbf{k}, \quad t = \sqrt{3}$

58. $\mathbf{r}(t) = (e^t \cos t)\mathbf{i} + (e^t \sin t)\mathbf{j} + e^t\mathbf{k}, \quad t = \ln 2$

59. $\mathbf{r}(t) = (t - \sin t)\mathbf{i} + (1 - \cos t)\mathbf{j} + \sqrt{-t}\,\mathbf{k}, \quad t = -3\pi$

60. $\mathbf{r}(t) = (3t - t^2)\mathbf{i} + (3t^2)\mathbf{j} + (3t + t^3)\mathbf{k}, \quad t = 1$

11.5

Planetary Motion and Satellites

In this section, we derive Kepler's laws of planetary motion from Newton's laws of motion and gravitation and discuss the orbits of Earth satellites. The derivation of Kepler's laws from Newton's is one of the triumphs of calculus. It draws on almost everything we have studied so far, including the algebra and geometry of vectors in space, the calculus of vector functions, the solutions of differential equations and initial value problems, and the polar coordinate description of conic sections.

Vector Equations for Motion in Polar and Cylindrical Coordinates

When a particle moves along a curve in the polar coordinate plane, we express its position, velocity, and acceleration in terms of the moving unit vectors

$$\mathbf{u}_r = (\cos\theta)\mathbf{i} + (\sin\theta)\mathbf{j}, \qquad \mathbf{u}_\theta = -(\sin\theta)\mathbf{i} + (\cos\theta)\mathbf{j}, \tag{1}$$

shown in Fig. 11.35. The vector $\mathbf{u}_r$ points along the position vector $\overrightarrow{OP}$, so $\mathbf{r} = r\mathbf{u}_r$. The vector $\mathbf{u}_\theta$, orthogonal to $\mathbf{u}_r$, points in the direction of increasing θ.

We find from (1) that

$$\frac{d\mathbf{u}_r}{d\theta} = -(\sin\theta)\mathbf{i} + (\cos\theta)\mathbf{j} = \mathbf{u}_\theta$$

$$\frac{d\mathbf{u}_\theta}{d\theta} = -(\cos\theta)\mathbf{i} - (\sin\theta)\mathbf{j} = -\mathbf{u}_r. \tag{2}$$

When we differentiate $\mathbf{u}_r$ and $\mathbf{u}_\theta$ with respect to t to find how they change with time, the Chain Rule gives

$$\dot{\mathbf{u}}_r = \frac{d\mathbf{u}_r}{d\theta}\dot{\theta} = \dot{\theta}\,\mathbf{u}_\theta, \qquad \dot{\mathbf{u}}_\theta = \frac{d\mathbf{u}_\theta}{d\theta}\dot{\theta} = -\dot{\theta}\,\mathbf{u}_r. \tag{3}$$

Hence,

$$\mathbf{v} = \dot{\mathbf{r}} = \frac{d}{dt}\left(r\,\mathbf{u}_r\right) = \dot{r}\,\mathbf{u}_r + r\dot{\mathbf{u}}_r = \dot{r}\,\mathbf{u}_r + r\dot{\theta}\,\mathbf{u}_\theta. \tag{4}$$

See Fig. 11.36.

The acceleration is

$$\mathbf{a} = \dot{\mathbf{v}} = (\ddot{r}\,\mathbf{u}_r + \dot{r}\,\dot{\mathbf{u}}_r) + (\dot{r}\dot{\theta}\,\mathbf{u}_\theta + r\ddot{\theta}\,\mathbf{u}_\theta + r\dot{\theta}\,\dot{\mathbf{u}}_\theta). \tag{5}$$

When Eqs. (3) are used to evaluate $\dot{\mathbf{u}}_r$ and $\dot{\mathbf{u}}_\theta$ and the components are separated, the equation for acceleration becomes

$$\mathbf{a} = (\ddot{r} - r\dot{\theta}^2)\,\mathbf{u}_r + (r\ddot{\theta} + 2\dot{r}\dot{\theta})\,\mathbf{u}_\theta. \tag{6}$$

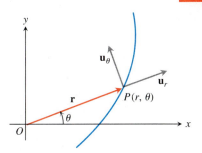

11.35 The length of $\mathbf{r}$ is the positive polar coordinate r of the point P. Thus, $\mathbf{u}_r$, which is $\mathbf{r}/|\mathbf{r}|$, is also $\mathbf{r}/r$. Equations (1) express $\mathbf{u}_r$ and $\mathbf{u}_\theta$ in terms of $\mathbf{i}$ and $\mathbf{j}$.

11.36 In polar coordinates, the velocity vector is

$$\mathbf{v} = \dot{r}\,\mathbf{u}_r + r\dot{\theta}\,\mathbf{u}_\theta.$$

As in the previous section, we use Newton's dot notation for time derivatives to keep the formulas as simple as we can: $\dot{\mathbf{u}}_r$ means $d\mathbf{u}_r/dt$, $\dot{\theta}$ means $d\theta/dt$, and so on.

To extend these equations of motion to space, we add $z\mathbf{k}$ to the right-hand side of the equation $\mathbf{r} = r\mathbf{u}_r$. Then, in cylindrical coordinates,

Notice that $|\mathbf{r}| \neq r$ if $z \neq 0$.

$$\mathbf{r} = r\,\mathbf{u}_r + z\,\mathbf{k}$$
$$\mathbf{v} = \dot{r}\,\mathbf{u}_r + r\dot{\theta}\,\mathbf{u}_\theta + \dot{z}\,\mathbf{k} \tag{7}$$
$$\mathbf{a} = (\ddot{r} - r\dot{\theta}^2)\,\mathbf{u}_r + (r\ddot{\theta} + 2\dot{r}\dot{\theta})\mathbf{u}_\theta + \ddot{z}\,\mathbf{k}.$$

The vectors $\mathbf{u}_r$, $\mathbf{u}_\theta$, and $\mathbf{k}$ make a right-handed frame (Fig. 11.37) in which

$$\mathbf{u}_r \times \mathbf{u}_\theta = \mathbf{k}, \qquad \mathbf{u}_\theta \times \mathbf{k} = \mathbf{u}_r, \qquad \mathbf{k} \times \mathbf{u}_r = \mathbf{u}_\theta. \tag{8}$$

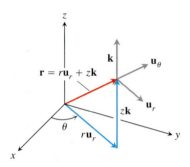

11.37 Position vector and basic unit vectors in cylindrical coordinates.

Planets Move in Planes

Newton's Law of Gravitation says that if $\mathbf{r}$ is the radius vector from the center of a sun of mass M to the center of a planet of mass m, then the force $\mathbf{F}$ of the gravitational attraction between the planet and sun is

$$\mathbf{F} = -\frac{GmM}{|\mathbf{r}|^2}\frac{\mathbf{r}}{|\mathbf{r}|} \tag{9}$$

(Fig. 11.38). The number G is the (universal) **gravitational constant.** If we measure mass in kilograms, force in newtons, and distance in meters, G is about $6.6720 \times 10^{-11}\ \mathrm{Nm^2kg^{-2}}$.

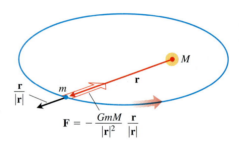

$$\mathbf{F} = -\frac{GmM}{|\mathbf{r}|^2}\frac{\mathbf{r}}{|\mathbf{r}|}$$

11.38 The force of gravity is directed along the line joining the centers of mass.

Combining Eq. (9) with Newton's second law, $\mathbf{F} = m\ddot{\mathbf{r}}$, for the force acting on the planet gives

$$m\ddot{\mathbf{r}} = -\frac{GmM}{|\mathbf{r}|^2}\frac{\mathbf{r}}{|\mathbf{r}|},$$
$$\ddot{\mathbf{r}} = -\frac{GM}{|\mathbf{r}|^2}\frac{\mathbf{r}}{|\mathbf{r}|}. \tag{10}$$

The planet is accelerated toward the sun's center at all times.

Equation (10) says that $\ddot{\mathbf{r}}$ is a scalar multiple of $\mathbf{r}$, so that

$$\mathbf{r} \times \ddot{\mathbf{r}} = \mathbf{0}. \tag{11}$$

A routine calculation shows $\mathbf{r} \times \ddot{\mathbf{r}}$ to be the derivative of $\mathbf{r} \times \dot{\mathbf{r}}$:

$$\frac{d}{dt}(\mathbf{r} \times \dot{\mathbf{r}}) = \underbrace{\dot{\mathbf{r}} \times \dot{\mathbf{r}}}_{0} + \mathbf{r} \times \ddot{\mathbf{r}} = \mathbf{r} \times \ddot{\mathbf{r}}. \tag{12}$$

Hence Eq. (11) is equivalent to

$$\frac{d}{dt}(\mathbf{r} \times \dot{\mathbf{r}}) = \mathbf{0}, \tag{13}$$

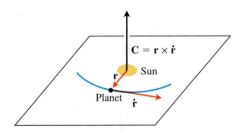

11.39 A planet that obeys Newton's laws of gravitation and motion travels in the plane through the sun's center of mass perpendicular to $\mathbf{C} = \mathbf{r} \times \dot{\mathbf{r}}$.

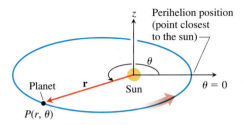

11.40 The coordinate system for planetary motion. The motion is counterclockwise when viewed from above, as it is here, and $\dot{\theta} > 0$.

which integrates to

$$\mathbf{r} \times \dot{\mathbf{r}} = \mathbf{C} \tag{14}$$

for some constant vector $\mathbf{C}$.

Equation (14) tells us that $\mathbf{r}$ and $\dot{\mathbf{r}}$ always lie in a plane perpendicular to $\mathbf{C}$. Hence, the planet moves in a fixed plane through the center of its sun (Fig. 11.39).

Coordinates and Initial Conditions

We now introduce cylindrical coordinates in a way that places the origin at the sun's center of mass and makes the plane of the planet's motion the polar coordinate plane. This makes $\mathbf{r}$ the planet's polar coordinate position vector and makes $|\mathbf{r}|$ equal to r and $\mathbf{r}/|\mathbf{r}|$ equal to $\mathbf{u}_r$. We also position the z-axis in a way that makes $\mathbf{k}$ the direction of $\mathbf{C}$. Thus, $\mathbf{k}$ has the same right-hand relation to $\mathbf{r} \times \dot{\mathbf{r}}$ that $\mathbf{C}$ does, and the planet's motion is counterclockwise when viewed from the positive z-axis. This makes θ increase with t, so that $\dot{\theta} > 0$ for all t. Finally, we rotate the polar coordinate plane about the z-axis, if necessary, to make the initial ray coincide with the direction $\mathbf{r}$ has when the planet is closest to the sun. This runs the ray through the planet's **perihelion** position (Fig. 11.40).

If we measure time so that $t = 0$ at perihelion, we have the following initial conditions for the planet's motion.

1. $r = r_0$, the minimum radius, when $t = 0$
2. $\dot{r} = 0$ when $t = 0$ (because r has a minimum value then)
3. $\theta = 0$ when $t = 0$
4. $|\mathbf{v}| = v_0$ when $t = 0$

Since

$$
\begin{aligned}
v_0 &= |\mathbf{v}|_{t=0} \\
&= \left| \dot{r}\,\mathbf{u}_r + r\dot{\theta}\,\mathbf{u}_\theta \right|_{t=0} && \text{Eq. (4)} \\
&= \left| r\dot{\theta}\,\mathbf{u}_\theta \right|_{t=0} && \dot{r} = 0 \text{ when } t = 0 \\
&= \left(\left| r\dot{\theta} \right| |\mathbf{u}_\theta| \right)_{t=0} && \\
&= \left| r\dot{\theta} \right|_{t=0} && |\mathbf{u}_\theta| = 1 \\
&= (r\dot{\theta})_{t=0}, && r \text{ and } \dot{\theta} \text{ both positive}
\end{aligned}
$$

we also know that

5. $r\dot{\theta} = v_0$ when $t = 0$.

Statement of Kepler's First Law (The Conic Section Law)

Kepler's first law says that a planet's path is a conic section with the sun at one focus. The eccentricity of the conic is

$$e = \frac{r_0 v_0^2}{GM} - 1 \tag{15}$$

and the polar equation is

$$r = \frac{(1+e)r_0}{1 + e\cos\theta}. \tag{16}$$

The derivation uses Kepler's second law, so we will state and prove the second law before proving the first law.

Kepler's Second Law (The Equal Area Law)

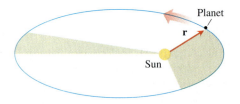

11.41 The line joining a planet to its sun sweeps over equal areas in equal times.

The German astronomer, mathematician, and physicist Johannes Kepler (1571–1630) was the first, and until Descartes the only, scientist to demand physical (as opposed to theological) explanations of celestial phenomena. His three laws of motion, the results of a lifetime of work, changed the course of astronomy forever and played a crucial role in the development of Newton's physics.

Kepler's second law says that the radius vector from the sun to a planet (the vector **r** in our model) sweeps out equal areas in equal times (Fig. 11.41). To derive the law, we use Eq. (4) to evaluate the cross product $\mathbf{C} = \mathbf{r} \times \dot{\mathbf{r}}$ from Eq. (14):

$$\mathbf{C} = \mathbf{r} \times \dot{\mathbf{r}} = \mathbf{r} \times \mathbf{v}$$

$$= r\,\mathbf{u}_r \times (\dot{r}\,\mathbf{u}_r + r\dot{\theta}\,\mathbf{u}_\theta) \qquad \text{Eq. (4)}$$

$$= r\dot{r}\,\underbrace{(\mathbf{u}_r \times \mathbf{u}_r)}_{\mathbf{0}} + r(r\dot{\theta})\,\underbrace{(\mathbf{u}_r \times \mathbf{u}_\theta)}_{\mathbf{k}} \tag{17}$$

$$= r(r\dot{\theta})\,\mathbf{k}.$$

Setting t equal to zero shows that

$$\mathbf{C} = [r(r\dot{\theta})]_{t=0}\,\mathbf{k} = r_0 v_0\,\mathbf{k}. \tag{18}$$

Substituting this value for **C** in Eq. (17) gives

$$r_0 v_0\,\mathbf{k} = r^2\dot{\theta}\,\mathbf{k}, \qquad \text{or} \qquad r^2\dot{\theta} = r_0 v_0. \tag{19}$$

This is where the area comes in. The area differential in polar coordinates is

$$dA = \frac{1}{2}r^2\,d\theta$$

(Section 9.9). Accordingly, dA/dt has the constant value

$$\frac{dA}{dt} = \frac{1}{2}r^2\dot{\theta} = \frac{1}{2}r_0 v_0, \tag{20}$$

which is Kepler's second law.

For Earth, r_0 is about 150,000,000 km, v_0 is about 30 km/sec, and dA/dt is about 2,250,000,000 km²/sec. Every time your heart beats, Earth advances 30 km along its orbit, and the radius joining Earth to the sun sweeps out 2,250,000,000 km² of area.

Proof of Kepler's First Law

To prove that a planet moves along a conic section with one focus at its sun, we need to express the planet's radius r as a function of θ. This requires a long sequence of calculations and some substitutions that are not altogether obvious.

We begin with the equation that comes from equating the coefficients of $\mathbf{u}_r = \mathbf{r}/|\mathbf{r}|$ in Eqs. (6) and (10):

$$\ddot{r} - r\dot{\theta}^2 = -\frac{GM}{r^2}. \tag{21}$$

We eliminate $\dot{\theta}$ temporarily by replacing it with $r_0 v_0/r^2$ from Eq. (19) and rearrange the resulting equation to get

$$\ddot{r} = \frac{r_0{}^2 v_0{}^2}{r^3} - \frac{GM}{r^2}. \tag{22}$$

We change this into a first order equation by a change of variable. With

$$p = \frac{dr}{dt}, \qquad \frac{d^2 r}{dt^2} = \frac{dp}{dt} = \frac{dp}{dr}\frac{dr}{dt} = p\frac{dp}{dr}, \qquad \textcolor{blue}{\text{Chain Rule}}$$

Eq. (22) becomes

$$p\frac{dp}{dr} = \frac{r_0{}^2 v_0{}^2}{r^3} - \frac{GM}{r^2}. \tag{23}$$

Multiplying through by 2 and integrating with respect to r gives

$$p^2 = (\dot{r})^2 = -\frac{r_0{}^2 v_0{}^2}{r^2} + \frac{2GM}{r} + C_1. \tag{24}$$

The initial conditions that $r = r_0$ and $\dot{r} = 0$ when $t = 0$ determine the value of C_1 to be

$$C_1 = v_0{}^2 - \frac{2GM}{r_0}.$$

Accordingly, Eq. (24), after a suitable rearrangement, becomes

$$\dot{r}^2 = v_0{}^2\left(1 - \frac{r_0{}^2}{r^2}\right) + 2GM\left(\frac{1}{r} - \frac{1}{r_0}\right). \tag{25}$$

The effect of going from Eq. (21) to Eq. (25) has been to replace a second order differential equation in r by a first order differential equation in r. Our goal is still to express r in terms of θ, so we now bring θ back into the picture. To accomplish this, we divide both sides of Eq. (25) by the squares of the corresponding sides of the equation $r^2\dot{\theta} = r_0 v_0$ (Eq. 19) and use the fact that $\dot{r}/\dot{\theta} = (dr/dt)/(d\theta/dt) = dr/d\theta$ to get

$$\frac{1}{r^4}\left(\frac{dr}{d\theta}\right)^2 = \frac{1}{r_0{}^2} - \frac{1}{r^2} + \frac{2GM}{r_0{}^2 v_0{}^2}\left(\frac{1}{r} - \frac{1}{r_0}\right)$$

$$= \frac{1}{r_0{}^2} - \frac{1}{r^2} + 2h\left(\frac{1}{r} - \frac{1}{r_0}\right). \qquad \textcolor{blue}{h = \frac{GM}{r_0{}^2 v_0{}^2}} \tag{26}$$

To simplify further, we substitute

$$u = \frac{1}{r}, \qquad u_0 = \frac{1}{r_0}, \qquad \frac{du}{d\theta} = -\frac{1}{r^2}\frac{dr}{d\theta}, \qquad \left(\frac{du}{d\theta}\right)^2 = \frac{1}{r^4}\left(\frac{dr}{d\theta}\right)^2,$$

obtaining

$$\left(\frac{du}{d\theta}\right)^2 = u_0{}^2 - u^2 + 2hu - 2hu_0 = (u_0 - h)^2 - (u - h)^2, \tag{27}$$

$$\frac{du}{d\theta} = \pm\sqrt{(u_0 - h)^2 - (u - h)^2}. \tag{28}$$

Which sign do we take? We know that $\dot{\theta} = r_0 v_0/r^2$ is positive. Also, r starts from a minimum value at $t = 0$, so it cannot immediately decrease, and $\dot{r} \geq 0$, at least for early positive values of t. Therefore,

$$\frac{dr}{d\theta} = \frac{\dot{r}}{\dot{\theta}} \geq 0 \qquad \text{and} \qquad \frac{du}{d\theta} = -\frac{1}{r^2}\frac{dr}{d\theta} \leq 0.$$

The correct sign for Eq. (28) is the negative sign. With this determined, we rearrange Eq. (28) and integrate both sides with respect to θ:

$$\frac{-1}{\sqrt{(u_0 - h)^2 - (u - h)^2}}\frac{du}{d\theta} = 1$$

$$\cos^{-1}\left(\frac{u - h}{u_0 - h}\right) = \theta + C_2. \tag{29}$$

The constant C_2 is zero because $u = u_0$ when $\theta = 0$ and $\cos^{-1}(1) = 0$. Therefore,

$$\frac{u - h}{u_0 - h} = \cos\theta$$

and

$$\frac{1}{r} = u = h + (u_0 - h)\cos\theta. \tag{30}$$

A few more algebraic maneuvers produce the final equation

$$r = \frac{(1 + e)r_0}{1 + e\cos\theta}, \tag{31}$$

where

$$e = \frac{1}{r_0 h} - 1 = \frac{r_0 v_0^2}{GM} - 1. \tag{32}$$

Together, Eqs. (31) and (32) say that the path of the planet is a conic section with one focus at the sun and with eccentricity $(r_0 v_0^2/GM) - 1$. This is the modern formulation of Kepler's first law.

Statement of Kepler's Third Law (The Time–Distance Law)

The time T it takes a planet to go around its sun once is the planet's **orbital period.** *Kepler's third law* says that T and the orbit's semimajor axis a are related by the equation

$$\frac{T^2}{a^3} = \frac{4\pi^2}{GM}. \tag{33}$$

Since the right-hand side of this equation is constant within a given solar system, the ratio of T^2 to a^3 is *the same for every planet in the system.*

Kepler's third law is the starting point for working out the size of our solar system. It allows the semimajor axis of each planetary orbit to be expressed in astronomical units, Earth's semimajor axis being one unit. The distance between any two planets at any time can then be predicted in astronomical units and all that remains is to find one of these distances in kilometers. This can be done by bouncing radar waves off Venus, for example. The astronomical unit is now known, after a series of such measurements, to be 149,597,870 km.

We derive Kepler's third law by combining two formulas for the area enclosed by the planet's elliptical orbit:

Formula 1:   Area $= \pi ab$ *The geometry formula in which a is the semimajor axis and b is the semiminor axis*

Formula 2:   Area $= \int_0^T dA$

$$= \int_0^T \frac{1}{2} r_0 v_0 \, dt \qquad \text{Eq. (20)}$$

$$= \frac{1}{2} T r_0 v_0.$$

Equating these gives

$$T = \frac{2\pi ab}{r_0 v_0} = \frac{2\pi a^2}{r_0 v_0} \sqrt{1 - e^2}. \qquad \text{\textit{For any ellipse,}}\ b = a\sqrt{1-e^2} \qquad (34)$$

It remains only to express a and e in terms of r_0, v_0, G, and M. Equation (32) does this for e. For a, we observe that setting θ equal to π in Eq. (31) gives

$$r_{\max} = r_0 \frac{1+e}{1-e}.$$

Hence,

$$2a = r_0 + r_{\max} = \frac{2r_0}{1-e} = \frac{2r_0 GM}{2GM - r_0 v_0{}^2}. \qquad (35)$$

Squaring both sides of Eq. (34) and substituting the results of Eqs. (32) and (35) now produces Kepler's third law (Exercise 14).

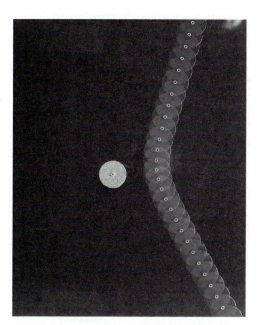

11.42 This multiflash photograph shows a body being deflected by an inverse square law force. It moves along a hyperbola.

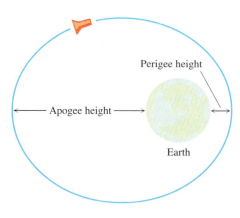

11.43 The orbit of an Earth satellite: $2a$ = diameter of earth + perigee height + apogee height.

Orbit Data

Although Kepler discovered his laws empirically and stated them only for the six planets known at the time, the modern derivations of Kepler's laws show that they apply to any body driven by a force that obeys an inverse square law. They apply to Halley's comet and the asteroid Icarus. They apply to the moon's orbit about Earth, and they applied to the orbit of the spacecraft *Apollo 8* about the moon. They also applied to the air puck shown in Fig. 11.42 being deflected by an inverse square law force—its path is a hyperbola. Charged particles fired at the nuclei of atoms scatter along hyperbolic paths.

Tables 11.1–11.3 give additional data for planetary orbits and for the orbits of seven of Earth's artificial satellites (Fig. 11.43). *Vanguard 1* sent back data that revealed differences between the levels of Earth's oceans and provided the first determination of the precise locations of some of the more isolated Pacific islands. The data also verified that the gravitation of the sun and moon would affect the orbits of Earth's satellites and that solar radiation could exert enough pressure to deform an orbit.

Syncom 3 is one of a series of U.S. Department of Defense telecommunications satellites. *Tiros 11* (for "television infrared observation satellite") is one of a series of weather satellites. *GOES 4* (for "geostationary operational environmental satellite") is one of a series of satellites designed to gather information about Earth's atmosphere. Its orbital period, 1436.2 minutes, is nearly the same as Earth's rotational period of 1436.1 minutes, and its orbit is nearly circular ($e = 0.0003$). *Intelsat 5* is a heavy-capacity commercial telecommunications satellite.

Table 11.1 Values of *a, e,* and *T* for the major planets

Planet	Semimajor axis $a^{\dagger}$	Eccentricity *e*	Period *T*
Mercury	57.95	0.2056	87.967 days
Venus	108.11	0.0068	224.701 days
Earth	149.57	0.0167	365.256 days
Mars	227.84	0.0934	1.8808 years
Jupiter	778.14	0.0484	11.8613 years
Saturn	1427.0	0.0543	29.4568 years
Uranus	2870.3	0.0460	84.0081 years
Neptune	4499.9	0.0082	164.784 years
Pluto	5909	0.2481	248.35 years

† Millions of kilometers

Table 11.2 Data on Earth's satellites

Name	Launch date	Time or expected time aloft	Mass at launch (kg)	Period (min)	Perigee height (km)	Apogee height (km)	Semimajor axis *a* (km)	Eccentricity
Sputnik 1	Oct. 1957	57.6 days	83.6	96.2	215	939	6,955	0.052
Vanguard 1	March 1958	300 years	1.47	138.5	649	4,340	8,872	0.208
Syncom 3	Aug. 1964	$>10^6$ years	39	1436.2	35,718	35,903	42,189	0.002
Skylab 4	Nov. 1973	84.06 days	13,980	93.11	422	437	6,808	0.001
Tiros 11	Oct. 1978	500 years	734	102.12	850	866	7,236	0.001
GOES 4	Sept. 1980	$>10^6$ years	627	1436.2	35,776	35,800	42,166	0.0003
Intelsat 5	Dec. 1980	$>10^6$ years	1,928	1417.67	35,143	35,707	41,803	0.007

Table 11.3 Numerical data

Gravitational constant: $G = 6.6720 \times 10^{-11} \ \text{Nm}^2\text{kg}^{-2}$
(When you use this value of *G* in a calculation, remember to express force in newtons, distance in meters, mass in kilograms, and time in seconds.)

Sun's mass: 1.99×10^{30} kg

Earth's mass: 5.975×10^{24} kg

Equatorial radius of Earth: 6378.533 km

Polar radius of Earth: 6356.912 km

Earth's rotational period: 1436.1 min

Earth's orbital period: 1 year = 365.256 days

Exercises 11.5

Reminder: When a calculation involves the gravitational constant G, express force in newtons, distance in meters, mass in kilograms, and time in seconds.

Calculator Exercises

1. Since the orbit of *Skylab 4* had a semimajor axis of $a = 6808$ km, Kepler's third law with M equal to the earth's mass should give the period. Calculate it. Compare your result with the value in Table 11.2.

2. Earth's distance from the sun at perihelion is approximately 149,577,000 km, and the eccentricity of the earth's orbit about the sun is 0.0167. Find the velocity v_0 of Earth in its orbit at perihelion. (Use Eq. 15.)

3. In July 1965, the USSR launched *Proton I,* weighing 12,200 kg (at launch), with a perigee height of 183 km, an apogee height of 589 km, and a period of 92.25 min. Using the relevant data for the mass of Earth and the gravitational constant G, find the semimajor axis a of the orbit from Eq. (33). Compare your answer with the number you get by adding the perigee and apogee heights to the diameter of the earth.

4. **a)** The *Viking 1* orbiter, which surveyed Mars from August 1975 to June 1976, had a period of 1639 min. Use this and the fact that the mass of Mars is 6.418×10^{23} kg to find the semimajor axis of the *Viking 1* orbit.

   b) The *Viking 1* orbiter was 1499 km from the surface of Mars at its closest point and 35,800 km from the surface at its farthest point. Use this information together with the value you obtained in part (a) to estimate the average diameter of Mars.

5. The *Viking 2* orbiter, which surveyed Mars from September 1975 to August 1976, moved in an ellipse whose semimajor axis was 22,030 km. What was the orbital period? (Express your answer in minutes.)

6. *Geosynchronous orbits.* Several satellites in the earth's equatorial plane have nearly circular orbits whose periods are the same as the earth's rotational period. Such orbits are called **geosynchronous** or **geostationary** because they hold the satellite over the same spot on Earth's surface.

   a) Approximately what is the semimajor axis of a geosynchronous orbit? Give reasons for your answer.

   b) About how high is a geosynchronous orbit above the Earth's surface?

   c) Which of the satellites in Table 11.2 have (nearly) geosynchronous orbits?

7. The mass of Mars is 6.418×10^{23} kg. If a satellite revolving about Mars is to hold a stationary orbit (have the same period as the period of Mars's rotation, which is 1477.4 min), what must the semimajor axis of its orbit be? Give reasons for your answer.

8. The period of the moon's rotation about Earth is 2.36055×10^6 sec. About how far away is the moon?

9. A satellite moves around Earth in a circular orbit. Express the satellite's speed as a function of the orbit's radius.

10. If T is measured in seconds and a in meters, what is the value of T^2/a^3 for planets in our solar system? for satellites orbiting Earth? for satellites orbiting the moon? (The moon's mass is 7.354×10^{22} kg.)

Noncalculator Exercises

11. For what values of v_0 in Eq. (15) is the orbit in Eq. (16) a circle? an ellipse? a parabola? a hyperbola?

12. Show that a planet in a circular orbit moves with a constant speed. (*Hint:* This is a consequence of one of Kepler's laws.)

13. Suppose that $\mathbf{r}$ is the position vector of a particle moving along a plane curve and dA/dt is the rate at which the vector sweeps out area. Without introducing coordinates, and assuming the necessary derivatives exist, give a geometric argument based on increments and limits for the validity of the equation

$$\frac{dA}{dt} = \frac{1}{2}|\mathbf{r} \times \dot{\mathbf{r}}|.$$

14. Complete the derivation of Kepler's third law (the part following Eq. 34).

15. Two planets, planet A and planet B, are orbiting their sun in circular orbits with A being the inner planet and B being farther away from the sun. Suppose the positions of A and B at time t are

$$\mathbf{r}_A (t) = 2\cos (2\pi t)\,\mathbf{i} + 2\sin (2\pi t)\,\mathbf{j}$$

and

$$\mathbf{r}_B (t) = 3\cos (\pi t)\,\mathbf{i} + 3\sin (\pi t)\,\mathbf{j},$$

respectively, where the sun is assumed to be located at the origin and distance is measured in astronomical units. (Notice that planet A moves faster than planet B.)

   The people on planet A regard their planet, not the sun, as the center of their planetary system (their solar system).

   a) Using planet A as the origin of a new coordinate system, give parametric equations for the location of planet B at time t. Write your answer in terms of $\cos (\pi t)$ and $\sin (\pi t)$.

   b) GRAPHER Using planet A as the origin, sketch a graph of the path of planet B.

   This exercise illustrates the difficulty that people before Kepler's time, with an earth-centered (planet A) view of our solar system, had in understanding the motions of the planets (i.e., planet $B =$ Mars). See D. G. Saari's article in *The American Mathematical Monthly,* Vol. 97, Feb. 1990, pp. 105–119.

16. Kepler discovered that the path of the earth around the sun is an ellipse with the sun at one of the foci. Let $\mathbf{r}(t)$ be the position vector from the center of the sun to the center of the earth at time t. Let $\mathbf{w}$ be the vector from the earth's South Pole to North Pole. It is known that $\mathbf{w}$ is constant and not orthogonal to the plane of the ellipse (the earth's axis is tilted). In terms of $\mathbf{r}(t)$ and $\mathbf{w}$, give the mathematical meaning of (i) perihelion, (ii) aphelion, (iii) equinox, (iv) summer solstice, (v) winter solstice.

CHAPTER 11 QUESTIONS TO GUIDE YOUR REVIEW

1. State the rules for differentiating and integrating vector functions. Give examples.

2. How do you define and calculate the velocity, speed, direction of motion, and acceleration of a body moving along a sufficiently differentiable space curve? Give an example.

3. What is special about the derivatives of vector functions of constant length? Give an example.

4. What are the vector and parametric equations for ideal projectile motion? How do you find a projectile's maximum height, flight time, and range? Give examples.

5. How do you define and calculate the length of a segment of a smooth space curve? Give an example. What mathematical assumptions are involved in the definition?

6. How do you measure distance along a smooth curve in space from a preselected base point? Give an example.

7. What is a differentiable curve's unit tangent vector? Give an example.

8. Define curvature, circle of curvature (osculating circle), center of curvature, and radius of curvature for twice-differentiable curves in the plane. Give examples. What curves have zero curvature? constant curvature?

9. What is a plane curve's principal normal vector? When is it defined? Which way does it point? Give an example.

10. How do you define $\mathbf{N}$ and κ for curves in space? How are these quantities related? Give examples.

11. What is a curve's binormal vector? Give an example. How is this vector related to the curve's torsion? Give an example.

12. What formulas are available for writing a moving body's acceleration as a sum of its tangential and normal components? Give an example. Why might one want to write the acceleration this way? What if the body moves at a constant speed? at a constant speed around a circle?

13. State Kepler's laws. To what do they apply?

CHAPTER 11 PRACTICE EXERCISES

Motion in a Cartesian Plane

In Exercises 1 and 2, graph the curves and sketch their velocity and acceleration vectors at the given values of t. Then write $\mathbf{a}$ in the form $\mathbf{a} = a_T \mathbf{T} + a_N \mathbf{N}$ without finding $\mathbf{T}$ and $\mathbf{N}$, and find the value of κ at the given values of t.

1. $\mathbf{r}(t) = (4 \cos t) \mathbf{i} + (\sqrt{2} \sin t) \mathbf{j}, \quad t = 0$ and $\pi/4$

2. $\mathbf{r}(t) = (\sqrt{3} \sec t) \mathbf{i} + (\sqrt{3} \tan t) \mathbf{j}, \quad t = 0$

3. The position of a particle in the plane at time t is

$$\mathbf{r} = \frac{1}{\sqrt{1 + t^2}} \mathbf{i} + \frac{t}{\sqrt{1 + t^2}} \mathbf{j}.$$

Find the particle's highest speed.

4. Suppose $\mathbf{r}(t) = (e^t \cos t) \mathbf{i} + (e^t \sin t) \mathbf{j}$. Show that the angle between $\mathbf{r}$ and $\mathbf{a}$ never changes. What *is* the angle?

5. At point P, the velocity and acceleration of a particle moving in the plane are $\mathbf{v} = 3 \mathbf{i} + 4 \mathbf{j}$ and $\mathbf{a} = 5 \mathbf{i} + 15 \mathbf{j}$. Find the curvature of the particle's path at P.

6. Find the point on the curve $y = e^x$ where the curvature is greatest.

7. A particle moves around the unit circle in the xy-plane. Its position at time t is $\mathbf{r} = x \mathbf{i} + y \mathbf{j}$, where x and y are differentiable functions of t. Find dy/dt if $\mathbf{v} \cdot \mathbf{i} = y$. Is the motion clockwise, or counterclockwise?

8. You send a message through a pneumatic tube that follows the

curve $9y = x^3$ (distance in meters). At the point (3, 3), $\mathbf{v} \cdot \mathbf{i} = 4$ and $\mathbf{a} \cdot \mathbf{i} = -2$. Find the values of $\mathbf{v} \cdot \mathbf{j}$ and $\mathbf{a} \cdot \mathbf{j}$ at (3, 3).

9. A particle moves in the plane so that its velocity and position vectors are always orthogonal. Show that the particle moves in a circle centered at the origin.

10. A circular wheel with radius 1 ft and center C rolls to the right along the x-axis at a half-turn per second. (See accompanying figure.) At time t seconds, the position vector of the point P on the wheel's circumference is

$$\mathbf{r} = (\pi t - \sin \pi t)\,\mathbf{i} + (1 - \cos \pi t)\,\mathbf{j}.$$

a) Sketch the curve traced by P during the interval $0 \le t \le 3$.
b) Find $\mathbf{v}$ and $\mathbf{a}$ at $t = 0, 1, 2$, and 3 and add these vectors to your sketch.
c) At any given time, what is the forward speed of the topmost point of the wheel? of C?

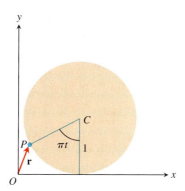

Projectile Motion and Motion in a Plane

11. *Shot put.* A shot leaves the thrower's hand 6.5 ft above the ground at a 45° angle at 44 ft/sec. Where is it 3 sec later?

12. *Javelin.* A javelin leaves the thrower's hand 7 ft above the ground at a 45° angle at 80 ft/sec. How high does it go?

13. A golf ball is hit with an initial speed v_0 at an angle α to the horizontal from a point that lies at the foot of a straight-sided hill that is inclined at an angle ϕ to the horizontal, where

$$0 < \phi < \alpha < \frac{\pi}{2}.$$

Show that the ball lands at a distance

$$\frac{2v_0{}^2 \cos \alpha}{g \cos^2 \phi} \sin(\alpha - \phi),$$

measured up the face of the hill. Hence, show that the greatest range that can be achieved for a given v_0 occurs when $\alpha = (\phi/2) + (\pi/4)$, i.e., when the initial velocity vector bisects the angle between the vertical and the hill.

14. *The Dictator.* The Civil War mortar Dictator weighed so much (17,120 lb) that it had to be mounted on a railroad car. It had a 13-in. bore and used a 20-lb powder charge to fire a 200-lb shell. The mortar was made by Mr. Charles Knapp in his ironworks

in Pittsburgh, Pennsylvania, and was used by the Union army in 1864 in the siege of Petersburg, Virginia. How far did it shoot? Here we have a difference of opinion. The ordnance manual claimed 4325 yd, while field officers claimed 4752 yd. Assuming a 45° firing angle, what muzzle speeds are involved here?

15. *The world's record for popping a champagne cork*

a) Until 1988, the world's record for popping a champagne cork was 109 ft. 6 in., once held by Captain Michael Hill of the British Royal Artillery (of course). Assuming Cpt. Hill held the bottle neck at ground level at a 45° angle, and the cork behaved like an ideal projectile, how fast was the cork going as it left the bottle?

b) A new world record of 177 ft. 9 in. was set on June 5, 1988, by Prof. Emeritus Heinrich of Rensselaer Polytechnic Institute, firing from 4 ft. above ground level at the Woodbury Vineyards Winery, New York. Assuming an ideal trajectory, what was the cork's initial speed?

16. *Javelin.* In Potsdam in 1988, Petra Felke of (then) East Germany set a women's world record by throwing a javelin 262 ft 5 in.

a) Assuming that Felke launched the javelin at a 40° angle to the horizontal 6.5 ft above the ground, what was the javelin's initial speed?

b) How high did the javelin go?

17. *Synchronous curves.* By eliminating α from the ideal projectile equations

$$x = (v_0 \cos \alpha)t, \qquad y = (v_0 \sin \alpha)t - \frac{1}{2}gt^2,$$

show that $x^2 + (y + gt^2/2)^2 = v_0{}^2 t^2$. This shows that projectiles launched simultaneously from the origin at the same initial speed will, at any given instant, all lie on the circle of radius $v_0 t$ centered at $(0, -gt^2/2)$, regardless of their launch angle. These circles are the *synchronous curves* of the launching.

18. Show that the radius of curvature of a twice-differentiable plane curve $\mathbf{r}(t) = f(t)\mathbf{i} + g(t)\mathbf{j}$ is given by the formula

$$\rho = \frac{\dot{x}^2 + \dot{y}^2}{\sqrt{\ddot{x}^2 + \ddot{y}^2 - \ddot{s}^2}}, \quad \text{where } \ddot{s} = \frac{d}{dt}\sqrt{\dot{x}^2 + \dot{y}^2}.$$

19. Express the curvature of the curve

$$\mathbf{r}(t) = \left(\int_0^t \cos\left(\frac{1}{2}\pi\theta^2\right) d\theta\right)\mathbf{i} + \left(\int_0^t \sin\left(\frac{1}{2}\pi\theta^2\right) d\theta\right)\mathbf{j}$$

as a function of the directed distance s measured along the curve from the origin (Fig. 11.44).

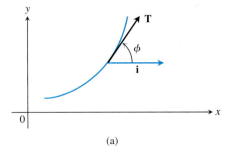

(a)

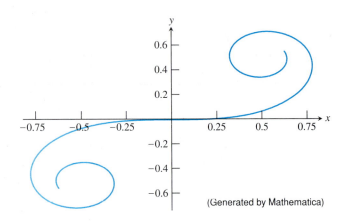

(Generated by Mathematica)

11.44 The curve in Exercise 19.

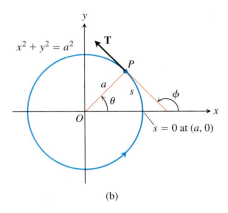

(b)

11.45 Figures for Exercise 20.

20. *An alternative definition of curvature in the plane.* An alternative definition gives the curvature of a sufficiently differentiable plane curve to be $|d\phi/ds|$, where ϕ is the angle between **T** and **i** (Fig. 11.45(a)). Figure 11.45(b) shows the distance s measured counterclockwise around the circle $x^2 + y^2 = a^2$ from the point $(a, 0)$ to a point P, along with the angle ϕ at P. Calculate the circle's curvature using the alternative definition. (*Hint:* $\phi = \theta + \pi/2$.)

Motion in Space

Find the lengths of the curves in Exercises 21 and 22.

21. $\mathbf{r}(t) = (2\cos t)\mathbf{i} + (2\sin t)\mathbf{j} + t^2\mathbf{k}, \quad 0 \le t \le \pi/4$

22. $\mathbf{r}(t) = (3\cos t)\mathbf{i} + (3\sin t)\mathbf{j} + 2t^{3/2}\mathbf{k}, \quad 0 \le t \le 3$

In Exercises 23–26, find **T**, **N**, **B**, κ, and τ at the given value of t.

23. $\mathbf{r}(t) = \frac{4}{9}(1+t)^{3/2}\mathbf{i} + \frac{4}{9}(1-t)^{3/2}\mathbf{j} + \frac{1}{3}t\mathbf{k}, \quad t = 0$

24. $\mathbf{r}(t) = (e^t \sin 2t)\mathbf{i} + (e^t \cos 2t)\mathbf{j} + 2e^t\mathbf{k}, \quad t = 0$

25. $\mathbf{r}(t) = t\mathbf{i} + \frac{1}{2}e^{2t}\mathbf{j}, \quad t = \ln 2$

26. $\mathbf{r}(t) = (3\cosh 2t)\mathbf{i} + (3\sinh 2t)\mathbf{j} + 6t\mathbf{k}, \quad t = \ln 2$

In Exercises 27 and 28, write **a** in the form $\mathbf{a} = a_T\mathbf{T} + a_N\mathbf{N}$ at $t = 0$ without finding **T** and **N**.

27. $\mathbf{r}(t) = (2 + 3t + 3t^2)\mathbf{i} + (4t + 4t^2)\mathbf{j} - (6\cos t)\mathbf{k}$

28. $\mathbf{r}(t) = (2 + t)\mathbf{i} + (t + 2t^2)\mathbf{j} + (1 + t^2)\mathbf{k}$

29. Find **T**, **N**, **B**, κ, and τ as functions of t if $\mathbf{r}(t) = (\sin t)\mathbf{i} + (\sqrt{2}\cos t)\mathbf{j} + (\sin t)\mathbf{k}$.

30. At what times in the interval $0 \le t \le \pi$ are the velocity and acceleration vectors of the motion $\mathbf{r}(t) = \mathbf{i} + (5\cos t)\mathbf{j} + (3\sin t)\mathbf{k}$ orthogonal?

31. The position of a particle moving in space at time $t \ge 0$ is

$$\mathbf{r}(t) = 2\mathbf{i} + \left(4\sin\frac{t}{2}\right)\mathbf{j} + \left(3 - \frac{t}{\pi}\right)\mathbf{k}.$$

Find the first time $\mathbf{r}$ is orthogonal to the vector $\mathbf{i} - \mathbf{j}$.

32. Find equations for the osculating, normal, and rectifying planes of the curve $\mathbf{r}(t) = t\mathbf{i} + t^2\mathbf{j} + t^3\mathbf{k}$ at the point $(1, 1, 1)$.

33. Find parametric equations for the line that is tangent to the curve $\mathbf{r}(t) = e^t\mathbf{i} + (\sin t)\mathbf{j} + \ln(1 - t)\mathbf{k}$ at $t = 0$.

34. Find parametric equations for the line tangent to the helix $\mathbf{r}(t) = (\sqrt{2}\cos t)\mathbf{i} + (\sqrt{2}\sin t)\mathbf{j} + t\mathbf{k}$ at the point where $t = \pi/4$.

35. *The view from Skylab 4.* What percentage of Earth's surface area could the astronauts see when *Skylab 4* was at its apogee height, 437 km above the surface? To find out, model the visible surface as the surface generated by revolving the circular arc *GT*, shown here, about the *y*-axis. Then carry out these steps:

1. Use similar triangles in the figure to show that $y_0/6380 = 6380/(6380 + 437)$. Solve for y_0.

2. To four significant digits, calculate the visible area as
$$VA = \int_{y_0}^{6380} 2\pi x \sqrt{1 + \left(\frac{dx}{dy}\right)^2}\, dy.$$

3. Express the result as a percentage of Earth's surface area.

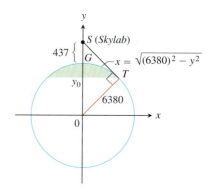

CHAPTER 11 ADDITIONAL EXERCISES–THEORY, EXAMPLES, APPLICATIONS

Applications

1. A straight river is 100 m wide. A rowboat leaves the far shore at time $t = 0$. The person in the boat rows at a rate of 20 m/min, always toward the near shore. The velocity of the river at (x, y) is $\mathbf{v} = \left(-\dfrac{1}{250}(y - 50)^2 + 10\right)\mathbf{i}$ m/min, $0 < y < 100$.

   a) Given that $\mathbf{r}(0) = 0\mathbf{i} + 100\mathbf{j}$, what is the position of the boat at time t?

   b) How far downstream will the boat land on the near shore?

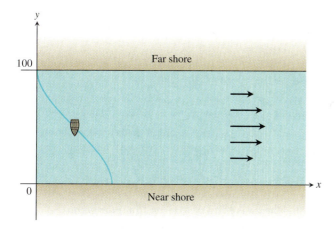

2. A straight river is 20 m wide. The velocity of the river at (x, y) is
$$\mathbf{v} = -\frac{3x(20 - x)}{100}\mathbf{j} \text{ m/min}, \quad 0 \le x \le 20.$$

A boat leaves the shore at $(0, 0)$ and travels through the water with a constant velocity. It arrives at the opposite shore at $(20, 0)$. The speed of the boat is always $\sqrt{20}$ m/min.

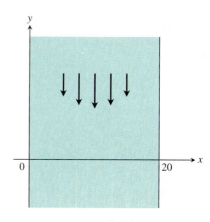

   a) Find the velocity of the boat.

   b) Find the location of the boat at time t.

   c) Sketch the path of the boat.

3. The line through the origin and the point $A(1, 1, 1)$ is the axis of rotation of a rigid body rotating with a constant angular speed of $3/2$ rad/sec. The rotation appears to be clockwise when we look toward the origin from A. Find the velocity $\mathbf{v}$ of the point of the body that is at the position $B\,(1, 3, 2)$.

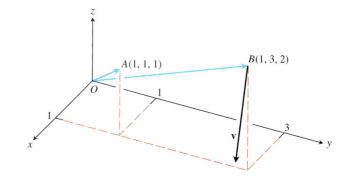

4. The curve

$$\mathbf{r}(t) = (2\sqrt{t}\cos t)\,\mathbf{i} + (3\sqrt{t}\sin t)\,\mathbf{j} + \sqrt{1-t}\,\mathbf{k}, \quad 0 \le t \le 1,$$

lies on a quadric surface. Describe the surface and find an equation for it.

5. A frictionless particle P, starting from rest at time $t = 0$ at the point $(a, 0, 0)$, slides down the helix

$$\mathbf{r}(\theta) = (a\cos\theta)\,\mathbf{i} + (a\sin\theta)\,\mathbf{j} + b\theta\,\mathbf{k} \quad (a, b > 0)$$

under the influence of gravity, as in Fig. 11.46. The θ in this equation is the cylindrical coordinate θ and the helix is the curve $r = a, z = b\theta, \theta \ge 0$, in cylindrical coordinates. We assume θ to be a differentiable function of t for the motion. The law of conservation of energy tells us that the particle's speed after it has fallen straight down a distance z is $\sqrt{2gz}$, where g is the constant acceleration of gravity.

a) Find the angular velocity $d\theta/dt$ when $\theta = 2\pi$.

b) Express the particle's θ- and z-coordinates as functions of t.

c) Express the tangential and normal components of the velocity $d\mathbf{r}/dt$ and acceleration $d^2\mathbf{r}/dt^2$ as functions of t. Does the acceleration have any nonzero component in the direction of the binormal vector $\mathbf{B}$?

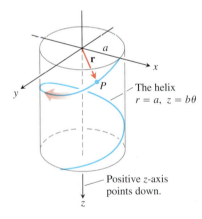

11.46 The circular helix in Exercise 5.

6. Suppose the curve in Exercise 5 is replaced by the conical helix $r = a\theta, z = b\theta$ shown in Fig. 11.47.

a) Express the angular velocity $d\theta/dt$ as a function of θ.

b) Express the distance the particle travels along the helix as a function of θ.

Polar Coordinate Systems and Motion in Space

7. Deduce from the orbit equation

$$r = \frac{(1+e)r_0}{1 + e\cos\theta}$$

that a planet is closest to its sun when $\theta = 0$ and show that $r = r_0$ at that time.

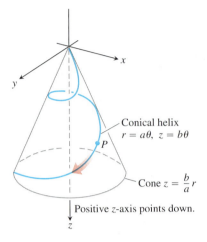

11.47 The conical helix in Exercise 6.

8. *A Kepler equation.* The problem of locating a planet in its orbit at a given time and date eventually leads to solving "Kepler" equations of the form

$$f(x) = x - 1 - \frac{1}{2}\sin x = 0.$$

a) Show that this particular equation has a solution between $x = 0$ and $x = 2$.

b) With your computer or calculator in radian mode, use Newton's method to find the solution to as many places as you can.

9. In Section 11.5, we found the velocity of a particle moving in the plane to be

$$\mathbf{v} = \dot{x}\,\mathbf{i} + \dot{y}\,\mathbf{j} = \dot{r}\,\mathbf{u}_r + r\dot{\theta}\,\mathbf{u}_\theta.$$

a) Express $\dot{x}$ and $\dot{y}$ in terms of $\dot{r}$ and $r\dot{\theta}$ by evaluating the dot products $\mathbf{v} \cdot \mathbf{i}$ and $\mathbf{v} \cdot \mathbf{j}$.

b) Express $\dot{r}$ and $r\dot{\theta}$ in terms of $\dot{x}$ and $\dot{y}$ by evaluating the dot products $\mathbf{v} \cdot \mathbf{u}_r$ and $\mathbf{v} \cdot \mathbf{u}_\theta$.

10. Express the curvature of a twice-differentiable curve $r = f(\theta)$ in the polar coordinate plane in terms of f and its derivatives.

11. A slender rod through the origin of the polar coordinate plane rotates (in the plane) about the origin at the rate of 3 rad/min. A beetle starting from the point $(2, 0)$ crawls along the rod toward the origin at the rate of 1 in./min.

a) Find the beetle's acceleration and velocity in polar form when it is halfway to (1 in. from) the origin.

b) CALCULATOR To the nearest tenth of an inch, what will be the length of the path the beetle has traveled by the time it reaches the origin?

12. *Conservation of angular momentum.* Let $\mathbf{r}(t)$ denote the position in space of a moving object at time t. Suppose the force acting on the object at time t is

$$\mathbf{F}(t) = -\frac{c}{|\mathbf{r}(t)|^3}\,\mathbf{r}(t),$$

where c is a constant. In physics the **angular momentum** of an object at time t is defined to be $\mathbf{L}(t) = \mathbf{r}(t) \times m\mathbf{v}(t)$, where m is the mass of the object and $\mathbf{v}(t)$ is the velocity. Prove that angular momentum is a conserved quantity; i.e., prove that $\mathbf{L}(t)$ is a constant vector, independent of time. Remember Newton's law $\mathbf{F} = m\mathbf{a}$. (This is a calculus problem, not a physics problem.)

Cylindrical Coordinate Systems

13. *Unit vectors for position and motion in cylindrical coordinates.* When the position of a particle moving in space is given in cylindrical coordinates, the unit vectors we use to describe its position and motion are

$$\mathbf{u}_r = (\cos\theta)\,\mathbf{i} + (\sin\theta)\,\mathbf{j}, \quad \mathbf{u}_\theta = -(\sin\theta)\,\mathbf{i} + (\cos\theta)\,\mathbf{j},$$

and $\mathbf{k}$ (Fig. 11.48). The particle's position vector is then $\mathbf{r} = r\,\mathbf{u}_r + z\,\mathbf{k}$, where r is the positive polar distance coordinate of the particle's position.

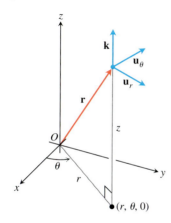

11.48 The unit vectors for describing motion in cylindrical coordinates (Exercise 13).

a) Show that $\mathbf{u}_r$, $\mathbf{u}_\theta$, and $\mathbf{k}$, in this order, form a right-handed frame of unit vectors.

b) Show that

$$\frac{d\mathbf{u}_r}{d\theta} = \mathbf{u}_\theta \quad \text{and} \quad \frac{d\mathbf{u}_\theta}{d\theta} = -\mathbf{u}_r.$$

c) Assuming that the necessary derivatives with respect to t exist, express $\mathbf{v} = \dot{\mathbf{r}}$ and $\mathbf{a} = \ddot{\mathbf{r}}$ in terms of $\mathbf{u}_r$, $\mathbf{u}_\theta$, $\mathbf{k}$, $\dot{r}$, and $\dot{\theta}$. (The dots indicate derivatives with respect to t: $\dot{\mathbf{r}}$ means $d\mathbf{r}/dt$, $\ddot{\mathbf{r}}$ means $d^2\mathbf{r}/dt^2$, and so on.) Section 11.5 derives these formulas and shows how the vectors mentioned here are used in describing planetary motion.

14. *Arc length in cylindrical coordinates*

a) Show that when you express $ds^2 = dx^2 + dy^2 + dz^2$ in terms of cylindrical coordinates, you get $ds^2 = dr^2 + r^2 d\theta^2 + dz^2$.

b) Interpret this result geometrically in terms of the edges and a diagonal of a box. Sketch the box.

c) Use the result in (a) to find the length of the curve $r = e^\theta$, $z = e^\theta$, $0 \le \theta \le \ln 8$.

Spherical Coordinate Systems

15. *Unit vectors for position and motion in spherical coordinates.* Hold two of the three spherical coordinates ρ, ϕ, θ of a point P in space constant while letting the remaining coordinate increase. Let $\mathbf{u}$, with a subscript corresponding to the increasing coordinate, be the unit vector that points in the direction in which P then starts to move. The three resulting unit vectors, $\mathbf{u}_\rho, \mathbf{u}_\phi$, and $\mathbf{u}_\theta$ at P are shown in Fig. 11.49.

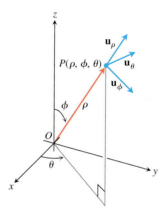

11.49 The unit vectors for describing motion in spherical coordinates (Exercise 15).

a) Express $\mathbf{u}_\rho$, $\mathbf{u}_\phi$, and $\mathbf{u}_\theta$ in terms of $\mathbf{i}, \mathbf{j}$, and $\mathbf{k}$.

b) Show that $\mathbf{u}_\rho \cdot \mathbf{u}_\phi = 0$.

c) Show that $\mathbf{u}_\theta = \mathbf{u}_\rho \times \mathbf{u}_\phi$.

d) Show that $\mathbf{u}_\rho$, $\mathbf{u}_\phi$, and $\mathbf{u}_\theta$, in that order, make a right-handed frame of mutually orthogonal vectors.

16. *Arc length in spherical coordinates*

a) Show that when you express $ds^2 = dx^2 + dy^2 + dz^2$ in terms of spherical coordinates, you get $ds^2 = d\rho^2 + \rho^2 d\phi^2 + \rho^2 \sin^2\phi\, d\theta^2$.

b) Interpret this result geometrically in terms of the edges and a diagonal of a "box" cut from a solid sphere. Sketch the box.

c) Use the result in (a) to find the length of the curve $\rho = 2e^\theta$, $\phi = \pi/6$, $0 \le \theta \le \ln 8$.

Multivariable Functions and Partial Derivatives

OVERVIEW   Functions with two or more independent variables appear more often in science than functions of a single variable, and their calculus is even richer. Their derivatives are more varied and more interesting because of the different ways in which the variables can interact. Their integrals lead to a greater variety of applications. The studies of probability, statistics, fluid dynamics, and electricity, to mention only a few, all lead in natural ways to functions of more than one variable. The mathematics of these functions is one of the finest achievements in science.

12.1 Functions of Several Variables

Many functions depend on more than one independent variable. The function $V = \pi r^2 h$ calculates the volume of a right circular cylinder from its radius and height. The function $f(x, y) = x^2 + y^2$ calculates the height of the paraboloid $z = x^2 + y^2$ above the point $P(x, y)$ from the two coordinates of P. In this section, we define functions of more than one independent variable and discuss ways to graph them.

Functions and Variables

Real-valued functions of several independent real variables are defined much the way you would imagine from the single-variable case. The domains are sets of ordered pairs (triples, quadruples, whatever) of real numbers, and the ranges are sets of real numbers of the kind we have worked with all along.

Definitions

Suppose D is a set of n-tuples of real numbers $(x_1, x_2, \ldots, x_n)$. A **real-valued function** f on D is a rule that assigns a real number

$$w = f(x_1, x_2, \ldots, x_n)$$

to each element in D. The set D is the function's **domain.** The set of w-values taken on by f is the function's **range.** The symbol w is the **dependent variable** of f, and f is said to be a function of the n **independent variables** x_1 to x_n. We also call the x's the function's **input variables** and call w the function's **output variable.**

If f is a function of two independent variables, we usually call the independent variables x and y and picture the domain of f as a region in the xy-plane. If f is a function of three independent variables, we call the variables x, y, and z and picture the domain as a region in space.

In applications, we tend to use letters that remind us of what the variables stand for. To say that the volume of a right circular cylinder is a function of its radius and height, we might write $V = f(r, h)$. To be more specific, we might replace the notation $f(r, h)$ by the formula that calculates the value of V from the values of r and h, and write $V = \pi r^2 h$. In either case, r and h would be the independent variables and V the dependent variable of the function.

As usual, we evaluate functions defined by formulas by substituting the values of the independent variables in the formula and calculating the corresponding value of the dependent variable.

EXAMPLE 1   The value of $f(x, y, z) = \sqrt{x^2 + y^2 + z^2}$ at the point $(3, 0, 4)$ is

$$f(3, 0, 4) = \sqrt{(3)^2 + (0)^2 + (4)^2} = \sqrt{25} = 5.$$ ❏

Domains

In defining functions of more than one variable, we follow the usual practice of excluding inputs that lead to complex numbers or division by zero. If $f(x, y) = \sqrt{y - x^2}$, y cannot be less than x^2. If $f(x, y) = 1/(xy)$, xy cannot be zero. The domains of functions are otherwise assumed to be the largest sets for which the defining rules generate real numbers.

EXAMPLE 2   *Functions of two variables*

Function	Domain	Range
$w = \sqrt{y - x^2}$	$y \geq x^2$	$[0, \infty)$
$w = \dfrac{1}{xy}$	$xy \neq 0$	$(-\infty, 0) \cup (0, \infty)$
$w = \sin xy$	Entire plane	$[-1, 1]$

❏

EXAMPLE 3   *Functions of three variables*

Function	Domain	Range
$w = \sqrt{x^2 + y^2 + z^2}$	Entire space	$[0, \infty)$
$w = \dfrac{1}{x^2 + y^2 + z^2}$	$(x, y, z) \neq (0, 0, 0)$	$(0, \infty)$
$w = xy \ln z$	Half-space $z > 0$	$(-\infty, \infty)$

❏

The domains of functions defined on portions of the plane can have interior points and boundary points just the way the domains of functions defined on intervals of the real line can.

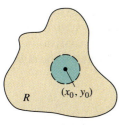

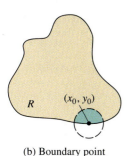

(a) Interior point

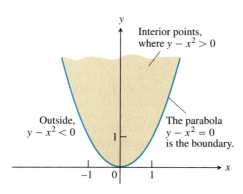

(b) Boundary point

12.1 Interior points and boundary points of a plane region R. An interior point is necessarily a point of R. A boundary point of R need not belong to R.

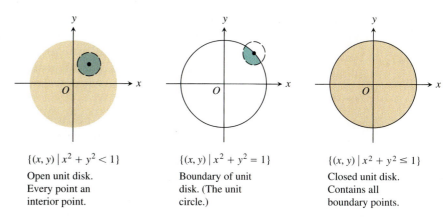

$\{(x, y) \mid x^2 + y^2 < 1\}$
Open unit disk. Every point an interior point.

$\{(x, y) \mid x^2 + y^2 = 1\}$
Boundary of unit disk. (The unit circle.)

$\{(x, y) \mid x^2 + y^2 \leq 1\}$
Closed unit disk. Contains all boundary points.

12.2 Interior points and boundary points of the unit disk in the plane.

Definitions

A point (x_0, y_0) in a region (set) R in the xy-plane is an **interior point** of R if it is the center of a disk that lies entirely in R (Fig. 12.1). A point (x_0, y_0) is a **boundary point** of R if every disk centered at (x_0, y_0) contains points that lie outside of R as well as points that lie in R. (The boundary point itself need not belong to R.)

The interior points of a region, as a set, make up the **interior** of the region. The region's boundary points make up its **boundary.** A region is **open** if it consists entirely of interior points. A region is **closed** if it contains all of its boundary points (Fig. 12.2).

As with intervals of real numbers, some regions in the plane are neither open nor closed. If you start with the open disk in Fig. 12.2 and add to it some but not all of its boundary points, the resulting set is neither open nor closed. The boundary points that *are* there keep the set from being open. The absence of the remaining boundary points keeps the set from being closed.

Definitions

A region in the plane is **bounded** if it lies inside a disk of fixed radius. A region is **unbounded** if it is not bounded.

EXAMPLE 4

Bounded sets in the plane: Line segments, triangles, interiors of triangles, rectangles, disks

Unbounded sets in the plane: Lines, coordinate axes, the graphs of functions defined on infinite intervals, quadrants, half-planes, the plane itself ❑

EXAMPLE 5   The domain of the function $f(x, y) = \sqrt{y - x^2}$ is closed and unbounded (Fig. 12.3). The parabola $y = x^2$ is the boundary of the domain. The points above the parabola make up the domain's interior. ❑

12.3 The domain of $f(x, y) = \sqrt{y - x^2}$ consists of the shaded region and its bounding parabola $y = x^2$.

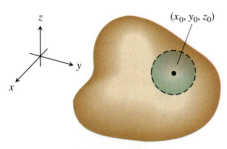

(a) Interior point

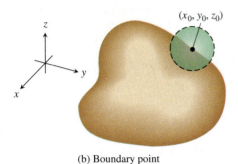

(b) Boundary point

12.4 Interior points and boundary points of a region in space.

The definitions of interior, boundary, open, closed, bounded, and unbounded for regions in space are similar to those for regions in the plane. To accommodate the extra dimension, we use balls instead of disks. A **closed ball** consists of the region of points inside a sphere together with the sphere. An **open ball** is the region of points inside a sphere without the bounding sphere.

Definitions

A point (x_0, y_0, z_0) in a region D in space is an **interior point** of D if it is the center of a ball that lies entirely in D (Fig. 12.4). A point (x_0, y_0, z_0) is a **boundary point** of D if every sphere centered at (x_0, y_0, z_0) encloses points that lie outside D as well as points that lie inside D. The **interior** of D is the set of interior points of D. The **boundary** of D is the set of boundary points of D.

A region D is **open** if it consists entirely of interior points. A region is **closed** if it contains its entire boundary.

EXAMPLE 6

Open sets in space:	Open balls; the open half-space $z > 0$; the first octant (bounding planes absent); space itself
Closed sets in space:	Lines; planes; closed balls; the closed half-space $z \geq 0$; the first octant together with its bounding planes; space itself
Neither open nor closed:	A closed ball with part of its bounding sphere removed; solid cube with a missing face, edge, or corner point

Graphs and Level Curves of Functions of Two Variables

There are two standard ways to picture the values of a function $f(x, y)$. One is to draw and label curves in the domain on which f has a constant value. The other is to sketch the surface $z = f(x, y)$ in space.

Definitions

The set of points in the plane where a function $f(x, y)$ has a constant value $f(x, y) = c$ is called a **level curve** of f. The set of all points $(x, y, f(x, y))$ in space, for (x, y) in the domain of f, is called the **graph** of f. The graph of f is also called the **surface $z = f(x, y)$**.

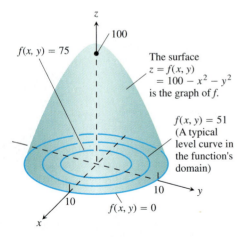

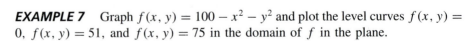

The surface
$z = f(x, y)$
$= 100 - x^2 - y^2$
is the graph of f.

$f(x, y) = 51$
(A typical level curve in the function's domain)

12.5 The graph and selected level curves of the function $f(x, y) = 100 - x^2 - y^2$.

EXAMPLE 7 Graph $f(x, y) = 100 - x^2 - y^2$ and plot the level curves $f(x, y) = 0$, $f(x, y) = 51$, and $f(x, y) = 75$ in the domain of f in the plane.

Solution The domain of f is the entire xy-plane, and the range of f is the set of real numbers less than or equal to 100. The graph is the paraboloid $z = 100 - x^2 - y^2$, a portion of which is shown in Fig. 12.5.

The level curve $f(x, y) = 0$ is the set of points in the xy-plane at which

$$f(x, y) = 100 - x^2 - y^2 = 0, \qquad \text{or} \qquad x^2 + y^2 = 100,$$

The contour line $f(x, y) = 100 - x^2 - y^2 = 75$ is the circle $x^2 + y^2 = 25$ in the plane $z = 75$.

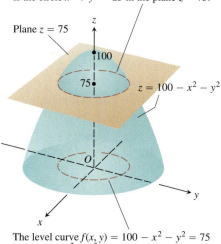

The level curve $f(x, y) = 100 - x^2 - y^2 = 75$ is the circle $x^2 + y^2 = 25$ in the xy-plane.

12.6 The graph of $f(x, y) = 100 - x^2 - y^2$ and its intersection with the plane $z = 75$.

which is the circle of radius 10 centered at the origin. Similarly, the level curves $f(x, y) = 51$ and $f(x, y) = 75$ (Fig. 12.5) are the circles

$$f(x, y) = 100 - x^2 - y^2 = 51, \quad \text{or} \quad x^2 + y^2 = 49,$$
$$f(x, y) = 100 - x^2 - y^2 = 75, \quad \text{or} \quad x^2 + y^2 = 25.$$

The level curve $f(x, y) = 100$ consists of the origin alone. (It is still a level curve.)

Contour Lines

The curve in space in which the plane $z = c$ cuts a surface $z = f(x, y)$ is made up of the points that represent the function value $f(x, y) = c$. It is called the **contour line** $f(x, y) = c$ to distinguish it from the level curve $f(x, y) = c$ in the domain of f. Figure 12.6 shows the contour line $f(x, y) = 75$ on the surface $z = 100 - x^2 - y^2$ defined by the function $f(x, y) = 100 - x^2 - y^2$. The contour line lies directly above the circle $x^2 + y^2 = 25$, which is the level curve $f(x, y) = 75$ in the function's domain.

Not everyone makes this distinction, however, and you may wish to call both kinds of curves by a single name and rely on context to convey which one you have in mind. On most maps, for example, the curves that represent constant elevation (height above sea level) are called contours, not level curves (Fig. 12.7).

Level Surfaces of Functions of Three Variables

In the plane, the points where a function of two independent variables has a constant value $f(x, y) = c$ make a curve in the function's domain. In space, the points where a function of three independent variables has a constant value $f(x, y, z) = c$ make a surface in the function's domain.

Definition

The set of points (x, y, z) in space where a function of three independent variables has a constant value $f(x, y, z) = c$ is called a **level surface** of f.

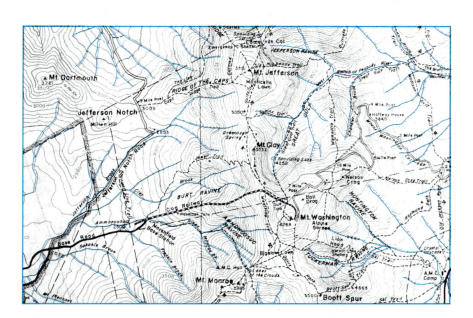

12.7 Contours on Mt. Washington in north central New Hampshire. The streams, which follow paths of steepest descent, run perpendicular to the contours. So does the Cog Railway.

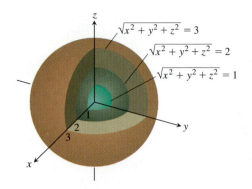

12.8 The level surfaces of $f(x, y, z) = \sqrt{x^2 + y^2 + z^2}$ are concentric spheres.

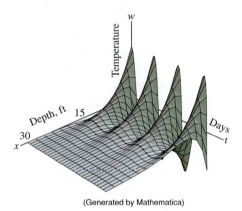

(Generated by Mathematica)

12.9 This computer-generated graph of

$$w = \cos(1.7 \times 10^{-2}t - 0.2x)e^{-0.2x}$$

shows the seasonal variation of the temperature below ground as a fraction of surface temperature. At $x = 15$ ft the variation is only 5% of the variation at the surface. At $x = 30$ ft the variation is less than 0.25% of the surface variation. (Adapted from art provided by Norton Starr for G. C. Berresford's "Differential Equations and Root Cellars," *The UMAP Journal,* Vol. 2, No. 3 [1981], pp. 53–75.)

EXAMPLE 8 Describe the level surfaces of the function

$$f(x, y, z) = \sqrt{x^2 + y^2 + z^2}.$$

Solution The value of f is the distance from the origin to the point (x, y, z). Each level surface $\sqrt{x^2 + y^2 + z^2} = c$, $c > 0$, is a sphere of radius c centered at the origin. Figure 12.8 shows a cutaway view of three of these spheres. The level surface $\sqrt{x^2 + y^2 + z^2} = 0$ consists of the origin alone.

We are not graphing the function here. The graph of the function, made up of the points $(x, y, z, \sqrt{x^2 + y^2 + z^2})$, lies in a four-variable space. Instead, we are looking at level surfaces in the function's domain.

The function's level surfaces show how the function's values change as we move through its domain. If we remain on a sphere of radius c centered at the origin, the function maintains a constant value, namely c. If we move from one sphere to another, the function's value changes. It increases if we move away from the origin and decreases if we move toward the origin. The way the function's values change depends on the direction we take. The dependence of change on direction is important. We will return to it in Section 12.7. ❑

Computer Graphing

The three-dimensional graphing programs for computers make it possible to graph functions of two variables with only a few keystrokes. We can often get information more quickly from a graph than from a formula.

EXAMPLE 9 Figure 12.9 shows a computer-generated graph of the function $w = \cos(1.7 \times 10^{-2}t - 0.2x)\,e^{-0.2x}$, where t is in days and x is in feet. The graph shows how the temperature beneath the earth's surface varies with time. The variation is given as a fraction of the variation at the surface. At a depth of 15 ft, the variation (change in vertical amplitude in the figure) is about 5 percent of the surface variation. At 30 ft, there is almost no variation during the year.

The graph also shows that the temperature 15 ft below the surface is about half a year out of phase with the surface temperature. When the temperature is lowest on the surface (late January, say) it is at its highest 15 ft below. Fifteen feet below the ground, the seasons are reversed. ❑

Exercises 12.1

Domain, Range, and Level Curves

In Exercises 1–12, (a) find the function's domain, (b) find the function's range, (c) describe the function's level curves, (d) find the boundary of the function's domain, (e) determine if the domain is an open region, a closed region, or neither, and (f) decide if the domain is bounded or unbounded.

1. $f(x, y) = y - x$

2. $f(x, y) = \sqrt{y - x}$

3. $f(x, y) = 4x^2 + 9y^2$

4. $f(x, y) = x^2 - y^2$

5. $f(x, y) = xy$

6. $f(x, y) = y/x^2$

7. $f(x, y) = \dfrac{1}{\sqrt{16 - x^2 - y^2}}$

8. $f(x, y) = \sqrt{9 - x^2 - y^2}$

9. $f(x, y) = \ln(x^2 + y^2)$

10. $f(x, y) = e^{-(x^2 + y^2)}$

11. $f(x, y) = \sin^{-1}(y - x)$

12. $f(x, y) = \tan^{-1}\left(\dfrac{y}{x}\right)$

Identifying Surfaces and Level Curves

Exercises 13–18 show level curves for the functions graphed in (a)–(f). Match each set of curves with the appropriate function.

13.

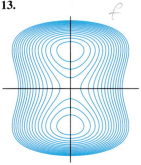

14.

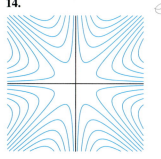

15.

16.

17.

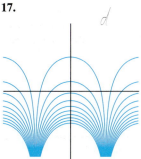

18.

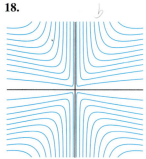

(a)

$z = (\cos x)(\cos y)\, e^{-\sqrt{x^2+y^2}/4}$

(b)

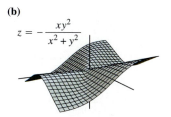

$z = -\dfrac{xy^2}{x^2+y^2}$

(c)

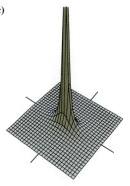

$z = \dfrac{1}{(4x^2+y^2)}$

(d)

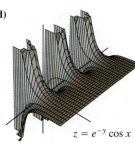

$z = e^{-y}\cos x$

(e)

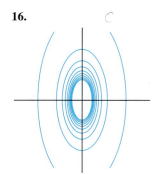

$z = \dfrac{xy(x^2-y^2)}{(x^2+y^2)}$

(f)

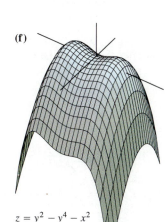

$z = y^2 - y^4 - x^2$

Identifying Functions of Two Variables

Display the values of the functions in Exercises 19–28 in two ways: (a) by sketching the surface $z = f(x, y)$ and (b) by drawing an assortment of level curves in the function's domain. Label each level curve with its function value.

19. $f(x, y) = y^2$ **20.** $f(x, y) = 4 - y^2$

21. $f(x, y) = x^2 + y^2$ **22.** $f(x, y) = \sqrt{x^2 + y^2}$

23. $f(x, y) = -(x^2 + y^2)$ **24.** $f(x, y) = 4 - x^2 - y^2$

25. $f(x, y) = 4x^2 + y^2$ **26.** $f(x, y) = 4x^2 + y^2 + 1$

27. $f(x, y) = 1 - |y|$ **28.** $f(x, y) = 1 - |x| - |y|$

Level Surfaces

In Exercises 29–36, sketch a typical level surface for the function.

29. $f(x, y, z) = x^2 + y^2 + z^2$

30. $f(x, y, z) = \ln(x^2 + y^2 + z^2)$

31. $f(x, y, z) = x + z$ **32.** $f(x, y, z) = z$

33. $f(x, y, z) = x^2 + y^2$ **34.** $f(x, y, z) = y^2 + z^2$

35. $f(x, y, z) = z - x^2 - y^2$

36. $f(x, y, z) = (x^2/25) + (y^2/16) + (z^2/9)$

Finding a Level Curve

In Exercises 37–40, find an equation for the level curve of the function $f(x, y)$ that passes through the given point.

37. $f(x, y) = 16 - x^2 - y^2$, $\left(2\sqrt{2}, \sqrt{2}\right)$

38. $f(x, y) = \sqrt{x^2 - 1}$, $(1, 0)$

39. $f(x, y) = \int_x^y \dfrac{dt}{1 + t^2}$, $\left(-\sqrt{2}, \sqrt{2}\right)$

40. $f(x, y) = \sum_{n=0}^{\infty} \left(\dfrac{x}{y}\right)^n$, $(1, 2)$

Finding a Level Surface

In Exercises 41–44, find an equation for the level surface of the function through the given point.

41. $f(x, y, z) = \sqrt{x - y} - \ln z$, $(3, -1, 1)$

42. $f(x, y, z) = \ln(x^2 + y + z^2)$, $(-1, 2, 1)$

43. $g(x, y, z) = \sum_{n=0}^{\infty} \dfrac{(x + y)^n}{n! \, z^n}$, $(\ln 2, \ln 4, 3)$

44. $g(x, y, z) = \int_x^y \dfrac{d\theta}{\sqrt{1 - \theta^2}} + \int_{\sqrt{2}}^z \dfrac{dt}{t\sqrt{t^2 - 1}}$, $(0, 1/2, 2)$

Theory and Examples

45. *The maximum value of a function on a line in space.* Does the function $f(x, y, z) = xyz$ have a maximum value on the line $x = 20 - t$, $y = t$, $z = 20$? If so, what is it? Give reasons for your answer. (*Hint:* Along the line, $w = f(x, y, z)$ is a differentiable function of t.)

46. *The minimum value of a function on a line in space.* Does the function $f(x, y, z) = xy - z$ have a minimum value on the line $x = t - 1$, $y = t - 2$, $z = t + 7$? If so, what is it? Give reasons for your answer. (*Hint:* Along the line, $w = f(x, y, z)$ is a differentiable function of t.)

47. *The Concorde's sonic booms.* The width w of the region in which people on the ground hear the *Concorde's* sonic boom directly, not reflected from a layer in the atmosphere, is a function of

$T =$ air temperature at ground level (in degrees Kelvin),

$h =$ the *Concorde's* altitude (in km),

$d =$ the vertical temperature gradient (temperature drop in degrees Kelvin per km).

The formula for w is

$$w = 4(Th/d)^{1/2}.$$

See Fig. 12.10.

The Washington-bound *Concorde* approaches the United States from Europe on a course that takes it south of Nantucket Island at an altitude of 16.8 km. If the surface temperature is 290 K and the vertical temperature gradient is 5 K/km, how many

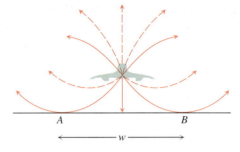

Sonic boom carpet

12.10 Sound waves from the *Concorde* bend as the temperature changes above and below the altitude at which the plane flies. The sonic boom carpet is the region on the ground that receives shock waves directly from the plane, not reflected from the atmosphere or diffracted along the ground. The carpet is determined by the grazing rays striking the ground from the point directly under the plane (Exercise 47).

kilometers south of Nantucket must the plane be flown to keep its sonic boom carpet away from the island? (From "Concorde Sonic Booms as an Atmospheric Probe" by N. K. Balachandra, W. L. Donn, and D. H. Rind, *Science*, July 1, 1977, Vol. 197, pp. 47–49).

48. As you know, the graph of a real-valued function of a single real variable is a set in a two-coordinate space. The graph of a real-valued function of two independent real variables is a set in a three-coordinate space. The graph of a real-valued function of three independent real variables is a set in a four-coordinate space. How would you define the graph of a real-valued function $f(x_1, x_2, x_3, x_4)$ of four independent real variables? How would you define the graph of a real-valued function $f(x_1, x_2, x_3, \ldots, x_n)$ of n independent real variables?

✹ CAS Explorations and Projects—Explicit Surfaces

Use a CAS to perform the following steps for each of the functions in Exercises 49–52.

a) Plot the surface over the given rectangle.

b) Plot several level curves in the rectangle.

c) Plot the level curve of f through the given point.

49. $f(x, y) = x \sin \dfrac{y}{2} + y \sin 2x$, $0 \le x \le 5\pi$, $0 \le y \le 5\pi$, $P(3\pi, 3\pi)$

50. $f(x, y) = (\sin x)(\cos y) \, e^{\sqrt{x^2 + y^2}/8}$, $0 \le x \le 5\pi$, $0 \le y \le 5\pi$, $P(4\pi, 4\pi)$

51. $f(x, y) = \sin(x + 2\cos y)$, $-2\pi \le x \le 2\pi$, $-2\pi \le y \le 2\pi$, $P(\pi, \pi)$

52. $f(x, y) = e^{(x^{0.1} - y)} \sin(x^2 + y^2)$, $0 \le x \le 2\pi$, $-2\pi \le y \le \pi$, $P(\pi, -\pi)$

✪ CAS Explorations and Projects—Implicit Surfaces

Use a CAS to plot the level surfaces in Exercises 53–56.

53. $4 \ln (x^2 + y^2 + z^2) = 1$ **54.** $x^2 + z^2 = 1$

55. $x + y^2 - 3z^2 = 1$

56. $\sin \left(\dfrac{x}{2} \right) - (\cos y) \sqrt{x^2 + z^2} = 2$

✪ CAS Explorations and Projects— Parametrized Surfaces

Just as you describe curves in the plane parametrically with a pair of equations $x = f(t)$, $y = g(t)$ defined on some parameter interval I, you can sometimes describe surfaces in space with a triple of equations $x = f(u, v)$, $y = g(u, v)$, $z = h(u, v)$ defined on some parameter rectangle $a \le u \le b, c \le v \le d$. Many computer algebra systems permit you to plot such surfaces in *parametric mode*. (Parametrized surfaces are discussed in detail in Section 14.6.) Use a CAS to plot the surfaces in Exercises 57–60. Also plot several level curves in the xy-plane.

57. $x = u \cos v$, $y = u \sin v$, $z = u$, $0 \le u \le 2$, $0 \le v \le 2\pi$

58. $x = u \cos v$, $y = u \sin v$, $z = v$, $0 \le u \le 2$, $0 \le v \le 2\pi$

59. $x = (2 + \cos u) \cos v$, $y = (2 + \cos u) \sin v$, $z = \sin u$,
$0 \le u \le 2\pi$, $0 \le v \le 2\pi$

60. $x = 2 \cos u \cos v$, $y = 2 \cos u \sin v$, $z = 2 \sin u$,
$0 \le u \le 2\pi$, $0 \le v \le \pi$

12.2 Limits and Continuity

This section treats limits and continuity for multivariable functions.

Limits

If the values of $f(x, y)$ lie arbitrarily close to a fixed real number L for all points (x, y) sufficiently close to a point (x_0, y_0), we say that f approaches the limit L as (x, y) approaches (x_0, y_0). This is similar to the informal definition for the limit of a function of a single variable. Notice, however, that if (x_0, y_0) lies in the interior of f's domain, (x, y) can approach (x_0, y_0) from any direction. The direction of approach can be an issue, as in some of the examples that follow.

> **Definition**
>
> We say that a function $f(x, y)$ approaches the **limit** L as (x, y) approaches (x_0, y_0), and write
>
> $$\lim_{(x,y) \to (x_0, y_0)} f(x, y) = L$$
>
> if, for every number $\epsilon > 0$, there exists a corresponding number $\delta > 0$ such that for all (x, y) in the domain of f,
>
> $$0 < \sqrt{(x - x_0)^2 + (y - y_0)^2} < \delta \quad \Rightarrow \quad |f(x, y) - L| < \epsilon. \quad (1)$$

The δ-ϵ requirement in the definition of limit is equivalent to the requirement that, given $\epsilon > 0$, there exists a corresponding $\delta > 0$ such that for all x,

$$0 < |x - x_0| < \delta \quad \text{and} \quad 0 < |y - y_0| < \delta \quad \Rightarrow \quad |f(x, y) - L| < \epsilon \quad (2)$$

(Exercise 59). Thus, in calculating limits we can think either in terms of distance in the plane or in terms of differences in coordinates.

The definition of limit applies to boundary points (x_0, y_0) as well as interior points of the domain of f. The only requirement is that the point (x, y) remain in the domain at all times.

It can be shown, as for functions of a single variable, that

$$\lim_{(x,y)\to(x_0,y_0)} x = x_0$$

$$\lim_{(x,y)\to(x_0,y_0)} y = y_0 \tag{3}$$

$$\lim_{(x,y)\to(x_0,y_0)} k = k. \quad \text{(Any number } k)$$

It can also be shown that the limit of the sum of two functions is the sum of their limits (when they both exist), with similar results for the limits of the differences, products, constant multiples, quotients, and powers.

Theorem 1

Properties of Limits of Functions of Two Variables

The following rules hold if

$$\lim_{(x,y)\to(x_0,y_0)} f(x,y) = L \quad \text{and} \quad \lim_{(x,y)\to(x_0,y_0)} g(x,y) = M.$$

1. *Sum Rule:* $\quad \lim [f(x,y) + g(x,y)] = L + M$

2. *Difference Rule:* $\quad \lim [f(x,y) - g(x,y)] = L - M$

3. *Product Rule:* $\quad \lim f(x,y) \cdot g(x,y) = L \cdot M$

4. *Constant Multiple Rule:* $\quad \lim kf(x,y) = kL \quad \text{(Any number } k)$

5. *Quotient Rule:* $\quad \lim \dfrac{f(x,y)}{g(x,y)} = \dfrac{L}{M} \quad \text{if} \quad M \neq 0.$

6. *Power Rule:* If m and n are integers, then
$$\lim [f(x,y)]^{m/n} = L^{m/n},$$
provided $L^{m/n}$ is a real number.

All limits are to be taken as $(x,y) \to (x_0, y_0)$, and L and M are to be real numbers.

When we apply Theorem 1 to the limits in Eqs. (3), we obtain the useful result that the limits of polynomials and rational functions as $(x,y) \to (x_0, y_0)$ can be calculated by evaluating the functions at (x_0, y_0). The only requirement is that the functions be defined at (x_0, y_0).

EXAMPLE 1

a) $\displaystyle \lim_{(x,y)\to(0,1)} \frac{x - xy + 3}{x^2 y + 5xy - y^3} = \frac{0 - (0)(1) + 3}{(0)^2(1) + 5(0)(1) - (1)^3} = -3$

b) $\displaystyle \lim_{(x,y)\to(3,-4)} \sqrt{x^2 + y^2} = \sqrt{(3)^2 + (-4)^2} = \sqrt{25} = 5$

EXAMPLE 2 Find

$$\lim_{(x,y)\to(0,0)} \frac{x^2 - xy}{\sqrt{x} - \sqrt{y}}.$$

Solution Since the denominator $\sqrt{x} - \sqrt{y}$ approaches 0 as $(x,y) \to (0,0)$, we cannot use the Quotient Rule from Theorem 1. However, if we multiply numerator and denominator by $\sqrt{x} + \sqrt{y}$, we produce an equivalent fraction whose limit we *can* find:

$$\lim_{(x,y)\to(0,0)} \frac{x^2 - xy}{\sqrt{x} - \sqrt{y}} = \lim_{(x,y)\to(0,0)} \frac{(x^2 - xy)(\sqrt{x} + \sqrt{y})}{(\sqrt{x} - \sqrt{y})(\sqrt{x} + \sqrt{y})}$$

$$= \lim_{(x,y)\to(0,0)} \frac{x(x - y)(\sqrt{x} + \sqrt{y})}{x - y} \qquad \text{Algebra}$$

$$= \lim_{(x,y)\to(0,0)} x(\sqrt{x} + \sqrt{y}) \qquad \begin{array}{l}\text{Cancel the}\\ \text{factor}\\ (x - y).\end{array}$$

$$= 0(\sqrt{0} + \sqrt{0}) = 0 \qquad \square$$

Continuity

As with functions of a single variable, continuity is defined in terms of limits.

> **Definitions**
>
> A function $f(x, y)$ is **continuous at the point** (x_0, y_0) if
>
> 1. f is defined at (x_0, y_0),
> 2. $\lim_{(x,y)\to(x_0,y_0)} f(x, y)$ exists,
> 3. $\lim_{(x,y)\to(x_0,y_0)} f(x, y) = f(x_0, y_0)$.
>
> A function is **continuous** if it is continuous at every point of its domain.

As with the definition of limit, the definition of continuity applies at boundary points as well as interior points of the domain of f. The only requirement is that the point (x, y) remain in the domain at all times.

As you may have guessed, one of the consequences of Theorem 1 is that algebraic combinations of continuous functions are continuous at every point at which all the functions involved are defined. This means that sums, differences, products, constant multiples, quotients, and powers of continuous functions are continuous where defined. In particular, polynomials and rational functions of two variables are continuous at every point at which they are defined.

If $z = f(x, y)$ is a continuous function of x and y, and $w = g(z)$ is a continuous function of z, then the composite $w = g(f(x, y))$ is continuous. Thus,

$$e^{x-y}, \qquad \cos\frac{xy}{x^2 + 1}, \qquad \ln(1 + x^2 y^2)$$

are continuous at every point (x, y).

As with functions of a single variable, the general rule is that composites of continuous functions are continuous. The only requirement is that each function be continuous where it is applied.

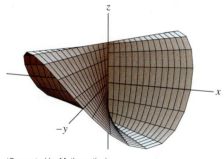

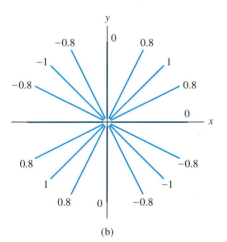

(Generated by Mathematica)

(a)

(b)

12.11 (a) The graph of

$$f(x, y) = \begin{cases} \dfrac{2xy}{x^2 + y^2}, & (x, y) \neq (0, 0) \\ 0, & (x, y) = (0, 0) \end{cases}$$

The function is continuous at every point except the origin. (b) The level curves of f.

EXAMPLE 3 Show that

$$f(x, y) = \begin{cases} \dfrac{2xy}{x^2 + y^2}, & (x, y) \neq (0, 0) \\ 0, & (x, y) = (0, 0) \end{cases}$$

is continuous at every point except the origin (Fig. 12.11).

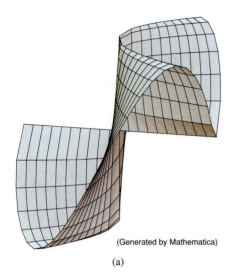

(Generated by Mathematica)

(a)

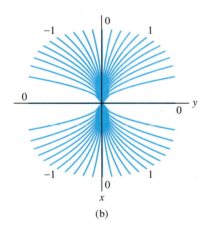

(b)

12.12 (a) The graph of $f(x, y) = 2x^2y/(x^4 + y^2)$. As the graph suggests and the level-curve values in (b) confirm, $\lim_{(x,y)\to(0,0)} f(x, y)$ does not exist.

Solution The function f is continuous at any point $(x, y) \neq (0, 0)$ because its values are then given by a rational function of x and y.

At $(0, 0)$ the value of f is defined, but f, we claim, has no limit as $(x, y) \to (0, 0)$. The reason is that different paths of approach to the origin can lead to different results, as we will now see.

For every value of m, the function f has a constant value on the "punctured" line $y = mx$, $x \neq 0$, because

$$f(x, y)\Big|_{y=mx} = \frac{2xy}{x^2 + y^2}\Big|_{y=mx} = \frac{2x(mx)}{x^2 + (mx)^2} = \frac{2mx^2}{x^2 + m^2x^2} = \frac{2m}{1 + m^2}.$$

Therefore, f has this number as its limit as (x, y) approaches $(0, 0)$ along the line:

$$\lim_{\substack{(x,y)\to(0,0)\\ \text{along } y=mx}} f(x, y) = \lim_{(x,y)\to(0,0)} \left[f(x, y)\Big|_{y=mx} \right] = \frac{2m}{1 + m^2}.$$

This limit changes with m. There is therefore no single number we may call the limit of f as (x, y) approaches the origin. The limit fails to exist, and the function is not continuous.

Example 3 illustrates an important point about limits of functions of two variables (or even more variables, for that matter). For a limit to exist at a point, the limit must be the same along every approach path. Therefore, if we ever find paths with different limits, we know the function has no limit at the point they approach.

> **The Two-Path Test for the Nonexistence of a Limit**
>
> If a function $f(x, y)$ has different limits along two different paths as (x, y) approaches (x_0, y_0), then $\lim_{(x,y)\to(x_0,y_0)} f(x, y)$ does not exist.

EXAMPLE 4   Show that the function

$$f(x, y) = \frac{2x^2y}{x^4 + y^2}$$

(Fig. 12.12) has no limit as (x, y) approaches $(0, 0)$.

Solution Along the curve $y = kx^2$, $x \neq 0$, the function has the constant value

$$f(x, y)\Big|_{y=kx^2} = \frac{2x^2y}{x^4 + y^2}\Big|_{y=kx^2} = \frac{2x^2(kx^2)}{x^4 + (kx^2)^2} = \frac{2kx^4}{x^4 + k^2x^4} = \frac{2k}{1 + k^2}.$$

Therefore,

$$\lim_{\substack{(x,y)\to(0,0)\\ \text{along } y=kx^2}} f(x, y) = \lim_{(x,y)\to(0,0)} \left[f(x, y)\Big|_{y=kx^2} \right] = \frac{2k}{1 + k^2}.$$

This limit varies with the path of approach. If (x, y) approaches $(0, 0)$ along the parabola $y = x^2$, for instance, $k = 1$ and the limit is 1. If (x, y) approaches $(0, 0)$ along the x-axis, $k = 0$ and the limit is 0. By the two-path test, f has no limit as (x, y) approaches $(0, 0)$.

The language here may seem contradictory. You might well ask, "What do you mean f has no limit as (x, y) approaches the origin—it has lots of limits." But that is the point. There is no single path-independent limit, and therefore, by the definition, $\lim_{(x,y)\to(0,0)} f(x, y)$ does not exist. It is our translating this formal statement into the more colloquial "has no limit" that creates the apparent contradiction. The mathematics is fine. The problem arises in how we talk about it. We need the formality to keep things straight. ❑

Functions of More Than Two Variables

The definitions of limit and continuity for functions of two variables and the conclusions about limits and continuity for sums, products, quotients, powers, and composites all extend to functions of three or more variables. Functions like

$$\ln (x + y + z) \qquad \text{and} \qquad \frac{y \sin z}{x - 1}$$

are continuous throughout their domains, and limits like

$$\lim_{P\to(1,0,-1)} \frac{e^{x+z}}{z^2 + \cos \sqrt{xy}} = \frac{e^{1-1}}{(-1)^2 + \cos 0} = \frac{1}{2},$$

where P denotes the point (x, y, z), may be found by direct substitution.

Exercises 12.2

Evaluating Limits

Find the limits in Exercises 1–12.

1. $\displaystyle\lim_{(x,y)\to(0,0)} \frac{3x^2 - y^2 + 5}{x^2 + y^2 + 2}$

2. $\displaystyle\lim_{(x,y)\to(0,4)} \frac{x}{\sqrt{y}}$

3. $\displaystyle\lim_{(x,y)\to(3,4)} \sqrt{x^2 + y^2 - 1}$

4. $\displaystyle\lim_{(x,y)\to(2,-3)} \left(\frac{1}{x} + \frac{1}{y}\right)^2$

5. $\displaystyle\lim_{(x,y)\to(0,\pi/4)} \sec x \tan y$

6. $\displaystyle\lim_{(x,y)\to(0,0)} \cos \frac{x^2 + y^3}{x + y + 1}$

7. $\displaystyle\lim_{(x,y)\to(0,\ln 2)} e^{x-y}$

8. $\displaystyle\lim_{(x,y)\to(1,1)} \ln |1 + x^2 y^2|$

9. $\displaystyle\lim_{(x,y)\to(0,0)} \frac{e^y \sin x}{x}$

10. $\displaystyle\lim_{(x,y)\to(1,1)} \cos \sqrt[3]{|xy| - 1}$

11. $\displaystyle\lim_{(x,y)\to(1,0)} \frac{x \sin y}{x^2 + 1}$

12. $\displaystyle\lim_{(x,y)\to(\pi/2,0)} \frac{\cos y + 1}{y - \sin x}$

Limits of Quotients

Find the limits in Exercises 13–20 by rewriting the fractions first.

13. $\displaystyle\lim_{\substack{(x,y)\to(1,1) \\ x\neq y}} \frac{x^2 - 2xy + y^2}{x - y}$

14. $\displaystyle\lim_{\substack{(x,y)\to(1,1) \\ x\neq y}} \frac{x^2 - y^2}{x - y}$

15. $\displaystyle\lim_{\substack{(x,y)\to(1,1) \\ x\neq 1}} \frac{xy - y - 2x + 2}{x - 1}$

16. $\displaystyle\lim_{\substack{(x,y)\to(2,-4) \\ y\neq-4,\ x\neq x^2}} \frac{y + 4}{x^2 y - xy + 4x^2 - 4x}$

17. $\displaystyle\lim_{\substack{(x,y)\to(0,0) \\ x\neq y}} \frac{x - y + 2\sqrt{x} - 2\sqrt{y}}{\sqrt{x} - \sqrt{y}}$

18. $\displaystyle\lim_{\substack{(x,y)\to(2,2) \\ x+y\neq4}} \frac{x + y - 4}{\sqrt{x + y} - 2}$

19. $\displaystyle\lim_{\substack{(x,y)\to(2,0) \\ 2x-y\neq4}} \frac{\sqrt{2x - y} - 2}{2x - y - 4}$

20. $\displaystyle\lim_{\substack{(x,y)\to(4,3) \\ x\neq y+1}} \frac{\sqrt{x} - \sqrt{y+1}}{x - y - 1}$

Limits with Three Variables

Find the limits in Exercises 21–26.

21. $\displaystyle\lim_{P\to(1,3,4)} \left(\frac{1}{x} + \frac{1}{y} + \frac{1}{z}\right)$

22. $\displaystyle\lim_{P\to(1,-1,-1)} \frac{2xy + yz}{x^2 + z^2}$

23. $\displaystyle\lim_{P\to(3,3,0)} (\sin^2 x + \cos^2 y + \sec^2 z)$

24. $\displaystyle\lim_{P\to(-1/4,\pi/2,2)} \tan^{-1} xyz$

25. $\displaystyle\lim_{P\to(\pi,0,3)} ze^{-2y} \cos 2x$

26. $\displaystyle\lim_{P\to(0,-2,0)} \ln \sqrt{x^2 + y^2 + z^2}$

Continuity in the Plane

At what points (x, y) in the plane are the functions in Exercises 27–30 continuous?

27. a) $f(x, y) = \sin(x + y)$ **b)** $f(x, y) = \ln(x^2 + y^2)$

28. a) $f(x, y) = \dfrac{x + y}{x - y}$ **b)** $f(x, y) = \dfrac{y}{x^2 + 1}$

29. a) $g(x, y) = \sin \dfrac{1}{xy}$ **b)** $g(x, y) = \dfrac{x + y}{2 + \cos x}$

30. a) $g(x, y) = \dfrac{x^2 + y^2}{x^2 - 3x + 2}$ **b)** $g(x, y) = \dfrac{1}{x^2 - y}$

Continuity in Space

At what points (x, y, z) in space are the functions in Exercises 31–34 continuous?

31. a) $f(x, y, z) = x^2 + y^2 - 2z^2$

    b) $f(x, y, z) = \sqrt{x^2 + y^2 - 1}$

32. a) $f(x, y, z) = \ln xyz$ **b)** $f(x, y, z) = e^{x+y} \cos z$

33. a) $h(x, y, z) = xy \sin \dfrac{1}{z}$ **b)** $h(x, y, z) = \dfrac{1}{x^2 + z^2 - 1}$

34. a) $h(x, y, z) = \dfrac{1}{|y| + |z|}$ **b)** $h(x, y, z) = \dfrac{1}{|xy| + |z|}$

No Limit at a Point

By considering different paths of approach, show that the functions in Exercises 35–42 have no limit as $(x, y) \to (0, 0)$.

35. $f(x, y) = -\dfrac{x}{\sqrt{x^2 + y^2}}$ **36.** $f(x, y) = \dfrac{x^4}{x^4 + y^2}$

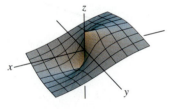

(Generated by Mathematica)   (Generated by Mathematica)

37. $f(x, y) = \dfrac{x^4 - y^2}{x^4 + y^2}$ **38.** $f(x, y) = \dfrac{xy}{|xy|}$

39. $g(x, y) = \dfrac{x - y}{x + y}$ **40.** $g(x, y) = \dfrac{x + y}{x - y}$

41. $h(x, y) = \dfrac{x^2 + y}{y}$ **42.** $h(x, y) = \dfrac{x^2}{x^2 - y}$

Theory and Examples

43. If $\lim_{(x,y)\to(x_0, y_0)} f(x, y) = L$, must f be defined at (x_0, y_0)? Give reasons for your answer.

44. If $f(x_0, y_0) = 3$, what can you say about

$$\lim_{(x,y)\to(x_0, y_0)} f(x, y)$$

if f is continuous at (x_0, y_0)? if f is not continuous at (x_0, y_0)? Give reasons for your answer.

The Sandwich Theorem for functions of two variables states that if $g(x, y) \le f(x, y) \le h(x, y)$ for all $(x, y) \ne (x_0, y_0)$ in a disk centered at (x_0, y_0) and if g and h have the same finite limit L as $(x, y) \to (x_0, y_0)$, then

$$\lim_{(x,y)\to(x_0, y_0)} f(x, y) = L.$$

Use this result to support your answers to the questions in Exercises 45–48.

45. Does knowing that

$$1 - \frac{x^2 y^2}{3} < \frac{\tan^{-1} xy}{xy} < 1$$

tell you anything about

$$\lim_{(x,y)\to(0,0)} \frac{\tan^{-1} xy}{xy}?$$

Give reasons for your answer.

46. Does knowing that

$$2|xy| - \frac{x^2 y^2}{6} < 4 - 4 \cos \sqrt{|xy|} < 2|xy|$$

tell you anything about

$$\lim_{(x,y)\to(0,0)} \frac{4 - 4 \cos \sqrt{|xy|}}{|xy|}?$$

Give reasons for your answer.

47. Does knowing that $|\sin(1/x)| \le 1$ tell you anything about

$$\lim_{(x,y)\to(0,0)} y \sin \frac{1}{x}?$$

Give reasons for your answer.

48. Does knowing that $|\cos(1/y)| \le 1$ tell you anything about

$$\lim_{(x,y)\to(0,0)} x \cos \frac{1}{y}?$$

Give reasons for your answer.

49. (*Continuation of Example 3.*)

    a) Reread Example 3. Then substitute $m = \tan \theta$ into the formula

$$\left. f(x, y) \right|_{y=mx} = \frac{2m}{1 + m^2}$$

    and simplify the result to show how the value of f varies with the line's angle of inclination.

    b) Use the formula you obtained in (a) to show that the limit of f as $(x, y) \to (0, 0)$ along the line $y = mx$ varies from -1 to 1 depending on the angle of approach.

50. Define $f(0, 0)$ in a way that extends

$$f(x, y) = xy \frac{x^2 - y^2}{x^2 + y^2}$$

to be continuous at the origin.

Changing to Polar Coordinates

If you cannot make any headway with $\lim_{(x,y)\to(0,0)} f(x, y)$ in rectangular coordinates, try changing to polar coordinates. Substitute $x = r \cos \theta$, $y = r \sin \theta$, and investigate the limit of the resulting expression as $r \to 0$. In other words, try to decide whether there exists a number L satisfying the following criterion:

Given $\epsilon > 0$, there exists a $\delta > 0$ such that for all r and θ,

$$|r| < \delta \quad \Rightarrow \quad |f(r, \theta) - L| < \epsilon. \tag{4}$$

If such an L exists, then

$$\lim_{(x,y)\to(0,0)} f(x, y) = \lim_{r\to 0} f(r, \theta) = L.$$

For instance,

$$\lim_{(x,y)\to(0,0)} \frac{x^3}{x^2 + y^2} = \lim_{r\to 0} \frac{r^3 \cos^3 \theta}{r^2} = \lim_{r\to 0} r \cos^3 \theta = 0.$$

To verify the last of these equalities, we need to show that (4) is satisfied with $f(r, \theta) = r \cos^3 \theta$ and $L = 0$. That is, we need to show that given any $\epsilon > 0$ there exists a $\delta > 0$ such that for all r and θ,

$$|r| < \delta \quad \Rightarrow \quad |r \cos^3 \theta - 0| < \epsilon.$$

Since

$$|r \cos^3 \theta| = |r|| \cos^3 \theta| \le |r| \cdot 1 = |r|,$$

the implication holds for all r and θ if we take $\delta = \epsilon$.

In contrast,

$$\frac{x^2}{x^2 + y^2} = \frac{r^2 \cos^2 \theta}{r^2} = \cos^2 \theta$$

takes on all values from 0 to 1 regardless of how small $|r|$ is, so that $\lim_{(x,y)\to(0,0)} x^2/(x^2 + y^2)$ does not exist.

In each of these instances, the existence or nonexistence of the limit as $r \to 0$ is fairly clear. Shifting to polar coordinates does not always help, however, and may even tempt us to false conclusions. For example, the limit may exist along every straight line (or ray) $\theta = $ constant and yet fail to exist in the broader sense. Example 4 illustrates this point. In polar coordinates, $f(x, y) = (2x^2y)/(x^4 + y^2)$ becomes

$$f(r \cos \theta, r \sin \theta) = \frac{r \cos \theta \sin 2\theta}{r^2 \cos^4 \theta + \sin^2 \theta}$$

for $r \ne 0$. If we hold θ constant and let $r \to 0$, the limit is 0. On the path $y = x^2$, however, we have $r \sin \theta = r^2 \cos^2 \theta$ and

$$f(r \cos \theta, r \sin \theta) = \frac{r \cos \theta \sin 2\theta}{r^2 \cos^4 \theta + (r \cos^2 \theta)^2}$$

$$= \frac{2r \cos^2 \theta \sin \theta}{2r^2 \cos^4 \theta} = \frac{r \sin \theta}{r^2 \cos^2 \theta} = 1.$$

In Exercises 51–56, find the limit of f as $(x, y) \to (0, 0)$ or show that the limit does not exist.

51. $f(x, y) = \dfrac{x^3 - xy^2}{x^2 + y^2}$

52. $f(x, y) = \cos\left(\dfrac{x^3 - y^3}{x^2 + y^2}\right)$

53. $f(x, y) = \dfrac{y^2}{x^2 + y^2}$

54. $f(x, y) = \dfrac{2x}{x^2 + x + y^2}$

55. $f(x, y) = \tan^{-1}\left(\dfrac{|x| + |y|}{x^2 + y^2}\right)$

56. $f(x, y) = \dfrac{x^2 - y^2}{x^2 + y^2}$

In Exercises 57 and 58, define $f(0, 0)$ in a way that extends f to be continuous at the origin.

57. $f(x, y) = \ln\left(\dfrac{3x^2 - x^2y^2 + 3y^2}{x^2 + y^2}\right)$

58. $f(x, y) = \dfrac{2xy^2}{x^2 + y^2}$

Using the δ-ε Definitions

59. Show that the δ-ϵ requirement in the definition of limit expressed in Eq. (1) is equivalent to the requirement expressed in Eq. (2).

60. Using the formal δ-ϵ definition of limit of a function $f(x, y)$ as $(x, y) \to (x_0, y_0)$ as a guide, state a formal definition for the limit of a function $g(x, y, z)$ as $(x, y, z) \to (x_0, y_0, z_0)$. What would be the analogous definition for a function $h(x, y, z, t)$ of four independent variables?

Each of Exercises 61–64 gives a function $f(x, y)$ and a positive number ϵ. In each exercise, either show that there exists a $\delta > 0$ such that for all (x, y),

$$\sqrt{x^2 + y^2} < \delta \quad \Rightarrow \quad |f(x, y) - f(0, 0)| < \epsilon$$

or show that there exists a $\delta > 0$ such that for all (x, y),

$$|x| < \delta \text{ and } |y| < \delta \quad \Rightarrow \quad |f(x, y) - f(0, 0)| < \epsilon.$$

Do either one or the other, whichever seems more convenient. There is no need to do both.

61. $f(x, y) = x^2 + y^2$, $\epsilon = 0.01$

62. $f(x, y) = y/(x^2 + 1)$, $\epsilon = 0.05$

63. $f(x, y) = (x + y)/(x^2 + 1)$, $\epsilon = 0.01$

64. $f(x, y) = (x + y)/(2 + \cos x)$, $\epsilon = 0.02$

Each of Exercises 65–68 gives a function $f(x, y, z)$ and a positive number ϵ. In each exercise, either show that there exists a $\delta > 0$ such that for all (x, y, z),

$$\sqrt{x^2 + y^2 + z^2} < \delta \quad \Rightarrow \quad |f(x, y, z) - f(0, 0, 0)| < \epsilon$$

or show that there exists a $\delta > 0$ such that for all (x, y, z),

$$|x| < \delta, \quad |y| < \delta, \quad \text{and}$$

$$|z| < \delta \quad \Rightarrow \quad |f(x, y, z) - f(0, 0, 0)| < \epsilon.$$

Do either one or the other, whichever seems more convenient. There is no need to do both.

65. $f(x, y, z) = x^2 + y^2 + z^2$, $\epsilon = 0.015$

66. $f(x, y, z) = xyz$, $\epsilon = 0.008$

67. $f(x, y, z) = \dfrac{x + y + z}{x^2 + y^2 + z^2 + 1}$, $\epsilon = 0.015$

68. $f(x, y, z) = \tan^2 x + \tan^2 y + \tan^2 z$, $\epsilon = 0.03$

69. Show that $f(x, y, z) = x + y - z$ is continuous at every point (x_0, y_0, z_0).

70. Show that $f(x, y, z) = x^2 + y^2 + z^2$ is continuous at the origin.

Partial Derivatives

When we hold all but one of the independent variables of a function constant and differentiate with respect to that one variable, we get a "partial" derivative. This section shows how partial derivatives arise and how to calculate partial derivatives by applying the rules for differentiating functions of a single variable.

Definitions and Notation

If (x_0, y_0) is a point in the domain of a function $f(x, y)$, the vertical plane $y = y_0$ will cut the surface $z = f(x, y)$ in the curve $z = f(x, y_0)$ (Fig. 12.13). This curve is the graph of the function $z = f(x, y_0)$ in the plane $y = y_0$. The horizontal coordinate in this plane is x; the vertical coordinate is z.

We define the partial derivative of f with respect to x at the point (x_0, y_0) as the ordinary derivative of $f(x, y_0)$ with respect to x at the point $x = x_0$.

Definition

The **partial derivative of $f(x, y)$ with respect to x** at the point (x_0, y_0) is

$$\frac{\partial f}{\partial x}\bigg|_{(x_0, y_0)} = \frac{d}{dx} f(x, y_0)\bigg|_{x=x_0} = \lim_{h \to 0} \frac{f(x_0 + h, y_0) - f(x_0, y_0)}{h}, \quad (1)$$

provided the limit exists. (Think of ∂ as a kind of d.)

The slope of the curve $z = f(x, y_0)$ at the point $P(x_0, y_0, f(x_0, y_0))$ in the plane $y = y_0$ is the value of the partial derivative of f with respect to x at (x_0, y_0). The tangent line to the curve at P is the line in the plane $y = y_0$ that passes through P with this slope. The partial derivative $\partial f/\partial x$ at (x_0, y_0) gives the rate of change of f with respect to x when y is held fixed at the value y_0. This is the rate of change of f in the direction of **i** at (x_0, y_0).

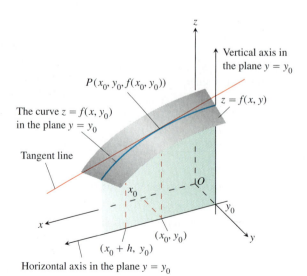

12.13 The intersection of the plane $y = y_0$ with the surface $z = f(x, y)$, viewed from a point above the first quadrant of the xy-plane.

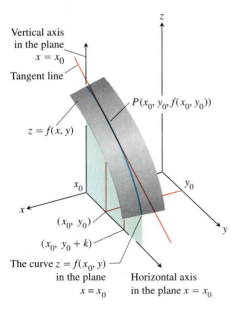

Vertical axis
in the plane
$x = x_0$

Tangent line

$P(x_0, y_0, f(x_0, y_0))$

$z = f(x, y)$

x_0

y_0

x

(x_0, y_0)

$(x_0, y_0 + k)$

The curve $z = f(x_0, y)$
in the plane   Horizontal axis
$x = x_0$ in the plane $x = x_0$

12.14 The intersection of the plane $x = x_0$ with the surface $z = f(x, y)$, viewed from above the first quadrant of the xy-plane.

The notation for a partial derivative depends on what we want to emphasize:

$\dfrac{\partial f}{\partial x}(x_0, y_0)$ or $f_x(x_0, y_0)$ — "Partial derivative of f with respect to x at (x_0, y_0)" or "f sub x at (x_0, y_0)." Convenient for stressing the point (x_0, y_0).

$\dfrac{\partial z}{\partial x}\Big|_{(x_0, y_0)}$ — "Partial derivative of z with respect to x at (x_0, y_0)." Common in science and engineering when you are dealing with variables and do not mention the function explicitly.

$f_x,\ \dfrac{\partial f}{\partial x},\ z_x,\ \text{or}\ \dfrac{\partial z}{\partial x}$ — "Partial derivative of f (or z) with respect to x." Convenient when you regard the partial derivative as a function in its own right.

The definition of the partial derivative of $f(x, y)$ with respect to y at a point (x_0, y_0) is similar to the definition of the partial derivative of f with respect to x. We hold x fixed at the value x_0 and take the ordinary derivative of $f(x_0, y)$ with respect to y at y_0.

Definition

The **partial derivative of $f(x, y)$ with respect to y** at the point (x_0, y_0) is

$$\frac{\partial f}{\partial y}\Big|_{(x_0, y_0)} = \frac{d}{dy} f(x_0, y)\Big|_{y = y_0}$$

$$= \lim_{h \to 0} \frac{f(x_0, y_0 + h) - f(x_0, y_0)}{h}, \qquad (2)$$

provided the limit exists.

The slope of the curve $z = f(x_0, y)$ at the point $P(x_0, y_0, f(x_0, y_0))$ in the vertical plane $x = x_0$ (Fig. 12.14) is the partial derivative of f with respect to y at (x_0, y_0). The tangent line to the curve at P is the line in the plane $x = x_0$ that passes through P with this slope. The partial derivative gives the rate of change of f with respect to y at (x_0, y_0) when x is held fixed at the value x_0. This is the rate of change of f in the direction of $\mathbf{j}$ at (x_0, y_0).

The partial derivative with respect to y is denoted the same way as the partial derivative with respect to x:

$$\frac{\partial f}{\partial y}(x_0, y_0), \qquad f_y(x_0, y_0), \qquad \frac{\partial f}{\partial y}, \qquad f_y.$$

Notice that we now have two tangent lines associated with the surface $z = f(x, y)$ at the point $P(x_0, y_0, f(x_0, y_0))$ (Fig. 12.15, on the following page). Is the plane they determine tangent to the surface at P? It would be nice if it were, but we have to learn more about partial derivatives before we can find out.

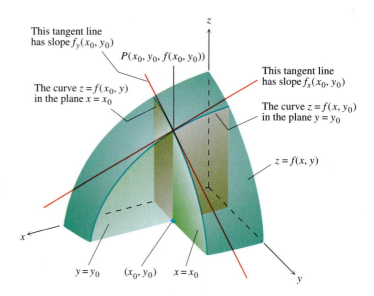

12.15 Figures 12.13 and 12.14 combined. The tangent lines at the point $(x_0, y_0, f(x_0, y_0))$ determine a plane that, in this picture at least, appears to be tangent to the surface.

Calculations

As Eq. (1) shows, we calculate $\partial f/\partial x$ by differentiating f with respect to x in the usual way while treating y as a constant. As Eq. (2) shows, we can calculate $\partial f/\partial y$ by differentiating f with respect to y in the usual way while holding x constant.

EXAMPLE 1   Find the values of $\partial f/\partial x$ and $\partial f/\partial y$ at the point $(4, -5)$ if

$$f(x, y) = x^2 + 3xy + y - 1.$$

Solution   To find $\partial f/\partial x$, we regard y as a constant and differentiate with respect to x:

$$\frac{\partial f}{\partial x} = \frac{\partial}{\partial x}(x^2 + 3xy + y - 1) = 2x + 3 \cdot 1 \cdot y + 0 - 0 = 2x + 3y.$$

The value of $\partial f/\partial x$ at $(4, -5)$ is $2(4) + 3(-5) = -7$.
To find $\partial f/\partial y$, we regard x as a constant and differentiate with respect to y:

$$\frac{\partial f}{\partial y} = \frac{\partial}{\partial y}(x^2 + 3xy + y - 1) = 0 + 3 \cdot x \cdot 1 + 1 - 0 = 3x + 1.$$

The value of $\partial f/\partial y$ at $(4, -5)$ is $3(4) + 1 = 13$.   ❑

EXAMPLE 2   Find $\partial f/\partial y$ if $f(x, y) = y \sin xy$.

Solution   We treat x as a constant and f as a product of y and $\sin xy$:

$$\frac{\partial f}{\partial y} = \frac{\partial}{\partial y}(y \sin xy) = y \frac{\partial}{\partial y} \sin xy + (\sin xy) \frac{\partial}{\partial y}(y)$$

$$= (y \cos xy) \frac{\partial}{\partial y}(xy) + \sin xy = xy \cos xy + \sin xy.$$ ❑

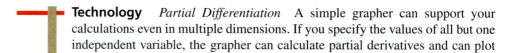

Technology   *Partial Differentiation*   A simple grapher can support your calculations even in multiple dimensions. If you specify the values of all but one independent variable, the grapher can calculate partial derivatives and can plot traces with respect to that remaining variable. Typically a Computer Algebra System can compute partial derivatives symbolically and numerically as easily as it can compute simple derivatives. Most systems use the same command to differentiate a function, regardless of the number of variables. (Simply specify the variable with which differentiation is to take place.)

EXAMPLE 3   Find f_x if $f(x, y) = \dfrac{2y}{y + \cos x}$.

Solution   We treat f as a quotient. With y held constant, we get

$$f_x = \frac{\partial}{\partial x}\left(\frac{2y}{y + \cos x}\right) = \frac{(y + \cos x)\dfrac{\partial}{\partial x}(2y) - 2y\dfrac{\partial}{\partial x}(y + \cos x)}{(y + \cos x)^2}$$

$$= \frac{(y + \cos x)(0) - 2y(-\sin x)}{(y + \cos x)^2} = \frac{2y\sin x}{(y + \cos x)^2}.$$

EXAMPLE 4   The plane $x = 1$ intersects the paraboloid $z = x^2 + y^2$ in a parabola. Find the slope of the tangent to the parabola at $(1, 2, 5)$ (Fig. 12.16).

Solution   The slope is the value of the partial derivative $\partial z/\partial y$ at $(1, 2)$:

$$\frac{\partial z}{\partial y}\bigg|_{(1,2)} = \frac{\partial}{\partial y}(x^2 + y^2)\bigg|_{(1,2)} = 2y\bigg|_{(1,2)} = 2(2) = 4.$$

As a check, we can treat the parabola as the graph of the single-variable function $z = (1)^2 + y^2 = 1 + y^2$ in the plane $x = 1$ and ask for the slope at $y = 2$. The slope, calculated now as an ordinary derivative, is

$$\frac{dz}{dy}\bigg|_{y=2} = \frac{d}{dy}(1 + y^2)\bigg|_{y=2} = 2y\bigg|_{y=2} = 4.$$

Implicit differentiation works for partial derivatives the way it works for ordinary derivatives.

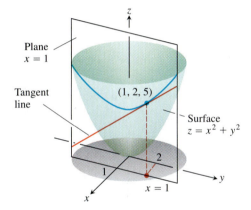

12.16 The tangent to the curve of intersection of the plane $x = 1$ and surface $z = x^2 + y^2$ at the point $(1, 2, 5)$ (Example 4).

EXAMPLE 5   Find $\partial z/\partial x$ if the equation

$$yz - \ln z = x + y$$

defines z as a function of the two independent variables x and y and the partial derivative exists.

Solution   We differentiate both sides of the equation with respect to x, holding y constant and treating z as a differentiable function of x:

$$\frac{\partial}{\partial x}(yz) - \frac{\partial}{\partial x}\ln z = \frac{\partial x}{\partial x} + \frac{\partial y}{\partial x}$$

$$y\frac{\partial z}{\partial x} - \frac{1}{z}\frac{\partial z}{\partial x} = 1 + 0 \qquad \text{\small With } y \text{ constant,}\ \frac{\partial}{\partial x}(yz) = y\frac{\partial z}{\partial x}.$$

$$\left(y - \frac{1}{z}\right)\frac{\partial z}{\partial x} = 1$$

$$\frac{\partial z}{\partial x} = \frac{z}{yz - 1}.$$

Functions of More Than Two Variables

The definitions of the partial derivatives of functions of more than two independent variables are like the definitions for functions of two variables. They are ordinary derivatives with respect to one variable, taken while the other independent variables are held constant.

EXAMPLE 6   If x, y, and z are independent variables and

$$f(x, y, z) = x \sin (y + 3z),$$

then

$$\frac{\partial f}{\partial z} = \frac{\partial}{\partial z} [x \sin (y + 3z)] = x \frac{\partial}{\partial z} \sin (y + 3z)$$

$$= x \cos (y + 3z) \frac{\partial}{\partial z} (y + 3z) = 3x \cos (y + 3z).$$ ❑

EXAMPLE 7   *Electrical resistors in parallel*

If resistors of R_1, R_2, and R_3 ohms are connected in parallel to make an R-ohm resistor, the value of R can be found from the equation

$$\frac{1}{R} = \frac{1}{R_1} + \frac{1}{R_2} + \frac{1}{R_3} \tag{3}$$

(Fig. 12.17). Find the value of $\partial R/\partial R_2$ when $R_1 = 30$, $R_2 = 45$, and $R_3 = 90$ ohms.

Solution   To find $\partial R/\partial R_2$, we treat R_1 and R_3 as constants and differentiate both sides of Eq. (3) with respect to R_2:

$$\frac{\partial}{\partial R_2} \left(\frac{1}{R} \right) = \frac{\partial}{\partial R_2} \left(\frac{1}{R_1} + \frac{1}{R_2} + \frac{1}{R_3} \right)$$

$$-\frac{1}{R^2} \frac{\partial R}{\partial R_2} = 0 - \frac{1}{R_2^2} + 0$$

$$\frac{\partial R}{\partial R_2} = \frac{R^2}{R_2^2} = \left(\frac{R}{R_2} \right)^2.$$

When $R_1 = 30$, $R_2 = 45$, and $R_3 = 90$,

$$\frac{1}{R} = \frac{1}{30} + \frac{1}{45} + \frac{1}{90} = \frac{3 + 2 + 1}{90} = \frac{6}{90} = \frac{1}{15},$$

so $R = 15$ and

$$\frac{\partial R}{\partial R_2} = \left(\frac{15}{45} \right)^2 = \left(\frac{1}{3} \right)^2 = \frac{1}{9}.$$ ❑

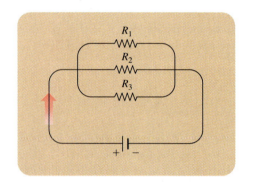

12.17 Resistors arranged this way are said to be connected in parallel (Example 7). Each resistor lets a portion of the current through. Their combined resistance R is calculated with the formula $\frac{1}{R} = \frac{1}{R_1} + \frac{1}{R_2} + \frac{1}{R_3}$.

The Relationship Between Continuity and the Existence of Partial Derivatives

A function $f(x, y)$ can have partial derivatives with respect to both x and y at a point without being continuous there. This is different from functions of a single variable, where the existence of a derivative implies continuity. However, if the partial derivatives of $f(x, y)$ exist and are continuous throughout a disk centered at (x_0, y_0), then f is continuous at (x_0, y_0), as we will see in the next section.

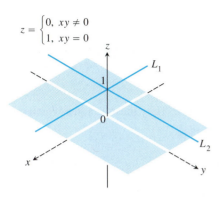

$$z = \begin{cases} 0, & xy \neq 0 \\ 1, & xy = 0 \end{cases}$$

12.18 The graph of

$$f(x, y) = \begin{cases} 0, & xy \neq 0 \\ 1, & xy = 0 \end{cases}$$

consists of the lines L_1 and L_2 and the four open quadrants of the *xy*-plane. The function has partial derivatives at the origin but is not continuous there.

EXAMPLE 8 The function

$$f(x, y) = \begin{cases} 0, & xy \neq 0 \\ 1, & xy = 0 \end{cases}$$

(Fig. 12.18) is not continuous at $(0, 0)$. The limit of f as (x, y) approaches $(0, 0)$ along the line $y = x$ is 0, but $f(0, 0) = 1$. The partial derivatives f_x and f_y, being the slopes of the horizontal lines L_1 and L_2 in Fig. 12.18, both exist at $(0, 0)$. ◻

Second Order Partial Derivatives

When we differentiate a function $f(x, y)$ twice, we produce its second order derivatives. These derivatives are usually denoted by

$\dfrac{\partial^2 f}{\partial x^2}$	"*d* squared *f d x* squared"	or	f_{xx} "*f* sub *x x*"
$\dfrac{\partial^2 f}{\partial y^2}$	"*d* squared *f d y* squared"		f_{yy} "*f* sub *y y*"
$\dfrac{\partial^2 f}{\partial x \partial y}$	"*d* squared *f d x d y*"		f_{yx} "*f* sub *y x*"
$\dfrac{\partial^2 f}{\partial y \partial x}$	"*d* squared *f d y d x*"		f_{xy} "*f* sub *x y*"

The defining equations are

$$\frac{\partial^2 f}{\partial x^2} = \frac{\partial}{\partial x}\left(\frac{\partial f}{\partial x}\right), \qquad \frac{\partial^2 f}{\partial x \, \partial y} = \frac{\partial}{\partial x}\left(\frac{\partial f}{\partial y}\right),$$

and so on. Notice the order in which the derivatives are taken:

$\dfrac{\partial^2 f}{\partial x \partial y}$ Differentiate first with respect to y, then with respect to x.

$f_{yx} = (f_y)_x$ Means the same thing.

EXAMPLE 9 If $f(x, y) = x \cos y + y e^x$, then

$$\frac{\partial f}{\partial x} = \cos y + y e^x$$

$$\frac{\partial^2 f}{\partial y \, \partial x} = \frac{\partial}{\partial y}\left(\frac{\partial f}{\partial x}\right) = -\sin y + e^x$$

$$\frac{\partial^2 f}{\partial x^2} = \frac{\partial}{\partial x}\left(\frac{\partial f}{\partial x}\right) = y e^x.$$

Also,

$$\frac{\partial f}{\partial y} = -x \sin y + e^x$$

$$\frac{\partial^2 f}{\partial x \, \partial y} = \frac{\partial}{\partial x}\left(\frac{\partial f}{\partial y}\right) = -\sin y + e^x$$

$$\frac{\partial^2 f}{\partial y^2} = \frac{\partial}{\partial y}\left(\frac{\partial f}{\partial y}\right) = -x \cos y.$$

◻

Euler's Theorem

You may have noticed that the "mixed" second order partial derivatives

$$\frac{\partial^2 f}{\partial y \, \partial x} \quad \text{and} \quad \frac{\partial^2 f}{\partial x \, \partial y}$$

in Example 9 were equal. This was not a coincidence. They must be equal whenever f, f_x, f_y, f_{xy}, and f_{yx} are continuous.

Theorem 2

Euler's Theorem (The Mixed Derivative Theorem)

If $f(x, y)$ and its partial derivatives f_x, f_y, f_{xy}, and f_{yx} are defined throughout an open region containing a point (a, b) and are all continuous at (a, b), then

$$f_{xy}(a, b) = f_{yx}(a, b). \tag{4}$$

You can find a proof of Theorem 2 in Appendix 9.

Theorem 2 says that to calculate a mixed second order derivative we may differentiate in either order. This can work to our advantage.

EXAMPLE 10   Find $\partial^2 w / \partial x \, \partial y$ if

$$w = xy + \frac{e^y}{y^2 + 1}.$$

Solution   The symbol $\partial^2 w / \partial x \, \partial y$ tells us to differentiate first with respect to y and then with respect to x. However, if we postpone the differentiation with respect to y and differentiate first with respect to x, we get the answer more quickly. In two steps,

$$\frac{\partial w}{\partial x} = y \quad \text{and} \quad \frac{\partial^2 w}{\partial y \, \partial x} = 1.$$

We are in for more work if we differentiate first with respect to y. (Just try it.)   ❑

Partial Derivatives of Still Higher Order

Although we will deal mostly with first and second order partial derivatives, because these appear the most frequently in applications, there is no theoretical limit to how many times we can differentiate a function as long as the derivatives involved exist. Thus we get third and fourth order derivatives denoted by symbols like

$$\frac{\partial^3 f}{\partial x \, \partial y^2} = f_{yyx},$$

$$\frac{\partial^4 f}{\partial x^2 \, \partial y^2} = f_{yyxx},$$

and so on. As with second order derivatives, the order of differentiation is immaterial as long as the derivatives through the order in question are continuous.

Exercises 12.3

Calculating First Order Partial Derivatives

In Exercises 1–22, find $\partial f/\partial x$ and $\partial f/\partial y$.

1. $f(x, y) = 2x^2 - 3y - 4$
2. $f(x, y) = x^2 - xy + y^2$

3. $f(x, y) = (x^2 - 1)(y + 2)$

4. $f(x, y) = 5xy - 7x^2 - y^2 + 3x - 6y + 2$

5. $f(x, y) = (xy - 1)^2$
6. $f(x, y) = (2x - 3y)^3$

7. $f(x, y) = \sqrt{x^2 + y^2}$
8. $f(x, y) = (x^3 + (y/2))^{2/3}$

9. $f(x, y) = 1/(x + y)$
10. $f(x, \grave{y}) = x/(x^2 + y^2)$

11. $f(x, y) = (x + y)/(xy - 1)$

12. $f(x, y) = \tan^{-1}(y/x)$
13. $f(x, y) = e^{(x+y+1)}$

14. $f(x, y) = e^{-x} \sin(x + y)$
15. $f(x, y) = \ln(x + y)$

16. $f(x, y) = e^{xy} \ln y$
17. $f(x, y) = \sin^2(x - 3y)$

18. $f(x, y) = \cos^2(3x - y^2)$
19. $f(x, y) = x^y$

20. $f(x, y) = \log_y x$

21. $f(x, y) = \displaystyle\int_x^y g(t)\, dt$ *(g continuous for all t)*

22. $f(x, y) = \displaystyle\sum_{n=0}^{\infty} (xy)^n \quad (|xy| < 1)$

In Exercises 23–34, find f_x, f_y, and f_z.

23. $f(x, y, z) = 1 + xy^2 - 2z^2$
24. $f(x, y, z) = xy + yz + xz$

25. $f(x, y, z) = x - \sqrt{y^2 + z^2}$

26. $f(x, y, z) = (x^2 + y^2 + z^2)^{-1/2}$

27. $f(x, y, z) = \sin^{-1}(xyz)$
28. $f(x, y, z) = \sec^{-1}(x + yz)$

29. $f(x, y, z) = \ln(x + 2y + 3z)$

30. $f(x, y, z) = yz \ln(xy)$
31. $f(x, y, z) = e^{-(x^2+y^2+z^2)}$

32. $f(x, y, z) = e^{-xyz}$

33. $f(x, y, z) = \tanh(x + 2y + 3z)$

34. $f(x, y, z) = \sinh(xy - z^2)$

In Exercises 35–40, find the partial derivative of the function with respect to each variable.

35. $f(t, \alpha) = \cos(2\pi t - \alpha)$
36. $g(u, v) = v^2 e^{(2u/v)}$

37. $h(\rho, \phi, \theta) = \rho \sin \phi \cos \theta$

38. $g(r, \theta, z) = r(1 - \cos \theta) - z$

39. *Work done by the heart.* (Section 3.7, Exercise 56)

$$W(P, V, \delta, v, g) = PV + \frac{V\delta v^2}{2g}$$

40. *Wilson lot size formula.* (Section 3.6, Exercise 57)

$$A(c, h, k, m, q) = \frac{km}{q} + cm + \frac{hq}{2}$$

Calculating Second Order Partial Derivatives

Find all the second order partial derivatives of the functions in Exercises 41–46.

41. $f(x, y) = x + y + xy$
42. $f(x, y) = \sin xy$

43. $g(x, y) = x^2 y + \cos y + y \sin x$

44. $h(x, y) = xe^y + y + 1$
45. $r(x, y) = \ln(x + y)$

46. $s(x, y) = \tan^{-1}(y/x)$

Mixed Partial Derivatives

In Exercises 47–50, verify that $w_{xy} = w_{yx}$.

47. $w = \ln(2x + 3y)$
48. $w = e^x + x \ln y + y \ln x$

49. $w = xy^2 + x^2 y^3 + x^3 y^4$
50. $w = x \sin y + y \sin x + xy$

51. Which order of differentiation will calculate f_{xy} faster: x first, or y first? Try to answer without writing anything down.

   a) $f(x, y) = x \sin y + e^y$
   b) $f(x, y) = 1/x$
   c) $f(x, y) = y + (x/y)$
   d) $f(x, y) = y + x^2 y + 4y^3 - \ln(y^2 + 1)$
   e) $f(x, y) = x^2 + 5xy + \sin x + 7e^x$
   f) $f(x, y) = x \ln xy$

52. The fifth order partial derivative $\partial^5 f/\partial x^2 \partial y^3$ is zero for each of the following functions. To show this as quickly as possible, which variable would you differentiate with respect to first: x, or y? Try to answer without writing anything down.

   a) $f(x, y) = y^2 x^4 e^x + 2$
   b) $f(x, y) = y^2 + y(\sin x - x^4)$
   c) $f(x, y) = x^2 + 5xy + \sin x + 7e^x$
   d) $f(x, y) = xe^{y^2/2}$

Using the Partial Derivative Definition

In Exercises 53 and 54, use the limit definition of partial derivative to compute the partial derivatives of the functions at the specified points.

53. $f(x, y) = 1 - x + y - 3x^2 y$, $\dfrac{\partial f}{\partial x}$ and $\dfrac{\partial f}{\partial y}$ at $(1, 2)$

54. $f(x, y) = 4 + 2x - 3y - xy^2$, $\dfrac{\partial f}{\partial x}$ and $\dfrac{\partial f}{\partial y}$ at $(-2, 1)$

55. Let $w = f(x, y, z)$ be a function of three independent variables, and write the formal definition of the partial derivative $\partial f/\partial z$ at (x_0, y_0, z_0). Use this definition to find $\partial f/\partial z$ at $(1, 2, 3)$ for $f(x, y, z) = x^2 yz^2$.

56. Let $w = f(x, y, z)$ be a function of three independent variables and write the formal definition of the partial derivative $\partial f/\partial y$ at (x_0, y_0, z_0). Use this definition to find $\partial f/\partial y$ at $(-1, 0, 3)$ for $f(x, y, z) = -2xy^2 + yz^2$.

Differentiating Implicitly

57. Find the value of $\partial z/\partial x$ at the point $(1, 1, 1)$ if the equation

$$xy + z^3 x - 2yz = 0$$

defines z as a function of the two independent variables x and y and the partial derivative exists.

58. Find the value of $\partial x/\partial z$ at the point $(1, -1, -3)$ if the equation

$$xz + y \ln x - x^2 + 4 = 0$$

defines x as a function of the two independent variables y and z and the partial derivative exists.

Exercises 59 and 60 are about the triangle shown here.

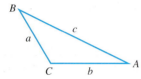

59. Express A implicitly as a function of $a, b,$ and c and calculate $\partial A/\partial a$ and $\partial A/\partial b$.

60. Express a implicitly as a function of $A, b,$ and B and calculate $\partial a/\partial A$ and $\partial a/\partial B$.

61. Express v_x in terms of u and v if the equations $x = v \ln u$ and $y = u \ln v$ define u and v as functions of the independent variables x and y, and if v_x exists. (*Hint:* Differentiate both equations with respect to x and solve for v_x with Cramer's rule.)

62. Find $\partial x/\partial u$ and $\partial y/\partial u$ if the equations $u = x^2 - y^2$ and $v = x^2 - y$ define x and y as functions of the independent variables u and v, and the partial derivatives exist. (See the hint in Exercise 61.) Then let $s = x^2 + y^2$ and find $\partial s/\partial u$.

Laplace Equations

The three-dimensional Laplace equation

$$\frac{\partial^2 f}{\partial x^2} + \frac{\partial^2 f}{\partial y^2} + \frac{\partial^2 f}{\partial z^2} = 0 \qquad (5)$$

is satisfied by steady-state temperature distributions $T = f(x, y, z)$ in space, by gravitational potentials, and by electrostatic potentials. The *two-dimensional Laplace equation*

$$\frac{\partial^2 f}{\partial x^2} + \frac{\partial^2 f}{\partial y^2} = 0, \qquad (6)$$

obtained by dropping the $\partial^2 f/\partial z^2$ term from Eq. (5), describes potentials and steady-state temperature distributions in a plane (Fig. 12.19).

Show that each function in Exercises 63–68 satisfies a Laplace equation.

63. $f(x, y, z) = x^2 + y^2 - 2z^2$

64. $f(x, y, z) = 2z^3 - 3(x^2 + y^2)z$

65. $f(x, y) = e^{-2y} \cos 2x$

66. $f(x, y) = \ln \sqrt{x^2 + y^2}$

67. $f(x, y, z) = (x^2 + y^2 + z^2)^{-1/2}$

68. $f(x, y, z) = e^{3x+4y} \cos 5z$

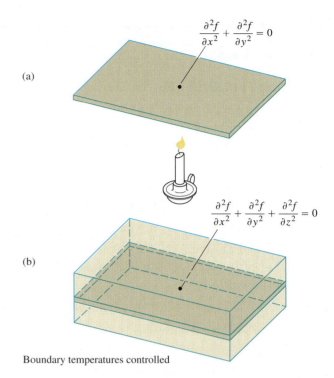

(a)

(b)

Boundary temperatures controlled

12.19 Steady-state temperature distributions in planes and solids satisfy Laplace equations. The plane (a) may be treated as a thin slice of the solid (b) perpendicular to the z-axis.

The Wave Equation

If we stand on an ocean shore and take a snapshot of the waves, the picture shows a regular pattern of peaks and valleys in an instant of time. We see periodic vertical motion in space, with respect to distance. If we stand in the water, we can feel the rise and fall of the water as the waves go by. We see periodic vertical motion in time. In physics, this beautiful symmetry is expressed by the *one-dimensional wave equation*

$$\frac{\partial^2 w}{\partial t^2} = c^2 \frac{\partial^2 w}{\partial x^2}, \qquad (7)$$

where w is the wave height, x is the distance variable, t is the time variable, and c is the velocity with which the waves are propagated.

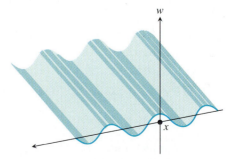

In our example, x is the distance across the ocean's surface, but in other applications x might be the distance along a vibrating string, distance through air (sound waves), or distance through space (light waves). The number c varies with the medium and type of wave.

Show that the functions in Exercises 69–75 are all solutions of the wave equation.

69. $w = \sin(x + ct)$

70. $w = \cos(2x + 2ct)$

71. $w = \sin(x + ct) + \cos(2x + 2ct)$

72. $w = \ln(2x + 2ct)$

73. $w = \tan(2x - 2ct)$

74. $w = 5\cos(3x + 3ct) + e^{x+ct}$

75. $w = f(u)$, where f is a differentiable function of u and $u = a(x + ct)$, where a is a constant.

| 12.4 | # Differentiability, Linearization, and Differentials |

In this section, we define differentiability and proceed from there to linearizations and differentials. The mathematical results of the section stem from the Increment Theorem. As we will see in the next section, this theorem also underlies the Chain Rule for multivariable functions.

Differentiability

Surprising as it may seem, the starting point for differentiability is not Fermat's difference quotient but rather the idea of increment. You may recall from our work with functions of a single variable that if $y = f(x)$ is differentiable at $x = x_0$, then the change in the value of f that results from changing x from x_0 to $x_0 + \Delta x$ is given by an equation of the form

$$\Delta y = f'(x_0)\Delta x + \epsilon\,\Delta x \tag{1}$$

in which $\epsilon \to 0$ as $\Delta x \to 0$. For functions of two variables, the analogous property becomes the definition of differentiability. The Increment Theorem (from advanced calculus) tells us when to expect the property to hold.

Theorem 3
The Increment Theorem for Functions of Two Variables

Suppose that the first partial derivatives of $f(x, y)$ are defined throughout an open region R containing the point (x_0, y_0) and that f_x and f_y are continuous at (x_0, y_0). Then the change

$$\Delta z = f(x_0 + \Delta x, y_0 + \Delta y) - f(x_0, y_0)$$

in the value of f that results from moving from (x_0, y_0) to another point $(x_0 + \Delta x, y_0 + \Delta y)$ in R satisfies an equation of the form

$$\Delta z = f_x(x_0, y_0)\Delta x + f_y(x_0, y_0)\Delta y + \epsilon_1\Delta x + \epsilon_2\Delta y, \tag{2}$$

in which $\epsilon_1, \epsilon_2 \to 0$ as $\Delta x, \Delta y \to 0$.

You will see where the epsilons come from if you read the proof in Appendix 10. You will also see that similar results hold for functions of more than two independent variables.

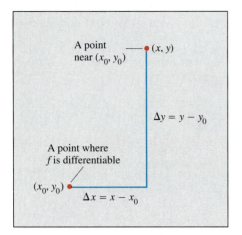

12.20 If f is differentiable at (x_0, y_0), then the value of f at any point (x, y) nearby is approximately $f(x_0, y_0) + f_x(x_0, y_0)\Delta x + f_y(x_0, y_0)\Delta y$.

As we can see from Theorems 3 and 4, a function $f(x, y)$ must be continuous at a point (x_0, y_0) if f_x and f_y are continuous throughout an open region containing (x_0, y_0). But remember that it is still possible for a function of two variables to be discontinuous at a point where its first partial derivatives exist, as we saw in Section 12.3, Example 8. Existence alone is not enough.

Definition

A function $f(x, y)$ is **differentiable at** $(\mathbf{x_0, y_0})$ if $f_x(x_0, y_0)$ and $f_y(x_0, y_0)$ exist and Eq. (2) holds for f at (x_0, y_0). We call f **differentiable** if it is differentiable at every point in its domain.

In light of this definition, we have the immediate corollary of Theorem 3 that a function is differentiable if its first partial derivatives are *continuous*.

Corollary of Theorem 3

If the partial derivatives f_x and f_y of a function $f(x, y)$ are continuous throughout an open region R, then f is differentiable at every point of R.

If we replace the Δz in Eq. (2) by the expression $f(x, y) - f(x_0, y_0)$ and rewrite the equation as

$$f(x, y) = f(x_0, y_0) + f_x(x_0, y_0)\Delta x + f_y(x_0, y_0)\Delta y + \epsilon_1 \Delta x + \epsilon_2 \Delta y, \quad \text{(3)}$$

we see that the right-hand side of the new equation approaches $f(x_0, y_0)$ as Δx and Δy approach 0. This tells us that a function $f(x, y)$ is continuous at every point where it is differentiable.

Theorem 4

If a function $f(x, y)$ is differentiable at (x_0, y_0), then f is continuous at (x_0, y_0).

How to Linearize a Function of Two Variables

Functions of two variables can be complicated, and we sometimes need to replace them with simpler ones that give the accuracy required for specific applications without being so hard to work with. We do this in a way that is similar to the way we find linear replacements for functions of a single variable (Section 3.7).

Suppose the function we wish to replace is $z = f(x, y)$ and that we want the replacement to be effective near a point (x_0, y_0) at which we know the values of $f, f_x,$ and f_y and at which f is differentiable. Since f is differentiable, Eq. (3) holds for f at (x_0, y_0). Therefore, if we move from (x_0, y_0) to any point (x, y) by increments $\Delta x = x - x_0$ and $\Delta y = y - y_0$ (Fig. 12.20), the new value of f will be

$$f(x, y) = f(x_0, y_0) + f_x(x_0, y_0)(x - x_0)$$
$$+ f_y(x_0, y_0)(y - y_0) + \epsilon_1 \Delta x + \epsilon_2 \Delta y, \qquad \begin{array}{l} \text{Eq. (3), with} \\ \Delta x = x - x_0 \\ \text{and } \Delta y = y - y_0 \end{array}$$

where $\epsilon_1, \epsilon_2 \to 0$ as $\Delta x, \Delta y \to 0$. If the increments Δx and Δy are small, the products $\epsilon_1 \Delta x$ and $\epsilon_2 \Delta y$ will eventually be smaller still and we will have

$$f(x, y) \approx \underbrace{f(x_0, y_0) + f_x(x_0, y_0)(x - x_0) + f_y(x_0, y_0)(y - y_0)}_{L(x, y)}.$$

In other words, as long as Δx and Δy are small, f will have approximately the same value as the linear function L. If f is hard to use, and our work can tolerate the error involved, we may safely replace f by L.

Definitions

The **linearization** of a function $f(x, y)$ at a point (x_0, y_0) where f is differentiable is the function

$$L(x, y) = f(x_0, y_0) + f_x(x_0, y_0)(x - x_0) + f_y(x_0, y_0)(y - y_0). \quad \textbf{(4)}$$

The approximation

$$f(x, y) \approx L(x, y)$$

is the **standard linear approximation** of f at (x_0, y_0).

In Section 12.8 we will see that the plane $z = L(x, y)$ is tangent to the surface $z = f(x, y)$ at the point (x_0, y_0). Thus, the linearization of a function of two variables is a tangent-*plane* approximation in the same way that the linearization of a function of a single variable is a tangent-*line* approximation.

EXAMPLE 1   Find the linearization of

$$f(x, y) = x^2 - xy + \frac{1}{2}y^2 + 3$$

at the point $(3, 2)$.

Solution   We evaluate Eq. (4) with

$$f(x_0, y_0) = \left(x^2 - xy + \frac{1}{2}y^2 + 3 \right)_{(3,2)} = 8,$$

$$f_x(x_0, y_0) = \frac{\partial}{\partial x} \left(x^2 - xy + \frac{1}{2}y^2 + 3 \right)_{(3,2)} = (2x - y)_{(3,2)} = 4,$$

$$f_y(x_0, y_0) = \frac{\partial}{\partial y} \left(x^2 - xy + \frac{1}{2}y^2 + 3 \right)_{(3,2)} = (-x + y)_{(3,2)} = -1,$$

getting

$$L(x, y) = f(x_0, y_0) + f_x(x_0, y_0)(x - x_0) + f_y(x_0, y_0)(y - y_0) \qquad \text{Eq. (4)}$$

$$= 8 + (4)(x - 3) + (-1)(y - 2) = 4x - y - 2.$$

The linearization of f at $(3, 2)$ is $L(x, y) = 4x - y - 2$.   ❑

How Accurate Is the Standard Linear Approximation?

To find the error in the approximation $f(x, y) \approx L(x, y)$, we use the second order partial derivatives of f. Suppose that the first and second order partial derivatives of f are continuous throughout an open set containing a closed rectangular region R centered at (x_0, y_0) and given by the inequalities

$$|x - x_0| \leq h, \qquad |y - y_0| \leq k$$

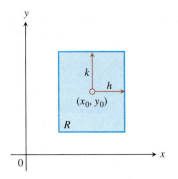

12.21 The rectangular region R: $|x - x_0| \le h$, $|y - y_0| \le k$ in the xy-plane. On this kind of region, we can find useful error bounds for our approximations.

(Fig. 12.21). Since R is closed and bounded, the second partial derivatives all take on absolute maximum values on R. If B is the largest of these values, then, as explained in Section 12.10, the error $E(x, y) = f(x, y) - L(x, y)$ in the standard linear approximation satisfies the inequality

$$|E(x, y)| \le \frac{1}{2} B \left(|x - x_0| + |y - y_0| \right)^2$$

throughout R.

When we use this inequality to estimate E, we usually cannot find the values of f_{xx}, f_{yy}, and f_{xy} that determine B and we have to settle for an upper bound or "worst-case" value instead. If M is any common upper bound for $|f_{xx}|$, $\left| f_{yy} \right|$, and $\left| f_{xy} \right|$ on R, then B will be less than or equal to M and we will know that

$$|E(x, y)| \le \frac{1}{2} M \left(|x - x_0| + |y - y_0| \right)^2 .$$

This is the inequality normally used in estimating E. When we need to make $|E(x, y)|$ small for a given M, we just make $|x - x_0|$ and $|y - y_0|$ small.

The Error in the Standard Linear Approximation

If f has continuous first and second partial derivatives throughout an open set containing a rectangle R centered at (x_0, y_0) and if M is any upper bound for the values of $|f_{xx}|$, $|f_{yy}|$, and $|f_{xy}|$ on R, then the error $E(x, y)$ incurred in replacing $f(x, y)$ on R by its linearization

$$L(x, y) = f(x_0, y_0) + f_x(x_0, y_0)(x - x_0) + f_y(x_0, y_0)(y - y_0)$$

satisfies the inequality

$$|E(x, y)| \le \frac{1}{2} M \left(|x - x_0| + |y - y_0| \right)^2 . \tag{5}$$

EXAMPLE 2 In Example 1, we found the linearization of

$$f(x, y) = x^2 - xy + \frac{1}{2}y^2 + 3$$

at $(3, 2)$ to be

$$L(x, y) = 4x - y - 2.$$

Find an upper bound for the error in the approximation $f(x, y) \approx L(x, y)$ over the rectangle

$$R: \quad |x - 3| \le 0.1, \quad |y - 2| \le 0.1.$$

Express the upper bound as a percentage of $f(3, 2)$, the value of f at the center of the rectangle.

Solution We use the inequality

$$|E(x, y)| \le \frac{1}{2} M \left(|x - x_0| + |y - y_0| \right)^2 . \qquad \text{Eq. (5)}$$

To find a suitable value for M, we calculate f_{xx}, f_{xy}, and f_{yy}, finding, after a routine differentiation, that all three derivatives are constant, with values

$$|f_{xx}| = |2| = 2, \qquad |f_{xy}| = |-1| = 1, \qquad |f_{yy}| = |1| = 1.$$

The largest of these is 2, so we may safely take M to be 2. With $(x_0, y_0) = (3, 2)$, we then know that, throughout R,

$$|E(x, y)| \leq \frac{1}{2}(2)(|x - 3| + |y - 2|)^2 = (|x - 3| + |y - 2|)^2.$$

Finally, since $|x - 3| \leq 0.1$ and $|y - 2| \leq 0.1$ on R, we have

$$|E(x, y)| \leq (0.1 + 0.1)^2 = 0.04.$$

As a percentage of $f(3, 2) = 8$, the error is no greater than

$$\frac{0.04}{8} \times 100 = 0.5\%.$$

As long as (x, y) stays in R, the approximation $f(x, y) \approx L(x, y)$ will be in error by no more than 0.04, which is 1/2% of the value of f at the center of R. ❑

Predicting Change with Differentials

Suppose we know the values of a differentiable function $f(x, y)$ and its first partial derivatives at a point (x_0, y_0) and we want to predict how much the value of f will change if we move to a point $(x_0 + \Delta x, y_0 + \Delta y)$ nearby. If Δx and Δy are small, f and its linearization at (x_0, y_0) will change by nearly the same amount, so the change in L will give a practical estimate of the change in f.

The change in f is

$$\Delta f = f(x_0 + \Delta x, y_0 + \Delta y) - f(x_0, y_0).$$

A straightforward calculation with Eq. (4), using the notation $x - x_0 = \Delta x$ and $y - y_0 = \Delta y$, shows that the corresponding change in L is

$$\Delta L = L(x_0 + \Delta x, y_0 + \Delta y) - L(x_0, y_0)$$
$$= f_x(x_0, y_0)\Delta x + f_y(x_0, y_0)\Delta y.$$

The formula for Δf is usually as hard to work with as the formula for f. The change in L, however, is just a known constant times Δx plus a known constant times Δy.

The change ΔL is usually described in the more suggestive notation

$$df = f_x(x_0, y_0)\,dx + f_y(x_0, y_0)\,dy,$$

in which df denotes the change in the linearization that results from the changes dx and dy in x and y. As usual, we call dx and dy differentials of x and y, and call df the corresponding differential of f.

Definition

If we move from (x_0, y_0) to a point $(x_0 + dx, y_0 + dy)$ nearby, the resulting differential in f is

$$df = f_x(x_0, y_0)\,dx + f_y(x_0, y_0)\,dy. \tag{6}$$

This change in the linearization of f is called the **total differential of f.**

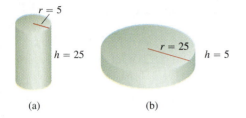

(a)   (b)

12.22 The volume of cylinder (a) is more sensitive to a small change in r than it is to an equally small change in h. The volume of cylinder (b) is more sensitive to small changes in h than it is to small changes in r.

Absolute change vs. relative change

If you measure a 20-volt potential with an error of 10 volts, your reading is probably too crude to be useful. You are off by 50%. But if you measure a 200,000-volt potential with an error of 10 volts, your reading is within 0.005% of the true value. An absolute error of 10 volts is significant in the first case but of no consequence in the second because the relative error is so small.

In other cases, a small relative error—say, traveling a few meters too far in a journey of hundreds of thousands of meters—can have spectacular consequences.

EXAMPLE 3   *Sensitivity to change*

Your company manufactures right circular cylindrical molasses storage tanks that are 25 ft high with a radius of 5 ft. How sensitive are the tanks' volumes to small variations in height and radius?

Solution   As a function of radius r and height h, the typical tank's volume is

$$V = \pi r^2 h.$$

The change in volume caused by small changes dr and dh in radius and height is approximately

$$dV = V_r(5, 25)\,dr + V_h(5, 25)\,dh \qquad \text{Eq. (6) with } f = V$$
$$\text{and } (x_0, y_0) = (5, 25)$$

$$= (2\pi rh)_{(5,25)}\,dr + (\pi r^2)_{(5,25)}\,dh$$

$$= 250\pi\,dr + 25\pi\,dh.$$

Thus, a 1-unit change in r will change V by about 250π units. A 1-unit change in h will change V by about 25π units. The tank's volume is 10 times more sensitive to a small change in r than it is to a small change of equal size in h. As a quality control engineer concerned with being sure the tanks have the correct volume, you would want to pay special attention to their radii.

In contrast, if the values of r and h are reversed to make $r = 25$ and $h = 5$, then the total differential in V becomes

$$dV = (2\pi rh)_{(25,5)}\,dr + (\pi r^2)_{(25,5)}\,dh = 250\pi\,dr + 625\pi\,dh.$$

Now the volume is more sensitive to changes in h than to changes in r (Fig. 12.22).

The general rule to be learned from this example is that functions are most sensitive to small changes in the variables that generate the largest partial derivatives. ❑

Absolute, Relative, and Percentage Change

When we move from (x_0, y_0) to a point nearby, we can describe the corresponding change in the value of a function $f(x, y)$ in three different ways.

	True	**Estimate**
Absolute change:	Δf	df
Relative change:	$\dfrac{\Delta f}{f(x_0, y_0)}$	$\dfrac{df}{f(x_0, y_0)}$
Percentage change:	$\dfrac{\Delta f}{f(x_0, y_0)} \times 100$	$\dfrac{df}{f(x_0, y_0)} \times 100$

EXAMPLE 4   Suppose that the variables r and h change from the initial values of $(r_0, h_0) = (1, 5)$ by the amounts $dr = 0.03$ and $dh = -0.1$. Estimate the resulting absolute, relative, and percentage changes in the values of the function $V = \pi r^2 h$.

Solution   To estimate the absolute change in V, we evaluate

$$dV = V_r(r_0, h_0)\,dr + V_h(r_0, h_0)\,dh$$

to get $dV = 2\pi r_0 h_0\,dr + \pi r_0^2\,dh = 2\pi(1)(5)(0.03) + \pi(1)^2(-0.1)$

$$= 0.3\pi - 0.1\pi = 0.2\pi.$$

We divide this by $V(r_0, h_0)$ to estimate the relative change:

$$\frac{dV}{V(r_0, h_0)} = \frac{0.2\pi}{\pi r_0^2 h_0} = \frac{0.2\pi}{\pi(1)^2(5)} = 0.04.$$

We multiply this by 100 to estimate the percentage change:

$$\frac{dV}{V(r_0, h_0)} \times 100 = 0.04 \times 100 = 4\%.$$

❑

EXAMPLE 5   The volume $V = \pi r^2 h$ of a right circular cylinder is to be calculated from measured values of r and h. Suppose that r is measured with an error of no more than 2% and h with an error of no more than 0.5%. Estimate the resulting possible percentage error in the calculation of V.

Solution   We are told that

$$\left| \frac{dr}{r} \times 100 \right| \le 2 \qquad \text{and} \qquad \left| \frac{dh}{h} \times 100 \right| \le 0.5.$$

Since

$$\frac{dV}{V} = \frac{2\pi r h \, dr + \pi r^2 dh}{\pi r^2 h} = \frac{2 \, dr}{r} + \frac{dh}{h},$$

we have

$$\left| \frac{dV}{V} \times 100 \right| = \left| 2\frac{dr}{r} \times 100 + \frac{dh}{h} \times 100 \right|$$

$$\le 2 \left| \frac{dr}{r} \times 100 \right| + \left| \frac{dh}{h} \times 100 \right| \le 2(2) + 0.5 = 4.5.$$

We estimate the error in the volume calculation to be at most 4.5%.   ❑

How accurately do we have to measure r and h to have a reasonable chance of calculating $V = \pi r^2 h$ with an error, say, of less than 2%? Questions like this are hard to answer because there is usually no single right answer. Since

$$\frac{dV}{V} = 2\frac{dr}{r} + \frac{dh}{h},$$

we see that dV/V is controlled by a combination of dr/r and dh/h. If we can measure h with great accuracy, we might come out all right even if we are sloppy about measuring r. On the other hand, our measurement of h might have so large a dh that the resulting dV/V would be too crude an estimate of $\Delta V/V$ to be useful even if dr were zero.

What we do in such cases is look for a reasonable square about the measured values (r_0, h_0) in which V will not vary by more than the allowed amount from $V_0 = \pi r_0^2 h_0$.

EXAMPLE 6   Find a reasonable square about the point $(r_0, h_0) = (5, 12)$ in which the value of $V = \pi r^2 h$ will not vary by more than ± 0.1.

Solution   We approximate the variation ΔV by the differential

$$dV = 2\pi r_0 h_0 \, dr + \pi r_0^2 dh = 2\pi(5)(12) \, dr + \pi(5)^2 dh = 120\pi \, dr + 25\pi \, dh.$$

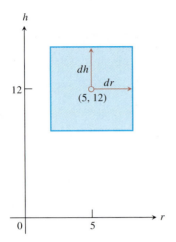

12.23 A small square about the point (5, 12) in the rh-plane (Example 6).

Since the region to which we are restricting our attention is a square (Fig. 12.23), we may set $dh = dr$ to get

$$dV = 120\pi\,dr + 25\pi\,dr = 145\pi\,dr.$$

We then ask, How small must we take dr to be sure that $|dV|$ is no larger than 0.1? To answer, we start with the inequality

$$|dV| \leq 0.1,$$

express dV in terms of dr,

$$|145\pi\,dr| \leq 0.1,$$

and find a corresponding upper bound for dr:

$$|dr| \leq \frac{0.1}{145\pi} \approx 2.1 \times 10^{-4}. \qquad \text{Rounding down to make sure } dr \text{ won't accidentally be too big}$$

With $dh = dr$, then, the square we want is described by the inequalities

$$|r - 5| \leq 2.1 \times 10^{-4}, \qquad |h - 12| \leq 2.1 \times 10^{-4}.$$

As long as (r, h) stays in this square, we may expect $|dV|$ to be less than or equal to 0.1 and we may expect $|\Delta V|$ to be approximately the same size.   ❑

Functions of More Than Two Variables

Analogous results hold for differentiable functions of more than two variables.

1. The **linearization** of $f(x, y, z)$ at a point $P_0(x_0, y_0, z_0)$ is

$$L(x, y, z) = f(P_0) + f_x(P_0)(x - x_0) + f_y(P_0)(y - y_0) + f_z(P_0)(z - z_0). \quad \text{(7)}$$

2. Suppose that R is a closed rectangular solid centered at P_0 and lying in an open region on which the second partial derivatives of f are continuous. Suppose also that $|f_{xx}|, |f_{yy}|, |f_{zz}|, |f_{xy}|, |f_{xz}|$, and $|f_{yz}|$ are all less than or equal to M throughout R. Then the **error** $E(x, y, z) = f(x, y, z) - L(x, y, z)$ in the approximation of f by L is bounded throughout R by the inequality

$$|E| \leq \frac{1}{2}M\left(|x - x_0| + |y - y_0| + |z - z_0|\right)^2. \quad \text{(8)}$$

3. If the second partial derivatives of f are continuous and if x, y, and z change from x_0, y_0, and z_0 by small amounts dx, dy, and dz, the **total differential**

$$df = f_x(P_0)\,dx + f_y(P_0)\,dy + f_z(P_0)\,dz$$

gives a good approximation of the resulting change in f.

EXAMPLE 7   Find the linearization $L(x, y, z)$ of

$$f(x, y, z) = x^2 - xy + 3\sin z$$

at the point $(x_0, y_0, z_0) = (2, 1, 0)$. Find an upper bound for the error incurred in replacing f by L on the rectangle

$$R: \quad |x - 2| \leq 0.01, \quad |y - 1| \leq 0.02, \quad |z| \leq 0.01.$$

Solution   A routine evaluation gives

$$f(2, 1, 0) = 2, \qquad f_x(2, 1, 0) = 3, \qquad f_y(2, 1, 0) = -2, \qquad f_z(2, 1, 0) = 3.$$

With these values, Eq. (7) becomes

$$L(x, y, z) = 2 + 3(x - 2) + (-2)(y - 1) + 3(z - 0) = 3x - 2y + 3z - 2.$$

Equation (8) gives an upper bound for the error incurred by replacing f by L on R. Since

$$f_{xx} = 2, \qquad f_{yy} = 0, \qquad f_{zz} = -3 \sin z,$$
$$f_{xy} = -1, \qquad f_{xz} = 0, \qquad f_{yz} = 0,$$

we may safely take M to be $\max | -3 \sin z| = 3$. Hence

$$|E| \leq \frac{1}{2}(3)(0.01 + 0.02 + 0.01)^2 = 0.0024.$$

The error will be no greater than 0.0024.   ❏

EXAMPLE 8   *Controlling sag in uniformly loaded beams*

A horizontal rectangular beam, supported at both ends, will sag when subjected to a uniform load (constant weight per linear foot). The amount S of sag (Fig. 12.24) is calculated with the formula

$$S = C \frac{px^4}{wh^3}.$$

In this equation,

$p = $ the load (newtons per meter of beam length),

$x = $ the length between supports (m),

$w = $ the width of the beam (m),

$h = $ the height of the beam (m),

$C = $ a constant that depends on the units of measurement and on the material from which the beam is made.

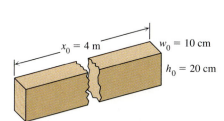

12.24 A beam supported at its two ends before and after loading. Example 8 shows how the sag S is related to the weight of the load and the dimensions of the beam.

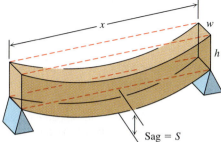

12.25 The dimensions of the beam in Example 8.

Find dS for a beam 4 m long, 10 cm wide, and 20 cm high that is subjected to a load of 100 N/m (Fig. 12.25). What conclusions can be drawn about the beam from the expression for dS?

Solution   Since S is a function of the four independent variables p, x, w, and h, its total differential dS is given by the equation

$$dS = S_p \, dp + S_x \, dx + S_w \, dw + S_h \, dh.$$

When we write this out for a particular set of values p_0, x_0, w_0, and h_0 and simplify the result, we find that

$$dS = S_0 \left(\frac{dp}{p_0} + \frac{4dx}{x_0} - \frac{dw}{w_0} - \frac{3dh}{h_0} \right),$$

where $S_0 = S(p_0, x_0, w_0, h_0) = Cp_0x_0^4/(w_0h_0^3)$.

If $p_0 = 100$ N/m, $x_0 = 4$ m, $w_0 = 0.1$ m, and $h_0 = 0.2$ m, then

$$dS = S_0 \left(\frac{dp}{100} + dx - 10dw - 15dh \right). \tag{9}$$

Here is what we can learn from Eq. (9). Since dp and dx appear with positive coefficients, increases in p and x will increase the sag. But dw and dh appear with negative coefficients, so increases in w and h will decrease the sag (make the beam stiffer). The sag is not very sensitive to changes in load because the coefficient of dp is 1/100. The magnitude of the coefficient of dh is greater than the magnitude of the coefficient of dw. Making the beam 1 cm higher will therefore decrease the sag more than making the beam 1 cm wider.   ❑

Exercises 12.4

Finding Linearizations

In Exercises 1–6, find the linearization $L(x, y)$ of the function at each point.

1. $f(x, y) = x^2 + y^2 + 1$ at (a) (0, 0), (b) (1, 1)

2. $f(x, y) = (x + y + 2)^2$ at (a) (0, 0), (b) (1, 2)

3. $f(x, y) = 3x - 4y + 5$ at (a) (0, 0), (b) (1, 1)

4. $f(x, y) = x^3 y^4$ at (a) (1, 1), (b) (0, 0)

5. $f(x, y) = e^x \cos y$ at (a) (0, 0), (b) (0, $\pi/2$)

6. $f(x, y) = e^{2y-x}$ at (a) (0, 0), (b) (1, 2)

Upper Bounds for Errors in Linear Approximations

In Exercises 7–12, find the linearization $L(x, y)$ of the function $f(x, y)$ at P_0. Then use inequality (5) to find an upper bound for the magnitude $|E|$ of the error in the approximation $f(x, y) \approx L(x, y)$ over the rectangle R.

7. $f(x, y) = x^2 - 3xy + 5$ at $P_0(2, 1)$,
   R: $|x - 2| \le 0.1$, $|y - 1| \le 0.1$

8. $f(x, y) = (1/2)x^2 + xy + (1/4)y^2 + 3x - 3y + 4$ at $P_0(2, 2)$,
   R: $|x - 2| \le 0.1$, $|y - 2| \le 0.1$

9. $f(x, y) = 1 + y + x \cos y$ at $P_0(0, 0)$,
   R: $|x| \le 0.2$, $|y| \le 0.2$
   (Use $|\cos y| \le 1$ and $|\sin y| \le 1$ in estimating E.)

10. $f(x, y) = xy^2 + y \cos(x - 1)$ at $P_0(1, 2)$,
    R: $|x - 1| \le 0.1$, $|y - 2| \le 0.1$

11. $f(x, y) = e^x \cos y$ at $P_0(0, 0)$,
    R: $|x| \le 0.1$, $|y| \le 0.1$
    (Use $e^x \le 1.11$ and $|\cos y| \le 1$ in estimating E.)

12. $f(x, y) = \ln x + \ln y$ at $P_0(1, 1)$,
    R: $|x - 1| \le 0.2$, $|y - 1| \le 0.2$

Sensitivity to Change. Estimates

13. You plan to calculate the area of a long, thin rectangle from measurements of its length and width. Which dimension should you measure more carefully? Give reasons for your answer.

14. a) Around the point (1, 0), is $f(x, y) = x^2(y + 1)$ more sensitive to changes in x, or to changes in y? Give reasons for your answer.

b) What ratio of dx to dy will make df equal zero at (1, 0)?

15. Suppose T is to be found from the formula $T = x(e^y + e^{-y})$ where x and y are found to be 2 and ln 2 with maximum possible errors of $|dx| = 0.1$ and $|dy| = 0.02$. Estimate the maximum possible error in the computed value of T.

16. About how accurately may $V = \pi r^2 h$ be calculated from measurements of r and h that are in error by 1%?

17. If $r = 5.0$ cm and $h = 12.0$ cm to the nearest millimeter, what should we expect the maximum percentage error in calculating $V = \pi r^2 h$ to be?

18. To estimate the volume of a cylinder of radius about 2 m and height about 3 m, about how accurately should the radius and height be measured so that the error in the volume estimate will not exceed 0.1 m^3? Assume that the possible error dr in measuring r is equal to the possible error dh in measuring h.

19. Give a reasonable square centered at (1, 1) over which the value of $f(x, y) = x^3 y^4$ will not vary by more than ± 0.1.

20. *Variation in electrical resistance.* The resistance R produced by wiring resistors of R_1 and R_2 ohms in parallel (Fig. 12.26) can be calculated from the formula

$$\frac{1}{R} = \frac{1}{R_1} + \frac{1}{R_2}.$$

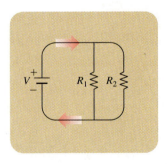

12.26 The circuit in Exercises 20 and 21.

a) Show that

$$dR = \left(\frac{R}{R_1}\right)^2 dR_1 + \left(\frac{R}{R_2}\right)^2 dR_2.$$

b) You have designed a two-resistor circuit like the one in Fig. 12.26 to have resistances of $R_1 = 100$ ohms and $R_2 = 400$

ohms, but there is always some variation in manufacturing and the resistors received by your firm will probably not have these exact values. Will the value of R be more sensitive to variation in R_1, or to variation in R_2? Give reasons for your answer.

21. (*Continuation of Exercise 20.*) In another circuit like the one in Fig. 12.26, you plan to change R_1 from 20 to 20.1 ohms and R_2 from 25 to 24.9 ohms. By about what percentage will this change R?

22. *Error carry-over in coordinate changes*

   a) If $x = 3 \pm 0.01$ and $y = 4 \pm 0.01$, as shown here, with approximately what accuracy can you calculate the polar coordinates r and θ of the point $P(x, y)$ from the formulas $r^2 = x^2 + y^2$ and $\theta = \tan^{-1}(y/x)$? Express your estimates as percentage changes of the values that r and θ have at the point $(x_0, y_0) = (3, 4)$.

   b) At the point $(x_0, y_0) = (3, 4)$, are the values of r and θ more sensitive to changes in x, or to changes in y? Give reasons for your answer.

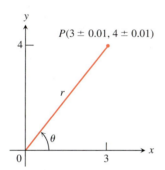

Functions of Three Variables

Find the linearizations $L(x, y, z)$ of the functions in Exercises 23–28 at the given points.

23. $f(x, y, z) = xy + yz + xz$ at

   a) (1, 1, 1)   **b)** (1, 0, 0)   **c)** (0, 0, 0)

24. $f(x, y, z) = x^2 + y^2 + z^2$ at

   a) (1, 1, 1)   **b)** (0, 1, 0)   **c)** (1, 0, 0)

25. $f(x, y, z) = \sqrt{x^2 + y^2 + z^2}$ at

   a) (1, 0, 0)   **b)** (1, 1, 0)   **c)** (1, 2, 2)

26. $f(x, y, z) = (\sin xy)/z$ at

   a) $(\pi/2, 1, 1)$ **b)** (2, 0, 1)

27. $f(x, y, z) = e^x + \cos(y + z)$ at

   a) (0, 0, 0)   **b)** $\left(0, \dfrac{\pi}{2}, 0\right)$ **c)** $\left(0, \dfrac{\pi}{4}, \dfrac{\pi}{4}\right)$

28. $f(x, y, z) = \tan^{-1}(xyz)$ at

   a) (1, 0, 0)   **b)** (1, 1, 0)   **c)** (1, 1, 1)

In Exercises 29–32, find the linearization $L(x, y, z)$ of the function $f(x, y, z)$ at P_0. Then use inequality (8) to find an upper bound for the magnitude of the error E in the approximation $f(x, y, z) \approx L(x, y, z)$ over the region R.

29. $f(x, y, z) = xz - 3yz + 2$ at $P_0(1, 1, 2)$
    R: $|x - 1| \le 0.01,$ $|y - 1| \le 0.01,$ $|z - 2| \le 0.02$

30. $f(x, y, z) = x^2 + xy + yz + (1/4)z^2$ at $P_0(1, 1, 2)$
    R: $|x - 1| \le 0.01,$ $|y - 1| \le 0.01,$ $|z - 2| \le 0.08$

31. $f(x, y, z) = xy + 2yz - 3xz$ at $P_0(1, 1, 0)$
    R: $|x - 1| \le 0.01,$ $|y - 1| \le 0.01,$ $|z| \le 0.01$

32. $f(x, y, z) = \sqrt{2} \cos x \sin(y + z)$ at $P_0(0, 0, \pi/4)$
    R: $|x| \le 0.01,$ $|y| \le 0.01,$ $|z - \pi/4| \le 0.01$

Theory and Examples

33. The beam of Example 8 is tipped on its side so that $h = 0.1$ m and $w = 0.2$ m.

   a) What is the value of dS now?

   b) Compare the sensitivity of the newly positioned beam to a small change in height with its sensitivity to an equally small change in width.

34. A standard 12-fl oz can of soda is essentially a cylinder of radius $r = 1$ in. and height $h = 5$ in.

   a) At these dimensions, how sensitive is the can's volume to a small change in the radius versus a small change in the height?

   b) Could you design a soda can that *appears* to hold more soda but in fact holds the same 12 fl oz? What might its dimensions be? (There is more than one correct answer.)

35. If $|a|$ is much greater than $|b|$, $|c|$, and $|d|$, to which of $a, b, c,$ and d is the value of the determinant

$$f(a, b, c, d) = \begin{vmatrix} a & b \\ c & d \end{vmatrix}$$

most sensitive? Give reasons for your answer.

36. Estimate how strongly simultaneous errors of 2% in $a, b,$ and c might affect the calculation of the product

$$p(a, b, c) = abc.$$

37. Estimate how much wood it takes to make a hollow rectangular box whose inside measurements are 5 ft long by 3 ft wide by 2 ft deep if the box is made of lumber 1/2-in. thick and the box has no top.

38. The area of a triangle is $(1/2)ab \sin C$, where a and b are the lengths of two sides of the triangle and C is the measure of the included angle. In surveying a triangular plot, you have measured

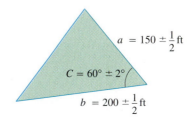

a, b, and C to be 150 ft, 200 ft, and 60°, respectively. By about how much could your area calculation be in error if your values of a and b are off by half a foot each and your measurement of C is off by 2°? See the figure. Remember to use radians.

39. Suppose that $u = xe^y + y \sin z$ and that x, y, and z can be measured with maximum possible errors of $\pm 0.2, \pm 0.6$, and $\pm \pi/180$, respectively. Estimate the maximum possible error in calculating u from the measured values $x = 2, y = \ln 3, z = \pi/2$.

40. *The Wilson lot size formula.* The Wilson lot size formula in economics says that the most economical quantity Q of goods (radios, shoes, brooms, whatever) for a store to order is given by the formula $Q = \sqrt{2KM/h}$, where K is the cost of placing the order, M is the number of items sold per week, and h is the weekly holding cost for each item (cost of space, utilities, security, and so on). To which of the variables K, M, and h is Q most sensitive near the point $(K_0, M_0, h_0) = (2, 20, 0.05)$? Give reasons for your answer.

41. Does a function $f(x, y)$ with continuous first partial derivatives throughout an open region R have to be continuous on R? Give reasons for your answer.

42. If a function $f(x, y)$ has continuous second partial derivatives throughout an open region R, must the first order partial derivatives of f be continuous on R? Give reasons for your answer.

12.5 The Chain Rule

When we are interested in the temperature $w = f(x, y, z)$ at points along a curve $x = g(t), y = h(t), z = k(t)$ in space, or in the pressure or density along a path through a gas or fluid, we may think of f as a function of the single variable t. For each value of t, the temperature at the point $(g(t), h(t), k(t))$ is the value of the composite function $f(g(t), h(t), k(t))$. If we then wish to know the rate at which f changes with respect to t along the path, we have only to differentiate this composite with respect to t, provided, of course, the derivative exists.

Sometimes we can find the derivative by substituting the formulas for g, h, and k into the formula for f and differentiating directly with respect to t. But we often have to work with functions whose formulas are too complicated for convenient substitution or for which formulas are not readily available. To find a function's derivatives under circumstances like these, we use the Chain Rule. The form the Chain Rule takes depends on how many variables are involved but, except for the presence of additional variables, it works just like the Chain Rule in Section 2.5.

The Chain Rule for Functions of Two Variables

In Section 2.5, we used the Chain Rule when $w = f(x)$ was a differentiable function of x and $x = g(t)$ was a differentiable function of t. This made w a differentiable function of t and the Chain Rule said that dw/dt could be calculated with the formula

$$\frac{dw}{dt} = \frac{dw}{dx}\frac{dx}{dt}.$$

The analogous formula for a function $w = f(x, y)$ is given in Theorem 5.

Theorem 5
Chain Rule for Functions of Two Independent Variables
If $w = f(x, y)$ is differentiable and x and y are differentiable functions of t, then w is a differentiable function of t and

$$\frac{dw}{dt} = \frac{\partial f}{\partial x}\frac{dx}{dt} + \frac{\partial f}{\partial y}\frac{dy}{dt}. \tag{1}$$

Proof The proof consists of showing that if x and y are differentiable at $t = t_0$, then w is differentiable at t_0 and

$$\left(\frac{dw}{dt}\right)_{t_0} = \left(\frac{\partial w}{\partial x}\right)_{P_0} \left(\frac{dx}{dt}\right)_{t_0} + \left(\frac{\partial w}{\partial y}\right)_{P_0} \left(\frac{dy}{dt}\right)_{t_0}, \qquad (2)$$

where $P_0 = (x(t_0), y(t_0))$.

Let Δx, Δy, and Δw be the increments that result from changing t from t_0 to $t_0 + \Delta t$. Since f is differentiable (remember the definition in Section 12.4),

$$\Delta w = \left(\frac{\partial w}{\partial x}\right)_{P_0} \Delta x + \left(\frac{\partial w}{\partial y}\right)_{P_0} \Delta y + \epsilon_1 \Delta x + \epsilon_2 \Delta y, \qquad (3)$$

where $\epsilon_1, \epsilon_2 \to 0$ as $\Delta x, \Delta y \to 0$. To find dw/dt, we divide Eq. (3) through by Δt and let Δt approach zero. The division gives

$$\frac{\Delta w}{\Delta t} = \left(\frac{\partial w}{\partial x}\right)_{P_0} \frac{\Delta x}{\Delta t} + \left(\frac{\partial w}{\partial y}\right)_{P_0} \frac{\Delta y}{\Delta t} + \epsilon_1 \frac{\Delta x}{\Delta t} + \epsilon_2 \frac{\Delta y}{\Delta t},$$

and letting Δt approach zero gives

$$\left(\frac{dw}{dt}\right)_{t_0} = \lim_{\Delta t \to 0} \frac{\Delta w}{\Delta t}$$

$$= \left(\frac{\partial w}{\partial x}\right)_{P_0} \left(\frac{dx}{dt}\right)_{t_0} + \left(\frac{\partial w}{\partial y}\right)_{P_0} \left(\frac{dy}{dt}\right)_{t_0} + 0 \cdot \left(\frac{dx}{dt}\right)_{t_0} + 0 \cdot \left(\frac{dy}{dt}\right)_{t_0}.$$

This establishes Eq. (2) and completes the proof. ❑

The way to remember the Chain Rule is to picture the diagram below. To find dw/dt, start at w and read down each route to t, multiplying derivatives along the way. Then add the products.

Chain Rule

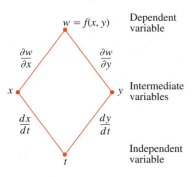

$$w = f(x, y) \quad \text{Dependent variable}$$
$$\frac{\partial w}{\partial x} \qquad \frac{\partial w}{\partial y}$$
$$x \qquad\qquad y \quad \text{Intermediate variables}$$
$$\frac{dx}{dt} \qquad \frac{dy}{dt}$$
$$t \quad \text{Independent variable}$$

$$\frac{dw}{dt} = \frac{\partial w}{\partial x}\frac{dx}{dt} + \frac{\partial w}{\partial y}\frac{dy}{dt}$$

The **tree diagram** in the margin provides a convenient way to remember the Chain Rule. From the diagram you see that when $t = t_0$ the derivatives dx/dt and dy/dt are evaluated at t_0. The value of t_0 then determines the value x_0 for the differentiable function x and the value y_0 for the differentiable function y. The partial derivatives $\partial w/\partial x$ and $\partial w/\partial y$ (which are themselves functions of x and y) are evaluated at the point $P_0(x_0, y_0)$ corresponding to t_0. The "true" independent variable is t, while x and y are *intermediate variables* (controlled by t) and w is the dependent variable.

A more precise notation for the Chain Rule shows how the various derivatives in Eq. (1) are evaluated:

$$\frac{dw}{dt}(t_0) = \frac{\partial f}{\partial x}(x_0, y_0) \cdot \frac{dx}{dt}(t_0) + \frac{\partial f}{\partial y}(x_0, y_0) \cdot \frac{dy}{dt}(t_0).$$

EXAMPLE 1 Use the Chain Rule to find the derivative of

$$w = xy$$

with respect to t along the path $x = \cos t$, $y = \sin t$. What is the derivative's value at $t = \pi/2$?

Solution We evaluate the right-hand side of Eq. (1) with $w = xy$, $x = \cos t$, and $y = \sin t$:

$$\frac{\partial w}{\partial x} = y = \sin t, \qquad \frac{\partial w}{\partial y} = x = \cos t, \qquad \frac{dx}{dt} = -\sin t, \qquad \frac{dy}{dt} = \cos t$$

$$\frac{dw}{dt} = \frac{\partial w}{\partial x}\frac{dx}{dt} + \frac{\partial w}{\partial y}\frac{dy}{dt} = (\sin t)(-\sin t) + (\cos t)(\cos t) \qquad \text{Eq. (1) with values from above}$$

$$= -\sin^2 t + \cos^2 t = \cos 2t.$$

Here we have three routes from w to t instead of two. But finding dw/dt is still the same. Read down each route, multiplying derivatives along the way; then add.

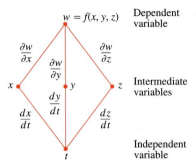

Chain Rule

$w = f(x, y, z)$ — Dependent variable

$\dfrac{\partial w}{\partial x}$ $\dfrac{\partial w}{\partial y}$ $\dfrac{\partial w}{\partial z}$

x y z — Intermediate variables

$\dfrac{dx}{dt}$ $\dfrac{dy}{dt}$ $\dfrac{dz}{dt}$

t — Independent variable

$$\frac{dw}{dt} = \frac{\partial w}{\partial x}\frac{dx}{dt} + \frac{\partial w}{\partial y}\frac{dy}{dt} + \frac{\partial w}{\partial z}\frac{dz}{dt}$$

The helix
$$\mathbf{r} = (\cos t)\mathbf{i} + (\sin t)\mathbf{j} + t\mathbf{k}$$

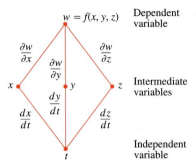

12.27 Example 2 shows how the values of $w = xy + z$ vary with t along this helix.

Notice in our calculation that we have substituted the functional expressions $x = \cos t$ and $y = \sin t$ in the partial derivatives $\partial w/\partial x$ and $\partial w/\partial y$. The resulting derivative dw/dt is then expressed in terms of the independent variable t (so the intermediate variables x and y do not appear).

In this example we can check the result with a more direct calculation. As a function of t,

$$w = xy = \cos t \sin t = \frac{1}{2}\sin 2t,$$

so

$$\frac{dw}{dt} = \frac{d}{dt}\left(\frac{1}{2}\sin 2t\right) = \frac{1}{2}\cdot 2\cos 2t = \cos 2t.$$

In either case,

$$\left(\frac{dw}{dt}\right)_{t=\pi/2} = \cos\left(2\cdot\frac{\pi}{2}\right) = \cos\pi = -1. \qquad \square$$

The Chain Rule for Functions of Three Variables

To get the Chain Rule for functions of three variables, we add a term to Eq. (1).

Chain Rule for Functions of Three Independent Variables

If $w = f(x, y, z)$ is differentiable and x, y, and z are differentiable functions of t, then w is a differentiable function of t and

$$\frac{dw}{dt} = \frac{\partial f}{\partial x}\frac{dx}{dt} + \frac{\partial f}{\partial y}\frac{dy}{dt} + \frac{\partial f}{\partial z}\frac{dz}{dt}. \qquad (4)$$

The derivation is identical with the derivation of Eq. (1) except that there are now three intermediate variables instead of two. The diagram we use for remembering the new equation is similar as well.

EXAMPLE 2 *Changes in a function's values along a helix*

Find dw/dt if

$$w = xy + z, \qquad x = \cos t, \qquad y = \sin t, \qquad z = t$$

(Fig. 12.27). What is the derivative's value at $t = 0$?

Solution

$$\frac{dw}{dt} = \frac{\partial w}{\partial x}\frac{dx}{dt} + \frac{\partial w}{\partial y}\frac{dy}{dt} + \frac{\partial w}{\partial z}\frac{dz}{dt} \qquad \text{Eq. (4)}$$

$$= (y)(-\sin t) + (x)(\cos t) + (1)(1)$$

$$= (\sin t)(-\sin t) + (\cos t)(\cos t) + 1 \qquad \text{Substitute for the intermediate variables.}$$

$$= -\sin^2 t + \cos^2 t + 1 = 1 + \cos 2t$$

$$\left(\frac{dw}{dt}\right)_{t=0} = 1 + \cos(0) = 2. \qquad \square$$

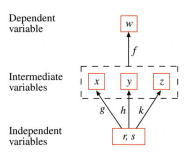

Dependent variable

Intermediate variables

Independent variables

$w = f(g(r, s), h(r, s), k(r, s))$

(a)

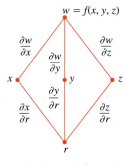

$\dfrac{\partial w}{\partial r} = \dfrac{\partial w}{\partial x}\dfrac{\partial x}{\partial r} + \dfrac{\partial w}{\partial y}\dfrac{\partial y}{\partial r} + \dfrac{\partial w}{\partial z}\dfrac{\partial z}{\partial r}$

(b)

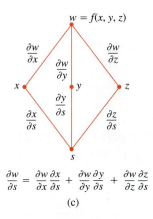

$\dfrac{\partial w}{\partial s} = \dfrac{\partial w}{\partial x}\dfrac{\partial x}{\partial s} + \dfrac{\partial w}{\partial y}\dfrac{\partial y}{\partial s} + \dfrac{\partial w}{\partial z}\dfrac{\partial z}{\partial s}$

(c)

12.28 Composite function and tree diagrams for Eqs. (5) and (6).

The Chain Rule for Functions Defined on Surfaces

If we are interested in the temperature $w = f(x, y, z)$ at points (x, y, z) on a globe in space, we might prefer to think of $x, y,$ and z as functions of the variables r and s that give the points' longitudes and latitudes. If $x = g(r, s)$, $y = h(r, s)$, and $z = k(r, s)$, we could then express the temperature as a function of r and s with the composite function

$$w = f(g(r, s), h(r, s), k(r, s)).$$

Under the right conditions, w would have partial derivatives with respect to both r and s that could be calculated in the following way.

Chain Rule for Two Independent Variables and Three Intermediate Variables

Suppose that $w = f(x, y, z)$, $x = g(r, s)$, $y = h(r, s)$, and $z = k(r, s)$. If all four functions are differentiable, then w has partial derivatives with respect to r and s, given by the formulas

$$\frac{\partial w}{\partial r} = \frac{\partial w}{\partial x}\frac{\partial x}{\partial r} + \frac{\partial w}{\partial y}\frac{\partial y}{\partial r} + \frac{\partial w}{\partial z}\frac{\partial z}{\partial r}, \quad (5)$$

$$\frac{\partial w}{\partial s} = \frac{\partial w}{\partial x}\frac{\partial x}{\partial s} + \frac{\partial w}{\partial y}\frac{\partial y}{\partial s} + \frac{\partial w}{\partial z}\frac{\partial z}{\partial s}. \quad (6)$$

Equation (5) can be derived from Eq. (4) by holding s fixed and setting r equal to t. Similarly, Eq. (6) can be derived by holding r fixed and setting s equal to t. The tree diagrams for Eqs. (5) and (6) are shown in Fig. 12.28.

EXAMPLE 3 Express $\partial w/\partial r$ and $\partial w/\partial s$ in terms of r and s if

$$w = x + 2y + z^2, \qquad x = \frac{r}{s}, \qquad y = r^2 + \ln s, \qquad z = 2r.$$

Solution
$$\frac{\partial w}{\partial r} = \frac{\partial w}{\partial x}\frac{\partial x}{\partial r} + \frac{\partial w}{\partial y}\frac{\partial y}{\partial r} + \frac{\partial w}{\partial z}\frac{\partial z}{\partial r} \qquad \text{Eq. (5)}$$

$$= (1)\left(\frac{1}{s}\right) + (2)(2r) + (2z)(2)$$

$$= \frac{1}{s} + 4r + (4r)(2) = \frac{1}{s} + 12r \qquad \text{Substitute for intermediate variable } z.$$

$$\frac{\partial w}{\partial s} = \frac{\partial w}{\partial x}\frac{\partial x}{\partial s} + \frac{\partial w}{\partial y}\frac{\partial y}{\partial s} + \frac{\partial w}{\partial z}\frac{\partial z}{\partial s} \qquad \text{Eq. (6)}$$

$$= (1)\left(-\frac{r}{s^2}\right) + (2)\left(\frac{1}{s}\right) + (2z)(0) = \frac{2}{s} - \frac{r}{s^2} \qquad \square$$

If f is a function of two variables instead of three, Eqs. (5) and (6) become one term shorter, because the intermediate variable z doesn't appear.

Chain Rule

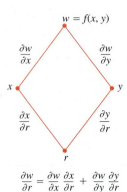

$$\frac{\partial w}{\partial r} = \frac{\partial w}{\partial x}\frac{\partial x}{\partial r} + \frac{\partial w}{\partial y}\frac{\partial y}{\partial r}$$

12.29 Tree diagram for the first of Eqs. (7).

Chain Rule

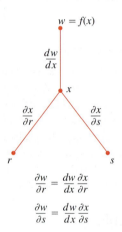

$$\frac{\partial w}{\partial r} = \frac{dw}{dx}\frac{\partial x}{\partial r}$$

$$\frac{\partial w}{\partial s} = \frac{dw}{dx}\frac{\partial x}{\partial s}$$

12.30 Tree diagram for Eqs. (8).

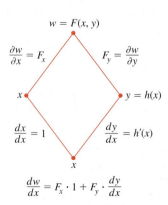

$$\frac{dw}{dx} = F_x \cdot 1 + F_y \cdot \frac{dy}{dx}$$

12.31 Tree diagram for Eq. (9).

If $w = f(x, y)$, $x = g(r, s)$, and $y = h(r, s)$, then

$$\frac{\partial w}{\partial r} = \frac{\partial w}{\partial x}\frac{\partial x}{\partial r} + \frac{\partial w}{\partial y}\frac{\partial y}{\partial r} \qquad \text{and} \qquad \frac{\partial w}{\partial s} = \frac{\partial w}{\partial x}\frac{\partial x}{\partial s} + \frac{\partial w}{\partial y}\frac{\partial y}{\partial s}. \qquad (7)$$

Figure 12.29 shows the tree diagram for the first of Eqs. (7). The diagram for the second equation is similar—just replace r with s.

EXAMPLE 4   Express $\partial w/\partial r$ and $\partial w/\partial s$ in terms of r and s if

$$w = x^2 + y^2, \qquad x = r - s, \qquad y = r + s.$$

Solution   We use Eqs. (7):

$$\frac{\partial w}{\partial r} = \frac{\partial w}{\partial x}\frac{\partial x}{\partial r} + \frac{\partial w}{\partial y}\frac{\partial y}{\partial r} \qquad\qquad \frac{\partial w}{\partial s} = \frac{\partial w}{\partial x}\frac{\partial x}{\partial s} + \frac{\partial w}{\partial y}\frac{\partial y}{\partial s}$$

$$= (2x)(1) + (2y)(1) \qquad\qquad = (2x)(-1) + (2y)(1)$$

$$= 2(r - s) + 2(r + s) \qquad\qquad = -2(r - s) + 2(r + s)$$

$$= 4r \qquad\qquad\qquad\qquad\qquad = 4s$$

Substitute for the intermediate variables.

If f is a function of x alone, Eqs. (5) and (6) simplify still further.

If $w = f(x)$ and $x = g(r, s)$, then

$$\frac{\partial w}{\partial r} = \frac{dw}{dx}\frac{\partial x}{\partial r} \qquad \text{and} \qquad \frac{\partial w}{\partial s} = \frac{dw}{dx}\frac{\partial x}{\partial s}. \qquad (8)$$

Here dw/dx is the ordinary (single-variable) derivative (Fig. 12.30).

Implicit Differentiation (Continued from Chapter 2)

Believe it or not, the two-variable Chain Rule in Eq. (1) leads to a formula that takes most of the work out of implicit differentiation. Suppose:

1.   The function $F(x, y)$ is differentiable and
2.   The equation $F(x, y) = 0$ defines y implicitly as a differentiable function of x, say $y = h(x)$.

Since $w = F(x, y) = 0$, the derivative dw/dx must be zero. Computing the derivative from the Chain Rule (tree diagram in Fig. 12.31), we find

$$0 = \frac{dw}{dx} = F_x \frac{dx}{dx} + F_y \frac{dy}{dx} \qquad \text{Eq. (1) with } t = x \text{ and } f = F$$

$$= F_x \cdot 1 + F_y \cdot \frac{dy}{dx}. \qquad (9)$$

If $F_y = \partial w / \partial y \neq 0$, we can solve Eq. (9) for dy/dx to get

$$\frac{dy}{dx} = -\frac{F_x}{F_y}.$$

Suppose that $F(x, y)$ is differentiable and that the equation $F(x, y) = 0$ defines y as a differentiable function of x. Then, at any point where $F_y \neq 0$,

$$\frac{dy}{dx} = -\frac{F_x}{F_y}. \tag{10}$$

EXAMPLE 5   Find dy/dx if $x^2 + \sin y - 2y = 0$.

Solution   Take $F(x, y) = x^2 + \sin y - 2y$. Then

$$\frac{dy}{dx} = -\frac{F_x}{F_y} = -\frac{2x}{\cos y - 2}. \qquad \text{Eq. (10)}$$

This calculation is significantly shorter than the single-variable calculation with which we found dy/dx in Section 2.6, Example 3.   ❑

Remembering the Different Forms of the Chain Rule

How are we to remember all the different forms of the Chain Rule? The answer is that there is no need to remember them all. The best thing to do is to draw the appropriate tree diagram by placing the dependent variable on top, the intermediate variables in the middle, and the selected independent variable at the bottom. To find the derivative of the dependent variable with respect to the selected independent variable, start at the dependent variable and read down each branch of the tree to the independent variable, calculating and multiplying the derivatives along the branch. Then add the products you found for the different branches. Let us summarize.

The Chain Rule for Functions of Many Variables

Suppose $w = f(x, y, \ldots, v)$ is a differentiable function of the variables $x, y, \ldots, v$ (a finite set) and the $x, y, \ldots, v$ are differentiable functions of $p, q, \ldots, t$ (another finite set). Then w is a differentiable function of the variables p through t and the partial derivatives of w with respect to these variables are given by equations of the form

$$\frac{\partial w}{\partial p} = \frac{\partial w}{\partial x}\frac{\partial x}{\partial p} + \frac{\partial w}{\partial y}\frac{\partial y}{\partial p} + \cdots + \frac{\partial w}{\partial v}\frac{\partial v}{\partial p}. \tag{11}$$

The other equations are obtained by replacing p by $q, \ldots, t$, one at a time.

One way to remember Eq. (11) is to think of the right-hand side as the dot product of two vectors with components

$$\underbrace{\left(\frac{\partial w}{\partial x}, \frac{\partial w}{\partial y}, \ldots, \frac{\partial w}{\partial v} \right)}_{\substack{\text{Derivatives of } w \text{ with} \\ \text{respect to the} \\ \text{intermediate variables}}} \quad \text{and} \quad \underbrace{\left(\frac{\partial x}{\partial p}, \frac{\partial y}{\partial p}, \ldots, \frac{\partial v}{\partial p} \right)}_{\substack{\text{Derivatives of the intermediate} \\ \text{variables with respect to the} \\ \text{selected independent variable}}}.$$

Exercises 12.5

Chain Rule: One Independent Variable

In Exercises 1–6, (a) express dw/dt as a function of t, both by using the Chain Rule and by expressing w in terms of t and differentiating directly with respect to t. Then (b) evaluate dw/dt at the given value of t.

1. $w = x^2 + y^2$, $\quad x = \cos t$, $\quad y = \sin t$; $\quad t = \pi$

2. $w = x^2 + y^2$, $\quad x = \cos t + \sin t$, $\quad y = \cos t - \sin t$; $\quad t = 0$

3. $w = \dfrac{x}{z} + \dfrac{y}{z}$, $\quad x = \cos^2 t$, $\quad y = \sin^2 t$, $\quad z = 1/t$; $\quad t = 3$

4. $w = \ln(x^2 + y^2 + z^2)$, $\quad x = \cos t$, $\quad y = \sin t$, $z = 4\sqrt{t}$; $\quad t = 3$

5. $w = 2ye^x - \ln z$, $\quad x = \ln(t^2 + 1)$, $\quad y = \tan^{-1} t$, $z = e^t$; $\quad t = 1$

6. $w = z - \sin xy$, $\quad x = t$, $\quad y = \ln t$, $\quad z = e^{t-1}$; $\quad t = 1$

Chain Rule: Two and Three Independent Variables

In Exercises 7 and 8, (a) express $\partial z/\partial r$ and $\partial z/\partial \theta$ as functions of r and θ both by using the Chain Rule and by expressing z directly in terms of r and θ before differentiating. Then (b) evaluate $\partial z/\partial r$ and $\partial z/\partial \theta$ at the given point (r, θ).

7. $z = 4e^x \ln y$, $\quad x = \ln(r \cos \theta)$, $\quad y = r \sin \theta$; $\quad (r, \theta) = (2, \pi/4)$

8. $z = \tan^{-1}(x/y)$, $\quad x = r \cos \theta$, $\quad y = r \sin \theta$; $\quad (r, \theta) = (1.3, \pi/6)$

In Exercises 9 and 10, (a) express $\partial w/\partial u$ and $\partial w/\partial v$ as functions of u and v both by using the Chain Rule and by expressing w directly in terms of u and v before differentiating. Then (b) evaluate $\partial w/\partial u$ and $\partial w/\partial v$ at the given point (u, v).

9. $w = xy + yz + xz$, $\quad x = u + v$, $\quad y = u - v$, $\quad z = uv$; $(u, v) = (1/2, 1)$

10. $w = \ln(x^2 + y^2 + z^2)$, $\quad x = ue^v \sin u$, $\quad y = ue^v \cos u$, $z = ue^v$; $\quad (u, v) = (-2, 0)$

In Exercises 11 and 12, (a) express $\partial u/\partial x$, $\partial u/\partial y$, and $\partial u/\partial z$ as functions of x, y, and z both by using the Chain Rule and by expressing u directly in terms of x, y, and z before differentiating. Then (b) evaluate $\partial u/\partial x$, $\partial u/\partial y$, and $\partial u/\partial z$ at the given point (x, y, z).

11. $u = \dfrac{p - q}{q - r}$, $\quad p = x + y + z$, $\quad q = x - y + z$, $r = x + y - z$; $\quad (x, y, z) = (\sqrt{3}, 2, 1)$

12. $u = e^{qr} \sin^{-1} p$, $\quad p = \sin x$, $\quad q = z^2 \ln y$, $\quad r = 1/z$; $(x, y, z) = (\pi/4, 1/2, -1/2)$

Using a Tree Diagram

In Exercises 13–24, draw a tree diagram and write a Chain Rule formula for each derivative.

13. $\dfrac{dz}{dt}$ for $z = f(x, y)$, $\quad x = g(t)$, $\quad y = h(t)$

14. $\dfrac{dz}{dt}$ for $z = f(u, v, w)$, $\quad u = g(t)$, $\quad v = h(t)$, $\quad w = k(t)$

15. $\dfrac{\partial w}{\partial u}$ and $\dfrac{\partial w}{\partial v}$ for $w = h(x, y, z)$, $\quad x = f(u, v)$, $\quad y = g(u, v)$, $z = k(u, v)$

16. $\dfrac{\partial w}{\partial x}$ and $\dfrac{\partial w}{\partial y}$ for $w = f(r, s, t)$, $\quad r = g(x, y)$, $\quad s = h(x, y)$, $t = k(x, y)$

17. $\dfrac{\partial w}{\partial u}$ and $\dfrac{\partial w}{\partial v}$ for $w = g(x, y)$, $\quad x = h(u, v)$, $\quad y = k(u, v)$

18. $\dfrac{\partial w}{\partial x}$ and $\dfrac{\partial w}{\partial y}$ for $w = g(u, v)$, $\quad u = h(x, y)$, $\quad v = k(x, y)$

19. $\dfrac{\partial z}{\partial t}$ and $\dfrac{\partial z}{\partial s}$ for $z = f(x, y)$, $\quad x = g(t, s)$, $\quad y = h(t, s)$

20. $\dfrac{\partial y}{\partial r}$ for $y = f(u)$, $\quad u = g(r, s)$

21. $\dfrac{\partial w}{\partial s}$ and $\dfrac{\partial w}{\partial t}$ for $w = g(u)$, $\quad u = h(s, t)$

22. $\dfrac{\partial w}{\partial p}$ for $w = f(x, y, z, v)$, $\quad x = g(p, q)$, $\quad y = h(p, q)$, $z = j(p, q)$, $\quad v = k(p, q)$

23. $\dfrac{\partial w}{\partial r}$ and $\dfrac{\partial w}{\partial s}$ for $w = f(x, y)$, $\quad x = g(r)$, $\quad y = h(s)$

24. $\dfrac{\partial w}{\partial s}$ for $w = g(x, y)$, $\quad x = h(r, s, t)$, $\quad y = k(r, s, t)$

Implicit Differentiation

Assuming that the equations in Exercises 25–28 define y as a differentiable function of x, use Eq. (10) to find the value of dy/dx at the given point.

25. $x^3 - 2y^2 + xy = 0$, $\quad (1, 1)$

26. $xy + y^2 - 3x - 3 = 0$, $\quad (-1, 1)$

27. $x^2 + xy + y^2 - 7 = 0$, $\quad (1, 2)$

28. $xe^y + \sin xy + y - \ln 2 = 0$, $\quad (0, \ln 2)$

Equation (10) can be generalized to functions of three variables and even more. The three-variable version goes like this:

If the equation $F(x, y, z) = 0$ determines z as a differentiable function of x and y, then, at points where $F_z \neq 0$,

$$\frac{\partial z}{\partial x} = -\frac{F_x}{F_z} \quad \text{and} \quad \frac{\partial z}{\partial y} = -\frac{F_y}{F_z}. \tag{12}$$

Use these equations to find the values of $\partial z/\partial x$ and $\partial z/\partial y$ at the points in Exercises 29–32.

29. $z^3 - xy + yz + y^3 - 2 = 0$, $\quad (1, 1, 1)$

30. $\dfrac{1}{x} + \dfrac{1}{y} + \dfrac{1}{z} - 1 = 0$, $(2, 3, 6)$

31. $\sin(x + y) + \sin(y + z) + \sin(x + z) = 0$, (π, π, π)

32. $xe^y + ye^z + 2\ln x - 2 - 3\ln 2 = 0$, $(1, \ln 2, \ln 3)$

Finding Specified Partial Derivatives

33. Find $\partial w/\partial r$ when $r = 1$, $s = -1$ if $w = (x + y + z)^2$, $x = r - s$, $y = \cos(r + s)$, $z = \sin(r + s)$.

34. Find $\partial w/\partial v$ when $u = -1$, $v = 2$ if $w = xy + \ln z$, $x = v^2/u$, $y = u + v$, $z = \cos u$.

35. Find $\partial w/\partial v$ when $u = 0$, $v = 0$ if $w = x^2 + (y/x)$, $x = u - 2v + 1$, $y = 2u + v - 2$.

36. Find $\partial z/\partial u$ when $u = 0$, $v = 1$ if $z = \sin xy + x \sin y$, $x = u^2 + v^2$, $y = uv$.

37. Find $\partial z/\partial u$ and $\partial z/\partial v$ when $u = \ln 2$, $v = 1$ if $z = 5\tan^{-1} x$ and $x = e^u + \ln v$.

38. Find $\partial z/\partial u$ and $\partial z/\partial v$ when $u = 1$ and $v = -2$ if $z = \ln q$ and $q = \sqrt{v} + 3\tan^{-1} u$.

Theory and Examples

39. *Changing voltage in a circuit.* The voltage V in a circuit that satisfies the law $V = IR$ is slowly dropping as the battery wears out. At the same time, the resistance R is increasing as the resistor heats up. Use the equation

$$\frac{dV}{dt} = \frac{\partial V}{\partial I}\frac{dI}{dt} + \frac{\partial V}{\partial R}\frac{dR}{dt}$$

to find how the current is changing at the instant when $R = 600$ ohms, $I = 0.04$ amp, $dR/dt = 0.5$ ohm/sec, and $dV/dt = -0.01$ volt/sec.

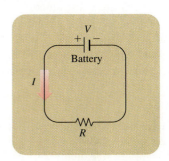

40. *Changing dimensions in a box.* The lengths a, b, and c of the edges of a rectangular box are changing with time. At the instant in question, $a = 1$ m, $b = 2$ m, $c = 3$ m, $da/dt = db/dt = 1$ m/sec, and $dc/dt = -3$ m/sec. At what rates are the box's volume V and surface area S changing at that instant? Are the box's interior diagonals increasing in length, or decreasing?

41. If $f(u, v, w)$ is differentiable and $u = x - y$, $v = y - z$, and $w = z - x$, show that

$$\frac{\partial f}{\partial x} + \frac{\partial f}{\partial y} + \frac{\partial f}{\partial z} = 0.$$

42. a) Show that if we substitute polar coordinates $x = r\cos\theta$ and $y = r\sin\theta$ in a differentiable function $w = f(x, y)$, then

$$\frac{\partial w}{\partial r} = f_x\cos\theta + f_y\sin\theta$$

and

$$\frac{1}{r}\frac{\partial w}{\partial \theta} = -f_x\sin\theta + f_y\cos\theta.$$

b) Solve the equations in (a) to express f_x and f_y in terms of $\partial w/\partial r$ and $\partial w/\partial\theta$.

c) Show that

$$(f_x)^2 + (f_y)^2 = \left(\frac{\partial w}{\partial r}\right)^2 + \frac{1}{r^2}\left(\frac{\partial w}{\partial \theta}\right)^2.$$

43. Show that if $w = f(u, v)$ satisfies the Laplace equation $f_{uu} + f_{vv} = 0$, and if $u = (x^2 - y^2)/2$ and $v = xy$, then w satisfies the Laplace equation $w_{xx} + w_{yy} = 0$.

44. Let $w = f(u) + g(v)$, where $u = x + iy$ and $v = x - iy$ and $i = \sqrt{-1}$. Show that w satisfies the Laplace equation $w_{xx} + w_{yy} = 0$ if all the necessary functions are differentiable.

Changes in Functions along Curves

45. Suppose that the partial derivatives of a function $f(x, y, z)$ at points on the helix $x = \cos t$, $y = \sin t$, $z = t$ are

$$f_x = \cos t, \quad f_y = \sin t, \quad f_z = t^2 + t - 2.$$

At what points on the curve, if any, can f take on extreme values?

46. Let $w = x^2 e^{2y}\cos 3z$. Find the value of dw/dt at the point $(1, \ln 2, 0)$ on the curve $x = \cos t$, $y = \ln(t + 2)$, $z = t$.

47. Let $T = f(x, y)$ be the temperature at the point (x, y) on the circle $x = \cos t$, $y = \sin t$, $0 \le t \le 2\pi$, and suppose that

$$\frac{\partial T}{\partial x} = 8x - 4y, \quad \frac{\partial T}{\partial y} = 8y - 4x.$$

a) Find where the maximum and minimum temperatures on the circle occur by examining the derivatives dT/dt and d^2T/dt^2.

b) Suppose $T = 4x^2 - 4xy + 4y^2$. Find the maximum and minimum values of T on the circle.

48. Let $T = g(x, y)$ be the temperature at the point (x, y) on the ellipse

$$x = 2\sqrt{2}\cos t, \quad y = \sqrt{2}\sin t, \quad 0 \le t \le 2\pi,$$

and suppose that

$$\frac{\partial T}{\partial x} = y, \quad \frac{\partial T}{\partial y} = x.$$

a) Locate the maximum and minimum temperatures on the ellipse by examining dT/dt and d^2T/dt^2.

b) Suppose that $T = xy - 2$. Find the maximum and minimum values of T on the ellipse.

Differentiating Integrals

Under mild continuity restrictions, it is true that if

$$F(x) = \int_a^b g(t, x)\, dt,$$

then $F'(x) = \int_a^b g_x(t, x)\, dt$. Using this fact and the Chain Rule, we can find the derivative of

$$F(x) = \int_a^{f(x)} g(t, x)\, dt$$

by letting

$$G(u, x) = \int_a^u g(t, x)\, dt,$$

where $u = f(x)$. Find the derivatives of the functions in Exercises 49 and 50.

49. $F(x) = \int_0^{x^2} \sqrt{t^4 + x^3}\, dt$

50. $F(x) = \int_{x^2}^1 \sqrt{t^3 + x^2}\, dt$

12.6

*Partial Derivatives with Constrained Variables

In finding partial derivatives of functions like $w = f(x, y)$, we have assumed x and y to be independent. But in many applications this is not the case. For example, the internal energy U of a gas may be expressed as a function $U = f(P, V, T)$ of pressure P, volume V, and temperature T. If the individual molecules of the gas do not interact, however, P, V, and T obey the ideal gas law

$$PV = nRT \qquad (n \text{ and } R \text{ constant})$$

and so fail to be independent. Finding partial derivatives in situations like these can be complicated. But it is better to face the complication now than to meet it for the first time while you are also trying to learn economics, engineering, or physics.

Decide Which Variables Are Dependent and Which Are Independent

If the variables in a function $w = f(x, y, z)$ are constrained by a relation like the one imposed on x, y, and z by the equation $z = x^2 + y^2$, the geometric meanings and the numerical values of the partial derivatives of f will depend on which variables are chosen to be dependent and which are chosen to be independent. To see how this choice can affect the outcome, we consider the calculation of $\partial w / \partial x$ when $w = x^2 + y^2 + z^2$ and $z = x^2 + y^2$.

EXAMPLE 1   Find $\partial w / \partial x$ if $w = x^2 + y^2 + z^2$ and $z = x^2 + y^2$.

Solution   We are given two equations in the four unknowns x, y, z, and w. Like many such systems, this one can be solved for two of the unknowns (the dependent variables) in terms of the others (the independent variables). In being asked for $\partial w / \partial x$, we are told that w is to be a dependent variable and x an independent variable. The possible choices for the other variables come down to

Dependent	Independent
w, z	x, y
w, y	x, z

In either case, we can express w explicitly in terms of the selected independent

---

*This section is based on notes written for MIT by Arthur P. Mattuck.

variables. We do this by using the second equation to eliminate the remaining dependent variable in the first equation.

In the first case, the remaining dependent variable is z. We eliminate it from the first equation by replacing it by $x^2 + y^2$. The resulting expression for w is

$$w = x^2 + y^2 + z^2 = x^2 + y^2 + (x^2 + y^2)^2$$
$$= x^2 + y^2 + x^4 + 2x^2y^2 + y^4$$

and

$$\frac{\partial w}{\partial x} = 2x + 4x^3 + 4xy^2. \tag{1}$$

This is the formula for $\partial w/\partial x$ when x and y are the independent variables.

In the second case, where the independent variables are x and z and the remaining dependent variable is y, we eliminate the dependent variable y in the expression for w by replacing y^2 by $z - x^2$. This gives

$$w = x^2 + y^2 + z^2 = x^2 + (z - x^2) + z^2 = z + z^2$$

and

$$\frac{\partial w}{\partial x} = 0. \tag{2}$$

This is the formula for $\partial w/\partial x$ when x and z are the independent variables.

The formulas for $\partial w/\partial x$ in Eqs. (1) and (2) are genuinely different. We cannot change either formula into the other by using the relation $z = x^2 + y^2$. There is not just one $\partial w/\partial x$, there are two, and we see that the original instruction to find $\partial w/\partial x$ was incomplete. *Which* $\partial w/\partial x$? we ask.

The geometric interpretations of Eqs. (1) and (2) help to explain why the equations differ. The function $w = x^2 + y^2 + z^2$ measures the square of the distance from the point (x, y, z) to the origin. The condition $z = x^2 + y^2$ says that the point (x, y, z) lies on the paraboloid of revolution shown in Fig. 12.32. What does it mean to calculate $\partial w/\partial x$ at a point $P(x, y, z)$ that can move only on this surface? What is the value of $\partial w/\partial x$ when the coordinates of P are, say, $(1, 0, 1)$?

If we take x and y to be independent, then we find $\partial w/\partial x$ by holding y fixed (at $y = 0$ in this case) and letting x vary. This means that P moves along the parabola $z = x^2$ in the xz-plane. As P moves on this parabola, w, which is the square of the distance from P to the origin, changes. We calculate $\partial w/\partial x$ in this case (our first solution above) to be

$$\frac{\partial w}{\partial x} = 2x + 4x^3 + 4xy^2.$$

At the point $P(1, 0, 1)$, the value of this derivative is

$$\frac{\partial w}{\partial x} = 2 + 4 + 0 = 6.$$

If we take x and z to be independent, then we find $\partial w/\partial x$ by holding z fixed while x varies. Since the z-coordinate of P is 1, varying x moves P along a circle in the plane $z = 1$. As P moves along this circle, its distance from the origin remains constant, and w, being the square of this distance, does not change. That is,

$$\frac{\partial w}{\partial x} = 0, \cdot$$

as we found in our second solution. ❑

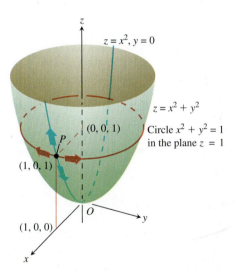

12.32 If P is constrained to lie on the paraboloid $z = x^2 + y^2$, the value of the partial derivative of $w = x^2 + y^2 + z^2$ with respect to x at P depends on the direction of motion (Example 1). (a) As x changes, with $y = 0$, P moves up or down the surface on the parabola $z = x^2$ in the xz-plane with $\partial w/\partial x = 2x + 4x^3$. (b) As x changes, with $z = 1$, P moves on the circle $x^2 + y^2 = 1$, $z = 1$, and $\partial w/\partial x = 0$.

Labels in figure: $z = x^2, y = 0$; $z = x^2 + y^2$; $(0, 0, 1)$; Circle $x^2 + y^2 = 1$ in the plane $z = 1$; P; $(1, 0, 1)$; $(1, 0, 0)$; O; x; y; z

How to Find $\partial w/\partial x$ When the Variables in $w = f(x, y, z)$ Are Constrained by Another Equation

As we saw in Example 1, a typical routine for finding $\partial w/\partial x$ when the variables in the function $w = f(x, y, z)$ are related by another equation has three steps. These steps apply to finding $\partial w/\partial y$ and $\partial w/\partial z$ as well.

Step 1 Decide which variables are to be dependent and which are to be independent. (In practice, the decision is based on the physical or theoretical context of our work. In the exercises at the end of this section, we say which variables are which.)

Step 2 Eliminate the other dependent variable(s) in the expression for w.

Step 3 Differentiate as usual.

If we cannot carry out step 2 after deciding which variables are dependent, we differentiate the equations as they are and try to solve for $\partial w/\partial x$ afterward. The next example shows how this is done.

EXAMPLE 2   Find $\partial w/\partial x$ at the point $(x, y, z) = (2, -1, 1)$ if

$$w = x^2 + y^2 + z^2, \qquad z^3 - xy + yz + y^3 = 1,$$

and x and y are the independent variables.

Solution   It is not convenient to eliminate z in the expression for w. We therefore differentiate both equations implicitly with respect to x, treating x and y as independent variables and w and z as dependent variables. This gives

$$\frac{\partial w}{\partial x} = 2x + 2z\frac{\partial z}{\partial x} \tag{3}$$

and

$$3z^2\frac{\partial z}{\partial x} - y + y\frac{\partial z}{\partial x} + 0 = 0. \tag{4}$$

These equations may now be combined to express $\partial w/\partial x$ in terms of x, y, and z. We solve Eq. (4) for $\partial z/\partial x$ to get

$$\frac{\partial z}{\partial x} = \frac{y}{y + 3z^2}$$

and substitute into Eq. (3) to get

$$\frac{\partial w}{\partial x} = 2x + \frac{2yz}{y + 3z^2}.$$

The value of this derivative at $(x, y, z) = (2, -1, 1)$ is

$$\left(\frac{\partial w}{\partial x}\right)_{(2, -1, 1)} = 2(2) + \frac{2(-1)(1)}{-1 + 3(1)^2} = 4 + \frac{-2}{2} = 3.$$

To show what variables are assumed to be independent in calculating a derivative, we can use the following notation:

$$\left(\frac{\partial w}{\partial x}\right)_y \qquad \partial w/\partial x \text{ with } x \text{ and } y \text{ independent}$$

$$\left(\frac{\partial f}{\partial y}\right)_{x,t} \qquad \partial f/\partial y \text{ with } y, x, \text{ and } t \text{ independent.}$$

EXAMPLE 3   Find $\left(\dfrac{\partial w}{\partial x}\right)_{y,z}$ if $w = x^2 + y - z + \sin t$ and $x + y = t$.

Solution

With x, y, z independent, we have

$$t = x + y, \qquad w = x^2 + y - z + \sin(x + y)$$

$$\left(\frac{\partial w}{\partial x}\right)_{y,z} = 2x + 0 - 0 + \cos(x + y)\frac{\partial}{\partial x}(x + y)$$

$$= 2x + \cos(x + y). \qquad \square$$

Arrow Diagrams

In solving problems like the one in Example 3, it often helps to start with an arrow diagram that shows how the variables and functions are related. If

$$w = x^2 + y - z + \sin t \qquad \text{and} \qquad x + y = t$$

and we are asked to find $\partial w/\partial x$ when $x, y,$ and z are independent, the appropriate diagram is one like this:

$$\begin{pmatrix} x \\ y \\ z \end{pmatrix} \quad \rightarrow \quad \begin{pmatrix} x \\ y \\ z \\ t \end{pmatrix} \quad \rightarrow \quad w \tag{5}$$

independent variables   intermediate variables and relations   dependent variable

$$x = x$$
$$y = y$$
$$z = z$$
$$t = x + y$$

The diagram shows the independent variables on the left, the intermediate variables and their relation to the independent variables in the middle, and the dependent variable on the right.

To find $\partial w/\partial x$, we first apply the four-variable form of the Chain Rule to w, getting

$$\frac{\partial w}{\partial x} = \frac{\partial w}{\partial x}\frac{\partial x}{\partial x} + \frac{\partial w}{\partial y}\frac{\partial y}{\partial x} + \frac{\partial w}{\partial z}\frac{\partial z}{\partial x} + \frac{\partial w}{\partial t}\frac{\partial t}{\partial x}. \tag{6}$$

We then use the formula for $w = x^2 + y - z + \sin t$ to evaluate the partial derivatives of w that appear on the right-hand side of Eq. (6). This gives

$$\frac{\partial w}{\partial x} = 2x\frac{\partial x}{\partial x} + (1)\frac{\partial y}{\partial x} + (-1)\frac{\partial z}{\partial x} + \cos t\frac{\partial t}{\partial x}$$

$$= 2x\frac{\partial x}{\partial x} + \frac{\partial y}{\partial x} - \frac{\partial z}{\partial x} + \cos t\frac{\partial t}{\partial x}. \tag{7}$$

To calculate the remaining partial derivatives, we apply what we know about the dependence and independence of the variables involved. As shown in the diagram (5), the variables x, y, and z are independent and $t = x + y$. Hence,

$$\frac{\partial x}{\partial x} = 1, \quad \frac{\partial y}{\partial x} = 0, \quad \frac{\partial z}{\partial x} = 0, \quad \frac{\partial t}{\partial x} = \frac{\partial}{\partial x}(x + y) = (1 + 0) = 1.$$

We substitute these values into Eq. (7) to find $\partial w / \partial x$:

$$\left(\frac{\partial w}{\partial x}\right)_{y,z} = 2x(1) + 0 - 0 + (\cos t)(1)$$

$$= 2x + \cos t$$

$$= 2x + \cos (x + y). \qquad \text{In terms of the independent variables}$$

Exercises 12.6

Finding Partial Derivatives with Constrained Variables

In Exercises 1–3, begin by drawing a diagram that shows the relations among the variables.

1. If $w = x^2 + y^2 + z^2$ and $z = x^2 + y^2$, find

   a) $\left(\dfrac{\partial w}{\partial y}\right)_z$ b) $\left(\dfrac{\partial w}{\partial z}\right)_x$ c) $\left(\dfrac{\partial w}{\partial z}\right)_y$

2. If $w = x^2 + y - z + \sin t$ and $x + y = t$, find

   a) $\left(\dfrac{\partial w}{\partial y}\right)_{x,z}$ b) $\left(\dfrac{\partial w}{\partial y}\right)_{z,t}$ c) $\left(\dfrac{\partial w}{\partial z}\right)_{x,y}$

   d) $\left(\dfrac{\partial w}{\partial z}\right)_{y,t}$ e) $\left(\dfrac{\partial w}{\partial t}\right)_{x,z}$ f) $\left(\dfrac{\partial w}{\partial t}\right)_{y,z}$

3. Let $U = f(P, V, T)$ be the internal energy of a gas that obeys the ideal gas law $PV = nRT$ (n and R constant). Find

   a) $\left(\dfrac{\partial U}{\partial P}\right)_V$ b) $\left(\dfrac{\partial U}{\partial T}\right)_V$

4. Find

   a) $\left(\dfrac{\partial w}{\partial x}\right)_y$ b) $\left(\dfrac{\partial w}{\partial z}\right)_y$

   at the point $(x, y, z) = (0, 1, \pi)$ if

   $$w = x^2 + y^2 + z^2 \quad \text{and} \quad y \sin z + z \sin x = 0.$$

5. Find

   a) $\left(\dfrac{\partial w}{\partial y}\right)_x$ b) $\left(\dfrac{\partial w}{\partial y}\right)_z$

   at the point $(w, x, y, z) = (4, 2, 1, -1)$ if

   $$w = x^2 y^2 + yz - z^3 \quad \text{and} \quad x^2 + y^2 + z^2 = 6.$$

6. Find $\left(\dfrac{\partial u}{\partial y}\right)_x$ at the point $(u, v) = (\sqrt{2}, 1)$ if $x = u^2 + v^2$ and $y = uv$.

7. Suppose that $x^2 + y^2 = r^2$ and $x = r \cos \theta$, as in polar coordinates. Find

   $$\left(\frac{\partial x}{\partial r}\right)_\theta \quad \text{and} \quad \left(\frac{\partial r}{\partial x}\right)_y.$$

8. Suppose that

   $$w = x^2 - y^2 + 4z + t \quad \text{and} \quad x + 2z + t = 25.$$

   Show that the equations

   $$\frac{\partial w}{\partial x} = 2x - 1 \quad \text{and} \quad \frac{\partial w}{\partial x} = 2x - 2$$

   each give $\partial w / \partial x$, depending on which variables are chosen to be dependent and which variables are chosen to be independent. Identify the independent variables in each case.

Partial Derivatives without Specific Formulas

9. Establish the fact, widely used in hydrodynamics, that if $f(x, y, z) = 0$, then

   $$\left(\frac{\partial x}{\partial y}\right)_z \left(\frac{\partial y}{\partial z}\right)_x \left(\frac{\partial z}{\partial x}\right)_y = -1.$$

   (*Hint:* Express all the derivatives in terms of the formal partial derivatives $\partial f / \partial x$, $\partial f / \partial y$, and $\partial f / \partial z$.)

10. If $z = x + f(u)$, where $u = xy$, show that

    $$x\frac{\partial z}{\partial x} - y\frac{\partial z}{\partial y} = x.$$

11. Suppose that the equation $g(x, y, z) = 0$ determines z as a differentiable function of the independent variables x and y and that $g_z \neq 0$. Show that

    $$\left(\frac{\partial z}{\partial y}\right)_x = -\frac{\partial g / \partial y}{\partial g / \partial z}.$$

12. Suppose that $f(x, y, z, w) = 0$ and $g(x, y, z, w) = 0$ determine z and w as differentiable functions of the independent variables

x and y, and suppose that

$$\frac{\partial f}{\partial z}\frac{\partial g}{\partial w} - \frac{\partial f}{\partial w}\frac{\partial g}{\partial z} \neq 0.$$

Show that

$$\left(\frac{\partial z}{\partial x}\right)_y = -\frac{\dfrac{\partial f}{\partial x}\dfrac{\partial g}{\partial w} - \dfrac{\partial f}{\partial w}\dfrac{\partial g}{\partial x}}{\dfrac{\partial f}{\partial z}\dfrac{\partial g}{\partial w} - \dfrac{\partial f}{\partial w}\dfrac{\partial g}{\partial z}}$$

and

$$\left(\frac{\partial w}{\partial y}\right)_x = -\frac{\dfrac{\partial f}{\partial z}\dfrac{\partial g}{\partial y} - \dfrac{\partial f}{\partial y}\dfrac{\partial g}{\partial z}}{\dfrac{\partial f}{\partial z}\dfrac{\partial g}{\partial w} - \dfrac{\partial f}{\partial w}\dfrac{\partial g}{\partial z}}.$$

12.7   Directional Derivatives, Gradient Vectors, and Tangent Planes

We know from Section 12.5 that if $f(x, y)$ is differentiable, then the rate at which f changes with respect to t along a differentiable curve $x = g(t)$, $y = h(t)$ is

$$\frac{df}{dt} = \frac{\partial f}{\partial x}\frac{dx}{dt} + \frac{\partial f}{\partial y}\frac{dy}{dt}.$$

At any point $P_0(x_0, y_0) = P_0(g(t_0), h(t_0))$, this equation gives the rate of change of f with respect to increasing t and therefore depends, among other things, on the direction of motion along the curve. This observation is particularly important when the curve is a straight line and t is the arc length parameter along the line measured from P_0 in the direction of a given unit vector $\mathbf{u}$. For then df/dt is the rate of change of f with respect to distance in its domain in the direction of $\mathbf{u}$. By varying $\mathbf{u}$, we find the rates at which f changes with respect to distance as we move through P_0 in different directions. These "directional derivatives" have useful interpretations in science and engineering as well as in mathematics. This section develops a formula for calculating them and proceeds from there to find equations for tangent planes and normal lines on surfaces in space.

Directional Derivatives in the Plane

Suppose that the function $f(x, y)$ is defined throughout a region R in the xy-plane, that $P_0(x_0, y_0)$ is a point in R, and that $\mathbf{u} = u_1\mathbf{i} + u_2\mathbf{j}$ is a unit vector. Then the equations

$$x = x_0 + su_1, \qquad y = y_0 + su_2$$

parametrize the line through P_0 parallel to $\mathbf{u}$. The parameter s measures arc length from P_0 in the direction of $\mathbf{u}$. We find the rate of change of f at P_0 in the direction of $\mathbf{u}$ by calculating df/ds at P_0 (Fig. 12.33):

Definition

The **derivative of f at $P_0(x_0, y_0)$ in the direction of the unit vector $\mathbf{u} = u_1\mathbf{i} + u_2\mathbf{j}$** is the number

$$\left(\frac{df}{ds}\right)_{\mathbf{u}, P_0} = \lim_{s \to 0} \frac{f(x_0 + su_1, y_0 + su_2) - f(x_0, y_0)}{s}, \tag{1}$$

provided the limit exists.

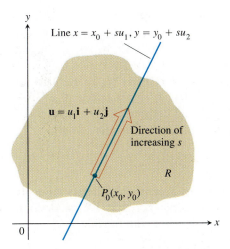

12.33 The rate of change of f in the direction of $\mathbf{u}$ at a point P_0 is the rate at which f changes along this line at P_0.

The directional derivative is also denoted by

$$(D_{\mathbf{u}}f)_{P_0}. \qquad \text{"The derivative of } f \text{ at } P_0 \text{ in the direction of } \mathbf{u}\text{"}$$

EXAMPLE 1   Find the derivative of

$$f(x,\, y) = x^2 + xy$$

at $P_0(1, 2)$ in the direction of the unit vector $\mathbf{u} = \left(1/\sqrt{2}\right)\mathbf{i} + \left(1/\sqrt{2}\right)\mathbf{j}$.

Solution

$$\left(\frac{df}{ds}\right)_{\mathbf{u},\, P_0} = \lim_{s\to 0}\frac{f(x_0 + su_1,\, y_0 + su_2) - f(x_0,\, y_0)}{s} \qquad \text{Eq. (1)}$$

$$= \lim_{s\to 0}\frac{f\left(1 + s\cdot\dfrac{1}{\sqrt{2}},\; 2 + s\cdot\dfrac{1}{\sqrt{2}}\right) - f(1, 2)}{s}$$

$$= \lim_{s\to 0}\frac{\left(1 + \dfrac{s}{\sqrt{2}}\right)^2 + \left(1 + \dfrac{s}{\sqrt{2}}\right)\left(2 + \dfrac{s}{\sqrt{2}}\right) - (1^2 + 1\cdot 2)}{s}$$

$$= \lim_{s\to 0}\frac{\left(1 + \dfrac{2s}{\sqrt{2}} + \dfrac{s^2}{2}\right) + \left(2 + \dfrac{3s}{\sqrt{2}} + \dfrac{s^2}{2}\right) - 3}{s}$$

$$= \lim_{s\to 0}\frac{\dfrac{5s}{\sqrt{2}} + s^2}{s} = \lim_{s\to 0}\left(\frac{5}{\sqrt{2}} + s\right) = \left(\frac{5}{\sqrt{2}} + 0\right) = \frac{5}{\sqrt{2}}.$$

The rate of change of $f(x,\, y) = x^2 + xy$ at $P_0(1, 2)$ in the direction $\mathbf{u} = \left(1/\sqrt{2}\right)\mathbf{i} + \left(1/\sqrt{2}\right)\mathbf{j}$ is $5/\sqrt{2}$.   ☐

Geometric Interpretation of the Directional Derivative

The equation $z = f(x,\, y)$ represents a surface S in space. If $z_0 = f(x_0,\, y_0)$, then the point $P(x_0,\, y_0,\, z_0)$ lies on S. The vertical plane that passes through P and $P_0(x_0,\, y_0)$ parallel to $\mathbf{u}$ intersects S in a curve C (Fig. 12.34). The rate of change of f in the direction of $\mathbf{u}$ is the slope of the tangent to C at P.

Notice that when $\mathbf{u} = \mathbf{i}$ the directional derivative at P_0 is $\partial f/\partial x$ evaluated at $(x_0,\, y_0)$. When $\mathbf{u} = \mathbf{j}$ the directional derivative at P_0 is $\partial f/\partial y$ evaluated at $(x_0,\, y_0)$. The directional derivative generalizes the two partial derivatives. We can now ask for the rate of change of f in any direction $\mathbf{u}$, not just the directions $\mathbf{i}$ and $\mathbf{j}$.

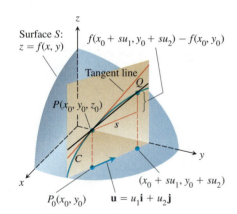

12.34 The slope of curve C at P_0 is

$$\lim_{Q\to P}\text{slope}(PQ)$$

$$= \lim_{s\to 0}\frac{f(x_0 + su_1,\, y_0 + su_2) - f(x_0,\, y_0)}{s}$$

$$= \left(\frac{df}{ds}\right)_{\mathbf{u}, P_0}.$$

Calculation

As you know, it is rarely convenient to calculate a derivative directly from its definition as a limit, and the directional derivative is no exception. We can develop a more efficient formula in the following way. We begin with the line

$$x = x_0 + su_1, \qquad y = y_0 + su_2, \qquad (2)$$

through $P_0(x_0,\, y_0)$, parametrized with the arc length parameter s increasing in the direction of the unit vector $\mathbf{u} = u_1 \mathbf{i} + u_2 \mathbf{j}$. Then

$$\left(\frac{df}{ds}\right)_{\mathbf{u}, P_0} = \left(\frac{\partial f}{\partial x}\right)_{P_0} \frac{dx}{ds} + \left(\frac{\partial f}{\partial y}\right)_{P_0} \frac{dy}{ds} \qquad \text{Chain Rule}$$

$$= \left(\frac{\partial f}{\partial x}\right)_{P_0} \cdot u_1 + \left(\frac{\partial f}{\partial y}\right)_{P_0} \cdot u_2 \qquad \begin{array}{l} \text{From Eqs. (2),} \\ dx/ds = u_1 \text{ and} \\ dy/ds = u_2 \end{array}$$

$$= \underbrace{\left[\left(\frac{\partial f}{\partial x}\right)_{P_0} \mathbf{i} + \left(\frac{\partial f}{\partial y}\right)_{P_0} \mathbf{j}\right]}_{\text{gradient of } f \text{ at } P_0} \cdot \underbrace{\left[u_1 \mathbf{i} + u_2 \mathbf{j}\right]}_{\text{direction } \mathbf{u}}. \qquad (3)$$

The notation ∇f is read "grad f" as well as "gradient of f" and "del f." The symbol ∇ by itself is read "del." Another notation for the gradient is grad f, read the way it is written.

Definition

The **gradient vector (gradient)** of $f(x, y)$ at a point $P_0(x_0, y_0)$ is the vector

$$\nabla f = \frac{\partial f}{\partial x} \mathbf{i} + \frac{\partial f}{\partial y} \mathbf{j}$$

obtained by evaluating the partial derivatives of f at P_0.

Equation (3) says that the derivative of f in the direction of $\mathbf{u}$ at P_0 is the dot product of $\mathbf{u}$ with the gradient of f at P_0.

Theorem 6

If the partial derivatives of $f(x, y)$ are defined at $P_0(x_0, y_0)$, then

$$\left(\frac{df}{ds}\right)_{\mathbf{u}, P_0} = (\nabla f)_{P_0} \cdot \mathbf{u}, \qquad (4)$$

the scalar product of the gradient f at P_0 and $\mathbf{u}$.

EXAMPLE 2   Find the derivative of $f(x, y) = xe^y + \cos(xy)$ at the point $(2, 0)$ in the direction of $\mathbf{A} = 3\mathbf{i} - 4\mathbf{j}$.

Solution   The direction of $\mathbf{A}$ is obtained by dividing $\mathbf{A}$ by its length:

$$\mathbf{u} = \frac{\mathbf{A}}{|\mathbf{A}|} = \frac{\mathbf{A}}{5} = \frac{3}{5}\mathbf{i} - \frac{4}{5}\mathbf{j}.$$

The partial derivatives of f at $(2, 0)$ are

$$f_x(2, 0) = (e^y - y \sin(xy))_{(2,0)} = e^0 - 0 = 1$$

$$f_y(2, 0) = (xe^y - x \sin(xy))_{(2,0)} = 2e^0 - 2 \cdot 0 = 2.$$

The gradient of f at $(2, 0)$ is

$$\nabla f\big|_{(2,0)} = f_x(2, 0)\mathbf{i} + f_y(2, 0)\mathbf{j} = \mathbf{i} + 2\mathbf{j}$$

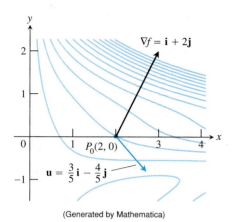

(Generated by Mathematica)

12.35 It is customary to picture ∇f as a vector in the domain of f. In the case of $f(x, y) = xe^y + \cos(xy)$, the domain is the entire plane. The rate at which f changes in the direction $\mathbf{u} = (3/5)\mathbf{i} - (4/5)\mathbf{j}$ is $\nabla f \cdot \mathbf{u} = -1$ (Example 2).

(Fig. 12.35). The derivative of f at $(2, 0)$ in the direction of $\mathbf{A}$ is therefore

$$(D_\mathbf{u} f)\big|_{(2,0)} = \nabla f\big|_{(2,0)} \cdot \mathbf{u} \qquad \text{Eq. (4)}$$

$$= (\mathbf{i} + 2\mathbf{j}) \cdot \left(\frac{3}{5}\mathbf{i} - \frac{4}{5}\mathbf{j}\right) = \frac{3}{5} - \frac{8}{5} = -1.$$

Properties of Directional Derivatives

Evaluating the dot product in the formula

$$D_\mathbf{u} f = \nabla f \cdot \mathbf{u} = |\nabla f||\mathbf{u}| \cos\theta = |\nabla f| \cos\theta$$

reveals the following properties.

Properties of the Directional Derivative $D_\mathbf{u} f = \nabla f \cdot \mathbf{u} = |\nabla f| \cos\theta$

1. The function f increases most rapidly when $\cos\theta = 1$, or when $\mathbf{u}$ is the direction of ∇f. That is, at each point P in its domain, f increases most rapidly in the direction of the gradient vector ∇f at P. The derivative in this direction is

$$D_\mathbf{u} f = |\nabla f| \cos(0) = |\nabla f|.$$

2. Similarly, f decreases most rapidly in the direction of $-\nabla f$. The derivative in this direction is $D_\mathbf{u} f = |\nabla f| \cos(\pi) = -|\nabla f|$.

3. Any direction $\mathbf{u}$ orthogonal to the gradient is a direction of zero change in f because θ then equals $\pi/2$ and

$$D_\mathbf{u} f = |\nabla f| \cos(\pi/2) = |\nabla f| \cdot 0 = 0.$$

As we will discuss later, these properties hold in three dimensions as well as two.

EXAMPLE 3   Find the directions in which $f(x, y) = (x^2/2) + (y^2/2)$ (a) increases most rapidly and (b) decreases most rapidly at the point $(1, 1)$. (c) What are the directions of zero change in f at $(1, 1)$?

Solution

a) The function increases most rapidly in the direction of ∇f at $(1, 1)$. The gradient is

$$(\nabla f)_{(1,1)} = (x\mathbf{i} + y\mathbf{j})_{(1,1)} = \mathbf{i} + \mathbf{j}.$$

Its direction is

$$\mathbf{u} = \frac{\mathbf{i} + \mathbf{j}}{|\mathbf{i} + \mathbf{j}|} = \frac{\mathbf{i} + \mathbf{j}}{\sqrt{(1)^2 + (1)^2}} = \frac{1}{\sqrt{2}}\mathbf{i} + \frac{1}{\sqrt{2}}\mathbf{j}.$$

b) The function decreases most rapidly in the direction of $-\nabla f$ at $(1, 1)$, which is

$$-\mathbf{u} = -\frac{1}{\sqrt{2}}\mathbf{i} - \frac{1}{\sqrt{2}}\mathbf{j}.$$

c) The directions of zero change at $(1, 1)$ are the directions orthogonal to ∇f:

$$\mathbf{n} = -\frac{1}{\sqrt{2}}\mathbf{i} + \frac{1}{\sqrt{2}}\mathbf{j} \quad \text{and} \quad -\mathbf{n} = \frac{1}{\sqrt{2}}\mathbf{i} - \frac{1}{\sqrt{2}}\mathbf{j}.$$

See Fig. 12.36. ☐

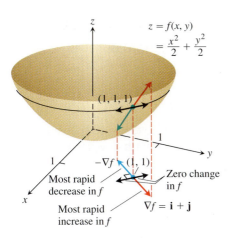

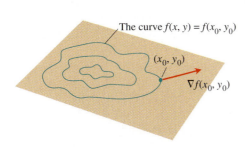

12.36 The direction in which $f(x, y) = (x^2/2) + (y^2/2)$ increases most rapidly at $(1, 1)$ is the direction of $\nabla f|_{(1,1)} = \mathbf{i} + \mathbf{j}$. It corresponds to the direction of steepest ascent on the surface at $(1, 1, 1)$.

Gradients and Tangents to Level Curves

If a differentiable function $f(x, y)$ has a constant value c along a smooth curve $\mathbf{r} = g(t)\mathbf{i} + h(t)\mathbf{j}$ (making the curve a level curve of f), then $f(g(t), h(t)) = c$. Differentiating both sides of this equation with respect to t leads to the equations

$$\frac{d}{dt}f(g(t), h(t)) = \frac{d}{dt}(c)$$

$$\frac{\partial f}{\partial x}\frac{dg}{dt} + \frac{\partial f}{\partial y}\frac{dh}{dt} = 0 \qquad \text{Chain Rule}$$

$$\underbrace{\left(\frac{\partial f}{\partial x}\mathbf{i} + \frac{\partial f}{\partial y}\mathbf{j}\right)}_{\nabla f} \cdot \underbrace{\left(\frac{dg}{dt}\mathbf{i} + \frac{dh}{dt}\mathbf{j}\right)}_{\frac{d\mathbf{r}}{dt}} = 0. \tag{5}$$

Equation (5) says that ∇f is normal to the tangent vector $d\mathbf{r}/dt$, so it is normal to the curve.

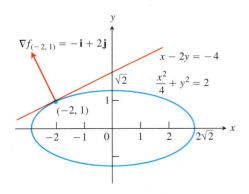

12.37 The gradient of a differentiable function of two variables at a point is always normal to the function's level curve through that point.

> At every point (x_0, y_0) in the domain of $f(x, y)$, the gradient of f is normal to the level curve through (x_0, y_0) (Fig. 12.37).

This observation enables us to find equations for tangent lines to level curves. They are the lines normal to the gradients. The line through a point $P_0(x_0, y_0)$ normal to a vector $\mathbf{N} = A\mathbf{i} + B\mathbf{j}$ has the equation

$$A(x - x_0) + B(y - y_0) = 0$$

(Exercise 59). If $\mathbf{N}$ is the gradient $(\nabla f)_{(x_0, y_0)} = f_x(x_0, y_0)\mathbf{i} + f_y(x_0, y_0)\mathbf{j}$, the equation becomes

$$f_x(x_0, y_0)(x - x_0) + f_y(x_0, y_0)(y - y_0) = 0. \tag{6}$$

EXAMPLE 4 Find an equation for the tangent to the ellipse

$$\frac{x^2}{4} + y^2 = 2$$

(Fig. 12.38) at the point $(-2, 1)$.

Solution The ellipse is a level curve of the function

$$f(x, y) = \frac{x^2}{4} + y^2.$$

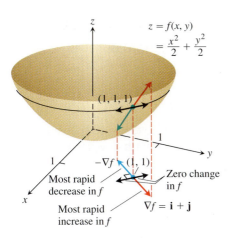

12.38 We can find the tangent to the ellipse $(x^2/4) + y^2 = 2$ by treating the ellipse as a level curve of the function $f(x, y) = (x^2/4) + y^2$ (Example 4).

The gradient of f at $(-2, 1)$ is

$$\nabla f\big|_{(-2,1)} = \left(\frac{x}{2}\mathbf{i} + 2y\mathbf{j}\right)_{(-2,1)} = -\mathbf{i} + 2\mathbf{j}.$$

The tangent is the line

$$(-1)(x + 2) + (2)(y - 1) = 0 \qquad \text{Eq. (6)}$$

$$x - 2y = -4. \qquad \square$$

Functions of Three Variables

We obtain three-variable formulas by adding the z-terms to the two-variable formulas. For a differentiable function $f(x, y, z)$ and a unit vector $\mathbf{u} = u_1\mathbf{i} + u_2\mathbf{j} + u_3\mathbf{k}$ in space, we have

$$\nabla f = \frac{\partial f}{\partial x}\mathbf{i} + \frac{\partial f}{\partial y}\mathbf{j} + \frac{\partial f}{\partial z}\mathbf{k}$$

and

$$D_{\mathbf{u}}f = \nabla f \cdot \mathbf{u} = \frac{\partial f}{\partial x}u_1 + \frac{\partial f}{\partial y}u_2 + \frac{\partial f}{\partial z}u_3.$$

The directional derivative can once again be written in the form

$$D_{\mathbf{u}}f = \nabla f \cdot \mathbf{u} = |\nabla f||u|\cos\theta = |\nabla f|\cos\theta,$$

so the properties listed earlier for functions of two variables continue to hold. At any given point, f increases most rapidly in the direction of ∇f and decreases most rapidly in the direction of $-\nabla f$. In any direction orthogonal to ∇f, the derivative is zero.

EXAMPLE 5

a) Find the derivative of $f(x, y, z) = x^3 - xy^2 - z$ at $P_0(1, 1, 0)$ in the direction of $\mathbf{A} = 2\mathbf{i} - 3\mathbf{j} + 6\mathbf{k}$.

b) In what directions does f change most rapidly at P_0, and what are the rates of change in these directions?

Solution

a) The direction of $\mathbf{A}$ is obtained by dividing $\mathbf{A}$ by its length:

$$|\mathbf{A}| = \sqrt{(2)^2 + (-3)^2 + (6)^2} = \sqrt{49} = 7$$

$$\mathbf{u} = \frac{\mathbf{A}}{|\mathbf{A}|} = \frac{2}{7}\mathbf{i} - \frac{3}{7}\mathbf{j} + \frac{6}{7}\mathbf{k}.$$

The partial derivatives of f at P_0 are

$$f_x = 3x^2 - y^2\big|_{(1,1,0)} = 2,$$

$$f_y = -2xy\big|_{(1,1,0)} = -2, \qquad f_z = -1\big|_{(1,1,0)} = -1.$$

The gradient of f at P_0 is

$$\nabla f\big|_{(1,1,0)} = 2\mathbf{i} - 2\mathbf{j} - \mathbf{k}.$$

The derivative of f at P_0 in the direction of $\mathbf{A}$ is therefore

$$(D_{\mathbf{u}}f)\big|_{(1,1,0)} = \nabla f\big|_{(1,1,0)} \cdot \mathbf{u} = (2\mathbf{i} - 2\mathbf{j} - \mathbf{k}) \cdot \left(\frac{2}{7}\mathbf{i} - \frac{3}{7}\mathbf{j} + \frac{6}{7}\mathbf{k}\right)$$

$$= \frac{4}{7} + \frac{6}{7} - \frac{6}{7} = \frac{4}{7}.$$

b) The function increases most rapidly in the direction of $\nabla f = 2\mathbf{i} - 2\mathbf{j} - \mathbf{k}$, and decreases most rapidly in the direction of $-\nabla f$. The rates of change in the directions are, respectively,

$$|\nabla f| = \sqrt{(2)^2 + (-2)^2 + (-1)^2} = \sqrt{9} = 3 \qquad \text{and} \qquad -|\nabla f| = -3. \quad \square$$

Equations for Tangent Planes and Normal Lines

If $\mathbf{r} = g(t)\mathbf{i} + h(t)\mathbf{j} + k(t)\mathbf{k}$ is a smooth curve on the level surface $f(x, y, z) = c$ of a differentiable function f, then $f(g(t), h(t), k(t)) = c$. Differentiating both sides of this equation with respect to t leads to

$$\frac{d}{dt} f(g(t), h(t), k(t)) = \frac{d}{dt}(c)$$

$$\frac{\partial f}{\partial x}\frac{dg}{dt} + \frac{\partial f}{\partial y}\frac{dh}{dt} + \frac{\partial f}{\partial z}\frac{dk}{dt} = 0 \qquad \text{Chain Rule}$$

$$\underbrace{\left(\frac{\partial f}{\partial x}\mathbf{i} + \frac{\partial f}{\partial y}\mathbf{j} + \frac{\partial f}{\partial z}\mathbf{k}\right)}_{\nabla f} \cdot \underbrace{\left(\frac{dg}{dt}\mathbf{i} + \frac{dh}{dt}\mathbf{j} + \frac{dk}{dt}\mathbf{k}\right)}_{d\mathbf{r}/dt} = 0. \tag{7}$$

At every point along the curve, ∇f is orthogonal to the curve's velocity vector.

Now let us restrict our attention to the curves that pass through P_0 (Fig. 12.39). All the velocity vectors at P_0 are orthogonal to ∇f at P_0, so the curves' tangent lines all lie in the plane through P_0 normal to ∇f. We call this plane the tangent plane of the surface at P_0. The line through P_0 perpendicular to the plane is the surface's normal line at P_0.

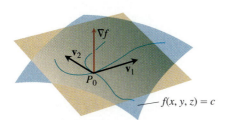

12.39 ∇f is orthogonal to the velocity vector of every smooth curve in the surface through P_0. The velocity vectors at P_0 therefore lie in a common plane, which we call the tangent plane at P_0.

Definitions

The **tangent plane** at the point $P_0(x_0, y_0, z_0)$ on the level surface $f(x, y, z) = c$ is the plane through P_0 normal to $\nabla f \big|_{P_0}$.

The **normal line** of the surface at P_0 is the line through P_0 parallel to $\nabla f \big|_{P_0}$.

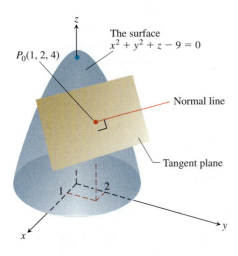

12.40 The tangent plane and normal line to the surface $x^2 + y^2 + z - 9 = 0$ at $P_0(1, 2, 4)$ (Example 6).

Thus, from Section 10.5, the tangent plane and normal line, respectively, have the following equations:

$$f_x(P_0)(x - x_0) + f_y(P_0)(y - y_0) + f_z(P_0)(z - z_0) = 0 \tag{8}$$

$$x = x_0 + f_x(P_0)t, \quad y = y_0 + f_y(P_0)t, \quad z = z_0 + f_z(P_0)t. \tag{9}$$

EXAMPLE 6 Find the tangent plane and normal line of the surface

$$f(x, y, z) = x^2 + y^2 + z - 9 = 0 \qquad \text{A circular paraboloid}$$

at the point $P_0(1, 2, 4)$.

Solution The surface is shown in Fig. 12.40.

The tangent plane is the plane through P_0 perpendicular to the gradient of f at P_0. The gradient is

$$\nabla f \big|_{P_0} = (2x\,\mathbf{i} + 2y\,\mathbf{j} + \mathbf{k})_{(1,2,4)} = 2\mathbf{i} + 4\mathbf{j} + \mathbf{k}.$$

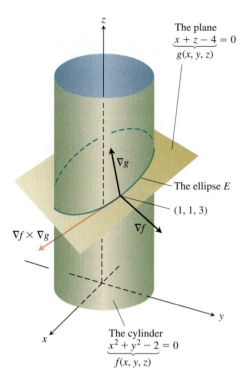

12.41 The cylinder $f(x, y, z) = x^2 + y^2 - 2$ $= 0$ and the plane $g(x, y, z) = x + z - 4 = 0$ intersect in an ellipse E (Example 7).

The plane is therefore the plane

$$2(x - 1) + 4(y - 2) + (z - 4) = 0, \qquad \text{or} \qquad 2x + 4y + z = 14.$$

The line normal to the surface at P_0 is

$$x = 1 + 2t, \qquad y = 2 + 4t, \qquad z = 4 + t. \qquad \square$$

EXAMPLE 7   The surfaces

$$f(x, y, z) = x^2 + y^2 - 2 = 0 \qquad \text{A cylinder}$$

and

$$g(x, y, z) = x + z - 4 = 0 \qquad \text{A plane}$$

meet in an ellipse E (Fig. 12.41). Find parametric equations for the line tangent to E at the point $P_0(1, 1, 3)$.

Solution   The tangent line is orthogonal to both ∇f and ∇g at P_0, and therefore parallel to $\mathbf{v} = \nabla f \times \nabla g$. The components of $\mathbf{v}$ and the coordinates of P_0 give us equations for the line. We have

$$\nabla f_{(1,1,3)} = (2x\,\mathbf{i} + 2y\,\mathbf{j})_{(1,1,3)} = 2\,\mathbf{i} + 2\,\mathbf{j}$$

$$\nabla g_{(1,1,3)} = (\mathbf{i} + \mathbf{k})_{(1,1,3)} = \mathbf{i} + \mathbf{k}$$

$$\mathbf{v} = (2\,\mathbf{i} + 2\,\mathbf{j}) \times (\mathbf{i} + \mathbf{k}) = \begin{vmatrix} \mathbf{i} & \mathbf{j} & \mathbf{k} \\ 2 & 2 & 0 \\ 1 & 0 & 1 \end{vmatrix} = 2\,\mathbf{i} - 2\,\mathbf{j} - 2\,\mathbf{k}.$$

The line is

$$x = 1 + 2t, \qquad y = 1 - 2t, \qquad z = 3 - 2t. \qquad \square$$

Planes Tangent to a Surface $z = f(x, y)$

To find an equation for the plane tangent to a surface $z = f(x, y)$ at a point $P_0(x_0, y_0, z_0)$ where $z_0 = f(x_0, y_0)$, we first observe that the equation $z = f(x, y)$ is equivalent to $f(x, y) - z = 0$. The surface $z = f(x, y)$ is therefore the zero level surface of the function $F(x, y, z) = f(x, y) - z$. The partial derivatives of F are

$$F_x = \frac{\partial}{\partial x}(f(x, y) - z) = f_x - 0 = f_x$$

$$F_y = \frac{\partial}{\partial y}(f(x, y) - z) = f_y - 0 = f_y$$

$$F_z = \frac{\partial}{\partial z}(f(x, y) - z) = 0 - 1 = -1.$$

The formula

$$F_x(P_0)(x - x_0) + F_y(P_0)(y - y_0) + F_z(P_0)(z - z_0) = 0 \qquad \text{Eq. (8) restated for } F(x, y, z)$$

for the plane tangent to the level surface at P_0 therefore reduces to

$$f_x(x_0, y_0)(x - x_0) + f_y(x_0, y_0)(y - y_0) - (z - z_0) = 0.$$

The plane tangent to the surface $z = f(x, y)$ at the point $P_0(x_0, y_0, z_0) = (x_0, y_0, f(x_0, y_0))$ is

$$f_x(x_0, y_0)(x - x_0) + f_y(x_0, y_0)(y - y_0) - (z - z_0) = 0. \qquad (10)$$

EXAMPLE 8 Find the plane tangent to the surface $z = x \cos y - ye^x$ at $(0, 0, 0)$.

Solution We calculate the partial derivatives of $f(x, y) = x \cos y - ye^x$ and use Eq. (10):

$$f_x(0, 0) = (\cos y - ye^x)_{(0,0)} = 1 - 0 \cdot 1 = 1$$

$$f_y(0, 0) = (-x \sin y - e^x)_{(0,0)} = 0 - 1 = -1.$$

The tangent plane is therefore

$$1 \cdot (x - 0) - 1 \cdot (y - 0) - (z - 0) = 0, \qquad \text{Eq. (10)}$$

or

$$x - y - z = 0. \qquad \Box$$

Increments and Distance

The directional derivative plays the role of an ordinary derivative when we want to estimate how much a function f changes if we move a small distance ds from a point P_0 to another point nearby. If f were a function of a single variable, we would have

$$df = f'(P_0)\, ds. \qquad \text{Ordinary derivative} \times \text{increment}$$

For a function of two or more variables, we use the formula

$$df = (\nabla f\big|_{P_0} \cdot \mathbf{u})\, ds, \qquad \text{Directional derivative} \times \text{increment}$$

where $\mathbf{u}$ is the direction of the motion away from P_0.

Estimating the Change in f in a Direction u

To estimate the change in the value of a function f when we move a small distance ds from a point P_0 in a particular direction $\mathbf{u}$, use the formula

$$df = \underbrace{(\nabla f\big|_{P_0} \cdot \mathbf{u})}_{\substack{\text{directional} \\ \text{derivative}}} \cdot \underbrace{ds}_{\substack{\text{distance} \\ \text{increment}}}$$

EXAMPLE 9 Estimate how much the value of

$$f(x, y, z) = xe^y + yz$$

will change if the point $P(x, y, z)$ moves 0.1 unit from $P_0(2, 0, 0)$ straight toward $P_1(4, 1, -2)$.

Solution We first find the derivative of f at P_0 in the direction of the vector

$$\overrightarrow{P_0 P_1} = 2\mathbf{i} + \mathbf{j} - 2\mathbf{k}.$$

The direction of this vector is

$$\mathbf{u} = \frac{\overrightarrow{P_0 P_1}}{|\overrightarrow{P_0 P_1}|} = \frac{\overrightarrow{P_0 P_1}}{3} = \frac{2}{3}\mathbf{i} + \frac{1}{3}\mathbf{j} - \frac{2}{3}\mathbf{k}.$$

The gradient of f at P_0 is

$$\nabla f\big|_{(2,0,0)} = (e^y\mathbf{i} + (xe^y + z)\mathbf{j} + y\,\mathbf{k})\big|_{(2,0,0)} = \mathbf{i} + 2\mathbf{j}.$$

Therefore,

$$\nabla f\big|_{P_0} \cdot \mathbf{u} = (\mathbf{i} + 2\mathbf{j}) \cdot \left(\frac{2}{3}\mathbf{i} + \frac{1}{3}\mathbf{j} - \frac{2}{3}\mathbf{k}\right) = \frac{2}{3} + \frac{2}{3} = \frac{4}{3}.$$

The change df in f that results from moving $ds = 0.1$ unit away from P_0 in the direction of $\mathbf{u}$ is approximately

$$df = (\nabla f\big|_{P_0} \cdot \mathbf{u})(ds) = \left(\frac{4}{3}\right)(0.1) \approx 0.13.$$

Algebra Rules for Gradients

If we know the gradients of two functions f and g, we automatically know the gradients of their constant multiples, sum, difference, product, and quotient.

These rules have the same form as the corresponding rules for derivatives, as they should (Exercise 65).

Algebra Rules for Gradients

1. *Constant Multiple Rule:* $\nabla(kf) = k\nabla f$ (any number k)

2. *Sum Rule:* $\nabla(f + g) = \nabla f + \nabla g$

3. *Difference Rule:* $\nabla(f - g) = \nabla f - \nabla g$

4. *Product Rule:* $\nabla(fg) = f\nabla g + g\nabla f$

5. *Quotient Rule:* $\nabla\left(\dfrac{f}{g}\right) = \dfrac{g\nabla f - f\nabla g}{g^2}$

EXAMPLE 10   We illustrate the rules with

$$f(x, y, z) = x - y \qquad g(x, y, z) = z$$

$$\nabla f = \mathbf{i} - \mathbf{j} \qquad\qquad \nabla g = \mathbf{k}.$$

We have:

1. $\nabla(2f) = \nabla(2x - 2y) = 2\mathbf{i} - 2\mathbf{j} = 2\nabla f$

2. $\nabla(f + g) = \nabla(x - y + z) = \mathbf{i} - \mathbf{j} + \mathbf{k} = \nabla f + \nabla g$

3. $\nabla(f - g) = \nabla(x - y - z) = \mathbf{i} - \mathbf{j} - \mathbf{k} = \nabla f - \nabla g$

4. $\nabla(fg) = \nabla(xz - yz) = z\mathbf{i} - z\mathbf{j} + (x - y)\mathbf{k} = g\nabla f + f\nabla g$

5. $\nabla\left(\dfrac{f}{g}\right) = \nabla\left(\dfrac{x-y}{z}\right) = \dfrac{\partial}{\partial x}\left(\dfrac{x-y}{z}\right)\mathbf{i} + \dfrac{\partial}{\partial y}\left(\dfrac{x-y}{z}\right)\mathbf{j} + \dfrac{\partial}{\partial z}\left(\dfrac{x-y}{z}\right)\mathbf{k}$

$= \dfrac{1}{z}\mathbf{i} - \dfrac{1}{z}\mathbf{j} + \dfrac{z\cdot 0 - (x-y)\cdot 1}{z^2}\mathbf{k}$

$= \dfrac{z\mathbf{i} - z\mathbf{j} - (x-y)\mathbf{k}}{z^2} = \dfrac{g\nabla f - f\nabla g}{g^2}$

Exercises 12.7

Calculating Gradients at Points

In Exercises 1–4, find the gradient of the function at the given point. Then sketch the gradient together with the level curve that passes through the point.

1. $f(x, y) = y - x$, $(2, 1)$

2. $f(x, y) = \ln(x^2 + y^2)$, $(1, 1)$

3. $g(x, y) = y - x^2$, $(-1, 0)$

4. $g(x, y) = \dfrac{x^2}{2} - \dfrac{y^2}{2}$, $(\sqrt{2}, 1)$

In Exercises 5–8, find ∇f at the given point.

5. $f(x, y, z) = x^2 + y^2 - 2z^2 + z\ln x$, $(1, 1, 1)$

6. $f(x, y, z) = 2z^3 - 3(x^2 + y^2)z + \tan^{-1} xz$, $(1, 1, 1)$

7. $f(x, y, z) = (x^2 + y^2 + z^2)^{-1/2} + \ln(xyz)$, $(-1, 2, -2)$

8. $f(x, y, z) = e^{x+y}\cos z + (y+1)\sin^{-1} x$, $(0, 0, \pi/6)$

Finding Directional Derivatives in the *xy*-Plane

In Exercises 9–16, find the derivative of the function at P_0 in the direction of **A**.

9. $f(x, y) = 2xy - 3y^2$, $P_0(5, 5)$, $\mathbf{A} = 4\mathbf{i} + 3\mathbf{j}$

10. $f(x, y) = 2x^2 + y^2$, $P_0(-1, 1)$, $\mathbf{A} = 3\mathbf{i} - 4\mathbf{j}$

11. $g(x, y) = x - (y^2/x) + \sqrt{3}\sec^{-1}(2xy)$, $P_0(1, 1)$,
$\mathbf{A} = 12\mathbf{i} + 5\mathbf{j}$

12. $h(x, y) = \tan^{-1}(y/x) + \sqrt{3}\sin^{-1}(xy/2)$, $P_0(1, 1)$,
$\mathbf{A} = 3\mathbf{i} - 2\mathbf{j}$

13. $f(x, y, z) = xy + yz + zx$, $P_0(1, -1, 2)$,
$\mathbf{A} = 3\mathbf{i} + 6\mathbf{j} - 2\mathbf{k}$

14. $f(x, y, z) = x^2 + 2y^2 - 3z^2$, $P_0(1, 1, 1)$, $\mathbf{A} = \mathbf{i} + \mathbf{j} + \mathbf{k}$

15. $g(x, y, z) = 3e^x \cos yz$, $P_0(0, 0, 0)$, $\mathbf{A} = 2\mathbf{i} + \mathbf{j} - 2\mathbf{k}$

16. $h(x, y, z) = \cos xy + e^{yz} + \ln zx$, $P_0(1, 0, 1/2)$,
$\mathbf{A} = \mathbf{i} + 2\mathbf{j} + 2\mathbf{k}$

Directions of Most Rapid Increase and Decrease

In Exercises 17–22, find the directions in which the functions increase and decrease most rapidly at P_0. Then find the derivatives of the functions in these directions.

17. $f(x, y) = x^2 + xy + y^2$, $P_0(-1, 1)$

18. $f(x, y) = x^2y + e^{xy}\sin y$, $P_0(1, 0)$

19. $f(x, y, z) = (x/y) - yz$, $P_0(4, 1, 1)$

20. $g(x, y, z) = xe^y + z^2$, $P_0(1, \ln 2, 1/2)$

21. $f(x, y, z) = \ln xy + \ln yz + \ln xz$, $P_0(1, 1, 1)$

22. $h(x, y, z) = \ln(x^2 + y^2 - 1) + y + 6z$, $P_0(1, 1, 0)$

Estimating Change

23. By about how much will
$$f(x, y, z) = \ln\sqrt{x^2 + y^2 + z^2}$$
change if the point $P(x, y, z)$ moves from $P_0(3, 4, 12)$ a distance of $ds = 0.1$ units in the direction of $3\mathbf{i} + 6\mathbf{j} - 2\mathbf{k}$?

24. By about how much will
$$f(x, y, z) = e^x \cos yz$$
change as the point $P(x, y, z)$ moves from the origin a distance of $ds = 0.1$ units in the direction of $2\mathbf{i} + 2\mathbf{j} - 2\mathbf{k}$?

25. By about how much will
$$g(x, y, z) = x + x\cos z - y\sin z + y$$
change if the point $P(x, y, z)$ moves from $P_0(2, -1, 0)$ a distance of $ds = 0.2$ units toward the point $P_1(0, 1, 2)$?

26. By about how much will
$$h(x, y, z) = \cos(\pi xy) + xz^2$$
change if the point $P(x, y, z)$ moves from $P_0(-1, -1, -1)$ a distance of $ds = 0.1$ units toward the origin?

Tangent Planes and Normal Lines to Surfaces

In Exercises 27–34, find equations for the (a) tangent plane and (b) normal line at the point P_0 on the given surface.

27. $x^2 + y^2 + z^2 = 3$, $P_0(1, 1, 1)$

28. $x^2 + y^2 - z^2 = 18$, $P_0(3, 5, -4)$

29. $2z - x^2 = 0$, $P_0(2, 0, 2)$

30. $x^2 + 2xy - y^2 + z^2 = 7$, $P_0(1, -1, 3)$

31. $\cos \pi x - x^2 y + e^{xz} + yz = 4$, $P_0(0, 1, 2)$

32. $x^2 - xy - y^2 - z = 0$, $P_0(1, 1, -1)$

33. $x + y + z = 1$, $P_0(0, 1, 0)$

34. $x^2 + y^2 - 2xy - x + 3y - z = -4$, $P_0(2, -3, 18)$

In Exercises 35–38, find an equation for the plane that is tangent to the given surface at the given point.

35. $z = \ln (x^2 + y^2)$, $(1, 0, 0)$ 36. $z = e^{-(x^2 + y^2)}$, $(0, 0, 1)$

37. $z = \sqrt{y - x}$, $(1, 2, 1)$ 38. $z = 4x^2 + y^2$, $(1, 1, 5)$

Tangent Lines to Curves

In Exercises 39–42, sketch the curve $f(x, y) = c$ together with ∇f and the tangent line at the given point. Then write an equation for the tangent line.

39. $x^2 + y^2 = 4$, $(\sqrt{2}, \sqrt{2})$

40. $x^2 - y = 1$, $(\sqrt{2}, 1)$

41. $xy = -4$, $(2, -2)$

42. $x^2 - xy + y^2 = 7$, $(-1, 2)$ (This is the curve in Section 2.6, Example 4.)

In Exercises 43–48, find parametric equations for the line tangent to the curve of intersection of the surfaces at the given point.

43. Surfaces: $x + y^2 + 2z = 4$, $x = 1$
    Point: $(1, 1, 1)$

44. Surfaces: $xyz = 1$, $x^2 + 2y^2 + 3z^2 = 6$
    Point: $(1, 1, 1)$

45. Surfaces: $x^2 + 2y + 2z = 4$, $y = 1$
    Point: $(1, 1, 1/2)$

46. Surfaces: $x + y^2 + z = 2$, $y = 1$
    Point: $(1/2, 1, 1/2)$

47. Surfaces: $x^3 + 3x^2 y^2 + y^3 + 4xy - z^2 = 0$,
                $x^2 + y^2 + z^2 = 11$
    Point: $(1, 1, 3)$

48. Surfaces: $x^2 + y^2 = 4$, $x^2 + y^2 - z = 0$
    Point: $(\sqrt{2}, \sqrt{2}, 4)$

Theory and Examples

49. In what directions is the derivative of $f(x, y) = xy + y^2$ at $P(3, 2)$ equal to zero?

50. In what two directions is the derivative of $f(x, y) = (x^2 - y^2)/(x^2 + y^2)$ at $P(1, 1)$ equal to zero?

51. Is there a direction **A** in which the rate of change of $f(x, y) = x^2 - 3xy + 4y^2$ at $P(1, 2)$ equals 14? Give reasons for your answer.

52. Is there a direction **A** in which the rate of change of the temperature function $T(x, y, z) = 2xy - yz$ (temperature in degrees Celsius, distance in feet) at $P(1, -1, 1)$ is $-3°$C/ft? Give reasons for your answer.

53. The derivative of $f(x, y)$ at $P_0(1, 2)$ in the direction of $\mathbf{i} + \mathbf{j}$ is $2\sqrt{2}$ and in the direction of $-2\mathbf{j}$ is -3. What is the derivative of f in the direction of $-\mathbf{i} - 2\mathbf{j}$? Give reasons for your answer.

54. The derivative of $f(x, y, z)$ at a point P is greatest in the direction of $\mathbf{A} = \mathbf{i} + \mathbf{j} - \mathbf{k}$. In this direction the value of the derivative is $2\sqrt{3}$.

    k)   What is ∇f at P? Give reasons for your answer.
    l)   What is the derivative of f at P in the direction of $\mathbf{i} + \mathbf{j}$?

55. *Temperature change along a circle.* Suppose that the Celsius temperature at the point (x, y) in the xy-plane is $T(x, y) = x \sin 2y$ and that distance in the xy-plane is measured in meters. A particle is moving *clockwise* around the circle of radius 1 m centered at the origin at the constant rate of 2 m/sec.

    a)   How fast is the temperature experienced by the particle changing in °C/m at the point $P(1/2, \sqrt{3}/2)$?
    b)   How fast is the temperature experienced by the particle changing in °C/sec at P?

56. *Change along the involute of a circle.* Find the derivative of $f(x, y) = x^2 + y^2$ in the direction of the unit tangent vector of the curve
$$\mathbf{r}(t) = (\cos t + t \sin t)\,\mathbf{i} + (\sin t - t \cos t)\,\mathbf{j}, \quad t > 0$$
(Fig. 12.42).

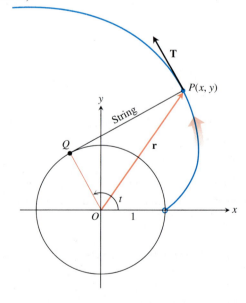

12.42 The involute of the unit circle from Section 11.3, Example 5. If you move out along the involute, covering distance along the curve at a constant rate, your distance from the origin will increase at a constant rate as well. (This is how to interpret the result of your calculation in Exercise 56.)

57. *Change along a helix.* Find the derivative of $f(x, y, z) = x^2 + y^2 + z^2$ in the direction of the unit tangent vector of the helix

$$\mathbf{r}(t) = (\cos t)\,\mathbf{i} + (\sin t)\,\mathbf{j} + t\,\mathbf{k}$$

at the points where $t = -\pi/4, 0$, and $\pi/4$. The function f gives the square of the distance from a point $P(x, y, z)$ on the helix to the origin. The derivatives calculated here give the rates at which the square of the distance is changing with respect to t as P moves through the points where $t = -\pi/4, 0$, and $\pi/4$.

58. The Celsius temperature in a region in space is given by $T(x, y, z) = 2x^2 - xyz$. A particle is moving in this region and its position at time t is given by $x = 2t^2, y = 3t, z = -t^2$, where time is measured in seconds and distance in meters.

a) How fast is the temperature experienced by the particle changing in °C/m when the particle is at the point $P(8, 6, -4)$?

b) How fast is the temperature experienced by the particle changing in °C/ sec at P?

59. Show that $A(x - x_0) + B(y - y_0) = 0$ is an equation for the line in the xy-plane through the point (x_0, y_0) normal to the vector $\mathbf{N} = A\,\mathbf{i} + B\,\mathbf{j}$.

60. *Normal curves and tangent curves.* A curve is *normal* to a surface $f(x, y, z) = c$ at a point of intersection if the curve's velocity vector is a scalar multiple of ∇f at the point. The curve is *tangent* to the surface at a point of intersection if its velocity vector is orthogonal to ∇f there.

a) Show that the curve

$$\mathbf{r}(t) = \sqrt{t}\,\mathbf{i} + \sqrt{t}\,\mathbf{j} - \frac{1}{4}(t + 3)\,\mathbf{k}$$

is normal to the surface $x^2 + y^2 - z = 3$ when $t = 1$.

b) Show that the curve

$$\mathbf{r}(t) = \sqrt{t}\,\mathbf{i} + \sqrt{t}\,\mathbf{j} + (2t - 1)\,\mathbf{k}$$

is tangent to the surface $x^2 + y^2 - z = 1$ when $t = 1$.

61. *Another way to see why gradients are normal to level curves.* Suppose that a differentiable function $f(x, y)$ has a constant value c along the differentiable curve $x = g(t), y = h(t)$ for all values of t. Differentiate both sides of the equation $f(g(t), h(t)) = c$ with respect to t to show that ∇f is normal to the curve's tangent vector at every point.

62. *The linearization of f(x, y) is a tangent-plane approximation.* Show that the tangent plane at the point $P_0(x_0, y_0, f(x_0, y_0))$ on the surface $z = f(x, y)$ defined by a differentiable function f is the plane

$$f_x(x_0, y_0)(x - x_0) + f_y(x_0, y_0)(y - y_0) - (z - f(x_0, y_0)) = 0$$

or

$$z = f(x_0, y_0) + f_x(x_0, y_0)(x - x_0) + f_y(x_0, y_0)(y - y_0).$$

Thus the tangent plane at P_0 is the graph of the linearization of f at P_0 (Fig. 12.43).

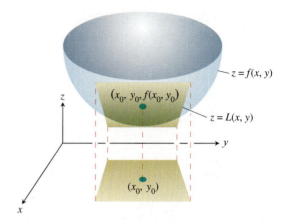

12.43 The graph of a function $z = f(x, y)$ and its linearization at a point (x_0, y_0). The plane defined by L is tangent to the surface at the point above the point (x_0, y_0). This furnishes a geometric explanation of why the values of L lie close to those of f in the immediate neighborhood of (x_0, y_0) (Exercise 62).

63. *Directional derivatives and scalar components.* How is the derivative of a differentiable function $f(x, y, z)$ at a point P_0 in the direction of a unit vector $\mathbf{u}$ related to the scalar component of $(\nabla f)_{P_0}$ in the direction of $\mathbf{u}$? Give reasons for your answer.

64. *Directional derivatives and partial derivatives.* Assuming that the necessary derivatives of $f(x, y, z)$ are defined, how are $D_{\mathbf{i}}f$, $D_{\mathbf{j}}f$, and $D_{\mathbf{k}}f$ related to f_x, f_y, and f_z? Give reasons for your answer.

65. *The algebra rules for gradients.* Given a constant k and the gradients

$$\nabla f = \frac{\partial f}{\partial x}\,\mathbf{i} + \frac{\partial f}{\partial y}\,\mathbf{j} + \frac{\partial f}{\partial z}\,\mathbf{k}$$

and

$$\nabla g = \frac{\partial g}{\partial x}\,\mathbf{i} + \frac{\partial g}{\partial y}\,\mathbf{j} + \frac{\partial g}{\partial z}\,\mathbf{k},$$

use the scalar equations

$$\frac{\partial}{\partial x}(kf) = k\frac{\partial f}{\partial x}, \quad \frac{\partial}{\partial x}(f \pm g) = \frac{\partial f}{\partial x} \pm \frac{\partial g}{\partial x},$$

$$\frac{\partial}{\partial x}(fg) = f\frac{\partial g}{\partial x} + g\frac{\partial f}{\partial x}, \quad \frac{\partial}{\partial x}\left(\frac{f}{g}\right) = \frac{g\frac{\partial f}{\partial x} - f\frac{\partial g}{\partial x}}{g^2},$$

and so on, to establish the following rules:

a) $\nabla(kf) = k\nabla f$

b) $\nabla(f + g) = \nabla f + \nabla g$

c) $\nabla(f - g) = \nabla f - \nabla g$

d) $\nabla(fg) = f\nabla g + g\nabla f$

e) $\nabla\left(\dfrac{f}{g}\right) = \dfrac{g\nabla f - f\nabla g}{g^2}$

12.8

Extreme Values and Saddle Points

Continuous functions defined on closed bounded regions in the xy-plane take on absolute maximum and minimum values on these domains (Figs. 12.44 and 12.45). It is important to be able to find these values and to know where they occur. We can often accomplish this by examining partial derivatives.

The Derivative Tests

To find the local extreme values of a function of a single variable, we look for points where the graph has a horizontal tangent line. At such points we then look for local maxima, local minima, and points of inflection. For a function $f(x, y)$ of two variables, we look for points where the surface $z = f(x, y)$ has a horizontal tangent *plane*. At such points we then look for local maxima, local minima, and saddle points (more about saddle points in a moment).

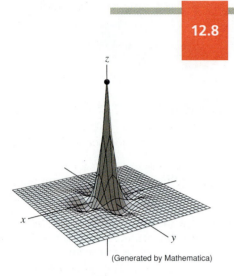

(Generated by Mathematica)

12.44 The function

$$z = (\cos x)(\cos y)e^{-\sqrt{x^2+y^2}}$$

has a maximum value of 1 and a minimum value of about -0.067 on the square region $|x| \le 3\pi/2$, $|y| \le 3\pi/2$.

> ### Definitions
>
> Let $f(x, y)$ be defined on a region R containing the point (a, b). Then
>
> 1. $f(a, b)$ is a **local maximum** value of f if $f(a, b) \ge f(x, y)$ for all domain points (x, y) in an open disk centered at (a, b).
> 2. $f(a, b)$ is a **local minimum** value of f if $f(a, b) \le f(x, y)$ for all domain points (x, y) in an open disk centered at (a, b).

Local maxima correspond to mountain peaks on the surface $z = f(x, y)$ and local minima correspond to valley bottoms (Fig. 12.46). At such points the tangent planes, when they exist, are horizontal. Local extrema are also called **relative extrema.**

As with functions of a single variable, the key to identifying the local extrema is a first derivative test.

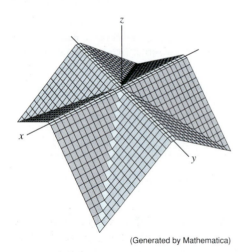

(Generated by Mathematica)

12.45 The "roof surface"

$$z = \frac{1}{2}\left(\,||x| - |y|| - |x| - |y|\,\right)$$

viewed from the point (10, 15, 20). The defining function has a maximum value of 0 and a minimum value of $-a$ on the square region $|x| \le a$, $|y| \le a$.

12.46 A local maximum is a mountain peak and a local minimum is a valley low.

> ### Theorem 7
> #### First Derivative Test for Local Extreme Values
>
> If $f(x, y)$ has a local maximum or minimum value at an interior point (a, b) of its domain, and if the first partial derivatives exist there, then $f_x(a, b) = 0$ and $f_y(a, b) = 0$.

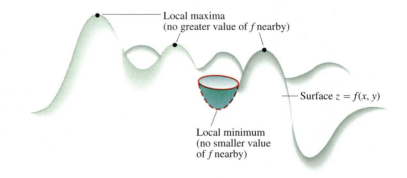

Local maxima
(no greater value of f nearby)

Surface $z = f(x, y)$

Local minimum
(no smaller value of f nearby)

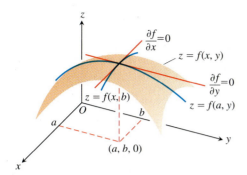

12.47 The maximum of f occurs at $x = a, y = b$.

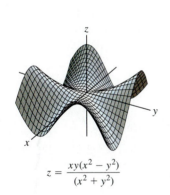

$$z = \frac{xy(x^2 - y^2)}{(x^2 + y^2)}$$

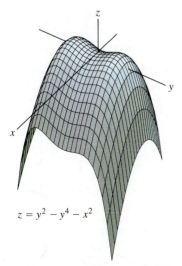

$z = y^2 - y^4 - x^2$

(Generated by Mathematica)

12.48 Saddle points at the origin.

Proof   Suppose that f has a local maximum value at an interior point (a, b) of its domain. Then

1. $x = a$ is an interior point of the domain of the curve $z = f(x, b)$ in which the plane $y = b$ cuts the surface $z = f(x, y)$ (Fig. 12.47).
2. The function $z = f(x, b)$ is a differentiable function of x at $x = a$ (the derivative is $f_x(a, b)$).
3. The function $z = f(x, b)$ has a local maximum value at $x = a$.
4. The value of the derivative of $z = f(x, b)$ at $x = a$ is therefore zero (Theorem 2, Section 3.1). Since this derivative is $f_x(a, b)$, we conclude that $f_x(a, b) = 0$.

A similar argument with the function $z = f(a, y)$ shows that $f_y(a, b) = 0$.

This proves the theorem for local maximum values. The proof for local minimum values is left as Exercise 48.   ❑

If we substitute the values $f_x(a, b) = 0$ and $f_y(a, b) = 0$ into the equation

$$f_x(a, b)(x - a) + f_y(a, b)(y - b) - (z - f(a, b)) = 0$$

for the tangent plane to the surface $z = f(x, y)$ at (a, b), the equation reduces to

$$0 \cdot (x - a) + 0 \cdot (y - b) - z + f(a, b) = 0$$

or

$$z = f(a, b).$$

Thus, Theorem 7 says that the surface does indeed have a horizontal tangent plane at a local extremum, provided there is a tangent plane there.

As in the single-variable case, Theorem 7 says that the only places a function $f(x, y)$ can ever have an extreme value are

1. Interior points where $f_x = f_y = 0$,
2. Interior points where one or both of f_x and f_y do not exist,
3. Boundary points of the function's domain.

Definition

An interior point of the domain of a function $f(x, y)$ where both f_x and f_y are zero or where one or both of f_x and f_y do not exist is a **critical point** of f.

Thus, the only points where a function $f(x, y)$ can assume extreme values are critical points and boundary points. As with differentiable functions of a single variable, not every critical point gives rise to a local extremum. A differentiable function of a single variable might have a point of inflection. A differentiable function of two variables might have a saddle point.

Definition

A differentiable function $f(x, y)$ has a **saddle point** at a critical point (a, b) if in every open disk centered at (a, b) there are domain points (x, y) where $f(x, y) > f(a, b)$ and domain points (x, y) where $f(x, y) < f(a, b)$. The corresponding point $(a, b, f(a, b))$ on the surface $z = f(x, y)$ is called a saddle point of the surface (Fig. 12.48).

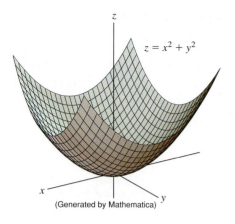

(Generated by Mathematica)

12.49 The graph of the function $f(x, y) = x^2 + y^2$ is the paraboloid $z = x^2 + y^2$. The function has only one critical point, the origin, which gives rise to a local minimum value of 0 (Example 1).

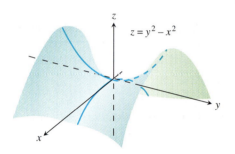

12.50 The origin is a saddle point of the function $f(x, y) = y^2 - x^2$. There are no local extreme values (Example 2).

EXAMPLE 1   Find the local extreme values of $f(x, y) = x^2 + y^2$.

Solution   The domain of f is the entire plane (so there are no boundary points) and the partial derivatives $f_x = 2x$ and $f_y = 2y$ exist everywhere. Therefore, local extreme values can occur only where

$$f_x = 2x = 0 \quad \text{and} \quad f_y = 2y = 0.$$

The only possibility is the origin, where the value of f is zero. Since f is never negative, we see that the origin gives a local minimum (Fig. 12.49). ❑

EXAMPLE 2   Find the local extreme values (if any) of $f(x, y) = y^2 - x^2$.

Solution   The domain of f is the entire plane (so there are no boundary points) and the partial derivatives $f_x = -2x$ and $f_y = 2y$ exist everywhere. Therefore, local extrema can occur only at the origin $(0, 0)$. However, along the positive x-axis f has the value $f(x, 0) = -x^2 < 0$; along the positive y-axis f has the value $f(0, y) = y^2 > 0$. Therefore every open disk in the xy-plane centered at $(0, 0)$ contains points where the function is positive and points where it is negative. The function has a saddle point at the origin (Fig. 12.50) instead of a local extreme value. We conclude that the function has no local extreme values. ❑

The fact that $f_x = f_y = 0$ at an interior point (a, b) of R does not tell us enough to be sure f has a local extreme value there. However, if f and its first and second partial derivatives are continuous on R, we may be able to learn the rest from the following theorem, proved in Section 12.10.

Theorem 8

Second Derivative Test for Local Extreme Values

Suppose $f(x, y)$ and its first and second partial derivatives are continuous throughout a disk centered at (a, b) and that $f_x(a, b) = f_y(a, b) = 0$. Then

i) f has a **local maximum** at (a, b) if $f_{xx} < 0$ and $f_{xx}f_{yy} - f_{xy}^2 > 0$ at (a, b);

ii) f has a **local minimum** at (a, b) if $f_{xx} > 0$ and $f_{xx}f_{yy} - f_{xy}^2 > 0$ at (a, b);

iii) f has a **saddle point** at (a, b) if $f_{xx}f_{yy} - f_{xy}^2 < 0$ at (a, b).

iv) **The test is inconclusive** at (a, b) if $f_{xx}f_{yy} - f_{xy}^2 = 0$ at (a, b). In this case, we must find some other way to determine the behavior of f at (a, b).

The expression $f_{xx}f_{yy} - f_{xy}^2$ is called the **discriminant** of f. It is sometimes easier to remember the determinant form,

$$f_{xx}f_{yy} - f_{xy}^2 = \begin{vmatrix} f_{xx} & f_{xy} \\ f_{xy} & f_{yy} \end{vmatrix}.$$

Theorem 8 says that if the discriminant is positive at the point (a, b), then the surface curves the same way in all directions: downwards if $f_{xx} < 0$, giving rise to a local maximum, and upwards if $f_{xx} > 0$, giving a local minimum. On the other hand, if the discriminant is negative at (a, b), then the surface curves up in some directions and down in others, so we have a saddle point.

EXAMPLE 3   Find the local extreme values of the function

$$f(x, y) = xy - x^2 - y^2 - 2x - 2y + 4.$$

Solution   The function is defined and differentiable for all x and y and its domain has no boundary points. The function therefore has extreme values only at the points where f_x and f_y are simultaneously zero. This leads to

$$f_x = y - 2x - 2 = 0, \qquad f_y = x - 2y - 2 = 0,$$

or

$$x = y = -2.$$

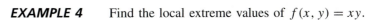

Therefore, the point $(-2, -2)$ is the only point where f may take on an extreme value. To see if it does so, we calculate

$$f_{xx} = -2, \qquad f_{yy} = -2, \qquad f_{xy} = 1.$$

The discriminant of f at $(a, b) = (-2, -2)$ is

$$f_{xx}f_{yy} - f_{xy}{}^2 = (-2)(-2) - (1)^2 = 4 - 1 = 3.$$

The combination

$$f_{xx} < 0 \qquad \text{and} \qquad f_{xx}f_{yy} - f_{xy}{}^2 > 0$$

tells us that f has a local maximum at $(-2, -2)$. The value of f at this point is $f(-2, -2) = 8$.   ❑

EXAMPLE 4   Find the local extreme values of $f(x, y) = xy$.

Solution   Since f is differentiable everywhere (Fig. 12.51), it can assume extreme values only where

$$f_x = y = 0 \qquad \text{and} \qquad f_y = x = 0.$$

Thus, the origin is the only point where f might have an extreme value. To see what happens there, we calculate

$$f_{xx} = 0, \qquad f_{yy} = 0, \qquad f_{xy} = 1.$$

The discriminant,

$$f_{xx}f_{yy} - f_{xy}{}^2 = -1,$$

is negative. Therefore the function has a saddle point at $(0, 0)$. We conclude that $f(x, y) = xy$ has no local extreme values.   ❑

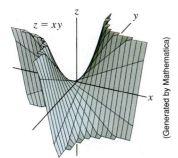

12.51 The surface $z = xy$ has a saddle point at the origin (Example 4).

(Generated by Mathematica)

Absolute Maxima and Minima on Closed Bounded Regions

We organize the search for the absolute extrema of a continuous function $f(x, y)$ on a closed and bounded region R into three steps.

Step 1: *List the interior points* of R where f may have local maxima and minima and evaluate f at these points. These are the points where $f_x = f_y = 0$ or where one or both of f_x and f_y fail to exist (the critical points of f).

Step 2: *List the boundary points* of R where f has local maxima and minima and evaluate f at these points. We will show how to do this shortly.

Step 3: *Look through the lists* for the maximum and minimum values of f. These will be the absolute maximum and minimum values of f on R. Since absolute maxima and minima are also local maxima and minima, the absolute maximum and minimum values of f already appear somewhere in the lists made in steps 1 and 2. We have only to glance at the lists to see what they are.

EXAMPLE 5   Find the absolute maximum and minimum values of

$$f(x, y) = 2 + 2x + 2y - x^2 - y^2$$

on the triangular plate in the first quadrant bounded by the lines $x = 0$, $y = 0$, $y = 9 - x$.

Solution   Since f is differentiable, the only places where f can assume these values are points inside the triangle (Fig. 12.52) where $f_x = f_y = 0$ and points on the boundary.

Interior points. For these we have

$$f_x = 2 - 2x = 0, \qquad f_y = 2 - 2y = 0,$$

yielding the single point $(x, y) = (1, 1)$. The value of f there is

$$f(1, 1) = 4.$$

Boundary points. We take the triangle one side at a time:

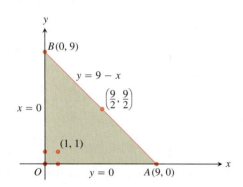

12.52 This triangular plate is the domain of the function in Example 5.

1. On the segment OA, $y = 0$. The function

   $$f(x, y) = f(x, 0) = 2 + 2x - x^2$$

   may now be regarded as a function of x defined on the closed interval $0 \le x \le 9$. Its extreme values (we know from Chapter 3) may occur at the endpoints

   $$x = 0 \qquad \text{where} \quad f(0, 0) = 2$$
   $$x = 9 \qquad \text{where} \quad f(9, 0) = 2 + 18 - 81 = -61$$

   and at the interior points where $f'(x, 0) = 2 - 2x = 0$. The only interior point where $f'(x, 0) = 0$ is $x = 1$, where

   $$f(x, 0) = f(1, 0) = 3.$$

2. On the segment OB, $x = 0$ and

   $$f(x, y) = f(0, y) = 2 + 2y - y^2.$$

   We know from the symmetry of f in x and y and from the analysis we just carried out that the candidates on this segment are

   $$f(0, 0) = 2, \qquad f(0, 9) = -61, \qquad f(0, 1) = 3.$$

3. We have already accounted for the values of f at the endpoints of AB, so we need only look at the interior points of AB. With $y = 9 - x$, we have

   $$f(x, y) = 2 + 2x + 2(9 - x) - x^2 - (9 - x)^2 = -61 + 18x - 2x^2.$$

   Setting $f'(x, 9 - x) = 18 - 4x = 0$ gives

   $$x = \frac{18}{4} = \frac{9}{2}.$$

At this value of x,

$$y = 9 - \frac{9}{2} = \frac{9}{2} \quad \text{and} \quad f(x, y) = f\left(\frac{9}{2}, \frac{9}{2}\right) = -\frac{41}{2}.$$

Summary. We list all the candidates: 4, 2, −61, 3, −(41/2). The maximum is 4, which f assumes at (1, 1). The minimum is −61, which f assumes at (0, 9) and (9, 0).

Conclusion

Despite the power of Theorem 7, we urge you to remember its limitations. It does not apply to boundary points of a function's domain, where it is possible for a function to have extreme values along with nonzero derivatives. And it does not apply to points where either f_x or f_y fails to exist.

Summary of Max-Min Tests

The extreme values of $f(x, y)$ can occur only at

(i) **boundary points** of the domain of f,

(ii) **critical points** (interior points where $f_x = f_y = 0$ or points where f_x or f_y fail to exist).

If the first and second order partial derivatives of f are continuous throughout a disk centered at a point (a, b), and $f_x(a, b) = f_y(a, b) = 0$, you may be able to classify $f(a, b)$ with the **second derivative test:**

(i) $f_{xx} < 0$ and $f_{xx}f_{yy} - f_{xy}^2 > 0$ at (a, b) $\Rightarrow$ **local maximum,**

(ii) $f_{xx} > 0$ and $f_{xx}f_{yy} - f_{xy}^2 > 0$ at (a, b) $\Rightarrow$ **local minimum,**

(iii) $f_{xx}f_{yy} - f_{xy}^2 < 0$ at (a, b) $\Rightarrow$ **saddle point,**

(iv) $f_{xx}f_{yy} - f_{xy}^2 = 0$ at (a, b) $\Rightarrow$ **test is inconclusive.**

Exercises 12.8

Finding Local Extrema

Find all the local maxima, local minima, and saddle points of the functions in Exercises 1–30.

1. $f(x, y) = x^2 + xy + y^2 + 3x - 3y + 4$

2. $f(x, y) = x^2 + 3xy + 3y^2 - 6x + 3y - 6$

3. $f(x, y) = 2xy - 5x^2 - 2y^2 + 4x + 4y - 4$

4. $f(x, y) = 2xy - 5x^2 - 2y^2 + 4x - 4$

5. $f(x, y) = x^2 + xy + 3x + 2y + 5$

6. $f(x, y) = y^2 + xy - 2x - 2y + 2$

7. $f(x, y) = 5xy - 7x^2 + 3x - 6y + 2$

8. $f(x, y) = 2xy - x^2 - 2y^2 + 3x + 4$

9. $f(x, y) = x^2 - 4xy + y^2 + 6y + 2$

10. $f(x, y) = 3x^2 + 6xy + 7y^2 - 2x + 4y$

11. $f(x, y) = 2x^2 + 3xy + 4y^2 - 5x + 2y$

12. $f(x, y) = 4x^2 - 6xy + 5y^2 - 20x + 26y$

13. $f(x, y) = x^2 - y^2 - 2x + 4y + 6$

14. $f(x, y) = x^2 - 2xy + 2y^2 - 2x + 2y + 1$

15. $f(x, y) = x^2 + 2xy$

16. $f(x, y) = 3 + 2x + 2y - 2x^2 - 2xy - y^2$

17. $f(x, y) = x^3 - y^3 - 2xy + 6$

18. $f(x, y) = x^3 + 3xy + y^3$

19. $f(x, y) = 6x^2 - 2x^3 + 3y^2 + 6xy$

20. $f(x, y) = 3y^2 - 2y^3 - 3x^2 + 6xy$

21. $f(x, y) = 9x^3 + y^3/3 - 4xy$

[handwritten: $f_x = 4y - 4x^3$]

22. $f(x, y) = 8x^3 + y^3 + 6xy$

[handwritten: $f_y = 4x - 4y^3$]

23. $f(x, y) = x^3 + y^3 + 3x^2 - 3y^2 - 8$

24. $f(x, y) = 2x^3 + 2y^3 - 9x^2 + 3y^2 - 12y$ *[handwritten: $f_{xx} = -12x^2$]*

25. $f(x, y) = 4xy - x^4 - y^4$

[handwritten: $f_{yy} = -12y^3$]

26. $f(x, y) = x^4 + y^4 + 4xy$

[handwritten: $f_{xy} = 4$]

27. $f(x, y) = \dfrac{1}{x^2 + y^2 - 1}$ **28.** $f(x, y) = \dfrac{1}{x} + xy + \dfrac{1}{y}$

29. $f(x, y) = y \sin x$ **30.** $f(x, y) = e^{2x} \cos y$

Finding Absolute Extrema

In Exercises 31–38, find the absolute maxima and minima of the functions on the given domains.

31. $f(x, y) = 2x^2 - 4x + y^2 - 4y + 1$ on the closed triangular plate bounded by the lines $x = 0$, $y = 2$, $y = 2x$ in the first quadrant

32. $D(x, y) = x^2 - xy + y^2 + 1$ on the closed triangular plate in the first quadrant bounded by the lines $x = 0$, $y = 4$, $y = x$

33. $f(x, y) = x^2 + y^2$ on the closed triangular plate bounded by the lines $x = 0$, $y = 0$, $y + 2x = 2$ in the first quadrant

34. $T(x, y) = x^2 + xy + y^2 - 6x$ on the rectangular plate $0 \le x \le 5$, $-3 \le y \le 3$

35. $T(x, y) = x^2 + xy + y^2 - 6x + 2$ on the rectangular plate $0 \le x \le 5$, $-3 \le y \le 0$

36. $f(x, y) = 48xy - 32x^3 - 24y^2$ on the rectangular plate $0 \le x \le 1$, $0 \le y \le 1$

37. $f(x, y) = (4x - x^2) \cos y$ on the rectangular plate $1 \le x \le 3$, $-\pi/4 \le y \le \pi/4$ (Fig. 12.53)

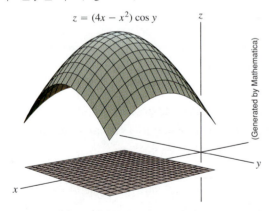

$z = (4x - x^2) \cos y$

12.53 The function and domain in Exercise 37.

38. $f(x, y) = 4x - 8xy + 2y + 1$ on the triangular plate bounded by the lines $x = 0$, $y = 0$, $x + y = 1$ in the first quadrant

39. Find two numbers a and b with $a \le b$ such that

$$\int_a^b (6 - x - x^2)\, dx$$

has its largest value.

[handwritten top right:
$4y - 4x^3 = 0$ $y = x^3$ $x - x^6 = 0$
$y - x^3 = 0$ $4x - 4x^6 = 0$ $x(1 - x^5) = 0$
$x = 0$
$x^5 = 1$
$x = \pm 1$
$y = 1, 0$
]

40. Find two numbers a and b with $a \le b$ such that

$$\int_a^b (24 - 2x - x^2)^{1/3}\, dx$$

has its largest value.

41. *Temperatures.* The flat circular plate in Fig. 12.54 has the shape of the region $x^2 + y^2 \le 1$. The plate, including the boundary where $x^2 + y^2 = 1$, is heated so that the temperature at the point (x, y) is

$$T(x, y) = x^2 + 2y^2 - x.$$

Find the temperatures at the hottest and coldest points on the plate.

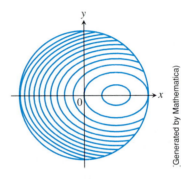

12.54 Curves of constant temperature are called isotherms. The figure shows isotherms of the temperature function $T(x, y) = x^2 + 2y^2 - x$ on the disk $x^2 + y^2 \le 1$ in the *xy*-plane. Exercise 41 asks you to locate the extreme temperatures.

42. Find the critical point of

$$f(x, y) = xy + 2x - \ln x^2 y$$

in the open first quadrant ($x > 0$, $y > 0$) and show that f takes on a minimum there (Fig. 12.55).

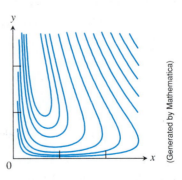

12.55 The function $f(x, y) = xy + 2x - \ln x^2 y$ (selected level curves shown here) takes on a minimum value somewhere in the open first quadrant $x > 0, y > 0$ (Exercise 42).

$2x - 4y = 0 \quad 2y - 4x = 0 \quad f_{xy} = -4 \quad (0,0)$
$x = 2y \quad 2y - 8y = 0 \quad -6y = 0$
$f_{xx} = 2 \quad f_{yy} = 2$

Theory and Examples

43. Find the maxima, minima, and saddle points of $f(x, y)$, if any, given that

a) $f_x = 2x - 4y$ and $f_y = 2y - 4x$

b) $f_x = 2x - 2$ and $f_y = 2y - 4$

c) $f_x = 9x^2 - 9$ and $f_y = 2y + 4$

Describe your reasoning in each case.

44. The discriminant $f_{xx}f_{yy} - f_{xy}^2$ is zero at the origin for each of the following functions, so the second derivative test fails there. Determine whether the function has a maximum, a minimum, or neither at the origin by imagining what the surface $z = f(x, y)$ looks like. Describe your reasoning in each case.

a) $f(x, y) = x^2 y^2$

b) $f(x, y) = 1 - x^2 y^2$

c) $f(x, y) = xy^2$

d) $f(x, y) = x^3 y^2$

e) $f(x, y) = x^3 y^3$

f) $f(x, y) = x^4 y^4$

45. Show that $(0, 0)$ is a critical point of $f(x, y) = x^2 + kxy + y^2$ no matter what value the constant k has. (*Hint:* Consider two cases: $k = 0$ and $k \neq 0$.)

46. For what values of the constant k does the second derivative test guarantee that $f(x, y) = x^2 + kxy + y^2$ will have a saddle point at $(0, 0)$? a local minimum at $(0, 0)$? For what values of k is the second derivative test inconclusive? Give reasons for your answers.

47. a) If $f_x(a, b) = f_y(a, b) = 0$, must f have a local maximum or minimum value at (a, b)? Give reasons for your answer.

b) Can you conclude anything about $f(a, b)$ if f and its first and second partial derivatives are continuous throughout a disk centered at (a, b) and $f_{xx}(a, b)$ and $f_{yy}(a, b)$ differ in sign? Give reasons for your answer.

48. Using the proof of Theorem 7 given in the text for the case in which f has a local maximum at (a, b), prove the theorem for the case in which f has a local minimum at (a, b).

49. Among all the points on the graph of $z = 10 - x^2 - y^2$ that lie above the plane $x + 2y + 3z = 0$, find the point farthest from the plane.

50. Find the point on the graph of $z = x^2 + y^2 + 10$ nearest the plane $x + 2y - z = 0$.

51. The function $f(x, y) = x + y$ fails to have an absolute maximum value in the closed first quadrant $x \geq 0$ and $y \geq 0$. Does this contradict the discussion on finding absolute extrema given in the text? Give reasons for your answer.

52. Consider the function $f(x, y) = x^2 + y^2 + 2xy - x - y + 1$ over the square $0 \leq x \leq 1$ and $0 \leq y \leq 1$.

a) Show that f has an absolute minimum along the line segment $2x + 2y = 1$ in this square. What *is* the absolute minimum value?

b) Find the absolute maximum value of f over the square.

Extreme Values on Parametrized Curves

To find the extreme values of a function $f(x, y)$ on a curve $x =$

$x(t), y = y(t)$, we treat f as a function of the single variable t and use the Chain Rule to find where df/dt is zero. As in any other single-variable case, the extreme values of f are then found among the values at the

a) critical points (points where df/dt is zero or fails to exist), and

b) endpoints of the parameter domain.

Find the absolute maximum and minimum values of the following functions on the given curves.

53. Functions:

a) $f(x, y) = x + y$

b) $g(x, y) = xy$

c) $h(x, y) = 2x^2 + y^2$

Curves:

i) The semicircle $x^2 + y^2 = 4$, $y \geq 0$

ii) The quarter circle $x^2 + y^2 = 4$, $x \geq 0$, $y \geq 0$

Use the parametric equations $x = 2 \cos t$, $y = 2 \sin t$.

54. Functions:

a) $f(x, y) = 2x + 3y$

b) $g(x, y) = xy$

c) $h(x, y) = x^2 + 3y^2$

Curves:

i) The semi-ellipse $(x^2/9) + (y^2/4) = 1$, $y \geq 0$

ii) The quarter ellipse $(x^2/9) + (y^2/4) = 1$, $x \geq 0$, $y \geq 0$

Use the parametric equations $x = 3 \cos t$, $y = 2 \sin t$.

55. Function: $f(x, y) = xy$

Curves:

i) The line $x = 2t$, $y = t + 1$

ii) The line segment $x = 2t$, $y = t + 1$, $-1 \leq t \leq 0$

iii) The line segment $x = 2t$, $y = t + 1$, $0 \leq t \leq 1$

56. Functions:

a) $f(x, y) = x^2 + y^2$

b) $g(x, y) = 1/(x^2 + y^2)$

Curves:

i) The line $x = t$, $y = 2 - 2t$

ii) The line segment $x = t$, $y = 2 - 2t$, $0 \leq t \leq 1$

Least Squares and Regression Lines

When we try to fit a line $y = mx + b$ to a set of numerical data points $(x_1, y_1), (x_2, y_2), \ldots, (x_n, y_n)$ (Fig. 12.56, on the following page), we usually choose the line that minimizes the sum of the squares of the vertical distances from the points to the line. In theory, this means finding the values of m and b that minimize the value of the function

$$w = (mx_1 + b - y_1)^2 + \cdots + (mx_n + b - y_n)^2. \tag{1}$$

The values of m and b that do this are found with the first and second derivative tests to be

$$m = \frac{\left(\sum x_k\right)\left(\sum y_k\right) - n \sum x_k y_k}{\left(\sum x_k\right)^2 - n \sum x_k^2}, \tag{2}$$

$$b = \frac{1}{n}\left(\sum y_k - m \sum x_k\right), \tag{3}$$

12.56 To fit a line to noncollinear points, we choose the line that minimizes the sum of the squares of the deviations.

12.57 The least squares line for the data in the example.

with all sums running from $k = 1$ to $k = n$. Many scientific calculators have these formulas built in, enabling you to find m and b with only a few key presses after you have entered the data.

The line $y = mx + b$ determined by these values of m and b is called the **least squares line, regression line,** or **trend line** for the data under study. Finding a least squares line lets you

1. summarize data with a simple expression,
2. predict values of y for other, experimentally untried values of x,
3. handle data analytically.

EXAMPLE   Find the least squares line for the points $(0, 1)$, $(1, 3)$, $(2, 2)$, $(3, 4)$, $(4, 5)$.

Solution   We organize the calculations in a table:

k	x_k	y_k	x_k^2	$x_k y_k$
1	0	1	0	0
2	1	3	1	3
3	2	2	4	4
4	3	4	9	12
5	4	5	16	20
Σ	10	15	30	39

Then we find

$$m = \frac{(10)(15) - 5(39)}{(10)^2 - 5(30)} = 0.9 \qquad \text{Eq. (2) with } n = 5 \text{ and data from the table}$$

and use the value of m to find

$$b = \frac{1}{5}(15 - (0.9)(10)) = 1.2. \qquad \text{Eq. (3) with } n = 5, m = 0.9$$

The least squares line is $y = 0.9x + 1.2$ (Fig. 12.57).   ❑

In Exercises 57–60, use Eqs. (2) and (3) to find the least squares line for each set of data points. Then use the linear equation you obtain to predict the value of y that would correspond to $x = 4$.

57. $(-1, 2)$, $(0, 1)$, $(3, -4)$ **58.** $(-2, 0)$, $(0, 2)$, $(2, 3)$

59. $(0, 0)$, $(1, 2)$, $(2, 3)$ **60.** $(0, 1)$, $(2, 2)$, $(3, 2)$

61. Write a linear equation for the effect of irrigation on the yield of alfalfa by fitting a least squares line to the data in Table 12.1 (from the University of California Experimental Station, *Bulletin* No. 450, p. 8). Plot the data and draw the line.

Table 12.1 Growth of alfalfa

x (total seasonal depth of water applied, in.)	y (average alfalfa yield, tons/acre)
12	5.27
18	5.68
24	6.25
30	7.21
36	8.20
42	8.71

62. *Craters of Mars.* One theory of crater formation suggests that the frequency of large craters should fall off as the square of the diameter (Marcus, *Science,* June 21, 1968, p. 1334). Pictures from *Mariner IV* show the frequencies listed in Table 12.2. Fit a line of the form $F = m(1/D^2) + b$ to the data. Plot the data and draw the line.

Table 12.2 Crater sizes on Mars

Diameter in km, D	$1/D^2$ (for left value of class interval)	Frequency, F
32–45	0.001	51
45–64	0.0005	22
64–90	0.00024	14
90–128	0.000123	4

Table 12.3 Compositions by Mozart

Köchel number, K	Year composed, y
1	1761
75	1771
155	1772
219	1775
271	1777
351	1780
425	1783
503	1786
575	1789
626	1791

Table 12.4 Sinkings of German submarines by U.S. during 16 consecutive months of WWII

Month	Guesses by U.S. (reported sinkings) x	Actual number y
1	3	3
2	2	2
3	4	6
4	2	3
5	5	4
6	5	3
7	9	11
8	12	9
9	8	10
10	13	16
11	14	13
12	3	5
13	4	6
14	13	19
15	10	15
16	16	15
	123	140

63. *Köchel numbers.* In 1862, the German musicologist Ludwig von Köchel made a chronological list of the musical works of Wolfgang Amadeus Mozart. This list is the source of the Köchel numbers, or "K numbers," that now accompany the titles of Mozart's pieces (Sinfonia Concertante in E-flat major, K.364, for example). Table 12.3 gives the Köchel numbers and composition dates (y) of ten of Mozart's works.

a) Plot y vs. K to show that y is close to being a linear function of K.

b) Find a least squares line $y = mK + b$ for the data and add the line to your plot in (a).

c) K.364 was composed in 1779. What date is predicted by the least squares line?

64. *Submarine sinkings.* The data in Table 12.4 show the results of a historical study of German submarines sunk by the U.S. Navy during 16 consecutive months of World War II. The data given for each month are the number of reported sinkings and the number of actual sinkings. The number of submarines sunk was slightly greater than the Navy's reports implied. Find a least squares line for estimating the number of actual sinkings from the number of reported sinkings.

CAS Explorations and Projects

In Exercises 65–70, you will explore functions to identify their local extrema. Use a CAS to perform the following steps:

a) Plot the function over the given rectangle.

b) Plot some level curves in the rectangle.

c) Calculate the function's first partial derivatives and use the CAS equation solver to find the critical points. How do the critical points relate to the level curves plotted in (b)? Which critical points, if any, appear to give a saddle point? Give reasons for your answer.

d) Calculate the function's second partial derivatives and find the discriminant $f_{xx}f_{yy} - f_{xy}^2$.

e) Using the max-min tests, classify the critical points found in (c). Are your findings consistent with your discussion in (c)?

65. $f(x, y) = x^2 + y^3 - 3xy, \quad -5 \le x \le 5, \quad -5 \le y \le 5$

66. $f(x, y) = x^3 - 3xy^2 + y^2, \quad -2 \le x \le 2, \quad -2 \le y \le 2$

67. $f(x, y) = x^4 + y^2 - 8x^2 - 6y + 16, \quad -3 \le x \le 3, \quad -6 \le y \le 6$

68. $f(x, y) = 2x^4 + y^4 - 2x^2 - 2y^2 + 3, \quad -3/2 \le x \le 3/2, \quad -3/2 \le y \le 3/2$

69. $f(x, y) = 5x^6 + 18x^5 - 30x^4 + 30xy^2 - 120x^3, \quad -4 \le x \le 3, \quad -2 \le y \le 2$

70. $f(x, y) = \begin{cases} x^5 \ln(x^2 + y^2), & (x, y) \ne (0, 0) \\ 0, & (x, y) = (0, 0) \end{cases}$, $-2 \le x \le 2, \quad -2 \le y \le 2$

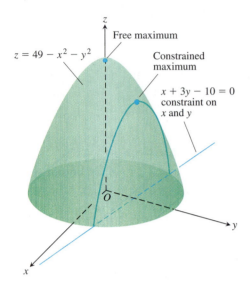

$z = 49 - x^2 - y^2$

Free maximum

Constrained maximum

$x + 3y - 10 = 0$ constraint on x and y

12.58 The function $f(x, y) = 49 - x^2 - y^2$, subject to the constraint $g(x, y) = x + 3y - 10 = 0$.

Lagrange Multipliers

As we saw in Section 12.8, we sometimes need to find the extreme values of a function whose domain is constrained to lie within some particular subset of the plane—a disk, for example, or a closed triangular region. But, as Fig. 12.58 suggests, a function may be subject to other kinds of constraints as well.

In this section, we explore a powerful method for finding extreme values of constrained functions: the method of *Lagrange multipliers*. Lagrange developed the method in 1755 to solve max-min problems in geometry. Today the method is important in economics, in engineering (where it is used in designing multistage rockets, for example), and in mathematics.

Constrained Maxima and Minima

EXAMPLE 1   Find the point $P(x, y, z)$ closest to the origin on the plane $2x + y - z - 5 = 0$.

Solution   The problem asks us to find the minimum value of the function

$$|\overrightarrow{OP}| = \sqrt{(x - 0)^2 + (y - 0)^2 + (z - 0)^2}$$
$$= \sqrt{x^2 + y^2 + z^2}$$

subject to the constraint that

$$2x + y - z - 5 = 0.$$

Since $|\overrightarrow{OP}|$ has a minimum value wherever the function

$$f(x, y, z) = x^2 + y^2 + z^2$$

has a minimum value, we may solve the problem by finding the minimum value of $f(x, y, z)$ subject to the constraint $2x + y - z - 5 = 0$. If we regard x and y as the independent variables in this equation and write z as

$$z = 2x + y - 5,$$

our problem reduces to one of finding the points (x, y) at which the function

$$h(x, y) = f(x, y, 2x + y - 5) = x^2 + y^2 + (2x + y - 5)^2$$

has its minimum value or values. Since the domain of h is the entire xy-plane, the first derivative test of Section 12.8 tells us that any minima that h might have must occur at points where

$$h_x = 2x + 2(2x + y - 5)(2) = 0, \qquad h_y = 2y + 2(2x + y - 5) = 0.$$

This leads to

$$10x + 4y = 20, \qquad 4x + 4y = 10,$$

and the solution

$$x = \frac{5}{3}, \qquad y = \frac{5}{6}.$$

We may apply a geometric argument together with the second derivative test to show that these values minimize h. The z-coordinate of the corresponding point on

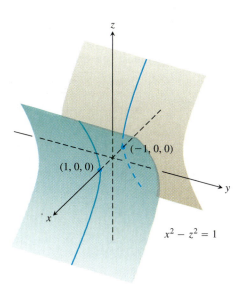

12.59 The hyperbolic cylinder $x^2 - z^2 - 1 = 0$ in Example 2.

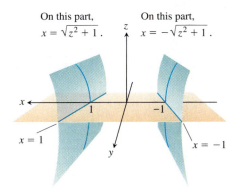

12.60 The region in the xy-plane from which the first two coordinates of the points (x, y, z) on the hyperbolic cylinder $x^2 - z^2 = 1$ are selected excludes the band $-1 < x < 1$ in the xy-plane.

the plane $z = 2x + y - 5$ is

$$z = 2 \left(\frac{5}{3} \right) + \frac{5}{6} - 5 = -\frac{5}{6}.$$

Therefore, the point we seek is

$$\text{Closest point:} \quad P \left(\frac{5}{3}, \frac{5}{6}, -\frac{5}{6} \right).$$

The distance from P to the origin is $5/\sqrt{6} \approx 2.04$.　　❑

Attempts to solve a constrained maximum or minimum problem by substitution, as we might call the method of Example 1, do not always go smoothly. This is one of the reasons for learning the new method of this section.

EXAMPLE 2　　Find the points closest to the origin on the hyperbolic cylinder $x^2 - z^2 - 1 = 0$.

Solution 1　　The cylinder is shown in Fig. 12.59. We seek the points on the cylinder closest to the origin. These are the points whose coordinates minimize the value of the function

$$f(x, y, z) = x^2 + y^2 + z^2 \qquad \text{Square of the distance}$$

subject to the constraint that $x^2 - z^2 - 1 = 0$. If we regard x and y as independent variables in the constraint equation, then

$$z^2 = x^2 - 1$$

and the values of $f(x, y, z) = x^2 + y^2 + z^2$ on the cylinder are given by the function

$$h(x, y) = x^2 + y^2 + (x^2 - 1) = 2x^2 + y^2 - 1.$$

To find the points on the cylinder whose coordinates minimize f, we look for the points in the xy-plane whose coordinates minimize h. The only extreme value of h occurs where

$$h_x = 4x = 0 \qquad \text{and} \qquad h_y = 2y = 0,$$

that is, at the point $(0, 0)$. But now we're in trouble—there are no points on the cylinder where both x and y are zero. What went wrong?

What happened was that the first derivative test found (as it should have) the point *in the domain of* h where h has a minimum value. We, on the other hand, want the points *on the cylinder* where h has a minimum value. While the domain of h is the entire xy-plane, the domain from which we can select the first two coordinates of the points (x, y, z) on the cylinder is restricted to the "shadow" of the cylinder on the xy-plane; it does not include the band between the lines $x = -1$ and $x = 1$ (Fig. 12.60).

We can avoid this problem if we treat y and z as independent variables (instead of x and y) and express x in terms of y and z as

$$x^2 = z^2 + 1.$$

With this substitution, $f(x, y, z) = x^2 + y^2 + z^2$ becomes

$$k(y, z) = (z^2 + 1) + y^2 + z^2 = 1 + y^2 + 2z^2$$

and we look for the points where k takes on its smallest value. The domain of

k in the yz-plane now matches the domain from which we select the y- and z-coordinates of the points (x, y, z) on the cylinder. Hence, the points that minimize k in the plane will have corresponding points on the cylinder. The smallest values of k occur where

$$k_y = 2y = 0 \qquad \text{and} \qquad k_z = 4z = 0,$$

or where $y = z = 0$. This leads to

$$x^2 = z^2 + 1 = 1, \qquad x = \pm 1.$$

The corresponding points on the cylinder are $(\pm 1, 0, 0)$. We can see from the inequality

$$k(y, z) = 1 + y^2 + 2z^2 \geq 1$$

that the points $(\pm 1, 0, 0)$ give a minimum value for k. We can also see that the minimum distance from the origin to a point on the cylinder is 1 unit.

Solution 2   Another way to find the points on the cylinder closest to the origin is to imagine a small sphere centered at the origin expanding like a soap bubble until it just touches the cylinder (Fig. 12.61). At each point of contact, the cylinder and sphere have the same tangent plane and normal line. Therefore, if the sphere and cylinder are represented as the level surfaces obtained by setting

$$f(x, y, z) = x^2 + y^2 + z^2 - a^2 \qquad \text{and} \qquad g(x, y, z) = x^2 - z^2 - 1$$

equal to 0, then the gradients ∇f and ∇g will be parallel where the surfaces touch. At any point of contact we should therefore be able to find a scalar λ ("lambda") such that

$$\nabla f = \lambda \nabla g,$$

or

$$2x\,\mathbf{i} + 2y\,\mathbf{j} + 2z\,\mathbf{k} = \lambda(2x\,\mathbf{i} - 2z\,\mathbf{k}).$$

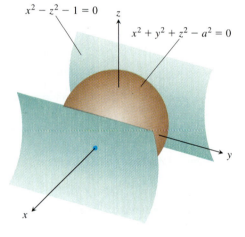

$x^2 - z^2 - 1 = 0$

$x^2 + y^2 + z^2 - a^2 = 0$

12.61 A sphere expanding like a soap bubble centered at the origin until it just touches the hyperbolic cylinder

$$x^2 - z^2 - 1 = 0.$$

See Solution 2 of Example 2.

Thus, the coordinates x, y, and z of any point of tangency will have to satisfy the three scalar equations

$$2x = 2\lambda x, \qquad 2y = 0, \qquad 2z = -2\lambda z. \tag{1}$$

For what values of λ will a point (x, y, z) whose coordinates satisfy the equations in (1) also lie on the surface $x^2 - z^2 - 1 = 0$? To answer this question, we use the fact that no point on the surface has a zero x-coordinate to conclude that $x \neq 0$ in the first equation in (1). This means that $2x = 2\lambda x$ only if

$$2 = 2\lambda, \qquad \text{or} \qquad \lambda = 1.$$

For $\lambda = 1$, the equation $2z = -2\lambda z$ becomes $2z = -2z$. If this equation is to be satisfied as well, z must be zero. Since $y = 0$ also (from the equation $2y = 0$), we conclude that the points we seek all have coordinates of the form

$$(x, 0, 0).$$

What points on the surface $x^2 - z^2 = 1$ have coordinates of this form? The points $(x, 0, 0)$ for which

$$x^2 - (0)^2 = 1, \qquad x^2 = 1, \qquad \text{or} \qquad x = \pm 1.$$

The points on the cylinder closest to the origin are the points $(\pm 1, 0, 0)$. ❏

The Method of Lagrange Multipliers

In Solution 2 of Example 2, we solved the problem by the **method of Lagrange multipliers.** In general terms, the method says that the extreme values of a function $f(x, y, z)$ whose variables are subject to a constraint $g(x, y, z) = 0$ are to be found on the surface $g = 0$ at the points where

$$\nabla f = \lambda \nabla g$$

for some scalar λ (called a **Lagrange multiplier**).

To explore the method further and see why it works, we first make the following observation, which we state as a theorem.

Theorem 9
The Orthogonal Gradient Theorem

Suppose that $f(x, y, z)$ is differentiable in a region whose interior contains a smooth curve

$$C: \quad \mathbf{r} = g(t)\,\mathbf{i} + h(t)\,\mathbf{j} + k(t)\,\mathbf{k}.$$

If P_0 is a point on C where f has a local maximum or minimum relative to its values on C, then ∇f is orthogonal to C at P_0.

Proof We show that ∇f is orthogonal to the curve's velocity vector at P_0. The values of f on C are given by the composite $f(g(t), h(t), k(t))$, whose derivative with respect to t is

$$\frac{df}{dt} = \frac{\partial f}{\partial x}\frac{dg}{dt} + \frac{\partial f}{\partial y}\frac{dh}{dt} + \frac{\partial f}{\partial z}\frac{dk}{dt} = \nabla f \cdot \mathbf{v}.$$

At any point P_0 where f has a local maximum or minimum relative to its values on the curve, $df/dt = 0$, so

$$\nabla f \cdot \mathbf{v} = 0. \qquad \square$$

By dropping the z-terms in Theorem 9, we obtain a similar result for functions of two variables.

Corollary of Theorem 9

At the points on a smooth curve $\mathbf{r} = g(t)\,\mathbf{i} + h(t)\,\mathbf{j}$ where a differentiable function $f(x, y)$ takes on its local maxima and minima relative to its values on the curve, $\nabla f \cdot \mathbf{v} = 0$.

Theorem 9 is the key to the method of Lagrange multipliers. Suppose that $f(x, y, z)$ and $g(x, y, z)$ are differentiable and that P_0 is a point on the surface $g(x, y, z) = 0$ where f has a local maximum or minimum value relative to its other values on the surface. Then f takes on a local maximum or minimum at P_0 relative to its values on every differentiable curve through P_0 on the surface $g(x, y, z) = 0$. Therefore, ∇f is orthogonal to the velocity vector of every such differentiable curve through P_0. But so is ∇g (because ∇g is orthogonal to the level surface $g = 0$, as we saw in Section 12.7). Therefore, at P_0, ∇f is some scalar multiple λ of ∇g.

> **The Method of Lagrange Multipliers**
>
> Suppose that $f(x, y, z)$ and $g(x, y, z)$ are differentiable. To find the local maximum and minimum values of f subject to the constraint $g(x, y, z) = 0$, find the values of x, y, z, and λ that simultaneously satisfy the equations
>
> $$\nabla f = \lambda \nabla g \qquad \text{and} \qquad g(x, y, z) = 0.$$
>
> For functions of two independent variables, the appropriate equations are
>
> $$\nabla f = \lambda \nabla g \qquad \text{and} \qquad g(x, y) = 0.$$

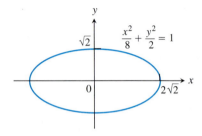

12.62 Example 3 shows how to find the largest and smallest values of the product xy on this ellipse.

EXAMPLE 3 Find the greatest and smallest values that the function

$$f(x, y) = xy$$

takes on the ellipse (Fig. 12.62)

$$\frac{x^2}{8} + \frac{y^2}{2} = 1.$$

Solution We want the extreme values of $f(x, y) = xy$ subject to the constraint

$$g(x, y) = \frac{x^2}{8} + \frac{y^2}{2} \quad 1 = 0.$$

To do so, we first find the values of x, y, and λ for which

$$\nabla f = \lambda \nabla g \qquad \text{and} \qquad g(x, y) = 0.$$

The gradient equation gives

$$y\mathbf{i} + x\mathbf{j} = \frac{\lambda}{4}x\mathbf{i} + \lambda y\mathbf{j},$$

from which we find

$$y = \frac{\lambda}{4}x, \qquad x = \lambda y, \qquad \text{and} \qquad y = \frac{\lambda}{4}(\lambda y) = \frac{\lambda^2}{4}y,$$

so that $y = 0$ or $\lambda = \pm 2$. We now consider these two cases.

Case 1: If $y = 0$, then $x = y = 0$. But $(0, 0)$ is not on the ellipse. Hence, $y \neq 0$.

Case 2: If $y \neq 0$, then $\lambda = \pm 2$ and $x = \pm 2y$. Substituting this in the equation $g(x, y) = 0$ gives

$$\frac{(\pm 2y)^2}{8} + \frac{y^2}{2} = 1, \qquad 4y^2 + 4y^2 = 8, \qquad \text{and} \qquad y = \pm 1.$$

The function $f(x, y) = xy$ therefore takes on its extreme values on the ellipse at the four points $(\pm 2, 1)$, $(\pm 2, -1)$. The extreme values are $xy = 2$ and $xy = -2$.

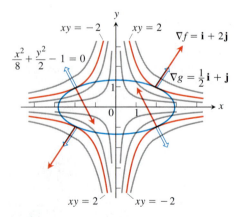

12.63 When subjected to the constraint $g(x, y) = x^2/8 + y^2/2 - 1 = 0$, the function $f(x, y) = xy$ takes on extreme values at the four points $(\pm 2, \pm 1)$. These are the points on the ellipse when ∇f (red) is a scalar multiple of ∇g (blue) (Example 3).

The Geometry of the Solution The level curves of the function $f(x, y) = xy$ are the hyperbolas $xy = c$ (Fig. 12.63). The farther the hyperbolas lie from the origin, the larger the absolute value of f. We want to find the extreme values of $f(x, y)$, given that the point (x, y) also lies on the ellipse $x^2 + 4y^2 = 8$. Which hyperbolas intersecting the ellipse lie farthest from the origin? The hyperbolas that just graze the ellipse, the ones that are tangent to it. At these points, any vector normal to the hyperbola is normal to the ellipse, so $\nabla f = y\mathbf{i} + x\mathbf{j}$ is a multiple ($\lambda = \pm 2$) of $\nabla g = (x/4)\mathbf{i} + y\mathbf{j}$. At the point $(2, 1)$, for example,

$$\nabla f = \mathbf{i} + 2\mathbf{j}, \qquad \nabla g = \frac{1}{2}\mathbf{i} + \mathbf{j}, \qquad \text{and} \qquad \nabla f = 2\nabla g.$$

At the point $(-2, 1)$,

$$\nabla f = \mathbf{i} - 2\mathbf{j}, \qquad \nabla g = -\frac{1}{2}\mathbf{i} + \mathbf{j}, \qquad \text{and} \qquad \nabla f = -2\nabla g. \qquad \square$$

EXAMPLE 4   Find the maximum and minimum values of the function $f(x, y) = 3x + 4y$ on the circle $x^2 + y^2 = 1$.

Solution   We model this as a Lagrange multiplier problem with

$$f(x, y) = 3x + 4y, \qquad g(x, y) = x^2 + y^2 - 1$$

and look for the values of x, y, and λ that satisfy the equations

$$\nabla f = \lambda \nabla g : \qquad 3\mathbf{i} + 4\mathbf{j} = 2x\lambda\mathbf{i} + 2y\lambda\mathbf{j},$$

$$g(x, y) = 0 : \qquad x^2 + y^2 - 1 = 0.$$

The gradient equation implies that $\lambda \neq 0$ and gives

$$x = \frac{3}{2\lambda}, \qquad y = \frac{2}{\lambda}.$$

These equations tell us, among other things, that x and y have the same sign. With these values for x and y, the equation $g(x, y) = 0$ gives

$$\left(\frac{3}{2\lambda}\right)^2 + \left(\frac{2}{\lambda}\right)^2 - 1 = 0,$$

so

$$\frac{9}{4\lambda^2} + \frac{4}{\lambda^2} = 1, \qquad 9 + 16 = 4\lambda^2, \qquad 4\lambda^2 = 25, \qquad \text{and} \qquad \lambda = \pm\frac{5}{2}.$$

Thus,

$$x = \frac{3}{2\lambda} = \pm\frac{3}{5}, \qquad y = \frac{2}{\lambda} = \pm\frac{4}{5},$$

and $f(x, y) = 3x + 4y$ has extreme values at $(x, y) = \pm(3/5, 4/5)$.

By calculating the value of $3x + 4y$ at the points $\pm(3/5, 4/5)$, we see that its maximum and minimum values on the circle $x^2 + y^2 = 1$ are

$$3\left(\frac{3}{5}\right) + 4\left(\frac{4}{5}\right) = \frac{25}{5} = 5 \qquad \text{and} \qquad 3\left(-\frac{3}{5}\right) + 4\left(-\frac{4}{5}\right) = -\frac{25}{5} = -5.$$

The Geometry of the Solution   (Fig. 12.64) The level curves of $f(x, y) = 3x + 4y$ are the lines $3x + 4y = c$. The farther the lines lie from the origin, the larger the absolute value of f. We want to find the extreme values of $f(x, y)$ given that the point (x, y) also lies on the circle $x^2 + y^2 = 1$. Which lines intersecting the circle lie farthest from the origin? The lines tangent to the circle. At the points of tangency, any vector normal to the line is normal to the circle, so the gradient $\nabla f = 3\mathbf{i} + 4\mathbf{j}$ is a multiple ($\lambda = \pm 5/2$) of the gradient $\nabla g = 2x\mathbf{i} + 2y\mathbf{j}$. At the point $(3/5, 4/5)$, for example,

$$\nabla f = 3\mathbf{i} + 4\mathbf{j}, \qquad \nabla g = \frac{6}{5}\mathbf{i} + \frac{8}{5}\mathbf{j}, \qquad \text{and} \qquad \nabla f = \frac{5}{2}\nabla g. \qquad \square$$

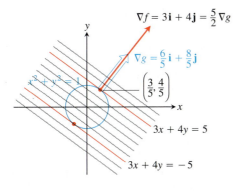

12.64 The function $f(x, y) = 3x + 4y$ takes on its largest value on the unit circle $g(x, y) = x^2 + y^2 - 1 = 0$ at the point $(3/5, 4/5)$ and its smallest value at the point $(-3/5, -4/5)$ (Example 4). At each of these points, ∇f is a scalar multiple of ∇g. The figure shows the gradients at the first point but not the second.

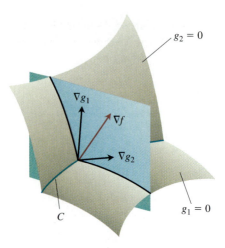

12.65 The vectors ∇g_1 and ∇g_2 lie in a plane perpendicular to the curve C because ∇g_1 is normal to the surface $g_1 = 0$ and ∇g_2 is normal to the surface $g_2 = 0$.

Lagrange Multipliers with Two Constraints

Many problems require us to find the extreme values of a differentiable function $f(x, y, z)$ whose variables are subject to two constraints. If the constraints are

$$g_1(x, y, z) = 0 \quad \text{and} \quad g_2(x, y, z) = 0$$

and g_1 and g_2 are differentiable, with ∇g_1 not parallel to ∇g_2, we find the constrained local maxima and minima of f by introducing two Lagrange multipliers λ and μ (mu, pronounced "mew"). That is, we locate the points $P(x, y, z)$ where f takes on its constrained extreme values by finding the values of $x, y, z, \lambda,$ and μ that simultaneously satisfy the equations

$$\nabla f = \lambda \nabla g_1 + \mu \nabla g_2, \quad g_1(x, y, z) = 0, \quad g_2(x, y, z) = 0. \tag{2}$$

The equations in (2) have a nice geometric interpretation. The surfaces $g_1 = 0$ and $g_2 = 0$ (usually) intersect in a smooth curve, say C (Fig. 12.65), and along this curve we seek the points where f has local maximum and minimum values relative to its other values on the curve. These are the points where ∇f is normal to C, as we saw in Theorem 9. But ∇g_1 and ∇g_2 are also normal to C at these points because C lies in the surfaces $g_1 = 0$ and $g_2 = 0$. Therefore ∇f lies in the plane determined by ∇g_1 and ∇g_2, which means that $\nabla f = \lambda \nabla g_1 + \mu \nabla g_2$ for some λ and μ. Since the points we seek also lie in both surfaces, their coordinates must satisfy the equations $g_1(x, y, z) = 0$ and $g_2(x, y, z) = 0$, which are the remaining requirements in Eqs. (2).

EXAMPLE 5   The plane $x + y + z = 1$ cuts the cylinder $x^2 + y^2 = 1$ in an ellipse (Fig. 12.66). Find the points on the ellipse that lie closest to and farthest from the origin.

Solution   We find the extreme values of

$$f(x, y, z) = x^2 + y^2 + z^2$$

(the square of the distance from (x, y, z) to the origin) subject to the constraints

$$g_1(x, y, z) = x^2 + y^2 - 1 = 0 \tag{3}$$

$$g_2(x, y, z) = x + y + z - 1 = 0. \tag{4}$$

The gradient equation in (2) then gives

$$\nabla f = \lambda \nabla g_1 + \mu \nabla g_2 \qquad \text{Eq. (2)}$$

$$2x\,\mathbf{i} + 2y\,\mathbf{j} + 2z\,\mathbf{k} = \lambda(2x\,\mathbf{i} + 2y\,\mathbf{j}) + \mu\,(\mathbf{i} + \mathbf{j} + \mathbf{k})$$

$$2x\,\mathbf{i} + 2y\,\mathbf{j} + 2z\,\mathbf{k} = (2\lambda x + \mu)\,\mathbf{i} + (2\lambda y + \mu)\,\mathbf{j} + \mu\,\mathbf{k}$$

or

$$2x = 2\lambda x + \mu, \quad 2y = 2\lambda y + \mu, \quad 2z = \mu. \tag{5}$$

The scalar equations in (5) yield

$$2x = 2\lambda x + 2z \quad \Rightarrow \quad (1 - \lambda)x = z,$$
$$2y = 2\lambda y + 2z \quad \Rightarrow \quad (1 - \lambda)y = z. \tag{6}$$

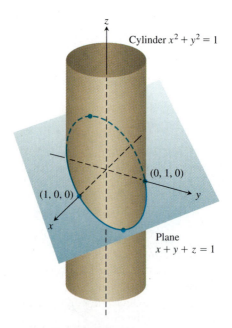

12.66 On the ellipse where the plane and cylinder meet, what are the points closest to and farthest from the origin (Example 5)?

Equations (6) are satisfied simultaneously if either $\lambda = 1$ and $z = 0$ or $\lambda \neq 1$ and $x = y = z/(1 - \lambda)$.

If $z = 0$, then solving Eqs. (3) and (4) simultaneously to find the corresponding points on the ellipse gives the two points $(1, 0, 0)$ and $(0, 1, 0)$. This makes sense when you look at Fig. 12.66.

If $x = y$, then Eqs. (3) and (4) give

$$x^2 + x^2 - 1 = 0 \qquad\qquad x + x + z - 1 = 0$$

$$2x^2 = 1 \qquad\qquad z = 1 - 2x$$

$$x = \pm\frac{\sqrt{2}}{2} \qquad\qquad z = 1 \mp \sqrt{2}.$$

The corresponding points on the ellipse are

$$P_1 = \left(\frac{\sqrt{2}}{2}, \frac{\sqrt{2}}{2}, 1 - \sqrt{2}\right) \qquad \text{and} \qquad P_2 = \left(-\frac{\sqrt{2}}{2}, -\frac{\sqrt{2}}{2}, 1 + \sqrt{2}\right).$$

But here we need to be careful. While P_1 and P_2 both give local maxima of f on the ellipse, P_2 is farther from the origin than P_1.

The points on the ellipse closest to the origin are $(1, 0, 0)$ and $(0, 1, 0)$. The point on the ellipse farthest from the origin is P_2. ❑

Exercises 12.9

Two Independent Variables with One Constraint

1. Find the points on the ellipse $x^2 + 2y^2 = 1$ where $f(x, y) = xy$ has its extreme values.

2. Find the extreme values of $f(x, y) = xy$ subject to the constraint $g(x, y) = x^2 + y^2 - 10 = 0$.

3. Find the maximum value of $f(x, y) = 49 - x^2 - y^2$ on the line $x + 3y = 10$ (Fig. 12.58).

4. Find the local extreme values of $f(x, y) = x^2 y$ on the line $x + y = 3$.

5. Find the points on the curve $xy^2 = 54$ nearest the origin.

6. Find the points on the curve $x^2 y = 2$ nearest the origin.

7. Use the method of Lagrange multipliers to find

   a) the minimum value of $x + y$, subject to the constraints $xy = 16, x > 0, y > 0$;

   b) the maximum value of xy, subject to the constraint $x + y = 16$.

   Comment on the geometry of each solution.

8. Find the points on the curve $x^2 + xy + y^2 = 1$ in the xy-plane that are nearest to and farthest from the origin.

9. Find the dimensions of the closed right circular cylindrical can of smallest surface area whose volume is 16π cm^3.

10. Find the radius and height of the open right circular cylinder of largest surface area that can be inscribed in a sphere of radius a. What *is* the largest surface area?

11. Use the method of Lagrange multipliers to find the dimensions of the rectangle of greatest area that can be inscribed in the ellipse $x^2/16 + y^2/9 = 1$ with sides parallel to the coordinate axes.

12. Find the dimensions of the rectangle of largest perimeter that can be inscribed in the ellipse $x^2/a^2 + y^2/b^2 = 1$ with sides parallel to the coordinate axes. What *is* the largest perimeter?

13. Find the maximum and minimum values of $x^2 + y^2$ subject to the constraint $x^2 - 2x + y^2 - 4y = 0$.

14. Find the maximum and minimum values of $3x - y + 6$ subject to the constraint $x^2 + y^2 = 4$.

15. The temperature at a point (x, y) on a metal plate is $T(x, y) = 4x^2 - 4xy + y^2$. An ant on the plate walks around the circle of radius 5 centered at the origin. What are the highest and lowest temperatures encountered by the ant?

16. Your firm has been asked to design a storage tank for liquid petroleum gas. The customer's specifications call for a cylindrical tank with hemispherical ends, and the tank is to hold 8000 m^3 of gas. The customer also wants to use the smallest amount of material possible in building the tank. What radius and height do you recommend for the cylindrical portion of the tank?

Three Independent Variables with One Constraint

17. Find the point on the plane $x + 2y + 3z = 13$ closest to the point $(1, 1, 1)$.

18. Find the point on the sphere $x^2 + y^2 + z^2 = 4$ which is farthest from the point $(1, -1, 1)$.

19. Find the minimum distance from the surface $x^2 + y^2 - z^2 = 1$ to the origin.

20. Find the point on the surface $z = xy + 1$ nearest the origin.

21. Find the points on the surface $z^2 = xy + 4$ closest to the origin.

22. Find the point(s) on the surface $xyz = 1$ closest to the origin.

23. Find the maximum and minimum values of

$$f(x, y, z) = x - 2y + 5z$$

on the sphere $x^2 + y^2 + z^2 = 30$.

24. Find the points on the sphere $x^2 + y^2 + z^2 = 25$ where $f(x, y, z) = x + 2y + 3z$ has its maximum and minimum values.

25. Find three real numbers whose sum is 9 and the sum of whose squares is as small as possible.

26. Find the largest product the positive numbers $x, y,$ and z can have if $x + y + z^2 = 16$.

27. Find the dimensions of the closed rectangular box with maximum volume that can be inscribed in the unit sphere.

28. Find the volume of the largest closed rectangular box in the first octant having three faces in the coordinate planes and a vertex on the plane $x/a + y/b + z/c = 1$, where $a > 0, b > 0,$ and $c > 0$.

29. A space probe in the shape of the ellipsoid

$$4x^2 + y^2 + 4z^2 = 16$$

enters the earth's atmosphere and its surface begins to heat. After one hour, the temperature at the point (x, y, z) on the probe's surface is

$$T(x, y, z) = 8x^2 + 4yz - 16z + 600.$$

Find the hottest point on the probe's surface.

30. Suppose that the Celsius temperature at the point (x, y, z) on the sphere $x^2 + y^2 + z^2 = 1$ is $T = 400xyz^2$. Locate the highest and lowest temperatures on the sphere.

31. *An example from economics.* In economics, the usefulness or *utility* of amounts x and y of two capital goods G_1 and G_2 is sometimes measured by a function $U(x, y)$. For example, G_1 and G_2 might be two chemicals a pharmaceutical company needs to have on hand and $U(x, y)$ the gain from manufacturing a product whose synthesis requires different amounts of the chemicals depending on the process used. If G_1 costs a dollars per kilogram, G_2 costs b dollars per kilogram, and the total amount allocated for the purchase of G_1 and G_2 together is c dollars, then the company's managers want to maximize $U(x, y)$ given that $ax + by = c$. Thus, they need to solve a typical Lagrange multiplier problem.

Suppose that

$$U(x, y) = xy + 2x$$

and that the equation $ax + by = c$ simplifies to

$$2x + y = 30.$$

Find the maximum value of U and the corresponding values of x and y subject to this latter constraint.

32. You are in charge of erecting a radio telescope on a newly discovered planet. To minimize interference, you want to place it where the magnetic field of the planet is weakest. The planet is spherical, with a radius of 6 units. Based on a coordinate system whose origin is at the center of the planet, the strength of the magnetic field is given by $M(x, y, z) = 6x - y^2 + xz + 60$. Where should you locate the radio telescope?

Lagrange Multipliers with Two Constraints

33. Maximize the function $f(x, y, z) = x^2 + 2y - z^2$ subject to the constraints $2x - y = 0$ and $y + z = 0$.

34. Minimize the function $f(x, y, z) = x^2 + y^2 + z^2$ subject to the constraints $x + 2y + 3z = 6$ and $x + 3y + 9z = 9$.

35. Find the point closest to the origin on the line of intersection of the planes $y + 2z = 12$ and $x + y = 6$.

36. Find the maximum value that $f(x, y, z) = x^2 + 2y - z^2$ can have on the line of intersection of the planes $2x - y = 0$ and $y + z = 0$.

37. Find the extreme values of $f(x, y, z) = x^2yz + 1$ on the intersection of the plane $z = 1$ with the sphere $x^2 + y^2 + z^2 = 10$.

38. a) Find the maximum value of $w = xyz$ on the line of intersection of the two planes $x + y + z = 40$ and $x + y - z = 0$.

b) Give a geometric argument to support your claim that you have found a maximum, and not a minimum, value of w.

39. Find the extreme values of the function $f(x, y, z) = xy + z^2$ on the circle in which the plane $y - x = 0$ intersects the sphere $x^2 + y^2 + z^2 = 4$.

40. Find the point closest to the origin on the curve of intersection of the plane $2y + 4z = 5$ and the cone $z^2 = 4x^2 + 4y^2$.

Theory and Examples

41. *The condition $\nabla f = \lambda \nabla g$ is not sufficient.* While $\nabla f = \lambda \nabla g$ is a necessary condition for the occurrence of an extreme value of $f(x, y)$ subject to the condition $g(x, y) = 0$, it does not in itself guarantee that one exists. As a case in point, try using the method of Lagrange multipliers to find a maximum value of $f(x, y) = x + y$ subject to the constraint that $xy = 16$. The method will identify the two points $(4, 4)$ and $(-4, -4)$ as candidates for the location of extreme values. Yet the sum $(x + y)$ has no maximum value on the hyperbola $xy = 16$. The farther you go from the origin on this hyperbola in the first quadrant, the larger the sum $f(x, y) = x + y$ becomes.

42. *A least squares plane.* The plane $z = Ax + By + C$ is to be "fitted" to the following points (x_k, y_k, z_k):

$$(0, 0, 0), \quad (0, 1, 1), \quad (1, 1, 1), \quad (1, 0, -1).$$

Find the values of $A, B,$ and C that minimize the sum

$$\sum_{k=1}^{4}(Ax_k + By_k + C - z_k)^2,$$

the sum of the squares of the deviations.

43. a) Show that the maximum value of $a^2b^2c^2$ on a sphere of radius r centered at the origin of a Cartesian abc-coordinate system is $(r^2/3)^3$.

b) Using part (a), show that for nonnegative numbers a, b, and c,

$$(abc)^{1/3} \le \frac{a+b+c}{3}.$$

That is, the *geometric mean* of three numbers is less than or equal to the *arithmetic mean*.

44. Let $a_1, a_2, \ldots, a_n$ be n positive numbers. Find the maximum of $\sum_{i=1}^{n} a_i x_i$ subject to the constraint $\sum_{i=1}^{n} x_i{}^2 = 1$.

✳ CAS Explorations and Projects

In Exercises 45–50, use a CAS to perform the following steps implementing the method of Lagrange multipliers for finding constrained extrema:

a) Form the function $h = f - \lambda_1 g_1 - \lambda_2 g_2$, where f is the function to optimize subject to the constraints $g_1 = 0$ and $g_2 = 0$.

b) Determine all the first partial derivatives of h, including the partials with respect to λ_1 and λ_2, and set them equal to 0.

c) Solve the system of equations found in (b) for all the unknowns, including λ_1 and λ_2.

d) Evaluate f at each of the solution points found in (c) and select the extreme value subject to the constraints asked for in the exercise.

45. Minimize $f(x, y, z) = xy + yz$ subject to the constraints $x^2 + y^2 - 2 = 0$ and $x^2 + z^2 - 2 = 0$.

46. Minimize $f(x, y, z) = xyz$ subject to the constraints $x^2 + y^2 - 1 = 0$ and $x - z = 0$.

47. Maximize $f(x, y, z) = x^2 + y^2 + z^2$ subject to the constraints $2y + 4z - 5 = 0$ and $4x^2 + 4y^2 - z^2 = 0$.

48. Minimize $f(x, y, z) = x^2 + y^2 + z^2$ subject to the constraints $x^2 - xy + y^2 - z^2 - 1 = 0$ and $x^2 + y^2 - 1 = 0$.

49. Minimize $f(x, y, z, w) = x^2 + y^2 + z^2 + w^2$ subject to the constraints $2x - y + z - w - 1 = 0$ and $x + y - z + w - 1 = 0$.

50. Determine the distance from the line $y = x + 1$ to the parabola $y^2 = x$. (*Hint:* Let (x, y) be a point on the line and (w, z) a point on the parabola. You want to minimize $(x - w)^2 + (y - z)^2$.)

Taylor's Formula

This section uses Taylor's formula (Section 8.10) to derive the second derivative test for local extreme values (Section 12.8) and the error formula for linearizations of functions of two independent variables (Section 12.4, Eq. 5). The use of Taylor's formula in these derivations leads to an extension of the formula that provides polynomial approximations of all orders for functions of two independent variables.

The Derivation of the Second Derivative Test

Let $f(x, y)$ have continuous partial derivatives in an open region R containing a point $P(a, b)$ where $f_x = f_y = 0$ (Fig. 12.67). Let h and k be increments small enough to put the point $S(a + h, b + k)$ and the line segment joining it to P inside R. We parametrize the segment PS as

$$x = a + th, \qquad y = b + tk, \qquad 0 \le t \le 1.$$

If $F(t) = f(a + th, b + tk)$, the Chain Rule gives

$$F'(t) = f_x \frac{dx}{dt} + f_y \frac{dy}{dt} = h f_x + k f_y.$$

Since f_x and f_y are differentiable (they have continuous partial derivatives), F' is a differentiable function of t and

$$F'' = \frac{\partial F'}{\partial x}\frac{dx}{dt} + \frac{\partial F'}{\partial y}\frac{dy}{dt} = \frac{\partial}{\partial x}\left(h f_x + k f_y\right)\cdot h + \frac{\partial}{\partial y}\left(h f_x + k f_y\right)\cdot k$$

$$= h^2 f_{xx} + 2hk f_{xy} + k^2 f_{yy}. \qquad f_{xy} = f_{yx}$$

Since F and F' are continuous on $[0, 1]$ and F' is differentiable on $(0, 1)$, we can

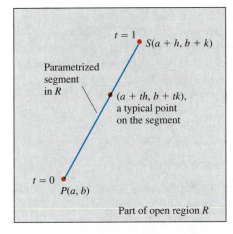

12.67 We begin the derivation of the second derivative test at $P(a, b)$ by parametrizing a typical line segment from P to a point S nearby.

In the figure:
$t = 1$, $S(a + h, b + k)$
Parametrized segment in R
$(a + th, b + tk)$, a typical point on the segment
$t = 0$, $P(a, b)$
Part of open region R

apply Taylor's formula with $n = 2$ and $a = 0$ to obtain

$$F(1) = F(0) + F'(0)(1-0) + F''(c)\frac{(1-0)^2}{2}$$

$$F(1) = F(0) + F'(0) + \frac{1}{2}F''(c) \tag{1}$$

for some c between 0 and 1. Writing Eq. (1) in terms of f gives

$$f(a+h, b+k) = f(a, b) + hf_x(a, b) + kf_y(a, b)$$

$$+ \frac{1}{2}\left(h^2 f_{xx} + 2hk f_{xy} + k^2 f_{yy}\right)\Big|_{(a+ch, b+ck)}. \tag{2}$$

Since $f_x(a, b) = f_y(a, b) = 0$, this reduces to

$$f(a+h, b+k) - f(a, b) = \frac{1}{2}\left(h^2 f_{xx} + 2hk f_{xy} + k^2 f_{yy}\right)\Big|_{(a+ch, b+ck)}. \tag{3}$$

The presence of an extremum of f at (a, b) is determined by the sign of $f(a+h, b+k) - f(a, b)$. By Eq. (3), this is the same as the sign of

$$Q(c) = (h^2 f_{xx} + 2hk f_{xy} + k^2 f_{yy})|_{(a+ch, b+ck)}.$$

Now, if $Q(0) \neq 0$, the sign of $Q(c)$ will be the same as the sign of $Q(0)$ for sufficiently small values of h and k. We can predict the sign of

$$Q(0) = h^2 f_{xx}(a, b) + 2hk f_{xy}(a, b) + k^2 f_{yy}(a, b) \tag{4}$$

from the signs of f_{xx} and $f_{xx} f_{yy} - f_{xy}^2$ at (a, b). Multiply both sides of Eq. (3) by f_{xx} and rearrange the right-hand side to get

$$f_{xx} Q(0) = (hf_{xx} + kf_{xy})^2 + (f_{xx} f_{yy} - f_{xy}^2)k^2. \tag{5}$$

From Eq. (5) we see that

1. If $f_{xx} < 0$ and $f_{xx} f_{yy} - f_{xy}^2 > 0$ at (a, b), then $Q(0) < 0$ for all sufficiently small nonzero values of h and k, and f has a *local maximum* value at (a, b).
2. If $f_{xx} > 0$ and $f_{xx} f_{yy} - f_{xy}^2 > 0$ at (a, b), then $Q(0) > 0$ for all sufficiently small nonzero values of h and k, and f has a *local minimum* value at (a, b).
3. If $f_{xx} f_{yy} - f_{xy}^2 < 0$ at (a, b), there are combinations of arbitrarily small nonzero values of h and k for which $Q(0) > 0$, and other values for which $Q(0) < 0$. Arbitrarily close to the point $P_0(a, b, f(a, b))$ on the surface $z = f(x, y)$ there are points above P_0 and points below P_0, so f has a *saddle point* at (a, b).
4. If $f_{xx} f_{yy} - f_{xy}^2 = 0$, another test is needed. The possibility that $Q(0)$ equals zero prevents us from drawing conclusions about the sign of $Q(c)$.

The Error Formula for Linear Approximations

We want to show that the difference $E(x, y)$ between the values of a function $f(x, y)$ and its linearization $L(x, y)$ at (x_0, y_0) satisfies the inequality

$$|E(x, y)| \leq \frac{1}{2}B(|x - x_0| + |y - y_0|)^2.$$

The function f is assumed to have continuous second partial derivatives throughout an open set containing a closed rectangular region R centered at (x_0, y_0).

The number B is the largest value that any of $|f_{xx}|$, $|f_{yy}|$, and $|f_{xy}|$ take on R.

The inequality we want comes from Eq. (2). We substitute x_0 and y_0 for a and b, and $x - x_0$ and $y - y_0$ for h and k, respectively, and rearrange the result as

$$f(x, y) = \underbrace{f(x_0, y_0) + f_x(x_0, y_0)(x - x_0) + f_y(x_0, y_0)(y - y_0)}_{\text{linearization } L(x, y)}$$

$$+ \underbrace{\frac{1}{2}\left((x - x_0)^2 f_{xx} + 2(x - x_0)(y - y_0)f_{xy} + (y - y_0)^2 f_{yy}\right)\Big|_{(x_0+c(x-x_0), y_0+c(y-y_0))}}_{\text{error } E(x, y)}.$$

This remarkable equation reveals that

$$|E| \le \frac{1}{2}\left(|x - x_0|^2|f_{xx}| + 2|x - x_0||y - y_0||f_{xy}| + |y - y_0|^2|f_{yy}|\right).$$

Hence, if B is an upper bound for the values of $|f_{xx}|$, $|f_{xy}|$, and $|f_{yy}|$ on R,

$$|E| \le \frac{1}{2}\left(|x - x_0|^2 B + 2|x - x_0||y - y_0|B + |y - y_0|^2 B\right)$$

$$\le \frac{1}{2}B\left(|x - x_0| + |y - y_0|\right)^2.$$

Taylor's Formula for Functions of Two Variables

The formulas derived earlier for F' and F'' can be obtained by applying to $f(x, y)$ the operators

$$\left(h\frac{\partial}{\partial x} + k\frac{\partial}{\partial y}\right) \quad \text{and} \quad \left(h\frac{\partial}{\partial x} + k\frac{\partial}{\partial y}\right)^2 = h^2\frac{\partial^2}{\partial x^2} + 2hk\frac{\partial^2}{\partial x\,\partial y} + k^2\frac{\partial^2}{\partial y^2}.$$

These are the first two instances of a more general formula,

$$F^{(n)}(t) = \frac{d^n}{dt^n}F(t) = \left(h\frac{\partial}{\partial x} + k\frac{\partial}{\partial y}\right)^n f(x, y), \tag{6}$$

which says that applying d^n/dt^n to $F(t)$ gives the same result as applying the operator

$$\left(h\frac{\partial}{\partial x} + k\frac{\partial}{\partial y}\right)^n$$

to $f(x, y)$ after expanding it by the binomial theorem.

If partial derivatives of f through order $n + 1$ are continuous throughout a rectangular region centered at (a, b), we may extend the Taylor formula for $F(t)$ to

$$F(t) = F(0) + F'(0)t + \frac{F''(0)}{2!}t^2 + \cdots + \frac{F^{(n)}(0)}{n!}t^n + \text{remainder},$$

and take $t = 1$ to obtain

$$F(1) = F(0) + F'(0) + \frac{F''(0)}{2!} + \cdots + \frac{F^{(n)}(0)}{n!} + \text{remainder}.$$

When we replace the first n derivatives on the right of this last series by their equivalent expressions from Eq. (6) evaluated at $t = 0$ and add the appropriate remainder term, we arrive at the following formula.

Taylor's Formula for $f(x, y)$ at the Point (a, b)

Suppose $f(x, y)$ and its partial derivatives through order $n + 1$ are continuous throughout an open rectangular region R centered at a point (a, b). Then, throughout R,

$$f(a + h, b + k) = f(a, b) + (hf_x + kf_y)|_{(a,b)} + \frac{1}{2!}(h^2 f_{xx} + 2hk f_{xy} + k^2 f_{yy})|_{(a,b)}$$

$$+ \frac{1}{3!}(h^3 f_{xxx} + 3h^2 k f_{xxy} + 3hk^2 f_{xyy} + k^3 f_{yyy})|_{(a,b)} + \cdots + \frac{1}{n!}\left(h\frac{\partial}{\partial x} + k\frac{\partial}{\partial y}\right)^n f\bigg|_{(a,b)}$$

$$+ \frac{1}{(n+1)!}\left(h\frac{\partial}{\partial x} + k\frac{\partial}{\partial y}\right)^{n+1} f\bigg|_{(a+ch, b+ck)}. \tag{7}$$

The first n derivative terms are evaluated at (a, b). The last term is evaluated at some point $(a + ch, b + ck)$ on the line segment joining (a, b) and $(a + h, b + k)$.

If $(a, b) = (0, 0)$ and we treat h and k as independent variables (denoting them now by x and y), then Eq. (7) assumes the following simpler form.

Taylor's Formula for $f(x, y)$ at the Origin

$$f(x, y) = f(0, 0) + xf_x + yf_y + \frac{1}{2!}(x^2 f_{xx} + 2xy f_{xy} + y^2 f_{yy})$$

$$+ \frac{1}{3!}(x^3 f_{xxx} + 3x^2 y f_{xxy} + 3xy^2 f_{xyy} + y^3 f_{yyy}) + \cdots + \frac{1}{n!}\left(x\frac{\partial}{\partial x} + y\frac{\partial}{\partial y}\right)^n f$$

$$+ \frac{1}{(n+1)!}\left(x\frac{\partial}{\partial x} + y\frac{\partial}{\partial y}\right)^{n+1} f\bigg|_{(cx, cy)} \tag{8}$$

The first n derivative terms are evaluated at $(0, 0)$. The last term is evaluated at a point on the line segment joining the origin and (x, y).

Taylor's formula provides polynomial approximations of two-variable functions. The first n derivative terms give the polynomial; the last term gives the approximation error. The first three terms of Taylor's formula give the function's linearization. To improve on the linearization, we add higher power terms.

EXAMPLE 1   Find a quadratic $f(x, y) = \sin x \sin y$ near the origin. How accurate is the approximation if $|x| \leq 0.1$ and $|y| \leq 0.1$?

Solution   We take $n = 2$ in Eq. (8):

$$f(x, y) = f(0, 0) + (xf_x + yf_y) + \frac{1}{2}(x^2 f_{xx} + 2xy f_{xy} + y^2 f_{yy})$$

$$+ \frac{1}{6}(x^3 f_{xxx} + 3x^2 y f_{xxy} + 3xy^2 f_{xyy} + y^3 f_{yyy})_{(cx, cy)}$$

with

$$f(0, 0) = \sin x \sin y|_{(0,0)} = 0, \qquad f_{xx}(0, 0) = -\sin x \sin y|_{(0,0)} = 0,$$
$$f_x(0, 0) = \cos x \sin y|_{(0,0)} = 0, \qquad f_{xy}(0, 0) = \cos x \cos y|_{(0,0)} = 1,$$
$$f_y(0, 0) = \sin x \cos y|_{(0,0)} = 0, \qquad f_{yy}(0, 0) = -\sin x \sin y|_{(0,0)} = 0,$$

we have

$$\sin x \sin y \approx 0 + 0 + 0 + \frac{1}{2}(x^2(0) + 2xy(1) + y^2(0)),$$

$$\sin x \sin y \approx xy.$$

The error in the approximation is

$$E(x, y) = \frac{1}{6}(x^3 f_{xxx} + 3x^2 y f_{xxy} + 3xy^2 f_{xyy} + y^3 f_{yyy})|_{(cx, cy)}.$$

The third derivatives never exceed 1 in absolute value because they are products of sines and cosines. Also, $|x| \leq 0.1$ and $|y| \leq 0.1$. Hence

$$|E(x, y)| \leq \frac{1}{6}((0.1)^3 + 3(0.1)^3 + 3(0.1)^3 + (0.1)^3) \leq \frac{8}{6}(0.1)^3 \leq 0.00134$$

(rounded up). The error will not exceed 0.00134 if $|x| \leq 0.1$ and $|y| \leq 0.1$. ❑

Exercises 12.10

Finding Quadratic and Cubic Approximations

In Exercises 1–10, use Taylor's formula for $f(x, y)$ at the origin to find quadratic and cubic approximations of f near the origin.

1. $f(x, y) = x e^y$
2. $f(x, y) = e^x \cos y$
3. $f(x, y) = y \sin x$
4. $f(x, y) = \sin x \cos y$
5. $f(x, y) = e^x \ln(1 + y)$
6. $f(x, y) = \ln(2x + y + 1)$
7. $f(x, y) = \sin(x^2 + y^2)$
8. $f(x, y) = \cos(x^2 + y^2)$

9. $f(x, y) = \dfrac{1}{1 - x - y}$
10. $f(x, y) = \dfrac{1}{1 - x - y + xy}$

11. Use Taylor's formula to find a quadratic approximation of $f(x, y) = \cos x \cos y$ at the origin. Estimate the error in the approximation if $|x| \leq 0.1$ and $|y| \leq 0.1$.

12. Use Taylor's formula to find a quadratic approximation of $e^x \sin y$ at the origin. Estimate the error in the approximation if $|x| \leq 0.1$ and $|y| \leq 0.1$.

CHAPTER   **12**   QUESTIONS TO GUIDE YOUR REVIEW

---

1. What is a real-valued function of two independent variables? three independent variables? Give examples.

2. What does it mean for sets in the plane or in space to be open? closed? Give examples. Give examples of sets that are neither open nor closed.

3. How can you display the values of a function $f(x, y)$ of two independent variables graphically? How do you do the same for a function $f(x, y, z)$ of three independent variables?

4. What does it mean for a function $f(x, y)$ to have limit L as $(x, y) \to (x_0, y_0)$? What are the basic properties of limits of functions of two independent variables?

5. When is a function of two (three) independent variables continuous at a point in its domain? Give examples of functions that are continuous at some points but not others.

6. What can be said about algebraic combinations and composites of continuous functions?

7. Explain the two-path test for nonexistence of limits.

8. How are the partial derivatives $\partial f/\partial x$ and $\partial f/\partial y$ of a function $f(x, y)$ defined? How are they interpreted and calculated?

9. How does the relation between first partial derivatives and continuity of functions of two independent variables differ from the relation between first derivatives and continuity for real-valued functions of a single independent variable? Give an example.

10. What is Euler's theorem for mixed second order partial derivatives? How can it help in calculating partial derivatives of second and higher orders? Give examples.

11. What does it mean for a function $f(x, y)$ to be differentiable? What does the Increment Theorem say about differentiability?

12. How can you sometimes decide from examining f_x and f_y that a function $f(x, y)$ is differentiable? What is the relation between the differentiability of f and the continuity of f at a point?

13. How do you linearize a function $f(x, y)$ of two independent variables at a point (x_0, y_0)? Why might you want to do this? How do you linearize a function of three independent variables?

14. What can you say about the accuracy of linear approximations of functions of two (three) independent variables?

15. If (x, y) moves from (x_0, y_0) to a point $(x_0 + dx, y_0 + dy)$ nearby, how can you estimate the resulting change in the value of a differentiable function $f(x, y)$? Give an example.

16. What is the Chain Rule? What form does it take for functions of two independent variables? three independent variables? functions defined on surfaces? How do you diagram these different forms? Give examples. What pattern enables one to remember all the different forms?

17. What is the derivative of a function $f(x, y)$ at a point P_0 in the direction of a unit vector $\mathbf{u}$? What rate does it describe? What geometric interpretation does it have? Give examples.

18. What is the gradient vector of a function $f(x, y)$? How is it related to the function's directional derivatives? State the analogous results for functions of three independent variables.

19. How do you find the tangent line at a point on a level curve of a differentiable function $f(x, y)$? How do you find the tangent plane and normal line at a point on a level surface of a differentiable function $f(x, y, z)$? Give examples.

20. How can you use directional derivatives to estimate change?

21. How do you define local maxima, local minima, and saddle points for a differentiable function $f(x, y)$? Give examples.

22. What derivative tests are available for determining the local extreme values of a function $f(x, y)$? How do they enable you to narrow your search for these values? Give examples.

23. How do you find the extrema of a continuous function $f(x, y)$ on a closed bounded region of the xy-plane? Give an example.

24. Describe the method of Lagrange multipliers and give examples.

25. How does Taylor's formula for a function $f(x, y)$ generate polynomial approximations and error estimates?

CHAPTER 12 PRACTICE EXERCISES

Domain, Range, and Level Curves

In Exercises 1–4, find the domain and range of the given function and identify its level curves. Sketch a typical level curve.

1. $f(x, y) = 9x^2 + y^2$ **2.** $f(x, y) = e^{x+y}$

3. $g(x, y) = 1/xy$ **4.** $g(x, y) = \sqrt{x^2 - y}$

In Exercise 5–8, find the domain and range of the given function and identify its level surfaces. Sketch a typical level surface.

5. $f(x, y, z) = x^2 + y^2 - z$

6. $g(x, y, z) = x^2 + 4y^2 + 9z^2$

7. $h(x, y, z) = \dfrac{1}{x^2 + y^2 + z^2}$

8. $k(x, y, z) = \dfrac{1}{x^2 + y^2 + z^2 + 1}$

Evaluating Limits

Find the limits in Exercises 9–14.

9. $\displaystyle\lim_{(x,y)\to(\pi,\ln 2)} e^y \cos x$ **10.** $\displaystyle\lim_{(x,y)\to(0,0)} \dfrac{2+y}{x+\cos y}$

11. $\displaystyle\lim_{\substack{(x,y)\to(1,1) \\ x\neq y}} \dfrac{x-y}{x^2-y^2}$ **12.** $\displaystyle\lim_{(x,y)\to(1,1)} \dfrac{x^3 y^3 - 1}{xy - 1}$

13. $\displaystyle\lim_{P\to(1,-1,e)} \ln|x+y+z|$

14. $\displaystyle\lim_{P\to(1,-1,-1)} \tan^{-1}(x+y+z)$

By considering different paths of approach, show that the limits in Exercises 15 and 16 do not exist.

15. $\displaystyle\lim_{\substack{(x,y)\to(0,0) \\ y\neq x^2}} \dfrac{y}{x^2 - y}$ **16.** $\displaystyle\lim_{\substack{(x,y)\to(0,0) \\ xy\neq 0}} \dfrac{x^2 + y^2}{xy}$

17. a) Let $f(x, y) = (x^2 - y^2)/(x^2 + y^2)$ for $(x, y) \neq (0, 0)$. Is it possible to define $f(0, 0)$ in a way that makes f continuous at the origin? Why?

b) Let

$$f(x, y) = \begin{cases} \dfrac{\sin(x-y)}{|x| + |y|}, & |x| + |y| \neq 0, \\ 0, & (x, y) = (0, 0). \end{cases}$$

Is f continuous at the origin? Why?

18. Let

$$f(r, \theta) = \begin{cases} \dfrac{\sin 6r}{6r}, & r \neq 0, \\ 1, & r = 0, \end{cases}$$

where r and θ are polar coordinates. Find

a) $\displaystyle\lim_{r\to 0} f(r, \theta)$ **b)** $f_r(0, 0)$ **c)** $f_\theta(r, \theta), \quad r \neq 0$

$z = f(r, \theta)$

(Generated by Mathematica)

Partial Derivatives

In Exercises 19–24, find the partial derivative of the function with respect to each variable.

19. $g(r, \theta) = r \cos \theta + r \sin \theta$

20. $f(x, y) = \frac{1}{2} \ln (x^2 + y^2) + \tan^{-1} \frac{y}{x}$

21. $f(R_1, R_2, R_3) = \frac{1}{R_1} + \frac{1}{R_2} + \frac{1}{R_3}$

22. $h(x, y, z) = \sin (2\pi x + y - 3z)$

23. $P(n, R, T, V) = \frac{nRT}{V}$ (the Ideal Gas Law)

24. $f(r, l, T, w) = \frac{1}{2rl} \sqrt{\frac{T}{\pi w}}$

Second Order Partials

Find the second order partial derivatives of the functions in Exercises 25–28.

25. $g(x, y) = y + \frac{x}{y}$

26. $g(x, y) = e^x + y \sin x$

27. $f(x, y) = x + xy - 5x^3 + \ln (x^2 + 1)$

28. $f(x, y) = y^2 - 3xy + \cos y + 7e^y$

Linearizations

In Exercises 29 and 30, find the linearization $L(x, y)$ of the function $f(x, y)$ at the point P_0. Then find an upper bound for the magnitude of the error E in the approximation $f(x, y) \approx L(x, y)$ over the rectangle R.

29. $f(x, y) = \sin x \cos y, \quad P_0(\pi/4, \pi/4)$

$R: \quad \left| x - \frac{\pi}{4} \right| \le 0.1, \quad \left| y - \frac{\pi}{4} \right| \le 0.1$

30. $f(x, y) = xy - 3y^2 + 2, \quad P_0(1, 1)$

$R: \quad |x - 1| \le 0.1, \quad |y - 1| \le 0.2$

Find the linearizations of the functions in Exercises 31 and 32 at the given points.

31. $f(x, y, z) = xy + 2yz - 3xz$ at $(1, 0, 0)$ and $(1, 1, 0)$

32. $f(x, y, z) = \sqrt{2} \cos x \sin (y + z)$ at $(0, 0, \pi/4)$ and $(\pi/4, \pi/4, 0)$

Estimates and Sensitivity to Change

33. You plan to calculate the volume inside a stretch of pipeline that is about 36 in. in diameter and 1 mi long. With which measurement should you be more careful—the length, or the diameter? Why?

34. Near the point $(1, 2)$, is $f(x, y) = x^2 - xy + y^2 - 3$ more sensitive to changes in x, or to changes in y? How do you know?

35. Suppose that the current I (amperes) in an electrical circuit is related to the voltage V (volts) and the resistance R (ohms) by the equation $I = V/R$. If the voltage drops from 24 to 23 volts and the resistance drops from 100 to 80 ohms, will I increase, or decrease? By about how much? Express the changes in V and R and the estimated change in I as percentages of their original values.

36. If $a = 10$ cm and $b = 16$ cm to the nearest millimeter, what should you expect the maximum percentage error to be in the calculated area $A = \pi ab$ of the ellipse $x^2/a^2 + y^2/b^2 = 1$?

37. Let $y = uv$ and $z = u + v$, where u and v are positive independent variables.

a) If u is measured with an error of 2% and v with an error of 3%, about what is the percentage error in the calculated value of y?

b) Show that the percentage error in the calculated value of z is less than the percentage error in the value of y.

38. *Cardiac index.* To make different people comparable in studies of cardiac output (Section 2.7, Exercise 25), researchers divide the measured cardiac output by the body surface area to find the *cardiac index C:*

$$C = \frac{\text{cardiac output}}{\text{body surface area}}.$$

The body surface area B is calculated with the formula

$$B = 71.84 w^{0.425} h^{0.725},$$

which gives B in square centimeters when w is measured in kilograms and h in centimeters. You are about to calculate the cardiac index of a person with the following measurements:

Cardiac output:	7 L/min
Weight:	70 kg
Height:	180 cm

Which will have a greater effect on the calculation, a 1-kg error in measuring the weight, or a 1-cm error in measuring the height?

Chain Rule Calculations

39. Find dw/dt at $t = 0$ if $w = \sin (xy + \pi)$, $x = e^t$, and $y = \ln (t + 1)$.

40. Find dw/dt at $t = 1$ if $w = xe^y + y \sin z - \cos z$, $x = 2\sqrt{t}$, $y = t - 1 + \ln t$, $z = \pi t$.

41. Find $\partial w / \partial r$ and $\partial w / \partial s$ when $r = \pi$ and $s = 0$ if $w = \sin(2x - y)$, $x = r + \sin s$, $y = rs$.

42. Find $\partial w / \partial u$ and $\partial w / \partial v$ when $u = v = 0$ if $w = \ln \sqrt{1 + x^2} - \tan^{-1} x$ and $x = 2e^u \cos v$.

43. Find the value of the derivative of $f(x, y, z) = xy + yz + xz$ with respect to t on the curve $x = \cos t$, $y = \sin t$, $z = \cos 2t$ at $t = 1$.

44. Show that if $w = f(s)$ is any differentiable function of s and if $s = y + 5x$, then

$$\frac{\partial w}{\partial x} - 5\frac{\partial w}{\partial y} = 0.$$

Implicit Differentiation

Assuming that the equations in Exercises 45 and 46 define y as a differentiable function of x, find the value of dy/dx at point P.

45. $1 - x - y^2 - \sin xy = 0$, $\quad P(0, 1)$

46. $2xy + e^{x+y} - 2 = 0$, $\quad P(0, \ln 2)$

Partial Derivatives with Constrained Variables

In Exercises 47 and 48, begin by drawing a diagram that shows the relations among the variables.

47. If $w = x^2 e^{yz}$ and $z = x^2 - y^2$, find

a) $\left(\dfrac{\partial w}{\partial y} \right)_z$ **b)** $\left(\dfrac{\partial w}{\partial z} \right)_x$ **c)** $\left(\dfrac{\partial w}{\partial z} \right)_y$

48. Let $U = f(P, V, T)$ be the internal energy of a gas that obeys the ideal gas law $PV = nRT$ (n and R constant). Find

a) $\left(\dfrac{\partial U}{\partial T} \right)_P$ **b)** $\left(\dfrac{\partial U}{\partial V} \right)_T$

Directional Derivatives

In Exercises 49–52, find the directions in which f increases and decreases most rapidly at P_0 and find the derivative of f in each direction. Also, find the derivative of f at P_0 in the direction of the vector $\mathbf{A}$.

49. $f(x, y) = \cos x \cos y$, $\quad P_0(\pi/4, \pi/4)$, $\quad \mathbf{A} = 3\mathbf{i} + 4\mathbf{j}$

50. $f(x, y) = x^2 e^{-2y}$, $\quad P_0(1, 0)$, $\quad \mathbf{A} = \mathbf{i} + \mathbf{j}$

51. $f(x, y, z) = \ln(2x + 3y + 6z)$, $\quad P_0(-1, -1, 1)$, $\mathbf{A} = 2\mathbf{i} + 3\mathbf{j} + 6\mathbf{k}$

52. $f(x, y, z) = x^2 + 3xy - z^2 + 2y + z + 4$, $P_0(0, 0, 0)$, $\mathbf{A} = \mathbf{i} + \mathbf{j} + \mathbf{k}$

53. Find the derivative of $f(x, y, z) = xyz$ in the direction of the velocity vector of the helix

$$\mathbf{r}(t) = (\cos 3t)\mathbf{i} + (\sin 3t)\mathbf{j} + 3t\,\mathbf{k}$$

at $t = \pi/3$.

54. What is the largest value that the directional derivative of $f(x, y, z) = xyz$ can have at the point $(1, 1, 1)$?

55. At the point $(1, 2)$ the function $f(x, y)$ has a derivative of 2 in the direction toward $(2, 2)$ and derivative of -2 in the direction toward $(1, 1)$.

a) Find $f_x(1, 2)$ and $f_y(1, 2)$.

b) Find the derivative of f at $(1, 2)$ in the direction toward the point $(4, 6)$.

56. Which of the following statements are true if $f(x, y)$ is differentiable at (x_0, y_0)?

a) If $\mathbf{u}$ is a unit vector, the derivative of f at (x_0, y_0) in the direction of $\mathbf{u}$ is $(f_x(x_0, y_0)\mathbf{i} + f_y(x_0, y_0)\mathbf{j}) \cdot \mathbf{u}$.

b) The derivative of f at (x_0, y_0) in the direction of $\mathbf{u}$ is a vector.

c) The directional derivative of f at (x_0, y_0) has its greatest value in the direction of ∇f.

d) At (x_0, y_0), vector ∇f is normal to the curve $f(x, y) = f(x_0, y_0)$.

Gradients, Tangent Planes, and Normal Lines

In Exercises 57 and 58, sketch the surface $f(x, y, z) = c$ together with ∇f at the given points.

57. $x^2 + y + z^2 = 0$; $\quad (0, -1, \pm 1)$, $\quad (0, 0, 0)$

58. $y^2 + z^2 = 4$; $\quad (2, \pm 2, 0)$, $\quad (2, 0, \pm 2)$

In Exercises 59 and 60, find an equation for the plane tangent to the level surface $f(x, y, z) = c$ at the point P_0. Also, find parametric equations for the line that is normal to the surface at P_0.

59. $x^2 - y - 5z = 0$, $\quad P_0(2, -1, 1)$

60. $x^2 + y^2 + z = 4$, $\quad P_0(1, 1, 2)$

In Exercises 61 and 62, find an equation for the plane tangent to the surface $z = f(x, y)$ at the given point

61. $z = \ln(x^2 + y^2)$, $\quad (0, 1, 0)$

62. $z = 1/(x^2 + y^2)$, $\quad (1, 1, 1/2)$

In Exercises 63 and 64, find equations for the lines that are tangent and normal to the level curve $f(x, y) = c$ at the point P_0. Then sketch the lines and level curve together with ∇f at P_0.

63. $y - \sin x = 1$, $\quad P_0(\pi, 1)$

64. $\dfrac{y^2}{2} - \dfrac{x^2}{2} = \dfrac{3}{2}$, $\quad P_0(1, 2)$

Tangent Lines to Curves

In Exercises 65 and 66, find parametric equations for the line that is tangent to the curve of intersection of the surfaces at the given point.

65. Surfaces: $x^2 + 2y + 2z = 4$, $\quad y = 1$
Point: $(1, 1, 1/2)$

66. Surfaces: $x + y^2 + z = 2$, $\quad y = 1$
Point: $(1/2, 1, 1/2)$

Local Extrema

Test the functions in Exercises 67–72 for local maxima and minima and saddle points. Find each function's values at these points.

67. $f(x, y) = x^2 - xy + y^2 + 2x + 2y - 4$

68. $f(x, y) = 5x^2 + 4xy - 2y^2 + 4x - 4y$

69. $f(x, y) = 2x^3 + 3xy + 2y^3$

70. $f(x, y) = x^3 + y^3 - 3xy + 15$

71. $f(x, y) = x^3 + y^3 + 3x^2 - 3y^2$

72. $f(x, y) = x^4 - 8x^2 + 3y^2 - 6y$

Absolute Extrema

In Exercises 73–80, find the absolute maximum and minimum values of f on the region R.

73. $f(x, y) = x^2 + xy + y^2 - 3x + 3y$
R: The triangular region cut from the first quadrant by the line $x + y = 4$

74. $f(x, y) = x^2 - y^2 - 2x + 4y + 1$
R: The rectangular region in the first quadrant bounded by the coordinate axes and the lines $x = 4$ and $y = 2$

75. $f(x, y) = y^2 - xy - 3y + 2x$
R: The square region enclosed by the lines $x = \pm 2$ and $y = \pm 2$

76. $f(x, y) = 2x + 2y - x^2 - y^2$
R: The square bounded by the coordinate axes and the lines $x = 2, y = 2$ in the first quadrant

77. $f(x, y) = x^2 - y^2 - 2x + 4y$
R: The triangular region bounded below by the x-axis, above by the line $y = x + 2$, and on the right by the line $x = 2$

78. $f(x, y) = 4xy - x^4 - y^4 + 16$
R: The triangular region bounded below by the line $y = -2$, above by the line $y = x$, and on the right by the line $x = 2$

79. $f(x, y) = x^3 + y^3 + 3x^2 - 3y^2$
R: The square region enclosed by the lines $x = \pm 1$ and $y = \pm 1$

80. $f(x, y) = x^3 + 3xy + y^3 + 1$
R: The square region enclosed by the lines $x = \pm 1$ and $y = \pm 1$

Lagrange Multipliers

81. Find the extreme values of $f(x, y) = x^3 + y^2$ on the circle $x^2 + y^2 = 1$.

82. Find the extreme values of $f(x, y) = xy$ on the circle $x^2 + y^2 = 1$.

83. Find the extreme values of $f(x, y) = x^2 + 3y^2 + 2y$ on the unit disk $x^2 + y^2 \le 1$.

84. Find the extreme values of $f(x, y) = x^2 + y^2 - 3x - xy$ on the disk $x^2 + y^2 \le 9$.

85. Find the extreme values of $f(x, y, z) = x - y + z$ on the unit sphere $x^2 + y^2 + z^2 = 1$.

86. Find the points on the surface $z^2 - xy = 4$ closest to the origin.

87. A closed rectangular box is to have volume V cm^3. The cost of the material used in the box is a cents/cm^2 for top and bottom, b cents/cm^2 for front and back, and c cents/cm^2 for the remaining sides. What dimensions minimize the total cost of materials?

88. Find the plane $x/a + y/b + z/c = 1$ that passes through the point $(2, 1, 2)$ and cuts off the least volume from the first octant.

89. Find the extreme values of $f(x, y, z) = x(y + z)$ on the curve of intersection of the right circular cylinder $x^2 + y^2 = 1$ and the hyperbolic cylinder $xz = 1$.

90. Find the point closest to the origin on the curve of intersection of the plane $x + y + z = 1$ and the cone $z^2 = 2x^2 + 2y^2$.

Theory and Examples

91. Let $w = f(r, \theta), r = \sqrt{x^2 + y^2}$, and $\theta = \tan^{-1}(y/x)$. Find $\partial w/\partial x$ and $\partial w/\partial y$ and express your answers in terms of r and θ.

92. Let $z = f(u, v), u = ax + by$, and $v = ax - by$. Express z_x and z_y in terms of f_u, f_v, and the constants a and b.

93. If a and b are constants, $w = u^3 + \tanh u + \cos u$, and $u = ax + by$, show that

$$a \frac{\partial w}{\partial y} = b \frac{\partial w}{\partial x}.$$

94. If $w = \ln(x^2 + y^2 + 2z)$, $x = r + s$, $y = r - s$, and $z = 2rs$, find w_r and w_s by the Chain Rule. Then check your answer another way.

95. The equations $e^u \cos v - x = 0$ and $e^u \sin v - y = 0$ define u and v as differentiable functions of x and y. Show that the angle between the vectors

$$\frac{\partial u}{\partial x}\mathbf{i} + \frac{\partial u}{\partial y}\mathbf{j} \quad \text{and} \quad \frac{\partial v}{\partial x}\mathbf{i} + \frac{\partial v}{\partial y}\mathbf{j}$$

is constant.

96. Introducing polar coordinates $x = r \cos \theta$ and $y = r \sin \theta$ changes $f(x, y)$ to $g(r, \theta)$. Find the value of $\partial^2 g/\partial \theta^2$ at the point $(r, \theta) = (2, \pi/2)$, given that

$$\frac{\partial f}{\partial x} = \frac{\partial f}{\partial y} = \frac{\partial^2 f}{\partial x^2} = \frac{\partial^2 f}{\partial y^2} = 1$$

at that point.

97. Find the points on the surface

$$(y + z)^2 + (z - x)^2 = 16$$

where the normal line is parallel to the yz-plane.

98. Find the points on the surface

$$xy + yz + zx - x - z^2 = 0$$

where the tangent plane is parallel to the xy-plane.

99. Suppose that $\nabla f(x, y, z)$ is always parallel to the position vector $x\,\mathbf{i} + y\,\mathbf{j} + z\,\mathbf{k}$. Show that $f(0, 0, a) = f(0, 0, -a)$ for any a.

100. Show that the directional derivative of

$$f(x, y, z) = \sqrt{x^2 + y^2 + z^2}$$

at the origin equals 1 in any direction but that f has no gradient vector at the origin.

101. Show that the line normal to the surface $xy + z = 2$ at the point $(1, 1, 1)$ passes through the origin.

102. a) Sketch the surface $x^2 - y^2 + z^2 = 4$.

b) Find a vector normal to the surface at $(2, -3, 3)$. Add the vector to your sketch.

c) Find the equations for the tangent plane and normal line at $(2, -3, 3)$.

CHAPTER 12   ADDITIONAL EXERCISES–THEORY, EXAMPLES, APPLICATIONS

Partial Derivatives

1. If you did Exercise 50 in Section 12.2, you know that the function

$$f(x, y) = \begin{cases} xy\dfrac{x^2 - y^2}{x^2 + y^2}, & (x, y) \neq (0, 0) \\ 0, & (x, y) = (0, 0) \end{cases}$$

(see the accompanying figure) is continuous at $(0, 0)$. Find $f_{xy}(0, 0)$ and $f_{yx}(0, 0)$.

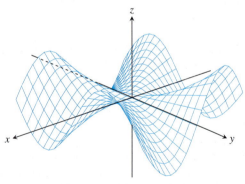

(Generated by Mathematica)

2. Find a function $w = f(x, y)$ whose first partial derivatives are $\partial w/\partial x = 1 + e^x \cos y$ and $\partial w/\partial y = 2y - e^x \sin y$, and whose value at the point $(\ln 2, 0)$ is $\ln 2$.

3. *A proof of Leibniz's rule.* Leibniz's rule says that if f is continuous on $[a, b]$ and if $u(x)$ and $v(x)$ are differentiable functions of x whose values lie in $[a, b]$, then

$$\frac{d}{dx} \int_{u(x)}^{v(x)} f(t)\, dt = f(v(x))\frac{dv}{dx} - f(u(x))\frac{du}{dx}.$$

Prove the rule by setting

$$g(u, v) = \int_u^v f(t)\, dt, \quad u = u(x), \quad v = v(x)$$

and calculating dg/dx with the Chain Rule.

4. Suppose that f is a twice-differentiable function of r, that $r = \sqrt{x^2 + y^2 + z^2}$, and that

$$f_{xx} + f_{yy} + f_{zz} = 0.$$

Show that for some constants a and b,

$$f(r) = \frac{a}{r} + b.$$

5. *Homogeneous functions.* A function $f(x, y)$ is *homogeneous of degree n* (n a nonnegative integer) if $f(tx, ty) = t^n f(x, y)$ for all $t, x,$ and y. For such a function (sufficiently differentiable), prove that

a) $x\dfrac{\partial f}{\partial x} + y\dfrac{\partial f}{\partial y} = nf(x, y)$

b) $x^2\left(\dfrac{\partial^2 f}{\partial x^2}\right) + 2xy\left(\dfrac{\partial^2 f}{\partial x \partial y}\right) + y^2\left(\dfrac{\partial^2 f}{\partial y^2}\right) = n(n-1)f.$

6. *Spherical coordinates.* Let $\mathbf{r} = x\,\mathbf{i} + y\,\mathbf{j} + z\,\mathbf{k}$. Express $x, y,$ and z as functions of the spherical coordinates $\rho, \phi,$ and θ and calculate $\partial\mathbf{r}/\partial\rho$, $\partial\mathbf{r}/\partial\phi$, and $\partial\mathbf{r}/\partial\theta$. Then express these derivatives in terms of the unit vectors

$$\mathbf{u}_\rho = (\sin\phi\,\cos\theta)\,\mathbf{i} + (\sin\phi\,\sin\theta)\,\mathbf{j} + (\cos\phi)\,\mathbf{k}$$

$$\mathbf{u}_\phi = (\cos\phi\,\cos\theta)\,\mathbf{i} + (\cos\phi\,\sin\theta)\,\mathbf{j} - (\sin\phi)\,\mathbf{k}$$

$$\mathbf{u}_\theta = -(\sin\theta)\,\mathbf{i} + (\cos\theta)\,\mathbf{j}.$$

Gradients and Tangents

7. Let $\mathbf{r} = x\,\mathbf{i} + y\,\mathbf{j} + z\,\mathbf{k}$ and let $r = |\mathbf{r}|$.

a) Show that $\nabla r = \mathbf{r}/r$.

b) Show that $\nabla(r^n) = nr^{n-2}\mathbf{r}$.

c) Find a function whose gradient equals $\mathbf{r}$.

d) Show that $\mathbf{r} \cdot d\mathbf{r} = r\, dr$.

e) Show that $\nabla(\mathbf{A} \cdot \mathbf{r}) = \mathbf{A}$ for any constant vector $\mathbf{A}$.

8. Suppose that a differentiable function $f(x, y)$ has the constant value c along the differentiable curve $x = g(t), y = h(t)$; that is,

$$f(g(t), h(t)) = c$$

for all values of t. Differentiate both sides of this equation with respect to t to show that ∇f is orthogonal to the curve's tangent vector at every point on the curve.

9. Show that the curve

$$\mathbf{r}(t) = (\ln t)\,\mathbf{i} + (t \ln t)\,\mathbf{j} + t\,\mathbf{k}$$

is tangent to the surface

$$xz^2 - yz + \cos xy = 1$$

at $(0, 0, 1)$.

10. Show that the curve

$$\mathbf{r}(t) = \left(\frac{t^3}{4} - 2\right)\mathbf{i} + \left(\frac{4}{t} - 3\right)\mathbf{j} + \cos(t - 2)\mathbf{k}$$

is tangent to the surface

$$x^3 + y^3 + z^3 - xyz = 0$$

at $(0, -1, 1)$.

11. *The gradient in cylindrical coordinates.* Suppose cylindrical coordinates r, θ, z are introduced into a function $w = f(x, y, z)$ to yield $w = F(r, \theta, z)$. Show that

$$\nabla w = \frac{\partial w}{\partial r}\mathbf{u}_r + \frac{1}{r}\frac{\partial w}{\partial \theta}\mathbf{u}_\theta + \frac{\partial w}{\partial z}\mathbf{k}, \qquad (1)$$

where

$$\mathbf{u}_r = (\cos\theta)\mathbf{i} + (\sin\theta)\mathbf{j}$$

$$\mathbf{u}_\theta = -(\sin\theta)\mathbf{i} + (\cos\theta)\mathbf{j}.$$

(*Hint:* Express the right-hand side of Eq. (1) in terms of $\mathbf{i}$, $\mathbf{j}$, and $\mathbf{k}$ and use the Chain Rule to express the components of $\mathbf{i}$, $\mathbf{j}$, and $\mathbf{k}$ in rectangular coordinates.)

12. *The gradient in spherical coordinates.* Suppose spherical coordinates ρ, ϕ, θ are introduced into a function $w = f(x, y, z)$ to yield a function $w = F(\rho, \phi, \theta)$. Show that

$$\nabla w = \frac{\partial w}{\partial \rho}\mathbf{u}_\rho + \frac{1}{\rho}\frac{\partial w}{\partial \phi}\mathbf{u}_\phi + \frac{1}{\rho\sin\phi}\frac{\partial w}{\partial \theta}\mathbf{u}_\theta, \qquad (2)$$

where

$$\mathbf{u}_\rho = (\sin\phi\cos\theta)\mathbf{i} + (\sin\phi\sin\theta)\mathbf{j} + (\cos\phi)\mathbf{k}$$

$$\mathbf{u}_\phi = (\cos\phi\cos\theta)\mathbf{i} + (\cos\phi\sin\theta)\mathbf{j} - (\sin\phi)\mathbf{k}$$

$$\mathbf{u}_\theta = -(\sin\theta)\mathbf{i} + (\cos\theta)\mathbf{j}.$$

(*Hint:* Express the right-hand side of Eq. (2) in terms of $\mathbf{i}$, $\mathbf{j}$, and $\mathbf{k}$ and use the Chain Rule to express the components of $\mathbf{i}$, $\mathbf{j}$, and $\mathbf{k}$ in rectangular coordinates.)

Extreme Values

13. Show that the only possible maxima and minima of z on the surface $z = x^3 + y^3 - 9xy + 27$ occur at $(0, 0)$ and $(3, 3)$. Show that neither a maximum nor a minimum occurs at $(0, 0)$. Determine whether z has a maximum or a minimum at $(3, 3)$.

14. Find the maximum value of $f(x, y) = 6xye^{-(2x+3y)}$ in the closed first quadrant (includes the nonnegative axes).

15. Find the minimum volume for a region bounded by the planes $x = 0, y = 0, z = 0$ and a plane tangent to the ellipsoid

$$\frac{x^2}{a^2} + \frac{y^2}{b^2} + \frac{z^2}{c^2} = 1$$

at a point in the first octant.

16. By minimizing the function $f(x, y, u, v) = (x - u)^2 + (y - v)^2$ subject to the constraints $y = x + 1$ and $u = v^2$, find the minimum distance in the xy-plane from the line $y = x + 1$ to the parabola $y^2 = x$.

Theory and Examples

17. Prove the following theorem: If $f(x, y)$ is defined in an open region R of the xy-plane, and if f_x and f_y are bounded on R, then $f(x, y)$ is continuous on R. (The assumption of boundedness is essential.)

18. Suppose $\mathbf{r}(t) = g(t)\mathbf{i} + h(t)\mathbf{j} + k(t)\mathbf{k}$ is a smooth curve in the domain of a differentiable function $f(x, y, z)$. Describe the relation between df/dt, ∇f, and $\mathbf{v} = d\mathbf{r}/dt$. What can be said about ∇f and $\mathbf{v}$ at interior points of the curve where f has extreme values relative to its other values on the curve? Give reasons for your answer.

19. Suppose that f and g are functions of x and y such that

$$\frac{\partial f}{\partial y} = \frac{\partial g}{\partial x} \quad \text{and} \quad \frac{\partial f}{\partial x} = \frac{\partial g}{\partial y},$$

and suppose that

$$\frac{\partial f}{\partial x} = 0, \quad f(1, 2) = g(1, 2) = 5, \quad \text{and} \quad f(0, 0) = 4.$$

Find $f(x, y)$ and $g(x, y)$.

20. We know that if $f(x, y)$ is a function of two variables and if $\mathbf{u} = a\mathbf{i} + b\mathbf{j}$ is a unit vector, then $D_\mathbf{u} f(x, y) = f_x(x, y)a + f_y(x, y)b$ is the rate of change of $f(x, y)$ at (x, y) in the direction of $\mathbf{u}$. Give a similar formula for the rate of change *of the rate of change* of $f(x, y)$ at (x, y) in the direction $\mathbf{u}$.

21. *Path of a heat-seeking particle.* A heat-seeking particle has the property that at any point (x, y) in the plane it moves in the direction of maximum temperature increase. If the temperature at (x, y) is $T(x, y) = -e^{-2y}\cos x$, find an equation $y = f(x)$ for the path of a heat-seeking particle at the point $(\pi/4, 0)$.

22. A particle traveling in a straight line with constant velocity $\mathbf{i} + \mathbf{j} - 5\mathbf{k}$ passes through the point $(0, 0, 30)$ and hits the surface $z = 2x^2 + 3y^2$. The particle ricochets off the surface, the angle of reflection being equal to the angle of incidence. Assuming no loss of speed, what is the velocity of the particle after the ricochet? Simplify your answer.

23. Let S be the surface that is the graph of $f(x, y) = 10 - x^2 - y^2$. Suppose the temperature in space at each point (x, y, z) is $T(x, y, z) = x^2y + y^2z + 4x + 14y + z$.

   a) Among all of the possible directions tangential to the surface S at the point $(0, 0, 10)$, which direction will make the rate of change of temperature at $(0, 0, 10)$ a maximum?

   b) Which direction tangential to S at the point $(1, 1, 8)$ will make the rate of change of temperature a maximum?

24. On a flat surface of land, geologists drilled a borehole straight down and hit a mineral deposit at 1000 ft. They drilled a second borehole 100 ft to the north of the first and hit the mineral deposit at 950 ft. A third borehole 100 ft east of the first borehole struck the mineral deposit at 1025 ft. The geologists have reasons to believe that the mineral deposit is in the shape of a dome and for the sake of economy they would like to find where the deposit is closest to the surface. Assuming the surface to be the xy-plane, in what direction from the first borehole would you suggest the geologists drill their fourth borehole?

The One-Dimensional Heat Equation

If $w(x, t)$ represents the temperature at position x at time t in a uniform conducting rod with perfectly insulated sides (see the accompanying figure), then the partial derivatives w_{xx} and w_t satisfy a differential equation of the form

$$w_{xx} = \frac{1}{c^2} w_t. \tag{3}$$

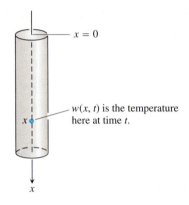

$x = 0$

$w(x, t)$ is the temperature here at time t.

x

x

This equation is called the **one-dimensional heat equation.** The value of the positive constant c^2 is determined by the material from which

the rod is made. It has been determined experimentally for a broad range of materials. For a given application one finds the appropriate value in a table. For dry soil, for example, $c^2 = 0.19$ ft^2/day.

In chemistry and biochemistry, the heat equation is known as the **diffusion equation.** In this context, $w(x, t)$ represents the concentration of a dissolved substance, a salt for instance, diffusing along a tube filled with liquid. The value of $w(x, t)$ is the concentration at point x at time t. In other applications, $w(x, t)$ represents the diffusion of a gas down a long, thin pipe.

In electrical engineering, the heat equation appears in the forms

$$v_{xx} = RCv_t \tag{4}$$

and

$$i_{xx} = RCi_t, \tag{5}$$

which are known as the **telegraph equations.** These equations describe the voltage v and the flow of current i in a coaxial cable or in any other cable in which leakage and inductance are negligible. The functions and constants in these equations are

$$v(x, t) = \text{voltage at point } x \text{ at time } t$$

$$R = \text{resistance per unit length}$$

$$C = \text{capacitance to ground per unit of cable length}$$

$$i(x, t) = \text{current at point } x \text{ at time } t.$$

25. Find all solutions of the one-dimensional heat equation of the form $w = e^{rt} \sin \pi x$, where r is a constant.

26. Find all solutions of the one-dimensional heat equation that have the form $w = e^{rt} \sin kx$ and satisfy the conditions that $w(0, t) = 0$ and $w(L, t) = 0$. What happens to these solutions as $t \to \infty$?

Multiple Integrals

OVERVIEW The problems we can solve by integrating functions of two and three variables are similar to the problems solved by single-variable integration, but more general. As in the previous chapter, we can perform the necessary calculations by drawing on our experience with functions of a single variable.

13.1 Double Integrals

We now show how to integrate a continuous function $f(x, y)$ over a bounded region in the xy-plane. There are many similarities between the "double" integrals we define here and the "single" integrals we defined in Chapter 4 for functions of a single variable. Every double integral can be evaluated in stages, using the single-integration methods already at our command.

Double Integrals over Rectangles

Suppose that $f(x, y)$ is defined on a rectangular region R given by

$$R: \quad a \leq x \leq b, \quad c \leq y \leq d.$$

We imagine R to be covered by a network of lines parallel to the x- and y-axes (Fig. 13.1). These lines divide R into small pieces of area $\Delta A = \Delta x \Delta y$. We number these in some order $\Delta A_1, \Delta A_2, \ldots, \Delta A_n$, choose a point (x_k, y_k) in each piece ΔA_k, and form the sum

$$S_n = \sum_{k=1}^{n} f(x_k, y_k) \, \Delta A_k. \tag{1}$$

If f is continuous throughout R, then, as we refine the mesh width to make both Δx and Δy go to zero, the sums in (1) approach a limit called the **double integral** of f over R. The notation for it is

$$\iint\limits_{R} f(x, y) \, dA \qquad \text{or} \qquad \iint\limits_{R} f(x, y) \, dx \, dy.$$

Thus,

$$\iint\limits_{R} f(x, y) \, dA = \lim_{\Delta A \to 0} \sum_{k=1}^{n} f(x_k, y_k) \Delta A_k. \tag{2}$$

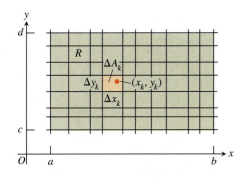

13.1 Rectangular grid partitioning the region R into small rectangles of area $\Delta A_k = \Delta x_k \Delta y_k$.

As with functions of a single variable, the sums approach this limit no matter how the intervals $[a, b]$ and $[c, d]$ that determine R are partitioned, as long as the norms of the partitions both go to zero. The limit in (2) is also independent of the order in which the areas ΔA_k are numbered and independent of the choice of the point (x_k, y_k) within each ΔA_k. The values of the individual approximating sums S_n depend on these choices, but the sums approach the same limit in the end. The proof of the existence and uniqueness of this limit for a continuous function f is given in more advanced texts. The continuity of f is a sufficient condition for the existence of the double integral, but not a necessary one. The limit in question exists for many discontinuous functions as well.

Properties of Double Integrals

Like single integrals, double integrals of continuous functions have algebraic properties that are useful in computations and applications.

1. $\displaystyle\iint\limits_{R} kf(x, y)\, dA = k \iint\limits_{R} f(x, y)\, dA \qquad$ (any number k)

2. $\displaystyle\iint\limits_{R} (f(x, y) \pm g(x, y))\, dA = \iint\limits_{R} f(x, y)\, dA \pm \iint\limits_{R} g(x, y)\, dA$

3. $\displaystyle\iint\limits_{R} f(x, y)\, dA \geq 0 \quad$ if $\quad f(x, y) \geq 0$ on R

4. $\displaystyle\iint\limits_{R} f(x, y)\, dA \geq \iint\limits_{R} g(x, y)\, dA \quad$ if $\quad f(x, y) \geq g(x, y)$ on R

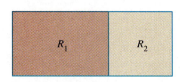

$$\iint\limits_{R_1 \cup R_2} f(x, y)\, dA = \iint\limits_{R_1} f(x, y)\, dA + \iint\limits_{R_2} f(x, y)\, dA$$

13.2 Double integrals have the same kind of domain additivity property that single integrals have.

These are like the single-integral properties in Section 4.5. There is also an additivity property:

5. $\displaystyle\iint\limits_{R} f(x, y)\, dA = \iint\limits_{R_1} f(x, y)\, dA + \iint\limits_{R_2} f(x, y)\, dA.$

It holds when R is the union of two nonoverlapping rectangles R_1 and R_2 (Fig. 13.2). Again, we omit the proof.

Double Integrals as Volumes

When $f(x, y)$ is positive, we may interpret the double integral of f over a rectangular region R as the volume of the solid prism bounded below by R and above by the surface $z = f(x, y)$ (Fig. 13.3). Each term $f(x_k, y_k)\, \Delta A_k$ in the sum $S_n = \sum f(x_k, y_k)\, \Delta A_k$ is the volume of a vertical rectangular prism that approximates the volume of the portion of the solid that stands directly above the base ΔA_k. The sum S_n thus approximates what we want to call the total volume of the solid. We *define* this volume to be

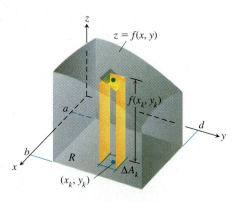

13.3 Approximating solids with rectangular prisms leads us to define the volumes of more general prisms as double integrals. The volume of the prism shown here is the double integral of $f(x, y)$ over the base region R.

$$\text{Volume} = \lim S_n = \iint\limits_{R} f(x, y)\, dA. \qquad (3)$$

As you might expect, this more general method of calculating volume agrees with the methods in Chapter 5, but we will not prove this here.

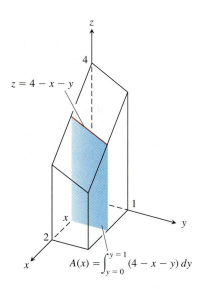

13.4 To obtain the cross-section area $A(x)$, we hold x fixed and integrate with respect to y.

Fubini's Theorem for Calculating Double Integrals

Suppose we wish to calculate the volume under the plane $z = 4 - x - y$ over the rectangular region R: $0 \le x \le 2$, $0 \le y \le 1$ in the xy-plane. If we apply the method of slicing from Section 5.2, with slices perpendicular to the x-axis (Fig. 13.4), then the volume is

$$\int_{x=0}^{x=2} A(x) \, dx, \tag{4}$$

where $A(x)$ is the cross-section area at x. For each value of x we may calculate $A(x)$ as the integral

$$A(x) = \int_{y=0}^{y=1} (4 - x - y) \, dy, \tag{5}$$

which is the area under the curve $z = 4 - x - y$ in the plane of the cross section at x. In calculating $A(x)$, x is held fixed and the integration takes place with respect to y. Combining (4) and (5), we see that the volume of the entire solid is

$$\text{Volume} = \int_{x=0}^{x=2} A(x) \, dx = \int_{x=0}^{x=2} \left(\int_{y=0}^{y=1} (4 - x - y) \, dy \right) dx$$

$$= \int_{x=0}^{x=2} \left[4y - xy - \frac{y^2}{2} \right]_{y=0}^{y=1} dx = \int_{x=0}^{x=2} \left(\frac{7}{2} - x \right) dx = \left[\frac{7}{2} x - \frac{x^2}{2} \right]_0^2 = 5. \tag{6}$$

If we had just wanted to write instructions for calculating the volume, without carrying out any of the integrations, we could write

$$\text{Volume} = \int_0^2 \int_0^1 (4 - x - y) \, dy \, dx.$$

The expression on the right, called an **iterated** or **repeated integral,** says that the volume is obtained by integrating $4 - x - y$ with respect to y from $y = 0$ to $y = 1$, holding x fixed, and then integrating the resulting expression in x with respect to x from $x = 0$ to $x = 2$.

What would have happened if we had calculated the volume by slicing with planes perpendicular to the y-axis (Fig. 13.5)? As a function of y, the typical cross-section area is

$$A(y) = \int_{x=0}^{x=2} (4 - x - y) \, dx = \left[4x - \frac{x^2}{2} - xy \right]_{x=0}^{x=2} = 6 - 2y. \tag{7}$$

The volume of the entire solid is therefore

$$\text{Volume} = \int_{y=0}^{y=1} A(y) \, dy = \int_{y=0}^{y=1} (6 - 2y) \, dy = \left[6y - y^2 \right]_0^1 = 5,$$

in agreement with our earlier calculation.

Again, we may give instructions for calculating the volume as an iterated integral by writing

$$\text{Volume} = \int_0^1 \int_0^2 (4 - x - y) \, dx \, dy.$$

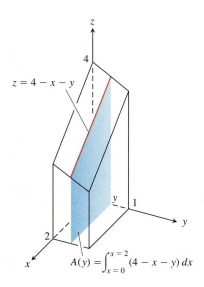

13.5 To obtain the cross-section area $A(y)$, we hold y fixed and integrate with respect to x.

The expression on the right says we can find the volume by integrating $4 - x - y$ with respect to x from $x = 0$ to $x = 2$ (as in Eq. 7) and integrating the result with respect to y from $y = 0$ to $y = 1$. In this iterated integral the order of integration is first x and then y, the reverse of the order in Eq. (6).

What do these two volume calculations with iterated integrals have to do with the double integral

$$\iint_R (4 - x - y) \, dA$$

over the rectangle $R : 0 \le x \le 2, \; 0 \le y \le 1$? The answer is that they both give the value of the double integral. A theorem published in 1907 by Guido Fubini (1879–1943) says that the double integral of any continuous function over a rectangle can be calculated as an iterated integral in either order of integration. (Fubini proved his theorem in greater generality, but this is how it translates into what we're doing now.)

Theorem 1

Fubini's Theorem (First Form)

If $f(x, y)$ is continuous on the rectangular region $R : a \le x \le b, \; c \le y \le d$, then

$$\iint_R f(x, y) \, dA = \int_c^d \int_a^b f(x, y) \, dx \, dy = \int_a^b \int_c^d f(x, y) \, dy \, dx.$$

Fubini's theorem says that double integrals over rectangles can be calculated as iterated integrals. This means we can evaluate a double integral by integrating with respect to one variable at a time.

Fubini's theorem also says that we may calculate the double integral by integrating in *either* order, a genuine convenience, as we will see in Example 3. In particular, when we calculate a volume by slicing, we may use either planes perpendicular to the x-axis or planes perpendicular to the y-axis.

EXAMPLE 1 Calculate $\iint_R f(x, y) \, dA$ for

$$f(x, y) = 1 - 6x^2 y \quad \text{and} \quad R : \; 0 \le x \le 2, \; -1 \le y \le 1.$$

Solution By Fubini's theorem,

$$\iint_R f(x, y) \, dA = \int_{-1}^1 \int_0^2 (1 - 6x^2 y) \, dx \, dy = \int_{-1}^1 \left[x - 2x^3 y \right]_{x=0}^{x=2} dy$$

$$= \int_{-1}^1 (2 - 16y) \, dy = \left[2y - 8y^2 \right]_{-1}^1 = 4.$$

Reversing the order of integration gives the same answer:

$$\int_0^2 \int_{-1}^1 (1 - 6x^2 y) \, dy \, dx = \int_0^2 \left[y - 3x^2 y^2 \right]_{y=-1}^{y=1} dx$$

$$= \int_0^2 \left[(1 - 3x^2) - (-1 - 3x^2) \right] dx = \int_0^2 2 \, dx = 4.$$

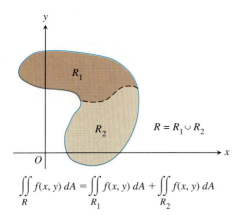

13.6 A rectangular grid partitioning a bounded nonrectangular region into cells.

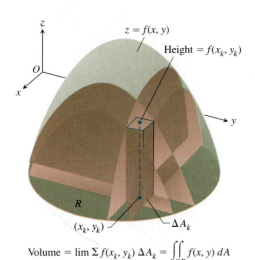

$$\iint_R f(x, y)\, dA = \iint_{R_1} f(x, y)\, dA + \iint_{R_2} f(x, y)\, dA$$

13.7 The additivity property for rectangular regions holds for regions bounded by continuous curves.

Technology *Multiple Integration* Most Computer Algebra Systems can calcuate both multiple and iterated integrals. The typical procedure is to apply the CAS integrate command in nested iterations according to the order of integration you specify:

Integral	Typical CAS Formulation
$\displaystyle\iint x^2 y\, dx\, dy$	int(int(x ^ 2 * y, x), y);
$\displaystyle\int_{-\pi/3}^{\pi/4} \int_0^1 x \cos y\, dx\, dy$	int(int(x* cos(y), x = 0 . . 1), y = −Pi/3 . . Pi/4);

If a CAS cannot produce an exact value for a definite integral, it can usually find an approximate value numerically.

Double Integrals over Bounded Nonrectangular Regions

To define the double integral of a function $f(x, y)$ over a bounded nonrectangular region, like the one shown in Fig. 13.6, we again imagine R to be covered by a rectangular grid, but we include in the partial sum only the small pieces of area $\Delta A = \Delta x \Delta y$ that lie entirely within the region (shaded in the figure). We number the pieces in some order, choose an arbitrary point (x_k, y_k) in each ΔA_k, and form the sum

$$S_n = \sum_{k=1}^{n} f(x_k, y_k)\, \Delta A_k.$$

The only difference between this sum and the one in Eq. (1) for rectangular regions is that now the areas ΔA_k may not cover all of R. But as the mesh becomes increasingly fine and the number of terms in S_n increases, more and more of R is included. If f is continuous and the boundary of R is made from the graphs of a finite number of continuous functions of x and/or continuous functions of y joined end to end, then the sums S_n will have a limit as the norms of the partitions that define the rectangular grid independently approach zero. We call the limit the **double integral** of f over R:

$$\iint_R f(x, y)\, dA = \lim_{\Delta A \to 0} \sum f(x_k, y_k)\, \Delta A_k.$$

This limit may also exist under less restrictive circumstances.

Double integrals of continuous functions over nonrectangular regions have the same algebraic properties as integrals over rectangular regions. The domain additivity property corresponding to property 5 says that if R is decomposed into nonoverlapping regions R_1 and R_2 with boundaries that are again made of a finite number of line segments or smooth curves (see Fig. 13.7 for an example), then

$$\iint_R f(x, y)\, dA = \iint_{R_1} f(x, y)\, dA + \iint_{R_2} f(x, y)\, dA.$$

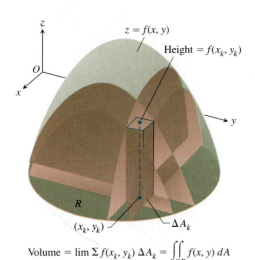

Volume $= \lim \Sigma f(x_k, y_k)\, \Delta A_k = \iint_R f(x, y)\, dA$

13.8 We define the volumes of solids with curved bases the same way we define the volumes of solids with rectangular bases.

If $f(x, y)$ is positive and continuous over R (Fig. 13.8), we define the volume of the solid region between R and the surface $z = f(x, y)$ to be $\iint_R f(x, y)\, dA$, as before.

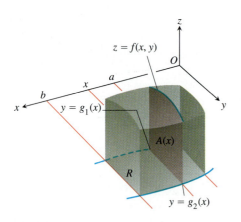

13.9 The area of the vertical slice shown here is

$$A(x) = \int_{g_1(x)}^{g_2(x)} f(x, y)\, dy.$$

To calculate the volume of the solid, we integrate this area from $x = a$ to $x = b$.

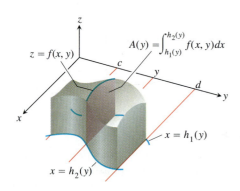

13.10 The volume of the solid shown here is

$$\int_c^d A(y)\, dy = \int_c^d \int_{h_1(y)}^{h_2(y)} f(x, y)\, dx\, dy.$$

If R is a region like the one shown in the xy-plane in Fig. 13.9, bounded "above" and "below" by the curves $y = g_2(x)$ and $y = g_1(x)$ and on the sides by the lines $x = a, x = b$, we may again calculate the volume by the method of slicing. We first calculate the cross-section area

$$A(x) = \int_{y=g_1(x)}^{y=g_2(x)} f(x, y)\, dy$$

and then integrate $A(x)$ from $x = a$ to $x = b$ to get the volume as an iterated integral:

$$V = \int_a^b A(x)\, dx = \int_a^b \int_{g_1(x)}^{g_2(x)} f(x, y)\, dy\, dx. \tag{8}$$

Similarly, if R is a region like the one shown in Fig. 13.10, bounded by the curves $x = h_2(y)$ and $x = h_1(y)$ and the lines $y = c$ and $y = d$, then the volume calculated by slicing is given by the iterated integral

$$\text{Volume} = \int_c^d \int_{h_1(y)}^{h_2(y)} f(x, y)\, dx\, dy. \tag{9}$$

The fact that the iterated integrals in Eqs. (8) and (9) both give the volume that we defined to be the double integral of f over R is a consequence of the following stronger form of Fubini's theorem.

Theorem 2

Fubini's Theorem (Stronger Form)

Let $f(x, y)$ be continuous on a region R.

1. If R is defined by $a \le x \le b$, $g_1(x) \le y \le g_2(x)$, with g_1 and g_2 continuous on $[a, b]$, then

$$\iint_R f(x, y)\, dA = \int_a^b \int_{g_1(x)}^{g_2(x)} f(x, y)\, dy\, dx.$$

2. If R is defined by $c \le y \le d$, $h_1(y) \le x \le h_2(y)$, with h_1 and h_2 continuous on $[c, d]$, then

$$\iint_R f(x, y)\, dA = \int_c^d \int_{h_1(y)}^{h_2(y)} f(x, y)\, dx\, dy.$$

EXAMPLE 2 Find the volume of the prism whose base is the triangle in the xy-plane bounded by the x-axis and the lines $y = x$ and $x = 1$ and whose top lies in the plane

$$z = f(x, y) = 3 - x - y.$$

Solution See Fig. 13.11. For any x between 0 and 1, y may vary from $y = 0$ to $y = x$ (Fig. 13.11b). Hence,

$$V = \int_0^1 \int_0^x (3 - x - y)\, dy\, dx = \int_0^1 \left[3y - xy - \frac{y^2}{2} \right]_{y=0}^{y=x} dx$$

$$= \int_0^1 \left(3x - \frac{3x^2}{2} \right) dx = \left[\frac{3x^2}{2} - \frac{x^3}{2} \right]_{x=0}^{x=1} = 1.$$

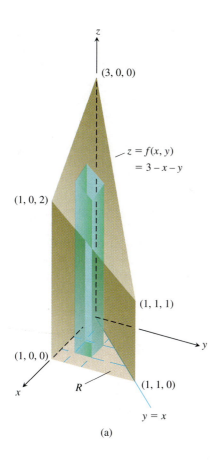

(a)

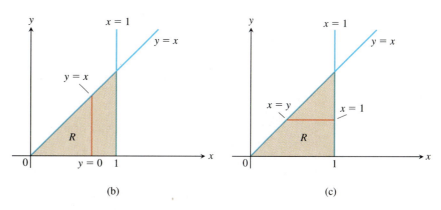

(b)

(c)

13.11 (a) Prism with a triangular base in the xy-plane. The volume of this prism is defined as a double integral over R. To evaluate it as an iterated integral, we may integrate first with respect to y and then with respect to x, or the other way around (Example 2). (b) Integration limits of

$$\int_{x=0}^{x=1} \int_{y=0}^{y=x} f(x, y)\, dy\, dx.$$

If we integrate first with respect to y, we integrate along a vertical line through R and then integrate from left to right to include all the vertical lines in R. (c) Integration limits of

$$\int_{y=0}^{y=1} \int_{x=y}^{x=1} f(x, y)\, dx\, dy.$$

If we integrate first with respect to x, we integrate along a horizontal line through R and then integrate from bottom to top to include all the horizontal lines in R.

When the order of integration is reversed (Fig. 13.11c), the integral for the volume is

$$V = \int_0^1 \int_y^1 (3 - x - y)\, dx\, dy = \int_0^1 \left[3x - \frac{x^2}{2} - xy \right]_{x=y}^{x=1} dy$$

$$= \int_0^1 \left(3 - \frac{1}{2} - y - 3y + \frac{y^2}{2} + y^2 \right) dy$$

$$= \int_0^1 \left(\frac{5}{2} - 4y + \frac{3}{2}y^2 \right) dy = \left[\frac{5}{2}y - 2y^2 + \frac{y^3}{2} \right]_{y=0}^{y=1} = 1.$$

The two integrals are equal, as they should be. ◻

While Fubini's theorem assures us that a double integral may be calculated as an iterated integral in either order of integration, the value of one integral may be easier to find than the value of the other. The next example shows how this can happen.

EXAMPLE 3   Calculate

$$\iint\limits_R \frac{\sin x}{x}\, dA,$$

where R is the triangle in the xy-plane bounded by the x-axis, the line $y = x$, and the line $x = 1$.

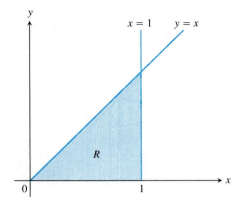

13.12 The region of integration in Example 3.

Solution The region of integration is shown in Fig. 13.12. If we integrate first with respect to y and then with respect to x, we find

$$\int_0^1 \left(\int_0^x \frac{\sin x}{x} \, dy \right) dx = \int_0^1 \left(y \, \frac{\sin x}{x} \right]_{y=0}^{y=x} \right) dx = \int_0^1 \sin x \, dx$$

$$= -\cos(1) + 1 \approx 0.46.$$

If we reverse the order of integration and attempt to calculate

$$\int_0^1 \int_y^1 \frac{\sin x}{x} \, dx \, dy,$$

we are stopped by the fact that $\int ((\sin x)/x) \, dx$ cannot be expressed in terms of elementary functions.

There is no general rule for predicting which order of integration will be the good one in circumstances like these, so don't worry about how to start your integrations. Just forge ahead and if the order you first choose doesn't work, try the other. ❑

Finding the Limits of Integration

The hardest part of evaluating a double integral can be finding the limits of integration. Fortunately, there is a good procedure to follow.

Procedure for Finding Limits of Integration

A. To evaluate $\iint_R f(x, y) \, dA$ over a region R, integrating first with respect to y and then with respect to x, take the following steps:

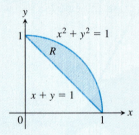
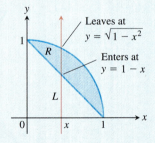
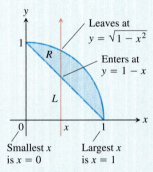

1. *A sketch.* Sketch the region of integration and label the bounding curves.

2. *The y-limits of integration.* Imagine a vertical line L cutting through R in the direction of increasing y. Mark the y-values where L enters and leaves. These are the y-limits of integration.

3. *The x-limits of integration.* Choose x-limits that include all the vertical lines through R. The integral is

$$\iint_R f(x, y) \, dA =$$

$$\int_{x=0}^{x=1} \int_{y=1-x}^{y=\sqrt{1-x^2}} f(x, y) \, dy \, dx.$$

B. To evaluate the same double integral as an iterated integral with the order of integration reversed, use horizontal lines instead of vertical lines. The integral is

$$\iint\limits_{R} f(x, y)\, dA = \int_0^1 \int_{1-y}^{\sqrt{1-y^2}} f(x, y)\, dx\, dy.$$

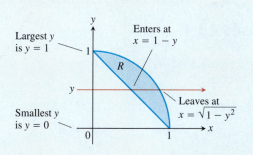

Largest y is $y = 1$

Enters at $x = 1 - y$

R

Leaves at $x = \sqrt{1 - y^2}$

Smallest y is $y = 0$

EXAMPLE 4   Sketch the region of integration for the integral

$$\int_0^2 \int_{x^2}^{2x} (4x + 2)\, dy\, dx$$

and write an equivalent integral with the order of integration reversed.

Solution   The region of integration is given by the inequalities $x^2 \le y \le 2x$ and $0 \le x \le 2$. It is therefore the region bounded by the curves $y = x^2$ and $y = 2x$ between $x = 0$ and $x = 2$ (Fig. 13.13a).

To find limits for integrating in the reverse order, we imagine a horizontal line passing from left to right through the region. It enters at $x = y/2$ and leaves at $x = \sqrt{y}$. To include all such lines, we let y run from $y = 0$ to $y = 4$ (Fig. 13.13b). The integral is

$$\int_0^4 \int_{y/2}^{\sqrt{y}} (4x + 2)\, dx\, dy.$$

The common value of these integrals is 8.

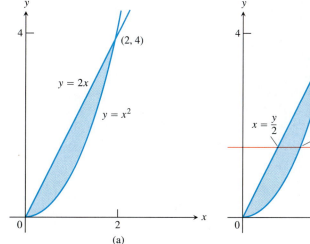
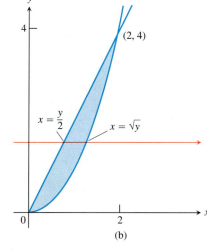

13.13 Figure for Example 4.

Exercises 13.1

Finding Regions of Integration and Double Integrals

In Exercises 1–10, sketch the region of integration and evaluate the integral.

1. $\displaystyle\int_0^3 \int_0^2 (4 - y^2)\, dy\, dx$

2. $\displaystyle\int_0^3 \int_{-2}^0 (x^2 y - 2xy)\, dy\, dx$

3. $\displaystyle\int_{-1}^0 \int_{-1}^1 (x + y + 1)\, dx\, dy$

4. $\displaystyle\int_\pi^{2\pi} \int_0^\pi (\sin x + \cos y)\, dx\, dy$

5. $\displaystyle\int_0^\pi \int_0^x x \sin y\, dy\, dx$

6. $\displaystyle\int_0^\pi \int_0^{\sin x} y\, dy\, dx$

7. $\displaystyle\int_1^{\ln 8} \int_0^{\ln y} e^{x+y}\, dx\, dy$

8. $\displaystyle\int_1^2 \int_y^{y^2} dx\, dy$

9. $\displaystyle\int_0^1 \int_0^{y^2} 3y^3 e^{xy}\, dx\, dy$

10. $\displaystyle\int_1^4 \int_0^{\sqrt{x}} \frac{3}{2} e^{y/\sqrt{x}}\, dy\, dx$

In Exercises 11–16, integrate f over the given region.

11. $f(x, y) = x/y$ over the region in the first quadrant bounded by the lines $y = x$, $y = 2x$, $x = 1$, $x = 2$

12. $f(x, y) = 1/(xy)$ over the square $1 \le x \le 2, 1 \le y \le 2$

13. $f(x, y) = x^2 + y^2$ over the triangular region with vertices $(0, 0)$, $(1, 0)$, and $(0, 1)$

14. $f(x, y) = y \cos xy$ over the rectangle $0 \le x \le \pi, 0 \le y \le 1$

15. $f(u, v) = v - \sqrt{u}$ over the triangular region cut from the first quadrant of the uv-plane by the line $u + v = 1$

16. $f(s, t) = e^s \ln t$ over the region in the first quadrant of the st-plane that lies above the curve $s = \ln t$ from $t = 1$ to $t = 2$

Each of Exercises 17–20 gives an integral over a region in a Cartesian coordinate plane. Sketch the region and evaluate the integral.

17. $\displaystyle\int_{-2}^0 \int_v^{-v} 2\, dp\, dv$ (the pv-plane)

18. $\displaystyle\int_0^1 \int_0^{\sqrt{1-s^2}} 8t\, dt\, ds$ (the st-plane)

19. $\displaystyle\int_{-\pi/3}^{\pi/3} \int_0^{\sec t} 3 \cos t\, du\, dt$ (the tu-plane)

20. $\displaystyle\int_0^3 \int_{-2}^{4-2u} \frac{4 - 2u}{v^2}\, dv\, du$ (the uv-plane)

Reversing the Order of Integration

In Exercises 21–30, sketch the region of integration and write an equivalent double integral with the order of integration reversed.

21. $\displaystyle\int_0^1 \int_2^{4-2x} dy\, dx$

22. $\displaystyle\int_0^2 \int_{y-2}^0 dx\, dy$

23. $\displaystyle\int_0^1 \int_y^{\sqrt{y}} dx\, dy$

24. $\displaystyle\int_0^1 \int_{1-x}^{1-x^2} dy\, dx$

25. $\displaystyle\int_0^1 \int_1^{e^x} dy\, dx$

26. $\displaystyle\int_0^{\ln 2} \int_{e^y}^2 dx\, dy$

27. $\displaystyle\int_0^{3/2} \int_0^{9-4x^2} 16x\, dy\, dx$

28. $\displaystyle\int_0^2 \int_0^{4-y^2} y\, dx\, dy$

29. $\displaystyle\int_0^1 \int_{-\sqrt{1-y^2}}^{\sqrt{1-y^2}} 3y\, dx\, dy$

30. $\displaystyle\int_0^2 \int_{-\sqrt{4-x^2}}^{\sqrt{4-x^2}} 6x\, dy\, dx$

Evaluating Double Integrals

In Exercises 31–40, sketch the region of integration, determine the order of integration, and evaluate the integral.

31. $\displaystyle\int_0^\pi \int_x^\pi \frac{\sin y}{y}\, dy\, dx$

32. $\displaystyle\int_0^2 \int_x^2 2y^2 \sin xy\, dy\, dx$

33. $\displaystyle\int_0^1 \int_y^1 x^2 e^{xy}\, dx\, dy$

34. $\displaystyle\int_0^2 \int_0^{4-x^2} \frac{xe^{2y}}{4 - y}\, dy\, dx$

35. $\displaystyle\int_0^{2\sqrt{\ln 3}} \int_{y/2}^{\sqrt{\ln 3}} e^{x^2}\, dx\, dy$

36. $\displaystyle\int_0^3 \int_{\sqrt{x/3}}^1 e^{y^3}\, dy\, dx$

37. $\displaystyle\int_0^{1/16} \int_{y^{1/4}}^{1/2} \cos(16\pi x^5)\, dx\, dy$

38. $\displaystyle\int_0^8 \int_{\sqrt[3]{x}}^2 \frac{dy\, dx}{y^4 + 1}$

39. $\displaystyle\iint_R (y - 2x^2)\, dA$ where R is the region inside the square $|x| + |y| = 1$

40. $\displaystyle\iint_R xy\, dA$ where R is the region bounded by the lines $y = x$, $y = 2x$, and $x + y = 2$

Volume Beneath a Surface $z = f(x, y)$

41. Find the volume of the region that lies under the paraboloid $z = x^2 + y^2$ and above the triangle enclosed by the lines $y = x$, $x = 0$, and $x + y = 2$ in the xy-plane.

42. Find the volume of the solid that is bounded above by the cylinder $z = x^2$ and below by the region enclosed by the parabola $y = 2 - x^2$ and the line $y = x$ in the xy-plane.

43. Find the volume of the solid whose base is the region in the xy-plane that is bounded by the parabola $y = 4 - x^2$ and the line $y = 3x$, while the top of the solid is bounded by the plane $z = x + 4$.

44. Find the volume of the solid in the first octant bounded by the coordinate planes, the cylinder $x^2 + y^2 = 4$, and the plane $z + y = 3$.

45. Find the volume of the solid in the first octant bounded by the coordinate planes, the plane $x = 3$, and the parabolic cylinder $z = 4 - y^2$.

46. Find the volume of the solid cut from the first octant by the surface $z = 4 - x^2 - y$.

47. Find the volume of the wedge cut from the first octant by the cylinder $z = 12 - 3y^2$ and the plane $x + y = 2$.

48. Find the volume of the solid cut from the square column $|x| + |y| \le 1$ by the planes $z = 0$ and $3x + z = 3$.

49. Find the volume of the solid that is bounded on the front and back by the planes $x = 2$ and $x = 1$, on the sides by the cylinders $y = \pm 1/x$, and above and below by the planes $z = x + 1$ and $z = 0$.

50. Find the volume of the solid that is bounded on the front and back by the planes $x = \pm \pi/3$, on the sides by the cylinders $y = \pm \sec x$, above by the cylinder $z = 1 + y^2$, and below by the xy-plane.

Integrals over Unbounded Regions

Evaluate the improper integrals in Exercises 51–54 as iterated integrals.

51. $\displaystyle \int_1^\infty \int_{e^{-x}}^1 \frac{1}{x^3 y} \, dy \, dx$

52. $\displaystyle \int_{-1}^1 \int_{-1/\sqrt{1-x^2}}^{1/\sqrt{1-x^2}} (2y + 1) \, dy \, dx$

53. $\displaystyle \int_{-\infty}^\infty \int_{-\infty}^\infty \frac{1}{(x^2 + 1)(y^2 + 1)} \, dx \, dy$

54. $\displaystyle \int_0^\infty \int_0^\infty x e^{-(x+2y)} \, dx \, dy$

Approximating Double Integrals

In Exercises 55 and 56, approximate the double integral of $f(x, y)$ over the region R partitioned by the given vertical lines $x = a$ and horizontal lines $y = c$. In each subrectangle use (x_k, y_k) as indicated for your approximation.

$$\iint_R f(x, y) \, dA \approx \sum_{k=1}^n f(x_k, y_k) \, \Delta A_k$$

55. $f(x, y) = x + y$ over the region R bounded above by the semicircle $y = \sqrt{1 - x^2}$ and below by the x-axis, using the partition $x = -1, -1/2, 0, 1/4, 1/2, 1$ and $y = 0, 1/2, 1$ with (x_k, y_k) the lower left corner in the kth subrectangle (provided the subrectangle lies within R)

56. $f(x, y) = x + 2y$ over the region R inside the circle $(x - 2)^2 + (y - 3)^2 = 1$ using the partition $x = 1, 3/2, 2, 5/2, 3$ and $y = 2, 5/2, 3, 7/2, 4$ with (x_k, y_k) the center (centroid) in the kth subrectangle (provided it lies within R)

Theory and Examples

57. Integrate $f(x, y) = \sqrt{4 - x^2}$ over the smaller sector cut from the disk $x^2 + y^2 \le 4$ by the rays $\theta = \pi/6$ and $\theta = \pi/2$.

58. Integrate $f(x, y) = 1/[(x^2 - x)(y - 1)^{2/3}]$ over the infinite rectangle $2 \le x < \infty$, $0 \le y \le 2$.

59. A solid right (noncircular) cylinder has its base R in the xy-plane and is bounded above by the paraboloid $z = x^2 + y^2$. The cylinder's volume is

$$V = \int_0^1 \int_0^y (x^2 + y^2) \, dx \, dy + \int_1^2 \int_0^{2-y} (x^2 + y^2) \, dx \, dy.$$

Sketch the base region R and express the cylinder's volume as a single iterated integral with the order of integration reversed. Then evaluate the integral to find the volume.

60. Evaluate the integral

$$\int_0^2 (\tan^{-1} \pi x - \tan^{-1} x) \, dx.$$

(*Hint:* Write the integrand as an integral.)

61. What region R in the xy-plane maximizes the value of

$$\iint_R (4 - x^2 - 2y^2) \, dA?$$

Give reasons for your answer.

62. What region R in the xy-plane minimizes the value of

$$\iint_R (x^2 + y^2 - 9) \, dA?$$

Give reasons for your answer.

63. Is it all right to evaluate the integral of a continuous function $f(x, y)$ over a rectangular region in the xy-plane and get different answers depending on the order of integration? Give reasons for your answer.

64. How would you evaluate the double integral of a continuous function $f(x, y)$ over the region R in the xy-plane enclosed by the triangle with vertices $(0, 1)$, $(2, 0)$, and $(1, 2)$? Give reasons for your answer.

65. Prove that

$$\int_{-\infty}^\infty \int_{-\infty}^\infty e^{-x^2 - y^2} dx \, dy = \lim_{b \to \infty} \int_{-b}^b \int_{-b}^b e^{-x^2 - y^2} dx \, dy$$

$$= 4 \left(\int_0^\infty e^{-x^2} dx \right)^2.$$

66. Evaluate the improper integral $\displaystyle \int_0^1 \int_0^3 \frac{x^2}{(y - 1)^{2/3}} \, dy \, dx$.

✳ Numerical Evaluation

Use a double-integral evaluator to estimate the values of the integrals in Exercises 67–70.

67. $\displaystyle \int_1^3 \int_1^x \frac{1}{xy} \, dy \, dx$

68. $\displaystyle \int_0^1 \int_0^1 e^{-(x^2 + y^2)} \, dy \, dx$

69. $\displaystyle \int_0^1 \int_0^1 \tan^{-1} xy \, dy \, dx$

70. $\displaystyle \int_{-1}^1 \int_0^{\sqrt{1-x^2}} 3\sqrt{1 - x^2 - y^2} \, dy \, dx$

Areas, Moments, and Centers of Mass

In this section we show how to use double integrals to define and calculate the areas of bounded regions in the plane and the masses, moments, centers of mass, and radii of gyration of thin plates covering these regions. The calculations are similar to the ones in Chapter 5, but now we can handle a greater variety of shapes.

Areas of Bounded Regions in the Plane

If we take $f(x, y) = 1$ in the definition of the double integral over a region R in the preceding section, the partial sums reduce to

$$S_n = \sum_{k=1}^{n} f(x_k, y_k) \Delta A_k = \sum_{k=1}^{n} \Delta A_k. \tag{1}$$

This approximates what we would like to call the area of R. As Δx and Δy approach zero, the coverage of R by the ΔA_k's (Fig. 13.14) becomes increasingly complete, and we define the area of R to be the limit

$$\text{Area} = \lim_{n \to \infty} \sum_{k=1}^{n} \Delta A_k = \iint_R dA. \tag{2}$$

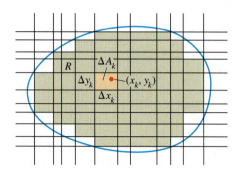

13.14 The first step in defining the area of a region is to partition the interior of the region into cells.

Definition

The **area** of a closed, bounded plane region R is

$$A = \iint_R dA. \tag{3}$$

As with the other definitions in this chapter, the definition here applies to a greater variety of regions than does the earlier single-variable definition of area, but it agrees with the earlier definition on regions to which they both apply.

To evaluate the integral in (3), we integrate the constant function $f(x, y) = 1$ over R.

EXAMPLE 1 Find the area of the region R bounded by $y = x$ and $y = x^2$ in the first quadrant.

Solution We sketch the region (Fig. 13.15) and calculate the area as

$$A = \int_0^1 \int_{x^2}^{x} dy\, dx = \int_0^1 \left[y \right]_{x^2}^{x} dx = \int_0^1 (x - x^2)\, dx = \left[\frac{x^2}{2} - \frac{x^3}{3} \right]_0^1 = \frac{1}{6}. \quad \square$$

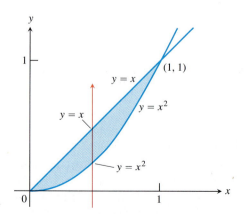

13.15 The area of the region between the parabola and the line in Example 1 is

$$\int_0^1 \int_{x^2}^{x} dy\, dx.$$

EXAMPLE 2 Find the area of the region R enclosed by the parabola $y = x^2$ and the line $y = x + 2$.

Solution If we divide R into the regions R_1 and R_2 shown in Fig. 13.16(a), we may

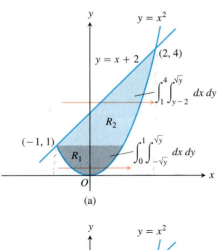

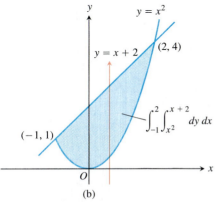

13.16 Calculating this area takes (a) two double integrals if the first integration is with respect to x, but (b) only one if the first integration is with respect to y (Example 2).

Global warming

The "global warming" controversy deals with whether the average air temperature over the surface of the earth is increasing.

---

calculate the area as

$$A = \iint_{R_1} dA + \iint_{R_2} dA = \int_0^1 \int_{-\sqrt{y}}^{\sqrt{y}} dx\,dy + \int_1^4 \int_{y-2}^{\sqrt{y}} dx\,dy.$$

On the other hand, reversing the order of integration (Fig. 13.16b) gives

$$A = \int_{-1}^2 \int_{x^2}^{x+2} dy\,dx.$$

This result is simpler and is the only one we would bother to write down in practice. The area is

$$A = \int_{-1}^2 \left[y \right]_{x^2}^{x+2} dx = \int_{-1}^2 (x + 2 - x^2)\,dx = \left[\frac{x^2}{2} + 2x - \frac{x^3}{3} \right]_{-1}^2 = \frac{9}{2}. \qquad \square$$

Average Value

The average value of an integrable function of a single variable on a closed interval is the integral of the function over the interval divided by the length of the interval. For an integrable function of two variables defined on a closed and bounded region that has a measurable area, the average value is the integral over the region divided by the area of the region. If f is the function and R the region, then

$$\textbf{Average value of } f \textbf{ over } R = \frac{1}{\text{area of } R} \iint_R f\,dA. \qquad (4)$$

If f is the area density of a thin plate covering R, then the double integral of f over R divided by the area of R is the plate's average density in units of mass per unit area. If $f(x, y)$ is the distance from the point (x, y) to a fixed point P, then the average value of f over R is the average distance of points in R from P.

EXAMPLE 3   Find the average value of $f(x, y) = x \cos xy$ over the rectangle $R: 0 \le x \le \pi, \ 0 \le y \le 1$.

Solution   The value of the integral of f over R is

$$\int_0^\pi \int_0^1 x \cos xy\,dy\,dx = \int_0^\pi \left[\sin xy \right]_{y=0}^{y=1} dx$$

$$= \int_0^\pi (\sin x - 0)\,dx = -\cos x \Big]_0^\pi = 1 + 1 = 2.$$

The area of R is π. The average value of f over R is $2/\pi$. $\qquad \square$

First and Second Moments and Centers of Mass

To find the moments and centers of mass of thin sheets and plates, we use formulas similar to those in Chapter 5. The main difference is that now, with double integrals, we can accommodate a greater variety of shapes and density functions. The formulas are given in Table 13.1, on the following page. The examples that follow show how the formulas are used.

The mathematical difference between the **first moments** M_x and M_y and the **moments of inertia,** or **second moments,** I_x and I_y is that the second moments use the *squares* of the "lever-arm" distances x and y.

Table 13.1 Mass and moment formulas for thin plates covering regions in the *xy*-plane

---

Density: $\quad \delta(x, y)$

Mass: $\quad M = \iint \delta(x, y)\, dA$ $\qquad$ **First moments:** $\quad M_x = \iint y\delta(x, y)\, dA, \quad M_y = \iint x\delta(x, y)\, dA$

Center of mass: $\quad \bar{x} = \dfrac{M_y}{M}, \quad \bar{y} = \dfrac{M_x}{M}$

Moments of inertia (second moments):

$\quad$ About the *x*-axis: $\quad I_x = \iint y^2\, \delta(x, y)\, dA$ $\qquad$ About the origin $\quad I_0 = \iint (x^2 + y^2)\, \delta(x, y)\, dA = I_x + I_y$
(polar moment):

$\quad$ About the *y*-axis: $\quad I_y = \iint x^2\, \delta(x, y)\, dA$

$\quad$ About a line *L*: $\quad I_L = \iint r^2(x, y)\delta(x, y)\, dA$, where $r(x, y)$ = distance from (x, y) to L

Radii of gyration: $\quad$ About the *x*-axis: $\quad R_x = \sqrt{I_x/M}$

$\qquad\qquad\qquad\quad$ About the *y*-axis: $\quad R_y = \sqrt{I_y/M}$

$\qquad\qquad\qquad\quad$ About the origin: $\quad R_0 = \sqrt{I_0/M}$

The moment I_0 is also called the **polar moment** of inertia about the origin. It is calculated by integrating the density $\delta(x, y)$ (mass per unit area) times $r^2 = x^2 + y^2$, the square of the distance from a representative point (x, y) to the origin. Notice that $I_0 = I_x + I_y$; once we find two, we get the third automatically. (The moment I_0 is sometimes called I_z, for moment of inertia about the *z*-axis. The identity $I_z = I_x + I_y$ is then called the **Perpendicular Axis Theorem.**)

The **radius of gyration** R_x is defined by the equation

$$I_x = MR_x{}^2.$$

It tells how far from the *x*-axis the entire mass of the plate might be concentrated to give the same I_x. The radius of gyration gives a convenient way to express the moment of inertia in terms of a mass and a length. The radii R_y and R_0 are defined in a similar way, with

$$I_y = MR_y{}^2 \qquad \text{and} \qquad I_0 = MR_0{}^2.$$

We take square roots to get the formulas in Table 13.1.

Why the interest in moments of inertia? A body's first moments tell us about balance and about the torque the body exerts about different axes in a gravitational field. But if the body is a rotating shaft, we are more likely to be interested in how much energy is stored in the shaft or about how much energy it will take to accelerate the shaft to a particular angular velocity. This is where the second moment or moment of inertia comes in.

Think of partitioning the shaft into small blocks of mass Δm_k and let r_k denote the distance from the *k*th block's center of mass to the axis of rotation (Fig. 13.17). If the shaft rotates at an angular velocity of $\omega = d\theta/dt$ radians per second, the block's center of mass will trace its orbit at a linear speed of

$$v_k = \frac{d}{dt}\left(r_k\theta\right) = r_k\frac{d\theta}{dt} = r_k\,\omega. \tag{5}$$

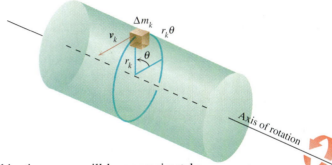

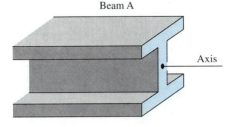

Beam A

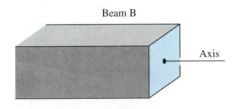

Beam B

13.17 To find an integral for the amount of energy stored in a rotating shaft, we first imagine the shaft to be partitioned into small blocks. Each block has its own kinetic energy. We add the contributions of the individual blocks to find the kinetic energy of the shaft.

The block's kinetic energy will be approximately

$$\frac{1}{2}\Delta m_k v_k{}^2 = \frac{1}{2}\Delta m_k (r_k\,\omega)^2 = \frac{1}{2}\,\omega^2 r_k{}^2\,\Delta m_k. \tag{6}$$

The kinetic energy of the shaft will be approximately

$$\sum \frac{1}{2}\,\omega^2 r_k{}^2 \Delta m_k. \tag{7}$$

The integral approached by these sums as the shaft is partitioned into smaller and smaller blocks gives the shaft's kinetic energy:

$$\mathrm{KE}_{\mathrm{shaft}} = \int \frac{1}{2}\omega^2 r^2\,dm = \frac{1}{2}\omega^2 \int r^2\,dm. \tag{8}$$

The factor

$$I = \int r^2\,dm \tag{9}$$

is the moment of inertia of the shaft about its axis of rotation, and we see from Eq. (8) that the shaft's kinetic energy is

$$\mathrm{KE}_{\mathrm{shaft}} = \frac{1}{2}I\omega^2. \tag{10}$$

To start a shaft of inertial moment I rotating at an angular velocity ω, we need to provide a kinetic energy of $\mathrm{KE} = (1/2)I\omega^2$. To stop the shaft, we have to take this amount of energy back out. To start a locomotive with mass m moving at a linear velocity v, we need to provide a kinetic energy of $\mathrm{KE} = (1/2)\,mv^2$. To stop the locomotive, we have to remove this amount of energy. The shaft's moment of inertia is analogous to the locomotive's mass. What makes the locomotive hard to start or stop is its mass. What makes the shaft hard to start or stop is its moment of inertia. The moment of inertia takes into account not only the mass but also its distribution.

The moment of inertia also plays a role in determining how much a horizontal metal beam will bend under a load. The stiffness of the beam is a constant times I, the polar moment of inertia of a typical cross section of the beam perpendicular to the beam's longitudinal axis. The greater the value of I, the stiffer the beam and the less it will bend under a given load. That is why we use I beams instead of beams whose cross sections are square. The flanges at the top and bottom of the beam hold most of the beam's mass away from the longitudinal axis to maximize the value of I (Fig. 13.18).

If you want to see the moment of inertia at work, try the following experiment. Tape two coins to the ends of a pencil and twiddle the pencil about the center of mass. The moment of inertia accounts for the resistance you feel each time you change the direction of motion. Now move the coins an equal distance toward the

13.18 The greater the polar moment of inertia of the cross section of a beam about the beam's longitudinal axis, the stiffer the beam. Beams A and B have the same cross-section area, but A is stiffer.

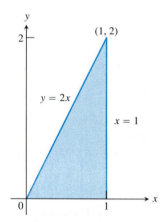

13.19 The triangular region covered by the plate in Example 4.

center of mass and twiddle the pencil again. The system has the same mass and the same center of mass but now offers less resistance to the changes in motion. The moment of inertia has been reduced. The moment of inertia is what gives a baseball bat, golf club, or tennis racket its "feel." Tennis rackets that weigh the same, look the same, and have identical centers of mass will feel different and behave differently if their masses are not distributed the same way.

EXAMPLE 4   A thin plate covers the triangular region bounded by the x-axis and the lines $x = 1$ and $y = 2x$ in the first quadrant. The plate's density at the point (x, y) is $\delta(x, y) = 6x + 6y + 6$. Find the plate's mass, first moments, center of mass, moments of inertia, and radii of gyration about the coordinate axes.

Solution   We sketch the plate and put in enough detail to determine the limits of integration for the integrals we have to evaluate (Fig. 13.19).

The plate's mass is

$$M = \int_0^1 \int_0^{2x} \delta(x, y)\, dy\, dx = \int_0^1 \int_0^{2x} (6x + 6y + 6)\, dy\, dx$$

$$= \int_0^1 \left[6xy + 3y^2 + 6y \right]_{y=0}^{y=2x} dx$$

$$= \int_0^1 (24x^2 + 12x)\, dx = \left[8x^3 + 6x^2 \right]_0^1 = 14.$$

The first moment about the x-axis is

$$M_x = \int_0^1 \int_0^{2x} y\delta(x, y)\, dy\, dx = \int_0^1 \int_0^{2x} (6xy + 6y^2 + 6y)\, dy\, dx$$

$$= \int_0^1 \left[3xy^2 + 2y^3 + 3y^2 \right]_{y=0}^{y=2x} dx = \int_0^1 (28x^3 + 12x^2)\, dx$$

$$= \left[7x^4 + 4x^3 \right]_0^1 = 11.$$

A similar calculation gives

$$M_y = \int_0^1 \int_0^{2x} x\delta(x, y)\, dy\, dx = 10.$$

The coordinates of the center of mass are therefore

$$\bar{x} = \frac{M_y}{M} = \frac{10}{14} = \frac{5}{7}, \qquad \bar{y} = \frac{M_x}{M} = \frac{11}{14}.$$

The moment of inertia about the x-axis is

$$I_x = \int_0^1 \int_0^{2x} y^2\delta(x, y)\, dy\, dx = \int_0^1 \int_0^{2x} (6xy^2 + 6y^3 + 6y^2)\, dy\, dx$$

$$= \int_0^1 \left[2xy^3 + \frac{3}{2}y^4 + 2y^3 \right]_{y=0}^{y=2x} dx = \int_0^1 (40x^4 + 16x^3)\, dx$$

$$= \left[8x^5 + 4x^4 \right]_0^1 = 12.$$

Similarly, the moment of inertia about the y-axis is

$$I_y = \int_0^1 \int_0^{2x} x^2 \delta(x, y)\, dy\, dx = \frac{39}{5}.$$

Since we know I_x and I_y, we do not need to evaluate an integral to find I_0; we can use the equation $I_0 = I_x + I_y$ instead:

$$I_0 = 12 + \frac{39}{5} = \frac{60 + 39}{5} = \frac{99}{5}.$$

The three radii of gyration are

$$R_x = \sqrt{I_x/M} = \sqrt{12/14} = \sqrt{6/7}$$

$$R_y = \sqrt{I_y/M} = \sqrt{\left(\frac{39}{5}\right)/14} = \sqrt{39/70}$$

$$R_0 = \sqrt{I_0/M} = \sqrt{\left(\frac{99}{5}\right)/14} = \sqrt{99/70}.$$

Centroids of Geometric Figures

When the density of an object is constant, it cancels out of the numerator and denominator of the formulas for $\bar{x}$ and $\bar{y}$. As far as $\bar{x}$ and $\bar{y}$ are concerned, δ might as well be 1. Thus, when δ is constant, the location of the center of mass becomes a feature of the object's shape and not of the material of which it is made. In such cases, engineers may call the center of mass the **centroid** of the shape. To find a centroid, we set δ equal to 1 and proceed to find $\bar{x}$ and $\bar{y}$ as before, by dividing first moments by masses.

EXAMPLE 5   Find the centroid of the region in the first quadrant that is bounded above by the line $y = x$ and below by the parabola $y = x^2$.

Solution   We sketch the region and include enough detail to determine the limits of integration (Fig. 13.20). We then set δ equal to 1 and evaluate the appropriate formulas from Table 13.1:

$$M = \int_0^1 \int_{x^2}^x 1\, dy\, dx = \int_0^1 \Big[y \Big]_{y=x^2}^{y=x} dx = \int_0^1 (x - x^2)\, dx = \left[\frac{x^2}{2} - \frac{x^3}{3} \right]_0^1 = \frac{1}{6}$$

$$M_x = \int_0^1 \int_{x^2}^x y\, dy\, dx = \int_0^1 \left[\frac{y^2}{2} \right]_{y=x^2}^{y=x} dx$$

$$= \int_0^1 \left(\frac{x^2}{2} - \frac{x^4}{2} \right) dx = \left[\frac{x^3}{6} - \frac{x^5}{10} \right]_0^1 = \frac{1}{15}$$

$$M_y = \int_0^1 \int_{x^2}^x x\, dy\, dx = \int_0^1 \Big[xy \Big]_{y=x^2}^{y=x} dx = \int_0^1 (x^2 - x^3)\, dx = \left[\frac{x^3}{3} - \frac{x^4}{4} \right]_0^1 = \frac{1}{12}.$$

From these values of M, M_x, and M_y, we find

$$\bar{x} = \frac{M_y}{M} = \frac{1/12}{1/6} = \frac{1}{2} \quad \text{and} \quad \bar{y} = \frac{M_x}{M} = \frac{1/15}{1/6} = \frac{2}{5}.$$

The centroid is the point $\left(\frac{1}{2}, \frac{2}{5} \right)$.

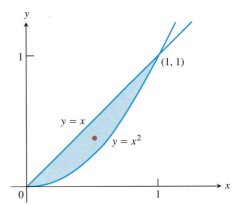

13.20 Example 5 finds the centroid of the region shown here.

Exercises 13.2

Area by Double Integration

In Exercises 1–8, sketch the region bounded by the given lines and curves. Then express the region's area as an iterated double integral and evaluate the integral.

1. The coordinate axes and the line $x + y = 2$

2. The lines $x = 0$, $y = 2x$, and $y = 4$

3. The parabola $x = -y^2$ and the line $y = x + 2$

4. The parabola $x = y - y^2$ and the line $y = -x$

5. The curve $y = e^x$ and the lines $y = 0$, $x = 0$, and $x = \ln 2$

6. The curves $y = \ln x$ and $y = 2 \ln x$ and the line $x = e$, in the first quadrant

7. The parabolas $x = y^2$ and $x = 2y - y^2$

8. The parabolas $x = y^2 - 1$ and $x = 2y^2 - 2$

The integrals and sums of integrals in Exercises 9–14 give the areas of regions in the xy-plane. Sketch each region, label each bounding curve with its equation, and give the coordinates of the points where the curves intersect. Then find the area of the region.

9. $\displaystyle\int_0^6 \int_{y^2/3}^{2y} dx\, dy$

10. $\displaystyle\int_0^3 \int_{-x}^{x(2-x)} dy\, dx$

11. $\displaystyle\int_0^{\pi/4} \int_{\sin x}^{\cos x} dy\, dx$

12. $\displaystyle\int_{-1}^2 \int_{y^2}^{y+2} dx\, dy$

13. $\displaystyle\int_{-1}^0 \int_{-2x}^{1-x} dy\, dx + \int_0^2 \int_{-x/2}^{1-x} dy\, dx$

14. $\displaystyle\int_0^2 \int_{x^2-4}^0 dy\, dx + \int_0^4 \int_0^{\sqrt{x}} dy\, dx$

Average Values

15. Find the average value of $f(x, y) = \sin(x + y)$ over

    a) the rectangle $0 \le x \le \pi$, $0 \le y \le \pi$,
    b) the rectangle $0 \le x \le \pi$, $0 \le y \le \pi/2$.

16. Which do you think will be larger, the average value of $f(x, y) = xy$ over the square $0 \le x \le 1$, $0 \le y \le 1$, or the average value of f over the quarter circle $x^2 + y^2 \le 1$ in the first quadrant? Calculate them to find out.

17. Find the average height of the paraboloid $z = x^2 + y^2$ over the square $0 \le x \le 2$, $0 \le y \le 2$.

18. Find the average value of $f(x, y) = 1/(xy)$ over the square $\ln 2 \le x \le 2 \ln 2$, $\ln 2 \le y \le 2 \ln 2$.

Constant Density

19. Find the center of mass of a thin plate of density $\delta = 3$ bounded by the lines $x = 0$, $y = x$, and the parabola $y = 2 - x^2$ in the first quadrant.

20. Find the moments of inertia and radii of gyration about the coordinate axes of a thin rectangular plate of constant density δ bounded by the lines $x = 3$ and $y = 3$ in the first quadrant.

21. Find the centroid of the region in the first quadrant bounded by the x-axis, the parabola $y^2 = 2x$, and the line $x + y = 4$.

22. Find the centroid of the triangular region cut from the first quadrant by the line $x + y = 3$.

23. Find the centroid of the semicircular region bounded by the x-axis and the curve $y = \sqrt{1 - x^2}$.

24. The area of the region in the first quadrant bounded by the parabola $y = 6x - x^2$ and the line $y = x$ is 125/6 square units. Find the centroid.

25. Find the centroid of the region cut from the first quadrant by the circle $x^2 + y^2 = a^2$.

26. Find the moment of inertia about the x-axis of a thin plate of density $\delta = 1$ bounded by the circle $x^2 + y^2 = 4$. Then use your result to find I_y and I_0 for the plate.

27. Find the centroid of the region between the x-axis and the arch $y = \sin x$, $0 \le x \le \pi$.

28. Find the moment of inertia with respect to the y-axis of a thin sheet of constant density $\delta = 1$ bounded by the curve $y = (\sin^2 x)/x^2$ and the interval $\pi \le x \le 2\pi$ of the x-axis.

29. *The centroid of an infinite region.* Find the centroid of the infinite region in the second quadrant enclosed by the coordinate axes and the curve $y = e^x$. (Use improper integrals in the mass-moment formulas.)

30. *The first moment of an infinite plate.* Find the first moment about the y-axis of a thin plate of density $\delta(x, y) = 1$ covering the infinite region under the curve $y = e^{-x^2/2}$ in the first quadrant.

Variable Density

31. Find the moment of inertia and radius of gyration about the x-axis of a thin plate bounded by the parabola $x = y - y^2$ and the line $x + y = 0$ if $\delta(x, y) = x + y$.

32. Find the mass of a thin plate occupying the smaller region cut from the ellipse $x^2 + 4y^2 = 12$ by the parabola $x = 4y^2$ if $\delta(x, y) = 5x$.

33. Find the center of mass of a thin triangular plate bounded by the y-axis and the lines $y = x$ and $y = 2 - x$ if $\delta(x, y) = 6x + 3y + 3$.

34. Find the center of mass and moment of inertia about the x-axis of a thin plate bounded by the curves $x = y^2$ and $x = 2y - y^2$ if the density at the point (x, y) is $\delta(x, y) = y + 1$.

35. Find the center of mass and the moment of inertia and radius of gyration about the y-axis of a thin rectangular plate cut from the first quadrant by the lines $x = 6$ and $y = 1$ if $\delta(x, y) = x + y + 1$.

36. Find the center of mass and the moment of inertia and radius of gyration about the y-axis of a thin plate bounded by the line $y = 1$ and the parabola $y = x^2$ if the density is $\delta(x, y) = y + 1$.

37. Find the center of mass and the moment of inertia and radius of gyration about the y-axis of a thin plate bounded by the x-axis, the lines $x = \pm 1$, and the parabola $y = x^2$ if $\delta(x, y) = 7y + 1$.

38. Find the center of mass and the moment of inertia and radius of gyration about the x-axis of a thin rectangular plate bounded by the lines $x = 0$, $x = 20$, $y = -1$, and $y = 1$ if $\delta(x, y) = 1 + (x/20)$.

39. Find the center of mass, the moments of inertia and radii of gyration about the coordinate axes, and the polar moment of inertia and radius of gyration of a thin triangular plate bounded by the lines $y = x$, $y = -x$, and $y = 1$ if $\delta(x, y) = y + 1$.

40. Repeat Exercise 39 for $\delta(x, y) = 3x^2 + 1$.

Theory and Examples

41. If $f(x, y) = (10{,}000\, e^y)/(1 + |x|/2)$ represents the "population density" of a certain bacteria on the xy-plane, where x and y are measured in centimeters, find the total population of bacteria within the rectangle $-5 \le x \le 5$ and $-2 \le y \le 0$.

42. If $f(x, y) = 100\,(y + 1)$ represents the population density of a planar region on Earth, where x and y are measured in miles, find the number of people in the region bounded by the curves $x = y^2$ and $x = 2y - y^2$.

43. *Appliance design.* When we design an appliance, one of the concerns is how hard the appliance will be to tip over. When tipped, it will right itself as long as its center of mass lies on the correct side of the *fulcrum*, the point on which the appliance is riding as it tips. Suppose the profile of an appliance of approximately constant density is parabolic, like an old-fashioned radio. It fills the region $0 \le y \le a(1 - x^2)$, $-1 \le x \le 1$, in the xy-plane (Fig. 13.21). What values of a will guarantee that the appliance will have to be tipped more than $45°$ to fall over?

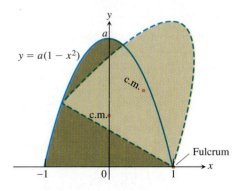

13.21 The profile of the appliance in Exercise 43.

44. *Minimizing a moment of inertia.* A rectangular plate of constant density $\delta(x, y) = 1$ occupies the region bounded by the lines $x = 4$ and $y = 2$ in the first quadrant. The moment of inertia I_a of the rectangle about the line $y = a$ is given by the integral

$$I_a = \int_0^4 \int_0^2 (y - a)^2 \, dy\, dx.$$

Find the value of a that minimizes I_a.

45. Find the centroid of the infinite region in the xy-plane bounded by the curves $y = 1/\sqrt{1 - x^2}$, $y = -1/\sqrt{1 - x^2}$, and the lines $x = 0$, $x = 1$.

46. Find the radius of gyration of a slender rod of constant linear density δ gm/cm and length L cm with respect to an axis

a) through the rod's center of mass perpendicular to the rod's axis;

b) perpendicular to the rod's axis at one end of the rod.

47. A thin plate of constant density δ occupies the region R in the xy-plane bounded by the curves $x = y^2$ and $x = 2y - y^2$ (see Exercise 34).

a) Find δ such that the plate has the same mass as the plate in Exercise 34.

b) Compare the value of δ found in part (a) with the average value of $\delta(x, y) = y + 1$ over R.

48. According to the *Texas Almanac*, Texas has 254 counties and a National Weather Service station in each county. Assume that at time t_0 each of the 254 weather stations recorded the local temperature. Find a formula that would give a reasonable approximation to the average temperature in Texas at time t_0. Your answer should involve information that is readily available in the *Texas Almanac*.

The Parallel Axis Theorem

Let $L_{c.m.}$ be a line in the xy-plane that runs through the center of mass of a thin plate of mass m covering a region in the plane. Let L be a line in the plane parallel to and h units away from $L_{c.m.}$. The **Parallel Axis Theorem** says that under these conditions the moments of inertia I_L and $I_{c.m.}$ of the plate about L and $L_{c.m.}$ satisfy the equation

$$I_L = I_{c.m.} + mh^2. \tag{1}$$

This equation gives a quick way to calculate one moment when the other moment and the mass are known.

49. *Proof of the Parallel Axis Theorem*

a) Show that the first moment of a thin flat plate about any line in the plane of the plate through the plate's center of mass is zero. (*Hint:* Place the center of mass at the origin with the line along the y-axis. What does the formula $\bar{x} = M_y/M$ then tell you?)

b) Use the result in (a) to derive the Parallel Axis Theorem. Assume that the plane is coordinatized in a way that makes $L_{c.m.}$ the y-axis and L the line $x = h$. Then expand the integrand of the integral for I_L to rewrite the integral as the sum of integrals whose values you recognize.

50. a) Use the Parallel Axis Theorem and the results of Example 4 to find the moments of inertia of the plate in Example 4 about the vertical and horizontal lines through the plate's center of mass.

b) Use the results in (a) to find the plate's moments of inertia about the lines $x = 1$ and $y = 2$.

Pappus's Formula

In addition to stating the centroid theorems in Section 5.10, Pappus knew that the centroid of the union of two nonoverlapping plane regions lies on the line segment joining their individual centroids. More specifically, suppose that m_1 and m_2 are the masses of thin plates P_1 and P_2 that cover nonoverlapping regions in the xy-plane. Let $\mathbf{c}_1$ and $\mathbf{c}_2$ be the vectors from the origin to the respective centers of mass of P_1 and P_2. Then the center of mass of the union $P_1 \cup P_2$ of the two plates is determined by the vector

$$\mathbf{c} = \frac{m_1\mathbf{c}_1 + m_2\mathbf{c}_2}{m_1 + m_2}. \qquad (2)$$

Equation (2) is known as **Pappus's formula.** For more than two nonoverlapping plates, as long as their number is finite, the formula generalizes to

$$\mathbf{c} = \frac{m_1\mathbf{c}_1 + m_2\mathbf{c}_2 + \cdots + m_n\mathbf{c}_n}{m_1 + m_2 + \cdots + m_n}. \qquad (3)$$

This formula is especially useful for finding the centroid of a plate of irregular shape that is made up of pieces of constant density whose centroids we know from geometry. We find the centroid of each piece and apply Eq. (3) to find the centroid of the plate.

51. Derive Pappus's formula (Eq. 2). (*Hint:* Sketch the plates as regions in the first quadrant and label their centers of mass as $(\bar{x}_1, \bar{y}_1)$ and $(\bar{x}_2, \bar{y}_2)$. What are the moments of $P_1 \cup P_2$ about the coordinate axes?)

52. Use Eq. (2) and mathematical induction to show that Eq. (3) holds for any positive integer $n > 2$.

53. Let A, B, and C be the shapes indicated in Fig. 13.22(a). Use Pappus's formula to find the centroid of

a) $A \cup B$ **b)** $A \cup C$ **c)** $B \cup C$
d) $A \cup B \cup C$

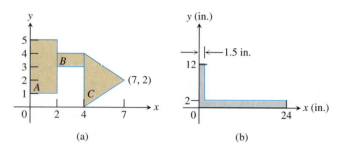

13.22 The figures for Exercises 53 and 54.

54. Locate the center of mass of the carpenter's square in Fig. 13.22(b).

55. An isosceles triangle T has base $2a$ and altitude h. The base lies along the diameter of a semicircular disk D of radius a so that the two together make a shape resembling an ice cream cone. What relation must hold between a and h to place the centroid of $T \cup D$ on the common boundary of T and D? inside T?

56. An isosceles triangle T of altitude h has as its base one side of a square Q whose edges have length s. (The square and triangle do not overlap.) What relation must hold between h and s to place the centroid of $T \cup Q$ on the base of the triangle? Compare your answer with the answer to Exercise 55.

13.3

Double Integrals in Polar Form

Integrals are sometimes easier to evaluate if we change to polar coordinates. This section shows how to accomplish the change and how to evaluate integrals over regions whose boundaries are given by polar equations.

Integrals in Polar Coordinates

When we defined the double integral of a function over a region R in the xy-plane, we began by cutting R into rectangles whose sides were parallel to the coordinate axes. These were the natural shapes to use because their sides have either constant x-values or constant y-values. In polar coordinates, the natural shape is a "polar rectangle" whose sides have constant r- and θ-values.

Suppose that a function $f(r, \theta)$ is defined over a region R that is bounded by the rays $\theta = \alpha$ and $\theta = \beta$ and by the continuous curves $r = g_1(\theta)$ and $r = g_2(\theta)$. Suppose also that $0 \leq g_1(\theta) \leq g_2(\theta) \leq a$ for every value of θ between α and β. Then R lies in a fan-shaped region Q defined by the inequalities $0 \leq r \leq a$ and $\alpha \leq \theta \leq \beta$. See Fig. 13.23.

13.23 The region R: $g_1(\theta) \le r \le g_2(\theta)$, $\alpha \le \theta \le \beta$ is contained in the fan-shaped region Q: $0 \le r \le a$, $\alpha \le \theta \le \beta$. The partition of Q by circular arcs and rays induces a partition of R.

We cover Q by a grid of circular arcs and rays. The arcs are cut from circles centered at the origin, with radii Δr, $2\Delta r$, ..., $m\Delta r$, where $\Delta r = a/m$. The rays are given by

$$\theta = \alpha, \quad \theta = \alpha + \Delta\theta, \quad \theta = \alpha + 2\Delta\theta, \quad \dots, \quad \theta = \alpha + m'\Delta\theta = \beta,$$

where $\Delta\theta = (\beta - \alpha)/m'$. The arcs and rays partition Q into small patches called "polar rectangles."

We number the polar rectangles that lie inside R (the order does not matter), calling their areas ΔA_1, ΔA_2, ..., ΔA_n.

We let (r_k, θ_k) be the center of the polar rectangle whose area is ΔA_k. By "center" we mean the point that lies halfway between the circular arcs on the ray that bisects the arcs. We then form the sum

$$S_n = \sum_{k=1}^{n} f(r_k, \theta_k)\Delta A_k. \tag{1}$$

If f is continuous throughout R, this sum will approach a limit as we refine the grid to make Δr and $\Delta\theta$ go to zero. The limit is called the double integral of f over R. In symbols,

$$\lim_{n \to \infty} S_n = \iint\limits_{R} f(r, \theta)\, dA.$$

To evaluate this limit, we first have to write the sum S_n in a way that expresses ΔA_k in terms of Δr and $\Delta\theta$. The radius of the inner arc bounding ΔA_k is $r_k - (\Delta r/2)$ (Fig. 13.24). The radius of the outer arc is $r_k + (\Delta r/2)$. The areas of the circular sectors subtended by these arcs at the origin are

$$\begin{array}{ll} \text{Inner} & \dfrac{1}{2}\left(r_k - \dfrac{\Delta r}{2}\right)^2 \Delta\theta \qquad \text{Outer} \qquad \dfrac{1}{2}\left(r_k + \dfrac{\Delta r}{2}\right)^2 \Delta\theta. \\ \text{radius:} & \hspace{4cm} \text{radius:} \end{array} \tag{2}$$

Therefore,

$$\Delta A_k = \text{Area of large sector} - \text{Area of small sector}$$

$$= \frac{\Delta\theta}{2}\left[\left(r_k + \frac{\Delta r}{2}\right)^2 - \left(r_k - \frac{\Delta r}{2}\right)^2\right] = \frac{\Delta\theta}{2}(2r_k\Delta r) = r_k\Delta r\Delta\theta.$$

Combining this result with Eq. (1) gives

$$S_n = \sum_{k=1}^{n} f(r_k, \theta_k)r_k\Delta r\Delta\theta. \tag{3}$$

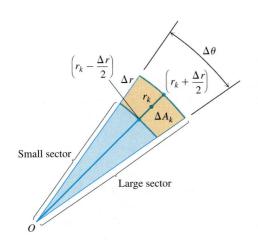

13.24 The observation that

$$\Delta A_k = \left(\begin{array}{c}\text{area of}\\\text{large sector}\end{array}\right) - \left(\begin{array}{c}\text{area of}\\\text{small sector}\end{array}\right)$$

leads to the formula $\Delta A_k = r_k\Delta r\Delta\theta$. The text explains why.

A version of Fubini's theorem now says that the limit approached by these sums can be evaluated by repeated single integrations with respect to r and θ as

$$\iint_R f(r, \theta)\, dA = \int_{\theta=\alpha}^{\theta=\beta} \int_{r=g_1(\theta)}^{r=g_2(\theta)} f(r, \theta)\, r\, dr\, d\theta. \tag{4}$$

Limits of Integration

The procedure for finding limits of integration in rectangular coordinates also works for polar coordinates.

How to Integrate in Polar Coordinates

To evaluate $\iint_R f(r, \theta)\, dA$ over a region R in polar coordinates, integrating first with respect to r and then with respect to θ, take the following steps.

1. *A sketch.* Sketch the region and label the bounding curves.

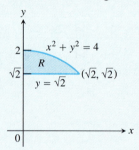

2. *The r-limits of integration.* Imagine a ray L from the origin cutting through R in the direction of increasing r. Mark the r-values where L enters and leaves R. These are the r-limits of integration. They usually depend on the angle θ that L makes with the positive x-axis.

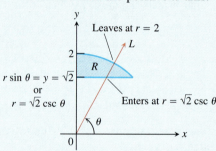

3. *The θ-limits of integration.* Find the smallest and largest θ-values that bound R. These are the θ-limits of integration.

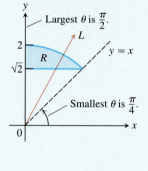

The integral is

$$\iint_R f(r, \theta)\, dA = \int_{\theta=\pi/4}^{\theta=\pi/2} \int_{r=\sqrt{2}\csc\theta}^{r=2} f(r, \theta)\, r\, dr\, d\theta.$$

EXAMPLE 1   Find the limits of integration for integrating $f(r, \theta)$ over the region R that lies inside the cardioid $r = 1 + \cos\theta$ and outside the circle $r = 1$.

Solution

Step 1: A sketch. We sketch the region and label the bounding curves (Fig. 13.25).

Step 2: The r-limits of integration. A typical ray from the origin enters R where $r = 1$ and leaves where $r = 1 + \cos\theta$.

Step 3: The θ-limits of integration. The rays from the origin that intersect R run from

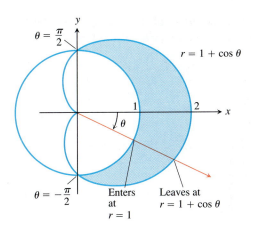

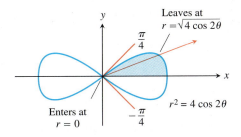

13.25 The sketch for Example 1.

13.26 To integrate over the shaded region, we run r from 0 to $\sqrt{4\cos 2\theta}$ and θ from 0 to $\pi/4$ (Example 2).

$\theta = -\pi/2$ to $\theta = \pi/2$. The integral is

$$\int_{-\pi/2}^{\pi/2} \int_{1}^{1+\cos\theta} f(r, \theta)\, r\, dr\, d\theta.$$

If $f(r, \theta)$ is the constant function whose value is 1, then the integral of f over R is the area of R.

Area in Polar Coordinates

The area of a closed and bounded region R in the polar coordinate plane is

$$A = \iint_R r\, dr\, d\theta. \tag{5}$$

As you might expect, this formula for area is consistent with all earlier formulas, although we will not prove the fact.

EXAMPLE 2   Find the area enclosed by the lemniscate $r^2 = 4\cos 2\theta$.

Solution   We graph the lemniscate to determine the limits of integration (Fig. 13.26) and see that the total area is 4 times the first-quadrant portion.

$$A = 4\int_0^{\pi/4} \int_0^{\sqrt{4\cos 2\theta}} r\, dr\, d\theta = 4\int_0^{\pi/4} \left[\frac{r^2}{2}\right]_{r=0}^{r=\sqrt{4\cos 2\theta}} d\theta$$

$$= 4\int_0^{\pi/4} 2\cos 2\theta\, d\theta = 4\sin 2\theta \Big]_0^{\pi/4} = 4.$$

Changing Cartesian Integrals into Polar Integrals

The procedure for changing a Cartesian integral $\iint_R f(x, y)\, dx\, dy$ into a polar integral has two steps.

Step 1: Substitute $x = r\cos\theta$ and $y = r\sin\theta$, and replace $dx\, dy$ by $r\, dr\, d\theta$ in the Cartesian integral.

Step 2: Supply polar limits of integration for the boundary of R.

The Cartesian integral then becomes

$$\iint_R f(x, y)\, dx\, dy = \iint_G f(r\cos\theta, r\sin\theta)\, r\, dr\, d\theta, \tag{6}$$

where G denotes the region of integration in polar coordinates. This is like the substitution method in Chapter 4 except that there are now two variables to substitute for instead of one. Notice that $dx\, dy$ is not replaced by $dr\, d\theta$ but by $r\, dr\, d\theta$. We will see why in Section 13.7.

EXAMPLE 3   Find the polar moment of inertia about the origin of a thin plate of density $\delta(x, y) = 1$ bounded by the quarter circle $x^2 + y^2 = 1$ in the first quadrant.

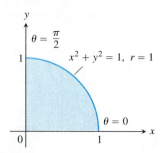

13.27 In polar coordinates, this region is described by simple inequalities:

$$0 \le r \le 1 \quad \text{and} \quad 0 \le \theta \le \pi/2$$

(Example 3).

Solution We sketch the plate to determine the limits of integration (Fig. 13.27). In Cartesian coordinates, the polar moment is the value of the integral

$$\int_0^1 \int_0^{\sqrt{1-x^2}} (x^2 + y^2)\, dy\, dx.$$

Integration with respect to y gives

$$\int_0^1 \left(x^2\sqrt{1 - x^2} + \frac{(1 - x^2)^{3/2}}{3} \right) dx,$$

an integral difficult to evaluate without tables.

Things go better if we change the original integral to polar coordinates. Substituting $x = r\cos\theta$, $y = r\sin\theta$, and replacing $dx\, dy$ by $r\, dr\, d\theta$, we get

$$\int_0^1 \int_0^{\sqrt{1-x^2}} (x^2 + y^2)\, dy\, dx = \int_0^{\pi/2} \int_0^1 (r^2)\, r\, dr\, d\theta$$

$$= \int_0^{\pi/2} \left[\frac{r^4}{4} \right]_{r=0}^{r=1} d\theta = \int_0^{\pi/2} \frac{1}{4}\, d\theta = \frac{\pi}{8}.$$

Why is the polar coordinate transformation so effective? One reason is that $x^2 + y^2$ simplifies to r^2. Another is that the limits of integration become constants. ❏

EXAMPLE 4 Evaluate

$$\iint_R e^{x^2+y^2}\, dy\, dx,$$

where R is the semicircular region bounded by the x-axis and the curve $y = \sqrt{1 - x^2}$ (Fig. 13.28).

Solution In Cartesian coordinates, the integral in question is a nonelementary integral and there is no direct way to integrate $e^{x^2+y^2}$ with respect to either x or y. Yet this integral and others like it are important in mathematics—in statistics, for example—and we must find a way to evaluate it. Polar coordinates save the day. Substituting $x = r\cos\theta$, $y = r\sin\theta$, and replacing $dy\, dx$ by $r\, dr\, d\theta$ enables us to evaluate the integral as

$$\iint_R e^{x^2+y^2}\, dy\, dx = \int_0^{\pi} \int_0^1 e^{r^2} r\, dr\, d\theta = \int_0^{\pi} \left[\frac{1}{2}e^{r^2} \right]_0^1 d\theta$$

$$= \int_0^{\pi} \frac{1}{2}(e - 1)\, d\theta = \frac{\pi}{2}(e - 1).$$

The r in the $r\, dr\, d\theta$ was just what we needed to integrate e^{r^2}. Without it we would have been stuck, as we were at the beginning. ❏

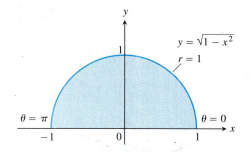

13.28 The semicircular region in Example 4 is the region

$$0 \le r \le 1, \quad 0 \le \theta \le \pi.$$

Exercises 13.3

Evaluating Polar Integrals

In Exercises 1–16, change the Cartesian integral into an equivalent polar integral. Then evaluate the polar integral.

1. $\displaystyle\int_{-1}^1 \int_0^{\sqrt{1-x^2}} dy\, dx$

2. $\displaystyle\int_{-1}^1 \int_{-\sqrt{1-x^2}}^{\sqrt{1-x^2}} dy\, dx$

3. $\displaystyle\int_0^1 \int_0^{\sqrt{1-y^2}} (x^2 + y^2)\, dx\, dy$

4. $\displaystyle\int_{-1}^1 \int_{-\sqrt{1-y^2}}^{\sqrt{1-y^2}} (x^2 + y^2)\, dx\, dy$

5. $\displaystyle\int_{-a}^a \int_{-\sqrt{a^2-x^2}}^{\sqrt{a^2-x^2}} dy\, dx$

6. $\displaystyle\int_0^2 \int_0^{\sqrt{4-y^2}} (x^2 + y^2)\, dx\, dy$

7. $\displaystyle\int_0^6 \int_0^y x\, dx\, dy$ $r^2\cos\theta\, dr\, d\theta$

8. $\displaystyle\int_0^2 \int_0^x y\, dy\, dx$

9. $\displaystyle\int_{-1}^0 \int_{-\sqrt{1-x^2}}^0 \frac{2}{1 + \sqrt{x^2 + y^2}}\, dy\, dx$

10. $\displaystyle\int_{-1}^1 \int_{-\sqrt{1-y^2}}^0 \frac{4\sqrt{x^2 + y^2}}{1 + x^2 + y^2}\, dx\, dy$

11. $\displaystyle\int_0^{\ln 2} \int_0^{\sqrt{(\ln 2)^2 - y^2}} e^{\sqrt{x^2+y^2}}\, dx\, dy$

12. $\displaystyle\int_0^1 \int_0^{\sqrt{1-x^2}} e^{-(x^2+y^2)}\, dy\, dx$

13. $\displaystyle\int_0^2 \int_0^{\sqrt{1-(x-1)^2}} \frac{x+y}{x^2 + y^2}\, dy\, dx$

14. $\displaystyle\int_0^2 \int_{-\sqrt{1-(y-1)^2}}^0 xy^2\, dx\, dy$

15. $\displaystyle\int_{-1}^1 \int_{-\sqrt{1-y^2}}^{\sqrt{1-y^2}} \ln(x^2 + y^2 + 1)\, dx\, dy$

16. $\displaystyle\int_{-1}^1 \int_{-\sqrt{1-x^2}}^{\sqrt{1-x^2}} \frac{2}{(1 + x^2 + y^2)^2}\, dy\, dx$

Finding Area in Polar Coordinates

17. Find the area of the region cut from the first quadrant by the curve $r = 2(2 - \sin 2\theta)^{1/2}$.

18. Find the area of the region that lies inside the cardioid $r = 1 + \cos\theta$ and outside the circle $r = 1$.

19. Find the area enclosed by one leaf of the rose $r = 12\cos 3\theta$.

20. Find the area of the region enclosed by the positive x-axis and spiral $r = 4\theta/3, 0 \le \theta \le 2\pi$. The region looks like a snail shell.

21. Find the area of the region cut from the first quadrant by the cardioid $r = 1 + \sin\theta$.

22. Find the area of the region common to the interiors of the cardioids $r = 1 + \cos\theta$ and $r = 1 - \cos\theta$.

Masses and Moments

23. Find the first moment about the x-axis of a thin plate of constant

density $\delta(x, y) = 3$, bounded below by the x-axis and above by the cardioid $r = 1 - \cos\theta$.

24. Find the moment of inertia about the x-axis and the polar moment of inertia about the origin of a thin disk bounded by the circle $x^2 + y^2 = a^2$ if the disk's density at the point (x, y) is $\delta(x, y) = k(x^2 + y^2), k$ a constant.

25. Find the mass of a thin plate covering the region outside the circle $r = 3$ and inside the circle $r = 6\sin\theta$ if the plate's density function is $\delta(x, y) = 1/r$.

26. Find the polar moment of inertia about the origin of a thin plate covering the region that lies inside the cardioid $r = 1 - \cos\theta$ and outside the circle $r = 1$ if the plate's density function is $\delta(x, y) = 1/r^2$.

27. Find the centroid of the region enclosed by the cardioid $r = 1 + \cos\theta$.

28. Find the polar moment of inertia about the origin of a thin plate enclosed by the cardioid $r = 1 + \cos\theta$ if the plate's density function is $\delta(x, y) = 1$.

Average Values

29. Find the average height of the hemisphere $z = \sqrt{a^2 - x^2 - y^2}$ above the disk $x^2 + y^2 \le a^2$ in the xy-plane.

30. Find the average height of the (single) cone $z = \sqrt{x^2 + y^2}$ above the disk $x^2 + y^2 \le a^2$ in the xy-plane.

31. Find the average distance from a point $P(x, y)$ in the disk $x^2 + y^2 \le a^2$ to the origin.

32. Find the average value of the *square* of the distance from the point $P(x, y)$ in the disk $x^2 + y^2 \le 1$ to the boundary point $A(1, 0)$.

Theory and Examples

33. Integrate $f(x, y) = [\ln(x^2 + y^2)]/\sqrt{x^2 + y^2}$ over the region $1 \le x^2 + y^2 \le e$.

34. Integrate $f(x, y) = [\ln(x^2 + y^2)]/(x^2 + y^2)$ over the region $1 \le x^2 + y^2 \le e^2$.

35. The region that lies inside the cardioid $r = 1 + \cos\theta$ and outside the circle $r = 1$ is the base of a solid right cylinder. The top of the cylinder lies in the plane $z = x$. Find the cylinder's volume.

36. The region enclosed by the lemniscate $r^2 = 2\cos 2\theta$ is the base of a solid right cylinder whose top is bounded by the sphere $z = \sqrt{2 - r^2}$. Find the cylinder's volume.

37. a) The usual way to evaluate the improper integral $I = \int_0^\infty e^{-x^2}\, dx$ is first to calculate its square:

$$I^2 = \left(\int_0^\infty e^{-x^2}\, dx\right)\left(\int_0^\infty e^{-y^2}\, dy\right) = \int_0^\infty \int_0^\infty e^{-(x^2+y^2)}\, dx\, dy.$$

Evaluate the last integral using polar coordinates and solve the resulting equation for I.

b) (*Continuation of Section 7.6, Exercise 92.*) Evaluate

$$\lim_{x\to\infty} \text{erf}(x) = \lim_{x\to\infty} \int_0^x \frac{2e^{-t^2}}{\sqrt{\pi}}\, dt.$$

38. Evaluate the integral

$$\int_0^\infty \int_0^\infty \frac{1}{(1+x^2+y^2)^2} \, dx \, dy.$$

39. Integrate the function $f(x, y) = 1/(1 - x^2 - y^2)$ over the disk $x^2 + y^2 \le 3/4$. Does the integral of $f(x, y)$ over the disk $x^2 + y^2 \le 1$ exist? Give reasons for your answer.

40. Use the double integral in polar coordinates to derive the formula

$$A = \int_\alpha^\beta \frac{1}{2} r^2 \, d\theta$$

for the area of the fan-shaped region between the origin and polar curve $r = f(\theta), \alpha \le \theta \le \beta$.

41. Let P_0 be a point inside a circle of radius a and let h denote the distance from P_0 to the center of the circle. Let d denote the distance from an arbitrary point P to P_0. Find the average value of d^2 over the region enclosed by the circle. (*Hint:* Simplify your work by placing the center of the circle at the origin and P_0 on the x-axis.)

42. Suppose that the area of a region in the polar coordinate plane is

$$A = \int_{\pi/4}^{3\pi/4} \int_{\csc \theta}^{2 \sin \theta} r \, dr \, d\theta.$$

a) Sketch the region and find its area.

b) Use one of Pappus's theorems together with the centroid information in Exercise 26 of Section 5.10 to find the volume of the solid generated by revolving the region about the x-axis.

CAS Explorations and Projects

In Exercises 43–46, use a CAS to change the Cartesian integrals into an equivalent polar integral and evaluate the polar integral. Perform the following steps in each exercise.

a) Plot the Cartesian region of integration in the xy-plane.

b) Change each boundary curve of the Cartesian region in (a) to its polar representation by solving its Cartesian equation for r and θ.

c) Using the results in (b), plot the polar region of integration in the $r\theta$-plane.

d) Change the integrand from Cartesian to polar coordinates. Determine the limits of integration from your plot in (c) and evaluate the polar integral using the CAS integration utility.

43. $\displaystyle\int_0^1 \int_x^1 \frac{y}{x^2+y^2} \, dy \, dx$

44. $\displaystyle\int_0^1 \int_0^{x/2} \frac{x}{x^2+y^2} \, dy \, dx$

45. $\displaystyle\int_0^1 \int_{-y/3}^{y/3} \frac{y}{\sqrt{x^2+y^2}} \, dx \, dy$

46. $\displaystyle\int_0^1 \int_y^{2-y} \sqrt{x+y} \, dx \, dy$

13.4	# Triple Integrals in Rectangular Coordinates

We use triple integrals to find the volumes of three-dimensional shapes, the masses and moments of solids, and the average values of functions of three variables. In Chapter 14, we will also see how these integrals arise in the studies of vector fields and fluid flow.

Triple Integrals

If $F(x, y, z)$ is a function defined on a closed bounded region D in space—the region occupied by a solid ball, for example, or a lump of clay—then the integral of F over D may be defined in the following way. We partition a rectangular region containing D into rectangular cells by planes parallel to the coordinate planes (Fig. 13.29). We number the cells that lie inside D from 1 to n in some order, a typical cell having dimensions Δx_k by Δy_k by Δz_k and volume ΔV_k. We choose a point (x_k, y_k, z_k) in each cell and form the sum

$$S_n = \sum_{k=1}^n F(x_k, y_k, z_k) \Delta V_k. \tag{1}$$

If F is continuous and the bounding surface of D is made of smooth surfaces joined along continuous curves, then as Δx_k, Δy_k, and Δz_k approach zero independently the sums S_n approach a limit

$$\lim_{n \to \infty} S_n = \iiint_D F(x, y, z) \, dV. \tag{2}$$

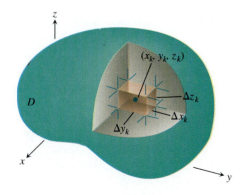

13.29 Partitioning a solid with rectangular cells of volume ΔV_k.

We call this limit the **triple integral of F over D.** The limit also exists for some discontinuous functions.

Properties of Triple Integrals

Triple integrals have the same algebraic properties as double and single integrals. If $F = F(x, y, z)$ and $G = G(x, y, z)$ are continuous, then

1. $\displaystyle\iiint\limits_{D} kF \, dV = k \iiint\limits_{D} F \, dV$ (any number k)

2. $\displaystyle\iiint\limits_{D} (F \pm G) \, dV = \iiint\limits_{D} F \, dV \pm \iiint\limits_{D} G \, dV$

3. $\displaystyle\iiint\limits_{D} F \, dV \geq 0$ if $F \geq 0$ on D

4. $\displaystyle\iiint\limits_{D} F \, dV \geq \iiint\limits_{D} G \, dV$ if $F \geq G$ on D.

Triple integrals also have an additivity property, used in physics and engineering as well as in mathematics. If the domain D of a continuous function F is partitioned by smooth surfaces into a finite number of nonoverlapping cells $D_1, D_2, \ldots, D_n$, then

5. $\displaystyle\iiint\limits_{D} F \, dV = \iiint\limits_{D_1} F \, dV + \iiint\limits_{D_2} F \, dV + \cdots + \iiint\limits_{D_n} F \, dV.$

Volume of a Region in Space

If F is the constant function whose value is 1, then the sums in Eq. (1) reduce to

$$S_n = \sum F(x_k, y_k, z_k) \Delta V_k = \sum 1 \cdot \Delta V_k = \sum \Delta V_k. \qquad (3)$$

As Δx, Δy, and Δz approach zero, the cells ΔV_k become smaller and more numerous and fill up more and more of D. We therefore define the volume of D to be the triple integral

$$\lim_{n \to \infty} \sum_{k=1}^{n} \Delta V_k = \iiint\limits_{D} dV.$$

Definition
The **volume** of a closed, bounded region D in space is

$$V = \iiint\limits_{D} dV. \qquad (4)$$

As we will see in a moment, this integral enables us to calculate the volumes of solids enclosed by curved surfaces.

Evaluation

We seldom evaluate a triple integral from its definition as a limit. Instead, we apply a three-dimensional version of Fubini's theorem to evaluate it by repeated single integrations. As with double integrals, there is a geometric procedure for finding the limits of integration.

How to Find Limits of Integration in Triple Integrals

To evaluate

$$\iiint_D F(x, y, z)\, dV$$

over a region D, integrating first with respect to z, then with respect to y, finally with x, take the following steps.

1. A *sketch.* Sketch the region D along with its "shadow" R (vertical projection) in the xy-plane. Label the upper and lower bounding surfaces of D and the upper and lower bounding curves of R.

2. *The z-limits of integration.* Draw a line M passing through a typical point (x, y) in R parallel to the z-axis. As z increases, M enters D at $z = f_1(x, y)$ and leaves at $z = f_2(x, y)$. These are the z-limits of integration.

3. *The y-limits of integration.* Draw a line L through (x, y) parallel to the y-axis. As y increases, L enters R at $y = g_1(x)$ and leaves at $y = g_2(x)$. These are the y-limits of integration.

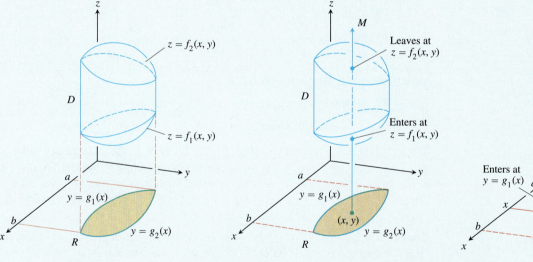
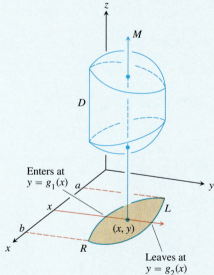

4. *The x-limits of integration.* Choose x-limits that include all lines through R parallel to the y-axis ($x = a$ and $x = b$ in the preceding figure). These are the x-limits of integration. The integral is

$$\int_{x=a}^{x=b} \int_{y=g_1(x)}^{y=g_2(x)} \int_{z=f_1(x,y)}^{z=f_2(x,y)} F(x, y, z)\, dz\, dy\, dx.$$

Follow similar procedures if you change the order of integration. The "shadow" of region D lies in the plane of the last two variables with respect to which the iterated integration takes place.

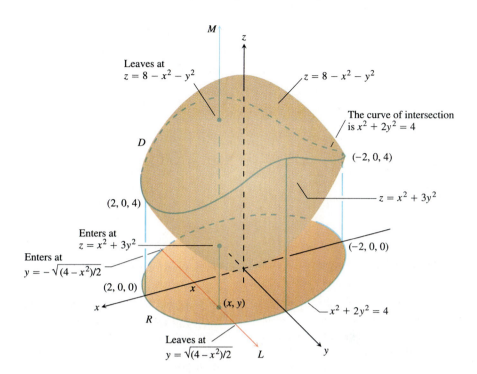

13.30 The volume of the region enclosed by these two paraboloids is calculated in Example 1.

EXAMPLE 1   Find the volume of the region D enclosed by the surfaces $z = x^2 + 3y^2$ and $z = 8 - x^2 - y^2$.

Solution   The volume is

$$V = \iiint\limits_{D} dz\,dy\,dx,$$

the integral of $F(x, y, z) = 1$ over D. To find the limits of integration for evaluating the integral, we take these steps.

Step 1: *A sketch.* The surfaces (Fig. 13.30) intersect on the elliptical cylinder $x^2 + 3y^2 = 8 - x^2 - y^2$ or $x^2 + 2y^2 = 4$. The boundary of the region R, the projection of D onto the xy-plane, is an ellipse with the same equation: $x^2 + 2y^2 = 4$. The "upper" boundary of R is the curve $y = \sqrt{(4 - x^2)/2}$. The lower boundary is the curve $y = -\sqrt{(4 - x^2)/2}$.

Step 2: *The z-limits of integration.* The line M passing through a typical point (x, y) in R parallel to the z-axis enters D at $z = x^2 + 3y^2$ and leaves at $z = 8 - x^2 - y^2$.

Step 3: *The y-limits of integration.* The line L through (x, y) parallel to the y-axis enters R at $y = -\sqrt{(4 - x^2)/2}$ and leaves at $y = \sqrt{(4 - x^2)/2}$.

Step 4: *The x-limits of integration.* As L sweeps across R, the value of x varies from $x = -2$ at $(-2, 0, 0)$ to $x = 2$ at $(2, 0, 0)$. The volume of D is

$$V = \iiint\limits_{D} dz\,dy\,dx$$

$$= \int_{-2}^{2} \int_{-\sqrt{(4-x^2)/2}}^{\sqrt{(4-x^2)/2}} \int_{x^2+3y^2}^{8-x^2-y^2} dz\,dy\,dx$$

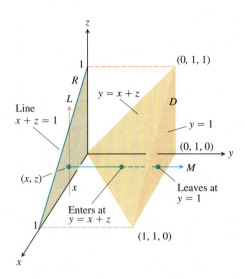

13.31 The tetrahedron in Example 2.

$$= \int_{-2}^{2} \int_{-\sqrt{(4-x^2)/2}}^{\sqrt{(4-x^2)/2}} (8 - 2x^2 - 4y^2)\, dy\, dx$$

$$= \int_{-2}^{2} \left[(8 - 2x^2)y - \frac{4}{3}y^3 \right]_{y=-\sqrt{(4-x^2)/2}}^{y=\sqrt{(4-x^2)/2}} dx$$

$$= \int_{-2}^{2} \left(2(8 - 2x^2)\sqrt{\frac{4-x^2}{2}} - \frac{8}{3}\left(\frac{4-x^2}{2}\right)^{3/2} \right) dx$$

$$= \int_{-2}^{2} \left[8\left(\frac{4-x^2}{2}\right)^{3/2} - \frac{8}{3}\left(\frac{4-x^2}{2}\right)^{3/2} \right] dx = \frac{4\sqrt{2}}{3} \int_{-2}^{2} (4 - x^2)^{3/2}\, dx$$

$$= 8\pi\sqrt{2}. \qquad \text{After integration with the substitution } x = 2\sin u \qquad \square$$

In the next example, we project D onto the xz-plane instead of the xy-plane.

EXAMPLE 2   Set up the limits of integration for evaluating the triple integral of a function $F(x, y, z)$ over the tetrahedron D with vertices $(0, 0, 0)$, $(1, 1, 0)$, $(0, 1, 0)$, and $(0, 1, 1)$.

Solution

Step 1: *A sketch.* We sketch D along with its "shadow" R in the xz-plane (Fig. 13.31). The upper (right-hand) bounding surface of D lies in the plane $y = 1$. The lower (left-hand) bounding surface lies in the plane $y = x + z$. The upper boundary of R is the line $z = 1 - x$. The lower boundary is the line $z = 0$.

Step 2: *The y-limits of integration.* The line through a typical point (x, z) in R parallel to the y-axis enters D at $y = x + z$ and leaves at $y = 1$.

Step 3: *The z-limits of integration.* The line L through (x, z) parallel to the z-axis enters R at $z = 0$ and leaves at $z = 1 - x$.

Step 4: *The x-limits of integration.* As L sweeps across R, the value of x varies from $x = 0$ to $x = 1$. The integral is

$$\int_{0}^{1} \int_{0}^{1-x} \int_{x+z}^{1} F(x, y, z)\, dy\, dz\, dx. \qquad \square$$

As we know, there are sometimes (but not always) two different orders in which the single integrations for evaluating a double integral may be worked. For triple integrals, there could be as many as *six*.

EXAMPLE 3   Each of the following integrals gives the volume of the solid shown in Fig. 13.32.

a) $\displaystyle\int_{0}^{1} \int_{0}^{1-z} \int_{0}^{2} dx\, dy\, dz$ **b)** $\displaystyle\int_{0}^{1} \int_{0}^{1-y} \int_{0}^{2} dx\, dz\, dy$

c) $\displaystyle\int_{0}^{1} \int_{0}^{2} \int_{0}^{1-z} dy\, dx\, dz$ **d)** $\displaystyle\int_{0}^{2} \int_{0}^{1} \int_{0}^{1-z} dy\, dz\, dx$

e) $\displaystyle\int_{0}^{1} \int_{0}^{2} \int_{0}^{1-y} dz\, dx\, dy$ **f)** $\displaystyle\int_{0}^{2} \int_{0}^{1} \int_{0}^{1-y} dz\, dy\, dx$ $\square$

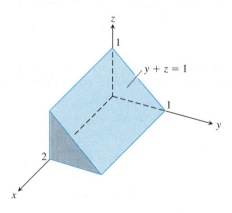

13.32 Example 3 gives six different iterated triple integrals for the volume of this prism.

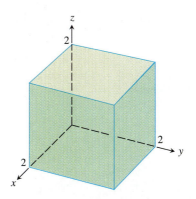

13.33 The region of integration in Example 4.

Average Value of a Function in Space

The average value of a function F over a region D in space is defined by the formula

$$\text{Average value of } F \text{ over } D = \frac{1}{\text{volume of } D} \iiint_D F \, dV. \qquad (5)$$

For example, if $F(x, y, z) = \sqrt{x^2 + y^2 + z^2}$, then the average value of F over D is the average distance of points in D from the origin. If $F(x, y, z)$ is the density of a solid that occupies a region D in space, then the average value of F over D is the average density of the solid in units of mass per unit volume.

EXAMPLE 4   Find the average value of $F(x, y, z) = xyz$ over the cube bounded by the coordinate planes and the planes $x = 2$, $y = 2$, and $z = 2$ in the first octant.

Solution   We sketch the cube with enough detail to show the limits of integration (Fig. 13.33). We then use Eq. (5) to calculate the average value of F over the cube. The volume of the cube is $(2)(2)(2) = 8$. The value of the integral of F over the cube is

$$\int_0^2 \int_0^2 \int_0^2 xyz \, dx \, dy \, dz = \int_0^2 \int_0^2 \left[\frac{x^2}{2} yz \right]_{x=0}^{x=2} dy \, dz = \int_0^2 \int_0^2 2yz \, dy \, dz$$

$$= \int_0^2 \left[y^2 z \right]_{y=0}^{y=2} dz = \int_0^2 4z \, dz = \left[2z^2 \right]_0^2 = 8.$$

With these values, Eq. (5) gives

$$\text{Average value of } xyz \text{ over the cube} = \frac{1}{\text{volume}} \iiint_{\text{cube}} xyz \, dV = \left(\frac{1}{8} \right)(8) = 1.$$

In evaluating the integral, we chose the order dx, dy, dz, but any of the other five possible orders would have done as well. ◻

Exercises 13.4

Evaluating Triple Integrals in Different Iterations

1. Find the common value of the integrals in Example 3.

2. Write six different iterated triple integrals for the volume of the rectangular solid in the first octant bounded by the coordinate planes and the planes $x = 1$, $y = 2$, and $z = 3$. Evaluate one of the integrals.

3. Write six different iterated triple integrals for the volume of the tetrahedron cut from the first octant by the plane $6x + 3y + 2z = 6$. Evaluate one of the integrals.

4. Write six different iterated triple integrals for the volume of the region in the first octant enclosed by the cylinder $x^2 + z^2 = 4$ and the plane $y = 3$. Evaluate one of the integrals.

5. Let D be the region bounded by the paraboloids $z = 8 - x^2 - y^2$ and $z = x^2 + y^2$. Write six different triple iterated integrals for the volume of D. Evaluate one of the integrals.

6. Let D be the region bounded by the paraboloid $z = x^2 + y^2$ and the plane $z = 2y$. Write triple iterated integrals in the order $dz \, dx \, dy$ and $dz \, dy \, dx$ that give the volume of D. Do not evaluate either integral.

Evaluating Triple Iterated Integrals

Evaluate the integrals in Exercises 7–20.

7. $\displaystyle\int_0^1 \int_0^1 \int_0^1 (x^2 + y^2 + z^2) \, dz \, dy \, dx$

8. $\int_0^{\sqrt{2}} \int_0^{3y} \int_{x^2+3y^2}^{8-x^2-y^2} dz\, dx\, dy$ **9.** $\int_1^e \int_1^e \int_1^e \frac{1}{xyz} dx\, dy\, dz$

10. $\int_0^1 \int_0^{3-3x} \int_0^{3-3x-y} dz\, dy\, dx$

11. $\int_0^1 \int_0^\pi \int_0^\pi y \sin z\, dx\, dy\, dz$

12. $\int_{-1}^1 \int_{-1}^1 \int_{-1}^1 (x+y+z)\, dy\, dx\, dz$

13. $\int_0^3 \int_0^{\sqrt{9-x^2}} \int_0^{\sqrt{9-x^2}} dz\, dy\, dx$

14. $\int_0^2 \int_{-\sqrt{4-y^2}}^{\sqrt{4-y^2}} \int_0^{2x+y} dz\, dx\, dy$

15. $\int_0^1 \int_0^{2-x} \int_0^{2-x-y} dz\, dy\, dx$

16. $\int_0^1 \int_0^{1-x^2} \int_3^{4-x^2-y} x\, dz\, dy\, dx$

17. $\int_0^\pi \int_0^\pi \int_0^\pi \cos(u+v+w)\, du\, dv\, dw$ (*uvw*-space)

18. $\int_1^e \int_1^e \int_1^e \ln r \ln s \ln t\, dt\, dr\, ds$ (*rst*-space)

19. $\int_0^{\pi/4} \int_0^{\ln \sec v} \int_{-\infty}^{2t} e^x\, dx\, dt\, dv$ (*tvx*-space)

20. $\int_0^7 \int_0^2 \int_0^{\sqrt{4-q^2}} \frac{q}{r+1} dp\, dq\, dr$ (*pqr*-space)

Volumes Using Triple Integrals

21. Here is the region of integration of the integral

$$\int_{-1}^1 \int_{x^2}^1 \int_0^{1-y} dz\, dy\, dx.$$

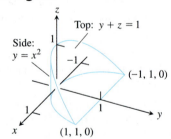

Top: $y+z=1$

Side: $y=x^2$

$(-1, 1, 0)$

$(1, 1, 0)$

Rewrite the integral as an equivalent iterated integral in the order

a) $dy\, dz\, dx$ **b)** $dy\, dx\, dz$
c) $dx\, dy\, dz$ **d)** $dx\, dz\, dy$
e) $dz\, dx\, dy$

22. Here is the region of integration of the integral

$$\int_0^1 \int_{-1}^0 \int_0^{y^2} dz\, dy\, dx.$$

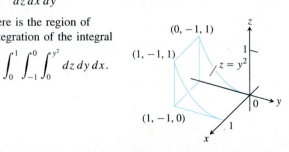

$(0, -1, 1)$

$(1, -1, 1)$

$z=y^2$

$(1, -1, 0)$

Rewrite the integral as an equivalent iterated integral in the order

a) $dy\, dz\, dx$ **b)** $dy\, dx\, dz$
c) $dx\, dy\, dz$ **d)** $dx\, dz\, dy$
e) $dz\, dx\, dy$

Find the volumes of the regions in Exercises 23–36.

23. The region between the cylinder $z=y^2$ and the xy-plane that is bounded by the planes $x=0, x=1,$ $y=-1, y=1$

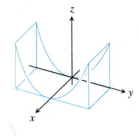

24. The region in the first octant bounded by the coordinate planes and the planes $x+z=1, y+2z=2$

25. The region in the first octant bounded by the coordinate planes, the plane $y+z=2,$ and the cylinder $x=4-y^2$

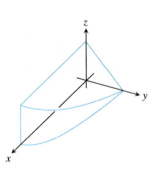

26. The wedge cut from the cylinder $x^2+y^2=1$ by the planes $z=-y$ and $z=0$

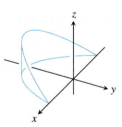

27. The tetrahedron in the first octant bounded by the coordinate planes and the plane $x+y/2+z/3=1$

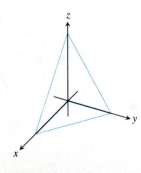

28. The region in the first octant bounded by the coordinate planes, the plane $y = 1 - x$, and the surface $z = \cos(\pi x/2)$, $0 \le x \le 1$

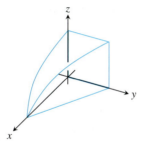

29. The region common to the interiors of the cylinders $x^2 + y^2 = 1$ and $x^2 + z^2 = 1$ (Fig. 13.34)

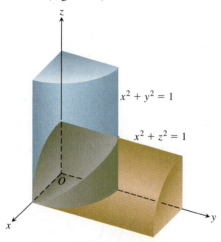

$x^2 + y^2 = 1$

$x^2 + z^2 = 1$

13.34 One-eighth of the region common to the cylinders $x^2 + y^2 = 1$ and $x^2 + z^2 = 1$ in Exercise 29.

30. The region in the first octant bounded by the coordinate planes and the surface $z = 4 - x^2 - y$

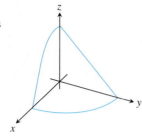

31. The region in the first octant bounded by the coordinate planes, the plane $x + y = 4$, and the cylinder $y^2 + 4z^2 = 16$

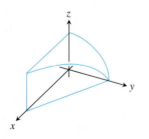

32. The region cut from the cylinder $x^2 + y^2 = 4$ by the plane $z = 0$ and the plane $x + z = 3$

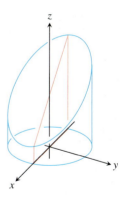

33. The region between the planes $x + y + 2z = 2$ and $2x + 2y + z = 4$ in the first octant

34. The finite region bounded by the planes $z = x$, $x + z = 8$, $z = y$, $y = 8$, and $z = 0$.

35. The region cut from the solid elliptical cylinder $x^2 + 4y^2 \le 4$ by the xy-plane and the plane $z = x + 2$

36. The region bounded in back by the plane $x = 0$, on the front and sides by the parabolic cylinder $x = 1 - y^2$, on the top by the paraboloid $z = x^2 + y^2$, and on the bottom by the xy-plane

Average Values

In Exercises 37–40, find the average value of $F(x, y, z)$ over the given region.

37. $F(x, y, z) = x^2 + 9$ over the cube in the first octant bounded by the coordinate planes and the planes $x = 2$, $y = 2$, and $z = 2$

38. $F(x, y, z) = x + y - z$ over the rectangular solid in the first octant bounded by the coordinate planes and the planes $x = 1$, $y = 1$, and $z = 2$

39. $F(x, y, z) = x^2 + y^2 + z^2$ over the cube in the first octant bounded by the coordinate planes and the planes $x = 1$, $y = 1$, and $z = 1$

40. $F(x, y, z) = xyz$ over the cube in the first octant bounded by the coordinate planes and the planes $x = 2$, $y = 2$, and $z = 2$

Changing the Order of Integration

Evaluate the integrals in Exercises 41–44 by changing the order of integration in an appropriate way.

41. $\displaystyle\int_0^4 \int_0^1 \int_{2y}^2 \frac{4\cos(x^2)}{2\sqrt{z}}\,dx\,dy\,dz$

42. $\displaystyle\int_0^1 \int_0^1 \int_{x^2}^1 12xz\,e^{zy^2}\,dy\,dx\,dz$

43. $\displaystyle\int_0^1 \int_{\sqrt[3]{z}}^1 \int_0^{\ln 3} \frac{\pi e^{2x} \sin \pi y^2}{y^2}\,dx\,dy\,dz$

44. $\displaystyle\int_0^2 \int_0^{4-x^2} \int_0^x \frac{\sin 2z}{4 - z}\,dy\,dz\,dx$

Theory and Examples

45. Solve for a:
$$\int_0^1 \int_0^{4-a-x^2} \int_a^{4-x^2-y} dz\, dy\, dx = \frac{4}{15}.$$

46. For what value of c is the volume of the ellipsoid $x^2 + (y/2)^2 + (z/c)^2 = 1$ equal to 8π?

47. What domain D in space minimizes the value of the integral
$$\iiint_D (4x^2 + 4y^2 + z^2 - 4)\, dV?$$

Give reasons for your answer.

48. What domain D in space maximizes the value of the integral
$$\iiint_D (1 - x^2 - y^2 - z^2)\, dV?$$

Give reasons for your answer.

CAS Explorations and Projects

In Exercises 49–52, use a CAS integration utility to evaluate the triple integral of the given function over the specified solid region.

49. $F(x, y, z) = x^2 y^2 z$ over the solid cylinder bounded by $x^2 + y^2 = 1$ and the planes $z = 0$ and $z = 1$.

50. $F(x, y, z) = |xyz|$ over the solid bounded below by the paraboloid $z = x^2 + y^2$ and above by the plane $z = 1$.

51. $F(x, y, z) = \dfrac{z}{(x^2 + y^2 + z^2)^{3/2}}$ over the solid bounded below by the cone $z = \sqrt{x^2 + y^2}$ and above by the plane $z = 1$.

52. $F(x, y, z) = x^4 + y^2 + z^2$ over the solid sphere $x^2 + y^2 + z^2 \le 1$.

13.5	# Masses and Moments in Three Dimensions

This section shows how to calculate the masses and moments of three-dimensional objects in Cartesian coordinates. The formulas are similar to those for two-dimensional objects. For calculations in spherical and cylindrical coordinates, see Section 13.6.

Masses and Moments

If $\delta(x, y, z)$ is the density of an object occupying a region D in space (mass per unit volume), the integral of δ over D gives the mass of the object. To see why, imagine partitioning the object into n mass elements like the one in Fig. 13.35. The object's mass is the limit

$$M = \lim_{n\to\infty} \sum_{k=1}^n \Delta m_k = \lim_{n\to\infty} \sum_{k=1}^n \delta(x_k, y_k, z_k)\Delta V_k = \iiint_D \delta(x, y, z)\, dV. \quad (1)$$

If $r(x, y, z)$ is the distance from the point (x, y, z) in D to a line L, then the moment of inertia of the mass $\Delta m_k = \delta(x_k, y_k, z_k)\Delta V_k$ about the line L (shown in Fig. 13.35) is approximately $\Delta I_k = r^2(x_k, y_k, z_k)\Delta m_k$. The moment of inertia of the entire object about L is

$$I_L = \lim_{n\to\infty} \sum_{k=1}^n \Delta I_k = \lim_{n\to\infty} \sum_{k=1}^n r^2(x_k, y_k, z_k)\,\delta(x_k, y_k, z_k)\Delta V_k = \iiint_D r^2\delta\, dV.$$

If L is the x-axis, then $r^2 = y^2 + z^2$ (Fig. 13.36) and

$$I_x = \iiint_D (y^2 + z^2)\,\delta\, dV.$$

Similarly,

$$I_y = \iiint_D (x^2 + z^2)\,\delta\, dV \qquad \text{and} \qquad I_z = \iiint_D (x^2 + y^2)\,\delta\, dV.$$

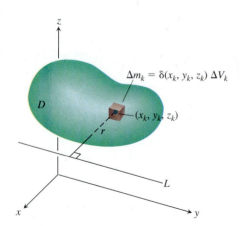

$\Delta m_k = \delta(x_k, y_k, z_k)\,\Delta V_k$

(x_k, y_k, z_k)

13.35 To define an object's mass and moment of inertia about a line, we first imagine it to be partitioned into a finite number of mass elements Δm_k.

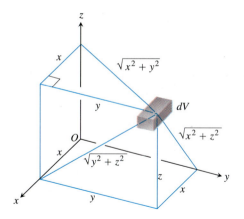

13.36 Distances from dV to the coordinate planes and axes.

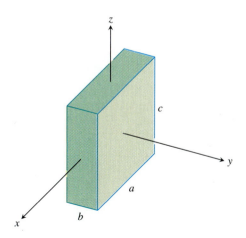

13.37 Example 1 calculates I_x, I_y, and I_z for the block shown here. The origin lies at the center of the block.

These and other useful formulas are summarized in Table 13.2.

Table 13.2 Mass and moment formulas for objects in space

Mass: $\quad M = \iiint\limits_{D} \delta \, dV \qquad (\delta = \text{density})$

First moments about the coordinate planes:

$$M_{yz} = \iiint\limits_{D} x \, \delta \, dV, \qquad M_{xz} = \iiint\limits_{D} y \, \delta \, dV, \qquad M_{xy} = \iiint\limits_{D} z \, \delta \, dV$$

Center of mass:

$$\bar{x} = \frac{M_{yz}}{M}, \qquad \bar{y} = \frac{M_{xz}}{M}, \qquad \bar{z} = \frac{M_{xy}}{M}$$

Moments of inertia (second moments):

$$I_x = \iiint (y^2 + z^2) \, \delta \, dV$$
$$I_y = \iiint (x^2 + z^2) \, \delta \, dV$$
$$I_z = \iiint (x^2 + y^2) \, \delta \, dV$$

Moment of inertia about a line L:

$$I_L = \iiint r^2 \, \delta \, dV \qquad (r(x, y, z) = \text{distance from points } (x, y, z) \text{ to line } L)$$

Radius of gyration about a line L:

$$R_L = \sqrt{I_L / M}$$

EXAMPLE 1 Find I_x, I_y, I_z for the rectangular solid of constant density δ shown in Fig. 13.37.

Solution The preceding formula for I_x gives

$$I_x = \int_{-c/2}^{c/2} \int_{-b/2}^{b/2} \int_{-a/2}^{a/2} (y^2 + z^2) \, \delta \, dx \, dy \, dz. \tag{2}$$

We can avoid some of the work of integration by observing that $(y^2 + z^2) \, \delta$ is an even function of x, y, and z and therefore

$$I_x = 8 \int_{0}^{c/2} \int_{0}^{b/2} \int_{0}^{a/2} (y^2 + z^2) \, \delta \, dx \, dy \, dz = 4a\delta \int_{0}^{c/2} \int_{0}^{b/2} (y^2 + z^2) \, dy \, dz$$

$$= 4a\delta \int_{0}^{c/2} \left[\frac{y^3}{3} + z^2 y \right]_{y=0}^{y=b/2} dz$$

$$= 4a\delta \int_{0}^{c/2} \left(\frac{b^3}{24} + \frac{z^2 b}{2} \right) dz$$

$$= 4a\delta \left(\frac{b^3 c}{48} + \frac{c^3 b}{48} \right) = \frac{abc\delta}{12} (b^2 + c^2) = \frac{M}{12} (b^2 + c^2).$$

Similarly,

$$I_y = \frac{M}{12} (a^2 + c^2) \qquad \text{and} \qquad I_z = \frac{M}{12} (a^2 + b^2).$$

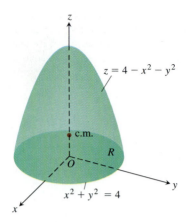

13.38 Example 2 finds the center of mass of this solid.

EXAMPLE 2 Find the center of mass of a solid of constant density δ bounded below by the disk $R: x^2 + y^2 \leq 4$ in the plane $z = 0$ and above by the paraboloid $z = 4 - x^2 - y^2$ (Fig. 13.38).

Solution By symmetry, $\bar{x} = \bar{y} = 0$. To find $\bar{z}$, we first calculate

$$M_{xy} = \iiint_R \int_{z=0}^{z=4-x^2-y^2} z\,\delta\,dz\,dy\,dx = \iint_R \left[\frac{z^2}{2}\right]_{z=0}^{z=4-x^2-y^2} \delta\,dy\,dx$$

$$= \frac{\delta}{2} \iint_R (4 - x^2 - y^2)^2\,dy\,dx$$

$$= \frac{\delta}{2} \int_0^{2\pi} \int_0^2 (4 - r^2)^2\,r\,dr\,d\theta \qquad \text{Polar coordinates}$$

$$= \frac{\delta}{2} \int_0^{2\pi} \left[-\frac{1}{6}(4 - r^2)^3\right]_{r=0}^{r=2} d\theta = \frac{16\delta}{3} \int_0^{2\pi} d\theta = \frac{32\pi\delta}{3}.$$

A similar calculation gives

$$M = \iiint_R \int_0^{4-x^2-y^2} \delta\,dz\,dy\,dx = 8\pi\,\delta.$$

Therefore $\bar{z} = (M_{xy}/M) = 4/3$, and the center of mass is $(\bar{x}, \bar{y}, \bar{z}) = (0, 0, 4/3)$.

When the density of a solid object is constant (as in Examples 1 and 2), the center of mass is called the **centroid** of the object (as was the case for two-dimensional shapes in Section 13.2).

Exercises 13.5

Constant Density

The solids in Exercises 1–12 all have constant density $\delta = 1$.

1. Evaluate the integral for I_x in Eq. (2) directly to show that the shortcut in Example 1 gives the same answer. Use the results in Example 1 to find the radius of gyration of the rectangular solid about each coordinate axis.

2. The coordinate axes in the figure to the right run through the centroid of a solid wedge parallel to the labeled edges. Find I_x, I_y, and I_z if $a = b = 6$ and $c = 4$.

3. Find the moments of inertia of the rectangular solid shown here with respect to its edges by calculating I_x, I_y, and I_z.

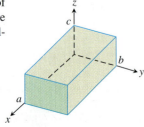

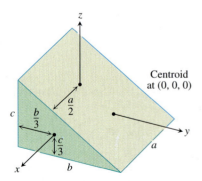

Figure for Exercise 2

4. **a)** Find the centroid and the moments of inertia I_x, I_y, and I_z of the tetrahedron whose vertices are the points $(0, 0, 0)$, $(1, 0, 0)$, $(0, 1, 0)$, and $(0, 0, 1)$.

   b) Find the radius of gyration of the tetrahedron about the x-axis. Compare it with the distance from the centroid to the x-axis.

5. A solid "trough" of constant density is bounded below by the surface $z = 4y^2$, above by the plane $z = 4$, and on the ends by the planes $x = 1$ and $x = -1$. Find the center of mass and the moments of inertia with respect to the three axes.

6. A solid of constant density is bounded below by the plane $z = 0$, on the sides by the elliptic cylinder $x^2 + 4y^2 = 4$, and above by the plane $z = 2 - x$ (see the figure).

a) Find $\bar{x}$ and $\bar{y}$.

b) Evaluate the integral

$$M_{xy} = \int_{-2}^{2} \int_{-(1/2)\sqrt{4-x^2}}^{(1/2)\sqrt{4-x^2}} \int_{0}^{2-x} z \, dz \, dy \, dx,$$

using integral tables to carry out the final integration with respect to x. Then divide M_{xy} by M to verify that $\bar{z} = 5/4$.

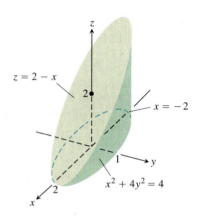

7. a) Find the center of mass of a solid of constant density bounded below by the paraboloid $z = x^2 + y^2$ and above by the plane $z = 4$.

b) Find the plane $z = c$ that divides the solid into two parts of equal volume. This plane does not pass through the center of mass.

8. A solid cube, 2 units on a side, is bounded by the planes $x = \pm 1$, $z = \pm 1$, $y = 3$, and $y = 5$. Find the center of mass and the moments of inertia and radii of gyration about the coordinate axes.

9. A wedge like the one in Exercise 2 has $a = 4$, $b = 6$, and $c = 3$. Make a quick sketch to check for yourself that the square of the distance from a typical point (x, y, z) of the wedge to the line $L: z = 0$, $y = 6$ is $r^2 = (y - 6)^2 + z^2$. Then calculate the moment of inertia and radius of gyration of the wedge about L.

10. A wedge like the one in Exercise 2 has $a = 4$, $b = 6$, and $c = 3$. Make a quick sketch to check for yourself that the square of the distance from a typical point (x, y, z) of the wedge to the line $L: x = 4$, $y = 0$ is $r^2 = (x - 4)^2 + y^2$. Then calculate the moment of inertia and radius of gyration of the wedge about L.

11. A solid like the one in Exercise 3 has $a = 4$, $b = 2$, and $c = 1$. Make a quick sketch to check for yourself that the square of the distance between a typical point (x, y, z) of the solid and the line $L: y = 2$, $z = 0$ is $r^2 = (y - 2)^2 + z^2$. Then find the moment of inertia and radius of gyration of the solid about L.

12. A solid like the one in Exercise 3 has $a = 4$, $b = 2$, and $c = 1$. Make a quick sketch to check for yourself that the square of the distance between a typical point (x, y, z) of the solid and the line $L: x = 4$, $y = 0$ is $r^2 = (x - 4)^2 + y^2$. Then find the moment of inertia and radius of gyration of the solid about L.

Variable Density

In Exercises 13 and 14, find (a) the mass of the solid and (b) the center of mass.

13. A solid region in the first octant is bounded by the coordinate planes and the plane $x + y + z = 2$. The density of the solid is $\delta(x, y, z) = 2x$.

14. A solid in the first octant is bounded by the planes $y = 0$ and $z = 0$ and by the surfaces $z = 4 - x^2$ and $x = y^2$ (see the figure). Its density function is $\delta(x, y, z) = kxy$.

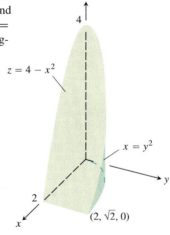

In Exercises 15 and 16, find

a) the mass of the solid
b) the center of mass
c) the moments of inertia about the coordinate axes
d) the radii of gyration about the coordinate axes.

15. A solid cube in the first octant is bounded by the coordinate planes and by the planes $x = 1$, $y = 1$, and $z = 1$. The density of the cube is $\delta(x, y, z) = x + y + z + 1$.

16. A wedge like the one in Exercise 2 has dimensions $a = 2$, $b = 6$, and $c = 3$. The density is $\delta(x, y, z) = x + 1$. Notice that if the density is constant, the center of mass will be $(0, 0, 0)$.

17. Find the mass of the solid bounded by the planes $x + z = 1$, $x - z = -1$, $y = 0$ and the surface $y = \sqrt{z}$. The density of the solid is $\delta(x, y, z) = 2y + 5$.

18. Find the mass of the solid region bounded by the parabolic surfaces $z = 16 - 2x^2 - 2y^2$ and $z = 2x^2 + 2y^2$ if the density of the solid is $\delta(x, y, z) = \sqrt{x^2 + y^2}$.

Work

In Exercises 19 and 20, calculate the following.

a) The amount of work done by (constant) gravity g in moving the liquid filled in the container to the xy-plane (*Hint:* Partition the liquid in the container into small volume elements ΔV_i and find the work done (approximately) by gravity on each element.

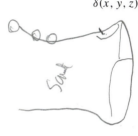

Summation and passage to the limit gives a triple integral to evaluate.)

b) The work done by gravity in moving the center of mass down to the xy-plane

19. The container is a cubical box in the first octant bounded by the coordinate planes and the planes $x = 1$, $y = 1$, and $z = 1$. The density of the liquid filling the box is $\delta(x, y, z) = x + y + z + 1$ (refer to Exercise 15).

20. The container is in the shape of the region bounded by $y = 0$, $z = 0$, $z = 4 - x^2$, and $x = y^2$. The density of the liquid filling the region is $\delta(x, y, z) = kxy$ (see Exercise 14).

The Parallel Axis Theorem

The Parallel Axis Theorem (Exercises 13.2) holds in three dimensions as well as in two. Let $L_{\text{c.m.}}$ be a line through the center of mass of a body of mass m and let L be a parallel line h units away from $L_{\text{c.m.}}$. The **Parallel Axis Theorem** says that the moments of inertia $I_{\text{c.m.}}$ and I_L of the body about $L_{\text{c.m.}}$ and L satisfy the equation

$$I_L = I_{\text{c.m.}} + mh^2. \tag{1}$$

As in the two-dimensional case, the theorem gives a quick way to calculate one moment when the other moment and the mass are known.

21. *Proof of the Parallel Axis Theorem*

a) Show that the first moment of a body in space about any plane through the body's center of mass is zero. (*Hint:* Place the body's center of mass at the origin and let the plane be the yz-plane. What does the formula $\bar{x} = M_{yz}/M$ then tell you?)

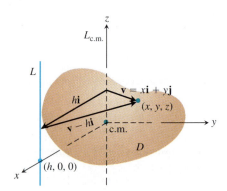

b) To prove the Parallel Axis Theorem, place the body with its center of mass at the origin, with the line $L_{\text{c.m.}}$ along the z-axis and the line L perpendicular to the xy-plane at the point $(h, 0, 0)$. Let D be the region of space occupied by the body. Then, in the notation of the figure,

$$I_L = \iiint\limits_D |\mathbf{v} - h\mathbf{i}|^2 \, dm. \tag{2}$$

Expand the integrand in this integral and complete the proof.

22. The moment of inertia about a diameter of a solid sphere of constant density and radius a is $(2/5)ma^2$, where m is the mass

of the sphere. Find the moment of inertia about a line tangent to the sphere.

23. The moment of inertia of the solid in Exercise 3 about the z-axis is $I_z = abc(a^2 + b^2)/3$.

a) Use Eq. (1) to find the moment of inertia and radius of gyration of the solid about the line parallel to the z-axis through the solid's center of mass.

b) Use Eq. (1) and the result in (a) to find the moment of inertia and radius of gyration of the solid about the line $x = 0$, $y = 2b$.

24. If $a = b = 6$ and $c = 4$, the moment of inertia of the solid wedge in Exercise 2 about the x-axis is $I_x = 208$. Find the moment of inertia of the wedge about the line $y = 4$, $z = -4/3$ (the edge of the wedge's narrow end).

Pappus's Formula

Pappus's formula (Exercises 13.2) holds in three dimensions as well as in two. Suppose that bodies B_1 and B_2 of mass m_1 and m_2, respectively, occupy nonoverlapping regions in space and that $\mathbf{c}_1$ and $\mathbf{c}_2$ are the vectors from the origin to the bodies' respective centers of mass. Then the center of mass of the union $B_1 \cup B_2$ of the two bodies is determined by the vector

$$\mathbf{c} = \frac{m_1 \mathbf{c}_1 + m_2 \mathbf{c}_2}{m_1 + m_2}. \tag{3}$$

As before, this formula is called **Pappus's formula.** As in the two-dimensional case, the formula generalizes to

$$\mathbf{c} = \frac{m_1 \mathbf{c}_1 + m_2 \mathbf{c}_2 + \cdots + m_n \mathbf{c}_n}{m_1 + m_2 + \cdots + m_n} \tag{4}$$

for n bodies.

25. Derive Pappus's formula (Eq. 3). (*Hint:* Sketch B_1 and B_2 as nonoverlapping regions in the first octant and label their centers of mass $(\bar{x}_1, \bar{y}_1, \bar{z}_1)$ and $(\bar{x}_2, \bar{y}_2, \bar{z}_2)$. Express the moments of $B_1 \cup B_2$ about the coordinate planes in terms of the masses m_1 and m_2 and the coordinates of these centers.)

26. The figure below shows a solid made from three rectangular solids of constant density $\delta = 1$. Use Pappus's formula to find the center of mass of

a) $A \cup B$ **b)** $A \cup C$

c) $B \cup C$ **d)** $A \cup B \cup C.$

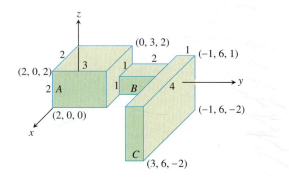

27. a) Suppose that a solid right circular cone C of base radius a and altitude h is constructed on the circular base of a solid hemisphere S of radius a so that the union of the two solids resembles an ice cream cone. The centroid of a solid cone lies one-fourth of the way from the base toward the vertex. The centroid of a solid hemisphere lies three-eighths of the way from the base to the top. What relation must hold between h and a to place the centroid of $C \cup S$ in the common base of the two solids?

b) If you have not already done so, answer the analogous ques-

tion about a triangle and a semicircle (Section 13.2, Exercise 55). The answers are not the same.

28. A solid pyramid P with height h and four congruent sides is built with its base as one face of a solid cube C whose edges have length s. The centroid of a solid pyramid lies one-fourth of the way from the base toward the vertex. What relation must hold between h and s to place the centroid of $P \cup C$ in the base of the pyramid? Compare your answer with the answer to Exercise 27. Also compare it to the answer to Exercise 56 in Section 13.2.

13.6

Triple Integrals in Cylindrical and Spherical Coordinates

When a calculation in physics, engineering, or geometry involves a cylinder, cone, or sphere, we can often simplify our work by using cylindrical or spherical coordinates.

Cylindrical Coordinates

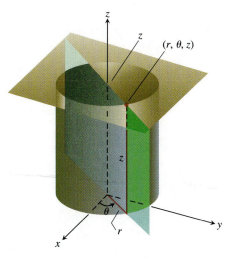

Cylindrical coordinates (Fig. 13.39) are good for describing cylinders whose axes run along the z-axis and planes that either contain the z-axis or lie perpendicular to the z-axis. As we saw in Section 10.7, surfaces like these have equations of constant coordinate value:

$$r = 4 \qquad \text{Cylinder, radius 4, axis the } z\text{-axis}$$

$$\theta = \frac{\pi}{3} \qquad \text{Plane containing the } z\text{-axis}$$

$$z = 2 \qquad \text{Plane perpendicular to the } z\text{-axis}$$

The volume element for subdividing a region in space with cylindrical coordinates is

$$dV = dz\, r\, dr\, d\theta \tag{1}$$

(Fig. 13.40). Triple integrals in cylindrical coordinates are then evaluated as iterated integrals, as in the following example.

13.39 Cylindrical coordinates and typical surfaces of constant coordinate value.

EXAMPLE 1 Find the limits of integration in cylindrical coordinates for integrating a function $f(r, \theta, z)$ over the region D bounded below by the plane $z = 0$, laterally by the circular cylinder $x^2 + (y - 1)^2 = 1$, and above by the paraboloid $z = x^2 + y^2$.

Solution

Step 1: *A sketch* (Fig. 13.41). The base of D is also the region's projection R on the xy-plane. The boundary of R is the circle $x^2 + (y - 1)^2 = 1$. Its polar coordinate equation is

$$x^2 + (y - 1)^2 = 1$$

$$x^2 + y^2 - 2y + 1 = 1$$

$$r^2 - 2r \sin \theta = 0$$

$$r = 2 \sin \theta.$$

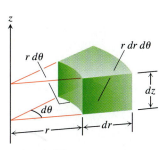

13.40 The volume element in cylindrical coordinates is $dV = dz\, r\, dr\, d\theta$.

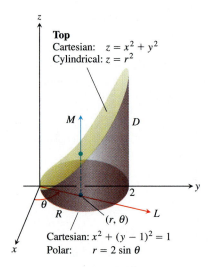

13.41 The figure for Example 1.

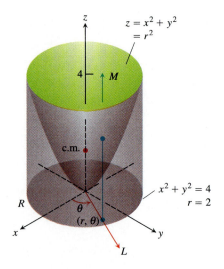

13.42 Example 2 shows how to find the centroid of this solid.

Step 2: *The z-limits of integration.* A line M through a typical point (r, θ) in R parallel to the z-axis enters D at $z = 0$ and leaves at $z = x^2 + y^2 = r^2$.

Step 3: *The r-limits of integration.* A ray L through (r, θ) from the origin enters R at $r = 0$ and leaves at $r = 2 \sin \theta$.

Step 4: *The θ-limits of integration.* As L sweeps across R, the angle θ it makes with the positive x-axis runs from $\theta = 0$ to $\theta = \pi$. The integral is

$$\iiint_D f(r, \theta, z) \, dV = \int_0^\pi \int_0^{2 \sin \theta} \int_0^{r^2} f(r, \theta, z) \, dz \, r \, dr \, d\theta.$$ ❑

Example 1 illustrates a good procedure for finding limits of integration in cylindrical coordinates. The procedure is summarized in the box on the following page.

EXAMPLE 2   Find the centroid ($\delta = 1$) of the solid enclosed by the cylinder $x^2 + y^2 = 4$, bounded above by the paraboloid $z = x^2 + y^2$ and below by the xy-plane.

Solution   We sketch the solid, bounded above by the paraboloid $z = r^2$ and below by the plane $z = 0$ (Fig. 13.42). Its base R is the disk $|r| \leq 2$ in the xy-plane.

The solid's centroid $(\overline{x}, \overline{y}, \overline{z})$ lies on its axis of symmetry, here the z-axis. This makes $\overline{x} = \overline{y} = 0$. To find $\overline{z}$, we divide the first moment M_{xy} by the mass M.

To find the limits of integration for the mass and moment integrals, we continue with the four basic steps. We completed step 1 with our initial sketch. The remaining steps give the limits of integration.

Step 2: *The z-limits.* A line M through a typical point (r, θ) in the base parallel to the z-axis enters the solid at $z = 0$ and leaves at $z = r^2$.

Step 3: *The r-limits.* A ray L through (r, θ) from the origin enters R at $r = 0$ and leaves at $r = 2$.

Step 4: *The θ-limits.* As L sweeps over the base like a clock hand, the angle θ it makes with the positive x-axis runs from $\theta = 0$ to $\theta = 2\pi$. The value of M_{xy} is

$$M_{xy} = \int_0^{2\pi} \int_0^2 \int_0^{r^2} z \, dz \, r \, dr \, d\theta = \int_0^{2\pi} \int_0^2 \left[\frac{z^2}{2} \right]_0^{r^2} r \, dr \, d\theta$$

$$= \int_0^{2\pi} \int_0^2 \frac{r^5}{2} \, dr \, d\theta = \int_0^{2\pi} \left[\frac{r^6}{12} \right]_0^2 d\theta = \int_0^{2\pi} \frac{16}{3} \, d\theta = \frac{32\pi}{3}.$$

The value of M is

$$M = \int_0^{2\pi} \int_0^2 \int_0^{r^2} dz \, r \, dr \, d\theta = \int_0^{2\pi} \int_0^2 \left[z \right]_0^{r^2} r \, dr \, d\theta$$

$$= \int_0^{2\pi} \int_0^2 r^3 \, dr \, d\theta = \int_0^{2\pi} \left[\frac{r^4}{4} \right]_0^2 d\theta = \int_0^{2\pi} 4 \, d\theta = 8\pi.$$

Therefore,

$$\overline{z} = \frac{M_{xy}}{M} = \frac{32\pi}{3} \frac{1}{8\pi} = \frac{4}{3},$$

and the centroid is $(0, 0, 4/3)$. Notice that the centroid lies outside the solid. ❑

How to Integrate in Cylindrical Coordinates

To evaluate

$$\iiint_D f(r, \theta, z) \, dV$$

over a region D in space in cylindrical coordinates, integrating first with respect to z, then with respect to r, and finally with respect to θ, take the following steps.

1. *A sketch.* Sketch the region D along with its projection R on the xy-plane. Label the surfaces and curves that bound D and R.

2. *The z-limits of integration.* Draw a line M through a typical point (r, θ) of R parallel to the z-axis. As z increases, M enters D at $z = g_1(r, \theta)$ and leaves at $z = g_2(r, \theta)$. These are the z-limits of integration.

3. *The r-limits of integration.* Draw a ray L through (r, θ) from the origin. The ray enters R at $r = h_1(\theta)$ and leaves at $r = h_2(\theta)$. These are the r-limits of integration.

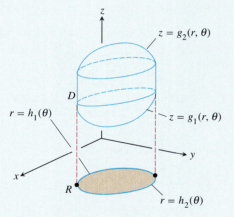
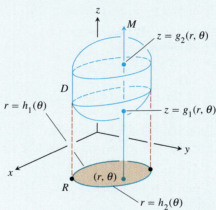
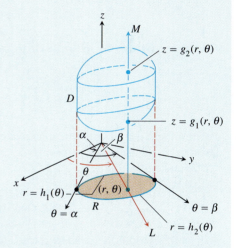

4. *The θ-limits of integration.* As L sweeps across R, the angle θ it makes with the positive x-axis runs from $\theta = \alpha$ to $\theta = \beta$. These are the θ-limits of integration. The integral is

$$\iiint_D f(r, \theta, z) \, dV = \int_{\theta=\alpha}^{\theta=\beta} \int_{r=h_1(\theta)}^{r=h_2(\theta)} \int_{z=g_1(r,\theta)}^{z=g_2(r,\theta)} f(r, \theta, z) \, dz \, r \, dr \, d\theta. \tag{2}$$

Spherical Coordinates

Spherical coordinates (Fig. 13.43, on the following page) are good for describing spheres centered at the origin, half-planes hinged along the z-axis, and single-napped cones whose vertices lie at the origin and whose axes lie along the z-axis. Surfaces like these have equations of constant coordinate value:

$$\rho = 4 \qquad \text{Sphere, radius 4, center at origin}$$

$$\phi = \frac{\pi}{3} \qquad \text{Cone opening up from the origin, making an angle of } \pi/3 \text{ radians with the positive } z\text{-axis}$$

$$\theta = \frac{\pi}{3} \qquad \text{Half-plane, hinged along the } z\text{-axis, making an angle of } \pi/3 \text{ radians with the positive } x\text{-axis}$$

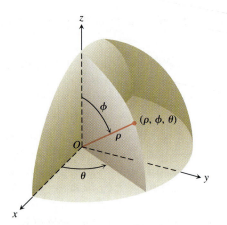

13.43 Spherical coordinates are measured with a distance and two angles.

The volume element in spherical coordinates is the volume of a **spherical wedge** defined by the differentials $d\rho$, $d\phi$, and $d\theta$ (Fig. 13.44). The wedge is approximately a rectangular box with one side a circular arc of length $\rho\, d\phi$, another side a circular arc of length $\rho \sin \phi\, d\theta$, and thickness $d\rho$. Therefore the volume element in spherical coordinates is

$$dV = \rho^2 \sin \phi\, d\rho\, d\phi\, d\theta, \tag{3}$$

and triple integrals take the form

$$\iiint F(\rho, \phi, \theta)\, dV = \iiint F(\rho, \phi, \theta)\, \rho^2 \sin \phi\, d\rho\, d\phi\, d\theta. \tag{4}$$

To evaluate these integrals, we usually integrate first with respect to ρ. The procedure for finding the limits of integration is shown in the following box. We restrict our attention to integrating over domains that are solids of revolution about the z-axis (or portions thereof) and for which the limits for θ and ϕ are constant.

How to Integrate in Spherical Coordinates

To evaluate

$$\iiint_D f(\rho, \phi, \theta)\, dV$$

over a region D in space in spherical coordinates, integrating first with respect to ρ, then with respect to ϕ, and finally with respect to θ, take the following steps.

1. *A sketch.* Sketch the region D along with its projection R on the xy-plane. Label the surfaces that bound D.
2. *The ρ-limits of integration.* Draw a ray M from the origin making an angle ϕ with the positive z-axis. Also draw the projection of M on the xy-plane (call the projection L). The ray L makes an angle θ with the positive x-axis. As ρ increases, M enters D at $\rho = g_1(\phi, \theta)$ and leaves at $\rho = g_2(\phi, \theta)$. These are the ρ-limits of integration.
3. *The ϕ-limits of integration.* For any given θ, the angle ϕ that M makes with the z-axis runs from $\phi = \phi_{min}$ to $\phi = \phi_{max}$. These are the ϕ-limits of integration.
4. *The θ-limits of integration.* The ray L sweeps over R as θ runs from α to β. These are the θ-limits of integration. The integral is

$$\iiint_D f(\rho, \phi, \theta)\, dV = \int_{\theta=\alpha}^{\theta=\beta} \int_{\phi=\phi_{min}}^{\phi=\phi_{max}} \int_{\rho=g_1(\phi,\theta)}^{\rho=g_2(\phi,\theta)} f(\rho, \phi, \theta)\, \rho^2 \sin \phi\, d\rho\, d\phi\, d\theta. \tag{5}$$

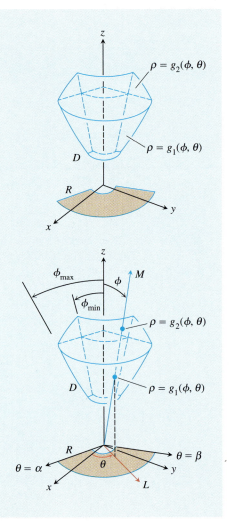

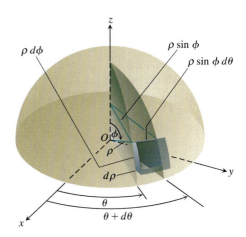

13.44 The volume element in spherical coordinates is

$$dV = d\rho \cdot \rho\, d\phi \cdot \rho \sin\phi\, d\theta$$
$$= \rho^2 \sin\phi\, d\rho\, d\phi\, d\theta.$$

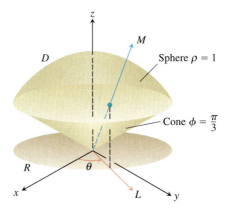

13.45 The solid in Example 3.

EXAMPLE 3   Find the volume of the upper region D cut from the solid sphere $\rho \le 1$ by the cone $\phi = \pi/3$.

Solution   The volume is $V = \iiint_D \rho^2 \sin\phi\, d\rho\, d\phi\, d\theta$, the integral of $f(\rho, \phi, \theta) = 1$ over D.

To find the limits of integration for evaluating the integral, we take the following steps.

Step 1: *A sketch.* We sketch D and its projection R on the xy-plane (Fig. 13.45).

Step 2: *The ρ-limits of integration.* We draw a ray M from the origin making an angle ϕ with the positive z-axis. We also draw L, the projection of M on the xy-plane, along with the angle θ that L makes with the positive x-axis. Ray M enters D at $\rho = 0$ and leaves at $\rho = 1$.

Step 3: *The ϕ-limits of integration.* The cone $\phi = \pi/3$ makes an angle of $\pi/3$ with the positive z-axis. For any given θ, the angle ϕ can run from $\phi = 0$ to $\phi = \pi/3$.

Step 4: *The θ-limits of integration.* The ray L sweeps over R as θ runs from 0 to 2π. The volume is

$$V = \iiint_D \rho^2 \sin\phi\, d\rho\, d\phi\, d\theta = \int_0^{2\pi} \int_0^{\pi/3} \int_0^1 \rho^2 \sin\phi\, d\rho\, d\phi\, d\theta$$

$$= \int_0^{2\pi} \int_0^{\pi/3} \left[\frac{\rho^3}{3}\right]_0^1 \sin\phi\, d\phi\, d\theta = \int_0^{2\pi} \int_0^{\pi/3} \frac{1}{3} \sin\phi\, d\phi\, d\theta$$

$$= \int_0^{2\pi} \left[-\frac{1}{3} \cos\phi\right]_0^{\pi/3} d\theta = \int_0^{2\pi} \left(-\frac{1}{6} + \frac{1}{3}\right) d\theta = \frac{1}{6}(2\pi) = \frac{\pi}{3}. \qquad \square$$

EXAMPLE 4   A solid of constant density $\delta = 1$ occupies the region D in Example 3. Find the solid's moment of inertia about the z-axis.

Solution   In rectangular coordinates, the moment is

$$I_z = \iiint (x^2 + y^2)\, dV.$$

In spherical coordinates, $x^2 + y^2 = (\rho \sin\phi \cos\theta)^2 + (\rho \sin\phi \sin\theta)^2 = \rho^2 \sin^2\phi$. Hence,

$$I_z = \iiint (\rho^2 \sin^2\phi)\, \rho^2 \sin\phi\, d\rho\, d\phi\, d\theta = \iiint \rho^4 \sin^3\phi\, d\rho\, d\phi\, d\theta.$$

For the region in Example 3, this becomes

$$I_z = \int_0^{2\pi} \int_0^{\pi/3} \int_0^1 \rho^4 \sin^3\phi\, d\rho\, d\phi\, d\theta = \int_0^{2\pi} \int_0^{\pi/3} \left[\frac{\rho^5}{5}\right]_0^1 \sin^3\phi\, d\phi\, d\theta$$

$$= \frac{1}{5} \int_0^{2\pi} \int_0^{\pi/3} (1 - \cos^2\phi) \sin\phi\, d\phi\, d\theta = \frac{1}{5} \int_0^{2\pi} \left[-\cos\phi + \frac{\cos^3\phi}{3}\right]_0^{\pi/3} d\theta$$

$$= \frac{1}{5} \int_0^{2\pi} \left(-\frac{1}{2} + 1 + \frac{1}{24} - \frac{1}{3}\right) d\theta = \frac{1}{5} \int_0^{2\pi} \frac{5}{24}\, d\theta = \frac{1}{24}(2\pi) = \frac{\pi}{12}. \qquad \square$$

Coordinate Conversion Formulas (from Section 10.8)

Cylindrical to Rectangular	Spherical to Rectangular	Spherical to Cylindrical
$x = r \cos \theta$	$x = \rho \sin \phi \cos \theta$	$r = \rho \sin \phi$
$y = r \sin \theta$	$y = \rho \sin \phi \sin \theta$	$z = \rho \cos \phi$
$z = z$	$z = \rho \cos \phi$	$\theta = \theta$

Corresponding volume elements

$$dV = dx\, dy\, dz$$
$$= dz\, r\, dr\, d\theta$$
$$= \rho^2 \sin \phi\, d\rho\, d\phi\, d\theta$$

Exercises 13.6

Cylindrical Coordinates

Evaluate the cylindrical coordinate integrals in Exercises 1–6.

1. $\displaystyle \int_0^{2\pi} \int_0^1 \int_r^{\sqrt{2-r^2}} dz\, r\, dr\, d\theta$

2. $\displaystyle \int_0^{2\pi} \int_0^3 \int_{r^2/3}^{\sqrt{18-r^2}} dz\, r\, dr\, d\theta$

3. $\displaystyle \int_0^{2\pi} \int_0^{\theta/2\pi} \int_0^{3+24r^2} dz\, r\, dr\, d\theta$

4. $\displaystyle \int_0^{\pi} \int_0^{\theta/\pi} \int_{-\sqrt{4-r^2}}^{3\sqrt{4-r^2}} z\, dz\, r\, dr\, d\theta$

5. $\displaystyle \int_0^{2\pi} \int_0^1 \int_r^{1/\sqrt{2-r^2}} 3\, dz\, r\, dr\, d\theta$

6. $\displaystyle \int_0^{2\pi} \int_0^1 \int_{-1/2}^{1/2} (r^2 \sin^2 \theta + z^2)\, dz\, r\, dr\, d\theta$

The integrals we have seen so far suggest that there are preferred orders of integration for cylindrical coordinates, but other orders usually work well and are occasionally easier to evaluate. Evaluate the integrals in Exercises 7–10.

7. $\displaystyle \int_0^{2\pi} \int_0^3 \int_0^{z/3} r^3\, dr\, dz\, d\theta$

8. $\displaystyle \int_{-1}^1 \int_0^{2\pi} \int_0^{1+\cos \theta} 4r\, dr\, d\theta\, dz$

9. $\displaystyle \int_0^1 \int_0^{\sqrt{z}} \int_0^{2\pi} (r^2 \cos^2 \theta + z^2)\, r\, d\theta\, dr\, dz$

10. $\displaystyle \int_0^2 \int_{r-2}^{\sqrt{4-r^2}} \int_0^{2\pi} (r \sin \theta + 1)\, r\, d\theta\, dz\, dr$

11. Let D be the region bounded below by the plane $z = 0$, above by the sphere $x^2 + y^2 + z^2 = 4$, and on the sides by the cylinder $x^2 + y^2 = 1$. Set up the triple integrals in cylindrical coordinates that give the volume of D using the following orders of integration.

a) $dz\, dr\, d\theta$
b) $dr\, dz\, d\theta$
c) $d\theta\, dz\, dr$

12. Let D be the region bounded below by the cone $z = \sqrt{x^2 + y^2}$ and above by the paraboloid $z = 2 - x^2 - y^2$. Set up the triple integrals in cylindrical coordinates that give the volume of D using the following orders of integration.

a) $dz\, dr\, d\theta$
b) $dr\, dz\, d\theta$
c) $d\theta\, dz\, dr$

13. Give the limits of integration for evaluating the integral

$$\iiint f(r, \theta, z)\ dz\, r\, dr\, d\theta$$

as an iterated integral over the region that is bounded below by the plane $z = 0$, on the side by the cylinder $r = \cos \theta$, and on top by the paraboloid $z = 3r^2$.

14. Convert the integral

$$\int_{-1}^1 \int_0^{\sqrt{1-y^2}} \int_0^x (x^2 + y^2)\, dz\, dx\, dy$$

to an equivalent integral in cylindrical coordinates and evaluate the result.

In Exercises 15–20, set up the iterated integral for evaluating $\iiint_D f(r, \theta, z)\, dz\, r\, dr\, d\theta$ over the given region D.

15. D is the right circular cylinder whose base is the circle $r = 2 \sin \theta$ in the xy-plane and whose top lies in the plane $z = 4 - y$.

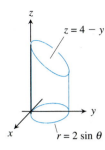

16. D is the right circular cylinder whose base is the circle $r = 3 \cos \theta$ and whose top lies in the plane $z = 5 - x$.

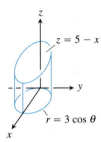

17. D is the solid right cylinder whose base is the region in the xy-plane that lies inside the cardioid $r = 1 + \cos \theta$ and outside the circle $r = 1$ and whose top lies in the plane $z = 4$.

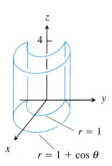

18. D is the solid right cylinder whose base is the region between the circles $r = \cos \theta$ and $r = 2 \cos \theta$, and whose top lies in the plane $z = 3 - y$.

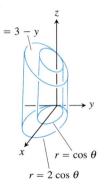

19. D is the prism whose base is the triangle in the xy-plane bounded by the x-axis and the lines $y = x$ and $x = 1$ and whose top lies in the plane $z = 2 - y$.

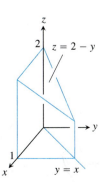

20. D is the prism whose base is the triangle in the xy-plane bounded by the y-axis and the lines $y = x$ and $y = 1$ and whose top lies in the plane $z = 2 - x$.

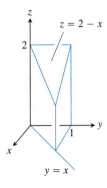

Spherical Coordinates

Evaluate the spherical coordinate integrals in Exercises 21–26.

21. $\displaystyle\int_0^{\pi} \int_0^{\pi} \int_0^{2 \sin \phi} \rho^2 \sin \phi \, d\rho \, d\phi \, d\theta$

22. $\displaystyle\int_0^{2\pi} \int_0^{\pi/4} \int_0^{2} (\rho \cos \phi) \rho^2 \sin \phi \, d\rho \, d\phi \, d\theta$

23. $\displaystyle\int_0^{2\pi} \int_0^{\pi} \int_0^{(1-\cos \phi)/2} \rho^2 \sin \phi \, d\rho \, d\phi \, d\theta$

24. $\displaystyle\int_0^{3\pi/2} \int_0^{\pi} \int_0^{1} 5\rho^3 \sin^3 \phi \, d\rho \, d\phi \, d\theta$

25. $\displaystyle\int_0^{2\pi} \int_0^{\pi/3} \int_{\sec \phi}^{2} 3\rho^2 \sin \phi \, d\rho \, d\phi \, d\theta$

26. $\displaystyle\int_0^{2\pi} \int_0^{\pi/4} \int_0^{\sec \phi} (\rho \cos \phi) \rho^2 \sin \phi \, d\rho \, d\phi \, d\theta$

The previous integrals suggest there are preferred orders of integration for spherical coordinates, but other orders are possible and occasionally easier to evaluate. Evaluate the integrals in Exercises 27–30.

27. $\displaystyle\int_0^{2} \int_{-\pi}^{0} \int_{\pi/4}^{\pi/2} \rho^3 \sin 2\phi \, d\phi \, d\theta \, d\rho$

28. $\displaystyle\int_{\pi/6}^{\pi/3} \int_{\csc \phi}^{2 \csc \phi} \int_0^{2\pi} \rho^2 \sin \phi \, d\theta \, d\rho \, d\phi$

29. $\displaystyle\int_0^{1} \int_0^{\pi} \int_0^{\pi/4} 12\rho \sin^3 \phi \, d\phi \, d\theta \, d\rho$

30. $\displaystyle\int_{\pi/6}^{\pi/2} \int_{-\pi/2}^{\pi/2} \int_{\csc \phi}^{2} 5\rho^4 \sin^3 \phi \, d\rho \, d\theta \, d\phi$

31. Let D be the region in Exercise 11. Set up the triple integrals in spherical coordinates that give the volume of D using the following orders of integration.

  a) $d\rho \, d\phi \, d\theta$ **b)** $d\phi \, d\rho \, d\theta$

32. Let D be the region bounded below by the cone $z = \sqrt{x^2 + y^2}$ and above by the plane $z = 1$. Set up the triple integrals in spherical coordinates that give the volume of D using the following orders of integration.

  a) $d\rho \, d\phi \, d\theta$ **b)** $d\phi \, d\rho \, d\theta$

In Exercises 33–38, (a) find the spherical coordinate limits for the integral that calculates the volume of the given solid, and (b) then evaluate the integral.

33. The solid between the sphere $\rho = \cos \phi$ and the hemisphere $\rho = 2, z \geq 0$

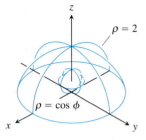

34. The solid bounded below by the hemisphere $\rho = 1, z \geq 0$, and above by the cardioid of revolution $\rho = 1 + \cos \phi$

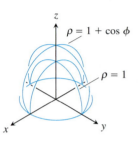

35. The solid enclosed by the cardioid of revolution $\rho = 1 - \cos \phi$

36. The upper portion cut from the solid in Exercise 35 by the xy-plane

37. The solid bounded below by the sphere $\rho = 2 \cos \phi$ and above by the cone $z = \sqrt{x^2 + y^2}$

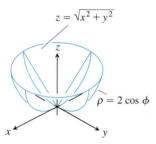

38. The solid bounded below by the xy-plane, on the sides by the sphere $\rho = 2$, and above by the cone $\phi = \pi/3$

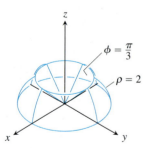

Rectangular, Cylindrical, and Spherical Coordinates

39. Set up triple integrals for the volume of the sphere $\rho = 2$ in (a) spherical, (b) cylindrical, and (c) rectangular coordinates.

40. Let D be the region in the first octant that is bounded below by the cone $\phi = \pi/4$ and above by the sphere $\rho = 3$. Express the volume of D as an iterated triple integral in (a) cylindrical and (b) spherical coordinates. Then (c) find V.

41. Let D be the smaller cap cut from a solid ball of radius 2 units by a plane 1 unit from the center of the sphere. Express the volume of D as an iterated triple integral in (a) spherical, (b) cylindrical, and (c) rectangular coordinates. Then (d) find the volume by evaluating one of the three triple integrals.

42. Express the moment of inertia I_z of the solid hemisphere $x^2 + y^2 + z^2 \leq 1, z \geq 0$, as an iterated integral in (a) cylindrical and (b) spherical coordinates. Then (c) find I_z.

Volumes

Find the volumes of the solids in Exercises 43–48.

43.

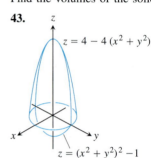

44.

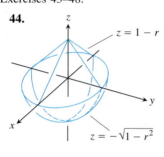

45.

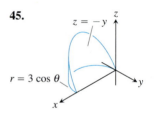

46.

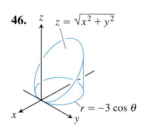

47.

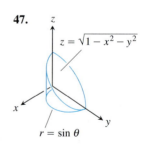

48.

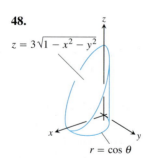

49. Find the volume of the portion of the solid sphere $\rho \leq a$ that lies between the cones $\phi = \pi/3$ and $\phi = 2\pi/3$.

50. Find the volume of the region cut from the solid sphere $\rho \leq a$ by the half-planes $\theta = 0$ and $\theta = \pi/6$ in the first octant.

51. Find the volume of the smaller region cut from the solid sphere $\rho \leq 2$ by the plane $z = 1$.

52. Find the volume of the solid enclosed by the cone $z = \sqrt{x^2 + y^2}$ between the planes $z = 1$ and $z = 2$.

53. Find the volume of the region bounded below by the plane $z = 0$, laterally by the cylinder $x^2 + y^2 = 1$, and above by the paraboloid $z = x^2 + y^2$.

54. Find the volume of the region bounded below by the paraboloid $z = x^2 + y^2$, laterally by the cylinder $x^2 + y^2 = 1$, and above by the paraboloid $z = x^2 + y^2 + 1$.

55. Find the volume of the solid cut from the thick-walled cylinder $1 \le x^2 + y^2 \le 2$ by the cones $z = \pm\sqrt{x^2 + y^2}$.

56. Find the volume of the region that lies inside the sphere $x^2 + y^2 + z^2 = 2$ and outside the cylinder $x^2 + y^2 = 1$.

57. Find the volume of the region enclosed by the cylinder $x^2 + y^2 = 4$ and the planes $z = 0$ and $y + z = 4$.

58. Find the volume of the region enclosed by the cylinder $x^2 + y^2 = 4$ and the planes $z = 0$ and $x + y + z = 4$.

59. Find the volume of the region bounded above by the paraboloid $z = 5 - x^2 - y^2$ and below by the paraboloid $z = 4x^2 + 4y^2$.

60. Find the volume of the region bounded above by the paraboloid $z = 9 - x^2 - y^2$, below by the xy-plane, and lying *outside* the cylinder $x^2 + y^2 = 1$.

61. Find the volume of the region cut from the solid cylinder $x^2 + y^2 \le 1$ by the sphere $x^2 + y^2 + z^2 = 4$.

62. Find the volume of the region bounded above by the sphere $x^2 + y^2 + z^2 = 2$ and below by the paraboloid $z = x^2 + y^2$.

Average Values

63. Find the average value of the function $f(r, \theta, z) = r$ over the region bounded by the cylinder $r = 1$ between the planes $z = -1$ and $z = 1$.

64. Find the average value of the function $f(r, \theta, z) = r$ over the solid ball bounded by the sphere $r^2 + z^2 = 1$. (This is the sphere $x^2 + y^2 + z^2 = 1$.)

65. Find the average value of the function $f(\rho, \phi, \theta) = \rho$ over the solid ball $\rho \le 1$.

66. Find the average value of the function $f(\rho, \phi, \theta) = \rho \cos \phi$ over the solid upper ball $\rho \le 1$, $0 \le \phi \le \pi/2$.

Masses, Moments, and Centroids

67. A solid of constant density is bounded below by the plane $z = 0$, above by the cone $z = r$, $r \ge 0$, and on the sides by the cylinder $r = 1$. Find the center of mass.

68. Find the centroid of the region in the first octant that is bounded above by the cone $z = \sqrt{x^2 + y^2}$, below by the plane $z = 0$, and on the sides by the cylinder $x^2 + y^2 = 4$ and the planes $x = 0$ and $y = 0$.

69. Find the centroid of the solid in Exercise 38.

70. Find the centroid of the solid bounded above by the sphere $\rho = a$ and below by the cone $\phi = \pi/4$.

71. Find the centroid of the region that is bounded above by the surface $z = \sqrt{r}$, on the sides by the cylinder $r = 4$, and below by the xy-plane.

72. Find the centroid of the region cut from the solid ball $r^2 + z^2 \le 1$ by the half-planes $\theta = -\pi/3, r \ge 0$, and $\theta = \pi/3, r \ge 0$.

73. Find the moment of inertia and radius of gyration about the z-axis of a thick-walled right circular cylinder bounded on the inside by the cylinder $r = 1$, on the outside by the cylinder $r = 2$, and on the top and bottom by the planes $z = 4$ and $z = 0$. (Take $\delta = 1$.)

74. Find the moment of inertia of a solid circular cylinder of radius 1 and height 2 (a) about the axis of the cylinder, (b) about a line through the centroid perpendicular to the axis of the cylinder. (Take $\delta = 1$.)

75. Find the moment of inertia of a right circular cone of base radius 1 and height 1 about an axis through the vertex parallel to the base. (Take $\delta = 1$.)

76. Find the moment of inertia of a solid sphere of radius a about a diameter. (Take $\delta = 1$.)

77. Find the moment of inertia of a right circular cone of base radius a and height h about its axis. (*Hint:* Place the cone with its vertex at the origin and its axis along the z-axis.)

78. A solid is bounded on the top by the paraboloid $z = r^2$, on the bottom by the plane $z = 0$, and on the sides by the cylinder $r = 1$. Find the center of mass and the moment of inertia and radius of gyration about the z-axis if the density is (a) $\delta(r, \theta, z) = z$; (b) $\delta(r, \theta, z) = r$.

79. A solid is bounded below by the cone $z = \sqrt{x^2 + y^2}$ and above by the plane $z = 1$. Find the center of mass and the moment of inertia and radius of gyration about the z-axis if the density is (a) $\delta(r, \theta, z) = z$; (b) $\delta(r, \theta, z) = z^2$.

80. A solid ball is bounded by the sphere $\rho = a$. Find the moment of inertia and radius of gyration about the z-axis if the density is

a) $\delta(\rho, \phi, \theta) = \rho^2$,   **b)** $\delta(\rho, \phi, \theta) = r = \rho \sin \phi$.

81. Show that the centroid of the solid semi-ellipsoid of revolution $(r^2/a^2) + (z^2/h^2) \le 1, z \ge 0$, lies on the z-axis three-eighths of the way from the base to the top. The special case $h = a$ gives a solid hemisphere. Thus the centroid of a solid hemisphere lies on the axis of symmetry three-eighths of the way from the base to the top.

82. Show that the centroid of a solid right circular cone is one-fourth of the way from the base to the vertex. (In general, the centroid of a solid cone or pyramid is one-fourth of the way from the centroid of the base to the vertex.)

83. A solid right circular cylinder is bounded by the cylinder $r = a$ and the planes $z = 0$ and $z = h$, $h > 0$. Find the center of mass and the moment of inertia and radius of gyration about the z-axis if the density is $\delta(r, \theta, z) = z + 1$.

84. A spherical planet of radius R has an atmosphere whose density is $\mu = \mu_0 e^{-ch}$, where h is the altitude above the surface of the planet, μ_0 is the density at sea level, and c is a positive constant. Find the mass of the planet's atmosphere.

85. A planet is in the shape of a sphere of radius R and total mass M with spherically symmetric density distribution that increases linearly as one approaches its center. What is the density at the center of this planet if the density at its edge (surface) is taken to be zero?

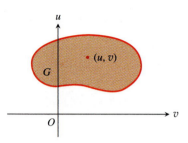

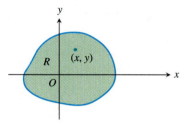

13.46 The equations $x = g(u, v)$ and $y = h(u, v)$ allow us to change an integral over a region R in the xy-plane into an integral over a region G in the uv-plane.

Notice the "Reversed" Order

The transforming equations $x = g(u, v)$ and $y = h(u, v)$ go from G to R, but we use them to change an integral over R into an integral over G.

13.7 Substitutions in Multiple Integrals

This section shows how to evaluate multiple integrals by substitution. As in single integration, the goal of substitution is to replace complicated integrals by ones that are easier to evaluate. Substitutions accomplish this by simplifying the integrand, the limits of integration, or both.

Substitutions in Double Integrals

The polar coordinate substitution of Section 13.3 is a special case of a more general substitution method for double integrals, a method that pictures changes in variables as transformations of regions.

Suppose that a region G in the uv-plane is transformed one-to-one into the region R in the xy-plane by equations of the form

$$x = g(u, v), \qquad y = h(u, v),$$

as suggested in Fig. 13.46. We call R the **image** of G under the transformation, and G the **preimage** of R. Any function $f(x, y)$ defined on R can be thought of as a function $f(g(u, v), h(u, v))$ defined on G as well. How is the integral of $f(x, y)$ over R related to the integral of $f(g(u, v), h(u, v))$ over G?

The answer is: If g, h, and f have continuous partial derivatives and $J(u, v)$ (to be discussed in a moment) is zero only at isolated points, if at all, then

$$\iint\limits_{R} f(x, y)\, dx\, dy = \iint\limits_{G} f(g(u, v), h(u, v))\, |J(u, v)|\, du\, dv. \qquad (1)$$

The factor $J(u, v)$, whose absolute value appears in Eq. (1), is the *Jacobian* of the coordinate transformation, named after the mathematician Carl Jacobi.

Definition

The **Jacobian determinant** or **Jacobian** of the coordinate transformation $x = g(u, v), y = h(u, v)$ is

$$J(u, v) = \begin{vmatrix} \dfrac{\partial x}{\partial u} & \dfrac{\partial x}{\partial v} \\[2mm] \dfrac{\partial y}{\partial u} & \dfrac{\partial y}{\partial v} \end{vmatrix} = \frac{\partial x}{\partial u}\frac{\partial y}{\partial v} - \frac{\partial y}{\partial u}\frac{\partial x}{\partial v}. \qquad (2)$$

The Jacobian is also denoted by

$$J(u, v) = \frac{\partial(x, y)}{\partial(u, v)}$$

to help remember how the determinant in Eq. (2) is constructed from the partial derivatives of x and y. The derivation of Eq. (1) is intricate and properly belongs to a course in advanced calculus. We will not give the derivation here.

For polar coordinates, we have r and θ in place of u and v. With $x = r\cos\theta$

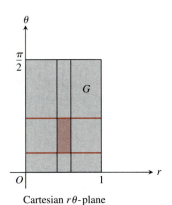

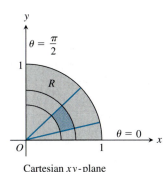

13.47 The equations $x = r \cos \theta, y = r \sin \theta$ transform G into R.

and $y = r \sin \theta$, the Jacobian is

$$J(r, \theta) = \begin{vmatrix} \dfrac{\partial x}{\partial r} & \dfrac{\partial x}{\partial \theta} \\[2mm] \dfrac{\partial y}{\partial r} & \dfrac{\partial y}{\partial \theta} \end{vmatrix} = \begin{vmatrix} \cos \theta & -r \sin \theta \\ \sin \theta & r \cos \theta \end{vmatrix} = r(\cos^2 \theta + \sin^2 \theta) = r.$$

Hence, Eq. (1) becomes

$$\iint_R f(x, y)\, dx\, dy = \iint_G f(r \cos \theta, r \sin \theta) \, |r|\, dr\, d\theta$$

$$= \iint_G f(r \cos \theta, r \sin \theta)\, r\, dr\, d\theta, \qquad \text{If } r \geq 0 \quad (3)$$

which is Eq. (6) in Section 13.3.

Figure 13.47 shows how the equations $x = r \cos \theta$, $y = r \sin \theta$ transform the rectangle G: $0 \leq r \leq 1$, $0 \leq \theta \leq \pi/2$ into the quarter circle R bounded by $x^2 + y^2 = 1$ in the first quadrant of the xy-plane.

Notice that the integral on the right-hand side of Eq. (3) is not the integral of $f(r \cos \theta, r \sin \theta)$ over a region in the polar coordinate plane. It is the integral of the product of $f(r \cos \theta, r \sin \theta)$ and r over a region G in the *Cartesian $r\theta$-plane*.

Here is an example of another substitution.

EXAMPLE 1   Evaluate

$$\int_0^4 \int_{x=y/2}^{x=(y/2)+1} \frac{2x - y}{2}\, dx\, dy$$

by applying the transformation

$$u = \frac{2x - y}{2}, \qquad v = \frac{y}{2} \qquad (4)$$

and integrating over an appropriate region in the uv-plane.

Solution   We sketch the region R of integration in the xy-plane and identify its boundaries (Fig. 13.48).

To apply Eq. (1), we need to find the corresponding uv-region G and the Jacobian of the transformation. To find them, we first solve Eqs. (4) for x and y in terms of u and v. Routine algebra gives

$$x = u + v, \qquad y = 2v. \qquad (5)$$

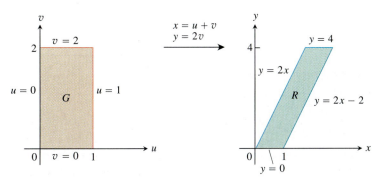

13.48 The equations $x = u + v$ and $y = 2v$ transform G into R. Reversing the transformation by the equations $u = (2x - y)/2$ and $v = y/2$ transforms R into G. See Example 1.

We then find the boundaries of G by substituting these expressions into the equations for the boundaries of R (Fig. 13.48).

xy-equations for the boundary of R	Corresponding uv-equations for the boundary of G	Simplified uv-equations
$x = y/2$	$u + v = 2v/2 = v$	$u = 0$
$x = (y/2) + 1$	$u + v = (2v/2) + 1 = v + 1$	$u = 1$
$y = 0$	$2v = 0$	$v = 0$
$y = 4$	$2v = 4$	$v = 2$

The Jacobian of the transformation (again from Eqs. 5) is

$$J(u, v) = \begin{vmatrix} \dfrac{\partial x}{\partial u} & \dfrac{\partial x}{\partial v} \\[2mm] \dfrac{\partial y}{\partial u} & \dfrac{\partial y}{\partial v} \end{vmatrix} = \begin{vmatrix} \dfrac{\partial}{\partial u}(u + v) & \dfrac{\partial}{\partial v}(u + v) \\[2mm] \dfrac{\partial}{\partial u}(2v) & \dfrac{\partial}{\partial v}(2v) \end{vmatrix} = \begin{vmatrix} 1 & 1 \\ 0 & 2 \end{vmatrix} = 2.$$

We now have everything we need to apply Eq. (1):

$$\int_0^4 \int_{x=y/2}^{x=(y/2)+1} \frac{2x - y}{2}\, dx\, dy = \int_{v=0}^{v=2} \int_{u=0}^{u=1} u\,|J(u, v)|\, du\, dv$$

$$= \int_0^2 \int_0^1 (u)(2)\, du\, dv = \int_0^2 \left[u^2 \right]_0^1 dv = \int_0^2 dv = 2. \qquad \square$$

EXAMPLE 2   Evaluate

$$\int_0^1 \int_0^{1-x} \sqrt{x + y}\,(y - 2x)^2\, dy\, dx.$$

Solution   We sketch the region R of integration in the xy-plane and identify its boundaries (Fig. 13.49). The integrand suggests the transformation $u = x + y$ and $v = y - 2x$. Routine algebra produces x and y as functions of u and v:

$$x = \frac{u}{3} - \frac{v}{3}, \qquad y = \frac{2u}{3} + \frac{v}{3}. \qquad (6)$$

From Eqs. (6) we can find the boundaries of the uv-region G (Fig. 13.49).

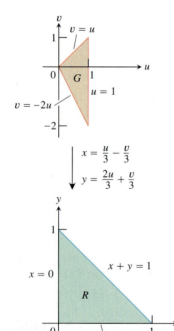

13.49 The equations $x = (u/3) - (v/3)$ and $y = (2u/3) + (v/3)$ transform G into R. Reversing the transformation by the equations $u = x + y$ and $v = y - 2x$ transforms R into G. See Example 2.

xy-equations for the boundary of R	Corresponding uv-equations for the boundary of G	Simplified uv-equations
$x + y = 1$	$\left(\dfrac{u}{3} - \dfrac{v}{3}\right) + \left(\dfrac{2u}{3} + \dfrac{v}{3}\right) = 1$	$u = 1$
$x = 0$	$\dfrac{u}{3} - \dfrac{v}{3} = 0$	$v = u$
$y = 0$	$\dfrac{2u}{3} + \dfrac{v}{3} = 0$	$v = -2u$

The Jacobian of the transformation in Eq. (6) is

$$J(u, v) = \begin{vmatrix} \dfrac{\partial x}{\partial u} & \dfrac{\partial x}{\partial v} \\[2mm] \dfrac{\partial y}{\partial u} & \dfrac{\partial y}{\partial v} \end{vmatrix} = \begin{vmatrix} \dfrac{1}{3} & -\dfrac{1}{3} \\[2mm] \dfrac{2}{3} & \dfrac{1}{3} \end{vmatrix} = \dfrac{1}{3}.$$

Applying Eq. (1), we evaluate the integral:

$$\int_0^1 \int_0^{1-x} \sqrt{x+y}\,(y-2x)^2\,dy\,dx = \int_{u=0}^{u=1} \int_{v=-2u}^{v=u} u^{1/2} v^2\,|J(u, v)|\,dv\,du$$

$$= \int_0^1 \int_{-2u}^{u} u^{1/2} v^2 \left(\frac{1}{3}\right) dv\,du = \frac{1}{3} \int_0^1 u^{1/2} \left(\frac{1}{3} v^3\right)\Big|_{v=-2u}^{v=u} du$$

$$= \frac{1}{9} \int_0^1 u^{1/2}(u^3 + 8u^3)\,du = \int_0^1 u^{7/2}\,du = \frac{2}{9} u^{9/2}\Big|_0^1 = \frac{2}{9}. \qquad \square$$

Substitutions in Triple Integrals

The cylindrical and spherical coordinate substitutions are special cases of a substitution method that pictures changes of variables in triple integrals as transformations of three-dimensional regions. The method is like the method for double integrals except that now we work in three dimensions instead of two.

Suppose that a region G in uvw-space is transformed one-to-one into the region D in xyz-space by differentiable equations of the form

$$x = g(u, v, w), \qquad y = h(u, v, w), \qquad z = k(u, v, w),$$

as suggested in Fig. 13.50. Then any function $F(x, y, z)$ defined on D can be thought of as a function

$$F(g(u, v, w), h(u, v, w), k(u, v, w)) = H(u, v, w)$$

defined on G. If g, h, and k have continuous first partial derivatives, then the integral of $F(x, y, z)$ over D is related to the integral of $H(u, v, w)$ over G by the equation

$$\iiint_D F(x, y, z)\,dx\,dy\,dz = \iiint_G H(u, v, w)\,|J(u, v, w)|\,du\,dv\,dw. \qquad (7)$$

The factor $J(u, v, w)$, whose absolute value appears in this equation, is the **Jacobian determinant**

$$J(u, v, w) = \begin{vmatrix} \dfrac{\partial x}{\partial u} & \dfrac{\partial x}{\partial v} & \dfrac{\partial x}{\partial w} \\[2mm] \dfrac{\partial y}{\partial u} & \dfrac{\partial y}{\partial v} & \dfrac{\partial y}{\partial w} \\[2mm] \dfrac{\partial z}{\partial u} & \dfrac{\partial z}{\partial v} & \dfrac{\partial z}{\partial w} \end{vmatrix} = \dfrac{\partial(x, y, z)}{\partial(u, v, w)}. \qquad (8)$$

As in the two-dimensional case, the derivation of the change-of-variable formula in Eq. (7) is complicated and we will not go into it here.

For cylindrical coordinates, r, θ, and z take the place of u, v, and w. The transformation from Cartesian $r\theta z$-space to Cartesian xyz-space is given by the equations

$$x = r \cos \theta, \qquad y = r \sin \theta, \qquad z = z.$$

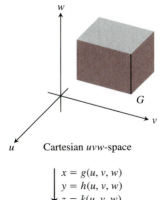

Cartesian uvw-space

$$\begin{aligned} x &= g(u, v, w) \\ y &= h(u, v, w) \\ z &= k(u, v, w) \end{aligned}$$

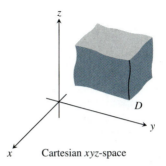

Cartesian xyz-space

13.50 The equations $x = g(u, v, w)$, $y = h(u, v, w)$, and $z = k(u, v, w)$ allow us to change an integral over a region D in Cartesian xyz-space into an integral over a region G in Cartesian uvw-space.

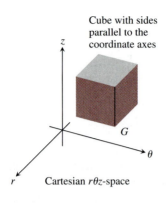

Cube with sides parallel to the coordinate axes

G

Cartesian $r\theta z$-space

$x = r \cos \theta$
$y = r \sin \theta$
$z = z$

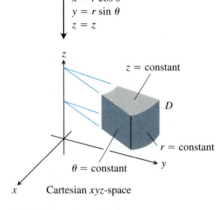

$z = $ constant

D

$r = $ constant

$\theta = $ constant

Cartesian xyz-space

13.51 The equations $x = r \cos \theta$, $y = r \sin \theta$, and $z = z$ transform G into D.

(Fig. 13.51). The Jacobian of the transformation is

$$J(r, \theta, z) = \begin{vmatrix} \dfrac{\partial x}{\partial r} & \dfrac{\partial x}{\partial \theta} & \dfrac{\partial x}{\partial z} \\ \dfrac{\partial y}{\partial r} & \dfrac{\partial y}{\partial \theta} & \dfrac{\partial y}{\partial z} \\ \dfrac{\partial z}{\partial r} & \dfrac{\partial z}{\partial \theta} & \dfrac{\partial z}{\partial z} \end{vmatrix} = \begin{vmatrix} \cos \theta & -r \sin \theta & 0 \\ \sin \theta & r \cos \theta & 0 \\ 0 & 0 & 1 \end{vmatrix}$$

$$= r \cos^2 \theta + r \sin^2 \theta = r.$$

The corresponding version of Eq. (7) is

$$\iiint_D F(x, y, z) \, dx \, dy \, dz = \iiint_G H(r, \theta, z) \, |r| \, dr \, d\theta \, dz. \qquad (9)$$

We can drop the absolute value signs whenever $r \geq 0$.

For spherical coordinates, ρ, ϕ, and θ take the place of u, v, and w. The transformation from Cartesian $\rho\phi\theta$-space to Cartesian xyz-space is given by

$$x = \rho \sin \phi \cos \theta, \qquad y = \rho \sin \phi \sin \theta, \qquad z = \rho \cos \phi$$

(Fig. 13.52). The Jacobian of the transformation is

$$J(\rho, \phi, \theta) = \begin{vmatrix} \dfrac{\partial x}{\partial \rho} & \dfrac{\partial x}{\partial \phi} & \dfrac{\partial x}{\partial \theta} \\ \dfrac{\partial y}{\partial \rho} & \dfrac{\partial y}{\partial \phi} & \dfrac{\partial y}{\partial \theta} \\ \dfrac{\partial z}{\partial \rho} & \dfrac{\partial z}{\partial \phi} & \dfrac{\partial z}{\partial \theta} \end{vmatrix} = \rho^2 \sin \phi \qquad (10)$$

(Exercise 17). The corresponding version of Eq. (7) is

$$\iiint_D F(x, y, z) \, dx \, dy \, dz = \iiint_G H(\rho, \phi, \theta) |\rho^2 \sin \phi| \, d\rho \, d\phi \, d\theta. \qquad (11)$$

We can drop the absolute value signs because $\sin \phi$ is never negative.

Here is an example of another substitution.

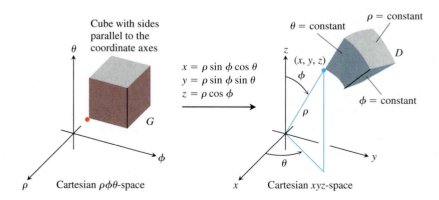

13.52 The equations $x = \rho \sin \phi \cos \theta$, $y = \rho \sin \phi \sin \theta$, and $z = \rho \cos \phi$ transform G into D.

Cube with sides parallel to the coordinate axes

θ

$x = \rho \sin \phi \cos \theta$
$y = \rho \sin \phi \sin \theta$
$z = \rho \cos \phi$

G

ϕ

ρ Cartesian $\rho\phi\theta$-space

$\theta = $ constant

$\rho = $ constant

(x, y, z)

ϕ

D

$\phi = $ constant

ρ

θ

Cartesian xyz-space

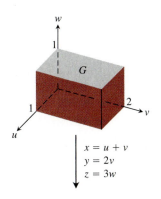

$x = u + v$
$y = 2v$
$z = 3w$

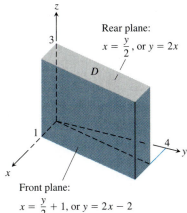

Rear plane:
$x = \dfrac{y}{2}$, or $y = 2x$

D

Front plane:
$x = \dfrac{y}{2} + 1$, or $y = 2x - 2$

13.53 The equations $x = u + v$, $y = 2v$, and $z = 3w$ transform G into D. Reversing the transformation by the equations $u = (2x - y)/2$, $v = y/2$, and $w = z/3$ transforms D into G. See Example 3.

Carl Gustav Jacob Jacobi

Jacobi (1804–1851), one of nineteenth-century Germany's most accomplished scientists, developed the theory of determinants and transformations into a powerful tool for evaluating multiple integrals and solving differential equations. He also applied transformation methods to study nonelementary integrals like the ones that arise in the calculation of arc length. Like Euler, Jacobi was a prolific writer and an even more prolific calculator and worked in a variety of mathematical and applied fields.

EXAMPLE 3   Evaluate

$$\int_0^3 \int_0^4 \int_{x=y/2}^{x=(y/2)+1} \left(\frac{2x - y}{2} + \frac{z}{3} \right) dx\, dy\, dz$$

by applying the transformation

$$u = (2x - y)/2, \qquad v = y/2, \qquad w = z/3 \qquad (12)$$

and integrating over an appropriate region in uvw-space.

Solution   We sketch the region D of integration in xyz-space and identify its boundaries (Fig. 13.53). In this case, the bounding surfaces are planes.

To apply Eq. (7), we need to find the corresponding uvw-region G and the Jacobian of the transformation. To find them, we first solve Eqs. (12) for x, y, and z in terms of u, v, and w. Routine algebra gives

$$x = u + v, \qquad y = 2v, \qquad z = 3w. \qquad (13)$$

We then find the boundaries of G by substituting these expressions into the equations for the boundaries of D:

xyz-equations for the boundary of D	Corresponding uvw-equations for the boundary of G	Simplified uvw-equations
$x = y/2$	$u + v = 2v/2 = v$	$u = 0$
$x = (y/2) + 1$	$u + v = (2v/2) + 1 = v + 1$	$u = 1$
$y = 0$	$2v = 0$	$v = 0$
$y = 4$	$2v = 4$	$v = 2$
$z = 0$	$3w = 0$	$w = 0$
$z = 3$	$3w = 3$	$w = 1$

The Jacobian of the transformation, again from Eqs. (13), is

$$J(u, v, w) = \begin{vmatrix} \dfrac{\partial x}{\partial u} & \dfrac{\partial x}{\partial v} & \dfrac{\partial x}{\partial w} \\ \dfrac{\partial y}{\partial u} & \dfrac{\partial y}{\partial v} & \dfrac{\partial y}{\partial w} \\ \dfrac{\partial z}{\partial u} & \dfrac{\partial z}{\partial v} & \dfrac{\partial z}{\partial w} \end{vmatrix} = \begin{vmatrix} 1 & 1 & 0 \\ 0 & 2 & 0 \\ 0 & 0 & 3 \end{vmatrix} = 6.$$

We now have everything we need to apply Eq. (7):

$$\int_0^3 \int_0^4 \int_{x=y/2}^{x=(y/2)+1} \left(\frac{2x - y}{2} + \frac{z}{3} \right) dx\, dy\, dz$$

$$= \int_0^1 \int_0^2 \int_0^1 (u + w)\, |J(u, v, w)|\, du\, dv\, dw$$

$$= \int_0^1 \int_0^2 \int_0^1 (u + w)(6)\, du\, dv\, dw = 6 \int_0^1 \int_0^2 \left[\frac{u^2}{2} + uw \right]_0^1 dv\, dw$$

$$= 6 \int_0^1 \int_0^2 \left(\frac{1}{2} + w \right) dv\, dw = 6 \int_0^1 \left[\frac{v}{2} + vw \right]_0^2 dw = 6 \int_0^1 (1 + 2w)\, dw$$

$$= 6 \left[w + w^2 \right]_0^1 = 6(2) = 12.$$

Exercises 13.7

Transformations of Coordinates

1. a) Solve the system

$$u = x - y, \qquad v = 2x + y$$

for x and y in terms of u and v. Then find the value of the Jacobian $\partial(x, y)/\partial(u, v)$.

b) Find the image under the transformation $u = x - y, v = 2x + y$ of the triangular region with vertices $(0, 0)$, $(1, 1)$, and $(1, -2)$ in the xy-plane. Sketch the transformed region in the uv-plane.

2. a) Solve the system

$$u = x + 2y, \qquad v = x - y$$

for x and y in terms of u and v. Then find the value of the Jacobian $\partial(x, y)/\partial(u, v)$.

b) Find the image under the transformation $u = x + 2y, v = x - y$ of the triangular region in the xy-plane bounded by the lines $y = 0$, $y = x$, and $x + 2y = 2$. Sketch the transformed region in the uv-plane.

3. a) Solve the system

$$u = 3x + 2y, \qquad v = x + 4y$$

for x and y in terms of u and v. Then find the value of the Jacobian $\partial(x, y)/\partial(u, v)$.

b) Find the image under the transformation $u = 3x + 2y, v = x + 4y$ of the triangular region in the xy-plane bounded by the x-axis, the y-axis, and the line $x + y = 1$. Sketch the transformed region in the uv-plane.

4. a) Solve the system

$$u = 2x - 3y, \qquad v = -x + y$$

for x and y in terms of u and v. Then find the value of the Jacobian $\partial(x, y)/\partial(u, v)$.

b) Find the image under the transformation $u = 2x - 3y, v = -x + y$ of the parallelogram R in the xy-plane with boundaries $x = -3$, $x = 0$, $y = x$, and $y = x + 1$. Sketch the transformed region in the uv-plane.

5. Find the Jacobian $\partial(x, y)/\partial(u, v)$ for the transformation

a) $x = u \cos v, \quad y = u \sin v$

b) $x = u \sin v, \quad y = u \cos v$.

6. Find the Jacobian $\partial(x, y, z)/\partial(u, v, w)$ of the transformation

a) $x = u \cos v, \quad y = u \sin v, \quad z = w$

b) $x = 2u - 1, \quad y = 3v - 4, \quad z = \dfrac{1}{2}(w - 4)$.

Double Integrals

7. Evaluate the integral

$$\int_0^4 \int_{x=y/2}^{x=(y/2)+1} \frac{2x - y}{2} dx \, dy$$

from Example 1 directly by integration with respect to x and y to confirm that its value is 2.

8. Use the transformation in Exercise 1 to evaluate the integral

$$\iint_R (2x^2 - xy - y^2) \, dx \, dy$$

for the region R in the first quadrant bounded by the lines $y = -2x + 4$, $y = -2x + 7$, $y = x - 2$, and $y = x + 1$.

9. Use the transformation in Exercise 3 to evaluate the integral

$$\iint_R (3x^2 + 14xy + 8y^2) \, dx \, dy$$

for the region R in the first quadrant bounded by the lines $y = -\dfrac{3}{2}x + 1$, $y = -\dfrac{3}{2}x + 3$, $y = -\dfrac{1}{4}x$, and $y = -\dfrac{1}{4}x + 1$.

10. Use the transformation and parallelogram R in Exercise 4 to evaluate the integral

$$\iint_R 2(x - y) \, dx \, dy.$$

11. Let R be the region in the first quadrant of the xy-plane bounded by the hyperbolas $xy = 1$, $xy = 9$ and the lines $y = x$, $y = 4x$. Use the transformation $x = u/v$, $y = uv$ with $u > 0$ and $v > 0$ to rewrite

$$\iint_R \left(\sqrt{\frac{y}{x}} + \sqrt{xy} \right) dx \, dy$$

as an integral over an appropriate region G in the uv-plane. Then evaluate the uv-integral over G.

12. a) Find the Jacobian of the transformation $x = u$, $y = uv$, and sketch the region $G: 1 \le u \le 2, 1 \le uv \le 2$ in the uv-plane.

b) Then use Eq. (1) to transform the integral

$$\int_1^2 \int_1^2 \frac{y}{x} \, dy \, dx$$

into an integral over G, and evaluate both integrals.

13. A thin plate of constant density covers the region bounded by the ellipse $x^2/a^2 + y^2/b^2 = 1$, $a > 0$, $b > 0$, in the xy-plane. Find the first moment of the plate about the origin. (*Hint:* Use the transformation $x = ar \cos \theta$, $y = br \sin \theta$.)

14. The area πab of the ellipse $x^2/a^2 + y^2/b^2 = 1$ can be found by integrating the function $f(x, y) = 1$ over the region bounded by the ellipse in the xy-plane. Evaluating the integral directly requires a trigonometric substitution. An easier way to evaluate the integral is to use the transformation $x = au, y = bv$ and evaluate the transformed integral over the disk $G: u^2 + v^2 \le 1$ in the uv-plane. Find the area this way.

15. Use the transformation in Exercise 2 to evaluate the integral

$$\int_0^{2/3} \int_y^{2-2y} (x + 2y) e^{(y-x)} \, dx \, dy$$

by first writing it as an integral over a region G in the uv-plane.

16. Use the transformation $x = u + (1/2)v$, $y = v$ to evaluate the integral

$$\int_0^2 \int_{y/2}^{(y+4)/2} y^3 (2x - y) e^{(2x-y)^2} \, dx \, dy$$

by first writing it as an integral over a region G in the uv-plane.

Triple Integrals

17. Evaluate the determinant in Eq. (10) to show that the Jacobian of the transformation from Cartesian $\rho\phi\theta$-space to Cartesian xyz-space is $\rho^2 \sin \phi$.

18. Evaluate the integral in Example 3 by integrating with respect to x, y, and z.

19. Find the volume of the ellipsoid

$$\frac{x^2}{a^2} + \frac{y^2}{b^2} + \frac{z^2}{c^2} = 1.$$

(*Hint:* Let $x = au$, $y = bv$, and $z = cw$. Then find the volume of an appropriate region in uvw-space.)

20. Evaluate

$$\iiint |xyz| \, dx \, dy \, dz$$

over the solid ellipsoid

$$\frac{x^2}{a^2} + \frac{y^2}{b^2} + \frac{z^2}{c^2} \le 1.$$

(*Hint:* Let $x = au$, $y = bv$, and $z = cw$. Then integrate over an appropriate region in uvw-space.)

21. Let D be the region in xyz-space defined by the inequalities

$$1 \le x \le 2, \qquad 0 \le xy \le 2, \qquad 0 \le z \le 1.$$

Evaluate

$$\iiint_D (x^2 y + 3 xyz) \, dx \, dy \, dz$$

by applying the transformation

$$u = x, \qquad v = xy, \qquad w = 3z$$

and integrating over an appropriate region G in uvw-space.

22. Assuming the result that the center of mass of a solid hemisphere lies on the axis of symmetry three-eighths of the way from the base toward the top, show, by transforming the appropriate integrals, that the center of mass of a solid semi-ellipsoid $(x^2/a^2) + (y^2/b^2) + (z^2/c^2) \le 1$, $z \ge 0$, lies on the z-axis three-eighths of the way from the base toward the top. (You can do this without evaluating any of the integrals.)

Single Integrals

23. *Substitutions in single integrals.* How can substitutions in single definite integrals be viewed as transformations of regions? What is the Jacobian in such a case? Illustrate with an example.

CHAPTER **13** QUESTIONS TO GUIDE YOUR REVIEW

---

1. Define the double integral of a function of two variables over a bounded region in the coordinate plane.

2. How are double integrals evaluated as iterated integrals? Does the order of integration matter? How are the limits of integration determined? Give examples.

3. How are double integrals used to calculate areas, average values, masses, moments, centers of mass, and radii of gyration? Give examples.

4. How can you change a double integral in rectangular coordinates into a double integral in polar coordinates? Why might it be worthwhile to do so? Give an example.

5. Define the triple integral of a function $f(x, y, z)$ over a bounded region in space.

6. How are triple integrals in rectangular coordinates evaluated? How are the limits of integration determined? Give an example.

7. How are triple integrals in rectangular coordinates used to calculate volumes, average values, masses, moments, centers of mass, and radii of gyration? Give examples.

8. How are triple integrals defined in cylindrical and spherical coordinates? Why might one prefer working in one of these coordinate systems to working in rectangular coordinates?

9. How are triple integrals in cylindrical and spherical coordinates evaluated? How are the limits of integration found? Give examples.

10. How are substitutions in double integrals pictured as transformations of two-dimensional regions? Give a sample calculation.

11. How are substitutions in triple integrals pictured as transformations of three-dimensional regions? Give a sample calculation.

Planar Regions of Integration

In Exercises 1–4, sketch the region of integration and evaluate the double integral.

1. $\displaystyle\int_{1}^{10}\int_{0}^{1/y} y e^{xy}\, dx\, dy$

2. $\displaystyle\int_{0}^{1}\int_{0}^{x^3} e^{y/x}\, dy\, dx$

3. $\displaystyle\int_{0}^{3/2}\int_{-\sqrt{9-4t^2}}^{\sqrt{9-4t^2}} t\, ds\, dt$

4. $\displaystyle\int_{0}^{1}\int_{\sqrt{y}}^{2-\sqrt{y}} x y\, dx\, dy$

Reversing the Order of Integration

In Exercises 5–8, sketch the region of integration and write an equivalent integral with the order of integration reversed. Then evaluate both integrals.

5. $\displaystyle\int_{0}^{4}\int_{-\sqrt{4-y}}^{(y-4)/2} dx\, dy$

6. $\displaystyle\int_{0}^{1}\int_{x^2}^{x} \sqrt{x}\, dy\, dx$

7. $\displaystyle\int_{0}^{3/2}\int_{-\sqrt{9-4y^2}}^{\sqrt{9-4y^2}} y\, dx\, dy$

8. $\displaystyle\int_{0}^{2}\int_{0}^{4-x^2} 2x\, dy\, dx$

Evaluating Double Integrals

Evaluate the integrals in Exercises 9–12.

9. $\displaystyle\int_{0}^{1}\int_{2y}^{2} 4 \cos(x^2)\, dx\, dy$

10. $\displaystyle\int_{0}^{2}\int_{y/2}^{1} e^{x^2}\, dx\, dy$

11. $\displaystyle\int_{0}^{8}\int_{\sqrt[3]{x}}^{2} \frac{dy\, dx}{y^4+1}$

12. $\displaystyle\int_{0}^{1}\int_{\sqrt[3]{y}}^{1} \frac{2\pi \sin \pi x^2}{x^2}\, dx\, dy$

Areas and Volumes

13. Find the area of the region enclosed by the line $y = 2x + 4$ and the parabola $y = 4 - x^2$ in the xy-plane.

14. Find the area of the "triangular" region in the xy-plane that is bounded on the right by the parabola $y = x^2$, on the left by the line $x + y = 2$, and above by the line $y = 4$.

15. Find the volume under the paraboloid $z = x^2 + y^2$ above the triangle enclosed by the lines $y = x, x = 0$, and $x + y = 2$ in the xy-plane.

16. Find the volume under the parabolic cylinder $z = x^2$ above the region enclosed by the parabola $y = 6 - x^2$ and the line $y = x$ in the xy-plane.

Average Values

Find the average value of $f(x, y) = xy$ over the regions in Exercises 17 and 18.

17. The square bounded by the lines $x = 1, y = 1$ in the first quadrant

18. The quarter circle $x^2 + y^2 \le 1$ in the first quadrant

Masses and Moments

19. Find the centroid of the "triangular" region bounded by the lines $x = 2, y = 2$ and the hyperbola $xy = 2$ in the xy-plane.

20. Find the centroid of the region between the parabola $x + y^2 - 2y = 0$ and the line $x + 2y = 0$ in the xy-plane.

21. Find the polar moment of inertia about the origin of a thin triangular plate of constant density $\delta = 3$, bounded by the y-axis and the lines $y = 2x$ and $y = 4$ in the xy-plane.

22. Find the polar moment of inertia about the center of a thin rectangular sheet of constant density $\delta = 1$ bounded by the lines

a) $x = \pm 2, \quad y = \pm 1$ in the xy-plane

b) $x = \pm a, \quad y = \pm b$ in the xy-plane.

(*Hint:* Find I_x. Then use the formula for I_x to find I_y and add the two to find I_0.)

23. Find the moment of inertia and radius of gyration about the x-axis of a thin plate of constant density δ covering the triangle with vertices $(0, 0)$, $(3, 0)$, and $(3, 2)$ in the xy-plane.

24. Find the center of mass and the moments of inertia and radii of gyration about the coordinate axes of a thin plate bounded by the line $y = x$ and the parabola $y = x^2$ in the xy-plane if the density is $\delta(x, y) = x + 1$.

25. Find the mass and first moments about the coordinate axes of a thin square plate bounded by the lines $x = \pm 1, y = \pm 1$ in the xy-plane if the density is $\delta(x, y) = x^2 + y^2 + 1/3$.

26. Find the moment of inertia and radius of gyration about the x-axis of a thin triangular plate of constant density δ whose base lies along the interval $[0, b]$ on the x-axis and whose vertex lies on the line $y = h$ above the x-axis. As you will see, it does not matter where on the line this vertex lies. All such triangles have the same moment of inertia and radius of gyration.

Polar Coordinates

Evaluate the integrals in Exercises 27 and 28 by changing to polar coordinates.

27. $\displaystyle\int_{-1}^{1}\int_{-\sqrt{1-x^2}}^{\sqrt{1-x^2}} \frac{2\, dy\, dx}{(1 + x^2 + y^2)^2}$

28. $\displaystyle\int_{-1}^{1}\int_{-\sqrt{1-y^2}}^{\sqrt{1-y^2}} \ln(x^2 + y^2 + 1)\, dx\, dy$

29. Find the centroid of the region in the polar coordinate plane defined by the inequalities $0 \le r \le 3$ and $-\pi/3 \le \theta \le \pi/3$.

30. Find the centroid of the region in the first quadrant bounded by the rays $\theta = 0$ and $\theta = \pi/2$ and the circles $r = 1$ and $r = 3$.

31. a) Find the centroid of the region in the polar coordinate plane that lies inside the cardioid $r = 1 + \cos\theta$ and outside the circle $r = 1$.

b) CALCULATOR Sketch the region and show the centroid in your sketch.

32. a) Find the centroid of the plane region defined by the polar coordinate inequalities $0 \le r \le a$, $-\alpha \le \theta \le \alpha$ $(0 < \alpha \le \pi)$. How does the centroid move as $\alpha \to \pi^-$?

b) CALCULATOR Sketch the region for $\alpha = 5\pi/6$ and show the centroid in your sketch.

33. Integrate the function $f(x, y) = 1/(1 + x^2 + y^2)^2$ over the region enclosed by one loop of the lemniscate $(x^2 + y^2)^2 - (x^2 - y^2) = 0$.

34. Integrate $f(x, y) = 1/(1 + x^2 + y^2)^2$ over

a) the triangle with vertices $(0, 0)$, $(1, 0)$, $(1, \sqrt{3})$;

b) the first quadrant of the xy-plane.

Triple Integrals in Cartesian Coordinates

Evaluate the integrals in Exercises 35–38.

35. $\displaystyle\int_0^\pi \int_0^\pi \int_0^\pi \cos(x + y + z)\, dx\, dy\, dz$

36. $\displaystyle\int_{\ln 6}^{\ln 7} \int_0^{\ln 2} \int_{\ln 4}^{\ln 5} e^{(x+y+z)}\, dz\, dy\, dx$

37. $\displaystyle\int_0^1 \int_0^{x^2} \int_0^{x+y} (2x - y - z)\, dz\, dy\, dx$

38. $\displaystyle\int_1^e \int_1^x \int_0^z \frac{2y}{z^3}\, dy\, dz\, dx$

39. Find the volume of the wedge-shaped region enclosed on the side by the cylinder $x = -\cos y$, $-\pi/2 \le y \le \pi/2$, on the top by the plane $z = -2x$, and below by the xy-plane.

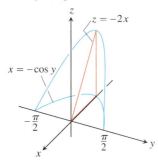

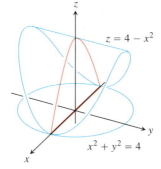

40. Find the volume of the solid that is bounded above by the cylinder $z = 4 - x^2$, on the sides by the cylinder $x^2 + y^2 = 4$, and below by the xy-plane.

41. Find the average value of $f(x, y, z) = 30xz\sqrt{x^2 + y}$ over the rectangular solid in the first octant bounded by the coordinate planes and the planes $x = 1$, $y = 3$, $z = 1$.

42. Find the average value of ρ over the solid sphere $\rho \le a$ (spherical coordinates).

Cylindrical and Spherical Coordinates

43. Convert

$$\int_0^{2\pi} \int_0^{\sqrt{2}} \int_r^{\sqrt{4-r^2}} 3\, dz\, r\, dr\, d\theta, \quad r \ge 0$$

to (a) rectangular coordinates with the order of integration $dz\, dx\, dy$, and (b) spherical coordinates. Then (c) evaluate one of the integrals.

44. (a) Convert to cylindrical coordinates. Then (b) evaluate the new integral.

$$\int_0^1 \int_{-\sqrt{1-x^2}}^{\sqrt{1-x^2}} \int_{-(x^2+y^2)}^{(x^2+y^2)} 21xy^2\, dz\, dy\, dx$$

45. (a) Convert to spherical coordinates. Then (b) evaluate the new integral.

$$\int_{-1}^1 \int_{-\sqrt{1-x^2}}^{\sqrt{1-x^2}} \int_{\sqrt{x^2+y^2}}^1 dz\, dy\, dx$$

46. Write an iterated triple integral for the integral of $f(x, y, z) = 6 + 4y$ over the region in the first octant bounded by the cone $z = \sqrt{x^2 + y^2}$, the cylinder $x^2 + y^2 = 1$, and the coordinate planes in (a) rectangular coordinates, (b) cylindrical coordinates, (c) spherical coordinates. Then (d) find the integral of f by evaluating one of the triple integrals.

47. Set up an integral in rectangular coordinates equivalent to the integral

$$\int_0^{\pi/2} \int_1^{\sqrt{3}} \int_1^{\sqrt{4-r^2}} r^3 \sin\theta \cos\theta \, z^2\, dz\, dr\, d\theta.$$

Arrange the order of integration to be z first, then y, then x.

48. The volume of a solid is

$$\int_0^2 \int_0^{\sqrt{2x-x^2}} \int_{-\sqrt{4-x^2-y^2}}^{\sqrt{4-x^2-y^2}} dz\, dy\, dx.$$

a) Describe the solid by giving equations for the surfaces that form its boundary.

b) Convert the integral to cylindrical coordinates but do not evaluate the integral.

49. Let D be the smaller spherical cap cut from a solid ball of radius 2 by a plane 1 unit from the center of the sphere. Express the volume of D as an iterated triple integral in (a) rectangular, (b) cylindrical, and (c) spherical coordinates. *Do not evaluate the integrals.*

50. Express the moment of inertia I_z of the solid hemisphere bounded below by the plane $z = 0$ and above by the sphere $x^2 + y^2 + z^2 = 1$ as an iterated integral in (a) rectangular, (b) cylindrical, and (c) spherical coordinates. *Do not evaluate the integrals.*

51. *Spherical vs. cylindrical coordinates.* Triple integrals involving spherical shapes do not always require spherical coordinates for convenient evaluation. Some calculations may be accomplished more easily with cylindrical coordinates. As a case in point, find the volume of the region bounded above by the sphere $x^2 + y^2 + z^2 = 8$ and below by the plane $z = 2$ by using (a) cylindrical coordinates, (b) spherical coordinates.

52. Find the moment of inertia about the z-axis of a solid of constant density $\delta = 1$ that is bounded above by the sphere $\rho = 2$ and below by the cone $\phi = \pi/3$ (spherical coordinates).

53. Find the moment of inertia of a solid of constant density δ bounded by two concentric spheres of radii a and b ($a < b$) about a diameter.

54. Find the moment of inertia about the z-axis of a solid of density $\delta = 1$ enclosed by the spherical coordinate surface $\rho = 1 - \cos \phi$.

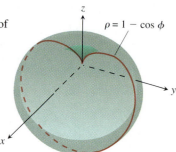

$\rho = 1 - \cos \phi$

Substitutions

55. Show that if $u = x - y$ and $v = y$, then

$$\int_0^\infty \int_0^x e^{-sx} f(x - y, y)\, dy\, dx = \int_0^\infty \int_0^\infty e^{-s(u+v)} f(u, v)\, du\, dv.$$

56. What relationship must hold between the constants a, b, and c to make

$$\int_{-\infty}^\infty \int_{-\infty}^\infty e^{-(ax^2 + 2bxy + cy^2)}\, dx\, dy = 1?$$

(*Hint:* Let $s = \alpha x + \beta y$ and $t = \gamma x + \delta y$, where $(\alpha\delta - \beta\gamma)^2 = ac - b^2$. Then $ax^2 + 2bxy + cy^2 = s^2 + t^2$.)

CHAPTER 13 ADDITIONAL EXERCISES–THEORY, EXAMPLES, APPLICATIONS

Volumes

1. The base of a sand pile covers the region in the xy-plane that is bounded by the parabola $x^2 + y = 6$ and the line $y = x$. The height of the sand above the point (x, y) is x^2. Express the volume of sand as (a) a double integral, (b) a triple integral. Then (c) find the volume.

2. A hemispherical bowl of radius 5 cm is filled with water to within 3 cm of the top. Find the volume of water in the bowl.

3. Find the volume of the portion of the solid cylinder $x^2 + y^2 \leq 1$ that lies between the planes $z = 0$ and $x + y + z = 2$.

4. Find the volume of the region bounded above by the sphere $x^2 + y^2 + z^2 = 2$ and below by the paraboloid $z = x^2 + y^2$.

5. Find the volume of the region bounded above by the paraboloid $z = 3 - x^2 - y^2$ and below by the paraboloid $z = 2x^2 + 2y^2$.

6. Find the volume of the region enclosed by the spherical coordinate surface $\rho = 2 \sin \phi$ (Fig. 13.54).

7. A circular cylindrical hole is bored through a solid sphere, the axis of the hole being a diameter of the sphere. The volume of the remaining solid is

$$V = 2 \int_0^{2\pi} \int_0^{\sqrt{3}} \int_1^{\sqrt{4-z^2}} r\, dr\, dz\, d\theta.$$

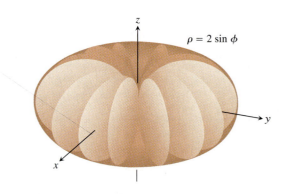

$\rho = 2 \sin \phi$

13.54 The surface in Exercise 6.

a) Find the radius of the hole and the radius of the sphere.
b) Evaluate the integral.

8. Find the volume of material cut from the solid sphere $r^2 + z^2 \leq 9$ by the cylinder $r = 3 \sin \theta$.

9. Find the volume of the region enclosed by the surfaces $z = x^2 + y^2$ and $z = (x^2 + y^2 + 1)/2$.

10. Find the volume of the region in the first octant that lies between the cylinders $r = 1$ and $r = 2$ and that is bounded below by the xy-plane and above by the surface $z = xy$.

Changing the Order of Integration

In Exercises 11 and 12, sketch the region of integration and write an equivalent iterated integral with the order of integration reversed.

11. $\displaystyle\int_0^1 \int_{x^2}^x f(x, y)\, dy\, dx$

12. $\displaystyle\int_0^4 \int_y^{2\sqrt{y}} f(x, y)\, dx\, dy$

13. Evaluate the integral

$$\int_0^\infty \frac{e^{-ax} - e^{-bx}}{x}\, dx.$$

(*Hint:* Use the relation

$$\frac{e^{-ax} - e^{-bx}}{x} = \int_a^b e^{-xy}\, dy$$

to form a double integral and evaluate the integral by changing the order of integration.)

14. a) Show, by changing to polar coordinates, that

$$\int_0^{a \sin \beta} \int_{y \cot \beta}^{\sqrt{a^2 - y^2}} \ln(x^2 + y^2)\, dx\, dy = a^2 \beta \left(\ln a - \frac{1}{2} \right),$$

where $a > 0$ and $0 < \beta < \pi/2$.

b) Rewrite the Cartesian integral with the order of integration reversed.

15. By changing the order of integration, show that the following double integral can be reduced to a single integral:

$$\int_0^x \int_0^u e^{m(x-t)} f(t)\, dt\, du = \int_0^x (x - t) e^{m(x-t)} f(t)\, dt.$$

Similarly, it can be shown that

$$\int_0^x \int_0^v \int_0^u e^{m(x-t)} f(t)\, dt\, du\, dv = \int_0^x \frac{(x-t)^2}{2} e^{m(x-t)} f(t)\, dt.$$

16. Sometimes a multiple integral with variable limits can be changed into one with constant limits. By changing the order of integration, show that

$$\int_0^1 f(x) \left(\int_0^x g(x - y) f(y)\, dy \right) dx$$

$$= \int_0^1 f(y) \left(\int_y^1 g(x - y) f(x)\, dx \right) dy$$

$$= \frac{1}{2} \int_0^1 \int_0^1 g(|x - y|) f(x)\, f(y)\, dx\, dy.$$

Masses and Moments

17. A thin plate of constant density is to occupy the triangular region in the first quadrant of the xy-plane having vertices $(0, 0)$, $(a, 0)$, and $(a, 1/a)$. What value of a will minimize the plate's polar moment of inertia about the origin?

18. Find the polar moment of inertia about the origin of a thin triangular plate of constant density $\delta = 3$ bounded by the y-axis and the lines $y = 2x$ and $y = 4$ in the xy-plane.

19. Find the centroid of the region in the polar coordinate plane

that lies inside the cardioid $r = 1 + \cos \theta$ and outside the circle $r = 1$.

20. Find the centroid of the boomerang-shaped region between the parabolas $y^2 = -4(x - 1)$ and $y^2 = -2(x - 2)$ in the xy-plane.

21. The counterweight of a flywheel of constant density 1 has the form of the smaller segment cut from a circle of radius a by a chord at a distance b from the center ($b < a$). Find the mass of the counterweight and its polar moment of inertia about the center of the wheel.

22. Find the radii of gyration about the x- and y-axes of a thin plate of density $\delta = 1$ enclosed by one loop of the lemniscate $r^2 = 2a^2 \cos 2\theta$.

23. A solid is bounded on the top by the paraboloid $z = r^2$, on the bottom by the plane $z = 0$, and on the sides by the cylinder $r = 1$. Find the center of mass and the moment of inertia and radius of gyration about the z-axis if the density is (a) $\delta(r, \theta, z) = z$; (b) $\delta(r, \theta, z) = r$.

24. A solid is bounded below by the cone $z = \sqrt{x^2 + y^2}$ and above by the plane $z = 1$. Find the center of mass and the moment of inertia and radius of gyration about the z-axis if the density is (a) $\delta(r, \theta, z) = z$; (b) $\delta(r, \theta, z) = z^2$.

25. Use spherical coordinates to find the centroid of a solid hemisphere of radius a.

26. Find the moment of inertia and radius of gyration of a solid sphere of radius a and density $\delta = 1$ about a diameter of the sphere.

Theory and Applications

27. Evaluate

$$\int_0^a \int_0^b e^{\max(b^2 x^2, \, a^2 y^2)}\, dy\, dx,$$

where a and b are positive numbers and

$$\max(b^2 x^2, \, a^2 y^2) = \begin{cases} b^2 x^2 & \text{if } b^2 x^2 \ge a^2 y^2 \\ a^2 y^2 & \text{if } b^2 x^2 < a^2 y^2. \end{cases}$$

28. Show that

$$\iint \frac{\partial^2 F(x, y)}{\partial x\, \partial y}\, dx\, dy$$

over the rectangle $x_0 \le x \le x_1$, $y_0 \le y \le y_1$, is

$$F(x_1, y_1) - F(x_0, y_1) - F(x_1, y_0) + F(x_0, y_0).$$

29. Suppose that $f(x, y)$ can be written as a product $f(x, y) = F(x)G(y)$ of a function of x and a function of y. Then the integral of f over the rectangle $R: a \le x \le b, c \le y \le d$ can be evaluated as a product as well, by the formula

$$\iint_R f(x, y)\, dA = \left(\int_a^b F(x)\, dx \right) \left(\int_c^d G(y)\, dy \right). \tag{1}$$

The argument is that

$$\iint\limits_{R} f(x, y)\, dA = \int_{c}^{d} \left(\int_{a}^{b} F(x) G(y)\, dx \right) dy \qquad \text{(i)}$$

$$= \int_{c}^{d} \left(G(y) \int_{a}^{b} F(x)\, dx \right) dy \qquad \text{(ii)}$$

$$= \int_{c}^{d} \left(\int_{a}^{b} F(x)\, dx \right) G(y)\, dy \qquad \text{(iii)}$$

$$= \left(\int_{a}^{b} F(x)\, dx \right) \int_{c}^{d} G(y)\, dy. \qquad \text{(iv)}$$

a) Give reasons for steps (i)–(iv).

When it applies, Eq. (1) can be a time saver. Use it to evaluate the following integrals.

b) $\displaystyle \int_{0}^{\ln 2} \int_{0}^{\pi/2} e^{x} \cos y\, dy\, dx$

c) $\displaystyle \int_{1}^{2} \int_{-1}^{1} \frac{x}{y^{2}}\, dx\, dy$

30. Let $D_{\mathbf{u}} f$ denote the derivative of $f(x, y) = (x^{2} + y^{2})/2$ in the direction of the unit vector $\mathbf{u} = u_{1}\mathbf{i} + u_{2}\mathbf{j}$.

a) Find the average value of $D_{\mathbf{u}} f$ over the triangular region cut from the first quadrant by the line $x + y = 1$.

b) Show in general that the average value of $D_{\mathbf{u}} f$ over a region in the xy-plane is the value of $D_{\mathbf{u}} f$ at the centroid of the region.

31. *The value of $\Gamma(1/2)$.* As we saw in Additional Exercises 49 and 50 in Chapter 7, the gamma function,

$$\Gamma(x) = \int_{0}^{\infty} t^{x-1} e^{-t}\, dt,$$

extends the factorial function from the nonnegative integers to other real values. Of particular interest in the theory of differential equations is the number

$$\Gamma\left(\frac{1}{2}\right) = \int_{0}^{\infty} t^{(1/2)-1} e^{-t}\, dt = \int_{0}^{\infty} \frac{e^{-t}}{\sqrt{t}}\, dt. \qquad \text{(2)}$$

a) If you have not yet done Exercise 37 in Section 13.3, do it now to show that

$$I = \int_{0}^{\infty} e^{-y^{2}}\, dy = \frac{\sqrt{\pi}}{2}.$$

b) Substitute $y = \sqrt{t}$ in Eq. (2) to show that $\Gamma(1/2) = 2I = \sqrt{\pi}$.

32. The electrical charge distribution on a circular plate of radius R meters is $\sigma(r, \theta) = kr(1 - \sin \theta)$ coulomb/m^{2} (k a constant). Integrate σ over the plate to find the total charge Q.

33. *A parabolic rain gauge.* A bowl is in the shape of the graph of $z = x^{2} + y^{2}$ from $z = 0$ to $z = 10$ in. You plan to calibrate the bowl to make it into a rain gauge. What height in the bowl would correspond to 1 in. of rain? 3 in. of rain?

34. *Water in a satellite dish.* A parabolic satellite dish is 2 m wide and 1/2 m deep. Its axis of symmetry is tilted 30 degrees from the vertical.

a) Set up, but do not evaluate, a triple integral in rectangular coordinates that gives the amount of water the satellite dish will hold. (*Hint:* Put your coordinate system so that the satellite dish is in "standard position" and the plane of the water level is slanted.) (*Caution:* The limits of integration are not "nice.")

b) What would be the smallest tilt of the satellite dish so that it holds no water?

35. *Cylindrical shells.* In Section 5.4, we learned how to find the volume of a solid of revolution using the shell method, namely if the region between the curve $y = f(x)$ and the x-axis from a to b ($0 < a < b$) is revolved about the y-axis the volume of the resulting solid is $\int_{a}^{b} 2\pi x\, f(x)\, dx$. Prove that finding volumes by using triple integrals gives the same result. (*Hint:* Use cylindrical coordinates with the roles of y and z changed.)

36. *An infinite half-cylinder.* Let D be the interior of the infinite right circular half-cylinder of radius 1 with its single-end face suspended 1 unit above the origin and its axis the ray from $(0, 0, 1)$ to ∞. Use cylindrical coordinates to evaluate

$$\iiint\limits_{D} z(r^{2} + z^{2})^{-5/2}\, dV.$$

37. *Hypervolume.* We have learned that $\int_{a}^{b} 1\, dx$ is the length of the interval $[a, b]$ on the number line (one-dimensional space), $\iint_{R} 1\, dA$ is the area of region R in the xy-plane (two-dimensional space), and $\iiint_{D} 1\, dV$ is the volume of the region D in three-dimensional space (xyz-space). We could continue: If Q is a region in 4-space ($xyzw$-space), then $\iiiint_{Q} 1\, dV$ is the "hypervolume" of Q. Use your generalizing abilities and a Cartesian coordinate system of 4-space to find the hypervolume inside the unit 4-sphere $x^{2} + y^{2} + z^{2} + w^{2} = 1$.

CHAPTER 14

Integration in Vector Fields

OVERVIEW   This chapter treats integration in vector fields. The mathematics in this chapter is the mathematics that is used to describe the properties of electromagnetism, explain the flow of heat in stars, and calculate the work it takes to put a satellite in orbit.

14.1 Line Integrals

When a curve $\mathbf{r}(t) = g(t)\,\mathbf{i} + h(t)\,\mathbf{j} + k(t)\,\mathbf{k}$, $a \leq t \leq b$, passes through the domain of a function $f(x, y, z)$ in space, the values of f along the curve are given by the composite function $f(g(t), h(t), k(t))$. If we integrate this composite with respect to arc length from $t = a$ to $t = b$, we calculate the so-called line integral of f along the curve. Despite the three-dimensional geometry, the line integral is an ordinary integral of a real-valued function over an interval of real numbers.

The importance of line integrals lies in their application. These are the integrals with which we calculate the work done by variable forces along paths in space and the rates at which fluids flow along curves and across boundaries.

Definitions and Notation

Suppose that $f(x, y, z)$ is a function whose domain contains the curve $\mathbf{r}(t) = g(t)\,\mathbf{i} + h(t)\,\mathbf{j} + k(t)\,\mathbf{k}$, $a \leq t \leq b$. We partition the curve into a finite number of subarcs (Fig. 14.1). The typical subarc has length Δs_k. In each subarc we choose a point (x_k, y_k, z_k) and form the sum

$$S_n = \sum_{k=1}^{n} f(x_k, y_k, z_k)\,\Delta s_k. \tag{1}$$

If f is continuous and the functions g, h, and k have continuous first derivatives, then the sums in (1) approach a limit as n increases, and the lengths Δs_k approach zero. We call this limit the **integral of f over the curve from a to b.** If the curve is denoted by a single letter, C for example, the notation for the integral is

$$\int_C f(x, y, z)\,ds \qquad \text{``The integral of } f \text{ over } C\text{''} \tag{2}$$

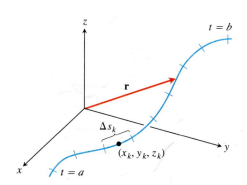

14.1 The curve $\mathbf{r} = g(t)\,\mathbf{i} + h(t)\,\mathbf{j} + k(t)\,\mathbf{k}$, partitioned into small arcs from $t = a$ to $t = b$. The length of a typical subarc is Δs_k.

1061

Evaluation for Smooth Curves

If $\mathbf{r}(t)$ is smooth for $a \le t \le b$ ($\mathbf{v} = d\mathbf{r}/dt$ is continuous and never $\mathbf{0}$), we can use the equation

$$s(t) = \int_a^t |\mathbf{v}(\tau)| \, d\tau \qquad \text{Eq. (4) of Section 11.3, with } t_0 = a$$

to express ds in Eq. (2) as $ds = |\mathbf{v}(t)| \, dt$. A theorem from advanced calculus says that we can then evaluate the integral of f over C as

$$\int_C f(x, y, z) \, ds = \int_a^b f(g(t), h(t), k(t)) |\mathbf{v}(t)| \, dt.$$

This formula will evaluate the integral correctly no matter what parametrization we use, as long as the parametrization is smooth.

How to Evaluate a Line Integral

To integrate a continuous function $f(x, y, z)$ over a curve C:

1. Find a smooth parametrization of C,

$$\mathbf{r}(t) = g(t)\,\mathbf{i} + h(t)\,\mathbf{j} + k(t)\,\mathbf{k}, \qquad a \le t \le b.$$

2. Evaluate the integral as

$$\int_C f(x, y, z) \, ds = \int_a^b f(g(t), h(t), k(t)) \, |\mathbf{v}(t)| \, dt. \qquad (3)$$

Notice that if f has the constant value 1, then the integral of f over C gives the length of C.

EXAMPLE 1   Integrate $f(x, y, z) = x - 3y^2 + z$ over the line segment C joining the origin and the point $(1, 1, 1)$ (Fig. 14.2).

Solution   We choose the simplest parametrization we can think of:

$$\mathbf{r}(t) = t\,\mathbf{i} + t\,\mathbf{j} + t\,\mathbf{k}, \qquad 0 \le t \le 1.$$

The components have continuous first derivatives and $|\mathbf{v}(t)| = \sqrt{1^2 + 1^2 + 1^2} = \sqrt{3}$ is never 0, so the parametrization is smooth. The integral of f over C is

$$\int_C f(x, y, z) \, ds = \int_0^1 f(t, t, t) \left(\sqrt{3}\right) dt \qquad \text{Eq. (3)}$$

$$= \int_0^1 (t - 3t^2 + t) \sqrt{3} \, dt$$

$$= \sqrt{3} \int_0^1 (2t - 3t^2) \, dt = \sqrt{3} \left[t^2 - t^3 \right]_0^1 = 0. \qquad \square$$

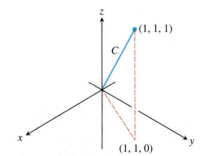

14.2 The integration path in Example 1.

Additivity

Line integrals have the useful property that if a curve C is made by joining a finite number of curves $C_1, C_2, \ldots, C_n$ end to end, then the integral of a function over

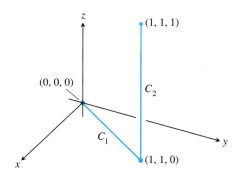

14.3 The path of integration in Example 2.

C is the sum of the integrals over the curves that make it up:

$$\int_C f\,ds = \int_{C_1} f\,ds + \int_{C_2} f\,ds + \cdots + \int_{C_n} f\,ds. \tag{4}$$

EXAMPLE 2   Figure 14.3 shows another path from the origin to $(1, 1, 1)$, the union of line segments C_1 and C_2. Integrate $f(x, y, z) = x - 3y^2 + z$ over $C_1 \cup C_2$.

Solution   We choose the simplest parametrizations for C_1 and C_2 we can think of, checking the lengths of the velocity vectors as we go along:

$$C_1: \quad \mathbf{r}(t) = t\,\mathbf{i} + t\,\mathbf{j}, \quad 0 \le t \le 1; \quad |\mathbf{v}| = \sqrt{1^2 + 1^2} = \sqrt{2}$$

$$C_2: \quad \mathbf{r}(t) = \mathbf{i} + \mathbf{j} + t\,\mathbf{k}, \quad 0 \le t \le 1; \quad |\mathbf{v}| = \sqrt{0^2 + 0^2 + 1^2} = 1.$$

With these parametrizations we find that

$$\int_{C_1 \cup C_2} f(x, y, z)\,ds = \int_{C_1} f(x, y, z)\,ds + \int_{C_2} f(x, y, z)\,ds \qquad \text{Eq. (4)}$$

$$= \int_0^1 f(t, t, 0)\,\sqrt{2}\,dt + \int_0^1 f(1, 1, t)(1)\,dt \qquad \text{Eq. (3)}$$

$$= \int_0^1 (t - 3t^2 + 0)\,\sqrt{2}\,dt + \int_0^1 (1 - 3 + t)(1)\,dt$$

$$= \sqrt{2}\left[\frac{t^2}{2} - t^3\right]_0^1 + \left[\frac{t^2}{2} - 2t\right]_0^1 = -\frac{\sqrt{2}}{2} - \frac{3}{2}. \qquad \square$$

Notice three things about the integrations in Examples 1 and 2. First, as soon as the components of the appropriate curve were substituted into the formula for f, the integration became a standard integration with respect to t. Second, the integral of f over $C_1 \cup C_2$ was obtained by integrating f over each section of the path and adding the results. Third, the integrals of f over C and $C_1 \cup C_2$ had different values. For most functions, the value of the integral along a path joining two points changes if you change the path between them. For some functions, however, the value remains the same, as we will see in Section 14.3.

Mass and Moment Calculations

We treat coil springs and wires like masses distributed along smooth curves in space. The distribution is described by a continuous density function $\delta(x, y, z)$ (mass per unit length). The spring's or wire's mass, center-of-mass, and moments are then calculated with the formulas in Table 14.1, on the following page. The formulas also apply to thin rods.

EXAMPLE 3   A coil spring lies along the helix

$$\mathbf{r}(t) = (\cos 4t)\,\mathbf{i} + (\sin 4t)\,\mathbf{j} + t\,\mathbf{k}, \qquad 0 \le t \le 2\pi.$$

The spring's density is a constant, $\delta = 1$. Find the spring's mass and center of mass, and its moment of inertia and radius of gyration about the z-axis.

Solution   We sketch the spring (Fig. 14.4). Because of the symmetries involved, the center of mass lies at the point $(0, 0, \pi)$ on the z-axis.

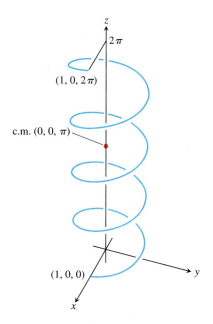

14.4 The helical spring in Example 3.

Table 14.1 Mass and moment formulas for coil springs, thin rods, and wires lying along a smooth curve C in space

Mass: $\quad M = \displaystyle\int_C \delta(x, y, z) \, ds$

First moments about the coordinate planes:

$$M_{yz} = \int_C x \, \delta \, ds, \qquad M_{xz} = \int_C y \, \delta \, ds, \qquad M_{xy} = \int_C z \, \delta \, ds$$

Coordinates of the center of mass:

$$\bar{x} = M_{yz}/M, \qquad \bar{y} = M_{xz}/M, \qquad \bar{z} = M_{xy}/M$$

Moments of inertia:

$$I_x = \int_C (y^2 + z^2) \, \delta \, ds, \qquad I_y = \int_C (x^2 + z^2) \, \delta \, ds$$

$$I_z = \int_C (x^2 + y^2) \, \delta \, ds, \qquad I_L = \int_C r^2 \, \delta \, ds$$

$$r(x, y, z) = \text{distance from point } (x, y, z) \text{ to line } L$$

Radius of gyration about a line L: $\quad R_L = \sqrt{I_L/M}$

For the remaining calculations, we first find $|\mathbf{v}(t)|$:

$$|\mathbf{v}(t)| = \sqrt{\left(\frac{dx}{dt}\right)^2 + \left(\frac{dy}{dt}\right)^2 + \left(\frac{dz}{dt}\right)^2}$$

$$= \sqrt{(-4 \sin 4t)^2 + (4 \cos 4t)^2 + 1} = \sqrt{17}.$$

We then evaluate the formulas from Table 14.1 using Eq. (3):

$$M = \int_{\text{Helix}} \delta \, ds = \int_0^{2\pi} (1) \sqrt{17} \, dt = 2\pi \sqrt{17}$$

$$I_z = \int_{\text{Helix}} (x^2 + y^2) \delta \, ds = \int_0^{2\pi} (\cos^2 4t + \sin^2 4t)(1)\sqrt{17} \, dt$$

$$= \int_0^{2\pi} \sqrt{17} \, dt = 2\pi \sqrt{17}$$

$$R_z = \sqrt{I_z/M} = \sqrt{2\pi \sqrt{17}/(2\pi \sqrt{17})} = 1.$$

Notice that the radius of gyration about the z-axis is the radius of the cylinder around which the helix winds. ◻

EXAMPLE 4 A slender metal arch, denser at the bottom than top, lies along the semicircle $y^2 + z^2 = 1$, $z \geq 0$, in the yz-plane (Fig. 14.5). Find the center of the arch's mass if the density at the point (x, y, z) on the arch is $\delta(x, y, z) = 2 - z$.

Solution We know that $\bar{x} = 0$ and $\bar{y} = 0$ because the arch lies in the yz-plane with its mass distributed symmetrically about the z-axis. To find $\bar{z}$, we parametrize the circle as

$$\mathbf{r}(t) = (\cos t) \mathbf{j} + (\sin t) \mathbf{k}, \qquad 0 \leq t \leq \pi.$$

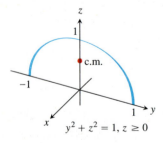

14.5 Example 4 shows how to find the center of mass of a circular arch of variable density.

For this parametrization,

$$|\mathbf{v}(t)| = \sqrt{\left(\frac{dx}{dt}\right)^2 + \left(\frac{dy}{dt}\right)^2 + \left(\frac{dz}{dt}\right)^2} = \sqrt{(0)^2 + (-\sin t)^2 + (\cos t)^2} = 1.$$

The formulas in Table 14.1 then give

$$M = \int_C \delta \, ds = \int_C (2 - z) \, ds = \int_0^\pi (2 - \sin t) \, dt = 2\pi - 2$$

$$M_{xy} = \int_C z \delta \, ds = \int_C z(2 - z) \, ds = \int_0^\pi (\sin t)(2 - \sin t) \, dt$$

$$= \int_0^\pi (2 \sin t - \sin^2 t) \, dt = \frac{8 - \pi}{2}$$

$$\bar{z} = \frac{M_{xy}}{M} = \frac{8 - \pi}{2} \cdot \frac{1}{2\pi - 2} = \frac{8 - \pi}{4\pi - 4} \approx 0.57.$$

With $\bar{z}$ to the nearest hundredth, the center of mass is $(0, 0, 0.57)$.

Exercises 14.1

Graphs of Vector Equations

Match the vector equations in Exercises 1–8 with the graphs in Fig. 14.6.

(a)

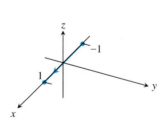

(b)

(c)

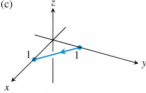

(d)

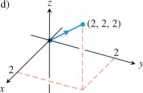

(e)

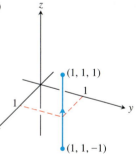

(f)

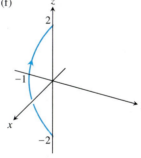

(g)

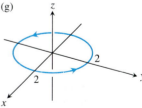

(h)

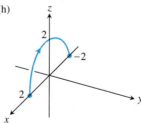

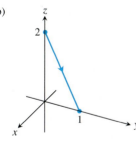

14.6 The graphs for Exercises 1–8.

1. $\mathbf{r}(t) = t\mathbf{i} + (1 - t)\mathbf{j}, \quad 0 \le t \le 1$

2. $\mathbf{r}(t) = \mathbf{i} + \mathbf{j} + t\mathbf{k}, \quad -1 \le t \le 1$

3. $\mathbf{r}(t) = (2 \cos t)\mathbf{i} + (2 \sin t)\mathbf{j}, \quad 0 \le t \le 2\pi$

4. $\mathbf{r}(t) = t\mathbf{i}, \quad -1 \le t \le 1$

5. $\mathbf{r}(t) = t\mathbf{i} + t\mathbf{j} + t\mathbf{k}, \quad 0 \le t \le 2$

6. $\mathbf{r}(t) = t\mathbf{j} + (2 - 2t)\mathbf{k}, \quad 0 \le t \le 1$

7. $\mathbf{r}(t) = (t^2 - 1)\mathbf{j} + 2t\mathbf{k}, \quad -1 \le t \le 1$

8. $\mathbf{r}(t) = (2 \cos t)\mathbf{i} + (2 \sin t)\mathbf{k}, \quad 0 \le t \le \pi$

Evaluating Line Integrals over Space Curves

9. Evaluate $\int_C (x + y)\,ds$ where C is the straight-line segment $x = t, y = (1 - t), z = 0$, from $(0, 1, 0)$ to $(1, 0, 0)$.

10. Evaluate $\int_C (x - y + z - 2)\,ds$ where C is the straight-line segment $x = t, y = (1 - t), z = 1$, from $(0, 1, 1)$ to $(1, 0, 1)$.

11. Evaluate $\int_C (xy + y + z)\,ds$ along the curve $\mathbf{r}(t) = 2t\,\mathbf{i} + t\,\mathbf{j} + (2 - 2t)\,\mathbf{k}, 0 \le t \le 1$.

12. Evaluate $\int_C \sqrt{x^2 + y^2}\,ds$ along the curve $\mathbf{r}(t) = (4 \cos t)\,\mathbf{i} + (4 \sin t)\,\mathbf{j} + 3t\,\mathbf{k}, -2\pi \le t \le 2\pi$.

13. Find the line integral of $f(x, y, z) = x + y + z$ over the straight-line segment from $(1, 2, 3)$ to $(0, -1, 1)$.

14. Find the line integral of $f(x, y, z) = \sqrt{3}/(x^2 + y^2 + z^2)$ over the curve $\mathbf{r}(t) = t\,\mathbf{i} + t\,\mathbf{j} + t\,\mathbf{k}, 1 \le t \le \infty$.

15. Integrate $f(x, y, z) = x + \sqrt{y} - z^2$ over the path from $(0, 0, 0)$ to $(1, 1, 1)$ (Fig. 14.7a) given by
    $C_1: \quad \mathbf{r}(t) = t\,\mathbf{i} + t^2\,\mathbf{j}, \quad 0 \le t \le 1$
    $C_2: \quad \mathbf{r}(t) = \mathbf{i} + \mathbf{j} + t\,\mathbf{k}, \quad 0 \le t \le 1$

16. Integrate $f(x, y, z) = x + \sqrt{y} - z^2$ over the path from $(0, 0, 0)$ to $(1, 1, 1)$ (Fig. 14.7b) given by
    $C_1: \quad \mathbf{r}(t) = t\,\mathbf{k}, \quad 0 \le t \le 1$
    $C_2: \quad \mathbf{r}(t) = t\,\mathbf{j} + \mathbf{k}, \quad 0 \le t \le 1$
    $C_3: \quad \mathbf{r}(t) = t\,\mathbf{i} + \mathbf{j} + \mathbf{k}, \quad 0 \le t \le 1$

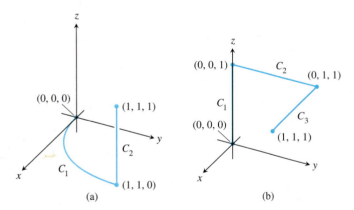

(a)   (b)

14.7 The paths of integration for Exercises 15 and 16.

17. Integrate $f(x, y, z) = (x + y + z)/(x^2 + y^2 + z^2)$ over the path $\mathbf{r}(t) = t\,\mathbf{i} + t\,\mathbf{j} + t\,\mathbf{k}, 0 < a \le t \le b$.

18. Integrate $f(x, y, z) = -\sqrt{x^2 + z^2}$ over the circle
    $$\mathbf{r}(t) = (a \cos t)\,\mathbf{j} + (a \sin t)\,\mathbf{k}, 0 \le t \le 2\pi.$$

Line Integrals over Plane Curves

In Exercises 19–22, integrate f over the given curve.

19. $f(x, y) = x^3/y, \quad C: \quad y = x^2/2, \quad 0 \le x \le 2$

20. $f(x, y) = (x + y^2)/\sqrt{1 + x^2}, \quad C: \quad y = x^2/2$ from $(1, 1/2)$ to $(0, 0)$

21. $f(x, y) = x + y, \quad C: \quad x^2 + y^2 = 4$ in the first quadrant from $(2, 0)$ to $(0, 2)$

22. $f(x, y) = x^2 - y, \quad C: \quad x^2 + y^2 = 4$ in the first quadrant from $(0, 2)$ to $(\sqrt{2}, \sqrt{2})$

Mass and Moments

23. Find the mass of a wire that lies along the curve $\mathbf{r}(t) = (t^2 - 1)\,\mathbf{j} + 2t\,\mathbf{k}, 0 \le t \le 1$, if the density is $\delta = (3/2)\,t$.

24. A wire of density $\delta(x, y, z) = 15\sqrt{y + 2}$ lies along the curve $\mathbf{r}(t) = (t^2 - 1)\,\mathbf{j} + 2t\,\mathbf{k}, -1 \le t \le 1$. Find its center of mass. Then sketch the curve and center of mass together.

25. Find the mass of a thin wire lying along the curve $\mathbf{r}(t) = \sqrt{2}t\,\mathbf{i} + \sqrt{2}t\,\mathbf{j} + (4 - t^2)\,\mathbf{k}, 0 \le t \le 1$, if the density is (a) $\delta = 3t$, (b) $\delta = 1$.

26. Find the center of mass of a thin wire lying along the curve $\mathbf{r}(t) = t\,\mathbf{i} + 2t\,\mathbf{j} + (2/3)t^{3/2}\,\mathbf{k}, 0 \le t \le 2$, if the density is $\delta = 3\sqrt{5 + t}$.

27. A circular wire hoop of constant density δ lies along the circle $x^2 + y^2 = a^2$ in the xy-plane. Find the hoop's moment of inertia and radius of gyration about the z-axis.

28. A slender rod of constant density lies along the line segment $\mathbf{r}(t) = t\,\mathbf{j} + (2 - 2t)\,\mathbf{k}, 0 \le t \le 1$, in the yz-plane. Find the moments of inertia and radii of gyration of the rod about the three coordinate axes.

29. A spring of constant density δ lies along the helix
    $$\mathbf{r}(t) = (\cos t)\,\mathbf{i} + (\sin t)\,\mathbf{j} + t\,\mathbf{k}, 0 \le t \le 2\pi.$$

    a) Find I_z and R_z.

    b) Suppose you have another spring of constant density δ that is twice as long as the spring in (a) and lies along the helix for $0 \le t \le 4\pi$. Do you expect I_z and R_z for the longer spring to be the same as those for the shorter one, or should they be different? Check your predictions by calculating I_z and R_z for the longer spring.

30. A wire of constant density $\delta = 1$ lies along the curve
    $$\mathbf{r}(t) = (t \cos t)\,\mathbf{i} + (t \sin t)\,\mathbf{j} + (2\sqrt{2}/3)\,t^{3/2}\,\mathbf{k}, \quad 0 \le t \le 1.$$

    Find $\bar{z}, I_z$, and R_z.

31. Find I_x and R_x for the arch in Example 4.

32. Find the center of mass, and the moments of inertia and radii of gyration about the coordinate axes of a thin wire lying along the curve
    $$\mathbf{r}(t) = t\,\mathbf{i} + \frac{2\sqrt{2}}{3}\,t^{3/2}\,\mathbf{j} + \frac{t^2}{2}\,\mathbf{k}, \quad 0 \le t \le 2,$$

    if the density is $\delta = 1/(t + 1)$.

⚙ CAS Explorations and Projects

In Exercises 33–36, use a CAS to perform the following steps to evaluate the line integrals:

a) Find $ds = |\mathbf{v}(t)|\,dt$ for the path $\mathbf{r}(t) = g(t)\,\mathbf{i} + h(t)\,\mathbf{j} + k(t)\,\mathbf{k}$.

b) Express the integrand $f(g(t), h(t), k(t))\,|\mathbf{v}(t)|$ as a function of the parameter t.

c) Evaluate $\int_C f\,ds$ using Eq. (3) in the text.

33. $f(x, y, z) = \sqrt{1 + 30x^2 + 10y}$; $\mathbf{r}(t) = t\,\mathbf{i} + t^2\,\mathbf{j} + 3t^2\,\mathbf{k}$,
$0 \le t \le 2$

34. $f(x, y, z) = \sqrt{1 + x^3 + 5y^3}$; $\mathbf{r}(t) = t\,\mathbf{i} + \dfrac{1}{3}t^2\,\mathbf{j} + \sqrt{t}\,\mathbf{k}$,
$0 \le t \le 2$

35. $f(x, y, z) = x\sqrt{y} - 3z^2$; $\mathbf{r}(t) = \cos 2t\,\mathbf{i} + \sin 2t\,\mathbf{j} + 5t\,\mathbf{k}$,
$0 \le t \le 2\pi$

36. $f(x, y, z) = \left(1 + \dfrac{9}{4}z^{1/3}\right)^{1/4}$; $\mathbf{r}(t) = \cos 2t\,\mathbf{i} + \sin 2t\,\mathbf{j} + t^{5/2}\,\mathbf{k}$,
$0 \le t \le 2\pi$

14.2 Vector Fields, Work, Circulation, and Flux

When we study physical phenomena that are represented by vectors, we replace integrals over closed intervals by integrals over paths through vector fields. We use such integrals to find the work done in moving an object along a path against a variable force (a vehicle sent into space against Earth's gravitational field) or to find the work done by a vector field in moving an object along a path through the field (the work done by an accelerator in raising the energy of a particle). We also use line integrals to find the rates at which fluids flow along and across curves.

Vector Fields

A **vector field** on a domain in the plane or in space is a function that assigns a vector to each point in the domain. A field of three-dimensional vectors might have a formula like

$$\mathbf{F}(x, y, z) = M(x, y, z)\,\mathbf{i} + N(x, y, z)\,\mathbf{j} + P(x, y, z)\,\mathbf{k}.$$

The field is **continuous** if the **component functions** M, N, and P are continuous, **differentiable** if M, N, and P are differentiable, and so on. A field of two-dimensional vectors might have a formula like

$$\mathbf{F}(x, y) = M(x, y)\,\mathbf{i} + N(x, y)\,\mathbf{j}.$$

If we attach a projectile's velocity vector to each point of the projectile's trajectory in the plane of motion, we have a two-dimensional field defined along the trajectory. If we attach the gradient vector of a scalar function to each point of a level surface of the function, we have a three-dimensional field on the surface. If we attach the velocity vector to each point of a flowing fluid, we have a three-dimensional field defined on a region in space. These and other fields are illustrated in Figs. 14.8–14.16. Some of the illustrations give formulas for the fields as well.

To sketch the fields that had formulas, we picked a representative selection of domain points and sketched the vectors attached to them. Notice the convention that the arrows representing the vectors are drawn with their tails, not their heads, at the points where the vector functions are evaluated. This is different from the way we drew the position vectors of the planets and projectiles in Chapter 11, with their tails at the origin and their heads at the planet's and projectile's locations.

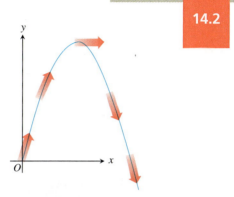

14.8 The velocity vectors $\mathbf{v}(t)$ of a projectile's motion make a vector field along the trajectory.

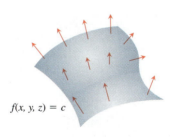

$f(x, y, z) = c$

14.9 The field of gradient vectors ∇f on a surface $f(x, y, z) = c$.

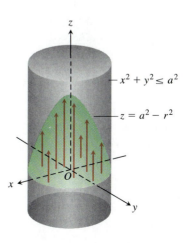

$x^2 + y^2 \le a^2$

$z = a^2 - r^2$

14.10 The flow of fluid in a long cylindrical pipe. The vectors $\mathbf{v} = (a^2 - r^2)\,\mathbf{k}$ inside the cylinder that have their bases in the xy-plane have their tips on the paraboloid $z = a^2 - r^2$.

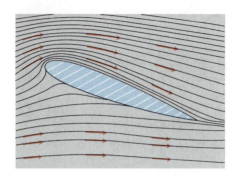

14.11 Velocity vectors of a flow around an airfoil in a wind tunnel. The streamlines were made visible by kerosene smoke. (Adapted from *NCFMF Book of Film Notes,* 1974, MIT Press with Education Development Center, Inc., Newton, Massachusetts.)

14.12 Streamlines in a contracting channel. The water speeds up as the channel narrows and the velocity vectors increase in length. (Adapted from *NCFMF Book of Film Notes,* 1974, MIT Press with Education Development Center, Inc., Newton, Massachusetts.)

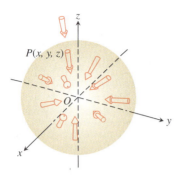

14.13 Vectors in the gravitational field

$$\mathbf{F} = -\frac{GM(x\,\mathbf{i} + y\,\mathbf{j} + z\,\mathbf{k})}{(x^2 + y^2 + z^2)^{3/2}}.$$

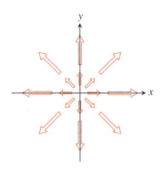

14.14 The radial field $\mathbf{F} = x\,\mathbf{i} + y\,\mathbf{j}$ of position vectors of points in the plane. Notice the convention that an arrow is drawn with its tail, not its head, at the point where $\mathbf{F}$ is evaluated.

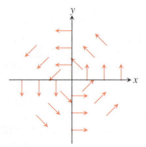

14.15 The circumferential or "spin" field of unit vectors

$$\mathbf{F} = (-y\,\mathbf{i} + x\,\mathbf{j})/(x^2 + y^2)^{1/2}$$

in the plane. The field is not defined at the origin.

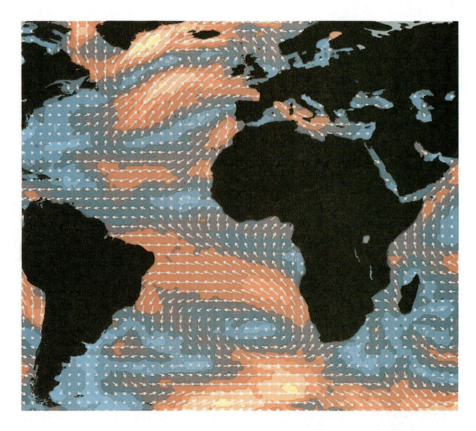

WIND SPEED, M/S

0   2   4   6   8   10   12   14   16+

14.16 NASA's *Seasat* used radar during a 3-day period in September 1978 to take 350,000 wind measurements over the world's oceans. The arrows show wind direction; their length and the color contouring indicate speed. Notice the heavy storm south of Greenland.

33. $f(x, y, z) = \sqrt{1 + 30x^2 + 10y}$; $\quad \mathbf{r}(t) = t\,\mathbf{i} + t^2\,\mathbf{j} + 3t^2\,\mathbf{k}$,
$\quad 0 \le t \le 2$

34. $f(x, y, z) = \sqrt{1 + x^3 + 5y^3}$; $\quad \mathbf{r}(t) = t\,\mathbf{i} + \frac{1}{3}t^2\,\mathbf{j} + \sqrt{t}\,\mathbf{k}$,
$\quad 0 \le t \le 2$

35. $f(x, y, z) = x\sqrt{y} - 3z^2$; $\quad \mathbf{r}(t) = \cos 2t\,\mathbf{i} + \sin 2t\,\mathbf{j} + 5t\,\mathbf{k}$,
$\quad 0 \le t \le 2\pi$

36. $f(x, y, z) = \left(1 + \frac{9}{4}z^{1/3}\right)^{1/4}$; $\quad \mathbf{r}(t) = \cos 2t\,\mathbf{i} + \sin 2t\,\mathbf{j} + t^{5/2}\,\mathbf{k}$,
$\quad 0 \le t \le 2\pi$

Vector Fields, Work, Circulation, and Flux

When we study physical phenomena that are represented by vectors, we replace integrals over closed intervals by integrals over paths through vector fields. We use such integrals to find the work done in moving an object along a path against a variable force (a vehicle sent into space against Earth's gravitational field) or to find the work done by a vector field in moving an object along a path through the field (the work done by an accelerator in raising the energy of a particle). We also use line integrals to find the rates at which fluids flow along and across curves.

Vector Fields

A **vector field** on a domain in the plane or in space is a function that assigns a vector to each point in the domain. A field of three-dimensional vectors might have a formula like

$$\mathbf{F}(x, y, z) = M(x, y, z)\,\mathbf{i} + N(x, y, z)\,\mathbf{j} + P(x, y, z)\,\mathbf{k}.$$

The field is **continuous** if the **component functions** M, N, and P are continuous, **differentiable** if M, N, and P are differentiable, and so on. A field of two-dimensional vectors might have a formula like

$$\mathbf{F}(x, y) = M(x, y)\,\mathbf{i} + N(x, y)\,\mathbf{j}.$$

If we attach a projectile's velocity vector to each point of the projectile's trajectory in the plane of motion, we have a two-dimensional field defined along the trajectory. If we attach the gradient vector of a scalar function to each point of a level surface of the function, we have a three-dimensional field on the surface. If we attach the velocity vector to each point of a flowing fluid, we have a three-dimensional field defined on a region in space. These and other fields are illustrated in Figs. 14.8–14.16. Some of the illustrations give formulas for the fields as well.

To sketch the fields that had formulas, we picked a representative selection of domain points and sketched the vectors attached to them. Notice the convention that the arrows representing the vectors are drawn with their tails, not their heads, at the points where the vector functions are evaluated. This is different from the way we drew the position vectors of the planets and projectiles in Chapter 11, with their tails at the origin and their heads at the planet's and projectile's locations.

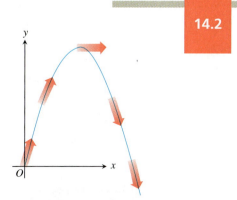

14.8 The velocity vectors $\mathbf{v}(t)$ of a projectile's motion make a vector field along the trajectory.

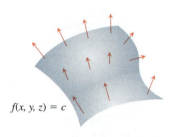

$f(x, y, z) = c$

14.9 The field of gradient vectors ∇f on a surface $f(x, y, z) = c$.

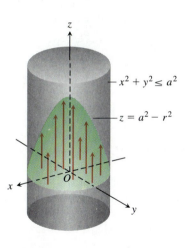

14.10 The flow of fluid in a long cylindrical pipe. The vectors $\mathbf{v} = (a^2 - r^2)\,\mathbf{k}$ inside the cylinder that have their bases in the xy-plane have their tips on the paraboloid $z = a^2 - r^2$.

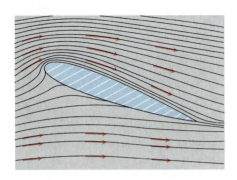

14.11 Velocity vectors of a flow around an airfoil in a wind tunnel. The streamlines were made visible by kerosene smoke. (Adapted from *NCFMF Book of Film Notes,* 1974, MIT Press with Education Development Center, Inc., Newton, Massachusetts.)

14.12 Streamlines in a contracting channel. The water speeds up as the channel narrows and the velocity vectors increase in length. (Adapted from *NCFMF Book of Film Notes,* 1974, MIT Press with Education Development Center, Inc., Newton, Massachusetts.)

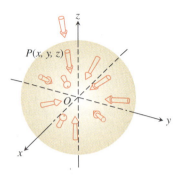

14.13 Vectors in the gravitational field

$$\mathbf{F} = -\frac{GM(x\,\mathbf{i} + y\,\mathbf{j} + z\,\mathbf{k})}{(x^2 + y^2 + z^2)^{3/2}}.$$

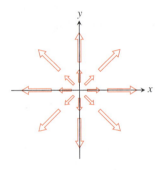

14.14 The radial field $\mathbf{F} = x\,\mathbf{i} + y\,\mathbf{j}$ of position vectors of points in the plane. Notice the convention that an arrow is drawn with its tail, not its head, at the point where $\mathbf{F}$ is evaluated.

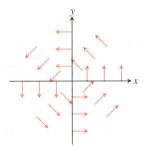

14.15 The circumferential or "spin" field of unit vectors

$$\mathbf{F} = (-y\,\mathbf{i} + x\,\mathbf{j})/(x^2 + y^2)^{1/2}$$

in the plane. The field is not defined at the origin.

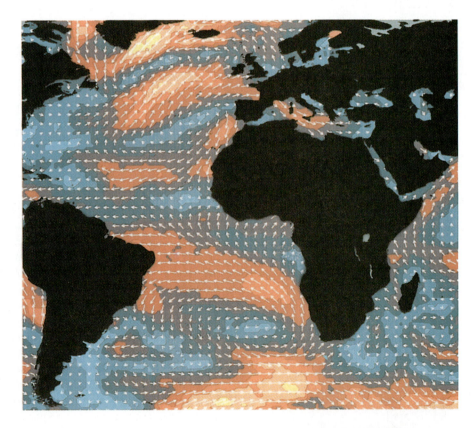

WIND SPEED, M/S

0 2 4 6 8 10 12 14 16+

14.16 NASA's *Seasat* used radar during a 3-day period in September 1978 to take 350,000 wind measurements over the world's oceans. The arrows show wind direction; their length and the color contouring indicate speed. Notice the heavy storm south of Greenland.

Gradient Fields

> **Definition**
>
> The **gradient field** of a differentiable function $f(x, y, z)$ is the field of gradient vectors
>
> $$\nabla f = \frac{\partial f}{\partial x}\mathbf{i} + \frac{\partial f}{\partial y}\mathbf{j} + \frac{\partial f}{\partial z}\mathbf{k}.$$

EXAMPLE 1   Find the gradient field of $f(x, y, z) = xyz$.

Solution   The gradient field of f is the field $\mathbf{F} = \nabla f = yz\,\mathbf{i} + xz\,\mathbf{j} + xy\,\mathbf{k}.$ ☐

As we will see in Section 14.3, gradient fields are of special importance in engineering, mathematics, and physics.

The Work Done by a Force over a Curve in Space

Suppose that the vector field $\mathbf{F} = M(x, y, z)\,\mathbf{i} + N(x, y, z)\,\mathbf{j} + P(x, y, z)\,\mathbf{k}$ represents a force throughout a region in space (it might be the force of gravity or an electromagnetic force of some kind) and that

$$\mathbf{r}(t) = g(t)\,\mathbf{i} + h(t)\,\mathbf{j} + k(t)\,\mathbf{k}, \quad a \le t \le b,$$

is a smooth curve in the region. Then the integral of $\mathbf{F} \cdot \mathbf{T}$, the scalar component of $\mathbf{F}$ in the direction of the curve's unit tangent vector, over the curve is called the work done by $\mathbf{F}$ over the curve from a to b (Fig. 14.17).

> **Definition**
>
> The **work** done by a force $\mathbf{F} = M(x, y, z)\,\mathbf{i} + N(x, y, z)\,\mathbf{j} + P(x, y, z)\,\mathbf{k}$ over a smooth curve $\mathbf{r}(t) = g(t)\,\mathbf{i} + h(t)\,\mathbf{j} + k(t)\,\mathbf{k}$ from $t = a$ to $t = b$ is
>
> $$W = \int_{t=a}^{t=b} \mathbf{F} \cdot \mathbf{T}\, ds. \tag{1}$$

We motivate Eq. (1) with the same kind of reasoning we used in Section 5.8 to derive the formula $W = \int_a^b F(x)\, dx$ for the work done by a continuous force of magnitude $F(x)$ directed along an interval of the x-axis. We divide the curve into short segments, apply the constant-force-times-distance formula for work to approximate the work over each curved segment, add the results to approximate the work over the entire curve, and calculate the work as the limit of the approximating sums as the segments become shorter and more numerous. To find exactly what the limiting integral should be, we partition the parameter interval $I = [a, b]$ in the usual way and choose a point c_k in each subinterval $[t_k, t_{k+1}]$. The partition of I determines ("induces," we say) a partition of the curve, with the point P_k being the tip of the position vector $\mathbf{r}$ at $t = t_k$ and Δs_k being the length of the curve segment $P_k P_{k+1}$ (Fig. 14.18). If $\mathbf{F}_k$ denotes the value of $\mathbf{F}$ at the point on the curve corresponding to $t = c_k$, and $\mathbf{T}_k$ denotes the curve's tangent vector at this point, then $\mathbf{F}_k \cdot \mathbf{T}_k$ is the scalar component of $\mathbf{F}$ in the direction of $\mathbf{T}$ at $t = c_k$

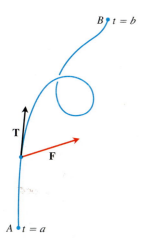

14.17 The work done by a continuous field $\mathbf{F}$ over a smooth path $\mathbf{r} = g(t)\,\mathbf{i} + h(t)\,\mathbf{j} + k(t)\,\mathbf{k}$ from A to B is the integral of $\mathbf{F} \cdot \mathbf{T}$ over the path from $t = a$ to $t = b$.

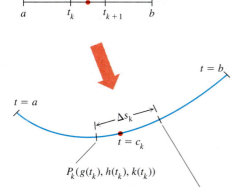

14.18 Each partition of a parameter interval $a \le t \le b$ induces a partition of the curve $\mathbf{r} = g(t)\,\mathbf{i} + h(t)\,\mathbf{j} + k(t)\,\mathbf{k}$.

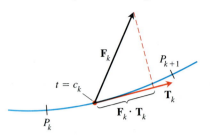

14.19 An enlarged view of the curve segment $P_k P_{k+1}$ in Fig. 14.18, showing the force vector and unit tangent vector at the point on the curve where $t = c_k$.

(Fig. 14.19). The work done by $\mathbf{F}$ along the curve segment $P_k P_{k+1}$ will be approximately

$$\left(\begin{array}{c} \text{force component in} \\ \text{direction of motion} \end{array} \right) \times \left(\begin{array}{c} \text{distance} \\ \text{applied} \end{array} \right) = \mathbf{F}_k \cdot \mathbf{T}_k \Delta s_k.$$

The work done by $\mathbf{F}$ along the curve from $t = a$ to $t = b$ will be approximately

$$\sum_{k=1}^{n} \mathbf{F}_k \cdot \mathbf{T}_k \Delta s_k.$$

As the norm of the partition of $[a, b]$ approaches zero, the norm of the induced partition of the curve approaches zero and these sums approach the line integral

$$\int_{t=a}^{t=b} \mathbf{F} \cdot \mathbf{T} \, ds.$$

The sign of the number we calculate with this integral depends on the direction in which the curve is traversed as t increases. If we reverse the direction of motion, we reverse the direction of $\mathbf{T}$ and change the sign of $\mathbf{F} \cdot \mathbf{T}$ and its integral.

Notation and Evaluation

Table 14.2 shows six ways to write the work integral in Eq. (1).

Table 14.2 Different ways to write the work integral

$\mathbf{W} = \displaystyle\int_{t=a}^{t=b} \mathbf{F} \cdot \mathbf{T} \, ds$	The definition
$= \displaystyle\int_{t=a}^{t=b} \mathbf{F} \cdot d\mathbf{r}$	Compact differential form
$= \displaystyle\int_{a}^{b} \mathbf{F} \cdot \dfrac{d\mathbf{r}}{dt} dt$	Expanded to include dt; emphasizes the parameter t and velocity vector $d\mathbf{r}/dt$
$= \displaystyle\int_{a}^{b} \left(M \dfrac{dg}{dt} + N \dfrac{dh}{dt} + P \dfrac{dk}{dt} \right) dt$	Emphasizes the component functions
$= \displaystyle\int_{a}^{b} \left(M \dfrac{dx}{dt} + N \dfrac{dy}{dt} + P \dfrac{dz}{dt} \right) dt$	Abbreviates the components of $\mathbf{r}$
$= \displaystyle\int_{a}^{b} M \, dx + N \, dy + P \, dz$	dt's canceled; the most common form

Despite their variety, the formulas in Table 14.2 are all evaluated the same way.

How to Evaluate a Work Integral

To evaluate the work integral, take these steps:

1. Evaluate $\mathbf{F}$ on the curve as a function of the parameter t.
2. Find $d\mathbf{r}/dt$.
3. Dot $\mathbf{F}$ with $d\mathbf{r}/dt$.
4. Integrate from $t = a$ to $t = b$.

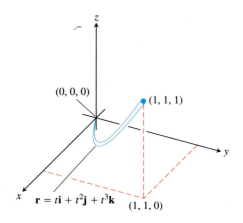

14.20 The curve in Example 2.

EXAMPLE 2   Find the work done by $\mathbf{F} = (y - x^2)\mathbf{i} + (z - y^2)\mathbf{j} + (x - z^2)\mathbf{k}$ over the curve $\mathbf{r}(t) = t\mathbf{i} + t^2\mathbf{j} + t^3\mathbf{k}$, $0 \le t \le 1$, from (0, 0, 0) to (1, 1, 1) (Fig. 14.20).

Solution

Step 1: *Evaluate* $\mathbf{F}$ *on the curve.*

$$\mathbf{F} = (y - x^2)\mathbf{i} + (z - y^2)\mathbf{j} + (x - z^2)\mathbf{k}$$

$$= \underbrace{(t^2 - t^2)}_{0}\mathbf{i} + (t^3 - t^4)\mathbf{j} + (t - t^6)\mathbf{k}$$

Step 2: *Find* $d\mathbf{r}/dt$.

$$\frac{d\mathbf{r}}{dt} = \frac{d}{dt}(t\mathbf{i} + t^2\mathbf{j} + t^3\mathbf{k}) = \mathbf{i} + 2t\mathbf{j} + 3t^2\mathbf{k}$$

Step 3: *Dot* $\mathbf{F}$ *with* $d\mathbf{r}/dt$.

$$\mathbf{F} \cdot \frac{d\mathbf{r}}{dt} = [(t^3 - t^4)\mathbf{j} + (t - t^6)\mathbf{k}] \cdot (\mathbf{i} + 2t\mathbf{j} + 3t^2\mathbf{k})$$

$$= (t^3 - t^4)(2t) + (t - t^6)(3t^2) = 2t^4 - 2t^5 + 3t^3 - 3t^8$$

Step 4: *Integrate from* $t = 0$ *to* $t = 1$.

$$\text{Work} = \int_0^1 (2t^4 - 2t^5 + 3t^3 - 3t^8)\,dt$$

$$= \left[\frac{2}{5}t^5 - \frac{2}{6}t^6 + \frac{3}{4}t^4 - \frac{3}{9}t^9\right]_0^1 = \frac{29}{60}$$

Flow Integrals and Circulation

Instead of being a force field, suppose that $\mathbf{F} = M\mathbf{i} + N\mathbf{j} + P\mathbf{k}$ represents the velocity field of a fluid flowing through a region in space (a tidal basin or the turbine chamber of a hydroelectric generator, for example). Under these circumstances, the integral of $\mathbf{F} \cdot \mathbf{T}$ along a curve in the region gives the fluid's flow along the curve.

> **Definitions**
>
> If $\mathbf{r}(t) = g(t)\mathbf{i} + h(t)\mathbf{j} + k(t)\mathbf{k}$, $a \le t \le b$, is a smooth curve in the domain of a continuous velocity field $\mathbf{F} = M(x, y, z)\mathbf{i} + N(x, y, z)\mathbf{j} + P(x, y, z)\mathbf{k}$, the **flow** along the curve from $t = a$ to $t = b$ is the integral of $\mathbf{F} \cdot \mathbf{T}$ over the curve from a to b:
>
> $$\text{Flow} = \int_a^b \mathbf{F} \cdot \mathbf{T}\,ds. \qquad (2)$$
>
> The integral in this case is called a **flow integral.** If the curve is a closed loop, the flow is called the **circulation** around the curve.

We evaluate flow integrals the same way we evaluate work integrals.

EXAMPLE 3   A fluid's velocity field is $\mathbf{F} = x\mathbf{i} + z\mathbf{j} + y\mathbf{k}$. Find the flow along the helix $\mathbf{r}(t) = (\cos t)\mathbf{i} + (\sin t)\mathbf{j} + t\mathbf{k}$, $0 \le t \le \pi/2$.

Solution

Step 1: *Evaluate* **F** *on the curve.*

$$\mathbf{F} = x\,\mathbf{i} + z\,\mathbf{j} + y\,\mathbf{k} = (\cos t)\,\mathbf{i} + t\,\mathbf{j} + (\sin t)\,\mathbf{k}$$

Step 2: *Find* $d\mathbf{r}/dt$.

$$\frac{d\mathbf{r}}{dt} = (-\sin t)\,\mathbf{i} + (\cos t)\,\mathbf{j} + \mathbf{k}$$

Step 3: *Find* $\mathbf{F} \cdot (d\mathbf{r}/dt)$.

$$\mathbf{F} \cdot \frac{d\mathbf{r}}{dt} = (\cos t)(-\sin t) + (t)(\cos t) + (\sin t)(1)$$

$$= -\sin t \cos t + t \cos t + \sin t$$

Step 4: *Integrate from* $t = a$ *to* $t = b$.

$$\text{Flow} = \int_{t=a}^{t=b} \mathbf{F} \cdot \frac{d\mathbf{r}}{dt}\,dt = \int_0^{\pi/2} (-\sin t \cos t + t \cos t + \sin t)\,dt$$

$$= \left[\frac{\cos^2 t}{2} + t \sin t\right]_0^{\pi/2} = \left(0 + \frac{\pi}{2}\right) - \left(\frac{1}{2} + 0\right) = \frac{\pi}{2} - \frac{1}{2} \qquad \square$$

EXAMPLE 4   Find the circulation of the field $\mathbf{F} = (x - y)\,\mathbf{i} + x\,\mathbf{j}$ around the circle $\mathbf{r}(t) = (\cos t)\,\mathbf{i} + (\sin t)\,\mathbf{j}$, $0 \le t \le 2\pi$.

Solution

1. On the circle, $\mathbf{F} = (x - y)\,\mathbf{i} + x\,\mathbf{j} = (\cos t - \sin t)\,\mathbf{i} + (\cos t)\,\mathbf{j}$.

2. $\dfrac{d\mathbf{r}}{dt} = (-\sin t)\,\mathbf{i} + (\cos t)\,\mathbf{j}$

3. $\mathbf{F} \cdot \dfrac{d\mathbf{r}}{dt} = -\sin t \cos t + \underbrace{\sin^2 t + \cos^2 t}_{1}$

4. $\text{Circulation} = \displaystyle\int_0^{2\pi} \mathbf{F} \cdot \frac{d\mathbf{r}}{dt}\,dt = \int_0^{2\pi} (1 - \sin t \cos t)\,dt$

$$= \left[t - \frac{\sin^2 t}{2}\right]_0^{2\pi} = 2\pi. \qquad \square$$

Flux Across a Plane Curve

To find the rate at which a fluid is entering or leaving a region enclosed by a smooth curve C in the xy-plane, we calculate the line integral over C of $\mathbf{F} \cdot \mathbf{n}$, the scalar component of the fluid's velocity field in the direction of the curve's outward-pointing normal vector. The value of this integral is called the flux of $\mathbf{F}$ across C. *Flux* is Latin for *flow*, but many flux calculations involve no motion at all. If $\mathbf{F}$ were an electric field or a magnetic field, for instance, the integral of $\mathbf{F} \cdot \mathbf{n}$ would still be called the flux of the field across C.

Definition

If C is a smooth closed curve in the domain of a continuous vector field $\mathbf{F} = M(x, y)\,\mathbf{i} + N(x, y)\,\mathbf{j}$ in the plane, and if $\mathbf{n}$ is the outward-pointing unit normal vector on C, the **flux** of $\mathbf{F}$ across C is given by the following line integral:

$$\text{Flux of } \mathbf{F} \text{ across } C = \int_C \mathbf{F} \cdot \mathbf{n} \, ds. \tag{3}$$

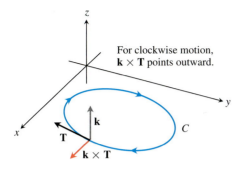

For clockwise motion, $\mathbf{k} \times \mathbf{T}$ points outward.

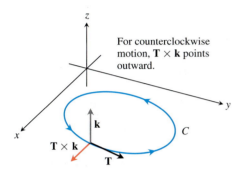

For counterclockwise motion, $\mathbf{T} \times \mathbf{k}$ points outward.

14.21 To find an outward unit normal vector for a smooth curve C in the xy-plane that is traversed counterclockwise as t increases, we take $\mathbf{n} = \mathbf{T} \times \mathbf{k}$.

Notice the difference between flux and circulation. The flux of $\mathbf{F}$ across C is the line integral with respect to arc length of $\mathbf{F} \cdot \mathbf{n}$, the scalar component of $\mathbf{F}$ in the direction of the outward normal. The circulation of $\mathbf{F}$ around C is the line integral with respect to arc length of $\mathbf{F} \cdot \mathbf{T}$, the scalar component of $\mathbf{F}$ in the direction of the unit tangent vector. Flux is the integral of the normal component of $\mathbf{F}$; circulation is the integral of the tangential component of $\mathbf{F}$.

To evaluate the integral in (3), we begin with a parametrization

$$x = g(t), \qquad y = h(t), \qquad a \le t \le b,$$

that traces the curve C exactly once as t increases from a to b. We can find the outward unit normal vector $\mathbf{n}$ by crossing the curve's unit tangent vector $\mathbf{T}$ with the vector $\mathbf{k}$. But which order do we choose, $\mathbf{T} \times \mathbf{k}$ or $\mathbf{k} \times \mathbf{T}$? Which one points outward? It depends on which way C is traversed as the parameter t increases. If the motion is clockwise, then $\mathbf{k} \times \mathbf{T}$ points outward; if the motion is counterclockwise, then $\mathbf{T} \times \mathbf{k}$ points outward (Fig. 14.21). The usual choice is $\mathbf{n} = \mathbf{T} \times \mathbf{k}$, the choice that assumes counterclockwise motion. Thus, while the value of the arc length integral in the definition of flux in Eq. (3) does not depend on which way C is traversed, the formulas we are about to derive for evaluating the integral in Eq. (3) will assume counterclockwise motion.

In terms of components,

$$\mathbf{n} = \mathbf{T} \times \mathbf{k} = \left(\frac{dx}{ds} \mathbf{i} + \frac{dy}{ds} \mathbf{j} \right) \times \mathbf{k} = \frac{dy}{ds} \mathbf{i} - \frac{dx}{ds} \mathbf{j}.$$

If $\mathbf{F} = M(x, y) \mathbf{i} + N(x, y) \mathbf{j}$, then

$$\mathbf{F} \cdot \mathbf{n} = M(x, y) \frac{dy}{ds} - N(x, y) \frac{dx}{ds}.$$

Hence,

$$\int_C \mathbf{F} \cdot \mathbf{n} \, ds = \int_C \left(M \frac{dy}{ds} - N \frac{dx}{ds} \right) ds = \oint_C M \, dy - N \, dx.$$

We put a directed circle $\circlearrowleft$ on the last integral as a reminder that the integration around the closed curve C is to be in the counterclockwise direction. To evaluate this integral, we express $M, dy, N,$ and dx in terms of t and integrate from $t = a$ to $t = b$. We do not need to know either $\mathbf{n}$ or ds to find the flux.

The Formula for Calculating Flux Across a Smooth Closed Plane Curve

$$(\text{Flux of } \mathbf{F} = M \mathbf{i} + N \mathbf{j} \text{ across } C) = \oint_C M \, dy - N \, dx \tag{4}$$

The integral can be evaluated from any smooth parametrization $x = g(t)$, $y = h(t), a \le t \le b$, that traces C counterclockwise exactly once.

EXAMPLE 5   Find the flux of $\mathbf{F} = (x - y)\mathbf{i} + x\mathbf{j}$ across the circle $x^2 + y^2 = 1$ in the xy-plane.

Solution   The parametrization $\mathbf{r}(t) = (\cos t)\mathbf{i} + (\sin t)\mathbf{j}$, $0 \le t \le 2\pi$, traces the circle counterclockwise exactly once. We can therefore use this parametrization in Eq. (4). With

$$M = x - y = \cos t - \sin t, \qquad dy = d(\sin t) = \cos t \, dt$$

$$N = x = \cos t, \qquad dx = d(\cos t) = -\sin t \, dt,$$

we find

$$\text{Flux} = \int_C M \, dy - N \, dx = \int_0^{2\pi} (\cos^2 t - \sin t \cos t + \cos t \, \sin t) \, dt \qquad \text{Eq. (4)}$$

$$= \int_0^{2\pi} \cos^2 t \, dt = \int_0^{2\pi} \frac{1 + \cos 2t}{2} \, dt = \left[\frac{t}{2} + \frac{\sin 2t}{4}\right]_0^{2\pi} = \pi.$$

The flux of $\mathbf{F}$ across the circle is π. Since the answer is positive, the net flow across the curve is outward. A net inward flow would have given a negative flux. ❏

Exercises 14.2

Vector and Gradient Fields

Find the gradient fields of the functions in Exercises 1–4.

1. $f(x, y, z) = (x^2 + y^2 + z^2)^{-1/2}$

2. $f(x, y, z) = \ln \sqrt{x^2 + y^2 + z^2}$

3. $g(x, y, z) = e^z - \ln(x^2 + y^2)$

4. $g(x, y, z) = xy + yz + xz$

5. Give a formula $\mathbf{F} = M(x, y)\mathbf{i} + N(x, y)\mathbf{j}$ for the vector field in the plane that has the property that $\mathbf{F}$ points toward the origin with magnitude inversely proportional to the square of the distance from (x, y) to the origin. (The field is not defined at $(0, 0)$.)

6. Give a formula $\mathbf{F} = M(x, y)\mathbf{i} + N(x, y)\mathbf{j}$ for the vector field in the plane that has the properties that $\mathbf{F} = \mathbf{0}$ at $(0, 0)$ and that at any other point (a, b), $\mathbf{F}$ is tangent to the circle $x^2 + y^2 = a^2 + b^2$ and points in the clockwise direction with magnitude $|\mathbf{F}| = \sqrt{a^2 + b^2}$.

Work

In Exercises 7–12, find the work done by force $\mathbf{F}$ from $(0, 0, 0)$ to $(1, 1, 1)$ over each of the following paths (Fig. 14.22):

a) The straight-line path C_1: $\mathbf{r}(t) = t\mathbf{i} + t\mathbf{j} + t\mathbf{k}$, $0 \le t \le 1$

b) The curved path C_2: $\mathbf{r}(t) = t\mathbf{i} + t^2\mathbf{j} + t^4\mathbf{k}$, $0 \le t \le 1$

c) The path $C_3 \cup C_4$ consisting of the line segment from $(0, 0, 0)$ to $(1, 1, 0)$ followed by the segment from $(1, 1, 0)$ to $(1, 1, 1)$

7. $\mathbf{F} = 3y\mathbf{i} + 2x\mathbf{j} + 4z\mathbf{k}$

8. $\mathbf{F} = [1/(x^2 + 1)]\mathbf{j}$

9. $\mathbf{F} = \sqrt{z}\mathbf{i} - 2x\mathbf{j} + \sqrt{y}\mathbf{k}$

10. $\mathbf{F} = xy\mathbf{i} + yz\mathbf{j} + xz\mathbf{k}$

11. $\mathbf{F} = (3x^2 - 3x)\mathbf{i} + 3z\mathbf{j} + \mathbf{k}$

12. $\mathbf{F} = (y + z)\mathbf{i} + (z + x)\mathbf{j} + (x + y)\mathbf{k}$

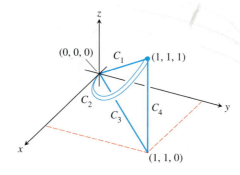

14.22 The paths from $(0, 0, 0)$ to $(1, 1, 1)$.

In Exercises 13–16, find the work done by $\mathbf{F}$ over the curve in the direction of increasing t.

13. $\mathbf{F} = xy\mathbf{i} + y\mathbf{j} - yz\mathbf{k}$
$\mathbf{r}(t) = t\mathbf{i} + t^2\mathbf{j} + t\mathbf{k}$, $0 \le t \le 1$

14. $\mathbf{F} = 2y\mathbf{i} + 3x\mathbf{j} + (x + y)\mathbf{k}$
$\mathbf{r}(t) = (\cos t)\mathbf{i} + (\sin t)\mathbf{j} + (t/6)\mathbf{k}$, $0 \le t \le 2\pi$

15. $\mathbf{F} = z\mathbf{i} + x\mathbf{j} + y\mathbf{k}$
$\mathbf{r}(t) = (\sin t)\mathbf{i} + (\cos t)\mathbf{j} + t\mathbf{k}$, $0 \le t \le 2\pi$

16. $\mathbf{F} = 6z\mathbf{i} + y^2\mathbf{j} + 12x\mathbf{k}$
$\mathbf{r}(t) = (\sin t)\mathbf{i} + (\cos t)\mathbf{j} + (t/6)\mathbf{k}$, $0 \le t \le 2\pi$

Line Integrals and Vector Fields in the Plane

17. Evaluate $\int_C xy \, dx + (x + y) \, dy$ along the curve $y = x^2$ from $(-1, 1)$ to $(2, 4)$.

18. Evaluate $\int_C (x - y)\,dx + (x + y)\,dy$ counterclockwise around the triangle with vertices $(0, 0)$, $(1, 0)$, and $(0, 1)$.

19. Evaluate $\int_C \mathbf{F} \cdot \mathbf{T}\,ds$ for the vector field $\mathbf{F} = x^2\mathbf{i} - y\mathbf{j}$ along the curve $x = y^2$ from $(4, 2)$ to $(1, -1)$.

20. Evaluate $\int_C \mathbf{F} \cdot d\mathbf{r}$ for the vector field $\mathbf{F} = y\mathbf{i} - x\mathbf{j}$ counterclockwise along the unit circle $x^2 + y^2 = 1$ from $(1, 0)$ to $(0, 1)$.

21. Find the work done by the force $\mathbf{F} = xy\mathbf{i} + (y - x)\mathbf{j}$ over the straight line from $(1, 1)$ to $(2, 3)$.

22. Find the work done by the gradient of $f(x, y) = (x + y)^2$ counterclockwise around the circle $x^2 + y^2 = 4$ from $(2, 0)$ to itself.

23. Find the circulation and flux of the fields

$$\mathbf{F}_1 = x\mathbf{i} + y\mathbf{j} \quad \text{and} \quad \mathbf{F}_2 = -y\mathbf{i} + x\mathbf{j}$$

around and across each of the following curves.

a) The circle $\mathbf{r}(t) = (\cos t)\mathbf{i} + (\sin t)\mathbf{j}, \quad 0 \le t \le 2\pi$

b) The ellipse $\mathbf{r}(t) = (\cos t)\mathbf{i} + (4 \sin t)\mathbf{j}, \quad 0 \le t \le 2\pi$

24. Find the flux of the fields

$$\mathbf{F}_1 = 2x\mathbf{i} - 3y\mathbf{j} \quad \text{and} \quad \mathbf{F}_2 = 2x\mathbf{i} + (x - y)\mathbf{j}$$

across the circle

$$\mathbf{r}(t) = (a \cos t)\mathbf{i} + (a \sin t)\mathbf{j}, \quad 0 \le t \le 2\pi.$$

In Exercises 25–28, find the circulation and flux of the field $\mathbf{F}$ around and across the closed semicircular path that consists of the semicircular arch $\mathbf{r}_1(t) = (a \cos t)\mathbf{i} + (a \sin t)\mathbf{j}, 0 \le t \le \pi$, followed by the line segment $\mathbf{r}_2(t) = t\mathbf{i}, -a \le t \le a$.

25. $\mathbf{F} = x\mathbf{i} + y\mathbf{j}$

26. $\mathbf{F} = x^2\mathbf{i} + y^2\mathbf{j}$

27. $\mathbf{F} = -y\mathbf{i} + x\mathbf{j}$

28. $\mathbf{F} = -y^2\mathbf{i} + x^2\mathbf{j}$

29. Evaluate the flow integral of the velocity field $\mathbf{F} = (x + y)\mathbf{i} - (x^2 + y^2)\mathbf{j}$ along each of the following paths from $(1, 0)$ to $(-1, 0)$ in the xy-plane.

a) The upper half of the circle $x^2 + y^2 = 1$

b) The line segment from $(1, 0)$ to $(-1, 0)$

c) The line segment from $(1, 0)$ to $(0, -1)$ followed by the line segment from $(0, -1)$ to $(-1, 0)$

30. Find the flux of the field $\mathbf{F}$ in Exercise 29 outward across the triangle with vertices $(1, 0)$, $(0, 1)$, $(-1, 0)$.

Sketching and Finding Fields in the Plane

31. Draw the spin field

$$\mathbf{F} = -\frac{y}{\sqrt{x^2 + y^2}}\mathbf{i} + \frac{x}{\sqrt{x^2 + y^2}}\mathbf{j}$$

(see Fig. 14.15) along with its horizontal and vertical components at a representative assortment of points on the circle $x^2 + y^2 = 4$.

32. Draw the radial field

$$\mathbf{F} = x\mathbf{i} + y\mathbf{j}$$

(see Fig. 14.14) along with its horizontal and vertical components at a representative assortment of points on the circle $x^2 + y^2 = 1$.

33. a) Find a field $\mathbf{G} = P(x, y)\mathbf{i} + Q(x, y)\mathbf{j}$ in the xy-plane with the property that at any point $(a, b) \ne (0, 0)$, $\mathbf{G}$ is a vector of magnitude $\sqrt{a^2 + b^2}$ tangent to the circle $x^2 + y^2 = a^2 + b^2$ and pointing in the counterclockwise direction. (The field is undefined at $(0, 0)$.)

b) How is $\mathbf{G}$ related to the spin field $\mathbf{F}$ in Fig. 14.15?

34. a) Find a field $\mathbf{G} = P(x, y)\mathbf{i} + Q(x, y)\mathbf{j}$ in the xy-plane with the property that at any point $(a, b) \ne (0, 0)$, $\mathbf{G}$ is a unit vector tangent to the circle $x^2 + y^2 = a^2 + b^2$ and pointing in the clockwise direction.

b) How is $\mathbf{G}$ related to the spin field $\mathbf{F}$ in Fig. 14.15?

35. Find a field $\mathbf{F} = M(x, y)\mathbf{i} + N(x, y)\mathbf{j}$ in the xy-plane with the property that at each point $(x, y) \ne (0, 0)$, $\mathbf{F}$ is a unit vector pointing toward the origin. (The field is undefined at $(0, 0)$.)

36. Find a field $\mathbf{F} = M(x, y)\mathbf{i} + N(x, y)\mathbf{j}$ in the xy-plane with the property that at each point $(x, y) \ne (0, 0)$, $\mathbf{F}$ points toward the origin and $|\mathbf{F}|$ is (a) the distance from (x, y) to the origin, (b) inversely proportional to the distance from (x, y) to the origin. (The field is undefined at $(0, 0)$.)

Flow Integrals in Space

In Exercises 37–40, $\mathbf{F}$ is the velocity field of a fluid flowing through a region in space. Find the flow along the given curve in the direction of increasing t.

37. $\mathbf{F} = -4xy\mathbf{i} + 8y\mathbf{j} + 2\mathbf{k}$
$\mathbf{r}(t) = t\mathbf{i} + t^2\mathbf{j} + \mathbf{k}, \quad 0 \le t \le 2$

38. $\mathbf{F} = x^2\mathbf{i} + yz\mathbf{j} + y^2\mathbf{k}$
$\mathbf{r}(t) = 3t\mathbf{j} + 4t\mathbf{k}, \quad 0 \le t \le 1$

39. $\mathbf{F} = (x - z)\mathbf{i} + x\mathbf{k}$
$\mathbf{r}(t) = (\cos t)\mathbf{i} + (\sin t)\mathbf{k}, \quad 0 \le t \le \pi$

40. $\mathbf{F} = -y\mathbf{i} + x\mathbf{j} + 2\mathbf{k}$
$\mathbf{r}(t) = (-2\cos t)\mathbf{i} + (2\sin t)\mathbf{j} + 2t\mathbf{k}, \quad 0 \le t \le 2\pi$

41. Find the circulation of $\mathbf{F} = 2x\mathbf{i} + 2z\mathbf{j} + 2y\mathbf{k}$ around the closed path consisting of the following three curves traversed in the direction of increasing t:

C_1: $\mathbf{r}(t) = (\cos t)\mathbf{i} + (\sin t)\mathbf{j} + t\mathbf{k}, \quad 0 \le t \le \pi/2$

C_2: $\mathbf{r}(t) = \mathbf{j} + (\pi/2)(1 - t)\mathbf{k}, \quad 0 \le t \le 1$

C_3: $\mathbf{r}(t) = t\mathbf{i} + (1 - t)\mathbf{j}, \quad 0 \le t \le 1$

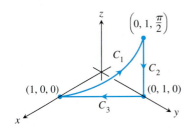

42. Let C be the ellipse in which the plane $2x + 3y - z = 0$ meets the cylinder $x^2 + y^2 = 12$. Show, without evaluating either line integral directly, that the circulation of the field $\mathbf{F} = x\,\mathbf{i} + y\,\mathbf{j} + z\,\mathbf{k}$ around C in either direction is zero.

43. The field $\mathbf{F} = xy\,\mathbf{i} + y\,\mathbf{j} - yz\,\mathbf{k}$ is the velocity field of a flow in space. Find the flow from $(0, 0, 0)$ to $(1, 1, 1)$ along the curve of intersection of the cylinder $y = x^2$ and the plane $z = x$. (*Hint:* Use $t = x$ as the parameter.)

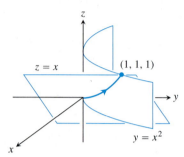

44. Find the flow of the field $\mathbf{F} = \nabla(xy^2z^3)$

a) once around the curve C in Exercise 42, clockwise as viewed from above.

b) along the line segment from $(1, 1, 1)$ to $(2, 1, -1)$.

Theory and Examples

45. Suppose $f(t)$ is differentiable and positive for $a \le t \le b$. Let C be the path $\mathbf{r}(t) = t\,\mathbf{i} + f(t)\,\mathbf{j}$, $a \le t \le b$, and $\mathbf{F} = y\,\mathbf{i}$. Is there any relation between the value of the work integral

$$\int_C \mathbf{F} \cdot d\mathbf{r}$$

and the area of the region bounded by the t-axis, the graph of f, and the lines $t = a$ and $t = b$? Give reasons for your answer.

46. A particle moves along the smooth curve $y = f(x)$ from $(a, f(a))$ to $(b, f(b))$. The force moving the particle has constant magnitude k and always points away from the origin. Show that the work done by the force is

$$\int_C \mathbf{F} \cdot \mathbf{T}\,ds = k\left[(b^2 + (f(b))^2)^{1/2} - (a^2 + (f(a))^2)^{1/2}\right].$$

✲ CAS Explorations and Projects

In Exercises 47–52, use a CAS to perform the following steps for finding the work done by force $\mathbf{F}$ over the given path:

a) Find $d\mathbf{r}$ for the path $\mathbf{r}(t) = g(t)\,\mathbf{i} + h(t)\,\mathbf{j} + k(t)\,\mathbf{k}$.

b) Evaluate the force $\mathbf{F}$ along the path.

c) Evaluate $\displaystyle\int_C \mathbf{F} \cdot d\mathbf{r}$.

47. $\mathbf{F} = xy^6\,\mathbf{i} + 3x(xy^5 + 2)\,\mathbf{j}$; $\mathbf{r}(t) = 2\cos t\,\mathbf{i} + \sin t\,\mathbf{j}, 0 \le t \le 2\pi$

48. $\mathbf{F} = \dfrac{3}{1+x^2}\,\mathbf{i} + \dfrac{2}{1+y^2}\,\mathbf{j}$; $\mathbf{r}(t) = \cos t\,\mathbf{i} + \sin t\,\mathbf{j}$, $0 \le t \le \pi$

49. $\mathbf{F} = (y + yz\cos xyz)\,\mathbf{i} + (x^2 + xz\cos xyz)\,\mathbf{j} + (z + xy\cos xyz)\,\mathbf{k}$; $\mathbf{r}(t) = 2\cos t\,\mathbf{i} + 3\sin t\,\mathbf{j} + \mathbf{k}$, $0 \le t \le 2\pi$

50. $\mathbf{F} = 2xy\,\mathbf{i} - y^2\,\mathbf{j} + ze^x\,\mathbf{k}$; $\mathbf{r}(t) = -t\,\mathbf{i} + \sqrt{t}\,\mathbf{j} + 3t\,\mathbf{k}$, $1 \le t \le 4$

51. $\mathbf{F} = (2y + \sin x)\,\mathbf{i} + (z^2 + (1/3)\cos y)\,\mathbf{j} + x^4\,\mathbf{k}$; $\mathbf{r}(t) = \sin t\,\mathbf{i} + \cos t\,\mathbf{j} + \sin 2t\,\mathbf{k}$, $-\pi/2 \le t \le \pi/2$

52. $\mathbf{F} = (x^2y)\,\mathbf{i} + \dfrac{1}{3}x^3\,\mathbf{j} + xy\,\mathbf{k}$; $\mathbf{r}(t) = \cos t\,\mathbf{i} + \sin t\,\mathbf{j} + (2\sin^2(t) - 1)\,\mathbf{k}$, $0 \le t \le 2\pi$

14.3

Path Independence, Potential Functions, and Conservative Fields

In gravitational and electric fields, the amount of work it takes to move a mass or a charge from one point to another depends only on the object's initial and final positions and not on the path taken in between. This section discusses the notion of path independence of work integrals and describes the remarkable properties of fields in which work integrals are path independent.

Path Independence

If A and B are two points in an open region D in space, the work $\int \mathbf{F} \cdot d\mathbf{r}$ done in moving a particle from A to B by a field $\mathbf{F}$ defined on D usually depends on the path taken. For some special fields, however, the integral's value is the same for all paths from A to B. If this is true for all points A and B in D, we say that the integral $\int \mathbf{F} \cdot d\mathbf{r}$ is path independent in D and that $\mathbf{F}$ is conservative on D.

Definitions

Let **F** be a field defined on an open region D in space, and suppose that for any two points A and B in D the work $\int_A^B \mathbf{F} \cdot d\mathbf{r}$ done in moving from A to B is the same over all paths from A to B. Then the integral $\int \mathbf{F} \cdot d\mathbf{r}$ is **path independent in D** and the field **F** is **conservative on D.**

The word *conservative* comes from physics, where it refers to fields in which the principle of conservation of energy holds (it does, in conservative fields).

Under conditions normally met in practice, a field **F** is conservative if and only if it is the gradient field of a scalar function f; that is, if and only if $\mathbf{F} = \nabla f$ for some f. The function f is then called a potential function for **F**.

Definition

If **F** is a field defined on D and $\mathbf{F} = \nabla f$ for some scalar function f on D, then f is called a **potential function** for **F**.

An electric potential is a scalar function whose gradient field is an electric field. A gravitational potential is a scalar function whose gradient field is a gravitational field, and so on. As we will see, once we have found a potential function f for a field **F,** we can evaluate all the work integrals in the domain of **F** by

$$\int_A^B \mathbf{F} \cdot d\mathbf{r} = \int_A^B \nabla f \cdot d\mathbf{r} = f(B) - f(A). \qquad (1)$$

If you think of ∇f for functions of several variables as being something like the derivative f' for functions of a single variable, then you see that Eq. (1) is the vector calculus analogue of the Fundamental Theorem of Calculus formula

$$\int_a^b f'(x) \, dx = f(b) - f(a).$$

Conservative fields have other remarkable properties we will study as we go along. For example, saying that **F** is conservative on D is equivalent to saying that the integral of **F** around every closed path in D is zero. Naturally, we need to impose conditions on the curves, fields, and domains to make Eq. (1) and its implications hold.

We assume that all curves are **piecewise smooth,** i.e., made up of finitely many smooth pieces connected end to end, as discussed in Section 11.1. We also assume that the components of **F** have continuous first partial derivatives. When $\mathbf{F} = \nabla f$, this continuity requirement guarantees that the mixed second derivatives of the potential function f are equal, a result we will find revealing in studying conservative fields **F.**

We assume D to be an *open* region in space. This means that every point in D is the center of a ball that lies entirely in D. We also assume D to be **connected,** which in an open region means that every point can be connected to every other point by a smooth curve that lies in the region.

Line Integrals in Conservative Fields

The following result provides a convenient way to evaluate a line integral in a conservative field. The result establishes that the value of the integral depends only on the endpoints and not on the specific path joining them.

> **Theorem 1**
> **The Fundamental Theorem of Line Integrals**
> 1. Let $\mathbf{F} = M\mathbf{i} + N\mathbf{j} + P\mathbf{k}$ be a vector field whose components are continuous throughout an open connected region D in space. Then there exists a differentiable function f such that
>
> $$\mathbf{F} = \nabla f = \frac{\partial f}{\partial x}\mathbf{i} + \frac{\partial f}{\partial y}\mathbf{j} + \frac{\partial f}{\partial z}\mathbf{k}$$
>
> if and only if for all points A and B in D the value of $\int_A^B \mathbf{F} \cdot d\mathbf{r}$ is independent of the path joining A to B in D.
> 2. If the integral is independent of the path from A to B, its value is
>
> $$\int_A^B \mathbf{F} \cdot d\mathbf{r} = f(B) - f(A).$$

Proof That $\mathbf{F} = \nabla f$ Implies Path Independence of the Integral   Suppose that A and B are two points in D and that C: $\mathbf{r}(t) = g(t)\mathbf{i} + h(t)\mathbf{j} + k(t)\mathbf{k}$, $a \leq t \leq b$, is a smooth curve in D joining A and B. Along the curve, f is a differentiable function of t and

$$\frac{df}{dt} = \frac{\partial f}{\partial x}\frac{dx}{dt} + \frac{\partial f}{\partial y}\frac{dy}{dt} + \frac{\partial f}{\partial z}\frac{dz}{dt} \qquad \text{Chain Rule}$$

$$= \nabla f \cdot \left(\frac{dx}{dt}\mathbf{i} + \frac{dy}{dt}\mathbf{j} + \frac{dz}{dt}\mathbf{k}\right) = \nabla f \cdot \frac{d\mathbf{r}}{dt} = \mathbf{F} \cdot \frac{d\mathbf{r}}{dt}. \qquad \text{Because } \mathbf{F} = \nabla f \qquad (2)$$

Therefore, $\displaystyle \int_C \mathbf{F} \cdot d\mathbf{r} = \int_{t=a}^{t=b} \mathbf{F} \cdot \frac{d\mathbf{r}}{dt}\, dt = \int_a^b \frac{df}{dt}\, dt \qquad \text{Eq. (2)}$

$$= f(g(t), h(t), k(t)) \Big]_a^b = f(B) - f(A).$$

Thus, the value of the work integral depends only on the values of f at A and B and not on the path in between. This proves Part 2 as well as the forward implication in Part 1. We omit the more technical proof of the reverse implication.   ❏

EXAMPLE 1   Find the work done by the conservative field

$$\mathbf{F} = yz\mathbf{i} + xz\mathbf{j} + xy\mathbf{k} = \nabla(xyz)$$

along any smooth curve C joining the point $(-1, 3, 9)$ to $(1, 6, -4)$.

Solution   With $f(x, y, z) = xyz$, we have

$$\int_A^B \mathbf{F} \cdot d\mathbf{r} = \int_A^B \nabla f \cdot d\mathbf{r} \qquad \mathbf{F} = \nabla f$$

$$= f(B) - f(A) \qquad \text{Fundamental Theorem, Part 2}$$

$$= xyz\Big|_{(1,6,-4)} - xyz\Big|_{(-1,3,9)}$$

$$= (1)(6)(-4) - (-1)(3)(9)$$

$$= -24 + 27 = 3.$$

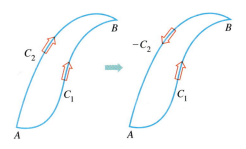

14.23 If we have two paths from A to B, one of them can be reversed to make a loop.

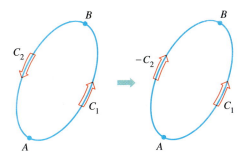

14.24 If A and B lie on a loop, we can reverse part of the loop to make two paths from A to B.

Theorem 2

The following statements are equivalent:

1. $\int \mathbf{F} \cdot d\mathbf{r} = 0$ around every closed loop in D.
2. The field $\mathbf{F}$ is conservative on D.

Proof That **(1)** $\Rightarrow$ **(2)**   We want to show that for any two points A and B in D the integral of $\mathbf{F} \cdot d\mathbf{r}$ has the same value over any two paths C_1 and C_2 from A to B. We reverse the direction on C_2 to make a path $-C_2$ from B to A (Fig. 14.23). Together, C_1 and $-C_2$ make a closed loop C, and

$$\int_{C_1} \mathbf{F} \cdot d\mathbf{r} - \int_{C_2} \mathbf{F} \cdot d\mathbf{r} = \int_{C_1} \mathbf{F} \cdot d\mathbf{r} + \int_{-C_2} \mathbf{F} \cdot d\mathbf{r} = \int_{C} \mathbf{F} \cdot d\mathbf{r} = 0.$$

Thus the integrals over C_1 and C_2 give the same value.

Proof That **(2)** $\Rightarrow$ **(1)**   We want to show that the integral of $\mathbf{F} \cdot d\mathbf{r}$ is zero over any closed loop C. We pick two points A and B on C and use them to break C into two pieces: C_1 from A and B followed by C_2 from B back to A (Fig. 14.24). Then

$$\oint_{C} \mathbf{F} \cdot d\mathbf{r} = \int_{C_1} \mathbf{F} \cdot d\mathbf{r} + \int_{C_2} \mathbf{F} \cdot d\mathbf{r} = \int_{A}^{B} \mathbf{F} \cdot d\mathbf{r} - \int_{A}^{B} \mathbf{F} \cdot d\mathbf{r} = 0. \qquad \square$$

The following diagram summarizes the results of Theorems 1 and 2.

$$\underset{\mathbf{F} = \nabla f \text{ on } D}{} \quad \overset{\text{Theorem 1}}{\Leftrightarrow} \quad \underset{\substack{\mathbf{F} \text{ conservative} \\ \text{on } D}}{} \quad \overset{\text{Theorem 2}}{\Leftrightarrow} \quad \underset{\substack{\oint_{C} \mathbf{F} \cdot d\mathbf{r} = 0 \\ \text{over any closed} \\ \text{path in } D}}{}$$

Now that we see how convenient it is to evaluate line integrals in conservative fields, two questions remain:

1. How do we know when a given field $\mathbf{F}$ is conservative?
2. If $\mathbf{F}$ is in fact conservative, how do we find a potential function f (so that $\mathbf{F} = \nabla f$)?

Finding Potentials for Conservative Fields

The test for being conservative is this:

The Component Test for Conservative Fields

Let $\mathbf{F} = M(x, y, z)\,\mathbf{i} + N(x, y, z)\,\mathbf{j} + P(x, y, z)\,\mathbf{k}$ be a field whose component functions have continuous first partial derivatives. Then, $\mathbf{F}$ is conservative if and only if

$$\frac{\partial P}{\partial y} = \frac{\partial N}{\partial z}, \qquad \frac{\partial M}{\partial z} = \frac{\partial P}{\partial x}, \qquad \text{and} \qquad \frac{\partial N}{\partial x} = \frac{\partial M}{\partial y}. \qquad (3)$$

Proof We show that Eqs. (3) must hold if **F** is conservative. There is a potential function f such that

$$\mathbf{F} = M\,\mathbf{i} + N\,\mathbf{j} + P\,\mathbf{k} = \frac{\partial f}{\partial x}\mathbf{i} + \frac{\partial f}{\partial y}\mathbf{j} + \frac{\partial f}{\partial z}\mathbf{k}.$$

Hence

$$\frac{\partial P}{\partial y} = \frac{\partial}{\partial y}\left(\frac{\partial f}{\partial z}\right) = \frac{\partial^2 f}{\partial y\,\partial z}$$

$$= \frac{\partial^2 f}{\partial z\,\partial y} \qquad \text{\color{blue}Continuity implies that the mixed partial derivatives are equal.}$$

$$= \frac{\partial}{\partial z}\left(\frac{\partial f}{\partial y}\right) = \frac{\partial N}{\partial z}.$$

The other two equations in (3) are proved similarly.

The second half of the proof, that Eqs. (3) imply that **F** is conservative, is a consequence of Stokes's theorem, taken up in Section 14.7. ❏

When we know that **F** is conservative, we usually want to find a potential function for **F**. This requires solving the equation $\nabla f = \mathbf{F}$ or

$$\frac{\partial f}{\partial x}\mathbf{i} + \frac{\partial f}{\partial y}\mathbf{j} + \frac{\partial f}{\partial z}\mathbf{k} = M\,\mathbf{i} + N\,\mathbf{j} + P\,\mathbf{k}$$

for f. We accomplish this by integrating the three equations

$$\frac{\partial f}{\partial x} = M, \qquad \frac{\partial f}{\partial y} = N, \qquad \frac{\partial f}{\partial z} = P.$$

EXAMPLE 2 Show that $\mathbf{F} = (e^x \cos y + yz)\,\mathbf{i} + (xz - e^x \sin y)\,\mathbf{j} + (xy + z)\,\mathbf{k}$ is conservative and find a potential function for it.

Solution We apply the test in Eqs. (3) to

$$M = e^x \cos y + yz, \qquad N = xz - e^x \sin y, \qquad P = xy + z$$

and calculate

$$\frac{\partial P}{\partial y} = x = \frac{\partial N}{\partial z}, \qquad \frac{\partial M}{\partial z} = y = \frac{\partial P}{\partial x}, \qquad \frac{\partial N}{\partial x} = -e^x \sin y + z = \frac{\partial M}{\partial y}.$$

Together, these equalities tell us that there is a function f with $\nabla f = \mathbf{F}$.

We find f by integrating the equations

$$\frac{\partial f}{\partial x} = e^x \cos y + yz, \qquad \frac{\partial f}{\partial y} = xz - e^x \sin y, \qquad \frac{\partial f}{\partial z} = xy + z. \qquad (4)$$

We integrate the first equation with respect to x, holding y and z fixed, to get

$$f(x, y, z) = e^x \cos y + xyz + g(y, z).$$

We write the constant of integration as a function of y and z because its value may change if y and z change. We then calculate $\partial f/\partial y$ from this equation and match it with the expression for $\partial f/\partial y$ in Eq. (4). This gives

$$-e^x \sin y + xz + \frac{\partial g}{\partial y} = xz - e^x \sin y,$$

so $\partial g/\partial y = 0$. Therefore, g is a function of z alone, and

$$f(x, y, z) = e^x \cos y + xyz + h(z).$$

We now calculate $\partial f/\partial z$ from this equation and match it to the formula for $\partial f/\partial z$ in Eq. (4). This gives

$$xy + \frac{dh}{dz} = xy + z, \qquad \text{or} \qquad \frac{dh}{dz} = z,$$

so

$$h(z) = \frac{z^2}{2} + C.$$

Hence,

$$f(x, y, z) = e^x \cos y + xyz + \frac{z^2}{2} + C.$$

We have found infinitely many potential functions for **F,** one for each value of C. ❏

EXAMPLE 3   Show that $\mathbf{F} = (2x - 3)\,\mathbf{i} - z\,\mathbf{j} + (\cos z)\,\mathbf{k}$ is not conservative.

Solution   We apply the component test in Eqs. (3) and find right away that

$$\frac{\partial P}{\partial y} = \frac{\partial}{\partial y}(\cos z) = 0, \qquad \frac{\partial N}{\partial z} = \frac{\partial}{\partial z}(-z) = -1.$$

The two are unequal, so **F** is not conservative. No further testing is required. ❏

Exact Differential Forms

As we will see in the next section and again later on, it is often convenient to express work and circulation integrals in the "differential" form

$$\int_A^B M\,dx + N\,dy + P\,dz$$

mentioned in Section 14.2. Such integrals are relatively easy to evaluate if $M\,dx + N\,dy + P\,dz$ is the differential of a function f. For then

$$\int_A^B M\,dx + N\,dy + P\,dz = \int_A^B \frac{\partial f}{\partial x}\,dx + \frac{\partial f}{\partial y}\,dy + \frac{\partial f}{\partial z}\,dz$$

$$= \int_A^B \nabla f \cdot d\mathbf{r}$$

$$= f(B) - f(A). \qquad \text{Theorem 1}$$

Thus

$$\int_A^B df = f(B) - f(A),$$

just as with differentiable functions of a single variable.

> **Definitions**
>
> The form $M(x, y, z)\,dx + N(x, y, z)\,dy + P(x, y, z)\,dz$ is called a **differential form.** A differential form is **exact** on a domain D in space if
>
> $$M\,dx + N\,dy + P\,dz = \frac{\partial f}{\partial x}\,dx + \frac{\partial f}{\partial y}\,dy + \frac{\partial f}{\partial z}\,dz = df$$
>
> for some (scalar) function f throughout D.

Notice that if $M\,dx + N\,dy + P\,dz = df$ on D, then $\mathbf{F} = M\mathbf{i} + N\mathbf{j} + P\mathbf{k}$ is the gradient field of f on D. Conversely, if $\mathbf{F} = \nabla f$, then the form $M\,dx + N\,dy + P\,dz$ is exact. The test for the form's being exact is therefore the same as the test for $\mathbf{F}$'s being conservative.

The Test for Exactness of $M\,dx + N\,dy + P\,dz$

The differential form $M\,dx + N\,dy + P\,dz$ is exact if and only if

$$\frac{\partial P}{\partial y} = \frac{\partial N}{\partial z}, \qquad \frac{\partial M}{\partial z} = \frac{\partial P}{\partial x}, \qquad \text{and} \qquad \frac{\partial N}{\partial x} = \frac{\partial M}{\partial y}. \tag{5}$$

This is equivalent to saying that the field $\mathbf{F} = M\mathbf{i} + N\mathbf{j} + P\mathbf{k}$ is conservative.

EXAMPLE 4 Show that $y\,dx + x\,dy + 4\,dz$ is exact, and evaluate the integral

$$\int_{(1,1,1)}^{(2,3,-1)} y\,dx + x\,dy + 4\,dz$$

over the line segment from $(1, 1, 1)$ to $(2, 3, -1)$.

Solution We let $M = y$, $N = x$, $P = 4$ and apply the test of Eq. (5):

$$\frac{\partial P}{\partial y} = 0 = \frac{\partial N}{\partial z}, \qquad \frac{\partial M}{\partial z} = 0 = \frac{\partial P}{\partial x}, \qquad \frac{\partial N}{\partial x} = 1 = \frac{\partial M}{\partial y}.$$

These equalities tell us that $y\,dx + x\,dy + 4\,dz$ is exact, so

$$y\,dx + x\,dy + 4\,dz = df$$

for some function f, and the integral's value is $f(2, 3, -1) - f(1, 1, 1)$.

We find f up to a constant by integrating the equations

$$\frac{\partial f}{\partial x} = y, \qquad \frac{\partial f}{\partial y} = x, \qquad \frac{\partial f}{\partial z} = 4. \tag{6}$$

From the first equation we get

$$f(x, y, z) = xy + g(y, z).$$

The second equation tells us that

$$\frac{\partial f}{\partial y} = x + \frac{\partial g}{\partial y} = x, \qquad \text{or} \qquad \frac{\partial g}{\partial y} = 0.$$

Hence, g is a function of z alone, and

$$f(x, y, z) = xy + h(z).$$

The third of Eqs. (6) tells us that

$$\frac{\partial f}{\partial z} = 0 + \frac{dh}{dz} = 4, \qquad \text{or} \qquad h(z) = 4z + C.$$

Therefore, $\qquad\qquad\qquad f(x, y, z) = xy + 4z + C.$

The value of the integral is

$$f(2, 3, -1) - f(1, 1, 1) = 2 + C - (5 + C) = -3. \qquad \square$$

Exercises 14.3

Testing for Conservative Fields

Which fields in Exercises 1–6 are conservative, and which are not?

1. $\mathbf{F} = yz\,\mathbf{i} + xz\,\mathbf{j} + xy\,\mathbf{k}$

2. $\mathbf{F} = (y \sin z)\,\mathbf{i} + (x \sin z)\,\mathbf{j} + (xy \cos z)\,\mathbf{k}$

3. $\mathbf{F} = y\,\mathbf{i} + (x + z)\,\mathbf{j} - y\,\mathbf{k}$

4. $\mathbf{F} = -y\,\mathbf{i} + x\,\mathbf{j}$

5. $\mathbf{F} = (z + y)\,\mathbf{i} + z\,\mathbf{j} + (y + x)\,\mathbf{k}$

6. $\mathbf{F} = (e^x \cos y)\,\mathbf{i} - (e^x \sin y)\,\mathbf{j} + z\,\mathbf{k}$

Finding Potential Functions

In Exercises 7–12, find a potential function f for the field $\mathbf{F}$.

7. $\mathbf{F} = 2x\,\mathbf{i} + 3y\,\mathbf{j} + 4z\,\mathbf{k}$

8. $\mathbf{F} = (y + z)\,\mathbf{i} + (x + z)\,\mathbf{j} + (x + y)\,\mathbf{k}$

9. $\mathbf{F} = e^{y+2z}\,(\mathbf{i} + x\,\mathbf{j} + 2x\,\mathbf{k})$

10. $\mathbf{F} = (y \sin z)\,\mathbf{i} + (x \sin z)\,\mathbf{j} + (xy \cos z)\,\mathbf{k}$

11. $\mathbf{F} = (\ln x + \sec^2(x + y))\,\mathbf{i} +$
$$\left(\sec^2(x + y) + \frac{y}{y^2 + z^2}\right)\mathbf{j} + \frac{z}{y^2 + z^2}\,\mathbf{k}$$

12. $\mathbf{F} = \dfrac{y}{1 + x^2 y^2}\,\mathbf{i} + \left(\dfrac{x}{1 + x^2 y^2} + \dfrac{z}{\sqrt{1 - y^2 z^2}}\right)\mathbf{j} +$
$$\left(\frac{y}{\sqrt{1 - y^2 z^2}} + \frac{1}{z}\right)\mathbf{k}$$

Evaluating Line Integrals

In Exercises 13–22, show that the differential forms in the integrals are exact. Then evaluate the integrals.

13. $\displaystyle\int_{(0,0,0)}^{(2,3,-6)} 2x\,dx + 2y\,dy + 2z\,dz$

14. $\displaystyle\int_{(1,1,2)}^{(3,5,0)} yz\,dx + xz\,dy + xy\,dz$

15. $\displaystyle\int_{(0,0,0)}^{(1,2,3)} 2xy\,dx + (x^2 - z^2)\,dy - 2yz\,dz$

16. $\displaystyle\int_{(0,0,0)}^{(3,3,1)} 2x\,dx - y^2\,dy - \frac{4}{1 + z^2}\,dz$

17. $\displaystyle\int_{(1,0,0)}^{(0,1,1)} \sin y \cos x\,dx + \cos y \sin x\,dy + dz$

18. $\displaystyle\int_{(0,2,1)}^{(1,\pi/2,2)} 2 \cos y\,dx + \left(\frac{1}{y} - 2x \sin y\right)dy + \frac{1}{z}\,dz$

19. $\displaystyle\int_{(1,1,1)}^{(1,2,3)} 3x^2\,dx + \frac{z^2}{y}\,dy + 2z \ln y\,dz$

20. $\displaystyle\int_{(1,2,1)}^{(2,1,1)} (2x \ln y - yz)\,dx + \left(\frac{x^2}{y} - xz\right)dy - xy\,dz$

21. $\displaystyle\int_{(1,1,1)}^{(2,2,2)} \frac{1}{y}\,dx + \left(\frac{1}{z} - \frac{x}{y^2}\right)dy - \frac{y}{z^2}\,dz$

22. $\displaystyle\int_{(-1,-1,-1)}^{(2,2,2)} \frac{2x\,dx + 2y\,dy + 2z\,dz}{x^2 + y^2 + z^2}$

23. Evaluate the integral
$$\int_{(1,1,1)}^{(2,3,-1)} y\,dx + x\,dy + 4\,dz$$

from Example 4 by finding parametric equations for the line segment from $(1, 1, 1)$ to $(2, 3, -1)$ and evaluating the line integral of $\mathbf{F} = y\,\mathbf{i} + x\,\mathbf{j} + 4\,\mathbf{k}$ along the segment. Since $\mathbf{F}$ is conservative, the integral is independent of the path.

24. Evaluate $\displaystyle\int_C x^2\,dx + yz\,dy + (y^2/2)\,dz$

along the line segment C joining $(0, 0, 0)$ to $(0, 3, 4)$.

Theory, Applications, and Examples

Show that the values of the integrals in Exercises 25 and 26 do not depend on the path taken from A to B.

25. $\displaystyle\int_A^B z^2\,dx + 2y\,dy + 2xz\,dz$

26. $\displaystyle\int_A^B \frac{x\,dx + y\,dy + z\,dz}{\sqrt{x^2 + y^2 + z^2}}$

In Exercises 27 and 28, express $\mathbf{F}$ in the form ∇f.

27. $\mathbf{F} = \dfrac{2x}{y}\,\mathbf{i} + \left(\dfrac{1 - x^2}{y^2}\right)\mathbf{j}$

28. $\mathbf{F} = (e^x \ln y)\,\mathbf{i} + \left(\dfrac{e^x}{y} + \sin z\right)\mathbf{j} + (y \cos z)\,\mathbf{k}$

29. Find the work done by $\mathbf{F} = (x^2 + y)\,\mathbf{i} + (y^2 + x)\,\mathbf{j} + ze^z\,\mathbf{k}$ over the following paths from $(1, 0, 0)$ to $(1, 0, 1)$.

 a) The line segment $x = 1, y = 0, 0 \leq z \leq 1$

 b) The helix $\mathbf{r}(t) = (\cos t)\,\mathbf{i} + (\sin t)\,\mathbf{j} + (t/2\pi)\,\mathbf{k}, 0 \leq t \leq 2\pi$

 c) The x-axis from $(1, 0, 0)$ to $(0, 0, 0)$ followed by the parabola $z = x^2, y = 0$ from $(0, 0, 0)$ to $(1, 0, 1)$

30. Find the work done by $\mathbf{F} = e^{yz}\,\mathbf{i} + (xz\,e^{yz} + z \cos y)\,\mathbf{j} + (xy\,e^{yz} + \sin y)\,\mathbf{k}$ over the following paths from $(1, 0, 1)$ to $(1, \pi/2, 0)$.

 a) The line segment $x = 1, y = \pi t/2, z = 1 - t, 0 \leq t \leq 1$

 b) The line segment from $(1, 0, 1)$ to the origin followed by the line segment from the origin to $(1, \pi/2, 0)$

 c) The line segment from $(1, 0, 1)$ to $(1, 0, 0)$, followed by the x-axis from $(1, 0, 0)$ to the origin, followed by the parabola $y = \pi x^2/2, z = 0$

31. Let $\mathbf{F} = \nabla(x^3 y^2)$ and let C be the path in the xy-plane from $(-1, 1)$ to $(1, 1)$ that consists of the line segment from $(-1, 1)$ to $(0, 0)$ followed by the line segment from $(0, 0)$ to $(1, 1)$. Evaluate $\int_C \mathbf{F} \cdot d\mathbf{r}$ in two ways:

a) Find parametrizations for the segments that make up C and evaluate the integral;

b) Use the fact that $f(x, y) = x^3 y^2$ is a potential function for $\mathbf{F}$.

32. Evaluate $\int_C 2x \cos y \, dx - x^2 \sin y \, dy$ along the following paths C in the xy-plane.

a) The parabola $y = (x - 1)^2$ from $(1, 0)$ to $(0, 1)$

b) The line segment from $(-1, \pi)$ to $(1, 0)$

c) The x-axis from $(-1, 0)$ to $(1, 0)$

d) The astroid $\mathbf{r}(t) = (\cos^3 t)\mathbf{i} + (\sin^3 t)\mathbf{j}$, $0 \le t \le 2\pi$, counterclockwise from $(1, 0)$ back to $(1, 0)$.

33. Find a potential function for the gravitational field

$$\mathbf{F} = -GmM \frac{x\mathbf{i} + y\mathbf{j} + z\mathbf{k}}{(x^2 + y^2 + z^2)^{3/2}} \qquad (G, m, \text{ and } M \text{ are constants}).$$

34. (*Continuation of Exercise 33.*) Let P_1 and P_2 be points at distances s_1 and s_2 from the origin. Show that the work done by the gravitational field in Exercise 33 in moving a particle from P_1 to P_2 is the quantity

$$GmM \left(\frac{1}{s_2} - \frac{1}{s_1} \right).$$

35. a) How are the constants $a, b,$ and c related if the following differential form is exact?

$$(ay^2 + 2czx) \, dx + y(bx + cz) \, dy + (ay^2 + cx^2) \, dz$$

b) For what values of b and c will

$$\mathbf{F} = (y^2 + 2czx)\mathbf{i} + y(bx + cz)\mathbf{j} + (y^2 + cx^2)\mathbf{k}$$

be a gradient field?

36. Suppose that $\mathbf{F} = \nabla f$ is a conservative vector field and

$$g(x, y, z) = \int_{(0,0,0)}^{(x,y,z)} \mathbf{F} \cdot d\mathbf{r}.$$

Show that $\nabla g = \mathbf{F}$.

37. You have been asked to find the path along which a force field $\mathbf{F}$ will perform the least work in moving a particle between two locations. A quick calculation on your part shows $\mathbf{F}$ to be conservative. How should you respond? Give reasons for your answer.

38. By experiment, you find that a force field $\mathbf{F}$ performs only half as much work in moving an object along path C_1 from A to B as it does in moving the object along path C_2 from A to B. What can you conclude about $\mathbf{F}$? Give reasons for your answer.

14.4

Green's Theorem in the Plane

We now come to a theorem that can be used to describe the relationship between the way an incompressible fluid flows along or across the boundary of a plane region and the way it moves inside the region. The connection between the fluid's boundary behavior and its internal behavior is made possible by the notions of divergence and curl. The divergence of a fluid's velocity field measures the rate at which fluid is being piped into or out of the region at any given point. The curl measures the fluid's rate of rotation at each point.

Green's theorem states that, under conditions usually met in practice, the outward flux of a vector field across the boundary of a plane region equals the double integral of the divergence of the field over the interior of the region. In another form, it states that the counterclockwise circulation of a field around the boundary of a region equals the double integral of the curl of the field over the region.

Green's theorem is one of the great theorems of calculus. It is deep and surprising and has far-reaching consequences. In pure mathematics, it ranks in importance with the Fundamental Theorem of Calculus. In applied mathematics, the generalizations of Green's theorem to three dimensions provide the foundation for theorems about electricity, magnetism, and fluid flow.

We talk in terms of velocity fields of fluid flows because fluid flows are easy to picture. We would like you to be aware, however, that Green's theorem applies to any vector field satisfying certain mathematical conditions. It does not depend for its validity on the field's having a particular physical interpretation.

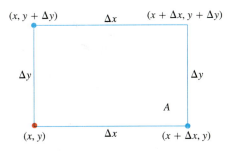

$(x, y + \Delta y)$ $\quad \Delta x \quad$ $(x + \Delta x, y + \Delta y)$

Δy $\qquad$ Δy

A

(x, y) $\quad \Delta x \quad$ $(x + \Delta x, y)$

14.25 The rectangle for defining the flux density (divergence) of a vector field at a point (x, y).

Flux Density at a Point: Divergence

We need two new ideas for Green's theorem. The first is the idea of the flux density of a vector field at a point, which in mathematics is called the *divergence* of the vector field. We obtain it in the following way.

Suppose that $\mathbf{F}(x, y) = M(x, y)\mathbf{i} + N(x, y)\mathbf{j}$ is the velocity field of a fluid flow in the plane and that the first partial derivatives of M and N are continuous at each point of a region R. Let (x, y) be a point in R and let A be a small rectangle with one corner at (x, y) that, along with its interior, lies entirely in R (Fig. 14.25). The sides of the rectangle, parallel to the coordinate axes, have lengths of Δx and Δy. The rate at which fluid leaves the rectangle across the bottom edge is approximately

$$\mathbf{F}(x, y) \cdot (-\mathbf{j})\Delta x = -N(x, y)\Delta x. \tag{1}$$

This is the scalar component of the velocity at (x, y) in the direction of the outward normal times the length of the segment. If the velocity is in meters per second, for example, the exit rate will be in meters per second times meters or square meters per second. The rates at which the fluid crosses the other three sides in the directions of their outward normals can be estimated in a similar way. All told, we have

Top: $\quad \mathbf{F}(x, y + \Delta y) \cdot \mathbf{j}\, \Delta x = N(x, y + \Delta y)\Delta x$
Bottom: $\quad \mathbf{F}(x, y) \cdot (-\mathbf{j})\, \Delta x = -N(x, y)\Delta x$
Right: $\quad \mathbf{F}(x + \Delta x, y) \cdot \mathbf{i}\, \Delta y = M(x + \Delta x, y)\, \Delta y$
Left: $\quad \mathbf{F}(x, y) \cdot (-\mathbf{i})\, \Delta y = -M(x, y)\, \Delta y.$

$$\tag{2}$$

Combining opposite pairs gives

Top and bottom: $\quad (N(x, y + \Delta y) - N(x, y))\, \Delta x \approx \left(\dfrac{\partial N}{\partial y}\Delta y \right) \Delta x \tag{3}$

Right and left: $\quad (M(x + \Delta x, y) - M(x, y))\, \Delta y \approx \left(\dfrac{\partial M}{\partial x}\Delta x \right) \Delta y. \tag{4}$

Adding (3) and (4) gives

$$\text{Flux across rectangle boundary} \approx \left(\dfrac{\partial M}{\partial x} + \dfrac{\partial N}{\partial y} \right) \Delta x\, \Delta y.$$

We now divide by $\Delta x\, \Delta y$ to estimate the total flux per unit area or flux density for the rectangle:

$$\dfrac{\text{Flux across rectangle boundary}}{\text{Rectangle area}} \approx \left(\dfrac{\partial M}{\partial x} + \dfrac{\partial N}{\partial y} \right).$$

Finally, we let Δx and Δy approach zero to define what we call the *flux density* of $\mathbf{F}$ at the point (x, y).

In mathematics, we call the flux density the *divergence* of $\mathbf{F}$. The symbol for it is div $\mathbf{F}$, pronounced "divergence of $\mathbf{F}$" or "div $\mathbf{F}$."

Definition

The **flux density** or **divergence** of a vector field $\mathbf{F} = M\mathbf{i} + N\mathbf{j}$ at the point (x, y) is

$$\text{div } \mathbf{F} = \dfrac{\partial M}{\partial x} + \dfrac{\partial N}{\partial y}. \tag{5}$$

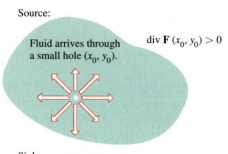

Source:

Fluid arrives through a small hole (x_0, y_0).

$\text{div } \mathbf{F}\,(x_0, y_0) > 0$

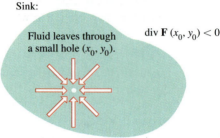

Sink:

Fluid leaves through a small hole (x_0, y_0).

$\text{div } \mathbf{F}\,(x_0, y_0) < 0$

14.26 In the flow of an incompressible fluid across a plane region, the divergence is positive at a "source," a point where fluid enters the system, and negative at a "sink," a point where the fluid leaves the system.

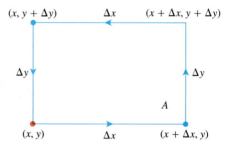

14.27 The rectangle for defining the circulation density (curl) of a vector field at a point (x, y).

Intuitively, if water were flowing into a region through a small hole at the point (x_0, y_0), the lines of flow would diverge there (hence the name) and, since water would be flowing out of a small rectangle about (x_0, y_0), the divergence of $\mathbf{F}$ at (x_0, y_0) would be positive. If the water were draining out instead of flowing in, the divergence would be negative. See Fig. 14.26.

EXAMPLE 1 Find the divergence of $\mathbf{F}(x, y) = (x^2 - y)\mathbf{i} + (xy - y^2)\mathbf{j}$.

Solution We use the formula in Eq. (5):

$$\text{div } \mathbf{F} = \frac{\partial M}{\partial x} + \frac{\partial N}{\partial y} = \frac{\partial}{\partial x}(x^2 - y) + \frac{\partial}{\partial y}(xy - y^2)$$

$$= 2x + x - 2y = 3x - 2y. \qquad \square$$

Circulation Density at a Point: The Curl

The second of the two new ideas we need for Green's theorem is the idea of circulation density of a vector field $\mathbf{F}$ at a point, which in mathematics is called the *curl* of $\mathbf{F}$. To obtain it, we return to the velocity field

$$\mathbf{F}(x, y) = M(x, y)\mathbf{i} + N(x, y)\mathbf{j}$$

and the rectangle A. The rectangle is redrawn here as Fig. 14.27.

The counterclockwise circulation of $\mathbf{F}$ around the boundary of A is the sum of flow rates along the sides. For the bottom edge, the flow rate is approximately

$$\mathbf{F}(x, y) \cdot \mathbf{i}\, \Delta x = M(x, y)\, \Delta x. \tag{6}$$

This is the scalar component of the velocity $\mathbf{F}(x, y)$ in the direction of the tangent vector $\mathbf{i}$ times the length of the segment. The rates of flow along the other sides in the counterclockwise direction are expressed in a similar way. In all, we have

Top: $\quad \mathbf{F}(x, y + \Delta y) \cdot (-\mathbf{i})\, \Delta x = -M(x, y + \Delta y)\, \Delta x$

Bottom: $\quad \mathbf{F}(x, y) \cdot \mathbf{i}\, \Delta x = M(x, y)\, \Delta x$

Right: $\quad \mathbf{F}(x + \Delta x, y) \cdot \mathbf{j}\, \Delta y = N(x + \Delta x, y)\, \Delta y$

Left: $\quad \mathbf{F}(x, y) \cdot (-\mathbf{j})\, \Delta y = -N(x, y)\, \Delta y.$
$\tag{7}$

We add opposite pairs to get

Top and bottom:

$$-(M(x, y + \Delta y) - M(x, y))\, \Delta x \approx -\left(\frac{\partial M}{\partial y}\Delta y\right)\Delta x \tag{8}$$

Right and left:

$$(N(x + \Delta x, y) - N(x, y))\, \Delta y \approx \left(\frac{\partial N}{\partial x}\Delta x\right)\Delta y. \tag{9}$$

Adding (8) and (9) and dividing by $\Delta x \Delta y$ gives an estimate of the circulation density for the rectangle:

$$\frac{\text{Circulation around rectangle}}{\text{Rectangle area}} \approx \frac{\partial N}{\partial x} - \frac{\partial M}{\partial y}.$$

Finally, we let Δx and Δy approach zero to define what we call the *circulation density* of $\mathbf{F}$ at the point (x, y).

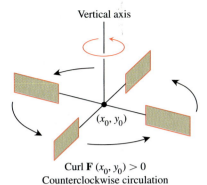

Vertical axis

Curl **F** $(x_0, y_0) > 0$
Counterclockwise circulation

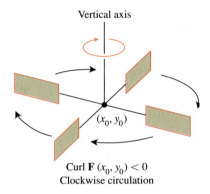

Vertical axis

Curl **F** $(x_0, y_0) < 0$
Clockwise circulation

14.28 In the flow of an incompressible fluid over a plane region, the curl measures the rate of the fluid's rotation at a point. The curl is positive at points where the rotation is counterclockwise and negative where the rotation is clockwise.

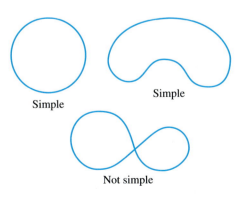

Simple

Simple

Not simple

14.29 In proving Green's theorem, we distinguish between two kinds of closed curves, simple and not simple. Simple curves do not cross themselves. A circle is simple but a figure 8 is not.

Definition

The **circulation density** or **curl** of a vector field $\mathbf{F} = M\,\mathbf{i} + N\,\mathbf{j}$ at the point (x, y) is

$$\operatorname{curl} \mathbf{F} = \frac{\partial N}{\partial x} - \frac{\partial M}{\partial y}. \tag{10}$$

If water is moving about a region in the xy-plane in a thin layer, then the circulation, or curl, at a point (x_0, y_0) gives a way to measure how fast and in what direction a small paddle wheel will spin if it is put into the water at (x_0, y_0) with its axis perpendicular to the plane (Fig. 14.28).

EXAMPLE 2   Find the curl of the vector field

$$\mathbf{F}(x, y) = (x^2 - y)\,\mathbf{i} + (xy - y^2)\,\mathbf{j}.$$

Solution   We use the formula in Eq. (10):

$$\operatorname{curl} \mathbf{F} = \frac{\partial N}{\partial x} - \frac{\partial M}{\partial y} = \frac{\partial}{\partial x}(xy - y^2) - \frac{\partial}{\partial y}(x^2 - y) = y + 1. \quad \square$$

Green's Theorem in the Plane

In one form, Green's theorem says that under suitable conditions the outward flux of a vector field across a simple closed curve in the plane (Fig. 14.29) equals the double integral of the divergence of the field over the region enclosed by the curve. Recall the formulas for flux in Eqs. (3) and (4) in Section 14.2.

Theorem 3
Green's Theorem (Flux-Divergence or Normal Form)

The outward flux of a field $\mathbf{F} = M\,\mathbf{i} + N\,\mathbf{j}$ across a simple closed curve C equals the double integral of div $\mathbf{F}$ over the region R enclosed by C.

$$\underbrace{\oint_C \mathbf{F} \cdot \mathbf{n}\,ds = \oint_C M\,dy - N\,dx}_{\text{outward flux}} = \underbrace{\iint_R \left(\frac{\partial M}{\partial x} + \frac{\partial N}{\partial y} \right) dx\,dy}_{\text{divergence integral}} \tag{11}$$

In another form, Green's theorem says that the counterclockwise circulation of a vector field around a simple closed curve is the double integral of the curl of the field over the region enclosed by the curve.

Theorem 4
Green's Theorem (Circulation-Curl or Tangential Form)

The counterclockwise circulation of a field $\mathbf{F} = M\,\mathbf{i} + N\,\mathbf{j}$ around a simple closed curve C in the plane equals the double integral of curl $\mathbf{F}$ over the

(continued)

For a two-dimensional field $\mathbf{F} = M\,\mathbf{i} + N\,\mathbf{j}$, the integral in Eq. (2), Section 14.2, for circulation takes the equivalent form

$$\oint_C \mathbf{F} \cdot \mathbf{T}\,ds = \oint_C M\,dx + N\,dy.$$

region R enclosed by C.

$$\underset{\substack{\text{counterclockwise}\\\text{circulation}}}{\oint_C \mathbf{F} \cdot \mathbf{T}\,ds} = \oint_C M\,dx + N\,dy = \underset{\text{curl integral}}{\iint_R \left(\frac{\partial N}{\partial x} - \frac{\partial M}{\partial y}\right) dx\,dy} \qquad (12)$$

The two forms of Green's theorem are equivalent. Applying Eq. (11) to the field $\mathbf{G}_1 = N\,\mathbf{i} - M\,\mathbf{j}$ gives Eq. (12), and applying Eq. (12) to $\mathbf{G}_2 = -N\,\mathbf{i} + M\,\mathbf{j}$ gives Eq. (11).

We need two kinds of assumptions for Green's theorem to hold. First, we need conditions on M and N to ensure the existence of the integrals. The usual assumptions are that M, N, and their first partial derivatives are continuous at every point of some open region containing C and R. Second, we need geometric conditions on the curve C. It must be simple, closed, and made up of pieces along which we can integrate M and N. The usual assumptions are that C is piecewise smooth. The proof we give for Green's theorem, however, assumes things about the shape of R as well. You can find proofs that are less restrictive in more advanced texts. First let's look at some examples.

EXAMPLE 3   Verify both forms of Green's theorem for the field

$$\mathbf{F}(x, y) = (x - y)\,\mathbf{i} + x\,\mathbf{j}$$

and the region R bounded by the unit circle

$$C: \qquad \mathbf{r}(t) = (\cos t)\,\mathbf{i} + (\sin t)\,\mathbf{j}, \qquad 0 \le t \le 2\pi.$$

Solution   We first express all functions, derivatives, and differentials in terms of t:

$$M = \cos t - \sin t, \qquad dx = d(\cos t) = -\sin t\,dt,$$

$$N = \cos t, \qquad dy = d(\sin t) = \cos t\,dt,$$

$$\frac{\partial M}{\partial x} = 1, \qquad \frac{\partial M}{\partial y} = -1, \qquad \frac{\partial N}{\partial x} = 1, \qquad \frac{\partial N}{\partial y} = 0.$$

The two sides of Eq. (11):

$$\oint_C M\,dy - N\,dx = \int_{t=0}^{t=2\pi} (\cos t - \sin t)(\cos t\,dt) - (\cos t)(-\sin t\,dt)$$

$$= \int_0^{2\pi} \cos^2 t\,dt = \pi$$

$$\iint_R \left(\frac{\partial M}{\partial x} + \frac{\partial N}{\partial y}\right) dx\,dy = \iint_R (1 + 0)\,dx\,dy$$

$$= \iint_R dx\,dy = \text{area of unit circle} = \pi.$$

The two sides of Eq. (12):

$$\oint_C M\,dx + N\,dy = \int_{t=0}^{t=2\pi} (\cos t - \sin t)(-\sin t\,dt) + (\cos t)(\cos t\,dt)$$

$$= \int_0^{2\pi} (-\sin t \cos t + 1)\,dt = 2\pi$$

$$\iint_R \left(\frac{\partial N}{\partial x} - \frac{\partial M}{\partial y}\right) dx\,dy = \iint_R (1 - (-1))\,dx\,dy = 2\iint_R dx\,dy = 2\pi.$$

Using Green's Theorem to Evaluate Line Integrals

If we construct a closed curve C by piecing a number of different curves end to end, the process of evaluating a line integral over C can be lengthy because there are so many different integrals to evaluate. However, if C bounds a region R to which Green's theorem applies, we can use Green's theorem to change the line integral around C into one double integral over R.

EXAMPLE 4   Evaluate the integral

$$\oint_C xy\,dy - y^2\,dx,$$

where C is the square cut from the first quadrant by the lines $x = 1$ and $y = 1$.

Solution   We can use either form of Green's theorem to change the line integral into a double integral over the square.

1. *With Eq. (11):* Taking $M = xy$, $N = y^2$, and C and R as the square's boundary and interior gives

$$\oint_C xy\,dy - y^2\,dx = \iint_R (y + 2y)\,dx\,dy = \int_0^1 \int_0^1 3y\,dx\,dy$$

$$= \int_0^1 \left[3xy\right]_{x=0}^{x=1} dy = \int_0^1 3y\,dy = \frac{3}{2}y^2\Big]_0^1 = \frac{3}{2}.$$

2. *With Eq. (12):* Taking $M = -y^2$ and $N = xy$ gives the same result:

$$\oint_C -y^2\,dx + xy\,dy = \iint_R (y - (-2y))\,dx\,dy = \frac{3}{2}.$$

The Green of Green's Theorem

The Green of Green's theorem was George Green (1793–1841), a self-taught scientist in Nottingham, England. Green's work on the mathematical foundations of gravitation, electricity, and magnetism was published privately in 1828 in a short book entitled *An Essay on the Application of Mathematical Analysis to Electricity and Magnetism.* The book sold all of fifty-two copies (fewer than one hundred were printed), the copies going mostly to Green's patrons and personal friends. A few weeks before Green's death in 1841, William Thomson noticed a reference to Green's book and in 1845 was finally able to locate a copy. Excited by what he read, Thomson shared Green's ideas with other scientists and had the book republished in a series of journal articles. Green's mathematics provided the foundation on which Thomson, Stokes, Rayleigh, and Maxwell built the present-day theory of electromagnetism.

EXAMPLE 5   Calculate the outward flux of the field $\mathbf{F}(x, y) = x\,\mathbf{i} + y^2\,\mathbf{j}$ across the square bounded by the lines $x = \pm 1$ and $y = \pm 1$.

Solution   Calculating the flux with a line integral would take four integrations, one for each side of the square. With Green's theorem, we can change the line integral to one double integral. With $M = x$, $N = y^2$, C the square, and R the square's interior, we have

$$\text{Flux} = \oint_C \mathbf{F} \cdot \mathbf{n} \, ds = \oint_C M \, dy - N \, dx$$

$$= \iint_R \left(\frac{\partial M}{\partial x} + \frac{\partial N}{\partial y} \right) dx \, dy \qquad \text{Green's theorem}$$

$$= \int_{-1}^{1} \int_{-1}^{1} (1 + 2y) \, dx \, dy = \int_{-1}^{1} \left[x + 2xy \right]_{x=-1}^{x=1} dy$$

$$= \int_{-1}^{1} (2 + 4y) \, dy = \left[2y + 2y^2 \right]_{-1}^{1} = 4.$$

A Proof of Green's Theorem (Special Regions)

Let C be a smooth simple closed curve in the xy-plane with the property that lines parallel to the axes cut it in no more than two points. Let R be the region enclosed by C and suppose that M, N, and their first partial derivatives are continuous at every point of some open region containing C and R. We want to prove the circulation-curl form of Green's theorem,

$$\oint_C M \, dx + N \, dy = \iint_R \left(\frac{\partial N}{\partial x} - \frac{\partial M}{\partial y} \right) dx \, dy. \qquad (13)$$

Figure 14.30 shows C made up of two directed parts:

$$C_1: \quad y = f_1(x), \quad a \le x \le b, \qquad C_2: \quad y = f_2(x), \quad b \ge x \ge a.$$

For any x between a and b, we can integrate $\partial M / \partial y$ with respect to y from $y = f_1(x)$ to $y = f_2(x)$ and obtain

$$\int_{f_1(x)}^{f_2(x)} \frac{\partial M}{\partial y} \, dy = M(x, y) \Big]_{y=f_1(x)}^{y=f_2(x)} = M(x, f_2(x)) - M(x, f_1(x)). \qquad (14)$$

We can then integrate this with respect to x from a to b:

$$\int_a^b \int_{f_1(x)}^{f_2(x)} \frac{\partial M}{\partial y} \, dy \, dx = \int_a^b [M(x, f_2(x)) - M(x, f_1(x))] \, dx$$

$$= -\int_b^a M(x, f_2(x)) \, dx - \int_a^b M(x, f_1(x)) \, dx$$

$$= -\int_{C_2} M \, dx - \int_{C_1} M \, dx$$

$$= -\oint_C M \, dx.$$

Therefore

$$\oint_C M \, dx = \iint_R \left(-\frac{\partial M}{\partial y} \right) dx \, dy. \qquad (15)$$

Equation (15) is half the result we need for Eq. (13). We derive the other half by integrating $\partial N / \partial x$ first with respect to x and then with respect to y, as suggested by Fig. 14.31. This shows the curve C of Fig. 14.30 decomposed into the two

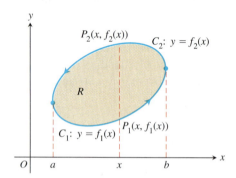

14.30 The boundary curve C is made up of C_1, the graph of $y = f_1(x)$, and C_2, the graph of $y = f_2(x)$.

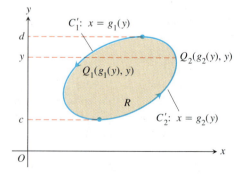

14.31 The boundary curve C is made up of C_1', the graph of $x = g_1(y)$, and C_2', the graph of $x = g_2(y)$.

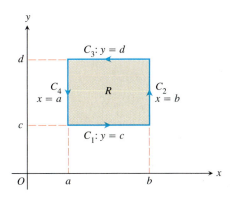

14.32 To prove Green's theorem for a rectangle, we divide the boundary into four directed line segments.

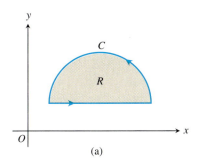

(a)

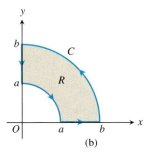

(b)

14.33 Other regions to which Green's theorem applies.

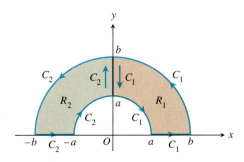

14.34 A region R that combines regions R_1 and R_2.

directed parts C_1': $x = g_1(y)$, $d \geq y \geq c$ and C_2': $x = g_2(y)$, $c \leq y \leq d$. The result of this double integration is

$$\oint_C N \, dy = \iint_R \frac{\partial N}{\partial x} \, dx \, dy. \tag{16}$$

Combining Eqs. (15) and (16) gives Eq. (13). This concludes the proof. ❑

Extending the Proof to Other Regions

The argument we just gave does not apply directly to the rectangular region in Fig. 14.32 because the lines $x = a$, $x = b$, $y = c$, and $y = d$ meet the region's boundary in more than two points. However, if we divide the boundary C into four directed line segments,

$$C_1: \quad y = c, \quad a \leq x \leq b, \qquad C_2: \quad x = b, \quad c \leq y \leq d,$$
$$C_3: \quad y = d, \quad b \geq x \geq a, \qquad C_4: \quad x = a, \quad d \geq y \geq c,$$

we can modify the argument in the following way.

Proceeding as in the proof of Eq. (16), we have

$$\int_c^d \int_a^b \frac{\partial N}{\partial x} \, dx \, dy = \int_c^d (N(b, y) - N(a, y)) \, dy$$

$$= \int_c^d N(b, y) \, dy + \int_d^c N(a, y) \, dy \tag{17}$$

$$= \int_{C_2} N \, dy + \int_{C_4} N \, dy.$$

Because y is constant along C_1 and C_3, $\int_{C_1} N \, dy = \int_{C_3} N \, dy = 0$, so we can add $\int_{C_1} N \, dy + \int_{C_3} N \, dy$ to the right-hand side of Eq. (17) without changing the equality. Doing so, we have

$$\int_c^d \int_a^b \frac{\partial N}{\partial x} \, dx \, dy = \oint_C N \, dy. \tag{18}$$

Similarly, we can show that

$$\int_a^b \int_c^d \frac{\partial M}{\partial y} \, dy \, dx = -\oint_C M \, dx. \tag{19}$$

Subtracting Eq. (19) from Eq. (18), we again arrive at

$$\oint_C M \, dx + N \, dy = \iint_R \left(\frac{\partial N}{\partial x} - \frac{\partial M}{\partial y} \right) dx \, dy.$$

Regions like those in Fig. 14.33 can be handled with no greater difficulty. Equation (13) still applies. It also applies to the horseshoe-shaped region R shown in Fig. 14.34, as we see by putting together the regions R_1 and R_2 and their boundaries. Green's theorem applies to C_1, R_1 and to C_2, R_2, yielding

$$\int_{C_1} M \, dx + N \, dy = \iint_{R_1} \left(\frac{\partial N}{\partial x} - \frac{\partial M}{\partial y} \right) dx \, dy$$

$$\int_{C_2} M \, dx + N \, dy = \iint_{R_2} \left(\frac{\partial N}{\partial x} - \frac{\partial M}{\partial y} \right) dx \, dy.$$

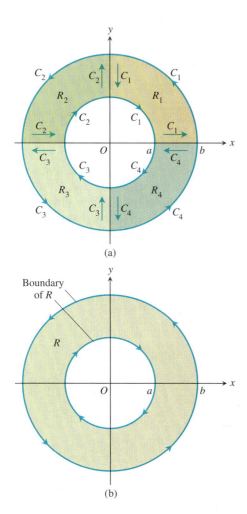

(a)

(b)

14.35 The annular region R combines four smaller regions. In polar coordinates, $r = a$ for the inner circle, $r = b$ for the outer circle, and $a \leq r \leq b$ for the region itself.

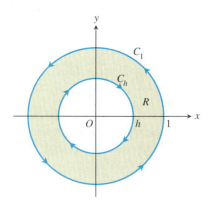

14.36 Green's theorem may be applied to the annular region R by integrating along the boundaries as shown (Example 6).

When we add these two equations, the line integral along the y-axis from b to a for C_1 cancels the integral over the same segment but in the opposite direction for C_2. Hence

$$\oint_C M\,dx + N\,dy = \iint_R \left(\frac{\partial N}{\partial x} - \frac{\partial M}{\partial y} \right) dx\,dy,$$

where C consists of the two segments of the x-axis from $-b$ to $-a$ and from a to b and of the two semicircles, and where R is the region inside C.

The device of adding line integrals over separate boundaries to build up an integral over a single boundary can be extended to any finite number of subregions. In Fig. 14.35(a), let C_1 be the boundary, oriented counterclockwise, of the region R_1 in the first quadrant. Similarly for the other three quadrants: C_i is the boundary of the region R_i, $i = 1, 2, 3, 4$. By Green's theorem,

$$\oint_{C_i} M\,dx + N\,dy = \iint_{R_i} \left(\frac{\partial N}{\partial x} - \frac{\partial M}{\partial y} \right) dx\,dy. \tag{20}$$

We add Eqs. (20) for $i = 1, 2, 3, 4$, and get (Fig. 14.35b):

$$\oint_{r=b} (M\,dx + N\,dy) + \oint_{r=a} (M\,dx + N\,dy) = \iint_{a \leq r \leq b} \left(\frac{\partial N}{\partial x} - \frac{\partial M}{\partial y} \right) dx\,dy. \tag{21}$$

Equation (21) says that the double integral of $(\partial N/\partial x) - (\partial M/\partial y)$ over the annular ring R equals the line integral of $M\,dx + N\,dy$ over the complete boundary of R in the direction that keeps R on our left as we progress (Fig. 14.35b).

EXAMPLE 6 Verify the circulation form of Green's theorem (Eq. 12) on the annular ring R: $h^2 \leq x^2 + y^2 \leq 1, 0 < h < 1$ (Fig. 14.36), if

$$M = \frac{-y}{x^2 + y^2}, \qquad N = \frac{x}{x^2 + y^2}.$$

Solution The boundary of R consists of the circle

$$C_1: \qquad x = \cos t, \quad y = \sin t, \quad 0 \leq t \leq 2\pi,$$

traversed counterclockwise as t increases, and the circle

$$C_h: \qquad x = h\cos\theta, \quad y = -h\sin\theta, \quad 0 \leq \theta \leq 2\pi,$$

traversed clockwise as θ increases. The functions M and N and their partial derivatives are continuous throughout R. Moreover,

$$\frac{\partial M}{\partial y} = \frac{(x^2 + y^2)(-1) + y(2y)}{(x^2 + y^2)^2}$$

$$= \frac{y^2 - x^2}{(x^2 + y^2)^2} = \frac{\partial N}{\partial x},$$

so

$$\iint_R \left(\frac{\partial N}{\partial x} - \frac{\partial M}{\partial y} \right) dx\,dy = \iint_R 0\,dx\,dy = 0.$$

The integral of $M\,dx + N\,dy$ over the boundary of R is

$$\int_C M\,dx + N\,dy = \oint_{C_1} \frac{x\,dy - y\,dx}{x^2 + y^2} + \oint_{C_h} \frac{x\,dy - y\,dx}{x^2 + y^2}$$

$$= \int_0^{2\pi} (\cos^2 t + \sin^2 t)\,dt - \int_0^{2\pi} \frac{h^2(\cos^2 \theta + \sin^2 \theta)}{h^2}\,d\theta$$

$$= 2\pi - 2\pi = 0. \qquad \square$$

The functions M and N in Example 6 are discontinuous at $(0, 0)$, so we cannot apply Green's theorem to the circle C_1 and the region inside it. We must exclude the origin. We do so by excluding the points inside C_h.

We could replace the circle C_1 in Example 6 by an ellipse or any other simple closed curve K surrounding C_h (Fig. 14.37). The result would still be

$$\oint_K (M\,dx + N\,dy) + \oint_{C_h} (M\,dx + N\,dy) = \iint_R \left(\frac{\partial N}{\partial x} - \frac{\partial M}{\partial y} \right) dy\,dx = 0,$$

which leads to the surprising conclusion that

$$\oint_K (M\,dx + N\,dy) = 2\pi$$

for any such curve K. We can explain this result by changing to polar coordinates. With

$$x = r \cos \theta \qquad\qquad y = r \sin \theta$$

$$dx = -r \sin \theta\,d\theta + \cos \theta\,dr, \qquad dy = r \cos \theta\,d\theta + \sin \theta\,dr,$$

we have

$$\frac{x\,dy - y\,dx}{x^2 + y^2} = \frac{r^2(\cos^2 \theta + \sin^2 \theta)\,d\theta}{r^2} = d\theta,$$

and θ increases by 2π as we traverse K once counterclockwise.

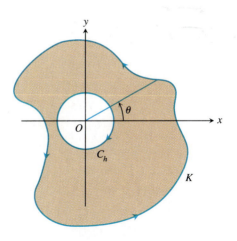

14.37 The region bounded by the circle C_h and the curve K.

Exercises 14.4

Verifying Green's Theorem

In Exercises 1–4, verify Green's theorem by evaluating both sides of Eqs. (11) and (12) for the field $\mathbf{F} = M\,\mathbf{i} + N\,\mathbf{j}$. Take the domains of integration in each case to be the disk $R: x^2 + y^2 \le a^2$ and its bounding circle $C: \mathbf{r} = (a \cos t)\,\mathbf{i} + (a \sin t)\,\mathbf{j},\ 0 \le t \le 2\pi$.

1. $\mathbf{F} = -y\,\mathbf{i} + x\,\mathbf{j}$ 2. $\mathbf{F} = y\,\mathbf{i}$

3. $\mathbf{F} = 2x\,\mathbf{i} - 3y\,\mathbf{j}$ 4. $\mathbf{F} = -x^2 y\,\mathbf{i} + xy^2\,\mathbf{j}$

Counterclockwise Circulation and Outward Flux

In Exercises 5–10, use Green's theorem to find the counterclockwise circulation and outward flux for the field $\mathbf{F}$ and curve C.

5. $\mathbf{F} = (x - y)\,\mathbf{i} + (y - x)\,\mathbf{j}$
   C: The square bounded by $x = 0, x = 1, y = 0, y = 1$

6. $\mathbf{F} = (x^2 + 4y)\,\mathbf{i} + (x + y^2)\,\mathbf{j}$
   C: The square bounded by $x = 0, x = 1, y = 0, y = 1$

7. $\mathbf{F} = (y^2 - x^2)\,\mathbf{i} + (x^2 + y^2)\,\mathbf{j}$
   C: The triangle bounded by $y = 0, x = 3$, and $y = x$

8. $\mathbf{F} = (x + y)\,\mathbf{i} - (x^2 + y^2)\,\mathbf{j}$
   C: The triangle bounded by $y = 0, x = 1$, and $y = x$

9. $\mathbf{F} = (x + e^x \sin y)\,\mathbf{i} + (x + e^x \cos y)\,\mathbf{j}$
   C: The right-hand loop of the lemniscate $r^2 = \cos 2\theta$

10. $\mathbf{F} = \left(\tan^{-1} \dfrac{y}{x} \right) \mathbf{i} + \ln (x^2 + y^2)\,\mathbf{j}$

    C: The boundary of the region defined by the polar coordinate inequalities $1 \le r \le 2, 0 \le \theta \le \pi$

11. Find the counterclockwise circulation and outward flux of the

field $\mathbf{F} = xy\,\mathbf{i} + y^2\,\mathbf{j}$ around and over the boundary of the region enclosed by the curves $y = x^2$ and $y = x$ in the first quadrant.

12. Find the counterclockwise circulation and the outward flux of the field $\mathbf{F} = (-\sin y)\,\mathbf{i} + (x\cos y)\,\mathbf{j}$ around and over the square cut from the first quadrant by the lines $x = \pi/2$ and $y = \pi/2$.

13. Find the outward flux of the field

$$\mathbf{F} = \left(3xy - \frac{x}{1+y^2}\right)\mathbf{i} + (e^x + \tan^{-1} y)\,\mathbf{j}$$

across the cardioid $r = a(1 + \cos\theta), a > 0$.

14. Find the counterclockwise circulation of $\mathbf{F} = (y + e^x \ln y)\,\mathbf{i} + (e^x/y)\,\mathbf{j}$ around the boundary of the region that is bounded above by the curve $y = 3 - x^2$ and below by the curve $y = x^4 + 1$.

Work

In Exercises 15 and 16, find the work done by $\mathbf{F}$ in moving a particle once counterclockwise around the given curve.

15. $\mathbf{F} = 2xy^3\,\mathbf{i} + 4x^2 y^2\,\mathbf{j}$
    C: The boundary of the "triangular" region in the first quadrant enclosed by the x-axis, the line $x = 1$, and the curve $y = x^3$

16. $\mathbf{F} = (4x - 2y)\,\mathbf{i} + (2x - 4y)\,\mathbf{j}$
    C: The circle $(x - 2)^2 + (y - 2)^2 = 4$

Evaluating Line Integrals in the Plane

Apply Green's theorem to evaluate the integrals in Exercises 17–20.

17. $\oint_C (y^2\,dx + x^2\,dy)$
    C: The triangle bounded by $x = 0, x + y = 1, y = 0$

18. $\oint_C (3y\,dx + 2x\,dy)$
    C: The boundary of $0 \le x \le \pi, 0 \le y \le \sin x$

19. $\oint_C (6y + x)\,dx + (y + 2x)\,dy$
    C: The circle $(x - 2)^2 + (y - 3)^2 = 4$

20. $\oint_C (2x + y^2)\,dx + (2xy + 3y)\,dy$
    C: Any simple closed curve in the plane for which Green's theorem holds

Calculating Area with Green's Theorem

If a simple closed curve C in the plane and the region R it encloses satisfy the hypotheses of Green's theorem, the area of R is given by:

> **Green's Theorem Area Formula**
>
> $$\text{Area of } R = \frac{1}{2}\oint_C x\,dy - y\,dx \qquad (22)$$

The reason is that by Eq. (11), run backward,

$$\text{Area of } R = \iint_R dy\,dx = \iint_R \left(\frac{1}{2} + \frac{1}{2}\right) dy\,dx$$

$$= \oint_C \frac{1}{2}x\,dy - \frac{1}{2}y\,dx.$$

Use the Green's theorem area formula (Eq. 22) to find the areas of the regions enclosed by the curves in Exercises 21–24.

21. The circle $\mathbf{r}(t) = (a\cos t)\,\mathbf{i} + (a\sin t)\,\mathbf{j}, \quad 0 \le t \le 2\pi$

22. The ellipse $\mathbf{r}(t) = (a\cos t)\,\mathbf{i} + (b\sin t)\,\mathbf{j}, \quad 0 \le t \le 2\pi$

23. The astroid (Fig. 9.42) $\mathbf{r}(t) = (\cos^3 t)\,\mathbf{i} + (\sin^3 t)\,\mathbf{j}, \quad 0 \le t \le 2\pi$

24. The curve (Fig. 9.75) $\mathbf{r}(t) = t^2\,\mathbf{i} + ((t^3/3) - t)\,\mathbf{j},$
    $-\sqrt{3} \le t \le \sqrt{3}$

Theory and Examples

25. Let C be the boundary of a region on which Green's theorem holds. Use Green's theorem to calculate

    a) $\oint_C f(x)\,dx + g(y)\,dy,$

    b) $\oint_C ky\,dx + hx\,dy \quad (k \text{ and } h \text{ constants}).$

26. Show that the value of

    $$\oint_C xy^2\,dx + (x^2 y + 2x)\,dy$$

    around any square depends only on the area of the square and not on its location in the plane.

27. What is special about the integral

    $$\oint_C 4x^3 y\,dx + x^4\,dy?$$

    Give reasons for your answer.

28. What is special about the integral

    $$\oint_C -y^3\,dx + x^3\,dy?$$

    Give reasons for your answer.

29. Show that if R is a region in the plane bounded by a piecewise smooth simple closed curve C, then

    $$\text{Area of } R = \oint_C x\,dy = -\oint_C y\,dx.$$

30. Suppose that a nonnegative function $y = f(x)$ has a continuous first derivative on $[a, b]$. Let C be the boundary of the region in the xy-plane that is bounded below by the x-axis, above by the graph of f, and on the sides by the lines $x = a$ and $x = b$. Show

that

$$\int_a^b f(x)\,dx = -\oint_C y\,dx.$$

31. Let A be the area and $\bar{x}$ the x-coordinate of the centroid of a region R that is bounded by a piecewise smooth simple closed curve C in the xy-plane. Show that

$$\frac{1}{2}\oint_C x^2\,dy = -\oint_C xy\,dx = \frac{1}{3}\oint_C x^2\,dy - xy\,dx = A\bar{x}.$$

32. Let I_y be the moment of inertia about the y-axis of the region in Exercise 31. Show that

$$\frac{1}{3}\oint_C x^3\,dy = -\oint_C x^2 y\,dx = \frac{1}{4}\oint_C x^3\,dy - x^2 y\,dx = I_y.$$

33. *Green's theorem and Laplace's equation.* Assuming that all the necessary derivatives exist and are continuous, show that if $f(x, y)$ satisfies the Laplace equation

$$\frac{\partial^2 f}{\partial x^2} + \frac{\partial^2 f}{\partial y^2} = 0,$$

then

$$\oint_C \frac{\partial f}{\partial y}\,dx - \frac{\partial f}{\partial x}\,dy = 0$$

for all closed curves C to which Green's theorem applies. (The converse is also true: If the line integral is always zero, then f satisfies the Laplace equation.)

34. Among all smooth simple closed curves in the plane, oriented counterclockwise, find the one along which the work done by

$$\mathbf{F} = \left(\frac{1}{4}x^2 y + \frac{1}{3}y^3\right)\mathbf{i} + x\,\mathbf{j}$$

is greatest. (*Hint:* Where is curl $\mathbf{F}$ positive?)

35. Green's theorem holds for a region R with any finite number of holes as long as the bounding curves are smooth, simple, and closed and we integrate over each component of the boundary in the direction that keeps R on our immediate left as we go along (Fig. 14.38).

a) Let $f(x, y) = \ln(x^2 + y^2)$ and let C be the circle $x^2 + y^2 = a^2$. Evaluate the flux integral

$$\oint_C \nabla f \cdot \mathbf{n}\,ds.$$

b) Let K be an arbitrary smooth simple closed curve in the plane that does not pass through $(0, 0)$. Use Green's theorem to show that

$$\oint_K \nabla f \cdot \mathbf{n}\,ds$$

has two possible values, depending on whether $(0, 0)$ lies inside K or outside K.

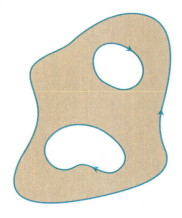

14.38 Green's theorem holds for regions with more than one hole (Exercise 35).

36. *Bendixson's criterion.* The **streamlines** of a planar fluid flow are the smooth curves traced by the fluid's individual particles. The vectors $\mathbf{F} = M(x, y)\mathbf{i} + N(x, y)\mathbf{j}$ of the flow's velocity field are the tangent vectors of the streamlines. Show that if the flow takes place over a *simply connected* region R (no holes or missing points) and that if $M_x + N_y \neq 0$ throughout R, then none of the streamlines in R is closed. In other words, no particle of fluid ever has a closed trajectory in R. The criterion $M_x + N_y \neq 0$ is called **Bendixson's criterion** for the nonexistence of closed trajectories.

37. Establish Eq. (16) to finish the proof of the special case of Green's theorem.

38. Establish Eq. (19) to complete the argument for the extension of Green's theorem.

39. Can anything be said about the curl of a conservative two-dimensional vector field? Give reasons for your answer.

40. Does Green's theorem give any information about the circulation of a conservative field? Does this agree with anything else you know? Give reasons for your answer.

CAS Explorations and Projects

In Exercises 41–44, use a CAS and Green's theorem to find the counterclockwise circulation of the field $\mathbf{F}$ around the simple closed curve C. Perform the following CAS steps:

a) Plot C in the xy-plane.

b) Determine the integrand $\dfrac{\partial N}{\partial x} - \dfrac{\partial M}{\partial y}$ for the curl form of Green's theorem.

c) Determine the (double integral) limits of integration from your plot in (a) and evaluate the curl integral for the circulation.

41. $\mathbf{F} = (2x - y)\mathbf{i} + (x + 3y)\mathbf{j}$, C: The ellipse $x^2 + 4y^2 = 4$

42. $\mathbf{F} = (2x^3 - y^3)\mathbf{i} + (x^3 + y^3)\mathbf{j}$, C: The ellipse $\dfrac{x^2}{4} + \dfrac{y^2}{9} = 1$

43. $\mathbf{F} = x^{-1}e^y\mathbf{i} + (e^y \ln x + 2x)\mathbf{j}$,
C: The boundary of the region defined by $y = 1 + x^4$ (below) and $y = 2$ (above)

44. $\mathbf{F} = x\,e^y\mathbf{i} + 4x^2 \ln y\,\mathbf{j}$, C: The triangle with vertices $(0, 0)$, $(2, 0)$, and $(0, 4)$

Surface Area and Surface Integrals

We know how to integrate a function over a flat region in a plane, but what if the function is defined over a curved surface? How do we calculate its integral then? The trick to evaluating one of these so-called surface integrals is to rewrite it as a double integral over a region in a coordinate plane beneath the surface (Fig. 14.39). In Sections 14.7 and 14.8 we will see how surface integrals provide just what we need to generalize the two forms of Green's theorem to three dimensions.

The Definition of Surface Area

Figure 14.40 shows a surface S lying above its "shadow" region R in a plane beneath it. The surface is defined by the equation $f(x, y, z) = c$. If the surface is **smooth** (∇f is continuous and never vanishes on S), we can define and calculate its area as a double integral over R.

The first step in defining the area of S is to partition the region R into small rectangles ΔA_k of the kind we would use if we were defining an integral over R. Directly above each ΔA_k lies a patch of surface $\Delta \sigma_k$ that we may approximate with a portion ΔP_k of the tangent plane. To be specific, we suppose that ΔP_k is a portion of the plane that is tangent to the surface at the point $T_k(x_k, y_k, z_k)$ directly above the back corner C_k of ΔA_k. If the tangent plane is parallel to R, then ΔP_k will be congruent to ΔA_k. Otherwise, it will be a parallelogram whose area is somewhat larger than the area of ΔA_k.

Figure 14.41 gives a magnified view of $\Delta \sigma_k$ and ΔP_k, showing the gradient vector $\nabla f(x_k, y_k, z_k)$ at T_k and a unit vector $\mathbf{p}$ that is normal to R. The figure also shows the angle γ_k between ∇f and $\mathbf{p}$. The other vectors in the picture, $\mathbf{u}_k$ and $\mathbf{v}_k$, lie along the edges of the patch ΔP_k in the tangent plane. Thus, both $\mathbf{u}_k \times \mathbf{v}_k$ and ∇f are normal to the tangent plane.

We now need the fact from advanced vector geometry that $|(\mathbf{u}_k \times \mathbf{v}_k) \cdot \mathbf{p}|$ is the area of the projection of the parallelogram determined by $\mathbf{u}_k$ and $\mathbf{v}_k$ onto any plane whose normal is $\mathbf{p}$. In our case, this translates into the statement

$$|(\mathbf{u}_k \times \mathbf{v}_k) \cdot \mathbf{p}| = \Delta A_k. \tag{1}$$

Now, $|\mathbf{u}_k \times \mathbf{v}_k|$ itself is the area ΔP_k (standard fact about cross products) so Eq. (1) becomes

$$\underbrace{|\mathbf{u}_k \times \mathbf{v}_k|}_{\Delta P_k} \; \underbrace{|\mathbf{p}|}_{1} \; \underbrace{|\cos (\text{angle between } \mathbf{u}_k \times \mathbf{v}_k \text{ and } \mathbf{p})|}_{\substack{\text{same as } |\cos \gamma_k| \text{ because} \\ \nabla f \text{ and } \mathbf{u}_k \times \mathbf{v}_k \text{ are both} \\ \text{normal to the tangent plane}}} = \Delta A_k \tag{2}$$

or

$$\Delta P_k |\cos \gamma_k| = \Delta A_k$$

or

$$\Delta P_k = \frac{\Delta A_k}{|\cos \gamma_k|},$$

provided $\cos \gamma_k \neq 0$. We will have $\cos \gamma_k \neq 0$ as long as ∇f is not parallel to the ground plane and $\nabla f \cdot \mathbf{p} \neq 0$.

Since the patches ΔP_k approximate the surface patches $\Delta \sigma_k$ that fit together to make S, the sum

$$\sum \Delta P_k = \sum \frac{\Delta A_k}{|\cos \gamma_k|} \tag{3}$$

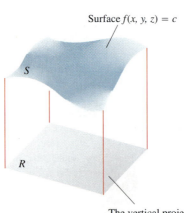

Surface $f(x, y, z) = c$

The vertical projection or "shadow" of S on a coordinate plane

14.39 As we will soon see, the integral of a function $g(x, y, z)$ over a surface S in space can be calculated by evaluating a related double integral over the vertical projection or "shadow" of S on a coordinate plane.

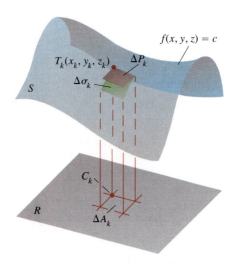

$f(x, y, z) = c$

14.40 A surface S and its vertical projection onto a plane beneath it. You can think of R as the shadow of S on the plane. The tangent plate ΔP_k approximates the surface patch $\Delta \sigma_k$ above ΔA_k.

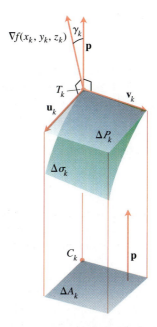

14.41 Magnified view from the preceding figure. The vector $\mathbf{u}_k \times \mathbf{v}_k$ (not shown) is parallel to the vector ∇f because both vectors are normal to the plane of ΔP_k.

looks like an approximation of what we might like to call the surface area of S. It also looks as if the approximation would improve if we refined the partition of R. In fact, the sums on the right-hand side of Eq. (3) are approximating sums for the double integral

$$\iint_R \frac{1}{|\cos \gamma|} \, dA. \tag{4}$$

We therefore define the **area** of S to be the value of this integral whenever it exists.

A Practical Formula

For any surface $f(x, y, z) = c$, we have $|\nabla f \cdot \mathbf{p}| = |\nabla f||\mathbf{p}||\cos \gamma|$, so

$$\frac{1}{|\cos \gamma|} = \frac{|\nabla f|}{|\nabla f \cdot \mathbf{p}|}.$$

This combines with Eq. (4) to give a practical formula for area.

The Formula for Surface Area

The area of the surface $f(x, y, z) = c$ over a closed and bounded plane region R is

$$\text{Surface area} = \iint_R \frac{|\nabla f|}{|\nabla f \cdot \mathbf{p}|} \, dA, \tag{5}$$

where $\mathbf{p}$ is a unit vector normal to R and $\nabla f \cdot \mathbf{p} \neq 0$.

Thus, the area is the double integral over R of the magnitude of ∇f divided by the magnitude of the scalar component of ∇f normal to R.

We reached Eq. (5) under the assumption that $\nabla f \cdot \mathbf{p} \neq 0$ throughout R and that ∇f is continuous. Whenever the integral exists, however, we define its value to be the area of the portion of the surface $f(x, y, z) = c$ that lies over R.

EXAMPLE 1 Find the area of the surface cut from the bottom of the paraboloid $x^2 + y^2 - z = 0$ by the plane $z = 4$.

Solution We sketch the surface S and the region R below it in the xy-plane (Fig. 14.42). The surface S is part of the level surface $f(x, y, z) = x^2 + y^2 - z = 0$, and R is the disk $x^2 + y^2 \leq 4$ in the xy-plane. To get a unit vector normal to the plane of R, we can take $\mathbf{p} = \mathbf{k}$.

At any point (x, y, z) on the surface, we have

$$f(x, y, z) = x^2 + y^2 - z$$

$$\nabla f = 2x\,\mathbf{i} + 2y\,\mathbf{j} - \mathbf{k}$$

$$|\nabla f| = \sqrt{(2x)^2 + (2y)^2 + (-1)^2}$$

$$= \sqrt{4x^2 + 4y^2 + 1}$$

$$|\nabla f \cdot \mathbf{p}| = |\nabla f \cdot \mathbf{k}| = |-1| = 1.$$

14.42 The area of this parabolic surface is calculated in Example 1.

In the region $R, dA = dx \, dy$. Therefore,

$$\text{Surface area} = \iint\limits_{R} \frac{|\nabla f|}{|\nabla f \cdot \mathbf{p}|} \, dA \qquad \text{Eq. (5)}$$

$$= \iint\limits_{x^2+y^2\leq 4} \sqrt{4x^2 + 4y^2 + 1} \, dx \, dy$$

$$= \int_0^{2\pi} \int_0^2 \sqrt{4r^2 + 1} \, r \, dr \, d\theta \qquad \text{Polar coordinates}$$

$$= \int_0^{2\pi} \left[\frac{1}{12}(4r^2 + 1)^{3/2} \right]_0^2 d\theta$$

$$= \int_0^{2\pi} \frac{1}{12}(17^{3/2} - 1) \, d\theta = \frac{\pi}{6}(17\sqrt{17} - 1). \qquad \square$$

EXAMPLE 2   Find the area of the cap cut from the hemisphere $x^2 + y^2 + z^2 = 2, z \geq 0$, by the cylinder $x^2 + y^2 = 1$ (Fig. 14.43).

Solution   The cap S is part of the level surface $f(x, y, z) = x^2 + y^2 + z^2 = 2$. It projects one-to-one onto the disk $R: x^2 + y^2 \leq 1$ in the xy-plane. The vector $\mathbf{p} = \mathbf{k}$ is normal to the plane of R.

At any point on the surface,

$$f(x, y, z) = x^2 + y^2 + z^2$$

$$\nabla f = 2x \, \mathbf{i} + 2y \, \mathbf{j} + 2z \, \mathbf{k}$$

$$|\nabla f| = 2\sqrt{x^2 + y^2 + z^2} = 2\sqrt{2} \qquad \begin{array}{l}\text{Because } x^2 + y^2 + z^2 = 2 \\ \text{at points of } S\end{array}$$

$$|\nabla f \cdot \mathbf{p}| = |\nabla f \cdot \mathbf{k}| = |2z| = 2z.$$

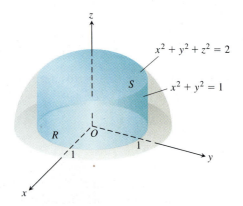

14.43 The cap cut from the hemisphere by the cylinder projects vertically onto the disk $R: x^2 + y^2 \leq 1$ (Example 2).

Therefore,

$$\text{Surface area} = \iint\limits_{R} \frac{|\nabla f|}{|\nabla f \cdot \mathbf{p}|} \, dA = \iint\limits_{R} \frac{2\sqrt{2}}{2z} \, dA = \sqrt{2} \iint\limits_{R} \frac{dA}{z}. \qquad (6)$$

What do we do about the z?

Since z is the z-coordinate of a point on the sphere, we can express it in terms of x and y as

$$z = \sqrt{2 - x^2 - y^2}.$$

We continue the work of Eq. (6) with this substitution:

$$\text{Surface area} = \sqrt{2} \iint\limits_{R} \frac{dA}{z} = \sqrt{2} \iint\limits_{x^2+y^2\leq 1} \frac{dA}{\sqrt{2 - x^2 - y^2}}$$

$$= \sqrt{2} \int_0^{2\pi} \int_0^1 \frac{r \, dr \, d\theta}{\sqrt{2 - r^2}} \qquad \text{Polar coordinates}$$

$$= \sqrt{2} \int_0^{2\pi} \left[-(2 - r^2)^{1/2} \right]_{r=0}^{r=1} d\theta$$

$$= \sqrt{2} \int_0^{2\pi} (\sqrt{2} - 1) \, d\theta = 2\pi(2 - \sqrt{2}). \qquad \square$$

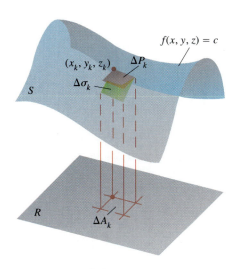

$f(x, y, z) = c$

(x_k, y_k, z_k) ΔP_k

$\Delta \sigma_k$

S

R ΔA_k

14.44 If we know how an electrical charge is distributed over a surface, we can find the total charge with a suitably modified surface integral.

Surface Integrals

We now show how to integrate a function over a surface, using the ideas just developed for calculating surface area.

Suppose, for example, that we have an electrical charge distributed over a surface $f(x, y, z) = c$ like the one shown in Fig. 14.44 and that the function $g(x, y, z)$ gives the charge per unit area (charge density) at each point on S. Then we may calculate the total charge on S as an integral in the following way.

We partition the shadow region R on the ground plane beneath the surface into small rectangles of the kind we would use if we were defining the surface area of S. Then directly above each ΔA_k lies a patch of surface $\Delta \sigma_k$ that we approximate with a parallelogram-shaped portion of tangent plane, ΔP_k.

Up to this point the construction proceeds as in the definition of surface area, but now we take one additional step: We evaluate g at (x_k, y_k, z_k) and then approximate the total charge on the surface patch $\Delta \sigma_k$ by the product $g(x_k, y_k, z_k) \Delta P_k$. The rationale is that when the partition of R is sufficiently fine, the value of g throughout $\Delta \sigma_k$ is nearly constant and ΔP_k is nearly the same as $\Delta \sigma_k$. The total charge over S is then approximated by the sum

$$\text{Total charge} \approx \sum g(x_k, y_k, z_k) \Delta P_k = \sum g(x_k, y_k, z_k) \frac{\Delta A_k}{|\cos \gamma_k|}. \tag{7}$$

If f, the function defining the surface S, and its first partial derivatives are continuous, and if g is continuous over S, then the sums on the right-hand side of Eq. (7) approach the limit

$$\iint\limits_R g(x, y, z) \frac{dA}{|\cos \gamma|} = \iint\limits_R g(x, y, z) \frac{|\nabla f|}{|\nabla f \cdot \mathbf{p}|} dA \tag{8}$$

as the partition of R is refined in the usual way. This limit is called the integral of g over the surface S and is calculated as a double integral over R. The value of the integral is the total charge on the surface S.

As you might expect, the formula in Eq. (8) defines the integral of *any* function g over the surface S as long as the integral exists.

Definitions

If R is the shadow region of a surface S defined by the equation $f(x, y, z) = c$, and g is a continuous function defined at the points of S, then the **integral of g over S** is the integral

$$\iint\limits_R g(x, y, z) \frac{|\nabla f|}{|\nabla f \cdot \mathbf{p}|} dA, \tag{9}$$

where $\mathbf{p}$ is a unit vector normal to R and $\nabla f \cdot \mathbf{p} \neq 0$. The integral itself is called a **surface integral.**

The integral in (9) takes on different meanings in different applications. If g has the constant value 1, the integral gives the area of S. If g gives the mass density of a thin shell of material modeled by S, the integral gives the mass of the shell.

Algebraic Properties: The Surface Area Differential

We can abbreviate the integral in (9) by writing $d\sigma$ for $(|\nabla f| / |\nabla f \cdot \mathbf{p}|) \, dA$.

The Surface Area Differential and the Differential Form for Surface Integrals

$$d\sigma = \frac{|\nabla f|}{|\nabla f \cdot \mathbf{p}|} \, dA \qquad \qquad \iint_S g \, d\sigma \qquad \qquad (10)$$

surface area
differential

differential formula
for surface integrals

Surface integrals behave like other double integrals, the integral of the sum of two functions being the sum of their integrals and so on. The domain additivity property takes the form

$$\iint_S g \, d\sigma = \iint_{S_1} g \, d\sigma + \iint_{S_2} g \, d\sigma + \cdots + \iint_{S_n} g \, d\sigma.$$

The idea is that if S is partitioned by smooth curves into a finite number of nonoverlapping smooth patches (i.e., if S is **piecewise smooth**), then the integral over S is the sum of the integrals over the patches. Thus, the integral of a function over the surface of a cube is the sum of the integrals over the faces of the cube. We integrate over a turtle shell of welded plates by integrating one plate at a time and adding the results.

EXAMPLE 3   Integrate $g(x, y, z) = xyz$ over the surface of the cube cut from the first octant by the planes $x = 1$, $y = 1$, and $z = 1$ (Fig. 14.45).

Solution   We integrate xyz over each of the six sides and add the results. Since $xyz = 0$ on the sides that lie in the coordinate planes, the integral over the surface of the cube reduces to

$$\iint_{\substack{\text{cube} \\ \text{surface}}} xyz \, d\sigma = \iint_{\text{side } A} xyz \, d\sigma + \iint_{\text{side } B} xyz \, d\sigma + \iint_{\text{side } C} xyz \, d\sigma.$$

Side A is the surface $f(x, y, z) = z = 1$ over the square region R_{xy}: $0 \le x \le 1$, $0 \le y \le 1$, in the xy-plane. For this surface and region,

$$\mathbf{p} = \mathbf{k}, \qquad \nabla f = \mathbf{k}, \qquad |\nabla f| = 1, \qquad |\nabla f \cdot \mathbf{p}| = |\mathbf{k} \cdot \mathbf{k}| = 1,$$

$$d\sigma = \frac{|\nabla f|}{|\nabla f \cdot \mathbf{p}|} \, dA = \frac{1}{1} \, dx \, dy = dx \, dy,$$

$$xyz = xy(1) = xy,$$

and

$$\iint_{\text{side } A} xyz \, d\sigma = \iint_{R_{xy}} xy \, dx \, dy = \int_0^1 \int_0^1 xy \, dx \, dy = \int_0^1 \frac{y}{2} \, dy = \frac{1}{4}.$$

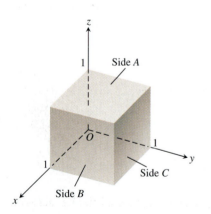

14.45 To integrate a function over the surface of a cube, we integrate over each face and add the results (Example 3).

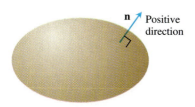

n Positive direction

14.46 Smooth closed surfaces in space are orientable. The outward unit normal vector defines the positive direction at each point.

14.47 To make a Möbius band, take a rectangular strip of paper *abcd*, give the end *bc* a single twist, and paste the ends of the strip together to match *a* with *c* and *b* with *d*. The Möbius band is a nonorientable or one-sided surface.

Symmetry tells us that the integrals of xyz over sides B and C are also 1/4. Hence,

$$\iint\limits_{\substack{\text{cube} \\ \text{surface}}} xyz\, d\sigma = \frac{1}{4} + \frac{1}{4} + \frac{1}{4} = \frac{3}{4}.$$

Orientation

We call a smooth surface S **orientable** or **two-sided** if it is possible to define a field **n** of unit normal vectors on S that varies continuously with position. Any patch or subportion of an orientable surface is orientable. Spheres and other smooth closed surfaces in space (smooth surfaces that enclose solids) are orientable. By convention, we choose **n** on a closed surface to point outward.

Once **n** has been chosen, we say that we have **oriented** the surface, and we call the surface together with its normal field an **oriented surface.** The vector **n** at any point is called the **positive direction** at that point (Fig. 14.46).

The Möbius band in Fig. 14.47 is not orientable. No matter where you start to construct a continuous unit normal field (shown as the shaft of a thumbtack in the figure), moving the vector continuously around the surface in the manner shown will return it to the starting point with a direction opposite to the one it had when it started out. The vector at that point cannot point both ways and yet it must if the field is to be continuous. We conclude that no such field exists.

The Surface Integral for Flux

Suppose that **F** is a continuous vector field defined over an oriented surface S and that **n** is the chosen unit normal field on the surface. We call the integral of $\mathbf{F} \cdot \mathbf{n}$ over S the flux across S in the positive direction. Thus, the flux is the integral over S of the scalar component of **F** in the direction of **n.**

> **Definition**
>
> The **flux** of a three-dimensional vector field **F** across an oriented surface S in the direction of **n** is given by the formula
>
> $$\text{Flux} = \iint\limits_{S} \mathbf{F} \cdot \mathbf{n}\, d\sigma. \tag{11}$$

The definition is analogous to the flux of a two-dimensional field **F** across a plane curve C. In the plane (Section 14.2), the flux is

$$\int_C \mathbf{F} \cdot \mathbf{n}\, ds,$$

the integral of the scalar component of **F** normal to the curve.

If **F** is the velocity field of a three-dimensional fluid flow, the flux of **F** across S is the net rate at which fluid is crossing S in the chosen positive direction. We will discuss such flows in more detail in Section 14.7.

If **S** is part of a level surface $g(x, y, z) = c$, then **n** may be taken to be one of the two fields

$$\mathbf{n} = \pm \frac{\nabla g}{|\nabla g|}, \tag{12}$$

depending on which one gives the preferred direction. The corresponding flux is

$$\text{Flux} = \iint_S \mathbf{F} \cdot \mathbf{n} \, d\sigma \qquad \text{Eq. (11)}$$

$$= \iint_R \left(\mathbf{F} \cdot \frac{\pm \nabla g}{|\nabla g|} \right) \frac{|\nabla g|}{|\nabla g \cdot \mathbf{p}|} \, dA \qquad \text{Eqs. (12) and (10)}$$

$$= \iint_R \mathbf{F} \cdot \frac{\pm \nabla g}{|\nabla g \cdot \mathbf{p}|} \, dA. \qquad (13)$$

EXAMPLE 4   Find the flux of $\mathbf{F} = yz \mathbf{j} + z^2 \mathbf{k}$ outward through the surface S cut from the cylinder $y^2 + z^2 = 1$, $z \geq 0$, by the planes $x = 0$ and $x = 1$.

Solution   The outward normal field on S (Fig. 14.48) may be calculated from the gradient of $g(x, y, z) = y^2 + z^2$ to be

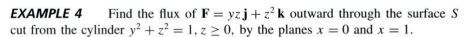

$$\mathbf{n} = +\frac{\nabla g}{|\nabla g|} = \frac{2y \mathbf{j} + 2z \mathbf{k}}{\sqrt{4y^2 + 4z^2}} = \frac{2y \mathbf{j} + 2z \mathbf{k}}{2\sqrt{1}} = y \mathbf{j} + z \mathbf{k}.$$

With $\mathbf{p} = \mathbf{k}$, we also have

$$d\sigma = \frac{|\nabla g|}{|\nabla g \cdot \mathbf{k}|} \, dA = \frac{2}{|2z|} \, dA = \frac{1}{z} \, dA.$$

We can drop the absolute value bars because $z \geq 0$ on S.

The value of $\mathbf{F} \cdot \mathbf{n}$ on the surface is given by the formula

$$\mathbf{F} \cdot \mathbf{n} = (yz \mathbf{j} + z^2 \mathbf{k}) \cdot (y \mathbf{j} + z \mathbf{k})$$

$$= y^2 z + z^3 = z(y^2 + z^2)$$

$$= z. \qquad \qquad y^2 + z^2 = 1 \text{ on } S$$

Therefore, the flux of $\mathbf{F}$ outward through S is

$$\iint_S \mathbf{F} \cdot \mathbf{n} \, d\sigma = \iint_S (z) \left(\frac{1}{z} \, dA \right) = \iint_{R_{xy}} dA = \text{area}(R_{xy}) = 2.$$

Moments and Masses of Thin Shells

Thin shells of material like bowls, metal drums, and domes are modeled with surfaces. Their moments and masses are calculated with the formulas in Table 14.3.

EXAMPLE 5   Find the center of mass of a thin hemispherical shell of radius a and constant density δ.

Solution   We model the shell with the hemisphere

$$f(x, y, z) = x^2 + y^2 + z^2 = a^2, \qquad z \geq 0$$

(Fig. 14.49). The symmetry of the surface about the z-axis tells us that $\bar{x} = \bar{y} = 0$. It remains only to find $\bar{z}$ from the formula $\bar{z} = M_{xy}/M$.

The mass of the shell is

$$M = \iint_S \delta \, d\sigma = \delta \iint_S d\sigma = (\delta)(\text{area of } S) = 2\pi a^2 \delta.$$

14.48 Example 4 calculates the flux of a vector field outward through this surface. The area of the shadow region R_{xy} is 2.

14.49 The center of mass of a thin hemispherical shell of constant density lies on the axis of symmetry halfway from the base to the top (Example 5).

Table 14.3 Mass and moment formulas for very thin shells

Mass: $M = \iint\limits_{S} \delta(x, y, z)\, d\sigma$ $(\delta(x, y, z) = $ density at (x, y, z), mass per unit area)

First moments about the coordinate planes:

$$M_{yz} = \iint\limits_{S} x\delta\, d\sigma, \qquad M_{xz} = \iint\limits_{S} y\delta\, d\sigma, \qquad M_{xy} = \iint\limits_{S} z\delta\, d\sigma$$

Coordinates of center of mass:

$$\overline{x} = M_{yz}/M, \qquad \overline{y} = M_{xz}/M, \qquad \overline{z} = M_{xy}/M$$

Moments of inertia:

$$I_x = \iint\limits_{S} (y^2 + z^2)\, \delta\, d\sigma, \qquad I_y = \iint\limits_{S} (x^2 + z^2)\, \delta\, d\sigma,$$

$$I_z = \iint\limits_{S} (x^2 + y^2)\, \delta\, d\sigma, \qquad I_L = \iint\limits_{S} r^2 \delta\, d\sigma,$$

$r(x, y, z) = $ distance from point (x, y, z) to line L

Radius of gyration about a line L: $R_L = \sqrt{I_L/M}$

To evaluate the integral for M_{xy}, we take $\mathbf{p} = \mathbf{k}$ and calculate

$$|\nabla f| = |2x\,\mathbf{i} + 2y\,\mathbf{j} + 2z\,\mathbf{k}| = 2\sqrt{x^2 + y^2 + z^2} = 2a$$

$$|\nabla f \cdot \mathbf{p}| = |\nabla f \cdot \mathbf{k}| = |2z| = 2z$$

$$d\sigma = \frac{|\nabla f|}{|\nabla f \cdot \mathbf{p}|}\, dA = \frac{a}{z}\, dA.$$

Then

$$M_{xy} = \iint\limits_{S} z\delta\, d\sigma = \delta \iint\limits_{R} z\frac{a}{z}\, dA = \delta a \iint\limits_{R} dA = \delta a(\pi a^2) = \delta \pi a^3$$

$$\overline{z} = \frac{M_{xy}}{M} = \frac{\pi a^3 \delta}{2\pi a^2 \delta} = \frac{a}{2}.$$

The shell's center of mass is the point $(0, 0, a/2)$.

Exercises 14.5

Surface Area

1. Find the area of the surface cut from the paraboloid $x^2 + y^2 - z = 0$ by the plane $z = 2$.

2. Find the area of the band cut from the paraboloid $x^2 + y^2 - z = 0$ by the planes $z = 2$ and $z = 6$.

3. Find the area of the region cut from the plane $x + 2y + 2z = 5$ by the cylinder whose walls are $x = y^2$ and $x = 2 - y^2$.

4. Find the area of the portion of the surface $x^2 - 2z = 0$ that lies above the triangle bounded by the lines $x = \sqrt{3}$, $y = 0$, and $y = x$ in the xy-plane.

5. Find the area of the surface $x^2 - 2y - 2z = 0$ that lies above the triangle bounded by the lines $x = 2$, $y = 0$, and $y = 3x$ in the xy-plane.

6. Find the area of the cap cut from the sphere $x^2 + y^2 + z^2 = 2$ by the cone $z = \sqrt{x^2 + y^2}$.

7. Find the area of the ellipse cut from the plane $z = cx$ by the cylinder $x^2 + y^2 = 1$.

8. Find the area of the upper portion of the cylinder $x^2 + z^2 = 1$ that lies between the planes $x = \pm 1/2$ and $y = \pm 1/2$.

9. Find the area of the portion of the paraboloid $x = 4 - y^2 - z^2$ that lies above the ring $1 \le y^2 + z^2 \le 4$ in the yz-plane.

10. Find the area of the surface cut from the paraboloid $x^2 + y + z^2 = 2$ by the plane $y = 0$.

11. Find the area of the surface $x^2 - 2\ln x + \sqrt{15}y - z = 0$ above the square $R: 1 \le x \le 2, 0 \le y \le 1$, in the xy-plane.

12. Find the area of the surface $2x^{3/2} + 2y^{3/2} - 3z = 0$ above the square $R: 0 \le x \le 1, 0 \le y \le 1$, in the xy-plane.

Surface Integrals

13. Integrate $g(x, y, z) = x + y + z$ over the surface of the cube cut from the first octant by the planes $x = a$, $y = a$, $z = a$.

14. Integrate $g(x, y, z) = y + z$ over the surface of the wedge in the first octant bounded by the coordinate planes and the planes $x = 2$ and $y + z = 1$.

15. Integrate $g(x, y, z) = xyz$ over the surface of the rectangular solid cut from the first octant by the planes $x = a$, $y = b$, and $z = c$.

16. Integrate $g(x, y, z) = xyz$ over the surface of the rectangular solid bounded by the planes $x = \pm a$, $y = \pm b$, and $z = \pm c$.

17. Integrate $g(x, y, z) = x + y + z$ over the portion of the plane $2x + 2y + z = 2$ that lies in the first octant.

18. Integrate $g(x, y, z) = x\sqrt{y^2 + 4}$ over the surface cut from the parabolic cylinder $y^2 + 4z = 16$ by the planes $x = 0, x = 1$, and $z = 0$.

Flux Across a Surface

In Exercises 19 and 20, find the flux of the field **F** across the portion of the given surface in the specified direction.

19. $\mathbf{F}(x, y, z) = -\mathbf{i} + 2\mathbf{j} + 3\mathbf{k}$
    S: rectangular surface $z = 0$, $0 \le x \le 2$, $0 \le y \le 3$, direction $\mathbf{k}$

20. $\mathbf{F}(x, y, z) = yx^2\mathbf{i} - 2\mathbf{j} + xz\mathbf{k}$
    S: rectangular surface $y = 0$, $-1 \le x \le 2$, $2 \le z \le 7$, direction $-\mathbf{j}$

In Exercises 21–26, find the flux of the field **F** across the portion of the sphere $x^2 + y^2 + z^2 = a^2$ in the first octant in the direction away from the origin.

21. $\mathbf{F}(x, y, z) = z\mathbf{k}$

22. $\mathbf{F}(x, y, z) = -y\mathbf{i} + x\mathbf{j}$

23. $\mathbf{F}(x, y, z) = y\mathbf{i} - x\mathbf{j} + \mathbf{k}$

24. $\mathbf{F}(x, y, z) = zx\mathbf{i} + zy\mathbf{j} + z^2\mathbf{k}$

25. $\mathbf{F}(x, y, z) = x\mathbf{i} + y\mathbf{j} + z\mathbf{k}$

26. $\mathbf{F}(x, y, z) = \dfrac{x\mathbf{i} + y\mathbf{j} + z\mathbf{k}}{\sqrt{x^2 + y^2 + z^2}}$

27. Find the flux of the field $\mathbf{F}(x, y, z) = z^2\mathbf{i} + x\mathbf{j} - 3z\mathbf{k}$ upward through the surface cut from the parabolic cylinder $z = 4 - y^2$ by the planes $x = 0, x = 1$, and $z = 0$.

28. Find the flux of the field $\mathbf{F}(x, y, z) = 4x\mathbf{i} + 4y\mathbf{j} + 2\mathbf{k}$ outward (away from the z-axis) through the surface cut from the bottom of the paraboloid $z = x^2 + y^2$ by the plane $z = 1$.

29. Let S be the portion of the cylinder $y = e^x$ in the first octant that projects parallel to the x-axis onto the rectangle $R_{yz}: 1 \le y \le 2$, $0 \le z \le 1$ in the yz-plane (Fig. 14.50). Let **n** be the unit vector normal to S that points away from the yz-plane. Find the flux of the field $\mathbf{F}(x, y, z) = -2\mathbf{i} + 2y\mathbf{j} + z\mathbf{k}$ across S in the direction of **n**.

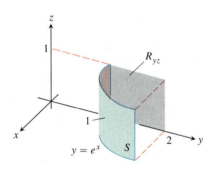

14.50 The surface and region in Exercise 29.

30. Let S be the portion of the cylinder $y = \ln x$ in the first octant whose projection parallel to the y-axis onto the xz-plane is the rectangle $R_{xz}: 1 \le x \le e, 0 \le z \le 1$. Let **n** be the unit vector normal to S that points away from the xz-plane. Find the flux of $\mathbf{F} = 2y\mathbf{j} + z\mathbf{k}$ through S in the direction of **n**.

31. Find the outward flux of the field $\mathbf{F} = 2xy\mathbf{i} + 2yz\mathbf{j} + 2xz\mathbf{k}$ across the surface of the cube cut from the first octant by the planes $x = a$, $y = a$, $z = a$.

32. Find the outward flux of the field $\mathbf{F} = xz\mathbf{i} + yz\mathbf{j} + \mathbf{k}$ across the surface of the upper cap cut from the solid sphere $x^2 + y^2 + z^2 \le 25$ by the plane $z = 3$.

Moments and Masses

33. Find the centroid of the portion of the sphere $x^2 + y^2 + z^2 = a^2$ that lies in the first octant.

34. Find the centroid of the surface cut from the cylinder $y^2 + z^2 = 9$, $z \ge 0$, by the planes $x = 0$ and $x = 3$ (resembles the surface in Example 4).

35. Find the center of mass and the moment of inertia and radius of gyration about the z-axis of a thin shell of constant density δ cut from the cone $x^2 + y^2 - z^2 = 0$ by the planes $z = 1$ and $z = 2$.

36. Find the moment of inertia about the z-axis of a thin shell of constant density δ cut from the cone $4x^2 + 4y^2 - z^2 = 0$, $z \geq 0$, by the circular cylinder $x^2 + y^2 = 2x$ (Fig. 14.51).

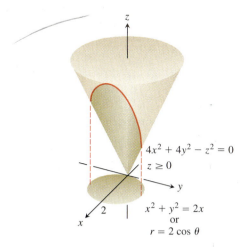

$$4x^2 + 4y^2 - z^2 = 0$$
$$z \geq 0$$

$$x^2 + y^2 = 2x$$
or
$$r = 2 \cos \theta$$

14.51 The surface in Exercise 36.

37. a) Find the moment of inertia about a diameter of a thin spherical shell of radius a and constant density δ. (Work with a hemispherical shell and double the result.)

b) Use the Parallel Axis Theorem (Exercises 13.5) and the result in (a) to find the moment of inertia about a line tangent to the shell.

38. a) Find the centroid of the lateral surface of a solid cone of base radius a and height h (cone surface minus the base).

b) Use Pappus's formula (Exercises 13.5) and the result in (a) to find the centroid of the complete surface of a solid cone (side plus base).

c) A cone of radius a and height h is joined to a hemisphere of radius a to make a surface S that resembles an ice cream cone. Use Pappus's formula and the results in (a) and Example 5 to find the centroid of S. How high does the cone have to be to place the centroid in the plane shared by the bases of the hemisphere and cone?

Special Formulas for Surface Area

If S is the surface defined by a function $z = f(x, y)$ that has continuous first partial derivatives throughout a region R_{xy} in the xy-plane (Fig. 14.52), then S is also the level surface $F(x, y, z) = 0$ of the function $F(x, y, z) = f(x, y) - z$. Taking the unit normal to R_{xy} to be $\mathbf{p} = \mathbf{k}$ then gives

$$|\nabla F| = |f_x \mathbf{i} + f_y \mathbf{j} - \mathbf{k}| = \sqrt{f_x{}^2 + f_y{}^2 + 1},$$
$$|\nabla F \cdot \mathbf{p}| = |(f_x \mathbf{i} + f_y \mathbf{j} - \mathbf{k}) \cdot \mathbf{k}| = |-1| = 1,$$

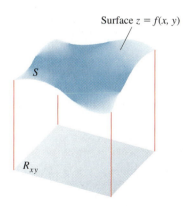

Surface $z = f(x, y)$

14.52 For a surface $z = f(x, y)$, the surface area formula in Eq. (5) takes the form

$$A = \iint_{R_{xy}} \sqrt{f_x{}^2 + f_y{}^2 + 1} \, dx \, dy.$$

and

$$A = \iint_{R_{xy}} \frac{|\nabla F|}{|\nabla F \cdot \mathbf{p}|} \, dA = \iint_{R_{xy}} \sqrt{f_x{}^2 + f_y{}^2 + 1} \, dx \, dy. \quad (14)$$

Similarly, the area of a smooth surface $x = f(y, z)$ over a region R_{yz} in the yz-plane is

$$A = \iint_{R_{yz}} \sqrt{f_y{}^2 + f_z{}^2 + 1} \, dy \, dz, \quad (15)$$

and the area of a smooth $y = f(x, z)$ over a region R_{xz} in the xz-plane is

$$A = \iint_{R_{xz}} \sqrt{f_x{}^2 + f_z{}^2 + 1} \, dx \, dz. \quad (16)$$

Use Eqs. (14)–(16) to find the areas of the surfaces in Exercises 39–44.

39. The surface cut from the bottom of the paraboloid $z = x^2 + y^2$ by the plane $z = 3$

40. The surface cut from the "nose" of the paraboloid $x = 1 - y^2 - z^2$ by the yz-plane

41. The portion of the cone $z = \sqrt{x^2 + y^2}$ that lies over the region between the circle $x^2 + y^2 = 1$ and the ellipse $9x^2 + 4y^2 = 36$ in the xy-plane. (*Hint:* Use formulas from geometry to find the area of the region.)

42. The triangle cut from the plane $2x + 6y + 3z = 6$ by the bounding planes of the first octant. Calculate the area three ways, once with each area formula

43. The surface in the first octant cut from the cylinder $y = (2/3)z^{3/2}$ by the planes $x = 1$ and $y = 16/3$

44. The portion of the plane $y + z = 4$ that lies above the region cut from the first quadrant of the xz-plane by the parabola $x = 4 - z^2$

14.6 Parametrized Surfaces

We have defined curves in the plane in three different ways:

Explicit form:	$y = f(x)$
Implicit form:	$F(x, y) = 0$
Parametric vector form:	$\mathbf{r}(t) = f(t)\mathbf{i} + g(t)\mathbf{j}, \quad a \leq t \leq b.$

We have analogous definitions of surfaces in space:

Explicit form:	$z = f(x, y)$
Implicit form:	$F(x, y, z) = 0.$

There is also a parametric form that gives the position of a point on the surface as a vector function of two variables. The present section extends the investigation of surface area and surface integrals to surfaces described parametrically.

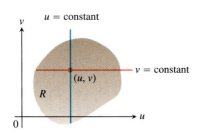

Parametrizations of Surfaces

Let

$$\mathbf{r}(u, v) = f(u, v)\mathbf{i} + g(u, v)\mathbf{j} + h(u, v)\mathbf{k} \tag{1}$$

be a continuous vector function that is defined on a region R in the uv-plane and one-to-one on the interior of R (Fig. 14.53). We call the range of $\mathbf{r}$ the **surface** S defined or traced by $\mathbf{r}$, and Eq. (1) together with the domain R constitute a **parametrization** of the surface. The variables u and v are the **parameters,** and R is the **parameter domain.** To simplify our discussion, we will take R to be a rectangle defined by inequalities of the form $a \leq u \leq b, c \leq v \leq d$. The requirement that $\mathbf{r}$ be one-to-one on the interior of R ensures that S does not cross itself. Notice that Eq. (1) is the vector equivalent of *three* parametric equations:

$$x = f(u, v), \qquad y = g(u, v), \qquad z = h(u, v).$$

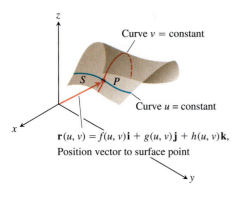

14.53 A parametrized surface.

EXAMPLE 1 Find a parametrization of the cone

$$z = \sqrt{x^2 + y^2}, \qquad 0 \leq z \leq 1.$$

Solution Here, cylindrical coordinates provide everything we need. A typical point (x, y, z) on the cone (Fig. 14.54) has $x = r \cos \theta$, $y = r \sin \theta$, and $z = \sqrt{x^2 + y^2} = r$, with $0 \leq r \leq 1$ and $0 \leq \theta \leq 2\pi$. Taking $u = r$ and $v = \theta$ in Eq. (1) gives the parametrization

$$\mathbf{r}(r, \theta) = (r \cos \theta)\mathbf{i} + (r \sin \theta)\mathbf{j} + r\mathbf{k}, \qquad 0 \leq r \leq 1, \quad 0 \leq \theta \leq 2\pi. \quad \square$$

EXAMPLE 2 Find a parametrization of the sphere $x^2 + y^2 + z^2 = a^2$.

Solution Spherical coordinates provide what we need. A typical point (x, y, z) on the sphere (Fig. 14.55) has $x = a \sin \phi \cos \theta$, $y = a \sin \phi \sin \theta$, and $z = a \cos \phi$, $0 \leq \phi \leq \pi$, $0 \leq \theta \leq 2\pi$. Taking $u = \phi$ and $v = \theta$ in Eq. (1) gives the parametrization

$$\mathbf{r}(\phi, \theta) = (a \sin \phi \cos \theta)\mathbf{i} + (a \sin \phi \sin \theta)\mathbf{j} + (a \cos \phi)\mathbf{k},$$

$$0 \leq \phi \leq \pi, \quad 0 \leq \theta \leq 2\pi. \quad \square$$

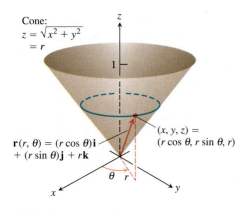

14.54 The cone in Example 1.

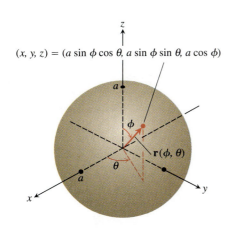

$(x, y, z) = (a \sin \phi \cos \theta, a \sin \phi \sin \theta, a \cos \phi)$

14.55 The sphere in Example 2.

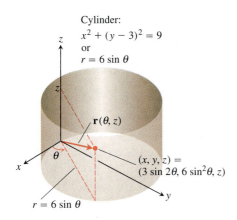

Cylinder:
$x^2 + (y - 3)^2 = 9$
or
$r = 6 \sin \theta$

$\mathbf{r}(\theta, z)$

$(x, y, z) = (3 \sin 2\theta, 6 \sin^2 \theta, z)$

$r = 6 \sin \theta$

14.56 The cylinder in Example 3.

EXAMPLE 3   Find a parametrization of the cylinder

$$x^2 + (y - 3)^2 = 9, \qquad 0 \le z \le 5.$$

Solution   In cylindrical coordinates, a point (x, y, z) has $x = r \cos \theta$, $y = r \sin \theta$, and $z = z$. For points on the cylinder $x^2 + (y - 3)^2 = 9$ (Fig. 14.56), $r = 6 \sin \theta$, $0 \le \theta \le \pi$ (Section 10.7, Example 5). A typical point on the cylinder therefore has

$$x = r \cos \theta = 6 \sin \theta \cos \theta = 3 \sin 2\theta$$

$$y = r \sin \theta = 6 \sin^2 \theta$$

$$z = z.$$

Taking $u = \theta$ and $v = z$ in Eq. (1) gives the parametrization

$$\mathbf{r}(\theta, z) = (3 \sin 2\theta)\,\mathbf{i} + (6 \sin^2 \theta)\,\mathbf{j} + z\,\mathbf{k}, \qquad 0 \le \theta \le \pi, \quad 0 \le z \le 5. \quad \square$$

Surface Area

Our goal is to find a double integral for calculating the area of a curved surface S based on the parametrization

$$\mathbf{r}(u, v) = f(u, v)\,\mathbf{i} + g(u, v)\,\mathbf{j} + h(u, v)\,\mathbf{k}, \qquad a \le u \le b, \quad c \le v \le d.$$

We need to assume that S is smooth enough for the construction we are about to carry out. The definition of smoothness involves the partial derivatives of $\mathbf{r}$ with respect to u and v:

$$\mathbf{r}_u = \frac{\partial \mathbf{r}}{\partial u} = \frac{\partial f}{\partial u}\mathbf{i} + \frac{\partial g}{\partial u}\mathbf{j} + \frac{\partial h}{\partial u}\mathbf{k}$$

$$\mathbf{r}_v = \frac{\partial \mathbf{r}}{\partial v} = \frac{\partial f}{\partial v}\mathbf{i} + \frac{\partial g}{\partial v}\mathbf{j} + \frac{\partial h}{\partial v}\mathbf{k}.$$

Definition

A parametrized surface $\mathbf{r}(u, v) = f(u, v)\,\mathbf{i} + g(u, v)\,\mathbf{j} + h(u, v)\,\mathbf{k}$ is **smooth** if $\mathbf{r}_u$ and $\mathbf{r}_v$ are continuous and $\mathbf{r}_u \times \mathbf{r}_v$ is never zero on the parameter domain.

Now consider a small rectangle ΔA_{uv} in R with sides on the lines $u = u_0$, $u = u_0 + \Delta u$, $v = v_0$, and $v = v_0 + \Delta v$ (Fig. 14.57, on the following page). Each side of ΔA_{uv} maps to a curve on the surface S, and together these four curves bound a "curved area element" $\Delta \sigma_{uv}$. In the notation of the figure, the side $v = v_0$ maps to curve C_1, the side $u = u_0$ maps to C_2, and their common vertex (u_0, v_0) maps to P_0. Figure 14.58 (on the following page) shows an enlarged view of $\Delta \sigma_{uv}$. The vector $\mathbf{r}_u(u_0, v_0)$ is tangent to C_1 at P_0. Likewise, $\mathbf{r}_v(u_0, v_0)$ is tangent to C_2 at P_0. The cross product $\mathbf{r}_u \times \mathbf{r}_v$ is normal to the surface at P_0. (Here is where we begin to use the assumption that S is smooth. We want to be sure that $\mathbf{r}_u \times \mathbf{r}_v \ne \mathbf{0}$.)

We next approximate the surface element $\Delta \sigma_{uv}$ by the parallelogram on the tangent plane whose sides are determined by the vectors $\Delta u\,\mathbf{r}_u$ and $\Delta v\,\mathbf{r}_v$ (Fig. 14.59, on the following page). The area of this parallelogram is

$$|\Delta u\,\mathbf{r}_u \times \Delta v\,\mathbf{r}_v| = |\mathbf{r}_u \times \mathbf{r}_v|\Delta u\,\Delta v. \tag{2}$$

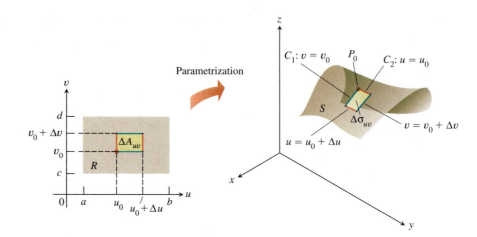

Parametrization

14.57 A rectangular area element ΔA_{uv} in the uv-plane maps onto a curved area element $\Delta\sigma_{uv}$ on S.

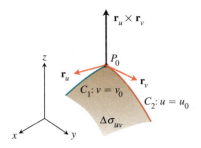

14.58 A magnified view of a surface area element $\Delta\sigma_{uv}$.

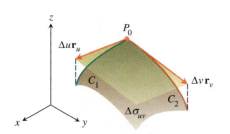

14.59 The parallelogram determined by the vectors $\Delta u\,\mathbf{r}_u$ and $\Delta v\,\mathbf{r}_v$ approximates the surface area element $\Delta\sigma_{uv}$.

A partition of the region R in the uv-plane by rectangular regions ΔA_{uv} generates a partition of the surface S into surface area elements $\Delta\sigma_{uv}$. We approximate the area of each surface element $\Delta\sigma_{uv}$ by the parallelogram area in Eq. (2) and sum these areas together to obtain an approximation of the area of S:

$$\sum_u \sum_v |\mathbf{r}_u \times \mathbf{r}_v|\,\Delta u\,\Delta v. \tag{3}$$

As Δu and Δv approach zero independently, the continuity of $\mathbf{r}_u$ and $\mathbf{r}_v$ guarantees that the sum in Eq. (3) approaches the double integral $\int_c^d \int_a^b |\mathbf{r}_u \times \mathbf{r}_v|\,du\,dv$. This double integral gives the area of the surface S.

Parametric Formula for the Area of a Smooth Surface

The area of the smooth surface

$$\mathbf{r}(u, v) = f(u, v)\,\mathbf{i} + g(u, v)\,\mathbf{j} + h(u, v)\,\mathbf{k}, \qquad a \le u \le b, \quad c \le v \le d$$

is

$$A = \int_c^d \int_a^b |\mathbf{r}_u \times \mathbf{r}_v|\,du\,dv. \tag{4}$$

As in Section 14.5, we can abbreviate the integral in (4) by writing $d\sigma$ for $|\mathbf{r}_u \times \mathbf{r}_v|\,du\,dv$.

Surface Area Differential and the Differential Formula for Surface Area

$$d\sigma = |\mathbf{r}_u \times \mathbf{r}_v|\,du\,dv \qquad\qquad \iint_S d\sigma \tag{5}$$

surface area differential

differential formula for surface area

EXAMPLE 4   Find the surface area of the cone in Example 1 (Fig. 14.54).

Solution   In Example 1 we found the parametrization

$$\mathbf{r}(r, \theta) = (r \cos \theta)\,\mathbf{i} + (r \sin \theta)\,\mathbf{j} + r\,\mathbf{k}, \qquad 0 \le r \le 1, \qquad 0 \le \theta \le 2\pi.$$

To apply Eq. (4) we first find $\mathbf{r}_r \times \mathbf{r}_\theta$:

$$\mathbf{r}_r \times \mathbf{r}_\theta = \begin{vmatrix} \mathbf{i} & \mathbf{j} & \mathbf{k} \\ \cos \theta & \sin \theta & 1 \\ -r \sin \theta & r \cos \theta & 0 \end{vmatrix}$$

$$= -(r \cos \theta)\,\mathbf{i} - (r \sin \theta)\,\mathbf{j} + \underbrace{(r \cos^2 \theta + r \sin^2 \theta)}_{r}\,\mathbf{k}.$$

Thus, $|\mathbf{r}_r \times \mathbf{r}_\theta| = \sqrt{r^2 \cos^2 \theta + r^2 \sin^2 \theta + r^2} = \sqrt{2r^2} = \sqrt{2}\,r.$ The area of the cone is

$$A = \int_0^{2\pi} \int_0^1 |\mathbf{r}_r \times \mathbf{r}_\theta|\, dr\, d\theta \qquad \text{Eq. (4) with } u = r,\, v = \theta$$

$$= \int_0^{2\pi} \int_0^1 \sqrt{2}\,r\, dr\, d\theta = \int_0^{2\pi} \frac{\sqrt{2}}{2}\, d\theta = \frac{\sqrt{2}}{2}\,(2\pi) = \pi\sqrt{2}. \qquad \square$$

EXAMPLE 5   Find the surface area of a sphere of radius a.

Solution   We use the parametrization from Example 2:

$$\mathbf{r}(\phi, \theta) = (a \sin \phi \cos \theta)\,\mathbf{i} + (a \sin \phi \sin \theta)\,\mathbf{j} + (a \cos \phi)\,\mathbf{k},$$

$$0 \le \phi \le \pi, \quad 0 \le \theta \le 2\pi.$$

For $\mathbf{r}_\phi \times \mathbf{r}_\theta$ we get

$$\mathbf{r}_\phi \times \mathbf{r}_\theta = \begin{vmatrix} \mathbf{i} & \mathbf{j} & \mathbf{k} \\ a \cos \phi \cos \theta & a \cos \phi \sin \theta & -a \sin \phi \\ -a \sin \phi \sin \theta & a \sin \phi \cos \theta & 0 \end{vmatrix}$$

$$= (a^2 \sin^2 \phi \cos \theta)\,\mathbf{i} + (a^2 \sin^2 \phi \sin \theta)\,\mathbf{j} + (a^2 \sin \phi \cos \phi)\,\mathbf{k}.$$

Thus,

$$|\mathbf{r}_\phi \times \mathbf{r}_\theta| = \sqrt{a^4 \sin^4 \phi \cos^2 \theta + a^4 \sin^4 \phi \sin^2 \theta + a^4 \sin^2 \phi \cos^2 \phi}$$

$$= \sqrt{a^4 \sin^4 \phi + a^4 \sin^2 \phi \cos^2 \phi} = \sqrt{a^4 \sin^2 \phi\,(\sin^2 \phi + \cos^2 \phi)}$$

$$= a^2 \sqrt{\sin^2 \phi} = a^2 \sin \phi,$$

since $\sin \phi \ge 0$ for $0 \le \phi \le \pi$. Therefore the area of the sphere is

$$A = \int_0^{2\pi} \int_0^\pi a^2 \sin \phi\, d\phi\, d\theta$$

$$= \int_0^{2\pi} \left[-a^2 \cos \phi \right]_0^\pi d\theta = \int_0^{2\pi} 2a^2\, d\theta = 4\pi a^2. \qquad \square$$

Surface Integrals

Having found the formula for calculating the area of a parametrized surface, we can now integrate a function over the surface using the parametrized form.

Definition

If S is a smooth surface defined parametrically as $\mathbf{r}(u, v) = f(u, v)\mathbf{i} + g(u, v)\mathbf{j} + h(u, v)\mathbf{k}$, $a \leq u \leq b$, $c \leq v \leq d$, and $G(x, y, z)$ is a continuous function defined on S, then the **integral of G over S** is

$$\iint\limits_{S} G(x, y, z)\, d\sigma = \int_{c}^{d} \int_{a}^{b} G(f(u, v), g(u, v), h(u, v))|\mathbf{r}_u \times \mathbf{r}_v|\, du\, dv.$$

EXAMPLE 6   Integrate $G(x, y, z) = x^2$ over the cone $z = \sqrt{x^2 + y^2}$, $0 \leq z \leq 1$.

Solution   Continuing the work in Examples 1 and 4, we have $|\mathbf{r}_r \times \mathbf{r}_\theta| = \sqrt{2}\, r$ and

$$\iint\limits_{S} x^2\, d\sigma = \int_{0}^{2\pi} \int_{0}^{1} (r^2 \cos^2 \theta)(\sqrt{2}\, r)\, dr\, d\theta \qquad x = r \cos \theta$$

$$= \sqrt{2} \int_{0}^{2\pi} \int_{0}^{1} r^3 \cos^2 \theta\, dr\, d\theta$$

$$= \frac{\sqrt{2}}{4} \int_{0}^{2\pi} \cos^2 \theta\, d\theta = \frac{\sqrt{2}}{4} \left[\frac{\theta}{2} + \frac{1}{4} \sin 2\theta \right]_{0}^{2\pi} = \frac{\pi \sqrt{2}}{4}.$$

EXAMPLE 7   Find the flux of $\mathbf{F} = yz\,\mathbf{i} + x\,\mathbf{j} - z^2\,\mathbf{k}$ outward through the parabolic cylinder $y = x^2$, $0 \leq x \leq 1$, $0 \leq z \leq 4$ (Fig. 14.60).

Solution   On the surface we have $x = x$, $y = x^2$, and $z = z$, so we automatically have the parametrization $\mathbf{r}(x, z) = x\,\mathbf{i} + x^2\,\mathbf{j} + z\,\mathbf{k}$, $0 \leq x \leq 1$, $0 \leq z \leq 4$. The cross product of tangent vectors is

$$\mathbf{r}_x \times \mathbf{r}_z = \begin{vmatrix} \mathbf{i} & \mathbf{j} & \mathbf{k} \\ 1 & 2x & 0 \\ 0 & 0 & 1 \end{vmatrix} = 2x\,\mathbf{i} - \mathbf{j}.$$

The unit normal pointing outward from the surface is

$$\mathbf{n} = \frac{\mathbf{r}_x \times \mathbf{r}_z}{|\mathbf{r}_x \times \mathbf{r}_z|} = \frac{2x\,\mathbf{i} - \mathbf{j}}{\sqrt{4x^2 + 1}}.$$

On the surface, $y = x^2$, so the vector field is

$$\mathbf{F} = yz\,\mathbf{i} + x\,\mathbf{j} - z^2\,\mathbf{k} = x^2 z\,\mathbf{i} + x\,\mathbf{j} - z^2\,\mathbf{k}.$$

Thus,

$$\mathbf{F} \cdot \mathbf{n} = \frac{1}{\sqrt{4x^2 + 1}} \left((x^2 z)(2x) + (x)(-1) + (-z^2)(0) \right)$$

$$= \frac{2x^3 z - x}{\sqrt{4x^2 + 1}}.$$

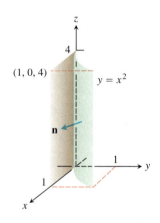

14.60 The parabolic surface in Example 7.

The flux of **F** outward through the surface is

$$\iint\limits_{S} \mathbf{F} \cdot \mathbf{n}\, d\sigma = \int_{0}^{4} \int_{0}^{1} \frac{2x^3z - x}{\sqrt{4x^2 + 1}}\, |\mathbf{r}_x \times \mathbf{r}_z|\, dx\, dz$$

$$= \int_{0}^{4} \int_{0}^{1} \frac{2x^3z - x}{\sqrt{4x^2 + 1}} \sqrt{4x^2 + 1}\, dx\, dz$$

$$= \int_{0}^{4} \int_{0}^{1} (2x^3z - x)\, dx\, dz = \int_{0}^{4} \left[\frac{1}{2}x^4z - \frac{1}{2}x^2 \right]_{x=0}^{x=1}\, dz$$

$$= \int_{0}^{4} \frac{1}{2}(z - 1)\, dz = \frac{1}{4}(z - 1)^2 \bigg]_{0}^{4}$$

$$= \frac{1}{4}(9) - \frac{1}{4}(1) = 2.$$

❑

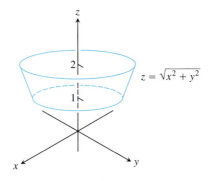

14.61 The cone frustum in Example 8.

EXAMPLE 8   Find the center of mass of a thin shell of constant density δ cut from the cone $z = \sqrt{x^2 + y^2}$ by the planes $z = 1$ and $z = 2$ (Fig. 14.61).

Solution   The symmetry of the surface about the z-axis tells us that $\bar{x} = \bar{y} = 0$. We find $\bar{z} = M_{xy}/M$. Working as in Examples 1 and 4 we have

$$\mathbf{r}(r, \theta) = r \cos\theta\, \mathbf{i} + r \sin\theta\, \mathbf{j} + r\, \mathbf{k}, \qquad 1 \le r \le 2, \quad 0 \le \theta \le 2\pi,$$

and

$$|\mathbf{r}_r \times \mathbf{r}_\theta| = \sqrt{2}\, r.$$

Therefore,

$$M = \iint\limits_{S} \delta\, d\sigma = \int_{0}^{2\pi} \int_{1}^{2} \delta\sqrt{2}\, r\, dr\, d\theta$$

$$= \delta\sqrt{2} \int_{0}^{2\pi} \left[\frac{r^2}{2} \right]_{1}^{2}\, d\theta = \delta\sqrt{2} \int_{0}^{2\pi} \left(2 - \frac{1}{2} \right)\, d\theta$$

$$= \delta\sqrt{2} \left[\frac{3\theta}{2} \right]_{0}^{2\pi} = 3\pi\, \delta\sqrt{2}$$

$$M_{xy} = \iint\limits_{S} \delta z\, d\sigma = \int_{0}^{2\pi} \int_{1}^{2} \delta r \sqrt{2}\, r\, dr\, d\theta$$

$$= \delta\sqrt{2} \int_{0}^{2\pi} \int_{1}^{2} r^2\, dr\, d\theta = \delta\sqrt{2} \int_{0}^{2\pi} \left[\frac{r^3}{3} \right]_{1}^{2}\, d\theta$$

$$= \delta\sqrt{2} \int_{0}^{2\pi} \frac{7}{3}\, d\theta = \frac{14}{3}\pi\, \delta\sqrt{2}$$

$$\bar{z} = \frac{M_{xy}}{M} = \frac{14\pi\, \delta\sqrt{2}}{3(3\pi\, \delta\sqrt{2})} = \frac{14}{9}.$$

The shell's center of mass is the point $(0, 0, 14/9)$.

❑

Exercises 14.6

Finding Parametrizations for Surfaces

In Exercises 1–16, find a parametrization of the surface. (There are many correct ways to do these, so your answers may not be the same as those in the back of the book.)

1. The paraboloid $z = x^2 + y^2, \quad z \le 4$

2. The paraboloid $z = 9 - x^2 - y^2, \quad z \ge 0$

3. The first-octant portion of the cone $z = \sqrt{x^2 + y^2}/2$ between the planes $z = 0$ and $z = 3$

4. The portion of the cone $z = 2\sqrt{x^2 + y^2}$ between the planes $z = 2$ and $z = 4$

5. The cap cut from the sphere $x^2 + y^2 + z^2 = 9$ by the cone $z = \sqrt{x^2 + y^2}$

6. The portion of the sphere $x^2 + y^2 + z^2 = 4$ in the first octant between the xy-plane and the cone $z = \sqrt{x^2 + y^2}$

7. The portion of the sphere $x^2 + y^2 + z^2 = 3$ between the planes $z = \sqrt{3}/2$ and $z = -\sqrt{3}/2$

8. The upper portion cut from the sphere $x^2 + y^2 + z^2 = 8$ by the plane $z = -2$

9. The surface cut from the parabolic cylinder $z = 4 - y^2$ by the planes $x = 0, x = 2,$ and $z = 0$

10. The surface cut from the parabolic cylinder $y = x^2$ by the planes $z = 0, z = 3,$ and $y = 2$

11. The portion of the cylinder $y^2 + z^2 = 9$ between the planes $x = 0$ and $x = 3$

12. The portion of the cylinder $x^2 + z^2 = 4$ above the xy-plane between the planes $y = -2$ and $y = 2$

13. The portion of the plane $x + y + z = 1$
    a)   inside the cylinder $x^2 + y^2 = 9$
    b)   inside the cylinder $y^2 + z^2 = 9$

14. The portion of the plane $x - y + 2z = 2$
    a)   inside the cylinder $x^2 + z^2 = 3$
    b)   inside the cylinder $y^2 + z^2 = 2$

15. The portion of the cylinder $(x - 2)^2 + z^2 = 4$ between the planes $y = 0$ and $y = 3$

16. The portion of the cylinder $y^2 + (z - 5)^2 = 25$ between the planes $x = 0$ and $x = 10$

17. The portion of the plane $y + 2z = 2$ inside the cylinder $x^2 + y^2 = 1$

18. The portion of the plane $z = -x$ inside the cylinder $x^2 + y^2 = 4$

19. The portion of the cone $z = 2\sqrt{x^2 + y^2}$ between the planes $z = 2$ and $z = 6$

20. The portion of the cone $z = \sqrt{x^2 + y^2}/3$ between the planes $z = 1$ and $z = 4/3$

21. The portion of the cylinder $x^2 + y^2 = 1$ between the planes $z = 1$ and $z = 4$

22. The portion of the cylinder $x^2 + z^2 = 10$ between the planes $y = -1$ and $y = 1$

23. The cap cut from the paraboloid $z = 2 - x^2 - y^2$ by the cone $z = \sqrt{x^2 + y^2}$

24. The portion of the paraboloid $z = x^2 + y^2$ between the planes $z = 1$ and $z = 4$

25. The lower portion cut from the sphere $x^2 + y^2 + z^2 = 2$ by the cone $z = \sqrt{x^2 + y^2}$

26. The portion of the sphere $x^2 + y^2 + z^2 = 4$ between the planes $z = -1$ and $z = \sqrt{3}$

Parametrized Surface Integrals

In Exercises 27–34, integrate the given function over the given surface.

27. $G(x, y, z) = x$, over the parabolic cylinder $y = x^2, 0 \le x \le 2, 0 \le z \le 3$

28. $G(x, y, z) = z$, over the cylindrical surface $y^2 + z^2 = 4, z \ge 0, 1 \le x \le 4$

29. $G(x, y, z) = x^2$, over the unit sphere $x^2 + y^2 + z^2 = 1$

30. $G(x, y, z) = z^2$, over the hemisphere $x^2 + y^2 + z^2 = a^2, z \ge 0$

31. $F(x, y, z) = z$, over the portion of the plane $x + y + z = 4$ that lies above the square $0 \le x \le 1, 0 \le y \le 1$, in the xy-plane

32. $F(x, y, z) = z - x$, over the cone $z = \sqrt{x^2 + y^2}, 0 \le z \le 1$

33. $H(x, y, z) = x^2\sqrt{5 - 4z}$, over the parabolic dome $z = 1 - x^2 - y^2, z \ge 0$

34. $H(x, y, z) = yz$, over the part of the sphere $x^2 + y^2 + z^2 = 4$ that lies above the cone $z = \sqrt{x^2 + y^2}$

Areas of Parametrized Surfaces

In Exercises 17–26, use a parametrization to express the area of the surface as a double integral. Then evaluate the integral. (There are many correct ways to set up the integrals, so your integrals may not be the same as those in the back of the book. They should have the same values, however.)

Flux Across Parametrized Surfaces

In Exercises 35–44, use a parametrization to find the flux $\iint_S \mathbf{F} \cdot \mathbf{n} \, d\sigma$ across the surface in the given direction.

35. $\mathbf{F} = z^2\mathbf{i} + x\mathbf{j} - 3z\mathbf{k}$ outward (normal away from the x-axis) through the surface cut from the parabolic cylinder $z = 4 - y^2$ by the planes $x = 0, x = 1,$ and $z = 0$

36. $\mathbf{F} = x^2 \mathbf{j} - xz \mathbf{k}$ outward (normal away from the yz-plane) through the surface cut from the parabolic cylinder $y = x^2$, $-1 \le x \le 1$, by the planes $z = 0$ and $z = 2$

37. $\mathbf{F} = z \mathbf{k}$ across the portion of the sphere $x^2 + y^2 + z^2 = a^2$ in the first octant in the direction away from the origin

38. $\mathbf{F} = x \mathbf{i} + y \mathbf{j} + z \mathbf{k}$ across the sphere $x^2 + y^2 + z^2 = a^2$ in the direction away from the origin

39. $\mathbf{F} = 2xy \mathbf{i} + 2yz \mathbf{j} + 2xz \mathbf{k}$ upward across the portion of the plane $x + y + z = 2a$ that lies above the square $0 \le x \le a, 0 \le y \le a$, in the xy-plane

40. $\mathbf{F} = x \mathbf{i} + y \mathbf{j} + z \mathbf{k}$ outward through the portion of the cylinder $x^2 + y^2 = 1$ cut by the planes $z = 0$ and $z = a$

41. $\mathbf{F} = xy \mathbf{i} - z \mathbf{k}$ outward (normal away from the z-axis) through the cone $z = \sqrt{x^2 + y^2}, 0 \le z \le 1$

42. $\mathbf{F} = y^2 \mathbf{i} + xz \mathbf{j} - \mathbf{k}$ outward (normal away from the z-axis) through the cone $z = 2\sqrt{x^2 + y^2}, 0 \le z \le 2$

43. $\mathbf{F} = -x \mathbf{i} - y \mathbf{j} + z^2 \mathbf{k}$ outward (normal away from the z-axis) through the portion of the cone $z = \sqrt{x^2 + y^2}$ between the planes $z = 1$ and $z = 2$

44. $\mathbf{F} = 4x \mathbf{i} + 4y \mathbf{j} + 2 \mathbf{k}$ outward (normal away from the z-axis) through the surface cut from the bottom of the paraboloid $z = x^2 + y^2$ by the plane $z = 1$

Moments and Masses

45. Find the centroid of the portion of the sphere $x^2 + y^2 + z^2 = a^2$ that lies in the first octant.

46. Find the center of mass and the moment of inertia and radius of gyration about the z-axis of a thin shell of constant density δ cut from the cone $x^2 + y^2 - z^2 = 0$ by the planes $z = 1$ and $z = 2$.

47. Find the moment of inertia about the z-axis of a thin spherical shell $x^2 + y^2 + z^2 = a^2$ of constant density δ.

48. Find the moment of inertia about the z-axis of a thin conical shell $z = \sqrt{x^2 + y^2}, 0 \le z \le 1$, of constant density δ.

Tangent Planes to Parametrized Surfaces

The tangent plane at a point $P_0 (f (u_0, v_0), g (u_0, v_0), h (u_0, v_0))$ on a parametrized surface $\mathbf{r} (u, v) = f (u, v) \mathbf{i} + g (u, v) \mathbf{j} + h (u, v) \mathbf{k}$ is the plane through P_0 normal to the vector $\mathbf{r}_u (u_0, v_0) \times \mathbf{r}_v (u_0, v_0)$, which is the cross product of the tangent vectors $\mathbf{r}_u (u_0, v_0)$ and $\mathbf{r}_v (u_0, v_0)$ at P_0. In Exercises 49–52, find an equation for the plane that is tangent to the surface at the given point P_0. Then find a Cartesian equation for the surface and sketch the surface and tangent plane together.

49. The cone $\mathbf{r} (r, \theta) = (r \cos \theta) \mathbf{i} + (r \sin \theta) \mathbf{j} + r \mathbf{k}$, $r \ge 0, 0 \le \theta \le 2\pi$ at the point $P_0 \left(\sqrt{2}, \sqrt{2}, 2 \right)$ corresponding to $(r, \theta) = (2, \pi/4)$.

50. The hemisphere surface

$$\mathbf{r} (\phi, \theta) = (4 \sin \phi \cos \theta) \mathbf{i} + (4 \sin \phi \sin \theta) \mathbf{j} + (4 \cos \phi) \mathbf{k},$$

$0 \le \phi \le \pi/2, 0 \le \theta \le 2\pi$, at the point $P_0 \left(\sqrt{2}, \sqrt{2}, 2\sqrt{3} \right)$ corresponding to $(\phi, \theta) = (\pi/6, \pi/4)$

51. The circular cylinder $\mathbf{r} (\theta, z) = (3 \sin 2\theta) \mathbf{i} + (6 \sin^2 \theta) \mathbf{j} + z \mathbf{k}$, $0 \le \theta \le \pi$, at the point $P_0 \left(\dfrac{3\sqrt{3}}{2}, \dfrac{9}{2}, 0 \right)$ corresponding to (θ, z) = $(\pi/3, 0)$ (See Example 3.)

52. The parabolic cylinder surface $\mathbf{r} (x, y) = x \mathbf{i} + y \mathbf{j} - x^2 \mathbf{k}, -\infty < x < \infty, -\infty < y < \infty$, at the point $P_0 (1, 2, -1)$ corresponding to $(x, y) = (1, 2)$

Further Examples of Parametrizations

53. a) A *torus of revolution* (doughnut) is the surface obtained by rotating a circle C in the xz-plane about the z-axis in space. If the radius of C is $r > 0$ and the center is $(R, 0, 0)$, show that a parametrization of the torus is

$$\mathbf{r} (u, v) = ((R + r \cos u) \cos v) \mathbf{i}$$
$$+ ((R + r \cos u) \sin v) \mathbf{j} + (r \sin u) \mathbf{k},$$

where $0 \le u \le 2\pi$ and $0 \le v \le 2\pi$ are the angles in Fig. 14.62.

b) Show that the surface area of the torus is $A = 4\pi^2 R r$.

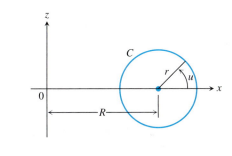

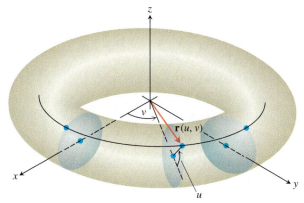

14.62 The torus surface in Exercise 53.

54. *Parametrization of a surface of revolution.* Suppose the parametrized curve C: $(f (u), g (u))$ is revolved about the x-axis, where $g (u) > 0$ for $a \le u \le b$.

a) Show that

$$\mathbf{r} (u, v) = f (u) \mathbf{i} + (g (u) \cos v) \mathbf{j} + (g (u) \sin v) \mathbf{k}$$

is a parametrization of the resulting surface of revolution, where $0 \leq v \leq 2\pi$ is the angle from the xy-plane to the point $\mathbf{r}(u, v)$ on the surface. (See the accompanying figure.) Notice that $f(u)$ measures distance *along* the axis of revolution and $g(u)$ measures distance *from* the axis of revolution.

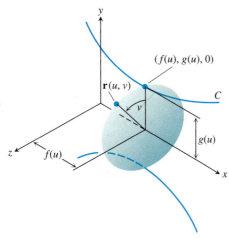

b) Find a parametrization for the surface obtained by revolving the curve $x = y^2$, $y \geq 0$, about the x-axis.

55. a) Recall the parametrization $x = a \cos \theta$, $y = b \sin \theta$, $0 \leq \theta \leq 2\pi$ for the ellipse $(x^2/a^2) + (y^2/b^2) = 1$ (Section 9.4, Example 5). Using the angles θ and ϕ as defined in spherical coordinates, show that

$$\mathbf{r}(\theta, \phi) = (a \cos \theta \cos \phi)\,\mathbf{i}$$
$$+ (b \sin \theta \cos \phi)\,\mathbf{j} + (c \sin \phi)\,\mathbf{k}$$

is a parametrization of the ellipsoid $(x^2/a^2) + (y^2/b^2) + (z^2/c^2) = 1$.

b) Write an integral for the surface area of the ellipsoid, but do not evaluate the integral.

56. a) Find a parametrization for the hyperboloid of one sheet $x^2 + y^2 - z^2 = 1$ in terms of the angle θ associated with the circle $x^2 + y^2 = r^2$ and the hyperbolic parameter u associated with the hyperbolic function $r^2 - z^2 = 1$. (See Section 6.10. Exercise 86.)

b) Generalize the result in (a) to the hyperboloid $(x^2/a^2) + (y^2/b^2) - (z^2/c^2) = 1$.

57. (*Continuation of Exercise 56.*) Find a Cartesian equation for the plane tangent to the hyperboloid $x^2 + y^2 - z^2 = 25$ at the point $(x_0, y_0, 0)$, where $x_0^2 + y_0^2 = 25$.

58. Find a parametrization of the hyperboloid of two sheets $(z^2/c^2) - (x^2/a^2) - (y^2/b^2) = 1$.

14.7 Stokes's Theorem

As we saw in Section 14.4, the circulation density or curl of a two-dimensional field $\mathbf{F} = M\,\mathbf{i} + N\,\mathbf{j}$ at a point (x, y) is described by the scalar quantity $(\partial N/\partial x - \partial M/\partial y)$. In three dimensions, the circulation around a point P in a plane is described with a vector. This vector is normal to the plane of the circulation (Fig. 14.63) and points in the direction that gives it a right-hand relation to the circulation line. The length of the vector gives the rate of the fluid's rotation, which usually varies as the circulation plane is tilted about P. It turns out that the vector of greatest circulation in a flow with velocity field $\mathbf{F} = M\,\mathbf{i} + N\,\mathbf{j} + P\,\mathbf{k}$ is

$$\operatorname{curl} \mathbf{F} = \left(\frac{\partial P}{\partial y} - \frac{\partial N}{\partial z}\right)\mathbf{i} + \left(\frac{\partial M}{\partial z} - \frac{\partial P}{\partial x}\right)\mathbf{j} + \left(\frac{\partial N}{\partial x} - \frac{\partial M}{\partial y}\right)\mathbf{k}. \qquad (1)$$

We get this information from Stokes's theorem, the generalization of the circulation-curl form of Green's theorem to space.

Del Notation

The formula for curl $\mathbf{F}$ in Eq. (1) is usually written using the symbolic operator

$$\nabla = \mathbf{i}\,\frac{\partial}{\partial x} + \mathbf{j}\,\frac{\partial}{\partial y} + \mathbf{k}\,\frac{\partial}{\partial z}. \qquad (2)$$

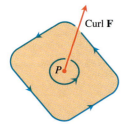

14.63 The circulation vector at a point P in a plane in a three-dimensional fluid flow. Notice its right-hand relation to the circulation line.

George Gabriel Stokes

Sir George Gabriel Stokes (1819–1903), one of the most influential scientific figures of his century, was Lucasian Professor of Mathematics at Cambridge University from 1849 until his death in 1903. His theoretical and experimental investigations covered hydrodynamics, elasticity, light, gravity, sound, heat, meteorology, and solar physics. He left electricity and magnetism to his friend William Thomson, Baron Kelvin of Largs. It is another one of those delightful quirks of history that the theorem we call Stokes's theorem isn't his theorem at all. He learned of it from Thomson in 1850 and a few years later included it among the questions on an examination he wrote for the Smith Prize. It has been known as Stokes's theorem ever since. As usual, things have balanced out. Stokes was the original discoverer of the principles of spectrum analysis that we now credit to Bunsen and Kirchhoff.

(The symbol ∇ is pronounced "del.") The curl of $\mathbf{F}$ is $\nabla \times \mathbf{F}$:

$$\nabla \times \mathbf{F} = \begin{vmatrix} \mathbf{i} & \mathbf{j} & \mathbf{k} \\ \dfrac{\partial}{\partial x} & \dfrac{\partial}{\partial y} & \dfrac{\partial}{\partial z} \\ M & N & P \end{vmatrix}$$

$$= \left(\frac{\partial P}{\partial y} - \frac{\partial N}{\partial z} \right) \mathbf{i} + \left(\frac{\partial M}{\partial z} - \frac{\partial P}{\partial x} \right) \mathbf{j} + \left(\frac{\partial N}{\partial x} - \frac{\partial M}{\partial y} \right) \mathbf{k} \qquad (3)$$

$$= \text{curl } \mathbf{F}.$$

$$\boxed{\text{curl } \mathbf{F} = \nabla \times \mathbf{F}} \qquad (4)$$

EXAMPLE 1   Find the curl of $\mathbf{F} = (x^2 - y)\mathbf{i} + 4z\mathbf{j} + x^2\mathbf{k}.$

Solution

$$\text{curl } \mathbf{F} = \nabla \times \mathbf{F} \qquad \text{Eq. (4)}$$

$$= \begin{vmatrix} \mathbf{i} & \mathbf{j} & \mathbf{k} \\ \dfrac{\partial}{\partial x} & \dfrac{\partial}{\partial y} & \dfrac{\partial}{\partial z} \\ x^2 - y & 4z & x^2 \end{vmatrix}$$

$$= \left(\frac{\partial}{\partial y}(x^2) - \frac{\partial}{\partial z}(4z) \right) \mathbf{i} - \left(\frac{\partial}{\partial x}(x^2) - \frac{\partial}{\partial z}(x^2 - y) \right) \mathbf{j}$$

$$+ \left(\frac{\partial}{\partial x}(4z) - \frac{\partial}{\partial y}(x^2 - y) \right) \mathbf{k}$$

$$= (0 - 4)\mathbf{i} - (2x - 0)\mathbf{j} + (0 + 1)\mathbf{k}$$

$$= -4\mathbf{i} - 2x\mathbf{j} + \mathbf{k} \qquad \square$$

As we will see, the operator ∇ has a number of other applications. For instance, when applied to a scalar function $f(x, y, z)$, it gives the gradient of f:

$$\nabla f = \frac{\partial f}{\partial x}\mathbf{i} + \frac{\partial f}{\partial y}\mathbf{j} + \frac{\partial f}{\partial z}\mathbf{k}.$$

This may now be read as "del f" as well as "grad f."

Stokes's Theorem

Stokes's theorem says that, under conditions normally met in practice, the circulation of a vector field around the boundary of an oriented surface in space in the direction counterclockwise with respect to the surface's unit normal vector field $\mathbf{n}$ (Fig. 14.64) equals the integral of the normal component of the curl of the field over the surface.

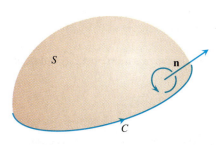

14.64 The orientation of the bounding curve C gives it a right-handed relation to the normal field **n**.

> ### Theorem 5
> ### Stokes's Theorem
>
> The circulation of $\mathbf{F} = M\mathbf{i} + N\mathbf{j} + P\mathbf{k}$ around the boundary C of an oriented surface S in the direction counterclockwise with respect to the surface's unit normal vector $\mathbf{n}$ equals the integral of $\nabla \times \mathbf{F} \cdot \mathbf{n}$ over S.
>
> $$\underset{\substack{\text{counterclockwise}\\ \text{circulation}}}{\oint_C \mathbf{F} \cdot d\mathbf{r}} = \underset{\text{curl integral}}{\iint_S \nabla \times \mathbf{F} \cdot \mathbf{n}\, d\sigma} \qquad (5)$$

Notice from Eq. (5) that if two different oriented surfaces S_1 and S_2 have the same boundary C, then their curl integrals are equal:

$$\iint_{S_1} \nabla \times \mathbf{F} \cdot \mathbf{n}_1\, d\sigma = \iint_{S_2} \nabla \times \mathbf{F} \cdot \mathbf{n}_2\, d\sigma.$$

Both curl integrals equal the counterclockwise circulation integral on the left side of Eq. (5) as long as the unit normal vectors $\mathbf{n}_1$ and $\mathbf{n}_2$ correctly orient the surfaces.

Naturally, we need some mathematical restrictions on $\mathbf{F}$, C, and S to ensure the existence of the integrals in Stokes's equation. The usual restrictions are that all the functions and derivatives involved be continuous.

If C is a curve in the xy-plane, oriented counterclockwise, and R is the region in the xy-plane bounded by C, then $d\sigma = dx\,dy$ and

$$(\nabla \times \mathbf{F}) \cdot \mathbf{n} = (\nabla \times \mathbf{F}) \cdot \mathbf{k} = \left(\frac{\partial N}{\partial x} - \frac{\partial M}{\partial y} \right). \qquad (6)$$

Green:

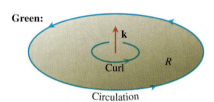

Stokes:

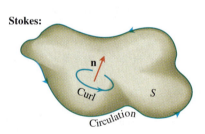

14.65 Green's theorem vs. Stokes's theorem.

Under these conditions, Stokes's equation becomes

$$\oint_C \mathbf{F} \cdot d\mathbf{r} = \iint_R \left(\frac{\partial N}{\partial x} - \frac{\partial M}{\partial y} \right) dx\,dy,$$

which is the circulation-curl form of the equation in Green's theorem. Conversely, by reversing these steps we can rewrite the circulation-curl form of Green's theorem for two-dimensional fields in del notation as

$$\oint_C \mathbf{F} \cdot d\mathbf{r} = \iint_R \nabla \times \mathbf{F} \cdot \mathbf{k}\, dA. \qquad (7)$$

See Fig. 14.65.

EXAMPLE 2   Evaluate Eq. (5) for the hemisphere S: $x^2 + y^2 + z^2 = 9$, $z \ge 0$, its bounding circle C: $x^2 + y^2 = 9$, $z = 0$, and the field $\mathbf{F} = y\mathbf{i} - x\mathbf{j}$.

Solution   We calculate the counterclockwise circulation around C (as viewed from above) using the parametrization $\mathbf{r}(\theta) = (3\cos\theta)\mathbf{i} + (3\sin\theta)\mathbf{j}$, $0 \le \theta \le 2\pi$:

$$d\mathbf{r} = (-3\sin\theta\, d\theta)\mathbf{i} + (3\cos\theta\, d\theta)\mathbf{j}$$

$$\mathbf{F} = y\mathbf{i} - x\mathbf{j} = (3\sin\theta)\mathbf{i} - (3\cos\theta)\mathbf{j}$$

$$\mathbf{F} \cdot d\mathbf{r} = -9\sin^2\theta\, d\theta - 9\cos^2\theta\, d\theta = -9\, d\theta$$

$$\oint_C \mathbf{F} \cdot d\mathbf{r} = \int_0^{2\pi} -9\, d\theta = -18\pi.$$

For the curl integral of **F**, we have

$$\nabla \times \mathbf{F} = \left(\frac{\partial P}{\partial y} - \frac{\partial N}{\partial z} \right) \mathbf{i} + \left(\frac{\partial M}{\partial z} - \frac{\partial P}{\partial x} \right) \mathbf{j} + \left(\frac{\partial N}{\partial x} - \frac{\partial M}{\partial y} \right) \mathbf{k}$$

$$= (0 - 0)\,\mathbf{i} + (0 - 0)\,\mathbf{j} + (-1 - 1)\,\mathbf{k} = -2\,\mathbf{k}$$

$$\mathbf{n} = \frac{x\,\mathbf{i} + y\,\mathbf{j} + z\,\mathbf{k}}{\sqrt{x^2 + y^2 + z^2}} = \frac{x\,\mathbf{i} + y\,\mathbf{j} + z\,\mathbf{k}}{3} \qquad \text{Outer unit normal}$$

$$d\sigma = \frac{3}{z}\,dA \qquad \text{Section 14.5, Example 5, with } a = 3$$

$$\nabla \times \mathbf{F} \cdot \mathbf{n}\,d\sigma = -\frac{2z}{3}\frac{3}{z}\,dA = -2\,dA$$

and

$$\iint_S \nabla \times \mathbf{F} \cdot \mathbf{n}\,d\sigma = \iint_{x^2 + y^2 \le 9} -2\,dA = -18\pi.$$

The circulation around the circle equals the integral of the curl over the hemisphere, as it should. ◻

EXAMPLE 3 Find the circulation of the field $\mathbf{F} = (x^2 - y)\,\mathbf{i} + 4z\,\mathbf{j} + x^2\,\mathbf{k}$ around the curve C in which the plane $z = 2$ meets the cone $z = \sqrt{x^2 + y^2}$, counterclockwise as viewed from above (Fig. 14.66).

Solution Stokes's theorem enables us to find the circulation by integrating over the surface of the cone. Traversing C in the counterclockwise direction viewed from above corresponds to taking the *inner* normal **n** to the cone (which has a positive z-component).

We parametrize the cone as

$$\mathbf{r}(r, \theta) = (r \cos \theta)\,\mathbf{i} + (r \sin \theta)\,\mathbf{j} + r\,\mathbf{k}, \qquad 0 \le r \le 2, \quad 0 \le \theta \le 2\pi.$$

We then have

$$\mathbf{n} = \frac{\mathbf{r}_r \times \mathbf{r}_\theta}{|\mathbf{r}_r \times \mathbf{r}_\theta|} = \frac{-(r \cos \theta)\,\mathbf{i} - (r \sin \theta)\,\mathbf{j} + r\,\mathbf{k}}{r\sqrt{2}} \qquad \text{Section 14.6, Example 4}$$

$$= \frac{1}{\sqrt{2}}(-(\cos \theta)\,\mathbf{i} - (\sin \theta)\,\mathbf{j} + \mathbf{k})$$

$$d\sigma = r\sqrt{2}\,dr\,d\theta \qquad \text{Section 14.6, Example 4}$$

$$\nabla \times \mathbf{F} = -4\,\mathbf{i} - 2x\,\mathbf{j} + \mathbf{k} \qquad \text{Example 1}$$

$$= -4\,\mathbf{i} - 2r \cos \theta\,\mathbf{j} + \mathbf{k}. \qquad x = r \cos \theta$$

Accordingly,

$$\nabla \times \mathbf{F} \cdot \mathbf{n} = \frac{1}{\sqrt{2}}(4 \cos \theta + 2r \cos \theta \sin \theta + 1)$$

$$= \frac{1}{\sqrt{2}}(4 \cos \theta + r \sin 2\theta + 1)$$

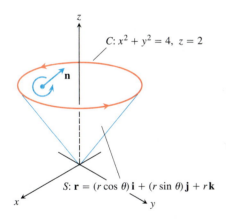

$C: x^2 + y^2 = 4,\ z = 2$

$S: \mathbf{r} = (r \cos \theta)\,\mathbf{i} + (r \sin \theta)\,\mathbf{j} + r\,\mathbf{k}$

14.66 The curve C and cone S in Example 3.

and the circulation is

$$\oint_C \mathbf{F} \cdot d\mathbf{r} = \iint_S \nabla \times \mathbf{F} \cdot \mathbf{n}\, d\sigma \qquad \text{Stokes's theorem}$$

$$= \int_0^{2\pi} \int_0^2 \frac{1}{\sqrt{2}} \left(4\cos\theta + r\sin 2\theta + 1 \right) \left(r\sqrt{2}\, dr\, d\theta \right) = 4\pi.$$

An Interpretation of $\nabla \times \mathbf{F}$

Suppose that $\mathbf{v}(x, y, z)$ is the velocity of a moving fluid whose density at (x, y, z) is $\delta(x, y, z)$, and let $\mathbf{F} = \delta\mathbf{v}$. Then

$$\oint_C \mathbf{F} \cdot d\mathbf{r}$$

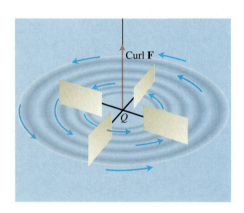

is the circulation of the fluid around the closed curve C. By Stokes's theorem, the circulation is equal to the flux of $\nabla \times \mathbf{F}$ through a surface S spanning C:

$$\oint_C \mathbf{F} \cdot d\mathbf{r} = \iint_S \nabla \times \mathbf{F} \cdot \mathbf{n}\, d\sigma.$$

Suppose we fix a point Q in the domain of $\mathbf{F}$ and a direction $\mathbf{u}$ at Q. Let C be a circle of radius ρ, with center at Q, whose plane is normal to $\mathbf{u}$. If $\nabla \times \mathbf{F}$ is continuous at Q, then the average value of the $\mathbf{u}$-component of $\nabla \times \mathbf{F}$ over the circular disk S bounded by C approaches the $\mathbf{u}$-component of $\nabla \times \mathbf{F}$ at Q as $\rho \to 0$:

$$(\nabla \times \mathbf{F} \cdot \mathbf{u})_Q = \lim_{\rho \to 0} \frac{1}{\pi\rho^2} \iint_S \nabla \times \mathbf{F} \cdot \mathbf{u}\, d\sigma. \tag{8}$$

14.67 The paddle wheel interpretation of curl **F**.

If we replace the double integral in Eq. (8) by the circulation, we get

$$(\nabla \times \mathbf{F} \cdot \mathbf{u})_Q = \lim_{\rho \to 0} \frac{1}{\pi\rho^2} \oint_C \mathbf{F} \cdot d\mathbf{r}. \tag{9}$$

The left-hand side of Eq. (9) has its maximum value when $\mathbf{u}$ is the direction of $\nabla \times \mathbf{F}$. When ρ is small, the limit on the right-hand side of Eq. (9) is approximately

$$\frac{1}{\pi\rho^2} \oint_C \mathbf{F} \cdot d\mathbf{r},$$

which is the circulation around C divided by the area of the disk (circulation density). Suppose that a small paddle wheel of radius ρ is introduced into the fluid at Q, with its axle directed along $\mathbf{u}$. The circulation of the fluid around C will affect the rate of spin of the paddle wheel. The wheel will spin fastest when the circulation integral is maximized; therefore it will spin fastest when the axle of the paddle wheel points in the direction of $\nabla \times \mathbf{F}$ (Fig. 14.67).

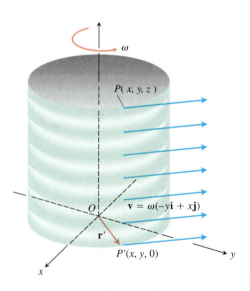

14.68 A steady rotational flow parallel to the xy-plane, with constant angular velocity ω in the positive (counterclockwise) direction.

EXAMPLE 4 A fluid of constant density δ rotates around the z-axis with velocity $\mathbf{v} = \omega(-y\,\mathbf{i} + x\,\mathbf{j})$, where ω is a positive constant called the *angular velocity* of the rotation (Fig. 14.68). If $\mathbf{F} = \delta\mathbf{v}$, find $\nabla \times \mathbf{F}$ and relate it to the circulation density.

Solution With $\mathbf{F} = \delta\mathbf{v} = -\delta\omega y\,\mathbf{i} + \delta\omega x\,\mathbf{j}$,

$$\nabla \times \mathbf{F} = \left(\frac{\partial P}{\partial y} - \frac{\partial N}{\partial z} \right) \mathbf{i} + \left(\frac{\partial M}{\partial z} - \frac{\partial P}{\partial x} \right) \mathbf{j} + \left(\frac{\partial N}{\partial x} - \frac{\partial M}{\partial y} \right) \mathbf{k}$$

$$= (0 - 0)\,\mathbf{i} + (0 - 0)\,\mathbf{j} + (\delta\omega - (-\delta\omega))\,\mathbf{k} = 2\,\delta\omega\,\mathbf{k}.$$

By Stokes's theorem, the circulation of **F** around a circle C of radius ρ bounding a disk S in a plane normal to $\nabla \times \mathbf{F}$, say the xy-plane, is

$$\oint_C \mathbf{F} \cdot d\mathbf{r} = \iint_S \nabla \times \mathbf{F} \cdot \mathbf{n}\, d\sigma = \iint_S 2\,\delta\omega\, \mathbf{k} \cdot \mathbf{k}\, dx\, dy = (2\,\delta\omega)(\pi\rho^2).$$

Thus,

$$(\nabla \times \mathbf{F}) \cdot \mathbf{k} = 2\,\delta\omega = \frac{1}{\pi\rho^2} \oint_C \mathbf{F} \cdot d\mathbf{r},$$

in agreement with Eq. (9) with $\mathbf{u} = \mathbf{k}$.

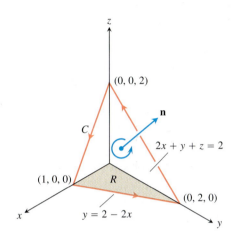

14.69 The planar surface in Example 5.

EXAMPLE 5   Use Stokes's theorem to evaluate $\int_C \mathbf{F} \cdot d\mathbf{r}$, if $\mathbf{F} = xz\,\mathbf{i} + xy\,\mathbf{j} + 3xz\,\mathbf{k}$ and C is the boundary of the portion of the plane $2x + y + z = 2$ in the first octant, traversed counterclockwise as viewed from above (Fig. 14.69).

Solution   The plane is the level surface $f(x, y, z) = 2$ of the function $f(x, y, z) = 2x + y + z$. The unit normal vector

$$\mathbf{n} = \frac{\nabla f}{|\nabla f|} = \frac{(2\mathbf{i} + \mathbf{j} + \mathbf{k})}{|2\mathbf{i} + \mathbf{j} + \mathbf{k}|} = \frac{1}{\sqrt{6}}(2\mathbf{i} + \mathbf{j} + \mathbf{k})$$

is consistent with the counterclockwise motion around C. To apply Stokes's theorem, we find

$$\operatorname{curl} \mathbf{F} = \nabla \times \mathbf{F} = \begin{vmatrix} \mathbf{i} & \mathbf{j} & \mathbf{k} \\ \dfrac{\partial}{\partial x} & \dfrac{\partial}{\partial y} & \dfrac{\partial}{\partial z} \\ xz & xy & 3xz \end{vmatrix} = (x - 3z)\,\mathbf{j} + y\,\mathbf{k}.$$

On the plane, z equals $2 - 2x - y$, so

$$\nabla \times \mathbf{F} = (x - 3(2 - 2x - y))\,\mathbf{j} + y\,\mathbf{k} = (7x + 3y - 6)\,\mathbf{j} + y\,\mathbf{k}$$

and

$$\nabla \times \mathbf{F} \cdot \mathbf{n} = \frac{1}{\sqrt{6}}(7x + 3y - 6 + y) = \frac{1}{\sqrt{6}}(7x + 4y - 6).$$

The surface area element is

$$d\sigma = \frac{|\nabla f|}{|\nabla f \cdot \mathbf{k}|}\, dA = \frac{\sqrt{6}}{1}\, dx\, dy.$$

The circulation is $\oint_C \mathbf{F} \cdot d\mathbf{r} = \iint_S \nabla \times \mathbf{F} \cdot \mathbf{n}\, d\sigma$ Stokes's theorem

$$= \int_0^1 \int_0^{2-2x} \frac{1}{\sqrt{6}}(7x + 4y - 6)\,\sqrt{6}\, dy\, dx$$

$$= \int_0^1 \int_0^{2-2x} (7x + 4y - 6)\, dy\, dx = -1.$$

Proof of Stokes's Theorem for Polyhedral Surfaces

Let S be a polyhedral surface consisting of a finite number of plane regions. (Think of one of Buckminster Fuller's geodesic domes.) We apply Green's theorem to each

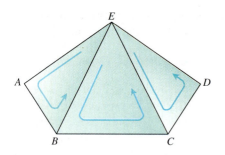

14.70 Part of a polyhedral surface.

separate panel of S. There are two types of panels:

1. those that are surrounded on all sides by other panels and
2. those that have one or more edges that are not adjacent to other panels.

The boundary Δ of S consists of those edges of the type 2 panels that are not adjacent to other panels. In Fig. 14.70, the triangles EAB, BCE, and CDE represent a part of S, with $ABCD$ part of the boundary Δ. Applying Green's theorem to the three triangles in turn and adding the results, we get

$$\left(\oint_{EAB} + \oint_{BCE} + \oint_{CDE} \right) \mathbf{F} \cdot d\mathbf{r} = \left(\iint_{EAB} + \iint_{BCE} + \iint_{CDE} \right) \nabla \times \mathbf{F} \cdot \mathbf{n}\, d\sigma. \quad (10)$$

The three line integrals on the left-hand side of Eq. (10) combine into a single line integral taken around the periphery $ABCDE$ because the integrals along interior segments cancel in pairs. For example, the integral along segment BE in triangle ABE is opposite in sign to the integral along the same segment in triangle EBC. Similarly for segment CE. Hence (10) reduces to

$$\oint_{ABCDE} \mathbf{F} \cdot d\mathbf{r} = \iint_{ABCDE} \nabla \times \mathbf{F} \cdot \mathbf{n}\, d\sigma.$$

When we apply Green's theorem to all the panels and add the results, we get

$$\oint_{\Delta} \mathbf{F} \cdot d\mathbf{r} = \iint_{S} \nabla \times \mathbf{F} \cdot \mathbf{n}\, d\sigma. \quad (11)$$

This is Stokes's theorem for a polyhedral surface S. You can find proofs for more general surfaces in advanced calculus texts. ❏

Stokes's Theorem for Surfaces with Holes

Stokes's theorem can be extended to an oriented surface S that has one or more holes (Fig. 14.71), in a way analogous to the extension of Green's theorem: The surface integral over S of the normal component of $\nabla \times \mathbf{F}$ equals the sum of the line integrals around all the boundary curves of the tangential component of $\mathbf{F}$, where the curves are to be traced in the direction induced by the orientation of S.

An Important Identity

The following identity arises frequently in mathematics and the physical sciences.

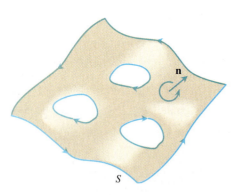

14.71 Stokes's theorem also holds for oriented surfaces with holes.

$$\text{curl grad } f = \mathbf{0} \qquad \text{or} \qquad \nabla \times \nabla f = \mathbf{0} \qquad (12)$$

This identity holds for any function $f(x, y, z)$ whose second partial derivatives are continuous. The proof goes like this:

$$\nabla \times \nabla f = \begin{vmatrix} \mathbf{i} & \mathbf{j} & \mathbf{k} \\ \dfrac{\partial}{\partial x} & \dfrac{\partial}{\partial y} & \dfrac{\partial}{\partial z} \\ \dfrac{\partial f}{\partial x} & \dfrac{\partial f}{\partial y} & \dfrac{\partial f}{\partial z} \end{vmatrix} = (f_{zy} - f_{yz})\mathbf{i} - (f_{zx} - f_{xz})\mathbf{j} + (f_{yx} - f_{xy})\mathbf{k}.$$

Connected and simply connected.

Connected but not simply connected.

Connected and simply connected.

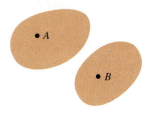

Simply connected but not connected.
No path from A to B lies entirely in the region.

14.72 Connectivity and simple connectivity are not the same. Neither implies the other, as these pictures of plane regions illustrate. To make three-dimensional regions with these properties, thicken the plane regions into cylinders.

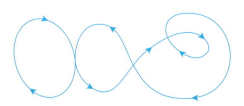

14.73 In a simply connected open region in space, differentiable curves that cross themselves can be divided into loops to which Stokes's theorem applies.

If the second partial derivatives are continuous, the mixed second derivatives in parentheses are equal (Euler's theorem, Section 12.3) and the vector is zero.

Conservative Fields and Stokes's Theorem

In Section 14.3, we found that saying that a field **F** is conservative in an open region D in space is equivalent to saying that the integral of **F** around every closed loop in D is zero. This, in turn, is equivalent in *simply connected* open regions to saying that $\nabla \times \mathbf{F} = \mathbf{0}$. A region D is **simply connected** if every closed path in D can be contracted to a point in D without ever leaving D. If D consisted of space with a line removed, for example, D would not be simply connected. There would be no way to contract a loop around the line to a point without leaving D. On the other hand, space itself *is* simply connected (Fig. 14.72).

Theorem 6
If $\nabla \times \mathbf{F} = \mathbf{0}$ at every point of a simply connected open region D in space, then on any piecewise smooth closed path C in D,

$$\oint_C \mathbf{F} \cdot d\mathbf{r} = 0.$$

Sketch of a Proof   Theorem 6 is usually proved in two steps. The first step is for simple closed curves. A theorem from topology, a branch of advanced mathematics, states that every differentiable simple closed curve C in a simply connected open region D is the boundary of a smooth two-sided surface S that also lies in D. Hence, by Stokes's theorem,

$$\oint_C \mathbf{F} \cdot d\mathbf{r} = \iint_S \nabla \times \mathbf{F} \cdot \mathbf{n} \, d\sigma = 0.$$

The second step is for curves that cross themselves, like the one in Fig. 14.73. The idea is to break these into simple loops spanned by orientable surfaces, apply Stokes's theorem one loop at a time, and add the results.   ❑

The following diagram summarizes the results for conservative fields defined on connected, simply connected open regions.

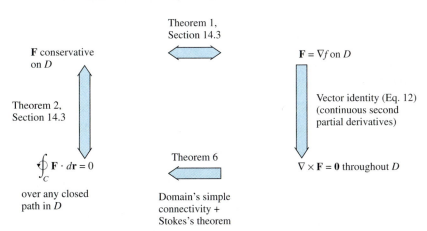

Exercises 14.7

Using Stokes's Theorem to Calculate Circulation

In Exercises 1–6, use the surface integral in Stokes's theorem to calculate the circulation of the field **F** around the curve C in the indicated direction.

1. $\mathbf{F} = x^2\mathbf{i} + 2x\mathbf{j} + z^2\mathbf{k}$
   C: The ellipse $4x^2 + y^2 = 4$ in the xy-plane, counterclockwise when viewed from above

2. $\mathbf{F} = 2y\mathbf{i} + 3x\mathbf{j} - z^2\mathbf{k}$
   C: The circle $x^2 + y^2 = 9$ in the xy-plane, counterclockwise when viewed from above

3. $\mathbf{F} = y\mathbf{i} + xz\mathbf{j} + x^2\mathbf{k}$
   C: The boundary of the triangle cut from the plane $x + y + z = 1$ by the first octant, counterclockwise when viewed from above

4. $\mathbf{F} = (y^2 + z^2)\mathbf{i} + (x^2 + z^2)\mathbf{j} + (x^2 + y^2)\mathbf{k}$
   C: The boundary of the triangle cut from the plane $x + y + z = 1$ by the first octant, counterclockwise when viewed from above

5. $\mathbf{F} = (y^2 + z^2)\mathbf{i} + (x^2 + y^2)\mathbf{j} + (x^2 + y^2)\mathbf{k}$
   C: The square bounded by the lines $x = \pm 1$ and $y = \pm 1$ in the xy-plane, counterclockwise when viewed from above

6. $\mathbf{F} = x^2 y^3 \mathbf{i} + \mathbf{j} + z\mathbf{k}$
   C: The intersection of the cylinder $x^2 + y^2 = 4$ and the hemisphere $x^2 + y^2 + z^2 = 16$, $z \geq 0$

Flux of the Curl

7. Let **n** be the outer unit normal of the elliptical shell
$$S: \quad 4x^2 + 9y^2 + 36z^2 = 36, \quad z \geq 0,$$
and let
$$\mathbf{F} = y\mathbf{i} + x^2\mathbf{j} + (x^2 + y^4)^{3/2} \sin e^{\sqrt{xyz}}\,\mathbf{k}.$$
Find the value of
$$\iint_S \nabla \times \mathbf{F} \cdot \mathbf{n}\,d\sigma.$$

(*Hint:* One parametrization of the ellipse at the base of the shell is $x = 3 \cos t$, $y = 2 \sin t$, $0 \leq t \leq 2\pi$.)

8. Let **n** be the outer unit normal (normal away from the origin) of the parabolic shell
$$S: \quad 4x^2 + y + z^2 = 4, \quad y \geq 0$$
and let
$$\mathbf{F} = \left(-z + \frac{1}{2+x}\right)\mathbf{i} + (\tan^{-1} y)\mathbf{j} + \left(x + \frac{1}{4+z}\right)\mathbf{k}.$$
Find the value of
$$\iint_S \nabla \times \mathbf{F} \cdot \mathbf{n}\,d\sigma.$$

9. Let S be the cylinder $x^2 + y^2 = a^2$, $0 \leq z \leq h$, together with its top, $x^2 + y^2 \leq a^2$, $z = h$. Let $\mathbf{F} = -y\mathbf{i} + x\mathbf{j} + x^2\mathbf{k}$. Use Stokes's theorem to calculate the flux of $\nabla \times \mathbf{F}$ outward through S.

10. Evaluate
$$\iint_S \nabla \times (y\mathbf{i}) \cdot \mathbf{n}\,d\sigma,$$
where S is the hemisphere $x^2 + y^2 + z^2 = 1$, $z \geq 0$.

11. Show that
$$\iint_S \nabla \times \mathbf{F} \cdot \mathbf{n}\,d\sigma$$
has the same value for all oriented surfaces S that span C and that induce the same positive direction on C.

12. Let **F** be a differentiable vector field defined on a region containing a smooth closed oriented surface S and its interior. Let **n** be the unit normal vector field on S. Suppose that S is the union of two surfaces S_1 and S_2 joined along a smooth simple closed curve C. Can anything be said about
$$\iint_S \nabla \times \mathbf{F} \cdot \mathbf{n}\,d\sigma?$$
Give reasons for your answer.

Stokes's Theorem for Parametrized Surfaces

In Exercises 13–18, use the surface integral in Stokes's theorem to calculate the flux of the curl of the field **F** across the surface S in the direction of the outward unit normal **n**.

13. $\mathbf{F} = 2z\mathbf{i} + 3x\mathbf{j} + 5y\mathbf{k}$
   S: $\mathbf{r}(r, \theta) = (r \cos \theta)\mathbf{i} + (r \sin \theta)\mathbf{j} + (4 - r^2)\mathbf{k}$, $0 \leq r \leq 2$, $0 \leq \theta \leq 2\pi$

14. $\mathbf{F} = (y - z)\mathbf{i} + (z - x)\mathbf{j} + (x + z)\mathbf{k}$
   S: $\mathbf{r}(r, \theta) = (r \cos \theta)\mathbf{i} + (r \sin \theta)\mathbf{j} + (9 - r^2)\mathbf{k}$, $0 \leq r \leq 3$, $0 \leq \theta \leq 2\pi$

15. $\mathbf{F} = x^2 y\mathbf{i} + 2y^3 z\mathbf{j} + 3z\mathbf{k}$
   S: $\mathbf{r}(r, \theta) = (r \cos \theta)\mathbf{i} + (r \sin \theta)\mathbf{j} + r\mathbf{k}$, $0 \leq r \leq 1$, $0 \leq \theta \leq 2\pi$

16. $\mathbf{F} = (x - y)\mathbf{i} + (y - z)\mathbf{j} + (z - x)\mathbf{k}$
   S: $\mathbf{r}(r, \theta) = (r \cos \theta)\mathbf{i} + (r \sin \theta)\mathbf{j} + (5 - r)\mathbf{k}$, $0 \leq r \leq 5$, $0 \leq \theta \leq 2\pi$

17. $\mathbf{F} = 3y\mathbf{i} + (5 - 2x)\mathbf{j} + (z^2 - 2)\mathbf{k}$
   S: $\mathbf{r}(\phi, \theta) = (\sqrt{3} \sin \phi \cos \theta)\mathbf{i} + (\sqrt{3} \sin \phi \sin \theta)\mathbf{j} + (\sqrt{3} \cos \phi)\mathbf{k}$, $0 \leq \phi \leq \pi/2$, $0 \leq \theta \leq 2\pi$

18. $\mathbf{F} = y^2\mathbf{i} + z^2\mathbf{j} + x\mathbf{k}$
   S: $\mathbf{r}(\phi, \theta) = (2 \sin \phi \cos \theta)\mathbf{i} + (2 \sin \phi \sin \theta)\mathbf{j} + (2 \cos \phi)\mathbf{k}$, $0 \leq \phi \leq \pi/2$, $0 \leq \theta \leq 2\pi$

Theory and Examples

19. Use the identity $\nabla \times \nabla f = 0$ (Eq. 12 in the text) and Stokes's theorem to show that the circulations of the following fields around the boundary of any smooth orientable surface in space are zero.

a) $\mathbf{F} = 2x\,\mathbf{i} + 2y\,\mathbf{j} + 2z\,\mathbf{k}$

b) $\mathbf{F} = \nabla(xy^2z^3)$

c) $\mathbf{F} = \nabla \times (x\,\mathbf{i} + y\,\mathbf{j} + z\,\mathbf{k})$

d) $\mathbf{F} = \nabla f$

20. Let $f(x, y, z) = (x^2 + y^2 + z^2)^{-1/2}$. Show that the clockwise circulation of the field $\mathbf{F} = \nabla f$ around the circle $x^2 + y^2 = a^2$ in the xy-plane is zero

a) by taking $\mathbf{r} = (a \cos t)\,\mathbf{i} + (a \sin t)\,\mathbf{j}, 0 \leq t \leq 2\pi$, and integrating $\mathbf{F} \cdot d\mathbf{r}$ over the circle, and

b) by applying Stokes's theorem.

21. Let C be a simple closed smooth curve in the plane $2x + 2y + z = 2$, oriented as shown here. Show that

$$\oint_C 2y\,dx + 3z\,dy - x\,dz$$

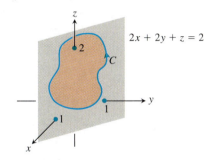

depends only on the area of the region enclosed by C and not on the position or shape of C.

22. Show that if $\mathbf{F} = x\,\mathbf{i} + y\,\mathbf{j} + z\,\mathbf{k}$, then $\nabla \times \mathbf{F} = 0$.

23. Find a vector field with twice-differentiable components whose curl is $x\,\mathbf{i} + y\,\mathbf{j} + z\,\mathbf{k}$ or prove that no such field exists.

24. Does Stokes's theorem say anything special about circulation in a field whose curl is zero? Give reasons for your answer.

25. Let R be a region in the xy-plane that is bounded by a piecewise smooth simple closed curve C, and suppose that the moments of inertia of R about the x- and y-axes are known to be I_x and I_y. Evaluate the integral

$$\oint_C \nabla(r^4) \cdot \mathbf{n}\,ds,$$

where $r = \sqrt{x^2 + y^2}$, in terms of I_x and I_y.

26. Show that the curl of

$$\mathbf{F} = \frac{-y}{x^2 + y^2}\,\mathbf{i} + \frac{x}{x^2 + y^2}\,\mathbf{j} + z\,\mathbf{k}$$

is zero but that

$$\oint_C \mathbf{F} \cdot d\mathbf{r}$$

is not zero if C is the circle $x^2 + y^2 = 1$ in the xy-plane. (Theorem 6 does not apply here because the domain of $\mathbf{F}$ is not simply connected. The field $\mathbf{F}$ is not defined along the z-axis so there is no way to contract C to a point without leaving the domain of $\mathbf{F}$.)

14.8

The Divergence Theorem and a Unified Theory

The divergence form of Green's theorem in the plane states that the net outward flux of a vector field across a simple closed curve can be calculated by integrating the divergence of the field over the region enclosed by the curve. The corresponding theorem in three dimensions, called the Divergence Theorem, states that the net outward flux of a vector field across a closed surface in space can be calculated by integrating the divergence of the field over the region enclosed by the surface. In this section, we prove the Divergence Theorem and show how it simplifies the calculation of flux. We also derive Gauss's law for flux in an electric field and the continuity equation of hydrodynamics. Finally, we unify the chapter's vector integral theorems into a single fundamental theorem.

Divergence in Three Dimensions

The **divergence** of a vector field $\mathbf{F} = M(x, y, z)\,\mathbf{i} + N(x, y, z)\,\mathbf{j} + P(x, y, z)\,\mathbf{k}$ is the scaler function

$$\text{div }\mathbf{F} = \nabla \cdot \mathbf{F} = \frac{\partial M}{\partial x} + \frac{\partial N}{\partial y} + \frac{\partial P}{\partial z}. \tag{1}$$

The Divergence Theorem

Mikhail Vassilievich Ostrogradsky (1801–1862) was the first mathematician to publish a proof of the Divergence Theorem. Upon being denied his degree at Kharkhov University by the minister for religious affairs and national education (for atheism), Ostrogradsky left Russia for Paris in 1822, attracted by the presence of Laplace, Legendre, Fourier, Poisson, and Cauchy. While working on the theory of heat in the mid-1820s, he formulated the Divergence Theorem as a tool for converting volume integrals to surface integrals.

Carl Friedrich Gauss (1777–1855) had already proved the theorem while working on the theory of gravitation, but his notebooks were not to be published until many years later. (The theorem is sometimes called Gauss's theorem.) The list of Gauss's accomplishments in science and mathematics is truly astonishing, ranging from the invention of the electric telegraph (with Wilhelm Weber in 1833) to the development of a wonderfully accurate theory of planetary orbits and to work in non-Euclidean geometry that later became fundamental to Einstein's general theory of relativity.

The symbol "div **F**" is read as "divergence of **F**" or "div **F**." The notation $\nabla \cdot \mathbf{F}$ is read "del dot **F**."

Div **F** has the same physical interpretation in three dimensions that it does in two. If **F** is the velocity field of a fluid flow, the value of div **F** at a point (x, y, z) is the rate at which fluid is being piped in or drained away at (x, y, z). The divergence is the flux per unit volume or flux density at the point.

EXAMPLE 1   Find the divergence of $\mathbf{F} = 2xz\,\mathbf{i} - xy\,\mathbf{j} - z\,\mathbf{k}$.

Solution   The divergence of **F** is

$$\nabla \cdot \mathbf{F} = \frac{\partial}{\partial x}(2xz) + \frac{\partial}{\partial y}(-xy) + \frac{\partial}{\partial z}(-z) = 2z - x - 1.$$

The Divergence Theorem

The Divergence Theorem says that under suitable conditions the outward flux of a vector field across a closed surface (oriented outward) equals the triple integral of the divergence of the field over the region enclosed by the surface.

> **Theorem 7**
>
> **The Divergence Theorem**
>
> The flux of a vector field $\mathbf{F} = M\,\mathbf{i} + N\,\mathbf{j} + P\,\mathbf{k}$ across a closed oriented surface S in the direction of the surface's outward unit normal field **n** equals the integral of $\nabla \cdot \mathbf{F}$ over the region D enclosed by the surface:
>
> $$\underset{S}{\iint} \mathbf{F} \cdot \mathbf{n}\, d\sigma = \underset{D}{\iiint} \nabla \cdot \mathbf{F}\, dV. \qquad (2)$$
>
> $\underset{\text{flux}}{\underset{\text{outward}}{}}$ $\underset{\text{integral}}{\underset{\text{divergence}}{}}$

EXAMPLE 2   Evaluate both sides of Eq. (2) for the field $\mathbf{F} = x\,\mathbf{i} + y\,\mathbf{j} + z\,\mathbf{k}$ over the sphere $x^2 + y^2 + z^2 = a^2$.

Solution   The outer unit normal to S, calculated from the gradient of $f(x, y, z) = x^2 + y^2 + z^2 - a^2$, is

$$\mathbf{n} = \frac{2(x\,\mathbf{i} + y\,\mathbf{j} + z\,\mathbf{k})}{\sqrt{4(x^2 + y^2 + z^2)}} = \frac{x\,\mathbf{i} + y\,\mathbf{j} + z\,\mathbf{k}}{a}.$$

Hence

$$\mathbf{F} \cdot \mathbf{n}\, d\sigma = \frac{x^2 + y^2 + z^2}{a}\, d\sigma = \frac{a^2}{a}\, d\sigma = a\, d\sigma$$

because $x^2 + y^2 + z^2 = a^2$ on the surface. Therefore

$$\underset{S}{\iint} \mathbf{F} \cdot \mathbf{n}\, d\sigma = \underset{S}{\iint} a\, d\sigma = a \underset{S}{\iint} d\sigma = a\,(4\pi a^2) = 4\pi a^3.$$

The divergence of $\mathbf{F}$ is

$$\nabla \cdot \mathbf{F} = \frac{\partial}{\partial x}(x) + \frac{\partial}{\partial y}(y) + \frac{\partial}{\partial z}(z) = 3,$$

so

$$\iiint_D \nabla \cdot \mathbf{F}\, dV = \iiint_D 3\, dV = 3\left(\frac{4}{3}\pi a^3\right) = 4\pi a^3.$$

EXAMPLE 3   Find the flux of $\mathbf{F} = xy\,\mathbf{i} + yz\,\mathbf{j} + xz\,\mathbf{k}$ outward through the surface of the cube cut from the first octant by the planes $x = 1$, $y = 1$, and $z = 1$.

Solution   Instead of calculating the flux as a sum of six separate integrals, one for each face of the cube, we can calculate the flux by integrating the divergence

$$\nabla \cdot \mathbf{F} = \frac{\partial}{\partial x}(xy) + \frac{\partial}{\partial y}(yz) + \frac{\partial}{\partial z}(xz) = y + z + x$$

over the cube's interior:

$$\text{Flux} = \iint_{\substack{\text{cube} \\ \text{surface}}} \mathbf{F} \cdot \mathbf{n}\, d\sigma = \iiint_{\substack{\text{cube} \\ \text{interior}}} \nabla \cdot \mathbf{F}\, dV \qquad \text{\color{teal}The Divergence Theorem}$$

$$= \int_0^1 \int_0^1 \int_0^1 (x + y + z)\, dx\, dy\, dz = \frac{3}{2}. \qquad \text{\color{teal}Routine integration}$$

Proof of the Divergence Theorem (Special Regions)

To prove the Divergence Theorem, we assume that the components of $\mathbf{F}$ have continuous first partial derivatives. We also assume that D is a convex region with no holes or bubbles, such as a solid sphere, cube, or ellipsoid, and that S is a piecewise smooth surface. In addition, we assume that any line perpendicular to the xy-plane at an interior point of the region R_{xy} that is the projection of D on the xy-plane intersects the surface S in exactly two points, producing surfaces

$$S_1: \qquad z = f_1(x, y), \quad (x, y) \text{ in } R_{xy}$$

$$S_2: \qquad z = f_2(x, y), \quad (x, y) \text{ in } R_{xy},$$

with $f_1 \le f_2$. We make similar assumptions about the projection of D onto the other coordinate planes. See Fig. 14.74.

The components of the unit normal vector $\mathbf{n} = n_1\,\mathbf{i} + n_2\,\mathbf{j} + n_3\,\mathbf{k}$ are the cosines of the angles α, β, and γ that $\mathbf{n}$ makes with $\mathbf{i}$, $\mathbf{j}$, and $\mathbf{k}$ (Fig. 14.75). This is true because all the vectors involved are unit vectors. We have

$$n_1 = \mathbf{n} \cdot \mathbf{i} = |\mathbf{n}|\,|\mathbf{i}| \cos \alpha = \cos \alpha$$

$$n_2 = \mathbf{n} \cdot \mathbf{j} = |\mathbf{n}|\,|\mathbf{j}| \cos \beta = \cos \beta$$

$$n_3 = \mathbf{n} \cdot \mathbf{k} = |\mathbf{n}|\,|\mathbf{k}| \cos \gamma = \cos \gamma.$$

Thus,

$$\mathbf{n} = (\cos \alpha)\,\mathbf{i} + (\cos \beta)\,\mathbf{j} + (\cos \gamma)\,\mathbf{k}$$

and

$$\mathbf{F} \cdot \mathbf{n} = M \cos \alpha + N \cos \beta + P \cos \gamma.$$

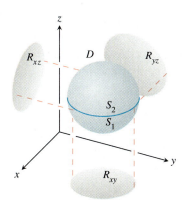

14.74 We first prove the Divergence Theorem for the kind of three-dimensional region shown here. We then extend the theorem to other regions.

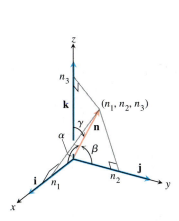

14.75 The scalar components of a unit normal vector $\mathbf{n}$ are the cosines of the angles α, β, and γ that it makes with $\mathbf{i}$, $\mathbf{j}$, and $\mathbf{k}$.

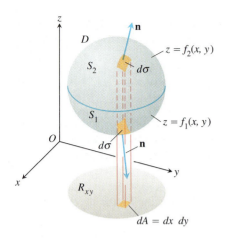

14.76 The three-dimensional region D enclosed by the surfaces S_1 and S_2 shown here projects vertically onto a two-dimensional region R_{xy} in the xy-plane.

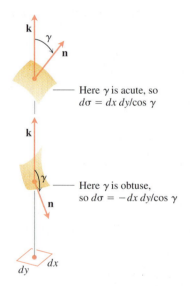

14.77 An enlarged view of the area patches in Fig. 14.76. The relations $d\sigma = \pm\, dx\, dy/\cos \gamma$ are derived in Section 14.5.

In component form, the Divergence Theorem states that

$$\iint\limits_{S} (M \cos \alpha + N \cos \beta + P \cos \gamma)\, d\sigma = \iiint\limits_{D} \left(\frac{\partial M}{\partial x} + \frac{\partial N}{\partial y} + \frac{\partial P}{\partial z} \right) dx\, dy\, dz.$$

We prove the theorem by proving the three following equalities:

$$\iint\limits_{S} M \cos \alpha\, d\sigma = \iiint\limits_{D} \frac{\partial M}{\partial x}\, dx\, dy\, dz \qquad (3)$$

$$\iint\limits_{S} N \cos \beta\, d\sigma = \iiint\limits_{D} \frac{\partial N}{\partial y}\, dx\, dy\, dz \qquad (4)$$

$$\iint\limits_{S} P \cos \gamma\, d\sigma = \iiint\limits_{D} \frac{\partial P}{\partial z}\, dx\, dy\, dz \qquad (5)$$

We prove Eq. (5) by converting the surface integral on the left to a double integral over the projection R_{xy} of D on the xy-plane (Fig. 14.76). The surface S consists of an upper part S_2 whose equation is $z = f_2(x, y)$ and a lower part S_1 whose equation is $z = f_1(x, y)$. On S_2, the outer normal $\mathbf{n}$ has a positive $\mathbf{k}$-component and

$$\cos \gamma\, d\sigma = dx\, dy \quad \text{because} \quad d\sigma = \frac{dA}{|\cos \gamma|} = \frac{dx\, dy}{\cos \gamma}.$$

See Fig. 14.77. On S_1, the outer normal $\mathbf{n}$ has a negative $\mathbf{k}$-component and

$$\cos \gamma\, d\sigma = -dx\, dy.$$

Therefore,

$$\iint\limits_{S} P \cos \gamma\, d\sigma = \iint\limits_{S_2} P \cos \gamma\, d\sigma + \iint\limits_{S_1} P \cos \gamma\, d\sigma$$

$$= \iint\limits_{R_{xy}} P\,(x, y, f_2(x, y))\, dx\, dy - \iint\limits_{R_{xy}} P\,(x, y, f_1(x, y))\, dx\, dy$$

$$= \iint\limits_{R_{xy}} [P(x, y, f_2(x, y)) - P(x, y, f_1(x, y))]\, dx\, dy$$

$$= \iint\limits_{R_{xy}} \left[\int_{f_1 (x,y)}^{f_2(x,y)} \frac{\partial P}{\partial z}\, dz \right] dx\, dy = \iiint\limits_{D} \frac{\partial P}{\partial z}\, dz\, dx\, dy.$$

This proves Eq. (5).

The proofs for Eqs. (3) and (4) follow the same pattern; or just permute x, y, z; M, N, P; α, β, γ, in order, and get those results from Eq. (5).

The Divergence Theorem for Other Regions

The Divergence Theorem can be extended to regions that can be partitioned into a finite number of simple regions of the type just discussed and to regions that can be defined as limits of simpler regions in certain ways. For example, suppose that D is the region between two concentric spheres and that $\mathbf{F}$ has continuously differentiable components throughout D and on the bounding surfaces. Split D by an equatorial plane and apply the Divergence Theorem to each half separately. The

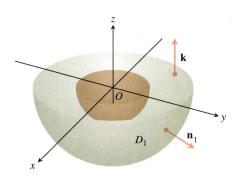

14.78 The lower half of the solid region between two concentric spheres.

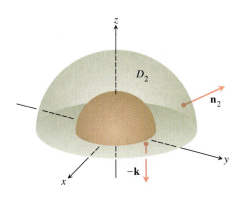

14.79 The upper half of the solid region between two concentric spheres.

bottom half, D_1, is shown in Fig. 14.78. The surface that bounds D_1 consists of an outer hemisphere, a plane washer-shaped base, and an inner hemisphere. The Divergence Theorem says that

$$\iint_{S_1} \mathbf{F} \cdot \mathbf{n}_1 \, d\sigma_1 = \iiint_{D_1} \nabla \cdot \mathbf{F} \, dV_1. \tag{6}$$

The unit normal $\mathbf{n}_1$ that points outward from D_1 points away from the origin along the outer surface, equals $\mathbf{k}$ along the flat base, and points toward the origin along the inner surface. Next apply the Divergence Theorem to D_2, as shown in Fig. 14.79:

$$\iint_{S_2} \mathbf{F} \cdot \mathbf{n}_2 \, d\sigma_2 = \iiint_{D_2} \nabla \cdot \mathbf{F} \, dV_2. \tag{7}$$

As we follow $\mathbf{n}_2$ over S_2, pointing outward from D_2, we see that $\mathbf{n}_2$ equals $-\mathbf{k}$ along the washer-shaped base in the xy-plane, points away from the origin on the outer sphere, and points toward the origin on the inner sphere. When we add Eqs. (6) and (7), the integrals over the flat base cancel because of the opposite signs of $\mathbf{n}_1$ and $\mathbf{n}_2$. We thus arrive at the result

$$\iint_S \mathbf{F} \cdot \mathbf{n} \, d\sigma = \iiint_D \nabla \cdot \mathbf{F} \, dV,$$

with D the region between the spheres, S the boundary of D consisting of two spheres, and $\mathbf{n}$ the unit normal to S directed outward from D.

EXAMPLE 4   Find the net outward flux of the field

$$\mathbf{F} = \frac{x\,\mathbf{i} + y\,\mathbf{j} + z\,\mathbf{k}}{\rho^3}, \qquad \rho = \sqrt{x^2 + y^2 + z^2}$$

across the boundary of the region D: $0 < a^2 \le x^2 + y^2 + z^2 \le b^2$.

Solution   The flux can be calculated by integrating $\nabla \cdot \mathbf{F}$ over D. We have

$$\frac{\partial \rho}{\partial x} = \frac{1}{2}(x^2 + y^2 + z^2)^{-1/2}(2x) = \frac{x}{\rho}$$

and

$$\frac{\partial M}{\partial x} = \frac{\partial}{\partial x}(x\rho^{-3}) = \rho^{-3} - 3x\rho^{-4}\frac{\partial \rho}{\partial x} = \frac{1}{\rho^3} - \frac{3x^2}{\rho^5}.$$

Similarly,

$$\frac{\partial N}{\partial y} = \frac{1}{\rho^3} - \frac{3y^2}{\rho^5} \quad \text{and} \quad \frac{\partial P}{\partial z} = \frac{1}{\rho^3} - \frac{3z^2}{\rho^5}.$$

Hence,

$$\text{div } \mathbf{F} = \frac{3}{\rho^3} - \frac{3}{\rho^5}(x^2 + y^2 + z^2) = \frac{3}{\rho^3} - \frac{3\rho^2}{\rho^5} = 0$$

and

$$\iiint_D \nabla \cdot \mathbf{F} \, dV = 0.$$

So the integral of $\nabla \cdot \mathbf{F}$ over D is zero and the net outward flux across the boundary of D is zero. But there is more to learn from this example. The flux leaving D across the inner sphere S_a is the negative of the flux leaving D across the outer sphere S_b (because the sum of these fluxes is zero). This means that the flux of $\mathbf{F}$ across S_a in the direction away from the origin equals the flux of $\mathbf{F}$

across S_b in the direction away from the origin. Thus, the flux of $\mathbf{F}$ across a sphere centered at the origin is independent of the radius of the sphere. What is this flux?

To find it, we evaluate the flux integral directly. The outward unit normal on the sphere of radius a is

$$\mathbf{n} = \frac{x\,\mathbf{i} + y\,\mathbf{j} + z\,\mathbf{k}}{\sqrt{x^2 + y^2 + z^2}} = \frac{x\,\mathbf{i} + y\,\mathbf{j} + z\,\mathbf{k}}{a}.$$

Hence, on the sphere,

$$\mathbf{F} \cdot \mathbf{n} = \frac{x\,\mathbf{i} + y\,\mathbf{j} + z\,\mathbf{k}}{a^3} \cdot \frac{x\,\mathbf{i} + y\,\mathbf{j} + z\,\mathbf{k}}{a} = \frac{x^2 + y^2 + z^2}{a^4} = \frac{a^2}{a^4} = \frac{1}{a^2}$$

and

$$\iint_{S_a} \mathbf{F} \cdot \mathbf{n}\, d\sigma = \frac{1}{a^2} \iint_{S_a} d\sigma = \frac{1}{a^2}\,(4\pi a^2) = 4\pi.$$

The outward flux of $\mathbf{F}$ across any sphere centered at the origin is 4π. ◻

Gauss's Law—One of the Four Great Laws of Electromagnetic Theory

There is more to be learned from Example 4. In electromagnetic theory, the electric field created by a point charge q located at the origin is the inverse square field

$$\mathbf{E}(x, y, z) = \frac{1}{4\pi\,\epsilon_0}\frac{q}{|\mathbf{r}|^2}\left(\frac{\mathbf{r}}{|\mathbf{r}|}\right) = \frac{q}{4\pi\,\epsilon_0}\frac{\mathbf{r}}{|\mathbf{r}|^3} = \frac{q}{4\pi\,\epsilon_0}\frac{x\,\mathbf{i} + y\,\mathbf{j} + z\,\mathbf{k}}{\rho^3}$$

where ϵ_0 is a physical constant, $\mathbf{r}$ is the position vector of the point (x, y, z), and $\rho = |\mathbf{r}| = \sqrt{x^2 + y^2 + z^2}$. In the notation of Example 4,

$$\mathbf{E} = \frac{q}{4\pi\,\epsilon_0}\,\mathbf{F}.$$

The calculations in Example 4 show that the outward flux of $\mathbf{E}$ across any sphere centered at the origin is q/ϵ_0. But this result is not confined to spheres. The outward flux of $\mathbf{E}$ across any closed surface S that encloses the origin (and to which the Divergence Theorem applies) is also q/ϵ_0. To see why, we have only to imagine a large sphere S_a centered at the origin and enclosing the surface S. Since

$$\nabla \cdot \mathbf{E} = \nabla \cdot \frac{q}{4\pi\,\epsilon_0}\,\mathbf{F} = \frac{q}{4\pi\,\epsilon_0}\,\nabla \cdot \mathbf{F} = 0$$

when $\rho > 0$, the integral of $\nabla \cdot \mathbf{E}$ over the region D between S and S_a is zero. Hence, by the Divergence Theorem,

$$\iint_{\substack{\text{boundary} \\ \text{of } D}} \mathbf{E} \cdot \mathbf{n}\, d\sigma = 0,$$

and the flux of $\mathbf{E}$ across S in the direction away from the origin must be the same as the flux of $\mathbf{E}$ across S_a in the direction away from the origin, which is $4\pi q$. This statement, called *Gauss's law*, also applies to charge distributions that are more general than the one assumed here, as you will see in nearly any physics text.

$$\text{Gauss's Law:} \qquad \iint_S \mathbf{E} \cdot \mathbf{n}\, d\sigma = \frac{q}{\epsilon_0}$$

The Continuity Equation of Hydrodynamics

Let D be a region in space bounded by a closed oriented surface S. If $\mathbf{v}(x, y, z)$ is the velocity field of a fluid flowing smoothly through D, $\delta = \delta(t, x, y, z)$ is the fluid's density at (x, y, z) at time t, and $\mathbf{F} = \delta\mathbf{v}$, then the **continuity equation** of hydrodynamics states that

$$\nabla \cdot \mathbf{F} + \frac{\partial \delta}{\partial t} = 0.$$

If the functions involved have continuous first partial derivatives, the equation evolves naturally from the Divergence Theorem, as we will now see.

First, the integral

$$\iint\limits_{S} \mathbf{F} \cdot \mathbf{n} \, d\sigma$$

is the rate at which mass leaves D across S (leaves because $\mathbf{n}$ is the outer normal). To see why, consider a patch of area $\Delta\sigma$ on the surface (Fig. 14.80). In a short time interval Δt, the volume ΔV of fluid that flows across the patch is approximately equal to the volume of a cylinder with base area $\Delta\sigma$ and height $(\mathbf{v}\,\Delta t) \cdot \mathbf{n}$, where $\mathbf{v}$ is a velocity vector rooted at a point of the patch:

$$\Delta V \approx \mathbf{v} \cdot \mathbf{n} \, \Delta\sigma \, \Delta t.$$

The mass of this volume of fluid is about

$$\Delta m \approx \delta\mathbf{v} \cdot \mathbf{n} \, \Delta\sigma \, \Delta t,$$

so the rate at which mass is flowing out of D across the patch is about

$$\frac{\Delta m}{\Delta t} \approx \delta\mathbf{v} \cdot \mathbf{n} \, \Delta\sigma.$$

This leads to the approximation

$$\frac{\Sigma \, \Delta m}{\Delta t} \approx \sum \delta\mathbf{v} \cdot \mathbf{n} \, \Delta\sigma$$

as an estimate of the average rate at which mass flows across S. Finally, letting $\Delta\sigma \to 0$ and $\Delta t \to 0$ gives the instantaneous rate at which mass leaves D across S as

$$\frac{dm}{dt} = \iint\limits_{S} \delta\mathbf{v} \cdot \mathbf{n} \, d\sigma,$$

which for our particular flow is

$$\frac{dm}{dt} = \iint\limits_{S} \mathbf{F} \cdot \mathbf{n} \, d\sigma,$$

Now let B be a solid sphere centered at a point Q in the flow. The average value of $\nabla \cdot \mathbf{F}$ over B is

$$\frac{1}{\text{volume of } B} \iiint\limits_{B} \nabla \cdot \mathbf{F} \, dV.$$

It is a consequence of the continuity of the divergence that $\nabla \cdot \mathbf{F}$ actually takes on

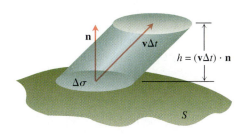

14.80 The fluid that flows upward through the patch $\Delta\sigma$ in a short time Δt fills a "cylinder" whose volume is approximately base $\times$ height $= \mathbf{v} \cdot \mathbf{n} \, \Delta\sigma \, \Delta t$.

this value at some point P in B. Thus,

$$(\nabla \cdot \mathbf{F})_P = \frac{1}{\text{volume of } B} \iiint_B \nabla \cdot \mathbf{F} \, dV = \frac{\iint_S \mathbf{F} \cdot \mathbf{n} \, d\sigma}{\text{volume of } B}$$

$$= \frac{\text{rate at which mass leaves } B \text{ across its surface } S}{\text{volume of } B}. \tag{8}$$

The fraction on the right describes decrease in mass per unit volume.

Now let the radius of B approach zero while the center Q stays fixed. The left-hand side of Eq. (8) converges to $(\nabla \cdot \mathbf{F})_Q$, the right side to $(-\partial\delta/\partial t)_Q$. The equality of these two limits is the continuity equation

$$\nabla \cdot \mathbf{F} = -\frac{\partial\delta}{\partial t}.$$

The continuity equation "explains" $\nabla \cdot \mathbf{F}$: The divergence of $\mathbf{F}$ at a point is the rate at which the density of the fluid is decreasing there.

The Divergence Theorem

$$\iint_S \mathbf{F} \cdot \mathbf{n} \, d\sigma = \iiint_D \nabla \cdot \mathbf{F} \, dV$$

now says that the net decrease in density of the fluid in region D is accounted for by the mass transported across the surface S. In a way, the theorem is a statement about conservation of mass.

Unifying the Integral Theorems

If we think of a two-dimensional field $\mathbf{F} = M(x, y)\mathbf{i} + N(x, y)\mathbf{j}$ as a three-dimensional field whose $\mathbf{k}$-component is zero, then $\nabla \cdot \mathbf{F} = (\partial M/\partial x) + (\partial N/\partial y)$ and the normal form of Green's theorem can be written as

$$\oint_C \mathbf{F} \cdot \mathbf{n} \, ds = \iint_R \left(\frac{\partial M}{\partial x} + \frac{\partial N}{\partial y} \right) dx \, dy = \iint_R \nabla \cdot \mathbf{F} \, dA.$$

Similarly, $\nabla \times \mathbf{F} \cdot \mathbf{k} = (\partial N/\partial x) - (\partial M/\partial y)$, so the tangential form of Green's theorem can be written as

$$\oint_C \mathbf{F} \cdot d\mathbf{r} = \iint_R \left(\frac{\partial N}{\partial x} - \frac{\partial M}{\partial y} \right) dx \, dy = \iint_R \nabla \times \mathbf{F} \cdot \mathbf{k} \, dA.$$

With the equations of Green's theorem now in del notation, we can see their relationships to the equations in Stokes's theorem and the Divergence Theorem.

Green's Theorem and Its Generalization to Three Dimensions

Normal form of
Green's theorem:

$$\oint_C \mathbf{F} \cdot \mathbf{n} \, ds = \iint_R \nabla \cdot \mathbf{F} \, dA$$

Divergence Theorem:

$$\iint_S \mathbf{F} \cdot \mathbf{n} \, d\sigma = \iiint_D \nabla \cdot \mathbf{F} \, dV$$

Tangential form of
Green's theorem:
$$\oint_C \mathbf{F} \cdot d\mathbf{r} = \iint_R \nabla \times \mathbf{F} \cdot \mathbf{k} \, dA$$

Stokes's theorem:
$$\oint_C \mathbf{F} \cdot d\mathbf{r} = \iint_S \nabla \times \mathbf{F} \cdot \mathbf{n} \, d\sigma$$

Notice how Stokes's theorem generalizes the tangential (curl) form of Green's theorem from a flat surface in the plane to a surface in three-dimensional space. In each case, the integral of the normal component of curl $\mathbf{F}$ over the interior of the surface equals the circulation of $\mathbf{F}$ around the boundary.

Likewise, the Divergence Theorem generalizes the normal (flux) form of Green's theorem from a two-dimensional region in the plane to a three-dimensional region in space. In each case, the integral of $\nabla \cdot \mathbf{F}$ over the interior of the region equals the total flux of the field across the boundary.

There is still more to be learned here. All of these results can be thought of as forms of a *single fundamental theorem*. Think back to the Fundamental Theorem of Calculus in Section 4.7. It says that if $f(x)$ is differentiable on $[a, b]$ then

$$\int_a^b \frac{df}{dx} \, dx = f(b) - f(a).$$

$$\mathbf{n} = -\mathbf{i} \qquad \mathbf{n} = \mathbf{i}$$
$$a \qquad b \qquad \longrightarrow x$$

14.81 The outward unit normals at the boundary of $[a, b]$ in one-dimensional space.

If we let $\mathbf{F} = f(x)\mathbf{i}$ throughout $[a, b]$, then $(df/dx) = \nabla \cdot \mathbf{F}$. If we define the unit vector $\mathbf{n}$ normal to the boundary of $[a, b]$ to be $\mathbf{i}$ at b and $-\mathbf{i}$ at a (Fig. 14.81) then

$$f(b) - f(a) = f(b)\mathbf{i} \cdot (\mathbf{i}) + f(a)\mathbf{i} \cdot (-\mathbf{i})$$

$$= \mathbf{F}(b) \cdot \mathbf{n} + \mathbf{F}(a) \cdot \mathbf{n}$$

$$= \text{total outward flux of } \mathbf{F} \text{ across the boundary of } [a, b].$$

The Fundamental Theorem now says that

$$\int_{[a, b]} \nabla \cdot \mathbf{F} \, dx = \text{total outward flux of } \mathbf{F} \text{ across the boundary.}$$

The Fundamental Theorem of Calculus, the flux form of Green's theorem, and the Divergence Theorem all say that the integral of the differential operator $\nabla\cdot$ operating on a field $\mathbf{F}$ over a region equals the sum of the normal field components over the boundary of the region.

Stokes's theorem and the circulation form of Green's theorem say that, when things are properly oriented, the integral of the normal component of the curl operating on a field equals the sum of the tangential field components on the boundary of the surface.

The beauty of these interpretations is the observance of a marvelous underlying principle, which we might state as follows.

> The integral of a differential operator acting on a field over a region equals the sum of the field components appropriate to the operator over the boundary of the region.

Exercises 14.8

Calculating Divergence

In Exercises 1–4, find the divergence of the field.

1. The spin field in Fig. 14.15.

2. The radial field in Fig. 14.14.

3. The gravitational field in Fig. 14.13.

4. The velocity field in Fig. 14.10.

Using the Divergence Theorem to Calculate Outward Flux

In Exercises 5–16, use the Divergence Theorem to find the outward flux of **F** across the boundary of the region D.

5. $\mathbf{F} = (y - x)\mathbf{i} + (z - y)\mathbf{j} + (y - x)\mathbf{k}$
   D: The cube bounded by the planes $x = \pm 1, y = \pm 1,$ and $z = \pm 1$

6. $\mathbf{F} = x^2\mathbf{i} + y^2\mathbf{j} + z^2\mathbf{k}$
   a) D: The cube cut from the first octant by the planes $x = 1, y = 1,$ and $z = 1$
   b) D: The cube bounded by the planes $x = \pm 1, y = \pm 1,$ and $z = \pm 1$
   c) D: The region cut from the solid cylinder $x^2 + y^2 \le 4$ by the planes $z = 0$ and $z = 1$

7. $\mathbf{F} = y\mathbf{i} + xy\mathbf{j} - z\mathbf{k}$
   D: The region inside the solid cylinder $x^2 + y^2 \le 4$ between the plane $z = 0$ and the paraboloid $z = x^2 + y^2$

8. $\mathbf{F} = x^2\mathbf{i} + xz\mathbf{j} + 3z\mathbf{k}$
   D: the solid sphere $x^2 + y^2 + z^2 \le 4$

9. $\mathbf{F} = x^2\mathbf{i} - 2xy\mathbf{j} + 3xz\mathbf{k}$
   D: The region cut from the first octant by the sphere $x^2 + y^2 + z^2 = 4$

10. $\mathbf{F} = (6x^2 + 2xy)\mathbf{i} + (2y + x^2z)\mathbf{j} + 4x^2y^3\mathbf{k}$
    D: The region cut from the first octant by the cylinder $x^2 + y^2 = 4$ and the plane $z = 3$

11. $\mathbf{F} = 2xz\mathbf{i} - xy\mathbf{j} - z^2\mathbf{k}$
    D: The wedge cut from the first octant by the plane $y + z = 4$ and the elliptical cylinder $4x^2 + y^2 = 16$

12. $\mathbf{F} = x^3\mathbf{i} + y^3\mathbf{j} + z^3\mathbf{k}$
    D: The solid sphere $x^2 + y^2 + z^2 \le a^2$

13. $\mathbf{F} = \sqrt{x^2 + y^2 + z^2}\,(x\mathbf{i} + y\mathbf{j} + z\mathbf{k})$
    D: The region $1 \le x^2 + y^2 + z^2 \le 2$

14. $\mathbf{F} = (x\mathbf{i} + y\mathbf{j} + z\mathbf{k})/\sqrt{x^2 + y^2 + z^2}$
    D: The region $1 \le x^2 + y^2 + z^2 \le 4$

15. $\mathbf{F} = (5x^3 + 12xy^2)\mathbf{i} + (y^3 + e^y \sin z)\mathbf{j} + (5z^3 + e^y \cos z)\mathbf{k}$
    D: The solid region between the spheres $x^2 + y^2 + z^2 = 1$ and $x^2 + y^2 + z^2 = 2$

16. $\mathbf{F} = \ln(x^2 + y^2)\mathbf{i} - \left(\dfrac{2z}{x}\tan^{-1}\dfrac{y}{x}\right)\mathbf{j} + z\sqrt{x^2 + y^2}\,\mathbf{k}$
    D: The thick-walled cylinder $1 \le x^2 + y^2 \le 2, \quad -1 \le z \le 2$

Properties of Curl and Divergence

17. div (curl **G**) = 0
    a) Show that if the necessary partial derivatives of the components of the field $\mathbf{G} = M\mathbf{i} + N\mathbf{j} + P\mathbf{k}$ are continuous, then $\nabla \cdot \nabla \times \mathbf{G} = 0$.
    b) What, if anything, can you conclude about the flux of the field $\nabla \times \mathbf{G}$ across a closed surface? Give reasons for your answer.

18. Let $\mathbf{F}_1$ and $\mathbf{F}_2$ be differentiable vector fields, and let a and b be arbitrary real constants. Verify the following identities.
    a) $\nabla \cdot (a\mathbf{F}_1 + b\mathbf{F}_2) = a\nabla \cdot \mathbf{F}_1 + b\nabla \cdot \mathbf{F}_2$
    b) $\nabla \times (a\mathbf{F}_1 + b\mathbf{F}_2) = a\nabla \times \mathbf{F}_1 + b\nabla \times \mathbf{F}_2$
    c) $\nabla \cdot (\mathbf{F}_1 \times \mathbf{F}_2) = \mathbf{F}_2 \cdot \nabla \times \mathbf{F}_1 - \mathbf{F}_1 \cdot \nabla \times \mathbf{F}_2$

19. Let **F** be a differentiable vector field and let $g(x, y, z)$ be a differentiable scalar function. Verify the following identities.
    a) $\nabla \cdot (g\mathbf{F}) = g\nabla \cdot \mathbf{F} + \nabla g \cdot \mathbf{F}$
    b) $\nabla \times (g\mathbf{F}) = g\nabla \times \mathbf{F} + \nabla g \times \mathbf{F}$

20. If $\mathbf{F} = M\mathbf{i} + N\mathbf{j} + P\mathbf{k}$ is a differentiable vector field, we define the notation $\mathbf{F} \cdot \nabla$ to mean

    $$M\frac{\partial}{\partial x} + N\frac{\partial}{\partial y} + P\frac{\partial}{\partial z}.$$

    For differentiable vector fields $\mathbf{F}_1$ and $\mathbf{F}_2$ verify the following identities.
    a) $\nabla \times (\mathbf{F}_1 \times \mathbf{F}_2) = (\mathbf{F}_2 \cdot \nabla)\mathbf{F}_1 - (\mathbf{F}_1 \cdot \nabla)\mathbf{F}_2$ $+ (\nabla \cdot \mathbf{F}_2)\mathbf{F}_1 - (\nabla \cdot \mathbf{F}_1)\mathbf{F}_2$
    b) $\nabla(\mathbf{F}_1 \cdot \mathbf{F}_2) = (\mathbf{F}_1 \cdot \nabla)\mathbf{F}_2 + (\mathbf{F}_2 \cdot \nabla)\mathbf{F}_1$ $+ \mathbf{F}_1 \times (\nabla \times \mathbf{F}_2) + \mathbf{F}_2 \times (\nabla \times \mathbf{F}_1)$

Theory and Examples

21. Let **F** be a field whose components have continuous first partial derivatives throughout a portion of space containing a region D bounded by a smooth closed surface S. If $|\mathbf{F}| \le 1$, can any bound be placed on the size of

    $$\iiint_D \nabla \cdot \mathbf{F}\, dV?$$

    Give reasons for your answer.

22. The base of the closed cubelike surface shown here is the unit square in the xy-plane. The four sides lie in the planes $x = 0, x = 1, y = 0,$ and $y = 1$. The top is an arbitrary smooth surface whose identity is unknown. Let $\mathbf{F} = x\mathbf{i} - 2y\mathbf{j} + (z + 3)\mathbf{k}$, and suppose the outward flux of **F** through side A is 1 and through side B is -3. Can you conclude anything about the outward flux through the top? Give reasons for your answer.

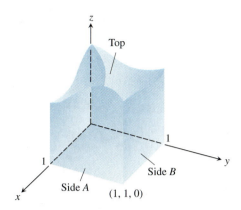

23. a) Show that the flux of the position vector field $\mathbf{F} = x\,\mathbf{i} + y\,\mathbf{j} + z\,\mathbf{k}$ outward through a smooth closed surface S is three times the volume of the region enclosed by the surface.

b) Let $\mathbf{n}$ be the outward unit normal vector field on S. Show that it is not possible for $\mathbf{F}$ to be orthogonal to $\mathbf{n}$ at every point of S.

24. Among all rectangular solids defined by the inequalities $0 \le x \le a, 0 \le y \le b, 0 \le z \le 1$, find the one for which the total flux of $\mathbf{F} = (-x^2 - 4xy)\,\mathbf{i} - 6yz\,\mathbf{j} + 12z\,\mathbf{k}$ outward through the six sides is greatest. What *is* the greatest flux?

25. Let $\mathbf{F} = x\,\mathbf{i} + y\,\mathbf{j} + z\,\mathbf{k}$ and suppose that the surface S and region D satisfy the hypotheses of the Divergence Theorem. Show that the volume of D is given by the formula

$$\text{Volume of } D = \frac{1}{3} \iint_S \mathbf{F} \cdot \mathbf{n}\,d\sigma.$$

26. Show that the outward flux of a constant vector field $\mathbf{F} = \mathbf{C}$ across any closed surface to which the Divergence Theorem applies is zero.

27. *Harmonic functions.* A function $f(x, y, z)$ is said to be **harmonic** in a region D in space if it satisfies the Laplace equation

$$\nabla^2 f = \nabla \cdot \nabla f = \frac{\partial^2 f}{\partial x^2} + \frac{\partial^2 f}{\partial y^2} + \frac{\partial^2 f}{\partial z^2} = 0$$

throughout D.

a) Suppose that f is harmonic throughout a bounded region D enclosed by a smooth surface S and that $\mathbf{n}$ is the chosen unit normal vector on S. Show that the integral over S of $\nabla f \cdot \mathbf{n}$, the derivative of f in the direction of $\mathbf{n}$, is zero.

b) Show that if f is harmonic on D, then

$$\iint_S f \nabla f \cdot \mathbf{n}\,d\sigma = \iiint_D |\nabla f|^2\,dV.$$

28. Let S be the surface of the portion of the solid sphere $x^2 + y^2 + z^2 \le a^2$ that lies in the first octant and let $f(x, y, z) = \ln \sqrt{x^2 + y^2 + z^2}$. Calculate

$$\iint_S \nabla f \cdot \mathbf{n}\,d\sigma.$$

$(\nabla f \cdot \mathbf{n}$ is the derivative of f in the direction of $\mathbf{n}$.)

29. *Green's first formula.* Suppose that f and g are scalar functions with continuous first- and second-order partial derivatives throughout a closed region D that is bounded by a piecewise smooth surface S. Show that

$$\iint_S f \nabla g \cdot \mathbf{n}\,d\sigma = \iiint_D (f \nabla^2 g + \nabla f \cdot \nabla g)\,dV. \qquad (9)$$

Equation (9) is **Green's first formula.** (*Hint:* Apply the Divergence Theorem to the field $\mathbf{F} = f \nabla g$.)

30. *Green's second formula.* (*Continuation of Exercise 29*). Interchange f and g in Eq. (9) to obtain a similar formula. Then subtract this formula from Eq. (9) to show that

$$\iint_S (f \nabla g - g \nabla f) \cdot \mathbf{n}\,d\sigma$$

$$= \iiint_D (f \nabla^2 g - g \nabla^2 f)\,dV. \qquad (10)$$

This equation is **Green's second formula.**

31. *Conservation of mass.* Let $\mathbf{v}(t, x, y, z)$ be a continuously differentiable vector field over the region D in space and let $p(t, x, y, z)$ be a continuously differentiable scalar function. The variable t represents the time domain. The Law of Conservation of Mass asserts that

$$\frac{d}{dt} \iiint_D p(t, x, y, z)\,dV = -\iint_S p\mathbf{v} \cdot \mathbf{n}\,d\sigma,$$

where S is the surface enclosing D.

a) Give a physical interpretation of the conservation of mass law if $\mathbf{v}$ is a velocity flow field and p represents the density of the fluid at point (x, y, z) at time t.

b) Use the Divergence Theorem and Leibniz's rule,

$$\frac{d}{dt} \iiint_D p(t, x, y, z)\,dV = \iiint_D \frac{\partial p}{\partial t}\,dV,$$

to show that the Law of Conservation of Mass is equivalent to the continuity equation,

$$\nabla \cdot p\mathbf{v} + \frac{\partial p}{\partial t} = 0.$$

(In the first term $\nabla \cdot p\mathbf{v}$ the variable t is held fixed and in the second term $\partial p / \partial t$ it is assumed that the point (x, y, z) in D is held fixed.)

32. *General diffusion equation.* Let $T(t, x, y, z)$ be a function with continuous second derivatives giving the temperature at time t at the point (x, y, z) of a solid occupying a region D in space. If the solid's specific heat and mass density are denoted by the constants c and ρ respectively, the quantity $c \rho T$ is called the solid's **heat energy per unit volume.**

a) Explain why $-\nabla T$ points in the direction of heat flow.

b) Let $-k\nabla T$ denote the **energy flux vector.** (Here the constant k is called the **conductivity.**) Assuming the Law of

Conservation of Mass with $-k\nabla T = \mathbf{v}$ and $c\rho T = p$ in Exercise 31, derive the diffusion (heat) equation

$$\frac{\partial T}{\partial t} = K\nabla^2 T,$$

where $K = k/(c\rho) > 0$ is the *diffusivity* constant. (Notice

that if $T(t, x)$ represents the temperature at time t at position x in a uniform conducting rod with perfectly insulated sides, then $\nabla^2 T = \partial^2 T/\partial x^2$ and the diffusion equation reduces to the one-dimensional heat equation in the Chapter 12 Additional Exercises.)

CHAPTER 14 QUESTIONS TO GUIDE YOUR REVIEW

1. What are line integrals? How are they evaluated? Give examples.

2. How can you use line integrals to find the centers of mass of springs? Explain.

3. What is a vector field? a gradient field? Give examples.

4. How do you calculate the work done by a force in moving a particle along a curve? Give an example.

5. What are flow, circulation, and flux?

6. What is special about path independent fields?

7. How can you tell when a field is conservative?

8. What is a potential function? Show by example how to find a potential function for a conservative field.

9. What is a differential form? What does it mean for such a form to be exact? How do you test for exactness? Give examples.

10. What is the divergence of a vector field? How can you interpret it?

11. What is the curl of a vector field? How can you interpret it?

12. What is Green's theorem? How can you interpret it?

13. How do you calculate the area of a curved surface in space? Give an example.

14. What is an oriented surface? How do you calculate the flux of a three-dimensional vector field across an oriented surface? Give an example.

15. What are surface integrals? What can you calculate with them? Give an example.

16. What is a parametrized surface? How do you find the area of such a surface? Give examples.

17. How do you integrate a function over a parametrized surface? Give an example.

18. What is Stokes's theorem? How can you interpret it?

19. Summarize the chapter's results on conservative fields.

20. What is the Divergence Theorem? How can you interpret it?

21. How does the Divergence Theorem generalize Green's theorem?

22. How does Stokes's theorem generalize Green's theorem?

23. How can Green's theorem, Stokes's theorem, and the Divergence Theorem be thought of as forms of a single fundamental theorem?

CHAPTER 14 PRACTICE EXERCISES

Evaluating Line Integrals

1. Figure 14.82 shows two polygonal paths in space joining the origin to the point $(1, 1, 1)$. Integrate $f(x, y, z) = 2x - 3y^2 - 2z + 3$ over each path.

2. Figure 14.83 shows three polygonal paths joining the origin to the point $(1, 1, 1)$. Integrate $f(x, y, z) = x^2 + y - z$ over each path.

3. Integrate $f(x, y, z) = \sqrt{x^2 + z^2}$ over the circle

$$\mathbf{r}(t) = (a\cos t)\mathbf{j} + (a\sin t)\mathbf{k}, \quad 0 \le t \le 2\pi.$$

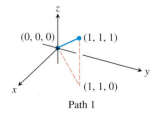

Path 1

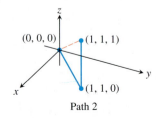

Path 2

14.82 The paths in Exercise 1.

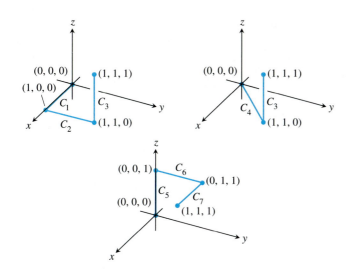

14.83 The paths in Exercise 2.

4. Integrate $f(x, y, z) = \sqrt{x^2 + z^2}$ over the involute curve
$$\mathbf{r}(t) = (\cos t + t \sin t)\mathbf{i} + (\sin t - t \cos t)\mathbf{j}, \ 0 \le t \le \sqrt{3}.$$
(See Fig. 11.20.)

Evaluate the integrals in Exercises 5 and 6.

5. $\displaystyle\int_{(-1,1,1)}^{(4,-3,0)} \frac{dx + dy + dz}{\sqrt{x + y + z}}$

6. $\displaystyle\int_{(1,1,1)}^{(10,3,3)} dx - \sqrt{\frac{z}{y}}\, dy - \sqrt{\frac{y}{z}}\, dz$

7. Integrate $\mathbf{F} = -(y \sin z)\mathbf{i} + (x \sin z)\mathbf{j} + (xy \cos z)\mathbf{k}$ around the circle cut from the sphere $x^2 + y^2 + z^2 = 5$ by the plane $z = -1$, clockwise as viewed from above.

8. Integrate $\mathbf{F} = 3x^2 y\,\mathbf{i} + (x^3 + 1)\mathbf{j} + 9z^2\,\mathbf{k}$ around the circle cut from the sphere $x^2 + y^2 + z^2 = 9$ by the plane $x = 2$.

Evaluate the line integrals in Exercises 9 and 10.

9. $\displaystyle\int_C 8x \sin y\, dx - 8y \cos x\, dy$

C is the square cut from the first quadrant by the lines $x = \pi/2$ and $y = \pi/2$.

10. $\displaystyle\int_C y^2\, dx + x^2\, dy$

C is the circle $x^2 + y^2 = 4$.

Evaluating Surface Integrals

11. Find the area of the elliptical region cut from the plane $x + y + z = 1$ by the cylinder $x^2 + y^2 = 1$.

12. Find the area of the cap cut from the paraboloid $y^2 + z^2 = 3x$ by the plane $x = 1$.

13. Find the area of the cap cut from the top of the sphere $x^2 + y^2 + z^2 = 1$ by the plane $z = \sqrt{2}/2$.

14. a) Find the area of the surface cut from the hemisphere $x^2 + y^2 + z^2 = 4$, $z \ge 0$, by the cylinder $x^2 + y^2 = 2x$.

b) Find the area of the portion of the cylinder that lies inside the hemisphere. (*Hint:* Project onto the xz-plane. Or evaluate the integral $\int h\, ds$, where h is the altitude of the cylinder and ds is the element of arc length on the circle $x^2 + y^2 = 2x$ in the xy-plane.)

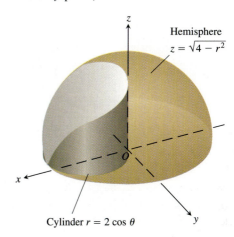

15. Find the area of the triangle in which the plane $(x/a) + (y/b) + (z/c) = 1 \ (a, b, c > 0)$ intersects the first octant. Check your answer with an appropriate vector calculation.

16. Integrate

a) $g(x, y, z) = \dfrac{yz}{\sqrt{4y^2 + 1}}$ **b)** $g(x, y, z) = \dfrac{z}{\sqrt{4y^2 + 1}}$

over the surface cut from the parabolic cylinder $y^2 - z = 1$ by the planes $x = 0$, $x = 3$, and $z = 0$.

17. Integrate $g(x, y, z) = x^4 y(y^2 + z^2)$ over the portion of the cylinder $y^2 + z^2 = 25$ that lies in the first octant between the planes $x = 0$ and $x = 1$ and above the plane $z = 3$.

18. CALCULATOR The state of Wyoming is bounded by the meridians $111° \, 3'$ and $104° \, 3'$ west longitude and by the circles $41°$ and $45°$ north latitude. Assuming that the earth is a sphere of radius $R = 3959$ mi, find the area of Wyoming.

Parametrized Surfaces

Find the parametrizations for the surfaces in Exercises 19–24. (There are many ways to do these, so your answers may not be the same as those in the back of the book.)

19. The portion of the sphere $x^2 + y^2 + z^2 = 36$ between the planes $z = -3$ and $z = 3\sqrt{3}$

20. The portion of the paraboloid $z = -(x^2 + y^2)/2$ above the plane $z = -2$

21. The cone $z = 1 + \sqrt{x^2 + y^2}$, $z \le 3$

22. The portion of the plane $4x + 2y + 4z = 12$ that lies above the square $0 \le x \le 2$, $0 \le y \le 2$ in the first quadrant

23. The portion of the paraboloid $y = 2(x^2 + z^2)$, $y \le 2$, that lies above the xy-plane

24. The portion of the hemisphere $x^2 + y^2 + z^2 = 10$, $y \geq 0$, in the first octant

25. Find the area of the surface

$$\mathbf{r}(u, v) = (u + v)\mathbf{i} + (u - v)\mathbf{j} + v\mathbf{k}, \quad 0 \leq u \leq 1, \quad 0 \leq v \leq 1.$$

26. Integrate $f(x, y, z) = xy - z^2$ over the surface in Exercise 25.

27. Find the surface area of the helicoid

$$\mathbf{r}(r, \theta) = r \cos \theta \, \mathbf{i} + r \sin \theta \, \mathbf{j} + \theta \mathbf{k}, \quad 0 \leq \theta \leq 2\pi \text{ and } 0 \leq r \leq 1,$$

in the accompanying figure.

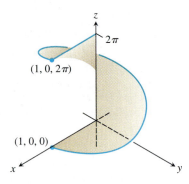

28. Evaluate the integral $\iint_S \sqrt{x^2 + y^2 + 1} \, d\sigma$, where S is the helicoid in Exercise 27.

Conservative Fields

Which of the fields in Exercises 29–32 are conservative, and which are not?

29. $\mathbf{F} = x\mathbf{i} + y\mathbf{j} + z\mathbf{k}$

30. $\mathbf{F} = (x\mathbf{i} + y\mathbf{j} + z\mathbf{k})/(x^2 + y^2 + z^2)^{3/2}$

31. $\mathbf{F} = x e^y \mathbf{i} + y e^z \mathbf{j} + z e^x \mathbf{k}$

32. $\mathbf{F} = (\mathbf{i} + z\mathbf{j} + y\mathbf{k})/(x + yz)$

Find potential functions for the fields in Exercises 33 and 34.

33. $\mathbf{F} = 2\mathbf{i} + (2y + z)\mathbf{j} + (y + 1)\mathbf{k}$

34. $\mathbf{F} = (z \cos xz)\mathbf{i} + e^y \mathbf{j} + (x \cos xz)\mathbf{k}$

Work and Circulation

In Exercises 35 and 36, find the work done by each field along the paths from $(0, 0, 0)$ to $(1, 1, 1)$ in Fig. 14.82.

35. $\mathbf{F} = 2xy\mathbf{i} + \mathbf{j} + x^2\mathbf{k}$ **36.** $\mathbf{F} = 2xy\mathbf{i} + x^2\mathbf{j} + \mathbf{k}$

37. Find the work done by

$$\mathbf{F} = \frac{x\mathbf{i} + y\mathbf{j}}{(x^2 + y^2)^{3/2}}$$

over the plane curve $\mathbf{r}(t) = (e^t \cos t)\mathbf{i} + (e^t \sin t)\mathbf{j}$ from the point $(1, 0)$ to the point $(e^{2\pi}, 0)$ in two ways:

a) by using the parametrization of the curve to evaluate the work integral.

b) by evaluating a potential function for $\mathbf{F}$.

38. Find the flow of the field $\mathbf{F} = \nabla(x^2 z e^y)$

a) once around the ellipse C in which the plane $x + y + z = 1$ intersects the cylinder $x^2 + z^2 = 25$, clockwise as viewed from the positive y-axis.

b) along the curved boundary of the helicoid in Exercise 27 from $(1, 0, 0)$ to $(1, 0, 2\pi)$.

39. Suppose $\mathbf{F}(x, y) = (x + y)\mathbf{i} - (x^2 + y^2)\mathbf{j}$ is the velocity field of a fluid flowing across the xy-plane. Find the flow along each of the following paths from $(1, 0)$ to $(-1, 0)$.

a) The upper half of the circle $x^2 + y^2 = 1$

b) The line segment from $(1, 0)$ to $(-1, 0)$

c) The line segment from $(1, 0)$ to $(0, -1)$ followed by the line segment from $(0, -1)$ to $(-1, 0)$

40. Find the circulation of $\mathbf{F} = 2x\mathbf{i} + 2z\mathbf{j} + 2y\mathbf{k}$ along the closed path consisting of the helix $\mathbf{r}_1(t) = (\cos t)\mathbf{i} + (\sin t)\mathbf{j} + t\mathbf{k}$, $0 \leq t \leq \pi/2$, followed by the line segments $\mathbf{r}_2(t) = \mathbf{j} + (\pi/2)(1 - t)\mathbf{k}, 0 \leq t \leq 1$, and $\mathbf{r}_3(t) = t\mathbf{i} + (1 - t)\mathbf{j}, 0 \leq t \leq 1$.

In Exercises 41 and 42, use the surface integral in Stokes's theorem to find the circulation of the field $\mathbf{F}$ around the curve C in the indicated direction.

41. $\mathbf{F} = y^2\mathbf{i} - y\mathbf{j} + 3z^2\mathbf{k}$
C: The ellipse in which the plane $2x + 6y - 3z = 6$ meets the cylinder $x^2 + y^2 = 1$, counterclockwise as viewed from above

42. $\mathbf{F} = (x^2 + y)\mathbf{i} + (x + y)\mathbf{j} + (4y^2 - z)\mathbf{k}$
C: The circle in which the plane $z = -y$ meets the sphere $x^2 + y^2 + z^2 = 4$, counterclockwise as viewed from above

Mass and Moments

43. Find the mass of a thin wire lying along the curve $\mathbf{r}(t) = \sqrt{2}t\mathbf{i} + \sqrt{2}t\mathbf{j} + (4 - t^2)\mathbf{k}$, $0 \leq t \leq 1$, if the density at t is (a) $\delta = 3t$, (b) $\delta = 1$.

44. Find the center of mass of a thin wire lying along the curve $\mathbf{r}(t) = t\mathbf{i} + 2t\mathbf{j} + (2/3)t^{2/3}\mathbf{k}$, $0 \leq t \leq 2$, if the density at t is $\delta = 3\sqrt{5 + t}$.

45. Find the center of mass and the moments of inertia and radii of gyration about the coordinate axes of a thin wire lying along the curve

$$\mathbf{r}(t) = t\mathbf{i} + \frac{2\sqrt{2}}{3}t^{3/2}\mathbf{j} + \frac{t^2}{2}\mathbf{k}, \quad 0 \leq t \leq 2,$$

if the density at t is $\delta = 1/(t + 1)$.

46. A slender metal arch lies along the semicircle $y = \sqrt{a^2 - x^2}$ in the xy-plane. The density at the point (x, y) on the arch is $\delta(x, y) = 2a - y$. Find the center of mass.

47. A wire of constant density $\delta = 1$ lies along the curve $\mathbf{r}(t) = (e^t \cos t)\mathbf{i} + (e^t \sin t)\mathbf{j} + e^t\mathbf{k}, 0 \leq t \leq \ln 2$. Find $\bar{z}, I_z,$ and R_z.

48. Find the mass and center of mass of a wire of constant density δ that lies along the helix $\mathbf{r}(t) = (2 \sin t)\mathbf{i} + (2 \cos t)\mathbf{j} + 3t\mathbf{k}, 0 \leq t \leq 2\pi$.

49. Find I_z, R_z, and the center of mass of a thin shell of density $\delta(x, y, z) = z$ cut from the upper portion of the sphere $x^2 + y^2 + z^2 = 25$ by the plane $z = 3$.

50. Find the moment of inertia about the z-axis of the surface of the cube cut from the first octant by the planes $x = 1$, $y = 1$, and $z = 1$ if the density is $\delta = 1$.

Flux Across a Plane Curve or Surface

Use Green's theorem to find the counterclockwise circulation and outward flux for the fields and curves in Exercises 51 and 52.

51. $\mathbf{F} = (2xy + x)\mathbf{i} + (xy - y)\mathbf{j}$
   C: The square bounded by $x = 0$, $x = 1$, $y = 0$, $y = 1$

52. $\mathbf{F} = (y - 6x^2)\mathbf{i} + (x + y^2)\mathbf{j}$
   C: The triangle made by the lines $y = 0$, $y = x$, and $x = 1$

53. Show that

$$\oint_C \ln x \, \sin y \, dy - \frac{\cos y}{x} \, dx = 0$$

for any closed curve C to which Green's theorem applies.

54. a) Show that the outward flux of the position vector field $\mathbf{F} = x\mathbf{i} + y\mathbf{j}$ across any closed curve to which Green's theorem applies is twice the area of the region enclosed by the curve.

   b) Let $\mathbf{n}$ be the outward unit normal vector to a closed curve to which Green's theorem applies. Show that it is not possible for $\mathbf{F} = x\mathbf{i} + y\mathbf{j}$ to be orthogonal to $\mathbf{n}$ at every point of C.

In Exercises 55–58, find the outward flux of $\mathbf{F}$ across the boundary of D.

55. $\mathbf{F} = 2xy\mathbf{i} + 2yz\mathbf{j} + 2xz\mathbf{k}$
   D: The cube cut from the first octant by the planes $x = 1$, $y = 1$, $z = 1$

56. $\mathbf{F} = xz\mathbf{i} + yz\mathbf{j} + \mathbf{k}$
   D: The entire surface of the upper cap cut from the solid sphere $x^2 + y^2 + z^2 \le 25$ by the plane $z = 3$

57. $\mathbf{F} = -2x\mathbf{i} - 3y\mathbf{j} + z\mathbf{k}$
   D: The upper region cut from the solid sphere $x^2 + y^2 + z^2 \le 2$ by the paraboloid $z = x^2 + y^2$

58. $\mathbf{F} = (6x + y)\mathbf{i} - (x + z)\mathbf{j} + 4yz\mathbf{k}$
   D: The region in the first octant bounded by the cone $z = \sqrt{x^2 + y^2}$, the cylinder $x^2 + y^2 = 1$, and the coordinate planes

59. Let S be the surface that is bounded on the left by the hemisphere $x^2 + y^2 + z^2 = a^2$, $y \le 0$, in the middle by the cylinder $x^2 + z^2 = a^2$, $0 \le y \le a$, and on the right by the plane $y = a$. Find the flux of the field $\mathbf{F} = y\mathbf{i} + z\mathbf{j} + x\mathbf{k}$ outward across S.

60. Find the outward flux of the field $\mathbf{F} = 3xz^2\mathbf{i} + y\mathbf{j} - z^3\mathbf{k}$ across the surface of the solid in the first octant that is bounded by the cylinder $x^2 + 4y^2 = 16$ and the planes $y = 2z$, $x = 0$, and $z = 0$.

61. Use the Divergence Theorem to find the flux of $\mathbf{F} = xy^2\mathbf{i} + x^2y\mathbf{j} + y\mathbf{k}$ outward through the region enclosed by the cylinder $x^2 + y^2 = 1$ and the planes $z = 1$ and $z = -1$.

62. Find the flux of $\mathbf{F} = (3z + 1)\mathbf{k}$ upward across the hemisphere $x^2 + y^2 + z^2 = a^2$, $z \ge 0$ (a) with the Divergence Theorem, (b) by evaluating the flux integral directly.

CHAPTER | **14** | ADDITIONAL EXERCISES–THEORY, EXAMPLES, APPLICATIONS

Finding Areas with Green's Theorem

Use the Green's theorem area formula, Eq. (22) in Exercises 14.4, to find the areas of the regions enclosed by the curves in Exercises 1–4.

1. The limaçon $x = 2\cos t - \cos 2t$, $y = 2\sin t - \sin 2t$, $0 \le t \le 2\pi$

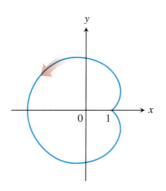

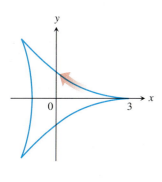

2. The deltoid $x = 2\cos t + \cos 2t$, $y = 2\sin t - \sin 2t$, $0 \le t \le 2\pi$

3. The eight curve $x = (1/2)\sin 2t$, $y = \sin t$, $0 \le t \le \pi$ (one loop)

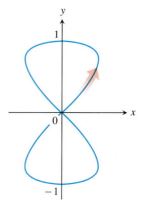

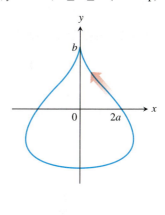

4. The teardrop $x = 2a\cos t - a\sin 2t$, $y = b\sin t$, $0 \le t \le 2\pi$

Theory and Applications

5. a) Give an example of a vector field $\mathbf{F}(x, y, z)$ that has value $\mathbf{0}$ at only one point and such that curl $\mathbf{F}$ is nonzero everywhere. Be sure to identify the point and compute the curl.

b) Give an example of a vector field $\mathbf{F}(x, y, z)$ that has value $\mathbf{0}$ on precisely one line and such that curl $\mathbf{F}$ is nonzero everywhere. Be sure to identify the line and compute the curl.

c) Give an example of a vector field $\mathbf{F}(x, y, z)$ that has value $\mathbf{0}$ on a surface and such that curl $\mathbf{F}$ is nonzero everywhere. Be sure to identify the surface and compute the curl.

6. Find all points (a, b, c) on the sphere $x^2 + y^2 + z^2 = R^2$ where the vector field $\mathbf{F} = yz^2\,\mathbf{i} + xz^2\,\mathbf{j} + 2xyz\,\mathbf{k}$ is normal to the surface and $\mathbf{F}(a, b, c) \neq \mathbf{0}$.

7. Find the mass of a spherical shell of radius R such that at each point (x, y, z) on the surface the mass density $\delta(x, y, z)$ is its distance to some fixed point (a, b, c) on the surface.

8. Find the mass of a helicoid

$$\mathbf{r}(r, \theta) = (r \cos \theta)\,\mathbf{i} + (r \sin \theta)\,\mathbf{j} + \theta\,\mathbf{k},$$

$0 \le r \le 1$, $0 \le \theta \le 2\pi$, if the density function is $\delta(x, y, z) = 2\sqrt{x^2 + y^2}$. See Practice Exercise 27 for a figure.

9. Among all rectangular regions $0 \le x \le a$, $0 \le y \le b$, find the one for which the total outward flux of $\mathbf{F} = (x^2 + 4xy)\,\mathbf{i} - 6y\,\mathbf{j}$ across the four sides is least. What *is* the least flux?

10. Find an equation for the plane through the origin such that the circulation of the flow field $\mathbf{F} = z\,\mathbf{i} + x\,\mathbf{j} + y\,\mathbf{k}$ around the circle of intersection of the plane with the sphere $x^2 + y^2 + z^2 = 4$ is a maximum.

11. A string lies along the circle $x^2 + y^2 = 4$ from $(2, 0)$ to $(0, 2)$ in the first quadrant. The density of the string is $\rho(x, y) = xy$.

a) Partition the string into a finite number of subarcs to show that the work done by gravity to move the string straight down to the x-axis is given by

$$\text{Work} = \lim_{n \to \infty} \sum_{k=1}^{n} g\,x_k\,y_k{}^2\Delta s_k = \int_C g\,xy^2\,ds,$$

where g is the gravitational constant.

b) Find the total work done by evaluating the line integral in part (a).

c) Show that the total work done equals the work required to move the string's center of mass $(\bar{x}, \bar{y})$ straight down to the x-axis.

12. A thin sheet lies along the portion of the plane $x + y + z = 1$ in the first octant. The density of the sheet is $\delta(x, y, z) = xy$.

a) Partition the sheet into a finite number of subpieces to show that the work done by gravity to move the sheet straight down to the xy-plane is given by

$$\text{Work} = \lim_{n \to \infty} \sum_{k=1}^{n} g\,x_k\,y_k\,z_k\Delta\sigma_k = \iint_S g\,xyz\,d\sigma,$$

where g is the gravitational constant.

b) Find the total work done by evaluating the surface integral in part (a).

c) Show that the total work done equals the work required to move the sheet's center of mass $(\bar{x}, \bar{y}, \bar{z})$ straight down to the xy-plane.

13. *Archimedes' principle.* If an object such as a ball is placed in a liquid, it will either sink to the bottom, float, or sink a certain distance and remain suspended in the liquid. Suppose a fluid has constant weight density w and that the fluid's surface coincides with the plane $z = 4$. A spherical ball remains suspended in the fluid and occupies the region $x^2 + y^2 + (z - 2)^2 \le 1$.

a) Show that the surface integral giving the magnitude of the total force on the ball due to the fluid's pressure is

$$\text{Force} = \lim_{n \to \infty} \sum_{k=1}^{n} w\,(4 - z_k)\,\Delta\sigma_k = \iint_S w\,(4 - z)\,d\sigma.$$

b) Since the ball is not moving, it is being held up by the buoyant force of the liquid. Show that the magnitude of the buoyant force on the sphere is

$$\text{Buoyant force} = \iint_S w\,(z - 4)\,\mathbf{k} \cdot \mathbf{n}\,d\sigma,$$

where $\mathbf{n}$ is the outer unit normal at (x, y, z). This illustrates Archimedes' principle that the magnitude of the buoyant force on a submerged solid equals the weight of the displaced fluid.

c) Use the Divergence Theorem to find the magnitude of the buoyant force in part (b).

14. *Fluid force on a curved surface.* A cone in the shape of the surface $z = \sqrt{x^2 + y^2}$, $0 \le z \le 2$, is filled with a liquid of constant weight density w. Assuming the xy-plane is "ground level," show that the total force on the portion of the cone from $z = 1$ to $z = 2$ due to liquid pressure is the surface integral

$$F = \iint_S w\,(2 - z)\,d\sigma.$$

Evaluate the integral.

15. *Faraday's law.* If $\mathbf{E}(t, x, y, z)$ and $\mathbf{B}(t, x, y, z)$ represent the electric and magnetic fields at point (x, y, z) at time t, a basic principle of electromagnetic theory says that $\nabla \times \mathbf{E} = -\partial\mathbf{B}/\partial t$. In this expression $\nabla \times \mathbf{E}$ is computed with t held fixed and $\partial\mathbf{B}/\partial t$ is calculated with (x, y, z) fixed. Use Stokes's theorem to derive Faraday's law

$$\oint_C \mathbf{E} \cdot d\mathbf{r} = -\frac{\partial}{\partial t} \iint_S \mathbf{B} \cdot \mathbf{n}\,\sigma,$$

where C represents a wire loop through which current flows counterclockwise with respect to the surface's unit normal $\mathbf{n}$, giving rise to the voltage

$$\oint_C \mathbf{E} \cdot d\mathbf{r}$$

around C. The surface integral on the right side of the equation is called the **magnetic flux,** and S is any oriented surface with boundary C.

16. Let

$$\mathbf{F} = -\frac{GmM}{|\mathbf{r}|^3}\,\mathbf{r}$$

be the gravitational force field defined for $\mathbf{r} \neq \mathbf{0}$. Use Gauss's law in Section 14.8 to show that there is no continuously differentiable vector field $\mathbf{H}$ satisfying $\mathbf{F} = \nabla \times \mathbf{H}$.

17. If $f(x, y, z)$ and $g(x, y, z)$ are continuously differentiable scalar functions defined over the oriented surface S with boundary curve C, prove that

$$\iint\limits_{S} (\nabla f \times \nabla g) \cdot \mathbf{n}\, d\sigma = \oint_{C} f\,\nabla g \cdot d\mathbf{r}.$$

18. Suppose that $\nabla \cdot \mathbf{F}_1 = \nabla \cdot \mathbf{F}_2$ and $\nabla \times \mathbf{F}_1 = \nabla \times \mathbf{F}_2$ over a region D enclosed by the oriented surface S with outward unit normal $\mathbf{n}$ and that $\mathbf{F}_1 \cdot \mathbf{n} = \mathbf{F}_2 \cdot \mathbf{n}$ on S. Prove that $\mathbf{F}_1 = \mathbf{F}_2$ throughout D.

19. Prove or disprove that if $\nabla \cdot \mathbf{F} = 0$ and $\nabla \times \mathbf{F} = \mathbf{0}$, then $\mathbf{F} = \mathbf{0}$.

20. Let S be an oriented surface parametrized by $\mathbf{r}(u, v)$. Define the notation $d\sigma = \mathbf{r}_u\, du \times \mathbf{r}_v\, dv$ so that $d\sigma$ is a vector normal to the surface. Also, the magnitude $d\sigma = |d\sigma|$ is the element of surface area (by Eq. 5 in Section 14.6). Derive the identity

$$d\sigma = (EG - F^2)^{1/2}\, du\, dv$$

where

$$E = |\mathbf{r}_u|^2, \quad F = \mathbf{r}_u \cdot \mathbf{r}_v, \quad \text{and} \quad G = |\mathbf{r}_v|^2.$$

21. Show that the volume V of a region D in space enclosed by the oriented surface S with outward normal $\mathbf{n}$ satisfies the identity

$$V = \frac{1}{3} \iint\limits_{S} \mathbf{r} \cdot \mathbf{n}\, d\sigma,$$

where $\mathbf{r}$ is the position vector of the point (x, y, z) in D.

Appendices

Mathematical Induction

Many formulas, like

$$1 + 2 + \cdots + n = \frac{n(n+1)}{2},$$

can be shown to hold for every positive integer n by applying an axiom called the *mathematical induction principle.* A proof that uses this axiom is called a *proof by mathematical induction* or a *proof by induction.*

The steps in proving a formula by induction are the following.

Step 1: Check that the formula holds for $n = 1$.

Step 2: Prove that if the formula holds for any positive integer $n = k$, then it also holds for the next integer, $n = k + 1$.

Once these steps are completed (the axiom says), we know that the formula holds for all positive integers n. By step 1 it holds for $n = 1$. By step 2 it holds for $n = 2$, and therefore by step 2 also for $n = 3$, and by step 2 again for $n = 4$, and so on. If the first domino falls, and the kth domino always knocks over the $(k + 1)$st when it falls, all the dominoes fall.

From another point of view, suppose we have a sequence of statements S_1, $S_2, \ldots, S_n, \ldots$, one for each positive integer. Suppose we can show that assuming any one of the statements to be true implies that the next statement in line is true. Suppose that we can also show that S_1 is true. Then we may conclude that the statements are true from S_1 on.

EXAMPLE 1   Show that for every positive integer n,

$$1 + 2 + \cdots + n = \frac{n(n+1)}{2}.$$

Solution   We accomplish the proof by carrying out the two steps above.

Step 1: The formula holds for $n = 1$ because

$$1 = \frac{1(1+1)}{2}.$$

Step 2: If the formula holds for $n = k$, does it also hold for $n = k + 1$? The answer is yes, and here's why: If

$$1 + 2 + \cdots + k = \frac{k(k+1)}{2},$$

A-1

then

$$1 + 2 + \cdots + k + (k + 1) = \frac{k(k + 1)}{2} + (k + 1) = \frac{k^2 + k + 2k + 2}{2}$$

$$= \frac{(k + 1)(k + 2)}{2} = \frac{(k + 1)((k + 1) + 1)}{2}.$$

The last expression in this string of equalities is the expression $n(n + 1)/2$ for $n = (k + 1)$.

The mathematical induction principle now guarantees the original formula for all positive integers n. Notice that all *we* have to do is carry out steps 1 and 2. The mathematical induction principle does the rest. ❑

EXAMPLE 2   Show that for all positive integers n,

$$\frac{1}{2^1} + \frac{1}{2^2} + \cdots + \frac{1}{2^n} = 1 - \frac{1}{2^n}.$$

Solution   We accomplish the proof by carrying out the two steps of mathematical induction.

Step 1: The formula holds for $n = 1$ because

$$\frac{1}{2^1} = 1 - \frac{1}{2^1}.$$

Step 2: If

$$\frac{1}{2^1} + \frac{1}{2^2} + \cdots + \frac{1}{2^k} = 1 - \frac{1}{2^k},$$

then

$$\frac{1}{2^1} + \frac{1}{2^2} + \cdots + \frac{1}{2^k} + \frac{1}{2^{k+1}} = 1 - \frac{1}{2^k} + \frac{1}{2^{k+1}} = 1 - \frac{1 \cdot 2}{2^k \cdot 2} + \frac{1}{2^{k+1}}$$

$$= 1 - \frac{2}{2^{k+1}} + \frac{1}{2^{k+1}} = 1 - \frac{1}{2^{k+1}}.$$

Thus, the original formula holds for $n = (k + 1)$ whenever it holds for $n = k$.

With these steps verified, the mathematical induction principle now guarantees the formula for every positive integer n. ❑

Other Starting Integers

Instead of starting at $n = 1$, some induction arguments start at another integer. The steps for such an argument are as follows.

Step 1: Check that the formula holds for $n = n_1$ (the first appropriate integer).

Step 2: Prove that if the formula holds for any integer $n = k \geq n_1$, then it also holds for $n = (k + 1)$.

Once these steps are completed, the mathematical induction principle guarantees the formula for all $n \geq n_1$.

EXAMPLE 3   Show that $n! > 3^n$ if n is large enough.

Solution   How large is large enough? We experiment:

n	1	2	3	4	5	6	7
$n!$	1	2	6	24	120	720	5040
3^n	3	9	27	81	243	729	2187

It looks as if $n! > 3^n$ for $n \geq 7$. To be sure, we apply mathematical induction. We take $n_1 = 7$ in step 1 and try for step 2.

Suppose $k! > 3^k$ for some $k \geq 7$. Then

$$(k+1)! = (k+1)(k!) > (k+1)3^k > 7 \cdot 3^k > 3^{k+1}.$$

Thus, for $k \geq 7$,

$$k! > 3^k \quad \Rightarrow \quad (k+1)! > 3^{k+1}.$$

The mathematical induction principle now guarantees $n! \geq 3^n$ for all $n \geq 7$.   ❏

Exercises A.1

1. Assuming that the triangle inequality $|a + b| \leq |a| + |b|$ holds for any two numbers a and b, show that

$$|x_1 + x_2 + \cdots + x_n| \leq |x_1| + |x_2| + \cdots + |x_n|$$

for any n numbers.

2. Show that if $r \neq 1$, then

$$1 + r + r^2 + \cdots + r^n = \frac{1 - r^{n+1}}{1 - r}$$

for every positive integer n.

3. Use the Product Rule, $\dfrac{d}{dx}(uv) = u\dfrac{dv}{dx} + v\dfrac{du}{dx}$,

and the fact that $\dfrac{d}{dx}(x) = 1$

to show that $\dfrac{d}{dx}(x^n) = nx^{n-1}$

for every positive integer n.

4. Suppose that a function $f(x)$ has the property that $f(x_1x_2) = f(x_1) + f(x_2)$ for any two positive numbers x_1 and x_2. Show that

$$f(x_1x_2 \cdots x_n) = f(x_1) + f(x_2) + \cdots + f(x_n)$$

for the product of any n positive numbers $x_1, x_2 \ldots, x_n$.

5. Show that

$$\frac{2}{3^1} + \frac{2}{3^2} + \cdots + \frac{2}{3^n} = 1 - \frac{1}{3^n}$$

for all positive integers n.

6. Show that $n! > n^3$ if n is large enough.

7. Show that $2^n > n^2$ if n is large enough.

8. Show that $2^n \geq 1/8$ for $n \geq -3$.

9. *Sums of squares.* Show that the sum of the squares of the first n positive integers is

$$\frac{n\left(n + \dfrac{1}{2}\right)(n + 1)}{3}.$$

10. *Sums of cubes.* Show that the sum of the cubes of the first n positive integers is $(n(n + 1)/2)^2$.

11. *Rules for finite sums.* Show that the following finite sum rules hold for every positive integer n.

a) $\displaystyle\sum_{k=1}^{n}(a_k + b_k) = \sum_{k=1}^{n}a_k + \sum_{k=1}^{n}b_k$

b) $\displaystyle\sum_{k=1}^{n}(a_k - b_k) = \sum_{k=1}^{n}a_k - \sum_{k=1}^{n}b_k$

c) $\displaystyle\sum_{k=1}^{n}ca_k = c \cdot \sum_{k=1}^{n}a_k$ (Any number c)

d) $\displaystyle\sum_{k=1}^{n}a_k = n \cdot c$ (if a_k has the constant value c)

12. Show that $|x^n| = |x|^n$ for every positive integer n and every real number x.

A.2 Proofs of Limit Theorems in Section 1.2

This appendix proves Theorem 1, Parts 2–5, and Theorem 4 from Section 1.2.

Theorem 1

Properties of Limits

The following rules hold if $\lim_{x \to c} f(x) = L$ and $\lim_{x \to c} g(x) = M$ (L and M real numbers).

1. *Sum Rule:* $\qquad\qquad\qquad \lim_{x \to c} [f(x) + g(x)] = L + M$

2. *Difference Rule:* $\qquad\qquad \lim_{x \to c} [f(x) - g(x)] = L - M$

3. *Product Rule:* $\qquad\qquad\; \lim_{x \to c} f(x) \cdot g(x) = L \cdot M$

4. *Constant Multiple Rule:* $\quad \lim_{x \to c} k f(x) = kL \qquad$ (any number k)

5. *Quotient Rule:* $\qquad\qquad\; \lim_{x \to c} \dfrac{f(x)}{g(x)} = \dfrac{L}{M}, \qquad$ if $M \neq 0$

6. *Power Rule:* $\qquad\qquad\quad$ If m and n are integers, then

$$\lim_{x \to c} [f(x)]^{m/n} = L^{m/n}$$

provided $L^{m/n}$ is a real number.

We proved the Sum Rule in Section 1.3 and the Power Rule is proved in more advanced texts. We obtain the Difference Rule by replacing $g(x)$ by $-g(x)$ and M by $-M$ in the Sum Rule. The Constant Multiple Rule is the special case $g(x) = k$ of the Product Rule. This leaves only the Product and Quotient Rules.

Proof of the Limit Product Rule We show that for any $\epsilon > 0$ there exists a $\delta > 0$ such that for all x in the intersection D of the domains of f and g,

$$0 < |x - c| < \delta \quad \Rightarrow \quad |f(x) g(x) - LM| < \epsilon.$$

Suppose then that ϵ is a positive number, and write $f(x)$ and $g(x)$ as

$$f(x) = L + (f(x) - L), \qquad g(x) = M + (g(x) - M).$$

Multiply these expressions together and subtract LM:

$$
\begin{aligned}
f(x) \cdot g(x) - LM &= (L + (f(x) - L))(M + (g(x) - M)) - LM \\
&= LM + L(g(x) - M) + M(f(x) - L) \\
&\quad + (f(x) - L)(g(x) - M) - LM \\
&= L(g(x) - M) + M(f(x) - L) + (f(x) - L)(g(x) - M).
\end{aligned}
\tag{1}
$$

Since f and g have limits L and M as $x \to c$, there exist positive numbers $\delta_1, \delta_2, \delta_3$, and δ_4 such that for all x in D

$$0 < |x - c| < \delta_1 \implies |f(x) - L| < \sqrt{\epsilon/3}$$

$$0 < |x - c| < \delta_2 \implies |g(x) - M| < \sqrt{\epsilon/3}$$

$$0 < |x - c| < \delta_3 \implies |f(x) - L| < \epsilon/(3(1 + |M|))$$

$$0 < |x - c| < \delta_4 \implies |g(x) - M| < \epsilon/(3(1 + |L|))$$

$$(2)$$

If we take δ to be the smallest numbers δ_1 through δ_4, the inequalities on the right-hand side of (2) will hold simultaneously for $0 < |x - c| < \delta$. Therefore, for all x in D, $0 < |x - c| < \delta$ implies

$$|f(x) \cdot g(x) - LM|$$

$$\leq |L||g(x) - M| + |M||f(x) - L| + |f(x) - L||g(x) - M|$$

$$\leq (1 + |L|)|g(x) - M| + (1 + |M|)|f(x) - L| + |f(x) - L||g(x) - M|$$

$$\leq \frac{\epsilon}{3} + \frac{\epsilon}{3} + \sqrt{\frac{\epsilon}{3}}\sqrt{\frac{\epsilon}{3}} = \epsilon.$$

Triangle inequality applied to Eq. (1)

Values from (2)

This completes the proof of the Limit Product Rule.   ❏

Proof of the Limit Quotient Rule   We show that $\lim_{x \to c}(1/g(x)) = 1/M$. We can then conclude that

$$\lim_{x \to c} \frac{f(x)}{g(x)} = \lim_{x \to c} \left(f(x) \cdot \frac{1}{g(x)} \right) = \lim_{x \to c} f(x) \cdot \lim_{x \to c} g(x) = L \cdot \frac{1}{M} = \frac{L}{M}$$

by the Limit Product Rule.

Let $\epsilon > 0$ be given. To show that $\lim_{x \to c}(1/g(x)) = 1/M$, we need to show that there exists a $\delta > 0$ such that for all x

$$0 < |x - c| < \delta \implies \left| \frac{1}{g(x)} - \frac{1}{M} \right| < \epsilon.$$

Since $|M| > 0$, there exists a positive number δ_1 such that for all x

$$0 < |x - c| < \delta_1 \implies |g(x) - M| < \frac{M}{2}.$$

$$(3)$$

For any numbers A and B it can be shown that $|A| - |B| \leq |A - B|$ and $|B| - |A| \leq |A - B|$, from which it follows that $||A| - |B|| \leq |A - B|$. With $A = g(x)$ and $B = M$, this becomes

$$||g(x)| - |M|| \leq |g(x) - M|,$$

which can be combined with the inequality on the right in (3) to get, in turn,

$$||g(x)| - |M|| < \frac{|M|}{2}$$

$$-\frac{|M|}{2} < |g(x)| - |M| < \frac{|M|}{2}$$

$$\frac{|M|}{2} < |g(x)| < \frac{3|M|}{2}$$

$$|M| < 2|g(x)| < 3|M|$$

$$\frac{1}{|g(x)|} < \frac{2}{|M|} < \frac{3}{|g(x)|}$$

$$(4)$$

Therefore, $0 < |x - c| < \delta_1$ implies that

$$
\left| \frac{1}{g(x)} - \frac{1}{M} \right| = \left| \frac{M - g(x)}{Mg(x)} \right| \le \frac{1}{|M|} \cdot \frac{1}{|g(x)|} \cdot |M - g(x)|
$$

$$
< \frac{1}{|M|} \cdot \frac{2}{|M|} \cdot |M - g(x)|. \qquad \text{Inequality (4)}
$$

(5)

Since $(1/2)|M|^2 \epsilon > 0$, there exists a number $\delta_2 > 0$ such that for all x

$$
0 < |x - c| < \delta_2 \quad \Rightarrow \quad |M - g(x)| < \frac{\epsilon}{2}|M|^2. \tag{6}
$$

If we take δ to be the smaller of δ_1 and δ_2, the conclusions in (5) and (6) both hold for all x such that $0 < |x - c| < \delta$. Combining these conclusions gives

$$
0 < |x - c| < \delta \quad \Rightarrow \quad \left| \frac{1}{g(x)} - \frac{1}{M} \right| < \epsilon.
$$

This concludes the proof of the Limit Quotient Rule. ❑

Theorem 4

The Sandwich Theorem

Suppose that $g(x) \le f(x) \le h(x)$ for all x in some open interval containing c, except possibly at $x = c$ itself. Suppose also that $\lim_{x \to c} g(x) = \lim_{x \to c} h(x) = L$. Then $\lim_{x \to c} f(x) = L$.

Proof for Right-hand Limits Suppose $\lim_{x \to c^+} g(x) = \lim_{x \to c^+} h(x) = L$. Then for any $\epsilon > 0$ there exists a $\delta > 0$ such that for all x the inequality $c < x < c + \delta$ implies

$$
L - \epsilon < g(x) < L + \epsilon \qquad \text{and} \qquad L - \epsilon < h(x) < L + \epsilon. \tag{7}
$$

These inequalities combine with the inequality $g(x) \le f(x) \le h(x)$ to give

$$
L - \epsilon < g(x) \le f(x) \le h(x) < L + \epsilon,
$$

$$
L - \epsilon < f(x) < L + \epsilon, \tag{8}
$$

$$
-\epsilon < f(x) - L < \epsilon.
$$

Therefore, for all x, the inequality $c < x < c + \delta$ implies $|f(x) - L| < \epsilon$. ❑

Proof for Left-hand Limits Suppose $\lim_{x \to c^-} g(x) = \lim_{x \to c^-} h(x) = L$. Then for any $\epsilon > 0$ there exists a $\delta > 0$ such that for all x the inequality $c - \delta < x < c$ implies

$$
L - \epsilon < g(x) < L + \epsilon \qquad \text{and} \qquad L - \epsilon < h(x) < L + \epsilon. \tag{9}
$$

We conclude as before that for all x, $c - \delta < x < c$ implies $|f(x) - L| < \epsilon$. ❑

Proof for Two-sided Limits If $\lim_{x \to c} g(x) = \lim_{x \to c} h(x) = L$, then $g(x)$ and $h(x)$ both approach L as $x \to c^+$ and as $x \to c^-$; so $\lim_{x \to c^+} f(x) = L$ and $\lim_{x \to c^-} f(x) = L$. Hence $\lim_{x \to c} f(x)$ exists and equals L. ❑

Exercises A.2

1. Suppose that functions $f_1(x)$, $f_2(x)$, and $f_3(x)$ have limits L_1, L_2, and L_3, respectively, as $x \to c$. Show that their sum has limit $L_1 + L_2 + L_3$. Use mathematical induction (Appendix 1) to generalize this result to the sum of any finite number of functions.

2. Use mathematical induction and the Limit Product Rule in Theorem 1 to show that if functions $f_1(x)$, $f_2(x)$, ..., $f_n(x)$ have limits L_1, L_2, ..., L_n as $x \to c$, then
   $$\lim_{x \to c} f_1(x) f_2(x) \cdot \cdots \cdot f_n(x) = L_1 \cdot L_2 \cdot \cdots \cdot L_n.$$

3. Use the fact that $\lim_{x \to c} x = c$ and the result of Exercise 2 to show that $\lim_{x \to c} x^n = c^n$ for any integer $n > 1$.

4. *Limits of polynomials.* Use the fact that $\lim_{x \to c}(k) = k$ for any number k together with the results of Exercises 1 and 3 to show that $\lim_{x \to c} f(x) = f(c)$ for any polynomial function
   $$f(x) = a_0 x^n + a_1 x^{n-1} + \cdots + a_{n-1}x + a_n.$$

5. *Limits of rational functions.* Use Theorem 1 and the result of Exercise 4 to show that if $f(x)$ and $g(x)$ are polynomial functions and $g(c) \neq 0$, then
   $$\lim_{x \to c} \frac{f(x)}{g(x)} = \frac{f(c)}{g(c)}.$$

6. *Composites of continuous functions.* Figure A.1 gives the diagram for a proof that the composite of two continuous functions is continuous. Reconstruct the proof from the diagram. The statement to be proved is this: If f is continuous at $x = c$ and g is continuous at $f(c)$, then $g \circ f$ is continuous at c.

   Assume that c is an interior point of the domain of f and that $f(c)$ is an interior point of the domain of g. This will make the limits involved two-sided. (The arguments for the cases that involve one-sided limits are similar.)

A.1 The diagram for a proof that the composite of two continuous functions is continuous. The continuity of composites holds for any finite number of functions. The only requirement is that each function be continuous where it is applied. In the figure, *f* is to be continuous at *c* and *g* at *f(c)*.

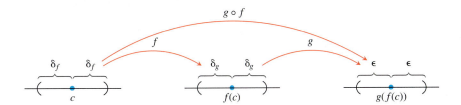

Complex Numbers

Complex numbers are expressions of the form $a + ib$, where a and b are real numbers and i is a symbol for $\sqrt{-1}$. Unfortunately, the words "real" and "imaginary" have connotations that somehow place $\sqrt{-1}$ in a less favorable position in our minds than $\sqrt{2}$. As a matter of fact, a good deal of imagination, in the sense of *inventiveness,* has been required to construct the *real* number system, which forms the basis of the calculus. In this appendix we review the various stages of this invention. The further invention of a complex number system will then not seem so strange.

The Development of the Real Numbers

The earliest stage of number development was the recognition of the **counting numbers** 1, 2, 3, . . . , which we now call the **natural numbers** or the **positive integers.** Certain simple arithmetic operations can be performed with these numbers without getting outside the system. That is, the system of positive integers is **closed** under the operations of addition and multiplication. By this we mean that if m and n are any positive integers, then

$$m + n = p \qquad \text{and} \qquad mn = q \tag{1}$$

are also positive integers. Given the two positive integers on the left-hand side of either equation in (1), we can find the corresponding positive integer on the right. More than this, we can sometimes specify the positive integers m and p and find a positive integer n such that $m + n = p$. For instance, $3 + n = 7$ can be solved when the only numbers we know are the positive integers. But the equation $7 + n = 3$ cannot be solved unless the number system is enlarged.

The number zero and the negative integers were invented to solve equations like $7 + n = 3$. In a civilization that recognizes all the **integers**

$$\ldots, -3, -2, -1, 0, 1, 2, 3, \ldots, \tag{2}$$

an educated person can always find the missing integer that solves the equation $m + n = p$ when given the other two integers in the equation.

Suppose our educated people also know how to multiply any two of the integers in (2). If, in Eqs. (1), they are given m and q, they discover that sometimes they can find n and sometimes they cannot. If their imagination is still in good working order, they may be inspired to invent still more numbers and introduce fractions, which are just ordered pairs m/n of integers m and n. The number zero has special properties that may bother them for a while, but they ultimately discover that it is handy to have all ratios of integers m/n, excluding only those having zero in the denominator. This system, called the set of **rational numbers,** is now rich enough for them to perform the so-called **rational operations** of arithmetic:

1. a) addition   **2.** a) multiplication
   b) subtraction   b) division

on any two numbers in the system, *except that they cannot divide by zero.*

The geometry of the unit square (Fig. A.2) and the Pythagorean theorem showed that they could construct a geometric line segment that, in terms of some basic unit of length, has length equal to $\sqrt{2}$. Thus they could solve the equation

$$x^2 = 2$$

by a geometric construction. But then they discovered that the line segment representing $\sqrt{2}$ and the line segment representing the unit of length 1 were incommensurable quantities. This means that the ratio $\sqrt{2}/1$ cannot be expressed as the ratio of two *integer* multiples of some other, presumably more fundamental, unit of length. That is, our educated people could not find a rational number solution of the equation $x^2 = 2$.

There *is* no rational number whose square is 2. To see why, suppose that there were such a rational number. Then we could find integers p and q with no common factor other than 1, and such that

$$p^2 = 2q^2. \tag{3}$$

Since p and q are integers, p must be even; otherwise its product with itself would be odd. In symbols, $p = 2p_1$, where p_1 is an integer. This leads to $2p_1^2 = q^2$, which says q must be even, say $q = 2q_1$, where q_1 is an integer. This makes 2 a factor of both p and q, contrary to our choice of p and q as integers with no common factor other than 1. Hence there is no rational number whose square is 2.

Although our educated people could not find a rational solution of the equation $x^2 = 2$, they could get a sequence of rational numbers

$$\frac{1}{1}, \frac{7}{5}, \frac{41}{29}, \frac{239}{169}, \ldots, \tag{4}$$

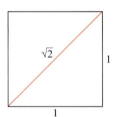

A.2 With a straightedge and compass, it is possible to construct a segment of irrational length.

whose squares form a sequence

$$\frac{1}{1}, \frac{49}{25}, \frac{1681}{841}, \frac{57,121}{28,561}, \ldots, \tag{5}$$

that converges to 2 as its limit. This time their imagination suggested that they needed the concept of a limit of a sequence of rational numbers. If we accept the fact that an increasing sequence that is bounded from above always approaches a limit and observe that the sequence in (4) has these properties, then we want it to have a limit L. This would also mean, from (5), that $L^2 = 2$, and hence L is *not* one of our rational numbers. If to the rational numbers we further add the limits of all bounded increasing sequences of rational numbers, we arrive at the system of all "real" numbers. The word *real* is placed in quotes because there is nothing that is either "more real" or "less real" about this system than there is about any other mathematical system.

The Complex Numbers

Imagination was called upon at many stages during the development of the real number system. In fact, the art of invention was needed at least three times in constructing the systems we have discussed so far:

1. The *first invented* system: the set of *all integers* as constructed from the counting numbers.
2. The *second invented* system: the set of *rational numbers* m/n as constructed from the integers.
3. The *third invented* system: the set of all *real numbers* x as constructed from the rational numbers.

These invented systems form a hierarchy in which each system contains the previous system. Each system is also richer than its predecessor in that it permits additional operations to be performed without going outside the system:

1. In the system of all integers, we can solve all equations of the form

$$x + a = 0, \tag{6}$$

where a can be any integer.
2. In the system of all rational numbers, we can solve all equations of the form

$$ax + b = 0, \tag{7}$$

provided a and b are rational numbers and $a \neq 0$.
3. In the system of all real numbers, we can solve all of the equations in (6) and (7) and, in addition, all quadratic equations

$$ax^2 + bx + c = 0 \quad \text{having} \quad a \neq 0 \quad \text{and} \quad b^2 - 4ac \geq 0. \tag{8}$$

You are probably familiar with the formula that gives the solutions of (8), namely,

$$x = \frac{-b \pm \sqrt{b^2 - 4ac}}{2a}, \tag{9}$$

and are familiar with the further fact that when the discriminant, $d = b^2 - 4ac$, is negative, the solutions in (9) do *not* belong to any of the systems discussed above. In fact, the very simple quadratic equation

$$x^2 + 1 = 0$$

is impossible to solve if the only number systems that can be used are the three invented systems mentioned so far.

Thus we come to the *fourth invented* system, the set of all complex numbers $a + ib$. We could dispense entirely with the symbol i and use a notation like (a, b). We would then speak simply of a pair of real numbers a and b. Since, under algebraic operations, the numbers a and b are treated somewhat differently, it is essential to keep the *order* straight. We therefore might say that the **complex number system** consists of the set of all ordered pairs of real numbers (a, b), together with the rules by which they are to be equated, added, multiplied, and so on, listed below. We will use both the (a, b) notation and the notation $a + ib$ in the discussion that follows. We call a the **real part** and b the **imaginary part** of the complex number (a, b).

We make the following definitions.

Equality

$a + ib = c + id$

if and only if

$a = c$ and $b = d$

Two complex numbers (a, b) and (c, d) are *equal* if and only if $a = c$ and $b = d$.

Addition

$(a + ib) + (c + id)$
$= (a + c) + i(b + d)$

The sum of the two complex numbers (a, b) and (c, d) is the complex number $(a + c, b + d)$.

Multiplication

$(a + ib)(c + id)$
$= (ac - bd) + i(ad + bc)$

The product of two complex numbers (a, b) and (c, d) is the complex number $(ac - bd, ad + bc)$.

$c(a + ib) = ac + i(bc)$

The product of a real number c and the complex number (a, b) is the complex number (ac, bc).

The set of all complex numbers (a, b) in which the second number b is zero has all the properties of the set of real numbers a. For example, addition and multiplication of $(a, 0)$ and $(c, 0)$ give

$$(a, 0) + (c, 0) = (a + c, 0),$$

$$(a, 0) \cdot (c, 0) = (ac, 0),$$

which are numbers of the same type with imaginary part equal to zero. Also, if we multiply a "real number" $(a, 0)$ and the complex number (c, d), we get

$$(a, 0) \cdot (c, d) = (ac, ad) = a(c, d).$$

In particular, the complex number $(0, 0)$ plays the role of zero in the complex number system, and the complex number $(1, 0)$ plays the role of unity.

The number pair $(0, 1)$, which has real part equal to zero and imaginary part equal to one, has the property that its square,

$$(0, 1)(0, 1) = (-1, 0),$$

has real part equal to minus one and imaginary part equal to zero. Therefore, in the system of complex numbers (a, b), there is a number $x = (0, 1)$ whose square can be added to unity $= (1, 0)$ to produce zero $= (0, 0)$; that is,

$$(0, 1)^2 + (1, 0) = (0, 0).$$

The equation

$$x^2 + 1 = 0$$

therefore has a solution $x = (0, 1)$ in this new number system.

You are probably more familiar with the $a + ib$ notation than you are with the notation (a, b). And since the laws of algebra for the ordered pairs enable us to write

$$(a, b) = (a, 0) + (0, b) = a(1, 0) + b(0, 1),$$

while $(1, 0)$ behaves like unity and $(0, 1)$ behaves like a square root of minus one, we need not hesitate to write $a + ib$ in place of (a, b). The i associated with b is like a tracer element that tags the imaginary part of $a + ib$. We can pass at will from the realm of ordered pairs (a, b) to the realm of expressions $a + ib$, and conversely. But there is nothing less "real" about the symbol $(0, 1) = i$ than there is about the symbol $(1, 0) = 1$, once we have learned the laws of algebra in the complex number system (a, b).

To reduce any rational combination of complex numbers to a single complex number, we apply the laws of elementary algebra, replacing i^2 wherever it appears by -1. Of course, we cannot divide by the complex number $(0, 0) = 0 + i\,0$. But if $a + ib \neq 0$, then we may carry out a division as follows:

$$\frac{c + id}{a + ib} = \frac{(c + id)(a - ib)}{(a + ib)(a - ib)} = \frac{(ac + bd) + i(ad - bc)}{a^2 + b^2}.$$

The result is a complex number $x + iy$ with

$$x = \frac{ac + bd}{a^2 + b^2}, \qquad y = \frac{ad - bc}{a^2 + b^2},$$

and $a^2 + b^2 \neq 0$, since $a + ib = (a, b) \neq (0, 0)$.

The number $a - ib$ that is used as multiplier to clear the i from the denominator is called the **complex conjugate** of $a + ib$. It is customary to use $\bar{z}$ (read "z bar") to denote the complex conjugate of z; thus

$$z = a + ib, \qquad \bar{z} = a - ib.$$

Multiplying the numerator and denominator of the fraction $(c + id)/(a + ib)$ by the complex conjugate of the denominator will always replace the denominator by a real number.

EXAMPLE 1

a) $(2 + 3i) + (6 - 2i) = (2 + 6) + (3 - 2)i = 8 + i$

b) $(2 + 3i) - (6 - 2i) = (2 - 6) + (3 - (-2))i = -4 + 5i$

c) $(2 + 3i)(6 - 2i) = (2)(6) + (2)(-2i) + (3i)(6) + (3i)(-2i)$

$$= 12 - 4i + 18i - 6i^2 = 12 + 14i + 6 = 18 + 14i$$

d) $\dfrac{2 + 3i}{6 - 2i} = \dfrac{2 + 3i}{6 - 2i}\,\dfrac{6 + 2i}{6 + 2i}$

$$= \frac{12 + 4i + 18i + 6i^2}{36 + 12i - 12i - 4i^2}$$

$$= \frac{6 + 22i}{40} = \frac{3}{20} + \frac{11}{20}i$$

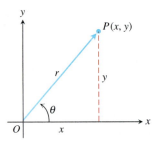

A.3 This Argand diagram represents $z = x + iy$ both as a point $P(x, y)$ and as a vector $\overrightarrow{OP}$.

Argand Diagrams

There are two geometric representations of the complex number $z = x + iy$:

a)   as the point $P(x, y)$ in the xy-plane and

b)   as the vector $\overrightarrow{OP}$ from the origin to P.

In each representation, the x-axis is called the **real axis** and the y-axis is the **imaginary axis.** Both representations are **Argand diagrams** for $x + iy$ (Fig. A.3).

In terms of the polar coordinates of x and y, we have

$$x = r \cos \theta, \qquad y = r \sin \theta,$$

and

$$z = x + iy = r(\cos \theta + i \sin \theta). \tag{10}$$

We define the **absolute value** of a complex number $x + iy$ to be the length r of a vector $\overrightarrow{OP}$ from the origin to $P(x, y)$. We denote the absolute value by vertical bars, thus:

$$|x + iy| = \sqrt{x^2 + y^2}.$$

If we always choose the polar coordinates r and θ so that r is nonnegative, then

$$r = |x + iy|.$$

The polar angle θ is called the **argument** of z and is written $\theta = \arg z$. Of course, any integer multiple of 2π may be added to θ to produce another appropriate angle.

The following equation gives a useful formula connecting a complex number z, its conjugate $\bar{z}$, and its absolute value $|z|$, namely,

$$z \cdot \bar{z} = |z|^2.$$

Euler's Formula, Products, and Quotients

The identity

$$e^{i\theta} = \cos \theta + i \sin \theta,$$

called **Euler's formula,** enables us to rewrite Eq. (10) as

$$z = r e^{i\theta}.$$

This, in turn, leads to the following rules for calculating products, quotients, powers, and roots of complex numbers. It also leads to Argand diagrams for $e^{i\theta}$. Since $\cos \theta + i \sin \theta$ is what we get from Eq. (10) by taking $r = 1$, we can say that $e^{i\theta}$ is represented by a unit vector that makes an angle θ with the positive x-axis, as shown in Fig. A.4.

Products   To multiply two complex numbers, we multiply their absolute values and add their angles. Let

$$z_1 = r_1 e^{i\theta_1}, \qquad z_2 = r_2 e^{i\theta_2}, \tag{11}$$

so that

$$|z_1| = r_1, \quad \arg z_1 = \theta_1; \qquad |z_2| = r_2, \quad \arg z_2 = \theta_2.$$

Then

$$z_1 z_2 = r_1 e^{i\theta_1} \cdot r_2 e^{i\theta_2} = r_1 r_2 e^{i(\theta_1 + \theta_2)}$$

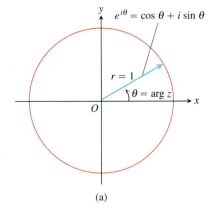

(a)

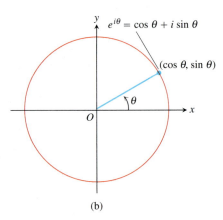

(b)

A.4 Argand diagrams for $e^{i\theta} = \cos \theta + i \sin \theta$ (a) as a vector, (b) as a point.

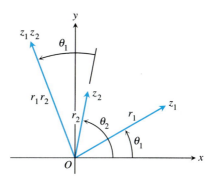

A.5 When z_1 and z_2 are multiplied, $|z_1 z_2| = r_1 \cdot r_2$ and $\arg (z_1 z_2) = \theta_1 + \theta_2$.

---

exp (A) stands for e^A.

---

and hence

$$|z_1 z_2| = r_1 r_2 = |z_1| \cdot |z_2|,$$

$$\arg (z_1 z_2) = \theta_1 + \theta_2 = \arg z_1 + \arg z_2. \tag{12}$$

Thus the product of two complex numbers is represented by a vector whose length is the product of the lengths of the two factors and whose argument is the sum of their arguments (Fig. A.5). In particular, a vector may be rotated counterclockwise through an angle θ by multiplying it by $e^{i\theta}$. Multiplication by i rotates $90°$, by -1 rotates $180°$, by $-i$ rotates $270°$, etc.

EXAMPLE 2   Let $z_1 = 1 + i, z_2 = \sqrt{3} - i$. We plot these complex numbers in an Argand diagram (Fig. A.6) from which we read off the polar representations

$$z_1 = \sqrt{2} e^{i\pi/4}, \qquad z_2 = 2 e^{-i\pi/6}.$$

Then

$$z_1 z_2 = 2\sqrt{2} \exp \left(\frac{i\pi}{4} - \frac{i\pi}{6} \right) = 2\sqrt{2} \exp \left(\frac{i\pi}{12} \right)$$

$$= 2\sqrt{2} \left(\cos \frac{\pi}{12} + i \sin \frac{\pi}{12} \right) \approx 2.73 + 0.73i.$$

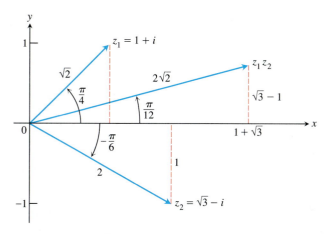

A.6 To multiply two complex numbers, multiply their absolute values and add their arguments.   ❏

Quotients

Suppose $r_2 \neq 0$ in Eq. (11). Then

$$\frac{z_1}{z_2} = \frac{r_1 e^{i\theta_1}}{r_2 e^{i\theta_2}} = \frac{r_1}{r_2} e^{i(\theta_1 - \theta_2)}.$$

Hence

$$\left| \frac{z_1}{z_2} \right| = \frac{r_1}{r_2} = \frac{|z_1|}{|z_2|} \quad \text{and} \quad \arg \left(\frac{z_1}{z_2} \right) = \theta_1 - \theta_2 = \arg z_1 - \arg z_2.$$

That is, we divide lengths and subtract angles.

EXAMPLE 3   Let $z_1 = 1 + i$ and $z_2 = \sqrt{3} - i$, as in Example 2. Then

$$\frac{1+i}{\sqrt{3}-i} = \frac{\sqrt{2}e^{i\pi/4}}{2e^{-i\pi/6}} = \frac{\sqrt{2}}{2}e^{5\pi i/12} \approx 0.707\left(\cos\frac{5\pi}{12} + i\,\sin\frac{5\pi}{12}\right)$$

$$\approx 0.183 + 0.\overline{683}i.$$ ❑

Powers

If n is a positive integer, we may apply the product formulas in (12) to find

$$z^n = z \cdot z \cdot \cdots \cdot z. \qquad \textcolor{blue}{n\ \text{factors}}$$

With $z = re^{i\theta}$, we obtain

$$z^n = (re^{i\theta})^n = r^n e^{i(\theta+\theta+\cdots+\theta)} \qquad \textcolor{blue}{n\ \text{summands}}$$
$$= r^n e^{in\theta}. \tag{13}$$

The length $r = |z|$ is raised to the nth power and the angle $\theta = \arg z$ is multiplied by n.

If we take $r = 1$ in Eq. (13), we obtain De Moivre's theorem.

De Moivre's Theorem

$$(\cos\theta + i\,\sin\theta)^n = \cos n\theta + i\,\sin n\theta. \tag{14}$$

If we expand the left-hand side of De Moivre's equation (Eq. 14) by the binomial theorem and reduce it to the form $a + ib$, we obtain formulas for $\cos n\theta$ and $\sin n\theta$ as polynomials of degree n in $\cos\theta$ and $\sin\theta$.

EXAMPLE 4   If $n = 3$ in Eq. (14), we have

$$(\cos\theta + i\,\sin\theta)^3 = \cos 3\theta + i\,\sin 3\theta.$$

The left-hand side of this equation is

$$\cos^3\theta + 3i\,\cos^2\theta\,\sin\theta - 3\cos\theta\,\sin^2\theta - i\,\sin^3\theta.$$

The real part of this must equal $\cos 3\theta$ and the imaginary part must equal $\sin 3\theta$. Therefore,

$$\cos 3\theta = \cos^3\theta - 3\cos\theta\,\sin^2\theta,$$
$$\sin 3\theta = 3\cos^2\theta\,\sin\theta - \sin^3\theta.$$ ❑

Roots   If $z = re^{i\theta}$ is a complex number different from zero and n is a positive integer, then there are precisely n different complex numbers $w_0, w_1, \ldots, w_{n-1}$, that are nth roots of z. To see why, let $w = \rho e^{i\alpha}$ be an nth root of $z = re^{i\theta}$, so that

$$w^n = z$$

or

$$\rho^n e^{in\alpha} = re^{i\theta}.$$

Then

$$\rho = \sqrt[n]{r}$$

is the real, positive nth root of r. As regards the angle, although we cannot say that $n\alpha$ and θ must be equal, we can say that they may differ only by an integer multiple of 2π. That is,

$$n\alpha = \theta + 2k\pi, \qquad k = 0, \pm 1, \pm 2, \ldots.$$

Therefore,

$$\alpha = \frac{\theta}{n} + k\frac{2\pi}{n}.$$

Hence all nth roots of $z = re^{i\theta}$ are given by

$$\sqrt[n]{re^{i\theta}} = \sqrt[n]{r}\,\exp\,i\left(\frac{\theta}{n} + k\frac{2\pi}{n}\right), \qquad k = 0, \pm 1, \pm 2, \ldots. \qquad (15)$$

There might appear to be infinitely many different answers corresponding to the infinitely many possible values of k. But $k = n + m$ gives the same answer as $k = m$ in Eq. (15). Thus we need only take n consecutive values for k to obtain all the different nth roots of z. For convenience, we take

$$k = 0, 1, 2, \ldots, n - 1.$$

All the nth roots of $re^{i\theta}$ lie on a circle centered at the origin O and having radius equal to the real, positive nth root of r. One of them has argument $\alpha = \theta/n$. The others are uniformly spaced around the circle, each being separated from its neighbors by an angle equal to $2\pi/n$. Figure A.7 illustrates the placement of the three cube roots, w_0, w_1, w_2, of the complex number $z = re^{i\theta}$.

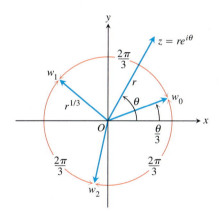

A.7 The three cube roots of $z = re^{i\theta}$.

EXAMPLE 5   Find the four fourth roots of -16.

Solution   As our first step, we plot the number -16 in an Argand diagram (Fig. A.8) and determine its polar representation $re^{i\theta}$. Here, $z = -16, r = +16$, and $\theta = \pi$. One of the fourth roots of $16e^{i\pi}$ is $2e^{i\pi/4}$. We obtain others by successive additions of $2\pi/4 = \pi/2$ to the argument of this first one. Hence

$$\sqrt[4]{16\exp i\pi} = 2\,\exp\,i\left(\frac{\pi}{4}, \frac{3\pi}{4}, \frac{5\pi}{4}, \frac{7\pi}{4}\right),$$

and the four roots are

$$w_0 = 2\left[\cos\frac{\pi}{4} + i\,\sin\frac{\pi}{4}\right] = \sqrt{2}(1 + i),$$

$$w_1 = 2\left[\cos\frac{3\pi}{4} + i\,\sin\frac{3\pi}{4}\right] = \sqrt{2}(-1 + i),$$

$$w_2 = 2\left[\cos\frac{5\pi}{4} + i\,\sin\frac{5\pi}{4}\right] = \sqrt{2}(-1 - i),$$

$$w_3 = 2\left[\cos\frac{7\pi}{4} + i\,\sin\frac{7\pi}{4}\right] = \sqrt{2}(1 - i).$$

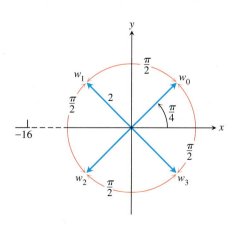

A.8 The four fourth roots of -16.

The Fundamental Theorem of Algebra   One may well say that the invention of $\sqrt{-1}$ is all well and good and leads to a number system that is richer than the real number system alone; but where will this process end? Are we also going to invent still more systems so as to obtain $\sqrt[4]{-1}$, $\sqrt[6]{-1}$, and so on? By now it should be clear that this is not necessary. These numbers are already expressible in terms of the complex number system $a + ib$. In fact, the Fundamental Theorem of Algebra says that with the introduction of the complex numbers we now have enough numbers to factor every polynomial into a product of linear factors and hence enough numbers to solve every possible polynomial equation.

The Fundamental Theorem of Algebra

Every polynomial equation of the form

$$a_0 z^n + a_1 z^{n-1} + a_2 z^{n-2} + \cdots + a_{n-1} z + a_n = 0,$$

in which the coefficients $a_0, a_1, \ldots, a_n$ are any complex numbers, whose degree n is greater than or equal to one, and whose leading coefficient a_0 is not zero, has exactly n roots in the complex number system, provided each multiple root of multiplicity m is counted as m roots.

A proof of this theorem can be found in almost any text on the theory of functions of a complex variable.

Exercises A.3

Operations with Complex Numbers

1. *How computers multiply complex numbers*
   Find $(a, b) \cdot (c, d) = (ac - bd, ad + bc)$.

   a) $(2, 3) \cdot (4, -2)$ b) $(2, -1) \cdot (-2, 3)$
   c) $(-1, -2) \cdot (2, 1)$

   (This is how complex numbers are multiplied by computers.)

2. Solve the following equations for the real numbers, x and y.

   a) $(3 + 4i)^2 - 2(x - iy) = x + iy$

   b) $\left(\dfrac{1+i}{1-i}\right)^2 + \dfrac{1}{x+iy} = 1 + i$

   c) $(3 - 2i)(x + iy) = 2(x - 2iy) + 2i - 1$

Graphing and Geometry

3. How may the following complex numbers be obtained from $z = x + iy$ geometrically? Sketch.

   a) $\bar{z}$ b) $\overline{(-z)}$
   c) $-z$ d) $1/z$

4. Show that the distance between the two points z_1 and z_2 in an Argand diagram is equal to $|z_1 - z_2|$.

In Exercises 5–10, graph the points $z = x + iy$ that satisfy the given conditions.

5. a) $|z| = 2$ b) $|z| < 2$ c) $|z| > 2$

6. $|z - 1| = 2$ 7. $|z + 1| = 1$

8. $|z + 1| = |z - 1|$ 9. $|z + i| = |z - 1|$

10. $|z + 1| \geq |z|$

Express the complex numbers in Exercises 11–14 in the form $re^{i\theta}$, with $r \geq 0$ and $-\pi < \theta \leq \pi$. Draw an Argand diagram for each calculation.

11. $(1 + \sqrt{-3})^2$ 12. $\dfrac{1+i}{1-i}$

13. $\dfrac{1 + i\sqrt{3}}{1 - i\sqrt{3}}$ 14. $(2 + 3i)(1 - 2i)$

Theory and Examples

15. Show with an Argand diagram that the law for adding complex numbers is the same as the parallelogram law for adding vectors.

16. Show that the conjugate of the sum (product, or quotient) of two complex numbers z_1 and z_2 is the same as the sum (product, or quotient) of their conjugates.

17. *Complex roots of polynomials with real coefficients come in complex-conjugate pairs.*

    a) Extend the results of Exercise 16 to show that $f(\bar{z}) = \overline{f(z)}$

if

$$f(z) = a_0 z^n + a_1 z^{n-1} + \cdots + a_{n-1} z + a_n$$

is a polynomial with real coefficients $a_0, \ldots, a_n$.

b) If z is a root of the equation $f(z) = 0$, where $f(z)$ is a polynomial with real coefficients as in part (a), show that the conjugate $\bar{z}$ is also a root of the equation. (*Hint:* Let $f(z) = u + iv = 0$; then both u and v are zero. Now use the fact that $f(\bar{z}) = \overline{f(z)} = u - iv$.)

18. Show that $|\bar{z}| = |z|$.

19. If z and $\bar{z}$ are equal, what can you say about the location of the point z in the complex plane?

20. Let $Re(z)$ denote the real part of z and $Im(z)$ the imaginary part. Show that the following relations hold for any complex numbers z, z_1, and z_2.

a) $z + \bar{z} = 2Re(z)$ **b)** $z - \bar{z} = 2i\,Im(z)$

c) $|Re(z)| \le |z|$

d) $|z_1 + z_2|^2 = |z_1|^2 + |z_2|^2 + 2Re(z_1\bar{z}_2)$
e) $|z_1 + z_2| \le |z_1| + |z_2|$

Use De Moivre's theorem to express the trigonometric functions in Exercises 21 and 22 in terms of $\cos\theta$ and $\sin\theta$.

21. $\cos 4\theta$ **22.** $\sin 4\theta$

Roots

23. Find the three cube roots of 1.

24. Find the two square roots of i.

25. Find the three cube roots of $-8i$.

26. Find the six sixth roots of 64.

27. Find the four solutions of the equation $z^4 - 2z^2 + 4 = 0$.

28. Find the six solutions of the equation $z^6 + 2z^3 + 2 = 0$.

29. Find all solutions of the equation $x^4 + 4x^2 + 16 = 0$.

30. Solve the equation $x^4 + 1 = 0$.

A.4 Simpson's One-Third Rule

Simpson's rule for approximating $\int_a^b f(x)\,dx$ is based on approximating the graph of f with parabolic arcs.

The area of the shaded region under the parabola in Fig. A.9 is

$$\text{Area} = \frac{h}{3}(y_0 + 4y_1 + y_2).$$

This formula is known as Simpson's one-third rule.

We can derive the formula as follows. To simplify the algebra, we use the coordinate system in Fig. A.10. The area under the parabola is the same no matter where the y-axis is, as long as we preserve the vertical scale. The parabola has an equation of the form $y = Ax^2 + Bx + C$, so the area under it from $x = -h$ to

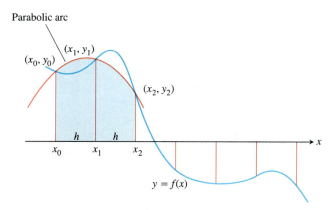

A.9 Simpson's rule approximates short stretches of curve with parabolic arcs.

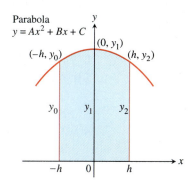

A.10 By integrating from $-h$ to h, the shaded area is found to be

$$\frac{h}{3}(y_0 + 4y_1 + y_2).$$

$x = h$ is

$$\text{Area} = \int_{-h}^{h} (Ax^2 + Bx + C)\, dx = \left[\frac{Ax^3}{3} + \frac{Bx^2}{2} + Cx \right]_{-h}^{h}$$

$$= \frac{2Ah^3}{3} + 2Ch$$

$$= \frac{h}{3}(2Ah^2 + 6C).$$

Since the curve passes through $(-h, y_0)$, $(0, y_1)$, and (h, y_2), we also have

$$y_0 = Ah^2 - Bh + C, \qquad y_1 = C, \qquad y_2 = Ah^2 + Bh + C.$$

From these equations we obtain

$$C = y_1,$$

$$Ah^2 - Bh = y_0 - y_1,$$

$$Ah^2 + Bh = y_2 - y_1,$$

$$2Ah^2 = y_0 + y_2 - 2y_1.$$

These substitutions for C and $2Ah^2$ give

$$\text{Area} = \frac{h}{3}(2Ah^2 + 6C) = \frac{h}{3}((y_0 + y_2 - 2y_1) + 6y_1) = \frac{h}{3}(y_0 + 4y_1 + y_2).$$

A.5

Cauchy's Mean Value Theorem and the Stronger Form of l'Hôpital's Rule

This appendix proves the finite-limit case of the stronger form of l'Hôpital's Rule (Section 6.6, Theorem 3).

L'Hôpital's Rule (Stronger Form)

Suppose that

$$f(x_0) = g(x_0) = 0$$

and that the functions f and g are both differentiable on an open interval (a, b) that contains the point x_0. Suppose also that $g' \neq 0$ at every point in (a, b) except possibly x_0. Then

$$\lim_{x \to x_0} \frac{f(x)}{g(x)} = \lim_{x \to x_0} \frac{f'(x)}{g'(x)}, \tag{1}$$

provided the limit on the right exists.

The proof of the stronger form of l'Hôpital's rule is based on Cauchy's Mean Value Theorem, a mean value theorem that involves two functions instead of one. We prove Cauchy's theorem first and then show how it leads to l'Hôpital's rule.

Cauchy's Mean Value Theorem

Suppose that functions f and g are continuous on $[a, b]$ and differentiable throughout (a, b) and suppose also that $g' \neq 0$ throughout (a, b). Then there exists a number c in (a, b) at which

$$\frac{f'(c)}{g'(c)} = \frac{f(b) - f(a)}{g(b) - g(a)}. \tag{2}$$

The ordinary Mean Value Theorem (Section 3.2, Theorem 4) is the case $g(x) = x$.

Proof of Cauchy's Mean Value Theorem   We apply the Mean Value Theorem of Section 3.2 twice. First we use it to show that $g(a) \neq g(b)$. For if $g(b)$ did equal $g(a)$, then the Mean Value Theorem would give

$$g'(c) = \frac{g(b) - g(a)}{b - a} = 0$$

for some c between a and b. This cannot happen because $g'(x) \neq 0$ in (a, b).

We next apply the Mean Value Theorem to the function

$$F(x) = f(x) - f(a) - \frac{f(b) - f(a)}{g(b) - g(a)} [g(x) - g(a)].$$

This function is continuous and differentiable where f and g are, and $F(b) = F(a) = 0$. Therefore there is a number c between a and b for which $F'(c) = 0$. In terms of f and g this says

$$F'(c) = f'(c) - \frac{f(b) - f(a)}{g(b) - g(a)} [g'(c)] = 0,$$

or

$$\frac{f'(c)}{g'(c)} = \frac{f(b) - f(a)}{g(b) - g(a)},$$

which is Eq. (2). ❑

Proof of the Stronger Form of l'Hôpital's Rule   We first establish Eq. (1) for the case $x \to x_0^+$. The method needs almost no change to apply to $x \to x_0^-$, and the combination of these two cases establishes the result.

Suppose that x lies to the right of x_0. Then $g'(x) \neq 0$ and we can apply Cauchy's Mean Value Theorem to the closed interval from x_0 to x. This produces a number c between x_0 and x such that

$$\frac{f'(c)}{g'(c)} = \frac{f(x) - f(x_0)}{g(x) - g(x_0)}.$$

But $f(x_0) = g(x_0) = 0$, so

$$\frac{f'(c)}{g'(c)} = \frac{f(x)}{g(x)}.$$

As x approaches x_0, c approaches x_0 because it lies between x and x_0. Therefore,

$$\lim_{x \to x_0^+} \frac{f(x)}{g(x)} = \lim_{c \to x_0^+} \frac{f'(c)}{g'(c)} = \lim_{x \to x_0^+} \frac{f'(x)}{g'(x)}.$$

This establishes l'Hôpital's rule for the case where x approaches x_0 from above. The case where x approaches x_0 from below is proved by applying Cauchy's Mean Value Theorem to the closed interval $[x, x_0], x < x_0$. ❑

A.6 Limits That Arise Frequently

This appendix verifies limits (4)–(6) in Section 8.2, Table 1.

Limit 4: If $|x| < 1$, $\displaystyle\lim_{n\to\infty} x^n = 0$ We need to show that to each $\epsilon > 0$ there corresponds an integer N so large that $|x^n| < \epsilon$ for all n greater than N. Since $\epsilon^{1/n} \to 1$, while $|x| < 1$, there exists an integer N for which $\epsilon^{1/N} > |x|$. In other words,

$$|x^N| = |x|^N < \epsilon. \tag{1}$$

This is the integer we seek because, if $|x| < 1$, then

$$|x^n| < |x^N| \text{ for all } n > N. \tag{2}$$

Combining (1) and (2) produces $|x^n| < \epsilon$ for all $n > N$, concluding the proof.

Limit 5: For any number x, $\displaystyle\lim_{n\to\infty} \left(1 + \frac{x}{n}\right)^n = e^x$ Let

$$a_n = \left(1 + \frac{x}{n}\right)^n.$$

Then

$$\ln a_n = \ln \left(1 + \frac{x}{n}\right)^n = n \ln \left(1 + \frac{x}{n}\right) \to x,$$

as we can see by the following application of l'Hôpital's rule, in which we differentiate with respect to n:

$$\lim_{n\to\infty} n \ln \left(1 + \frac{x}{n}\right) = \lim_{n\to\infty} \frac{\ln(1 + x/n)}{1/n}$$

$$= \lim_{n\to\infty} \frac{\left(\dfrac{1}{1 + x/n}\right) \cdot \left(-\dfrac{x}{n^2}\right)}{-1/n^2} = \lim_{n\to\infty} \frac{x}{1 + x/n} = x.$$

Apply Theorem 4, Section 8.2, with $f(x) = e^x$ to conclude that

$$\left(1 + \frac{x}{n}\right)^n = a_n = e^{\ln a_n} \to e^x.$$

Limit 6: For any number x, $\displaystyle\lim_{n\to\infty} \frac{x^n}{n!} = 0$ Since

$$-\frac{|x|^n}{n!} \le \frac{x^n}{n!} \le \frac{|x|^n}{n!},$$

all we need to show is that $|x|^n/n! \to 0$. We can then apply the Sandwich Theorem for Sequences (Section 8.2, Theorem 3) to conclude that $x^n/n! \to 0$.

The first step in showing that $|x|^n/n! \to 0$ is to choose an integer $M > |x|$, so that $(|x|/M) < 1$. By Limit 4, just proved, we then have $(|x|/M)^n \to 0$. We then restrict our attention to values of $n > M$. For these values of n, we can write

$$\frac{|x|^n}{n!} = \frac{|x|^n}{1 \cdot 2 \cdot \cdots \cdot M \cdot \underbrace{(M+1)(M+2) \cdot \cdots \cdot n}_{(n-M) \text{ factors}}}$$

$$\leq \frac{|x|^n}{M! M^{n-M}} = \frac{|x|^n M^M}{M! M^n} = \frac{M^M}{M!} \left(\frac{|x|}{M}\right)^n.$$

Thus,

$$0 \leq \frac{|x|^n}{n!} \leq \frac{M^M}{M!} \left(\frac{|x|}{M}\right)^n.$$

Now, the constant $M^M/M!$ does not change as n increases. Thus the Sandwich Theorem tell us that $|x|^n/n! \to 0$ because $(|x|/M)^n \to 0$.

| A.7 | # The Distributive Law for Vector Cross Products |

In this appendix we prove the distributive law

$$\mathbf{A} \times (\mathbf{B} + \mathbf{C}) = \mathbf{A} \times \mathbf{B} + \mathbf{A} \times \mathbf{C} \tag{1}$$

from Eq. (6) in Section 10.4.

Proof To derive Eq. (1), we construct $\mathbf{A} \times \mathbf{B}$ a new way. We draw $\mathbf{A}$ and $\mathbf{B}$ from the common point O and construct a plane M perpendicular to $\mathbf{A}$ at O (Fig. A.11). We then project $\mathbf{B}$ orthogonally onto M, yielding a vector $\mathbf{B}'$ with length $|\mathbf{B}| \sin \theta$. We rotate $\mathbf{B}'$ 90° about $\mathbf{A}$ in the positive sense to produce a vector $\mathbf{B}''$. Finally, we multiply $\mathbf{B}''$ by the length of $\mathbf{A}$. The resulting vector $|\mathbf{A}|\mathbf{B}''$ is equal to $\mathbf{A} \times \mathbf{B}$ since $\mathbf{B}''$ has the same direction as $\mathbf{A} \times \mathbf{B}$ by its construction (Fig. A.11) and

$$|\mathbf{A}||\mathbf{B}''| = |\mathbf{A}||\mathbf{B}'| = |\mathbf{A}||\mathbf{B}| \sin \theta = |\mathbf{A} \times \mathbf{B}|.$$

Now each of these three operations, namely,

1. projection onto M,
2. rotation about $\mathbf{A}$ through 90°,
3. multiplication by the scalar $|\mathbf{A}|$,

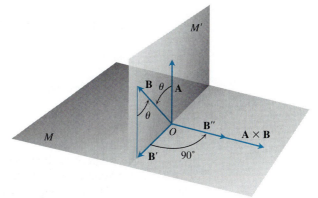

A.11 As explained in the text, $\mathbf{A} \times \mathbf{B} = |\mathbf{A}|\mathbf{B}''$.

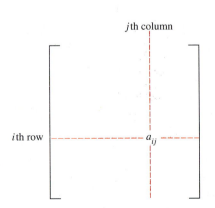

A.12 The vectors, **B, C, B + C**, and their projections onto a plane perpendicular to **A**.

when applied to a triangle whose plane is not parallel to **A,** will produce another triangle. If we start with the triangle whose sides are **B, C,** and **B + C** (Fig. A.12) and apply these three steps, we successively obtain

1. a triangle whose sides are **B′, C′,** and **(B + C)′** satisfying the vector equation

$$\mathbf{B'} + \mathbf{C'} = \mathbf{(B + C)'};$$

2. a triangle whose sides are **B″, C″,** and **(B + C)″** satisfying the vector equation

$$\mathbf{B''} + \mathbf{C''} = \mathbf{(B + C)''}$$

    (the double prime on each vector has the same meaning as in Fig. A.11); and, finally,

3. a triangle whose sides are **|A|B″, |A|C″,** and **|A|(B + C)″** satisfying the vector equation

$$|\mathbf{A}|\mathbf{B''} + |\mathbf{A}|\mathbf{C''} = |\mathbf{A}|\mathbf{(B + C)''}. \tag{2}$$

Substituting $|\mathbf{A}|\mathbf{B''} = \mathbf{A} \times \mathbf{B}, |\mathbf{A}|\mathbf{C''} = \mathbf{A} \times \mathbf{C},$ and $|\mathbf{A}|\mathbf{(B + C)''} = \mathbf{A} \times \mathbf{(B + C)}$ from our discussion above into Eq. (2) gives

$$\mathbf{A} \times \mathbf{B} + \mathbf{A} \times \mathbf{C} = \mathbf{A} \times \mathbf{(B + C)},$$

which is the law we wanted to establish.   ❏

A.8   Determinants and Cramer's Rule

A rectangular array of numbers like

$$A = \begin{bmatrix} 2 & 1 & 3 \\ 1 & 0 & -2 \end{bmatrix}$$

is called a **matrix.** We call A a 2 by 3 matrix because it has two rows and three columns. An m by n matrix has m rows and n columns, and the **entry** or **element** (number) in the ith row and jth column is denoted by a_{ij}. The matrix

$$A = \begin{bmatrix} 2 & 1 & 3 \\ 1 & 0 & -2 \end{bmatrix}$$

has

$$a_{11} = 2, \qquad a_{12} = 1, \qquad a_{13} = 3,$$

$$a_{21} = 1, \qquad a_{22} = 0, \qquad a_{23} = -2.$$

A matrix with the same number of rows as columns is a **square matrix.** It is a **matrix of order *n*** if the number of rows and columns is n.

With each square matrix A we associate a number det A or $|a_{ij}|$, called the **determinant** of A, calculated from the entries of A in the following way. For $n = 1$ and $n = 2$, we define

The vertical bars in the notation $|a_{ij}|$ do not mean absolute value.

$$\det [a] = a, \tag{1}$$

$$\det \begin{bmatrix} a_{11} & a_{12} \\ a_{21} & a_{22} \end{bmatrix} = a_{11}a_{22} - a_{21}a_{12}. \tag{2}$$

For a matrix of order 3, we define

$$\det A = \det \begin{bmatrix} a_{11} & a_{12} & a_{13} \\ a_{21} & a_{22} & a_{23} \\ a_{31} & a_{32} & a_{33} \end{bmatrix} = \begin{array}{l} \text{Sum of all signed products} \\ \text{of the form } \pm a_{1i}a_{2j}a_{3k}, \end{array} \tag{3}$$

where i, j, k is a permutation of 1, 2, 3 in some order. There are $3! = 6$ such permutations, so there are six terms in the sum. The sign is positive when the index of the permutation is even and negative when the index is odd.

Definition
Index of a Permutation

Given any permutation of the numbers $1, 2, 3, \ldots, n$, denote the permutation by $i_1, i_2, i_3, \ldots, i_n$. In this arrangement, some of the numbers following i_1 may be less than i_1, and the number of these is called the **number of inversions** in the arrangement pertaining to i_1. Likewise, there are a number of inversions pertaining to each of the other i's; it is the number of indices that come after that particular i in the arrangement and are less than it. The **index** of the permutation is the sum of all of the numbers of inversions pertaining to the separate indices.

EXAMPLE 1   For $n = 5$, the permutation

$$5 \quad 3 \quad 1 \quad 2 \quad 4$$

has 4 inversions pertaining to the first element, 5, 2 inversions pertaining to the second element, 3, and no further inversions, so the index is $4 + 2 = 6$.   ❑

The following table shows the permutations of 1, 2, 3, the index of each permutation, and the signed product in the determinant of Eq. (3).

Permutation	Index	Signed product
1 2 3	0	$+a_{11}a_{22}a_{33}$
1 3 2	1	$-a_{11}a_{23}a_{32}$
2 1 3	1	$-a_{12}a_{21}a_{33}$
2 3 1	2	$+a_{12}a_{23}a_{31}$
3 1 2	2	$+a_{13}a_{21}a_{32}$
3 2 1	3	$-a_{13}a_{22}a_{31}$

$$\tag{4}$$

The sum of the six signed products is

$$a_{11}(a_{22}a_{33} - a_{23}a_{32}) - a_{12}(a_{21}a_{33} - a_{23}a_{31}) + a_{13}(a_{21}a_{32} - a_{22}a_{31})$$

$$= a_{11} \begin{vmatrix} a_{22} & a_{23} \\ a_{32} & a_{33} \end{vmatrix} - a_{12} \begin{vmatrix} a_{21} & a_{23} \\ a_{31} & a_{33} \end{vmatrix} + a_{13} \begin{vmatrix} a_{21} & a_{22} \\ a_{31} & a_{32} \end{vmatrix} = \begin{vmatrix} a_{11} & a_{12} & a_{13} \\ a_{21} & a_{22} & a_{23} \\ a_{31} & a_{32} & a_{33} \end{vmatrix}. \quad (5)$$

The formula

$$\begin{vmatrix} a_{11} & a_{12} & a_{13} \\ a_{21} & a_{22} & a_{23} \\ a_{31} & a_{32} & a_{33} \end{vmatrix} = a_{11} \begin{vmatrix} a_{22} & a_{23} \\ a_{32} & a_{33} \end{vmatrix} - a_{12} \begin{vmatrix} a_{21} & a_{23} \\ a_{31} & a_{33} \end{vmatrix} + a_{13} \begin{vmatrix} a_{21} & a_{22} \\ a_{31} & a_{32} \end{vmatrix} \quad (6)$$

reduces the calculation of a 3 by 3 determinant to the calculation of three 2 by 2 determinants.

Many people prefer to remember the following scheme for calculating the six signed products in the determinant of a 3 by 3 matrix:

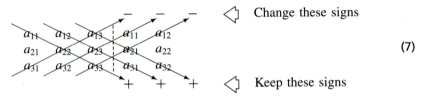

(7)

Minors and Cofactors

The second order determinants on the right-hand side of Eq. (6) are called the **minors** (short for "minor determinants") of the entries they multiply. Thus,

$$\begin{vmatrix} a_{22} & a_{23} \\ a_{32} & a_{33} \end{vmatrix} \text{ is the minor of } a_{11}, \qquad \begin{vmatrix} a_{21} & a_{23} \\ a_{31} & a_{33} \end{vmatrix} \text{ is the minor of } a_{12},$$

and so on. The minor of the element a_{ij} in a matrix A is the determinant of the matrix that remains after we delete the row and column containing a_{ij}:

$$\begin{vmatrix} a_{11} & a_{12} & a_{13} \\ a_{21} & a_{22} & a_{23} \\ a_{31} & a_{32} & a_{33} \end{vmatrix}. \qquad \text{The minor of } a_{22} \text{ is } \begin{vmatrix} a_{11} & a_{13} \\ a_{31} & a_{33} \end{vmatrix}.$$

$$\begin{vmatrix} a_{11} & a_{12} & a_{13} \\ a_{21} & a_{22} & a_{23} \\ a_{31} & a_{32} & a_{33} \end{vmatrix}. \qquad \text{The minor of } a_{23} \text{ is } \begin{vmatrix} a_{11} & a_{12} \\ a_{31} & a_{32} \end{vmatrix}.$$

The **cofactor** A_{ij} of a_{ij} is $(-1)^{i+j}$ times the minor of a_{ij}. Thus,

$$A_{22} = (-1)^{2+2} \begin{vmatrix} a_{11} & a_{13} \\ a_{31} & a_{33} \end{vmatrix} = \begin{vmatrix} a_{11} & a_{13} \\ a_{31} & a_{33} \end{vmatrix},$$

$$A_{23} = (-1)^{2+3} \begin{vmatrix} a_{11} & a_{12} \\ a_{31} & a_{32} \end{vmatrix} = - \begin{vmatrix} a_{11} & a_{12} \\ a_{31} & a_{32} \end{vmatrix}.$$

The factor $(-1)^{i+j}$ changes the sign of the minor when $i + j$ is odd. There is a checkerboard pattern for remembering these changes:

$$\begin{matrix} + & - & + \\ - & + & - \\ + & - & + \end{matrix}$$

In the upper left corner, $i = 1$, $j = 1$ and $(-1)^{1+1} = +1$. In going from any cell to an adjacent cell in the same row or column, we change i by 1 or j by 1, but not both, so we change the exponent from even to odd or from odd to even, which changes the sign from $+$ to $-$ or from $-$ to $+$.

When we rewrite Eq. (6) in terms of cofactors we get

$$\det A = a_{11}A_{11} + a_{12}A_{12} + a_{13}A_{13}. \tag{8}$$

EXAMPLE 2   Find the determinant of

$$A = \begin{bmatrix} 2 & 1 & 3 \\ 3 & -1 & -2 \\ 2 & 3 & 1 \end{bmatrix}.$$

Solution 1   The cofactors are

$$A_{11} = (-1)^{1+1} \begin{vmatrix} -1 & -2 \\ 3 & 1 \end{vmatrix}, \qquad A_{12} = (-1)^{1+2} \begin{vmatrix} 3 & -2 \\ 2 & 1 \end{vmatrix},$$

$$A_{13} = (-1)^{1+3} \begin{vmatrix} 3 & -1 \\ 2 & 3 \end{vmatrix}.$$

To find $\det A$, we multiply each element of the first row of A by its cofactor and add:

$$\det A = 2 \begin{vmatrix} -1 & -2 \\ 3 & 1 \end{vmatrix} + (-1) \begin{vmatrix} 3 & -2 \\ 2 & 1 \end{vmatrix} + 3 \begin{vmatrix} 3 & -1 \\ 2 & 3 \end{vmatrix}$$

$$= 2(-1+6) - 1(3+4) + 3(9+2) = 10 - 7 + 33 = 36.$$

Solution 2   From (7) we find

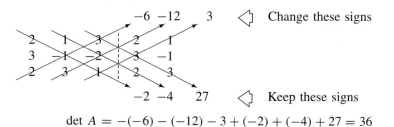

$$\det A = -(-6) - (-12) - 3 + (-2) + (-4) + 27 = 36$$

Expanding by Columns or by Other Rows

The determinant of a square matrix can be calculated from the cofactors of any row or any column.

If we were to expand the determinant in Example 2 by cofactors according to elements of its third column, say, we would get

$$+3 \begin{vmatrix} 3 & -1 \\ 2 & 3 \end{vmatrix} - (-2) \begin{vmatrix} 2 & 1 \\ 2 & 3 \end{vmatrix} + 1 \begin{vmatrix} 2 & 1 \\ 3 & -1 \end{vmatrix}$$

$$= 3(9+2) + 2(6-2) + 1(-2-3) = 33 + 8 - 5 = 36.$$

Useful Facts About Determinants

Fact 1: If two rows (or columns) are identical, the determinant is zero.

Fact 2: Interchanging two rows (or columns) changes the sign of the determinant.

Fact 3: The determinant is the sum of the products of the elements of the ith row (or column) by their cofactors, for any i.

Fact 4: The determinant of the transpose of a matrix is the same as the determinant of the original matrix. (The **transpose** of a matrix is obtained by writing the rows as columns.)

Fact 5: Multiplying each element of some row (or column) by a constant c multiplies the determinant by c.

Fact 6: If all elements above the main diagonal (or all below it) are zero, the determinant is the product of the elements on the main diagonal. (The **main diagonal** is the diagonal from upper left to lower right.)

EXAMPLE 3

$$\begin{vmatrix} 3 & 4 & 7 \\ 0 & -2 & 5 \\ 0 & 0 & 5 \end{vmatrix} = (3)(-2)(5) = -30$$

Fact 7: If the elements of any row are multiplied by the cofactors of the corresponding elements of a different row and these products are summed, the sum is zero.

EXAMPLE 4   If A_{11}, A_{12}, A_{13} are the cofactors of the elements of the first row of $A = (a_{ij})$, then the sums

$$a_{21}A_{11} + a_{22}A_{12} + a_{23}A_{13}$$

(elements of second row times cofactors of elements of first row) and

$$a_{31}A_{11} + a_{32}A_{12} + a_{33}A_{13}$$

are both zero.

Fact 8: If the elements of any column are multiplied by the cofactors of the corresponding elements of a different column and these products are summed, the sum is zero.

Fact 9: If each element of a row is multiplied by a constant c and the results added to a different row, the determinant is not changed. A similar result holds for columns.

EXAMPLE 5   If we start with

$$A = \begin{bmatrix} 2 & 1 & 3 \\ 3 & -1 & -2 \\ 2 & 3 & 1 \end{bmatrix}$$

and add -2 times row 1 to row 2 (subtract 2 times row 1 from row 2), we get

$$B = \begin{bmatrix} 2 & 1 & 3 \\ -1 & -3 & -8 \\ 2 & 3 & 1 \end{bmatrix}.$$

Since $\det A = 36$ (Example 2), we should find that $\det B = 36$ as well. Indeed we do, as the following calculation shows:

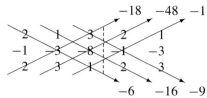

$$\det B = -(-18) - (-48) - (-1) + (-6) + (-16) + (-9)$$
$$= 18 + 48 + 1 - 6 - 16 - 9 = 67 - 31 = 36.$$

EXAMPLE 6   Evaluate the fourth order determinant

$$D = \begin{vmatrix} 1 & -2 & 3 & 1 \\ 2 & 1 & 0 & 2 \\ -1 & 2 & 1 & -2 \\ 0 & 1 & 2 & 1 \end{vmatrix}.$$

Solution   We subtract 2 times row 1 from row 2 and add row 1 to row 3 to get

$$D = \begin{vmatrix} 1 & -2 & 3 & 1 \\ 0 & 5 & -6 & 0 \\ 0 & 0 & 4 & -1 \\ 0 & 1 & 2 & 1 \end{vmatrix}.$$

We then multiply the elements of the first column by their cofactors to get

$$D = \begin{vmatrix} 5 & -6 & 0 \\ 0 & 4 & -1 \\ 1 & 2 & 1 \end{vmatrix} = 5(4 + 2) - (-6)(0 + 1) + 0 = 36.$$

Cramer's Rule

If the determinant $D = \det A = \begin{vmatrix} a_{11} & a_{12} \\ a_{21} & a_{22} \end{vmatrix} = 0$, the system

$$a_{11}x + a_{12}y = b_1,$$

$$a_{21}x + a_{22}y = b_2 \tag{9}$$

has either infinitely many solutions or no solution at all. The system

$$x + y = 0,$$

$$2x + 2y = 0$$

whose determinant is

$$D = \begin{vmatrix} 1 & 1 \\ 2 & 2 \end{vmatrix} = 2 - 2 = 0$$

has infinitely many solutions. We can find an x to match any given y. The system

$$x + y = 0,$$

$$2x + 2y = 2$$

has no solution. If $x + y = 0$, then $2x + 2y = 2(x + y)$ cannot be 2.

If $D \neq 0$, the system (9) has a unique solution, and Cramer's rule states that it may be found from the formulas

$$x = \frac{\begin{vmatrix} b_1 & a_{12} \\ b_2 & a_{22} \end{vmatrix}}{D}, \qquad y = \frac{\begin{vmatrix} a_{11} & b_1 \\ a_{21} & b_2 \end{vmatrix}}{D}. \tag{10}$$

The numerator in the formula for x comes from replacing the first column in A (the x-column) by the column of constants b_1 and b_2 (the b-column). Replacing the y-column by the b-column gives the numerator of the y-solution.

EXAMPLE 7 Solve the system

$$3x - y = 9,$$

$$x + 2y = -4.$$

Solution We use Eqs. (10). The determinant of the coefficient matrix is

$$D = \begin{vmatrix} 3 & -1 \\ 1 & 2 \end{vmatrix} = 6 + 1 = 7.$$

Hence,

$$x = \frac{\begin{vmatrix} 9 & -1 \\ -4 & 2 \end{vmatrix}}{D} = \frac{18 - 4}{7} = \frac{14}{7} = 2,$$

$$y = \frac{\begin{vmatrix} 3 & 9 \\ 1 & -4 \end{vmatrix}}{D} = \frac{-12 - 9}{7} = \frac{-21}{7} = -3.$$

Systems of three equations in three unknowns work the same way. If

$$D = \det A = \begin{vmatrix} a_{11} & a_{12} & a_{13} \\ a_{21} & a_{22} & a_{23} \\ a_{31} & a_{32} & a_{33} \end{vmatrix} = 0,$$

the system

$$a_{11}x + a_{12}y + a_{13}z = b_1,$$

$$a_{21}x + a_{22}y + a_{23}z = b_2, \tag{11}$$

$$a_{31}x + a_{32}y + a_{33}z = b_3$$

has either infinitely many solutions or no solution at all. If $D \neq 0$, the system has a unique solution, given by Cramer's rule:

$$x = \frac{1}{D} \begin{vmatrix} b_1 & a_{12} & a_{13} \\ b_2 & a_{22} & a_{23} \\ b_3 & a_{32} & a_{33} \end{vmatrix}, \qquad y = \frac{1}{D} \begin{vmatrix} a_{11} & b_1 & a_{13} \\ a_{21} & b_2 & a_{23} \\ a_{31} & b_3 & a_{33} \end{vmatrix}.$$

$$z = \frac{1}{D} \begin{vmatrix} a_{11} & a_{12} & b_1 \\ a_{21} & a_{22} & b_2 \\ a_{31} & a_{32} & b_3 \end{vmatrix}.$$

The pattern continues in higher dimensions.

Exercises A.8

Evaluating Determinants

Evaluate the following determinants.

1. $\begin{vmatrix} 2 & 3 & 1 \\ 4 & 5 & 2 \\ 1 & 2 & 3 \end{vmatrix}$

2. $\begin{vmatrix} 2 & -1 & -2 \\ -1 & 2 & 1 \\ 3 & 0 & -3 \end{vmatrix}$

3. $\begin{vmatrix} 1 & 2 & 3 & 4 \\ 0 & 1 & 2 & 3 \\ 0 & 0 & 2 & 1 \\ 0 & 0 & 3 & 2 \end{vmatrix}$

4. $\begin{vmatrix} 1 & -1 & 2 & 3 \\ 2 & 1 & 2 & 6 \\ 1 & 0 & 2 & 3 \\ -2 & 2 & 0 & -5 \end{vmatrix}$

Evaluate the following determinants by expanding according to the cofactors of (a) the third row and (b) the second column.

5. $\begin{vmatrix} 2 & -1 & 2 \\ 1 & 0 & 3 \\ 0 & 2 & 1 \end{vmatrix}$

6. $\begin{vmatrix} 1 & 0 & -1 \\ 0 & 2 & -2 \\ 2 & 0 & 1 \end{vmatrix}$

7. $\begin{vmatrix} 1 & 1 & 0 & 0 \\ 0 & 0 & -2 & 1 \\ 0 & -1 & 0 & 7 \\ 3 & 0 & 2 & 1 \end{vmatrix}$

8. $\begin{vmatrix} 0 & 1 & 0 & 0 \\ 0 & 1 & 1 & 0 \\ 1 & 1 & 1 & 1 \\ 1 & 1 & 0 & 0 \end{vmatrix}$

Systems of Equations

Solve the following systems of equations by Cramer's rule.

9. $x + 8y = 4$
$3x - y = -13$

10. $2x + 3y = 5$
$3x - y = 2$

11. $4x - 3y = 6$
$3x - 2y = 5$

12. $x + y + z = 2$
$2x - y + z = 0$
$x + 2y - z = 4$

13. $2x + y - z = 2$
$x - y + z = 7$
$2x + 2y + z = 4$

14. $2x - 4y = 6$
$x + y + z = 1$
$5y + 7z = 10$

15. $x \quad - z = 3$
$2y - 2z = 2$
$2x \quad + z = 3$

16. $x_1 + x_2 - x_3 + x_4 = 2$
$x_1 - x_2 + x_3 + x_4 = -1$
$x_1 + x_2 + x_3 - x_4 = 2$
$x_1 \quad + x_3 + x_4 = -1$

Theory and Examples

17. Find values of h and k for which the system
$$2x + hy = 8,$$
$$x + 3y = k$$
has (a) infinitely many solutions, (b) no solution at all.

18. For what value of x will
$$\begin{vmatrix} x & x & 1 \\ 2 & 0 & 5 \\ 6 & 7 & 1 \end{vmatrix} = 0?$$

19. Suppose $u, v,$ and w are twice-differentiable functions of x that satisfy the relation $au + bv + cw = 0$, where $a, b,$ and c are constants, not all zero. Show that
$$\begin{vmatrix} u & v & w \\ u' & v' & w' \\ u'' & v'' & w'' \end{vmatrix} = 0.$$

20. *Partial fractions.* Expanding the quotient
$$\frac{ax + b}{(x - r_1)(x - r_2)}$$
by partial fractions calls for finding the values of C and D that make the equation
$$\frac{ax + b}{(x - r_1)(x - r_2)} = \frac{C}{x - r_1} + \frac{D}{x - r_2}$$
hold for all x.

a) Find a system of linear equations that determines C and D.

b) Under what circumstances does the system of equations in part (a) have a unique solution? That is, when is the determinant of the coefficient matrix of the system different from zero?

A.9

Euler's Theorem and the Increment Theorem

This appendix derives Euler's Theorem (Theorem 2, Section 12.3) and the Increment Theorem for Functions of Two Variables (Theorem 3, Section 12.4). Euler first published his theorem in 1734, in a series of papers he wrote on hydrodynamics.

Euler's Theorem

If $f(x, y)$ and its partial derivatives $f_x, f_y, f_{xy},$ and f_{yx} are defined throughout an open region containing a point (a, b) and are all continuous at (a, b), then $f_{xy}(a, b) = f_{yx}(a, b)$.

Proof The equality of $f_{xy}(a, b)$ and $f_{yx}(a, b)$ can be established by four applications of the Mean Value Theorem (Theorem 4, Section 3.2). By hypothesis, the point (a, b) lies in the interior of a rectangle R in the xy-plane on which f, f_x, f_y, f_{xy}, and f_{yx} are all defined. We let h and k be numbers such that the point $(a + h, b + k)$ also lies in the rectangle R, and we consider the difference

$$\Delta = F(a + h) - F(a), \tag{1}$$

where

$$F(x) = f(x, b + k) - f(x, b). \tag{2}$$

We apply the Mean Value Theorem to F (which is continuous because it is differentiable), and Eq. (1) becomes

$$\Delta = hF'(c_1), \tag{3}$$

where c_1 lies between a and $a + h$. From Eq. (2),

$$F'(x) = f_x(x, b + k) - f_x(x, b),$$

so Eq. (3) becomes

$$\Delta = h[f_x(c_1, b + k) - f_x(c_1, b)]. \tag{4}$$

Now we apply the Mean Value Theorem to the function $g(y) = f_x(c_1, y)$ and have

$$g(b + k) - g(b) = kg'(d_1),$$

or

$$f_x(c_1, b + k) - f_x(c_1, b) = kf_{xy}(c_1, d_1),$$

for some d_1 between b and $b + k$. By substituting this into Eq. (4), we get

$$\Delta = hkf_{xy}(c_1, d_1), \tag{5}$$

for some point (c_1, d_1) in the rectangle R' whose vertices are the four points (a, b), $(a + h, b)$, $(a + h, b + k)$, and $(a, b + k)$. (See Fig. A.13.)

By substituting from Eq. (2) into Eq. (1), we may also write

$$\Delta = f(a + h, b + k) - f(a + h, b) - f(a, b + k) + f(a, b)$$
$$= [f(a + h, b + k) - f(a, b + k)] - [f(a + h, b) - f(a, b)] \tag{6}$$
$$= \phi(b + k) - \phi(b),$$

where
$$\phi(y) = f(a + h, y) - f(a, y). \tag{7}$$

The Mean Value Theorem applied to Eq. (6) now gives

$$\Delta = k\phi'(d_2), \tag{8}$$

for some d_2 between b and $b + k$. By Eq. (7),

$$\phi'(y) = f_y(a + h, y) - f_y(a, y). \tag{9}$$

Substituting from Eq. (9) into Eq. (8) gives

$$\Delta = k[f_y(a + h, d_2) - f_y(a, d_2)].$$

Finally, we apply the Mean Value Theorem to the expression in brackets and get

$$\Delta = khf_{yx}(c_2, d_2), \tag{10}$$

for some c_2 between a and $a + h$.

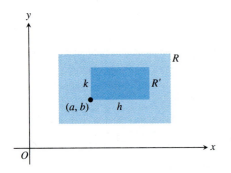

A.13 The key to proving $f_{xy}(a, b) = f_{yx}(a, b)$ is the fact that no matter how small R' is, f_{xy} and f_{yx} take on equal values somewhere inside R' (although not necessarily at the same point).

Together, Eqs. (5) and (10) show that

$$f_{xy}(c_1, d_1) = f_{yx}(c_2, d_2), \tag{11}$$

where (c_1, d_1) and (c_2, d_2) both lie in the rectangle R' (Fig. A.13). Equation (11) is not quite the result we want, since it says only that f_{xy} has the same value at (c_1, d_1) that f_{yx} has at (c_2, d_2). But the numbers h and k in our discussion may be made as small as we wish. The hypothesis that f_{xy} and f_{yx} are both continuous at (a, b) means that $f_{xy}(c_1, d_1) = f_{xy}(a, b) + \epsilon_1$ and $f_{yx}(c_2, d_2) = f_{yx}(a, b) + \epsilon_2$, where $\epsilon_1, \epsilon_2 \to 0$ as $h, k \to 0$. Hence, if we let h and $k \to 0$, we have $f_{xy}(a, b) = f_{yx}(a, b)$. ❏

The equality of $f_{xy}(a, b)$ and $f_{yx}(a, b)$ can be proved with hypotheses weaker than the ones we assumed. For example, it is enough for f, f_x, and f_y to exist in R and for f_{xy} to be continuous at (a, b). Then f_{yx} will exist at (a, b) and will equal f_{xy} at that point.

The Increment Theorem for Functions of Two Variables

Suppose that the first partial derivatives of $z = f(x, y)$ are defined throughout an open region R containing the point (x_0, y_0) and that f_x and f_y are continuous at (x_0, y_0). Then the change $\Delta z = f(x_0 + \Delta x, y_0 + \Delta y) - f(x_0, y_0)$ in the value of f that results from moving from (x_0, y_0) to another point $(x_0 + \Delta x, y_0 + \Delta y)$ in R satisfies an equation of the form

$$\Delta z = f_x(x_0, y_0)\Delta x + f_y(x_0, y_0)\Delta y + \epsilon_1 \Delta x + \epsilon_2 \Delta y,$$

in which $\epsilon_1, \epsilon_2 \to 0$ as $\Delta x, \Delta y \to 0$.

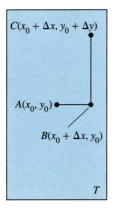

A.14 The rectangular region T in the proof of the Increment Theorem. The figure is drawn for Δx and Δy positive, but either increment might be zero or negative.

Proof We work within a rectangle T centered at $A(x_0, y_0)$ and lying within R, and we assume that Δx and Δy are already so small that the line segment joining A to $B(x_0 + \Delta x, y_0)$ and the line segment joining B to $C(x_0 + \Delta x, y_0 + \Delta y)$ lie in the interior of T (Fig. A.14).

We may think of Δz as the sum $\Delta z = \Delta z_1 + \Delta z_2$ of two increments, where

$$\Delta z_1 = f(x_0 + \Delta x, y_0) - f(x_0, y_0)$$

is the change in the value of f from A to B and

$$\Delta z_2 = f(x_0 + \Delta x, y_0 + \Delta y) - f(x_0 + \Delta x, y_0)$$

is the change in the value of f from B to C (Fig. A.15, on the following page).

On the closed interval of x-values joining x_0 to $x_0 + \Delta x$, the function $F(x) = f(x, x_0)$ is a differentiable (and hence continuous) function of x, with derivative

$$F'(x) = f_x(x, y_0).$$

By the Mean Value Theorem (Theorem 4, Section 3.2), there is an x-value c between x_0 and $x_0 + \Delta x$ at which

$$F(x_0 + \Delta x) - F(x_0) = F'(c)\Delta x$$

or

$$f(x_0 + \Delta x, y_0) - f(x_0, y_0) = f_x(c, y_0)\Delta x$$

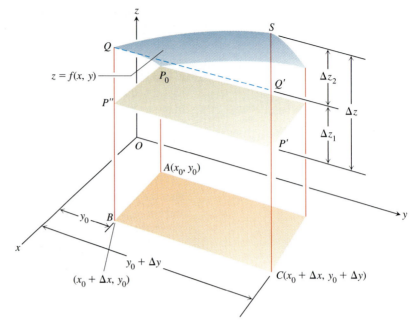

A.15 Part of the surface $z = f(x, y)$ near $P_0(x_0, y_0, f(x_0, y_0))$. The points P_0, P', and P'' have the same height $z_0 = f(x_0, y_0)$ above the xy-plane. The change in z is $\Delta z = P'S$. The change

$$\Delta z_1 = f(x_0 + \Delta x, y_0) - f(x_0, y_0),$$

shown as $P''Q = P'Q'$, is caused by changing x from x_0 to $x_0 + \Delta x$ while holding y equal to y_0. Then, with x held equal to $x_0 + \Delta x$,

$$\Delta z_2 = f(x_0 + \Delta x, y_0 + \Delta y)$$
$$-f(x_0 + \Delta x, y_0)$$

is the change in z caused by changing y from y_0 to $y_0 + \Delta y$. This is represented by $Q'S$. The total change in z is the sum of Δz_1 and Δz_2.

or

$$\Delta z_1 = f_x(c, y_0)\Delta x. \tag{12}$$

Similarly, $G(y) = f(x_0 + \Delta x, y)$ is a differentiable (and hence continuous) function of y on the closed y-interval joining y_0 and $y_0 + \Delta y$, with derivative

$$G'(y) = f_y(x_0 + \Delta x, y).$$

Hence there is a y-value d between y_0 and $y_0 + \Delta y$ at which

$$G(y_0 + \Delta y) - G(y_0) = G'(d)\Delta y$$

or

$$f(x_0 + \Delta x, y_0 + \Delta y) - f(x_0 + \Delta x, y) = f_y(x_0 + \Delta x, d)\Delta y$$

or

$$\Delta z_2 = f_y(x_0 + \Delta x, d)\Delta y. \tag{13}$$

Now, as Δx and $\Delta y \to 0$, we know $c \to x_0$ and $d \to y_0$. Therefore, since f_x and f_y are continuous at (x_0, y_0), the quantities

$$\epsilon_1 = f_x(c, y_0) - f_x(x_0, y_0),$$
$$\epsilon_2 = f_y(x_0 + \Delta x, d) - f_y(x_0, y_0) \tag{14}$$

both approach zero as Δx and $\Delta y \to 0$.

Finally,

$$\Delta z = \Delta z_1 + \Delta z_2$$
$$= f_x(c, y_0)\Delta x + f_y(x_0 + \Delta x, d)\,\Delta y \qquad \text{From (12) and (13)}$$
$$= [f_x(x_0, y_0) + \epsilon_1]\,\Delta x + [f_y(x_0, y_0) + \epsilon_2]\,\Delta y \qquad \text{From (14)}$$
$$= f_x(x_0, y_0)\,\Delta x + f_y(x_0, y_0)\,\Delta y + \epsilon_1\,\Delta x + \epsilon_2\,\Delta y,$$

where ϵ_1 and $\epsilon_2 \to 0$ as Δx and $\Delta y \to 0$. This is what we set out to prove.   ❑

Analogous results hold for functions of any finite number of independent variables. Suppose that the first partial derivatives of

$$w = f(x, y, z)$$

are defined throughout an open region containing the point (x_0, y_0, z_0) and that f_x, f_y, and f_z are continuous at (x_0, y_0). Then

$$\Delta w = f(x_0 + \Delta x, y_0 + \Delta y, z_0 + \Delta z) - f(x_0, y_0, z_0)$$
$$= f_x \Delta x + f_y \Delta y + f_z \Delta z + \epsilon_1 \Delta x + \epsilon_2 \Delta y + \epsilon_3 \Delta z, \tag{15}$$

where

$$\epsilon_1, \epsilon_2, \epsilon_3 \to 0 \quad \text{when} \quad \Delta x, \Delta y, \text{ and } \Delta z \to 0.$$

The partial derivatives f_x, f_y, f_z in this formula are to be evaluated at the point (x_0, y_0, z_0).

The result (15) can be proved by treating Δw as the sum of three increments,

$$\Delta w_1 = f(x_0 + \Delta x, y_0, z_0) - f(x_0, y_0, z_0), \tag{16}$$

$$\Delta w_2 = f(x_0 + \Delta x, y_0 + \Delta y, z_0) - f(x_0 + \Delta x, y_0, z_0), \tag{17}$$

$$\Delta w_3 = f(x_0 + \Delta x, y_0 + \Delta y, z_0 + \Delta z) - f(x_0 + \Delta x, y_0 + \Delta y, z_0), \tag{18}$$

and applying the Mean Value Theorem to each of these separately. Two coordinates remain constant and only one varies in each of these partial increments $\Delta w_1, \Delta w_2, \Delta w_3$. In (17), for example, only y varies, since x is held equal to $x_0 + \Delta x$ and z is held equal to z_0. Since $f(x_0 + \Delta x, y, z_0)$ is a continuous function of y with a derivative f_y, it is subject to the Mean Value Theorem, and we have

$$\Delta w_2 = f_y(x_0 + \Delta x, y_1, z_0)\Delta y$$

for some y_1 between y_0 and $y_0 + \Delta y$.

Answers

Section 8.1, pp. 619–622

1. $a_1 = 0$, $a_2 = -1/4$, $a_3 = -2/9$, $a_4 = -3/16$

3. $a_1 = 1$, $a_2 = -1/3$, $a_3 = 1/5$, $a_4 = -1/7$

5. $a_1 = 1/2$, $a_2 = 1/2$, $a_3 = 1/2$, $a_4 = 1/2$

7. $1, \dfrac{3}{2}, \dfrac{7}{4}, \dfrac{15}{8}, \dfrac{31}{16}, \dfrac{63}{32}, \dfrac{127}{64}, \dfrac{255}{128}, \dfrac{511}{256}, \dfrac{1023}{512}$

9. $2, 1, -\dfrac{1}{2}, -\dfrac{1}{4}, \dfrac{1}{8}, \dfrac{1}{16}, -\dfrac{1}{32}, -\dfrac{1}{64}, \dfrac{1}{128}, \dfrac{1}{256}$

11. $1, 1, 2, 3, 5, 8, 13, 21, 34, 55$ **13.** $a_n = (-1)^{n+1}$, $n \geq 1$

15. $a_n = (-1)^{n+1}(n)^2$, $n \geq 1$ **17.** $a_n = n^2 - 1$, $n \geq 1$

19. $a_n = 4n - 3$, $n \geq 1$ **21.** $a_n = \dfrac{1 + (-1)^{n+1}}{2}$, $n \geq 1$

23. $N = 692$, $a_n = \sqrt[n]{0.5}$, $L = 1$

25. $N = 65$, $a_n = (0.9)^n$, $L = 0$ **27.** b) $\sqrt{3}$

31. Nondecreasing, bounded   **33.** Not nondecreasing, bounded

35. Converges, nondecreasing sequence theorem

37. Converges, nondecreasing sequence theorem

39. Diverges, definition of divergence   **43.** Converges

45. Converges

Section 8.2, pp. 628–630

1. Converges, 2   **3.** Converges, -1 **5.** Converges, -5

7. Diverges   **9.** Diverges   **11.** Converges, 1/2

13. Converges, 0   **15.** Converges, $\sqrt{2}$ **17.** Converges, 1

19. Converges, 0   **21.** Converges, 0   **23.** Converges, 0

25. Converges, 1   **27.** Converges, e^7 **29.** Converges, 1

31. Converges, 1   **33.** Diverges   **35.** Converges, 4

37. Converges, 0   **39.** Diverges   **41.** Converges, e^{-1}

43. Converges, $e^{2/3}$ **45.** Converges, x $(x > 0)$ **47.** Converges, 0

49. Converges, 1   **51.** Converges, 1/2   **53.** Converges, $\pi/2$

55. Converges, 0   **57.** Converges, 0   **59.** Converges, 1/2

61. Converges, 0   **63.** $x_n = 2^{n-2}$

65. a) $f(x) = x^2 - 2$, $1.414213562 \approx \sqrt{2}$

b) $f(x) = \tan(x) - 1$, $0.7853981635 \approx \pi/4$

c) $f(x) = e^x$, diverges   **67.** b) 1   **75.** 1   **77.** -0.73908456

79. 0.853748068 **83.** -3

Section 8.3, pp. 638–640

1. $s_n = \dfrac{2(1 - (1/3)^n)}{1 - (1/3)}$, 3   **3.** $s_n = \dfrac{1 - (-1/2)^n}{1 - (-1/2)}$, 2/3

5. $s_n = \dfrac{1}{2} - \dfrac{1}{n+2}$, $\dfrac{1}{2}$ **7.** $1 - \dfrac{1}{4} + \dfrac{1}{16} - \dfrac{1}{64} + \cdots, \dfrac{4}{5}$

9. $\dfrac{7}{4} + \dfrac{7}{16} + \dfrac{7}{64} + \cdots, \dfrac{7}{3}$

11. $(5 + 1) + \left(\dfrac{5}{2} + \dfrac{1}{3}\right) + \left(\dfrac{5}{4} + \dfrac{1}{9}\right) + \left(\dfrac{5}{8} + \dfrac{1}{27}\right) + \cdots, \dfrac{23}{2}$

13. $(1 + 1) + \left(\dfrac{1}{2} - \dfrac{1}{5}\right) + \left(\dfrac{1}{4} + \dfrac{1}{25}\right) + \left(\dfrac{1}{8} - \dfrac{1}{125}\right) + \cdots, \dfrac{17}{6}$

15. 1   **17.** 5   **19.** Converges, 1   **21.** Converges, $-\dfrac{1}{\ln 2}$

23. Converges, $2 + \sqrt{2}$ **25.** Converges, 1   **27.** Diverges

29. Converges, $\dfrac{e^2}{e^2 - 1}$ **31.** Converges, 2/9   **33.** Converges, 3/2

35. Diverges   **37.** Diverges   **39.** Converges, $\dfrac{\pi}{\pi - e}$

41. $a = 1, r = -x$; converges to $1/(1 + x)$ for $|x| < 1$

43. $a = 3, r = (x - 1)/2$; converges to $6/(3 - x)$ for x in $(-1, 3)$

45. $|x| < \dfrac{1}{2}, \dfrac{1}{1 - 2x}$ **47.** $-2 < x < 0, \dfrac{1}{2 + x}$

49. $x \neq (2k + 1)\dfrac{\pi}{2}$, k an integer; $\dfrac{1}{1 - \sin x}$ **51.** 23/99

53. 7/9   **55.** 1/15   **57.** 41251/33300

59. a) $\displaystyle\sum_{n=-2}^{\infty}\frac{1}{(n+4)(n+5)}$ b) $\displaystyle\sum_{n=0}^{\infty}\frac{1}{(n+2)(n+3)}$

c) $\displaystyle\sum_{n=5}^{\infty}\frac{1}{(n-3)(n-2)}$ **61.** a) Answers may vary.

b) Answers may vary. c) Answers may vary. **69.** a) $r = 3/5$

b) $r = -3/10$ **71.** $|r| < 1,\ \dfrac{1+2r}{1-r^2}$ **73.** 28 m **75.** 8 m^2

77. a) $3\left(\dfrac{4}{3}\right)^{n-1}$

b) $A_n = A + \dfrac{1}{3}A + \dfrac{1}{3}\left(\dfrac{4}{9}\right)A + \cdots + \dfrac{1}{3}\left(\dfrac{4}{9}\right)^{n-2}A$,
$\displaystyle\lim_{n\to\infty}A_n = 2\sqrt{3}/5$

Section 8.4, pp. 643–644

1. Converges; geometric series, $r = \dfrac{1}{10} < 1$

3. Diverges; $\displaystyle\lim_{n\to\infty}\frac{n}{n+1} = 1 \neq 0$

5. Diverges; p-series, $p < 1$

7. Converges; geometric series, $r = \dfrac{1}{8} < 1$

9. Diverges; Integral Test

11. Converges; geometric series, $r = 2/3 < 1$

13. Diverges; Integral Test

15. Diverges; $\displaystyle\lim_{n\to\infty}\frac{2^n}{n+1} \neq 0$

17. Diverges; $\lim_{n\to\infty}\ (\sqrt{n}/\ln n) \neq 0$

19. Diverges; geometric series, $r = \dfrac{1}{\ln 2} > 1$

21. Converges; Integral Test

23. Diverges; nth-Term Test

25. Converges; Integral Test

27. Converges; Integral Test

29. Converges; Integral Test

31. $a = 1$ **33.** b) About 41.55

35. True

Section 8.5, p. 649

1. Diverges; limit comparison with $\sum(1/\sqrt{n})$

3. Converges; compare with $\sum(1/2^n)$ **5.** Diverges; nth-Term Test

7. Converges; $\left(\dfrac{n}{3n+1}\right)^n < \left(\dfrac{n}{3n}\right)^n = \left(\dfrac{1}{3}\right)^n$

9. Diverges; direct comparison with $\sum(1/n)$

11. Converges; limit comparison with $\sum(1/n^2)$

13. Diverges; limit comparison with $\sum(1/n)$

15. Diverges; limit comparison with $\sum(1/n)$

17. Diverges; Integral Test

19. Converges; compare with $\sum(1/n^{3/2})$

21. Converges; $\dfrac{1}{n2^n} \leq \dfrac{1}{2^n}$

23. Converges; $\dfrac{1}{3^{n-1}+1} < \dfrac{1}{3^{n-1}}$

25. Diverges; limit comparison with $\sum(1/n)$

27. Converges; compare with $\sum(1/n^2)$

29. Converges; $\dfrac{\tan^{-1}n}{n^{1.1}} < \dfrac{\pi/2}{n^{1.1}}$

31. Converges; compare with $\sum(1/n^2)$

33. Diverges; $3n > n\sqrt[n]{n} \Rightarrow \dfrac{1}{3n} < \dfrac{1}{n\sqrt[n]{n}} \Rightarrow \displaystyle\sum_{n=1}^{\infty}\frac{1}{n\sqrt[n]{n}}$ diverges

35. Converges; limit comparison with $\sum(1/n^2)$

Section 8.6, pp. 654–655

1. Converges; Ratio Test **3.** Diverges; Ratio Test

5. Converges; Ratio Test **7.** Converges; compare with $\sum(3/(1.25)^n)$

9. Diverges; $\displaystyle\lim_{n\to\infty}\left(1-\frac{3}{n}\right)^n = e^{-3} \neq 0$

11. Converges; compare with $\sum(1/n^2)$

13. Diverges; compare with $\sum(1/(2n))$

15. Diverges; compare with $\sum(1/n)$ **17.** Converges; Ratio Test

19. Converges; Ratio Test **21.** Converges; Ratio Test

23. Converges; Root Test **25.** Converges; compare with $\sum(1/n^2)$

27. Converges; Ratio Test **29.** Diverges; Ratio Test

31. Converges; Ratio Test **33.** Converges; Ratio Test

35. Diverges; $a_n = \left(\dfrac{1}{3}\right)^{(1/n!)} \to 1$ **37.** Converges; Ratio Test

39. Diverges; Root Test **41.** Converges; Root Test

43. Converges; Ratio Test **47.** Yes

Section 8.7, pp. 661–663

1. Converges by Theorem 8 **3.** Diverges; $a_n \nrightarrow 0$

5. Converges by Theorem 8 **7.** Diverges; $a_n \to 1/2$

9. Converges by Theorem 8

11. Converges absolutely. Series of absolute values is a convergent geometric series.

13. Converges conditionally. $1/\sqrt{n} \to 0$ but $\displaystyle\sum_{n=1}^{\infty}\frac{1}{\sqrt{n}}$ diverges.

15. Converges absolutely. Compare with $\sum_{n=1}^{\infty}(1/n^2)$.

17. Converges conditionally. $1/(n+3) \to 0$ but $\sum_{n=1}^{\infty}\dfrac{1}{n+3}$ diverges (compare with $\sum_{n=1}^{\infty}(1/n)$).

19. Diverges; $\dfrac{3+n}{5+n} \to 1$

21. Converges conditionally; $\left(\dfrac{1}{n^2} + \dfrac{1}{n}\right) \to 0$ but $(1+n)/n^2 > 1/n$

23. Converges absolutely; Root Test

25. Converges absolutely by Integral Test   **27.** Diverges; $a_n \nrightarrow 0$

29. Converges absolutely by the Ratio Test

31. Converges absolutely; $\dfrac{1}{n^2 + 2n + 1} < \dfrac{1}{n^2}$

33. Converges absolutely since $\left|\dfrac{\cos n\pi}{n\sqrt{n}}\right| = \left|\dfrac{(-1)^{n+1}}{n^{3/2}}\right| = \dfrac{1}{n^{3/2}}$
(convergent p–series)

35. Converges absolutely by Root Test   **37.** Diverges; $a_n \to \infty$

39. Converges conditionally; $\sqrt{n+1} - \sqrt{n} = 1/(\sqrt{n} + \sqrt{n+1}) \to$
0, but series of absolute values diverges (compare with $\sum(1/\sqrt{n})$)

41. Diverges, $a_n \to 1/2 \neq 0$

43. Converges absolutely; $\operatorname{sech} n = \dfrac{2}{e^n + e^{-n}} = \dfrac{2e^n}{e^{2n}+1} < \dfrac{2e^n}{e^{2n}}$

$= \dfrac{2}{e^n}$, a term from a convergent geometric series.

45. $|\text{Error}| < 0.2$ **47.** $|\text{Error}| < 2 \times 10^{-11}$ **49.** 0.54030

51. a) $a_n \geq a_{n+1}$ b) $-1/2$

Section 8.8, pp. 671–672

1. a) 1, $-1 < x < 1$ b) $-1 < x < 1$ c) none

3. a) $1/4$, $-1/2 < x < 0$ b) $-1/2 < x < 0$ c) none

5. a) 10, $-8 < x < 12$ b) $-8 < x < 12$ c) none

7. a) 1, $-1 < x < 1$ b) $-1 < x < 1$ c) none

9. a) 3, $[-3, 3]$ b) $[-3, 3]$ c) none   **11.** a) ∞, for all x

b) for all x c) none   **13.** a) ∞, for all x b) for all x c) none

15. a) 1, $-1 \leq x < 1$ b) $-1 < x < 1$ c) $x = -1$

17. a) 5, $-8 < x < 2$ b) $-8 < x < 2$ c) none

19. a) 3, $-3 < x < 3$ b) $-3 < x < 3$ c) none

21. a) 1, $-1 < x < 1$ b) $-1 < x < 1$ c) none

23. a) 0, $x = 0$ b) $x = 0$ c) none   **25.** a) 2, $-4 < x \leq 0$

b) $-4 < x < 0$ c) $x = 0$ **27.** a) 1, $-1 \leq x \leq 1$

b) $-1 \leq x \leq 1$ c) none   **29.** a) 1, $1 \leq x \leq 3/2$

b) $1 \leq x \leq 3/2$ c) none   **31.** a) 1, $(-1 - \pi) \leq x < (1 - \pi)$

b) $(-1 - \pi) < x < (1 - \pi)$ c) $x = -1 - \pi$

33. $-1 < x < 3$, $4/(3 + 2x - x^2)$ **35.** $0 < x < 16$, $2/(4 - \sqrt{x})$

37. $-\sqrt{2} < x < \sqrt{2}$, $3/(2 - x^2)$

39. $1 < x < 5$, $2/(x - 1)$, $1 < x < 5$, $-2/(x - 1)^2$

41. a) $\cos x = 1 - \dfrac{x^2}{2!} + \dfrac{x^4}{4!} - \dfrac{x^6}{6!} + \dfrac{x^8}{8!} - \dfrac{x^{10}}{10!} + \cdots$; converges for all x

b) and c) $2x - \dfrac{2^3 x^3}{3!} + \dfrac{2^5 x^5}{5!} - \dfrac{2^7 x^7}{7!} + \dfrac{2^9 x^9}{9!} - \dfrac{2^{11} x^{11}}{11!} + \cdots$

43. a) $\dfrac{x^2}{2} + \dfrac{x^4}{12} + \dfrac{x^6}{45} + \dfrac{17x^8}{2520} + \dfrac{31x^{10}}{14175}, -\dfrac{\pi}{2} < x < \dfrac{\pi}{2}$

b) $1 + x^2 + \dfrac{2x^4}{3} + \dfrac{17x^6}{45} + \dfrac{62x^8}{315} + \cdots, -\dfrac{\pi}{2} < x < \dfrac{\pi}{2}$

Section 8.9, pp. 677–678

1. $P_0(x) = 0$, $P_1(x) = x - 1$, $P_2(x) = (x - 1) - \dfrac{1}{2}(x - 1)^2$,

$P_3(x) = (x - 1) - \dfrac{1}{2}(x - 1)^2 + \dfrac{1}{3}(x - 1)^3$

3. $P_0(x) = \dfrac{1}{2}$, $P_1(x) = \dfrac{1}{2} - \dfrac{1}{4}(x - 2)$,

$P_2(x) = \dfrac{1}{2} - \dfrac{1}{4}(x - 2) + \dfrac{1}{8}(x - 2)^2$,

$P_3(x) = \dfrac{1}{2} - \dfrac{1}{4}(x - 2) + \dfrac{1}{8}(x - 2)^2 - \dfrac{1}{16}(x - 2)^3$

5. $P_0(x) = \dfrac{\sqrt{2}}{2}$, $P_1(x) = \dfrac{\sqrt{2}}{2} + \dfrac{\sqrt{2}}{2}\left(x - \dfrac{\pi}{4}\right)$,

$P_2(x) = \dfrac{\sqrt{2}}{2} + \dfrac{\sqrt{2}}{2}\left(x - \dfrac{\pi}{4}\right) - \dfrac{\sqrt{2}}{4}\left(x - \dfrac{\pi}{4}\right)^2$,

$P_3(x) = \dfrac{\sqrt{2}}{2} + \dfrac{\sqrt{2}}{2}\left(x - \dfrac{\pi}{4}\right) - \dfrac{\sqrt{2}}{4}\left(x - \dfrac{\pi}{4}\right)^2 - \dfrac{\sqrt{2}}{12}\left(x - \dfrac{\pi}{4}\right)^3$

7. $P_0(x) = 2$, $P_1(x) = 2 + \dfrac{1}{4}(x - 4)$,

$P_2(x) = 2 + \dfrac{1}{4}(x - 4) - \dfrac{1}{64}(x - 4)^2$,

$P_3(x) = 2 + \dfrac{1}{4}(x - 4) - \dfrac{1}{64}(x - 4)^2 + \dfrac{1}{512}(x - 4)^3$

9. $\displaystyle\sum_{n=0}^{\infty} \dfrac{(-x)^n}{n!} = 1 - x + \dfrac{x^2}{2!} - \dfrac{x^3}{3!} + \dfrac{x^4}{4!} - \cdots$

11. $\displaystyle\sum_{n=0}^{\infty} (-1)^n x^n = 1 - x + x^2 - x^3 + \cdots$

13. $\displaystyle\sum_{n=0}^{\infty} \dfrac{(-1)^n 3^{2n+1} x^{2n+1}}{(2n + 1)!}$ **15.** $7\displaystyle\sum_{n=0}^{\infty} \dfrac{(-1)^n x^{2n}}{(2n)!}$ **17.** $\displaystyle\sum_{n=0}^{\infty} \dfrac{x^{2n}}{(2n)!}$

19. $x^4 - 2x^3 - 5x + 4$ **21.** $8 + 10(x - 2) + 6(x - 2)^2 + (x - 2)^3$

23. $21 - 36(x + 2) + 25(x + 2)^2 - 8(x + 2)^3 + (x + 2)^4$

25. $\displaystyle\sum_{n=0}^{\infty} (-1)^n (n + 1)(x - 1)^n$ **27.** $\displaystyle\sum_{n=0}^{\infty} \dfrac{e^2}{n!}(x - 2)^n$

33. $L(x) = 0$, $Q(x) = -x^2/2$ **35.** $L(x) = 1$, $Q(x) = 1 + x^2/2$

37. $L(x) = x$, $Q(x) = x$

Section 8.10, pp. 686–688

1. $\displaystyle\sum_{n=0}^{\infty} \dfrac{(-5x)^n}{n!} = 1 - 5x + \dfrac{5^2 x^2}{2!} - \dfrac{5^3 x^3}{3!} + \cdots$

3. $\displaystyle\sum_{n=0}^{\infty} \dfrac{5(-1)^n(-x)^{2n+1}}{(2n + 1)!} = \displaystyle\sum_{n=0}^{\infty} \dfrac{5(-1)^{n+1} x^{2n+1}}{(2n + 1)!}$

$= -5x + \dfrac{5x^3}{3!} - \dfrac{5x^5}{5!} + \dfrac{5x^7}{7!} + \cdots$

5. $\displaystyle\sum_{n=0}^{\infty} \frac{(-1)^n x^n}{(2n)!}$ **7.** $\displaystyle\sum_{n=0}^{\infty} \frac{x^{n+1}}{n!} = x + x^2 + \frac{x^3}{2!} + \frac{x^4}{3!} + \frac{x^5}{4!} + \cdots$

9. $\displaystyle\sum_{n=2}^{\infty} \frac{(-1)^n x^{2n}}{(2n)!} = \frac{x^4}{4!} - \frac{x^6}{6!} + \frac{x^8}{8!} - \frac{x^{10}}{10!} + \cdots$

11. $\displaystyle x - \frac{\pi^2 x^3}{2!} + \frac{\pi^4 x^5}{4!} - \frac{\pi^6 x^7}{6!} + \cdots = \sum_{n=0}^{\infty} \frac{(-1)^n \pi^{2n} x^{2n+1}}{(2n)!}$

13. $\displaystyle 1 + \sum_{n=1}^{\infty} \frac{(-1)^n (2x)^{2n}}{2 \cdot (2n)!} = 1 - \frac{(2x)^2}{2 \cdot 2!} +$
$\displaystyle \frac{(2x)^4}{2 \cdot 4!} - \frac{(2x)^6}{2 \cdot 6!} + \frac{(2x)8}{2 \cdot 8!} - \cdots$

15. $\displaystyle\sum_{n=0}^{\infty} (2x)^{n+2} = 2^2 x^2 + 2^3 x^3 + 2^4 x^4 + \cdots$

17. $\displaystyle\sum_{n=1}^{\infty} nx^{n-1} = 1 + 2x + 3x^2 + 4x^3 + \cdots$

19. $|x| < (0.06)^{1/5} < 0.56968$

21. $|\text{Error}| < (10^{-3})^3/6 < 1.67 \times 10^{-10}, \quad -10^{-3} < x < 0$

23. $|\text{Error}| < (3^{0.1})(0.1)^3/6 < 1.87 \times 10^{-5}$ **25.** 0.000293653

27. $|x| < 0.02$ **31.** $\sin x, \; x = 0.1; \sin(0.1)$

33. $\tan^{-1} x, \; x = \pi/3; \sqrt{3}$

35. $\displaystyle e^x \sin x = x + x^2 + \frac{x^3}{3} - \frac{x^5}{30} - \frac{x^6}{90} \cdots$

43. a) $\displaystyle Q(x) = 1 + kx + \frac{k(k-1)}{2} x^2$ b) for $0 \le x < 100^{-1/3}$

49. a) -1 b) $(1/\sqrt{2})(1 + i)$ c) $-i$

53. $\displaystyle x + x^2 + \frac{1}{3} x^3 - \frac{1}{30} x^5 \cdots$; will converge for all x

Section 8.11, pp. 697–699

1. $\displaystyle 1 + \frac{x}{2} - \frac{x^2}{8} + \frac{x^3}{16}$ **3.** $\displaystyle 1 + \frac{1}{2} x + \frac{3}{8} x^2 + \frac{5}{16} x^3 + \cdots$

5. $\displaystyle 1 - x + \frac{3x^2}{4} - \frac{x^3}{2}$ **7.** $\displaystyle 1 - \frac{x^3}{2} + \frac{3x^6}{8} - \frac{5x^9}{16}$

9. $\displaystyle 1 + \frac{1}{2x} - \frac{1}{8x^2} + \frac{1}{16x^3}$

11. $(1 + x)^4 = 1 + 4x + 6x^2 + 4x^3 + x^4$

13. $(1 - 2x)^3 = 1 - 6x + 12x^2 - 8x^3$

15. $\displaystyle y = \sum_{n=0}^{\infty} \frac{(-1)^n}{n!} x^n = e^{-x}$ **17.** $\displaystyle y = \sum_{n=1}^{\infty} (x^n/n!) = e^x - 1$

19. $\displaystyle y = \sum_{n=2}^{\infty} (x^n/n!) = e^x - x - 1$ **21.** $\displaystyle y = \sum_{n=0}^{\infty} \frac{x^{2n}}{2^n n!} = e^{x^2/2}$

23. $\displaystyle y = \sum_{n=0}^{\infty} 2x^n = \frac{2}{1-x}$ **25.** $\displaystyle y = \sum_{n=0}^{\infty} \frac{x^{2n+1}}{(2n+1)!} = \sinh x$

27. $\displaystyle y = 2 + x - 2 \sum_{n=1}^{\infty} \frac{(-1)^{n+1} x^{2n}}{(2n)!}$

29. $\displaystyle y = \sum_{n=0}^{\infty} \frac{-2(x-2)^{2n+1}}{(2n+1)!}$

31. $\displaystyle y = a + bx + \frac{1}{6} x^3 - \frac{ax^4}{3 \cdot 4} - \frac{bx^5}{4 \cdot 5} - \frac{x^7}{6 \cdot 6 \cdot 7} + \frac{ax^8}{3 \cdot 4 \cdot 7 \cdot 8}$
$\displaystyle + \frac{bx^9}{4 \cdot 5 \cdot 8 \cdot 9} \cdots .$ For $n \ge 6, \; a_n = (n-2)(n-3) a_{n-4}.$

33. 0.00267 **35.** 0.1 **37.** $0.0999\ 44461\ 1$ **39.** $0.1000\ 01$

41. $1/(13 \cdot 6!) \approx 0.00011$ **43.** $\displaystyle \frac{t^3}{3} - \frac{t^7}{7 \cdot 3!} + \frac{t^{11}}{11 \cdot 5!}$

45. a) $\displaystyle \frac{x^2}{2} - \frac{x^4}{12}$

b) $\displaystyle \frac{x^2}{2} - \frac{x^4}{3 \cdot 4} + \frac{x^6}{5 \cdot 6} - \frac{x^8}{7 \cdot 8} + \cdots + (-1)^{15} \frac{x^{32}}{31 \cdot 32}$

47. $1/2$ **49.** $-1/24$ **51.** $1/3$ **53.** -1 **55.** 2

59. 500 terms

61. 3 terms

63. a) $\displaystyle x + \frac{x^3}{6} + \frac{3x^5}{40} + \frac{5x^7}{112}$, radius of convergence $= 1$

b) $\displaystyle \frac{\pi}{2} - x - \frac{x^3}{6} - \frac{3x^5}{40} - \frac{5x^7}{112}$ **65.** $1 - 2x + 3x^2 - 4x^3 + \cdots$

71. c) $3\pi/4$

Chapter 8 Practice Exercises, pp. 700–702

1. Converges to 1   **3.** Converges to -1 **5.** Diverges
7. Converges to 0   **9.** Converges to 1   **11.** Converges to e^{-5}
13. Converges to 3   **15.** Converges to $\ln 2$ **17.** Diverges
19. $1/6$ **21.** $3/2$ **23.** $e/(e-1)$ **25.** Diverges
27. Converges conditionally   **29.** Converges conditionally
31. Converges absolutely   **33.** Converges absolutely
35. Converges absolutely   **37.** Converges absolutely
39. Converges absolutely   **41.** a) $3, \; -7 \le x < -1$
b) $-7 < x < -1$ c) $x = -7$ **43.** a) $1/3, \; 0 \le x \le 2/3$
b) $0 \le x \le 2/3$ c) none   **45.** a) ∞, for all x b) for all x
c) none   **47.** a) $\sqrt{3}, \; -\sqrt{3} < x < \sqrt{3}$ b) $-\sqrt{3} < x < \sqrt{3}$
c) none   **49.** a) $e, \; (-e, e)$ b) $(-e, e)$ c) $\{ \}$

51. $\displaystyle \frac{1}{1+x}, \; \frac{1}{4}, \; \frac{4}{5}$ **53.** $\sin x, \; \pi, \; 0$ **55.** $e^x, \; \ln 2, \; 2$

57. $\displaystyle\sum_{n=0}^{\infty} 2^n x^n$ **59.** $\displaystyle\sum_{n=0}^{\infty} \frac{(-1)^n \pi^{2n+1} x^{2n+1}}{(2n+1)!}$ **61.** $\displaystyle\sum_{n=0}^{\infty} \frac{(-1)^n x^{5n}}{(2n)!}$

63. $\displaystyle\sum_{n=0}^{\infty} \frac{((\pi x)/2)^n}{n!}$

65. $\displaystyle 2 - \frac{(x+1)}{2 \cdot 1!} + \frac{3(x+1)^2}{2^3 \cdot 2!} + \frac{9(x+1)^3}{2^5 \cdot 3!} + \cdots$

67. $\displaystyle \frac{1}{4} - \frac{1}{4^2}(x-3) + \frac{1}{4^3}(x-3)^2 - \frac{1}{4^4}(x-3)^3$

69. $y = \sum_{n=0}^{\infty} \frac{(-1)^{n+1}}{n!} x^n = -e^{-x}$

71. $y = 3 \sum_{n=0}^{\infty} \frac{(-1)^n 2^n}{n!} x^n = 3e^{-2x}$

73. $y = -1 - x + 2 \sum_{n=2}^{\infty} (x^n/n!) = 2e^x - 3x - 3$

75. $y = 1 + x + 2 \sum_{n=0}^{\infty} (x^n/n!) = 2e^x - 1 - x$ **77.** 0.4849 17143 1

79. $\approx 0.4872\ 22358\ 3$ **81.** 7/2   **83.** 1/12   **85.** -2
87. $r = -3,\ s = 9/2$
89. b) $|\text{error}| < |\sin(1/42)| < 0.02381$; an underestimate because the remainder is positive

91. 2/3   **93.** $\ln\left(\frac{n+1}{2n}\right)$; the series converges to $\ln\left(\frac{1}{2}\right)$.

95. a) ∞ **b)** $a = 1,\quad b = 0$ **97.** It converges.

Chapter 8 Additional Exercises, pp. 703–707

1. Converges; Comparison Test   **3.** Diverges; nth Term Test
5. Converges; Comparison Test   **7.** Diverges; nth Term Test

9. With $a = \pi/3$, $\cos x = \frac{1}{2} - \frac{\sqrt{3}}{2}(x - \pi/3) - \frac{1}{4}(x - \pi/3)^2 +$

$\frac{\sqrt{3}}{12}(x - \pi/3)^3 + \cdots$

11. With $a = 0$, $e^x = 1 + x + \frac{x^2}{2!} + \frac{x^3}{3!} + \cdots$

13. With $a = 22\pi$, $\cos x = 1 - \frac{1}{2}(x - 22\pi)^2 + \frac{1}{4!}(x - 22\pi)^4 -$

$\frac{1}{6!}(x - 22\pi)^6 + \cdots$

15. Converges, limit $= b$ **17.** $\pi/2$ **23.** $b = \pm\frac{1}{5}$

25. $a = 2,\ L = -7/6$ **29. b)** Yes

35. a) $\sum_{n=1}^{\infty} nx^{n-1}$ **b)** 6   **c)** $1/q$

37. a) $R_n = C_0 e^{-kt_0}\left(1 - e^{-nkt_0}\right) / \left(1 - e^{-kt_0}\right)$,
$R = C_0\left(e^{-kt_0}\right) / \left(1 - e^{-kt_0}\right) = C_0 / \left(e^{kt_0} - 1\right)$
b) $R_1 = 1/e \approx 0.368$,
$R_{10} = R(1 - e^{-10}) \approx R(0.9999546) \approx 0.58195$;
$R \approx 0.58198;\ 0 < (R - R_{10})/R < 0.0001$ **c)** 7

CHAPTER 9

Section 9.1, pp. 719–722

1. $y^2 = 8x$, $F(2, 0)$, directrix: $x = -2$
3. $x^2 = -6y$, $F(0, -3/2)$, directrix: $y = 3/2$

5. $\frac{x^2}{4} - \frac{y^2}{9} = 1$, $F(\pm\sqrt{13}, 0)$, $V(\pm 2, 0)$,

asymptotes: $y = \pm\frac{3}{2}x$

7. $\frac{x^2}{2} + y^2 = 1$, $F(\pm 1, 0)$, $V(\pm\sqrt{2}, 0)$

9.
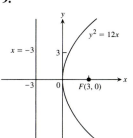

11.
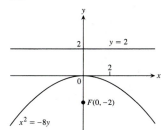

13.
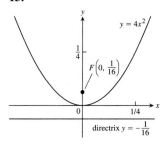

15.
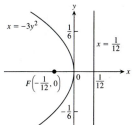

17.
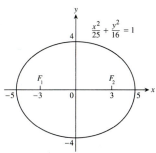

19.
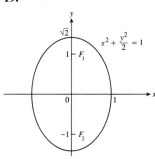

21.
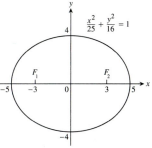

23.
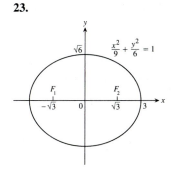

25. $\frac{x^2}{4} + \frac{y^2}{2} = 1$

27. Asymptotes: $y = \pm x$

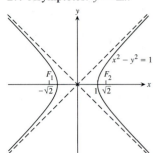

29. Asymptotes: $y = \pm x$

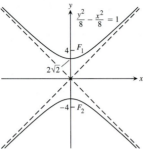

b)

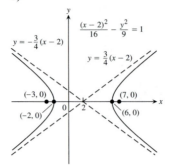

31. Asymptotes: $y = \pm 2x$

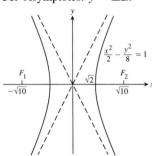

33. Asymptotes: $y = \pm \dfrac{x}{2}$

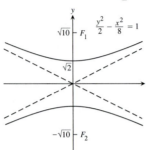

35. $y^2 - x^2 = 1$　　**37.** $\dfrac{x^2}{9} - \dfrac{y^2}{16} = 1$

39. a) Vertex: $(1, -2)$; focus: $(3, -2)$; directrix: $x = -1$

b)

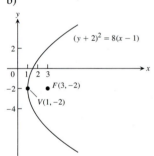

41. a) Foci: $(4 \pm \sqrt{7}, 3)$; vertices: $(8, 3)$ and $(0, 3)$; center: $(4, 3)$

b)

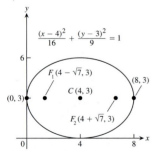

43. a) Center: $(2, 0)$; foci: $(7, 0)$ and $(-3, 0)$; vertices: $(6, 0)$ and $(-2, 0)$; asymptotes: $y = \pm \dfrac{3}{4}(x - 2)$

45. $(y + 3)^2 = 4(x + 2)$, $V(-2, -3)$, $F(-1, -3)$, directrix: $x = -3$

47. $(x - 1)^2 = 8(y + 7)$, $V(1, -7)$, $F(1, -5)$, directrix: $y = -9$

49. $\dfrac{(x + 2)^2}{6} + \dfrac{(y + 1)^2}{9} = 1$, $F(-2, \pm\sqrt{3} - 1)$, $V(-2, \pm 3 - 1)$, $C(-2, -1)$

51. $\dfrac{(x - 2)^2}{3} + \dfrac{(y - 3)^2}{2} = 1$, $F(3, 3)$ and $F(1, 3)$, $V(\pm\sqrt{3} + 2, 3)$, $C(2, 3)$

53. $\dfrac{(x - 2)^2}{4} - \dfrac{(y - 2)^2}{5} = 1$, $C(2, 2)$, $F(5, 2)$ and $F(-1, 2)$, $V(4, 2)$ and $V(0, 2)$; asymptotes: $(y - 2) = \pm\dfrac{\sqrt{5}}{2}(x - 2)$

55. $(y + 1)^2 - (x + 1)^2 = 1$, $C(-1, -1)$, $F(-1, \sqrt{2} - 1)$ and $F(-1, -\sqrt{2} - 1)$, $V(-1, 0)$ and $V(-1, -2)$; asymptotes: $(y + 1) = \pm(x + 1)$

57. $C(-2, 0)$, $a = 4$　　**59.** $V(-1, 1)$, $F(-1, 0)$

61. Ellipse: $\dfrac{(x + 2)^2}{5} + y^2 = 1$, $C(-2, 0)$, $F(0, 0)$ and $F(-4, 0)$, $V(\sqrt{5} - 2, 0)$ and $V(-\sqrt{5} - 2, 0)$

63. Ellipse: $\dfrac{(x - 1)^2}{2} + (y - 1)^2 = 1$, $C(1, 1)$, $F(2, 1)$ and $F(0, 1)$, $V(\sqrt{2} + 1, 1)$ and $V(-\sqrt{2} + 1, 1)$

65. Hyperbola: $(x - 1)^2 - (y - 2)^2 = 1$, $C(1, 2)$, $F(1 + \sqrt{2}, 2)$ and $F(1 - \sqrt{2}, 2)$, $V(2, 2)$ and $V(0, 2)$; asymptotes: $(y - 2) = \pm(x - 1)$

67. Hyperbola: $\dfrac{(y - 3)^2}{6} - \dfrac{x^2}{3} = 1$, $C(0, 3)$, $F(0, 6)$ and $F(0, 0)$, $V(0, \sqrt{6} + 3)$ and $V(0, -\sqrt{6} + 3)$; asymptotes: $y = \sqrt{2}x + 3$ or $y = -\sqrt{2}x + 3$

69.

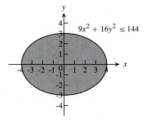

71.

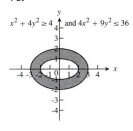

$x^2 + 4y^2 \geq 4$ and $4x^2 + 9y^2 \leq 36$

73.

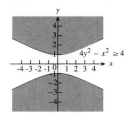

$4y^2 - x^2 \geq 4$

77. $3x^2 + 3y^2 - 7x - 7y + 4 = 0$
79. $(x + 2)^2 + (y - 1)^2 = 13$. The point is inside the circle.
81. b) $1 : 1$ **83.** Length $= 2\sqrt{2}$, width $= \sqrt{2}$, area $= 4$
85. 24π **87.** $(0, 16/(3\pi))$

Section 9.2, pp. 726–727

1. $e = 3/5$, $F(\pm 3, 0)$, $x = \pm 25/3$
3. $e = 1/\sqrt{2}$, $F(0, \pm 1)$, $y = \pm 2$
5. $e = 1/\sqrt{3}$, $F(0, \pm 1)$, $y = \pm 3$
7. $e = \sqrt{3}/3$, $F(\pm\sqrt{3}, 0)$, $x = \pm 3\sqrt{3}$

9. $\dfrac{x^2}{27} + \dfrac{y^2}{36} = 1$ **11.** $\dfrac{x^2}{4851} + \dfrac{y^2}{4900} = 1$

13. $e = \dfrac{\sqrt{5}}{3}$, $\dfrac{x^2}{9} + \dfrac{y^2}{4} = 1$ **15.** $e = 1/2$, $\dfrac{x^2}{64} + \dfrac{y^2}{48} = 1$

19. $\dfrac{(x-1)^2}{4} + \dfrac{(y-4)^2}{9} = 1$, $F(1, 4 \pm \sqrt{5})$, $e = \sqrt{5}/3$,
$y = 4 \pm (9\sqrt{5}/5)$
21. $a = 0$, $b = -4$, $c = 0$, $e = \sqrt{3}/2$
23. $e = \sqrt{2}$, $F(\pm\sqrt{2}, 0)$, $x = \pm 1/\sqrt{2}$
25. $e = \sqrt{2}$, $F(0, \pm 4)$, $y = \pm 2$
27. $e = \sqrt{5}$, $F(\pm\sqrt{10}, 0)$, $x = \pm 2/\sqrt{10}$

29. $e = \sqrt{5}$, $F(0, \pm\sqrt{10})$, $y = \pm 2/\sqrt{10}$ **31.** $y^2 - \dfrac{x^2}{8} = 1$

33. $x^2 - \dfrac{y^2}{8} = 1$ **35.** $e = \sqrt{2}$, $\dfrac{x^2}{8} - \dfrac{y^2}{8} = 1$

37. $e = 2$, $x^2 - \dfrac{y^2}{3} = 1$ **39.** $\dfrac{(y-6)^2}{36} - \dfrac{(x-1)^2}{45} = 1$

Section 9.3, pp. 733–734

1. Hyperbola   **3.** Ellipse   **5.** Parabola   **7.** Parabola
9. Hyperbola   **11.** Hyperbola   **13.** Ellipse   **15.** Ellipse
17. $x'^2 - y'^2 = 4$, hyperbola   **19.** $4x'^2 + 16y' = 0$, parabola
21. $y'^2 = 1$, parallel lines   **23.** $2\sqrt{2}x'^2 + 8\sqrt{2}y' = 0$, parabola
25. $4x'^2 + 2y'^2 = 19$, ellipse
27. $\sin\alpha = 1/\sqrt{5}$, $\cos\alpha = 2/\sqrt{5}$; or $\sin\alpha = -2/\sqrt{5}$,
$\cos\alpha = 1/\sqrt{5}$
29. $A' = 0.88$, $B' = 0.00$, $C' = 3.10$, $D' = 0.74$, $E' = -1.20$,
$F' = -3$, $0.88x'^2 + 3.10y'^2 + 0.74x' - 1.20y' - 3 = 0$, ellipse
31. $A' = 0.00$, $B' = 0.00$, $C' = 5.00$, $D' = 0$, $E' = 0$, $F' = -5$,
$5.00y'^2 - 5 = 0$ or $y' = \pm 1.00$, parallel lines

33. $A' = 5.05$, $B' = 0.00$, $C' = -0.05$, $D' = -5.07$, $E' = -6.18$,
$F' = -1$, $5.05x'^2 - 0.05y'^2 - 5.07x' - 6.18y' - 1 = 0$, hyperbola

35. a) $\dfrac{x'^2}{b^2} + \dfrac{y'^2}{a^2} = 1$ b) $\dfrac{y'^2}{a^2} - \dfrac{x'^2}{b^2} = 1$ c) $x'^2 + y'^2 = a^2$

d) $y' = -\dfrac{1}{m}x'$ e) $y' = -\dfrac{1}{m}x' + \dfrac{b}{m}$

37. a) $x'^2 - y'^2 = 2$ b) $x'^2 - y'^2 = 2a$ **43.** a) Parabola
45. a) Hyperbola
b)

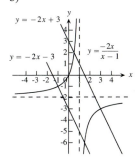

$y = -2x + 3$

$y = -2x - 3$

$y = \dfrac{-2x}{x - 1}$

c) $y = -2x - 3$, $y = -2x + 3$

Section 9.4, pp. 741–744

1.

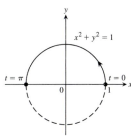

$x^2 + y^2 = 1$

$t = \pi$ $t = 0$

3.

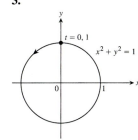

$t = 0, 1$

$x^2 + y^2 = 1$

5.

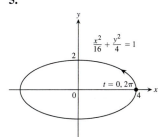

$\dfrac{x^2}{16} + \dfrac{y^2}{4} = 1$

$t = 0, 2\pi$

7.

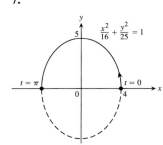

$\dfrac{x^2}{16} + \dfrac{y^2}{25} = 1$

$t = \pi$ $t = 0$

9.

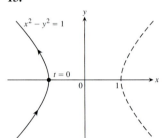

11.

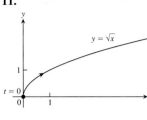

$$y = (a - b) \sin \theta - b \sin \left(\frac{a - b}{b} \theta \right)$$

35. $x = a \sin^2 t \tan t, \ y = a \sin^2 t$ **37.** $(1, 1)$

Section 9.5, pp. 749–751

1. $y = -x + 2\sqrt{2}, \ \dfrac{d^2y}{dx^2} = -\sqrt{2}$

3. $y = -\dfrac{1}{2}x + 2\sqrt{2}, \ \dfrac{d^2y}{dx^2} = -\dfrac{\sqrt{2}}{4}$ **5.** $y = x + \dfrac{1}{4}, \ \dfrac{d^2y}{dx^2} = -2$

7. $y = 2x - \sqrt{3}, \ \dfrac{d^2y}{dx^2} = -3\sqrt{3}$ **9.** $y = x - 4, \ \dfrac{d^2y}{dx^2} = \dfrac{1}{2}$

11. $y = \sqrt{3}x - \dfrac{\pi\sqrt{3}}{3} + 2, \ \dfrac{d^2y}{dx^2} = -4$ **13.** 0 **15.** -6 **17.** 4

19. 12 **21.** π^2 **23.** $8\pi^2$ **25.** $52\pi/3$ **27.** $3\pi\sqrt{5}$

29. a) $(\bar{x}, \bar{y}) = \left(\dfrac{12}{\pi} - \dfrac{24}{\pi^2}, \dfrac{24}{\pi^2} - 2 \right)$ b) Centroid: $(1.4, 0.4)$

31. a) $(\bar{x}, \bar{y}) = \left(\dfrac{1}{3}, \pi - \dfrac{4}{3} \right)$ **33.** a) π b) π **37.** $3\pi a^2$

39. $64\pi/3$ **41.** $\left(\dfrac{\sqrt{2}}{2}, 1 \right), \ y = 2x$ at $t = 0, \ y = -2x$ at $t = \pi$

13.

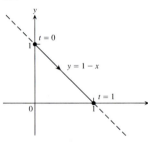

15.

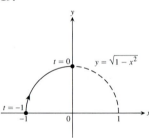

17.

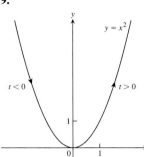

19.

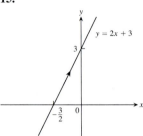

Section 9.6, pp. 755–756

1. a) e; b) g; c) h; d) f

3.

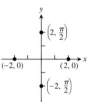

a) $\left(2, \dfrac{\pi}{2} + 2n\pi \right)$ and $\left(-2, \dfrac{\pi}{2} + (2n + 1)\pi \right)$, n an integer

b) $(2, 2n\pi)$ and $(-2, (2n + 1)\pi)$, n an integer

c) $\left(2, \dfrac{3\pi}{2} + 2n\pi \right)$ and $\left(-2, \dfrac{3\pi}{2} + (2n + 1)\pi \right)$, n an integer

d) $(2, (2n + 1)\pi)$ and $(-2, 2n\pi)$, n an integer

5. a) $(3, 0)$ b) $(-3, 0)$ c) $(-1, \sqrt{3})$ d) $(1, \sqrt{3})$,   e) $(3, 0)$

f) $(1, \sqrt{3})$ g) $(-3, 0)$ h) $(-1, \sqrt{3})$

21.

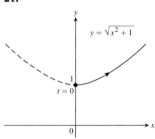

23.

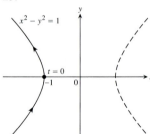

25. a) $x = a \cos t, \ y = -a \sin t, \ 0 \le t \le 2\pi$ b) $x = a \cos t,$
$y = a \sin t, \ 0 \le t \le 2\pi$ c) $x = a \cos t, \ y = -a \sin t,$
$0 \le t \le 4\pi$ d) $x = a \cos t, \ y = a \sin t, \ 0 \le t \le 4\pi$

27. $x = \dfrac{-at}{\sqrt{1 + t^2}}, \ y = \dfrac{a}{\sqrt{1 + t^2}}, \ -\infty < t < \infty$

29. $x = 2 \cot t, \ y = 2 \sin^2 t, \ 0 < t < \pi$

31. b) $x = x_1 t, \ y = y_1 t$ (answer not unique)   c) $x = -1 + t,$
$y = t$ (answer not unique)

33. $x = (a - b) \cos \theta + b \cos \left(\dfrac{a - b}{b} \theta \right),$

7.

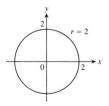

9.

11.

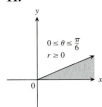

$0 \le \theta \le \frac{\pi}{6}$
$r \ge 0$

13.

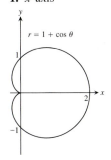

$\theta = \frac{\pi}{3}$
$-1 \le r \le 3$

$\frac{\pi}{3}$

Section 9.7, pp. 763–764

1. x-axis

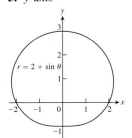

$r = 1 + \cos\theta$

3. y-axis

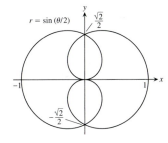

$r = 1 - \sin\theta$

15.

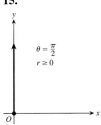

$\theta = \frac{\pi}{2}$
$r \ge 0$

17.

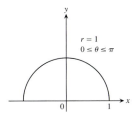

$r = 1$
$0 \le \theta \le \pi$

5. y-axis

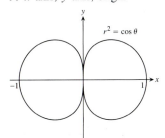

$r = 2 + \sin\theta$

7. x-axis

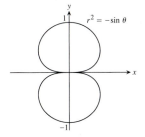

$r = \sin(\theta/2)$
$\frac{\sqrt{2}}{2}$
$-\frac{\sqrt{2}}{2}$

19.

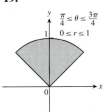

$\frac{\pi}{4} \le \theta \le \frac{3\pi}{4}$
$0 \le r \le 1$

21.

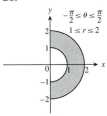

$-\frac{\pi}{2} \le \theta \le \frac{\pi}{2}$
$1 \le r \le 2$

9. x-axis, y-axis, origin

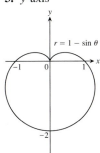

$r^2 = \cos\theta$

11. y-axis, x-axis, origin

$r^2 = -\sin\theta$

23. $x = 2$, vertical line through $(2, 0)$ **25.** $y = 0$, the x-axis
27. $y = 4$, horizontal line through $(0, 4)$
29. $x + y = 1$, line, $m = -1$, $b = 1$
31. $x^2 + y^2 = 1$, circle, $C(0, 0)$, radius 1
33. $y - 2x = 5$, line, $m = 2$, $b = 5$
35. $y^2 = x$, parabola, vertex $(0, 0)$, opens right
37. $y = e^x$, graph of natural exponential function
39. $x + y = \pm 1$, two straight lines of slope -1, y-intercepts $b = \pm 1$
41. $(x + 2)^2 + y^2 = 4$, circle, $C(-2, 0)$, radius 2
43. $x^2 + (y - 4)^2 = 16$, circle, $C(0, 4)$, radius 4
45. $(x - 1)^2 + (y - 1)^2 = 2$, circle, $C(1, 1)$, radius $\sqrt{2}$
47. $\sqrt{3}y + x = 4$ **49.** $r \cos\theta = 7$ **51.** $\theta = \pi/4$
53. $r = 2$ or $r = -2$ **55.** $4r^2 \cos^2\theta + 9r^2 \sin^2\theta = 36$
57. $r \sin^2\theta = 4 \cos\theta$ **59.** $r = 4 \sin\theta$
61. $r^2 = 6r \cos\theta - 2r \sin\theta - 6$ **63.** $(0, \theta)$, where θ is any angle

13. x-axis, y-axis, origin   **15.** Origin
17. The slope at $(-1, \pi/2)$ is -1, at $(-1, -\pi/2)$ is 1.

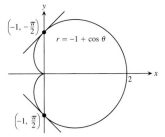

$\left(-1, -\frac{\pi}{2}\right)$

$r = -1 + \cos\theta$

$\left(-1, \frac{\pi}{2}\right)$

19. The slope at $(1, \pi/4)$ is -1, at $(-1, -\pi/4)$ is 1, at $(-1, 3\pi/4)$ is 1, at $(1, -3\pi/4)$ is -1.

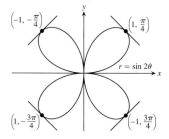

21. a)

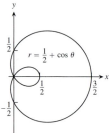

b)

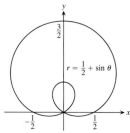

23. a)

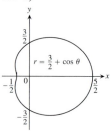

b)

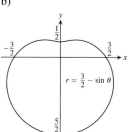

25.

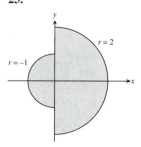

27.

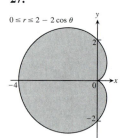

31. $(0,0)$, $(1, \pi/2)$, $(1, 3\pi/2)$
33. $(0,0)$, $(\sqrt{3}, \pi/3)$, $(-\sqrt{3}, -\pi/3)$
35. $(\sqrt{2}, \pm\pi/6)$, $(\sqrt{2}, \pm5\pi/6)$
37. $(1, \pi/12)$, $(1, 5\pi/12)$, $(1, 13\pi/12)$, $(1, 17\pi/12)$ **43. a**

51. $2y = \dfrac{2\sqrt{6}}{9}$

Section 9.8, pp. 768-770

1. $r \cos(\theta - \pi/6) = 5$, $y = -\sqrt{3}x + 10$
3. $r \cos(\theta - 4\pi/3) = 3$, $y = -(\sqrt{3}/3)x - 2\sqrt{3}$ **5.** $y = 2 - x$
7. $y = (\sqrt{3}/3)x + 2\sqrt{3}$ **9.** $r \cos\left(\theta - \dfrac{\pi}{4}\right) = 3$
11. $r \cos\left(\theta + \dfrac{\pi}{2}\right) = 5$ **13.** $r = 8 \cos\theta$ **15.** $r = 2\sqrt{2} \sin\theta$
17. $C(2, 0)$, radius $= 2$ **19.** $C(1, \pi)$, radius $= 1$
21. $(x - 6)^2 + y^2 = 36$, $r = 12 \cos\theta$
23. $x^2 + (y - 5)^2 = 25$, $r = 10 \sin\theta$
25. $(x + 1)^2 + y^2 = 1$, $r = -2 \cos\theta$
27. $x^2 + (y + 1/2)^2 = 1/4$, $r = -\sin\theta$ **29.** $r = 2/(1 + \cos\theta)$
31. $r = 30/(1 - 5\sin\theta)$ **33.** $r = 1/(2 + \cos\theta)$
35. $r = 10/(5 - \sin\theta)$
37.

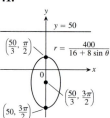

39.

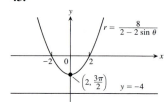

41.

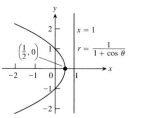

43.

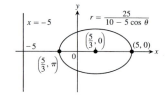

45.

57. b)

Planet	Perihelion	Aphelion
Mercury	0.3075 AU	0.4667 AU
Venus	0.7184 AU	0.7282 AU
Earth	0.9833 AU	1.0167 AU
Mars	1.3817 AU	1.6663 AU
Jupiter	4.9512 AU	5.4548 AU
Saturn	9.0210 AU	10.0570 AU
Uranus	18.2977 AU	20.0623 AU
Neptune	29.8135 AU	30.3065 AU
Pluto	29.6549 AU	49.2251 AU

59. a) $x^2 + (y - 2)^2 = 4$, $x = \sqrt{3}$

b)

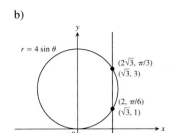

61. $r = 4/(1 + \cos \theta)$ **63.** b) The pins should be 2 in. apart.
65. $r = 2a \sin \theta$ (a circle) **67.** $r \cos (\theta - \alpha) = p$ (a line)

Section 9.9, pp. 775–777

1. 18π **3.** $\pi/8$ **5.** 2 **7.** $\dfrac{\pi}{2} - 1$ **9.** $5\pi - 8$

11. $3\sqrt{3} - \pi$ **13.** $\dfrac{\pi}{3} + \dfrac{\sqrt{3}}{2}$ **15.** $12\pi - 9\sqrt{3}$ **17.** a) $\dfrac{3}{2} - \dfrac{\pi}{4}$

19. $19/3$ **21.** 8 **23.** $3(\sqrt{2} + \ln(1 + \sqrt{2}))$ **25.** $\dfrac{\pi}{8} + \dfrac{3}{8}$

27. 2π **29.** $\pi\sqrt{2}$ **31.** $2\pi(2 - \sqrt{2})$ **37.** $\left(\dfrac{5}{6}a, 0\right)$

Chapter 9 Practice Exercises, pp. 778–782

1.

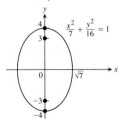

3.

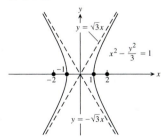

5. $e = 3/4$

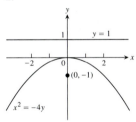

7. $e = 2$

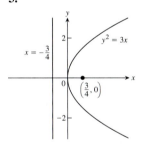

9. $(x - 2)^2 = -12(y - 3)$, $V(2, 3)$, $F(2, 0)$; directrix: $y = 6$
11. $\dfrac{(x + 3)^2}{9} + \dfrac{(y + 5)^2}{25} = 1$, $C(-3, -5)$, $V(-3, 0)$ and
$V(-3, -10)$, $F(-3, -1)$ and $F(-3, -9)$

13. $\dfrac{(y - 2\sqrt{2})^2}{8} - \dfrac{(x - 2)^2}{2} = 1$, $C(2, 2\sqrt{2})$, $V(2, 4\sqrt{2})$ and
$V(2, 0)$, $F(2, \sqrt{10} + 2\sqrt{2})$ and $F(2, -\sqrt{10} + 2\sqrt{2})$; asymptotes:
$y = 2x - 4 + 2\sqrt{2}$ and $y = -2x + 4 + 2\sqrt{2}$

15. Hyperbola: $\dfrac{(x - 2)^2}{4} - y^2 = 1$, $F(2 \pm \sqrt{5}, 0)$, $V(2 \pm 2, 0)$,
$C(2, 0)$; asymptotes: $y = \pm\dfrac{1}{2}(x - 2)$

17. Parabola: $(y - 1)^2 = -16(x + 3)$, $V(-3, 1)$, $F(-7, 1)$;
directrix: $x = 1$

19. Ellipse: $\dfrac{(x + 3)^2}{16} + \dfrac{(y - 2)^2}{9} = 1$, $F(\pm\sqrt{7} - 3, 2)$,
$V(\pm 4 - 3, 2)$, $C(-3, 2)$
21. Circle: $(x - 1)^2 + (y - 1)^2 = 2$, $C(1, 1)$, radius $= \sqrt{2}$
23. Ellipse **25.** Hyperbola **27.** Line
29. Ellipse, $5x'^2 + 3y'^2 = 30$ **31.** Hyperbola, $x'^2 - y'^2 = 2$
33. **35.**

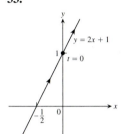 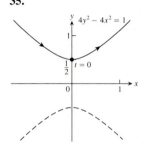

37.

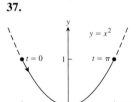

39. $x = 3 \cos t$, $y = 4 \sin t$, $0 \le t \le 2\pi$ **41.** $y = \dfrac{\sqrt{3}}{2}x + \dfrac{1}{4}$, $\dfrac{1}{4}$

43. $3 + \dfrac{\ln 2}{8}$ **45.** $76\pi/3$

47.

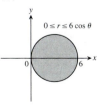

49. d **51.** l **53.** k **55.** i **57.** $(0, 0)$
59. $(0, 0)$, $(1, \pm\pi/2)$ **61.** The graphs coincide. **63.** $(\sqrt{2}, \pi/4)$
65. $y = x$, $y = -x$

67. At $(1, \pi/4) : r \cos(\theta - \pi/4) = 1$, At $(1, 3\pi/4)$:
$r \cos(\theta - 3\pi/4) = 1$, At $(1, 5\pi/4) : r \cos(\theta - 5\pi/4) = 1$,
At $(1, 7\pi/4) : r \cos(\theta - 7\pi/4) = 1$
69. $y = (\sqrt{3}/3)x - 4$ **71.** $x = 2$ **73.** $y = -3/2$
75. $x^2 + (y + 2)^2 = 4$ **77.** $(x - \sqrt{2})^2 + y^2 = 2$
79. $r = -5 \sin \theta$ **81.** $r = 3 \cos \theta$
83.   **85.**

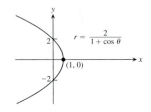   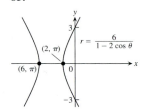

87. $r = \dfrac{4}{1 + 2 \cos \theta}$ **89.** $r = \dfrac{2}{2 + \sin \theta}$ **91.** $9\pi/2$
93. $2 + \pi/4$ **95.** 8 **97.** $\pi - 3$ **99.** $(2 - \sqrt{2})\pi$
101. a) 24π b) 16π **111.** $\pi/2$ **115.** $\left(2, \pm\dfrac{\pi}{3}\right)$, $\dfrac{\pi}{2}$
119. $\pi/2$ **121.** $\pi/4$

Chapter 9 Additional Exercises, pp. 783–786

1.

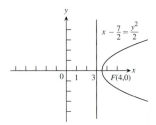

3. $3x^2 + 3y^2 - 8y + 4 = 0$ **5.** $(0, \pm 1)$

7. a) $\dfrac{(y - 1)^2}{16} - \dfrac{x^2}{48} = 1$ b) $\dfrac{16\left(y + \dfrac{3}{4}\right)^2}{25} - \dfrac{2x^2}{75} = 1$

17.   **19.**

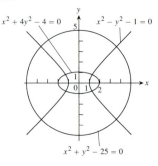   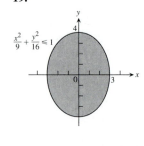

21.   **23.**

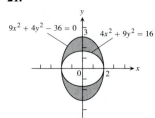   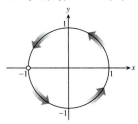

25. $x = (a + b) \cos \theta - b \cos\left(\dfrac{a + b}{b}\theta\right)$,

$y = (a + b) \sin \theta - b \sin\left(\dfrac{a + b}{b}\theta\right)$

27. $(-1, 0)$, $t = -1, 0, 1$

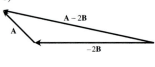

29. a) $r = e^{2\theta}$ b) $\dfrac{\sqrt{5}}{2}(e^{4\pi} - 1)$

31. $\dfrac{32\pi - 4\pi\sqrt{2}}{5}$ **33.** $r = \dfrac{4}{1 + 2 \cos \theta}$ **35.** $r = \dfrac{2}{2 + \sin \theta}$

37. a) $120°$ **39.** 1×10^7 mi. **41.** $e = \sqrt{2/3}$

43. Yes, a parabola **45.** a) $r = \dfrac{2a}{1 + \cos\left(\theta - \dfrac{\pi}{4}\right)}$

b) $r = \dfrac{8}{3 - \cos \theta}$ c) $r = \dfrac{3}{1 + 2 \sin \theta}$

CHAPTER 10

Section 10.1, pp. 794–795

1. a)   b)   c)

d)

3. $4\mathbf{i} + 5\mathbf{j}$ **5.** $(6 - (\sqrt{3}/\pi))\mathbf{i} - 20\mathbf{j}$

7. a) $\mathbf{w} = \mathbf{v} + \mathbf{u}$ b) $\mathbf{v} = \mathbf{w} - \mathbf{u}$

9.

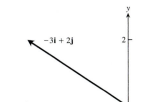

11.

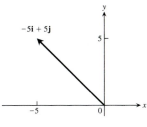

13.

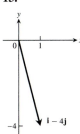

15.

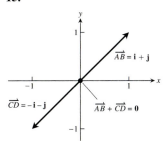

17. $(5, 8)$

19.

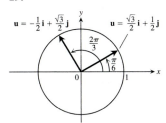

21.

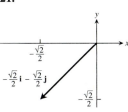

23. $\dfrac{3}{5}\mathbf{i} - \dfrac{4}{5}\mathbf{j}$

25. $\mathbf{u} = \dfrac{1}{\sqrt{17}}\mathbf{i} + \dfrac{4}{\sqrt{17}}\mathbf{j}, \quad -\mathbf{u} = -\dfrac{1}{\sqrt{17}}\mathbf{i} - \dfrac{4}{\sqrt{17}}\mathbf{j},$

$\mathbf{n} = \dfrac{4}{\sqrt{17}}\mathbf{i} - \dfrac{1}{\sqrt{17}}\mathbf{j}, \quad -\mathbf{n} = -\dfrac{4}{\sqrt{17}}\mathbf{i} + \dfrac{1}{\sqrt{17}}\mathbf{j}$

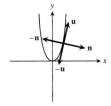

27. $\mathbf{u} = \dfrac{1}{\sqrt{5}}(2\mathbf{i} + \mathbf{j}), \quad -\mathbf{u} = \dfrac{1}{\sqrt{5}}(-2\mathbf{i} - \mathbf{j}), \quad \mathbf{n} = \dfrac{1}{\sqrt{5}}(-\mathbf{i} + 2\mathbf{j}),$

$-\mathbf{n} = \dfrac{1}{\sqrt{5}}(\mathbf{i} - 2\mathbf{j})$

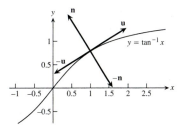

29. $\mathbf{u} = \dfrac{\pm 1}{5}(-4\mathbf{i} + 3\mathbf{j}), \quad \mathbf{v} = \dfrac{\pm 1}{5}(3\mathbf{i} + 4\mathbf{j})$

31. $\mathbf{u} = \dfrac{\pm 1}{2}(\mathbf{i} + \sqrt{3}\mathbf{j}), \quad \mathbf{v} = \dfrac{\pm 1}{2}(-\sqrt{3}\mathbf{i} + \mathbf{j})$

33. $13\left(\dfrac{5}{13}\mathbf{i} + \dfrac{12}{13}\mathbf{j}\right)$

35. $\dfrac{3}{5}\mathbf{i} - \dfrac{4}{5}\mathbf{j}$ and $-\dfrac{3}{5}\mathbf{i} + \dfrac{4}{5}\mathbf{j}$ **39.** $5\sqrt{3}\mathbf{i}, \; 5\mathbf{j}$

41. $\alpha = 3/2, \; \beta = 1/2$

43. a) $(5\cos 60°, 5\sin 60°) = \left(\dfrac{5}{2}, \dfrac{5\sqrt{3}}{2}\right)$

b) $(5\cos 60° + 10\cos 315°, \; 5\sin 60° + 10\sin 315°)$

$= \left(\dfrac{5 + \sqrt{2}}{2}, \dfrac{5\sqrt{3} - 10\sqrt{2}}{2}\right)$

45. The slope of $-\mathbf{v} = -a\mathbf{i} - b\mathbf{j}$ is $(-b)/(-a) = b/a$, the same as the slope of $\mathbf{v}$.

Section 10.2, pp. 804–806

1. The line through the point $(2, 3, 0)$ parallel to the z-axis

3. The x-axis   **5.** The circle $x^2 + y^2 = 4$ in the xy-plane

7. The circle $x^2 + z^2 = 4$ in the xz-plane

9. The circle $y^2 + z^2 = 1$ in the yz-plane

11. The circle $x^2 + y^2 = 16$ in the xy-plane

13. a) The first quadrant of the xy-plane

b) The fourth quadrant of the xy-plane

15. a) The ball of radius 1 centered at the origin

b) All points greater than 1 unit from the origin

17. a) The upper hemisphere of radius 1 centered at the origin

b) The solid upper hemisphere of radius 1 centered at the origin

19. a) $x = 3$ b) $y = -1$ c) $z = -2$ **21.** a) $z = 1$ b) $x = 3$

c) $y = -1$ **23.** a) $x^2 + (y - 2)^2 = 4, \; z = 0$

b) $(y - 2)^2 + z^2 = 4, \; x = 0$ c) $x^2 + z^2 = 4, \; y = 2$

25. a) $y = 3, \; z = -1$ b) $x = 1, \; z = -1$ c) $x = 1, \; y = 3$

27. $x^2 + y^2 + z^2 = 25, \; z = 3$ **29.** $0 \le z \le 1$ **31.** $z \le 0$

33. a) $(x - 1)^2 + (y - 1)^2 + (z - 1)^2 < 1$

b) $(x - 1)^2 + (y - 1)^2 + (z - 1)^2 > 1$ **35.** $3\left(\dfrac{2}{3}\mathbf{i} + \dfrac{1}{3}\mathbf{j} - \dfrac{2}{3}\mathbf{k}\right)$

37. $9\left(\dfrac{1}{9}\mathbf{i} + \dfrac{4}{9}\mathbf{j} - \dfrac{8}{9}\mathbf{k}\right)$ **39.** $5(\mathbf{k})$ **41.** $1\left(\dfrac{3}{5}\mathbf{i} + \dfrac{4}{5}\mathbf{k}\right)$

43. $\sqrt{\dfrac{1}{2}}\left(\dfrac{1}{\sqrt{3}}\mathbf{i} - \dfrac{1}{\sqrt{3}}\mathbf{j} - \dfrac{1}{\sqrt{3}}\mathbf{k}\right)$ **45.** a) $2\mathbf{i}$ b) $-\sqrt{3}\mathbf{k}$

c) $\dfrac{3}{10}\mathbf{j} + \dfrac{2}{5}\mathbf{k}$ d) $6\mathbf{i} - 2\mathbf{j} + 3\mathbf{k}$ **47.** $\dfrac{7}{13}(12\mathbf{i} - 5\mathbf{k})$

49. $-\dfrac{10}{7}\mathbf{i} + \dfrac{15}{7}\mathbf{j} - \dfrac{30}{7}\mathbf{k}$ **51.** a) 3 b) $\dfrac{2}{3}\mathbf{i} + \dfrac{2}{3}\mathbf{j} - \dfrac{1}{3}\mathbf{k}$

c) $(2, 2, 1/2)$ **53.** a) 7 b) $\dfrac{3}{7}\mathbf{i} - \dfrac{6}{7}\mathbf{j} + \dfrac{2}{7}\mathbf{k}$ c) $(5/2, 1, 6)$

55. a) $2\sqrt{3}$ b) $\dfrac{1}{\sqrt{3}}\mathbf{i} - \dfrac{1}{\sqrt{3}}\mathbf{j} - \dfrac{1}{\sqrt{3}}\mathbf{k}$ c) $(1, -1, -1)$

57. $A(4, -3, 5)$ **59.** $C(-2, 0, 2)$, $a = 2\sqrt{2}$

61. $C(\sqrt{2}, \sqrt{2}, -\sqrt{2})$, $a = \sqrt{2}$

63. $(x - 1)^2 + (y - 2)^2 + (z - 3)^2 = 14$

65. $(x + 2)^2 + y^2 + z^2 = 3$ **67.** $C(-2, 0, 2)$, $a = \sqrt{8}$

69. $C\left(-\dfrac{1}{4}, -\dfrac{1}{4}, -\dfrac{1}{4}\right)$, $a = \dfrac{5\sqrt{3}}{4}$

71. a) $\sqrt{y^2 + z^2}$ b) $\sqrt{x^2 + z^2}$ c) $\sqrt{x^2 + y^2}$

73. a) $\dfrac{3}{2}\mathbf{i} + \dfrac{3}{2}\mathbf{j} - 3\mathbf{k}$ b) $\mathbf{i} + \mathbf{j} - 2\mathbf{k}$ c) $(2, 2, 1)$

Section 10.3, pp. 812–814

1. a) $-25, 5, 5$ b) -1 c) -5 d) $-2\mathbf{i} + 4\mathbf{j} - \sqrt{5}\mathbf{k}$

3. a) $25, 15, 5$ b) $1/3$ c) $5/3$ d) $\dfrac{1}{9}(10\mathbf{i} + 11\mathbf{j} - 2\mathbf{k})$

5. a) $0, \sqrt{53}, 1$ b) 0 c) 0 d) 0 **7.** a) $2, \sqrt{34}, \sqrt{3}$

b) $\dfrac{2}{\sqrt{3}\sqrt{34}}$ c) $\dfrac{2}{\sqrt{34}}$ d) $\dfrac{1}{17}(5\mathbf{j} - 3\mathbf{k})$ **9.** a) $\sqrt{3} - \sqrt{2}, \sqrt{2}, 3$

b) $\dfrac{\sqrt{3} - \sqrt{2}}{3\sqrt{2}}$ c) $\dfrac{\sqrt{3} - \sqrt{2}}{\sqrt{2}}$ d) $\dfrac{\sqrt{3} - \sqrt{2}}{2}(-\mathbf{i} + \mathbf{j})$

11. $\left(\dfrac{3}{2}\mathbf{i} + \dfrac{3}{2}\mathbf{j}\right) + \left(-\dfrac{3}{2}\mathbf{i} + \dfrac{3}{2}\mathbf{j} + 4\mathbf{k}\right)$

13. $\left(\dfrac{14}{3}\mathbf{i} + \dfrac{28}{3}\mathbf{j} - \dfrac{14}{3}\mathbf{k}\right) + \left(\dfrac{10}{3}\mathbf{i} - \dfrac{16}{3}\mathbf{j} - \dfrac{22}{3}\mathbf{k}\right)$

15. The sum of two vectors of equal length is *always* orthogonal to their difference, as we can see from the equation

$$(\mathbf{v}_1 + \mathbf{v}_2) \cdot (\mathbf{v}_1 - \mathbf{v}_2) = \mathbf{v}_1 \cdot \mathbf{v}_1 + \mathbf{v}_2 \cdot \mathbf{v}_1 - \mathbf{v}_1 \cdot \mathbf{v}_2 - \mathbf{v}_2 \cdot \mathbf{v}_2$$

$$= |\mathbf{v}_1|^2 - |\mathbf{v}_2|^2.$$

21. $\tan^{-1}\sqrt{2}$ **23.** 0.75 rad   **25.** 1.77 rad

27. $\angle A \approx 1.24$ rad, $\angle B \approx 0.66$ rad, $\angle C \approx 1.24$ rad   **29.** 0.62 rad

31. a) Since $|\cos\theta| \leq 1$, we have

$$|\mathbf{u} \cdot \mathbf{v}| = |\mathbf{u}||\mathbf{v}||\cos\theta| \leq |\mathbf{u}||\mathbf{v}|(1) = |\mathbf{u}||\mathbf{v}|.$$

b) We have equality precisely when $|\cos\theta| = 1$ or when one or both of $\mathbf{u}$ and $\mathbf{v}$ are $\mathbf{0}$. In the case of nonzero vectors, we have equality when $\theta = 0$ or π, i.e., when the vectors are parallel.

33. a   **35.** a) $\sqrt{70}$ b) $\sqrt{568}$ **37.** 5 J   **39.** 3464.10 J

43. $x + 2y = 4$

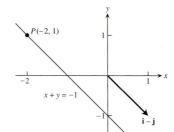

45. $-2x + y = -3$

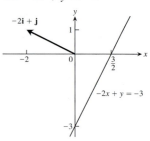

47. $x + y = -1$

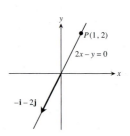

49. $2x - y = 0$

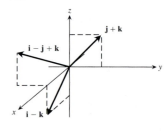

51. $\pi/4$ **53.** $\pi/6$ **55.** 0.14   **57.** $\pi/3$ and $2\pi/3$ at each point

59. At $(0, 0)$: $\pi/2$, at $(1, 1)$: $\pi/4$ and $3\pi/4$

Section 10.4, pp. 820–821

1. $|\mathbf{A} \times \mathbf{B}| = 3$, direction is $\dfrac{2}{3}\mathbf{i} + \dfrac{1}{3}\mathbf{j} + \dfrac{2}{3}\mathbf{k}$; $|\mathbf{B} \times \mathbf{A}| = 3$, direction

is $-\dfrac{2}{3}\mathbf{i} - \dfrac{1}{3}\mathbf{j} - \dfrac{2}{3}\mathbf{k}$

3. $|\mathbf{A} \times \mathbf{B}| = 0$, no direction; $|\mathbf{B} \times \mathbf{A}| = 0$, no direction

5. $|\mathbf{A} \times \mathbf{B}| = 6$, direction is $-\mathbf{k}$; $|\mathbf{B} \times \mathbf{A}| = 6$, direction is $\mathbf{k}$

7. $|\mathbf{A} \times \mathbf{B}| = 6\sqrt{5}$, direction is $\dfrac{1}{\sqrt{5}}\mathbf{i} - \dfrac{2}{\sqrt{5}}\mathbf{k}$; $|\mathbf{B} \times \mathbf{A}| = 6\sqrt{5}$,

direction is $-\dfrac{1}{\sqrt{5}}\mathbf{i} + \dfrac{2}{\sqrt{5}}\mathbf{k}$

9.

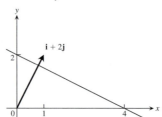

11.

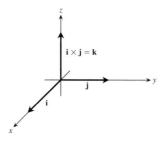

13.

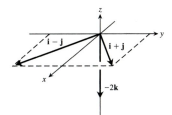

15. a) $2\sqrt{6}$ b) $\pm\dfrac{1}{\sqrt{6}}(2\mathbf{i}+\mathbf{j}+\mathbf{k})$ **17.** a) $\dfrac{\sqrt{2}}{2}$ b) $\pm\dfrac{1}{\sqrt{2}}(\mathbf{i}-\mathbf{j})$

19. a) None b) **A** and **C** **21.** $10\sqrt{3}$ ft·lb **23.** 8 **25.** 7

27. a) True b) Not always true c) True d) True
e) Not always true f) True g) True h) True

29. a) $\text{proj}_\mathbf{B}\,\mathbf{A}=\dfrac{\mathbf{A}\cdot\mathbf{B}}{\mathbf{B}\cdot\mathbf{B}}\,\mathbf{B}$ b) $\pm\mathbf{A}\times\mathbf{B}$ c) $\pm(\mathbf{A}\times\mathbf{B})\times\mathbf{C}$

d) $|(\mathbf{A}\times\mathbf{B})\cdot\mathbf{C}|$ **31.** a) Yes b) No c) Yes d) No

33. No, **B** need not equal **C**. For example, $\mathbf{i}+\mathbf{j}\neq-\mathbf{i}+\mathbf{j}$, but

$$\mathbf{i}\times(\mathbf{i}+\mathbf{j})=\mathbf{i}\times\mathbf{i}+\mathbf{i}\times\mathbf{j}=\mathbf{0}+\mathbf{k}=\mathbf{k}$$

$$\mathbf{i}\times(-\mathbf{i}+\mathbf{j})=-\mathbf{i}\times\mathbf{i}+\mathbf{i}\times\mathbf{j}=\mathbf{0}+\mathbf{k}=\mathbf{k}$$

35. 2 **37.** 13 **39.** 11/2 **41.** 25/2

43. If $\mathbf{A}=a_1\,\mathbf{i}+a_2\,\mathbf{j}$ and $\mathbf{B}=b_1\,\mathbf{i}+b_2\,\mathbf{j}$, then

$$\mathbf{A}\times\mathbf{B}=\begin{vmatrix}\mathbf{i}&\mathbf{j}&\mathbf{k}\\a_1&a_2&0\\b_1&b_2&0\end{vmatrix}=\begin{vmatrix}a_1&a_2\\b_1&b_2\end{vmatrix}\mathbf{k}$$

and the triangle's area is

$$\frac{1}{2}|\mathbf{A}\times\mathbf{B}|=\pm\frac{1}{2}\begin{vmatrix}a_1&a_2\\b_1&b_2\end{vmatrix}.$$

The applicable sign is $(+)$ if the acute angle from **A** to **B** runs counterclockwise in the xy-plane, and $(-)$ if it runs clockwise.

Section 10.5, pp. 827–829

1. $x=3+t,\ y=-4+t,\ z=-1+t$

3. $x=-2+5t,\ y=5t,\ z=3-5t$ **5.** $x=0,\ y=2t,\ z=t$

7. $x=1,\ y=1,\ z=1+t$ **9.** $x=t,\ y=-7+2t,\ z=2t$

11. $x=t,\ y=0,\ z=0$

13. $x=t,\ y=t,\ z=\dfrac{3}{2}t,\ 0\le t\le 1$

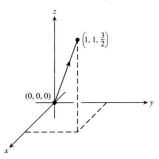

15. $x=1,\ y=1+t,\ z=0,\ -1\le t\le 0$

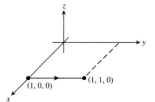

17. $x=0,\ y=1-2t,\ z=1,\ 0\le t\le 1$

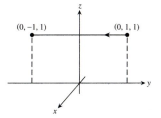

19. $x=2-2t,\ y=2t,\ z=2-2t,\ 0\le t\le 1$

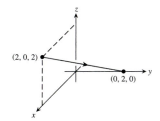

21. $3x-2y-z=-3$ **23.** $7x-5y-4z=6$

25. $x+3y+4z=34$ **27.** $(1,2,3),\ -20x+12y+z=7$

29. $y+z=3$ **31.** $x-y+z=0$ **33.** $2\sqrt{30}$ **35.** 0

37. $\dfrac{9\sqrt{42}}{7}$ **39.** 3 **41.** 19/5 **43.** 5/3 **45.** $9/\sqrt{41}$

47. $\pi/4$ **49.** 1.76 rad **51.** 0.82 rad **53.** $\left(\dfrac{3}{2},-\dfrac{3}{2},\dfrac{1}{2}\right)$

55. $(1,1,0)$ **57.** $x=1-t,\ y=1+t,\ z=-1$

59. $x=4,\ y=3+6t,\ z=1+3t$

61. $L1$ intersects $L2$; $L2$ is parallel to $L3$; $L1$ and $L3$ are skew.

63. $x=2+2t,\ y=-4-t,\ z=7+3t;\ x=-2-t,$
$y=-2+(1/2)t,\ z=1-(3/2)t$

65. $\left(0,-\dfrac{1}{2},-\dfrac{3}{2}\right),\ (-1,0,-3),\ (1,-1,0)$

69. Many possible answers. One possibility: $x+y=3$ and $2y+z=7$

71. $(x/a)+(y/b)+(z/c)=1$ describes all planes *except* those through the origin or parallel to a coordinate axis.

Section 10.6, pp. 839–841

1. d, ellipsoid **3.** a, cylinder **5.** l, hyperbolic paraboloid
7. b, cylinder **9.** k, hyperbolic paraboloid **11.** h, cone

13.

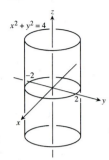

$x^2 + y^2 = 4$

15.

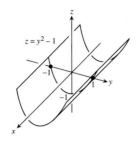

$z = y^2 - 1$

29.

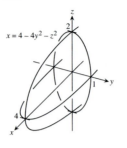

$x = 4 - 4y^2 - z^2$

31.

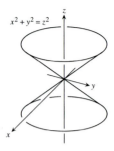

$x^2 + y^2 = z^2$

17.

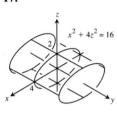

$x^2 + 4z^2 = 16$

19.

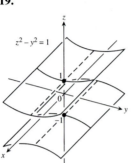

$z^2 - y^2 = 1$

33.

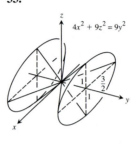

$4x^2 + 9z^2 = 9y^2$

35.

$x^2 + y^2 - z^2 = 1$

21.

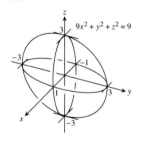

$9x^2 + y^2 + z^2 = 9$

23.

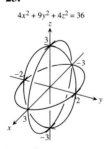

$4x^2 + 9y^2 + 4z^2 = 36$

37.

$$\frac{y^2}{4} + \frac{z^2}{9} - \frac{x^2}{4} = 1$$

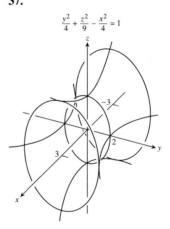

39.

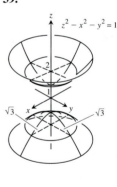

$z^2 - x^2 - y^2 = 1$

25.

$z = x^2 + 4y^2$

27.

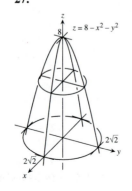

$z = 8 - x^2 - y^2$

41.

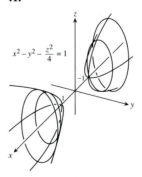

$x^2 - y^2 - \dfrac{z^2}{4} = 1$

43.

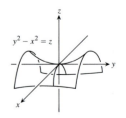

$y^2 - x^2 = z$

45.

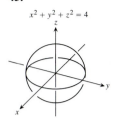

$x^2 + y^2 + z^2 = 4$

47.

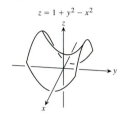

$z = 1 + y^2 - x^2$

65.

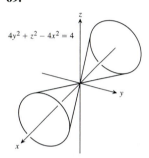

$z = -(x^2 + y^2)$

67.

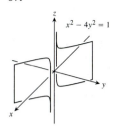

$x^2 - 4y^2 = 1$

49.

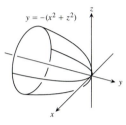

$y = -(x^2 + z^2)$

51.

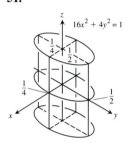

$16x^2 + 4y^2 = 1$

69.

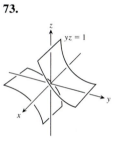

$4y^2 + z^2 - 4x^2 = 4$

71.

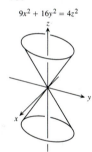

$x^2 + y^2 = z$

53.

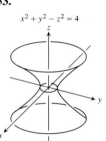

$x^2 + y^2 - z^2 = 4$

55.

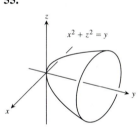

$x^2 + z^2 = y$

73.

$yz = 1$

75.

$9x^2 + 16y^2 = 4z^2$

57.

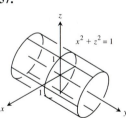

$x^2 + z^2 = 1$

59.

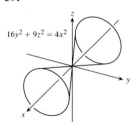

$16y^2 + 9z^2 = 4x^2$

77. a) $\dfrac{2\pi(9 - c^2)}{9}$ b) 8π c) $\dfrac{4\pi abc}{3}$

81. Vertex $(0, y_1, cy_1^2/b^2)$, focus $(0, y_1, c(y_1^2/b^2) - a^2/(4c))$

Section 10.7, pp. 846–847

	Rectangular	Cylindrical	Spherical
1.	$(0, 0, 0)$	$(0, 0, 0)$	$(0, 0, 0)$
3.	$(0, 1, 0)$	$(1, \pi/2, 0)$	$(1, \pi/2, \pi/2)$
5.	$(1, 0, 0)$	$(1, 0, 0)$	$(1, \pi/2, 0)$
7.	$(0, 1, 1)$	$(1, \pi/2, 1)$	$(\sqrt{2}, \pi/4, \pi/2)$
9.	$(0, -2\sqrt{2}, 0)$	$(2\sqrt{2}, 3\pi/2, 0)$	$(2\sqrt{2}, \pi/2, 3\pi/2)$

11. $x^2 + y^2 = 0$, $\theta = 0$ or $\theta = \pi$, the z-axis

13. $z = 0$, $\phi = \pi/2$, the xy-plane

15. $z = r$, $0 \le r \le 1$; $\phi = \pi/4$, $0 \le \rho \le \sqrt{2}$; a (finite) cone

17. $x = 0$, $\theta = \pi/2$, the yz-plane

19. $r^2 + z^2 = 4$, $\rho = 2$, sphere of radius 2 centered at the origin

21. $x^2 + y^2 + \left(z - \dfrac{5}{2}\right)^2 = \dfrac{25}{4}$, $r^2 + z^2 = 5z$, sphere of radius 5/2 centered at $(0, 0, 5/2)$ (rectangular)

61.

$9x^2 + 4y^2 + z^2 = 36$

63.

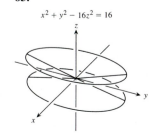

$x^2 + y^2 - 16z^2 = 16$

23. $y = 1$, $\rho \sin \phi \sin \theta = 1$, the plane $y = 1$

25. $z = \sqrt{2}$, the plane $z = \sqrt{2}$

27. $r^2 + z^2 = 2z$, $z \le 1$; $\rho = 2 \cos \phi$, $\pi/4 \le \phi \le \pi/2$; lower half (hemisphere) of the sphere of radius 1 centered at $(0, 0, 1)$ (rectangular)

29. $x^2 + y^2 + z^2 = 9$, $-3/2 \le z \le 3/2$; $r^2 + z^2 = 9$, $-3/2 \le z \le 3/2$; the portion of the sphere of radius 3 centered at the origin between the planes $z = -3/2$ and $z = 3/2$

31. $z = 4 - 4(x^2 + y^2)$, $0 \le z \le 4$; $\rho \cos \phi = 4 - 4\rho^2 \sin^2 \phi$, $0 \le \phi \le \pi/2$; the upper portion cut from the paraboloid $z = 4 - 4(x^2 + y^2)$ by the xy-plane

33. $z = -\sqrt{x^2 + y^2}$, $-1 \le z \le 0$; $z = -r$, $0 \le r \le 1$; cone, vertex at origin, base the circle $x^2 + y^2 = 1$ in the plane $z = -1$

35. $z + x^2 - y^2 = 0$ or $z = y^2 - x^2$, $\cos \phi + \rho \sin^2 \phi \cos 2\theta = 0$, hyperbolic paraboloid

37. $(2, 3, 1)$

39. Right circular cylinder parallel to the z-axis generated by the circle $r = -2 \sin \theta$ in the $r\theta$-plane

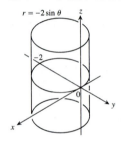

41. Cylinder of lines parallel to the z-axis generated by the cardioid $r = 1 - \cos \theta$ in the $r\theta$-plane

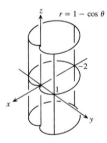

43. Cardioid of revolution symmetric about the y-axis, cusp at the origin pointing down

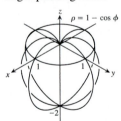

45. b) $\phi = \pi/2$

49. The surface's equation $r = f(z)$ tells us that the point $(r, \theta, z) = (f(z), \theta, z)$ will lie on the surface for all θ. In particular $(f(z), \theta +$ $\pi, z)$ lies on the surface whenever $(f(z), \theta, z)$ lies on the surface, so the surface is symmetric with respect to the z-axis.

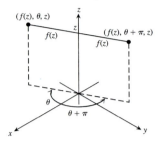

Chapter 10 Practice Exercises, pp. 848–851

1.

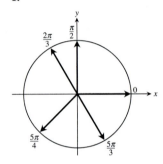

3. $2 \cdot \left(\dfrac{1}{\sqrt{2}} \mathbf{i} + \dfrac{1}{\sqrt{2}} \mathbf{j} \right)$ **5.** $7 \cdot \left(\dfrac{2}{7} \mathbf{i} - \dfrac{3}{7} \mathbf{j} + \dfrac{6}{7} \mathbf{k} \right)$

7. $\dfrac{8}{\sqrt{33}} \mathbf{i} - \dfrac{2}{\sqrt{33}} \mathbf{j} + \dfrac{8z}{\sqrt{33}} \mathbf{k}$

9. a) $\overrightarrow{BD} = \overrightarrow{AD} - \overrightarrow{AB}$

b) $\overrightarrow{AP} = \dfrac{1}{2} (\overrightarrow{AB} + \overrightarrow{AD})$

11.

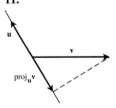

13. $|\mathbf{A}| = \sqrt{2}$, $|\mathbf{B}| = 3$, $\mathbf{A} \cdot \mathbf{B} = \mathbf{B} \cdot \mathbf{A} = 3$, $\mathbf{A} \times \mathbf{B} = -2\mathbf{i} + 2\mathbf{j} - \mathbf{k}$, $\mathbf{B} \times \mathbf{A} = 2\mathbf{i} - 2\mathbf{j} + \mathbf{k}$, $|\mathbf{A} \times \mathbf{B}| = 3$, $\theta = \pi/4$, $|\mathbf{B}| \cos \theta = 3/\sqrt{2}$, $\text{proj}_{\mathbf{A}} \mathbf{B} = (3/2)(\mathbf{i} + \mathbf{j})$

15. $\dfrac{4}{3} (2\mathbf{i} + \mathbf{j} - \mathbf{k}) - \dfrac{1}{3} (5\mathbf{i} + \mathbf{j} + 11\mathbf{k})$

17.

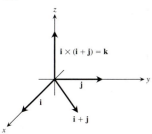

19. unit tangents $\pm\left(\dfrac{1}{\sqrt5}\mathbf{i}+\dfrac{2}{\sqrt5}\mathbf{j}\right)$, unit normals

$\pm\left(-\dfrac{2}{\sqrt5}\mathbf{i}+\dfrac{1}{\sqrt5}\mathbf{j}\right)$

23. $2\sqrt7$ **25.** a) $A=\sqrt{14}$ b) $V=1$ **29.** $\sqrt{78}/3$
31. $x=1-3t,\;y=2,\;z=3+7t$ **33.** $\sqrt2$
35. $2x+y-z=3$ **37.** $-9x+y+7z=4$

39. $\left(0,-\dfrac{1}{2},-\dfrac{3}{2}\right)$, $(-1,0,-3),\;(1,-1,0)$

41. $\pi/3$ **43.** $x=-5+5t,\;y=3-t,\;z=-3t$
45. b) $x=-12t,\;y=19/12+15t,\;z=1/6+6t$
47. Yes; $\mathbf{v}$ is parallel to the plane. **49.** 3 **51.** $-3\mathbf{j}+3\mathbf{k}$

53. $\dfrac{2}{\sqrt{35}}(5\mathbf{i}-\mathbf{j}-3\mathbf{k})$ **55.** $\left(\dfrac{11}{9},\dfrac{26}{9},\dfrac{7}{9}\right)$

57. $(1,-2,-1);\;x=1-5t,\;y=-2+3t,\;z=-1+4t$
59. $2x+7y+2z+10=0$ **61.** a) No b) no c) no d) no
e) yes **63.** $11/\sqrt{107}$
65.

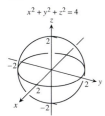

67.

69.

71.

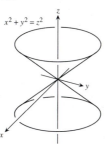

73.

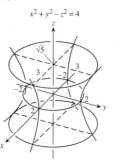

75.

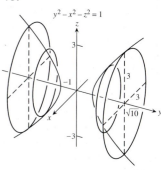

77. The y-axis in the xy-plane; the yz-plane in three dimensional space
79. The circle centered at $(0,0)$ with radius 2 in the xy-plane; the cylinder parallel to the z-axis in three dimensional space with the circle as a generating curve
81. The parabola $x=y^2$ in the xy-plane; the cylinder parallel to the z-axis in three dimensional space with the parabola as a generating curve
83. A cardioid in the $r\theta$-plane; a cylinder parallel with the z-axis in three dimensional space with the cardioid as a generating curve
85. A horizontal lemniscate of length $2\sqrt2$ in the $r\theta$-plane; the cylinder parallel to the z-axis in three dimensional space with the lemniscate as a generating curve
87. The sphere of radius 2 centered at the origin
89. The upper nappe of the cone having its vertex at the origin and making a $\pi/6$ angle with the z-axis
91. The upper hemisphere of the sphere of radius 1 centered at the origin

	Rectangular	Cylindrical	Spherical
93.	$(1,0,0)$	$(1,0,0)$	$(1,\pi/2,0)$
95.	$(0,1,1)$	$(1,\pi/2,1)$	$(\sqrt2,\pi/4,\pi/2)$
97.	$(-1,0,-1)$	$(1,\pi,-1)$	$(\sqrt2,3\pi/4,\pi)$

99. Cylindrical: $z=2$, spherical: $\rho\cos\phi=2$, a plane parallel with the xy-plane
101. Cylindrical: $r^2+z^2=-2z$, spherical: $\rho=-2\cos\phi$, sphere of radius 1 centered at $(0,0,-1)$ (rectangular)
103. Rectangular: $z=x^2+y^2$, spherical: $\rho=0$ or $\rho=\dfrac{\cos\phi}{\sin^2\phi}$ when $0<\phi<\pi/2$, a paraboloid symmetric to the z-axis, opening upward, vertex at the origin
105. Rectangular: $x^2+(y-7/2)^2=49/4$, spherical: $\rho\sin\phi=7\sin\theta$, cylinder parallel to the z-axis generated by the circle
107. Rectangular: $x^2+y^2+z^2=16$, cylindrical: $r^2+z^2=16$, sphere of radius 4 centered at the origin
109. Rectangular: $-\sqrt{x^2+y^2}=z$, cylindrical: $z=-r,r\ge0$, single cone making an angle of $3\pi/4$ with the positive z-axis, vertex at the origin

Chapter 10 Additional Exercises, pp. 851–853

1. $(26,23,-1/3)$ **3.** $\approx34{,}641$ J

17. b) $6/\sqrt{14}$ c) $2x - y + 2z = 8$
d) $x - 2y + z = 3 + 5\sqrt{6}$ and $x - 2y + z = 3 - 5\sqrt{6}$

23. $\mathbf{v} - 2\dfrac{\mathbf{v} \cdot \mathbf{z}}{|\mathbf{z}|^2}\mathbf{z}$ **25.** a) $|\mathbf{F}| = \dfrac{GMm}{d^2}\left(1 + \displaystyle\sum_{i=1}^{n}\dfrac{2}{(i^2 + 1)^{3/2}}\right)$
b) Yes

CHAPTER 11

Section 11.1, pp. 865–868

1. $y = x^2 - 2x$, $\mathbf{v} = \mathbf{i} + 2\mathbf{j}$, $\mathbf{a} = 2\mathbf{j}$

3. $y = \dfrac{2}{9}x^2$, $\mathbf{v} = 3\mathbf{i} + 4\mathbf{j}$, $\mathbf{a} = 3\mathbf{i} + 8\mathbf{j}$

5. $t = \dfrac{\pi}{4} : \mathbf{v} = \dfrac{\sqrt{2}}{2}\mathbf{i} - \dfrac{\sqrt{2}}{2}\mathbf{j}$, $\mathbf{a} = \dfrac{-\sqrt{2}}{2}\mathbf{i} - \dfrac{\sqrt{2}}{2}\mathbf{j}$;
$t = \pi/2 : \mathbf{v} = -\mathbf{j}$, $\mathbf{a} = -\mathbf{i}$

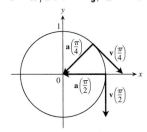

7. $t = \pi : \mathbf{v} = 2\mathbf{i}$, $\mathbf{a} = -\mathbf{j}$; $t = \dfrac{3\pi}{2} : \mathbf{v} = \mathbf{i} - \mathbf{j}$, $\mathbf{a} = -\mathbf{i}$

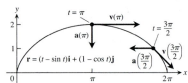

9. $\mathbf{v} = \mathbf{i} + 2t\mathbf{j} + 2\mathbf{k}$; $\mathbf{a} = 2\mathbf{j}$; speed: 3; direction: $\dfrac{1}{3}\mathbf{i} + \dfrac{2}{3}\mathbf{j} + \dfrac{2}{3}\mathbf{k}$;

$\mathbf{v}(1) = 3\left(\dfrac{1}{3}\mathbf{i} + \dfrac{2}{3}\mathbf{j} + \dfrac{2}{3}\mathbf{k}\right)$

11. $\mathbf{v} = (-2\sin t)\mathbf{i} + (3\cos t)\mathbf{j} + 4\mathbf{k}$;
$\mathbf{a} = (-2\cos t)\mathbf{i} - (3\sin t)\mathbf{j}$; speed: $2\sqrt{5}$;
direction: $(-1/\sqrt{5})\mathbf{i} + (2/\sqrt{5})\mathbf{k}$;
$\mathbf{v}(\pi/2) = 2\sqrt{5}[(-1/\sqrt{5})\mathbf{i} + (2/\sqrt{5})\mathbf{k}]$

13. $\mathbf{v} = \left(\dfrac{2}{t + 1}\right)\mathbf{i} + 2t\mathbf{j} + t\mathbf{k}$; $\mathbf{a} = \left(\dfrac{-2}{(t + 1)^2}\right)\mathbf{i} + 2\mathbf{j} + \mathbf{k}$;

speed: $\sqrt{6}$; direction: $\dfrac{1}{\sqrt{6}}\mathbf{i} + \dfrac{2}{\sqrt{6}}\mathbf{j} + \dfrac{1}{\sqrt{6}}\mathbf{k}$;

$\mathbf{v}(1) = \sqrt{6}\left(\dfrac{1}{\sqrt{6}}\mathbf{i} + \dfrac{2}{\sqrt{6}}\mathbf{j} + \dfrac{1}{\sqrt{6}}\mathbf{k}\right)$

15. $\pi/2$ **17.** $\pi/2$ **19.** $t = 0$, π, 2π

21. $(1/4)\mathbf{i} + 7\mathbf{j} + (3/2)\mathbf{k}$ **23.** $\left(\dfrac{\pi + 2\sqrt{2}}{2}\right)\mathbf{j} + 2\mathbf{k}$

25. $(\ln 4)\mathbf{i} + (\ln 4)\mathbf{j} + (\ln 2)\mathbf{k}$

27. $\mathbf{r}(t) = \left(\dfrac{-t^2}{2} + 1\right)\mathbf{i} + \left(\dfrac{-t^2}{2} + 2\right)\mathbf{j} + \left(\dfrac{-t^2}{2} + 3\right)\mathbf{k}$

29. $\mathbf{r}(t) = ((t + 1)^{3/2} - 1)\mathbf{i} + (-e^{-t} + 1)\mathbf{j} + (\ln(t + 1) + 1)\mathbf{k}$

31. $\mathbf{r}(t) = 8t\mathbf{i} + 8t\mathbf{j} + (-16t^2 + 100)\mathbf{k}$

33. $x = t$, $y = -1$, $z = 1 + t$

35. $x = at$, $y = a$, $z = 2\pi b + bt$

37. a) (i): It has constant speed 1 (ii): Yes (iii): Counterclockwise
(iv): Yes b) (i): It has constant speed 2 (ii): Yes (iii): Counterclockwise (iv): Yes c) (i): It has constant speed 1 (ii): Yes
(iii): Counterclockwise (iv): It starts at $(0, -1)$ instead of $(1, 0)$
d) (i): It has constant speed 1 (ii): Yes (iii): Clockwise (iv): Yes
e) (i): It has variable speed (ii): No (iii): Counterclockwise
(iv): Yes

39. $\mathbf{r}(t) = \left(\dfrac{3}{2}t^2 + \dfrac{6}{\sqrt{11}}t + 1\right)\mathbf{i} - \left(\dfrac{1}{2}t^2 + \dfrac{2}{\sqrt{11}}t - 2\right)\mathbf{j} +$

$\left(\dfrac{1}{2}t^2 + \dfrac{2}{\sqrt{11}}t + 3\right)\mathbf{k} = \left(\dfrac{1}{2}t^2 + \dfrac{2t}{\sqrt{11}}\right)(3\mathbf{i} - \mathbf{j} + \mathbf{k}) +$
$(\mathbf{i} + 2\mathbf{j} + 3\mathbf{k})$

41. $\mathbf{v} = 2\sqrt{5}\mathbf{i} + \sqrt{5}\mathbf{j}$

43. max $|\mathbf{v}| = 3$, min $|\mathbf{v}| = 2$, max $|\mathbf{a}| = 3$, min $|\mathbf{a}| = 2$

Section 11.2, pp. 873–876

1. 50 sec **3.** a) 72.2 sec, 25,510 m b) 4020 m c) 6378 m
5. $t \approx 2.135$ sec, $x \approx 66.42$ ft
7. $v_0 = 9.9$ m/sec, $\alpha = 18.4°$ or $71.6°$ **9.** 190 mph
11. The golf ball will clip the leaves at the top. **13.** 46.6 ft/sec
17. 141% **21.** 1.92 sec, 73.7 ft (approx.)

25. $\mathbf{v}(t) = -gt\mathbf{k} + \mathbf{v}_0$, $\mathbf{r}(t) = -\dfrac{1}{2}gt^2\mathbf{k} + \mathbf{v}_0 t$

Section 11.3, pp. 880–881

1. $\mathbf{T} = \left(-\dfrac{2}{3}\sin t\right)\mathbf{i} + \left(\dfrac{2}{3}\cos t\right)\mathbf{j} + \dfrac{\sqrt{5}}{3}\mathbf{k}$, 3π

3. $\mathbf{T} = \dfrac{1}{\sqrt{1 + t}}\mathbf{i} + \dfrac{\sqrt{t}}{\sqrt{1 + t}}\mathbf{k}$, $\dfrac{52}{3}$ **5.** $\mathbf{T} = -\cos t\mathbf{j} + \sin t\mathbf{k}$, $\dfrac{3}{2}$

7. $\mathbf{T} = \left(\dfrac{\cos t - t\sin t}{t + 1}\right)\mathbf{i} + \left(\dfrac{\sin t + t\cos t}{t + 1}\right)\mathbf{j} + \left(\dfrac{\sqrt{2}t^{1/2}}{t + 1}\right)\mathbf{k}$,

$\dfrac{\pi^2}{2} + \pi$

9. $(0, 5, 24\pi)$ **11.** $s(t) = 5t$, $L = \dfrac{5\pi}{2}$

13. $s(t) = \sqrt{3}e^t - \sqrt{3}$, $L = \dfrac{3\sqrt{3}}{4}$ **15.** $\sqrt{2} + \ln(1 + \sqrt{2})$

17. a) Cylinder is $x^2 + y^2 = 1$, plane is $x + z = 1$

b) and c)

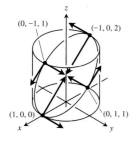

d) $L = \int_0^{2\pi} \sqrt{1 + \sin^2 t}\, dt$ e) $L \approx 7.64$

Section 11.4, pp. 890–893

1. $\mathbf{T} = (\cos t)\mathbf{i} - (\sin t)\mathbf{j}$, $\mathbf{N} = (-\sin t)\mathbf{i} - (\cos t)\mathbf{j}$, $\kappa = \cos t$

3. $\mathbf{T} = \dfrac{1}{\sqrt{1+t^2}}\mathbf{i} - \dfrac{t}{\sqrt{1+t^2}}\mathbf{j}$, $\mathbf{N} = \dfrac{-t}{\sqrt{1+t^2}}\mathbf{i} - \dfrac{1}{\sqrt{1+t^2}}\mathbf{j}$,

$\kappa = \dfrac{1}{2\left(\sqrt{1+t^2}\right)^3}$

5. $\mathbf{a} = \dfrac{2t}{\sqrt{1+t^2}}\mathbf{T} + \dfrac{2}{\sqrt{1+t^2}}\mathbf{N}$ 7. b) $\cos x$

9. b) $\mathbf{N} = \dfrac{-2e^{2t}}{\sqrt{1+4e^{4t}}}\mathbf{i} + \dfrac{1}{\sqrt{1+4e^{4t}}}\mathbf{j}$

c) $\mathbf{N} = -\dfrac{1}{2}\left(\sqrt{4 - t^2}\,\mathbf{i} + t\,\mathbf{j}\right)$

11. $\mathbf{T} = \dfrac{3\cos t}{5}\mathbf{i} - \dfrac{3\sin t}{5}\mathbf{j} + \dfrac{4}{5}\mathbf{k}$, $\mathbf{N} = (-\sin t)\mathbf{i} - (\cos t)\mathbf{j}$,

$\mathbf{B} = \left(\dfrac{4}{5}\cos t\right)\mathbf{i} - \left(\dfrac{4}{5}\sin t\right)\mathbf{j} - \dfrac{3}{5}\mathbf{k}$, $\kappa = \dfrac{3}{25}$, $\tau = -\dfrac{4}{25}$

13. $\mathbf{T} = \left(\dfrac{\cos t - \sin t}{\sqrt{2}}\right)\mathbf{i} + \left(\dfrac{\cos t + \sin t}{\sqrt{2}}\right)\mathbf{j}$,

$\mathbf{N} = \left(\dfrac{-\cos t - \sin t}{\sqrt{2}}\right)\mathbf{i} + \left(\dfrac{-\sin t + \cos t}{\sqrt{2}}\right)\mathbf{j}$,

$\mathbf{B} = \mathbf{k}, \kappa = \dfrac{1}{e^t\sqrt{2}}, \tau = 0$

15. $\mathbf{T} = \dfrac{t}{\sqrt{t^2 + 1}}\mathbf{i} + \dfrac{1}{\sqrt{t^2 + 1}}\mathbf{j}$, $\mathbf{N} = \dfrac{\mathbf{i}}{\sqrt{t^2 + 1}} - \dfrac{t\,\mathbf{j}}{\sqrt{t^2 + 1}}$,

$\mathbf{B} = -\mathbf{k}, \kappa = \dfrac{1}{t(t^2+1)^{3/2}}, \tau = 0$

17. $\mathbf{T} = \left(\operatorname{sech}\dfrac{t}{a}\right)\mathbf{i} + \left(\tanh\dfrac{t}{a}\right)\mathbf{j}$,

$\mathbf{N} = \left(-\tanh\dfrac{t}{a}\right)\mathbf{i} + \left(\operatorname{sech}\dfrac{t}{a}\right)\mathbf{j}$,

$\mathbf{B} = \mathbf{k}, \ \kappa = \dfrac{1}{a}\operatorname{sech}^2\dfrac{t}{a}, \ \tau = 0$

19. $\mathbf{a} = |a|\,\mathbf{N}$ 21. $\mathbf{a}(1) = \dfrac{4}{3}\mathbf{T} + \dfrac{2\sqrt{5}}{3}\mathbf{N}$ 23. $\mathbf{a}(0) = 2\,\mathbf{N}$

25. $\mathbf{r}\left(\dfrac{\pi}{4}\right) = \dfrac{\sqrt{2}}{2}\mathbf{i} + \dfrac{\sqrt{2}}{2}\mathbf{j} - \mathbf{k}$, $\mathbf{T}\left(\dfrac{\pi}{4}\right) = -\dfrac{\sqrt{2}}{2}\mathbf{i} + \dfrac{\sqrt{2}}{2}\mathbf{j}$,

$\mathbf{N}\left(\dfrac{\pi}{4}\right) = -\dfrac{\sqrt{2}}{2}\mathbf{i} - \dfrac{\sqrt{2}}{2}\mathbf{j}$, $\mathbf{B}\left(\dfrac{\pi}{4}\right) = \mathbf{k}$; osculating plane: $z = -1$;

normal plane: $-x + y = 0$; rectifying plane: $x + y = \sqrt{2}$

27. Yes. If the car is moving on a curved path ($\kappa \neq 0$), then $a_N = \kappa|\mathbf{v}|^2 \neq 0$ and $\mathbf{a} \neq \mathbf{0}$.

31. $|\mathbf{F}| = \kappa\left(m\left(\dfrac{ds}{dt}\right)^2\right)$ 35. $1/(2b)$ 39. a) $b - a$ b) π

45. $\kappa(x) = 2/(1 + 4x^2)^{3/2}$ 47. $\kappa(x) = |\sin x|/(1 + \cos^2 x)^{3/2}$

57. Components of $\mathbf{v}$: $-1.8701, 0.7089, 1.0000$
Components of $\mathbf{a}$: $-1.6960, -2.0307, 0$
Speed: 2.2361; Components of $\mathbf{T}$: $-0.8364, 0.3170, 0.4472$
Components of $\mathbf{N}$: $-0.4143, -0.8998, -0.1369$
Components of $\mathbf{B}$: $0.3590, -0.2998, 0.8839$; Curvature: 0.5060
Torsion: 0.2813; Tangential component of acceleration: 0.7746
Normal component of acceleration: 2.5298

59. Components of $\mathbf{v}$: $2.0000, 0, 0.1629$
Components of $\mathbf{a}$: $0, -1.0000, 0.0086$; Speed: 2.0066
Components of $\mathbf{T}$: $0.9967, 0, 0.0812$
Components of $\mathbf{N}$: $-0.0007, -1.0000, 0.0086$
Components of $\mathbf{B}$: $0.0812, -0.0086, -0.9967$; Curvature: 0.2484
Torsion: -0.0411; Tangential component of acceleration: 0.0007
Normal component of acceleration: 1.0000

Section 11.5, pp. 901–902

1. $T = 93.2$ min   3. $a = 6763$ km   5. $T = 1655$ min
7. $a = 20{,}430$ km   9. $|v| = 1.9966 \times 10^7\, r^{-1/2}$ m/sec

11. Circle: $v_0 = \sqrt{\dfrac{GM}{r_0}}$; ellipse: $\sqrt{\dfrac{GM}{r_0}} < v_0 < \sqrt{\dfrac{2GM}{r_0}}$;

parabola: $v_0 = \sqrt{\dfrac{2GM}{r_0}}$; hyperbola: $v_0 > \sqrt{\dfrac{2GM}{r_0}}$

15. a) $x(t) = 2 + (3 - 4\cos(\pi t))\cos(\pi t)$, $y(t) = (3 - 4\cos(\pi t))\sin(\pi t)$

Chapter 11 Practice Exercises, pp. 902–905

1. $\dfrac{x^2}{16} + \dfrac{y^2}{2} = 1$

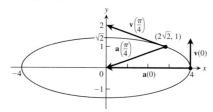

At $t = 0$: $a_T = 0$, $a_N = 4$, $\kappa = 2$;

At $t = \dfrac{\pi}{4}$: $a_T = \dfrac{7}{3}$, $a_N = \dfrac{4\sqrt{2}}{3}$, $\kappa = \dfrac{4\sqrt{2}}{27}$

3. $|\mathbf{v}|_{max} = 1$ **5.** $\kappa = 1/5$ **7.** $dy/dt = -x$; clockwise

11. Shot put is on the ground, about 66 ft, 5 in. from the stopboard.

15. a) 59.19 ft/sec   b) 74.58 ft/sec   **19.** $\kappa = \pi s$

21. Length $= \dfrac{\pi}{4}\sqrt{1 + \dfrac{\pi^2}{16}} + \ln\left(\dfrac{\pi}{4} + \sqrt{1 + \dfrac{\pi^2}{16}}\right)$

23. $\mathbf{T}(0) = \dfrac{2}{3}\mathbf{i} - \dfrac{2}{3}\mathbf{j} + \dfrac{1}{3}\mathbf{k}$; $\mathbf{N}(0) = \dfrac{1}{\sqrt{2}}\mathbf{i} + \dfrac{1}{\sqrt{2}}\mathbf{j}$;

$\mathbf{B}(0) = -\dfrac{1}{3\sqrt{2}}\mathbf{i} + \dfrac{1}{3\sqrt{2}}\mathbf{j} + \dfrac{4}{3\sqrt{2}}\mathbf{k}$; $\kappa = \dfrac{\sqrt{2}}{3}$; $\tau = \dfrac{1}{6}$

25. $\mathbf{T}(\ln 2) = \dfrac{1}{\sqrt{17}}\mathbf{i} + \dfrac{4}{\sqrt{17}}\mathbf{j}$; $\mathbf{N}(\ln 2) = -\dfrac{4}{\sqrt{17}}\mathbf{i} + \dfrac{1}{\sqrt{17}}\mathbf{j}$;

$\mathbf{B}(\ln 2) = \mathbf{k}$; $\kappa = \dfrac{8}{17\sqrt{17}}$; $\tau = 0$

27. $\mathbf{a}(0) = 10\,\mathbf{T} + 6\mathbf{N}$

29. $\mathbf{T} = \left(\dfrac{1}{\sqrt{2}}\cos t\right)\mathbf{i} - (\sin t)\mathbf{j} + \left(\dfrac{1}{\sqrt{2}}\cos t\right)\mathbf{k}$;

$\mathbf{N} = \left(-\dfrac{1}{\sqrt{2}}\sin t\right)\mathbf{i} - (\cos t)\mathbf{j} - \left(\dfrac{1}{\sqrt{2}}\sin t\right)\mathbf{k}$;

$\mathbf{B} = \dfrac{1}{\sqrt{2}}\mathbf{i} - \dfrac{1}{\sqrt{2}}\mathbf{k}$; $\kappa = \dfrac{1}{\sqrt{2}}$; $\tau = 0$

31. $\pi/3$ **33.** $x = 1 + t$, $y = t$, $z = -t$

35. 5971 km, 1.639×10^7 km², 3.21% visible

Chapter 11 Additional Exercises, pp. 905–907

1. a) $\mathbf{r}(t) = \left(-\dfrac{8}{15}t^3 + 4t^2\right)\mathbf{i} + (-20t + 100)\mathbf{j}$; b) $\dfrac{100}{3}$ m

3. $\mathbf{v} = \dfrac{\sqrt{3}}{2}\mathbf{i} + \dfrac{\sqrt{3}}{2}\mathbf{j} - \sqrt{3}\,\mathbf{k}$ **5.** a) $\dfrac{d\theta}{dt}\Big|_{\theta=2\pi} = 2\sqrt{\dfrac{\pi gb}{a^2+b^2}}$

b) $\theta = \dfrac{gbt^2}{2(a^2+b^2)}$, $z = \dfrac{gb^2t^2}{2(a^2+b^2)}$

c) $\mathbf{v}(t) = \dfrac{gbt}{\sqrt{a^2+b^2}}\,\mathbf{T}$; $\dfrac{d^2\mathbf{r}}{dt^2} = \dfrac{bg}{\sqrt{a^2+b^2}}\mathbf{T} + a\left(\dfrac{bgt}{a^2+b^2}\right)^2\mathbf{N}$

There is no component in the direction of $\mathbf{B}$.

9. a) $\dfrac{dx}{dt} = \dot{r}\cos\theta - r\dot{\theta}\sin\theta$, $\dfrac{dy}{dt} = \dot{r}\sin\theta + r\dot{\theta}\cos\theta$

b) $\dfrac{dr}{dt} = \dot{x}\cos\theta + \dot{y}\sin\theta$, $r\dfrac{d\theta}{dt} = -\dot{x}\sin\theta + \dot{y}\cos\theta$

11. a) $\mathbf{a}(1) = -9\mathbf{u}_r - 6\mathbf{u}_\theta$, $\mathbf{v}(1) = -\mathbf{u}_r + 3\mathbf{u}_\theta$ b) 6.5 in.

13. c) $\mathbf{v} = \dot{r}\,\mathbf{u}_r + r\dot{\theta}\,\mathbf{u}_\theta + \dot{z}\,\mathbf{k}$, $\mathbf{a} = (\ddot{r} - r\dot{\theta}^2)\,\mathbf{u}_r + (r\ddot{\theta} + 2\dot{r}\dot{\theta})\,\mathbf{u}_\theta + \ddot{z}\,\mathbf{k}$

15. a) $\mathbf{u}_\rho = \sin\phi\cos\theta\,\mathbf{i} + \sin\phi\sin\theta\,\mathbf{j} + \cos\phi\,\mathbf{k}$,

$\mathbf{u}_\phi = \cos\phi\cos\theta\,\mathbf{i} + \cos\phi\sin\theta\,\mathbf{j} - \sin\phi\,\mathbf{k}$,

$\mathbf{u}_\theta = -\sin\theta\,\mathbf{i} + \cos\theta\,\mathbf{j}$

CHAPTER 12

Section 12.1, pp. 914–917

1. a) All points in the xy-plane   b) All reals   c) The lines $y - x = c$ d) No boundary points   e) Both open and closed   f) Unbounded

3. a) All points in the xy-plane   b) $z \geq 0$ c) For $f(x, y) = 0$, the origin; for $f(x, y) \neq 0$, ellipses with the center $(0, 0)$, and major and minor axes, along the x- and y-axes, respectively   d) No boundary points   e) Both open and closed   f) Unbounded

5. a) All points in the xy-plane   b) All reals   c) For $f(x, y) = 0$, the x- and y-axes; for $f(x, y) \neq 0$, hyperbolas with the x- and y-axes as asymptotes   d) No boundary points   e) Both open and closed   f) Unbounded

7. a) All (x, y) satisfying $x^2 + y^2 < 16$ b) $z \geq 1/4$ c) Circles centered at the origin with radii $r < 4$ d) Boundary is the circle $x^2 + y^2 = 16$ e) Open   f) Bounded

9. a) $(x, y) \neq (0, 0)$ b) All reals   c) The circles with center $(0, 0)$ and radii $r > 0$ d) Boundary is the single point $(0, 0)$ e) Open   f) Unbounded

11. a) All (x, y) satisfying $-1 \leq y - x \leq 1$ b) $-\pi/2 \leq z \leq \pi/2$ c) Straight lines of the form $y - x = c$ where $-1 \leq c \leq 1$ d) Boundary is two straight lines $y = 1 + x$ and $y = -1 + x$ e) Closed   f) Unbounded

13. f   **15.** a   **17.** d

19. a)

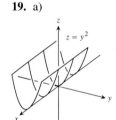

b)

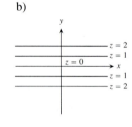

21. a)

b)

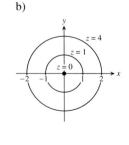

23. a)

$z = -(x^2 + y^2)$

b)

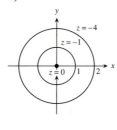

$z = -4$
$z = -1$
$z = 0$

25. a)

$z = 4x^2 + y^2$

b)

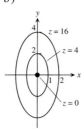

$z = 16$
$z = 4$
$z = 0$

27. a)

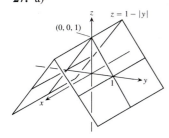

$z = 1 - |y|$
$(0, 0, 1)$

b)

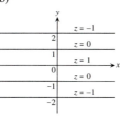

$z = -1$
$z = 0$
$z = 1$
$z = 0$
$z = -1$

29.

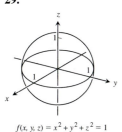

$f(x, y, z) = x^2 + y^2 + z^2 = 1$

31.

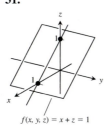

$f(x, y, z) = x + z = 1$

33.

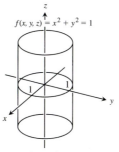

$f(x, y, z) = x^2 + y^2 = 1$

35.

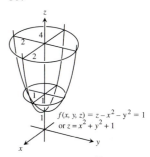

$f(x, y, z) = z - x^2 - y^2 = 1$
or $z = x^2 + y^2 + 1$

37. $x^2 + y^2 = 10$ **39.** $\tan^{-1} y - \tan^{-1} x = 2 \tan^{-1} \sqrt{2}$

41. $\sqrt{x - y} - \ln z = 2$ **43.** $\dfrac{x + y}{z} = \ln 2$ **45.** Yes, 2000

47. 63 km

Section 12.2, pp. 921–923

1. 5/2   **3.** $2\sqrt{6}$ **5.** 1   **7.** 1/2   **9.** 1   **11.** 0   **13.** 0
15. −1   **17.** 2   **19.** 1/4   **21.** 19/12   **23.** 2   **25.** 3
27. a) All (x, y) b) All (x, y) except $(0, 0)$
29. a) All (x, y) except where $x = 0$ or $y = 0$ b) All (x, y)
31. a) All (x, y, z) b) All (x, y, z) except the interior of the cylinder $x^2 + y^2 = 1$
33. a) All (x, y, z) with $z \neq 0$ b) All (x, y, z) with $x^2 + z^2 \neq 1$
35. Consider paths along $y = x, x > 0$, and along $y = x, x < 0$
37. Consider the paths $y = kx^2$, k a constant
39. Consider the paths $y = mx$, m a constant, $m \neq -1$
41. Consider the paths $y = kx^2$, k a constant, $k \neq 0$ **43.** No
45. The limit is 1   **47.** The limit is 0
49. a) $f(x, y)|_{y=mx} = \sin 2\theta$ where $\tan \theta = m$ **51.** 0
53. Does not exist   **55.** $\pi/2$ **57.** $f(0, 0) = \ln 3$ **61.** $\delta = 0.1$
63. $\delta = 0.005$ **65.** $\delta = \sqrt{0.015}$ **67.** $\delta = 0.005$

Section 12.3, pp. 931–933

1. $\dfrac{\partial f}{\partial x} = 4x,\quad \dfrac{\partial f}{\partial y} = -3$ **3.** $\dfrac{\partial f}{\partial x} = 2x(y + 2),\quad \dfrac{\partial f}{\partial y} = x^2 - 1$

5. $\dfrac{\partial f}{\partial x} = 2y(xy - 1),\quad \dfrac{\partial f}{\partial y} = 2x(xy - 1)$

7. $\dfrac{\partial f}{\partial x} = \dfrac{x}{\sqrt{x^2 + y^2}},\quad \dfrac{\partial f}{\partial y} = \dfrac{y}{\sqrt{x^2 + y^2}}$

9. $\dfrac{\partial f}{\partial x} = \dfrac{-1}{(x + y)^2},\quad \dfrac{\partial f}{\partial y} = \dfrac{-1}{(x + y)^2}$

11. $\dfrac{\partial f}{\partial x} = \dfrac{-y^2 - 1}{(xy - 1)^2},\quad \dfrac{\partial f}{\partial y} = \dfrac{-x^2 - 1}{(xy - 1)^2}$

13. $\dfrac{\partial f}{\partial x} = e^{x+y+1},\quad \dfrac{\partial f}{\partial y} = e^{x+y+1}$ **15.** $\dfrac{\partial f}{\partial x} = \dfrac{1}{x + y},\quad \dfrac{\partial f}{\partial y} = \dfrac{1}{x + y}$

17. $\dfrac{\partial f}{\partial x} = 2 \sin (x - 3y) \cos (x - 3y),$

$\dfrac{\partial f}{\partial y} = -6 \sin (x - 3y) \cos (x - 3y)$

19. $\dfrac{\partial f}{\partial x} = yx^{y-1}, \dfrac{\partial f}{\partial y} = x^y \ln x$　**21.** $\dfrac{\partial f}{\partial x} = -g(x), \dfrac{\partial f}{\partial y} = g(y)$

23. $f_x = y^2, \ f_y = 2xy, \ f_z = -4z$

25. $f_x = 1, \ f_y = -y(y^2 + z^2)^{-1/2}, \ f_z = -z(y^2 + z^2)^{-1/2}$

27. $f_x = \dfrac{yz}{\sqrt{1 - x^2 y^2 z^2}}, \ f_y = \dfrac{xz}{\sqrt{1 - x^2 y^2 z^2}}, \ f_z = \dfrac{xy}{\sqrt{1 - x^2 y^2 z^2}}$

29. $f_x = \dfrac{1}{x + 2y + 3z}, \ f_y = \dfrac{2}{x + 2y + 3z}, \ f_z = \dfrac{3}{x + 2y + 3z}$

31. $f_x = -2xe^{-(x^2+y^2+z^2)}, \ f_y = -2ye^{-(x^2+y^2+z^2)},$
$f_z = -2ze^{-(x^2+y^2+z^2)}$

33. $f_x = \operatorname{sech}^2(x + 2y + 3z), \ f_y = 2\operatorname{sech}^2(x + 2y + 3z),$
$f_z = 3\operatorname{sech}^2(x + 2y + 3z)$

35. $\dfrac{\partial f}{\partial t} = -2\pi \sin(2\pi t - \alpha), \ \dfrac{\partial f}{\partial \alpha} = \sin(2\pi t - \alpha)$

37. $\dfrac{\partial h}{\partial \rho} = \sin \phi \cos \theta, \ \dfrac{\partial h}{\partial \phi} = \rho \cos \phi \cos \theta, \ \dfrac{\partial h}{\partial \theta} = -\rho \sin \phi \sin \theta$

39. $W_P(P, V, \delta, v, g) = V, \ W_V(P, V, \delta, v, g) = P + \dfrac{\delta v^2}{2g},$
$W_P(P, V, \delta, v, g) = \dfrac{V v^2}{2g}, \ W_v(P, V, \delta, v, g) = \dfrac{V \delta v}{g},$
$W_g(P, V, \delta, v, g) = -\dfrac{V \delta v^2}{2g^2}$

41. $\dfrac{\partial f}{\partial x} = 1 + y, \ \dfrac{\partial f}{\partial y} = 1 + x, \ \dfrac{\partial^2 f}{\partial x^2} = 0, \ \dfrac{\partial^2 f}{\partial y^2} = 0,$
$\dfrac{\partial^2 f}{\partial y\, \partial x} = \dfrac{\partial^2 f}{\partial x\, \partial y} = 1$

43. $\dfrac{\partial g}{\partial x} = 2xy + y \cos x, \ \dfrac{\partial g}{\partial y} = x^2 - \sin y + \sin x,$
$\dfrac{\partial^2 g}{\partial x^2} = 2y - y \sin x, \ \dfrac{\partial^2 g}{\partial y^2} = -\cos y, \ \dfrac{\partial^2 g}{\partial y\, \partial x} = \dfrac{\partial^2 g}{\partial x\, \partial y} = 2x + \cos x$

45. $\dfrac{\partial r}{\partial x} = \dfrac{1}{x + y}, \ \dfrac{\partial r}{\partial y} = \dfrac{1}{x + y}, \ \dfrac{\partial^2 r}{\partial x^2} = \dfrac{-1}{(x + y)^2},$
$\dfrac{\partial^2 r}{\partial y^2} = \dfrac{-1}{(x + y)^2}, \ \dfrac{\partial^2 r}{\partial y\, \partial x} = \dfrac{\partial^2 r}{\partial x\, \partial y} = \dfrac{-1}{(x + y)^2}$

47. $\dfrac{\partial w}{\partial x} = \dfrac{2}{2x + 3y}, \ \dfrac{\partial w}{\partial y} = \dfrac{3}{2x + 3y}, \ \dfrac{\partial^2 w}{\partial y\, \partial x} = \dfrac{-6}{(2x + 3y)^2},$
$\dfrac{\partial^2 w}{\partial x\, \partial y} = \dfrac{-6}{(2x + 3y)^2}$

49. $\dfrac{\partial w}{\partial x} = y^2 + 2xy^3 + 3x^2y^4, \ \dfrac{\partial w}{\partial y} = 2xy + 3x^2y^2 + 4x^3y^3,$
$\dfrac{\partial^2 w}{\partial y\, \partial x} = 2y + 6xy^2 + 12x^2y^3, \ \dfrac{\partial^2 w}{\partial x\, \partial y} = 2y + 6xy^2 + 12x^2y^3$

51. a) x first **b)** y first **c)** x first **d)** x first **e)** y first
f) y first **53.** $f_x(1, 2) = -13, f_y(1, 2) = -2$ **55.** 12

57. -2 **59.** $\dfrac{\partial A}{\partial a} = \dfrac{a}{bc \sin A}, \ \dfrac{\partial A}{\partial b} = \dfrac{c \cos A - b}{bc \sin A}$

61. $v_x = \dfrac{\ln v}{(\ln u)(\ln v) - 1}$

Section 12.4, pp. 942–944

1. a) $L(x, y) = 1$ **b)** $L(x, y) = 2x + 2y - 1$
3. a) $L(x, y) = 3x - 4y + 5$ **b)** $L(x, y) = 3x - 4y + 5$
5. a) $L(x, y) = 1 + x$ **b)** $L(x, y) = -y + (\pi/2)$
7. $L(x, y) = 7 + x - 6y; 0.06$ **9.** $L(x, y) = x + y + 1; 0.08$
11. $L(x, y) = 1 + x; 0.0222$
13. Pay more attention to the smaller of the two dimensions. It will generate the larger partial derivative.
15. Maximum error (estimate) ≤ 0.31 in magnitude
17. Maximum percentage error $= \pm 4.83\%$
19. Let $|x - 1| \le 0.014, \ |y - 1| \le 0.014$ **21.** $\approx 0.1\%$
23. a) $L(x, y, z) = 2x + 2y + 2z - 3$ **b)** $L(x, y, z) = y + z$
c) $L(x, y, z) = 0$ **25. a)** $L(x, y, z) = x$
b) $L(x, y, z) = \dfrac{1}{\sqrt{2}} x + \dfrac{1}{\sqrt{2}} y$ **c)** $L(x, y, z) = \dfrac{1}{3}x + \dfrac{2}{3}y + \dfrac{2}{3}z$
27. a) $L(x, y, z) = 2 + x$ **b)** $L(x, y, z) = x - y - z + \dfrac{\pi}{2} + 1$
c) $L(x, y, z) = x - y - z + \dfrac{\pi}{2} + 1$
29. $L(x, y, z) = 2x - 6y - 2z + 6, \ 0.0024$
31. $L(x, y, z) = x + y - z - 1, 0.00135$
33. a) $S_0 \left(\dfrac{1}{100} dp + dx - 5dw - 30\, dh \right)$
b) More sensitive to a change in height
35. f is most sensitive to a change in d. **37.** $(47/24)$ ft^3
39. Magnitude of possible error ≤ 4.8 **41.** Yes

Section 12.5, pp. 950–952

1. $\dfrac{dw}{dt} = 0, \dfrac{dw}{dt}(\pi) = 0$ **3.** $\dfrac{dw}{dt} = 1, \dfrac{dw}{dt}(3) = 1$
5. $\dfrac{dw}{dt} = 4t \tan^{-1} t + 1, \dfrac{dw}{dt}(1) = \pi + 1$
7. a) $\dfrac{\partial z}{\partial r} = 4 \cos \theta \ln(r \sin \theta) + 4 \cos \theta,$
$\dfrac{\partial z}{\partial \theta} = -4r \sin \theta \ln(r \sin \theta) + \dfrac{4r \cos^2 \theta}{\sin \theta}$
b) $\dfrac{\partial z}{\partial r} = \sqrt{2}(\ln 2 + 2), \ \dfrac{\partial z}{\partial \theta} = -2\sqrt{2}\ln 2 + 4\sqrt{2}$
9. a) $\dfrac{\partial w}{\partial u} = 2u + 4uv, \ \dfrac{\partial w}{\partial v} = -2v + 2u^2$ **b)** $\dfrac{\partial w}{\partial u} = 3, \ \dfrac{\partial w}{\partial v} = -\dfrac{3}{2}$
11. a) $\dfrac{\partial u}{\partial x} = 0, \ \dfrac{\partial u}{\partial y} = \dfrac{z}{(z - y)^2}, \ \dfrac{\partial u}{\partial z} = \dfrac{-y}{(z - y)^2}$
b) $\dfrac{\partial u}{\partial x} = 0, \ \dfrac{\partial u}{\partial y} = 1, \ \dfrac{\partial u}{\partial z} = -2$
13. $\dfrac{dz}{dt} = \dfrac{\partial z}{\partial x}\dfrac{dx}{dt} + \dfrac{\partial z}{\partial y}\dfrac{dy}{dt}$

15. $\dfrac{\partial w}{\partial u} = \dfrac{\partial w}{\partial x}\dfrac{\partial x}{\partial u} + \dfrac{\partial w}{\partial y}\dfrac{\partial y}{\partial u} + \dfrac{\partial w}{\partial z}\dfrac{\partial z}{\partial u}$,

$\dfrac{\partial w}{\partial v} = \dfrac{\partial w}{\partial x}\dfrac{\partial x}{\partial v} + \dfrac{\partial w}{\partial y}\dfrac{\partial y}{\partial v} + \dfrac{\partial w}{\partial z}\dfrac{\partial z}{\partial v}$

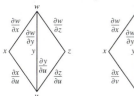

17. $\dfrac{\partial w}{\partial u} = \dfrac{\partial w}{\partial x}\dfrac{\partial x}{\partial u} + \dfrac{\partial w}{\partial y}\dfrac{\partial y}{\partial u}$, $\dfrac{\partial w}{\partial v} = \dfrac{\partial w}{\partial x}\dfrac{\partial x}{\partial v} + \dfrac{\partial w}{\partial y}\dfrac{\partial y}{\partial v}$

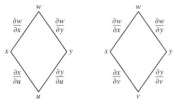

19. $\dfrac{\partial z}{\partial t} = \dfrac{\partial z}{\partial x}\dfrac{\partial x}{\partial t} + \dfrac{\partial z}{\partial y}\dfrac{\partial y}{\partial t}$, $\dfrac{\partial z}{\partial s} = \dfrac{\partial z}{\partial x}\dfrac{\partial x}{\partial s} + \dfrac{\partial z}{\partial y}\dfrac{\partial y}{\partial s}$

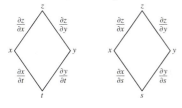

21. $\dfrac{\partial w}{\partial s} = \dfrac{dw}{du}\dfrac{\partial u}{\partial s}$, $\dfrac{\partial w}{\partial t} = \dfrac{dw}{du}\dfrac{\partial u}{\partial t}$

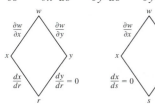

23. $\dfrac{\partial w}{\partial r} = \dfrac{\partial w}{\partial x}\dfrac{dx}{dr} + \dfrac{\partial w}{\partial y}\dfrac{dy}{dr} = \dfrac{\partial w}{\partial x}\dfrac{dx}{dr}$ since $\dfrac{dy}{dr} = 0$,

$\dfrac{\partial w}{\partial s} = \dfrac{\partial w}{\partial x}\dfrac{dx}{ds} + \dfrac{\partial w}{\partial y}\dfrac{dy}{ds} = \dfrac{\partial w}{\partial y}\dfrac{dy}{ds}$ since $\dfrac{dx}{ds} = 0$

25. $4/3$ **27.** $-4/5$ **29.** $\dfrac{\partial z}{\partial x} = \dfrac{1}{4}$, $\dfrac{\partial z}{\partial y} = -\dfrac{3}{4}$

31. $\dfrac{\partial z}{\partial x} = -1$, $\dfrac{\partial z}{\partial y} = -1$ **33.** 12 **35.** -7

37. $\dfrac{\partial z}{\partial u} = 2$, $\dfrac{\partial z}{\partial v} = 1$ **39.** -0.00005 amps/sec

45. $(\cos 1, \sin 1, 1)$ and $(\cos(-2), \sin(-2), -2)$

47. a) Maximum at $\left(-\dfrac{\sqrt{2}}{2}, \dfrac{\sqrt{2}}{2}\right)$ and $\left(\dfrac{\sqrt{2}}{2}, -\dfrac{\sqrt{2}}{2}\right)$, minimum at

$\left(\dfrac{\sqrt{2}}{2}, \dfrac{\sqrt{2}}{2}\right)$ and $\left(-\dfrac{\sqrt{2}}{2}, -\dfrac{\sqrt{2}}{2}\right)$

b) max $= 6$, min $= 2$ **49.** $2x\sqrt{x^8 + x^3} + \displaystyle\int_0^{x^2} \dfrac{3x^2}{2\sqrt{t^4 + x^3}}\, dt$

Section 12.6, pp. 956–957

1. a) 0 b) $1 + 2z$ c) $1 + 2z$ **3.** a) $\dfrac{\partial U}{\partial P} + \dfrac{\partial U}{\partial T}\left(\dfrac{V}{nR}\right)$

b) $\dfrac{\partial U}{\partial P}\left(\dfrac{nR}{V}\right) + \dfrac{\partial U}{\partial T}$ **5.** a) 5 b) 5 **7.** $\dfrac{x}{\sqrt{x^2 + y^2}}$

Section 12.7, pp. 967–969

1.

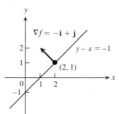

3.

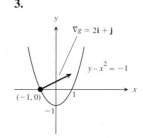

5. $\nabla f = 3\mathbf{i} + 2\mathbf{j} - 4\mathbf{k}$ **7.** $\nabla f = -\dfrac{26}{27}\mathbf{i} + \dfrac{23}{54}\mathbf{j} - \dfrac{23}{54}\mathbf{k}$ **9.** -4

11. $31/13$ **13.** 3 **15.** 2

17. $\mathbf{u} = -\dfrac{1}{\sqrt{2}}\mathbf{i} + \dfrac{1}{\sqrt{2}}\mathbf{j}$, $(D_{\mathbf{u}}f)_{P_0} = \sqrt{2}$; $-\mathbf{u} = \dfrac{1}{\sqrt{2}}\mathbf{i} - \dfrac{1}{\sqrt{2}}\mathbf{j}$,

$(D_{-\mathbf{u}}f)_{P_0} = -\sqrt{2}$

19. $\mathbf{u} = \dfrac{1}{3\sqrt{3}}\mathbf{i} - \dfrac{5}{3\sqrt{3}}\mathbf{j} - \dfrac{1}{3\sqrt{3}}\mathbf{k}$, $(D_{\mathbf{u}}f)_{P_0} = 3\sqrt{3}$;

$-\mathbf{u} = -\dfrac{1}{3\sqrt{3}}\mathbf{i} + \dfrac{5}{3\sqrt{3}}\mathbf{j} + \dfrac{1}{3\sqrt{3}}\mathbf{k}$, $(D_{-\mathbf{u}}f)_{P_0} = -3\sqrt{3}$

21. $\mathbf{u} = \dfrac{1}{\sqrt{3}}(\mathbf{i} + \mathbf{j} + \mathbf{k})$, $(D_{\mathbf{u}}f)_{P_0} = 2\sqrt{3}$; $-\mathbf{u} = -\dfrac{1}{\sqrt{3}}(\mathbf{i} + \mathbf{j} + \mathbf{k})$,

$(D_{-\mathbf{u}}f)_{P_0} = -2\sqrt{3}$

23. $df = \dfrac{9}{910} \approx 0.01$ **25.** $dg = 0$

27. Tangent: $x + y + z = 3$, normal line: $x = 1 + 2t$, $y = 1 + 2t$, $z = 1 + 2t$

29. Tangent: $2x - z - 2 = 0$, normal line: $x = 2 - 4t$, $y = 0$, $z = 2 + 2t$

31. Tangent: $2x + 2y + z - 4 = 0$, normal line: $x = 2t$, $y = 1 + 2t$, $z = 2 + t$

33. Tangent: $x + y + z - 1 = 0$, normal line: $x = t$, $y = 1 + t$, $z = t$

35. $2x - z - 2 = 0$ **37.** $x - y + 2z - 1 = 0$

39.

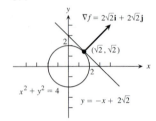

41.

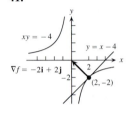

43. $x = 1$, $y = 1 + 2t$, $z = 1 - 2t$

45. $x = 1 - 2t$, $y = 1$, $z = \dfrac{1}{2} + 2t$

47. $x = 1 + 90t$, $y = 1 - 90t$, $z = 3$

49. $\mathbf{u} = \dfrac{7}{\sqrt{53}}\,\mathbf{i} - \dfrac{2}{\sqrt{53}}\,\mathbf{j}$, $-\mathbf{u} = -\dfrac{7}{\sqrt{53}}\,\mathbf{i} + \dfrac{2}{\sqrt{53}}\,\mathbf{j}$

51. No, the maximum rate of change is $\sqrt{185} < 14$.   **53.** $-\dfrac{7}{\sqrt{5}}$

55. a) $\dfrac{\sqrt{3}}{2}\sin\sqrt{3} - \dfrac{1}{2}\cos\sqrt{3} \approx 0.935°C/ft$

b) $\sqrt{3}\sin\sqrt{3} - \cos\sqrt{3} \approx 1.87°C/sec$

57. At $-\dfrac{\pi}{4}$, $-\dfrac{\pi}{2\sqrt{2}}$; at 0, 0; at $\dfrac{\pi}{4}$, $\dfrac{\pi}{2\sqrt{2}}$

Section 12.8, pp. 975–979

1. $f(-3, 3) = -5$, local minimum

3. $f\left(\dfrac{2}{3}, \dfrac{4}{3}\right) = 0$, local maximum   **5.** $f(-2, 1)$, saddle point

7. $f\left(\dfrac{6}{5}, \dfrac{69}{25}\right)$, saddle point   **9.** $f(2, 1)$, saddle point

11. $f(2, -1) = -6$, local minimum   **13.** $f(1, 2)$, saddle point

15. $f(0, 0)$, saddle point

17. $f(0, 0)$, saddle point; $f\left(-\dfrac{2}{3}, \dfrac{2}{3}\right) = \dfrac{170}{27}$, local maximum

19. $f(0, 0) = 0$, local minimum; $f(1, -1)$, saddle point

21. $f(0, 0)$, saddle point; $f\left(\dfrac{4}{9}, \dfrac{4}{3}\right) = -\dfrac{64}{81}$, local minimum

23. $f(0, 0)$, saddle point; $f(0, 2) = -12$, local minimum; $f(-2, 0) = -4$, local maximum; $f(-2, 2)$, saddle point

25. $f(0, 0)$, saddle point; $f(1, 1) = 2$, $f(-1, -1) = 2$, local maxima

27. $f(0, 0) = -1$, local maximum

29. $f(n\pi, 0)$, saddle point; $f(n\pi, 0) = 0$ for every n

31. Absolute maximum: 1 at $(0, 0)$; absolute minimum: -5 at $(1, 2)$

33. Absolute maximum: 4 at $(0, 2)$; absolute minimum: 0 at $(0, 0)$

35. Absolute maximum: 11 at $(0, -3)$; absolute minimum: -10 at $(4, -2)$

37. Absolute maximum: 4 at $(2, 0)$; absolute minimum: $\dfrac{3\sqrt{2}}{2}$ at $\left(3, -\dfrac{\pi}{4}\right)$, $\left(3, \dfrac{\pi}{4}\right)$, $\left(1, -\dfrac{\pi}{4}\right)$, and $\left(1, \dfrac{\pi}{4}\right)$ **39.** $a = -3$, $b = 2$

41. Hottest: $2\dfrac{1}{4}°$ at $\left(-\dfrac{1}{2}, \dfrac{\sqrt{3}}{2}\right)$ and $\left(-\dfrac{1}{2}, -\dfrac{\sqrt{3}}{2}\right)$; coldest: $-\dfrac{1}{4}°$ at $\left(\dfrac{1}{2}, 0\right)$

43. a) $f(0, 0)$, saddle point   b) $f(1, 2)$, local minimum   c) $f(1, -2)$, local minimum; $f(-1, -2)$, saddle point

49. $\left(\dfrac{1}{6}, \dfrac{1}{3}, \dfrac{355}{36}\right)$

53. a) On the semicircle, max $f = 2\sqrt{2}$ at $t = \pi/4$, min $f = -2$ at $t = \pi$. On the quarter circle, max $f = 2\sqrt{2}$ at $t = \pi/4$, min $f = 2$ at $t = 0, \pi/2$.

b) On the semicircle, max $g = 2$ at $t = \pi/4$, min $g = -2$ at $t = 3\pi/4$. On the quarter circle, max $g = 2$ at $t = \pi/4$, min $g = 0$ at $t = 0, \pi/2$.

c) On the semicircle, max $h = 8$ at $t = 0, \pi$; min $h = 4$ at $t = \pi/2$. On the quarter circle, max $h = 8$ at $t = 0$, min $h = 4$ at $t = \pi/2$.

55. i) min $f = -1/2$ at $t = -1/2$; no max   ii) max $f = 0$ at $t = -1, 0$; min $f = -1/2$ at $t = -1/2$ iii) max $f = 4$ at $t = 1$; min $f = 0$ at $t = 0$

57. $y = -\dfrac{20}{13}x + \dfrac{9}{13}$, $y|_{x=4} = -\dfrac{71}{13}$

59. $y = \dfrac{3}{2}x + \dfrac{1}{6}$, $y|_{x=4} = \dfrac{37}{6}$

61. $y = 0.122x + 3.58$

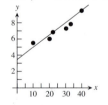

63. a)

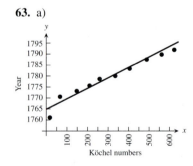

b) $y = 0.0427K + 1764.8$ c) 1780

Section 12.9, pp. 987–989

1. $\left(\pm\dfrac{1}{\sqrt{2}}, \dfrac{1}{2}\right)$, $\left(\pm\dfrac{1}{\sqrt{2}}, -\dfrac{1}{2}\right)$ **3.** 39   **5.** $(3, \pm3\sqrt{2})$ **7.** a) 8

b) 64   **9.** $r = 2$ cm, $h = 4$ cm   **11.** $l = 4\sqrt{2}$, $w = 3\sqrt{2}$

13. $f(0, 0) = 0$ is minimum, $f(2, 4) = 20$ is maximum

15. Minimum $= 0°$, maximum $= 125°$ **17.** $\left(\dfrac{3}{2}, 2, \dfrac{5}{2}\right)$ **19.** 1

21. $(0, 0, 2)$, $(0, 0, -2)$

23. $f(1, -2, 5) = 30$ is maximum, $f(-1, 2, -5) = -30$ is minimum

25. $3, 3, 3$ **27.** $\dfrac{2}{\sqrt{3}}$ by $\dfrac{2}{\sqrt{3}}$ by $\dfrac{2}{\sqrt{3}}$ units   **29.** $\left(\pm\dfrac{4}{3}, -\dfrac{4}{3}, -\dfrac{4}{3}\right)$

31. $U(8, 14) = \$128$ **33.** $f\left(\dfrac{2}{3}, \dfrac{4}{3}, -\dfrac{4}{3}\right) = \dfrac{4}{3}$ **35.** $(2, 4, 4)$

37. Maximum is $1 + 6\sqrt{3}$ at $(\pm\sqrt{6}, \sqrt{3}, 1)$, minimum is $1 - 6\sqrt{3}$ at $(\pm\sqrt{6}, -\sqrt{3}, 1)$

39. Maximum is 4 at $(0, 0, \pm 2)$, minimum is 2 at $(\pm\sqrt{2}, \pm\sqrt{2}, 0)$

Section 12.10, p. 993

1. Quadratic: $x + xy$; cubic: $x + xy + \dfrac{1}{2}xy^2$

3. Quadratic: xy; cubic: xy

5. Quadratic: $y + \dfrac{1}{2}(2xy - y^2)$;

cubic: $y + \dfrac{1}{2}(2xy - y^2) + \dfrac{1}{6}(3x^2y - 3xy^2 + 2y^3)$

7. Quadratic: $\dfrac{1}{2}(2x^2 + 2y^2) = x^2 + y^2$; cubic: $x^2 + y^2$

9. Quadratic: $1 + (x + y) + (x + y)^2$;
cubic: $1 + (x + y) + (x + y)^2 + (x + y)^3$

11. Quadratic: $1 - \dfrac{1}{2}x^2 - \dfrac{1}{2}y^2$, $E(x, y) \le 0.00134$

Chapter 12 Practice Exercises, pp. 994–998

1.

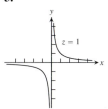

Domain: all points in the xy-plane; range: $z \ge 0$. Level curves are ellipses with major axis along the y-axis and minor axis along the x-axis.

3.

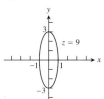

Domain: all (x, y) such that $x \neq 0$ and $y \neq 0$; range: $z \neq 0$. Level curves are hyperbolas with the x- and y-axes as asymptotes.

5.

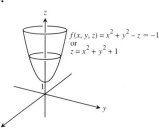

Domain: all (x, y, z) such that $(x, y, z) \neq (0, 0, 0)$; range: all real numbers. Level surfaces are paraboloids of revolution with the z-axis as axis.

7.

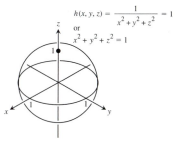

Domain: all (x, y, z) such that $(x, y, z) \neq (0, 0, 0)$; range: positive real numbers. Level surfaces are spheres with center $(0, 0, 0)$ and radius $r > 0$.

9. -2 **11.** $1/2$ **13.** 1 **15.** Let $y = kx^2$, $k \neq 1$

17. a) Does not exist   b) Not continuous at $(0, 0)$

19. $\dfrac{\partial g}{\partial r} = \cos\theta + \sin\theta$, $\dfrac{\partial g}{\partial\theta} = -r\sin\theta + r\cos\theta$

21. $\dfrac{\partial f}{\partial R_1} = -\dfrac{1}{R_1^2}$, $\dfrac{\partial f}{\partial R_2} = -\dfrac{1}{R_2^2}$, $\dfrac{\partial f}{\partial R_3} = -\dfrac{1}{R_3^2}$

23. $\dfrac{\partial P}{\partial n} = \dfrac{RT}{V}$, $\dfrac{\partial P}{\partial R} = \dfrac{nT}{V}$, $\dfrac{\partial P}{\partial T} = \dfrac{nR}{V}$, $\dfrac{\partial P}{\partial V} = -\dfrac{nRT}{V^2}$

25. $\dfrac{\partial^2 g}{\partial x^2} = 0$, $\dfrac{\partial^2 g}{\partial y^2} = \dfrac{2x}{y^3}$, $\dfrac{\partial^2 g}{\partial y\,\partial x} = \dfrac{\partial^2 g}{\partial x\,\partial y} = -\dfrac{1}{y^2}$

27. $\dfrac{\partial^2 f}{\partial x^2} = -30x + \dfrac{2 - 2x^2}{(x^2 + 1)^2}$, $\dfrac{\partial^2 f}{\partial y^2} = 0$, $\dfrac{\partial^2 f}{\partial y\,\partial x} = \dfrac{\partial^2 f}{\partial x\,\partial y} = 1$

29. Answers will depend on the upper bound used for $|f_{xx}|$, $|f_{xy}|$, $|f_{yy}|$. With $M = \sqrt{2}/2$, $|E| \le 0.0142$. With $M = 1$, $|E| \le 0.02$.

31. $L(x, y, z) = y - 3z$, $L(x, y, z) = x + y - z - 1$

33. Be more careful with the diameter.

35. $dl = 0.038$, % change in $V = -4.17\%$, % change in $R = -20\%$, % change in $l = 15.83\%$

37. a) 5% **39.** $\dfrac{dw}{dt}\bigg|_{t=0} = -1$

41. $\dfrac{\partial w}{\partial r}\bigg|_{(r,s)=(\pi,0)} = 2$, $\dfrac{\partial w}{\partial s}\bigg|_{(r,s)=(\pi,0)} = 2 - \pi$

43. $\dfrac{df}{dt}\bigg|_{t=1} = -(\sin 1 + \cos 2)\sin 1 + (\cos 1 + \cos 2)\cos 1 - 2(\sin 1 + \cos 1)\sin 2$

45. $\dfrac{dy}{dx}\bigg|_{(x,y)=(0,1)} = -1$

47. a) $(2y + x^2 z)e^{yz}$ b) $x^2 e^{yz}\left(y - \dfrac{z}{2y}\right)$ c) $(1 + x^2 y)e^{yz}$

49. Increases most rapidly in the direction $\mathbf{u} = -\dfrac{\sqrt{2}}{2}\mathbf{i} - \dfrac{\sqrt{2}}{2}\mathbf{j}$;

decreases most rapidly in the direction $-\mathbf{u} = \dfrac{\sqrt{2}}{2}\mathbf{i} + \dfrac{\sqrt{2}}{2}\mathbf{j}$;

$D_{\mathbf{u}}f = \dfrac{\sqrt{2}}{2}$, $D_{-\mathbf{u}}f = -\dfrac{\sqrt{2}}{2}$, $D_{\mathbf{u}_1}f = -\dfrac{7}{10}$

51. Increases most rapidly in the direction $\mathbf{u} = \dfrac{2}{7}\mathbf{i} + \dfrac{3}{7}\mathbf{j} + \dfrac{6}{7}\mathbf{k}$;

decreases most rapidly in the direction $-\mathbf{u} = -\dfrac{2}{7}\mathbf{i} - \dfrac{3}{7}\mathbf{j} - \dfrac{6}{7}\mathbf{k}$;
$D_{\mathbf{u}}f = 7$, $D_{-\mathbf{u}}f = -7$, $D_{\mathbf{u}_1}f = 7$
53. $\pi/\sqrt{2}$ **55.** a) $f_x(1,2) = f_y(1,2) = 2$ b) $14/5$
57.

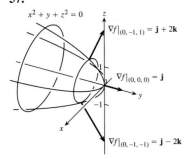

59. Tangent: $4x - y - 5z = 4$, normal line: $x = 2 + 4t$,
$y = -1 - t$, $z = 1 - 5t$
61. $2y - z - 2 = 0$
63. Tangent: $x + y = \pi + 1$, normal line: $y = x - \pi + 1$

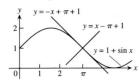

65. $x = 1 - 2t$, $y = 1$, $z = \dfrac{1}{2} + 2t$
67. Local minimum of -8 at $(-2, -2)$
69. Saddle point at $(0, 0)$, $f(0, 0) = 0$; local maximum of $1/4$ at
$\left(-\dfrac{1}{2}, -\dfrac{1}{2}\right)$
71. Saddle point at $(0, 0)$, $f(0, 0) = 0$; local minimum of -4 at
$(0, 2)$; local maximum of 4 at $(-2, 0)$; saddle point at $(-2, 2)$,
$f(-2, 2) = 0$
73. Absolute maximum: 28 at $(0, 4)$, absolute minimum: $-9/4$ at
$(3/2, 0)$
75. Absolute maximum: 18 at $(2, -2)$, absolute minimum: $-17/4$ at
$(-2, 1/2)$
77. Absolute maximum: 8 at $(-2, 0)$, absolute minimum: -1 at $(1, 0)$

79. Absolute maximum: 4 at $(1, 0)$, absolute minimum: -4 at $(0, -1)$
81. Absolute maximum: 1 at $(0, \pm1)$ and $(1, 0)$, absolute minimum:
-1 at $(-1, 0)$
83. Maximum: 5 at $(0, 1)$, minimum: $-1/3$ at $(0, -1/3)$
85. Maximum: $\sqrt{3}$ at $(1/\sqrt{3}, -1/\sqrt{3}, 1/\sqrt{3})$, minimum: $-\sqrt{3}$ at
$(-1/\sqrt{3}, 1/\sqrt{3}, -1/\sqrt{3})$
87. Width $= \left(\dfrac{c^2 V}{ab}\right)^{1/3}$, depth $= \left(\dfrac{b^2 V}{ac}\right)^{1/3}$, height $= \left(\dfrac{a^2 V}{bc}\right)^{1/3}$
89. Maximum $= 3/2$ at $(1/\sqrt{2}, 1/\sqrt{2}, \sqrt{2})$ and
$(-1/\sqrt{2}, -1/\sqrt{2}, -\sqrt{2})$, minimum $= 1/2$ at $(-1/\sqrt{2}, 1/\sqrt{2}, -\sqrt{2})$
and $(1/\sqrt{2}, -1/\sqrt{2}, \sqrt{2})$
91. $\dfrac{\partial w}{\partial x} = \cos\theta\,\dfrac{\partial w}{\partial r} - \dfrac{\sin\theta}{r}\,\dfrac{\partial w}{\partial \theta}$, $\dfrac{\partial w}{\partial y} = \sin\theta\,\dfrac{\partial w}{\partial r} + \dfrac{\cos\theta}{r}\,\dfrac{\partial w}{\partial \theta}$
97. $(t, -t \pm 4, t)$, t a real number

Chapter 12 Additional Exercises, pp. 998–1000

1. $f_{xy}(0, 0) = -1$, $f_{yx}(0, 0) = 1$ **7.** c) $r^2 = \dfrac{1}{2}(x^2 + y^2 + z^2)$

15. $V = \dfrac{\sqrt{3}abc}{2}$ **19.** $f(x, y) = \dfrac{y}{2} + 4$, $g(x, y) = \dfrac{x}{2} + \dfrac{9}{2}$

21. $y = 2\ln|\sin x| + \ln 2$ **23.** a) $\dfrac{1}{\sqrt{53}}(2\mathbf{i} + 7\mathbf{j})$

b) $\dfrac{-1}{\sqrt{29097}}(98\mathbf{i} - 127\mathbf{j} + 58\mathbf{k})$ **25.** $w = e^{-c^2\pi^2 t}\sin\pi x$

27. 0.213%

CHAPTER 13

Section 13.1, pp. 1010–1011

1. 16

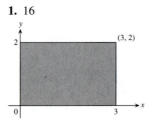

3. 1

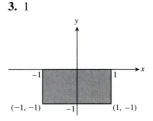

5. $\dfrac{\pi^2}{2} + 2$

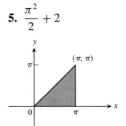

7. $8\ln 8 - 16 + e$

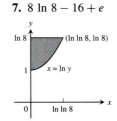

9. $e - 2$

11. $\dfrac{3}{2} \ln 2$ **13.** $1/6$ **15.** $-1/10$

17. 8

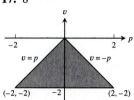

19. 2π

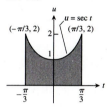

21. $\displaystyle\int_{2}^{4} \int_{0}^{(4-y)/2} dx\,dy$

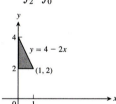

23. $\displaystyle\int_{0}^{1} \int_{x^2}^{x} dy\,dx$

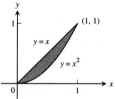

25. $\displaystyle\int_{1}^{e} \int_{\ln y}^{1} dx\,dy$

27. $\displaystyle\int_{0}^{9} \int_{0}^{(\sqrt{9-y})/2} 16x\,dx\,dy$

29. $\displaystyle\int_{-1}^{1} \int_{0}^{\sqrt{1-x^2}} 3y\,dy\,dx$

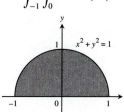

31. 2

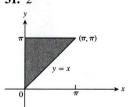

33. $\dfrac{e - 2}{2}$

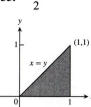

35. 2

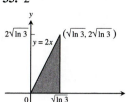

37. $\dfrac{1}{80\pi}$

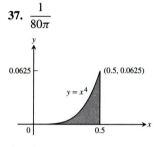

39. $-2/3$ **41.** $4/3$ **43.** $625/12$ **45.** 16 **47.** 20
49. $2(1 + \ln 2)$ **51.** 1 **53.** π^2 **55.** $-1/4$ **57.** $20\sqrt{3}/9$
59. $\displaystyle\int_{0}^{1} \int_{x}^{2-x} (x^2 + y^2)\,dy\,dx = \dfrac{4}{3}$

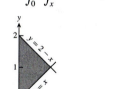

67. 0.603 **69.** 0.233

Section 13.2, pp. 1018–1020

1. $\displaystyle\int_{0}^{2} \int_{0}^{2-x} dy\,dx = 2$ or $\displaystyle\int_{0}^{2} \int_{0}^{2-y} dx\,dy = 2$

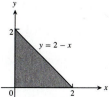

3. $\int_{-2}^{1}\int_{y-2}^{-y^2} dx\,dy = 9/2$

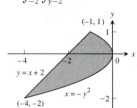

5. $\int_{0}^{\ln 2}\int_{0}^{e^x} dy\,dx = 1$

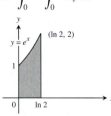

7. $\int_{0}^{1}\int_{y^2}^{2y-y^2} dx\,dy = 1/3$

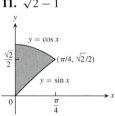

9. 12

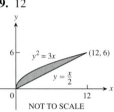

11. $\sqrt{2}-1$

13. 3/2

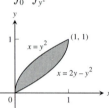

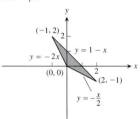

15. a) 0 b) $4/\pi^2$ **17.** 8/3 **19.** $\bar{x}=\dfrac{5}{14}$, $\bar{y}=\dfrac{38}{35}$

21. $\bar{x}=\dfrac{64}{35}$, $\bar{y}=\dfrac{5}{7}$ **23.** $\bar{x}=0$, $\bar{y}=\dfrac{4}{3\pi}$ **25.** $\bar{x}=\bar{y}=\dfrac{4a}{3\pi}$

27. $\bar{x}=\dfrac{\pi}{2}$, $\bar{y}=\dfrac{\pi}{8}$ **29.** $\bar{x}=-1$, $\bar{y}=\dfrac{1}{4}$

31. $I_x=\dfrac{64}{105}$, $R_x=2\sqrt{\dfrac{2}{7}}$ **33.** $\bar{x}=\dfrac{3}{8}$, $\bar{y}=\dfrac{17}{16}$

35. $\bar{x}=\dfrac{11}{3}$, $\bar{y}=\dfrac{14}{27}$, $I_y=432$, $R_y=4$

37. $\bar{x}=0$, $\bar{y}=\dfrac{13}{31}$, $I_y=\dfrac{7}{5}$, $R_y=\sqrt{\dfrac{21}{31}}$

39. $\bar{x}=0$, $\bar{y}=7/10$; $I_x=9/10$, $I_y=3/10$, $I_0=6/5$;

$R_x=\dfrac{3\sqrt{6}}{10}$, $R_y=\dfrac{3\sqrt{2}}{10}$, $R_0=\dfrac{3\sqrt{2}}{5}$

41. $40{,}000(1-e^{-2})\ln\left(\dfrac{7}{2}\right)\approx 43{,}329$

43. If $0 < a \le 5/2$, then the appliance will have to be tipped more than 45° to fall over.

45. $(\bar{x},\bar{y})=(2/\pi,0)$ **47.** a) 3/2 b) They are the same.

53. a) $\left(\dfrac{7}{5},\dfrac{31}{10}\right)$ b) $\left(\dfrac{19}{7},\dfrac{18}{7}\right)$ c) $\left(\dfrac{9}{2},\dfrac{19}{8}\right)$ d) $\left(\dfrac{11}{4},\dfrac{43}{16}\right)$

55. In order for c.m. to be on the common boundary, $h=a\sqrt{2}$. In order for c.m. to be inside T, $h > a\sqrt{2}$.

Section 13.3, pp. 1024–1026

1. $\pi/2$ **3.** $\pi/8$ **5.** πa^2 **7.** 36 **9.** $(1-\ln 2)\pi$

11. $(2\ln 2 - 1)(\pi/2)$ **13.** $\dfrac{\pi}{2}+1$ **15.** $\pi(\ln(4)-1)$

17. $2(\pi - 1)$ **19.** 12π **21.** $\dfrac{3\pi}{8}+1$ **23.** 4 **25.** $6\sqrt{3}-2\pi$

27. $\bar{x}=5/6$, $\bar{y}=0$ **29.** $2a/3$ **31.** $2a/3$ **33.** 2π

35. $\dfrac{4}{3}+\dfrac{5\pi}{8}$ **37.** a) $\sqrt{\pi}/2$ b) 1 **39.** $\pi\ln 4$, no

41. $\dfrac{1}{2}(a^2+2h^2)$

Section 13.4, pp. 1031–1034

1. 1

3. $\displaystyle\int_{0}^{1}\int_{0}^{2-2x}\int_{0}^{3-3x-3y/2} dz\,dy\,dx,\quad \int_{0}^{2}\int_{0}^{1-y/2}\int_{0}^{3-3x-3y/2} dz\,dx\,dy,$

$\displaystyle\int_{0}^{1}\int_{0}^{3-3x}\int_{0}^{2-2x-2z/3} dy\,dz\,dx,\quad \int_{0}^{3}\int_{0}^{1-z/3}\int_{0}^{2-2x-2z/3} dy\,dx\,dz,$

$\displaystyle\int_{0}^{2}\int_{0}^{3-3y/2}\int_{0}^{1-y/2-z/3} dx\,dz\,dy,\quad \int_{0}^{3}\int_{0}^{2-2z/3}\int_{0}^{1-y/2-z/3} dx\,dy\,dz.$

The value of all six integrals is 1.

5. $\displaystyle\int_{-2}^{2}\int_{-\sqrt{4-x^2}}^{\sqrt{4-x^2}}\int_{x^2+y^2}^{8-x^2-y^2} 1\,dz\,dy\,dx,$

$\displaystyle\int_{-2}^{2}\int_{-\sqrt{4-y^2}}^{\sqrt{4-y^2}}\int_{x^2+y^2}^{8-x^2-y^2} 1\,dz\,dx\,dy,$

$\displaystyle\int_{-2}^{2}\int_{4}^{8-y^2}\int_{-\sqrt{8-z-y^2}}^{\sqrt{8-z-y^2}} 1\,dx\,dz\,dy + \int_{-2}^{2}\int_{y^2}^{4}\int_{-\sqrt{z-y^2}}^{\sqrt{z-y^2}} 1\,dx\,dz\,dy,$

$\displaystyle\int_{4}^{8}\int_{-\sqrt{8-z}}^{\sqrt{8-z}}\int_{-\sqrt{8-z-y^2}}^{\sqrt{8-z-y^2}} 1\,dx\,dy\,dz + \int_{0}^{4}\int_{-\sqrt{z}}^{\sqrt{z}}\int_{-\sqrt{z-y^2}}^{\sqrt{z-y^2}} 1\,dx\,dy\,dz,$

$\displaystyle\int_{-2}^{2}\int_{4}^{8-x^2}\int_{-\sqrt{8-z-x^2}}^{\sqrt{8-z-x^2}} 1\,dy\,dz\,dx + \int_{-2}^{2}\int_{x^2}^{4}\int_{-\sqrt{z-x^2}}^{\sqrt{z-x^2}} 1\,dy\,dz\,dx,$

$\displaystyle\int_{4}^{8}\int_{-\sqrt{8-z}}^{\sqrt{8-z}}\int_{-\sqrt{8-z-x^2}}^{\sqrt{8-z-x^2}} 1\,dy\,dx\,dz + \int_{0}^{4}\int_{-\sqrt{z}}^{\sqrt{z}}\int_{-\sqrt{z-x^2}}^{\sqrt{z-x^2}} 1\,dy\,dx\,dz.$

The value of all six integrals is 16π.

7. 1 **9.** 1 **11.** $\dfrac{\pi^3}{2}(1-\cos 1)$ **13.** 18 **15.** 7/6 **17.** 0

19. $\dfrac{1}{2}-\dfrac{\pi}{8}$ **21.** a) $\displaystyle\int_{-1}^{1}\int_{0}^{1-x^2}\int_{x^2}^{1-z} dy\,dz\,dx$

b) $\displaystyle\int_{0}^{1}\int_{-\sqrt{1-z}}^{\sqrt{1-z}}\int_{x^2}^{1-z} dy\,dx\,dz$ c) $\displaystyle\int_{0}^{1}\int_{0}^{1-z}\int_{-\sqrt{y}}^{\sqrt{y}} dy\,dz\,dx$

d) $\int_0^1 \int_0^{1-y} \int_{-\sqrt{y}}^{\sqrt{y}} dx\, dz\, dy$ e) $\int_0^1 \int_{-\sqrt{y}}^{\sqrt{y}} \int_0^{1-y} dz\, dx\, dy$ **23.** 2/3

15. $\int_0^\pi \int_0^{2\sin\theta} \int_0^{4-r\sin\theta} f(r,\theta,z)\, dz\, r\, dr\, d\theta$

25. 20/3 **27.** 1 **29.** 16/3 **31.** $8\pi - \dfrac{32}{3}$ **33.** 2 **35.** 4π

17. $\int_{-\pi/2}^{\pi/2} \int_1^{1+\cos\theta} \int_0^4 f(r,\theta,z)\, dz\, r\, dr\, d\theta$

37. 31/3 **39.** 1 **41.** $2\sin 4$ **43.** 4 **45.** $a = 3$ or $a = \dfrac{13}{3}$

19. $\int_0^{\pi/4} \int_0^{\sec\theta} \int_0^{2-r\sin\theta} f(r,\theta,z)\, dz\, r\, dr\, d\theta$ **21.** π^2 **23.** $\pi/3$

Section 13.5, pp. 1036–1039

1. $R_x = \sqrt{\dfrac{b^2+c^2}{12}}$, $R_y = \sqrt{\dfrac{a^2+c^2}{12}}$, $R_z = \sqrt{\dfrac{a^2+b^2}{12}}$

25. 5π **27.** 2π **29.** $\left(\dfrac{8-5\sqrt{2}}{2}\right)\pi$

3. $I_x = \dfrac{M}{3}(b^2+c^2)$, $I_y = \dfrac{M}{3}(a^2+c^2)$, $I_z = \dfrac{M}{3}(a^2+b^2)$

31. a) $\int_0^{2\pi} \int_0^{\pi/6} \int_0^2 \rho^2 \sin\phi\, d\rho\, d\phi\, d\theta +$

$\int_0^{2\pi} \int_{\pi/6}^{\pi/2} \int_0^{\csc\phi} \rho^2 \sin\phi\, d\rho\, d\phi\, d\theta$

5. $\bar{x} = \bar{y} = 0$, $\bar{z} = \dfrac{12}{5}$, $I_x = \dfrac{7904}{105} \approx 75.28$, $I_y = \dfrac{4832}{63} \approx 76.70$,

$I_z = \dfrac{256}{45} \approx 5.69$

b) $\int_0^{2\pi} \int_1^2 \int_{\pi/6}^{\sin^{-1}(1/\rho)} \rho^2 \sin\phi\, d\phi\, d\rho\, d\theta +$

$\int_0^{2\pi} \int_0^2 \int_0^{\pi/6} \rho^2 \sin\phi\, d\phi\, d\rho\, d\theta$

7. a) $\bar{x} = \bar{y} = 0$, $\bar{z} = \dfrac{8}{3}$ b) $c = 2\sqrt{2}$

9. $I_L = 1386$, $R_L = \sqrt{\dfrac{77}{2}}$ **11.** $I_L = \dfrac{40}{3}$, $R_L = \sqrt{\dfrac{5}{3}}$ **13.** a) $\dfrac{4}{3}$

33. $\int_0^{2\pi} \int_0^{\pi/2} \int_{\cos\phi}^2 \rho^2 \sin\phi\, d\rho\, d\phi\, d\theta = \dfrac{31\pi}{6}$

b) $\bar{x} = \dfrac{4}{5}$, $\bar{y} = \bar{z} = \dfrac{2}{5}$ **15.** a) $\dfrac{5}{2}$ b) $\bar{x} = \bar{y} = \bar{z} = \dfrac{8}{15}$

35. $\int_0^{2\pi} \int_0^\pi \int_0^{1-\cos\phi} \rho^2 \sin\phi\, d\rho\, d\phi\, d\theta = \dfrac{8\pi}{3}$

c) $I_x = I_y = I_z = \dfrac{11}{6}$ d) $R_x = R_y = R_z = \sqrt{\dfrac{11}{15}}$ **17.** 3

37. $\int_0^{2\pi} \int_{\pi/4}^{\pi/2} \int_0^{2\cos\phi} \rho^2 \sin\phi\, d\rho\, d\phi\, d\theta = \dfrac{\pi}{3}$

19. a) $\dfrac{4}{3}g$ b) $\dfrac{4}{3}g$

39. a) $8\int_0^{\pi/2} \int_0^{\pi/2} \int_0^2 \rho^2 \sin\phi\, d\rho\, d\phi\, d\theta$

23. a) $I_{\text{c.m.}} = \dfrac{abc(a^2+b^2)}{12}$, $R_{\text{c.m.}} = \sqrt{\dfrac{a^2+b^2}{12}}$

b) $8\int_0^{\pi/2} \int_0^2 \int_0^{\sqrt{4-r^2}} r\, dz\, dr\, d\theta$

b) $I_L = \dfrac{abc(a^2+7b^2)}{3}$, $R_L = \sqrt{\dfrac{a^2+7b^2}{3}}$

c) $8\int_0^2 \int_0^{\sqrt{4-x^2}} \int_0^{\sqrt{4-x^2-y^2}} dz\, dy\, dx$

27. a) $h = a\sqrt{3}$ b) $h = a\sqrt{2}$

41. a) $\int_0^{2\pi} \int_0^{\pi/3} \int_{\sec\phi}^2 \rho^2 \sin\phi\, d\rho\, d\phi\, d\theta$

Section 13.6, pp. 1044–1047

1. $4\pi(\sqrt{2}-1)/3$ **3.** $17\pi/5$ **5.** $\pi(6\sqrt{2}-8)$ **7.** $3\pi/10$

b) $\int_0^{2\pi} \int_0^{\sqrt{3}} \int_1^{\sqrt{4-r^2}} r\, dz\, dr\, d\theta$

9. $\pi/3$ **11.** a) $\int_0^{2\pi} \int_0^1 \int_0^{\sqrt{4-r^2}} r\, dz\, dr\, d\theta$

c) $\int_{-\sqrt{3}}^{\sqrt{3}} \int_{-\sqrt{3-x^2}}^{\sqrt{3-x^2}} \int_1^{\sqrt{4-x^2-y^2}} dz\, dy\, dx$ d) $\dfrac{5\pi}{3}$ **43.** $8\pi/3$

b) $\int_0^{2\pi} \int_0^{\sqrt{3}} \int_0^1 r\, dr\, dz\, d\theta + \int_0^{2\pi} \int_{\sqrt{3}}^2 \int_0^{\sqrt{4-z^2}} r\, dr\, dz\, d\theta$

45. 9/4 **47.** $(3\pi - 4)/18$ **49.** $\dfrac{2\pi a^3}{3}$ **51.** $5\pi/3$ **53.** $\pi/2$

c) $\int_0^1 \int_0^{\sqrt{4-r^2}} \int_0^{2\pi} r\, d\theta\, dz\, dr$

55. $\dfrac{4(2\sqrt{2}-1)\pi}{3}$ **57.** 16π **59.** $5\pi/2$ **61.** $\dfrac{4\pi(8-3\sqrt{3})}{3}$

13. $\int_{-\pi/2}^{\pi/2} \int_0^{\cos\theta} \int_0^{3r^2} f(r,\theta,z)\, r\, dz\, dr\, d\theta$

63. 2/3 **65.** 3/4 **67.** $\bar{x} = \bar{y} = 0$, $\bar{z} = 3/8$
69. $(\bar{x}, \bar{y}, \bar{z}) = (0, 0, 3/8)$ **71.** $\bar{x} = \bar{y} = 0$, $\bar{z} = 5/6$

73. $I_z = 30\pi$, $R_z = \sqrt{\dfrac{5}{2}}$ **75.** $I_x = \pi/4$ **77.** $\dfrac{a^4 h\pi}{10}$

79. a) $(\bar{x}, \bar{y}, \bar{z}) = \left(0, 0, \dfrac{4}{5}\right)$, $I_z = \dfrac{\pi}{12}$, $R_z = \sqrt{\dfrac{1}{3}}$

b) $(\bar{x}, \bar{y}, \bar{z}) = \left(0, 0, \dfrac{5}{6}\right)$, $I_z = \dfrac{\pi}{14}$, $R_z = \sqrt{\dfrac{5}{14}}$

83. $(\bar{x}, \bar{y}, \bar{z}) = \left(0, 0, \dfrac{2h^2 + 3h}{3h + 6}\right)$, $I_z = \dfrac{\pi a^4 (h^2 + 2h)}{4}$, $R_z = \dfrac{a}{\sqrt{2}}$

85. $\dfrac{3M}{\pi R^3}$

Section 13.7, pp. 1054–1055

1. a) $x = \dfrac{u + v}{3}$, $y = \dfrac{v - 2u}{3}$; $\dfrac{1}{3}$

b) Triangular region with boundaries $u = 0$, $v = 0$, and $u + v = 3$

3. a) $x = \dfrac{1}{5}(2u - v)$, $y = \dfrac{1}{10}(3v - u)$; $\dfrac{1}{10}$

b) Triangular region with boundaries $3v = u$, $v = 2u$, and $3u + v = 10$

5. a) $\begin{vmatrix} \cos v & -u \sin v \\ \sin v & u \cos v \end{vmatrix} = u \cos^2 v + u \sin^2 v = u$

b) $\begin{vmatrix} \sin v & u \cos v \\ \cos v & -u \sin v \end{vmatrix} = -u \sin^2 v - u \cos^2 v = -u$

9. 64/5   **11.** $\displaystyle\int_1^2 \int_1^3 (u + v) \dfrac{2u}{v} \, du \, dv = 8 + \dfrac{52}{3} \ln 2$

13. $\dfrac{\pi ab(a^2 + b^2)}{4}$ **15.** $\dfrac{1}{3}\left(1 + \dfrac{3}{e^2}\right) \approx 0.4687$ **19.** $\dfrac{4\pi abc}{3}$

21. $\displaystyle\int_0^3 \int_0^2 \int_1^2 \left(\dfrac{v}{3} + \dfrac{vw}{3u}\right) du \, dv \, dw = 2 + \ln 8$

Chapter 13 Practice Exercises, pp. 1056–1058

1. $9e - 9$ **3.** $9/2$

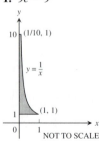

NOT TO SCALE

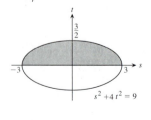

5. $\displaystyle\int_{-2}^0 \int_{2x+4}^{4-x^2} dy \, dx = \dfrac{4}{3}$

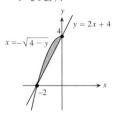

7. $\displaystyle\int_{-3}^3 \int_0^{(1/2)\sqrt{9-x^2}} y \, dy \, dx = \dfrac{9}{2}$

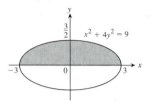

9. $\sin 4$ **11.** $\dfrac{\ln 17}{4}$ **13.** 4/3   **15.** 4/3   **17.** 1/4

19. $\bar{x} = \bar{y} = \dfrac{1}{2 - \ln 4}$ **21.** $I_0 = 104$ **23.** $I_x = 2\delta$, $R_x = \sqrt{\dfrac{2}{3}}$

25. $M = 4$, $M_x = 0$, $M_y = 0$ **27.** π **29.** $\bar{x} = \dfrac{3\sqrt{3}}{\pi}$, $\bar{y} = 0$

31. a) $\bar{x} = \dfrac{15\pi + 32}{6\pi + 48}$, $\bar{y} = 0$

b)

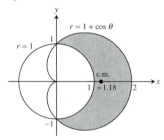

33. $\dfrac{\pi - 2}{4}$ **35.** 0   **37.** 8/35   **39.** $\pi/2$ **41.** $\dfrac{2(31 - 3^{5/2})}{3}$

43. a) $\displaystyle\int_{-\sqrt{2}}^{\sqrt{2}} \int_{-\sqrt{2-y^2}}^{\sqrt{2-y^2}} \int_{\sqrt{x^2+y^2}}^{\sqrt{4-x^2-y^2}} 3 \, dz \, dx \, dy$

b) $\displaystyle\int_0^{2\pi} \int_0^{\pi/4} \int_0^2 3\rho^2 \sin\phi \, d\rho \, d\phi \, d\theta$ c) $2\pi(8 - 4\sqrt{2})$

45. $\displaystyle\int_0^{2\pi} \int_0^{\pi/4} \int_0^{\sec\phi} \rho^2 \sin\phi \, d\rho \, d\phi \, d\theta = \dfrac{\pi}{3}$

47. $\displaystyle\int_0^1 \int_{\sqrt{1-x^2}}^{\sqrt{3-x^2}} \int_1^{\sqrt{4-x^2-y^2}} z^2 xy \, dz \, dy \, dx +$

$\displaystyle\int_1^{\sqrt{3}} \int_0^{\sqrt{3-x^2}} \int_1^{\sqrt{4-x^2-y^2}} z^2 xy \, dz \, dy \, dx$

49. a) $\displaystyle\int_{-\sqrt{3}}^{\sqrt{3}} \int_{-\sqrt{3-x^2}}^{\sqrt{3-x^2}} \int_1^{\sqrt{4-x^2-y^2}} dz \, dy \, dx$

b) $\displaystyle\int_0^{2\pi} \int_0^{\sqrt{3}} \int_1^{\sqrt{4-r^2}} r \, dz \, dr \, d\theta$

c) $\int_0^{2\pi} \int_0^{\pi/3} \int_{\sec \phi}^{2} \rho^2 \sin \phi \, d\rho \, d\phi \, d\theta$

51. a) $\dfrac{8\pi(4\sqrt{2}-5)}{3}$ b) $\dfrac{8\pi(4\sqrt{2}-5)}{3}$ **53.** $I_z = \dfrac{8\pi \delta(b^5 - a^5)}{15}$

Chapter 13 Additional Exercises, pp. 1058–1060

1. a) $\int_{-3}^{2} \int_{x}^{6-x^2} x^2 \, dy \, dx$ b) $\int_{-3}^{2} \int_{x}^{6-x^2} \int_{0}^{x^2} dz \, dy \, dx$ c) $\dfrac{125}{4}$

3. 2π **5.** $3\pi/2$ **7.** a) Hole radius $= 1$, sphere radius $= 2$
b) $4\sqrt{3}\pi$ **9.** $\pi/4$

11. $\int_0^1 \int_y^{\sqrt{y}} f(x, y) \, dx \, dy$

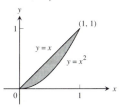

13. $\ln\left(\dfrac{b}{a}\right)$ **17.** $1/\sqrt[4]{3}$ **19.** $\bar{x} = \dfrac{15\pi + 32}{6\pi + 48}$, $\bar{y} = 0$

21. Mass $= a^2 \cos^{-1}\left(\dfrac{b}{a}\right) - b\sqrt{a^2 - b^2}$,

$I_0 = \dfrac{a^4}{2}\cos^{-1}\left(\dfrac{b}{a}\right) - \dfrac{b^3}{2}\sqrt{a^2 - b^2} - \dfrac{b^3}{6}(a^2 - b^2)^{3/2}$

23. a) $\bar{x} = \bar{y} = 0$, $\bar{z} = 1/2$; $I_z = \dfrac{\pi}{8}$, $R_z = \dfrac{\sqrt{3}}{2}$

b) $\bar{x} = \bar{y} = 0$, $\bar{z} = 5/14$; $I_z = \dfrac{2\pi}{7}$, $R_z = \sqrt{\dfrac{5}{7}}$

25. $\bar{x} = \bar{y} = 0$, $\bar{z} = \dfrac{3a}{8}$ **27.** $\dfrac{1}{ab}\left(e^{a^2 b^2} - 1\right)$ **29.** b) 1 c) 0

33. $h = \sqrt{20}$ in., $h = \sqrt{60}$ in.   **37.** $\dfrac{1}{2}\pi^2$

CHAPTER 14

Section 14.1, pp. 1065–1067

1. c   **3.** g   **5.** d   **7.** f   **9.** $\sqrt{2}$ **11.** 13/2   **13.** $3\sqrt{14}$

15. $(1/6)(5\sqrt{5} + 9)$ **17.** $\sqrt{3}\ln(b/a)$ **19.** $\dfrac{10\sqrt{5} - 2}{3}$ **21.** 8

23. $2\sqrt{2} - 1$ **25.** a) $4\sqrt{2} - 2$ b) $\sqrt{2} + \ln(1 + \sqrt{2})$
27. $I_z = 2\pi \delta a^3$, $R_z = a$ **29.** a) $I_z = 2\pi \sqrt{2} \delta$, $R_z = 1$
b) $I_z = 4\pi \sqrt{2} \delta$, $R_z = 1$ **31.** $I_x = 2\pi - 2$, $R_x = 1$

Section 14.2, pp. 1074–1076

1. $\nabla f = -(x\mathbf{i} + y\mathbf{j} + z\mathbf{k})(x^2 + y^2 + z^2)^{-3/2}$
3. $\nabla g = -(2x/(x^2 + y^2))\mathbf{i} - (2y/(x^2 + y^2))\mathbf{j} + e^z \mathbf{k}$
5. $\mathbf{F} = -\dfrac{kx}{(x^2 + y^2)^{3/2}}\mathbf{i} - \dfrac{ky}{(x^2 + y^2)^{3/2}}\mathbf{j}$, any $k > 0$

7. a) $\dfrac{9}{2}$ b) $\dfrac{13}{3}$ c) $\dfrac{9}{2}$ **9.** a) $\dfrac{1}{3}$ b) $-\dfrac{1}{5}$ c) 0   **11.** a) 2

b) $\dfrac{3}{2}$ c) $\dfrac{1}{2}$ **13.** 1/2   **15.** $-\pi$ **17.** 207/12   **19.** $-39/2$

21. 25/6   **23.** a) $\text{Circ}_1 = 0$, $\text{circ}_2 = 2\pi$, $\text{flux}_1 = 2\pi$, $\text{flux}_2 = 0$
b) $\text{Circ}_1 = 0$, $\text{circ}_2 = 8\pi$, $\text{flux}_1 = 8\pi$, $\text{flux}_2 = 0$
25. $\text{Circ} = 0$, $\text{flux} = a^2 \pi$ **27.** $\text{Circ} = a^2 \pi$, $\text{flux} = 0$
29. a) $-\pi/2$ b) 0   c) 1
31.

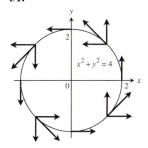

33. a) $\mathbf{G} = -y\mathbf{i} + x\mathbf{j}$ b) $\mathbf{G} = \sqrt{x^2 + y^2}\,\mathbf{F}$
35. $\mathbf{F} = -(x\mathbf{i} + y\mathbf{j})/\sqrt{x^2 + y^2}$ **37.** 48   **39.** π **41.** 0
43. 1/2

Section 14.3, pp. 1083–1084

1. Conservative   **3.** Not conservative   **5.** Not conservative

7. $f(x, y, z) = x^2 + \dfrac{3y^2}{2} + 2z^2 + C$ **9.** $f(x, y, z) = xe^{y+2z} + C$

11. $f(x, y, z) = x\ln x - x + \tan(x + y) + \dfrac{1}{2}\ln(y^2 + z^2) + C$

13. 49   **15.** -16 **17.** 1   **19.** $9\ln 2$ **21.** 0   **23.** -3

27. $\mathbf{F} = \nabla\left(\dfrac{x^2 - 1}{y}\right)$ **29.** a) 1 b) 1 c) 1   **31.** a) 2 b) 2

33. $f(x, y, z) = \dfrac{GmM}{(x^2 + y^2 + z^2)^{1/2}}$ **35.** a) $c = b = 2a$

b) $c = b = 2$
37. It does not matter what path you use. The work will be the same on any path because the field is conservative.

Section 14.4, pp. 1093–1095

1. Flux $= 0$, circ $= 2\pi a^2$ **3.** Flux $= -\pi a^2$, circ $= 0$
5. Flux $= 2$, circ $= 0$ **7.** Flux $= -9$, circ $= 9$
9. Flux $= 1/2$, circ $= 1/2$ **11.** Flux $= 1/5$, circ $= -1/12$

13. 0   **15.** 2/33   **17.** 0   **19.** -16π **21.** πa^2 **23.** $\dfrac{3}{8}\pi$

25. a) 0   b) $(h - k)(\text{area of the region})$ **35.** a) 0

Section 14.5, pp. 1103–1105

1. $\frac{13}{3}\pi$ **3.** 4   **5.** $6\sqrt{6} - 2\sqrt{2}$ **7.** $\pi\sqrt{c^2 + 1}$

9. $\frac{\pi}{6}(17\sqrt{17} - 5\sqrt{5})$ **11.** $3 + 2\ln 2$ **13.** $9a^3$

15. $\frac{abc}{4}(ab + ac + bc)$ **17.** 2   **19.** 18   **21.** $\pi a^3/6$

23. $\pi a^2/4$ **25.** $\pi a^3/2$ **27.** -32 **29.** -4 **31.** $3a^4$

33. $\left(\frac{a}{2}, \frac{a}{2}, \frac{a}{2}\right)$

35. $(\bar{x}, \bar{y}, \bar{z}) = \left(0, 0, \frac{14}{9}\right)$, $I_z = \frac{15\pi\sqrt{2}}{2}\delta$, $R_z = \frac{\sqrt{10}}{2}$

37. a) $\frac{8\pi}{3}a^4\delta$ b) $\frac{20\pi}{3}a^4\delta$ **39.** $\frac{\pi}{6}(13\sqrt{13} - 1)$ **41.** $5\pi\sqrt{2}$

43. $\frac{2}{3}(5\sqrt{5} - 1)$

Section 14.6, pp. 1112–1114

1. $\mathbf{r}(r, \theta) = (r\cos\theta)\mathbf{i} + (r\sin\theta)\mathbf{j} + r^2\mathbf{k}$, $0 \le r \le 2$, $0 \le \theta \le 2\pi$
3. $\mathbf{r}(r, \theta) = (r\cos\theta)\mathbf{i} + (r\sin\theta)\mathbf{j} + (r/2)\mathbf{k}$, $0 \le r \le 6$, $0 \le \theta \le \pi/2$
5. $\mathbf{r}(r, \theta) = (r\cos\theta)\mathbf{i} + (r\sin\theta)\mathbf{j} + \sqrt{9 - r^2}\mathbf{k}$, $0 \le r \le 3\sqrt{2}/2$, $0 \le \theta \le 2\pi$; Also: $\mathbf{r}(\phi, \theta) = (3\sin\phi\cos\theta)\mathbf{i} + (3\sin\phi\sin\theta)\mathbf{j} + (3\cos\phi)\mathbf{k}$, $0 \le \phi \le \pi/4, 0 \le \theta \le 2\pi$
7. $\mathbf{r}(\phi, \theta) = (\sqrt{3}\sin\phi\cos\theta)\mathbf{i} + (\sqrt{3}\sin\phi\sin\theta)\mathbf{j} + (\sqrt{3}\cos\phi)\mathbf{k}$, $\pi/3 \le \phi \le 2\pi/3$, $0 \le \theta \le 2\pi$
9. $\mathbf{r}(x, y) = x\mathbf{i} + y\mathbf{j} + (4 - y^2)\mathbf{k}$, $0 \le x \le 2$, $-2 \le y \le 2$
11. $\mathbf{r}(u, v) = u\mathbf{i} + (3\cos v)\mathbf{j} + (3\sin v)\mathbf{k}, 0 \le u \le 3, 0 \le v \le 2\pi$
13. a) $\mathbf{r}(r, \theta) = (r\cos\theta)\mathbf{i} + (r\sin\theta)\mathbf{j} + (1 - r\cos\theta - r\sin\theta)\mathbf{k}$, $0 \le r \le 3$, $0 \le \theta \le 2\pi$
b) $\mathbf{r}(u, v) = (1 - u\cos v - u\sin v)\mathbf{i} + (u\cos v)\mathbf{j} + (u\sin v)\mathbf{k}$, $0 \le u \le 3, 0 \le v \le 2\pi$
15. $\mathbf{r}(u, v) = (4\cos^2 v)\mathbf{i} + u\mathbf{j} + (4\cos v\sin v)\mathbf{k}$, $0 \le u \le 3$, $-(\pi/2) \le v \le (\pi/2)$; Another way: $\mathbf{r}(u, v) = (2 + 2\cos v)\mathbf{i} + u\mathbf{j} + (2\sin v)\mathbf{k}$, $0 \le u \le 3, 0 \le v \le 2\pi$

17. $\int_0^{2\pi}\int_0^1 \frac{\sqrt{5}}{2}r\,dr\,d\theta = \frac{\pi\sqrt{5}}{2}$

19. $\int_0^{2\pi}\int_1^3 r\sqrt{5}\,dr\,d\theta = 8\pi\sqrt{5}$

21. $\int_0^{2\pi}\int_1^4 1\,du\,dv = 6\pi$

23. $\int_0^{2\pi}\int_0^1 u\sqrt{4u^2 + 1}\,du\,dv = \frac{(5\sqrt{5} - 1)}{6}\pi$

25. $\int_0^{2\pi}\int_{\pi/4}^\pi 2\sin\phi\,d\phi\,d\theta = (4 + 2\sqrt{2})\pi$

27. $\iint_S x\,d\sigma = \int_0^3\int_0^2 u\sqrt{4u^2 + 1}\,du\,dv = \frac{17\sqrt{17} - 1}{4}$

29. $\iint_S x^2\,d\sigma = \int_0^{2\pi}\int_0^\pi \sin^3\phi\cos^2\theta\,d\phi\,d\theta = \frac{4\pi}{3}$

31. $\iint_S z\,d\sigma = \int_0^1\int_0^1 (4 - u - v)\sqrt{3}\,dv\,du = 3\sqrt{3}$
(for $x = u$, $y = v$)

33. $\iint_S x^2\sqrt{5 - 4z}\,d\sigma = \int_0^1\int_0^{2\pi} u^2\cos^2 v \cdot \sqrt{4u^2 + 1} \cdot$
$u\sqrt{4u^2 + 1}\,dv\,du = \int_0^1\int_0^{2\pi} u^3(4u^2 + 1)\cos^2 v\,dv\,du = \frac{11\pi}{12}$
35. -32 **37.** $\pi a^3/6$ **39.** $13a^4/6$ **41.** $2\pi/3$ **43.** $-73\pi/6$
45. $(a/2, a/2, a/2)$ **47.** $8\delta\pi a^4/3$
49.   **51.**

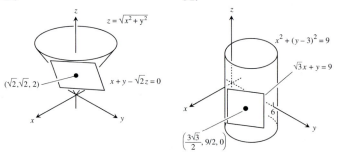

55. b) $A = \int_0^{2\pi}\int_0^\pi [a^2 b^2 \sin^2\phi\cos^2\phi + b^2 c^2\cos^4\phi\cos^2\theta + a^2 c^2\cos^4\phi\sin^2\theta]^{1/2}\,d\phi\,d\theta$

57. $x_0 x + y_0 y = 25$

Section 14.7, pp. 1122–1123

1. 4π **3.** $-5/6$ **5.** 0   **7.** -6π **9.** $2\pi a^2$ **13.** 12π

15. $-\frac{\pi}{4}$ **17.** -15π **25.** $16 I_y + 16 I_x$

Section 14.8, pp. 1132–1134

1. 0   **3.** 0   **5.** -16 **7.** -8π **9.** 3π **11.** $-40/3$
13. 12π **15.** $12\pi(4\sqrt{2} - 1)$
21. The integral's value never exceeds the surface area of S.

Chapter 14 Practice Exercises, pp. 1134–1137

1. Path 1: $2\sqrt{3}$, Path 2: $1 + 3\sqrt{2}$ **3.** $4a^2$ **5.** 0   **7.** $(0, 0, 0)$

9. 0   **11.** $\pi\sqrt{3}$ **13.** $2\pi\left(1 - \frac{1}{\sqrt{2}}\right)$ **15.** $\frac{abc}{2}\sqrt{\frac{1}{a^2} + \frac{1}{b^2} + \frac{1}{c^2}}$

17. 50
19. $\mathbf{r}(\phi, \theta) = (6\sin\phi\cos\theta)\mathbf{i} + (6\sin\phi\sin\theta)\mathbf{j} + (6\cos\phi)\mathbf{k}$, $(\pi/6) \le \phi \le 2\pi/3$, $0 \le \theta \le 2\pi$
21. $\mathbf{r}(r, \theta) = (r\cos\theta)\mathbf{i} + (r\sin\theta)\mathbf{j} + (1 + r)\mathbf{k}$, $0 \le r \le 2$, $0 \le \theta \le 2\pi$

23. $\mathbf{r}(u, v) = (u \cos v)\mathbf{i} + 2u^2\mathbf{j} + (u \sin v)\mathbf{k}$, $0 \le u \le 1$, $0 \le v \le \pi$

25. $\sqrt{6}$ **27.** $\pi[\sqrt{2} + \ln(1 + \sqrt{2})]$ **29.** Conservative

31. Not conservative   **33.** $f(x, y, z) = y^2 + yz + 2x + z$

35. Path 1: 2, Path 2: 8/3   **37.** a) $1 - e^{-2\pi}$ b) $1 - e^{-2\pi}$

39. a) $-\pi/2$ b) 0   c) 1   **41.** 0   **43.** a) $4\sqrt{2} - 2$
b) $\sqrt{2} + \ln(1 + \sqrt{2})$

45. $(\bar{x}, \bar{y}, \bar{z}) = \left(1, \dfrac{8}{15}, \dfrac{2}{3}\right)$; $I_x = \dfrac{232}{45}$, $I_y = \dfrac{64}{15}$, $I_z = \dfrac{56}{9}$;

$R_x = \sqrt{\dfrac{116}{45}}$, $R_y = \sqrt{\dfrac{32}{15}}$, $R_z = \sqrt{\dfrac{28}{9}}$

47. $\bar{z} = \dfrac{3}{2}$, $I_z = \dfrac{7\sqrt{3}}{3}$, $R_z = \sqrt{\dfrac{7}{3}}$

49. $(\bar{x}, \bar{y}, \bar{z}) = (0, 0, 49/12)$, $I_z = 640\pi$, $R_z = 2\sqrt{2}$

51. Flux: 3/2, Circ: $-1/2$ **55.** 3   **57.** $\dfrac{2\pi}{3}(7 - 8\sqrt{2})$ **59.** 0

61. π

Chapter 14 Additional Exercises, pp. 1137–1139

1. 6π **3.** 2/3   **5.** a) $\mathbf{F}(x, y, z) = z\mathbf{i} + x\mathbf{j} + y\mathbf{k}$
b) $\mathbf{F}(x, y, z) = z\mathbf{i} + y\mathbf{k}$ c) $\mathbf{F}(x, y, z) = z\mathbf{i}$ **7.** $\dfrac{16\pi R^3}{3}$

9. $a = 2, b = 1$. The minimum flux is -4.   **11.** b) $\dfrac{16}{3}g$
c) Work $= \left(\displaystyle\int_C g\,xy\,ds\right)$ $\bar{y} = g\displaystyle\int_C xy^2\,ds$ **13.** c) $\dfrac{4}{3}\pi w$

19. False if $\mathbf{F} = y\mathbf{i} + x\mathbf{j}$

APPENDICES

Appendix A.3, pp. A-16–A-17

1. a) $(14, 8)$ b) $(-1, 8)$ c) $(0, -5)$

3. a) By reflecting z across the real axis   b) By reflecting z across the imaginary axis   c) By reflecting z in the origin   d) By reflecting z in the real axis and then multiplying the length of the vector by $1/|z|^2$

5. a) Points on the circle $x^2 + y^2 = 4$ b) points inside the circle $x^2 + y^2 = 4$ c) points outside the circle $x^2 + y^2 = 4$

7. Points on a circle of radius 1, center $(-1, 0)$

9. Points on the line $y = -x$ **11.** $4\,e^{2\pi i/3}$ **13.** $1\,e^{2\pi i/3}$

21. $\cos^4\theta - 6\cos^2\theta\,\sin^2\theta + \sin^4\theta$ **23.** $1, -\dfrac{1}{2} \pm \dfrac{\sqrt{3}}{2}i$

25. $2i, -\sqrt{3} - i, \sqrt{3} - i$ **27.** $\dfrac{\sqrt{6}}{2} \pm \dfrac{\sqrt{2}}{2}i, -\dfrac{\sqrt{6}}{2} \pm \dfrac{\sqrt{2}}{2}i$

29. $1 \pm \sqrt{3}i, -1 \pm \sqrt{3}i$

Appendix A.8, p. A-29

1. -5 **3.** 1   **5.** -7 **7.** 38   **9.** $x = -4, y = 1$

11. $x = 3, y = 2$ **13.** $x = 3, y = -2, z = 2$

15. $x = 2, y = 0, z = -1$ **17.** a) $h = 6, k = 4$
b) $h = 6, k \ne 4$

Index

Note: Numbers in parentheses refer to exercises on the pages indicated.

A Brief Table of Integrals

1. $\int u\, dv = uv - \int v\, du$

2. $\int a^u\, du = \dfrac{a^u}{\ln a} + C, \quad a \neq 1, \quad a > 0$

3. $\int \cos u\, du = \sin u + C$

4. $\int \sin u\, du = -\cos u + C$

5. $\int (ax+b)^n\, dx = \dfrac{(ax+b)^{n+1}}{a(n+1)} + C, \quad n \neq -1$

6. $\int (ax+b)^{-1}\, dx = \dfrac{1}{a}\ln|ax+b| + C$

7. $\int x(ax+b)^n\, dx = \dfrac{(ax+b)^{n+1}}{a^2}\left[\dfrac{ax+b}{n+2} - \dfrac{b}{n+1}\right] + C, \quad n \neq -1, -2$

8. $\int x(ax+b)^{-1}\, dx = \dfrac{x}{a} - \dfrac{b}{a^2}\ln|ax+b| + C$

9. $\int x(ax+b)^{-2}\, dx = \dfrac{1}{a^2}\left[\ln|ax+b| + \dfrac{b}{ax+b}\right] + C$

10. $\int \dfrac{dx}{x(ax+b)} = \dfrac{1}{b}\ln\left|\dfrac{x}{ax+b}\right| + C$

11. $\int (\sqrt{ax+b})^n\, dx = \dfrac{2}{a}\dfrac{(\sqrt{ax+b})^{n+2}}{n+2} + C, \quad n \neq -2$

12. $\int \dfrac{\sqrt{ax+b}}{x}\, dx = 2\sqrt{ax+b} + b\int \dfrac{dx}{x\sqrt{ax+b}}$

13. a) $\int \dfrac{dx}{x\sqrt{ax-b}} = \dfrac{2}{\sqrt{b}}\tan^{-1}\sqrt{\dfrac{ax-b}{b}} + C$

b) $\int \dfrac{dx}{x\sqrt{ax+b}} = \dfrac{1}{\sqrt{b}}\ln\left|\dfrac{\sqrt{ax+b}-\sqrt{b}}{\sqrt{ax+b}+\sqrt{b}}\right| + C$

14. $\int \dfrac{\sqrt{ax+b}}{x^2}\, dx = -\dfrac{\sqrt{ax+b}}{x} + \dfrac{a}{2}\int \dfrac{dx}{x\sqrt{ax+b}} + C$

15. $\int \dfrac{dx}{x^2\sqrt{ax+b}} = -\dfrac{\sqrt{ax+b}}{bx} - \dfrac{a}{2b}\int \dfrac{dx}{x\sqrt{ax+b}} + C$

16. $\int \dfrac{dx}{a^2+x^2} = \dfrac{1}{a}\tan^{-1}\dfrac{x}{a} + C$

17. $\int \dfrac{dx}{(a^2+x^2)^2} = \dfrac{x}{2a^2(a^2+x^2)} + \dfrac{1}{2a^3}\tan^{-1}\dfrac{x}{a} + C$

18. $\int \dfrac{dx}{a^2-x^2} = \dfrac{1}{2a}\ln\left|\dfrac{x+a}{x-a}\right| + C$

19. $\int \dfrac{dx}{(a^2-x^2)^2} = \dfrac{x}{2a^2(a^2-x^2)} + \dfrac{1}{4a^3}\ln\left|\dfrac{x+a}{x-a}\right| + C$

20. $\int \dfrac{dx}{\sqrt{a^2+x^2}} = \sinh^{-1}\dfrac{x}{a} + C = \ln(x+\sqrt{a^2+x^2}) + C$

21. $\int \sqrt{a^2+x^2}\, dx = \dfrac{x}{2}\sqrt{a^2+x^2} +$

22. $\int x^2\sqrt{a^2+x^2}\, dx = \dfrac{x}{8}(a^2+2x^2)\sqrt{a^2+x^2} - \dfrac{a^4}{8}\ln(x+\sqrt{a^2+x^2}) + C$
$\qquad\qquad \dfrac{a^2}{2}\ln(x+\sqrt{a^2+x^2}) + C$

23. $\int \dfrac{\sqrt{a^2+x^2}}{x}\, dx = \sqrt{a^2+x^2} - a\ln\left|\dfrac{a+\sqrt{a^2+x^2}}{x}\right| + C$

24. $\int \dfrac{\sqrt{a^2+x^2}}{x^2}\, dx = \ln(x+\sqrt{a^2+x^2}) - \dfrac{\sqrt{a^2+x^2}}{x} + C$

25. $\int \dfrac{x^2}{\sqrt{a^2+x^2}}\, dx = -\dfrac{a^2}{2}\ln(x+\sqrt{a^2+x^2}) + \dfrac{x\sqrt{a^2+x^2}}{2} + C$

26. $\int \dfrac{dx}{x\sqrt{a^2+x^2}} = -\dfrac{1}{a}\ln\left|\dfrac{a+\sqrt{a^2+x^2}}{x}\right| + C$

27. $\int \dfrac{dx}{x^2\sqrt{a^2+x^2}} = -\dfrac{\sqrt{a^2+x^2}}{a^2x} + C$

28. $\int \dfrac{dx}{\sqrt{a^2-x^2}} = \sin^{-1}\dfrac{x}{a} + C$

29. $\int \sqrt{a^2-x^2}\, dx = \dfrac{x}{2}\sqrt{a^2-x^2} + \dfrac{a^2}{2}\sin^{-1}\dfrac{x}{a} + C$

30. $\displaystyle\int x^2\sqrt{a^2-x^2}\,dx = \frac{a^4}{8}\,\sin^{-1}\frac{x}{a} - \frac{1}{8}x\sqrt{a^2-x^2}(a^2-2x^2) + C$

31. $\displaystyle\int \frac{\sqrt{a^2-x^2}}{x}\,dx = \sqrt{a^2-x^2} - a\,\ln\left|\frac{a+\sqrt{a^2-x^2}}{x}\right| + C$ **32.** $\displaystyle\int \frac{\sqrt{a^2-x^2}}{x^2}\,dx = -\sin^{-1}\frac{x}{a} - \frac{\sqrt{a^2-x^2}}{x} + C$

33. $\displaystyle\int \frac{x^2}{\sqrt{a^2-x^2}}\,dx = \frac{a^2}{2}\,\sin^{-1}\frac{x}{a} - \frac{1}{2}x\sqrt{a^2-x^2} + C$ **34.** $\displaystyle\int \frac{dx}{x\sqrt{a^2-x^2}} = -\frac{1}{a}\,\ln\left|\frac{a+\sqrt{a^2-x^2}}{x}\right| + C$

35. $\displaystyle\int \frac{dx}{x^2\sqrt{a^2-x^2}} = -\frac{\sqrt{a^2-x^2}}{a^2x} + C$ **36.** $\displaystyle\int \frac{dx}{\sqrt{x^2-a^2}} = \cosh^{-1}\frac{x}{a} + C = \ln\left|x+\sqrt{x^2-a^2}\right| + C$

37. $\displaystyle\int \sqrt{x^2-a^2}\,dx = \frac{x}{2}\sqrt{x^2-a^2} - \frac{a^2}{2}\,\ln\left|x+\sqrt{x^2-a^2}\right| + C$

38. $\displaystyle\int (\sqrt{x^2-a^2})^n\,dx = \frac{x(\sqrt{x^2-a^2})^n}{n+1} - \frac{na^2}{n+1}\int(\sqrt{x^2-a^2})^{n-2}dx,\quad n\ne -1$

39. $\displaystyle\int \frac{dx}{(\sqrt{x^2-a^2})^n} = \frac{x(\sqrt{x^2-a^2})^{2-n}}{(2-n)a^2} - \frac{n-3}{(n-2)a^2}\int \frac{dx}{(\sqrt{x^2-a^2})^{n-2}},\quad n\ne 2$

40. $\displaystyle\int x(\sqrt{x^2-a^2})^n\,dx = \frac{(\sqrt{x^2-a^2})^{n+2}}{n+2} + C,\quad n\ne -2$

41. $\displaystyle\int x^2\sqrt{x^2-a^2}\,dx = \frac{x}{8}(2x^2-a^2)\sqrt{x^2-a^2} - \frac{a^4}{8}\,\ln\left|x+\sqrt{x^2-a^2}\right| + C$

42. $\displaystyle\int \frac{\sqrt{x^2-a^2}}{x}\,dx = \sqrt{x^2-a^2} - a\,\sec^{-1}\left|\frac{x}{a}\right| + C$ **43.** $\displaystyle\int \frac{\sqrt{x^2-a^2}}{x^2}\,dx = \ln\left|x+\sqrt{x^2-a^2}\right| - \frac{\sqrt{x^2-a^2}}{x} + C$

44. $\displaystyle\int \frac{x^2}{\sqrt{x^2-a^2}}\,dx = \frac{a^2}{2}\,\ln\left|x+\sqrt{x^2-a^2}\right| + \frac{x}{2}\sqrt{x^2-a^2} + C$

45. $\displaystyle\int \frac{dx}{x\sqrt{x^2-a^2}} = \frac{1}{a}\,\sec^{-1}\left|\frac{x}{a}\right| + C = \frac{1}{a}\,\cos^{-1}\left|\frac{a}{x}\right| + C$ **46.** $\displaystyle\int \frac{dx}{x^2\sqrt{x^2-a^2}} = \frac{\sqrt{x^2-a^2}}{a^2x} + C$

47. $\displaystyle\int \frac{dx}{\sqrt{2ax-x^2}} = \sin^{-1}\left(\frac{x-a}{a}\right) + C$

48. $\displaystyle\int \sqrt{2ax-x^2}\,dx = \frac{x-a}{2}\sqrt{2ax-x^2} + \frac{a^2}{2}\,\sin^{-1}\left(\frac{x-a}{a}\right) + C$

49. $\displaystyle\int (\sqrt{2ax-x^2})^n dx = \frac{(x-a)(\sqrt{2ax-x^2})^n}{n+1} + \frac{na^2}{n+1}\int(\sqrt{2ax-x^2})^{n-2}\,dx$

50. $\displaystyle\int \frac{dx}{(\sqrt{2ax-x^2})^n} = \frac{(x-a)(\sqrt{2ax-x^2})^{2-n}}{(n-2)a^2} + \frac{n-3}{(n-2)a^2}\int \frac{dx}{(\sqrt{2ax-x^2})^{n-2}}$

51. $\displaystyle\int x\sqrt{2ax-x^2}\,dx = \frac{(x+a)(2x-3a)\sqrt{2ax-x^2}}{6} + \frac{a^3}{2}\,\sin^{-1}\left(\frac{x-a}{a}\right) + C$

52. $\displaystyle\int \frac{\sqrt{2ax-x^2}}{x}\,dx = \sqrt{2ax-x^2} + a\,\sin^{-1}\left(\frac{x-a}{a}\right) + C$ **53.** $\displaystyle\int \frac{\sqrt{2ax-x^2}}{x^2}\,dx = -2\sqrt{\frac{2a-x}{x}} - \sin^{-1}\left(\frac{x-a}{a}\right) + C$

54. $\displaystyle\int \frac{x\,dx}{\sqrt{2ax-x^2}} = a\,\sin^{-1}\left(\frac{x-a}{a}\right) - \sqrt{2ax-x^2} + C$ **55.** $\displaystyle\int \frac{dx}{x\sqrt{2ax-x^2}} = -\frac{1}{a}\sqrt{\frac{2a-x}{x}} + C$

56. $\displaystyle\int \sin ax\, dx = -\frac{1}{a} \cos ax + C$

57. $\displaystyle\int \cos ax\, dx = \frac{1}{a} \sin ax + C$

58. $\displaystyle\int \sin^2 ax\, dx = \frac{x}{2} - \frac{\sin 2ax}{4a} + C$

59. $\displaystyle\int \cos^2 ax\, dx = \frac{x}{2} + \frac{\sin 2ax}{4a} + C$

60. $\displaystyle\int \sin^n ax\, dx = -\frac{\sin^{n-1} ax \cos ax}{na} + \frac{n-1}{n} \int \sin^{n-2} ax\, dx$

61. $\displaystyle\int \cos^n ax\, dx = \frac{\cos^{n-1} ax \sin ax}{na}$
$$+ \frac{n-1}{n} \int \cos^{n-2} ax\, dx$$

62. a) $\displaystyle\int \sin ax \cos bx\, dx = -\frac{\cos(a+b)x}{2(a+b)} - \frac{\cos(a-b)x}{2(a-b)} + C, \quad a^2 \neq b^2$

b) $\displaystyle\int \sin ax \sin bx\, dx = \frac{\sin(a-b)x}{2(a-b)} - \frac{\sin(a+b)x}{2(a+b)} + C, \quad a^2 \neq b^2$

c) $\displaystyle\int \cos ax \cos bx\, dx = \frac{\sin(a-b)x}{2(a-b)} + \frac{\sin(a+b)x}{2(a+b)} + C, \quad a^2 \neq b^2$

63. $\displaystyle\int \sin ax \cos ax\, dx = -\frac{\cos 2ax}{4a} + C$

64. $\displaystyle\int \sin^n ax \cos ax\, dx = \frac{\sin^{n+1} ax}{(n+1)a} + C, \quad n \neq -1$

65. $\displaystyle\int \frac{\cos ax}{\sin ax}\, dx = \frac{1}{a} \ln |\sin ax| + C$

66. $\displaystyle\int \cos^n ax \sin ax\, dx = -\frac{\cos^{n+1} ax}{(n+1)a} + C, \quad n \neq -1$

67. $\displaystyle\int \frac{\sin ax}{\cos ax}\, dx = -\frac{1}{a} \ln |\cos ax| + C$

68. $\displaystyle\int \sin^n ax \cos^m ax\, dx = -\frac{\sin^{n-1} ax \cos^{m+1} ax}{a(m+n)} + \frac{n-1}{m+n} \int \sin^{n-2} ax \cos^m ax\, dx, \quad n \neq -m \quad \text{(reduces } \sin^n ax)$

69. $\displaystyle\int \sin^n ax \cos^m ax\, dx = \frac{\sin^{n+1} ax \cos^{m-1} ax}{a(m+n)} + \frac{m-1}{m+n} \int \sin^n ax \cos^{m-2} ax\, dx, \quad m \neq -n \quad \text{(reduces } \cos^m ax)$

70. $\displaystyle\int \frac{dx}{b + c \sin ax} = \frac{-2}{a\sqrt{b^2 - c^2}} \tan^{-1}\left[\sqrt{\frac{b-c}{b+c}} \tan\left(\frac{\pi}{4} - \frac{ax}{2}\right)\right] + C, \quad b^2 > c^2$

71. $\displaystyle\int \frac{dx}{b + c \sin ax} = \frac{-1}{a\sqrt{c^2 - b^2}} \ln\left|\frac{c + b \sin ax + \sqrt{c^2 - b^2} \cos ax}{b + c \sin ax}\right| + C, \quad b^2 < c^2$

72. $\displaystyle\int \frac{dx}{1 + \sin ax} = -\frac{1}{a} \tan\left(\frac{\pi}{4} - \frac{ax}{2}\right) + C$

73. $\displaystyle\int \frac{dx}{1 - \sin ax} = \frac{1}{a} \tan\left(\frac{\pi}{4} + \frac{ax}{2}\right) + C$

74. $\displaystyle\int \frac{dx}{b + c \cos ax} = \frac{2}{a\sqrt{b^2 - c^2}} \tan^{-1}\left[\sqrt{\frac{b-c}{b+c}} \tan\frac{ax}{2}\right] + C, \quad b^2 > c^2$

75. $\displaystyle\int \frac{dx}{b + c \cos ax} = \frac{1}{a\sqrt{c^2 - b^2}} \ln\left|\frac{c + b \cos ax + \sqrt{c^2 - b^2} \sin ax}{b + c \cos ax}\right| + C, \quad b^2 < c^2$

76. $\displaystyle\int \frac{dx}{1 + \cos ax} = \frac{1}{a} \tan\frac{ax}{2} + C$

77. $\displaystyle\int \frac{dx}{1 - \cos ax} = -\frac{1}{a} \cot\frac{ax}{2} + C$

78. $\displaystyle\int x \sin ax\, dx = \frac{1}{a^2} \sin ax - \frac{x}{a} \cos ax + C$

79. $\displaystyle\int x \cos ax\, dx = \frac{1}{a^2} \cos ax + \frac{x}{a} \sin ax + C$

80. $\displaystyle\int x^n \sin ax \, dx = -\frac{x^n}{a} \cos ax + \frac{n}{a} \int x^{n-1} \cos ax \, dx$

81. $\displaystyle\int x^n \cos ax \, dx = \frac{x^n}{a} \sin ax - \frac{n}{a} \int x^{n-1} \sin ax \, dx$

82. $\displaystyle\int \tan ax \, dx = \frac{1}{a} \ln |\sec ax| + C$

83. $\displaystyle\int \cot ax \, dx = \frac{1}{a} \ln |\sin ax| + C$

84. $\displaystyle\int \tan^2 ax \, dx = \frac{1}{a} \tan ax - x + C$

85. $\displaystyle\int \cot^2 ax \, dx = -\frac{1}{a} \cot ax - x + C$

86. $\displaystyle\int \tan^n ax \, dx = \frac{\tan^{n-1} ax}{a(n-1)} - \int \tan^{n-2} ax \, dx, \quad n \neq 1$

87. $\displaystyle\int \cot^n ax \, dx = -\frac{\cot^{n-1} ax}{a(n-1)} - \int \cot^{n-2} ax \, dx, \quad n \neq 1$

88. $\displaystyle\int \sec ax \, dx = \frac{1}{a} \ln |\sec ax + \tan ax| + C$

89. $\displaystyle\int \csc ax \, dx = -\frac{1}{a} \ln |\csc ax + \cot ax| + C$

90. $\displaystyle\int \sec^2 ax \, dx = \frac{1}{a} \tan ax + C$

91. $\displaystyle\int \csc^2 ax \, dx = -\frac{1}{a} \cot ax + C$

92. $\displaystyle\int \sec^n ax \, dx = \frac{\sec^{n-2} ax \tan ax}{a(n-1)} + \frac{n-2}{n-1} \int \sec^{n-2} ax \, dx, \quad n \neq 1$

93. $\displaystyle\int \csc^n ax \, dx = -\frac{\csc^{n-2} ax \cot ax}{a(n-1)} + \frac{n-2}{n-1} \int \csc^{n-2} ax \, dx, \quad n \neq 1$

94. $\displaystyle\int \sec^n ax \tan ax \, dx = \frac{\sec^n ax}{na} + C, \quad n \neq 0$

95. $\displaystyle\int \csc^n ax \cot ax \, dx = -\frac{\csc^n ax}{na} + C, \quad n \neq 0$

96. $\displaystyle\int \sin^{-1} ax \, dx = x \sin^{-1} ax + \frac{1}{a}\sqrt{1 - a^2 x^2} + C$

97. $\displaystyle\int \cos^{-1} ax \, dx = x \cos^{-1} ax - \frac{1}{a}\sqrt{1 - a^2 x^2} + C$

98. $\displaystyle\int \tan^{-1} ax \, dx = x \tan^{-1} ax - \frac{1}{2a} \ln (1 + a^2 x^2) + C$

99. $\displaystyle\int x^n \sin^{-1} ax \, dx = \frac{x^{n+1}}{n+1} \sin^{-1} ax - \frac{a}{n+1} \int \frac{x^{n+1} \, dx}{\sqrt{1 - a^2 x^2}}, \quad n \neq -1$

100. $\displaystyle\int x^n \cos^{-1} ax \, dx = \frac{x^{n+1}}{n+1} \cos^{-1} ax + \frac{a}{n+1} \int \frac{x^{n+1} \, dx}{\sqrt{1 - a^2 x^2}}, \quad n \neq -1$

101. $\displaystyle\int x^n \tan^{-1} ax \, dx = \frac{x^{n+1}}{n+1} \tan^{-1} ax - \frac{a}{n+1} \int \frac{x^{n+1} \, dx}{1 + a^2 x^2}, \quad n \neq -1$

102. $\displaystyle\int e^{ax} dx = \frac{1}{a} e^{ax} + C$

103. $\displaystyle\int b^{ax} dx = \frac{1}{a} \frac{b^{ax}}{\ln b} + C, \quad b > 0, \, b \neq 1$

104. $\displaystyle\int x e^{ax} dx = \frac{e^{ax}}{a^2} (ax - 1) + C$

105. $\displaystyle\int x^n e^{ax} dx = \frac{1}{a} x^n e^{ax} - \frac{n}{a} \int x^{n-1} e^{ax} \, dx$

106. $\displaystyle\int x^n b^{ax} dx = \frac{x^n b^{ax}}{a \ln b} - \frac{n}{a \ln b} \int x^{n-1} b^{ax} \, dx, \quad b > 0, b \neq 1$

107. $\displaystyle\int e^{ax} \sin bx \, dx = \frac{e^{ax}}{a^2 + b^2} (a \sin bx - b \cos bx) + C$

108. $\displaystyle\int e^{ax} \cos bx \, dx = \frac{e^{ax}}{a^2 + b^2} (a \cos bx + b \sin bx) + C$

109. $\displaystyle\int \ln ax \, dx = x \ln ax - x + C$

110. $\displaystyle\int x^n (\ln ax)^m \, dx = \frac{x^{n+1} (\ln ax)^m}{n+1} - \frac{m}{n+1} \int x^n (\ln ax)^{m-1} dx, \quad n \neq -1$

111. $\displaystyle\int x^{-1} (\ln ax)^m \, dx = \frac{(\ln ax)^{m+1}}{m+1} + C, \quad m \neq -1$

112. $\displaystyle\int \frac{dx}{x \ln ax} = \ln |\ln ax| + C$

113. $\displaystyle\int \sinh ax\, dx = \frac{1}{a}\cosh ax + C$

114. $\displaystyle\int \cosh ax\, dx = \frac{1}{a}\sinh ax + C$

115. $\displaystyle\int \sinh^2 ax\, dx = \frac{\sinh 2ax}{4a} - \frac{x}{2} + C$

116. $\displaystyle\int \cosh^2 ax\, dx = \frac{\sinh 2ax}{4a} + \frac{x}{2} + C$

117. $\displaystyle\int \sinh^n ax\, dx = \frac{\sinh^{n-1} ax \cosh ax}{na} - \frac{n-1}{n}\int \sinh^{n-2} ax\, dx, \quad n \neq 0$

118. $\displaystyle\int \cosh^n ax\, dx = \frac{\cosh^{n-1} ax \sinh ax}{na} + \frac{n-1}{n}\int \cosh^{n-2} ax\, dx, \quad n \neq 0$

119. $\displaystyle\int x \sinh ax\, dx = \frac{x}{a}\cosh ax - \frac{1}{a^2}\sinh ax + C$

120. $\displaystyle\int x \cosh ax\, dx = \frac{x}{a}\sinh ax - \frac{1}{a^2}\cosh ax + C$

121. $\displaystyle\int x^n \sinh ax\, dx = \frac{x^n}{a}\cosh ax - \frac{n}{a}\int x^{n-1}\cosh ax\, dx$

122. $\displaystyle\int x^n \cosh ax\, dx = \frac{x^n}{a}\sinh ax - \frac{n}{a}\int x^{n-1}\sinh ax\, dx$

123. $\displaystyle\int \tanh ax\, dx = \frac{1}{a}\ln(\cosh ax) + C$

124. $\displaystyle\int \coth ax\, dx = \frac{1}{a}\ln|\sinh ax| + C$

125. $\displaystyle\int \tanh^2 ax\, dx = x - \frac{1}{a}\tanh ax + C$

126. $\displaystyle\int \coth^2 ax\, dx = x - \frac{1}{a}\coth ax + C$

127. $\displaystyle\int \tanh^n ax\, dx = -\frac{\tanh^{n-1} ax}{(n-1)a} + \int \tanh^{n-2} ax\, dx, \quad n \neq 1$

128. $\displaystyle\int \coth^n ax\, dx = -\frac{\coth^{n-1} ax}{(n-1)a} + \int \coth^{n-2} ax\, dx, \quad n \neq 1$

129. $\displaystyle\int \operatorname{sech} ax\, dx = \frac{1}{a}\sin^{-1}(\tanh ax) + C$

130. $\displaystyle\int \operatorname{csch} ax\, dx = \frac{1}{a}\ln\left|\tanh \frac{ax}{2}\right| + C$

131. $\displaystyle\int \operatorname{sech}^2 ax\, dx = \frac{1}{a}\tanh ax + C$

132. $\displaystyle\int \operatorname{csch}^2 ax\, dx = -\frac{1}{a}\coth ax + C$

133. $\displaystyle\int \operatorname{sech}^n ax\, dx = \frac{\operatorname{sech}^{n-2} ax \tanh ax}{(n-1)a} + \frac{n-2}{n-1}\int \operatorname{sech}^{n-2} ax\, dx, \quad n \neq 1$

134. $\displaystyle\int \operatorname{csch}^n ax\, dx = -\frac{\operatorname{csch}^{n-2} ax \coth ax}{(n-1)a} - \frac{n-2}{n-1}\int \operatorname{csch}^{n-2} ax\, dx, \quad n \neq 1$

135. $\displaystyle\int \operatorname{sech}^n ax \tanh ax\, dx = -\frac{\operatorname{sech}^n ax}{na} + C, \quad n \neq 0$

136. $\displaystyle\int \operatorname{csch}^n ax \coth ax\, dx = -\frac{\operatorname{csch}^n ax}{na} + C, \quad n \neq 0$

137. $\displaystyle\int e^{ax}\sinh bx\, dx = \frac{e^{ax}}{2}\left[\frac{e^{bx}}{a+b} - \frac{e^{-bx}}{a-b}\right] + C, \quad a^2 \neq b^2$

138. $\displaystyle\int e^{ax}\cosh bx\, dx = \frac{e^{ax}}{2}\left[\frac{e^{bx}}{a+b} + \frac{e^{-bx}}{a-b}\right] + C, \quad a^2 \neq b^2$

139. $\displaystyle\int_0^\infty x^{n-1}e^{-x}\, dx = \Gamma(n) = (n-1)!, \quad n > 0$

140. $\displaystyle\int_0^\infty e^{-ax^2}\, dx = \frac{1}{2}\sqrt{\frac{\pi}{a}}, \quad a > 0$

141. $\displaystyle\int_0^{\pi/2}\sin^n x\, dx = \int_0^{\pi/2}\cos^n x\, dx = \begin{cases} \dfrac{1\cdot 3\cdot 5\cdots(n-1)}{2\cdot 4\cdot 6\cdots n}\cdot\dfrac{\pi}{2}, & \text{if } n \text{ is an even integer} \geq 2 \\[2ex] \dfrac{2\cdot 4\cdot 6\cdots(n-1)}{3\cdot 5\cdot 7\cdots n}, & \text{if } n \text{ is an odd integer} \geq 3 \end{cases}$